S0-DUU-433

High Altitude Tropical Biogeography

High Altitude Tropical Biogeography

EDITED BY

FRANÇOIS VUILLEUMIER

AND

MAXIMINA MONASTERIO

Published by Oxford University Press and the American Museum of Natural History

New York Oxford

Oxford University Press

1986

Oxford University Press

Oxford New York Toronto
Delhi Bombay Calcutta Madras Karachi
Kuala Lumpur Singapore Hong Kong Tokyo
Nairobi Dar es Salaam Cape Town
Melbourne Auckland

and associated companies in
Beirut Berlin Ibadan Mexico City Nicosia

Published by Oxford University Press, Inc.,
200 Madison Avenue, New York, New York 10016

Library of Congress Cataloging in Publication Data
Main entry under title:
High altitude tropical biogeography.
Includes index. 1. Biogeography—Tropics. 2. Alpine regions—Tropics.
I. Vuilleumier, François, 1938- .II. Monasterio, Maximina.
QH84.5.H63 1986 574.5′264 85-4979
ISBN 0-19-503625-5

Published with support from the Ambrose Monell Foundation to the American Museum of Natural History

9 8 7 6 5 4 3 2 1

Printed in the United States of America
on acid-free paper

Preface

We offer in this book an overview of evolution and biogeography in the biomes of high tropical mountains of the world called páramo and puna (South America), afroalpine (Africa), and tropical-alpine (Malesia). The chapters present data gathered by a number of specialists during years of field work. More important, however, these chapters synthesize this knowledge and give interpretations of the distribution, origins, ecological requirements, and evolutionary adaptations of key plant and animal components (both extant and fossil) of high tropical montane biomes.

All too often, the cold montane tropics are overlooked by students of tropical biology, who focus on the warm lowlands, either the rain forests or the savannas. Several important books synthesize what we know of these lowland ecosystems, but to our knowledge no such book has been put together for the high mountains of the world, although several papers, reviews, and books deal with specific taxa or regional aspects of high altitude tropical biogeography (for instance, Sturm, 1978; Vuilleumier and Ewert, 1978; Salgado-Labouriau, 1979; Webber, 1979; Monasterio, 1980; Vuilleumier and Simberloff, 1980; Cleef, 1981; Nievergelt, 1981; and Hamilton, 1982[1]). The present book thus fills an important gap in the literature on tropical biology.

High tropical mountains are the least known of all tropical biomes or ecosystems. At high altitudes but low latitudes above the timberline, each day is summertime and winter reigns every night. These unique climatic conditions make the high tropical regions of the world a special setting for the unfolding of evolutionary adaptations. Furthermore, high tropical biota are like ecological islands in a sea of warm tropical climate and vegetation. Relative insularity has contributed to the development of the biota and to some of its striking evolutionary and biogeographic characteristics. But in a more practical way, as far as their scientific study is concerned, this isolation has been an important obstacle to the acquisition of knowledge. Thus, unlike the lowland tropics, the high montane tropics are less well understood, either as entire ecosystems, or as local functional units. Much of what is known so far is presented in this book, which can be viewed as a summary of the first phase of research. What remains to be learned, especially about the functioning of high tropical mountain biota as ecosystems will probably be best gathered through long-term local studies carried out by several scientists working together in a multidisciplinary team. In this second phase of research, such processes as nutrient cycling, energy flow, carbon balance, and population recruitment will be studied. Whereas we know a great deal about the evolution and biogeography of key components of the high tropical biota on each continent, as this book clearly demonstrates, very little long-term intercontinental research has been pursued. The second phase of work will also focus on this level of approach, and thus a close look at the difficult problem of convergence, which can only be hinted at today, will have to be taken.

Programs now being implemented by several international research and scientific organizations under the aegis of the Decade of the Tropics (Solbrig and Golley, 1983) will permit just such kinds of focus to be realized. Because it summarizes what we know, this book will offer active researchers in these programs a wealth of both information and speculation upon which to draw for their own studies at another level.

This book will also be useful to teachers who give courses in tropical biology and who want to include in their lectures and field demonstrations the spectrum of tropical habitats, not only the lowland ones. Graduate students in search of thesis projects should have no trouble finding topics after reading this book. Perhaps even seasoned workers will climb once again to the high-altitude realms of the tropics to try to disprove some of the hypotheses they will find in this volume.

Finally, this book will be helpful to conservationists. High-altitude tropical ecosystems are very fragile. Their preservation in the face of ever-expanding human activities will depend on their rational use. Thus we must secure areas of tropical high mountain landscapes for natural parks and reserves. But we must also ensure that pastoral use of high tropical grasslands, plantation of trees in areas at or even above the continuous timberline, utilization of hydraulic resources, and agriculture at the upper limit of crop production are all carried out

[1]Full references to works cited in the preface can be found at the end of Chapter 1.

according to a reasonably complete understanding of the ecosystems so managed and of their floristic and faunistic elements, as well as their historical antecedents.

One of the main components of the biota not represented directly in this book is in fact *Homo sapiens*. For several reasons, we have had to abandon our original plan to include chapters on man at very high altitudes in the tropics. However, since journal articles and chapters in books have appeared that deal with human adaptation and evolution in high tropical mountains, we feel that the interested reader can easily turn elsewhere for excellent and detailed information and speculation (e.g., Baker and Little, 1976; Brush, 1976; Baker, 1978; Ellenberg, 1979; Thomas, 1979; Price, 1981).

To achieve our aim to present as comprehensive a view of high tropical montane biota and of their evolutionary history throughout the world as is possible given our present level of knowledge, we chose chapter topics that are both of specific interest to the student of tropical and montane biota, and of more general interest to students of evolutionary or biogeographic patterns and processes. We tried to represent most geographic areas of the high tropical montane biota of the world by one or more chapters. However, we could not obtain a balanced representation correlated to the relative area of each region. Geographic coverage is thus uneven. Similarly, we included both plant and animal subjects, as well as living and fossil taxa, but we could not make our selection on the basis of representativeness alone. As a result, here again, taxonomic coverage is uneven. Finally, because the level of understanding of the topics selected varies according to the degree of work thus far carried out on various taxa or on various processes (adaptation, distribution, ecological requirements), the depth of each chapter will be found to vary. We make no apologies to the reader for these gaps. We too regret that they exist, for they make for a more patchy book than we would all want. But since they largely reflect the gaps in our knowledge, they should be an incentive for further work rather than a subject for criticism.

We selected authors who had done first-hand and long-term research on their respective subjects and who represented a truly international community of scholars who had devoted many years of their lives to a study of high tropical mountain biota or of one of their key components. We were very fortunate to enlist the collaboration of a distinguished group of scientists, and we feel privileged to have worked closely with them during the preparation of this book. We thank them all for their willingness to embark with us on such a venture, for their cheerfulness about our scientific and editorial requirements, for their patience, and above all for their hard work. Other important students of high tropical mountain biota are unfortunately not authors of chapters in this book. Some of these scientists had commitments that did not allow them to join us; or else we could not include their work for other reasons. One of our authors, Alfredo Barrera, died during the preparation of the book. We are greatly saddened to think that this fine Mexican scientist is no longer with us as part of our community of scholars. We also regret that his contribution was not sufficiently advanced at the time of his death that we could complete it and include it as a memorial tribute to his excellent work.

We would like to express our thanks to the authorities of the University of the Andes (Mérida, Venezuela) for help with fieldwork, facilities, and financial matters, and those of the American Museum of Natural History (New York, U.S.A.) for financial, logistical, and secretarial help. We greatly appreciate the many courtesies extended to us by the staff of Oxford University Press, especially Susan Meigs and William Curtis.

New York F.V.
Mérida M.M.
June 1985

Contents

Contributors

Cei, José M., Departamento de Ciencias Naturales, Universidad Nacional de Río Cuarto, Río Cuarto, Córdoba, Argentina

Cleef, Antoine M., Laboratory Hugo de Vries, University of Amsterdam, Sarphatistraat 221, 1018 BX Amsterdam, The Netherlands

Cuatrecasas, José, Department of Botany, Smithsonian Institution, Washington D.C. 20560

Descimon, Henri, Laboratoire de Biologie animale, Université de Provence, 3 Place Victor-Hugo, 13331 Marseille Cédex, France

Dorst, Jean, Zoologie, Mammifères et Oiseaux, Muséum National d'Histoire Naturelle, 55, rue de Buffon, 75005 Paris, France

Dowsett, R. J., Percy FitzPatrick Institute of African Ornithology, University of Cape Town, Rondebosch 7700, South Africa

Gill, Frank B., Department of Ornithology, The Academy of Natural Sciences of Philadelphia, 19th and the Parkway, Philadelphia, Pennsylvania 19103

Hedberg, Olov, Institute of Systematic Botony, P.O. Box 541, S-751 21 Uppsala, Sweden

Hoffstetter, Robert, Institut de Paléontologie, Muséum National d'Histoire Naturelle, 8, rue de Buffon, 75005 Paris, France

Holloway, Jeremy D., Commonwealth Institute of Entomology, British Museum (Natural History), Cromwell Road, London SW7 5BD, England

Lynch, John D., School of Biological Sciences, The University of Nebraska, Lincoln, Nebraska 68588

Mabberley, D. J., Departments of Botany and of Agricultural and Forest Sciences, University of Oxford, Oxford OX1 3PN, England

Monasterio, Maximina, Departamento de Biología, Ecología Vegetal, Facultad de Ciencias, Universidad de los Andes, Mérida, Venezuela

Reig, Osvaldo A., Departamento de Ciencias Biológicas, Facultad de Ciencias Exactas y Naturales, Universidad de Buenos Aires, Ciudad Universitaria, 1428 Nuñez, Buenos Aires, Argentina

Salgado-Labouriau, Maria Léa, Centro de Ecología, Instituto Venezolano de Investigaciones Científicas, Apartado 1827, Caracas 101, Venezuela

Sarmiento, Guillermo, Departamento de Biología, Ecología Vegetal, Facultad de Ciencias, Universidad de los Andes, Mérida, Venezuela

Simpson, Beryl B., Department of Botany, University of Texas, Austin, Texas 78712

Smith, Jeremy M. B., Department of Geography, University of New England, Armidale, New South Wales 235, Australia

Steyermark, Julian A., Missouri Botanical Garden, P.O. Box 229, St. Louis, Missouri 63166

van der Hammen, Thomas, Laboratory Hugo de Vries, University of Amsterdam, Sarphatistraat 221, 1018 BX Amsterdam, The Netherlands

Villwock, Wolfgang, Zoological Institute and Zoological Museum, Martin Luther King Platz 3, University of Hamburg, 2000 Hamburg 13, Federal Republic of Germany

Vuilleumier, François, Department of Ornithology, American Museum of Natural History, Central Park West at 79th Street, New York, New York 10024

Wing, Elizabeth S., The Florida State Museum, University of Florida, Gainesville, Florida 32611

Wolf, Larry L., Department of Biology, Syracuse University, Syracuse, New York 13210

High Altitude Tropical Biogeography

1

Introduction: High Tropical Mountain Biota of the World

MAXIMINA MONASTERIO AND FRANÇOIS VUILLEUMIER

The term "high tropical mountain biota of the world" means the flora and fauna, either integrated together into local communities or into ecosystems, or separated into some of its taxonomic components (birds, vascular plants), found at high altitudes above the upper altitudinal limit of continuous forest vegetation of both the Old and New World. If trees are defined on the basis of architectural models (see Hallé, Oldeman, and Tomlinson, 1978), then trees conforming to Corner's model of growth occur in the highest Andes. In the tropical mountains, furthermore, one must distinguish between timberline formed by continuous forest vegetation, and "timberline" that results from patches of trees (defined now on the basis of branching patterns) above continuous forest. In this book, we distinguish between forest and nonforest vegetation on the basis of the traditional concepts of tree and forest.

High montane ecosystems occurring between the Tropic of Cancer and the Tropic of Capricorn (Fig. 1–1) are known by the name given to the most distinctive local plant formations. Thus the names of páramo (northern Andes), puna (central Andes), afroalpine (East African mountains), or tropical-alpine (Malesia) appear regularly in this book. Unfortunately, no single name encompasses all the types of high tropical mountain vegetation above the continuous timberline. The afroalpine and the tropical-alpine correspond rather closely to the Andean páramo in vegetation physiognomy, but the puna of the

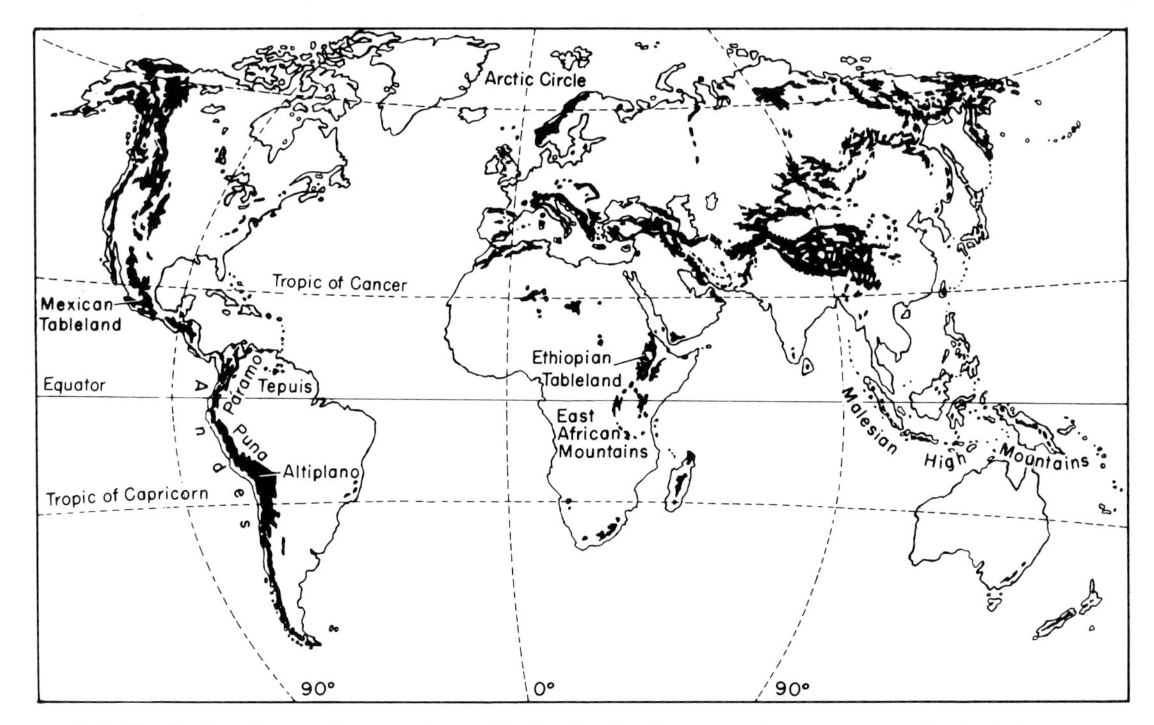

FIG. 1–1. Distribution of mountain vegetation worldwide, showing the extent of montane vegetation in the intertropical zone. Modified and simplified from world vegetation map, *Times* (London) *Atlas*, vol. I (1958), Plate 4, scale 1:65,000,000.

central Andes has no real equivalent in the Old World tropics; indeed, in some ways, the puna appears more similar to the steppes of the Tibetan Plateau. In spite of such differences in the aspect of vegetation and underlying climate (see Chap. 2) of various high tropical mountain areas of the world, it might be useful, for the sake of nomenclatural simplicity if for no other reason, to call all high tropical mountain biota above the timberline simply *páramo*. The addition of a geographic adjective, if necessary, would then be all that is needed to characterize the particular area that is being discussed, for example, Andean páramo, African páramo, Papuan páramo, and so forth.

We will use this simplified nomenclature in this introductory chapter, although the authors of chapters in this book have used the more traditional nomenclature.

Basically, páramos are high mountain grasslands, in general dominated physiognomically by the presence of bunch grass (Gramineae), but also containing other plants, especially shrubs or rosette trees of the family Compositae. Some páramos are almost pure dense grassland, but at the other extreme, especially at very high altitudes, other páramos are almost totally devoid of grasses and dominated instead by the peculiar giant rosettes of members of the family Compositae. Elsewhere mixtures predominate. In at least one case, Mt. Kinabalu in Borneo, there is no grassland, and in fact very little soil and vegetation, above the timberline.

Other plant groups are also important in the composition of páramo vegetation, especially (locally) some Bromeliaceae (particularly the genus *Puya*), some Campanulaceae (*Lobelia*), some Cruciferae (*Draba*), some Rosaceae (*Polylepis*), and some Cyatheaceae (*Cyathea*). The "ground" can be either a dense, mirelike, spongy mass of plant life (mosses and others) covering the soil entirely with a thick layer, or else the soil is essentially dry and barren, with sparse cushion and rosette plants, growing in the more sheltered areas. At very high altitudes, frequent or regular nightly frosts produce conspicuous frost polygons and striated ground, characteristics one usually associates with the arctic, not the tropics.

Páramos can thus appear either as lush environments, with many plants that look strange to a biologist from the temperate zone, or as very open, desert-like landscapes, with occasional oases of somewhat denser vegetation in thermally privileged, usually rocky, areas.

Montane rain forests or cloud forests are the vegetation formations encountered below páramos on wet mountains or along the wet slopes of some mountains. The ecotone between páramo and upper montane rain forest is usually a tangled thicket of epiphyte-laden shrubs or gnarled trees growing in impenetrable copses surrounded by tall grassland. Elsewhere the timberline is very sharp, making the transition from forest to páramo quite abrupt. Along dry slopes, transitions can be more subtle, since forest is usually absent at montane elevations and replaced instead by an open vegetation of shrubs and locally of cacti, more or less rapidly giving ground to the páramo plants.

Animal life in páramos is usually much poorer than in tropical environments at lower altitudes, whether wet or dry. There are fewer species of fewer taxonomic groups, but characteristically some groups (families, genera) dominate the communities. This is one of the features of insularity. It clearly looks as if the rigors of a high-altitude environment severely limited the ability of tropical groups to establish themselves permanently there. In fact, páramos are perhaps the tropical environment with the largest representation of temperate groups (this is also true for plants).

Thus the páramos are both "bridge and barrier": the first for temperate taxa, the second for tropical ones, although such an overly dichotomous view has obvious exceptions. By and large, páramo animals lack the extravagance of color and form assumed by their relatives in the lowland tropics: they are either smaller, or duller, or both. Often also, páramo animals appear scarce. Many of them lead hidden lives, in nooks and crannies of the vegetation or of the inanimate substrate.

The greatest extension of páramo *sensu lato* is in the Andes, where the dry and open formation called puna has a vast and continuous distribution in a region known as the altiplano (central and southern Peru, Bolivia, northern Chile, and northern Argentina); the páramo *sensu stricto* occurs in smaller and disjunct patches in Venezuela, Colombia, Ecuador, and northern Peru. In Africa, the approximate equivalent of the Andean altiplano is found on the Ethiopian tableland, whereas the African páramo, like the Andean páramo *sensu stricto*, is disjunct on the isolated summits of East African mountains. In Asia and the Pacific area (Malesian and Melanesian high mountains) all páramos are patchy and discontinuous, occurring either on single small or relatively small islands (like Borneo) or in patches on large, continentlike islands (New Guinea).

Although "alpinelike" in some of their characteristics, the páramo of high tropical mountains differs in important respects from alpine biota in North America, Europe, Asia, or New Zealand at high latitudes. Several of these differences derive from the fact that the oscillations of the main climatic variables (temperature, precipitation) at high altitudes in intertropical regions are quite distinct in their periodicity from those prevailing at temperate latitudes (see Chap. 2). A number of ecological and evolutionary consequences have resulted in adaptations unique to the taxa living in the páramo environment. Several of these adaptations will be discussed later in this book. High-altitude tropical areas are available year-round as habitats for both plants and animals. In the most favorable cases, year-round reproduction is possible and in some instances does occur. This constant or virtually constant availability, contrasted to the highly seasonal unavailability of alpine biota at temperate latitudes because of the unproductive and snowbound winter period, is likely to have been one of the factors that has permitted the páramo biota to be much richer in habitat types, and in species, than their temperate, alpine, counterparts. For these reasons we prefer to call tropical high mountain biota páramo rather than alpinelike. In our view, therefore, alpine is a term that should be restricted to the seasonal areas of high mountain biota at high latitudes, and páramo should include only the year-round available habitats of high tropical mountains.

The páramo biota discussed in this book are for the most part geologically very young. Phases of uplift in the main mountain systems in the intertropical regions of the world have "created" these ecosystems out of different vegetation types, whether savannalike or forestlike, at lower elevations. The appearance of the cold tropics in the late Cenozoic permitted immigration to and subsequent occupation of new vegetation types. Various plant and animal taxa were able to expand into this new area from other regions, either at lower elevations or at lower latitudes. Some of these taxa adapted to their new environment and even underwent extensive adaptive radiation there, as described in several chapters here. Differences in the ability of various taxa to disperse and colonize the proto-páramo must have been due in part to the intrinsic capacity for dispersal and adaptation to new environments, and partly to the geographic positions of the cold tropics on each continent. Thus in the Americas, the high mountains are located along the western margin of the tropical area. The Andes could thus have served both as a corridor between northern temperate and southern temperate (Patagonian or Fuegian) elements, and as a new area for colonists from lower altitudes along the tropical slopes. In Africa, the islandlike nature of the páramo on East African mountains can be contrasted with the plateaulike distribution of high tropical biota on the Ethiopian tableland. In the Indo-Pacific, the insular nature of both the páramo environment and of the lower altitudinal belts below, especially montane forest, and the linear arrangement of these vegetation types between continental Asia and insular New Guinea produced yet another background for colonization and establishment.

The study of páramo biota began a long time ago, at first of course with various phases of exploration and collection of botanical and zoological material. Jeannel (1950) described these early explorations in the high mountains of East Africa in a superb monograph, to cite only one reference. This type of work is still carried out today, but less intensively than in the past. Yet it must go on, for we still need to make thorough inventories of a number of plant and animal groups. We do know a great deal about the taxonomic composition and geographic distribution of many taxa of vascular plants in most high-altitude tropical areas, and a similar statement can be made for several animal groups, for instance, birds, some mammals, and to a lesser extent some Lepidoptera. But our knowledge of the basis facts of taxonomy, relationships, distribution, and ecological preferences of many other plant and animal groups is woefully incomplete. Among taxa in greatest need of such basic study we can cite lichens, fungi, earthworms, Collembola, Arachnida, many Insecta, especially perhaps Diptera and Hymenoptera. Not only are these groups important in themselves, but they are also a fundamental part of the communities as links in the food webs and trophic chains. Thus until we know what taxa we are dealing with we will not be able to answer ecological or evolutionary questions. Basic taxonomic work is urgently needed before more problem-oriented research can begin.

At the present state in our understanding of páramo biota, we can offer hypotheses and scenarios about the dispersion and origins of selected taxa of plants and animals, or of some plant or animal "communities" (the term is here used in the sense of taxocene, for instance, the "community" of birds of the Andean páramos). We have enough information to start verifying several ecological or evolutionary hypotheses about specific aspects of adaptations to high altitudes in the tropics, for example, plant breeding strategies and adaptive syndromes, or resource limitations in high-altitude hummingbirds. But we cannot yet make very profound commentaries about these habitats as ecosystems. For instance, we know very little about energy transfer and flow, evolution of biomass, trophic relationships, or the role of predation in regulating community structure. We can describe some of the parts of the páramo biota of the world, but we are not yet ready to integrate them into a functional whole.

Most chapters in this book, reflecting what we see as the state of the art in evolutionary and biogeographic research at the level of resolution approached here, deal more with ecology and evolution in broad time and space dimensions than with specific, or punctual and almost dimensionless, analyses of adaptations. The measurement of limiting resources, the definition of niche parameters, the fine-scale study of interspecific interactions (e.g., allelopathy), the quantitative analysis of predation or symbiosis are all areas for future work.

We believe that future research will have to be carried out at one or more of three levels. First, the inventory of taxa (at least key taxa in the functioning of páramo biota) will have to be improved. Only then will it be possible to pose questions at the second level, beginning with how. Finally, at the third level, long-term studies will have to be undertaken at many localities. All three levels are not necessarily independent of each other, of course. Some questions can indeed be asked before the inventory of a given group of plants or animals has begun in earnest, and can in fact direct the survey in fruitful directions. But the long-term research objectives will ultimately depend, in their feasibility, on the availability of local or regional centers of research located within the high tropical montane areas themselves. An example is the plant ecology group at the University of the Andes in Mérida, Venezuela.

To present the reader with a picture of the research carried out on páramo biota of the world we divided the book into five parts, all but the first preceded by a brief editorial introduction. Part I contains a single chapter, in which the peculiar aspects of the climate at high altitude in the tropics are described. This chapter is fundamental for the rest of the book because it gives the climatological background necessary for an understanding of the páramo biota, past and present.

The second part of the book consists of four chapters (two on plants and two on birds), analyzing the adaptive strategies of some of the most obvious or physiognomically dominant components of the biota. *Espeletia* in the Andes and *Dendrosenecio* in East African mountains are certainly among the most conspicuous plants of the páramo landscape. Some of their modes of life are now well known. The two chapters on birds are comparative. Hummingbirds in the New World and sunbirds in the Old are often thought of as ecologically equivalent, if not actually convergent taxa, and their adaptations are studied in a comparative framework. At a broader taxonomic, ecological, and biogeographic level, the avian associations of the high Andes are compared to those of the Ethiopian tableland in an effort to detect communitywide convergences or at least parallelisms in general form and function and in habitat selection. The hypotheses so generated should make much easier the task of testing specific details about parallel or convergent evolutionary pathways between two different páramo avifaunas or their components.

In part III the historical developments of high-altitude biota of the Andes are examined in four chapters treating floras (two chapters) and faunas, especially mammals (two chapters). These chapters present historical data and rely on the results of researches into fossil pollen or fossil bones, interpreted in the light of geologic and climatic evolution. The time span involved is relatively short, since it concerns chiefly the Pliocene and Pleistocene epochs, when the Andes reached their present height and became covered with the vegetation types and the habitats that characterize altitudes above the upper limit of continuous forest.

The fourth part of the book is devoted to studies of adaptive radiation and evolution at the species level, especially speciation, among some of the endemic taxa of the high tropical Andes and the Venezuelan Tepuis: *Polylepis* (the only genus of plants that forms forests or woodlands above the upper limit of continuous forests in the high Andes), the frog genus *Telmatobius* (with a wide range of adaptations to aquatic and semiaquatic habitats), the fish genus *Orestias* found in closed basins of the central Andes, and the very diversified rodents (an extraordinary instance of continental adaptive radiation). The panorama of taxa thus encompasses plants and animals, and among the latter both aquatic and terrestrial groups.

Finally, in part V, the geographic origins of the floras and faunas are examined in detail. A chapter on the African páramo flora is followed by five chapters on various animal groups, both invertebrates (butterflies) and vertebrates (birds). These chapters include Africa, Indo-Australia, and the Andes. In the absence of a fossil record for these taxa, their origins are inferred on the basis of present distributional patterns.

Because high tropical mountain biota are intermediate between the strikingly simple island biota and the extraordinarily complex tropical lowland biota, they constitute a superb natural laboratory for research into fundamental evolutionary and biogeographic processes. They also offer an opportunity to integrate basic research with the expanding needs of human population. The key components of high altitude tropical mountain life, namely insularity within a continental situation, variability along altitudinal gradients, and great diversity in the origins of floristic or faunistic elements, must form the basis of future research programs and must be fully understood before development proceeds further.

REFERENCES

Baker, P. T., ed. 1978. *The biology of high altitude peoples*. Cambridge: Cambridge University Press.

Baker, P. T., and M. A. Little, eds. 1976. *Man in the Andes: A multidisciplinary study of high altitude Quechua*. Stroudsburg, Pa.: Dowden, Hutchinson and Ross.

Brush, S. B. 1976. Man's use of an Andean ecosystem. *Human Ecology* 4: 147–166.

Cleef, A. M. 1981. *The vegetation of the páramos of the Colombian Cordillera Oriental*. Dissertationes Botanicae, 61. Vaduz: Cramer.

Ellenberg, H. 1979. Man's influence on tropical mountain ecosystems in South America. *J. Ecol.* 67: 401–416.

Hallé, F., R. A. A. Oldeman, and P. B. Tomlinson. 1978. *Tropical trees and forests. An architectural analysis*. Berlin: Springer-Verlag.

Hamilton, A. C. 1982. *Environmental history of East Africa. A study of the Quaternary*. New York: Academic Press.

Jeannel, R. 1950. *Hautes montagnes d'Afrique*. Publ. Muséum National Histoire Naturelle, Paris. Suppl. No. 1.

Monasterio, M., ed. 1980. *Estudios ecológicos en los páramos andinos*. Mérida: Ediciones Universidad de los Andes.

Nievergelt, B. 1981. *Ibexes in an African environment*. Ecological Studies 40. Berlin: Springer-Verlag.

Price, L. W. 1981. *Mountains and man*. Berkeley: University of California Press.

Salgado-Labouriau, M. L., ed. 1979. *El medio ambiente páramo*. Caracas: Ediciones Centro de Estudios Avanzados, IVIC.

Solbrig, O. T., and F. B. Golley. 1983. A decade of the tropics. *Biol. Int. Spec. Issue* No. 2: 1–15.

Sturm, H. 1978. *Zur Ökologie der andinen Paramoregion*. The Hague: Junk.

Thomas, R. B. 1979. Effects of change on high mountain human adaptive patterns. In P. J. Webber, ed., *High altitude geoecology*, pp. 139–188. Boulder, Colo.: Westview Press.

Vuilleumier, F., and D. Ewert. 1978. The distribution of birds in Venezuelan páramos. *Bull. Amer. Mus. Nat. Hist.* 162: 47–88.

Vuilleumier, F., and D. Simberloff. 1980. Ecology versus history as determinants of patchy and insular distributions in high Andean birds. In M. K. Hecht, W. C. Steere, and B. Wallace, eds., *Evolutionary biology*, vol. 12, pp. 235–379. New York: Plenum Press.

Webber, P. J., ed. 1979. *High altitude geoecology*. Boulder, Colo.: Westview Press.

I

THE CLIMATIC BACKGROUND

2

Ecological Features of Climate in High Tropical Mountains

GUILLERMO SARMIENTO

Climate and soil are the basis and substrate of all life, vegetal and animal. High tropical mountain climates, extremely peculiar in their combination of features, are examined here for the first time in intra- and intercontinental comparisons involving all major high tropical mountains of the world.

After defining tropical climates in general I will then discuss tropical mountain climates. This will be followed by detailed analysis of climatic variables pertinent to the establishment and maintenance of life on tropical mountains (the Andes, Mexico, East Africa, New Guinea). Finally I will summarize common patterns and ecologically crucial features, in order to set the stage for the rest of the book, where climate will be discussed only in a more local or regional fashion, or if appropriate to the theme of each chapter.

TROPICAL CLIMATES AND HIGH MOUNTAIN CLIMATES

Particular Features of Cyclic Changes in the Tropics

Cyclic changes in the tropics are best illustrated by comparing solar radiation along a latitudinal gradient from the equator (0°) to middle latitudes (50°) (Table 2–1). On the equator, the daily total amount of solar radiation at the equinoxes, when it reaches its annual maximum, is only about 13 per cent higher than the minimum amount of radiation intercepted at the solstices (Table 2–1, first line). At first this percentage increases slowly with latitude, together with solar declination at the solstices, so that at 23° 27′ (at the tropics) the annual range of extraterrestrial irradiation is still only about 60% of the winter solstice minimum (third line of Table 2–1). This percentage increases sharply outside the tropical belt, and reaches almost 400% at 50° latitude North and South (line four of Table 2–1). That is, at this middle latitude, the amount of solar radiation at the top of the atmosphere on midsummer day is nearly five times that of the daily value at the winter solstice. (At ground level, insolation at low latitudes, though variable locally, also maintains constantly high values throughout the year, in contrast with the strongly seasonal radiation climate of temperate zones.)

These figures help us to understand one of the most essential climatic characteristics opposing the intertropical regions to the rest of the earth's surface, in other words, the relative annual constancy in daily solar radiation, both in the upper atmosphere and at the earth's surface, and its well-known consequence, the low seasonal variability in mean air temperature. Therefore, tropical climates differ sharply from middle- and high-latitude climates in having very reduced month-to-month variation in both mean temperatures and day length, a fact that surely permit us

TABLE 2–1. Annual range of daily solar radiation at the top of the atmosphere at various latitudes.

Latitude	Daily max cal/cm²/day	Daily min cal/cm²/day	Maximum / Minimum	(Max − min) / min × 100
0°	930	820	1.13	13.4
10°N	905	760	1.19	19.1
23° 27′N	980	610	1.61	60.7
50°N	1030	210	4.90	390.5

Data from List (1971).

to consider tropical environments to be remarkably constant. (Seasons do not have the same meaning in the tropics as in higher latitudes, even when they can be recognized by rainfall patterns or by slight variations in temperature. Summer and winter will be employed here to mean the six-month period when the sun is over the corresponding or the opposite hemisphere.)

Climatic constancy has been utilized to circumscribe the area under tropical climates, either by defining as tropical those regions where differences in temperature between the warmest and the coldest months are smaller than the average daily range, or by applying the criterion of a maximum threshold in annual variation in temperature, usually 10°C. Either criterion produces results that do not differ significantly. Near the oceans, the limits of the tropical zones so defined do not depart much from the tropical parallels, but as one moves to the interior of the continents, increasing continentality and higher seasonality displace the limit of the tropical area toward lower latitudes. The area of tropical climates is thus narrowed.

In contrast to the small variation in daily temperature from month to month, the daily or circadian cycle is quite marked in nearly all tropical regions, making it the major environmental cyclic pulsation. The amplitude between the minimum night temperature and the maximum day temperature (which depends on many factors), is at least 3 times, and frequently more than 10 times, higher than the difference between the means of the coldest and warmest months. Climatic factors such as relative humidity show a regular daily variation inversely correlated with air temperature. As I will discuss later, cloudiness, winds and fogs also vary following a predictable daily pattern according to temperature and air flow induced by heating or cooling of the lower atmosphere.

Besides their year-round constancy in mean temperature, some tropical climates show distinctive patterns in the distribution of precipitation and consequently in relative humidity and soil water availability. Except in the wettest areas where rainfall is more or less evenly distributed throughout the year, or in arid regions (which have a permanent water deficiency), most tropical regions have seasons with heavy rainfall alternating with almost rainless ones. This is particularly true in areas under the direct influence of the trade winds, with their cyclic displacement across the thermal equator. These winds and the atmospheric circulation associated with them are responsible for the widespread occurrence of highly seasonal rainfall climates within tropical latitudes. By contrast, in regions under the influence of the equatorial trough, where air convection predominates, the two rainy seasons correspond with each passage of the sun across the equator, and the two dry seasons coincide with the solstices.

Irrespective of the causes of rainfall seasonality, many tropical areas have either two or four distinct and highly contrasted seasons, including rainless periods that extend from 1 to 6 or 7 months. This annual pulsation may induce slight but significant changes in the temperature regime. Since high cloudiness prevails during the rainy seasons, the total solar radiation at ground level decreases, whereas high relative humidity at night greatly diminishes the coldness due to longwave outgoing radiation from the ground and vegetation. These combined effects decrease the amplitude of daily temperature variation. The opposite conditions prevail during the rainless periods, when low cloudiness, clear skies, and dry atmosphere lead to higher day and lower night temperatures, that is, an increase in the amplitude of daily temperature fluctuations. In this way, seasonality in rainfall brings thermoperiodism, an annual cycle with dampened temperature oscillations and higher night minima during the wet seasons and with greater temperature fluctuations and lower night minima during the dry seasons. I describe later how this variability in daily oscillations of temperature combined with humidity seasons becomes a conspicuous feature of high-altitude climates.

Altitudinal gradients represent the most obvious axes of climatic and ecological variability in the tropical zone. In any large mountain mass, every type of temperature and humidity regime may be encountered along gradients from the warm lowlands to the nival summits. Indeed, the whole range of temperature and rainfall that exists over the nontropical parts of the planet may be found in tropical regions, the values at any tropical site being strongly dependent on that site's altitude above sea level. In wet tropical mountains air temperature decreases at an average rate of about 0.6°C per 100 m elevation, with slight variations according to local conditions. The altitudinal gradient in temperature determines the occurrence of the various thermal belts classically recognized by biogeographers in tropical mountains (van Steenis, 1935; Weberbauer, 1945; Cuatrecasas, 1958; Boughey, 1965).

Without entering into details of vertical zonation, I want to remind the reader that rainfall varies much more than temperature in tropical mountains since rainfall heavily depends on the precise geographic conditions of each mountain system. Lauer (1976) analyzed these patterns along the slopes of various tropical mountain regions. His data show that, as a general rule, the amount of precipitation increases from low altitudes to a maximum at a middle altitude, roughly corresponding to the occurrence of montane or cloud forests, and then decreases more or less steadily to the highest elevations. Later I will consider some of these patterns with particular reference to the upper belts.

Environmental Pulsations in High Tropical Mountains

The position of the tropical belt on the planet and the general circulation of air masses at low latitudes are responsible for the major features of tropical climates. Tropical highlands should therefore have the same kind of environmental rhythm as the lowlands. The relatively high variability of daily cycles, contrasting with the relatively low variability of yearly cycles, are thus the major features of tropical mountain climate. Having emphasized the overwhelming importance of 24-hour cycles, I now want to consider in more detail the specific features that distinguish tropical highlands from tropical lowlands. I will discuss, first the main patterns of temperature and precipitation, then some major trends of geographical variation in the various mountain systems and finally the climatic characteristics of a few well-known localities.

In the tropics high-altitude areas are the sole regions where low temperature predominates and is an important factor in plant, animal, and human life. According to the vertical lapse rate, at about 3000 m mean annual temperature is about 10°C. Average daily ranges remain below 15° in wet mountains up to the altitude of permanent snow (about 4700 m). As month-to-month variation is inconspicuous, the temperature of 10°C roughly corresponds to the climatic boundary between montane and páramo climates, that is, between G and H types in the Koeppen system of classification adapted to mountain climates (Andressen and Ponte, 1973). Irrespective of whether this classification is indeed applicable to tropical mountains, on wet slopes at least this boundary roughly coincides with the first appearance of freezing temperatures, in the form of a few days of frost that can occur at any time of the year.

Frost occurrence can surely be assumed to have great significance as an ecological boundary for tropical, warm-adapted life. On humid slopes, this frost boundary is located between 3000 and 3300 m and corresponds with the upper limit of montane forests, which are replaced by páramo formations (grassland and scrub) higher up. The frequency of frost increases slowly with higher altitudes, to about 100 frost days at 4500 m, and then more rapidly to attain the nival limit between 4700 and 4900 m (Fig. 2–1). On dry mountains, frost starts at somewhat lower elevations, but the limit of permanent snow may be as high as about 6000m. As one approaches the border of the tropical zone, as in the Mexican volcanoes, freezing temperatures begin at much lower altitudes and the altitudinal increase in number of frost days per year is more gradual.

Several outstanding environmental consequences on living organisms follow from these facts. First, temperature remains so low most of the time that adaptations to cold are necessary for survival. Second, perennial species must have mechanisms of frost resistance. Third, these adaptations have to be permanent, since there is no definite growth season in tropical mountains, unlike temperate mountains, and since even with only occasional frosts, the annual frost-free period may be quite short because freezing may occur at any time of the year. This is particularly true in the wettest climates, whereas in sites with a definite dry season, frost days are concentrated in this period. Fourth, as insolation is often fairly low, at least during the rainy periods, and as daylight temperatures remain low also, assimilation and growth in plants may be limited by insufficient light and suboptimal temperatures.

All tropical mountains share the characteristics mentioned above concerning means and oscillations in temperature, but they vary widely in their precipitation regime and hence in humidity conditions. These, in turn, determine the annual regime of frosts. As a general rule, highlands and adjacent windward lowlands have the same rainfall pattern. Thus, in regions influenced by the trade winds, mountain slopes show a clear unimodal seasonality in rainfall, whereas in highlands influenced by the equatorial trough, a distinctive bimodal pattern of precipitation is found. These contrasting precipitation regimes are unrelated to the total amount of rainfall of each locality.

During the rainy seasons an uneven distribution of rainfall is associated with less sunshine, more cloudiness, less total solar radiation, and less heat loss due to terrestrial long-wave radiation. The opposite characteristics prevail during the dry seasons. In this way, a seasonal regime of temperature exists that is obviously not compar-

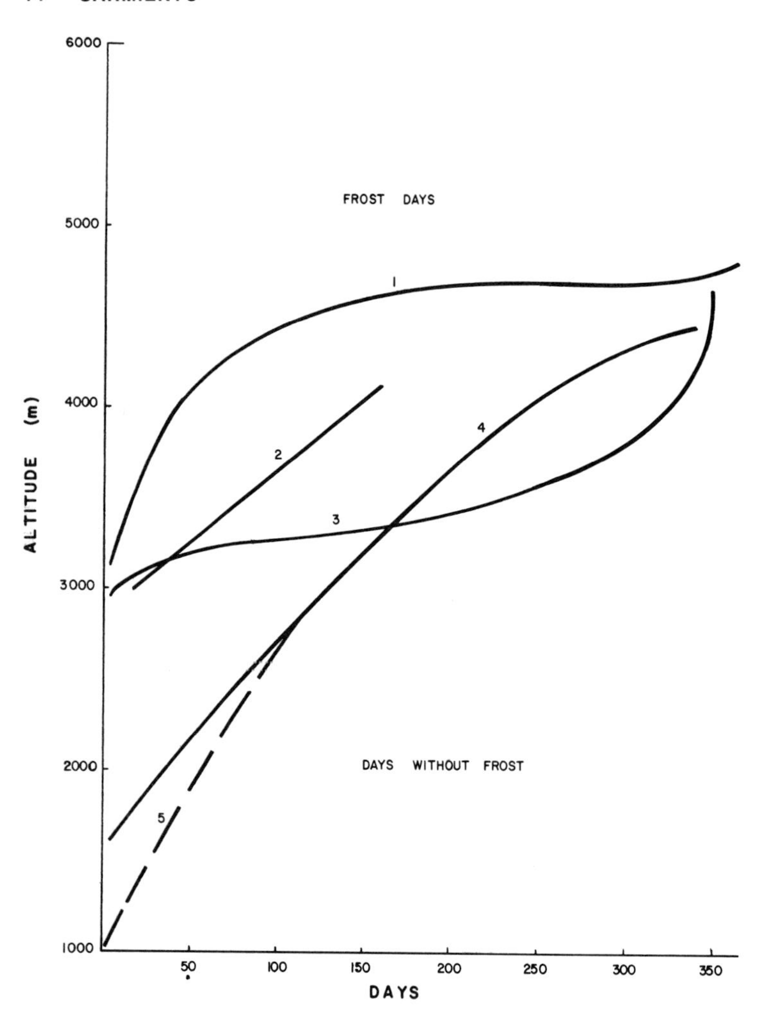

FIG. 2–1. Altitudinal variation in mean number of frost days per year in some American highlands. (1) Sierra Nevada de Mérida, Venezuela; (2) upper Chama Valley, Venezuela; (3) Volcano El Misti, Perú (from Troll, 1968); (4) the Mexican meseta; (5) Sierra Madre Oriental (after Lauer, 1973a, 1973b).

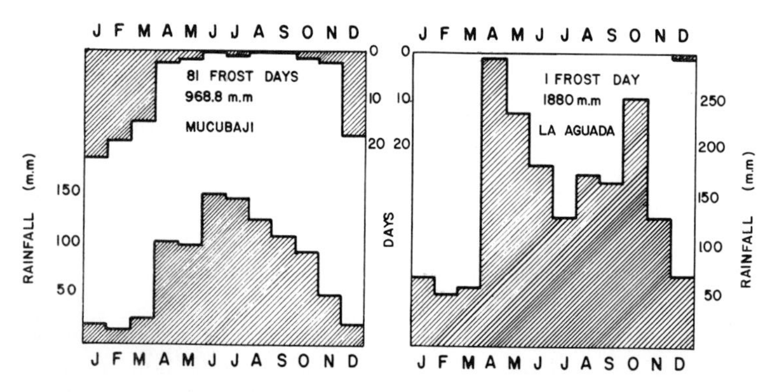

FIG. 2–2. Contrasting patterns of frost distribution in two localities of the Sierra Nevada de Mérida, Venezuela. Both sites are at a similar altitude but differ in amount and distribution of rainfall. Mucubají, the drier site, shows an annual average of 81 frost days, and La Aguada, with almost twice as much rainfall, has a mean of only 1 frost day per year. From Monasterio and Reyes, 1980.

able in intensity with the sharply seasonal mid-latitude climates, but that nevertheless shows a clear annual rhythm since frosts occur mostly during the less rainy periods. A comparison between two sites at similar elevations and only a few kilometers apart in the Venezuelan Andes illustrates this difference (Monasterio and Reyes, 1980). La Aguada has a bimodal rainfall pattern wherein the less rainy months receive more than 50 mm of precipitation, while Mucubají, with a unimodal pattern, has a four-month dry season with less than 100 mm rainfall. Figure 2–2 shows that the dry season at Mucubají is also a very definite frost season, when freezing temperatures occur almost nightly, whereas La Aguada has only very occasional frosts (only three in three years of records, all in December). These differences in frost regime are scarcely reflected in mean monthly temperatures, yet they have great ecological significance. In fact, even in sites with a very strongly seasonal climate, the difference in mean temperature between extreme months is less than 4°–5°C. At Mucubají, for instance, this difference is only 1.2°C; hence seasonality will become apparent only either in the mean minima or in the absolute minimum.

MAJOR GEOGRAPHIC TRENDS IN HIGH-ALTITUDE CLIMATES

After a consideration of the common features of all tropical high-altitude climates, I will provide details concerning the regional patterns of climatic variation within each major tropical highland region. My purpose is not to present a detailed geography of these climates. This would be unrealistic given the paucity of available meteorologic information. Instead, the aim of this section is to sketch the main patterns of climatic distribution and variation to serve as background for the discussions of biogeographic and evolutionary processes occurring in these environments. I start with the Andes, which is the major continuous cordillera crossing the tropics, and will continue with the more discontinuous highlands of Mexico and Central America, tropical Africa, and Malesia.

The Tropical Andes

The Andean cordilleras form a nearly continuous high-altitude chain in western South America, from the southern edge of the tropical zone at 17°–18°S in Bolivia northward to 10°N in western Venezuela (Fig. 2–3). The tropical Andes contain the largest extension of low-temperature areas within the tropical belt of the world. Not surprisingly, in such a huge geographic area that extends latitudinally thousands of kilometers, an amazing variety of climates exists along elevation and latitude gradients. Even restricting myself to the upper belts, I can only sketch a broad picture of environmental variability. As temperature closely depends on the elevation lapse rate, both means and ranges vary with altitude. But as night minima as well as frost frequency are correlated with the timing of rainfall the amount of rainfall and its annual distribution become the two principal climatic factors of ecological importance varying across and along the Andean cordilleras.

Considering altitudinal gradients first, they evidently differ from one latitude to another, and within each main chain, from one slope to its opposite. For example, in the Andes of Mérida (Fig. 2–4), four contrasting slopes have to be considered (Andressen and Ponte, 1973; Monasterio and Reyes, 1980). The southeast slope, facing the llanos, has a unimodal, two-seasonal or tropical rainfall pattern, with a more or less pronounced dry season during the winter, from December to March. Mucubají (Fig. 2–5) shows this type of rainfall pattern. Maximum annual rainfall (about 3000 mm) falls at altitudes between 400 and 800 m (Fig. 2–6). Precipitation gradually diminishes upward to 2200–2400 m, and then more steeply at higher altitudes to less than 800 mm near the summits at 4,200 m. The opposite slope, exposed to the northwest and facing Lake Maracaibo (Fig. 2–4), shows instead a bimodal, four-seasonal or equatorial pattern, with one main dry season in winter and one secondary minimum during the summer months, while the two annual peaks of precipitation correspond to May and

FIG. 2–3. The tropical Andes as delimited by the 2000-m contour. (1) Chita; (2) Aquitania; (3) Chiquinquirá; (4) El Granizo; (5) Chingaza; (6) Popayán; (7) Quito; (8) Izobamba; (9) Cajamarca; (10) Cerro de Pasco; (11) Huancayo; (12) Chuquibambilla; (13) La Paz; (14) Oruro.

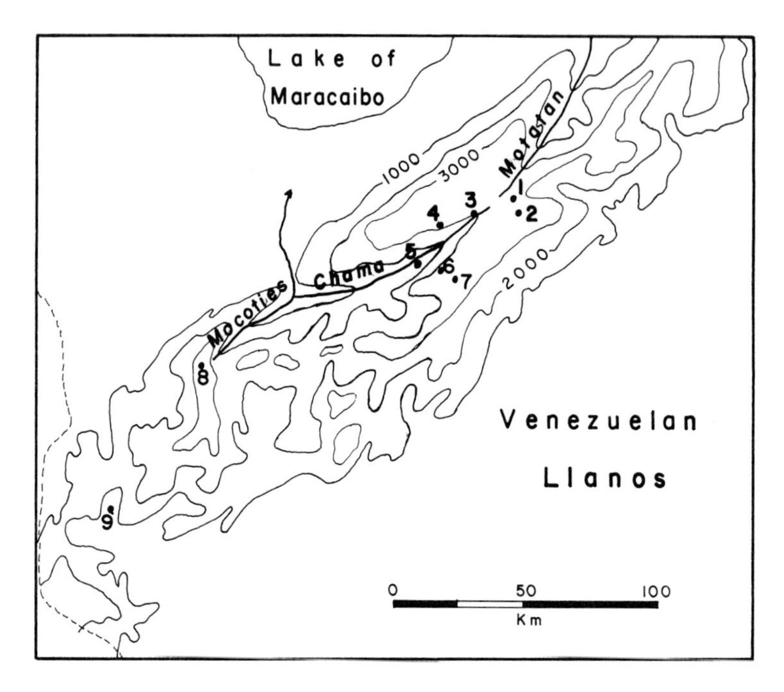

FIG. 2–4. The Venezuelan Andes as delimited by the 1000-m contour. (1) Pico El Aguila; (2) Mucubají; (3) Granja Mucuchíes; (4) La Culata; (5) Mérida; (6) La Aguada; (7) Pico Espejo; (8) Páramo La Negra; (9) San Cristobal.

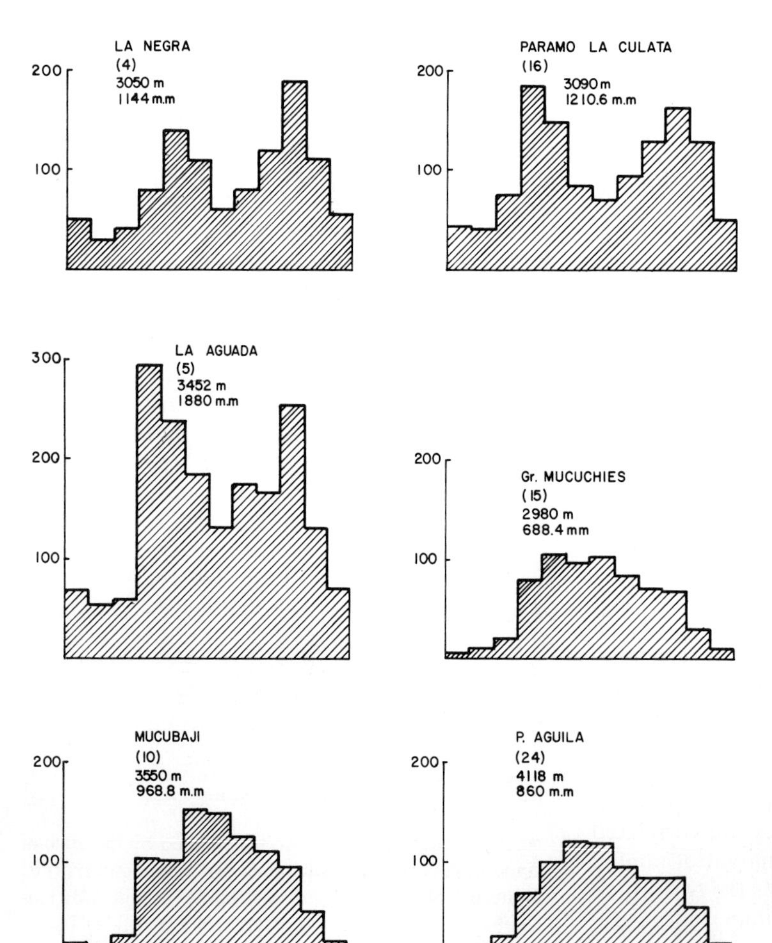

FIG. 2–5. Precipitation patterns in the Venezuelan Andes. Notice the bimodal pattern of La Negra, Páramo La Culata and La Aguada, and the unimodal distribution of precipitation at Granja Mucuchíes, Mucubají, and Pico El Aguila. In both cases the yearly minimum occurs from December to March, that is, in the Northern Hemisphere winter.

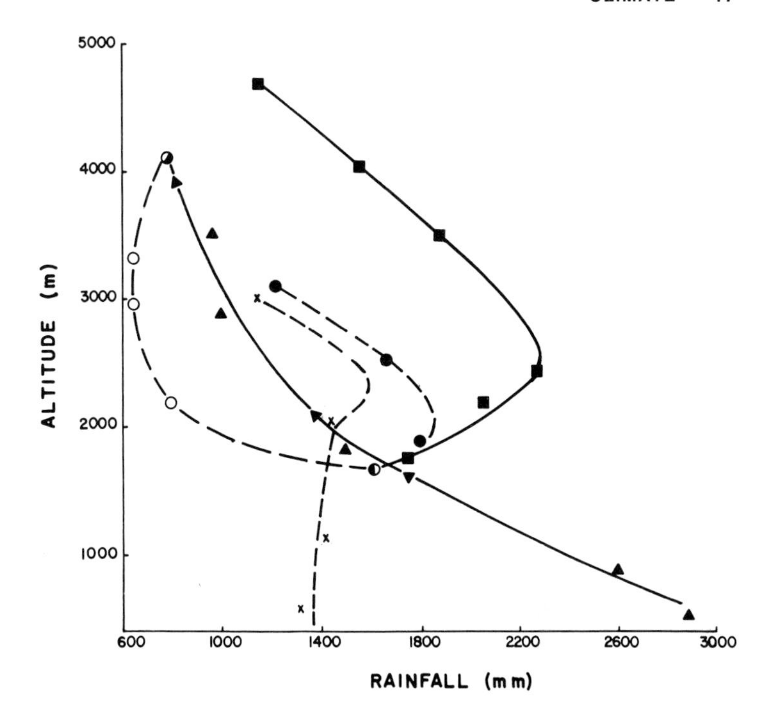

FIG. 2–6. Altitudinal gradients of precipitation in the Andes of Mérida, Venezuela. ▲—▲, SE slope, facing the llanos, where maximum rainfall occurs below 1000 m; ×—×, NW slope, facing the Lake of Maracaibo; ■—■, NW interior slope from Mérida to Pico Espejo; ●—●, SE interior slope from Mérida to Páramo La Culata; ○—○, upper Chama Valley from Mérida to Pico El Aguila.

October, respectively. Páramo La Negra (Fig. 2–5) shows this rainfall pattern. On the Maracaibo side of the cordillera rainfall is about 1400–1500 mm from the lowlands up all along the slope but reaches a small peak between 2000 and 2500 m, where cloud forest occurs (Fig. 2–6).

Furthermore, the Venezuelan Andes are deeply cut by long valleys parallel to the main chains occupying tectonic grabens: the Mocotíes, Chama, and Motatán valleys (Fig. 2–4). Thus, two interior versants appear besides the two external slopes just considered. Both interior versants have a bimodal rainfall distribution like the northwestern side, but the interior northwestern slope is much more rainy than the southeast one, with a maximum of nearly 2300 mm at an altitude of 2400 m in the wettest part of the Chama valley, in the Sierra Nevada National Park. La Aguada (Figs. 2–2 and 2–5), a somewhat higher station within this park, is still a very rainy locality. Precipitation decreases with increasing altitude to 1173 mm at the highest weather station in Venezuela, Pico Espejo at 4765 m, near the glacial tongue of the Pico Bolívar (Fig. 2–6). The southeastern interior versant facing the Sierra Nevada is significantly drier, with a maximum annual rainfall of 1800 mm at about 2000 m, decreasing to 1100 mm at the lower edge of the páramo (3000 m). This amount is nearly 1000 mm less than the opposite slope at the same elevation. Páramo La Culata (Fig. 2–5) is the highest weather station on this slope. As Monasterio and Reyes (1980) pointed out, these contrasting rainfall patterns, together with the vertical and horizontal variation in annual totals, heavily influence minimum temperatures and frost regimes that contribute to the wide diversity of high-altitude environments and to the richness of the biota of the high Venezuelan Andes.

Within the same area of the Andes of Mérida, it is worthwhile to refer to the peculiar situation met with in the upper Chama Valley between 2000 and 3500 m, where under the driest climate of the Venezuelan highlands a diversified agriculture flourishes: irrigation horticulture on alluvial soils, dry farming with potatoes and wheat as main crops on the slopes. In fact, annual rainfall varies from 800 to less than 600 mm with a unimodal pattern where less than 25 mm falls during the four month dry season (Fig. 2–6). Chavez (1962) analyzed the temperature regime in a locality within this area: Granja de Mucuchíes, at 2980 m, where annual rainfall reaches 688 mm (Fig. 2–5). Although the mean monthly temperature varies only from 10.6°C in January to 12.1°C in May, the mean minimum in January is 3.6°C whereas during the rainy season minima always remain above 7°C. In 12 years, an annual average of 16 days with frost was recorded, 15 of them during the dry season. These data corroborate the close association between rainfall pattern and minimum temperature that leads to a sharp decrease in temperature during the nights of the rainless period. In the same area of the Andes, at similar elevations or even 100 or 200 m higher, but with a wetter climate and a less severe

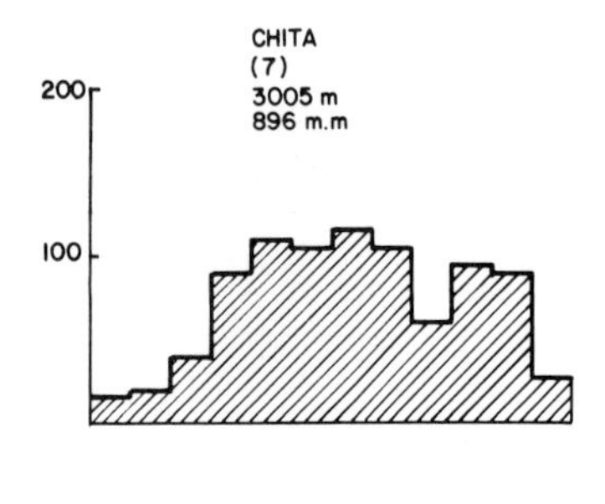

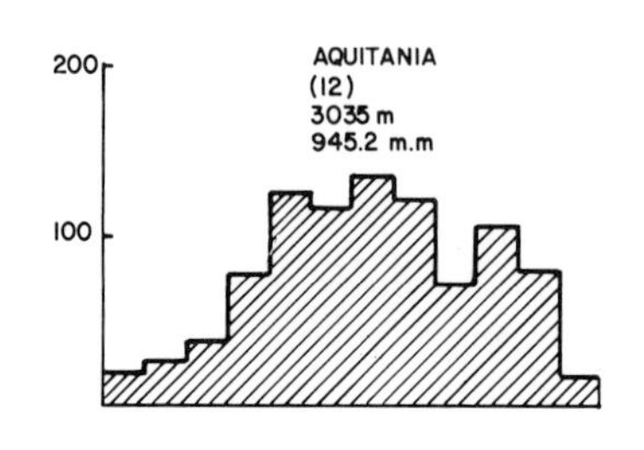

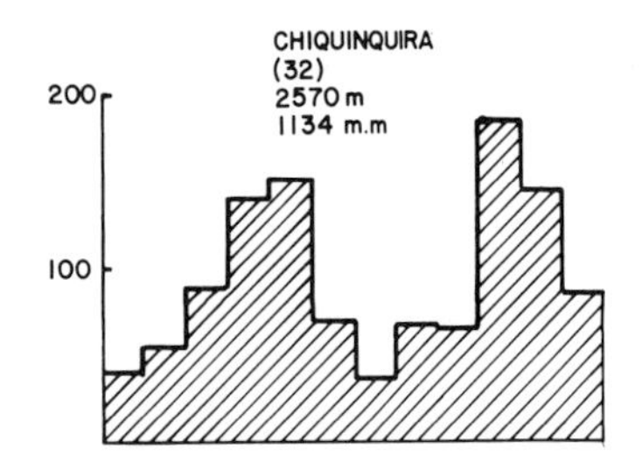

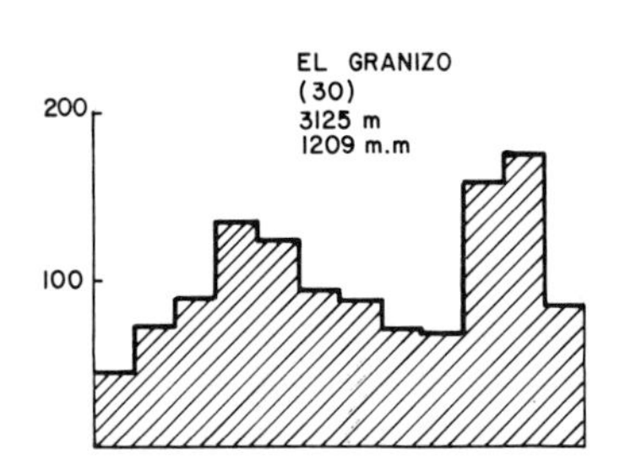

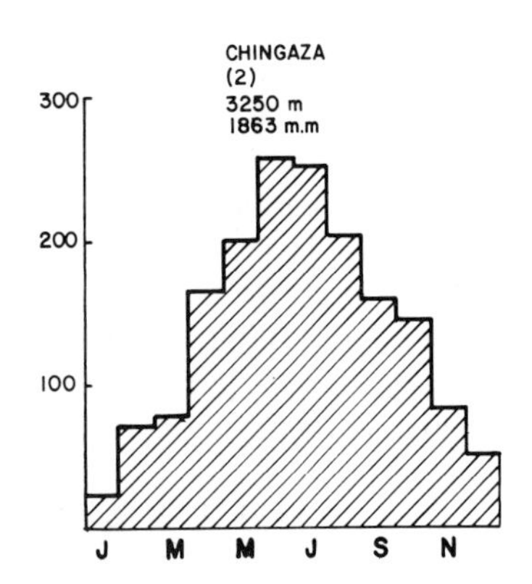

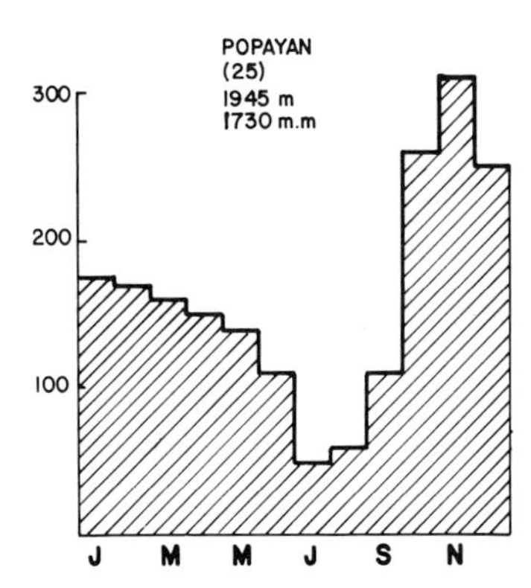

FIG. 2–7. Precipitation patterns in the Colombian Andes. Notice that the bimodal regime with an absolute minimum in the first months of the year, typical of the northernmost localities (Chita and Aquitania), changes to a bimodal pattern with two roughly equivalent dry seasons in the central part of the cordilleras (Chiquinquirá and El Granizo), while Chingaza, on the eastern slope of the Cordillera Oriental, has a unimodal pattern. Popayán, in the southern part of the Cordillera Central, also shows a unimodal pattern but with the minimum during the Southern Hemisphere winter.

drought, there is no frost and cloud forests normally occur. It is not evident what the original vegetation of the dry upper Chama Valley could have been, although the well-known abundance of hard tussock grasses along the roads and in some rangelands and the total absence of trees on the slopes strongly suggest a steppelike plant formation similar to the *pajonales* of the moist Peruvian Andes. This could be one of the rare situations in the generally moist northern Venezuelan Andes where the páramo does not represent the zonal vegetation above the forest, being replaced instead by a punalike grassland.

In the Colombian Andes, an equatorial rainfall regime predominates on the western slope of the Cordillera Oriental (Eastern Cordillera), with the main dry season from December to March and a second scarcely discernible minimum in September. Chita (1 in Fig. 2–3), on the western slope of the imposing Sierra Nevada del Cocuy, and Aquitania (2 in Fig. 2–3), some 100 km to the southwest, may serve to illustrate this pattern (Fig. 2–7). Here the altitudinal belt of maximum annual precipitation extends from 2000 to 3200 m.

The eastern slope of the Cordillera Oriental, facing the llanos, shows an almost biseasonal pattern, since the midyear low is hardly visible or does not appear at all. This windward side, where the air masses coming from the east discharge their humidity, constitutes one of the rainiest parts of the Colombian Andes, with heavy rainfall all the way from their base to their summits. Chingaza (Fig. 2–7) in the páramos to the east of Bogotá, is a representative site from these very wet highlands. Both in rainfall pattern and type of altitudinal gradient, the eastern slope resembles the southeastern side of the Venezuelan Andes, since maximum precipitation occurs at low elevation. Thus, Villavicencio in the piedmont of the Cordillera Oriental at 423 m receives more than 4000 mm of rain (Snow, 1976).

Below 6°N, as we approach the thermal equator, located in this part of South America between 4° and 5°N, the distribution of precipitation on the western slope of the Cordillera Oriental begins to change, stressing first the midyear minimum, so that both dry seasons become comparable in intensity and extension, to displace, farther south, the main dry season to the middle of the year, that is, to the Southern Hemisphere winter. Chiquinquirá in the Altiplano de Boyacá shows this type of bimodal pattern (Fig. 2–7). Weischet (1969) gives a profile through the Cordillera Oriental at 5°N, showing how cloud formation and rains occur all along the eastern slope as well as on both borders of the Sabana de Bogotá. On this plateau at altitudes between 2500 and 2800 m, the rainshadow effect of the eastern peaks decreases the annual totals to 700–1100 mm. However, the summits just bordering the Sabana are somewhat moister, like El Granizo at the Páramo de Cruz Verde (Fig. 2–7), just above Bogotá, while the windward upper side in this mountain area seems to be one of the wettest of all Colombian páramos, with annual totals from 1600 to 2900 mm, the latter figure recorded at Chuza at 3350 m (Cleef, 1981). Chingaza at 3250 m represents one of these moist sites (Fig. 2–7).

Guhl (1968) gives preliminary indications about the climate in the Páramo de Sumapaz, a huge massif in the Cordillera Oriental, immediately south of Bogotá. He concludes from his data that the windward eastern slope is the rainiest, with a maximum of 2368 mm recorded at Santa Rosa at 3400 m (four years). January is the driest month and June the rainiest, when a maximum of 599 mm was recorded in 1964. The lee side of this massif is less rainy, with precipitation from 1200 to 1400 mm between 3200 and 3500 m. Along this slope, the pattern becomes bimodal, April and October being the wettest months and January the driest. Farther downslope, this lee side becomes drier still; rainfall thus attains only 732 mm at El Hato (3100 m), although with the same bimodal pattern. Guhl (1968) remarks that the climatic equator seems to run across this mountain, since its northern part has a typical Northern Hemisphere rainfall regime while its southern areas approach the Southern Hemisphere pattern.

In the Cordillera Central and the Cordillera Occidental, precipitation follows a four-seasonal pattern too, with minimum rainfall during the winter months, but even in these two chains both dry seasons tend to be of similar intensity southward from 4° 30′N. The east slope of the Cordillera Central has a very moist cloud forest belt; thus El Paso at 3265 m receives 2296 mm of precipitation, but rainfall decreases again at higher elevations. The páramo therefore receives between 1100 and 1300 mm, that is about as much rain as falls on the opposite slope of the Cordillera Oriental. The highest recording station at El Ruiz (4200 m) registers 1011 mm of annual precipitation. Weischet (1969) also gives a transect of these two chains at 5°N where he shows the altitudinal gradient on the west slope of the Cordillera Central that faces the Cauca Valley, where rainfall decreases from a maximum of 2800 mm at 1400 m (Naranjal) to 2000 mm at 2700 m (Las Palomas), and to 1200 mm at 3250 m (Esperanza). But at about 4°N this west slope of the central chain becomes very rainy all the way to the top, with an annual rainfall up to 2200 mm in the páramos of the Departamentos del Valle and Cauca. Popayán (Fig. 2–7) is located in the southern part of this moister area. Rainfall becomes still more abundant in the lowlands of the Pacific slope of the Cordillera Occidental, where the highest precipitation in South America seems to occur, but unfortunately there are very few data on the climate of the western slope of the Western Andes.

In southern Colombia, below 2°N, the three major mountain chains converge into one single cordillera. In this area, rainfall distribution appears to be almost unimodal, June, July and August being the less rainy months, that is, the Southern Hemisphere winter. Farther south, the Ecuadorian highlands have a bimodal pattern, but also show a pronounced midyear minimum (Fig. 2–8). I want to point out here that the Colombian highlands are moister than other parts of the tropical Andes. Many areas receive more than 2000 mm of rain, which at these elevations certainly represents a very humid climate. But this is by no means the general situation throughout the Colombian Andes, since in most of the areas above 3000 m less rainy climates prevail, with annual totals in the range of 700–1300 mm.

At about 2°S in the Andes of Ecuador, a broad and dry intermontane valley gives way to a high altitude semiarid area comparable to the Chama Valley in Venezuela. The minimum rainfall in this area is recorded at Pachamama, at 3600 m, with only 382 mm. The landscape has a punalike physiognomy that contrasts with the rest of the Ecuadorian Andes (Johnson, 1976).

To summarize the main trends in the timing of rains in the high Andes from Venezuela to Ecuador, a bimodal pattern with the main dry season from December to March characterizes the environmental rhythmicity from Venezuela to central Colombia. The only exception is the slope facing the llanos, where the unimodal pattern typical of these lowlands extends to the highest mountain summits. At about 5°N in central Colombia, the double rainfall peaks occur in the three Andean chains, but the most pronounced dry season shifts from the first months of the year to midyear. Farther south, the secondary minimum that corresponds to the Northern Hemisphere winter tends to disappear and the climate becomes almost two-seasonal with a major dry season during the Southern Hemisphere winter, even though the region is in the Northern Hemisphere. That is, the Southern Hemisphere rainfall regime extends north of the equator to southern Colombia. This climatic trend is reinforced in Ecuador and culminates in a neat two-seasonal regime in the Peruvian Andes.

From central Ecuador to Bolivia, across the whole extension of the Peruvian Andes, the east–west rainfall gradient becomes by far the main axis of environmental variation, dividing the high Andes in quite distinct ecological zones that are roughly parallel to the chains (Troll, 1968). The Amazon-facing eastern slopes and related highlands conform to the wettest climatic zone, to which correspond, as zonal plant forma-

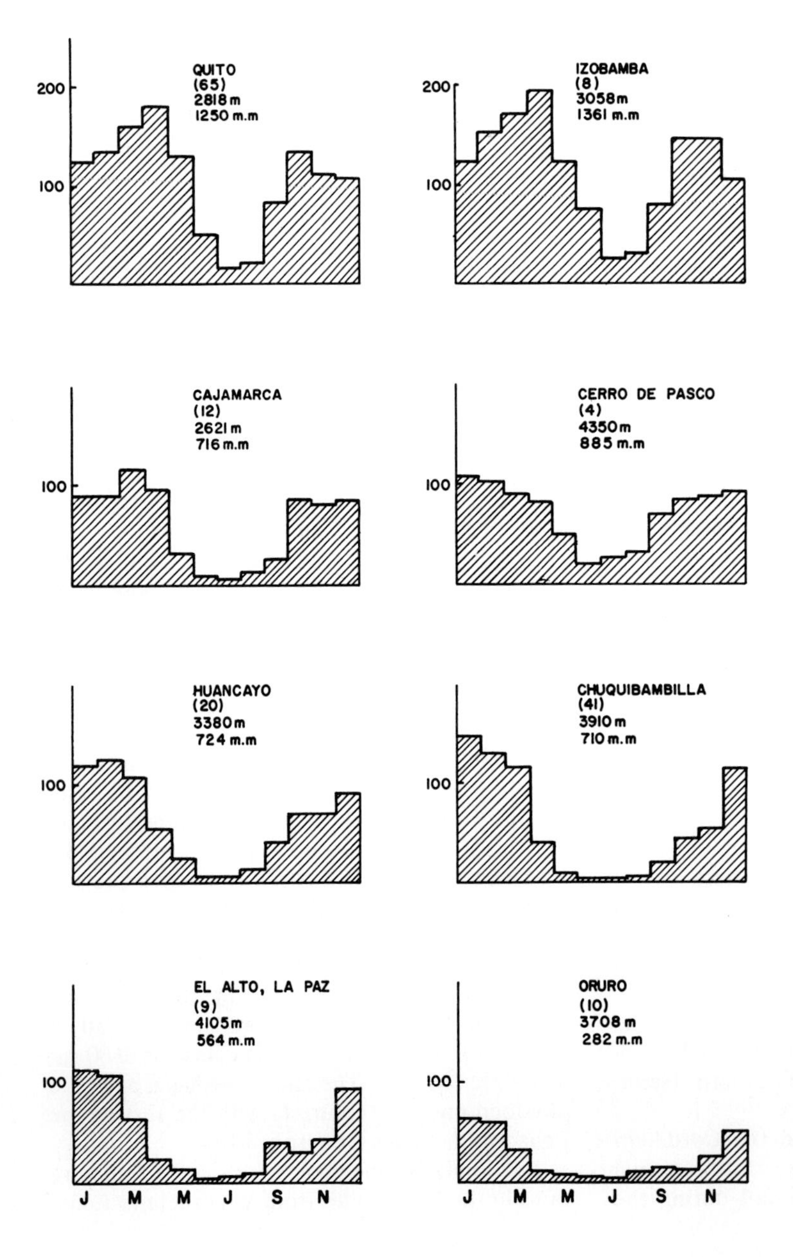

FIG. 2–8. Precipitation patterns in the Andes from Ecuador to Bolivia. The two Ecuadorian sites, Quito and Izobamba, have bimodal regimes with a midyear minimum. The remaining localities in Perú, (Cajamarca, Cerro de Pasco, Huancayo, Chuquibambilla) and Bolivia (El Alto, Oruro) all show a unimodal regime with a pronounced drought during the Southern Hemisphere winter.

tions, first the cloud forests or *yungas*, and above them the mountain grasslands or *pajonales*. High amounts of orographic rain can fall even at high altitudes on the rainside of these eastern chains, as at Talenga (10°S, 3995 m), east of a 5600 m range, with a rainfall of 1968 m (Johnson, 1976). Westward, the Peruvian altiplano or puna lies entirely in the rainshadow of these high chains. Its eastern part, the moist puna, still receives enough rainfall to maintain a rather closed grassy vegetation, while the central and western areas, together with the Pacific slope, are extremely dry and covered with the sparse vegetation of the shrubby dry puna. But disregarding this driest zone, the Peruvian altiplano has a unimodal rainfall regime with a strongly accentuated dry season during the Southern Hemisphere winter from June to August. Cajamarca, Cerro de Pasco, Huancayo and Chuquibambilla (Fig. 2–8), form a north–south sequence of 1500 km along the Peruvian Andes that may be representative of precipitation conditions in the moister part of the altiplano. The eastern border of the puna in northern Bolivia is still drier, La Paz receiving only 564 mm (Fig. 2–8), whereas farther south, the puna becomes exceedingly dry, Oruro receiving only 282 mm. Drought and cold increase southward and westward, reaching their extreme in the Atacama Desert at the boundary of Bolivia, Chile, and Argentina.

According to Johnson (1976), besides the major east–west variation across the central Andes, several climatic gradients are quite apparent along the Andes from Ecuador to Bolivia. The first is a gradient of southward-decreasing total rainfall, smooth from Ecuador to Peru, then more noticeably marked below 15°S. This change toward aridity is accompanied by a second climatic gradient of increasing rainfall seasonality, with a more intense and longer dry season. Thus, in the Andes of Ecuador, about 70% of the annual precipitation falls during the rainy season, whereas in southern Peru and in Bolivia, this figure reaches 80% and even 90%. A third gradient is the gradual occurrence of a cold season during the winter months, giving the climatic characteristics of subtropical latitudes. In Ecuador, the lowest published temperature up to 1969 was −6.8° at Río Pita (3860 m), but in the grassland area of the southern Peruvian altiplano minima of −15° have been recorded (Winterhalder and Thomas, 1978), and in the Bolivian puna a −30° record is not unusual. Further details will be given in the discussion of the climate of the Nuñoa area of southern Peru. A fourth gradient that appears along the Andes relates to rainfall reliability. In Peru and Bolivia, for instance, quite dry years do occur periodically. The coefficient of dispersion in yearly rainfall varies between 10% and 15% in Ecuador and central Peru, to increase to 27% in Cuzco, 19% in La Paz, and 28% in Oruro (Johnson, 1976).

It is not easy to locate the southern limit of the tropical Andes, partly because the eastern border of the Bolivian altiplano is not well known climatologically. It may be said, nevertheless, that between 17° and 18°S, the annual range of temperature sharply increases to 10°C while the daily range increases from a maximum of 25° in the southern Peruvian Andes to 30° or even 40° in the Bolivian puna. Under these conditions of severe winter temperatures, the tropical high-altitude biota give way to a cold-temperate one. An ecological corroboration of a definite change in the environment is the replacement of high-elevation grasslands (*pajonales*) more akin to the páramos of the northern Andes by plant formations of the dry and desert puna.

Climatic Trends in Other Tropical Highlands

Unlike in South America, where a nearly continuous zone of cool climates extends for thousands of kilometers along the tropical Andes, in all other tropical regions high summits may be envisaged as more or less disjunct archipelagos of small cold islands separated by extensive warm lowlands. This is the case in tropical Mexico and Central America, in Malesia, West Africa, and to a somewhat lesser extent in East Africa. I shall consider first Central America, then Africa, and finally Malesia.

Main Climatic Features of the Mexican Meseta and the Central American Highlands

Three mountain areas reach over 3000 m in Central America and tropical Mexico. These are the volcanoes of the eastern Mexican Meseta, namely, Ixtaccíhuatl (5285 m), Popocatépetl (5455 m), Pico de Orizaba (5675 m), and a few others that are several thousand meters higher than the average altitude of the plateau (2000–2500 m). In northern Central America, the Guatemalan highlands culminate in the Tacana (4064 m) and Tajumulco (4210 m) volcanoes. Finally, the most important mountain area in southern Central America is the Cordillera de Talamanca of Costa Rica, with its extensions into Panama. Several peaks in this range rise above the 3000-m level, culminating in Chirripó volcano

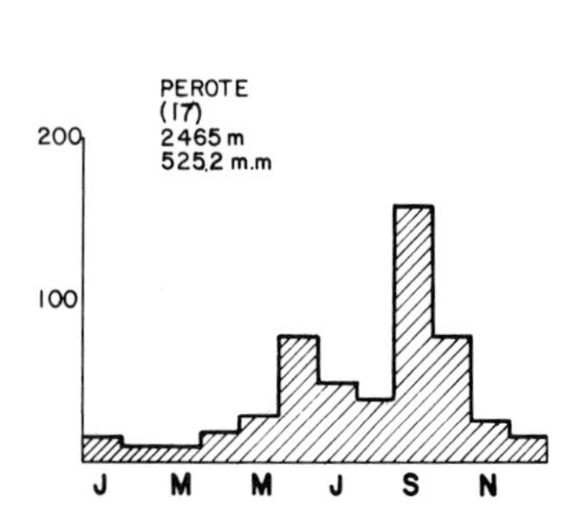

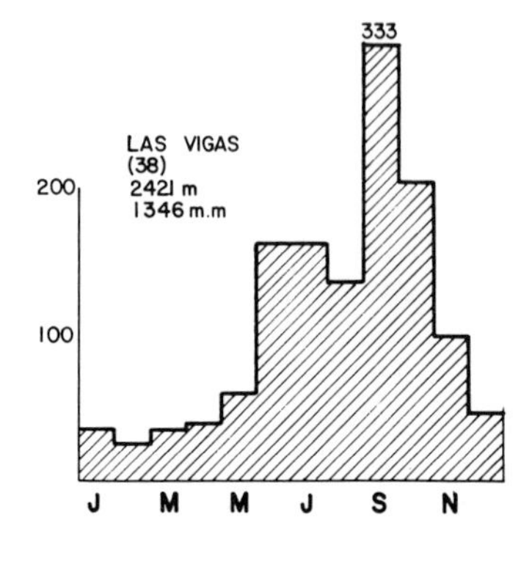

FIG. 2–9. Precipitation pattern in two Mexican localities: Perote, on the windward slope of the Sierra Madre Oriental and Las Vigas, on the eastern meseta, in the rainshadow of the chain. Data from Garcia (1970).

at 3820 m. The Central Cordillera of Costa Rica is hardly lower, its highest peak being the Irazú volcano (3422 m).

Lauer (1973) indicated that the eastern Mexican plateau bears only a rim-tropical character, with a tropical climate during the moist warm season and a strictly extratropical climate during the dry cool season. Cold continental air masses from North America (*nortes*) produce appreciable temperature decreases for several consecutive days during the winter. This influence is particularly noticeable on the northeastern slopes of the Sierra Madre Oriental. But in spite of these periodic intrusions of cold continental air, temperatures are higher over the Mexican meseta than in the free atmosphere over the Pacific and Gulf coasts at similar altitude. For instance the mean annual temperature at 3150 m over Tocubaya exceeds by about 3° that of the free atmosphere over Veracruz (Hastenrath, 1968). The causes of these higher temperatures are not yet clearly understood. The average absolute frost limit lies at about 1900 m on the meseta, while on the eastern slopes of the Sierra Madre Oriental, under more direct and frequent influence of the nortes, this limit probably lies 500 m lower. At 2500 m on the eastern meseta, with January temperatures at 10–12°, there are more than 50 frost days per year (Fig. 2–1), this number increasing at a rate of about 10 days of frost for every 100 m (Lauer, 1973). More details about temperature and rainfall will be given in a later section when I discuss the case of Pico de Orizaba, the highest volcano in this area.

Concerning altitudinal gradients in rainfall, a maximum up to 3000 mm or more is reached at the tropical montane forest belt from 2000 to 2300 m. On Ixtaccíhuatl at 3000 m, rainfall is about 1100–1200 mm, depending on exposure, the annual total decreasing to 800 mm at 4000 m in the pure grassland belt (*zacatonales*) (Lauer, 1978). Rainfall regime is tropical: the central volcanoes receive only 9% of annual rainfall during the winter, and at Orizaba, at the eastern edge of the meseta, the winter rains amount to just 12% of the annual total (Lauer, 1973). The rainshadow effect of the eastern high chains is quite noticeable. Two contrasting situations (Fig. 2–9) are met with, for instance, in Las Vigas (19° 38′N) and Perote (19° 38′N), at about the same altitude (approximately 2400 m), but the former site is on the windward slope of the Sierra Madre Oriental and Perote is on the meseta at the leeward side of Cofre de Perote, one of the highest peaks of this chain (4282 m). Table 2–2 shows the sharp decrease in precipitation in the meseta locality where the annual, January, and May temperatures are higher, both because of the aforementioned effect occurring on the meseta, and because of the far greater insolation reaching this dry locality.

Little is known about highland climates in Central America. The main generalization concerning these peaks is that they certainly are very

TABLE 2–2. Temperature and rainfall in two contrasting highland localities in southern México. January and May are the two extreme months concerning temperature.

	Temperature					Winter
	Mean	January	May	Annual range	Rainfall (mm)	rainfall (%)
Las Vigas (2421 m)	11.5	8.4	13.8	5.4	1346.4	7.2
Perote (2465 m)	12.7	9.5	14.9	6.4	525.2	6.7

Data from Garcia (1970).

rainy and cloudy. At the highest weather station in Costa Rica, Villa Mills at 3003 m in the Cordillera de Talamanca, more than 2500 mm has been recorded. Notwithstanding this high annual total, the regime is neatly bimodal, with four dry months (January to April), but during the rainy seasons rainfall peaks to more than 300 or even 400 mm per month (Weber, 1959).

A last point to remark on, related to the climate of these highlands, is the amazing lowering of the upper tree line from Mexico to Panama (Hastenrath, 1968). On the Mexican volcanoes, it lies at 4000 m, where *Pinus hartwegii* forest gives way to pure grasslands or *zacatonales*. At Mount Tajumulco in Guatemala, some 1000 km farther south, the upper tree line is at about 3800–3900 m. The climatic timberline is not reached in El Salvador, Honduras, or Nicaragua, but farther south, on the southernmost peaks of Central America, in Costa Rica and Panama, the upper tree line is found at about 3100 m only. This poleward rise in the altitude of timberline was associated by Hastenrath with a rise of the same order in isothermal surfaces, as I previously mentioned. But the northward drop in precipitation also seems to influence the elevation of both tree line and snow line.

East Africa, West Africa, and the Sahara

The Ethiopian plateau, with average altitudes over 2000 m, represents the largest highland area of tropical Africa, where several chains and peaks surpass 3000 and even 4000 m, to culminate in the Ras Dashan at 4620 m. As the climates at the highest altitudes in Ethiopia are scarcely known, I will refer mostly to the better known climates of the plateau (Brown and Cocheme, 1973). In East Africa rainfall increases with elevation up to an altitude between 2600 and 3400 m (Nieuwolt, 1974), so one may surmise that the high mountains of Ethiopia would have moister conditions than the lower plateau nearby. On the summits rainfall would probably decrease as is the rule on most mountains.

A major climatic gradient runs from the northeast ranges facing the Red Sea, with 400 mm or less rainfall, to the southwest plateau where annual totals attain 2500 mm. Consequently, the long dry period in the north and east decreases south and west to a minimum at about 7°N. Thus in Asmara (2325 m), in the northern highlands, annual precipitation barely reaches 550 mm and the dry period lasts ten months (September to June). Southward, Gondar (2120 m) has 1250 mm and seven dry months, Debra Markos (2313 m) 1350 mm and five dry months, whereas Addis Ababa (2370 m), 200 km to the southeast, has 1070 mm and six dry months. The most abundant rainfall is recorded in the southwestern plateau; thus Gore (2005 m) receives 2240 mm and shows only two dry months, January and February. But eastward, precipitation decreases again: Goba (2730 m), 300 km southeast of Addis Ababa, receives 730 mm and has four dry months. All these plateau areas exhibit a tropical rainfall regime with summer rains.

Less is known about temperature than rainfall in the Ethiopian highlands. At Addis Ababa, the coldest minima were recorded during the driest month, November, with a mean minimum of 4.2°. Rare cases of frost have been recorded in January (Brown and Cocheme, 1973). The warmest minima were noted during the rains. One may presume that the drier climates also have the greater daily and yearly temperature ranges and the lower limit of frost. In the southwestern highlands (Lauer, 1976), the cloud forest belt occurs between 2000 and 2500 m, where rainfall exceeds 2000 mm, decreasing gradually upward to 1600 mm at almost 3000 m, in the montane or "subalpine" *Hagenia-Hypericum* woodland. Few data are available on the climate of the páramo or afroalpine above this woodland formation, but Lauer (1976) indicated snowfall above 3800 m.

The other high mountains of East Africa, almost at equatorial latitudes, are either isolated volcanoes, like Mt. Kenya (5195 m), Kilimanjaro (5899 m), Elgon (4324 m), and a few others, or short volcanic chains on the Zaire-Uganda-Rwanda border, where the Ruwenzori (5119 m) and the Virunga (4500 m) have several summits above 3000 m. In spite of the very scarce records existing from the higher parts of many of these East African massifs, there appear to be substantial differences between them that are reflected in their vegetation zonation. Hedberg (1951) used this zonation to arrange these mountains in a series according to increasing dryness. Ruwenzori is the wettest mountain, followed by the Virunga volcanoes, Mt. Kenya, Aberdare, Elgon, Kilimanjaro, and Meru. With regard to timing of the rainfall seasons, all these highlands seem to have a bimodal regime with the main dry season in winter; that is, it changes from December to March north of the equator, to June through August in the southern mountains (Fig. 2–10). A more detailed consideration of the climate of Mt. Kenya and Mt. Kilimanjaro is offered in a later section.

It must be pointed out that exposure is a most important climatic factor in the high mountains of East Africa. Figure 2–11 shows the situation around Mt. Kenya. Similarly, on Kilimanjaro,

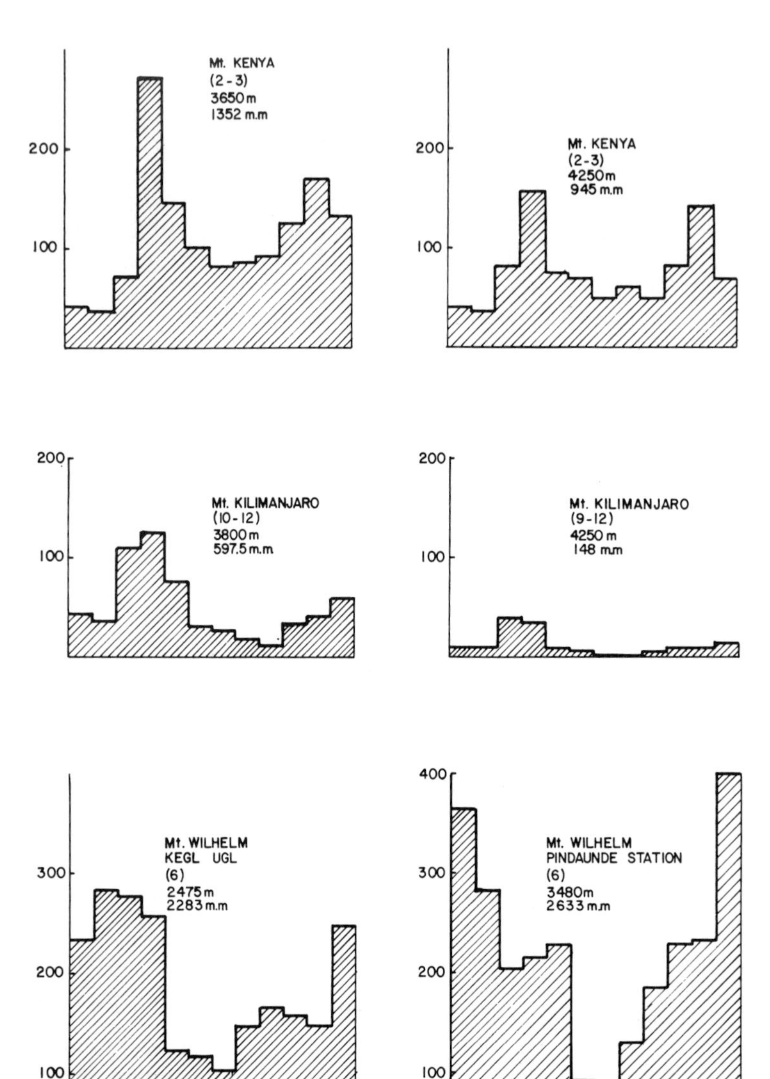

FIG. 2–10. Precipitation patterns in high African and Malesian mountains. Mt. Kenya has an equatorial regime, moist to its highest elevations, whereas Mt. Kilimanjaro, with the same regime, becomes increasingly dry toward its summits. Mt. Wilhelm is fairly humid throughout, though with a well-defined midyear drought. Data for Mt. Kenya and Mt. Kilimanjaro from Hedberg (1964); Mt. Wilhelm data from Hnatiuk et al. (1976).

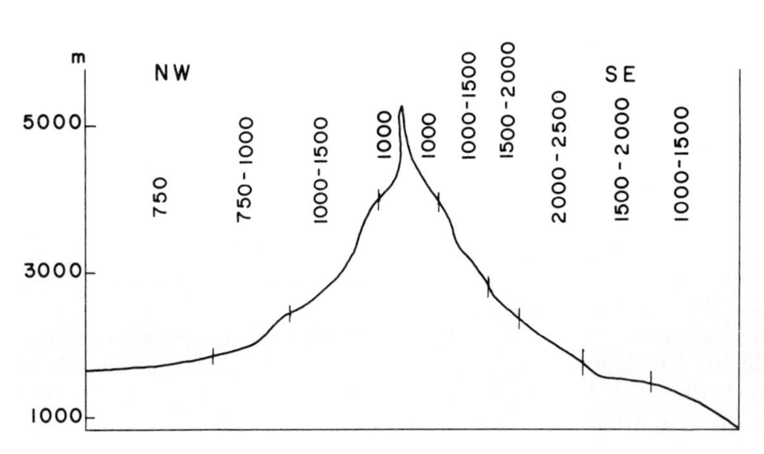

FIG. 2–11. Rainfall profile (in mm) in the Mt. Kenya area. Notice the strong rainshadow effect on the NW slope. Adapted from Thompson (1966).

the southern and southeastern sectors receive at least twice as much rain and have much more cloud during the rainy season, as the western and northern slopes at comparable altitudes (Hedberg, 1951). That is, according to the movement of the humid air masses, not only are the easternmost mountains the wettest, but in every single chain the southern and southeastern slopes are the rainiest.

On isolated Mt. Cameroon (4070 m) in West Africa, the upper belts are exceptionally wet (Lauer, 1976), with more than 8 m of precipitation at 1000 m and more than 2000 mm up to the highest peak. Indeed, this mountain, together with Mt. Kinabalu in Borneo, seems to be the rainiest tropical mountain in the world. Quite the opposite situation is met with in the Ahaggar and Tibesti, two isolated ranges reaching 3000 m in the middle of the Sahara. There, the summits receive only 100–200 mm rainfall (Yacono, 1968).

Malesia

Islands in Malay Archipelago (Java, Celebes, Borneo), and New Guinea all have peaks above 3000 m, but only Borneo and New Guinea have a tropical-alpine belt above the limit of upper continuous forest (Smith, 1977). The mountains of New Guinea are the largest and highest of the whole of Malesia. Their climate has been reviewed recently by Barry (1978a, 1980). Unfortunately, high-altitude meteorological records are lacking, and the only highland site consistently recorded for several years is the Australian National University research station on Mt. Wilhelm. This mountain will be taken as a case study representing high-altitude climates in this region. Besides these data, Allison and Bennett (1976) have given a preliminary characterization of the climate and microclimate of Mt. Carstensz, in West New Guinea, about 1000 km west of Mt. Wilhelm.

Barry (1979) remarked that Malesia is one of the most persistently cloudy regions in equatorial latitudes, a fact associated with the equatorial trough. Mainland New Guinea in particular is an area of maximum cloudiness in all months. Considering the New Guinea highlands as a whole, an east to west gradient of increasing precipitation and perhaps of narrowing temperature extremes seems to exist (Smith, 1980). Mt. Wilhelm (4510 m) has an annual precipitation of about 3000 mm or more throughout its summits (Hnatiuk, Smith and McVean, 1976), whereas Mt. Carstensz, the highest mountain in New Guinea (4804 m), would be rainier still and its temperature regime milder still. In fact, in three months of records, the mean maximum and mean minimum temperatures at 4250 m were 6.8° and 1.5°, respectively, with extremes of 9.3° and 0.1°. According to Allison and Bennett (1976), the most striking feature of the climate of Mt. Jaya, in the Carstensz area, is the lack of any seasonal differentiation. Mean temperatures have a very small annual range, and the diurnal temperature range is also very small: 3.4° at 3600 m, and 2.7° at 4250 m. Furthermore, precipitation and cloud cover probably have little seasonal variation also. This constancy is due to the high moisture content of the air, resulting from the closeness of a warm sea. The minimum daily relative humidity at 4251 m was above 70% in all recorded days.

In spite of the mild character of this highland climate, exceptional periods of frost have been recorded during equally abnormal droughts (Brown and Powell, 1974). In one such year, 1972, when monthly rainfall in most stations was less than 20% of normal values, six frost days were recorded at Tambul (2250 m) and ground frost occurred even down to 1600 m in some valleys. But records are not long enough to give a clear picture of the incidence of extreme temperatures, particularly the frequency of frosts.

Precipitation seasonality is almost absent in western Papua New Guinea toward the border with West New Guinea (Irian Jaya), but in the Eastern Highlands June and July show a relative decrease in rainfall (Fig. 2–10). This pattern also characterizes West New Guinea, but the equatorial pattern occurs only in a narrow sector along the southern edge of the central cordillera (Barry, 1980). Another interesting feature noted by this author is that the decrease of total precipitation with altitude at the highest elevations is less pronounced in the mountains of New Guinea than in many equatorial mountain areas.

Mt. Kinabalu (4101 m) in northern Borneo, is the highest mountain between Burma in Southeast Asia and New Guinea. On the basis of very short periods of observation, Smith (1977b, 1980) suggested that precipitation probably exceeds 3000 mm per year at 3350 m, and it has been estimated to be over 5000 mm at the summit. Rain falls throughout the year, with a wetter season from November to January. Ground frosts seem to be frequent but the absolute minimum temperature recorded at 3350 m was 0.6°. The number of frost days on this mountain is probably the lowest in Malesia. Forest is virtually continuous up to 3350 m, then begins an almost bare rocky summit area, where dwarf trees occur only on less steep slopes and in gullies and crevices up to 3950 m.

Although the climatic data available from the Malesian highlands are scanty, these mountains

TABLE 2–3. Climatic data from selected high Andean localities from Venezuela to Perú.

	Latitude	Altitude (m)	Rainfall (mm)	Evaporation (mm)	Temperature °C				Frost days	Insolation (h)
					Mean	Mean max warmest month	Mean min coldest month	Annual range		
Pico El Aguila, Venezuela	08°52′ N	4118	869	851	2.8	7.8	−1.0	2.7	156	–
Mucubají, Venezuela	08°48′ N	3550	969	763	5.4	11.5	−1.0	1.2	81	1638
Granja Mucuchíes, Venezuela	08°04′ N	2980	688	741	11.5	18.5	3.6	1.5	17	–
Quito, Ecuador	00°13′ S	2818	1250	–	13.0	23.0	7.0	0.3	0	2049
Izobamba, Ecuador	00°22′ S	3058	1361	–	11.5	19.0	4.0	0.8	A few	1895
Huancayo, Peru	12°07′ S	3380	724	–	11.8	20.5	0.5	3.6	62	2488
Chuquibambilla, Peru	14°47′ S	3910	830	–	7.1	18.0	−10.0	6.5	161	–

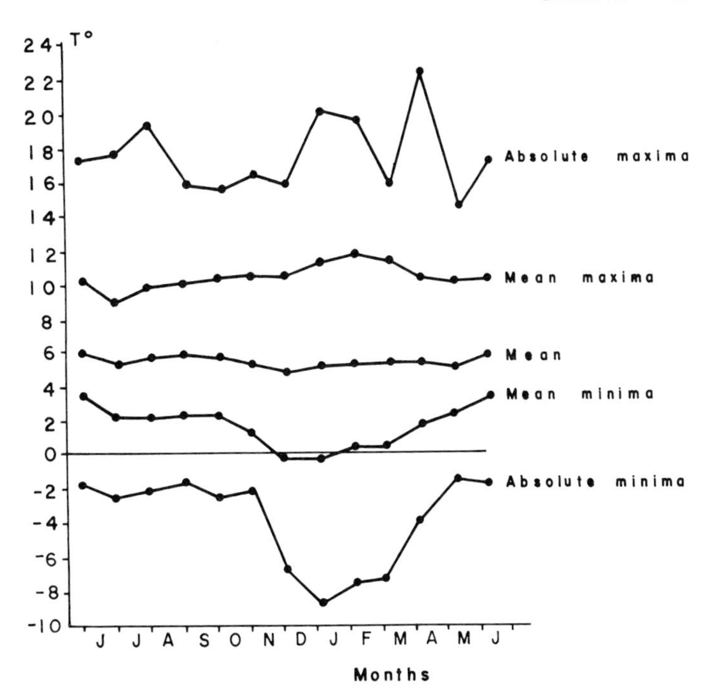

FIG. 2–12. Yearly pattern of temperature at Páramo de Mucubají, Venezuela. Notice the sharp decrease in the minima, especially absolute minima, during the dry season (December to March). From Azócar and Monasterio (1980a).

appear to be among the wettest in the tropics, with very heavy and almost evenly distributed precipitation. Smith (1980a) believes that these climatic conditions, less harsh than in other tropical ranges, may explain the exceptionally high altitude reached by tropical montane forests in these islands. Because the forest reaches such high altitudes, the páramo or tropical-alpine belt is quite reduced even on the highest massifs. It is practically nonexistent on Mt. Kinabalu, both because of a high upper tree limit and a conspicuous bare rocky summit.

SOME CASE STUDIES

Páramo de Mucubají, Venezuela

The meteorological station at the Páramo de Mucubají is located at 8°48′N and at an altitude of 3550 m in the Sierra Nevada de Mérida; it stands in a glacial valley about 900 m below the local summits of the Andes. The valley descends first to the northeast, but a few kilometers down it makes a 90° angle and opens south to the llanos. Mucubají shows the two-seasonal rainfall pattern characteristic of this side of the Venezuelan Andes. The whole area is covered by páramo vegetation, except for some patches of *Polylepis sericea* forest restricted to rocky slopes. Mucubají is at the upper limit of agriculture in the area (potatoes, wheat), although at this elevation field crops are cultivated only once every few years (Monasterio, 1980b).

My climatic analysis is based primarily on the detailed study of Azócar and Monasterio (1980a, 1980b). Two complementary aspects will be treated, namely, the annual variation and the daily cycles. Mean annual temperature is 5.4° (Table 2–3), with a range of only 1.2° between the coldest (December) and the warmest month (June). Monthly mean maxima range between 9° and 11.5° (Fig. 2–12), and mean minima vary from −1° to 3.5°. The highest recorded temperature in nine years was 22.2°, in March, and the lowest was −8.6° in January. The frost regime has already been discussed (Fig. 2–2). On average there are 81 frost days per year, varying between 56 and 96 days from year to year. These frost-change days are fairly concentrated in the driest period of the year, that is, from November to March, when 71 days occur. January has the greatest number, 22 days, and September is the only month totally free from freezing temperatures. Snow has never been recorded at this elevation.

Rainfall distribution is shown in Fig. 2–5. The ten year average is 968.8 mm, with an interannual variability of 13%. Between November and March, rains do not attain 30 mm per month. Relative humidity maintains itself between 60% and 70% during these dry months, but during the eight rainy months it is nearly always above 80%. Annual evaporation is 763.3 mm and calculated annual potential evapotranspiration is 540 mm.

Only during the dry season does monthly evaporation exceed precipitation.

Insolation records for seven years give a mean of 1637.7 sunny hours, that is, an average of almost 4.5 hours per day. During the dry season, insolation is above 200 hours per month (6.6 hours per day), but it decreases to less than 100 hours per month (3.3 hours per day) during the rainy season. Wind speed at 10 m above the ground ranges from 2.5 to 3.5 m/sec.

In summary the annual climatic regime at Mucubají is a fairly seasonal high-altitude tropical climate, with a four-month dry and cold season, when frosts are frequent, and a milder rainy season, with lower maxima but only sporadic freezing temperatures. However, there is no frost-free season, or at least it does not last more than one month. Insolation during the cloudy, rainy season is about half that of the sunny days of the dry season. Wind velocity is low, and does not seem to have a significant ecological effect on plant and animal life.

It is of interest to find other areas of the world with climatic conditions comparable to those of Mucubají. These places are far away from the tropics, in austral South America and its neighbouring islands. Thus in Ushuaia (Argentina) or San Isidro (Chile), at 54–55°S, mean annual temperature lies between 5° and 6°, and annual rainfall attains 800 mm. Similarly, in the Malvinas (Falkland) Islands, at 52°S, temperature and rainfall are within the same ranges. In these austral localities, yearly temperature fluctuation is greater than at Mucubají, although because of their oceanic nature it does not exceed 8°. Wind, however, does play a crucial ecological role, because it is strong and sustained.

Considering now the daily rhythms, I will discuss the average conditions in two contrasting months: January, indicative of the dry season, and August, representative of the rainy season. Azócar and Monasterio (1980b) recorded air temperature at 10 cm above the ground during one year, several hundred meters away from the meteorological station, but at almost the same elevation. In January, the average hourly temperatures ranged between −3° and 17° (Fig. 2–13). After the daily minimum at dawn, air warms quickly as the sun rises, the temperature increasing 12 degrees between 08.00 and 10.00 hours. Between 10.00 and 16.00 hours, under a clear sky, air temperature remains above 14°. The evening cooling is slower than was the morning warming. Average temperature during the sunlight hours was almost 10°. Night cooling is intense due to high reirradiation under a cloudless sky. Average night temperature is about 1°, but below-freezing temperatures persist for eight consecutive hours. Relative humidity (February) increases at night to 90%, while it reached its minimum at noon with values of 30% to 40%. Mean daily evaporation is 2.2 mm.

In August, average hourly temperature varied only between 3° and 8.5°. Because of the lack of direct isolation, either the morning warming or the evening cooling of the air proceeds slowly. The average daylight temperature is only 6.5°, but the mean night temperature exceeds 4°. As

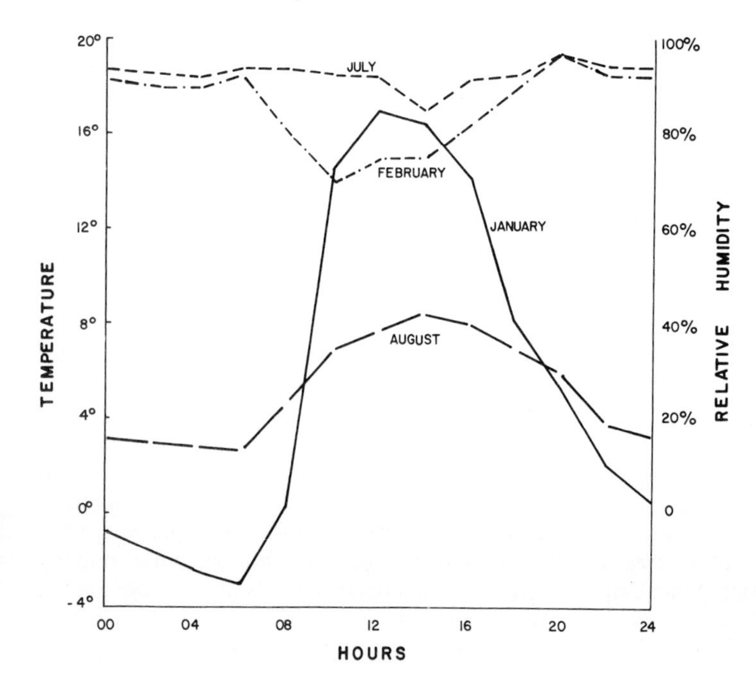

FIG. 2–13. Daily cycles of temperature (—— ; — — —) and relative humidity (– – – – – ; –·–·–·) at 10 cm above the ground in the Páramo de Mucubají, Venezuela. Averages from months of the dry season (January and February) and of the rainy season (July and August). Data from Azócar and Monasterio (1980b).

has already been said, the average number of frost-change days in August is less than one per year. Relative humidity (July) decreased from 90% or 100% at night to about 70% at noon. Mean daily evaporation is 1.8 mm.

During the wet season, high cloudiness and low insolation, together with fogs, mists, and rains, determine a fairly stable moist and cool climate. Both air temperature and relative humidity have low amplitudes. By contrast, differences in temperature and humidity between day and night become sharper during the dry season. In this four-month period, freezing temperatures were recorded during more than half of the nights. Climatic conditions in daylight hours may be comparable to those of clear winter days in many temperate areas.

Pico del Aguila, Venezuela

Pico del Aguila (8°52′N) lies at an altitude of 4118 m in the Andes of Mérida, 5 km as the crow flies from Mucubají. About 300 m higher than this mountain pass, the surrounding summits are known as páramos of Mucuchíes and Piedras Blancas. This high Andean area is characteristically covered with desert páramo vegetation (Monasterio, 1979, 1980a; Chap. 3). The upper limit of agriculture is found several hundred meters below. I refer to Monasterio (1979), who discussed the climate and microclimates of this region.

With an annual precipitation of 869 mm, the unimodal seasonal pattern leaves four dry months, December to March, with only 72 mm rainfall (Fig. 2–5). In 22 years, the interannual variability attained 12.5%. Snowfall is common, particularly during the wet season, but rarely persists more than a few hours. Mean annual temperature is 2.8°. Monthly maxima range between 4.7° and 7.8°, being highest in the dry months. Mean minima vary from −1.0° to 1.1° and are lowest in the dry season. Relative humidity remains close to 100% all day during the rainy months but decreases to minimum values of 20% to 40% on clear days. Annual evaporation, rather high given the prevailing low temperatures, amounts to 851 mm, varying from 111 mm in March to 47 mm in August. Incoming radiation under clear skies frequently reaches a peak of 1.6 or 1.7 cal/cm^2/min, but normally the rapid variations in cloudiness produce numerous sharp oscillations in radiation throughout the day. Monthly totals during the rainy season are in the range of 10–13 kcal/cm^2, increasing to 14–15 kcal/cm^2 in the dry months.

The most outstanding feature of the frost regime in the Pico del Aguila area is its large variation from year to year, which depends mostly on humidity and rainfall. In effect, given that mean temperatures are quite near 0°, a slight increase in the night cooling of the air may produce freezing temperatures. As cooling is directly related to cloud cover and rain, a period with precipitation below the mean has fewer rainy days, and thus becomes a relatively cool period with an increased number of frost nights. On the contrary, a rainier year than the average is also a year with many fewer frost days than the mean. In this way, even a decrease smaller than a couple of degrees in the night temperature may produce a freezing night, increasing dramatically the annual number of frost-change days. For this reason, the number of days of frost differs widely from year to year. Thus, one year with a precipitation 100 mm above the average may have only 100 frost days, whereas the following year with a rainfall 100 mm below the mean may have 200 frost days. The mean number of frost days thus does not have much ecological meaning, its variability being more significant for the survival of plant and animal populations. It should be noted that this frost regime refers to screen temperatures 1.5 m above the ground. The near ground conditions are much harsher and ground frost may be observed on most days throughout the year.

To show the daily cycle of temperature and humidity, I will consider two contrasting days, one in the dry, the other in the rainy season. The average daily cycles in two contrasting months are depicted in Fig. 2–14. A typical pattern during the dry months (for example, March) begins with a clear sky. Minimum temperature before dawn is about −1° and occasionally may be as low as −2°. Air warms rapidly between 08.00 and 10.00 hours, to attain maxima of about 7°–8° between 12.00 and 14.00 hours. Maximum temperatures rarely exceed 9°. The evening cooling is more gradual, so that during most light hours temperature persistently remains above 4°. Relative humidity peaks at 100% in the afternoon as the sky closes and fog may be formed. However, clouds tend to disappear after sunset, and the night is cloudless. Minimum relative humidity may be attained between 24.00 and 08.00 hours, when values as low as 20% or 30% are not infrequent. Evaporation in such a clear day may exceed 7 mm. As an example of daily weather in the rainy season during a typical rainy day in July, the minimum temperature is quite close to or slightly below zero, freezing depending mostly on rainfall duration. Maxima range from 4° to 5°, but the average temperature is 2°. Persistently high

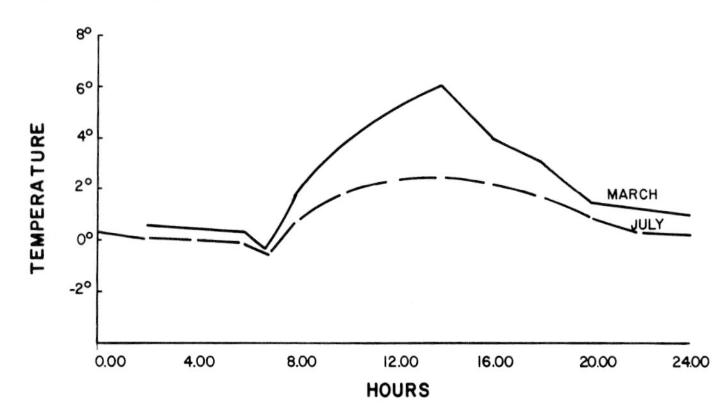

FIG. 2–14. Daily cycles of temperature at Pico El Aguila, Venezuela. Hourly means from a month of the dry season (March) and of the rainy season (July).

humidity is maintained all day. There is hardly any daily cycle, and fairly cold and humid conditions prevail the whole day. Evaporation on a rainy and constantly humid day may be lower than 0.5 mm.

Páramo de Monserrate, Colombia

I mentioned in a previous section that the leeward slope of the Cordillera Oriental above the Sabana de Bogotá has a bimodal rainfall pattern with annual totals of 1100 to 1300 mm. Fig. 2–7 shows the distribution of precipitation at one locality, El Granizo, at 3125 m, in a formerly forested zone now covered with low trees. The páramo begins 100 or 200 m above (Páramo de Monserrate). The midyear dry season is almost equivalent in length and intensity of drought to the first months of the year, which at this latitude correspond to the astronomical winter. Rainfall reliability seems to be somewhat higher than in other highland localities of northern South America, with 17 per cent variation in 30 years of records.

There are very few continuous records of temperatures. Bernal and Figueroa (1980) gave some air temperature data for the Páramo de Monserrate at 3025 m. They recorded temperatures at 100 cm above the ground for nine months. Their data show a mean of 11.5°, a mean maximum of 19.3°, and a mean minimum of 3.8°. January and February are the driest and coldest months, when the absolute minima reach −1° and a few frost days occur.

Bernal and Figueroa (1980) also measured daily cycles of temperature and humidity. Figure 2–15 shows their data for three days in Sep-

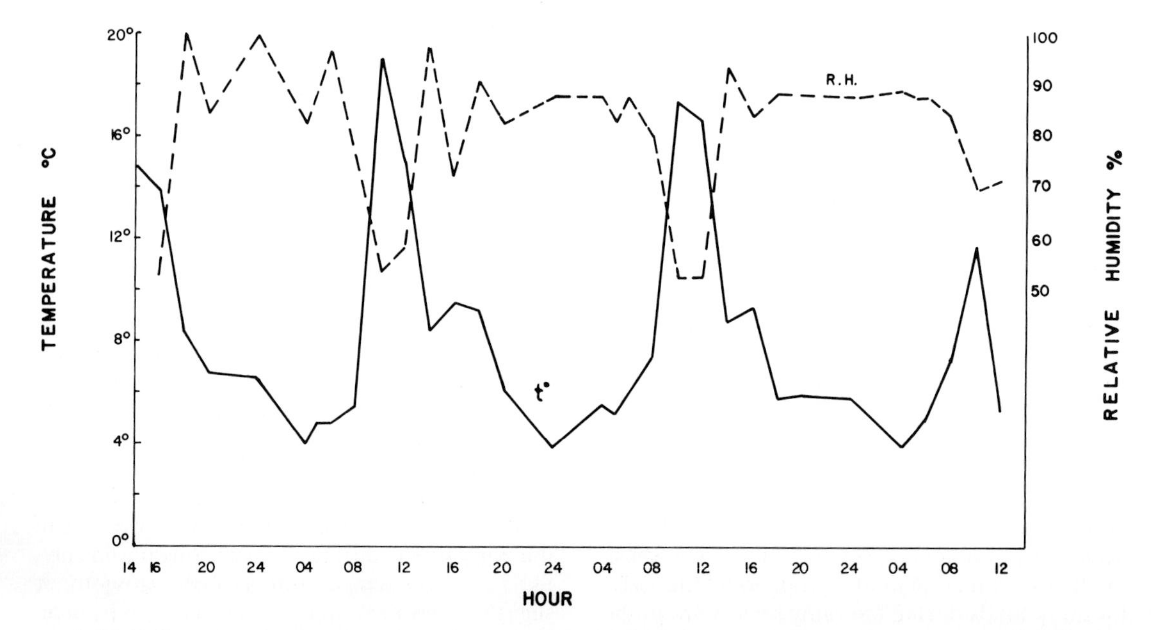

FIG. 2–15. Daily cycles of temperature (——) and relative humidity (— — —) during three days at the Páramo de Monserrate, Colombia. After Bernal and Figueroa (1980).

tember. Minima were above 4° while maxima reached 17° or 18°. Morning warming, as usual in the páramo, was remarkably rapid, whereas afternoon cooling, due to increased cloudiness and fog, began very early in the afternoon and proceeded much more slowly. Relative humidity attains high values, except in the morning when it decreases to 50% or 60%.

These Colombian sites seem to have a milder climate than the two sites in the Venezuelan Andes previously considered, and are closer to the Ecuadorian sites I will discuss in the next section. As was the case along the humid slopes of the Sierra Nevada de Mérida, such as La Aguada, the localities in the forest–páramo boundary either in Venezuela and Colombia or in Ecuador, have very few frost days occurring during the less rainy period of the year. Under these temperature and humidity conditions, the upper limit of cloud forest is exceptionally high for the Andes, since it may reach 3300 m or more. High cloudiness and low sunshine appear to be more crucial limiting factors than low temperature minima and frosts.

Quito-Izobamba Region, Ecuador

Quito (0°13′S, 2818 m) and Izobamba (0°22′S, 3058 m) illustrate climatic conditions in the equatorial Andes (Table 2–3) and permit one to compare two neighboring sites almost at the same altitude, but in two contrasting environments: montane forest and páramo. Quito (Fig. 2–8), situated in an area formerly covered by montane forest, has an average annual rainfall of 1250 mm, with a variation between 890 and 1366 mm in 65 years of records (Johnson, 1976). The pattern is equatorial with the main dry season in July and August, that is, during the Southern Hemisphere winter. As I have mentioned in an earlier section, this midyear drought becomes more pronounced as one proceeds southward through the Peruvian Andes.

Mean annual temperature is 13.0°, with a month-to-month variation of only 0.3° (30 years). Absolute extremes are 28° and 2°, and frost is totally unknown. Relative humidity remains above 70 per cent all year long, with medium to high cloudiness and very low winds. Insolation attains 2049 hours per year, with a seasonal variation from 131 hours per month in April to 212 hours in July. During the rainy seasons, daily temperatures fluctuate between 8° and 21°, while during the driest period of the year the mean daily range varies from 7° to 22°. These figures show how constant the climate is, with daily cycles of temperature and humidity almost similar throughout the year in spite of the seasonal differences in sunshine and rainfall.

Izobamba (Fig. 2–8), situated above the continuous forest line, has an overall climate quite similar to that of Quito, with the same equatorial rainfall pattern but a slightly higher annual total: 1361 mm. Mean annual temperature is 11.5° and yearly amplitude is less than 1°. The absolute maximum and minimum temperatures recorded are 0° (September) and 24° (December and January). September is the only month when occasional frosts may occur. Insolation attains 1895 hours per year, with seasonal variation from 118 hours (April) to 193 hours of sunshine per month (July and August). The mean temperature range in the dry seasons days is 15 degrees (4°–19°) while during the rainy seasons it decreases to 11 degrees (6°–17°).

The two sites thus have quite similar climatic conditions, with the differences due to a 230-m difference in altitude. These equatorial highlands have one of the most equable climates in the South American mountains, with fairly tenuous seasonal rhythmicity given by the rainfall seasons. But however similar the two localities may appear, they represent sharply contrasting ecological zones: montane forest and páramo. The major qualitative difference between the two environments seems to be the total absence of frost from the lower site and its occasional occurrence in the páramo locality, but it is hard to believe that this sporadic factor could have such an important ecological consequence. Nevertheless, the same feature has been repeatedly found on moist Andean slopes, where the frost line runs quite close to the upper forest line.

Altiplano of Southern Peru

The last case study in the Andes is used to discuss climatic features in the moist puna of southern Peru, where environmental conditions are more severe than in Venezuela or Colombia because of increased continentality and the transition toward the extratropical zone. Tosi (1957) analyzed the climate in the Mantaro Valley around Jauja and Huancayo in the central puna. Johnson (1976) gave a general outline of the climates in the Peruvian Andes, while Brooke Thomas and Winterhalder (1976), and Winterhalder and Brooke Thomas (1978) discussed in more detail the climate in the Chuquibambilla-Nuñoa region of southern Peru (see Figs. 2–3 and 2–8, and Table 2–3).

With an annual precipitation in the range of

700 mm to 850 mm, the eastern Peruvian puna, between 12° and 15°S, can be compared with the driest sites in the Venezuelan páramo. However, the altiplano has a more seasonal pattern with a unimodal regime. During the three or four dry months corresponding to the Southern Hemisphere winter, monthly precipitation is very low or even nil. As a consequence of this prolonged period of sharp drought, minimum temperature is much lower than in the driest páramo at comparable altitudes, so that freezing temperatures occur almost daily (Figs. 2–16, 2–19). In the rainy season, frost is less frequent: one to five days per month; thus Nuñoa, at almost 4000 m, has 161 frost days per year. But already at Huancayo (3380 m), with just 62 frost days per year, there is no predictable or extensive frost-free period. Moreover, not only do the number of frost-change days seem to be somewhat greater in the moist puna than in the driest páramos, but the minima may be up to 10 degrees lower.

Strong interannual variability in total rainfall characterizes the altiplano, with certain years well below the mean. At Chuquibambilla, for instance, during one such dry year, minimum temperatures in June and July frequently reached −15°. Although mean annual temperature in the moist puna may be similar to, or even several degrees higher than means in northern Andean localities at the same altitude, suggesting warmer conditions for the altiplano, the daily ranges here are much wider and the night minima lower. At Chuquibambilla, during the dry season, the mean daily range is 25 degrees, with mean minima of −10°. These amplitudes seem, however, to become somewhat smoother with altitude (Fig. 2–19).

Another factor contributes to make dry season temperatures still more severe. Antarctic air masses, after crossing southern South America, penetrate northward to near equatorial latitudes and regularly move through the altiplano. Under such conditions, temperature drops abruptly to −12° or −15°, with daily maxima of only 3° or 4°. This situation may persist for a few days.

Less total rainfall and clearer skies lead to higher annual insolation in the puna than in the páramos. Thus, Huancayo has an average of 2488 hours of sunshine per year, a value almost twice that of the sunshine recorded in the Andes of

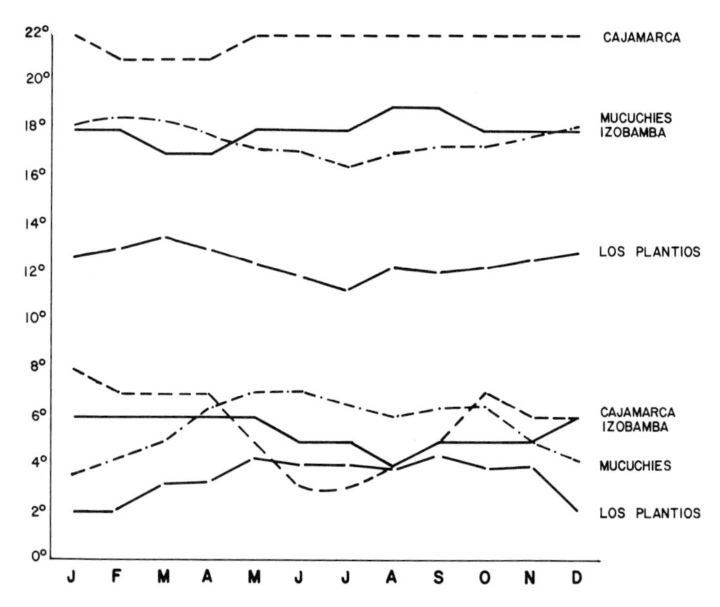

FIG. 2–16. Monthly maximum and minimum temperatures in four Andean localities at about the same elevation (about 3000 m), but differing in rainfall and latitude. Notice how the páramo site (Los Plantíos) has the lowest maxima and also the lowest minima, except during the two rainiest months (June and July). Mucuchíes, a dry site in the upper Chama Valley in Venezuela, and Izobamba, a páramo site in Ecuador, are intermediate in their maxima, whereas the Peruvian locality of Cajamarca has persistently the highest maxima but shows sharp contrast in the minima between the rainy and the dry season.

	LATITUDE	ALT. (m)	RAINFALL (mm)
Cajamarca	07° 08'S	2621	716
Los Plantíos	08° 49'N	2878	1003
Mucuchíes	08° 43'N	2980	688
Izobamba	00° 22'S	3058	1361

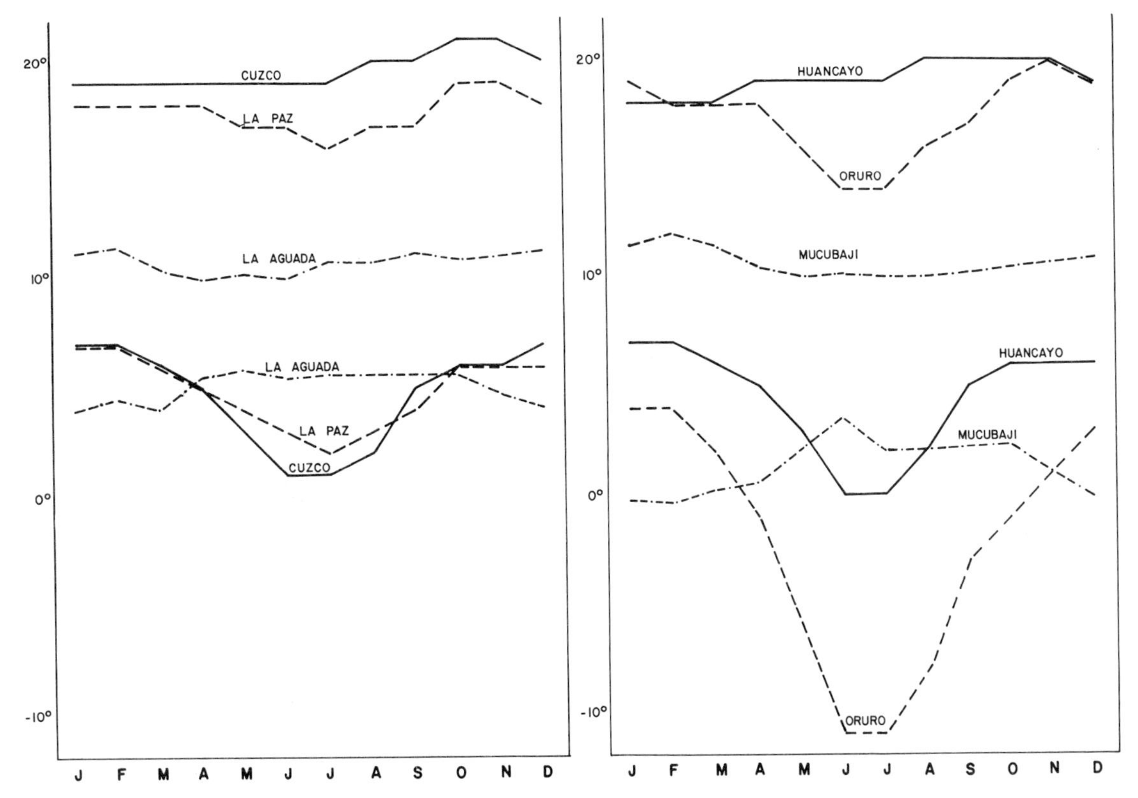

FIG. 2–17. Monthly maximum and minimum temperatures in six Andean localities at about the same elevation (about 3500 m), but differing in rainfall and latitude. Mucubají and La Aguada, in the Venezuelan páramo, have the lowest maxima and rather constant minima throughout the year, whereas Huancayo and Cuzco, on the Peruvian altiplano, and Oruro and La Paz, in the Bolivian puna, have higher maxima as well as a sharp decrease in minima during the winter months. At Oruro, the driest and southernmost locality, the winter decrease also affects the maxima.

	LATITUDE	ALT. (m)	RAINFALL (mm)
Mucubají	08° 48'N	3550	969
La Aguada	08° 35'N	3452	1811
Huancayo	12° 07'S	3380	724
Cuzco	13° 33'S	3312	750
La Paz	16° 30'S	3632	488
Oruro	17° 58'S	3708	282

northern South America. In the high valleys that deeply dissect the altiplano, such as the Mantaro Valley in the Huancayo area, winds may be important, particularly from September to November (Tosi, 1957). These winds have a daily pattern. From a light breeze in the morning, wind increases in the afternoon to average speeds of 5–12 km/hr, to die down later and end up in calm nights. The dessicating effects of wind necessitate the establishment of tree barriers around field crops.

To summarize and compare the environmental conditions prevailing in the Peruvian altiplano and those in the northern Andes (Figs. 2–16 to 2–19), I conclude that the moist eastern puna has a harsher climate induced both by the intensive drought of the midyear months and by the much lower temperatures of the dry season. At similar altitudes, puna localities have similar or higher means than páramo sites, but minima are many degrees lower. The dry season is much more severe in the puna than in the northern Andes and may well be called "winter," at least from the viewpoint of temperature, the puna being the sole tropical region in South America with a truly cold season. A complementary feature reinforcing the dry and cold character of the moist puna climate as compared with even the driest páramos is the higher unreliability of rainfall. Not unfrequently, in exceptionally dry and cold years temperatures reach the lowest values recorded in tropical mountains. It is clear that a major ecological gradient exists from the moist páramo of the Sierra Nevada de Mérida or of the eastern slope of the Colombian Cordillera Oriental, through the drier páramo, the moist puna, the dry puna, and finally to the desert puna of western Bolivia and northern Chile. The latter area lies outside the tropics and represents the most extreme type of highland environment in South America.

Pico de Orizaba, Mexico

The climate of the upper slopes of this great Mexican volcano (19°N, 5675 m) was analyzed by Lauer and Klaus (1975). They recorded several climatic elements at four elevations, from March 2 to 26 in 1974. Table 2–4 shows a typical daily cycle at each of the four sites for a sunny day and a rainy one, respectively. The altitudinal gradient in mean annual temperature is linear between 10° at 3000 m and 0° at 5000 m, whereas the number of frost days increases from 120 to 360 (Fig. 2–1). According to Lauer and Klaus (1975), a decrease with elevation in the maximum and minimum temperatures and in daily temperature range appears to be a valid generalization. It is interesting to notice that at timberline, formed by *Pinus hartwegii* forest at about 4000 m, there are about 200 frost days per year (Fig. 2–1). On the basis of data from the nearby Nevado de Toluca at 4120 m, these authors conclude that on these volcanoes freezing temperatures may occur throughout the year but with increased frequency during the dry winter months, whereas April and May are the months with fewer frost days. They also noticed the increase with elevation in the number

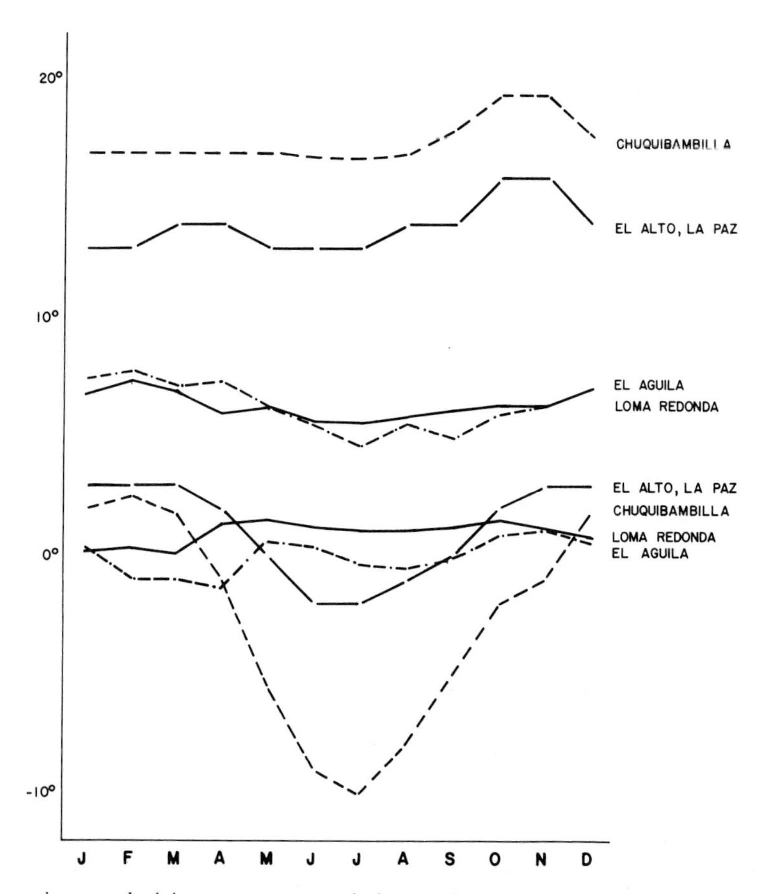

FIG. 2–18. Monthly maximum and minimum temperatures in four Andean localities at about the same elevation (about 4000 m) but differing in rainfall and latitude. The two sites in the Venezuelan páramo, El Aguila and Loma Redonda, have the lowest maxima and rather constant minima throughout the year. Chuquibambilla in the Peruvian puna and La Paz (El Alto) on the Bolivian altiplano, have higher maxima and a sharp decrease in the minima during the winter months.

	LATITUDE	ALT. (m)	RAINFALL (mm)
El Aguila	08° 51'N	4118	860
Loma Redonda	08° 35'N	4045	1553
Chuquibambilla	14° 47'S	3910	830
La Paz (El Alto)	16° 30'S	4105	564

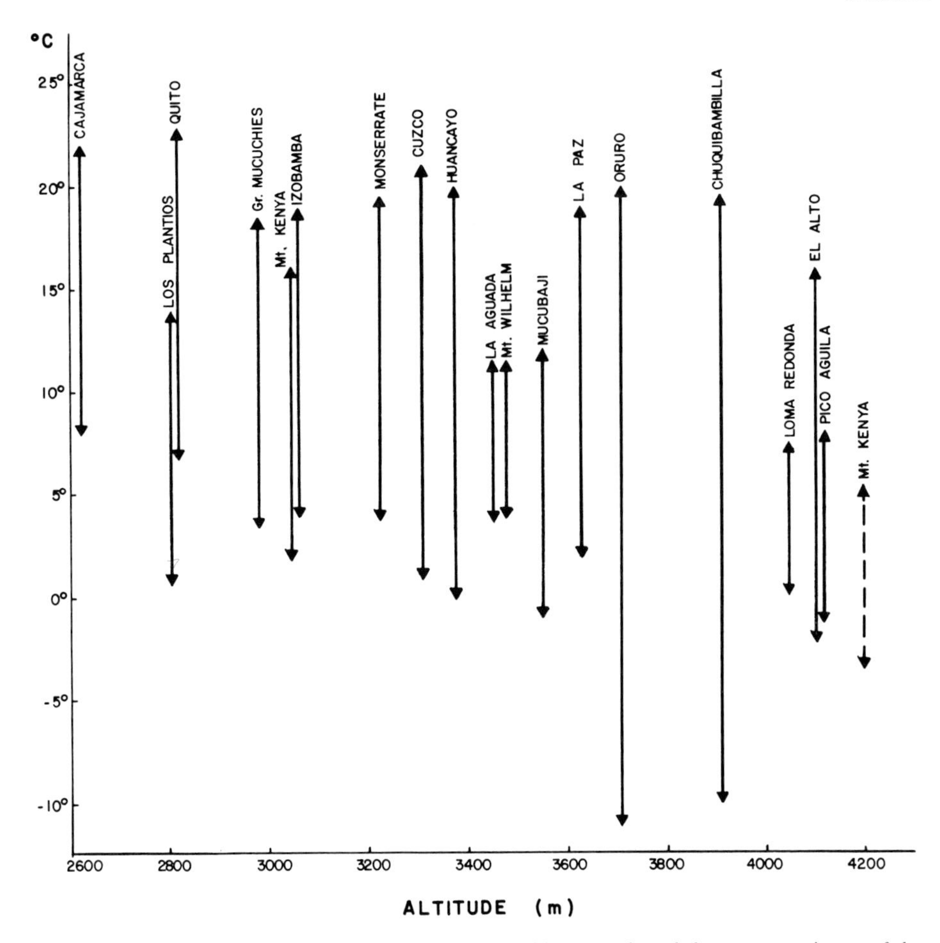

FIG. 2–19. Temperature range between the mean minimum of the coldest month and the mean maximum of the warmest month in various tropical highlands, arranged along an altitudinal gradient. Notice how the Venezuelan páramo sites, Los Plantíos, La Aguada, Mucubají, Loma Redonda, and Pico El Aguila, together with Mt. Kenya and Mt. Wilhelm, show the lowest maxima and the highest minima (i.e., the lowest annual ranges). By contrast, the puna sites, like Cuzco, Huancayo, Chuquibambilla, La Paz, Oruro, and El Alto, have the highest maxima and relatively lowest minima. The páramo sites in Colombia (Monserrate) and Ecuador (Izobamba) as well as the dry Venezuelan locality Granja Mucuchíes, show an intermediate pattern.

TABLE 2–4. Twenty-four-hour range in air temperature (°C) and relative humidity (%) along an elevational gradient on Pico de Orizaba, México.

Elevation	3480 m				3990 m				4250 m				4690 m			
	Temp.		RH		Temp.		RH		Temp.		RH		Temp.		RH	
	max	min	max	min	max	min	max	min	max	min	max	min	max	min	max	min
March 3	17°	−1°	30	15	16°	−5°	50	20	8°	−1°	30	15	5°	−6°	90	60
March 25	17°	5°	50	20	16°	3°	50	10	8°	0°	50	15	5°	−3°	90	30

Two days with contrasting weather have been taken as examples: March 3, a sunny day, and March 25, a cloudy and rainy day.

Data from Lauer and Klaus (1975).

of frost hours per day, from 1 hour or less at 3000 m to 24 hours at 5000 m in the glacial area.

As discussed in an earlier section, the climatic character of the Mexican highlands is transitional between that of tropical and that of temperate mountains. On Orizaba, the annual range of temperature between the warmest and coldest months is still low: 5.5° at 3000 m, decreasing to 2.5° above 4000 m (Lauer and Klaus, 1975). Thus, from the viewpoint of annual constancy, these mountains could be considered tropical. But the frequent incursions of cold air masses from North America, and the sharp decreases in temperature that accompany them, lead to sharp seasonal differentiation and particularly a dry and cold season during the winter months. The biota of the Mexican highlands show more affinities with boreal than with tropical floras and faunas. This may be seen either by the occurrence of conifers as the dominants of the upper forest belt, or by the dominance of boreal grasses in the mountain grasslands (*zacatonales*). A true páramo belt is lacking.

It would be enlightening to compare the Mexican volcanoes with Andean highlands at similar latitudes and elevations and with comparable annual rainfall. Unfortunately, a similar situation does not exist in South America, since at 19°S the Andes are much more arid than the mountains of southern Mexico. A comparable situation might be found in the eastern cordilleras of southern Peru and northern Bolivia between 14° and 17°S, but unfortunately these areas are not well known climatically. The only valid comparison that might be made, then, is between the respective plant formations. In both Mexico and Peru–Bolivia, hard tussock grasses predominate above the continuous forest, but in South America below this high-altitude ecological zone, tropical montane forest occurs. In Mexico and northern Central America, coniferous forest and mixed formations occupy the slopes between the timberline and 1800–1900 m, where a richer forest with more tropical elements appears.

This contrast in vegetation suggests a close correspondence with the different elevations where frost first occurs: near 3000 m in the Peruvian Andes, but less than 2000 m on the Mexican altiplano (Fig. 2–1). In southern Bolivia and northern Argentina, which lie outside the tropical climatic area, the eastern slopes of the mountains are moist enough to support a montane forest (though under a climate with frequent winter frosts). A coniferous forest belt, in this case a pure *Podocarpus* forest, is interposed between the subtropical rain forest on the lower slopes and the puna grassland on the summits, thus repeating the zonation on the Mexican volcanoes. One may suggest therefore, that the coniferous forest occupies an altitudinal zone where frost gradually increases from only a few frost days to more than half frost days per year. This ecoclimatic zone does not occur on mountains at low latitudes, where the frost gradient is abrupt. Thus this formation appears only at the two borders of the tropical altitude zone: in Mexico and Central America in the north, and in Bolivia and Argentina in the south.

Mt. Kenya, Kenya

The huge isolated massif of Mt. Kenya (5195 m), located on the equator, displays a variety of climates according to differences in altitude and exposure. The southeast slope is the rainiest, with a mean gradient of about 2 mm per meter of altitude between 1400 and 2200 m, where a maximum of 2500 mm is reached and maintained for another 1000 m, to decrease again at a rate of about 3 mm per meter, to the summit area, where the estimated amount is around 850 mm (Thompson, 1966). Northwesterly slopes have, in contrast, only about half this amount of rainfall. Figure 2–11 shows a rainfall profile adapted from Thompson (1966). On all sides of the mountain, the precipitation pattern is equatorial, but on the windward side the main dry season is June through August, and on the lee side and on the adjacent western plains, in the rainshadow of the massif, the driest months are January and February (Fig. 2–10).

Little is known about the temperature regime of Mt. Kenya. Coe (1967b) gave one-month records at three altitudes (Table 2–5). These figures show rather low absolute and mean minima, suggesting that the climate has a continental character. Furthermore, the decrease in temperature range with altitude clearly appears along this gradient. Coe (1967b) also remarked that one of the most conspicuous factors in the temperature regime is the great speed at which temperatures fluctuate near the ground. Under clear skies, the ground temperature rises sharply away from that of the air, but in a similar way, when the sky becomes clouded, it falls rapidly until air and ground temperatures are almost equal. This pattern is repeated constantly during the day. Hedberg (1964) also remarked on the rapid temperature changes frequent in the afroalpine climates, due largely to the thin atmosphere with low heat capacity and variable cloudiness. He noted that changes up to 10° in half an hour are not rare.

TABLE 2–5. Temperature data (in °C) on the slopes of Mt. Kenya, recorded between December 19, 1957, and January 17, 1958.

Altitude (m)	Mean	Mean max	Mean min	Average daily range	Max daily range	Min daily range	Absolute max	Absolute min
3048	7.4	16.2	1.7	14.5	20.5	7.8	19.5	−1.5
4191	2.0	5.5	−3.6	8.9	15.0	7.2	11.0	−6.7
4770	−1.9	1.2	−3.9	5.1	10.0	2.8	5.0	−8.3

From Coe (1967b).

In summary, Mt. Kenya is an equatorial mountain with rather wet, continental, highland climates, the dry seasons are short, and the number of humid months varies from 9.5 to 11 (Hedberg, 1964). The rainfall regime and amount of rainfall on this mountain closely resemble the environmental conditions of most of the Venezuelan Andes, as can be seen from the previous analysis of the climate of the páramos of Mucubají and Piedras Blancas.

Mt. Kilimanjaro, Tanzania

Mt. Kilimanjaro (3°S, 5899 m) is another imposing isolated volcano. At its highest elevations it displays ecological conditions that contrast sharply with those found on neighboring Mt. Kenya. Rainfall, for instance, decreases rapidly with altitude, resulting in such a barren landscape that the uppermost belt on this mountain has been considered an alpine desert (Salt, 1954). Mt. Kilimanjaro is one of the best known African mountains with regard to precipitation, because nine rain gauges have been in operation in the upper areas since the 1950s, allowing one to distinguish its striking climatic contrasts.

The rainfall regime is equatorial throughout, with the lowest monthly minima during the Southern Hemisphere winter, while the rainiest months are March and April (Fig. 2–10). Hedberg (1964), using the information available up to that time, showed the altitudinal variation in annual precipitation on the two contrasting slopes, the windward southeast one and the leeward west side (Fig. 2–20). On the southeastern slope, rainfall decreases from a maximum amount at 2200 m to a minimum at 4250 m, where just 203 mm has been recorded. Precipitation increases again to more than 500 mm to decrease finally to an estimated annual amount of 15 mm at the summit. On the west slope, precipitation decreases with a still steeper gradient from 935 mm at 4000 m to 84 mm at 4750 m. In accordance with the extreme dryness of the summit, the snow line is reported to be at 5400 m, about 800 m higher than on Mt. Kenya. The glaciers descend to 4500 m along the moister southeast slope, but they occur only at 5700 m on the drier northeast slope (Hedberg, 1964).

Although at 300 m there are eight to ten humid months, at high altitudes only two months, March and April, could be considered humid. But in spite of the rainless nature of the summits, Salt (1954) remarked that precipitation does occur in the form of dew and no doubt contributes to the maintenance of a favorable water budget in high-altitude plants. One may infer from Hedberg (1964) that plant formations on Kilimanjaro resemble more closely the puna of the drier eastern Peruvian altiplano than the páramos of the moister northern Andes, or than neighboring Mt. Kenya. Thus, the characteristic woodlands of giant rosette trees are quite reduced and localized in wet alluvial soils of the high valleys of Kilimanjaro, whereas, as in the puna, the slopes are covered by hard tussock grasses and dwarf shrubs.

As is normally the case in dry climates, inter-

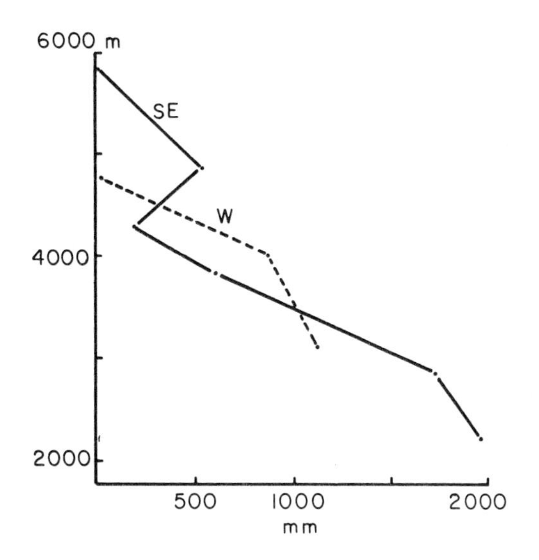

FIG. 2–20. Altitudinal variation in precipitation on the two opposite slopes of Mt. Kilimanjaro: the SE windward slope and the W rainshadow slope. Data from Hedberg (1964).

annual variability in precipitation is quite conspicuous on Mt. Kilimanjaro. On the basis of 6–12 years of records, the ratio between maximum and minimum rainfall attained the extreme value of 40 at 4750 m, thus indicating a tremendous unreliability that can be compared with that found in the Bolivian puna.

Very few data are available on temperature conditions at high altitudes. Klute (1920), as reported by Coe (1967a), recorded temperature at 4160 m during 46 consecutive days. Mean temperature was 1.8° with a daily amplitude of 10.1°. These few figures seem to confirm the rather extreme nature of the Kilimanjaro highlands, with wide daily ranges characteristic of climates with clear skies, low cloudiness, and high sunshine and radiation.

In spite of the much reduced extension of the afroalpine belt as compared with the páramo and puna of the South American Andes, a remarkable diversity of high-altitude environments and their corresponding biota is also apparent in Africa. This diversity constitutes an insular replica of the temperature and humidity gradients fully expressed in the Andes from the wettest and mildest páramos of Venezuela and Colombia, to the driest and most continental puna landscapes of the Peruvian and Bolivian Andes.

Mt. Wilhelm, New Guinea

Mt. Wilhelm (5°45′S, 4510 m) is one of the highest peaks in New Guinea. Hnatiuk, Smith, and McVean (1976) discussed the climate of Mt. Wilhelm on the basis of several years of records at the Australian National University Research Station at 3480 m, together with a few additional measurements at several other sites from 3215 m to 4400 m. A first singularity of this mountain is that the upper limit of natural forest is found at 3800–3900 m (Smith, 1975), that is, at least 600–700 m higher than in most other wet tropical mountains. Such a high timberline has been correlated with the rainy climate of this mountain: more than 3400 mm of rainfall at 3480 m and around 2900 mm at 4380 m. This factor may also cause the low upper altitudinal limit of agriculture, at about 2500 m.

Rainfall shows a pronounced single maximum during the Southern Hemisphere summer, but the two driest winter months each have more than 70 mm rainfall (Fig. 2–10). Snowfalls are frequent above 4000 m and may occur at all times of the year. Although there is a permanent water surplus, there may be short periods of water deficit during the dry season when up to 22 consecutive dry days have been recorded. Relative humidity is high, with minimum values between 40% and 80%.

Hnatiuk, Smith, and McVean (1976) distinguished three basic types of daily weather pattern whose distribution throughout the year reflects the seasonal cycle of rainfall. During the wet season, most days have fairly steady rain, thick fog, and little if any clear sky during the whole 24-hour cycle. Radiant energy input during such days may attain only 13% of total possible solar radiation. In the dry season, days tend to have clear sky either at night or during the day, with no or only light showers. On sunny days, up to 94% of total possible solar radiation may reach the land surface. A third weather pattern may occur at any time of year, when the day begins with a clear sky and often with ground frost or heavy dew. Cloud formation begins from the early morning on, and by about noon or in the early afternoon intermittent mist and showers occur. By sunset, clearing begins and in a couple of hours the sky again becomes cloudless.

Screen temperatures recorded at the research station give a yearly range of only 2.0°C, a mean maximum of 11.6°C, mean minimum of 4.0°C, with recorded extremes of 16.7° and −0.8°C. It is interesting to compare these figures with the extreme temperatures recorded over nine years at Mucubají, in the Venezuelan Andes, at similar elevation: 22.2°C and −8.6°C. The range is almost twice that of Mt. Wilhelm, suggesting what Hnatiuk, Smith, and McVean (1976) termed a quasi-oceanic climate. But as usual, the ground surface has a much more severe temperature regime than that shown by screen records. Thus, for instance, at 10 cm above a short grass turf, shielded from direct sunlight, mean maximum and mean minimum recorded during one year were 19.7°C and −1.1°C, respectively, with extremes of 29.4°C and −9.4°C. In correspondence with this temperature gradient upward from ground level, very few days with freezing temperatures have been recorded in the Stevenson screen at 1.5 m aboveground at 3480 m, but ground-level freezing temperatures appear to be quite common. Ground frost was recorded during more than 50% of nights in a year. It can occur any night of the year when skies remain clear. The number of frost days was not recorded. However, freezing temperatures seem to be common at ground level but rare at 1.5 m. Topography has an important influence on frost days, producing significant site-to-site variation during one year. There also appears to be great variability from year to year.

In summary, Mt. Wilhelm seems to have one of

the rainiest and wettest tropical high-altitude climates, with more than 3000 mm rainfall, ten rainy months, and two somewhat less rainy months, the driest one with 75 mm rainfall. As a consequence, the temperature climate is comparatively mild, with low maxima and high minima, reflected in the scarcity of freezing temperatures and frost days at elevations where these events are much more frequent in tropical mountains of South America or Africa. Both humidity and temperature constancy and narrow range impart a definite oceanic character to this island mountain, a situation that contrasts sharply with the continental mountains of East Africa and the South American Andes.

According to Smith (1980b), the major climatic limitation to life at high altitudes does not derive from low minimum temperatures but possibly from low daylight temperatures together with the very reduced solar radiation due to cloudiness, fog, and mist. These characteristics seem to favor montane forest over páramo-like formations and to disadvantage the growth of any type of crop.

TOPOCLIMATES AND MICROCLIMATES

The Influence of Aspect and Topography

In mountain systems with abrupt and irregular relief, horizontal surfaces are either reduced to small areas or are nearly inexistent. Under such circumstances, the concept of regional climate loses most of its value, since the weather conditions recorded at any given site represent, at best, a small area around that site. Several dozen meters away from the recording station, differences in elevation, in slope, or in topography may induce significant changes in climatic conditions. In this way, topoclimates become ecologically more meaningful than regional climates. Geiger (1966, 1969) and Barry and Van Wie (1974) review this subject in mountain areas in general, stressing the effects of three key factors: slope angle, slope aspect, and relative topographic position. The first two modify diurnal temperature and humidity through their action on insolation, while topography influences night climate through its action on downward cold air movement and the daily cycle of slope and valley winds.

The role of slope aspect is fairly different in tropical and in temperate situations. The orientation of mountain slopes with respect to the movement of air masses leads to a sharp contrast in the amount of precipitation at regional level between lee- and windward slope. At low latitudes slope aspect produces more contrasting topoclimatic differences between east and west slopes than between north- and south, as is normally the case at higher latitudes. Within the tropics, direct sunshine reaches either the north or the south-facing slopes, depending on the period of the year, thus erasing most of the differences between these two slopes. In contrast, the particular conditions of local air circulation in tropical mountains are responsible for differences in insolation between east and west-facing surfaces. The subsequent differences in temperature and humidity are often reflected in the structure and composition of biotic communities as well as in land use.

In many tropical mountains, the daily weather pattern determines that insolation on west-facing slopes be significantly reduced by cloudiness or fog during the afternoon, in contrast with east-facing slopes receiving early morning sunshine. Sites of easterly aspect receive greater direct insolation and are therefore drier, having higher maxima and lower minima. This pattern has been shown for instance on Mt. Wilhelm (Hnatiuk, Smith, and McVean, 1976; Smith, 1977a) and in the Venezuelan páramo (Azócar and Monasterio, 1979, 1980b).

On Mt. Wilhelm, Hnatiuk, Smith and McVean (1976) placed max–min thermometers 10 cm aboveground at 3480 m, 4020 m, and 4380 m, on slopes of opposing aspect but with similar plant cover and slope angle. The results of several weeks of recording show that minimum temperatures were generally lower, and maximum temperatures higher, on the east-facing slope. The differences between maximum temperatures may attain 10° or more. Hnatiuk et al. (1976) attribute higher maxima on east-facing slopes to a longer period of insolation due to a lower frequency of clouds in the morning than in the afternoon. Lower minimum temperatures may be due to drier soils, possibly because of greater evaporation under conditions of greater insolation. The differences may be confined, however, to the air layer below 20 cm and to the uppermost soil layers. Barry (1978b) measured soil temperatures on a 35° east-facing slope, and on a 25° west-facing slope about 30 m apart. Plant cover at both sites was tussock grass up to 30 cm high and sparse shrubs. The easterly slope averaged up to 10° warmer at 1 cm, in the morning before cloud had built up. But in the afternoon, the west-facing slope was just a few degrees warmer because cloud cover reduced the warming effect of direct sunshine. These results support the view that differences in east–west aspect could be

TABLE 2–6. The effect of slope aspect, topographic position, and vegetation on air temperature (in °C) at 10 cm and 150 cm aboveground in the Páramo de Mucubají (3700 m), Venezuela.

Slope	West-facing	East-facing	East-facing	Valley bottom
Vegetation	Páramo	Páramo	*Polylepis* forest	Páramo
10 cm				
Annual mean	3.2	6.1	5.4	4.9
Mean max	13.7	19.1	15.0	17.9
Mean min	−4.4	−2.3	0.3	−5.8
Frost days	230	115	39	235
150 cm				
Annual mean	3.0	5.4	5.3	3.8
Mean max	17.0	18.1	16.7	16.7
Mean min	−3.4	−1.2	−2.0	−3.4

Data from Azócar and Monasterio (1980b).

important ecological factors in many tropical mountains.

During one year Azócar and Monasterio (1980b) recorded air temperatures on slopes of contrasting aspect at the Páramo de Mucubají (3700 m) in the Venezuelan Andes. Both east- and west-facing slopes in this area have páramo vegetation with only minor floristic differences. Temperature was recorded at 10 cm and 150 cm above ground. The effect of aspect on air temperature is more evident at 10 cm, where the difference between annual means of east- and west-facing slopes reaches almost 3° (Table 2–6). Notice also how this warmer character of the easterly slope appears clearly reflected in the annual number of frost days. Frost is in fact half as frequent on the east-facing slope. That these differences in temperature were mainly due to differential insolation was shown by the fact that on days with clear mornings, the eastern aspect may receive up to 15% more sunshine than the opposite slope. Table 2–5 also indicates the influence of topographic position on site climate. The recording site in the valley bottom, at 3600 m, even though located 100 m lower than the two slope stations, does appear to be the coldest, as indicated by the lower mean minimum. Furthermore, the valley has three times more frost days than the average number of frost days recorded under screen at 150 cm in the nearby recording station at Mucubají, at the same altitude. The topoclimate of the valley is the coldest of the whole area probably because of the night inversion in temperatures due to downward flow of valley and slope winds.

The topoclimatic effect of slope angle has not been analyzed in tropical mountains, although it could be expected to have a significant influence. In contrast to the situation prevailing at middle latitudes (Geiger, 1966), in the tropics gently sloping surfaces ought to receive more direct sunshine than either horizontal or steeper slopes. However, no data are available to verify this suggestion.

The preceding examples illustrate the crucial importance of topoclimates as generators of particular microhabitats where living beings may escape from the full impact of adverse atmospheric conditions in high tropical mountains. All highland environments are therefore patchy and are differentiated into a rich and complex mosaic of microhabitats that often exhibit sharply contrasting patterns of temperature and humidity. This mosaic of environmental patterns must not only be studied from a static viewpoint. One must also remember that they may suffer astonishingly rapid changes in just a few minutes, leading to sudden alterations in radiation, humidity, wind, and temperature. The ecological consequences of this dynamism cannot be overemphasized.

The Microclimates of Plants and Plant Formations

Plant cover modifies the heat budget of a land surface by its influence on incoming and outgoing radiation, as well as by its direct modification of humidity and temperature below the canopy. These microclimatic influences have been well documented in tropical high mountains, and are particularly noticeable when different plant formations are compared with each other. This is the case, for instance, when forest and páramo formations growing at the same altitude in nearby sites are simultaneously monitored for weather variables.

Table 2–6 compares data on air temperature under the canopy of a *Polylepis sericea* grove at 3700 m in the Venezuelan Andes with the surrounding páramo dominated by rosettes of *Espeletia*, low shrubs, and herbs. Mean and maximum air temperatures are lower under the forest canopy than under the sparse cover of páramo plants, on a similar slope and at a similar elevation, but minimum temperatures are higher in the forest 10 cm above the ground. Notice also how the number of days of frost is greatly reduced inside the forest as compared to the sites with páramo vegetation. The forest canopy, even if it is low and light as at Mucubají, operates effectively as a screen for short- and and long-wave radiation. Temperature fluctuations are thus dampened and the daily climatic cycle is less contrasted.

Monasterio (1979) presented data from still higher elevations in the Andes concerning the microclimate of plants growing at the Páramo de Piedras Blancas at 4200 m (the climate in this area was described earlier in this chapter). The desert páramo formation, characteristic at this altitude, leaves the soil almost entirely devoid of plant cover, the giant rosettes of *Espeletia* being the major living component of this ecosystem. Daily cycles of air, soil, and plant temperatures were followed in the dry and in the rainy season. Under the cloudy skies and foggy days of the rainy season, vertical profiles of air temperature tend to flatten out; thus temperatures 10 cm aboveground were persistently about 1° higher than at 150 cm. But under the clear skies prevailing during the dry season, air temperature at 150 cm was several degrees higher than at 10 cm, at least throughout the night and the morning, while the reverse was true in the afternoon. These slight differences along a vertical gradient upward from the ground surface may help to explain the vertical structure of this ecosystem, with its green biomass located mostly well above ground level, between 1 and 3 m.

Species of *Espeletia* living at this high altitude are able to maintain the same temperature as that of the air throughout the 24-hour cycle. In fact, the difference between leaf and air temperatures never exceeds a few degrees, the leaves being slightly warmer during the daylight hours and cooler by night. Leaves can control their temperature in some way, but this behavior appears even more striking at the apical bud of the rosette, where the whole process of renewal of the assimilatory structures rests. The apical bud lies well protected among the dense envelope of developing leaves, thus allowing its temperature to be maintained several degrees higher than that of the air. The exception to this pattern is during the hours of more intense sunshine, in the morning, when the apical bud is cooler than the air (Monasterio, 1979). These features suggest that morphophysiological mechanisms operate to maintain a favorable heat budget in these plants submitted to circadian cold stress in the periglacial zone in the high tropical Andes.

Other interesting microclimatic data come from the upper limit of the *Pinus hartwegii* forest on Orizaba at 3990 m. At this altitude an open pine woodland intermingles with the mountain grassland (Lauer and Klaus, 1975). These authors recorded temperatures at various soil and plant surfaces during one daily cycle. At 150 cm above the ground needles of *P. hartwegii* maintain their temperature between 2° and 17° when air temperature ranges between −2° and 15.5°. Moreover within a tufted grass cushion of *Festuca* temperature oscillated between 0.5° and 27°. Clearly, living plant surfaces are able to maintain their temperature higher than the atmosphere near the ground, a feature that may be particularly important during frosts.

Other examples of the influence of plant cover on temperature conditions above different vegetation canopies and in various plant and soil surfaces have been reported from several tropical highlands, such as Mt. Wilhelm (Smith, 1977a; Barry, 1978b), Mt. Kilimanjaro (Coe, 1967a), and the Colombian páramo (Schnetter et al., 1976). All of these studies emphasize the fact that biological surfaces create and maintain their own boundary conditions at their interfaces with the lower atmosphere and that these biologically conditioned microclimates tend to favor survival under the limiting conditions imposed by low temperatures.

SUMMARY AND CONCLUSIONS

On the basis of this review of climatic conditions of tropical highlands, I wish to make some generalizations about the features constantly found in these regions, the range of variation in the major climatic parameters, the main peculiarities of some especially interesting situations, and finally, about the crucial ecological characteristics of climate for the upper limits of life or for survival.

All high tropical mountains share many important climatic features that make them quite specific environments for the colonization and the maintenance of plant, animal, and human populations. They are part of the world's circumtropical belt, and hence they show the relative annual constancy in incoming radiation, daylength, and temperature that generally distinguishes low from high latitudes. In this respect, they contrast sharply with the highly seasonal rhythmicity of higher latitudes, where life must cope with an unfavorable winter season and must profit from a rather short growth season to accomplish most of its activities. But tropical mountains differ from adjacent lowlands too, principally by the lower temperatures prevailing throughout the annual cycle. The climate of high tropical mountains makes these areas peculiar environments that are ecologically apart from both extratropical mountains and from tropical lowlands.

In spite of the many climatic features that distinguish tropical highlands from any other type of environment, the first generalization to be made is that it does not seem possible to refer to a single type of tropical high-altitude climate. Indeed, the diversity of weather types and regimes makes it necessary to distinguish several climates within the upper belts of tropical mountains.

A first gradient is latitudinal: annual ranges increase as one proceeds farther and farther away from the equator. At a given point, annual oscillation in temperature becomes so important as to delimit the tropical or thermically constant zone. This transition may be analyzed either in the central Andes or in the Mexican highlands, since these are the only two areas with a continuous range of high mountains from tropical to temperate latitudes. In both cases, some common features appear. Thus, for example, the influence of polar air masses is felt periodically during the winter, drastically lowering the temperature for several days. This naturally contributes to amplify differences in mean monthly temperatures.

A second gradient arises from the geographic position of each mountain massif, either with reference to distance from the ocean or with respect to the position of the chain in connection with the circulation of the lower atmosphere. Distance from the sea induces a differentiation between predominantly oceanic highlands and rather continental massifs. The New Guinea highlands afford a clear example of oceanic influence while the interior volcanoes of East Africa or the Peruvian and Bolivian altiplano represent continentality. The orientation of a given chain is important to induce differentiation between windward and leeward slopes. Examples were given of this feature, such as Mt. Kenya, Mt. Kilimanjaro, the different Andean chains in Colombia, and the contrasting Amazonian and Pacific slopes of the Peruvian Andes.

A third gradient of environmental variation is altitude. Altitudinal gradients of temperature and precipitation play a major role in the zonation of tropical mountains. Some conspicuous ecological boundaries related to these gradients are the upper continuous forest line, the timberline, the lower limit of periglacial climate, and the nival limit. The case of the upper limit of the tropical montane forest or cloud forest was considered in various situations. The most valid generalization seems to be that on moist slopes this boundary is closely related to frost frequency, since apparently even a few frost days per year are sufficient to replace the forest by páramo. In the case of the coniferous forest of the Mexican volcanoes, the situation is completely different, since the frost line delimits mixed forest from pine forest, whereas the latter vegetation formation extends well into the frost climate.

Periglacial climates, characterized by the actual periglacial sculpture of the land (Tricart, 1970, Hastenrath, 1973, 1977; Schubert, 1979, 1980), permit one to define two types of high-altitude ecosystems in the Venezuelan Andes, which Monasterio (1980a) named the Andean and the high Andean zones. However, it is not yet possible to indicate precisely what climatic features are at the root of this sharp geomorphological and ecological boundary. Finally, the nival zone occurs at quite difference altitudes in various chains, being related either to total rainfall or to the temperature lapse rate. Along the humid slopes of the Andes, the nival belt begins very close to the line of permanent frost, where every day in the year is either a frost day or a frost-change day. Thus, in the Sierra Nevada de Mérida, the glaciers descend to 4700 m. But in very dry mountains, as on the peaks that emerge from the puna, or on volcanoes such as Kilimanjaro, this limit may be found as high as 5500 or even 6000 m.

One of the more complex and less understood altitudinal gradients in tropical mountains is the increase in number of frost days with elevation. It differs widely in various mountain chains, being related to several concurrent factors such as latitude, amount and annual distribution of rainfall, and intensity of drought periods. On moist slopes, with short and weak dry periods, frost begins higher than on dry slopes, well above 3000 m in the lowest latitudes; upwards, the number of frost change days increases slowly up to 4500 m or more, to change suddenly at the 4700 m level at the nival limit (Fig. 2–1). Under these circumstances, the periglacial zone does not exist, since the change from half to all days in the year with freezing temperatures is too abrupt, collapsing thus the periglacial area between the páramo and the nival zones. Quite the opposite situation is found on the Mexican meseta, where the frost gradient is rather smooth from the level of first frost occurrence, at about 1600 m, up to 4000 m. The gradient then becomes steeper above this level, which precisely constitutes the pine forest–mountain grassland boundary (Fig. 2–1). On dry slopes of the northern Andes, such as the upper Chama Valley in Venezuela, frost behavior appears to be similar to that of the Mexican altiplano, but the curve is displaced toward higher elevations. Finally, on the high peaks that tower over the Peruvian puna, such as El Misti volcano, the curve relating the number of frost days per year to altitude appears to be entirely different from all others, because frost increase is abrupt at lower elevations, changing from a few frost days to about 200 frost days in the narrow range between 3200 and 3400 m. Above this level, the increase becomes gradual up to 4700 m. This zone of gradual increase corresponds in this area to the moist puna belt.

I wish to end with a brief consideration of the ecological impact of high-altitude climates on plant and animal populations. In this respect, some similarities and many differences between tropical and temperate mountains must be emphasized. In tropical highlands, strong winds or a seasonal snow cover do not operate on biological processes as they do in most temperate mountain areas. Moreover a frost-free growth season does not exist at any altitude above the frost line, since seasons are absent, and frost may occur at any time of the year (within the limitations imposed by the rainfall patterns discussed here).

The two factors that most authors consider to be highly significant filtering agents for maintenance and survival of life in tropical mountains are freezing temperatures at night and insufficient radiation and suboptimal temperatures during daylight hours. Adaptations to permit survival with the erratic or the normal occurrence of freezing imply serious modifications in form, function, behavior, or all three, which rather few tropical species have been able to accomplish. Furthermore, adaptations should also allow a favorable performance during the day, when quite often heat and light may be limiting factors and when sudden and hazardous changes do occur from day to day or even from hour to hour. The lack of well-marked annual thermal seasons does not imply lack of response of biological populations to temperature changes. Quite the contrary. Responses not only have to exist, but have to be instantaneous, and each species must be prepared to withstand as drastic changes in radiation, temperature, or humidity in a few minutes, as other populations suffer in contrasting annual seasons. In this way, a combination of opportunistic responses to sudden changes of extremely short periods and high frequency must accompany in their life strategies the capacity to permanently support low-temperature stress as well as the ability to maintain a favorable heat-and-carbon balance under conditions of limited energy supply.

ACKNOWLEDGMENTS

I want to express my warmest gratitude to the editors M. Monasterio and F. Vuilleumier for their kind request to collaborate with this chapter in their book. From M. Monasterio I also received many helpful suggestions that certainly improved the original typescript. I am grateful to J. M. B. Smith for his expert advice on the literature concerning Malesia.

REFERENCES

Allison, I., and J. Bennett 1976. Climate and microclimate. In *The Equatorial Glaciers of New Guinea*, G. S. Hope, J. A. Peterson, I. Allison, U. Radok, eds. pp. 61–81. Rotterdam: Balkema.

Andressen, R., and R. Ponte, 1973. *Estudio Integral de las Cuencas de los Ríos Chama y Capazón. Climatologiá e Hidrología*. Inst. Geogr. Conserv. Rec. Nat. Renov. Facultad Cienc. Forestales. Mérida: Universidad de Los Andes.

Azócar, A., and M. Monasterio, 1979. Variabilidad ambiental en el páramo de Mucubají. In *El Medio Ambiente Páramo*, M. L. Salgado-Labouriau ed., pp. 149–159. Caracas: Ediciones Centro de Estudios Avanzados.

——— 1980a. Caracterización ecológica del clima en el Páramo de Mucubají. In *Estudios Ecológicos en los Páramos Andinos*, M. Monasterio, ed., pp. 207–223. Mérida: Ediciones de la Universidad de Los Andes.

——— 1980b. Estudio de la variabilidad meso y microclimática en el Páramo de Mucubají. In *Estudios Ecológicos en los Páramos Andinos*, M. Monasterio, ed., pp. 225–262. Mérida: Ediciones de la Universidad de Los Andes.

Barry, R. G. 1978a. Aspects of the precipitation characteristics of the New Guinea mountains. *J. Trop. Geog.* 47: 13–30.

——— 1978b. Diurnal effects of topoclimate on an equatorial mountain. *14 Int. Tagung für Alpine Meteorologie* 72: 1–8.

——— 1979. High altitude climates. In *High altitude geoecology*, P. J. Webber, ed., pp. 55–74. AAAS Selected Symposium 12. Boulder: Westview Press.

——— 1980. Mountain climates of New Guinea. In *Alpine*

vegetation of New Guinea, P. Van Royen ed., pp. 74–109. Vaduz: Cramer.

Barry, R. G., and C. C. Van Wie, 1974. Topo- and microclimatology in alpine areas. In *Arctic and alpine environments*, J. D. Ives, R. G. Barry, eds., pp. 73–83. London: Methuen.

Bernal, A., and A. G. Figueroa, 1980. Estudio ecológico comparativo de la entomofauna en un bosque alto andino y un páramo localizado en la región de Monserrate, Bogotá. Bogotá: Universidad Nacional de Colombia, Facultad de Ciencias.

Boughey, A. S. 1965. Comparisons between the montane forest floras of North America, Africa and Asia. *Webbia* 19: 507–517.

Brooke Thomas, R., and B. P. Winterhalder, 1976. Physical and biotic environment of southern highland Peru. In *Man in the Andes*, P. T. Baker and M. A. Little, eds., pp. 21–59. Stroudsbourg, Pa.: Dowden, Hutchinson & Ross.

Brown, L. H., and J. Cocheme, 1973. A study of the agroclimatology of the highlands of Eastern Africa. WMO no. 339, Geneva: FAO-UNESCO-WMO.

Brown, M., and J. M. Powell, 1974. Frost and drought in the highlands of Papua New Guinea. *J. Trop. Geog.* 38: 1–6.

Chavez, L. F. 1962. Clima de las cuencas altas de los ríos Motatán, Chama y Santo Domingo. Caracas: Ministerio de Agricultura y Cría.

Cleef, A. M. 1981. The vegetation of the paramos of the Colombian Cordillera Oriental. Dissertationes Botanicae 61. Vaduz: Kramer.

Coe, M. J. 1967a. Microclimate and animal life in the equatorial mountains. *Zool. Africana* 4: 101–128.

——— 1967b. *The ecology of the alpine zone of Mount Kenya*. Monographiae Biologicae 17. The Hague: Junk.

Cuatrecasas, J. 1958. Aspectos de la vegetación natural de Colombia. *Rev. Acad. Colomb. Cienc. Exact. Fis. Nat.* 10: 221–264.

Flohn, H. 1974. Contribution to a comparative meteorology of mountain areas. In *Arctic and alpine environments*, J. D. Ives, and R. G. Barry, eds., pp. 55–71. London: Methuen.

Garcia, E. 1970. Los climas del Estado de Veracruz. *An. Inst. Biol. Univ. Nal. Auton. México* 41 *Ser. Botánica* 1: 3–42.

Geiger, R. 1966. *The climate near the ground.* Cambridge, Mass.: Harvard University Press.

——— 1969. Topoclimates. In *General climatology*, H. Flohn, ed., pp. 105–117. Vol. 2 of *World survey of climatology*, H. E. Landsberg, ed. Amsterdam: Elsevier.

Guhl, E. 1968. Los páramos circundantes de la sabana de Bogotá, su ecología y su importancia para el régimen hidrológico de la misma. In *Geo-ecology of the mountainous regions of the tropical Americas*, C. Troll, ed., pp. 195–212. Bonn: Dummlers.

Hastenrath, S. 1967. Rainfall distribution and regime in Central America. *Arch. Meteorol. Geophys. Bioklimatol.*, Ser. B. 15: 201–241.

——— 1968. Certain aspects of the three-dimensional distribution of climate and vegetation belts in the mountains of Central America and southern México. In *Geo-ecology of the mountainous regions of the tropical Americas*, C. Troll, ed., pp. 122–130. Bonn: Dummlers.

——— 1973. Observations on the periglacial morphology of Mts. Kenya and Kilimanjaro, East Africa. *Zeit. für Geomorph., N. F. Suppl.* 16: 161–179.

——— 1977. Observations on soil frost phenomena in the Peruvian Andes. *Zeit. für Geomorph., N. F.* 21: 357–362.

Hedberg, O. 1951. Vegetation belts of the East African mountains. *Svensk bot. Tidskr.* 45: 140–202.

——— 1964. *Features of afroalpine plant ecology*. Uppsala: Almqvist & Wiksells Boktryckeri.

Hnatiuk, R. I., J. M. B. Smith, and D. N. McVean, 1976. *The climate of Mt. Wilhelm*. Research School of Pacific Studies, Dept. of Biogeography & Geomorphology, Publ. BG/4. Canberra: Australian National University.

Johnson, A. M. 1976. The climate of Peru, Bolivia and Ecuador. In *Climates of Central and South America*, W. Schwerdtfeger, ed., pp. 147–218. Vol. 12 of *World survey of climatology*, H. E. Landsberg, ed. Amsterdam: Elsevier.

Klute, F. 1920. *Ergebnisse der Forschungen am Kilimandscharo 1912*. Berlin.

Lauer, W. 1973a. The altitudinal belts of the vegetation in the central Mexican highlands and their climatic conditions. *Arct. Alp. Res.* 5: A99–A113.

——— 1973b. Zusammenhange zwischen Klima und Vegetation am Ostabfall der Mexicanischen Meseta. *Erdkunde* 27: 192–213.

——— 1976. Zur Hygrischen Hohenstufung Tropischer Gebirge. In *Neotropische Oekosysteme*, F. Schmithüsen, ed., pp. 169–182. Biogeographica VII. The Hague: Junk.

——— 1978. Timberline studies in Central Mexico. *Arct. Alp. Res.* 10: 383–396.

——— 1979. La posición del páramo en la estructura del paisaje de los Andes tropicales. In *El Medio Ambiente Páramo*, M. L. Salgado-Labouriau, ed., pp. 29–45. Caracas: Ediciones Centro de Estudios Avanzados.

Lauer, W. and D. Klaus. 1975. Geoecological investigations on the timberline of Pico de Orizaba, México. *Arct. Alp. Res.* 7: 315–330.

List, R. J. 1971. *Smithsonian Meteorological Tables.* 6th ed. Washington, D.C.: Smithsonian Institution Press.

Monasterio, M. 1979. El páramo desértico en el altiandino de Venezuela. In *El Medio Ambiente Páramo*, M. L. Salgado Labouriau, ed., pp. 117–146. Caracas: Ediciones Centro de Estudios Avanzados.

——— 1980a. Las formaciones vegetales de los páramos de Venezuela. In *Estudios Ecológicos en los Páramos Andinos*, M. Monasterio, ed., pp. 93–158. Mérida: Ediciones de la Universidad de los Andes.

——— 1980b. El páramo de Mucubají dentro del cuadro general de los páramos venezolanos. In *Estudios Ecológicos en los Páramos Andinos*, M. Monasterio, ed., pp. 201–206. Mérida: Ediciones de la Universidad de los Andes.

Monasterio, M., and S. Reyes, 1980. Diversidad ambiental y variación de la vegetatión en los páramos de los

Andes Venezolanos. In *Estudios Ecológicos en los Páramos Andinos*, M. Monasterio, ed., pp. 47–91. Mérida: Ediciones de la Universidad de los Andes.

Nieuwolt, S. 1974. The influence of aspect and elevation on daily rainfall: Some examples from Tanzania. In *Agroclimatology of the Highlands of Eastern Africa.* Proceedings of the Technical Conference, Nairobi, October 1–5, 1973. WMO no. 389. Geneva: FAO-UNESCO-WMO.

Ortolani, M. 1965. Osservazioni sul clima delle Ande Centrali. *Rev. Geog. Ital.* 72: 217–235.

Salt, G. 1954. A contribution to the ecology of upper Kilimanjaro. *J. Ecol.* 42: 375–423.

Schnetter, R., G. Lozano-Contreras, M. L. Schnetter, and H. Cardoso. 1976. Estudios ecológicos en el páramo de Cruz Verde, Colombia. I. Ubicación geográfica, factores climáticos y edáficos. *Caldasia* 11 (54): 25–52.

Schubert, C. 1979. La zona del páramo: Morfología glacial y periglacial de los Andes de Venezuela. In *El Medio Ambiente Páramo*, M. L. Salgado Labouriau, ed., pp. 11–27. Caracas: Ediciones del Centro de Estudios Avanzados.

——— 1980. Aspectos geológicos de los Andes Venezolanos: historia, breve síntesis, el cuaternario y bibliografía. In *Estudios Ecológicos en los Páramos Andinos*, M. Monasterio, ed., pp. 29–46. Mérida: Ediciones de la Universidad de los Andes.

Smith, J. M. B. 1975. Mountain grasslands of New Guinea. *J. Biogeog.* 2: 27–44.

——— 1977a. Vegetation and microclimate of east- and west-facing slopes in the grasslands of Mt. Wilhelm, Papua New Guinea. *J. Ecol.* 65: 39–53.

——— 1977b. An ecological comparison of two tropical high mountains. *J. Trop. Geog.* 44: 71–80.

——— 1977c. Origins and ecology of the tropicalpine flora of Mt. Wilhelm, New Guinea. *Biol. J. Linn. Soc.* 9: 87–131.

——— 1980a. The vegetation of the summit zone of Mount Kinabalu. *New Phytol.* 84: 547–573.

——— 1980b. Ecology of the high mountains of New Guinea. In *The Alpine flora of New Guinea.* Vol. 1, *General Part,* P. Van Royen, ed. pp. 111–131. Vaduz: Cramer.

Snow, J. W. 1976. The climate of northern South America. In *Climates of Central and South America*, W. Schwerdtfeger, ed., pp. 295–403. Vol. 12, *World Survey of Climatology*, H. E. Landsberg, ed. Amsterdam: Elsevier.

Thompson, B. W. 1966. The mean annual rainfall of Mt. Kenya. *Weather* 21: 48–49.

Tosi, J. A. 1957. El clima y la ecología climática general de Huancayo, Peru. I.I.C.A. Publ. Miscelánea no. 11. Turrialba, Costa Rica.

Tricart, J. 1970. *Geomorphology of cold environments.* London: Macmillan.

Troll, C. 1968. The cordilleras of the tropical Americas. In *Geo-ecology of the mountainous regions of the tropical Americas*, C. Troll, ed., pp. 15–56. Bonn: Dummlers.

——— 1973. The upper timberlines in different climatic zones. *Arct. Alp. Res.* 5: A3–A18.

van Steenis, C. G. E. J. 1935. On the origin of the Malaysian mountain flora. Part 2. Altitudinal zones, general considerations and renewed statement of the problem. *Bull. Jard. Bot. Buitenz.* (Ser. 3) 13: 289–417.

——— 1968. Frost in the tropics. In *Recent advances in tropical ecology*, R. Misra, and B. Gopal, eds. pp. 154–167. Faribabad, India: Shri R. K. Jain.

Weberbauer, A. 1945. *El Mundo Vegetal de los Andes Peruanos.* Lima: Ministerio de Agricultura.

Weber, H. 1959. *Los Páramos de Costa Rica y su Concatenación Fitogeográfica con los Andes Suramericanos.* San José: Instituto Geográfico de Costa Rica.

Weischet, W. 1969. Klimatologische Regeln zur Verticalverteilung der Niederschlage in den Tropengebirgen. *Die Erde* 100(2–4): 287–306.

Winterhalder, B. P., and R. B. Thomas, 1978. Geoecology of southern highland Peru. A human adaptation perspective. *Inst. Arct. Alp. Res. Occ. Pap.* no. 27.

Yacono, D. 1968. Essai sur le climat de montagne au Sahara, l'Ahaggar. *Trav. Inst. Rech. Sahariennes*, 1.

II

ADAPTATIONS

According to Bock (1965), three concepts of adaptation can be defined. The first, universal adaptation, "best describes the essential interaction between living organisms and their environment." The second was called by Bock evolutionary adaptation, and "designates both a process and a state of being" at the level of the relationships between a given species and specific environmental conditions. The third, physiological adaptation, corresponds to "the ability of tissues to respond phenotypically to different environmental conditions during the life of an individual." There is clearly a hierarchy of levels of adaptation. Authors like Williams (1966) and Stern (1970) have emphasized the close relationship between adaptation and natural selection, a process that operates at several scales or on several planes. Lewontin (1970), for instance, described a hierarchical series of levels of selection, from molecules to species and communities, including individuals and populations.

We can thus organize, for tropical high montane biota, the general concept of adaptation into a hierarchical series of levels, including individual, populational, species, and community. In other words, research into such adaptations can be carried out separately at and within one or more of these levels, and furthermore, work can be done across the various levels. Plants of the high tropical mountains that seem to show convergent evolutionary responses (phenotypic in both morphological and physiological terms), such as *Espeletia* (Andean páramo) and *Dendrosenecio* (afroalpine) are especially good material for studies of adaptation. Mayr (1960) emphasized the suitability of birds generally for research on adaptation, singling out, among other taxa, hummingbirds, and among problems, that of convergence. Therefore it seemed to us that sunbirds of high African mountains and high-altitude hummingbirds in the high mountains of the New World tropics would be excellent material for an examination of adaptations, both physiological and ecological.

Recently, much emphasis has been given to communitywide adaptations and convergences (e.g., Cody and Mooney, 1978; Blondel et al., 1984). A broadly based comparison of avian communities of the high Andes and the Ethiopian tableland and afroalpine zone is the first attempt to analyze such adaptive convergences in birds of high tropical mountains.

We feel strongly that more work on adaptations in high tropical mountains should include intercontinental comparisons of ecologically equivalent taxa in climatically well-matched study sites.

REFERENCES

Blondel, J., F. Vuilleumier, L. F. Marcus, and E. Terouanne. 1984. Is there ecomorphological convergence among mediterranean bird communities of Chile, California and France? *Evol. Biol.* 18. M. K. Hecht, B. Wallace, and G. T. Prance, eds., pp. 141–213.

Bock, W. J. 1965. The role of adaptive mechanisms in the origin of higher levels of organization. *Syst. Zool.* 14: 272–278.

Cody, M. L., and H. A. Mooney. 1978. Convergence versus nonconvergence in mediterranean-climate ecosystems. *Ann. Rev. Ecol. Syst.* 9: 265–321.

Lewontin, R. C. 1970. The units of selection. *Ann. Rev. Ecol. Syst.* 1: 1–18.

Mayr, E. 1960. Chairman's introduction to the symposium on adaptive evolution. *Proc. XII Int. Ornithol. Congr.*, Vol. II: 495–498.

Stern, J. T. Jr. 1970. The meaning of 'adaptation' and its relation to the phenomenon of natural selection. *Evol. Biol.* 4. Th. Dobzhansky, M. K. Hecht, and W. C. Steere, eds., pp. 39–66.

Williams, G. C. 1966. *Adaptation and natural selection*. Princeton, N.J.: Princeton Univ. Press.

3

Adaptive Strategies of Espeletia *in the Andean Desert Páramo*

MAXIMINA MONASTERIO

THE DESERT PÁRAMO

The desert páramo, with its characteristic and spectacular giant rosettes of *Espeletia* (Fig. 3–1), is the most representative plant formation of the high Andean zone. It occurs intermingled with the periglacial desert that occupies the most fragile habitats, where scattered dwarf cushions, sessile rosettes, and lichens form a sparse cover that dots the prevailing bare ground.

The desert páramo is the most extreme environment colonized by *Espeletia* at the present time. The major area of this tropical, high Andean ecosystem is found in the Cordillera de Mérida, Venezuela, above 4000 m, where it reaches the very border of the glaciers at 4600 m. According to Tricart (1970), in the Andes of Mérida it is possible to fix at about 4000 m the lower limit of definite periglacial features. This same level marks the boundary between two clear-cut ecological zones: the Andean belt, downward, and the high Andean belt, toward the summits (Monasterio, 1980a).

The high Andean belt shows an insular pattern of distribution (Fig. 3–2), each "island" corresponding to one main sierra (Sierra Nevada, Sierra de Santo Domingo, Sierra de la Culata, etc.). Below 4000 m, the desert páramo is replaced by various páramo formations characterized by a wider diversity of life forms and by

FIG. 3–1. *E. timotensis* at the páramo of Pico del Aguila, Sierra de la Culata, 4200 m. Photo by M. Fariñas.

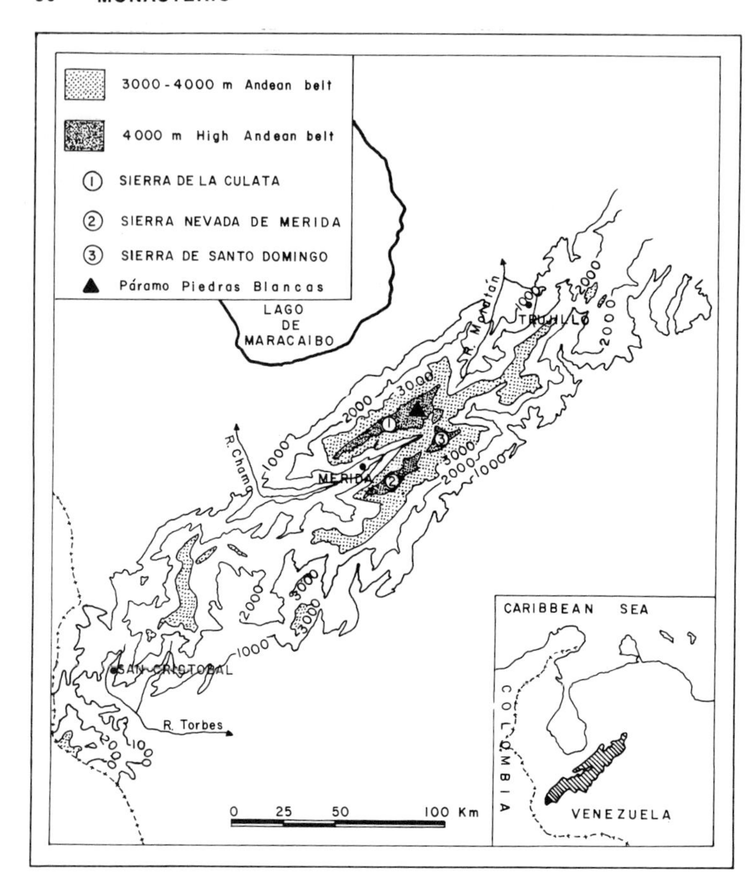

FIG. 3–2. Map of the surface corresponding to different altitudinal levels in the Cordillera de Mérida. At the 4000-m level, the highlands become discontinuous, giving rise to three separate areas or "islands."

the prevalence of small rosettes of *Espeletia* together with some branched forms (see Chap. 11).

The Periglacial Climate and Its Ecological Constraints

The desert páramo corresponds to the periglacial belt of low-latitude mountains where frost action and daily cycles of freeze–thaw promote specific geomorphogenetic processes and soil movements of profound ecological consequences. These climatic effects are reinforced by the open nature of the vegetation cover that leaves large areas of bare ground, favoring thus the alternation of rapid cooling and heating of the soil surface that lead to the daily cycles of freeze and thaw during most of the year. Soil particles gradually move down the slopes in the process of cryoturbation that induces such typical surface features as patterned ground in small-scale structures, due to their formation in daily cycles (Schubert, 1976).

The most common feature of patterned ground in the desert páramo is striped ground. Malagon (1982) analyzed this feature at Pico del Aguila, at 4118 m. He reported a mobile mantle of solifluction, about 7 cm thick, that migrates easily along the steep slopes prevailing in the highest areas. It is this moving mantle, porous and unstable, that receives the achenes of *Espeletia*, constituting its germinative niche. Seedlings thus have a low survival probability, since they can be swept away in the migrating solifluction mantle. Low temperature acts then indirectly on the survivorship of seedlings and juveniles, through the instability of the substrate rather than by its direct effect on plant tissues. We will return to this point, but it is worthwhile to emphasize now how the principal impact of the periglacial ecology occurs just at ground level (Fig. 3–3).

Besides the scarce cushions at ground level (Fig. 3–1), the desert páramo shows a conspicuous layer of giant rosettes that constitute an open stratum 1 to 3 m high. To understand this sharp contrast in the vertical distribution of the vegetation biomass, it is convenient to refer to microclimatic gradients along the soil and air in close contact with it. The main climatic features of the desert páramo within the framework of the high mountain climates have already been discussed by Sarmiento (Chap. 2), and therefore I will mention only a few additional environmental pulsations that may aid in the interpretation of the adaptive mechanisms of the giant rosettes of *Espeletia*.

Pico del Aguila, a representative site for the desert páramo, shows the annual constancy in mean daily temperature and in incoming radiation that characterizes low latitudes (Fig. 3–4), with a difference of only 2.7° between the mean temperature of the coldest and the warmest months: 1° and 3.7°, respectively. As was already pointed out, in high tropical mountains, daily cycles become the major cyclic environmental pulsations. However, rainfall seasonality also induces seasonal contrasts and amplifies daily cycles. At Pico del Aguila, wet and dry seasons are quite contrasted, since during the four months of the dry season (December to March) rainfall amounts to only 76 mm, that is, 8% of the annual total (Fig. 3–4). This contrasts sharply with the wettest months (May to August), which may be considered as perhumid, whereas the remaining months are intermediate between these two extremes. As a direct consequence of this rainfall pattern, the equatorial constancy is partially broken, introducing an annual pulsation that affects daily cycles inducing slight but noticeable changes in the solar radiation reaching the surface, as well as in air and soil temperatures and humidity. Two daily cycles, one in the moist, the other in the dry season, illustrate these facts (Table 3–1 and Figs. 3–5 and 3–6). Air temperatures were recorded hourly at 150 cm and 10 cm, as well as at the soil surface, and soil temperatures were recorded at 5, 10, 20 and 50 cm below the surface.

During the daily cycle in the dry season, air temperatures at 150 cm ranged from −2.5° to 11°, remaining at or below zero for 13 consecutive hours (8 P.M.–9 A.M.). At 10 cm, the pattern was similar but the daily oscillation was more pronounced: from −5° to 12°. During the night hours, air temperatures were consistently 2°–3° lower at 10 cm than at 150 cm. That is, from a thermal standpoint, the air level at 150 cm where most of the leaf and reproductive biomass of *Espeletia* is located, has more moderate conditions than the levels nearer to the ground. As far as soil temperature is concerned, an amplitude of 50° was recorded at the soil surface (bare ground), between a minimum of −10° and a maximum of 40°. As may be seen in Fig. 3–5, the ground surface remained at below freezing temperatures for 14 consecutive hours. The amplitude of daily oscillations dampened gradually with depth: 21° at 5 cm, 5° at 20 cm, and only 1° at 50 cm. Moreover, the freezing point was never

FIG. 3–3. A juvenile of *E. timotensis* surrounded by ice formed in the night and gradually melting in the morning. Páramo of Pico del Aguila, 4200 m. Photo by M. Fariñas.

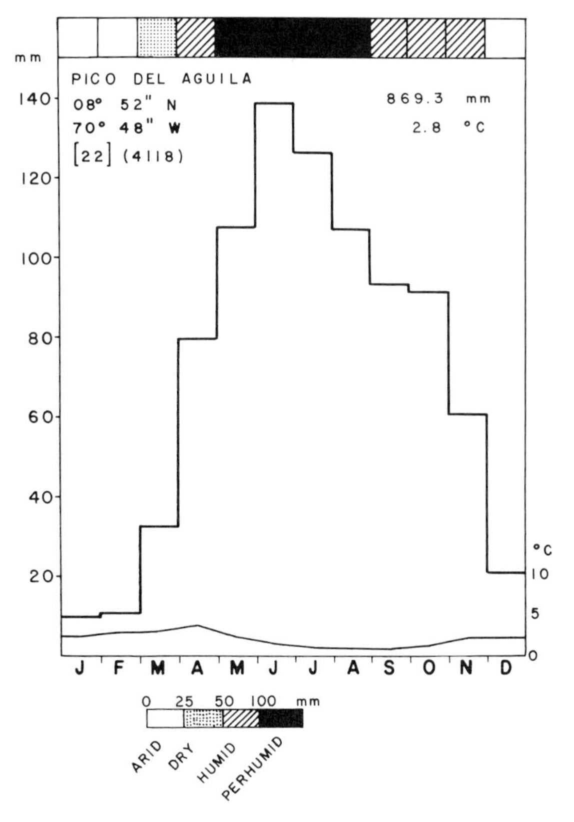

FIG. 3–4. Climate diagram for Pico del Aguila in the Sierra de la Culata, a typical desert páramo station. Periods of contrasting humidity conditions are shown from the arid to the perhumid.

TABLE 3–1. Microclimatic records in the Páramo de Piedras Blancas at 4200 m.

Micro-climatic factors (°C)		Air			Soil			*E. moritziana* (69 cm)			*E. timotensis* (165 cm)		
		150 cm	10 cm	Surface	−5	−20	−50	bud	ext. leaf	air	bud	ext. leaf	air
Mean temperature	January	2.7	2.3	7.0	5.4	6.5	6.0	5.5	2.4	2.4	5.4	4.3	2.5
	August	4.5	5	7.6	7.7	6.8	5.6	7.1	5.1	4.2	5.8	5.6	4.4
Temperature range	January	13.5	17.0	50.0	21.0	5.0	1.0	21.0	27.0	19.0	20.5	13.5	13.0
	August	10	11.1	25.5	9	3.9	1.4	16.3	14.5	10	11.6	15.2	12.1
Maximum temperature	January	11	12	40	19	9	6.5	19	21	16	19	11	11
	August	10	11.1	26	13	8.9	6.2	17.5	14	10	12.5	14.9	12
Minimum temperature	January	−2.5	−5.0	−10	−20	4	5.5	−2.0	−6.0	−3.0	−1.5	−2.5	−2.0
	August	0	0	0.5	4	5	4.8	1.2	−0.5	0	0.9	−0.3	−0.1

Note: Measurements were made hourly during two days, one in the dry season (January 12–13, 1978), the other in the rainy season (August 1–2, 1978). Extreme temperatures recorded at 10 cm above the ground were −14°C and 18°C in the dry season day, and −3°C and 15°C in the rainy season daily cycle.

reached at these two lowest levels. But we must keep in mind that most of the below–ground biomass of *Espeletia* occurs in the first 20 cm of soil.

Besides this daily pattern of air and soil temperature, in typical dry season days, the sky remains clear except for a few hours in the evening. Relative humidity varies from 20%–25% in the early morning to 70%–90% in the evening, whereas wind speed remains feeble all day.

During the wet season, the vertical gradient of temperature disappears; the minima are higher, whereas the maxima and the daily ranges are lower than in the dry period. Both cloudiness and relative humidity remain high for most of the day.

From the analysis of these two cycles, we may conclude that rather low air temperatures prevail all day and all year at the crown level of *Espeletia*. During the dry season, air and soil moisture are lower and daily ranges more pronounced, but the strong insulation may make the energy balance of the plant more favorable. By contrast, in the wet season moisture conditions improve and the minima are higher, but the lower incoming radiation and diurnal temperatures make the energy budget of plants less favorable. We see that neither season appears to be optimal, and thus that a given period cannot be considered the most favorable to all plant processes. In the wet season, for instance, pollination is negatively affected by the continuous rains and the frequent snowfalls, whereas growth is favored by the abundant water available in the soil.

Espeletia as a Colonizer of High Mountain Habitats

In previous works (Monasterio, 1979, 1980b) I analyzed several aspects of the desert páramo vegetation. This plant formation is structured by several species of giant rosettes of *Espeletia*. Since it occurs in a very open and discontinuous pattern (Figs. 3–1 and 3–7), the desert páramo leaves large areas of ground bare. Thus, it converges physiognomically toward other desert formations, particularly those of some mountain deserts in the Central Andes dominated by columnar cacti. In both cases we are dealing with cold deserts, though the desert páramo is much moister than the succulent deserts of the Central Andes.

The upper, woody layer in the desert páramo is formed by the giant rosettes of several species of *Espeletia*, while the ground layer is made of dwarf cushions (including *Mona*, *Azorella*, *Arenaria*, and *Calandrinia*), small rosettes (such as *Draba* and *Senecio*), and juveniles of *Espeletia*. As we have seen, the ground layer is subjected to hard microclimatic conditions. The adult rosettes of *Espeletia* in the upper layer live under a more stable thermal microclimate, since the crowns of the rosette trees are located well above the ground. A conspicuous feature of this plant formation is the maintenance of dead leaves attached to the trunks of *Espeletia* (Fig. 3–1), forming a thick mantle of slowly decomposing tissues that insulate the vascular system of the plant.

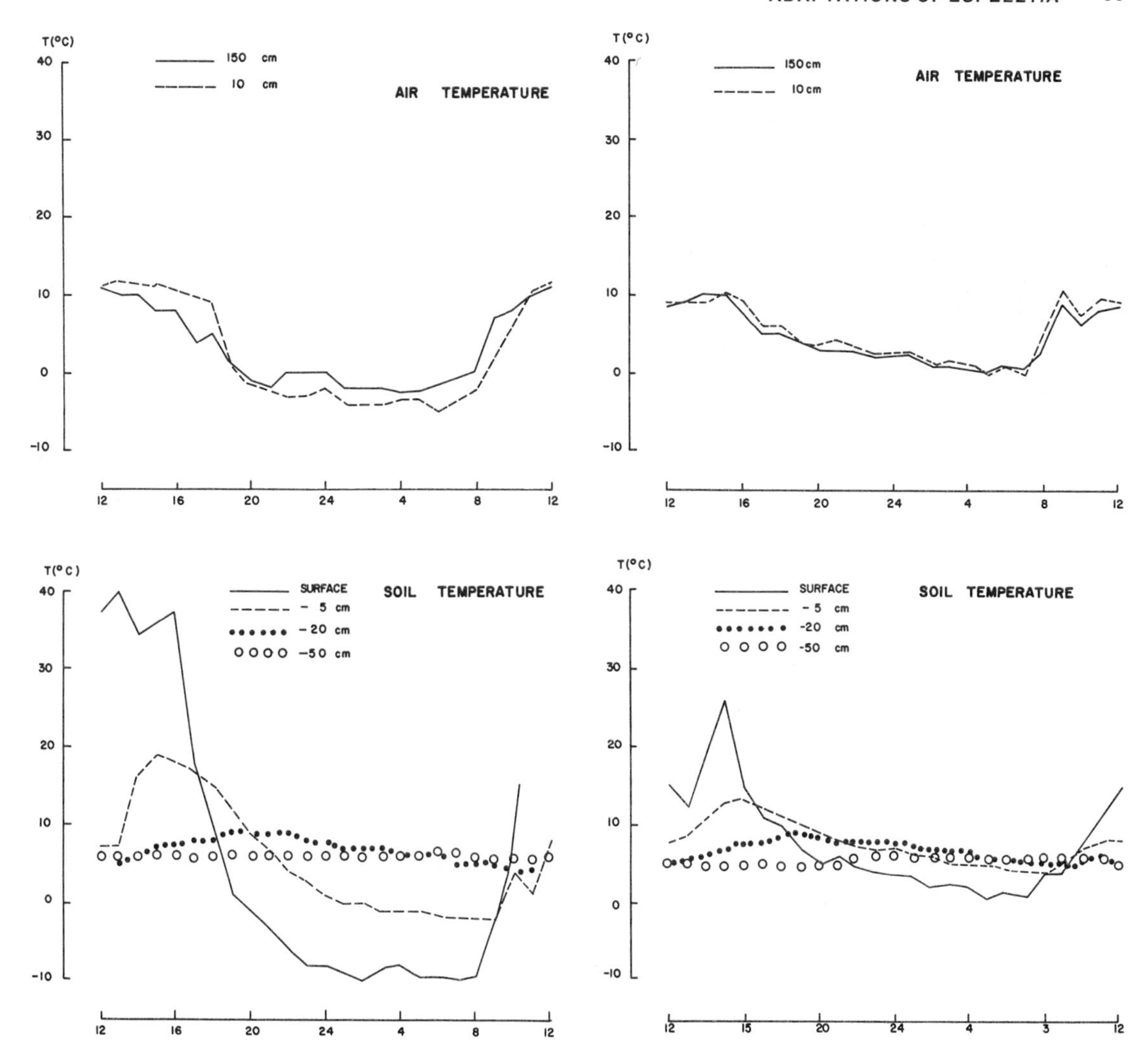

FIG. 3–5. *Top*: Daily course of air temperature 150 cm and 10 cm above ground for a 24-hour cycle in the dry season (January 12–13, 1978), at the Páramo de Piedras Blancas. *Bottom*: Daily course of soil temperature at various soil depths.

FIG. 3–6. *Top*: Daily course of air temperature 150 cm and 10 cm above ground for a 24-hour cycle in the wet season (August 1–2, 1978), at the Páramo de Piedras Blancas. *Bottom*: Daily course of soil temperature at various soil depths.

The total cover of *Espeletia* varies widely according to both altitude and habitat conditions, particularly the slope and the nature of parent materials. On gentle slopes, valley bottoms, and depressions, *Espeletia* may account for as much as 30% of ground cover. On steep slopes, instead, where most of the ground remains completely bare, the rosettes scarcely attain a cover of about 5%. The same factors affect the occurrence of the ground layer, which ranges in cover from 40% down to 2%.

Three species of *Espeletia* have colonized the periglacial habitats of the desert páramo in the Andes of Mérida: *E. timotensis* with its forma *lutescens* (see Cuatrecasas, Chap. 11), *E. spicata*, and *E. moritziana* (Cuatrecasas considers that these three species belong to his genus *Coespeletia*; see Chap. 11). Each of these species is exclusive of a given habitat where it coexists with a different set of ground layer species. Therefore, three different associations were distinguished in the desert páramo (Monasterio, 1979, 1980a). Given that each species population colonizes a different habitat and that the various types of habitats form a mosaic of small patches, any type of interspecific competition is avoided, with the possible exception of competition for biotic resources, such as pollinators, that might be

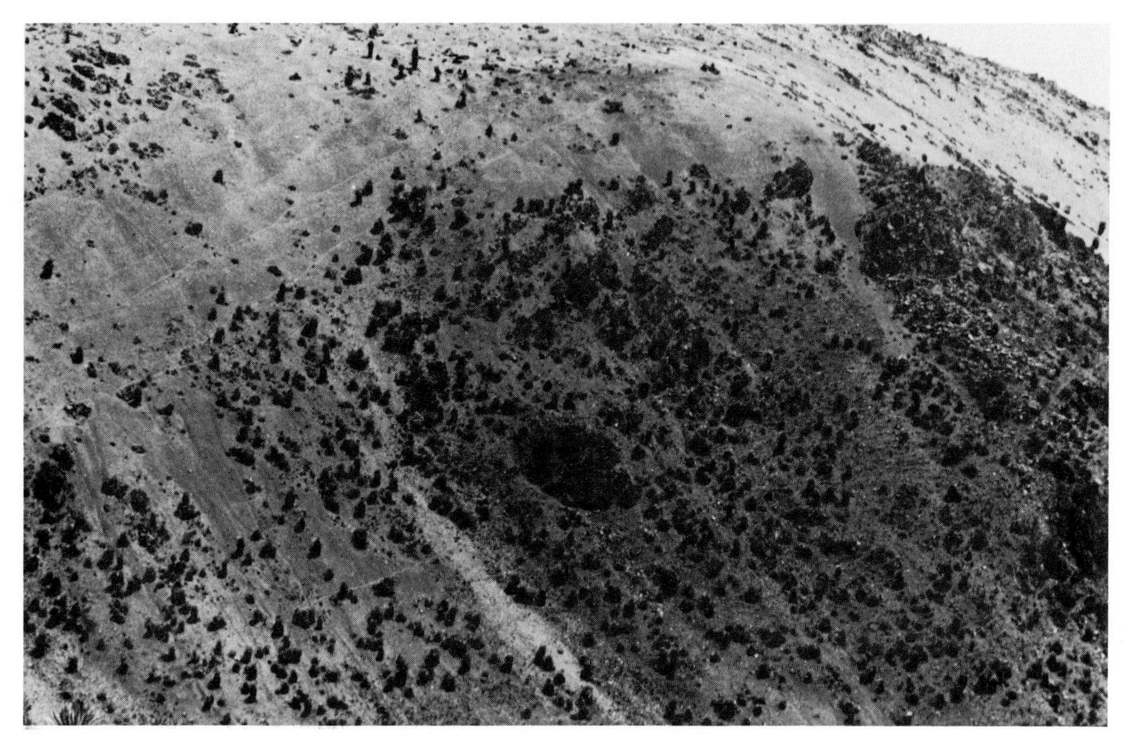

FIG. 3–7. A population of *E. timotensis* in the desert páramo. Páramo de Piedras Blancas, Sierra de la Culata, 4100–4300 m. Notice the distribution and cover of the rosettes. Our permanent plots are located at this site. Photo by Nuni Sarmiento.

shared by several of the adjacent species populations.

The desert páramo extends over a landscape modeled by late Pleistocene glacial events that left the characteristic features of glacial sculpture and depositions, such as glacial troughs, hanging valleys, roches moutonnées, cirques, horns, and till deposits. *E. timotensis* shows the widest ecological distribution and occurs between 4000 and 4500 m. Its habitats are the steep slopes covered by slope deposits, together with some erosion forms, like walls of cirques and valleys, covered with debris ranging in size from 2 to 20 cm, smaller sizes predominating. This is the species of *Espeletia* able to colonize the finest materials. Fine debris, together with the steep slopes, leads to great soil instability. However, when *E. timotensis* succeeds in colonizing this substrate, it acts as a stabilizing agent, slowing down the sliding of the solifluction mantle. The main soil pattern associated with these habitats is the striped ground.

E. timotensis shows true pioneer attributes, since it normally colonizes bare ground not previously occupied by other species. This means that a typical plant succession from lower plants to flowering species does not occur in this case. This pioneer behavior must be of great adaptive value in environments in which a sequence of glacial and interglacial events has continually changed ecological conditions. Besides, and in another temporal scale, that of present-day morphodynamics, the cryoturbation structures originating in the daily frost–defrost cycles hinder any possible stabilization of these high Andean habitats. The populations that colonize this environment play an important moderating role, fixing and wedging the mobile ground that thus forms small platforms or microterraces, as happens with *E. timotensis* on many of the steep slopes where it becomes established.

E. moritziana occurs on rocky sites, such as rocky summits, aretes, and horns, or over coarse periglacial debris formed by large boulders that carpet certain slopes in cirques and troughs (Fig. 3–8). Altitudinally, it ranges from 4100 to 4600 m, its upper limit surpassing that of *E. timotensis* because *E. moritziana* colonizes summits and mountain crests. The morphoclimatic effects of the periglacial climate are attenuated in the habitat of this species because rock interstices behave like thermal refuges by maintaining a more temperate microclimate that allows these populations to reach higher elevations.

E. spicata is found between 4000 and 4300 m, in habitats intermediate between those of the two

FIG. 3–8. *E. moritziana* growing on rocky crevices in the Páramo de Piedras Blancas, 4200 m. Photo by Carlos Estrada.

FIG. 3–9. *E. spicata* is normally found in intermediate habitats, covered by boulders and small angular debris. Páramo de Piedras Blancas, 4200 m. Photo by Carlos Estrada.

aforementioned species (Fig. 3–9), that is, on gentle slopes of cirques and valleys covered with medium-sized debris up to 50 cm in length.

Finally, I have to mention two additional species. In the desert páramo *E. semiglobulata* occurs in peat bogs between the steps of cirques and at the base of retreating glaciers, a perhumid environment where its decumbent trunks form giant cushions. *E. schultzii*, the most important species of *Espeletia* in the contiguous and lower Andean belt, reaches up to 4200 m in the desert páramo, where it occupies small areas having wetter conditions. This is not a giant rosette since it has only a small trunk. It is unable to colonize the extreme habitats of the high Andean belt, but its occurrence in small patches within the desert páramo is important, as we will see later, when the phenological niches of high Andean *Espeletia* are considered as a whole.

A Global Project Involving *Espeletia* of the Desert Páramo

As one approach to an understanding of the desert páramo ecosystem, my colleagues and I undertook a detailed analysis of the global strategies of the main species occurring in this belt: *E. timotensis*, *E. spicata*, and *E. moritziana*. Our study was carried out in the Páramo de Piedras Blancas (Fig. 3–2) between 4100 and 4300 m. The fieldwork began in 1971; intensive sampling in permanent plots started in 1977 and continues up to now.

Three main lines of research were followed. The first deals with the global strategy of these species, including aspects of demography, phenology, and growth. The second concerns the temperature balance and the water and photosynthetic budgets, and the third line covers reproductive aspects with particular emphasis on pollination. In this chapter I refer mainly to the first approach, mentioning very briefly some ecophysiological results that may aid in the interpretation of plant strategies.

As elements of the global adaptive strategy allowing the ecological success of *Espeletia* populations in the harsh environment of the desert páramo, we selected several events and processes of their life cycle. These include the architectural pattern, the vegetative and reproductive phenodynamics, leaf production and its rhythmicity, growth of leaves and reproductive structures, attainment of sexual maturity, reproductive effort, and seed rain. Field data were obtained from several hundred marked individuals in permanent plots. Several thousand additional individuals were sampled at regular intervals near the permanent plots. As the Páramo de Piedras Blancas can be reached after a two-hour drive from our laboratory in Mérida, we were able to make the intensive sampling required for this type of analysis. The results on populations dynamics will appear in another publication, and part of the ecophysiological aspects has already been published (Goldstein and Meinzer, 1983; Goldstein, Meinzer, and Monasterio, 1984).

THE DYNAMICS OF GROWTH AND REPRODUCTION IN *ESPELETIA*

In order to understand the adaptive strategies of *Espeletia* in the desert páramo, it seems essential to know the dynamics, capacity, and seasonal modalities of growth, as well as the part of the annual energy budget assigned to each organ and function. The phenological patterns and the reproductive behavior of their populations and individuals are also fundamental in the interpretation of their success as colonizers of the highest elevations of the cold tropics. To start with, we consider the architectural model of *Espeletia*, since it is the basis for an explanation of its dynamics of growth and reproduction.

Architectural Pattern

The three species of *Espeletia* that characterize the desert páramo belong to the Corner architectural model, defined by Hallé, Oldeman, and Tomlinson (1978) as monoaxial trees with indeterminate growth and lateral inflorescences. The Corner model corresponds to the life form *tall, polycarpic, pachycaulis, caulirosula* (Cuatrecasas, 1979), and to what Mabberley (1974) considered as "unbranched pachycaulis." Du Rietz (1931) employed the attractive name of "rosette tree." The advantage of architectural models over traditional life-form systems is that they imply and synthesize the dynamics of growth that expresses itself in a precise architectural pattern through the patterns of meristematic activity throughout the life cycle of the plant. In this sense, the ecological constraints of the desert páramo are reflected in the sequential growth patterns of the various plant parts as well as in their spatial distribution.

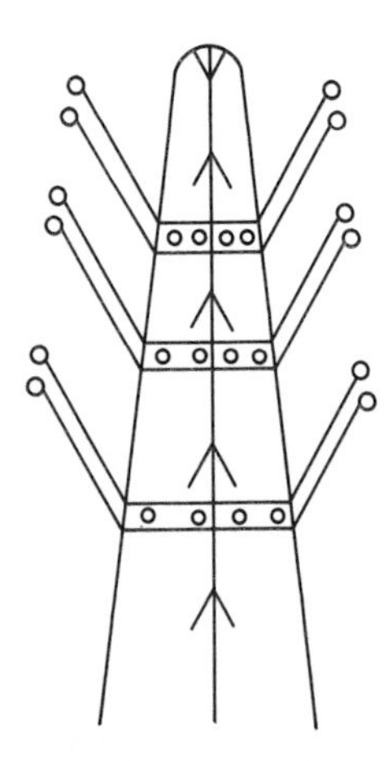

FIG. 3–10. Corner Model "monocaulous polycarpic tree," corresponding to all the species of *Espeletia* of the desert páramo. The stem has continuous growth and leaf production, but reproductive events occur at various frequencies throughout the adult plant life.

Specific Elements of the Corner Model in Desert Páramo Espeletia

In accordance with Corner's model, *E. timotensis*, *E. spicata*, and *E. moritziana* have a monocaulous structure built by a single shoot apical meristem responsible for the aboveground vegetative biomass (trunk and leaves). In contrast, the root system is built by various independent buds. The inflorescences are lateral, produced by

FIG. 3–11. The most conspicuous representative of the Corner Model is *E. timotensis*. Notice how the active leaves and the reproductive structures are located well above ground level. Photo by Nuni Sarmiento.

TABLE 3–2. Morphological and life cycle characteristics of *Espeletia* spp. in the desert páramo.

Species	Size adults (m)	Annual growth/ trunk (mm)	Estimated length of life cycle (years)	Mean number of leaves per rosette	Leaf size (length, width) (cm)	Length of inflor-escences (cm)	Average no. of heads/ inflor-escence	Head diameter (cm)
E. timotensis	2–3	15	170	240	50 × 3.5	110	11	3.5–4
E. spicata	1.5–2.5	20	130	590	35 × 2	80	22	2–2.5
E. moritziana	0.5–1	15	70	264	45 × 2	55	1	4–6

active buds in the axil of the young leaves of the rosette. Consequently, growth is indeterminate since it is not interrupted by reproduction, and the species are polycarpic. Monocarpic *Espeletia* differ from polycarpic ones in that their growth stops at the moment of reproduction, and they occur only below 4000 m, not being able to pass the barrier of the periglacial climate above 4000 m.

The three species have continuous growth, both in length of the trunk and in leaf production. The inflorescences, emergent from the axil of leaves, vary in their times of expansion at the specific, population, and individual levels (we will discuss this later). Corner's model of desert páramo *Espeletia* is exemplified in Figs. 3–10 and 3–11. Other elements associated with this architecture are presented in Table 3–2.

The growth in length of the unique axis and the production of leaves are continuous and synchronized as evidenced by the uniform pattern of leaf scars that may be observed on the denuded trunks. Phyllotaxis is typically spiral; the rosettes have spirally arranged leaves and congested persistent leaf scars on the trunk. Leaf abscission does not occur in these species; the leaves remain attached to the trunk and decompose very slowly during the life cycle of each individual, thus forming a case of marcescent leaves down to its base. Cuatrecases (Chap. 11) says: "The leaves of these species have enlarged imbricated sheaths arranged in multispiral circles at the end of the stem. The older, marcescent leaves attached to the stem embraced it with their closely imbricated sheaths." The trunk appears to be several times thicker than the real stem and acquires a columnar aspect. The adaptive significance of this bulky covering of dead leaves on the stem will be discussed later.

Biomass and Energy Allocation

The architectural pattern of *Espeletia* in the desert páramo reflects the allocation of their available energy to each plant organ or function. Consequently the architecture has great ecological importance since it controls the spatial distribution, both vertical and horizontal, of the energy incorporated into the ecosystem. Thus, architectural model and energy pattern are closely coupled.

It is of great interest to know how individuals growing under periglacial constraints allocate their resources to different functions, and to consider whether these allocation patterns constitute successful "decisions" that maximise their fitness in this environment. The proportion of the biomass allocated to aboveground and underground organs, to vegetative and reproductive structures, to assimilative organs, and to cryoprotection is particularly indicative in this context.

All three species of *Espeletia* in the desert páramo have the same pattern of resource distribution, only small features distinguishing each other. Figure 3–12 represents the case of a medium-sized individual of *E. timotensis*. The caloric values of the various plant parts appear in Table 3–3. Some main points to note are:

1. The aerial parts accumulate an overwhelming proportion of the total calories (96.6%) against only 3.4% represented by the underground biomass.
2. Within the aerial parts, the allocation of energy favors leaves (79.27%) over reproductive structures (inflorescences, heads, achenes: 11%), while only 4.4% is devoted to the stem.
3. Most of the energy stored in leaves corresponds to dead leaves (71.6%); green leaves instead accumulate 17.97% of the total stored energy.
4. The investment for rooting and support in these rosette trees is small—about 7% of their total energy. Given the unstable nature of the substrate, the root system has to be highly efficient to anchor the plant, and the stem is indeed very efficient to support

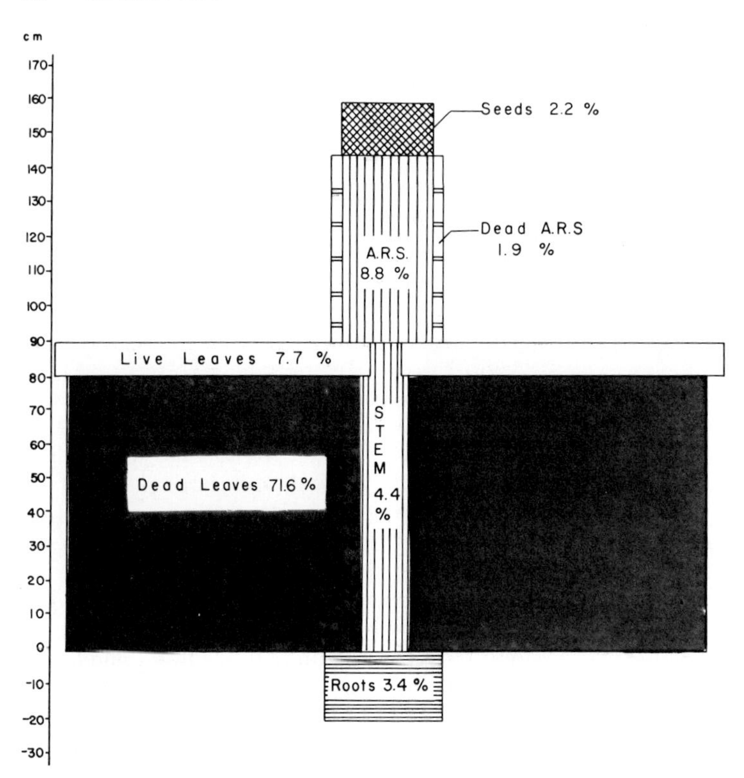

FIG. 3–12. Biomass allocation (%) in calories toward different plant organs in *E. timotensis*. Each figure is the mean of 10 replicates corresponding to 10 adult plants.

TABLE 3–3. Caloric values per gram of various plant structures in *E. timotensis*.

		Kcal/g
Live leaves		
Leaves of the bank		4.2
Juvenile leaves	(14– 28 cm^2)	4.1
	(28– 56 cm^2)	3.9
	(56– 84 cm^2)	4.5
Adult leaves	(84– 98 cm^2)	4.5
Mature leaves	(98–130 cm^2)	4.5
Pre-senescent leaves	(130–140 cm^2)	4.9
Senescent leaves		4.7
Dead leaves		4.4
Stem		
Cortex		4.7
Pith		4.3
Roots		4.9
Reproductive parts		
Inflorescences		4.4
Heads		4.7
Achenes		4.9
Maximum biomass per plant (g)	: 8,441.91	
Maximum biomass per plant (Kcal)	: 38,339.5	
Estimated annual leaf production (Kcal)	: 3,150.0	

Note: Each figure is the mean of 10 samples corresponding to 10 adult individuals.

the large mass of green and dead leaves and the reproductive structures as well.

5. *E. timotensis* clearly assigns the greater part of its energy to its photosynthetic structures, and most of it remains on the plant forming the cover of dead leaves around the trunk. However, this dead material, attached to the stem for the life span of the individual plant, does not represent a reservoir of energy to be used again, but it plays an important role in the transfer of nutrients from decaying leaves to actively growing tissues (Monasterio, 1980b). Besides this function as a nutrient sink, the cover of dead leaves isolates the living tissues of the stem, protecting them against low temperatures and regulating their water balance (Goldstein and Meinzer, 1983). Furthermore, the necromass is colonized by rich and diversified meso- and microfaunas that contribute to the cycling of nutrients in this ecosystem (Garay, Sarmiento-Monasterio, and Monasterio, 1982).

A first general conclusion that follows from this pattern of energy allocation is that the energy produced throughout the life cycle of the desert páramo *Espeletia* accumulates not just in their growing organs, such as stem and roots, but chiefly in the slowly decomposing cover of dead leaves that encloses the whole stem. The only energy dispersed by these plants is that forming part of the reproductive structures, mostly the achenes, since even the old inflorescences remain attached to the stem for a relatively long time.

The investment in reproduction appears to be rather high (11%), particularly if we consider that only one reproductive event was considered in our energy budget. A substantial part of the energy allocated to sexual reproduction goes to auxiliary structures (8.2%), but achenes represent only 2.2%. Thus most of the reproductive effort goes to structures that protect flowers or, like the axis of the inflorescences, dispose them at a favorable level either to facilitate pollination or to escape frost injury. The high investment in auxiliary reproductive structures gives a certain protection to all stages of the long reproductive process from buds to anthesis and seed dispersal. Figure 3–11 illustrates how the inflorescences of *E. timotensis* reach an air layer somewhat higher, and possibly with a more stable microclimate, than that of the rosette of green leaves.

Finally, it must be considered that the energy accumulated at any given moment in the life cycle of individuals of these species constitutes a fair approximation of the total energy allocated to vegetative plant parts throughout their life span. This is because the biomass allocation carries the whole history of the individual, since besides those parts with increasing biomass accumulation (e.g. the stem and roots), all leaves produced by an individual, as well as a part of the inflorescences, remain attached to the stem until its death. I will return to these aspects when dealing with reproduction.

Dynamics of Vegetative Growth

The rosette trees, ever-growing and with continuous activity, present the same aspect at any time of the year: the composition of the rosettes always looks uniform, and no ostensible changes in cover and biomass are noticeable at first sight. In this section, I analyze the rosette as a population of parts, considering the number of leaves existing at any given time, the rhythms of leaf initiation, expansion, and death in the dry and wet seasons, and the rates of leaf growth and turnover times of the rosette.

The principal problem is whether, in spite of their continuous growth, *E. timotensis*, *E. spicata*, and *E. moritziana* show a difference in rhythmicity along the seasons coupled with environmental changes, or whether an almost absolute constancy predominates, or else whether the dynamics of growth shows some endogenous rhythmicity.

The Demographic Structure of the Rosette

If the rosette is considered a whole composed of modular units, i.e., the leaves, a demographic approach can be applied to this ensemble. The knowledge of the demographic structure of leaves provides data useful to interpret the dynamics of leaf growth and the turnover time of the rosette.

The demographic approach applied to modular plant units is of recent use (Harper, 1977; Harper and Bell, 1979), but the concepts on which it is based are somewhat older (White, 1979). White (1979) considers the plant to be a metapopulation of parts.

In the cases of *E. timotensis*, *E. spicata*, and *E. moritziana*, which reproduce only by seeds, the modular units, the leaves, of a given individual all belong to one particular genet and in this sense they constitute a homogeneous ensemble.

The rosette with its spiral and compact arrangement of leaves presents at any given time the same sequence of developmental stages in its population of leaves. In its center (Figs. 3–13 and 3–14), a compact core of immature, not yet expanded, leaves, here named the *leaf bank* (foliar primordium and immature leaves), surrounds the apical meristem. Near or very near this center stand the recently expanded young leaves, and in sequence from the center to the periphery are all the other developmental stages, corresponding with increasing age to young, adult, mature, to senescent and finally dry leaves, thus gradually adding to the collar of dead leaves around the trunk.

Figure 3–15 represents the demographic structure of the rosette of *E. timotensis* on the basis of foliar area, which our data indicate has a 100% correlation with leaf length in this species. Two maxima appear in the frequency distribution, the first and higher one corresponding to the unexpanded units in the leaf bank (0–14 cm^2). This maximum is followed by a conspicuous inflection that corresponds to young leaves of intermediate sizes: 28–42, 42–56, 56–70, and 70–84 cm^2. Each successive age class has a smaller number of leaf units. The second maximum, which gives rise to a quite abrupt discontinuity in the graph, corresponds to adult leaves in the range of 84–126 cm^2, already in the final stages of their growth and development. The last class corresponds to leaves that are old but still alive. This increase in foliar

FIG. 3–13. Rosette of *E. timotensis*. In its center appears the leaf bank surrounding the single apical bud. Photo by Carlos Estrada.

FIG. 3–14. Rosette of *E. spicata*; various developmental leaf stages may be seen. Photo by Carlos Estrada.

area agrees with the disposition of leaves from the center to the top of the rosette. The distribution from center to periphery corresponds to sequential stages of development, and this arrangement agrees with the phyllotaxis in spiral.

The static vision of the rosette as represented in Fig. 3–15 needs for its interpretation the analysis of each leaf-area class in terms of its dynamics of growth and of the role of each leaf class in the productive process. Figure 3–16 indicates the dry weight of leaves per size class. In this example, the total dry weight of live leaves in the rosette attains 1665 g, while the leaf bank, with 174 units, attains 8.5 g, that is, 0.33% of the total weight. The leaf bank is not in contact with the external environment, their units do not photosynthesize, are yellow-white in color, and show only minimal growth, remaining in a state of dormancy or rest. For this reason we consider them to be a bank or reservoir of foliar resources. These very immature leaves behave as consumers and importers of resources from the rest of the plant.

In the external part of the rosette, the class of 14–28 cm^2 corresponds to freshly expanded leaves that still behave as consumers, their color being pale yellow. In the next several classes and up to 84 cm^2 of foliar area, a transitional and gradual process occurs that changes them from consumers to producers. But together with assimilation, leaf growth and elongation become the most conspicuous processes at this stage. The "residence time" in each of these classes is quite short.

The second maximum, 84–126 cm^2, corresponds to leaves at the peak of their assimilation process, the most efficient producers in the rosette, with gray-green color. Their growth slows while the residence time increases, and these assimilating leaves attain their optimum and become stable. As Fig. 3–16 shows, the dry

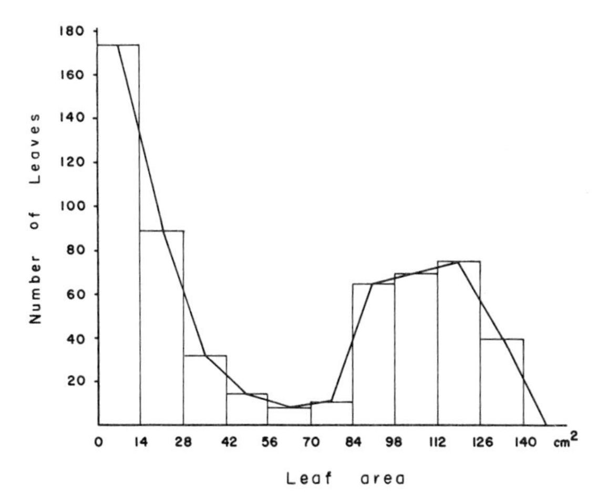

FIG. 3–15. Demographic structure of the rosette in *E. timotensis* showing the distribution of the population of leaves in leaf area classes.

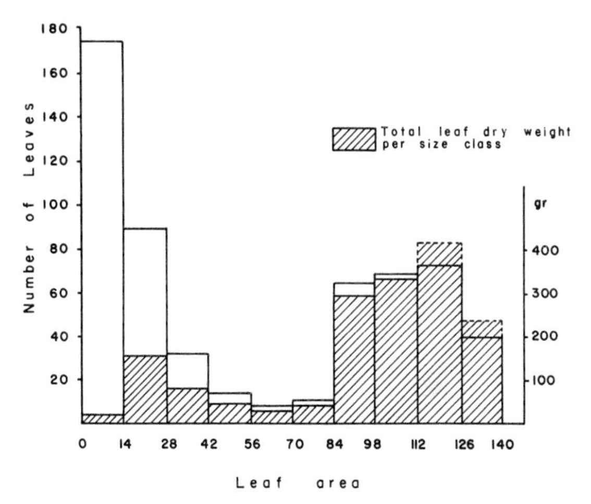

FIG. 3–16. Relationship between leaf area and leaf dry weight in *E. timotensis*.

weight of leaves in the classes from 84 to 126 cm^2 reaches 1050 g, which represents 63% of the total weight of the rosette. The last class reaches the top size and growth is then finished; the first signs of decay are already apparent. It is a fragile class, but assimilation remains active.

Leaf Life Cycle

As pointed out by Harper (1977), "a leaf has a life history, a changing pattern of behavior from birth, as a primordium on a meristem, to death from senescence or some environmental hazard." Figures 3–17 and 3–18 depict the leaf life cycles of *E. timotensis* and *E. spicata*, respectively. In the case of *E. timotensis* the measurement of leaf length of a cohort of leaves starts at the point of leaf expansion from the leaf bank. In the initial stages, the growth in length is only 1 mm/day, with a gradual increase up to a maximum rate of 2.5 mm/day. During the first six months, the growth rate curve shows a steep slope. During this period, the leaves expand up to 40 cm in length until they retain the maximum size of 50 cm. The maximal growth rates correspond to foliar areas between 45 and 100 cm^2, the minimal rates with areas up to their top size. Correspondence between length, area, and dry weight was obtained for about 500 leaves. In Fig. 3–17, at the start of the measurements, the recently expanded leaves have a dry weight of 0.68 g. After seven months of growth and assimilation, they reach the top weight of 6.29 g per leaf—an increase on the order of 6 g.

The top size and weight are reached seven months after the start of growth. After that, leaves remain in the apogee of their function for 14 or 15 consecutive months, and the first signs of senescence begin only 22 months after their start. The cohort studied since November 1978 showed two years later clear signs of withering but their basal parts were still active. Therefore, the leaf life cycle in *E. timotensis*, from the start of growth from the leaf bank and its incorporation into the external rosette, covers at least 20 months of optimal activity. But we must also consider the length of time they remain in the leaf bank, which we estimated to be 1.4 years. The total life cycle of the leaves in this species would then be over three years.

As each leaf develops, its position in the rosette changes. Thus, a spiral from the apical bud to the border of the rosette is traced during the leaf life cycle. Hence, the microenvironment of the leaf varies along this pathway, from the refuge conditions in the leaf bank, when leaves are totally dependent, to leaf expansion. This implies not only growth and assimilative autonomy, but also gradual changes in leaf angles that become greater as they approach senescence and then become incorporated into the collar of dead leaves. This angle has great significance, since it controls the incident light on the leaf surface, therefore influencing phtosynthesis and the temperature balance of the leaf. Following their spiral trajectory for two years, each leaf traverses a gradient of microenvironmental conditions within the rosette.

Leaf mortality is closely linked with leaf senescence, and each leaf dies when its life cycle is completed. The only mortality observed during field measurements was due to predation, mainly by insects, and it was quite low since insects produce some leaf damage but rarely leaf death. It

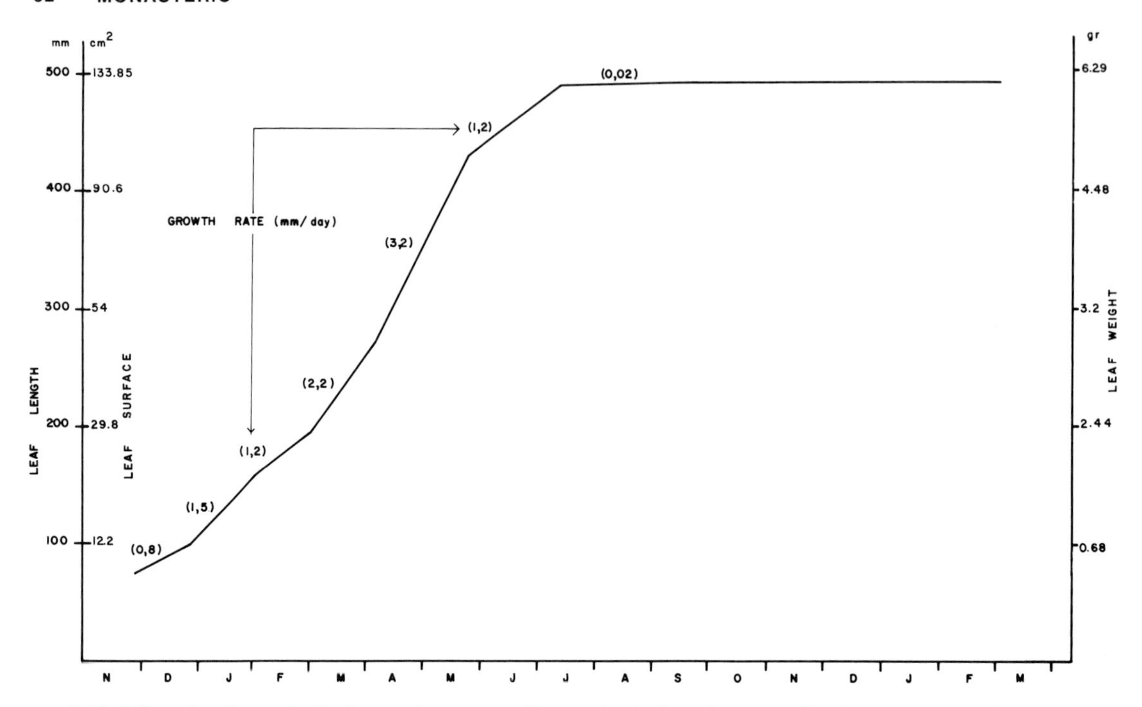

FIG. 3–17. Life cycle of leaves in *E. timotensis* corresponding to a leaf cohort that started in November 1978. Growth rates in each developmental stage are indicated as well as the correlation between leaf length, area, and dry weight.

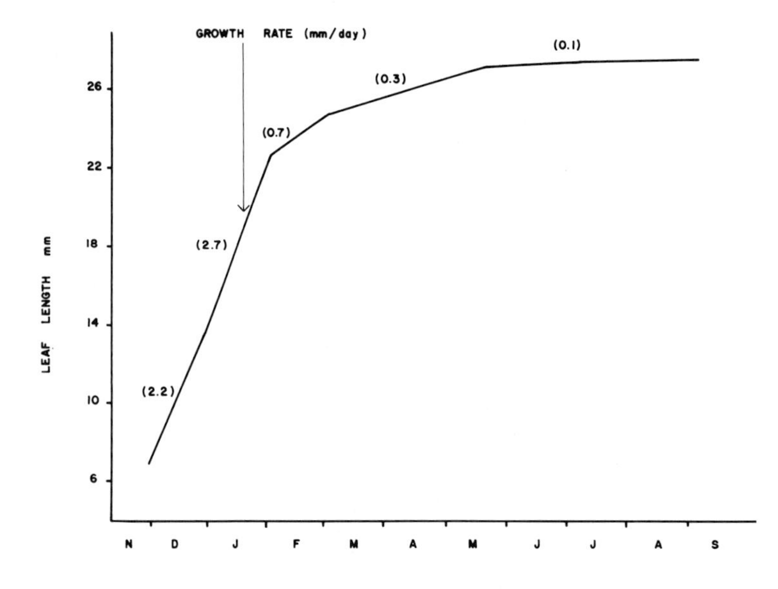

FIG. 3–18. Life cycle of leaves in *E. spicata* corresponding to a leaf cohort that started in November 1978. Growth rates in each developmental stage are indicated.

was not apparent whether any mortality was due to environmental hazards, caused either by the weight of the snow over the leaves during periods of heavy snowfall, or by frost for many consecutive days. However, during some exceptionally dry years, whole plants as well as certain leaves may die.

I refer now to the leaf cycle in *E. spicata* (Fig. 3–18). Its behavior is similar to that of *E. timotensis*, although development is more rapid during the first four months to reach the top size six months after their start. Nevertheless, the leaf span in the external rosette is shorter, being totally accomplished in 12 or 14 months. In *E. moritziana* it is less than one year.

Leaf Life Cycle and the Demographic Structure of the Rosette

The analysis of differential growth rates along the development of the leaves of *E. timotensis* allows one to interpret the demographic structure of the leaf population in the rosette. Why do leaf units accumulate in the leaf bank and also in adult size classes? Why are there so few leaves in other intermediate developmental stages?

Comparing Figs. 3–15 and 3–17, we see that low growth rates correspond with the accumulation of leaves in a given size-class, whereas a high growth rate, such as the one characterizing young leaves, brings about a rapid change toward the larger size classes.

Growth Rates in the Wet and Dry Seasons

Leaf growth seems to depend on different factors. Some are ontogenetic and relate to the developmental stage reached by each leaf, whereas others correspond to moisture seasons (Figs. 3–17 and 3–18). Maximal growth rates are found in young leaves: 14–30 cm in length in *E. timotensis*, 10–20 cm in *E. spicata*. In the latter species, maximum rates of 3.45 mm/day were recorded in leaves of 10–15 cm during the wet season, but this size class only attained 3.0 mm/day during the dry season. Similar results may be seen in Tables 3–4 and 3–5.

In the foliage of the ever-growing *Espeletia* of the desert páramo, growth takes place continuously, but not at constant rates. This implies that leaves incorporated from the leaf bank into the external rosette will receive different impulses for their initial development or for their maturation according to the season of the year. Nevertheless, given the extended leaf life cycle in these species, particularly in *E. timotensis*, each leaf, independent of its time of inception, will later be subjected to the influence of several dry and wet periods. Besides, its spiral trajectory will be conditioned by other leaves preceding or following it in the rosette. Observations by Estrada and Monasterio (unpublished) suggest that the duration of the leaf life cycle in these species is constant and unrelated to the starting period of each leaf as well as to the age of the individual to which they belong.

TABLE 3–4. Leaf growth rates of *E. timotensis* in the dry and rainy seasons for different size classes. Mean and standard error for 26 leaves in three plants.

Size class (cm)	Growth rate (mm/day)	
	Dry season	Wet season
6–13	1.51 ± 0.15	–
13–21	1.72 ± 0.16	–
21–29	1.93 ± 0.14	3.11 ± 0.15
29–36	1.83 ± 0.2	2.12 ± 0.29
36–43	0.56 ± 0.09	0.52 ± 0.12
43–50	0.11 ± 0.02	0.20 ± 0.04

TABLE 3–5. Leaf growth rates of *E. spicata* in the dry and rainy seasons for the different size classes. Mean and standard error for five plants.

Size class (cm)	Growth rate (mm/day)	
	Dry season	Wet season
5–10	2.55 ± 0.18	2.94 ± 0.12
11–15	3.00 ± 0.13	3.45 ± 0.09
16–20	2.78 ± 0.15	3.11 ± 0.12
21–25	1.83 ± 0.16	2.20 ± 0.11
26–30	0.40 ± 0.09	0.48 ± 0.06
31–35	0.05 ± 0.02	0.17 ± 0.03
36–40	−0.02 ± 0.03	–

Rosette Turnover Time

Long-term measurements of leaf expansion (three years) were made to determine the total number of leaves expanded per unit time during several different seasons of the year. Several individuals of each species were marked, and the total number of expanded leaves was counted at monthly intervals. We define rosette turnover time as the time taken to renew all leaves in the rosette.

As Fig. 3–19 indicates, the total number of leaves incorporated into the rosette in two consecutive years in *E. timotensis* is 240. The mean rate of leaf expansion in this species is 0.3 leaves per day; that is, a new leaf departs from the leaf bank toward the external rosette approximately every three days.

If the expansion rates during the wet and dry seasons are compared, it can be seen that their differences are not significant. Thus, of the two greatest expansion rates recorded, one was in the wet (0.41 leaves per day), the other in the dry season (0.49 leaves per day). Of two lowest values attained, one was in the dry (0.23 leaves per day), and the other in the wet season (0.21 leaves per day).

Certain anomalies in these expansion rates, however, deserve further comment. At the beginning of 1980 (Fig. 3–19), in the middle of the dry season, no leaf departure was recorded during a period of 35 days. But during this time, the leaf bank increased noticeably in all three observed species. A possible explanation for this is that normally the lowest temperatures occur in January, when there are frosts during many consecutive nights. But are there any mechanisms that might hinder the departure of leaves from the leaf bank? If the answer is yes, it could be the same type of mechanism as that of nyctaginasis (daily rhythm of opening and closing of the rosette), but in this case blocking leaf expansion. Anyway, the leaf bank appreciably increased during that cold and dry period, but remember that this is precisely the most stable and protected microenvironment within the rosette, and indeed the whole plant.

The mean number of expanded leaves per year was 120; individual variation is rather small. The rosette turnover time for *E. timotensis*, about 24 months, was the greatest determined in any of the three species. The dynamics of leaf production in *E. spicata* is shown in Fig. 3–20. In this species too, leaf expansion seems to be continuous, with an average rate of about one leaf per day. In total, 370 leaves are incorporated into the rosette yearly. These values are roughly double those obtained in *E. timotensis*; however, both leaf size and leaf weight are much higher in the latter species. Rosette turnover time is also shorter in *E. spicata*, about 13 or 14 months.

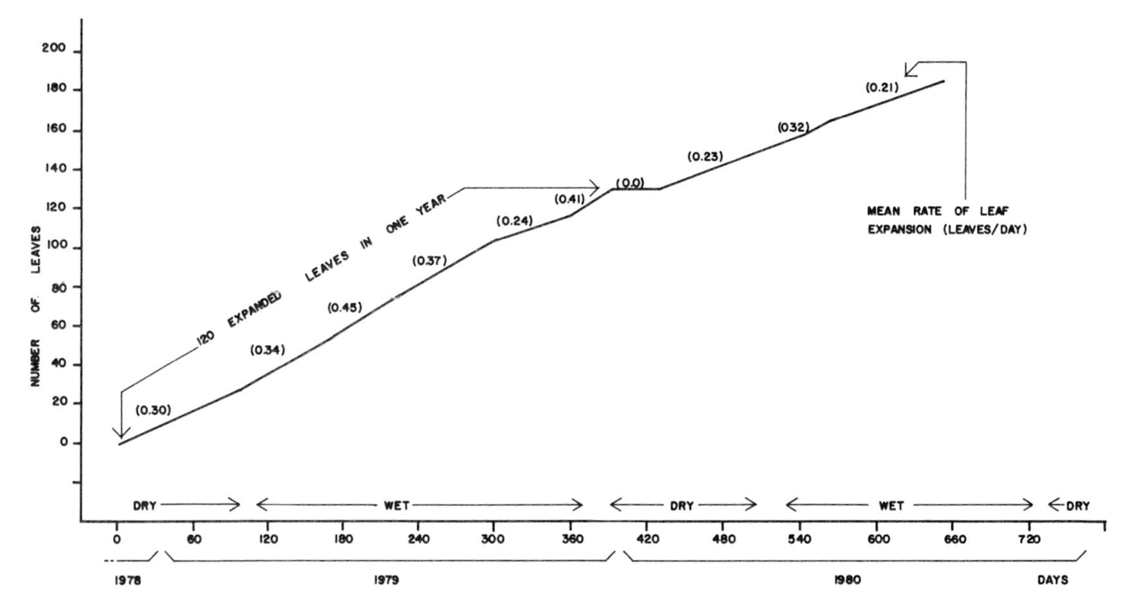

FIG. 3–19. Dynamics of leaf production in *E. timotensis* in contrasting seasons. The estimated rosette turnover time was 24 months.

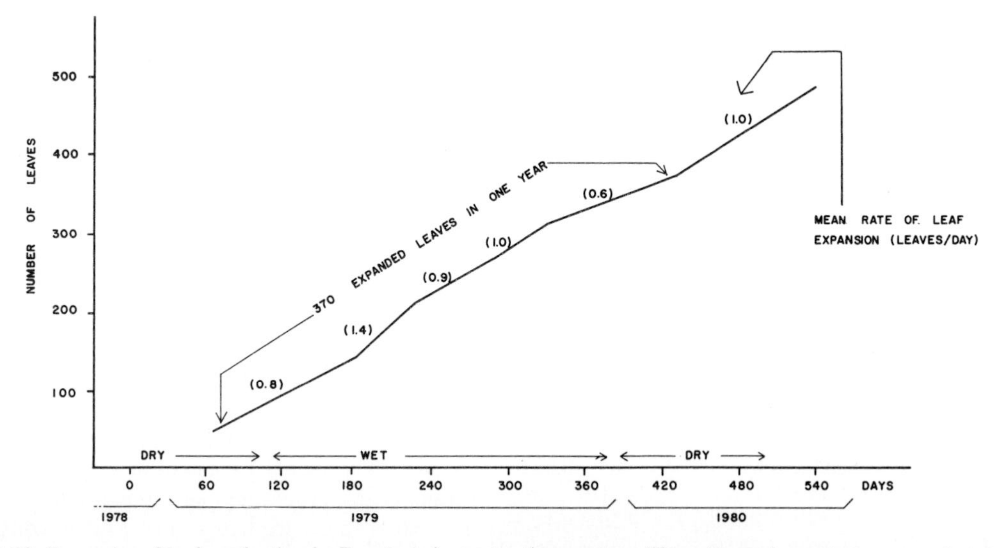

FIG. 3–20. Dynamics of leaf production in *E. spicata* in contrasting seasons. The estimated rosette turnover time was 14 months.

A comparison of the three species shows that *E. timotensis* has the longest turnover time (24 months), whereas in *E. spicata* and *E. moritziana* the respective values were 13–14 months and 12 months. Considering the life cycle of the three species, we see that *E. timotensis* has the longest life span followed by *E. spicata*, and *E. moritziana* (see Table 3–2). Thus, the longest-lived species seems to have the longest renewal rhythms, longevity being expressed not only by the length of the individual life cycle, but also by the longer turnover time of all organs, as I shall confirm later when considering reproductive aspects.

In comparative terms and as a point of reference, I want to consider the rosette turnover time in *E. schultzii*. Smith (1974) presented short-term measurements in populations of this species at various altitudes. At its upper altitudinal limit, about 4200 mm, *E. schultzii* shows a longer turnover time (12 months) than at 3600 m (7 months). Moreover, the various ecotypes along the altitudinal range of the species have different life spans, the most long-lived being at the highest altitude in the desert páramo.

Phenological Patterns

Reproduction is the most conspicuous process in the giant rosettes of the desert páramo. It implies a transitory rupture with the monocaulous form due to the appearance of numerous inflorescences emerging from the axils of the youngest leaves. Although reproduction is a long-term process in these polycarpic *Espeletia*, which reproduce only sexually, it does not interrupt leaf production or growth.

In the first part of this chapter I emphasized the spatial partition of the niche during the vegetative phases, each species of *Espeletia* having its own closely delimited habitat. This separation implies a sharp division of physical resources such as water, light, and nutrients. Apparently, there is no actual competition among these populations, since the niche is spatially partitioned. But what happens with the phenological niche? Are the reproductive processes synchronous in the various species or are the homologous phenological phases asynchronous. An analysis of the phenological pattern of each species may also provide a key to the degree of reproductive isolation among these closely related taxa.

I shall discuss in some detail the phenological pattern and the reproductive behavior of *E. timotensis*; the other *Espeletia* of the high Andean belt will be considered for comparative purposes only. Our phenological studies in the desert páramo started in 1971, and from 1977 on we gathered quantitative data on several hundred individuals in permanent plots.

The Phenological Pattern of E. timotensis

Our preliminary observations since 1971 suggested that this species has aperiodic reproductive behavior, because during several consecutive years its populations remained entirely vegetative. This was not the case with the other *Espeletia* in the desert páramo, which entered reproductive phases each year, as also pointed out by Smith (1974). In order to study this behavior more closely, we established permanent plots to follow reproduction at both the population and at the individual levels.

Apart from its aperiodic reproductive activity, when a given reproductive event erupted, the various phenophases followed a precise chronology. I give as an example the 1978–79 phenological cycle, for which I have data from 344 inflorescences that gave rise to 3677 heads.

The reproductive phenorhythm of *E. timotensis* may be summarized as follows (Fig. 3–21). The reproductive process starts in the middle of September with the emergence of the inflorescences, which are no more than 5 mm long and are scarcely perceptible hidden inside the rosette. From this starting point, a slow process of axis elongation and head maturation continues for almost one year. A maximal elongation rate of 4.98 mm/day was recorded in April–May when the axes were 21–30 mm long. When full grown, the inflorescences surpass 1 m (Fig. 3–21). A huge energetic investment is therefore directed toward the auxiliary reproductive structures whose building up takes about one year. In May, the first female flowers begin to open, followed by the normal sequence of phenophases in the processes of flowering and fruiting. A peak in the blooming of female flowers occurs in August, and a peak in pollination in September, but both processes go on for several consecutive months, from May to December. The maturing and dispersal of the achenes also continue for several months. Toward the end of June of the next year, most achenes have already dispersed, but a few remain on the heads.

In summary we have a period of 21 months from the starting point of reproduction in September 1978 to seed dispersal in June 1980. Thus, the reproductive schedule of this species does not follow an annual cycle but rather extends over a

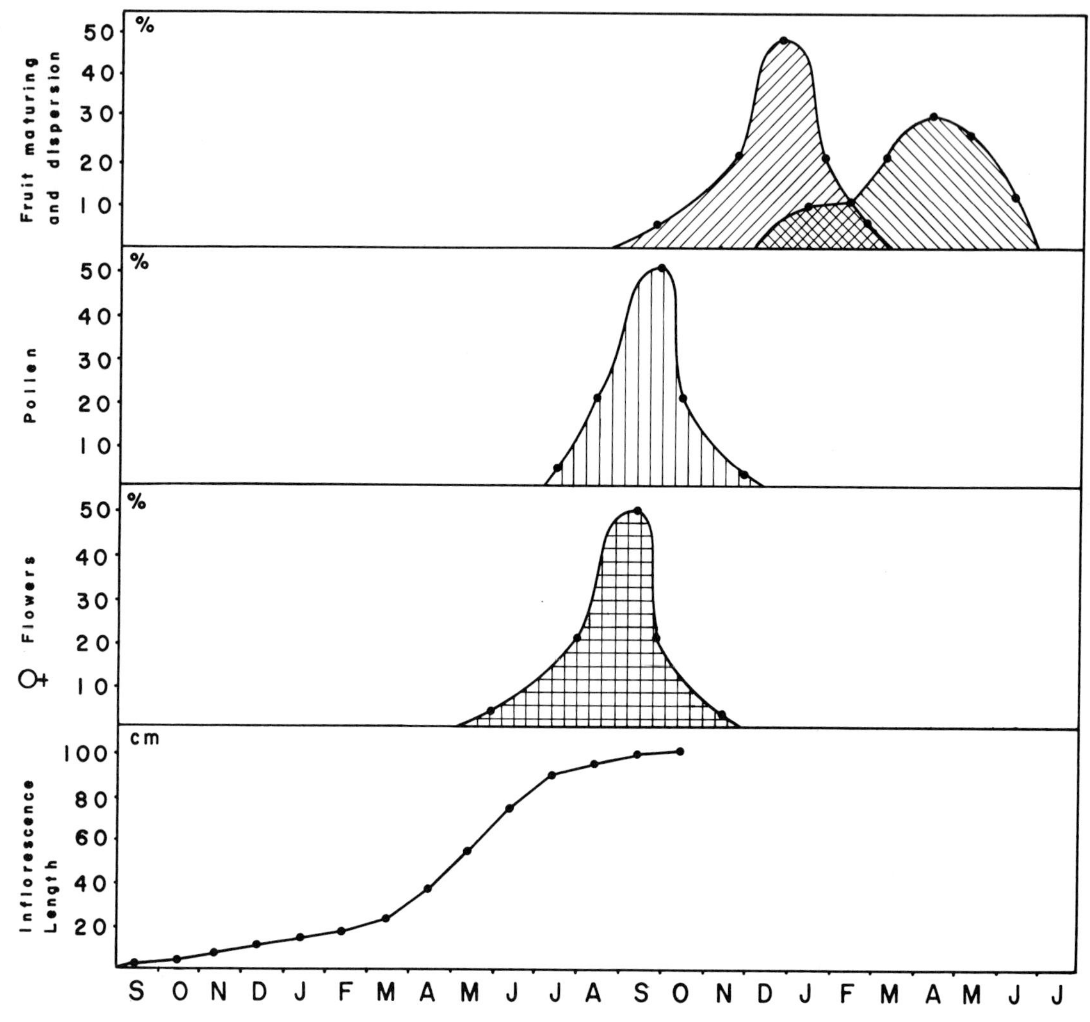

FIG. 3–21. Reproductive phenorhythm of *E. timotensis* in the Páramo de Piedras Blancas at 4200 m showing inflorescence elongation rate, production of female flowers, pollen, and fruit maturation and dispersal. This reproductive process started in September 1978 and the dispersal of achenes still continued in June 1980. Data based on 344 inflorescences that produced 3677 heads.

longer period, of the same order as the rosette turnover time.

I would like to remark on two aspects. The first is the duration of each phenophase for a relatively long period of time. This avoids the risks of losses due to environmental hazards and affords a greater probability of success for some key phases such as pollination and fertilization. The second aspect is the co-occurrence in a given population and in each individual of various phenological phases that may also contribute to avoid the risks of a total synchrony.

The reproductive process in *E. timotensis* extends over several moist seasons. The greatest energetic investment, that of building up the inflorescences, takes place mostly in the dry season, whereas pollination and fertilization occur in the wet season.

Reproductive Behavior of Individuals of E. timotensis

Long-term phenological observations were made on individuals of *E. timotensis* in permanent plots, in order to analyze reproductive behavior at the individual level as well as intrapopulation variability. Table 3–6 synthesizes seven years of observations (1976–82) on all adults in the permenent plots (58 individuals). Five different patterns were observed in this period. The first, shown by 25.8% of the population, is one in which no individual entered into reproduction during the seven

TABLE 3–6. Reproductive frequency in a population of 58 adults of *E. timotensis* in the period 1976–82.

Year							Number of reproductive events in 7 years	Number of individuals
1976	1977	1978	1979	1980	1981	1982		
1[a]	0	1[a]	0		1[a]	1[a]	4	2
1[b]		1[b]		1[b]		1[b]		
1[a]		1[a] 2[d]			1[a] 2[d]	2[d]		
1[b]	0		0	1[b]	1[b]		3	10
1[c]		1[c] 4[e]		4[e]	4[e]	1[c]		
		1[f]		1[f]		1[f]		
1[a]		1[a] 2[c]		2[c]				
	0		0			0	2	15
		8[b]		4[d]	8[b] 4[d]			
1	0	9	0	2	3	1	1	16
0	0	0	0	0	0	0	0	15
Total number of reproductive individuals per year								
7	0	31	0	15	24	7		

Note: Columns 1–7 show the individuals that started their reproductive cycle in each year; a letter (a to f) indicates those whose reproductive cycle is linked together in the same years. Five patterns are distinguished according to reproductive frequency. The number of plants showing each pattern is given in the last column.

consecutive years. A second pattern corresponds to those plants reproducing just once in that period (27.5% of the population); a third, to plants reproducing twice (25.8% of the population); a fourth, shown by individuals with three reproductive events (17%); and a fifth, shown by individuals with four reproductive events (3.4%).

The data in Table 3–6 also suggest relative synchrony in the reproductive behavior of individuals of this population. Thus, in 1977 and 1979 not a single plant bloomed. Although the sample is relatively small, this conclusion was supported by the additional sampling of 1500 plants each year outside the permanent plots. It appers then that the whole population of *E. timotensis* in the Páramo de Piedras Blancas fails to bloom in some years, as in 1977 and 1979. In contrast, in 1978 and 1981 a relatively large proportion of the population entered into reproduction (53% and 41% respectively).

On the other hand, individuals that flower more than twice in seven years seem to repeat their reproductive cycles together. This linkage implies highly variable pollen production and seed crops, with years when these resources are almost nil alternating with periods of abundant production.

Table 3–6 groups the different reproductive schedules into three main patterns that take into account rhythms and frequency of sexual reproduction (Fig. 3–22). The first one, (top of Fig. 3–22), corresponds to those plants in which variable periods set apart two successive reproductive events that never overlap. The minimum period to avoid overlapping would be two years, and the longest observed period is seven years. The second reproductive pattern (middle of Fig. 3–22), corresponds to those individuals that reproduced twice in seven years. Their behavior shows two successive reproductive events in two years. This led to an overlapping of phenological phases, since the last reproductive process started when the previous event was still far from being completed. This overlap, with the implied additional cost, was followed by several years without further reproduction. The third repro-

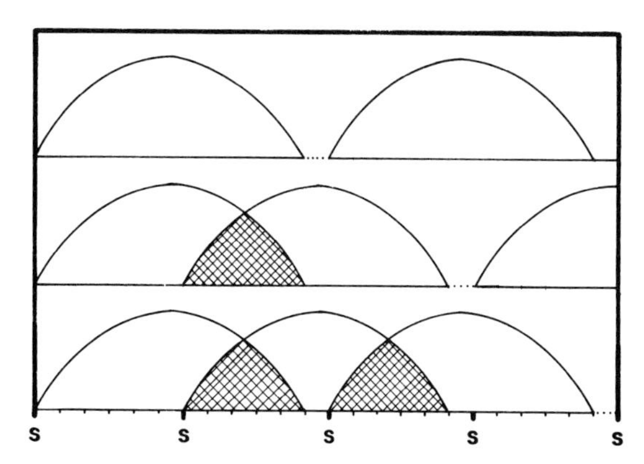

FIG. 3–22. Main reproductive patterns in *E. timotensis* during seven consecutive years. *Top:* Individuals without overlapping of reproductive events. *Middle:* Plants that reproduced twice in seven years (one overlap). *Bottom:* Plants that reproduced four times in seven years. S (September) indicates the start of reproduction.

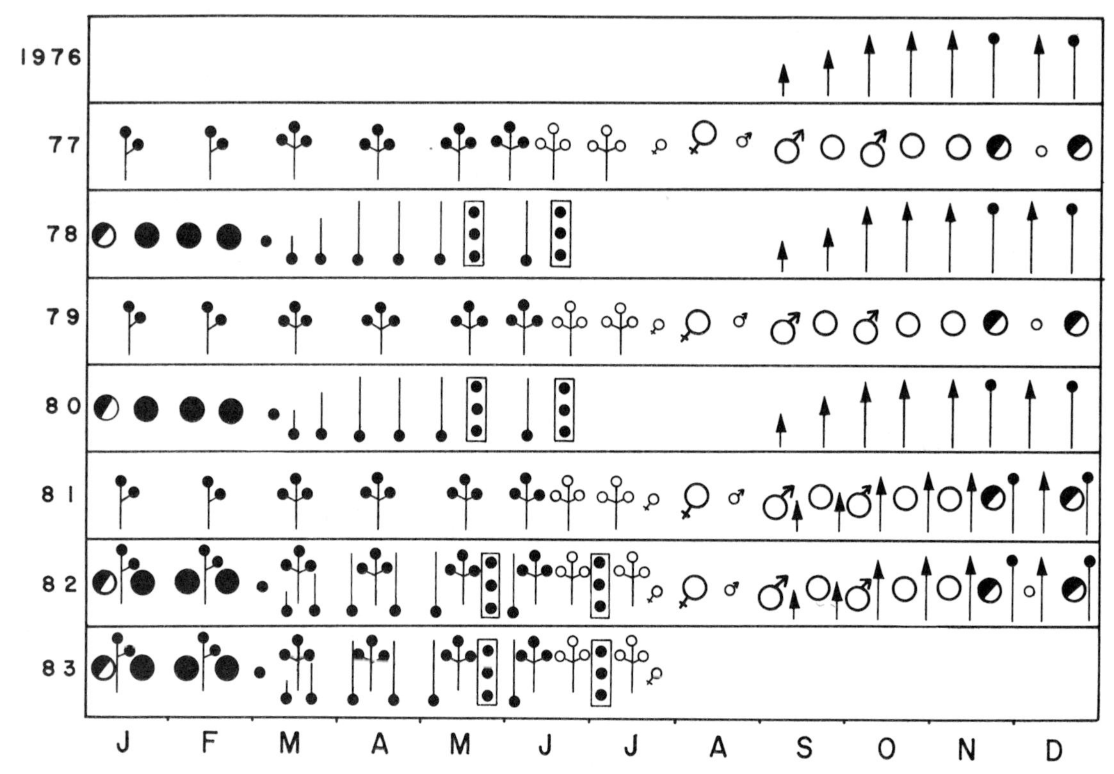

FIG. 3–23. Diversity of the phenological space in populations of *E. timotensis* (1976–82). A change is shown from an empty phenological space (first months of 1976) to one of maximal diversity in 1982 when several phenological phases coexisted.

Starting and growth of the inflorescences
Various stages in head formation and inflorescence elongation
Open heads, flower buds
Female flowers
Pollen
Maturing fruits
Maturing and mature fruits
Mature fruits
Fruit dispersal
Seed bank

ductive pattern (bottom of Fig. 3–22), corresponds to plants that flowered four ties in seven years. It is characterized by more than one overlap between successive reproductive events. This case might almost be considered as a pattern for continuous production of flowers, pollen, and fruit. As can be seen, these patterns range between two extremes: extended unfertility and multiple reproduction.

Diversity of Phenological Resources

The reproductive patterns exhibited by *E. timotensis* contribute to increase the phenological diversity of the plant community throughout the years. In seven years of records, the phenological space showed variable degrees of diversity (Fig. 3–23). Thus, up to September 1976 there was no reproductive activity in this population. At that time, a few plants started their reproductive cycles, which were completed in 1978. As I have shown, reproductive events were not started in 1977 or in 1979. From 1981 on, a certain overlap of reproductive events at the individual level began, with the consequent overlap at the population level, thus enriching the phenological diversity at any given time. A switch occurred then from an empty phenological space to one showing the greatest diversity. The consequences of these oscillations for pollinators (mostly arthropods) and granivores (mostly birds; see Vuilleumier and Ewert [1978, p. 78], especially *Phrygilus unicolor* in desert páramo), as well as for the renewal of the seed bank, can be easily visualized.

Phenological Niches: Interspecific Comparisons

In the desert páramo, reproductive processes do not acquire their real significance at the level of the whole ecosystem unless all the species of *Espeletia* can be considered together as populations producing phenological resources according to various interdigitated patterns. Figure 3–24 shows the temporal division of the phenological niche between the four main species of *Espeletia*: *E. timotensis*, *E. spicata*, *E. moritziana* and *E. schultzii*. The asynchrony between homologous phases in the first three species is readily apparent, whereas *E. schultzii* shows more continuous production. Apparently, the three typical species of the high Andean belt have an almost complete separation of their reproductive phenophases. This might permit the sharing of the same population of pollinators during the year. Even if more than one species attains the flowering phases at more or less the same time, the accomplishment of the successive flowering stages is not synchronous.

This way, the flowering behavior of each species suggests both a precise synchronization at the intrapopulation level and a conspicuous desynchronization of homologous phenophases among the different species. This behavior may have an adaptive value, since it allows the sequential utilization of one of the most limiting resources in this ecosystem: the populations of pollinators. Furthermore, this desynchronization contributes to the genetic isolation of the species of *Espeletia* in the desert páramo, promoting both their biological diversification and their ecological restriction to different habitats in the high Andean belt.

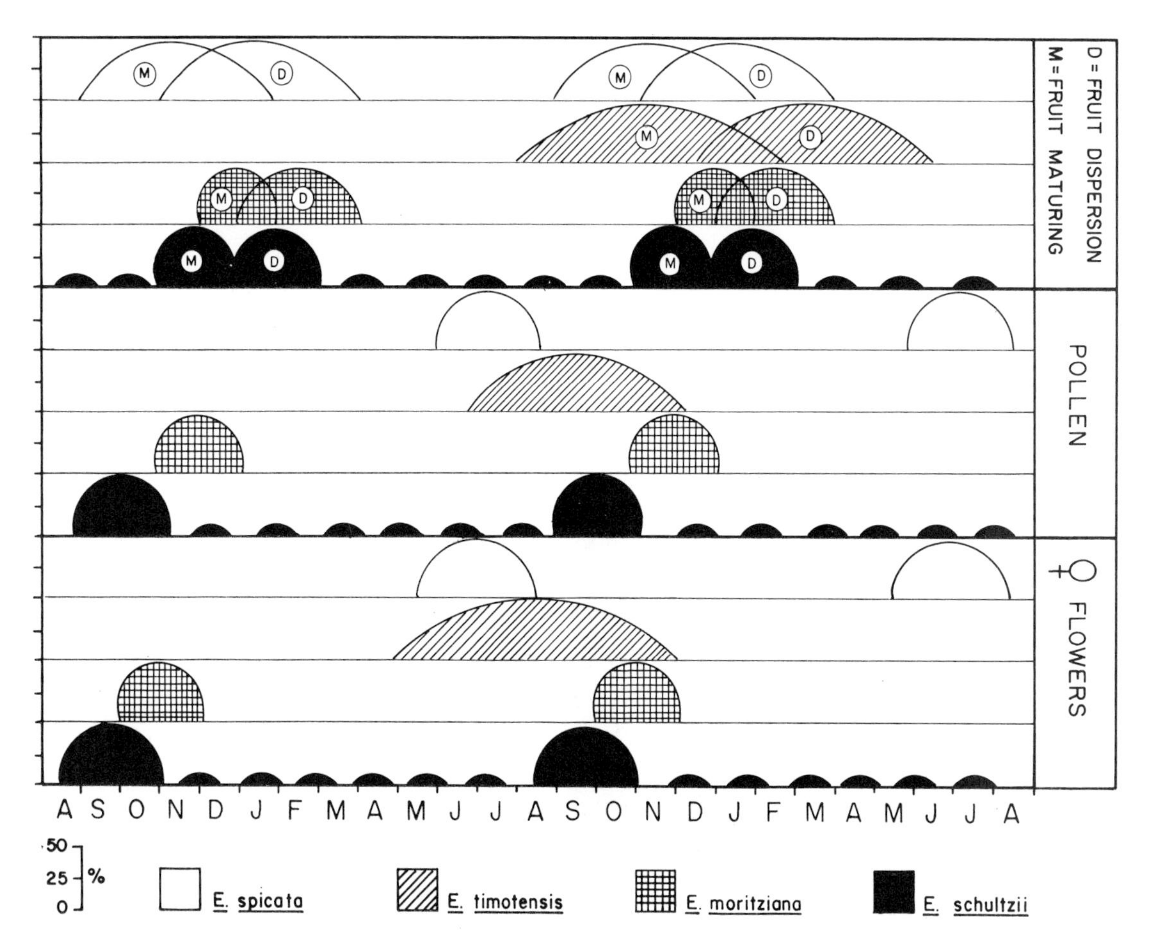

FIG. 3–24. Temporal partition of the phenological niche between the main species of the uppermost vegetation layer in the desert páramo: *E. timotensis, E. spicata,* and *E. moritziana.* Production of female flowers, pollen, and seeds is shown. Notice the desynchronization between these crucial phenophases in homologous species. *E. schultzii,* in contrast, has a continuous succession of reproductive phases throughout the year, although there is a major peak.

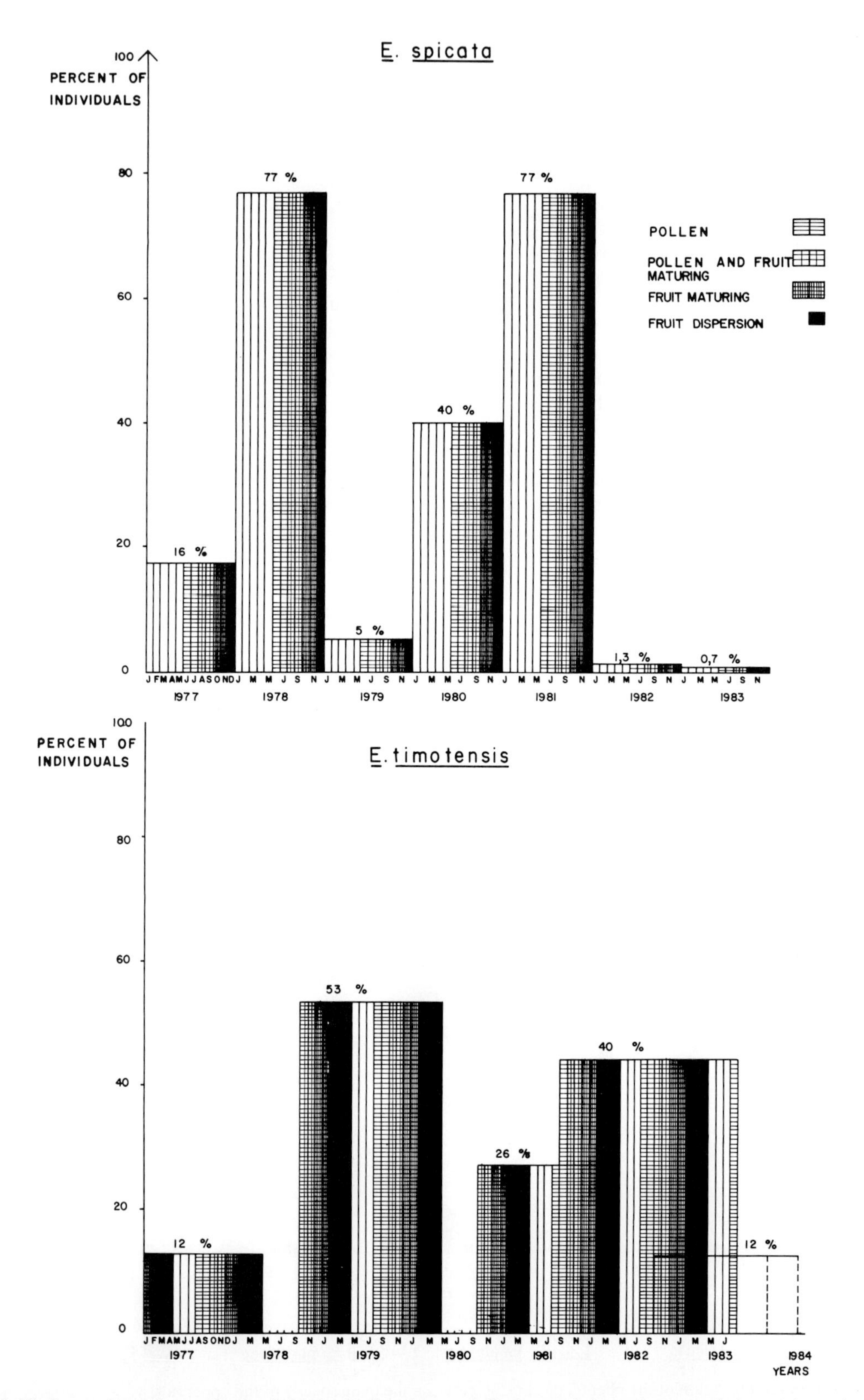

FIG. 3–25. Comparison of reproductive behavior in *E. spicata* and *E. timotensis* in seven years of observations. A clear desynchronization is readily apparent, since years with a higher proportion of reproductive plants in one species correspond with years of higher sterility in the other.

As indicated in Fig. 3–24, the period of pollen production extends from June–July in *E. spicata*, to August–October in *E. timotensis*, and to November–December in *E. moritziana*. *E. schultzii* produces small quantities all year, though flowering attains a peak in September–October.

If the reproductive behavior of *E. timotensis* and *E. spicata* is compared for the seven years with field records (Fig. 3–25), we see how the complementarity of resources also operates. In fact, those years with a greater proportion of reproducing individuals in *E. spicata* are those with a lower proportion of flowering in *E. timotensis*. This desynchronization at the specific level might also have an adaptive value since it allows a more regular distribution of resources through time. We must remember that in these desert páramo *Espeletia*, reproduction is a long-term, high-energy-demanding process and that most frequently reproductive cycles at the individual level appear well distributed along the years.

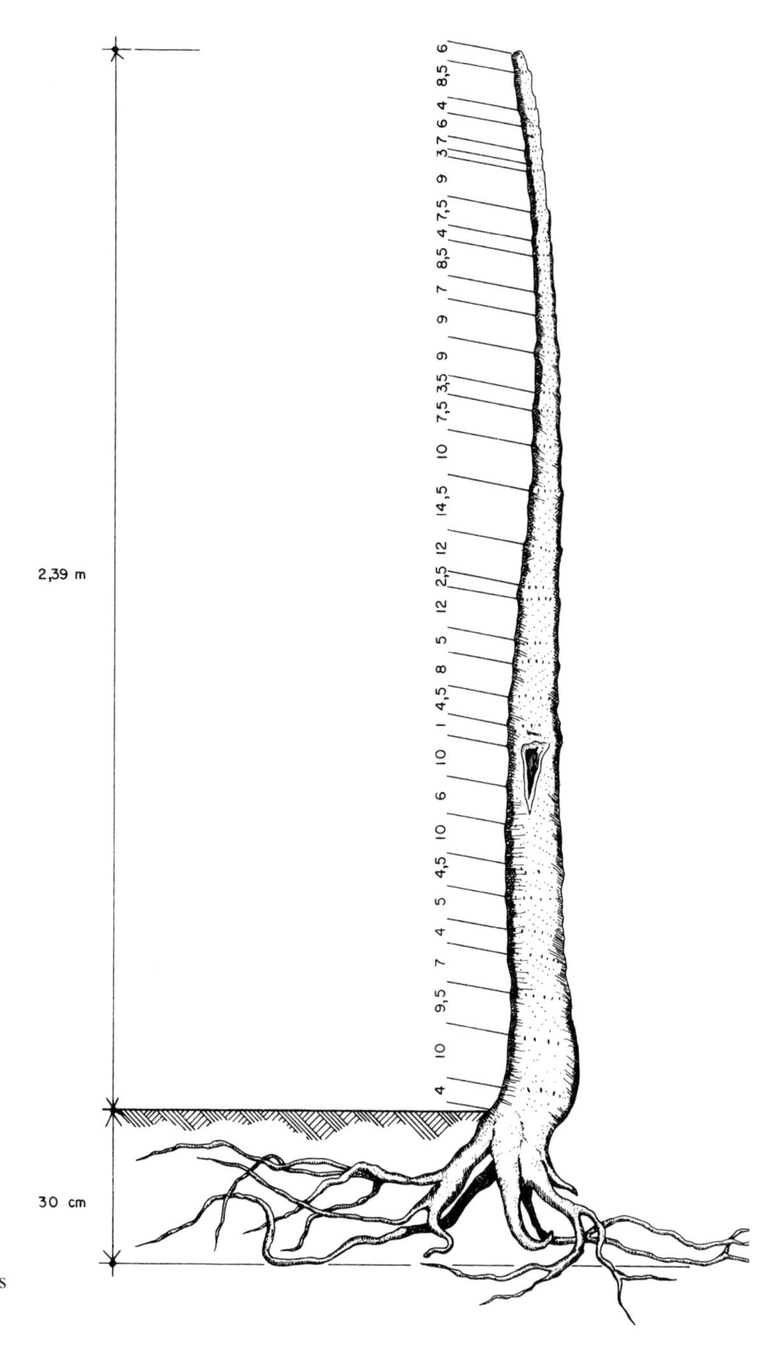

FIG. 3–26. Stem of an adult *Espeletia timotensis* where the scars of all reproductive events along its entire life cycle are new. The distance between consecutive flowering events is also shown.

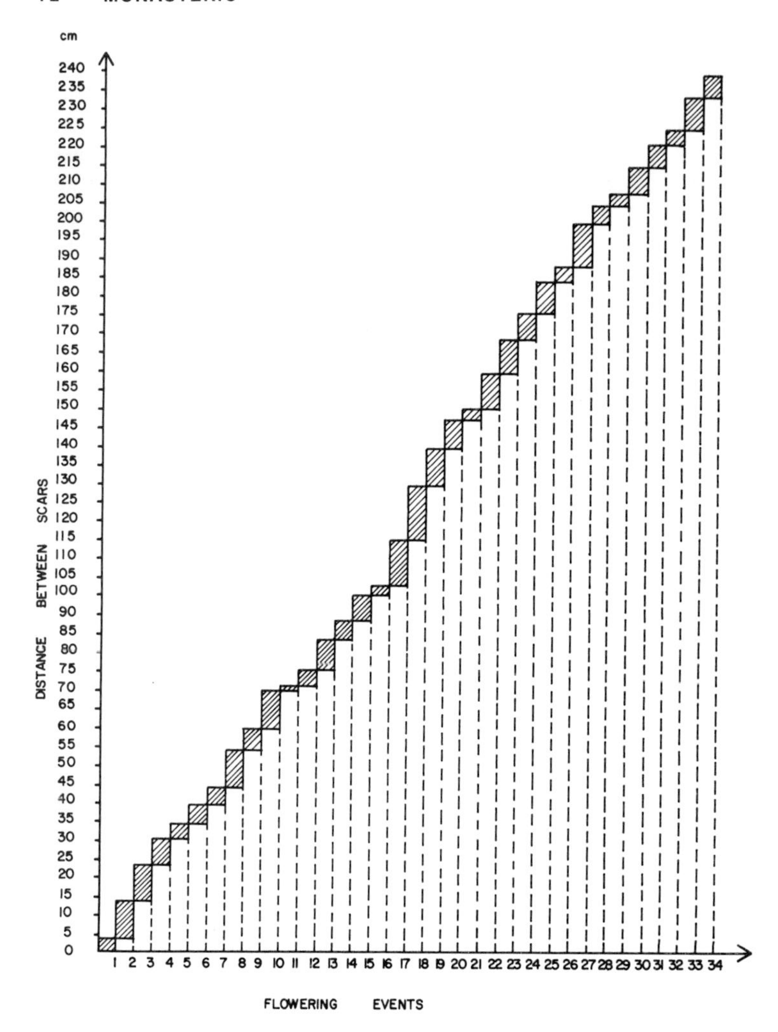

FIG. 3–27. Reproductive frequency by size or age, in the same plant shown in Fig. 3–26.

Reproductive Events and Production During the Life Cycle

In species with long cycles, such as these *Espeletia*, very long term observations, beyond the reach of a normal research program, would be necessary to disclose the reproductive behavior at the individual level. However, it is possible to obtain a picture of all reproductive cycles during the life span of individual plants by analyzing the scars left on the stem. Thus, Figure 3–26 depicts a dead adult plant whose trunk attained a length of 239 cm. On this trunk all the scars indicating reproductive events during its entire life cycle were marked, a total of 34, and the distance between each was measured. The reproductive rhythmicity appears clearly in Fig. 3–27. On the basis of these data, one may suggest, first, that sexual reproduction began at an early age, a conclusion also supported by other field observations; and second, that the intervals between reproductive cycles seem to be highly variable during the life span of the plant. If one considers that the stem of *E. timotensis* grows about 15 mm per year, by measuring distances between the scars we may conclude that in this particular plant, the first two reproductive cycles show a distance equivalent to two years of growth, then one reproductive event followed a period of about six or seven years without reproductive traces, followed again by a similar gap without flowering. Figure 3–27 suggests, third, that this individual has had periods of high fecundity, with quite rapid reproductive cycles, and other vegetative phases lasting up to 10 years.

In the context of the whole plant, it is interesting to know the total energy allocated by an individual to each plant part during its life cycle. I have already pointed out how the energy allocated in a given period of time represents in these *Espeletia* an estimate of the whole investment during the life cycle. However, it is possible to

TABLE 3–7. Reproductive events during the life cycle of an individual of *E. timotensis*.

Number of reproductive events	Distance between consecutive scars	Number of inflorescences	Estimated head number	Estimated achene number	Weight of achene yield (g)	Weight of inflorescence (g)	Total weight of reproductive structures (g)
1	–	3	33	9,647	5.4	76.3	81.7
2	4.0	2	22	6,358	4.7	50.90	55.6
3	13.0	20	220	63,580	47.68	509.0	556.68
4	7.0	25	275	79,475	59.6	636.25	695.85
5	3.0	10	110	31,790	23.8	254.5	278.3
6	8.5	29	319	92,191	69.1	738.05	807.15
7	8.0	8	88	25,432	19.0	203.60	226.60
8	7.5	19	209	60,401	45.3	485.55	530.85
9	9.0	10	110	31,790	23.8	254.5	278.3
10	6.0	18	198	57,222	42.9	458.10	501.0
11	8.0	24	264	76,296	57.2	600.80	658.0
12	11.5	26	286	82,654	61.9	661.70	723.60
13	11.5	29	319	92,191	69.1	738.05	807.15
14	11.0	35	385	101,265	75.94	890.75	966.69
15	6.0	26	286	82,654	61.9	661.70	723.60
16	4.5	12	132	37,148	41.86	305.40	347.26
17*	3.0	–	–	–	–	–	–
18*	1.5	–	–	–	–	–	–
Totals 18	124	290	3566	929,794	709.18	7525.15	8234.33

Note: Reproductive frequency is based on the distance between inflorescence scars on the stem. Each reproductive yield is estimated by counting the inflorescence scars. Data on head number per axis, seed number per head, and their respective weights, from Azócar and Monasterio (unpublished). *17 and 18 are events in the process of development.

obtain a more adequate picture, closer to the real allocation to different plant organs and functions. To do this, a living individual of *E. timotensis* with a trunk length of 121 cm was harvested. All the attached leaves, either living or dead, were counted (8500 leaves). Since we know that the blade or photosynthetic portion of an adult leaf has an average dry weight of 6.9 g, a figure of 61,200 g is obtained for the total leaf mass produced in its lifetime. The actual living leaf biomass, considering the leaf bank and the external rosette, amounted to 1700 g, corresponding to 420 leaves. This biomass constitutes the leaf production of more than two years. Given that the number of expanded leaves per year is about 120, this particular plant produced its standing mass of dead leaves in a period of 70 or 80 years. In this calculation the number of leaves produced by unit time and the size of the rosette were considered to become stable during the first two years of the plant's life (Estrada, personal communication).

Each reproductive event thus leaves clear traces on the stem. Furthermore, each inflorescence axis leaves its mark on the stem, thus allowing one to count the number of axes produced. We also know the mean number of heads per inflorescence and the mean number of flowers and achenes per head. On the basis of these data (Table 3–7) I arrived at a fair estimate of each reproductive crop during the life cycle of the individual, and hence of the total energy allocated to sexual reproduction. The harvested plant showed 16 different reproductive events. The initiation of two further reproductive cycles inside the rosette could also be observed. Table 3–7 indicates the distance between two successive events, the total number of floral axes, the estimated number of heads and achenes, and their corresponding total dry weight. The biomass allocated to each plant part or function throughout the life span of this particular individual is given in Table 3–7.

A first comparison to be made is between the vegetative and the reproductive investments. The ratio

$$\frac{\text{Reproductive biomass}}{\text{Vegetative biomass}} = \frac{8200 \text{ g}}{66{,}100 \text{ g}} = 0.12$$

and the reproductive effort:

$$\frac{\text{Reproductive biomass}}{\text{Total biomass}} = \frac{8200 \text{ g}}{74{,}300 \text{ g}} = 0.11$$

are both rather low. The ratio between below- and aboveground biomass is also low:

$$\frac{\text{Belowground biomass}}{\text{Aboveground biomass}} = \frac{784 \text{ g}}{74{,}300 \text{ g}} = 0.011$$

The biomass allocated to the various vegetative parts is:

Live leaves	1720 g	= 2.2% of the total biomass
Dead leaves	61,200 g	= 82.5% ,, ,, ,, ,,
Stem	2417 g	= 3.2% ,, ,, ,, ,,
bark	837.6 g	
pith	1567 g	
Roots	784 g	= 1.06% ,, ,, ,, ,,

Within the reproductive allocation we have:

Auxiliary reproductive structures 7525 g = 10% of the total biomass
Achenes 709 g = 0.9% of the total biomass

These data validate those previously considered on the energy allocated to plant parts at a given time. They confirm the minimal allocation to belowground organs, even if we keep in mind the errors in the measurement of root biomass due to harvest techniques. The low allocation to the stem is also striking, and it is interesting to notice how most of it corresponds to the pith, which as we will see later functions in these species as a water reservoir. The greatest allocation is to leaves; the total biomass allocated to reproduction appears relatively modest, and most of it is directed to the building of auxiliary structures. However, given the small size of the achenes in these *Espeletia*, the small biomass allocated to them throughout the life span of the plant corresponds to about one million achenes.

The strategy of energy allocation in *E. timotensis* might be described as continuous addition to the foliage in contrast with occasional or intermittent allocation to reproduction. Roots and stem do accumulate biomass, but apparently the belowground parts have shorter turnover times, being decomposed faster than the aerial organs. As a final observation, I want to emphasize that *E. timotensis* seems to be highly efficient in the capture and accumulation of energy, particularly if the very adverse environmental conditions under which it maintains itself in the desert páramo are considered.

SOME FUNCTIONAL ASPECTS

The giant caulescent rosettes of high altitude tropical regions have long been noted for their remarkable adaptations. The tall *Espeletia* of the desert páramo in the Venezuelan Andes have their equivalent in the genera *Senecio* (*Dendrose-*

necio) and *Lobelia* from equatorial Africa, which also contain several giant caulescent rosette species (Hedberg, 1964; Coe, 1967; Chap. 4). This striking morphological convergence in plants from disjunct tropical regions suggests that the giant rosette form represents an adaptive solution to high altitude tropical environments characterized by year-round low temperatures. These low temperatures make the soil water physiologically unavailable during the early morning hours, either because the soil water may be frozen, or because water uptake is impeded by freezing or near-freezing soil temperatures.

The Role of the Pith and Dead Leaves in the Water Balance

In the Andes as well as in Africa, the stems of giant rosette plants contain a voluminous parenchymatous pith that acts as a water source during periods of low water availability (Hedberg, 1964; Goldstein, Meinzer and Monasterio, 1984). The *Espeletia* that grow in the Andean páramos differ from each other not only in pith volume, but also in the ratio between the volume of the water reservoir and the transpiring surface (PV/LA) (Table 3–8). Species from higher and colder páramos have a higher PV/LA and can provide water to the transpirational stream for a longer period of time than species from lower páramos with smaller water storage capacities. The pith of the high Andean belt species can provide more than one and a half hours of water to satisfy the transpirational needs during the critical morning hours when soil temperture is low and evaporative demand relatively high.

Daily patterns of transpiration and leaf water potential show the buffering effect of the water stored in the pith (Fig. 3–28). The species from the desert páramo (4200 m) that show the highest storage capacity also exhibit small changes in leaf water potential, whereas the species from the wettest and lowest páramos (e.g., Páramo Batallón, 3100 m), with small pith reservoirs, exhibit pronounced changes in water potential under conditions of similar evaporative demand.

Variations in water storage capacity of the pith would also be expected during the life cycle of an individual. In *E. timotensis*, relative water storage capacity measured as PV/LA seems to increase rather linearly with plant height above 60 cm (Fig. 3–29). The importance of these changes in relative water storage is reflected in patterns of water balance in individuals of different sizes (Fig. 3–29). During the dry season, minimum leaf water potentials of adult individuals remain high, while minimum water potentials are significantly lower and wilting may occur in smaller plants (Goldstein and Meinzer, 1983; Goldstein et al., 1984). One could speculate that for this reason the risk of death is greater in small plants. Initial observations that tend to confirm these ideas show a high correlation between size, specific mortality, and water storage capacity of the pith for individuals up to 1 m tall.

The potential importance of the pith of the *Espeletia* of the desert páramo can be seen in the relationship between the plants and the bird *Cinclodes fuscus* (family Furnariidae). A nest of this species was found in September 1981 in the decaying pith of the fallen trunk of an individual of *Espeletia* (?*timotensis*) at 4100 m at Páramo de Piedras Blancas near the permanent plots (Vuilleumier, pers. comm.).

Another conspicuous feature shared by both Andean and African caulescent giant rosette species is the 10–30 cm thick layer of marcescent leaves surrounding the stem. This layer provides considerable temperature insulation and prevents stem temperatures from falling below 0°C during the night (Smith, 1979; Goldstein, Meinzer and Monasterio, 1984). Stem core temperatures for *E. timotensis* growing at 4200 m in the desert pár-

TABLE 3–8. Factors related to water budget in species of *Espeletia* in two contrasting páramos, one (Piedras Blancas) in the high Andean, the other (Batallón) in the Andean belt.

Páramo site	Mean temp. (°C)	Species	PV/LA (cm^3/cm^2)	$\triangle$M (g)	T (gh^{-1})	TH
Piedras Blancas (4200 m)	2.8	*E. timotensis*	0.105	176	70.7	2.5
		E. spicata	0.056	160	81.9	2.0
		E. moritziana	0.057	57	39.7	1.4
Batallón (3100 m)	9.3	*E. marcana*	0.038	68	55.3	1.6
		E. atropurpurea	0.018	9	16.3	0.6

Note: PV/LA, Pith volume/leaf area is a measure of relative capacitance; $\triangle$M, mass of water stored in the pith; T, transpiration rate; TH, transpiration hours needed to spend all the water stored in the pith; obtained by dividing $\triangle$M by T.
Source: Adapted from Goldstein et al., 1984.

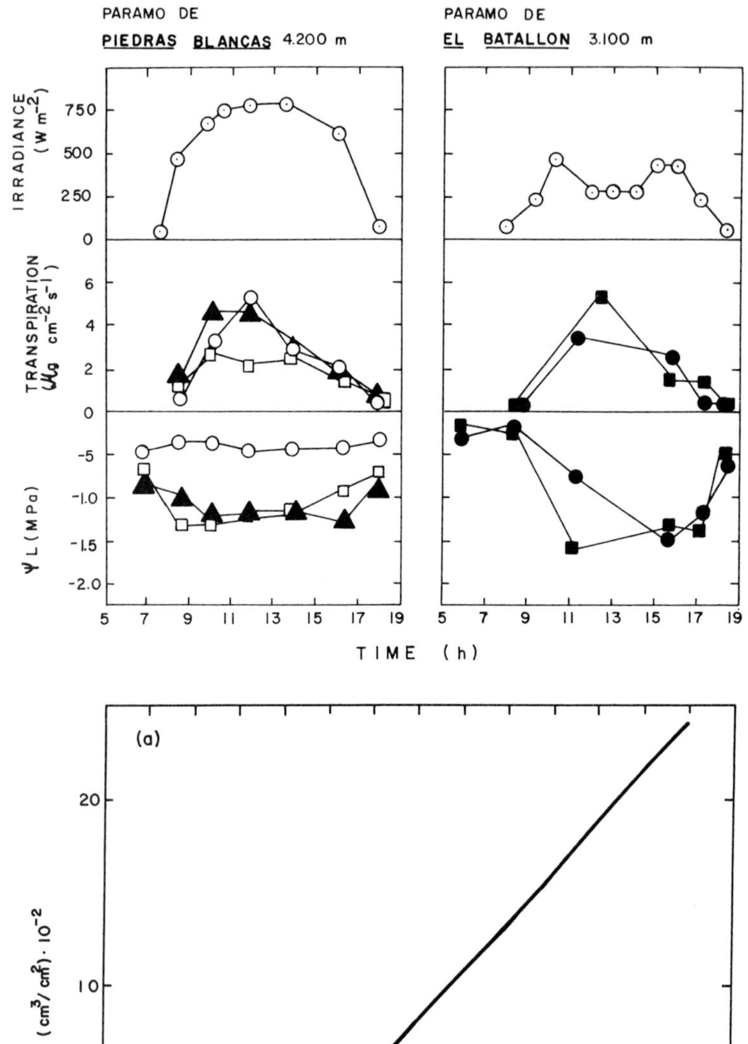

FIG. 3–28. Daily patterns of transpiration and leaf water potential in the giant rosettes of the Páramo de Piedras Blancas. *E. timotensis* (○), *E. moritziana* (▲), and *E. spicata* (□). In the Páramo El Batallón the acaulous rosettes of *E. marcana* (●) and *E. atropurpurea* (■) were studied. (After Goldstein et al., 1984.)

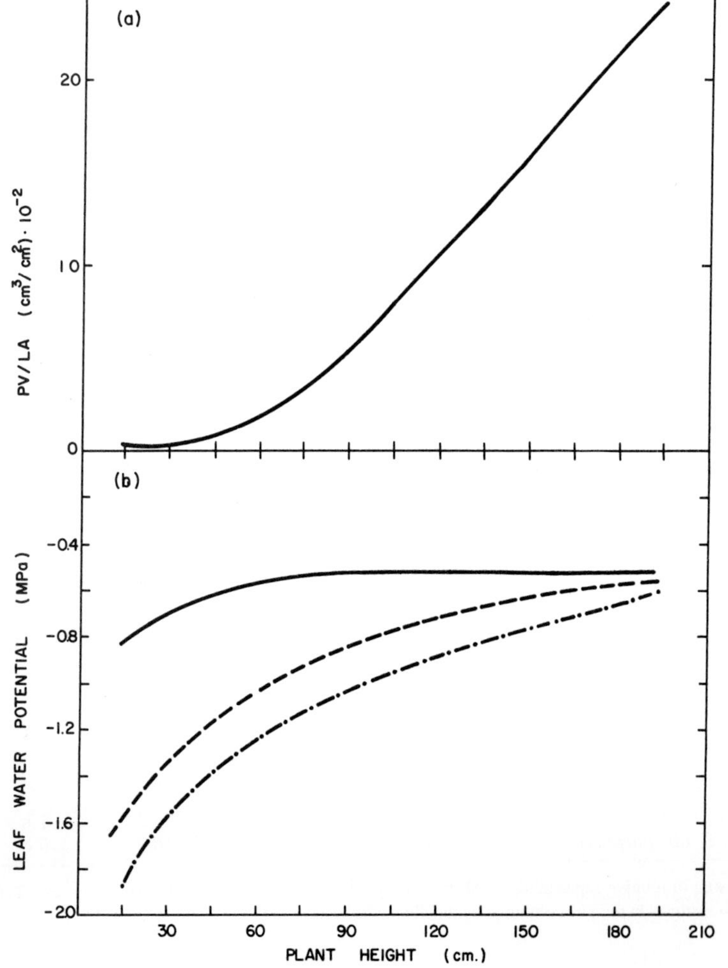

FIG. 3–29. (a) Variation in the capacity of water storage in the pith (PV/LA) in relation to size and age in *E. timotensis*. (b) Minimum leaf water potential reached in plants of *E. timotensis* of various sizes, in the wet (——) and in the dry (- - - -) season. (After Goldstein et al., 1984.)

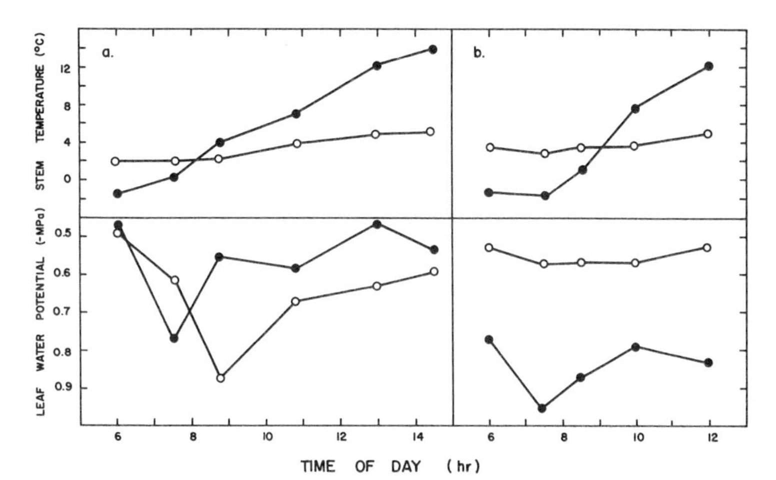

FIG. 3–30. Stem temperature and leaf water potential in *E. timotensis*. (○) Intact plants, (●) denuded plants (removal of attached dead leaves). (a) One day after the experimental removal of dead leaves. (b) Fifteen days afterward. (After Goldstein and Meinzer, 1983.)

amo remain above 2°C for much of the day, but exposed stems show pronounced temperature changes (Fig. 3–30). Thus, at least during the day, marcescent leaves enhanced chances of survival due to avoidance of direct stem freezing. In manipulative experiments carried out at 4200 m, leaf removal improved water balance during the following day as a consequence of higher stem temperatures and therefore improved water flow between the soil and/or the pith reservoir and the transpiring leaves. Two weeks later, however, leaf water potential of stripped plants was lower than that of intact plants, and symptoms of damage were evident. Hydraulic resistance on this day was four times greater in the stripped plants. All stripped individuals died within two months. The lethal effects of dead leaf removal could be attributed to one or more of the following causes: (1) inhibition of pith recharge by subfreezing stem temperatures, (2) embolism in stem xylem, and (3) frost injury to pith tissue. These results suggested that an insulating layer of marcescent leaves and the presence of an internal water reservoir closer to the rosette than the soil water are important adaptations for maintenance of a favorable water balance in high altitude tropical habitats where freezing temperatures occur regularly but last only a few hours.

Leaf Pubescence and Thermal Balance

Another conspicuous characteristic of many giant rosette plants, shared by the desert páramo *Espeletia*, is the presence of a thick pubescence layer in the active leaves. Field and laboratory measurements were used to develop an energy balance model for leaves of *E. timotensis* (Meinzer and Goldstein, 1985). Results of model simulation predict that the most important effect of the leaf hairs was on the boundary layer and on resistance to convective heat transfer, rather than on leaf absorbtion of solar radiation. Under clear-day conditions at 4200 m, the temperature of a pubescent leaf would be higher than that of a nonpubescent one in spite of the larger amount of solar radiation absorbed by the latter (Fig. 3–31). Thus, in contrast to many desert species in which leaf hairs reduce the radiant energy load and leaf temperatures by increasing reflectance, leaf hairs of *E. timotensis*, and probably of all the caulescent rosette species, result in greater daytime leaf temperatures by increasing the thickness of the boundary layer of unstirred air.

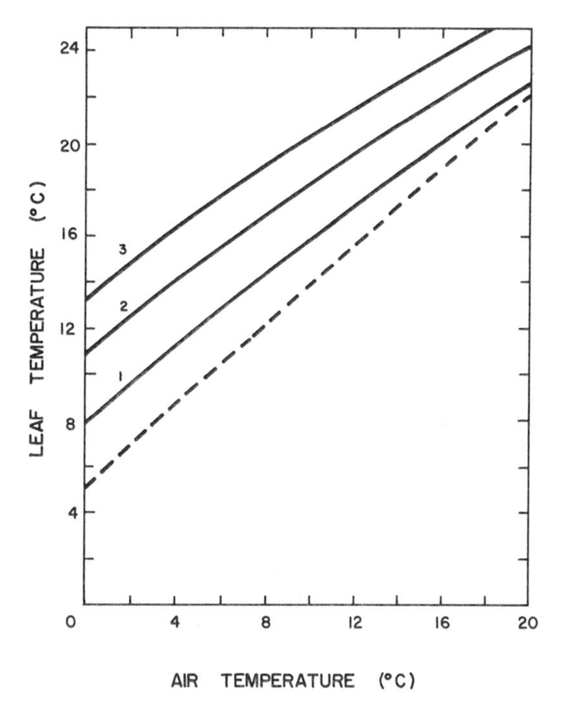

FIG. 3–31. Relation between simulated leaf temperature and air temperature in leaves of *E. timotensis* having 1, 2 and 3 cm of pubescence, and in glabrous leaves. (After Meinzer and Goldstein, 1985.)

Daytime leaf temperatures well above the low prevailing air temperatures would favor processes such as leaf growth and assimilate translocation. Leaf temperature can be increased without an accompanying increase in transpiration. This is important from an adaptive point of view because low temperatures severely limit water availability in the desert páramo.

Decomposition, Microfauna, and Nutrient Cycling

Variation in chemical composition with leaf age in the rosette leaves and in the standing leaf litter of *E. timotensis* and *E. schultzii* suggests a direct recycling of nutrients within the plant from dead leaves to growing tissues (Monasterio, 1980b). In this extreme environment where the periglacial climate hinders soil evolution (Malagon, 1982), this strategy would certainly favor greater efficiency in the use of nutrients, thus having clear adaptive advantages. Garay (1981) and Garay et al. (1982) analyzed the standing leaf litter of these two species, centering their attention on the relationships between nutrient stocks and the community of microarthropods, with the aim of understanding the role of microfauna in the processes of leaf decomposition and nutrient cycling.

Decomposition of the standing litter in *E. timotensis* follows two gradients, one toward the base of the stem, the other from outer to inner parts of the cover of dead leaves. Along both gradients, the proportion of amorphous organic matter increases. This organic matter appears to be formed almost exclusively of fecal pellets of arthropods (Garay, 1981). A high C:N ratio, induced by a low nitrogen content, in the dead leaves of the basal part of the stem also suggests a very low decomposition, and conversely, the weak decomposition of leaf sheaths may also be related to low nitrogen concentrations. Calcium and magnesium do not vary in the different parts of the standing litter, and potassium increases basipetally, probably due to the leaching of the upper leaves.

In *E. schultzii* (Diaz Rosales, 1983), the concentration of calcium and nitrogen is significantly higher than in *E. timotensis*, particularly in leaf sheaths, corresponding with higher decomposition rates. The protective function of the standing dead leaves thus would be less efficient in this predominantly Andean *Espeletia* than in *E. timotensis*, an exclusively high Andean species.

Garay (1981) reported a density of 130,000 microarthropods in the standing dead mass of a plant of *E. timotensis* with a trunk 110 cm high. Most of them occur toward the outer part of the standing litter and their abundance is highly correlated with nitrogen content.

From all these data we may conclude that leaf decomposition and its agents are of crucial importance to plants growing in soils with almost no chemical evolution and therefore with a very low content in available nutrients. In the long-lived species of *Espeletia*, like *E. timotensis*, nutrients once absorbed from the soil could be utilized repeatedly along numerous successive cycles of leafing and reproduction. Selection favoring longevity might therefore operate, since longevity appears correlated with better efficiency in nutrient use. On the other hand, a trade-off would be necessary between humification of the standing litter and its conservation as a protective cover against freezing temperatures. Thus, *E. schultzii*, with a more rapid decompositon and a shorter life span, is mostly restricted to the lower páramos and only occurs in protected sites in the high Andean belt.

Garay et al. (1982) also considered the decomposition-recycling system of the desert páramo *Espeletia* as giving rise to a strategy promoting the long-term maintenance of the colonized sites through the accumulation of organic matter, nutrients, and a rich fauna of decomposers in a single system that forms an integral part of the living plant.

Similar systems for more efficient cycling of minerals, but involving vertebrates instead of arthropods, may be found in ecologically similar areas elsewhere in the high tropical Andes. In the dry puna of Peru, where giant rosettes of *Puya raimondii* (Bromeliaceae) make up stands locally in the open grasslands (Vuilleumier, pers. comm.), a relationship between birds and plants has been postulated by Rees and Roe (1980). The sharply hooked leaves of *Puya raimondii* "catch" numerous birds which then die and remain as corpses for a long time in the rosette. Rees and Roe (1980) suggested that the dead birds, which decompose slowly in the dry climate of the puna, are a source of nutrients that would supplement the nutrient uptake by the root system.

CONCLUSIONS ON ADAPTIVE STRATEGIES

Several aspects of the adaptive behavior of the *Espeletia* in the desert páramo have been treated in this chapter, such as their dynamics of growth and reproduction, the patterns of biomass and energy allocation, the physioecological significance of the pith, dead leaves, and leaf pubescence, the fauna of the standing litter and its relation to decomposition and mineral cycling. All of these aspects constitute key elements in the interpretation of the global strategy of *E. timotensis*, *E. spicata*, and *E. moritziana*.

More than 50 species of *Espeletia*, having different life forms and architectural models, exist in the Venezuelan Andes, yet only five of them cross the barrier of the periglacial climate at 4000 m. All five follow the Corner architectural model. Some species of the lower páramos also belong to this model, but they are low rosettes, less than 1 m high. Thus, giant rosettes that follow the Corner model are restricted to the desert páramo.

Two of the high Andean species, *E. schultzii* and *E. semiglobulata*, occur in protected habitats, and their rosettes are lower. What characteristics of the three most widespread and typical desert páramo *Espeletia* differentiate them from other species and allow them to survive under the periglacial climate of the high summits, where, at 4600 m they reach the uppermost limit of flowering plants in the tropical Andes?

Two types of adaptation are fundamental to the ecological success of these desert páramo *Espeletia*: adaptations related to the maintenance of favorable temperature and water balance, and features that increase the ability of individuals to remain in the colonized sites after having passed the critical seedling phase. Among the most important features in this context we may consider the following:

1. Selection favors longevity. In effect, the three species of the desert páramo belonging to the Corner model have a long life span compared with that of the species with the same architectural model growing in the Andean belt below. The latter species have an average life span of about 50 years, but the high Andean belt species may live as long as 150 years. In an environment where low rate of seedling survival represents the major obstacle to the maintenance of plant populations, longevity acquires an undisputed adaptive value.
2. Long lifespan appears to be correlated with greater plant size and with the development of a conspicuous pith, which is as important in the water economy of these *Espeletia*, as it is for the covergent African species *Dendrosencio keniodendron* (Hedberg and Hedberg, 1979; see Chap. 4).
3. The internal recycling of nutrients from slowly decomposing standing dead litter, and the role of the protective cover of dead leaves in the insulation of the stem are particularly significant.
4. Only polycarpic species of *Espeletia* reach the desert páramo. These species not only reproduce many times during their life cycle, but they also attain sexual maturity quite early and increase in fecundity with age. Thus, an individual, after having surmounted the critical seedling phase, will produce about one million seeds during its lifetime. That is, for about 100 years it is continuously feeding the seed bank and so contributing to the success of its regeneration and to population stability.
5. The reproductive patterns show very little, if any, overlap among homologous phenophases in different species. Pollen is more or less continuously available to pollinators, thus ensuring that only sexually reproducing species are successfully pollinated.
6. Both growth and reproduction proceed gradually and continuously through small increments synchronized with the daily cycles of the equatorial or tropical mountain climate. However small these increments, the year-long persistence of the process leads to a considerable annual production. An adult rosette of *E. timotensis* produces an estimated 700 g dry weight per year.
7. A remarkable feature of desert páramo *Espeletia* is their strategy of consolidation and maintenance of their sites by increasing the efficient use of critical resources. This is obtained by the coherent ecological system around each living plant, formed by its live biomass, the standing dead litter, and the microfauna that humifies and slowly decomposes the organic matter, gradually liberating the sequestered nutrients. A pattern arises in which each plant constitutes a stable and organized micro-ecosystem surrounded by a desert of bare ground.

ACKNOWLEDGMENTS

This work was carried out with the active participation of Carlos Estrada, whom I wish to acknowledge for his permanent enthusiasm and recognized efficiency. Guillermo Goldstein and Frederick Meinzer were responsible for the ecophysiological part of this project and it was a highly profitable experience to work with them. Lina Sarmiento was irreplaceable in the delicate and time-consuming task of processing and computing the field data, where she was efficiently aided by Anairamiz Aranguren. Sarah Martin demonstrated an efficient capacity to type the manuscript in record time. I also want to acknowledge François Vuilleumier for his continuous support during all phases in the editing of this book. Finally, my most special thanks go to my colleague Guillermo Sarmiento for his collaboration in the crucial step of transposing my often diffuse English into a readable text.

REFERENCES

Coe, M. J. 1967. *The ecology of the alpine zone of Mount Kenya*. Monographiae Biologicae 17. The Hague: Junk.

Cuatrecasas, J. 1979. Growth forms of the Espeletiinae and their correlation to vegetative types in the high tropical Andes. In *Tropical botany*, K. Larsen and L. B. Holm–Nielsen, eds., pp. 397–410. London: Academic Press.

Diaz Rosales, H. 1983. Estudio de la comunidad de microarthrópodos en la hojarasca en pie de *Espeletia schultzii* WEDD en el Páramo Desértico. Facultad de Ciencias, Universidad de los Andes, Master's Thesis.

Du Rietz, G. E. 1931. Life forms of terrestrial flowering plants. *Acta Phytogeog. Suec.*, 3: 1–95.

Garay, I. 1981. Le peuplement de microarthropodes dans la litière sur pied de *Espeletia lutescens* et *Espeletia timotensis*. *Rev. Ecol. Biol. Sol.* 18: 209–219.

Garay, I., L. Sarmiento-Monasterio, and M. Monasterio 1982. Le páramo désertique: éléments biogènes, peuplements des microarthropodes et stratégies de survie de la végétation. In *Tendences nouvelles en biologie du sol,* Comptes rendus du VIIIe Colloque International de Zoologie du Sol, 1982. P. Lebrun, H. M. André, A. DeMedts, C. Grégoire-Wibo, and G. Wauthy, eds., pp. 127–134. Louvain-la-Nueve: Belgium.

Goldstein, G., and M. Meinzer, 1983. Influence of insulating dead leaves and low temperatures on water balance in an Andean giant rosette plant. *Plant Cell and Environment* 6: 649–656.

Goldstein, G., F. Meinzer and M. Monasterio, 1984. The role of capacitance in the water balance of Andean giant rosette species. *Plant Cell and Environment* 7: 179–186.

Goldstein, G., F. Meinzer and M. Monasterio, 1985. Physiological and mechanical factors in relation to size-dependent mortality in an Andean giant rosette species. *Oecol. Plant.* 6: 263–275.

Hallé, F., R. A. A. Oldeman, and P. B. Tomlinson, 1978. *Tropical trees and forests*. Berlin: Springer-Verlag.

Harper, J. 1977. *Population biology of plants*. London: Academic Press.

Harper, J., and A. Bell, 1979. The population dynamics of growth form in organisms with modular construction. In *Population dynamics*, R. L. Anderson, B. D. Turner, and L. R. Taylor, eds., pp. 29–52. Oxford: Blackwell Scientific Publications.

Hedberg, O. 1964. Features of afroalpine plant ecology. *Acta Phytogeog. Suec.* 49: 1–144.

Hedberg, I., and O. Hedberg, 1979. Tropical-alpine life-forms of vascular plants. *Oikos* 33: 297–307.

Mabberley, D. J. 1974. The pachycaul lobelias of Africa and St. Helena. *Kew Bull.* 29: 535–584.

Malagon, D. 1982. *Evolución de los suelos en el Páramo Andino*. Mérida: CIDIAT.

Meinzer, F., and G. Goldstein, 1985. Leaf pubescence and some of its consequences in an Andean giant rosette plant. *Ecology* 66: 512–520.

Monasterio, M. 1979. El Páramo Desértico en el altiandino de Venezuela. In *El Medio Ambiente Páramo*, M. L. Salgado-Labouriau, ed., pp. 117–146. Caracas: Ediciones Centro de Estudios Avanzados.

———. 1980a. Las formaciones vegetales de los páramos de Venezuela. In *Estudios Ecológicos en los Páramos Andinos*, M. Monasterio, ed., pp. 94–158. Mérida: Ediciones de la Universidad de Los Andes.

———. 1980b. Elementos para el análisis de la estrategia global en especies del Páramo Desértico. I. Demografía foliar y alocación de nutrientes en *Espeletia lutescens*. XXX Convención Anual de ASOVAC, Mérida, November 1980.

Monasterio, M., and S. Reyes, 1980. Diversidad ambiental y variación de la vegetación en los páramos de los Andes venezolanos. In *Estudios Ecológicos en los Páramos Andinos*, M. Monasterio, ed., pp. 47–91. Mérida: Ediciones de la Universidad de Los Andes.

Rees, W. E., and N. A. Roe, 1980. *Puya raimondii* (Pitcairnioideae, Bromeliaceae) and birds: An hypothesis on nutrient relationships. *Can. J. Bot.* 58: 1262–1268.

Smith, A. P. 1974. Population dynamics and life form of Espeletia in the Venezuelan Andes. Ph. D. thesis, Department of Botany, Duke University.

———. 1979. The function of dead leaves in *Espeletia schultzii* (Compositae), an Andean giant rosette plant. *Biotropica* 11: 43–47.

Schubert, C. 1976. Glaciación y morfología periglacial en los Andes venezolanos noroccidentales. *Bol. Soc. Venez. Ciencias Naturales* 32: 149–178.

Tricart, J. 1970. *Geomorphology of cold environments*. London: Macmillan Press.

Vuilleumier, F., and D. Ewert, 1978. The distribution of birds in Venezuelan páramos. *Bull. Am. Mus. Nat. Hist.* 162: 47–90.

White, J. 1979. The plant as a metapopulation. *Ann. Rev. Ecol. Syst.* 10: 109–145.

4

Adaptive Syndromes of the Afroalpine Species of Dendrosenecio

D. J. MABBERLEY

In the context of the African continent, the floristically impoverished vegetation of the upper slopes of the highest of the tropical mountains has an archipelago-like distribution, perched as it is above the similarly scattered but more widespread afromontane forest (White, 1983, p. 169). The afromontane forest is quite distinct from forests at lower altitudes but is rather homogeneous, "undifferentiated" in terms of African forests as a whole, and is separated, variously on different mountains, from the afroalpine vegetation by bamboo, *Hagenia* woodland, bushland, or thicket. The afroalpine vegetation itself is physiognomically very mixed (see Rehder et al., 1981), but because of the small stature of any shrubland within it, most conspicuous are the pachycaul species of *Dendrosenecio* (*Senecio, s. l.*, Compositae) and of *Lobelia* (Campanulaceae). Plants allied to both of these genera are found in vegetation types at lower altitudes, where they are rarely so conspicuous, however.

Hedberg has argued (Hedberg, 1964; Hedberg and Hedberg, 1979) that the pachycaul species of *Lobelia* and *Dendrosenecio* are the African representatives of the "giant rosette plants," the most conspicuous of the five major groupings of plant growth forms characteristic of tropical-alpine vegetation in both Africa and South America, viz., tussock grasses, acaulescent rosette plants, cushion plants, sclerophyllous shrubs, and giant rosette plants. He has argued further that such growth forms, or "syndromes" of morphological characters, represent adaptations to the extreme diurnal fluctuations characteristic of the climate of these mountains, for almost half of the vascular plant species fall into these five categories. Although the assemblage of the five growth forms together may be characteristic of the tropical-alpine vegetation, none of them is restricted to it. Nevertheless, examples of giant rosette plants in high montane areas of tropical South America include *Puya* (Bromeliaceae), *Espeletia*, *Culcitium* (Compositae), *Rumex* (Polygonaceae), *Draba* (Cruciferae), *Blechnum* (Blechnaceae), and *Lupinus* (Leguminosae); of the Himalayas, *Lobelia*, *Rheum* (Polygonaceae), *Arnebia* (Boraginaceae), and *Saussurea* (Compositae); of New Guinea, *Blechnum* and *Cyathea* (Cyatheaceae); of Hawaii, *Argyroxiphium* (Compositae); and of Africa *Lobelia*, *Dendrosenecio*, and *Carduus* (Compositae) (Mabberley, 1976; Smith, 1979). This pachycaul growth form is no exception to the others in being widespread elsewhere, being found throughout tropical regions, in rain forests, secondary vegetation, and desert regions besides above the "tree line" (Mabberley, 1979b). An account of its distribution in New Zealand is provided by Wardle (n.d.). Almost all the families listed here as having giant rosette plants in the tropical-alpine belt are represented by pachycaul plants of similar habit in other habitats.

In short, in the context of tropical plant life there is nothing "weird" about tropical-alpine pachycauls. The interesting questions concern what it is about the pachycaul habit that allows them to survive in the tropical-alpine belt, and why these pachycaul plants are able to live in the tropical-alpine environment when others cannot. In this chapter, I will attempt to answer these related questions by means of an analysis of *Dendrosenecio* and also *Lobelia* in high African mountains.

PACHYCAUL SPECIES OF *DENDROSENECIO*

Ecology

The four pachycaul species of *Dendrosenecio* are confined (Table 4–1, Fig. 4–1) to the nine highest massifs in Central and East Africa, viz. Mt. Kahuzi, Virunga volcanoes, Ruwenzori, Elgon, Cherangani Hills, the Aberdare Mountains, Mt. Kenya, Kilimanjaro, and Mt. Meru, which present a range of climate and geologic substrate from the wet metamorphic Ruwenzori in the west to

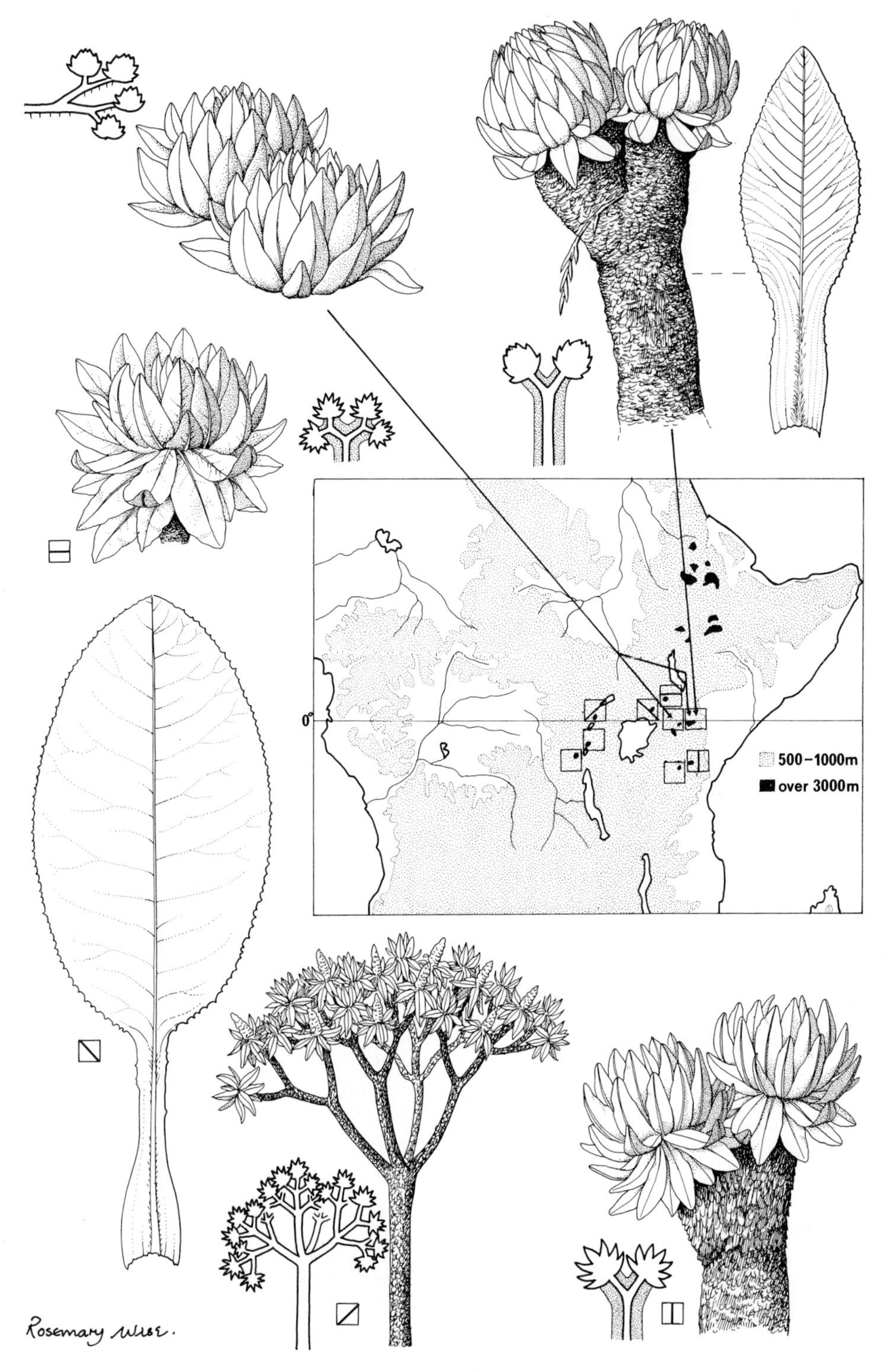

FIG. 4–1. Pachycaul *Dendrosenecio* life forms and distributions in tropical Africa. From *top right* counterclockwise: *D. keniodendron*; *D. keniensis* and *D. brassiciformis*; *D. johnstonii* subsp. *dalei*; *D. johnstonii* subsp. *elgonensis* (leaf), subsp. *adnivalis,* low altitude form; *D. johnstonii* subsp. *cottonii*. □ = distribution of *D. johnstonii*.

TABLE 4–1. Distribution of pachycaul *Dendrosenecio* spp. (see Appendix for nomenclature).

Species, subspecies and varieties	Massifs	Altitudinal range (m)
1. *D. johnstonii*		
a. subsp. *adnivalis*		
i. var. *adnivalis*	Mt. Kahuzi, Ruwenzori, Virunga Volanoes	2600–4500
ii. var. *friesiorum*	Ruwenzori	4000–4400
b. subsp. *cheranganiensis*	Cherangani Hills	2600–3400
c. subsp. *dalei*	Cherangani Hills	3050–3400
d. subsp. *elgonensis*	Mt. Elgon	2800–4250
e. subsp. *barbatipes*	Mt. Elgon	3650–4300
f. subsp. *battiscombei*	Aberdare Mountains, Mt. Kenya	2900–3800
g. subsp. *johnstonii*	Kilimanjaro, Mt. Meru	2450–4000
h. subsp. *cottonii*	Kilimanjaro	3700–4500
2. D. *keniodendron*	Mt. Kenya	3800–4650
3. D. *brassiciformis*	Aberdare Mountains	3000–3800
4. D. *keniensis* (*Senecio brassica*)	Mt. Kenya	3300–4500

the dry volcanic Meru and Kilimanjaro in the east. Most widespread is *D. johnstonii* (Fig. 4–2), represented on each massif by at least one subspecies separated ecogeographically from the others, whereas *D. keniodendron* (Fig. 4–3), a massive upright tree of the afroalpine belt, is restricted to Mt. Kenya and *D. keniensis* (Fig. 4–4) and *D. brassiciformis* (Fig. 4–5), dwarf or creeping species, are restricted to Mt. Kenya and the Aberdares, respectively. Of the subspecies of *D. johnstonii*, some, e.g., *johnstonii*, *battiscombei*, *cheranganiensis*, *elgonensis* and *adnivalis* are found in the thicket and bushland below the afroalpine belt proper, though the last extends into that belt, where, on different massifs, subspecies *cottonii* and *barbatipes* are found. *D. johnstonii* subsp. *dalei* is a dwarf tree of swampy ground in the Cheranganis.

It is true to say that, although a good deal of ecological work on the afroalpine taxa has been published, those plants from lower altitudes have been rather neglected. They all grow where groundwater is plentiful and thrive especially where this is slow-moving, as in gullies, at streamsides, and in seepage channels. The creeping *D. brassiciformis* is found from the ericaceous to the afroalpine belt in *Festuca* swamps and at streamsides in the Aberdares. *D. keniensis* is a conspicuous plant in the "vertical bog" of Mt. Kenya and is abundant in the lower moist regions of the valleys of the afroalpine belt while *D. keniodendron* is to be found most commonly on the higher and drier slopes.

D. keniodendron is found over a wide range of pH on Mt. Kenya, where Allt (1968), working in the Hausberg Valley, found it growing in areas with pH 5.0–6.25 (−7.5), and *D. keniensis* ("*Senecio brassica*") was found at pH 4.75–6.75. Progressing up the side of the valley was an increasing alkalinity and concomitant decrease in soil moisture, gradients reflected in the distribution of *D. keniodendron* throughout, but *D. keniensis* was restriced to the floor of the valley. Beck et al. (1981) have shown that *D. keniensis* and its associated flora are found in waterlogged but not flooded ground of the mountain wet gley or peaty gley type: *D. keniensis* with its pneumatophore-like root system is thus in contrast with the taprooted *D. keniodendron*, which grows on

FIG. 4–2. *Dendrosenecio johnstonii* subsp. *cheranganiensis*, Kenya, Cherangani Hills, Embobut Valley, 3240 m, August 1969.

FIG. 4–3. *D. keniodendron* with *Lobelia telekii,* Kenya, Mt. Kenya, Teleki Valley, January 10, 1971.

soil of the mountain brown soil–gley type. The species of *Dendrosenecio* appear to tolerate a certain amount of fire, which seems to be quite frequent, at least on Mt. Elgon and Mt. Kenya (Hedberg, 1964), as witnessed by their burnt-off frills of marcescent foliage. Although their root systems do not appear very strong ("so superficially rooted . . . that in spite of its [*D. johnstonii*] height and girth, I could pull it down with one hand"; Johnston, 1886, p. 268), the trees can grow on exposed ridges, swaying in the wind and losing their marcescent frills of leaves completely, for example *D. johnstonii* subsp. *battiscombei* observed near "The Twins," Aberdares, October 1970. In being the only woody plants of size in the afroalpine belt, *Dendrosenecio* species suffer from cold mountaineers' apparently relentless firewood forays, their marcescent frills are always dry, affording excellent tinder. They tolerate, and even luxuriate in, the equally relentless dunging activities of the hyrax (Hedberg, 1964). Above all, these massive plants survive the diur-

FIG. 4–4. *D. keniensis* (*Senecio brassica*), Kenya, Mt. Kenya, "vertical bog," 3500 m, January 10, 1971.

FIG. 4–5. *D. brassiciformis* (*Mabberley* 362), Kenya, Aberdares, Chebuswa, approximately 3300 m, October 31, 1970.

FIG. 4–6. Marcescent leaf bases and laminae of *D. johnstonii* subsp. *battiscombei,* Kenya, Aberdares, R. Gikururu, 2910 m, October 14, 1970.

nal extremes of the afroalpine environment, the *Tageszeitenklima* (Troll, 1947), which Hedberg (1957) has interpreted more euphoniously as "winter every night and summer every day."

On the wetter mountains, Ruwenzori and Virungas, epiphytes, notably mosses, liverworts, and ferns, may abound on the bark of the trunks, once the marcescent frills have been lost (Fig. 4–6); on drier mountains, lichens tend to be more common. Occasionally phanerogams such as *Poa schimperana* A. Rich., *Deschampsia flexuosa* (L.) Trin., *Sedum ruwenzoriense* Bak. f., *Senecio snowdenii* J. Hutch., *Cardamine obliqua* [Hochst. ex] A. Rich., and *Arabis alpina* L. as well as seedling *Dendrosenecio* spp. can be found as epiphytes. The lianoid *Galium ruwenzoriense* (Cortesi) Chiov. is a conspicuous feature of the *Dendrosenecio* woodlands of Ruwenzori and to a lesser extent the Cheranganis, scrambling over the trees to several meters. The marcescent leaves and woody leaf bases are a habitat for a range of bryophytes, lichens, and fungi including Uredinales (Jeannel, 1950). I have also collected numerous higher fungi from the base of *Dendrosenecio* trunks.

The marcescent frills of *D. keniodendron* provide a night shelter for the chironomid midges that breed in the buds of pachycaul *Lobelia* spp., and for many beetles and spiders (Coe, 1967). The leaves and inflorescences provide food for many insects.

Salt (1954) shot down an inflorescence of subsp. *cottonii* riddled with dipterous maggots on Kilimanjaro at 3600 m. Most spectacular, however, are the activities of the numerous weevils (Curculionidae), especially *Neoteripelus granulipennis* Hust. (*Seneciobius levenii* Auriv.) on Mt. Elgon. I have collected numerous weevils on many of the East African mountains, but they are yet to be identified. Hancock and Soundy (1931) and Scott (1935) have listed the Coleoptera collected from *Dendrosenecio* spp.; according to Bryk (1927), some weevils are found in the inflorescence.

High on Ruwenzori, Fishlock and Hancock (1932) noted geometrid moths, *Larentia* spp., among *Dendrosenecio* plants, and found cryptophagids, *Athelia ugandae* Bernh. (Staphilinidae), and the chironimid *Spaniotoma* (*Orthocladium*) sp. at 3600 m, as well as the weevils *Subleptospiris turbida* Mshll. between *Dendrosenecio* leaf bases. Hauman (1935) reported *Parasystates burgeoni* Mshll. eating leaves and *Pseudomesites senecionis* Mshll. in the bark of the Ruwenzori species.

The aforementioned authors have shown that in the Coleoptera restricted to the "giant plants," there is marked endemism paralleled by their host distribution patterns. Furthermore, flightlessness increases with altitude, probably associated with the afroalpine habitat "favouring" "cryptozoic" modes of life (Salt, 1954). Interestingly, there are species with well-developed wings including the cryptophagid *Micrambe senecionis* Scott associated with *D. johnstonii* subsp. *adnivalis* flowers on Ruwenzori; this species is a giant in its genus.

I have collected the grasshopper, *Parasphena cheranganica* Uvarov, on *D. johnstonii* subsp. *dalei*, *Psylla* sp. and other psyllids, some Acalypteratae (Diptera), and the hemipterous *Dindymus migratorius* Distant on subsp. *adnivalis*. I found several plants of *D. johnstonii* subsp. *battiscombei* and *D. brassiciformis* (Fig. 4–7) attacked by noctuid caterpillars that had chewed their way through the midribs and pith of the plants to the buds, causing them to collapse and die. I took several species of mollusk (*Vicarriihelix* and *Vitrina* spp.) from between the leaves of

FIG. 4–7. Young plant of *D. johnstonii* subsp. *battiscombei* (*Mabberley* 388), attacked by noctuid caterpillars, Kenya, Aberdares, "The Twins," 3420 m, October 29, 1970.

D. johnstonii subspp. *johnstonii* and *adnivalis* and *D. brassiciformis*. Like the insects, the snails use the leaf-rosettes as a shelter from cold at night and from desiccation by day (Coe, 1967).

The groove-toothed rat, *Otomys orestes orestes* Thomas lives at the base of *D. keniodendron* trees on Mt. Kenya; its hollows penetrate up from the ground into the marcescent leaves and leaf-bases (Coe, 1967). This tree is also the roost for the three resident passerine birds of the afroalpine belt of Mt. Kenya: the hill chat, *Cercomela sordida earnestii* Sharpe, the streaky seed-eater, *Serinus striolatus striolatus* (Rüppell) and the scarlet-tufted malachite sunbird, *Nectarinia johnstoni johnstoni* Shelley. This last bird collects the thick downy indumentum of the leaves of *D. keniensis* to line its nest (Coe, 1967).

The leaves of species of *Senecio* (*s.l.*) contain alkaloids toxic to vertebrates: they are antimitotic agents (Mattocks, 1972). Some animals are immune to the poisons, but the persistence of *D. keniensis*, *D. keniodendron* and the herbaceous *Senecio purtschelleri* Engl. around the otherwise almost denuded burrows of the hyrax, *Procavia johnstoni mackinderi* Thomas is proof that the toxins are repellent to some grazers. The unmolested plants, thriving on the hyrax droppings, camouflage the burrows, an interesting symbiosis (Coe, 1962).

The most remarkable exploitation of *Dendrosenecio* spp. is for water. The mountain gorilla of southwest Uganda rarely drinks (Schaller, 1963, p. 169) but snaps off the leaf-rosettes of *Dendrosenecio johnstonii*, bites off the noxious bark and wood and gnaws out the watery pith (p. 163), portions of which are characteristic of the gorilla's faeces (p. 88). The gorilla also takes the succulent leaf-bases (p. 361) for their water content.

Except as firewood, species of *Dendrosenecio* are not used by man. The local peoples rarely penetrate to high altitudes except to cross mountains such as Elgon and the Aberdares. There are few local names for the trees and they are little known to the people except the Bakonje porters of Ruwenzori, who regularly took mountaineers to the peaks. Occasionally they whittle the wood of old trees of *D. johnstonii* to make large cooking spoons. The wood is white, tough, and odor free. However, the species of *Dendrosenecio* are a tourist attraction in the Mountain National Parks, and as botanical "big game" (Hedberg, 1969) are an important resource in East Africa.

Phenology

Hauman (1935) calculated that a leaf of the afroalpine *D. johnstonii* var. *friesiorum* elongated from 22 to 35 mm in 20 days; at such a rate, a leaf would take one year to mature. By calculating the number of leaves produced during the ontogeny of a full-grown tree, he concluded that var. *friesiorum* would take 200 years to reach full size. Jeannel (1950, p. 208) calculated that a new leaf is added to a rosette every 20 days and that the trees grow about 1 m in 30 years, i.e. at a rate of about 3.3 cm a year. Hedberg (1969) has confirmed these calculations by revisiting a photographed tree of *D. keniodendron* in the Teleki Valley of Mt. Kenya after 19 years. It had grown approximately 45 cm, i.e., at a rate of around 2.5 cm a year, and would thus take about 200 years to mature. Beck et al. (1980) have shown that at 4100 m, this species produces some 50 new leaves per annum and a height increment of about 3 cm in unbranched specimens. After flowering, however, this increment may double (Beck et al., 1984) but then fall off again with increasing age of the tree. It seems likely that the forest forms with longer internodes and smaller leaf rosettes may grow more quickly, however.

Dendrosenecio groves undergo gregarious flowerings after long intervals, perhaps every

10–20 years. All branches of the tree may not flower at once, and trees may have single inflorescences sporadically. Flowering may occur at any time of year (Hedberg, 1957). *D. johnstonii* subsp. *adnivalis* ("*Senecio kahuzicus*") underwent a mass flowering in 1919 (Scaëtta, 1934), as did the populations on Mt. Elgon in 1935 (Synge, 1937, p. 69); subsp. *cheranganiensis* flowered similarly at 2700 m in July 1969 (Mabberley, 1971). A second population underwent mass flowering in 1966 and January 1972 (E. M. Tweedie, pers. comm.), and there was a mass-flowering of *D. johnstonii* subsp. *johnstonii* at 3000 m on Meru in December 1967 (D. Vesey Fitzgerald, Arusha National Park, pers. comm.). Data on this matter are scant, however, though *D. keniensis* is also known to have such mass flowerings (Coe, 1967, p. 23) and *D. keniodendron* flowered thus in 1916, 1922, 1929, 1957, 1974, and 1979 (Smith and Young, 1982). Ralph Tomlinson (letter to the author) reports that on Kilimanjaro, local mass flowerings of *D. johnstonii* subsp. *cottonii* seem to be associated with fire damage, for high percentages of trees recently affected were found to be in flower. During 1970–71, I saw no mass flowerings on Ruwenzori, Elgon, Cheranganis, Aberdares, Mt. Kenya, or Meru.

The inflorescence is a terminal pyramidal panicle that often exceeds a meter in length and is composed of many hundreds of capitula. The most massive inflorescences are found in plants flowering for the first time, in those with a massive stem, i.e., not the forest plants. After branching, the later inflorescences are smaller with fewer branches and capitula. Thus inflorescence initiation occurs on smaller branches as the plant ages. The capitula of the lower branches open first; the terminal capitula of the branch are usually followed by the lowest of the remaining capitula, and the remainder open last, working acropetally from there. The terminal capitulum of the inflorescence opens before the capitula of the small lateral branches immediately beneath it, the terminal part of the inflorescence behaving like a lateral branch. There may be in excess of 50 primary branches in the inflorescence, the basal ones bearing 20 or more secondary branches, of which the basal ones have 10 or more capitula, sometimes borne singly on peduncles, e.g., *D. brassiciformis*, or sometimes borne in pairs also, e.g., *D. johnstonii* subsp. *battiscombei*. The inflorescences are marcescent, remaining for many years between marcescent leaves of the stem, marking the interflowering distances on the trunk. This characteristic is common to many herbaceous "*Senecio*" spp., also, for example in the

FIG. 4–8. *Senecio roseiflorus*, Kenya, Aberdares, Malewa, approximately 3000 m, October 26, 1970.

afromontane flora, in *S. roseiflorus* R. E. Fr. (Fig. 4–8), *S. jacksonii* S. Moore, and *S. schweinfurthii* O. Hoffm.

According to Smith and Young (1982), individuals of *D. keniodendron* in flower are more likely to die and less likely to reproduce later than are individuals of the same size in the vegetative phase. If they do survive, however, they will produce more fruits than the previously vegetative ones because they will have more inflorescences due to the larger number of rosettes, which have arisen since the first flowering. Beck et al. (1984) found that of approximately 500,000 fruits per inflorescence, only 18% had seeds capable of germination and that less than 1% were left as young plants three years after germination.

The forest and swamp species have conspicuous and bright yellow capitula; those of the erect afroalpine species are dull yellow or orange, nodding, and although large, inconspicuous. There are no records in the literature of pollination having been observed in any *Dendrosenecio*. In the Cherangani Hills in Kenya, at 3300 m a flowering plant of *D. johnstonii* subsp. *dalei*, which is dwarf and therefore easy to study, and which has large ligulate capitula, was watched on December 12, 1970, and those insects actively visiting the flowers taken. Most of these insects have yet to be identified at the British Museum (Natural History). Numerous Coleoptera were seen walking over the extruded stamens and Diptera were observed visiting the flowers; one butterfly, *Issoria hanningtoni* Elwes, was also taken while visiting the flowers. In contrast, claims have been made for ornithophily (for instance Hauman, 1935), but on Mt. Kenya I observed the sunbirds and hill chats merely using the plants as perches. Their pollination of the pachycaul *Lobelia* spp. with large floral parts is better documented

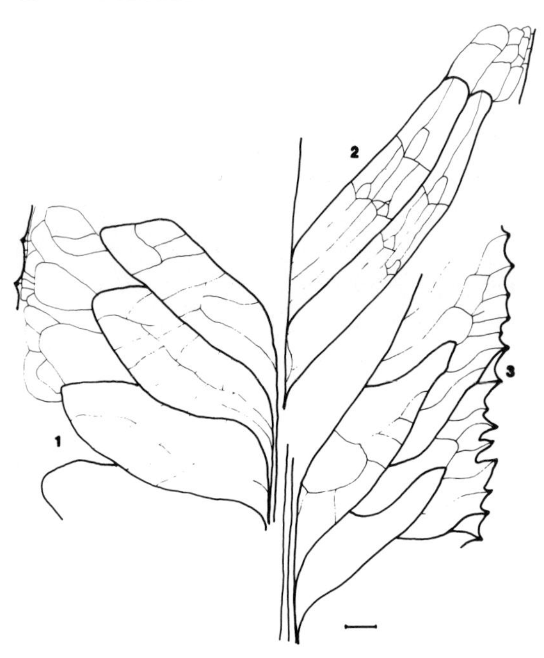

FIG. 4–9. *Dendrosenecio* leaf venation: (1) *D. johnstonii* subsp. *dalei* (*Mabberley* 505); (2) *D. keniensis* (*Mabberley* 408); (3) *D. johnstonii* subsp. *adnivalis* (*Mabberley* 521). Scale, 1 cm.

(Cheke, 1971; Young 1982); here sunbirds take insects as well as nectar. These insects are an important part of the sunbird diet (Coe, 1967; Cheke, 1971). It would appear that the flowers of *Dendrosenecio*, having little nectar, would be of little value to the birds. It might be more reasonable to suppose that the birds' relatively rare visits are associated with insect trapping. The inconspicuousness of the flowers of the species of the afroalpine belt, where sunbirds compose a high percentage of the vertebrate fauna (Coe, 1967), would support this idea. Furthermore their fetid "groundsel" odor, to be found in many Compositae, is attractive to the Diptera that pollinate many species in this family.

Ecological Morphology and Anatomy

The representatives of *D. johnstonii* found in the vegetation lying just below the afroalpine belt (Mabberley, 1973) are relatively thin stemmed, much branched, with thin bark compared with the afroalpine forms. They have little marcescent foliage, their leaf bases are scarcely persistent, and they have rather thin leaves that are sparsely pubescent, relatively wide, with a venation in which the costae (lateral veins) are not obviously looped together, and with prominent teeth and hydathodes (Fig. 4–9). Those in the afroalpine belt have thick marcescent collars of leaves held on by persistent leaf bases that remain adpressed to the trunk when the collars fall away. Their leaves are thick, relatively narrow, reflexed somewhat about the midrib, often markedly revolute at the margin, glossy adaxially and with a conspicuous hair cushion abaxially as well as a tomentum or dense pubescence; the costae are conspicuously looped together and the leaves are weakly toothed with few hydathodes. In the plants from lower altitudes, the leaves arise from lax rosettes that have small buds and few leaves, little slime around the bud, and weak diurnal movements. Their inflorescences are weakly pubescent, much branched and with few-flowered capitula, the outermost florets of which are conspicuously ligulate. Those at higher altitudes have tight rosettes with many leaves, and the buds are bathed in copious slime. They have conspicuous diurnal movements of the leaves closing over the bud at night. Their inflorescences are densely hairy, often less branched with large capitula of many florets, the outermost of which either have reflexed ligules or none. At the anatomical level, there is a decrease in size of epidermal cells with altitude, those of the adaxial surface being thicker-walled and more elaborately pitted; there is a decrease in size but an increase in number of airspaces in the spongy mesophyll, an increase in the number of stomata per unit area as well as more clearly defined palisade and spongy mesophyll layers (Hare, 1941; Mabberley, 1973), and an increase in secretory duct size and density of indumentum. Such a pattern of variation with altitude within one páramo species, *Espeletia schultzii* Wedd., has been demonstrated by Baruch (1979), who concluded that certain altitudinal ecotypes exist in that species. The variation in morphological and anatomical features in *Dendrosenecio* is rather continuous, so some of the specimens may be difficult to assign to one or another of the subspecies recognized on any one massif. The most striking variation in habit is not correlated with increasing altitude, however, but is associated with the occupation of swampy ground, viz. the dwarf subsp. *dalei* in the Cheranganis.

Compared with the pachycaul Compositae of the high tropical Andes, the range of life form exhibited by *Dendrosenecio* is very limited. According to the Cuatrecasas (1979, Chap. 11), the Espeletiinae of the Heliantheae in the Andes include plants which are hapaxanthic pachycauls (Holttum's Model, Hallé, Oldeman and Tomlinson, 1978), some sessile or with a tuber as well as with overground vegetative stems like the hapaxanthic *Lobelia* spp. of Africa (Mabberley,

1974b). Furthermore, there are unbranched pachycauls with lateral inflorescences and the same range of vegetative parts (Corner's Model, see Chap. 3). Sometimes these may be branched, but the branching trees with terminal inflorescences in the Espeletiinae are of two types, one of which, in *Tamania* and on a smaller scale in a species of *Ruilopezia*, corresponds to Leeuwenberg's Model, that is, terminal inflorescences and pseudodichotomous branching. This latter is the mode of branching in the pachycaul *Dendrosenecio* spp. (Fig. 4–1). *D johnstonii* subsp. *dalei* is merely a condensed form of this mode of branching and, except for its habit (Mabberley, 1971), indistinguishable from *D. johnstonii* subsp. *cheranganiensis* of the same massif. The candelabriform branching of these plants arises from the growth after the production of an inflorescence of one to eight lateral buds. This branching leads to a dwindling of the twig size and therefore leaves, one of "Corner's Rules" of Hallé et al. (1978), i.e. is apoxogenetic (Mabberley, 1973). In *D. brassiciformis* and especially in *D. keniensis*, however, the stem, with the same branching pattern, is prostrate, a *truncus superficialis*, lying along or at the surface of the swampy ground where these plants grow. Unlike the upright *D. johnstonii*, which has thick secondary xylem with growth rings, sufficient to make the large wooden cooking spoons, *D. keniensis* has little wood and the leaf traces pass into and up the pith first before passing out through to the petioles (Fig. 4–10). There is sclerenchyma in the cortex and conspicuous airspaces in the base of the petiole and cortex of the succulent roots (Mabberley, 1973). Further, the vessel-elements are about a quarter of the length (67 μ compared with 247–353 μ, Fig. 4–11).

FIG. 4–10. *Dendrosenecio* stem cross sections: (1) *D. johnstonii* subsp. *adnivalis* (*Mabberley* 522), scale, 1 mm; (2) same, secretory canal enlarged; (3) same, young plant; (4) *D. keniensis* (*Mabberley* 407), scale, 1 mm; (5) same as (4), portion of stele enlarged; (6) *Senecio vulgaris* L., to same scale as (3) and (4). Sc, sclereid; sec, secretory canal.

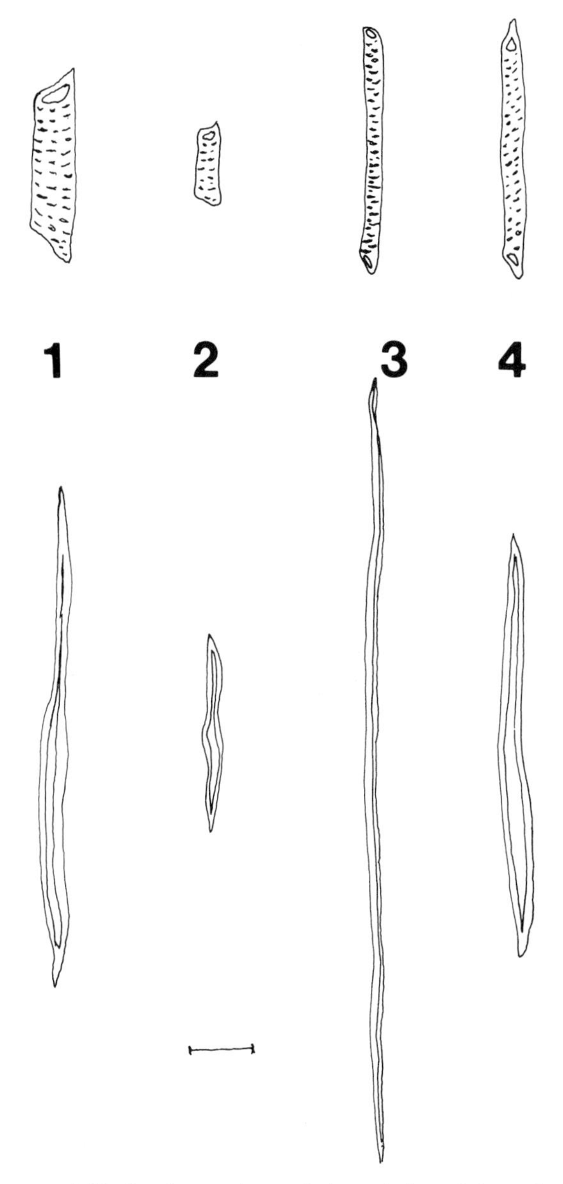

FIG. 4–11. *Dendrosenecio* wood elements (vessel elements *above*, fibers *below*): (1) *D. johnstonii* subsp. *battiscombei* (*Mabberley* 323); 2, *D. keniensis* (*Mabberley* 407); herbaceous species: (3) *S. vulgaris* L. (steps of Botany School, Cambridge); and (4) *S. snowdenii* J. Hutch. from East Africa. Scale, 100 µ.

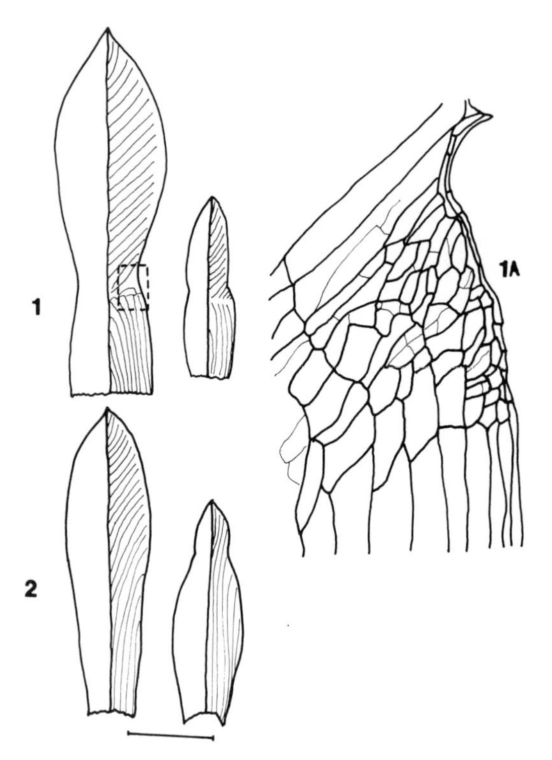

FIG. 4–12. Leaves and bracts of *Dendrosenecio:* (1) *D. keniensis* (*Mabberley* 408); (1A) enlarged; (2) *D. brassiciformis* (*Mabberley* 379). Scale, 10 cm.

The winged "petiole" found in some populations of *Dendrosenecio* is composed of thin winglike tissue webbing the leaf base with the lamina associated with the median and lateral strands of the developing venation. It resembles the bracts in morphology and anatomy: there are few hairs, glandular or otherwise; in the paludal species, *D. keniensis* and *D. brassiciformis* (Fig. 4–12), this feature has been used in keys to distinguish their leaves from those of other species, for this "petiole" appears distinct from the rest of the lamina. There is little diversity in size between the upper and lower epidermal cells compared with those of the lamina proper, and the individual cells are smaller, more elongated, and thinner-walled. There is little differentiation into palisade and mesophyll; each veinlet is accompanied by at least one secretory canal however. In decay the "wings" disappear and leave the midrib to become the persistent peg-like leaf base. The midrib consists of an arc of bundles embedded in parenchyma which begins to form early in the leaf development. The adaxial epidermis bears hairs which lie in the pool of slime which accumulates at the leaf base, and are sometimes coated in blue-green algae (?). The abaxial epidermis is similarly clothed with hairs. The internal parenchyma is composed of large cells, giving a bubbly appearance to the tissue that composes the bulk of the material which slowly rots to expose the bundles. The parenchyma of the abaxial half of the leaf base is composed of smaller more compacted cells. There is no obvious deposit, or change in the parenchyma of the leaf base to account for its persistence on the trunk of the tree for decades. However, note that the marcescent laminas forming the collars of dead matter on the trunks maintain a dry mircrohabitat beneath the living foliage. This is rarely moistened from above, for water accumulates between the leaf bases of the living leaves, as in *Ravenala* (Musaceae). That these cups are watertight is known only too well to those who have fallen into a plant of *D. keniensis* in the "vertical bog" of Mt. Kenya! As the leaves die, they cease their "sleep" movements and with the increase in diameter of the axis, the leaf bases become flattened against the axis, acting as support for the living leaves. A feature of the bundles which may assist preservation is their association with the secretory canals which harden as their contents crystallize in drying. Although these canals are found accompanying the smallest veins and are so common in the Compositae, no satisfactory explanation of their function has been given.

The hair-type of *Dendrosenecio* was shown by Hare (1941) to be waterabsorbing, and it fits the types of hair of other Compositae similarly proposed (Uphof, 1962). It is noteworthy that the longest hairs are found at the base of the leaves which are bathed in the cup of water and slime. The attenuating walls may suggest that the long hairs produce mucilage, however. Hare (1941) reported glandular hairs on the margin of the leaves of a Kilimanjaro population of *Dendrosenecio*; their structure corresponds to the mucilage producing type of Uphof (1962); these may be responsible for the slime so abundant in the buds. Such glands are found in many herbaceous species also; they are especially noticeable in *S. roseiflorus*, a very aromatic sticky plant of the bogs of Mt. Kenya and the Aberdare Mountains. In dry material the secretion of all species crystallizes as spicules around the "head" of the gland.

At the apex of each tooth at the leaf margin there is a hydathode or water pore. Near the base of the leaf where serrations are deeper, as if the leaf expanded before the characteristics of the more distal venation could develop, teeth may have lateral teeth themselves, each with a hydathode. The hydathodes occupy areas of the leaf-margin devoid of glands or other hairs; their structure is that of those of many plants of wet places (Hare, 1941). In general, my observations support these of Hare (1941), who considered the hydathodes a mechanism for extruding excess water, and the tomentum a protection against excess transpiration. Thus the tree might withstand a considerable range in atmospheric and ground conditions. I would suggest that the hairs at the base of the leaves, bathed in slime, may also be involved in water absorption, and that the slime may be involved in the leaf sleep movements characteristic of *Dendrosenecio*. The slime might aid the leaves to slide over one another or may be involved in the water exchange necessary to execute such movements. Alternatively, the movements may be growth phenomena; in basally growing leaves, there is a ready-made alternative to a pulvinus or other mechanism. (Leaves [basipetally growing] of cabbage rise at night [Cooke, 1881, p. 160], but others, e.g., *Siegesbeckia* [Compositae] droop [Darwin, 1880, p. 384].) Dolk (1931) has shown that the leaf movements in the compass plant, *Lactuca serriola* L. of the Compositae, are due to both asymmetric growth of the leaf base and to an epinastic movement. In the afroalpine flora, such movements are also found (Beck et al., 1982) in *Lobelia* and herbaceous *Helichrysum* and *Senecio* spp. of Mt. Kenya (Coe, 1967), and in *Carduus*. It is not known whether the movements are brought about by growth or turgor changes, although the degree of opening is temperature dependent (Coe, 1967). A parallel to *Dendrosenecio* is found in *Espeletia* of the Andean páramos (Hedberg, 1964). Slime, massive buds, and leaf movements are less noticeable in the forest species of both groups. Hedberg (1964) considered the slime to be an insulator for the bud, which is massive and compact, and Coe (1967) found small animals (insects and mollusks) sheltering in these cups, which rarely freeze solid, the slime acting as an antifreeze in the buds, sheltered by the leaves closed over them.

D. keniodendron may be readily distinguished from the other species by its leaves alone. These seem to have many more costae, which arises from the fact that during the development of the lamina, growth ceases at the margin leading to a rather toothless margin, but greatly increases near the midrib, such that the secondary linking veins are drawn out parallel to the costae, apparently increasing their number. In contrast, in *D. brassiciformis*, the leaf appears more phyllodic, for there is expansion between the lateral strands in the petiole compared with the rest of the leaf.

D. keniodendron is scarcely branched with a collar of marcescent leaves to the base: its habit is like that of the most extreme afroalpine forms of *D. johnstonii*. The marcescent collars are known to insulate the plant, and outside temperatures dropping as low as −5° C are not experienced within the collar, where they do not fall below freezing (Hedberg, 1964). Such marcescent collars are found in *Lobelia wollastonii* Bak. f. (Fig. 4–13) of Ruwenzori and the Virungas where

FIG. 4–13. *Lobelia wollastonii* (*Mabberley* 559), Uganda, Ruwenzori, Stuhlmann Pass, at approximately 4000 m, December 27, 1970.

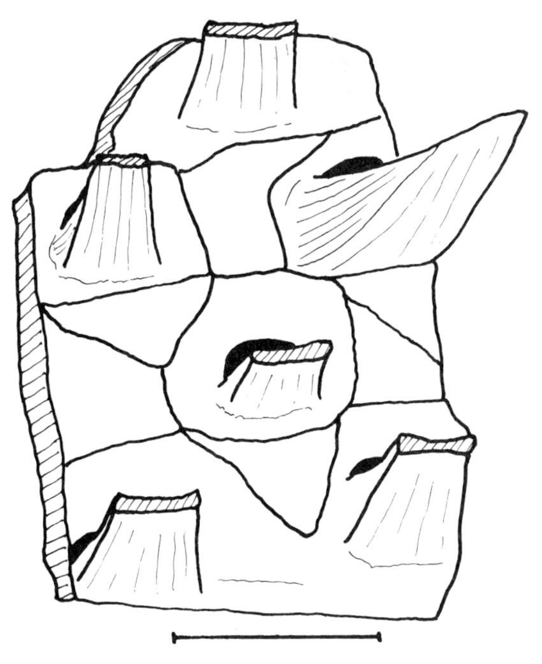

FIG. 4–14. Part of cork cylinder and leaf bases of dead *Lobelia wollastonii* (*Mabberley* 559) Scale, 1 cm.

FIG. 4–15. *Sambucus africana* Standley, Kenya, Aberdares, October 16, 1970.

FIG. 4–16. *Peucedanum kerstenii* Engler, Uganda, Ruwenzori, Bujuku Valley, December 27, 1970.

there is a corky process extending into the base of the leaf (Fig. 4–14) holding the old leaf to the stem (Mabberley, 1974b). This contrasts with the *Lobelia* spp. of lower altitudes and represents an example parallel to that in *Espeletia* spp. in the Andes (see Chap. 3). The removal of these leaves in *E. schulzii* is known to lead to the death of the plant, as demonstrated by Smith (1979). The marcescent collars in both *Dendrosenecio* and *Espeletia* harbor many invertebrates, when compared with the surrounding vegetation. Such marcescence is also found in *Sambucus* (Caprifoliaceae) and *Peucedanum* (Umbelliferae) in East Africa (Figs. 4–15 and 4–16); *Anaphalis* (Compositae), *Swertia* (Gentianaceae), and *Euphorbia* (Euphorbiaceae) in Java (van Steenis, 1972, Table 20); *Puya*, *Plantago* (Plantaginaceae), *Paepalanthus* (Eriocaulaceae), *Draba*, *Culcitium*, and *Blechnum* in the Andes; *Argyroxiphium* in Hawaii; and *Blechnum* and *Cyathea* in New Guinea. It must be pointed out that in many of these taxa marcescence may act as an insulator but that in *Cyathea* it is said merely to maintain relative humidity at the stem surface where there are delicate rootlets (Smith, 1979). Furthermore, marcescence is a common feature in many of these families represented. Indeed, its absence in some, as in the Compositae (Mabberley, 1976), is the more remarkable condition. Marcescence also occurs in taxa living in dry regions, as *Agave* (Agavaceae), *Yucca* and *Aloe* (Liliaceae), and *Washingtonia* (Palmae), where it may insulate against overheating.

The leaf movements in *D. keniodendron* are most marked and have been demonstrated to maintain the center of the plant above freezing, the temperature without dropping to −4°C, while the closed bud did not fall below 2°C (Hedberg, 1973). In *Espeletia schultzii* (Smith, 1974), death follows the artificial prevention of the closure of rosettes.

ORIGIN OF ECOLOGICAL SYNDROMES IN AFROALPINE SPECIES OF *DENDROSENECIO*

Within *Dendrosenecio johnstonii*, the most similar subspecies are found at lower altitudes, the most distinct in the afroalpine belt. It seems most probable therefore that the afroalpine forms have been derived in parallel on the different massifs. Further, there is a dwarf swamp form, subsp. *dalei*, which is very similar except in habit to the forest form, subsp. *cheranganiensis*. It is easier to argue, then, that the sparsely branched "hyperpachycaul" and the helophytes are secondary to the forest form with regular candelabrum branching (Mabberley, 1973, 1976; Nordenstam, 1978), a conclusion similar to that reached by Smith and Koch (1935) in their study of *Espeletia*. Within *D. johnstonii*, the populations on the wet mountains of the west are the least differentiated, though showing a wide variation in form of the leaves in particular (Mabberley, 1973). In the central area, Elgon and Cherangani, and in the east, Meru and Kilimanjaro, the variation is more clearcut into subspecies, but on Mt. Kenya and the Aberdares, the variation is such that three other distinct species are recognizable. These, the hyperpachycaul *D. keniodendron*, and the helophytes *D. brassiciformis* and *D. keniensis*, seem thus to have been derived from forest ancestors, and now appear to be fully fledged species.

FIG. 4–17. *Lobelia stuhlmannii* (*Mabberley* 532), Uganda, Ruwenzori, Nyamaleju, 3240 m, December 21, 1970.

COMPARISON WITH PACHYCAUL *LOBELIA* SPECIES IN AFRICA

Pachycaul *Lobelia* spp. are found in the tropics of the Old and the New Worlds. Most of them are found in forest or in disturbed habitats, such as roadsides. They are represented in Hawaii (Mabberley, 1974b) but are also characteristic of the African mountains, having a wider distribution, both geographically and altitudinally, than do the pachycaul *Dendrosenecio* spp. They are conspicuous plants of the montane forest and the afroalpine belt up to the snow line. They grow in damp and wet habitats, some species being helophytes. They are heliophilous, however, and when they occur at altitudes as low as 900 m, they are found in clearings, rocky gullies, and other semiopen sites (*L. longisepala* Engl.), or gullies and streamsides (*L. stricklandiae* Gilliland, *L. xongorolana* E. Wimmer). *Lobelia giberroa* Hemsl. is a common plant of secondary forest throughout eastern Africa from 1200–3000 m, springing up with such fast-growing trees as *Senecio mannii* Hook. f., and species of *Macaranga*, *Neoboutonia*, and *Vernonia*, but it also grows at high altitudes. *L. longisepala* is found in mixed rain forest, at lower altitudes nearer the coast (Usambaras) than inland (Ulugurus, Ngurus), where it is found in *Allanblackia* forest, perhaps an example of the *Massenerhebung* effect.

In Tanzania, the dominants of the "elfin" forests of the Ulugurus include *L. lukwangulensis* Engl., the crown of which disappears into the canopy of the *Allanblackia-Podocarpus* forest. In the Ukagurus, the endemic *L. sancta* Thulin is restricted to the steep slopes of the summit *Myrica-Syzygium-Podocarpus* forest. At higher altitudes, *L. stuhlmannii* Stuhlm. (*L. lanuriensis* De Wild., Ruwenzori, and Virungas, Fig. 4–17) and *L. bambuseti* R. E. and T. C. E. Fr. (Kenya and Aberdares, Fig. 4–18) are found in *Hagenia-Hypericum* woodland, bamboo, and ericaceous bush and thicket: *L. stuhlmannii* reaches the *Dendrosenecio* forests of the afroalpine belt. These associations have been described by Hedberg (1951). *L. petiolata* Hauman is found in the ericaceous thicket of Mt. Kahuzi but also at tracksides and in open habitats at lower altitudes.

In the highlands of East Africa, open habitats are provided by the seasonal swamps in the forest and in the surrounding cultivation. The creeping *L. aberdarica* R. E. and T. C. E. Fr. of the Kenya highlands and *L. mildbraedii* of Rwanda, Uganda and Tanzania occupy such habitats from 1800 m to the ericaceous thickets. At the upper edge of that vegetation and in the afroalpine belt such swampy habitats are occupied by the subspecies of *L. deckenii* (Asch.) Hemsl. (Fig. 4–19). In the afroalpine belt the sessile *L. telekii* Schweinf. (Fig. 4–3) is found on the rather dry moorland of Mt. Kenya (where, as an early colonizer of glacial soils, it germinates in the shelter of stones),

FIG. 4–18. *Lobelia bambuseti,* Kenya, Mt. Kenya, Naro Moro track, 3240 m, January 9, 1971.

FIG. 4–19. *Lobelia deckenii* subsp. *sattimae*, Kenya, Aberdares, "The Twins," 3240 m, October 29, 1970.

Aberdares and Elgon, and the stalked *L. wollastonii* (Fig. 4–13) in the wetter *Dendrosenecio* forests and *Helichrysum* scrub of Ruwenzori and the Virungas. The afroalpine belt of the Ethiopian highlands is famous for the enormous *L. rhynchopetalum* Hemsl., which extends down into the *Erica* thicket in the southern part of its range, where it is not such a tall plant.

In West Africa, *L. columnaris* Hook. f. grows in secondary grasslands and at forest edges from 1200 m upward on Fernando Pó and the highlands of the mainland; *L. barnsii* Exell grows in open situations on S. Tomé. The island populations grow at lower altitudes than those of the mainland, perhaps another manifestation of the *Massenerhebung* effect. Unlike the species of *Dendrosenecio*, the giant lobelias do not have stout woody trunks and are thus not sought out for firewood. They make good tinder when dry, however. The giant lobelias are unbranched and thus offer no crotches for epiphytes. Old plants may have a rich hepatic epiphytic flora, particularly on the wetter mountains, e.g., Ruwenzori and the Ulugurus, where *L. lukwangulensis* leaves are sometimes colonized by epiphyllae. This species was found to have the parasite *Loranthus lukwangulensis* Engler growing on it when first discovered by Stuhlmann (Engler, 1900).

Scott (1935) worked on the assemblages of Coleoptera restricted to the pachycaul *Lobelia* spp. He found a situation similar to that in *Dendrosenecio*, and *Euphorbia* of the Canary Islands, in that whole assemblages are characteristic of these plants. Some species, e.g., a silphid, spend their entire life cycle in a *Lobelia* plant as do also certain bibionid flies of *Lobelia* flowers (Coe, 1967). Interestingly, the species of insects associated with *L. columnaris* of West Africa, including nine species of Coleoptera (Paulian and Villiers, 1940), are considered by Scott (1952) to be a secondary association. This is in contrast with the species of East Africa where the distribution of the associated species of *Trechus* (Coleoptera) matches that of *Lobelia* (Scott, 1958). Many of the insects are flightless and "giant" within their own genera as are some of those associated with *Dendrosenecio*. Particular *Lobelia* species have been intensively studied. Scott (1958) recorded that the decaying stems of *L. rhynchopetalum* are the only sites for numerous Diptera. Noctuid moths are found in the inflorescences of this species and *L. giberroa*; numerous Hymenoptera visit the water between the leaves. Of the staphilinids, *Xylostiba scotti* Fagel is found in the flowers and *X. abyssinica* (Tottenham) under the bark in the Gughé Highland populations.

Hancock and Soundy (1931) reported chironomid midges under the decaying leaves of *L. deckenii* subsp. *elgonensis* (R. E. and T. C. E. Fr.) Mabb. Coe (1967) recorded these midges sheltered in the closed rosettes of the Mt. Kenya *L. deckenii* subsp. *keniensis* (R. E. and T. C. E. Fr.) Mabb. and larvae in the slimy water therein. The water never dries up, even in cultivation (McDouall, 1927) and does not freeze solid except at very low temperatures; the larvae are thus protected. *L. deckenii* subsp. *elgonensis* also yielded caterpillars and staphilinids to Hancock and Soundy (1931). Fishlock and Hancock (1932) recorded numerous insects from *L. wollastonii* on Ruwenzori. Particularly noticeable was the weevil *Pseudomesites lobeliae* Mshll. feeding on the fruit, and *Omalium algidum* Fauv. (Staphilinidae) previously recorded by Scott from *L. rhynchopetalum*. According to Jabbal and Harmsen (1969), the curculionid *Cessonus frigidus* Hust. is largely instrumental in the disintegration of the marcescent inflorescences of *L. deckenii* subsp. *keniensis*. In addition, I have collected noctuid caterpillars that burrow up the pith and into the inflorescence of *L. aberdarica*, *L. telekii*, and *L. deckenii* subsp. *sattimae* (R. E. and T. C. E. Fr.) Mabb. in the Aberdares.

Lobelia spp. are poisonous; the toxic principles are alkaloids, including lobeline. Edmunds (1904) showed that cats and dogs fed lobeline die from respiratory paralysis after primary respiratory stimulation and that frogs suffer a curare-type paralysis and strychnine-type convulsions, weakening of heart contractions, vomiting, and death. Steinberg and Volle (1972) have shown that lobeline in frogs interferes with the postjunctional action of "transmitter" at the nerve endings, which is needed like δ-tubocurarine.

As with *Senecio s.l.*, however, some herbivores are able to eat *Lobelia* spp. with impunity. *L. giberroa* has high concentrations of alkaloids (Hegnauer, 1966), and extracts of both this species and *L. mildbraedii* ("*L. suavibracteata*") injected into dogs have the same effect as standard lobeline (Tondeur and Charlier, 1950; Charlier and Tondeur, 1950), yet *L. giberroa* is browsed by elephants (Burtt, 1934). According to Hara and Kikuta (1958), lobeline depresses the blood pressure of herbivorous mammals, yet elevates it in carnivorous species. *L. giberroa* is gathered for goat-fodder (Dale, 1936, p. 154), a practice which I found to be particularly common in the dry season in the southern Aberdares, where large flocks are kept by the Kikuyu people.

L. deckenii subsp. *deckenii* is said to contain lobeline (Bally, 1938), but the closely related *L. deckenii* subsp. *keniensis* is browsed by hyrax and duiker (Coe, 1967), and I have observed eland browsing on similar *L. deckenii* subsp. *sattimae* in the swamps at 3600 m in the Aberdares. A remarkable example of *Lobelia* browsing is that of the rare bongo, studied by R. W. Wrangham (1970), who has kindly expanded the published account of observations in the Cheranganis (in correspondence of 2 Sept. 1970):

> It soon became clear that *L. giberroa* was so frequently eaten by them that it could be used as a marker of the presence of Bongo. . . . What Bongo did was to break off the stem at about three feet above the ground, and then eat the stem that they had broken off without eating the leaves. This left a pile of leaves . . . perhaps a dozen sometimes . . . near the plant. Since the pile was usually very tidy and often a few yards away from the plant, we imagined that the Bongo carried the broken stem to a place where it could eat in cover and comfort, and then stripped off the leaves. The mature leaves were scrupulously avoided. Two or three times we found leaves either on or off the plant which had a row of teeth-marks. But this was as much as the leaf was ever tried. The bunch of young leaves which form a shoot at the apex of the plant were often found to be missing, however, so it seems that the young leaves are tolerable. This shoot was sometimes missing from a plant whose stem had not been broken off. *L. giberroa* was certainly a favourite food of Bongo. The path taken by a group in the forest could be followed simply by finding where this plant had been eaten. This was particularly useful to us because the dryness of the sap of the broken stem told us roughly how long it was since Bongo had been in the area.

Dr. G. Jones told me that the seedlings of *L. rhynchopetalum* are grazed by *Arvicanthus* and *Tachyoryctes* (both rodents), but that they are undamaged if they attain 30 cm. He reported that it is also browsed by mountain nyalla in the Balé Province of Ethiopia.

The bracts of *L. deckenii* are often deeply shredded; this is a result of clawings by the sunbirds, particularly noticeable in subsp. *keniensis* from the activities of the Scarlet-tufted Malachite Sunbirds on Mt. Kenya, during visits for insects. The apex of those plants is flattened into a platform of small bracts; this is a much frequented perch of sunbirds and hill chats, and I have seen subsp. *elgonensis* similarly used by insect-catching chameleons in the Cheranganis; these perches make the chameleon vulnerable to raptors, however. The best-studied example of '*Lobelia*-living' is that of the Mountain Gorilla (Schaller, 1963). Above 3300 m, *Lobelia* with *Dendrosenecio* and *Peucedanum* forms the bulk of the gorilla's diet (p. 361); *L. giberroa* stem is eaten as are the root and leaf bases of *L. wollastonii*. The gorilla tackles the latter by breaking off the leaf rosette, eating the pith and leaf bases, and peeling the roots before eating the pith (p. 160). Stands of *L. giberroa* are highly favored as the foundations for the massive untidy nests of the gorilla (p. 172).

The inter-cicatrice distance on *Lobelia* stems is a reflection of the inter node distance. On some specimens of *L. giberroa* I found that there were two to four zones on the stem where the distance was very short, suggesting that there were two periods of relatively slow growth during the ontogeny of the stem. The adverse growing periods may have been periods of dry weather, the optimal growing periods being the rainy seasons. If the pattern reflects the climate, then the stems were only one or two years old because there are two rainy seasons in Tanzania. It has already been suggested by Greenway that the stems of *L. giberroa* are biennial (specimen 12331, East African Herbarium) and a similar maturation age is given for the Indian pachycaul, *L. nicotianifolia* [Roth ex] R. & S. ("*L. leschenaultiana*", Skottsberg, 1928). *L. stricklandiae* and *L. nicotianifolia* are treated as biennials in horticulture: seed of *L. columnaris* collected in West Africa (*Hepper* 1937) germinated and the resultant plants flowered at Kew a year later. *L. deckenii* is said to be short lived (Greenway, 1965) but lived for many years at Logan, Scotland, without flowering; however, no afroalpine species has flowered away from the tropics. *L. nubigena* Anth., the Bhutanese alpine, is also said to be short lived (Anthony, 1936). The short life cycle permits the giant lobelias to thrive in temporary secondary habitats. *L. giberroa* must grow at the remarkable rate of up to 5 m in two years; the enormous inflorescence of up to 4 m unfurling in as many months. The growth in height of the plant in the vegetative state almost reaches 1 cm per day! By contrast, the afroalpine *L. wollastonii* is said to grow for 10 to 40 years before flowering (Hauman, 1934); the Uluguru species *L. lukwangulensis* seems to grow at about the same rate, for a plant measured in January 1971 had grown about 20 cm by July 1972, i.e., at a rate of just over 1 cm per month.

As with *Dendrosenecio*, then, *Lobelia* is represented in the afroalpine belt by creeping species in wet habitats and erect or sessile species in drier ones. Their allies are in the forests and other vegetation types below. With increasing altitude, *Lobelia* spp. demonstrate an increase in the measure of pachycauly, a reduction in suckering, an elaboration of persistent leaf bases and marcescence (*L. wollastonii*) or the establishment of vegetative acaulescence, reduction in internode length and loss of lenticels, an increase in the number of leaves in buds, and leaves that are thicker, smaller, and narrower with fewer teeth and an apparent increase in the number of costae. Also observed are an elaboration of leaf movements and production of slime around the buds, the appearance of cortical laticifers, thicker leaf-cell walls, stomata restricted to the abaxial surface or equally distributed on both surfaces, a reduction of inflorescence branches, an increase in the degree of hollowness of the axis, an increase in pubescence, loss of bracteoles, less conspicuous pubescent flowers, smaller wingless seeds, and slower growth rate. These trends are seen in parallel in the mountains of the west through the series *L. stuhlmannii* and *L. wollastonii* and in the east through *L. bambuseti* and *L. telekii*. A similar pattern is seen in the Himalayas when the curious alpine *L. nubigena* is compared with the widespread *L. nicotianifolia* at lower altitudes.

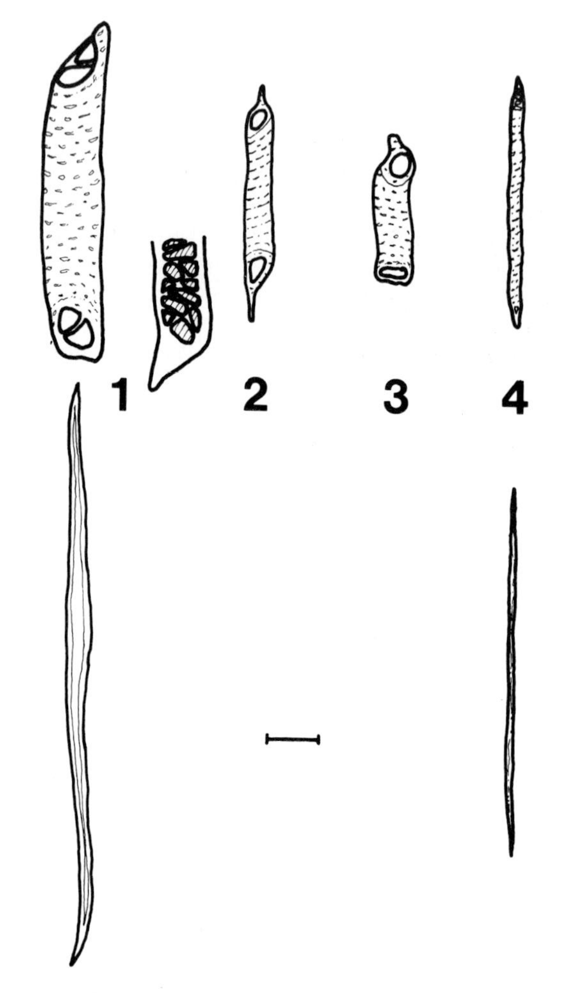

FIG. 4–20. *Lobelia* wood elements (vessel elements *above*, fibers *below*): (1) *L. giberroa* (*Mabberley* 317) with enlarged view of scalariform end plate of another element; (2) *L. wollastonii* (*Mobberley* 559: (3) *L. deckenii* subsp. *sattimae* (*Mabberley* 386) with (4) herbaceous *L. holstii* Engler of East Africa (*Mabberley* s.n.) for comparison. Scale, 100 μ.

The significance of the hollow inflorescence axis in *L. telekii* has been investigated by Krog et al. (1979). They showed that the inflorescence, which may attain 2 m, contains a slimy fluid that may reach 50 cm up the tube. The fluid becomes heated during the day but cools more slowly than the surrounding air after sunset. Further, ice is formed in vertical lamellae in the fluid. Due to the liberation of the heat of fusion of the freezing

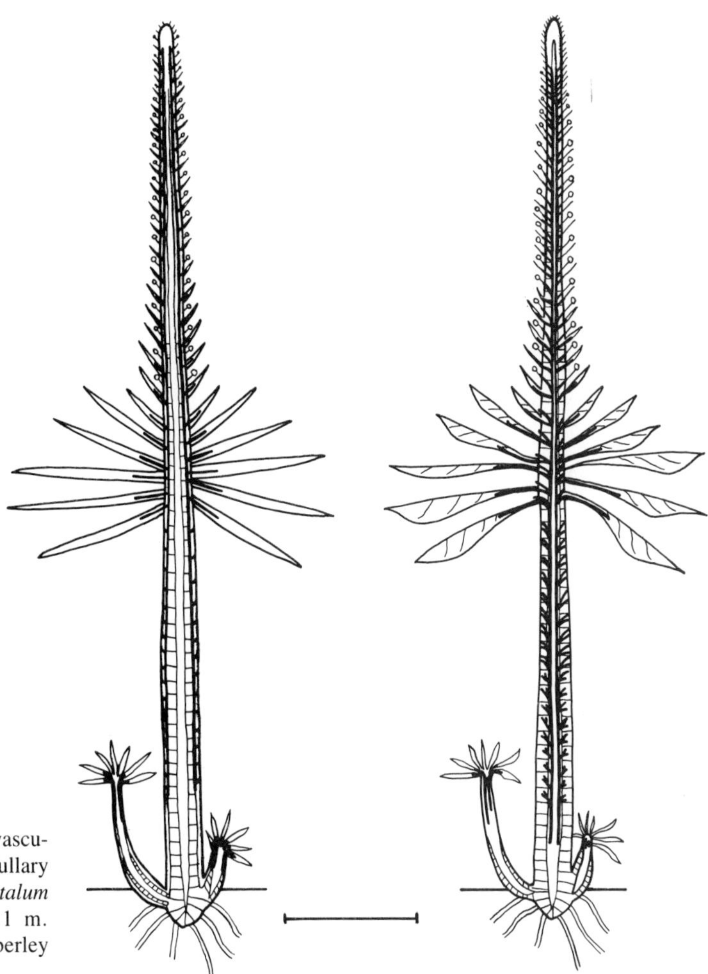

FIG. 4–21. *Lobelia* medullary and cortical vascular systems (V.S.) *Right: L. giberroa* (medullary system drawn heavily). *Left: L. rhynchopetalum* (cortical system drawn heavily). Scale, 1 m. Reproduced with permission from Mabberley (1974b).

water, the temperature of the unfrozen fraction of the fluid (around 98% by volume) is stabilized at the melting point of the fluid, ensuring maximal thermal protection of the plant. Furthermore, the freezing of the fluid will have a buffering effect on the air-filled chamber within the unfurling inflorescence, for the air within will circulate and cause internal heating at the top of the plant. The plant seems to induce freezing at as high a temperature as possible, a process known to be affected by the presence of nucleating agents, which in this case seem to be carbohydrates in the slime. The slime in the buds is of a consistency similar to that in the buds of *Dendrosenecio*. It appears to protect the bud from frost (Hedberg, 1964) and acts as shelter to invertebrates (Coe, 1967). The long hairs in the "cup" formed by the leaves may produce the slime or may be involved in absorbing water from the cup. At night the leaves fold over the bud in the afroalpine species, just as in *Dendrosenecio*.

Compared with the forest-living species, the swamp species have pneumatophore-like roots (Beck et al., 1981) and creeping stems with short vessel elements (Fig. 4–20) and weak wood development just as in *Dendrosenecio*. There is the appearance of the phyllodic leaf base as in *D. brassiciformis*, here associated with a new cortical system of vascular bundles (Figs. 4–21 and 4–22). The leaves are toothless and have fewer hydathodes; they undergo diurnal movements, elaborated at high altitudes, as in *Dendrosenecio*. The inflorescences are unbranched, hollow, and with brightly colored flowers, which at high altitudes (*L. deckenii*) have more or less unsplit corollas, visited by birds, which are said to be more efficient pollinators there (Cruden, 1972). Further, the bracts are larger and rather than being insulators, as in the afroalpines above, may act as efficient landing stages for the sunbird pollinators, which shred them with their claws. The seeds are winged, particularly in *L. deckenii*, helping to explain how that species is the most widespread of the afroalpine species, growing from Ruwenzori to Mt. Meru, divided into a number of rather poorly defined subspecies.

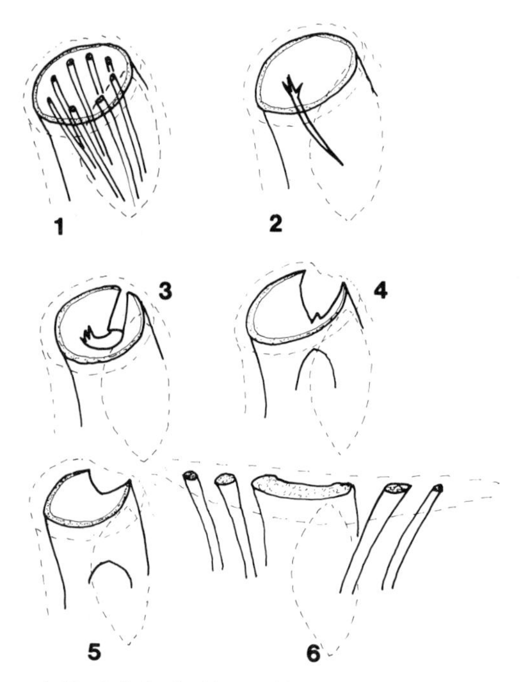

FIG. 4–22. *Lobelia* leaf base phloem, reconstructed from serial sections: (1) *L. giberroa* (*Mabberley* 317); (2–4) *L. nicotianifolia* (*Corner, s.n.*, Sri Lanka) and *L. stricklandiae* (*Mabberley* 1135); (5) *L. telekii*, *L. wollastonii*, *L. bambuseti*, *L. stuhlmannii*; (6) *L. aberdarica*, *L. deckenii*, *L. rhynchopetalum*. Dotted lines indicate the extent of petiole and leaf gap.

THE AFROALPINE PACHYCAULS IN A WIDER CONTEXT

In short, afroalpine hyperpachycaul and helophyte syndromes, apparently derived from forest ancestors similar to those in *Dendrosenecio*, are found in *Lobelia* also. The pachycaul construction lends itself admirably to occupation of the inhospitable afroalpine belt: particularly the massive construction allowing the damped cooling and heating in the diurnally fluctuating environment. Similarly, the pachycaul is suited to the diurnally fluctuating climate of deserts as seen in cacti, *Euphorbia*, and *Yucca* (Mabberley, 1976; Nobel, 1980). The deepseated cambium or its absence allows pachycauls to tolerate fire as in *Xanthorrhoea* and *Cyathea* (see Mabberley, 1976). The question then arises, why do not other pachycauls in the forests as Araliaceae, *Euphorbia* and Urticaceae have related species in the afroalpine belt? Clearly, besides the pachycaul attribute, these plants have something else. They are able to elaborate a hyperpachycaul construction, or an increasing primary growth, to enlarge the buds, to elaborate more hairs and slime and nyctinastic movements, to promote the retention of leaf bases and marcescence, as well as the expected thickening and narrowing of leaves. In so doing, the whole of the morphology of the plant is under increased apical control, with the suppression of lateral branches or suckers and the appearance of cortical systems associated with the streamlined phyllodic leaf. This all hinges on the mode of growth of organs in these groups, an essentially intercalary phenomenon, whether leaf or floral parts are concerned. This is what allows Campanulaceae and Compositae to become even more pachycaul just as can the monocot *Puya* (Bromeliaceae). The massiveness allows for relatively little circumference change with age, such that the leaf bases and marcescent leaves are not readily sloughed off in most pachycaul plants, just as in succulents in deserts the pachycaul habit may not merely present improved water storage, but also an uncracked photosynthetic surface in the form of the stem itself.

DISCUSSION: *SENECIO* AND THE ORIGIN OF HYPERPACHYCAUL AND HELOPHYTE SYNDROMES

In recent years, attempts have been made to bring order to the unwieldy genus *Senecio*, and to relate its diverse parts to the other genera of the Senecioneae. The great mass, some 1500 or so species, is still considered *Senecio sensu stricto* and a scheme for species referred to the genus on a worldwide scale is now available (Jeffrey, 1979). Thus the "Cacalioids," including *Lordhowea*, *Brachyglottis* and their allies, largely Gondwana groups, as well as *Ligularia*, *Doronicum* and their allies, a largely Eurasian assemblage, and the "Tephroseroids," a northern group, are excluded from the "Senecionoids." The latter have been divided into five major groupings, one of which, "Palustres," includes the pachycaul *Dendrosenecio* spp. and their herbaceous allies in one subgroup, the herbaceous *S. doria* L. of Europe and its allies in another, and the South American *S. hualtata* DC in a third. This relationship was posited by Mabberley (1973). The closest allies of *D. johnstonii* appear to be coarse herbaceous species of the montane forests and other habitats of tropical and southern Africa and Madagascar. The limits of the genus *Dendrosenecio* are being studied now at the Royal Botanic Gardens, Kew. Suffice it to say that the genus includes pachycaul trees as well as herbs, a situation similar to that in many other parts of the Senecioneae.

The cosmopolitan tribe Senecioneae comprises several thousand species of trees, shrubs, climbers, and even epiphytes and aquatics besides terrestrial herbs, both annual and perennial. Of the woody species, many of which are pachycaul, some are taxonomically quite isolated (such as *Lordhowea* of Lord Howe Island), but many are not as isolated as Stebbins (1977) would like us to believe. Besides *Dendrosenecio*, Africa provides another example in the *Kleinia* group, to which the pachycaul treelets *Senecio mannii* and *S. gigas* Vatke are referable (C Jeffrey, pers. comm.). This grouping includes the succulent *K. neriifolia* Haw. of the Canary Islands and a number of herbaceous and climbing species in the closely allied genus *Gynura*. Farther afield, the genus *Brachyglottis* in New Zealand includes herbaceous species that even hybridize with the woody ones (Mabberley, 1974a; Jeffrey, 1979). It is interesting to note that in all three of these groups, woody pachycaul species have branching patterns corresponding to Leeuwenberg's model. This also appears in *Pladaroxylon* (St. Helena), *Pittocaulon*, *Roldana*, and *Telanthophora* (tropical America), *Robinsonia* and *Symphochaete* (Juan Fernandez), and others (Mabberley 1974a). It also appears in Heliantheae (Cuatrecasas, 1979) and Astereae, Lactuceae, Cynareae, and Inuleae (Mabberley, 1979b). Are these pachycaul characters relicts as they seem to be in *Dendrosenecio*?

Pachycaul plants seem unfamiliar to botanists trained in temperate regions. They often make bad specimens and are difficult to file in herbaria (Corner, 1967a). In a serious study of evolution within genera with tropical representatives, however, they cannot be avoided, as seen classically in *Ficus* (Corner, 1967b) and also in Meliaceae (Mabberley, 1979a), in addition to Compositae and Campanulaceae as argued earlier. Some pachycauls are isolated taxonomically, but in large genera there is a complete gradation from the most pachycaul to the most leptocaul as in *Solanum* or *Euphorbia*. Examples of pachycauls associated with islands are provided by Mabberley (1979b)—to which may be added *Meryta latifolia* Seem. (Araliaceae, New Caledonia, where there are many other examples) and *Grias cauliflora* L. (Lecythidaceae, West Indies); I am indebted to Dr. H. Heine for bringing these to my notice. Both these latter and many other taxa mentioned in the earlier paper (Mabberley, 1979b) have pachycaul allies on continents where such plants are found in many habitats. To those may be added the woody Cyperaceae in the genus *Cephalocarpus* in America and the Malayan *Trigonostemon wetriifolius* Airy Shaw and Ng (Euphorbiaceae; Airy Shaw and Ng, 1978), while a linking series of genera is provided by the stout-budded *Hagenia* (Rosaceae) of the montane forests of eastern Africa, the pachycaul *Dendriopoterium* of the Canary Islands (Bramwell, 1980; Serebrykova and Petukhova, 1978) and the herbaceous *Sanguisorba* (*Poterium*) of the temperate zones. Growing in wet places and in the chalk grassland of Europe with *Sanguisorba* species are species of "*Senecio*" (*s.l.*) Have they travelled the same route? Do their allies live on subtropical islands and in tropical montane forests?

Syndrome and Strategy

The hyperpachycaul syndrome, apparently so well adapted to the tropical-alpine conditions, has clearly been derived in parallel manner from montane pachycaul plants in diverse genera. The evolution of such afroalpine plants with slower and more massive growth would seem to represent an increase in the "*K*-ness" rather than the "*r*-ness" in the fashionable jargon of plant "strategy." Similarly, Smith (1980) has shown how, within a single pachycaul species, *Espeletia schultzii* in the Venezuelan Andes, populations at higher altitudes are longer-lived and taller, unlike other plants where the general tendency would be to dwarfing (see also Chap. 3). More work on the populations of *Dendrosenecio* in the forests below the afroalpine belt is clearly now required.

The other syndrome, the helophyte, is also to be found in the afroalpine belt, again derived in parallel in *Lobelia* and *Dendrosenecio*. Turning back to the route from pseudodichotomously branching erect pachycaul tree through the *truncus superficialis* to the massive rhizome pre-adapted to storage and seasonal growth, the intriguing problem of apoxogenesis is got over in the creeping plant, for it dies at the base as it grows forward. Few rhizomatous plants have had their architecture scrutinized (Bell and Tomlinson, 1980) but it is known in *Alchemilla* (Kershaw, 1960) that the leaf size increases once the parental plant breaks up into separate plantlets. In other words, removal from the parent plant allows the build-up of apex size once more and therefore of leaf size. The helophyte syndrome, the maintenance of pachycaul massiveness below ground, is a strategy alternative to that shown by "woody" groups in the temperate zone. There, marcescence and massiveness are replaced by abscission and leptocauly. Within Compositae, this can be seen even in the tropical trees *Brachylaena* of Africa and Madagascar, *Vernonia* of Asia, and so on. In the Senecioneae, the Madagascan *S. hypargyreus* DC has a similar pattern: venation closed at an early stage, small buds and intermittent or seasonal growth with internodes—a syndrome clearly "preadapted" to the climate of the temperate zone.

Compositae and Campanulaceae in Africa and a restricted number of other families in the high montane tropics elsewhere are select as far as the tropical flora is concerned. The fact that these tropical plants should be the montane pachycaul representatives of families that are enormously successful in numbers of species in the temperate zones seems to argue for the overwhelming importance of the "phyletic heritage" as a constraint on the possible "adaptations" that may be selected, as so entertainingly argued from a zoological standpoint by Gould and Lewontin (1979). Here the constraints are architectural, the architecture of the pachycaul with intercalary meristems being most flexible in providing the starting point for hyperpachycaul and helophyte, the latter leading to the rhizome and the herbaceous hemicryptophytes and geophytes of the temperate zones. Clearly then, more thought should be given to "form" in the spirit of Grassé (1973) and Riedl (1977), so that the damaging dissection of organisms into "characters" and the consideration of the selective advantage of any one of these may be replaced by a more integrated approach. Indeed, explanations for the increasingly apparent widespread distribution of certain polymorphisms (see, e.g., Mabberley, 1980) may then become less tortuous. We may well, with Gould and Lewontin (1979), reread Charles Darwin and see what he actually said about natural selection as one, albeit a principal, mechanism of evolution.

ACKNOWLEDGMENTS

I would like to thank Mr. F. White for reading the manuscript and Vera Curtis-Smith for typing the final draft.

APPENDIX 4–A

The correct names for the giant groundsels in *Dendrosenecio* (Hedberg) Nordenstam are as follows (see Mabberley, 1973, and Nordenstam, 1978, for details):

1. ***D. johnstonii*** (*Oliver*) *Nordenstam* **emend.** Basionym: **Senecio johnstonii* Oliver in Johnston, Kilima—Njaro Exped.: 324 (1886). Type species. Eight ecogeographically separated subspecies including: *D.* ***johnstonii*** subsp. ***adnivalis*** (*Stapf*) *Mabb.*, ***comb. nova***. Basionym: *Senecio adnivalis* Stapf in *J. Linn. Soc. Bot. 37*: 521 (1906). This would appear to be the correct name for the plant until recently known as *S. johnstonii* subsp. *refractisquamatus*, according to the latest ruling on the priority of autonyms.
2. ***D. keniodendron*** (R. E. and T. C. E. Fr.) Nordenstam.
3. ****D. brassiciformis*** (*R. E. and T. C. E. Fr.*) *Mabb.*, ***comb. nova***. Basionym: *S. brassiciformis* R. E. and T. C. E. Fr. in *Svensk Bot. Tidskr*. 16: 338 (1922).
   With more available material, this now seems quite distinct from 3 in floral and vegetative characters as hinted at by Mabberley (1973).
4. ****D. keniensis*** (*Bak. f.*) *Mabb.* ***comb. nova***. Basionym: *S. keniensis* Bak. f. in *J. Bot., Lond*. 32: 140 (1894), *quoad inflorescentiam*. Synonym: *D. brassica* (R. E. and T. C. E. Fr.) Nordenstam. In accord with the International Code of Botanical Nomenclature, a lectotype has been selected from the mixed type collection (*Gregory. s. n.*, British Museum) of *S. keniensis*.

N.B. Hybrids between 2 and 3 are known (see Mabberley, 1973).

* Additions to *Index kewensis*

REFERENCES

Airy Shaw, H. K., and F. S. P. Ng. 1978. *Trigonostemon wetriifolius*, a new species from Endau-Rompin, South Peninsular Malaysia. *Malay. Forester* 41: 237–240.

Allt. G. 1968. pH and soil moisture in the substrate of giant senecios and lobelias on Mt. Kenya. *E. Afr. Wildl. J.* 6: 71–74.

Anthony, J. 1936. A remarkable alpine Lobelia from Bhutan. *Notes R. Bot. Gdn. Edinb.* 19: 175–176.

Bally, P. R. O. 1938. Heil- und Giftpflanzen der Eingeborenen von Tanganyika. *Fedde, Rep. Sp. nov. Reg. veg. Beih.* 102: 1–87.

Baruch, Z. 1979. Elevational differentiation in *Espeletia schulzii* (Compositae), a giant rosette plant of Venezuelan paramos. *Ecology* 60: 85–98.

Beck, E., H. Rehder, P. Pongratz, R. Scheibe, and M. Senser. 1981. Ecological analysis of the boundary between the afroalpine vegetation types "Dendrosenecio Woodlands" and "*Senecio brassica—Lobelia keniensis* community" on Mt. Kenya. *J. E. Africa Nat. Hist. Soc. Nat. Mus.* 172: 1–11.

Beck, E., R. Scheibe, M. Senser, and W. Müller. 1980.

Estimation of leaf and stem growth of unbranched *Senecio keniodendron* trees. *Flora* 170: 68–76.

Beck, E., I. Schlütter, R. Schreibe, and E.-D. Schulze. 1984. Growth rates and population rejuvenation of East African giant groundsels (*Dendrosenecio keniodendron*). *Flora* 175: 243–248.

Beck, E., M. Senser, R. Scheibe, H.-M. Steiger, and P. Pongratz. 1982. Frost avoidance and freezing tolerance in afroalpine "giant rosette" plants. *Pl. Cell Environ.* 5: 215–222.

Bell, A. D., and P. B. Tomlinson. 1980. Adaptive architecture in rhizomatous plants. *Bot. J. Linn. Soc.* 80: 125–160.

Bramwell, D. 1980. The endemic genera of Rosaceae (Poterieae) in Macaronesia. *Bot. Macar.* 6: 67–73.

Bryk, F. 1927. Ueber die Curculioniden-Fauna des Mount Elgongipfels. *Societ. Ent.* 42: 38–39.

Burtt, B. D. 1934. A botanical reconnaissance in the Birunga Volcanoes of Kigezi, Ruanda, Kivu. *Bull. Misc. Inf. Kew.* 1934: 145–165.

Charlier, R. and R. Tondeur. 1950. Alcaloïdes de *Lobelia suavibracteata* (Hauman). Extraction et étude pharmacologique. *Arch. Int. Pharmacodyn. Thér.* 83: 193–195.

Cheke, R. A. 1971. Feeding ecology and significance of interspecific territoriality of African montane sunbirds (Nectariniidae). *Rev. Zool. Bot. Afr.* 84: 50–64.

Coe, M. J. 1962. Notes on the habits of the Mount Kenya Hyrax (*Procavia johnstoni mackinderi* Thomas). *Proc. Zool. Soc. Lond.* 138: 639–654.

Coe, M. J. 1967. *The ecology of the alpine zone of Mount Kenya.* Monographiae Biologicae 17. The Hague: Junk.

Cooke, M. C. 1881. *Freaks and marvels of plant life.* London: Parker.

Corner, E. J. H. 1967a. On thinking big. *Phytomorphology* 17: 24–28.

Corner, E. J. H. 1967b. *Ficus* in the Solomon Islands and its bearing on the post-Jurassic history of Melanesia. *Phil. Trans. R. Soc. B* 253: 23–159.

Cruden, R. W. 1972. Pollinators in high-elevation ecosystems: relative effectiveness of birds and bees. *Science* 176: 1439–1440.

Cuatrecasas, J. 1979. Growth forms of the Espeletiinae and their correlation to vegetation types of the high tropical Andes. In *Tropical botany*, K. Larsen and L. B. Holm-Nielsen, eds., pp. 397–410. London: Academic Press.

Dale, I. R. 1936. *Trees and shrubs of Kenya Colony.* Nairobi: Government Printer.

Darwin, C. 1880. *The power of movement in plants.* London: Murray.

Dolk, H. E. 1931. The movements of leaves of the compass plant *Lactuca scariola. Am. J. Bot.* 18: 195–204.

Edmunds, C. W. 1904. On the action of lobeline. *Am. J. Physiol.* 11: 79–102.

Engler, A. 1900. Berichte über botanischen Ergebnisse der Nyassa-See-und Kinga-Gebirgs Expedition der Hermann- und Elise-geb. Heckmann-Wentzel-Stiftung. III. Die von W. Goetze und Dr Stuhlmann im Ulugurugebirge, sowie die von W. Goetze in der Kiseki und Khutu-Steppe und in Uhehe gesammelten Pflanzen. *Bot. Jahrb.* 28: 323–510.

Fishlock, C. W. L., and G. R. L. Hancock 1932. Notes on the flora and fauna of Ruwenzori with special reference to the Bujuku valley. *J. E. Africa Nat. Hist. Soc.* 44: 205–229.

Gould, S. J. and R. C. Lewontin. 1979. The spandrels of San Marco and the Panglossian paradigm: a critique of the adaptationist programme. *Proc. R. Soc. Lond. B* 205: 581–598.

Grassé, P. P. 1973. *L'évolution du vivant.* Paris: Albin Michel.

Greenway, P. J. 1965. The vegetation and flora of Mt. Kilimanjaro. *Tanganyika Notes Rec.* 64: 97–107.

Hallé, F., R. A. A. Oldeman, and P. B. Tomlinson. 1978. *Tropical trees and forests. An architectural analysis.* Berlin: Springer.

Hancock, G. L. R., and W. W. Soundy. 1931. Notes on the fauna and flora of Northern Bugishu and Masaba (Elgon). *J. E. Africa Nat. Hist. Soc.* 36: 165–183.

Hara, S., and Y. Kikuta. 1958. On the species difference of the appearance of blood pressure action of lobeline in animals. *Jap. J. Pharmacol.* 7: 104–108.

Hare, C. L. 1941. The arborescent senecios of Kilimanjaro: a study in ecological anatomy. *Trans. R. Soc. Edinburgh* 60: 355–371.

Hauman, L. 1934. Les "Lobelias" géants des montagnes du Congo Belge. *Mém. Inst. R. Colon. Belge, Sect. Sci. Nat. Méd. 8°* 2: 1–52.

Hauman, L. 1935. Les "Senecio" arborescents du Congo. *Rev. Zool. Bot. Afr.* 28: 1–76.

Hedberg, I., and O. Hedberg. 1979. Tropical-alpine forms of vascular plants. *Oikos* 33: 297–307.

Hedberg, O. 1951. Vegetation belts of the East African mountains. *Svensk Bot. Tidskr.* 45: 140–202.

Hedberg, O. 1957. Afro-alpine vascular plants. A taxonomic revision. *Symb. Bot. Upsal.* 15(1).

Hedberg, O. 1964. Features of afroalpine plant ecology. *Acta Phytogeogr. Suec.* 49: 1–144.

Hedberg, O. 1969. Growth of the East African giant senecios. *Nature* 222: 163.

Hedberg, O. 1973. Adaptive evolution in a tropical-alpine environment. In *Taxonomy and ecology*, V. H. Heywood, ed., pp. 71–92. London: Systematics Association.

Hegnauer, R. 1966. Lobeliaceae. *Chemotaxonomie der Pflanzen*, 4: 404–414. Basel and Stuttgart: Birkhauser.

Jabbal, I., and R. Harmsen 1969. Curculionidae (Weevils) of the alpine zone of Mt. Kenya. *J. E. Africa Nat. Hist. Soc.* 27: 141–154.

Jeannel, R. 1950. *Hautes montagnes d'Afrique.* Publ. Muséum National Histoire Naturelle, Paris. Suppl. 1.

Jeffrey, C. 1979. Generic and sectional limits in Senecio (Compositae): II. Evaluation of some recent studies. *Kew Bull.* 34: 49–58.

Johnston, H. H. 1886. *The Kilima-Njaro Expedition.* London: Kegan Paul.

Kershaw, K. A. 1960. Cyclic and pattern phenomena as exhibited by *Alchemilla alpina. J. Ecol.* 48: 442–453.

Krog, J. O., K. E. Zachariassen, B. Larsen, and O. Smidsrød. 1979. Thermal buffering in afro-alpine plants due to nucleating agent-induced water freezing. *Nature* 282: 300–301.

Mabberley, D. J. 1971. The Dendrosenecios of the Cherangani Hills. *Kew Bull.* 26: 33–36.

Mabberley, D. J. 1973. Evolution in the giant groundsels. *Kew Bull.* 28: 61–96.

Mabberley, D. J. 1974a. Branching in pachycaul Senecios: the Durian Theory and the evolution of angiospermous trees and herbs. *New Phytol.* 73: 967–975.

Mabberley, D. J. 1974b. The pachycaul lobelias of Africa and St. Helena. *Kew Bull.* 29: 535–584 [see also title page of vol. 30].

Mabberley, D. J. 1976. The origin of the afroalpine pachycaul flora and its implications. *Gard. Bull. Singapore* 29: 41–55.

Mabberley, D. J. 1979a. The species of *Chisocheton* (Meliaceae). *Bull. Br. Mus. (Nat. Hist.) Bot.* 6: 301–386.

Mabberley, D. J. 1979b. Pachycaul plants and islands. In *Plants and islands*, D. Bramwell, ed., pp. 259–277. London: Academic Press.

Mabberley, D. J. 1980. Review of J. Génermont, *Les mécanismes de l'évolution. New Phytol.* 84: 577.

McDouall, K. 1927. The gardens at Logan. *J. R. Hort. Soc.* 52: 1–14.

Mattocks, A. R. 1972. Toxicity and metabolism of Senecio alkaloids. In *Phytochemical ecology*, J. B. Harbourne, ed., pp. 179–200. London: Academic Press.

Nobel, P. S. 1980. Morphology, surface, temperatures, and northern limits of columnar cacti in the Sonoran Desert. *Ecology* 61: 1–7.

Nordenstam, B. 1978. Taxonomic studies in the tribe Senecioneae (Compositae). *Opera Bot.* 44: 1–83.

Paulian, R., and A. Villiers. 1940. Les coléoptères des Lobelias des montagnes du Cameroun. *Rev. Fr. Ent.* 7: 72–83.

Rehder, H., E. Beck, J. O. Kokwaro, and R. Scheibe. 1981. Vegetation analysis of the Upper Teleki Valley (Mount Kenya) and adjacent areas. *J. E. Afr. Nat. Hist. Soc. Nat. Mus.* 171: 1–8.

Riedl, R. 1977. A system-analytical approach to macroevolutionary phenomena. *Q. Rev. Biol.* 52: 351–370.

Salt, G. 1954. A contribution to the ecology of upper Kilimanjaro. *J. Ecol.* 42: 375–423.

Scaëtta, H. 1934. Le climat écologique de la dorsale Congo-Nil (Afrique centrale équatoriale). *Mém. Inst. R. Colon. Belge Sect. Sci. Nat. Méd. 4°* 3: 1–299.

Schaller, G. B. 1963. *The mountain gorilla: Ecology and behavior*. Chicago: University of Chicago Press.

Scott, H. 1935. Coleoptera associated with the giant lobelias and arborescent senecios of Eastern Africa. *J. Linn. Soc. (Zool.)* 39: 235–284.

———. 1952. Journey to the Gughé Highlands (Southern Ethiopia), 1948–49; biogeographical research at high altitudes. *Proc. Linn. Soc. Lond.* 163: 85–189.

———. 1958. Biogeographical research in High Simien (Northern Ethiopia), 1952–3. *Proc. Linn. Soc. Lond.* 170: 1–91.

Serebryakova, T. I., and L. V. Petukhova. 1978. "Architectural model" and life forms in some herbaceous Rosaceae (in Russian, English summary). *Byull. mosk. Obshch. Ispўt. Prir., Otd. Biol.* 6: 51–65.

Skottsberg, C. 1928. On some arborescent species of lobelia from tropical Asia. *Acta Hort. Goth.* 4: 1–26.

Smith, A. C., and M. F. Koch. 1935. The genus Espeletia: A study in phylogenetic taxonomy. *Brittonia* 1: 479–530.

Smith, A. P. 1974. Bud temperature in relation to nyctinastic leaf movement in an Andean giant rosette plant. *Biotropica* 6: 263–266.

———. 1979. Function of dead leaves in *Espeletia schultzii* (Compositae), an Andean caulescent rosette species. *Biotropica* 11: 43–47.

———. 1980. The paradox of plant height in an Andean giant rosette species. *J. Ecol.* 68: 63–73.

Smith, A. P. and T. P. Young. 1982. The cost of reproduction in *Senecio keniodendron*, a giant rosette species of Mt. Kenya. *Oecologia (Berl.)* 55: 243–247.

Stebbins, G. L. 1977. Development and comparative anatomy of the Compositae. In *The biology and chemistry of the Compositae*, Vol., 1, V. H. Heywood, J. B. Harbourne, and B. L. Turner, eds., pp. 91–109. London: Academic Press.

Steinberg, M. I., and R. L. Volle. 1972. A comparison of lobeline and nicotine at the frog neuromuscular junction. *Naunyn-Schmiedebergs Arch. Pharmak.* 272: 16–31.

Synge, P. M. 1937. *Mountains of the moon*. London: Drummond.

Tondeur, R., and R. Charlier. 1950. Alcaloïdes de *Lobelia giberroa* (Hemsl.): Extraction et étude pharmacologique. *Arch. Int. Pharmacodyn. Thér.* 83: 91–92.

Troll, C. 1947. Der asymmetrische Aufbau der Vegetationszonen und Vegetationstufen auf der Nord- und Südhalbkugel. *Ber. geobot. Inst. Rübel* 1947: 46–83.

van Steenis, C. G. G. J. 1972. *Mountain flora of Java*. Leiden: Brill.

Uphof, J. C. T. 1962. Plant hairs. *Handbuch der Pflanzenanatomie*, W. Zimmermann and P. G. Ozenda, eds., 2nd ed., 4(5). Berlin: Borntraeger.

Wardle, P. n.d. Ecological and geographical significance of some New Zealand growth forms. In *Geoecological relations between the southern temperate zone and the tropical mountains*, C. Troll and W. Lauer, eds., pp. 531–536. Wiesbaden: Steiner.

White, F. 1983. *The vegetation of Africa*. Paris: UNESCO.

Wrangham, R. W. 1970. Oxford Bongo Expedition 1968. *Bull. Oxf. Univ. Explor. Club.* 17: 138–50.

Young, T. P. 1982. Bird visitation, seed-set, and germination rates in two species of *Lobelia* on Mount Kenya. *Ecology* 63: 1983–1986.

5

Physiological and Ecological Adaptations of High Montane Sunbirds and Hummingbirds

LARRY L. WOLF AND FRANK B. GILL

High-montane environments appear inhospitable to small birds with high energy requirements, particularly so in the tropics where stark, cold, alpine-like or subalpine-like habitats contrast so sharply with the lush, productive, and warm lowlands. Reptiles and amphibians drop out as major faunal elements in the increasingly severe high altitude climates, and bird species diversity also tends to decline. Curiously, however, despite their susceptibility to energetic stress, small nectar-feeding birds are conspicuous, important members of high-elevation, tropical ecosystems.

In this chapter we explore possible adaptations to high montane living in New World hummingbirds (Trochilidae) and in Old World sunbirds (Nectariniidae). One problem is that, except for Carpenter's (1974, 1976a) studies of high Andean *Oreotrochilus*, little detailed information is available on tropical, high-altitude birds in general, and on nectarivores, in particular. As a result we will extrapolate liberally from studies of nectar feeders at montane elevations lower than the alpine-like systems of specific concern to other authors of this volume. To facilitate our task and to broaden our data base, we prefer not to be as precise as other chapters in this book in our definition of "high-montane" habitats or elevation. Few nectar-feeding birds restrict their activities to a narrow range of altitudes, climates and vegetation. Arbitrary altitude demarcations, therefore, would impose an artificial constraint on this review. In addition, at this stage, extrapolations from well-studied nectar-feeding bird systems permit us to draw some tentative conclusions about nectarivores in high-altitude tropical ecosystems which should serve as guidelines for much needed future research.

BIOGEOGRAPHY OF HIGH MONTANE TROPICAL NECTAR-FEEDING BIRDS

Approximately 10% of all birds use floral nectar as an energy source. In addition to hummingbirds (Trochilidae) and sunbirds (Nectariniidae), specialized nectar feeding occurs in the Australian honeyeater (Meliphagidae), in the flowerpeckers (Dicaeidae), Hawaiian honeycreepers (Drepanididae), some parrots (Loriidae), and the honeycreepers ("Coerebidae"). Some species of babblers (Timaliinae; e.g., *Myzornis*), white-eyes (Zosteropidae; e.g., *Z. olivacea*), and the enigmatic *Promerops* of South Africa have converged in morphology and habits with typical specialized nectar-feeding birds. Birds in many other families including wood warblers (Parulidae), weavers (Ploceidae), orioles (Icteridae), starlings (Sturnidae), Darwin's finches (Geospizidae), and white-eyes (Zosteropidae) exploit nectar opportunistically.

Each family of specialized nectar feeders has some high-altitude representatives (Table 5–1). In South America, 39% of Andean hummingbirds and honeycreepers (78 of 199 species) range into the temperate and páramo zones. They are a conspicuous species element (about 20%) of the high-altitude bird communities. Certain hummingbird genera, including *Oreotrochilus*, *Aglaeactis*, *Lafresnaya*, *Pterophanes*, *Coeligena*, *Heliangelus*, *Eriocnemis*, *Lesbia*, *Sappho*, *Ramphomicron*, *Metallura*, *Chalcostigma*, and *Oxypogon*, are primarily or entirely high-altitude taxa. No less than 12 species of hummingbirds and one "coerebid" are characteristic of the highest páramo elevations. Elsewhere, only a few species range into the páramo and moorland habitats at high elevations. Only three hummingbirds species do so in Central America, two sunbird species (especially *N. johnstoni*) in East Africa, and three honeyeaters and two lories in New Guinea. The high-altitude flowerpeckers in New Guinea, and perhaps also the three honeyeaters, do not depend on floral nectar for food, also using fruits and insects. In all areas, however, nectar-feeding birds thrive in montane habitats just below the tropical-alpine zone. Of 52 sunbird species in East Africa, 15 live above 2700 m, whereas 10 of 61 hummingbird species in Costa Rica and Panama do so. We can safely conclude that nec-

TABLE 5–1. Geography of high-altitude nectar-feeding birds.

	Central America Costa Rica, Panama[a]		South America Colombia, Ecuador Peru[b]		East Africa Uganda, Kenya, Tanzania[c]		New Guinea Irian Jaya, Papua New Guinea[d]	
	Total	Montane/páramo	Total	Montane/páramo	Total	Montane/páramo	Total	Montane/páramo
Trochilidae	61	10/3	169	67/12	–	–	–	–
"Coerebidae"	10	1/0	30	11/1	–	–	–	–
Nectariniidae	–	–	–	–	52	15/2	2	0
Meliphagidae	–	–	–	–	–	–	65	9/3
Dicaeidae	–	–	–	–	–	–	12	2/1
Loriidae	–	–	–	–	–	–	21	3/2
Total	71	11/3	199	78/13	52	15/2	100	14/6
(Percent in páramo)		(27)		(17)		(13)		(43)

Source: [a] R. S. Ridgely, *A guide to the birds of Panama.* Princeton, N.J.; Princeton University Press, 1976.

[b] R. Meyer de Schauensee. *A guide to the birds of South America.* Wynnewood, Pa.: Livingston, 1970.

[c] P. L. Britton, ed., *Birds of East Africa.* Nairobi: East Afr. Nat. Hist. Soc., 1980.

[d] A. L. Rand, and E. T. Gilliard, *Handbook of New Guinea birds.* Garden City, N.Y.: Natural History Press, 1968.

tar-feeding birds have successfully invaded the highest altitude ecosystems.

PHYSIOLOGICAL PROBLEMS AND ADAPTATIONS

Some basic physiological problems are shared by all birds that live at high elevations at any latitude. Air density is lower at higher elevations and temperatures, especially at night, are lower and more variable than at adjacent low elevations (see Chap. 2). These two characteristics of the physical environment influence the metabolic costs of living at high altitudes.

Effects of Low Air Density at High Elevations

The low air density affects activity in two ways. The low partial pressure of oxygen potentially makes metabolic oxygen needs more difficult to meet. Lower air density per se potentially increases flight effort, because less lift is generated on the wings of flying birds. In spite of these potential problems, House Sparrows (*Passer domesticus*) are capable of sustained flight at ambient pressures simulating an altitude of 6100 m (Tucker, 1968). Larger birds such as the Barheaded Goose (*Anser indicus*), vultures (*Gyps*), and choughs (*Pyrrhocorax*) have been reported flying as high as 11,000 m above sea level (Swan, 1970; Laybourne, 1974).

Oxygen Demand

The low partial pressure of oxygen characteristic of high altitudes potentially stresses respiratory support of normal brain and nervous system function of montane birds as it does in humans and in some other mammals (Schmidt-Nielsen, 1977). Some high montane mammals can cope successfully with the stresses imposed by low oxygen availability (e.g., Hainsworth, 1981). Llama (*Lama glama*) hemoglobin becomes more fully saturated with oxygen at lower partial pressures of oxygen than the hemoglobin of most other mammals. Humans may develop larger lung capacities and thus larger alveolar ventilation volumes at high elevations. However, the available data, based on low-altitude forms, suggest that no such major physiological adaptations are found in birds, despite their high levels of metabolism during flight (Schmidt-Nielsen, 1977).

The standard avian respiratory system may "preadapt" birds for high montane living by supplying adequate oxygen for active existence at high elevations (Schmidt-Nielsen 1977). Several avian respiratory features are involved. Air flow in the avian lung is unidirectional from back to front during both inspiration and expiration. The cross-current blood flow associated with this air movement permits high oxygen uptake, even when the difference in partial pressures of O_2 in blood and respiratory air is small (Piiper and Scheid, 1973).

Although the oxygen affinities of avian and mammalian blood are similar (Lutz et al., 1974), birds are more tolerant than mammals of low blood P_{CO_2}, consequent high blood pH (alkalosis), and the associated high ventilation rates. Not only is blood flow in the cerebral arteries maintained without severe, fatal constrictions (Maugh, 1979), but the oxygen affinity of the blood in the lungs increases at high elevations as a result of the increased pH, a reverse Bohr effect (Schmidt-Nielsen, 1977). Birds thus can live at high elevations with little or no respiratory penalty, apparently as a result of the high oxygen and ventilation demands of normal flight metabolism. In this sense birds are preadapted to high-altitude life.

The flights involved in long-distance, often nonstop migrations of some bird species probably have reduced the role of anaerobic metabolism and so probably have placed a selective premium on high rates of oxygen supply to the tissues. However, we must reemphasize that physiologists really have not yet studied high-altitude respiratory adaptations in tropical birds. Perhaps the best taxon to study would be hummingbirds, which have the highest per gram flight cost of any bird group.

Flight Costs

Low air densities at high elevations reduce lift (Greenewalt, 1975) and force a bird to work harder during flight, requiring even higher metabolism than expected for the same bird at lower altitudes (Berger, 1974). Theoretically, a bird either must flap faster, which requires more energy, or it must have greater wing surface area to produce compensatory lift as air density declines. Natural selection apparently has favored relatively longer wings at higher altitudes in some birds from both tropical and temperate latitudes (Moreau, 1957; Hamilton, 1961; Mayr, 1963; Gill, 1973).

The influence on flight metabolism of longer wings at high elevations can be investigated by considering the relative lift provided by wings. Wing disc loading is a measure of body weight supported by the wing surface or, more precisely, by the surface area swept out by the wings during one wing beat cycle (Greenewalt, 1960, 1975). Theoretical analyses and measurements in a wind tunnel indicate that flight costs will vary with wing disc loading, even with minor changes that occur as a result of feather abrasion and molt (see Epting and Casey, 1973; Epting, 1980, for measurements from hummingbirds). Flight costs also will vary with air density and with flight speed (Pennycuick, 1969; Tucker, 1974; Greenewalt, 1975).

We can consider specifically how hovering flight costs for hummingbirds vary with altitude. Hovering flight is equivalent to moving at zero flight speed and permits us to eliminate speed as a variable and yet make biologically meaningful comparisons of the energetic costs of one aspect of flight among species. In a sample of hummingbird species from the Peruvian Andes, the cost of hovering on a per gram basis did not vary appreciably with elevation (Feinsinger et al., 1979), partly because of high variance among species captured at the same elevation. Since air density varies with elevation, decreased wing disc loading must compensate for decreased lift from the air. Feinsinger et al. (1979) indicated that in their hummingbird sample body weights were not correlated with elevation and, therefore, that changes in wing length relative to body weight were responsible for adjustments in per gram flight costs.

Wing length must change predictably to keep per gram hovering costs the same as air density decreases with increasing elevation. Wing disc loading L_{wd}, is defined by Greenewalt (1975) as:

$$L_{wd} = W(\ell + 0.4\ell^{0.6})^{-2} \tag{1}$$

where W is the body weight (grams) and ℓ is the wing chord (cm). (For details on the derivation of the parameter values, the reader is referred to Greenewalt [1975] and to the use of Greenewalt's equations by Feinsinger et al. [1979].) The costs of hovering depend on the performance curve of an individual with respect to flight speed. The general shape of this curve is parabolic, although the specific curves vary with wing shape and body weight (Greenewalt, 1960, 1975; Pennycuick, 1969; Tucker, 1974). From the performance curve for a hummingbird we can obtain the cost of hovering from the following equation:

$$P_{hov} = 1.3 \times 10^3 \, (L_{wd}/p)^{\frac{1}{2}} \tag{2}$$

where P_{hov} has the units cal/g^{-1} · sec^{-1} and p is air density in g/cm^3 (Wolf and Hainsworth, 1978; Feinsinger et al., 1979). Air density decreases with elevation as

$$P_z = P_o\,(1 - 0.0065z/288)^{4.256} \tag{3}$$

where z is the elevation in meters (List, 1951).

If we assume costs per gram to be equal at all elevations and weight to be constant, then total hovering costs per unit time are equal at each elevation. We also set the wing length at sea level equal to unity (1). The necessary increase in wing length with elevation to meet these conditions is defined by combining equations (1), (2), and (3) to:

$$\ell_z + 0.404\ell_z^{\,0.6} = 1.404\,(1 - 0.0065z/288)^{-2.128} \tag{4}$$

This fairly complicated equation includes the elevation above sea level and fractional exponential functions of wing length at the higher elevation. However, the result is an almost linear increase in wing length relative to wing length at sea level. The wing length at 4500 m would have to be about 29% greater than at sea level (Fig. 5–1) to offset lower air density.

Reduced body weight is the other possible adaptation that could produce reduced wing disc loading at high elevations. If wing length is held constant, reduced wing disc loading requires a proportional weight reduction. Using the equations of Feinsinger et al. (1979), body weight must be reduced by a factor of P_h/P_ℓ to achieve identical costs per gram for a hovering hummingbird. Here P_h and P_ℓ are the air densities at the higher and lower elevations, respectively. A

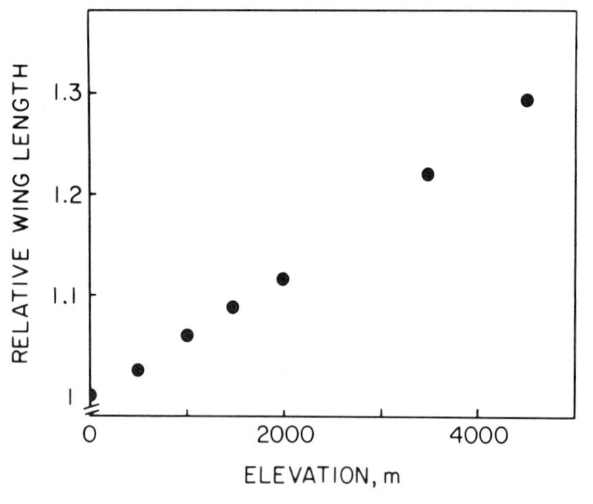

FIG. 5–1. The relationship between wing length and altitude for a bird in which body weight and wing disc loading are assumed to remain constant over an elevational gradient. We have scaled the increase relative to a wing length of 1.0 units at sea level.

hummingbird such as *Panterpe insignis*, which weighed 5.7 g at 3100 m, could weigh 6.8 g at 1400 m and have the same per gram costs. Conversely, the air density change from sea level to 4500 m would necessitate a 37% reduction in body weight.

Effects of Ambient Temperature

Standard Metabolism of Hummingbirds

Resting metabolic costs should increase with elevation, simply because of the generally lower ambient temperatures. Berger and Hart (1972) suggested that flight costs of hummingbirds might increase slightly with decreasing ambient temperatures, but this needs to be confirmed as we expect heat production during flight is sufficient to cover the slight increases in thermoregulatory costs. Hummingbirds tend to have slightly higher standard metabolic rates than would be predicted for nonpasserine birds of their body mass, but slightly lower than would be expected for equivalently sized passerines (Lasiewski and Dawson, 1967). Unfortunately, we have few data on standard metabolic costs for hummingbirds and even fewer for sunbirds. These minimal data indicate that the standard metabolism of high-altitude nectar feeders is not specialized (Wolf and Hainsworth, 1972; Wolf, Hainsworth, and Gill, 1975).

The habitats of extremely high elevations in East Africa and the Andes are inhabited by larger than average nectar-feeding birds (Coe, 1967; Carpenter, 1976a), e.g., *Nectarinia johnstoni* (13 g) in East Africa and *Oreotrochilus estella* (8.5 g) in Peru. Carpenter (1976a) suggested that large size generally is characteristic of hummingbirds at the highest elevations at which they occur, in keeping with Bergmann's rule. She related this to the relatively lower costs per gram of larger birds at all ambient temperatures, especially at lower ambient temperatures (Calder, 1974). However, since larger birds have higher total metabolic costs than smaller birds, and since her comparisons were inter- rather than intraspecific, her argument seems invalid.

The advantage of large size probably relates less to costs per se than to increased storage ability, rates of nectar uptake, and access to flowers through behavioral dominance. The critical requirement for high-altitude survival is the occurrence of sufficient food to support the high metabolic costs of cold conditions (e.g., Hainsworth and Wolf, 1972). Small forms are at an advantage due to their low total energy requirements, but larger forms have an advantage in terms of energy storage capacity and dominance.

Small birds, such as hummingbirds and sunbirds, do face potential problems from their low capacity to store energy relative to their metabolic costs. Storage capacity generally will increase linearly with body mass, while total maintenance costs increase approximately as the 0.75 power of mass (Calder, 1974; Brown, Calder, and Kodric-Brown, 1978). Thus, larger birds that have accumulated their maximum possible energy stores theoretically can survive longer while fasting at any particular temperature than can smaller birds.

Nevertheless, small species (less than 4 g in hummingbirds and less than 7 g in sunbirds) occur just below the highest elevations occupied by nectarivorous birds (among others, *Nectarinia mediocris* in Kenya, *Selasphorus flammula* in Costa Rica, and *Metallura tyrianthina* and *Lesbia nuna* in Peru). This suggests that size is not a critical adaptation, at least until one reaches extreme elevations.

Torpor in Hummingbirds

Most homeotherms, birds and mammals, maintain a nearly constant and relatively high body temperature, even when environmental temperatures drop much lower than their prevailing body temperature. Birds typically have body temperatures of 38–41°C (Calder, 1974), but we have relatively few measurements from individuals at high elevations. Body temperatures of homeotherms fluctuate through 2–4°C during a 24-hour cycle with the lowest temperatures occurring during sleep. Some homeotherms, especially very small species, can hibernate for long periods at body temperatures well below this normal range.

A few species of homeotherms, including hummingbirds, can survive periods of much reduced body temperature at low ambient temperatures during each 24-hour cycle. This drop in body temperature is accompanied by a coincident decrease in metabolism and has been called torpor. Roosting hummingbirds (*Oreotrochilus estella*) in the high Andes have been found torpid with nocturnal body temperatures as low as 5°C, although generally they had body temperatures from 6.5 to 10°C (Carpenter, 1974). Ambient temperatures in the caves where the birds roosted at night were as low as 3°C (although not as low as outside the caves), and body temperature remained about 1–2°C above that of the cave environment (Carpenter, 1976a).

The close correspondence between ambient and body temperature originally suggested to physiologists that hummingbirds abandoned

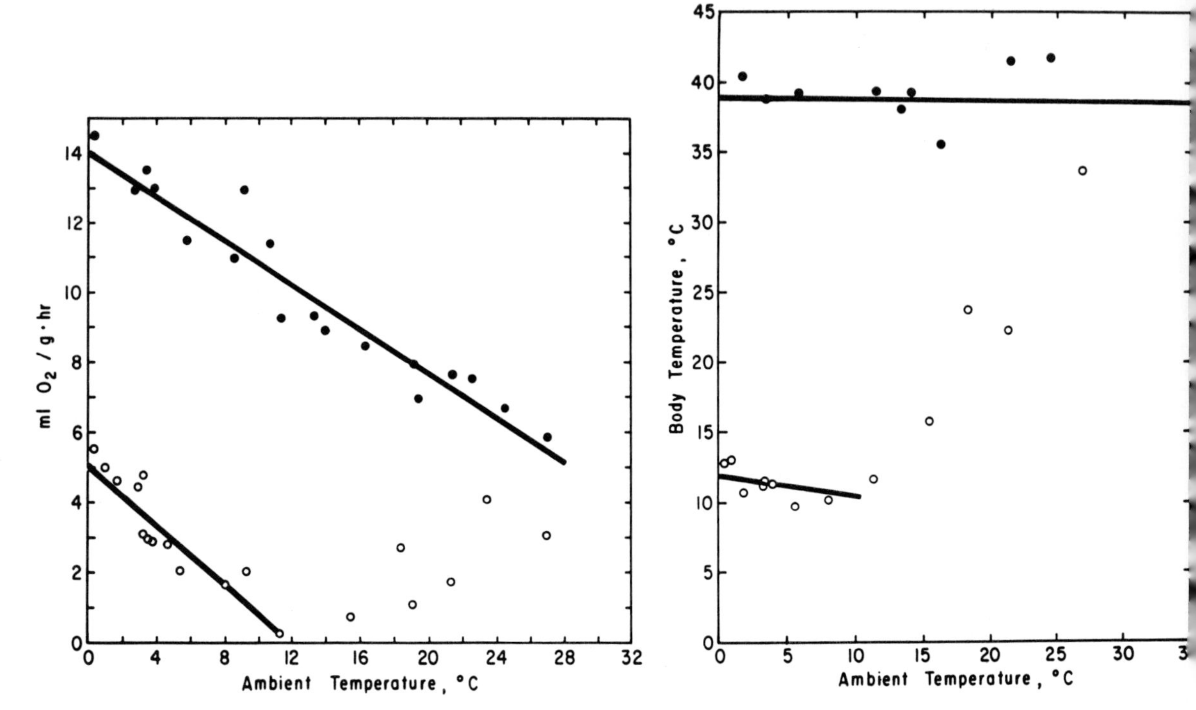

FIG. 5–2. Relationship between (*left*) metabolic rate per gram or (*right*) body temperature and ambient temperature for *Panterpe insignis*, a high montane hummingbird in Costa Rica. Circles are for torpid birds; dots are for non-torpid birds. From Wolf and Hainsworth (1972). Reprinted with permission from Comp. Biochem. Physiol, 41A, © Pergamon Press, 1972.

regulation of temperature via internal heat production (e.g., Kayser, 1965). However, recent work has shown that torpor is not an abandonment of regulation of body temperature, since a particular lower body temperature is maintained by increased metabolism if ambient temperatures fall sufficiently low (Fig. 5–2). Although this lower, regulated body temperature was not recognized by earlier workers on torpor in north temperate hummingbirds (e.g., Lasiewski, 1963), it apparently occurs in most or all hummingbird species, including those from low elevations (Hainsworth and Wolf, 1970, 1978; Wolf and Hainsworth, 1972; Carpenter, 1974). The occurrence of torpor per se among hummingbirds thus is not a specific physiological adaptation to increased metabolic costs or low nocturnal temperatures at high elevations, but seems to be directly related to the small size and the limited capacity of all hummingbirds to store excess energy.

One aspect of the ability to enter torpor, however, seems to be a direct adaptation to an elevational temperature gradient. The lower body temperature at which a hummingbird begins to regulate its temperature is a direct function of the normally prevailing, lower ambient temperatures in the few species that have been examined to date (Table 5–2; Hainsworth and Wolf, 1970, 1978; Wolf and Hainsworth, 1972; Carpenter, 1974; Hainsworth, Collins, and Wolf, 1977). The lowest reported regulated body temperature is about 6.5°C for *O. estella* in the laboratory and field (Carpenter, 1974, 1976a). This species regularly encounters temperatures near or below 0°C in its high Andean habitat. *Eulampis jugularis*, from the island of Dominica, Lesser Antilles, regulates its body temperature at about 18–20°C in torpor and normally does not encounter very low environmental temperatures (Hainsworth and Wolf, 1970).

The potential for saving energy might select for regular use of torpor to conserve metabolic reserves at night, but even *Oreotrochilus* does not enter torpor each night at high elevations between 14° and 16°S in the Andes. Carpenter (1976a) reported that about 10% of the roosting birds she examined during the rainy season (October–February) were torpid. About 60% of the birds were torpid in July and August, the dry season. Bouts of torpor were generally longer in the July to August period.

If torpor does not occur regularly as an energy-saving device for hummingbirds at night, then its

TABLE 5–2. Regulated body temperature in torpid hummingbirds and its relationship to body weight and environmental temperatures.

Species	Locality	Weight (g)	Regulated body temperature (°C)	Average lower ambient temperature (°C)
Archilochus alexandri[a]	Arizona	3.5	12–14	?
Calypte costae[b]	laboratory	3	20–22	19–24 (lab)
Panterpe insignis[c]	Costa Rica	5–6.5	10–12	4–6
Eugenes fulgens[c]	Costa Rica,	8–9.5	10–12	4–6
	Arizona[a]	7–8.5	12–14	?
Eulampis jugularis[d]	Dominica	7–10.9	18–20	18
Oreotrochilus estella[b]	Peru	8	5	5

[a] Hainsworth and Wolf (1978); [b] Carpenter (1976a); [c] Wolf and Hainsworth (1972); [d] Hainsworth and Wolf (1970).

occurrence might be under fairly strict control by the individual. The importance of high temperature for incubation would suggest that at least females should not regularly enter torpor during incubation. A nesting female Broad-tailed Hummingbird (*Selasphorus platycercus*) in the Colorado Rocky Mountains (3200 m) went into torpor during incubation and the early nesting period only when (daytime) feeding was curtailed, usually as a result of bad weather (Calder and Booser, 1973). Laboratory studies of hummingbirds suggest that they enter torpor only when energy reserves reach a critical, lower threshold (Hainsworth et al., 1977). The length of a torpor bout is regulated by when this nocturnal threshold is reached.

Sunbirds

The little information available on the physiology of sunbirds suggests they conform to the general predictive equations for passerines and have per-gram flight costs similar to those of hummingbirds (Wolf et al. 1975). Cheke (1971) suggested that some montane sunbirds may use nocturnal hypothermia to reduce energy requirements at low ambient temperatures, and perhaps when they have low energy reserves. The lower body temperatures achieved in his experiments generally were still quite high (25–30°C) and may have been little more than a slight extension of the range of normal temperature fluctuations over 24 hours exhibited by most homeotherms (e.g., King and Farner, 1961). With no additional evidence, it is also difficult to rule out the possibility that these captive birds were in poor health. In our limited laboratory studies of the metabolism of the Bronzy Sunbird (*Nectarinia kilimensis*), we found no evidence of such hypothermia, although body temperature decreased 2–3°C at low ambient temperatures (Fig. 5–3; Wolf et al., 1975). *Nectarinia johnstoni*, a species of the highest altitudes in East Africa, does "roost in deep holes, originally excavated by the Mountain Chat *Pinarochroa sordida* [= *Cercomela sordida*; Chap. 6] in the matted dead-leaf clusters of the Tree Groundsel (*Senecio keniodendron*" (= *Dendrosenecio keniodendron*; Chap. 4) (Williams, 1951, pp. 586–587), but no physiological studies have been made of this interesting species.

In sum, we know of no major physiological adaptations of nectarivores that relate specifically to high elevations. Most torpid hummingbirds probably regulate a lower body temperature and the specific temperature of regulation may be an adaptation to local temperatures. Other metabolic characteristics follow from predictive equations relating metabolism to body weight, independent of elevation. The physiological limitations imposed by high altitudes are likely to be reflected primarily in the ability of nectarivorous birds to meet their metabolic demands, which

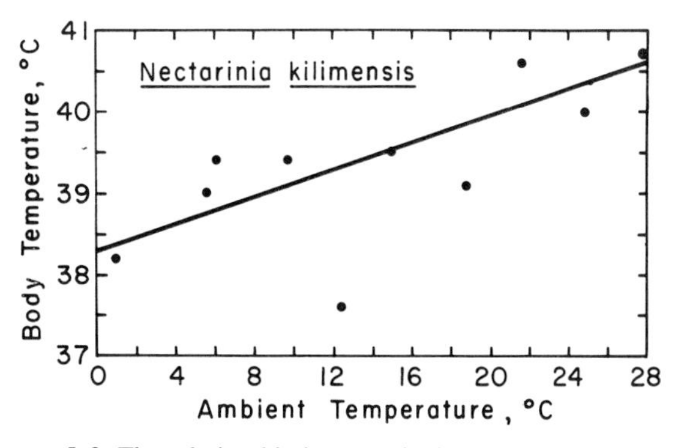

FIG. 5–3. The relationship between body temperature and ambient temperature in the Bronzy Sunbird, a montane species in East Africa. The conditions for the experiments were the same for this species and the hummingbird species shown in Fig. 5–2. From Wolf et al. (1975). Reprinted with permission from Ecology 56; © Ecological Society of America, 1975.

depend on the availability of energy, particularly nectar from flowers and perhaps also insects. If any physiological characteristics are essential for survival at high elevations they should determine which of the species that might invade the highlands actually occur there (e.g., perhaps mostly larger forms). The high-altitude species appear not to have made local physiological adjustments and their mobility often allows them to escape unsuitable conditions.

ECOLOGICAL ADAPTATIONS

Nectar as an Energy Resource

Sugars in floral nectar are the primary energy resources for hummingbirds and sunbirds at high elevations. Nectar is primarily a solution of sugars in water, but it may also contain small amounts of amino acids and lipids (Baker and Baker, 1975). The constituents of nectar in bird-visited flowers thus serve as an energy source with crucial nutrients, such as proteins and vitamins, coming from other sources. The most common additional food is insects that are captured either in the air or gleaned from vegetation (see, e.g., Snow, 1983). Insect catching among high-elevation nectarivores may account for up to 3% of the time budget of a bird that has regular access to a good supply of nectar-producing flowers (Wolf, 1975a; Wolf, Stiles, and Hainsworth, 1976) Some of the less specialized nectarivores, such as white-eyes, also may take some fruits (Gill, 1973).

The character of nectar as an energy resource changes along an elevational gradient in Costa Rica, the only tropical location where it has been systematically studied. High-elevation plant species tend to have lower nectar concentrations, lower nectar-production rates, and longer-lived flowers than those lower down (Hainsworth and Wolf, 1972; Wolf et al., 1976; Stiles, 1978, 1980).

The relative importance of birds as pollinators of plants (compared to insects) tends to increase with elevation in the tropics (Cruden, 1972), and probably also in western North America (F. L. Carpenter, per. comm.). This has been attributed to the endothermic physiology of birds, which permits them to be more predictable pollinators in periods of cold or inclement weather at high elevations (Cruden, 1972; Carpenter, 1976a,b; Brown et al., 1978).

Carpenter (1976a) reported that although hummingbird plants were in bloom most of the year at some localities in the puna of the Peruvian Andes, the flowers in the wet season were the ones that produced the most nectar. She suggested that the lack of flowering in the dry season was due to water stress. Elsewhere, however, most bird-pollinated plants bloom during the dry season. For example, most food plant species for hummingbirds in the second growth of the cut-over oak forest on the Cerro de la Muerte in Costa Rica (3100 m) bloomed during the dry season from February through April (Wolf et al., 1976).

The plant species on which many local sunbird populations depend in the Aberdare Mountains and adjacent sections of the Rift Valley of Kenya often occur in monospecific stands with relatively short blooming periods (Fig. 5–4). The sunbirds do not have the local temporal continuity in nectar availability found in the highlands of the New World tropics (e.g., Wolf et al., 1976). Although several plant species may bloom sequentially sunbirds often have to emigrate out of a local area to find flowers. The nectar resources are a spatiotemporal mosaic on the scale of kilometers and weeks. The required nomadic seasonal movements make permanent residency difficult. Elsewhere in the mountains of East Africa, some flowers may be available year round. Near Lake Kivu, in Zaire at 2100 m, staggered flowering of *Erythrina* and *Symphonia* trees sustains local populations of *Nectarinia kilimensis* and *N. purpureiventris*, respectively (Chapin, 1959).

To some extent, the present availability of nectar resources to nectar-feeding birds in the Amer-

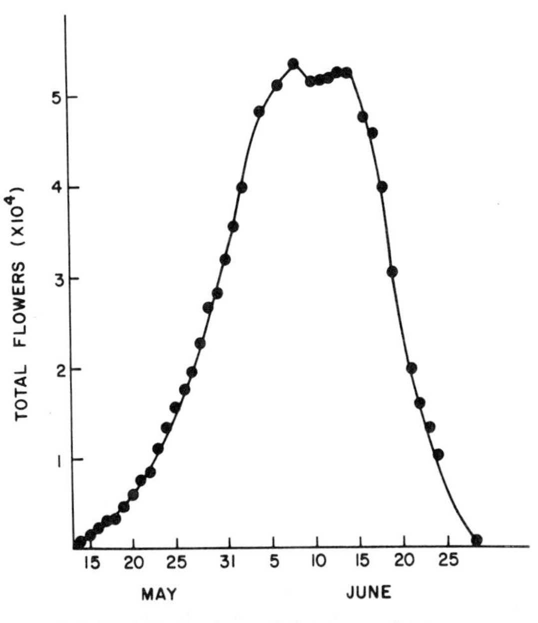

FIG. 5–4. The change in total flowers available as nectar sources for sunbirds foraging at a hedge of *Aloe kedongensis* near Gilgil, Kenya, 1975.

ican and African tropical highlands is determined by human activities throughout these montane areas. Human activity creates extensive areas of second-growth vegetation, the principal habitat for plant species that are bird pollinated in both regions. Most of these early successional plant species have increased dramatically in abundance following habitat degradation. The presence of the Green Violet-ear (*Colibri thalassinus*) in the hummingbird assemblage at 3100 m on the Cerro de la Muerte, Costa Rica, for instance, depends on the large populations of *Centropogon valerii* in local disturbances caused by logging and road-building activities. Although many other species of plants bloom at this locality, most are not visited with any regularity by *Colibri* (Wolf et al., 1976). In undisturbed habitat at a similar elevation on an adjacent mountain top, *Colibri*'s primary food plant is rare and *Colibri* itself is missing or very rare (Wolf et al., 1976).

Influences of disturbances notwithstanding, we feel that the following discussion is representative of the ecology of montane nectarivores in the Americas and Africa, since we are concerned with local responses to resources. The birds appear to be flexible in their responses to changing resource conditions, a flexibility produced by their close interaction with food supplies in successional habitats. Although the details of plant distributions and abundances have changed following human disturbance, the basic rules followed by the birds for exploiting these plants should not have changed.

Community Organization and Ecological Adaptations

Certain species of nectarivorous birds co-occur in time and space in assemblages or "communities." The species combinations and relative abundances will change through time at a particular montane locality. However, we can still inquire into the ecological rules that apparently govern which species co-occur and for how long. Many of these ecological rules are translated directly into behavioral interactions.

The organization of assemblages of montane nectar-feeding birds depends on the temporal and spatial patterns of nectar availability. Insects are necessary as nutrient sources, but probably are not sufficient to determine the presence and abundance of most species of nectarivorous birds, except perhaps at the highest altitudes of the páramo, puna, and afroalpine zones (Vuilleumier, pers. comm.). The juxtaposition of diurnal climatic extremes with the general wet and dry seasonal phenomena at all tropical elevations (see Chap. 2) may exacerbate, at least to some extent, the seasonal importance of nectar availability at high elevations. The question of species and morphological diversity of the birds in these assemblages revolves around the diversity of plant species that bloom simultaneously and also around the faunal and floral history of each area (Vuilleumier and Simberloff, 1980).

The temporal and spatial variation in nectar availability that occurs within an area makes it likely that altitudinal migration (in the sense of recurring seasonal movements) also will play a role in local species diversity (Feinsinger, 1980). Altitudinal migrations are an important aspect of life of nectarivores at high elevations (Wagner, 1945; Skutch, 1967; Carpenter, 1976a; Wolf, 1976; Wolf et al., 1976; Feinsinger, 1980). These migrations are analogous to the north–south migration patterns of temperate zone bird species.

The similarity of the community structure through time depends on the level, variance, and predictability of the resource base (Feinsinger, 1980). As the variance and unpredictability of the resources increase, addition and subtraction of species, via migrations, will be more common. The number of resident species probably is low among hummingbird assemblages at high elevations (Carpenter, 1976a; Wolf et al., 1976). At lower elevations, with more moderate climates (i.e., higher average temperatures and fewer day–night extremes) blooming seasons may be extended and superimposed, and the number of resident species of nectarivores increases (see, for example, Stiles, 1975; Feinsinger, 1980). At very high elevations, the single resident species is moderately larger (for instance *Oxypogon* in the Venezuelan páramos, *Chalcostigma* in the Colombian ones (Snow, 1983), *Oreotrochilus* in the Peruvian puna, *Panterpe* in the Talamanca Cordillera of Costa Rica and Panama) and able to dominate most of the smaller possible immigrants in encounters at contested food supplies.

Sunbird assemblages in montane East Africa (above 2000 m) often lack any resident species, reflecting the local and seasonal nature of floral nectar availability. Flowering at a single location usually involves only one common sunbird-pollinated plant species for two to three months (Fig. 5–4). Immigration and emigration of sunbirds as flower numbers change are conspicuous (Fig. 5–5).

Possible tendencies toward permanent residency involve species that are insectivorous during periods of low nectar availability. For example *Nectarinia tacazze* and *N. mediocris* are

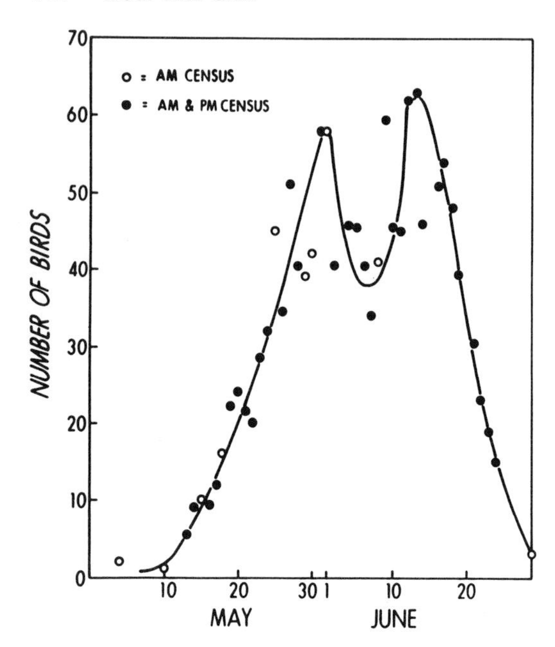

FIG. 5–5. Changes in the numbers of sunbirds foraging at the same hedge of *Aloe kedongensis* as in Fig. 5–4.

resident throughout the year in the *Hypericum-Hagenia* zone at 3000–3500 m on Mt. Kenya and the Aberdare Mountains in central Kenya. They feed on nectar in flowers of *Hypericum*, *Kniphofia*, and sometimes *Leonotis mollissima*. Some *Hypericum* shrubs may have flowers most of the year, but only a few at any time (pers. obs.). The two sunbird species also forage for insects throughout the year, especially in *Hagenia* trees. Curiously, they are subordinate to nonresident species like *N. reichenowi* and *N. famosa*, even though *N. tacazze* is larger than either of them. *N. purpureiventris* is resident year round at Lake Kivu (2100 m), a habit made possible by the sustained availability of flowering *Symphonia globulifera* trees (Chapin, 1959). *N. purpureiventris*, like *N. tacazze*, also eats many insects. In another example, *N. johnstoni* is resident above 4000 m on Mt. Kenya, Mt. Kilimanjaro, and in the Ruwenzori Mountains. Some aspects of its annual cycle are tied to flowering *Lobelia*, but at other times of the year it may be completely insectivorous (Williams, 1951; Coe, 1967; Young, pers. comm.). Similar patterns appear to exist in high Andean *Oxypogon* in Venezuela (Vuilleumier, pers. comm.), and in Colombia (Snow, 1983).

For birds that use the same basic resource pool, in this case the flower nectar, competition among individuals both within and among species is possible. This competition, which influences the ability of individuals to obtain energy from the environment, can take one of the two basic forms, both of which occur in assemblages of nectarivores (Gill, 1978; Wolf, 1978). The efficiency of nectar uptake, defined as energy intake minus cost per unit time, varies with the volume of nectar available in the flowers (Fig. 5–6). Nectar removed from a flower by a foraging bird is replaced rather slowly, so that subsequent birds visiting the same flower generally will be less efficient (Fig. 5–7; Gill and Wolf, 1979). This is an example of competition by use of resource (= exploitation competition) and generally will occur in most groups of nectarivores, unless the total bird population is very low. The second form of competition, interference competition, occurs if one individual aggressively defends a particular resource.

Carpenter (1978) has suggested that hummingbirds are influenced more strongly by interspecific competition, especially interference competition, than are high montane sunbirds. However, the hummingbird assemblages of the high mountains of the western United States (Gass, Angehr, and Centa, 1976; Kodric-Brown and Brown, 1978; Brown and Kodric-Brown, 1979; Gass, 1979; Carpenter, pers. comm.; Wolf, pers. obs.) resemble the montane sunbird systems of East Africa more closely than other hummingbird systems so far studied in the mountains of Central and South America. The hummingbird and sunbird assemblages involve transient species at rich, ephemeral flower resources with much interspecific aggression and high turnover of species and individuals. Several sympatric species of hummingbird flowers may bloom simultaneously in the western United States, but floral diversity is low compared to low-latitude hummingbird systems. The exception is in the high Peruvian Andes where, during the dry season, only a single

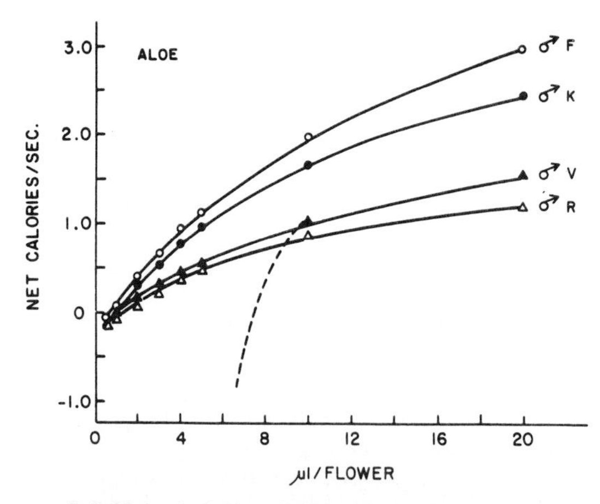

FIG. 5–6. Net calories obtained per second by several species of sunbirds foraging at *Aloe graminicola* flowers as a function of the nectar available per flower. F, *Nectarinia famosa*; K, *N. kilimensis*; V, *N. venusta*; R, *N. reichenowi*.

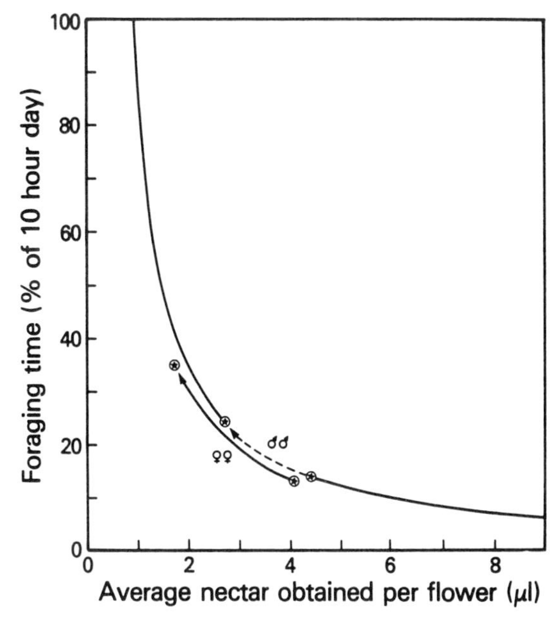

FIG. 5–7. Impact of nectar losses on foraging time budgets by nonterritorial *Nectarinia reichenowi* to competitors. Loss of nectar decreases the average obtained per flower visit and increases the number of flowers that a sunbird must visit to obtain a specified amount of energy each day, including the costs of the additional foraging. Because females are smaller, their costs are lower, and they can forage less each day than males at a particular nectar volume. Females tend to lose more nectar to dominant individuals, however, and must forage longer each day. From Gill and Wolf (1979).

principal plant species, *Chuquiraga spinosa*, is in bloom and only a single bird species, *Oreotrochilus estella*, remains year round (Carpenter 1976a). In the Ecuadorian Andes, Corley Smith (1969) suggested a close association between the closely related *O. chimborazo* and *Chuquiraga acutifolia*: "I have never found the bird without the shrub" (p. 18). See also Ortiz-Crepo and Bleiweiss, 1982.

Interspecific aggression and relative dominance among bird species in any assemblage will influence which species stay. In both sunbird and hummingbird assemblages there is a direct relationship between the size of a given species and that species's ability to win encounters with other species (Table 5–3; Wolf, 1970; Feinsinger, 1976; Wolf and Gill, pers. obs.). This implies a linear dominance hierarchy among species. However, minor nonlinearities exist. In the Talamanca Mountains of Costa Rica at 3100 m, *Panterpe insignis* can often displace the larger *Eugenes fulgens* from its territory, and in East Africa, *Nectarinia reichenowi* sometimes displaces the slightly larger *N. kilimensis* (Table 5–3).

The role of these dominance interactions,

TABLE 5–3. Dominance interactions among species-sex categories at a hedge of *Aloe* in May–June 1975.

	Nonterritorial individuals					
Chasee	KM	RM	KF	RF	FM	FF
Chaser						
KM	11	9	6	4	18	4
RM	4	3	3	4	7	7
KF			1		1	1
RF		2	1	4	23	6
FM					46	26
FF						1

Note: M, male; F, female; K, *Nectarinia kilimensis*; R, *N. reichenowi*; F, *N. famosa*.

including territoriality, in the organization of a particular assemblage depends on the level of nectar availability (Gill and Wolf, 1975; Carpenter and MacMillen, 1976; Wolf et al., 1976; Gill, 1978; Wolf, 1978). At high levels of nectar availability, measured as the standing crop in each flower, aggressive interactions decrease in frequency, territoriality breaks down, and competition is mostly exploitative. At intermediate levels of nectar availability, those species for which the net benefits (benefits minus costs) are sufficiently high will generally hold territories. In both sunbird and hummingbird systems, these are the larger species. The smaller, subordinate species cannot exclude the larger ones, making the net benefits of territoriality either low or negative.

In summary, assemblages of montane hummingbirds and sunbirds are governed by the same environmental variables and the same basic responses of the birds to these variables. Possible differences in the role of inter- and intraspecific competition within the hummingbird and sunbird systems depend on how closely the population sizes of each combination of species can track the changing resource levels. Low populations relative to resources will make exploitation competition more efficient than the energetically expensive interference competition (Gill, 1978; Wolf, 1978). In addition, montane hummingbird assemblages more often have multiple plant species in bloom simultaneously, providing niche refuges, and perhaps also lower temporal variation in blooming patterns of particular species. However, we emphasize that the same organizational rules apply equally well to low-elevation assemblages (Feinsinger, 1976; Feinsinger and Colwell, 1978; Brown and Kodric-Brown, 1979), although the relative importance of one or more variables may change among specific assemblages.

REPRODUCTIVE ADAPTATIONS

Timing

Reproduction in some tropical montane nectarivorous birds is seasonal and may be cued by photoperiodic changes (Carpenter, 1976a). Year-round breeding near the equator, as for example in *Oreotrochilus* (Corley Smith, 1969) and some sunbirds (*N. famosa*, *N. reichenowi*), however, suggests a lack of photoperiodic cueing of the breeding in some cases. Use of photoperiod as a cue to time breeding only reflects some other underlying, but correlated seasonal event important to successful breeding (Immelmann, 1971; Gwinner, 1975).

For most sunbirds and hummingbirds, the crucial environmental variable seems to be the availability of nectar in flowers. These flowers need to be ones exploited efficiently by the birds and common enough to provide energy needed during nesting (see Skead, 1967; Stiles, 1973; Carpenter, 1976a; Wolf and Wolf, 1976). In montane Guatemala, hummingbirds tend to nest during the dry season at a time when most bird-visited plants are flowering. In montane areas of central Mexico, the Green Violet-ear (*Colibri thalassinus*) may not breed during the rainy season (Wagner, 1945; Skutch, 1967). The major breeding season of *Oreotrochilus* at 12°S latitude in Peru, is from September to January, during the rainy season, when *Cajophora* spp. is in peak bloom (Carpenter, 1976a). In Costa Rica, at 3100 m, three hummingbird species (*Selasphorus flammula*, *Colibri thalassinus*, and *Eugenes fulgens*) breed primarily during the dry season when numerous plant species are in bloom (Wolf et al., 1976). Each bird species migrates up into the area to breed and then leaves for a portion of the year. The fourth, and only resident, species of hummingbird, *Panterpe insignis*, breeds during the wettest part of the rainy season, July to October, when the food plant at which it forages very efficiently (*Macleania glabra*) is in bloom (Wolf and Stiles, 1970). *Panterpe* does not breed when the other three species do. The three migrant species arrive and depart coincidentally with the changing availabiity of flowers at which they can forage efficiently (e.g., Hainsworth and Wolf, 1972).

At extreme elevations, harsh seasonality may prevent breeding at certain times of the year. Both Carpenter (1976a) and Corley Smith (1969) suggested that different species (or allospecies) of *Oreotrochilus* may have their breeding seasons limited by severe weather conditions. The hummingbirds in Carpenter's study area of southern Peru did breed during the warmest and least windy part of the year, which was also the wettest. It is possible that the poor nectar-producing quality of the primary dry season food plant, *Chuquiraga spinosa*, makes it energetically unfeasible for the birds to breed then. In Venezuelan páramos, wet season blooming of some plants, especially *Espeletia schultzii*, may be correlated with greater nectar productivity and the breeding of the hummingbird *Oxypogon guerinii*. There may also be a greater productivity of insects, which these birds consume (Vuilleumier, pers. comm.). The same might also be true in Colombian parámos (Snow, 1983).

As we noted earlier in our discussion of physiological characteristics, montane hummingbirds probably are capable of breeding even in very inclement conditions provided sufficient, efficiently exploitable energy is available. Severe climatic conditions reduce the probability of breeding indirectly through their effect on plant reproduction and directly through their effect on energy expenditures of the birds.

Sunbirds at 2000–2400 m in the Rift Valley of Kenya have been reported breeding in most months of the year (van Someren, 1956; Wolf and Wolf, 1976). The same appears true for *N. kilimensis* and *N. purpureiventris* at Lake Kivu (2100 m) (Chapin, 1959), and *N. johnstoni* on Mt. Kenya (4000 m) (Williams, 1951). But the situation is complicated slightly by the separation of monospecific patches of flowering plants in both time and space from patches of other plant species. We had *N. famosa*, *N. reichenowi*, *N. venusta*, and *N. kilimensis* all breeding in or near patches of flowering *Leonotis nepetifolia* (Gill and Wolf, pers. obs.). The ability of a particular pair of birds to breed in these situations depends on their having access to sufficient floral nectar to ensure an adequate energy supply. A resident pair of Malachite Sunbirds (*N. famosa*) began breeding just after a nonbreeding pair of Bronzy Sunbirds (*N. kilimensis*) left a patch of *Aloe graminicola* (Wolf and Wolf, 1976). Similarly, subordinate species, Malachite and Variable Sunbirds (*N. venusta*), bred in *Leonotis* patches either at the end of the bloom cycle when the larger, more dominant forms had left the area, or at times when dominant forms were absent (Wolf and Gill, pers. obs.). Such opportunistic breeding, especially by subordinate species of sunbirds, often relates to variations in the intensity of competition for floral nectar (Gill, 1978).

Coe (1967) has published the only suggestion that nectar availability is not the important timing device for breeding among sunbirds at high elevation. He reported that adults of *Nectarinia john-*

stoni breed up to at least 4300 m on Mt. Kenya, apparently are insectivorous, and may breed year round. More recent studies (Young, pers. comm.) indicate that *N. johnstoni* breeds twice a year, apparently taking advantage of two periods of high insect availability *plus* enhanced flowering of *Lobelia keniensis*.

Nests of several montane hummingbirds regularly are placed in caves, under earthen or rocky overhangs, or in other sheltered locations, including buildings (Carpenter, 1976a; Wolf and Stiles, per. obs; Vuilleumier, pers. comm.). But, more generally, high montane hummingbirds and sunbirds probably use a combination of nest insulation and nest placement in vegetation to reduce heat loss from birds and eggs in the nest during incubation and nestling development (Coe, 1967; Calder, 1973; Carpenter 1976a). The metabolic advantages of nest insulation characteristics accrue to both the adult female and the young. However, the relative importance of radiation heat loss to the sky, which is influenced by vegetation over the nest and by cup depth, obviously depends on the ambient temperatures (see Smith, Roberts, and Miller, 1974, for calculations on an incubating Anna Hummingbird [*Calypte anna*] in California). As the nocturnal temperature increases, the relative importance of other nest characteristics, such as insulation quality, increases.

Various authors have noted that hummingbird nests may be thicker at higher elevations (Wagner, 1955; Dorst, 1962; Snow, 1983; Vuilleumier, pers. comm.). Both *Oxypogon* in Venezuelan and Colombian páramos and *Oreotrochilus* in the Peruvian puna use woolly material from nearby plants to build enormous, well-insulated nests, that cannot be compared with the small structure of nests of lowland genera (Vuilleumier, pers. comm.; Snow, 1983). The model simulations by Smith et al. (1974) of an incubating female Anna Hummingbird suggests that insulation had the largest effect on metabolic costs. This result, in part, reflects the restrictive conditions of their simulations, particularly the relatively high nocturnal temperatures. However, the model does deal with characteristics of the nest, including thickness and surface area, that might be modified by the birds in relation to an elevational gradient.

A shaded nest at high elevations should reduce the incident heat load on the incubating female and unattended nestlings. Exposure of the nest might also be important at other times of the day. Dorst (1962) reported that *Oreotrochilus estella* nests in Peru tended to be oriented toward the east, perhaps to maximize early morning warming of the contents. Nest orientation, however, in populations of two species (or allospecies) of *Oreotrochilus* was random (Corley Smith, 1969; Carpenter, 1976a).

The cave sites and buildings used by nesting *Oreotrochilus* reduce the range of environmental temperature variation (Carpenter, 1976a). She suggested that these protected sites are partly responsible for the high nesting success of *Oreotrochilus* in her study area, but the high success also may relate to fewer potential predators at high elevations. Carpenter did report that nests in natural sites were significantly more successful than those on human-constructed sites. She offered no reasons for the difference, but the latter sites may be less buffered against climatic extremes.

In some populations of *Oreotrochilus estella*, nests may be grouped close together, especially within a single cave or rock outcrop (Dorst, 1962; Carpenter, 1976a). Closely spaced nests on one rock outcrop were defended only within 1.5 m of each nest and the females used distinct feeding areas on the nearby slopes (Carpenter, 1976a). There is generally little or no territoriality beyond the local nest site among hummingbirds at any elevation (e.g. Legg and Pitelka, 1956; Wolf and Wolf, 1971). Grouped nests might occur in any species of bird and probably depend on either clustered nest sites and/or patchy, yet extensive feeding areas that exceed the demands of a single nesting female or pair (Horn, 1968).

The bulky nest of sunbirds enclose the egg cavity except for the small side entrance. The nest of *N. johnstoni* on Mt. Kenya is an amazing, thickly insulated structure covered with strips of broad fibers of broad fibers of *Senecio brassica* (= *Dendrosenecio keniensis*; see Chap. 4) that act as a rain shield (Gill, pers. obs.). There could be some variation in the insulative quality of these nests, but the general structural architecture is independent of elevation, being found in most or all members of the family (Van Tyne and Berger, 1959). There is little information on this aspect of adaptation to high altitudes by sunbirds.

Pair Bonds and Parental Care

Montane sunbirds differ markedly from most hummingbirds in the type of pair bond formed during the breeding season, but there is no evidence in either family of long-term, nonbreeding pair bonds. However, the breeding pair bond difference dramatizes some of the basic differences between the two families and is not limited to montane species (Van Tyne and Berger, 1959;

Skead, 1967; Lack, 1968; Wolf and Stiles, 1970; Wolf and Wolf, 1976). Sunbirds apparently always form pair bonds and most of the evidence suggests strict monogamy (Skead, 1967; Wolf and Wolf, 1976). Hummingbirds as a group are notorious for their lack of pair formation during the breeding season (Van Tyne and Berger, 1959; Lack, 1968; Orians, 1969); a single species provides the only well-documented exception (Wolf and Stiles, 1970). *Panterpe insignis*, the Fiery-throated Hummingbird of the high mountains of Costa Rica and Panama, does form pair bonds during the breeding season, and can be polygynous as well as monogamous (Wolf and Stiles, 1970). This mating system contrasts strongly with other, promiscuous hummingbird species that breed in the same area, although not at the same time (discussed earlier).

The difference in mating systems between hummingbirds and sunbirds may be based on the characteristics of the flowers exploited by these birds during the breeding season. The sexes of most hummingbirds exploit different plant species, or the females use plants not sufficiently dense and abundant to be defendable by males. One common denominator among the hummingbird and sunbird species forming pair bonds is that they breed in conjunction with the blooming of a single plant species that generally is defendable by territorial males (Wolf and Stiles, 1970; Wolf and Wolf, 1976). This might provide a selective mechanism whereby females mate with males willing to share nectar supplies (Wolf, 1975b).

It has been suggested that the occurrence of only two eggs per clutch in hummingbirds permits the female to raise the young unaided (e.g., Orians, 1969). However, *Panterpe* also lays a two-egg clutch, as do most monogamous sunbirds (Van Tyne and Berger, 1959). Male aid in sunbirds probably is required to raise even two young in some circumstances, primarily through defense of the female's food supply while she is engaged more directly in caring for the nest and young (Wolf and Wolf, 1976). Many passerine bird species in the high Andes also produce only two eggs per clutch, probably as a response to direct difficulties of caring for more than two young successfully (Lack, 1968). The reasons for the limitation in these montane passerines may be somewhat different than for hummingbirds, which also produce two-egg clutches at low elevations.

Parental care is provided by the female alone in most hummingbird species. There are a few reports of male hummingbirds, including the montane species *Colibri coruscans*, incubating and/or feeding nestlings (Moore, 1947). However, these rare observations do not reflect the usual pattern of parental care for those particular species. Even in species that form a pair bond, the male probably does not contribute directly to the care of the nest or young (Wolf and Stiles, 1970; Stiles, pers. comm.). So far there is no evidence that the male *Panterpe* contributes more than defense of the food supply for the female and perhaps indirectly for the young to the extent that they receive nectar from the female.

Male sunbirds, while they are paired with a female during breeding, only feed the young. They do not help with incubation or nest building (Skead, 1967; Wolf and Wolf, 1976). The male often does accompany the female on trips outside the territory to gather nesting material, but this is interpreted as defense of the female against mating attempts by other males since the male does not gather or carry any nest material (Wolf and Wolf, 1976).

Hummingbird and sunbird young take relatively long to develop sufficiently to leave the nest, especially if compared to the incubation: nestling period ratio of other bird groups (Lack, 1968). At least in hummingbirds, the eggs are relatively small, suggesting that the young are hatched at an early stage of development and simply require longer to reach the fledgling stage. There also is rather large variability in the length of the nestling period for particular hummingbird species, e.g., the young of *Oreotrochilus estella* may be in the nest for 19–29 days (Dorst, 1962). This flexibility may permit young to fledge successfully more often in the colder and generally harsher climates of the high montane sites. However, this again is not a unique physiological adaptation to high altitudes.

SUMMARY

In this chapter we have suggested that tropical nectarivorous birds (hummingbirds and sunbirds) at high altitudes do not possess a suite of specific adaptations to the peculiar climatic and ecological conditions of such habitats (except perhaps for the large size of their nest). Rather they merely use adaptations generally characteristic of all such birds as they fit into a particular assemblage of nectarivores at high elevations. Some species are only short-term members of an assemblage; others may be permanent residents at high elevations. The presence and residency period of a particular species are correlated with the availability of nectar as an energy supply, hence to the flowering patterns of the plant species and presumably to the seasonal wet–dry cycles. The presence of a nectarivore also depends on the competitive interactions among the co-occurring individuals.

What is clear from this review is the need for detailed studies of truly high altitude nectarivores. Dorst's (1962), Corley Smith's (1969), Langner's (1973), and Carpenter's (1976a) studies of *Oreotrochilus* in the Andes and Young's unpublished studies of *Nectarinia johnstoni* on Mt. Kenya are important first steps. We hope this review will stimulate further efforts.

ACKNOWLEDGMENTS

The writing of this chapter was supported by the National Science Foundation (DEB 78–09060340, DEB 79–10799). We thank F. R. Hainsworth, E. Waltz, and the editors of this book for their comments on the manuscript. Their efforts should not be construed as agreement with all we have said, but they certainly tried hard to set us straight.

REFERENCES

Baker, H. G., and I. Baker. 1975. Studies of nectar-constitution and pollinator-plant coevolution. In *Coevolution of animals and plants*, L. E. Gilbert and P. H. Raven, eds., pp. 100–140. Austin: University of Texas Press,

Berger, M. 1974. Energiewechsel von Kolibris beim Schwirrflug unter Höhenbedingungen. *J. f. Ornithol.* 115: 273–288.

Berger, M. and J. S. Hart. 1972. Die Atmung beim Kolibri *Amazilia fimbriata* während des Schwirrfluges bei verschiedenen Umgebungstemperaturen. *J. Comp. Physiol.* 81: 363–380.

Britton, P. L., ed. 1980. *Birds of East Africa*. Nairobi: East Africa Natural History Society.

Brown, J. H., and A. Kodric-Brown. 1979. Convergence, competition and mimicry in a temperate community of hummingbird-pollinated flowers. *Ecology* 60: 1022–1035.

Brown, J. H., W. A. Calder, and A. Kodric-Brown. 1978. Correlates and consequences of body size in nectar-feeding birds. *Amer. Zool.* 18: 687–700.

Calder, W. A. 1973. Microhabitat selection during nesting of hummingbirds in the Rocky Mountains. *Ecology* 54: 127–134.

———. 1974. Consequences of body size for avian energetics. In *Avian energetics*, R. A. Paynter, Jr., ed., pp. 86–144. Cambridge, Mass.: Nuttall Ornithological Club.

Calder, W. A., and J. Booser. 1973. Hypothermia of broad-tailed hummingbirds during incubation in nature with ecological corelations. *Science* 180: 751–753.

Carpenter, F. L. 1974. Torpor in an Andean hummingbird: Its ecological significance. *Science* 183: 545–547.

———. 1976a. Ecology and evolution of an Andean hummingbird (*Oreotrochilus estella*). *Univ. Calif. Publ. Zool.* 106: 1–74.

———. 1976b. Plant-pollinator interaction in Hawaii: Pollination energetics of *Metrosideros collina* (Myrtaceae). *Ecology* 57: 1125–1144.

———. 1978. A spectrum of nectar-eater communities. *Amer. Zool.* 18: 809–819.

Carpenter, F. L., and R. E. MacMillen. 1976. Threshold model of feeding territoriality and test with a Hawaiian honeycreeper. *Science* 194: 639–642.

Chapin, J. P. 1959. Breeding cycles of *Nectarinia purpureiventris* and other Kivu birds. *Ostrich*, Supp. 3: 222–229.

Cheke, R. A. 1971. Temperature rhythms in African montane sunbirds. *Ibis* 113: 500–506.

Coe, M. J. 1967. *The ecology of the alpine zone of Mount Kenya*. Monographiae Biologicae 17. The Hague: Junk.

Corley Smith, G. T. C. 1969. A high altitude hummingbird on the volcano Cotopaxi. *Ibis* 111: 17–22.

Cruden, R. W. 1972. Pollinators in high-elevation ecosystems: Relative effectiveness of birds and bees. *Science* 176: 1439–1440.

Dorst, J. 1962. Nouvelles recherches biologiques sur les trochilidés des hautes Andes péruviennes (*Oreotrochilus estella*). *Oiseau Rev. Fr. Ornithol.* 32: 95–126.

Epting, R. J. 1980. Functional dependence of the power for hovering on wing disc loading in hummingbirds. *Physiol. Zool.* 53: 347–357.

Epting, R. J., and T. M. Casey. 1973. Power output and wing disc loading in hovering hummingbirds. *Amer. Natur.* 107: 761–765.

Feinsinger, P. 1976. Organization of a tropical guild of nectarivorous birds. *Ecol. Monogr.* 46: 257–291.

———. 1980. Asynchronous migration patterns and the coexistence of tropical hummingbirds. In *Migrant birds in the neotropics: Ecology, behavior, distribution, and conservation*, A. Keast and E. S. Morton, eds., pp. 411–419. Washington, D.C.: Smithsonian Institution Press.

Feinsinger, P., and R. K. Colwell. 1978. Community organization among neotropical nectar-feeding birds. *Amer. Zool.* 18: 779–795.

Feinsinger, P., R. K. Colwell, J. Terborgh, and S. B. Chaplin. 1979. Elevation and the morphology, flight energetics and foraging ecology of tropical hummingbirds. *Amer. Natur.* 113: 481–497.

Gass, C. L. 1979. Territory regulation, tenure and migration in rufous hummingbirds. *Can. J. Zool.* 57: 914–923.

Gass, C. L., G. Angehr, and J. Centa. 1976. Regulation of food supply by feeding territoriality in the rufous hummingbird. *Can. J. Zool.* 54: 2046–2054.

Gill, F. B. 1973. Intra-island variation in the Mascarene White-eye *Zosterops borbonica*. *Amer. Ornith. Union Monog.* no. 12.

———. 1978. Proximate costs of competition for nectar. *Amer. Zool.* 18: 753–763.

Gill, F. B., and L. L. Wolf. 1975. Economics of feeding

territoriality in the golden-winged sunbird. *Ecology* 56: 333–345.

———. 1979. Nectar loss by golden-winged sunbirds to competitors. *Auk* 96: 448–461.

Greenewalt, C. H. 1960. *Hummingbirds*. New York: Doubleday.

———. 1975. The flight of birds. *Trans. Amer. Philos. Soc.* 65(4): 1–67.

Gwinner, E. (1975) Circadian and circannual rhythms in birds. In *Avian biology*, Vol. 5, D. S. Farner and J. R. King, eds., pp. 221–274. New York: Academic Press.

Hainsworth, F. R. 1981. *Animal physiology: Adaptation in function*. Reading, Mass.: Addison-Wesley.

Hainsworth, F. R., and L. L. Wolf. 1970. Regulation of oxygen consumption and body temperature during torpor in a hummingbird, *Eulampis jugularis*. *Science* 168: 368–369.

———. 1972. Energetics of nectar extraction in a small, high altitude, tropical hummingbird. *J. Comp. Physiol.* 80: 377–387.

———. 1978. Regulation of metabolism during torpor in "temperate" zone hummingbirds. *Auk* 95: 197–199.

Hainsworth, F. R., B. G. Collins, and L. L. Wolf. 1977. The function of torpor in hummingbirds. *Physiol. Zool.* 50: 215–222.

Hamilton, T. H. 1961. The adaptive significance of intraspecific trends of variation in wing length and body size among bird species. *Evolution* 15: 180–195.

Horn, H. 1968. The adaptive significance of colonial nesting in the Brewer's blackbird (*Euphagus cyanocephalus*). *Ecology* 49: 682–694.

Immelmann, K. 1971. Ecological aspects of periodic reproduction. In *Avian biology*, Vol. 1, D. S. Farner and J. R. King, eds., pp. 342–376. New York: Academic Press.

Kayser, C. 1965. Hibernation. In *Physiological mammalogy*, Vol. 2, W. V. Mayer and R. G. Van Gelder, eds., pp. 179–296. New York: Academic Press.

King, J. R., and D. S. Farner. 1961. Energy metabolism, thermoregulation, and body temperature. In *Biology and comparative physiology of birds*, Vol. 2, A. J. Marshall, ed., pp. 215–288. New York: Academic Press.

Kodric-Brown, A., and J. H. Brown. 1978. Influence of economics, interspecific competition, and sexual dimorphism on territoriality of migrant rufous hummingbirds. *Ecology* 59: 285–296.

Lack, D. 1968. *Ecological adaptations for breeding in birds*. London: Methuen.

Langner, S. 1973. Zur Biologie des Hochlandkolibris *Oreotrochilus estella* in den Anden Boliviens. *Bonn. Zool. Beitr.* 24: 24–47.

Lasiewski, R. C. 1963. Oxygen consumption of torpid, resting, active, and flying hummingbirds. *Physiol. Zool.* 36: 122–140.

Lasiewski, R. C., and W. R. Dawson. 1967. A reexamination of the relation between standard metabolic rate and body weight in birds. *Condor* 69: 13–23.

Laybourne, R. C. 1974. Collision between a vulture and an aircraft at an altitude of 37000 feet. *Wilson Bull.* 86: 461–462.

Legg, K., and F. A. Pitelka. 1956. Ecologic overlap of Allen and Anna hummingbirds nesting at Santa Cruz, California. *Condor* 58: 393–405.

List, R. J. 1971. Smithsonian Meteorological Tables. 6th rev. ed. Washington, D.C.: Smithsonian Institution Press.

Lutz, P. L., I. S. Longmuir, J. V. Tuttle, and K. Schmidt-Nielsen. 1974. Oxygen affinity of bird blood. *Resp. Physiol.* 20: 325–330.

Maugh, T. H. 1979. Birds fly. Why can't I? *Science* 199: 1230.

Mayr, E. 1963. *Animal species and evolution*. Cambridge, Mass.: Harvard Univ. Press (Belknap Press).

Meyer de Schauensee, R. 1970. *A guide to the birds of South America*. Wynnewood, Pa.: Livingston.

Miller, R. S. 1967. Pattern and process in competition. *Adv. Ecol. Res.*, 4: 1–74.

———. 1969. Competition and species diversity. *Brookhaven Symp. Biol.* 22: 63–70.

Moore, R. T. 1947. Habits of male hummingbirds near their nests. *Wilson Bull.* 59: 21–25.

Moreau, R. D. 1957. Variation in the western Zosteropidae (Aves). *Bull. Brit. Mus. (Nat. Hist.) Zool.* 4: 312–433.

Orians, G. H. 1969. On the evolution of mating systems in birds and mammals. *Amer. Natur.* 103: 589–603.

Ortiz-Crespo, F. I., and R. Bleiweiss. 1982. The northern limit of the hummingbird genus *Oreotrochilus* in South America. *Auk* 99: 376–378.

Pennycuick, C. J. 1969. The mechanics of bird migration. *Ibis* 111: 525–556.

Piiper, J., and P. Scheid. 1973. Gas exchange in avian lungs: Models and experimental evidence. In *Comparative physiology: Locomotion, respiration, transport and blood*, L. Bolis, K. Schmidt-Nielsen, and P. Maddress, eds., pp. 161–185. Amsterdam: North-Holland.

Rand, A. L., and E. T. Gilliard. 1968. *Handbook of New Guinea Birds*. Garden City, N.Y.: Natural History Press.

Schmidt-Nielsen, K. 1977. The physiology of wild animals. *Proc. R. Soc. London B.* 199: 345–360.

Skead, C. J. 1967. *The sunbirds of southern Africa*. Cape Town: Balkema.

Skutch, A. 1967. The nesting seasons of Central American birds in relation to climate and food supply. *Ibis* 92: 185–222.

Smith, W. K., S. W. Roberts, and P. C. Miller. 1974. Calculating the nocturnal energy expenditure of an incubating Anna's hummingbird. *Condor* 76: 176–183.

Snow, D. W. 1983. The use of *Espeletia* by paramo hummingbirds in the Eastern Andes of Colombia, *Bull. Brit. Ornithol. Club* 103: 89–94.

Stiles, F. G. 1973. Food supply and the annual cycle of the Anna hummingbird. *Univ. Calif. Publ. Zool.* 97: 1–109.

———. 1975. Ecology, flowering phenology, and hummingbird pollination of some Costa Rican *Heliconia* species. *Ecology* 56: 285–301.

———. 1978. Ecological and evolutionary implications of bird pollination. *Amer. Zool.* 18: 715–727.

———. 1980. Ecological and evolutionary aspects of bird-flower coadaptations. *Proc. XVII Int. Ornithol. Congr.*, pp. 1173–1178. Berlin: Deutsche Ornithol. Gesellschaft.

Swan, L. W. 1970. Goose of the Himalayas. *Nat. Hist.* 1970: 68–75.

Tucker, V. A. 1968. Respiratory physiology of house sparrows in relation to high altitude flight. *J. Exp. Biol.* 48: 55–66.

———. 1974. Energetics of natural avian flight. In *Avian Energetics*, R. A. Paynter, Jr., ed., pp. 298–328. Cambridge, Mass.: Nuttall Ornithology Club.

van Someren, V. G. L. 1956. Days with birds: Studies of the habits of some East African species. *Fieldiana, Zool.* 38: 434–451.

Van Tyne, J., and A. J. Berger. 1959. *Fundamentals of ornithology*. New York: Wiley.

Vuilleumier, F., and D. Simberloff. 1980. Ecology versus history as determinants of patchy and insular distributions in high Andean birds. In *Evolutionary Biology*, Vol. 12, M. K. Hecht, W. C. Steere, and B. Wallace, eds., pp. 235–379. New York: Plenum.

Wagner, H. O. 1945. Notes on the life history of the Mexican violet-ear. *Wilson Bull.* 57: 165–187.

———. 1955. Einfluss der Poikilothermie bei Kolibris auf ihre Brutbiologie. *J. F. Ornithol.* 96: 361–388.

Williams, J. G. 1951. *Nectarinia johnstoni*: A revision of the species, together with data on plumages, moults, and habits. *Ibis* 93: 579–595.

Wolf, L. L. 1970. The impact of seasonal flowering on the biology of some tropical hummingbirds. *Condor* 72: 1–14.

———. 1975a. Energy intake and expenditures in a nectar-feeding sunbird. *Ecology* 56: 92–104.

———. 1975b. "Prostitution" behavior in a tropical hummingbird. *Condor* 77: 140–144.

———. 1976. Avifauna of the Cerro de la Muerte region, Costa Rica. *Amer. Mus. Novit.* no. 2606: 1–37.

———. 1978. Aggressive social organization in nectarivorous birds. *Amer. Zool.* 18: 765–778.

Wolf, L. L., and F. R. Hainsworth. 1972. Environmental influence on regulated body temperature in torpid hummingbirds. *Comp. Biochem. Physiol.* 41A: 167–173.

———. 1978. Energy: Expenditures and intakes. In *Chemical zoology*, Vol. 10, M. Florkin, B. T. Scheer, and A. H. Brush, eds., pp. 307–358. New York: Academic Press.

Wolf, L. L., and F. G. Stiles. 1970. Evolution of pair cooperation in a tropical hummingbird. *Evolution* 24: 759–773.

Wolf, L. L., and J. S. Wolf. 1971. Nesting of the purple-throated carib hummingbird. *Ibis* 113: 306–315.

———. 1976. Mating system and reproductive biology of malachite sunbirds. *Condor* 78: 27–39.

Wolf, L. L., F. R. Hainsworth, and F. B. Gill. 1975. Foraging efficiencies and time budgets in nectar feeding birds. *Ecology* 56: 117–128.

Wolf, L. L., F. G. Stiles, and F. R. Hainsworth. 1976. Ecological organization of a tropical, highland hummingbird community. *J. Anim. Ecol.* 45: 349–379.

6

Convergences in Bird Communities at High Altitudes in the Tropics (Especially the Andes and Africa) and at Temperate Latitudes (Tibet)

JEAN DORST AND FRANÇOIS VUILLEUMIER

Birds live in high mountains all over the world, some species commonly nesting as high as 5500 m or even higher. It is of great interest to compare the endproducts of evolution in high-altitude avifaunas in similar environments but in different regions. Such comparisons should be especially rewarding between the tropical Andes and tropical African mountains, where extensive high-altitude biomes (páramo and puna in the Andes; afroalpine in Africa; see Chap. 1) are physiognomically quite similar in spite of taxonomic differences in their floristic composition. These comparisons, furthermore, should be extended to temperate high mountains, especially the Himalaya, where ecological conditions on the high plateau of Tibet are similar, despite some clearcut differences, to those in the puna of the Andean altiplano.

Even though the avifaunas of each high mountain region in the tropics or the temperate zone have distinct taxonomic and geographic origins, striking similarities can be detected in general morphology, color and color pattern, and foraging or nesting behavior between pairs or trios of unrelated species, as well as among assemblages of unrelated species. Are these resemblances due to evolutionary convergence?

We suggest that the problem of convergence among pairs or trios of species, or among guilds (*sensu* Root, 1967; the so-called community level, see Orians and Paine, 1983, for a review), can be studied with greater ease in high tropical mountain biomes than in other tropical biomes because the high mountains are simple and similar structurally and young geologically. Similar environmental constraints may thus have acted within the same geologic time frame on different mountaintops in the intertropical region and in high mountains of temperate or subtropical regions such as the Himalaya, which could thus serve as a "control" for intertropical comparisons.

In this chapter we shall examine qualitative aspects of intercontinental convergences at two levels of biogeographic perception (Blondel and Choisy, 1983), regional and continental, and at two levels of evolutionary and ecologic integration, the species and the guild, after reviewing briefly some of the salient features of the climatic and environmental background for such convergences. Even though quantitative analyses of the different species and guilds involved would be desirable (see Cody and Mooney, 1978, and Blondel et al., 1984, for mediterranean bird communities), the comparative study of tropical high-altitude avian communities has not reached this stage of sophistication yet. For the time being therefore we must be content with relatively crude comparisons. The speculations that these comparisons will suggest to us should set the stage for quantitative work. Using our preliminary empirical descriptions students should be able to ask the appropriate questions and, after in-depth study in the field, should obtain answers that will test our ecological and evolutionary speculations concerning convergence among guilds.

We studied high-altitude avifaunas in Ethiopia and in Peru (J. D.), in the páramo and puna biomes at many localities throughout the tropical Andes (F. V.), in Mexico and southern Brazil (F. V.), and along the rim of the Tibetan plateau in Kashmir (F. V.) and India (J. D.). Although we have not studied páramo species of New Guinea, we have had experience with their congeners in Australia (J. D., F. V.). Between the two of us we have studied in the field the great majority of the genera and about three quarters of the species discussed in this chapter. We used the museum material of all these regions in the National

Museum, Paris (J. D.) and the American Museum, New York (F. V.). We thank our respective institutions for their support of our work. F. V. is very grateful to B. P. Hall for help with African avifaunas.

CLIMATIC AND ENVIRONMENTAL BACKGROUND

Ethiopia in this chapter refers to the high grassy or shrubby areas of the highlands (plateau and mountain slopes); East Africa refers to the afroalpine zone of Mts. Ruwenzori, Elgon, Kenya, and Kilimanjaro, and the Crater Highlands. Andean páramos are (usually) wet mountaintop grassland and shrubland in Venezuela, Colombia, Ecuador, and northern Peru; the puna is drier (steppe-like) grassy or shrubby vegetation in the high Andes of central and southern Peru, Bolivia, northwestern Argentina, and northern Chile; the altiplano is located within the puna biome (see next section). In accordance with Chapter 1, and for the sake of convenience we call páramo the high mountain grasslands of Mexico and of the Itatiaia massif in southeastern Brazil, as we do also the high areas above timberline (without grasslands) in Borneo (Mt. Kinabalu) and the grasslands above timberline in New Guinea.

Avian habitats reach up to well above 5000–5500 m in Tibet and the Andes. They are much less high (maximum about 4300–4700 m) in East Africa (highest summit, Mt. Kilimanjaro, 5899 m) and Mexico (highest summit, Pico de Orizaba, 5675). They are lower also in New Guinea (about 4000–4700 m; highest summit, Mt. Jaya, 4884 m), and even less so in Ethiopia (maximum about 4000–4200 m; highest summit in northern Ethiopia, Ras Dashan, 4620 m; in southern Ethiopia, Mount Batu, 4307 m) and in Borneo (Mt. Kinabalu, about 4101 m). Mt. Itatiaia is the lowest of all the páramo zones considered here (Agulhas Negras, 2797 m).

The climate of these regions has been described in detail in Chapter 2. We mention below therefore only some salient features, comparing especially the Andes with their African counterparts, since our avian comparisons will involve these two regions more than the others. The climate of the Tibetan plateau was reviewed by Vaurie (1972).

In the Andean páramos humidity and rainfall are usually high throughout the year; but in the puna a marked difference exists between a rather short wet season and a long dry season during which the daily temperature range is particularly wide. In the puna the length of the dry season increases roughly from north to south and from east to west (Troll, 1959). Similar daily and annual cycles in rainfall and humidity are characteristic of East Africa and Ethiopia. The páramos of East Africa (afroalpine zone) are humid all year round, due to mist and abundant rain, as in the northern Andes. At lower elevations on the Ethiopian plateau a well-defined rainy season contrasts strongly with the dry season, as in the Andean puna.

In Africa as in the Andes, mean annual temperatures are low at high altitudes. On the Ethiopian tableland as on the high plateaus of Peru, Bolivia, Argentina, and Chile above 3800 m, the daily range of temperature can be from −5°C to +15°C, particularly during the dry season. These plateaus are called the altiplano—the Spanish word for an area that is in fact a maze of valleys, hills, and vast flat expanses framed by tall mountain barriers; see Pearson (1951) for a pragmatic definition. High solar radiation during cloudless days, and low temperature during cloudless nights, together with wind, increase the thermal range and decrease the atmospheric humidity, thus creating cold tropical steppes and semideserts at high altitudes. The climate of the Tibetan plateau is similar to that of the puna, but temperatures are lower and winds much higher.

One of the most evident features of the altiplano landscape is the absence of trees, with a few notable exceptions. Indeed, most plants are very low and many are cushionlike (Cabrera, 1968). A wide-ranging plant association is the *pajonal de puna*, a steppe-like habitat of grasses which grow in separate tussocks. Another habitat is shrubsteppe, in the drier parts of the altiplano. Steep slopes and bottoms of narrow valleys, encased by cliffs and rocks, are covered by denser vegetation that locally may recall the páramo from farther north. Bushes and ligneous plants of several families constitute there a sclerophyllous matorral. In some well-protected and localized area *Polylepis* (Rosaceae) is the only plant that deserves the name of tree (see Chap. 12). Elsewhere, even more patchily distributed and limited in area are stands of the spectacular, giant Bromeliaceae *Puya raimondii* (Dorst, 1957). Open montane scrub on dry slopes and dense cloud forest on wet ones are the main habitats below the páramo and puna. Cloud forests are composed of many plant families that did not colonize the higher altitudes, though penetrating here and there along well-protected valleys up to about 3900–4000 m.

In Ethiopia and East Africa several plant associations can be observed along the altitudinal

TABLE 6–1. Approximate planar surface area of Andean páramo and puna, and of East African and Ethiopian afroalpine biomes.

Characteristics of region	Area (km^2) Africa	Area (km^2) Andes
Island-like	2,000 (afroalpine islands of East African mountains)	25,000 (páramos of the northern Andes)
Plateau-like	4,000 (Ethiopian tableland)	160,000 (puna of the central Andes, including the altiplano)
Totals	6,000	185,000

gradient (Jeannel, 1950; Hedberg, 1964). In Ethiopia for instance a zone from 3000 to 3200 m is covered by *Juniperus procera* forest, the so-called cedar tree of East Africa, mixed in some areas with *Podocarpus gracilior* and *Olea africana*. From 3200 m to 3400 m the dominant formation is a dense forest of *Hagenia abyssinica* ("Kosso"). From 3400 to 3550 m *Hypericum revolutum* dominates a forest interspersed with openings covered by grassland. Up to 3650 m and sometimes higher, the characteristic vegetation is dwarf forest dominated by giant heath (*Erica arborea*). Beyond the timberline we enter the páramo-like afroalpine zone, a vegetation including bushes, heathland, tussock grassland, and communities of rosette plants dominated by giant lobelias (*Lobelia* spp.) and arborescent groundsels (*Dendrosenecio* spp.) that recall the high Andean *Espeletia* spp. (see Chaps. 3 and 4). The giant *Lobelia rhynchopetalum* form copses at altitudes of about 4000 m that are reminiscent of the *Puya raimondii* stands of the Peruvian and Bolivian Andes.

Open habitats above the timberline are remarkably similar looking in the Andes and in East Africa or Ethiopia (compare for example Figs. 4 and 5 in Lauer, 1981; illustrations in Monasterio, 1980, for the Andes; in Jeannel, 1950, Hedberg, 1964, for East Africa). In New Guinea, tree ferns (*Cyathea*) assume life forms reminiscent of the giant rosette plants of the Andean or African páramos. In the Andes as in Ethiopia (but less in East Africa, and variably so in New Guinea) man has modified these high-altitude habitats for centuries, but it is diffucult to assess the importance of this activity on the biogeography of the biota, except for deforestation, which lowers the natural altitude of the continuous timberline and produces páramo-like situations at much lower altitudes than expected. Grazing by livestock or sheep of course is very destructive of the natural vegetation, and produces selective pressures on the habitat, the effects of which have barely been studied.

In many ways, the island-like nature of páramos is a feature shared by all these areas, in the Andes, East Africa, or New Guinea, but these vegetation types occupy very different surface areas on each continent. Similarly, one could compare the plateaus of the Andes with the Ethiopian highlands, but here again huge differences exist in total surface areas. These are illustrated in Table 6–1, comparing the two kinds of high mountain environments in the Andes and in Africa. The Andes are many times larger than their African equivalents.

Although the physiognomy of vegetation shows similarities in the Andes, in Africa, and in New Guinea, the plant taxa are often different (see Chaps. 7, 8, and 17). These gross structural similarities in vegetation, when viewed together with the similarities in climate, can be hypothesized to be propitious to convergences among the various consumer taxa, including the birds.

The Tibetan plateau, much higher and colder than any similar area in the tropics, is most comparable with the Andean altiplano. The part of Tibet called by Vaurie (1972) the Outer Plateau of Western Tibet appears to be most like the puna on the altiplano. Plates 4 and 5 in Vaurie (1972) recall the wet, grassy puna, and the dry, scrubby puna (as defined by Troll, 1959), respectively. Our avian comparisons involving Tibet and tropical high mountains will thus be made on the basis of Vaurie's (1972) species lists from Western Tibet.

We wish to stress that in spite of their apparent uniformity, and of the simplicity of their taxonomic composition, the páramos, the puna, the afroalpine, and the tropical-alpine of New Guinea (as well as the high Tibetan steppes) show great diversity in vegetation types, both on a local and on a regional scale. To convince oneself of this fact one needs only to look at illustrations, such as those in Jeannel (1950), Hedberg (1964), Dorst and Roux (1973a), and Nievergelt (1981) for Africa; those in Dorst (1956), Bourlière (1957), Cuatrecasas (1968), Cabrera (1968), Vuilleumier and Ewert (1978), and Monasterio (1980) for the Andes; and those in Archbold, Rand and Brass (1942) for New Guinea. Eventually, fine-grained field work on intercontinental convergences will have to take this variability into consideration.

COMPARISONS BETWEEN AFRICA AND THE ANDES

The Andean páramo and puna avifauna consists of 166 species in 84 genera and 32 families (Appendix 23–A, Chap. 23), and the afroalpine avifauna of Ethiopia and East Africa contains 53 species in 38 genera and 23 families (Appendix 6–A). Note that the list of afroalpine species in Appendix 6–A is somewhat different from the list of montane non-forest species in Appendix 22–A, because two different criteria were used in compiling them. Here all high-altitude species living above the timberline are included, regardless of whether they occur elsewhere, whereas in Chapter 22 are included only the species that are restricted to high altitudes. Furthermore, whereas only afroalpine taxa are included here, in Chapter 22 the altitudinal range of non-forest habitats is somewhat greater, and includes grasslands at lower elevations than those in the afroalpine zone (called moorland by Moreau, 1966).

In this section we compare the Andean and the African avifaunas, looking at patterns of resemblance and possible convergence, and examining differences as well. This analysis will be carried out first at the level of pairs or trios of species, and then at the level of guilds.

Table 6–2 groups most of the species of birds in Appendixes 23–A and 6–1 into trophic groups or guilds, arranged by the major habitat they occupy. This table was compiled on the basis of our own field work (Dorst, 1956, 1957, 1967; Dorst and Roux, 1973a,b; Vuilleumier and Ewert, 1978; Vuilleumier and Simberloff, 1980; also unpublished data from both of us), and according to published information (Chapin [1932], Brown [1965], Moreau [1966], Coe [1967], King [1973], and Lamprey [1974] for Africa; and Koepcke [1954], Bourlière [1957], Johnson [1965, 1967], Corley Smith [1969], Roe and Rees [1979], and Snow [1983] for the Andes).

TABLE 6–2. Assignment of African (Ethiopian plateau and afroalpine of East African mountains) and Andean birds (páramo and puna) to trophic groups or foraging guilds.

Habitat or vegetation type	Trophic group or foraging guild: Ethiopia and East Africa	Trophic group or foraging guild: Andes
Mixed, including cliffs, rocky slopes, heathland, grassland, and scrub	1. Predators	
	Buteo rufofuscus	*Circus cyaneus* (pá)
	Aquila rapax (E)	*Geranoaetus melanoleucus* (pá)
	A. verreauxii	*Buteo polyosoma* (pá)
	Falco tinnunculus (E)	*B. poecilochrous* (pu)
	F. biarmicus (E)	*Falco sparverius*
	F. peregrinus (E)	*F. femoralis*
	Bubo capensis	*Bubo virginianus*
		Athene cunicularia (pu)
	2. Scavengers (and omnivores)	
	Gypaetus barbatus	*Cathartes aura*
	Gyps rueppellii (E)	*Vultur gryphus*
	Corvus capensis (E)	*Polyborus carunculatus-megalopterus*
	C. albicollis (A)	
	C. crassirostris (E)	
	3. Aerial insectivores	
	Apus melba	*Caprimulgus longirostris*
	Apus niansae	*Notiochelidon murina* (pá)
	Apus aequatorialis (A)	*Petrochelidon andecola* (pu)
	Riparia cincta (E)	
	Hirundo fuligula (E)	
	Psalidoprocne pristoptera (E)	
Grassland, heathland and scrubland	4. Large ground seedeaters and/or insectivores	
	Francolinus castaneicollis (E)	*Nothura ornata* (pu)
	F. shelleyi (A)	*N. curvirostris* (pá)
	F. psilolaemus	*N. darwinii* (pu)
		Tinamotis pentlandii (pu)
		Gallinago stricklandii (pá)
		Attagis gayi
		Thinocorus orbignyianus (pu)

Habitat or vegetation type	Trophic group or foraging guild: Ethiopia and East Africa	Andes
		Oreopholus ruficollis (pu)
		Colaptes rupicola (pu)
	5. Small ground insectivores (and granivores)	
	Galerida malabarica (E)	*Asthenes modesta* (pu)
	Anthus novaeseelandiae (E)	*A. humilis* (pu)
	Macronyx flavicollis (E)	*A. wyatti*
	Lanius collaris (E)	*Upucerthia validirostris* (pu)
	Oenanthe lugens (E)	*Cinclodes fuscus*
	O. bottae (E)	*C. excelsior* (pá)
	Cercomela sordida	*Geositta punensis* (pu)
	Turdus litsipsirupa (E)	*G. cunicularia* (pu)
		G. tenuirostris (pu)
		Muscisaxicola maculirostris
		M. rufivertex (pu)
		M. juninensis (pu)
		M. alpina
		M. albifrons (pu)
		Mimus dorsalis (pu)
		Turdus chiguanco (pu)
		T. fuscater (pá)
		Sturnella magna (pá)
		Anthus furcatus (pu)
		A. hellmayri (pu)
		A. correndera (pu)
		A. bogotensis (pá)
	6. Small bush insectivores	
	Phylloscopus umbrovirens (A)	*Schizoeaca coryi* (pá)
	Bradypterus cinnamomeus (A)	*Leptasthenura aegithaloides* (pu)
	Cisticola brunnescens	*L. andicola*
	C. hunteri (A)	*Asthenes dorbignyi* (pu)
		A. flammulata
		Anairetes parulus
		Ochthoeca fumicolor (pá)
		O. oenanthoides (pu)
		Cistothorus platensis (pá)
		C. meridae (pá)
		C. apolinari (pá)
		Troglodytes aedon (pu)
	7. Nectarivores-insectivores (Andean ones also in rocky areas)	
	Nectarinia tacazze (E)	*Colibri coruscans*
	N. famosa	*Oreotrochilus chimborazo-estella*
	N. johnstoni (A)	*Patagona gigas* (pu)
	N. stuhlmanni (A)	*Ramphomicron dorsale* (pá)
		Chalcostigma olivaceum (pu)
		C. stanleyi (pá)
		C. heteropogon (pa)
		Oxypogon guerinii (pá)
	8. Small bush/ground seedeaters	
	Serinus canicollis	*Zonotrichia capensis* (pu)
	S. nigriceps (E)	*Phrygilus punensis* (pu)
	S. ankoberensis (E)	*P. fructiceti* (pu)
	S. striolatus	*P. unicolor*
	Passer griseus (E)	*P. plebejus*
		P. alaudinus (pu)
		Diuca speculifera (pu)
		Sicalis lutea (pu)
		S. uropygialis (pu)
		S. olivascens (pu)

Habitat or vegetation type	Trophic group or foraging guild: Ethiopia and East Africa	Trophic group or foraging guild: Andes
		Catamenia inornata
		C. homochroa (pá)
		Carduelis spinescens (pá)
		C. crassirostris (pu)
		C. atrata (pu)
		C. uropygialis (pu)
Cliffs and rocky slopes	9. Large insectivores/omnivores	
	Bostrychia carunulata (E) (breeding)	*Theristicus melanopis* (E) (breeding)
	10. Medium ground seedeaters	
	Columba albitorques (E)	*Metriopelia ceciliae* (pu)
		M. aymara (pu)
		M. melanoptera
	11. Medium-small insectivores/omnivores	
	Pyrhhocorax pyrrhocorax (E)	*Xolmis rufipennis* (pu)
	Onychognathus tenuirostris	*Agriornis montana*
	O. albirostris (E)	*A. andicola*
Wet meadows, moorlands, marshes, and bogs	12. Medium-large insectivores/omnivores	
	Bostrychia carunculata (E) (feeding)	*Theristicus melanopis* (feeding)
	Grus carunculatus (E)	*Plegadis ridgwayi* (pu)
	Rallus rougetii (E)	*Rallus sanguinolentus* (pu)
	Sarothrura affinis (A)	*Vanellus resplendens*
	Vanellus melanocephalus	*Gallinago nobilis* (pá)
	Gallinago nigripennis	*Gallinago andina* (pu)
	13. Large vegetarians	
	Cyanochen cyanopterus (E)	*Chloephaga melanoptera* (pu)
	14. Small insectivores	
		Cinclodes atacamensis (pu)
		Lessonia rufa (pu)
Lakes and lagoons	15. Large swimming vegetarians	
	Anas sparsa	*Anas flavirostris*
	A. undulata (E)	*A. specularioides* (pu)
		A. georgica
		A. versicolor (pu)
		Oxyura jamaicensis
		Gallinula chloropus
		Fulica ardesiaca
		F. gigantea (pu)
		F. cornuta (pu)
	16. Large swimming piscivores	
		Rollandia rolland
		R. microptera (pu)
		Podiceps occipitalis
		P. taczanowskii (pu)
		Phalacrocorax olivaceus
Rushes around lakes and lagoons	17. Small insectivores	
		Phleocryptes melanops (pu)
		Tachuris rubrigastra (pu)
		Agelaius thilius (pu)
Running streams	18. Large insectivores	
	Anas sparsa (also below afroalpine)	*Merganetta armata* (usually below páramo and puna)

Note: E, only in Ethiopian highlands; A, only in East African mountains; pá, only in páramos; pu, only in the puna.

Resemblances versus Differences in Pairs or Trios of Species

These species pairs or trios will be discussed in the order of the list of guilds or trophic groups of Table 6–2.

Scavengers (and Omnivores) (Guild no. 2)

The large vultures *Gyps* in Ethiopia and *Vultur* in the Andes, and the small scavenging corvids (*Corvus* spp.; Ethiopia, East Africa) and the scavenging falconid *Polyborus* (Andes), could be considered as cases of convergence. The two stocks of vultures are totally unrelated to each other, *Gyps* being an accipitrid, whereas *Vultur* (so-called New World vulture) may be related to storks; the crows and the caracara are also unrelated. The latter pair should probably be studied in detail. Such habits as flocking, following human beings or their domestic animals or roads, and varied diet should be compared on a quantitative basis.

Large Ground Seedeaters and/or Insectivores (Guild no. 4)

Francolins (Phasianidae) in Africa and tinamous (Tinamidae) in the Andes are similar kinds of game birds. In spite of their total lack of taxonomic relationships, they occupy similar habitats and seem to eat the same kinds of food. In both regions several species of different sizes may have evolved along similar lines. For example, the Grey-wing (*Francolinus psilolaemus*) may be the equivalent of the Andean *Nothura darwinii*. No comparative study between francolins and tinamous exists.

Small Ground Insectivores (and Granivores) (Guild no. 5)

Several pairs or trios of species may represent instances of convergence. They include the Ethiopian motacillid *Macronyx* and the Andean icterid *Sturnella*, the African rock-chats (*Cercomela*; turdids) and the Andean ground-tyrants (*Muscisaxicola*; tyrannids), and the African wheatears (*Oenanthe*; turdids) and the Andean miners (*Geositta*; furnariids). The African lark *Galerida* (alaudid) and the Andean *Geositta* or *Cinclodes* (furnariid) also show resemblances. These birds live in open grassy or scrubby habitats, with or without rocky outcrops depending on the species. Similarities exist in color pattern and general coloration (light grayish brown above and whitish below, light rump and dark tip of the tail in *Oenanthe-Geositta*), in general habits (locomotion, posture when on the ground, for instance *Cercomela-Muscisaxicola*), and in behavioral traits (up and down motions of the tail; several kinds of vocalizations; techniques of foraging for food). Vuilleumier (1971), Keast (1972), and Vaurie (1980) had commented earlier on some of the resemblances between the Old World Muscicapidae and the New World Furnariidae.

Several Andean ground-dwelling insectivores nest regularly or habitually in holes and burrows, which they sometimes excavate themselves (Dorst, 1962b). The African species either nest in more open situations or more simply in rock crevices or under stones. For instance, King (1973) reported a nest of *Cercomela sordida* in a "12-inch deep cleft between two rocks and under a third," a site reminiscent of the nest sites of several *Muscisaxicola* spp. (pers. obs.). Thus, while the tendency to underground nesting is present in members of the guild on both continents, it seems clearly more developed in the Andes. (Underground nesting is also either common or the rule in other Andean birds, see Dorst, 1962b; therefore this behavior may be more generally prevalent in the Andes than in Africa; several Tibetan birds also nest underground.)

Small Bush Insectivores (Guild no. 6)

The African warblers of the genera *Bradypterus* and *Cisticola* (sylviids) and the Andean wrens of the genus *Cistothorus* (troglodytid), show similarities in habitat preferences and general behavior but these remain to be studied comparatively. Alternatively, the African warblers might show convergence to small Andean furnariids of the genera *Asthenes* and *Schizoeaca*.

Small Bush/Ground Seedeaters (Guild no. 8)

Small sized ground-dwelling granivorous birds in Africa and the Andes show what may be convergences. Thus the various canaries or "siskins" (*Serinus* spp.) are ecologically and morphologically equivalent to several Andean siskins of the genus *Carduelis*. But since these two genera are both members of the cardueline finches, resemblances could be due to common descent rather than convergences of different stocks within the same family. Therefore the similarities between African *Serinus* and the Andean emberizine finches of the genera *Catamenia*, *Sicalis*, *Phrygilus*, *Diuca*, or *Zonotrichia*, might be more suggestive of convergence. The populations of some species of seedeaters reach high densities near human settements, especially sheep enclosures, where the Andean emberizine finch *Zonotrichia capensis* and the Ethiopian Black-headed Siskin (*Serinus nigriceps*) and the Gray-headed Sparrow (*Passer griseus*, Ploceidae) may be each other's ecological equivalents.

Large Insectivores/Omnivores (Guilds 9 and 12)

The Wattled Ibis (*Bostrychia carunculata*) of Ethiopia appears to be the counterpart of the Andean Buff-necked Ibis (*Theristicus melanopis*). Both species are comparable in their general behavior, diet (worms; insects, especially Coleoptera and Orthoptera collected in humid meadows and grassy steppes), and nesting habitat (crevices along cliffs or canyons). Since both species are members of the same family, however, their resemblances could be due to common ancestry rather than convergence.

Large Vegetarians (Guild no. 13)

The Blue-winged Goose of Ethiopia (*Cyanochen cyanopterus*) and the Andean Goose (*Chloephaga melanoptera*) are probably ecological equivalents. Both species are found in pairs or small groups searching in wet meadows for their mostly vegetal food. The Andean Goose is the only species of the group of southern South American Magellanic geese that lives at low latitudes but high altitudes. Both genera are related within the Anatidae, subfamily Tadorninae, as shown by the pattern of their chicks, some anatomical features, and aspects of their behavior, particularly vocalizations and sexual displays (see Johnsgard, 1965). Therefore, the characteristics they share may be due to phylogenetic relationships rather than convergent evolution.

Large Insectivores (Guild no. 18)

In Ethiopia the Black Duck (*Anas sparsa*) and in the Andes the Torrent Duck (*Merganetta armata*) live in mountain streams with fast-running water. Both species show convergences in their general behavior, resting on rocks in midstream or swimming with great ease in swift current. However, the Andean species is more specialized in morphology and anatomy, whereas the African species shows features of the more generalized Mallard duck. Resemblances include the possession of soft, almost rubberlike bills (though they are shaped differently), an adaptation for collecting food under stones, and an advantage in avoiding strikes against rocks (Johnsgard, 1965). The African species lives mainly on freshwater crabs, in addition to a few aquatic insects and larvae. The Andean species consumes mainly Plecoptera of the genus *Rheophila*.

Note that in mountain streams of New Guinea lives Salvadori's Teal or Duck (*Anas waigiuensis*; formerly placed in the monotypic genus *Salvadorina*), a species that may show behavioral similarities (convergences?) to the African and Andean ducks.

Resemblances versus Differences in Guilds

Predators (Guild no. 1)

This guild includes seven species in Africa and eight in the Andes. Thus species diversity appears to be about equal on the two continents. However, five of the seven African species are found in Ethiopia only, so the guild of predatory species is richer and more diverse in the Andes. The most common and widespread members of this guild and those most relevant for comparisons, are probably the buzzards (*Buteo*). Both the African species (*B. rufofuscus*) and the Andean ones (*B. polyosoma* or *B. poecilochrous*) have similar habits. They nest on small cliffs and prey on rodents in the same ways. Both the African species and the Andean ones show plumage polymorphism, including dark color phases (see Vaurie, 1962, for a detailed description of the Andean species). Although not as common as *Buteo* spp., owls of the genus *Bubo* are widespread in Africa and the Andes. Therefore, the predatory guild on both continents may be composed of about the same types of birds, basically a diurnal *Buteo* and a nocturnal *Bubo*. Other species may be present locally (for instance an eagle, *Aquila* in Ethiopia and the eaglelike *Geranoaetus* in the Andean páramos; and the kestrels *Falco tinnunculus* in Ethiopia, and *F. sparverius* in the Andes). Because of the taxonomic similarity between the members of the guild, resemblances are likely to be due to phylogenetic resemblances, rather than to convergences.

Table 6–3 shows that even though the guild has about the same number of species on both continents, the species/genus ratios are not similar. The greater ratio in Africa may mean that ecological niches are slightly more finely subdivided than in the Andes. If this were true, this might mean that prey diversity is somewhat greater in Africa.

Scavengers (Guild no. 2)

There are clearly more scavengers in Africa than in the Andes, not only in numbers of species (5 versus 3), but also in numbers of individuals per species. Scavengers are scarce in the Andes. The species/genus ratio (Table 6–3) reflects greater

TABLE 6–3. Numbers of species and genera and species/genus ratios in some avifaunas and some guilds of the high African mountains (including Ethiopia) and of the Andes.

	Number of Species	Number of Genera	Species/Genus Ratio
Avifauna			
African afroalpine (incl. Ethiopia)	53	38	1.39
Andean páramo and puna	166	84	1.98
Guild			
1. Predators			
Africa	7	4	1.75
Andes	8	6	1.33
2. Scavengers (and omnivores)			
Africa	5	3	1.67
Andes	3	3	1.00
4. Large ground seedeaters/insectivores			
Africa	3	1	3.00
Andes	9	7	1.29
5. Small ground insectivores/granivores			
Africa	8	7	1.14
Andes	22	9	2.44
7. Nectarivores-insectivores			
Africa	4	1	4.0
Andes	8	6	1.33
8. Small bush/ground seedeaters			
Africa	5	2	2.50
Andes	16	6	2.67
All guilds above pooled			
Africa	31	18	1.72
Andes	66	37	1.78

Note: Guild numbers as in Table 6–2.

diversity and suggests more ecological opportunities in Africa. The Lammergeier (*Gypaetus barbatus*) has no ecological equivalent in the Andes. Since the Turkey Vulture (*Cathartes aura*) is quite rare in the high Andes, scavengers there are reduced to the common caracara *Polyborus* and the rare Andean Condor *Vultur gryphus*.

This difference between Africa and the Andes may be related to the abundance and diversification of large herbivorous native mammals in Africa (see, e.g. Brown, 1969; Nievergelt, 1981), whereas their Andean counterparts are limited in both numbers of species and individuals. They are represented only by the rare camelids (vicuña, *Vicugna*, and guanaco, *Lama*) and the uncommon cervid *Hippocamelus*. The populations of these herbivores are not large enough to support a large number of individuals or species of scavengers. However, in the Pleistocene and the Late Cenozoic, the Andean altiplano did have many more large herbivorous mammals, now extinct (see lists in Chap. 9). Perhaps more scavenging birds existed then, and they subsequently became extinct. Their discovery in sites with fossil herbivorous mammals would be very exciting.

Aerial Insectivores (Guild no. 3)

This guild includes goatsuckers, swifts, and swallows, and is richer in Africa (6 species in 4 genera) than in the Andes, (3 species in 3 genera). Three of the 6 African species are localized to Ethiopia, and 1 to East Africa, and 2 Andean species are localized to the páramos (1) and the puna (1). This decreases the local level diversity on both continents, but still leaves the Andean diversity lower. In a general way, then, Africa seems to offer more ecological possibilities for aerial foragers (greater abundance and/or more different kinds of insects) than the Andes. Interestingly, there are no swifts in the guild in the Andes, whereas there are three in Africa. In contrast, there is one goatsucker in the Andes, but none in Africa. Only detailed studies will permit one to determine whether there is convergence among the taxonomically diverse members of the guild. These studies will have to take into account the morphology of the species involved, their foraging modes, and the kinds of insects they consume.

Large Ground Seedeaters and/or Insectivores (Guild no. 4)

We mentioned earlier (comparisons of pairs or trios of species) the possible convergences between *Francolinus* (Phasianidae, the only members of the guild in Africa) and the Tinamidae in the Andes. We note here that this guild includes a variety of other taxa in the Andes besides tinamous: a snipe, two seedsnipes, a plover, and a terrestrial woodpecker. One could argue that the snipe should be included in guild no. 12, which in the Andes contains *Gallinago andina* and *G. nobilis*, and in Africa *G. nigripennis*, but even if the African *G. nigripennis* were included in the present guild, the diversity of species in Africa would not reach that of the Andes. Although this guild is richer and much more diverse in the Andes than in Africa, the species/genus ratio is higher in Africa (3.00 versus 1.29; Table 6–3).

The taxa in guilds 4 and 12 could be pooled into a single guild of medium to large ground birds. This enlarged guild includes 9 species in 7 genera in Africa (a 1.29 species/genus ratio), versus 15 species in 11 genera in the Andes (a 1.36 species/genus ratio). The diversities judged by the species/genus ratios are now very similar, suggesting convergence in resource availability and/or niche subdivision, but there are more species in the Andes. Further study of this guild or guilds will first have to focus on the definition of the guild and what species to include in it before analyzing convergences.

Small Ground Insectivores (and Granivores) (Guild no. 5)

The species in this guild forage on the ground and eat primarily insects, although some species also consume seeds, either on a regular (e.g., *Galerida malabarica* in Africa) or occasional basis (e.g., *Turdus* spp. in both continents). Although relatively diverse in Africa (8 species in 7 genera, a 1.14 species/genus ratio; Table 6–3), the guild is much richer and more diverse in the Andes (22 species in 9 genera, a 2.44 species/genus ratio). Some clearcut convergences were described earlier at the level of pairs of species, and involve the following genera (African genus first/Andean one second): *Galerida/Geositta, Cinclodes*; *Macronyx/Sturnella*; *Oenanthe/Geositta, Muscisaxicola*; *Cercomela/Muscisaxicola*. Other resemblances are taxonomic, since the two regions share *Anthus* (1 species in Africa, 4 in the Andes) and *Turdus* (1 versus 2 species).

The small ground insectivores appear to converge in both regions, although the more diverse guild in the Andes suggests more resource availability. To date, besides anecdotal comments, such as those in Vuilleumier (1971), Keast (1972), and Vaurie (1980), no work has been done to substantiate the notion that small ground insectivores in Africa and the Andes show convergences at the guild or community level. This seems to us to be a good working hypothesis in need of either verification or rejection. Because the ground "adaptive zone" is very important in the open, high-altitude habitats discussed in this chapter and in this book, a detailed intercontinental study of small ground birds should certainly be attempted.

Small Bush Insectivores (Guild no. 6)

This guild includes 4 species in 3 genera in Africa, and 12 species in 7 genera in the Andes; the respective species/genus ratios are 1.33 and 1.71. Whereas some of the taxa involved might indeed converge (*Bradypterus* and *Cisticola* in Africa, *Asthenes* and *Cistothorus* in the Andes), others seem to be quite different morphologically and ecologically and to have no equivalents; for example *Phylloscopus* in Africa, and *Leptasthenura* and *Anairetes* in the Andes.

Bush habitats are more extensive in the Andean puna than in Africa, but bush-like habitats not necessarily formed by shrubs might serve as similar substrates. Hence the guild might be somewhat ill-defined at present, and other taxa might be excluded by other workers, for instance the species of *Ochthoeca*, which might be thought of as ground birds since they forage, at least occasionally, on the ground. They also forage on the wing around bushes, rather than in bushes, as the other members of the guilds do. However, their exclusion would only reduce the Andean part of the guild to 10 species in 6 genera, and to a 1.67 species/genus ratio, figures not very different from those obtained by including *Ochthoeca*.

Nectarivores-Insectivores (Guild no. 7)

Four species of sunbirds (Nectariniidae) in the genus *Nectarinia* are found in Africa, and 8 species of hummingbirds (Trochilidae) in 6 genera in the Andes. There are both parallels and differences between the African and the Andean members of the nectarivorous-insectivorous guild (see also Chap. 5). The resemblances include the use of some of the dominant plants which themselves appear to show convergences. The differences include the use of rocky habitats by Andean nectarivores only, and the greater diversity in the Andes than in Africa.

Páramo hummingbirds of the genera *Chalcostigma*, *Oxypogon*, and *Oreotrochilus* are comparable to afroalpine and high-altitude *Nectarinia* in that they consume insects as well as nectar, and are thus more eclectic in their diet than their counterparts at lower elevations, both in South America and in Africa. In the páramos of Venezuela (Vuilleumier, pers. obs.) and of Colombia (Snow, 1983), *Oxypogon guerinii* uses the dominant *Espeletia* spp. much as *N. johnstoni* uses the dominant giant *Lobelia* spp. in Africa (Coe, 1967). Other high Andean hummingbirds, especially *Oreotrochilus*, may be largely dependent on *Chuquiraga* spp. (Dorst, 1962a; Corley Smith, 1969; Carpenter, 1976; Ortiz-Crespo and Bleiweiss, 1982).

One major difference between African sunbirds and Andean hummingbirds is that the latter have become rupicolous, especially perhaps *Oreotrochilus*. They perch on the ground or rock faces, hunt insects along cliffs, and nest against rocky walls or under the roofs of some caves, or even in the middle of caves or mine tunnels in complete darkness (Dorst, 1962a; Langner, 1973; Carpenter, 1976).

In both the African mountains and the Andes nectarivores are very diverse at mid-montane altitudes but are rare and less diverse in the habitats above timberline. In Africa, only one species is found on each East African mountain (*Nectarinia stuhlmanni* on Mt. Ruwenzori, and *N. johnstoni* on other mountains, see Appendix 6–A). In the Ethiopian highlands, only two species live at high elevations, the Malachite Sunbird (*N. famosa*) and the Tacazze Sunbird (*N. tacazze*). The latter does not really live above timberline, however, but mainly in the canopy of the heath forest and in *Lobelia* and *Alchemilla* associations up to about 3400 m. In the páramos of Venezuela and Colombia, where two to four species may occur above timberline, one species is usually dominant over the others or more abundant than the others. But in most very high-altitude sites, usually only one species is found, an *Oxypogon* in páramos, and in the puna, *Oreotrochilus estella*.

Small Bush/Ground Seedeaters (Guild no. 8)

Fewer species of seedeaters (finchlike birds) live in Africa (5 species in 2 genera) than in the Andes (16 species in 6 genera). The species/genus ratio, however, is only slightly higher in the Andes (2.67) than in Africa (2.50) (Table 6–3). Not only are Andean finchlike birds speciose in the Andes, they are also numerically abundant, at least locally. The notable difference in species diversity of birds of this guild between Africa and the Andes could be correlated with the lesser extension of pure grasslands in Africa, especially in Ethiopia (where 2 of the 5 species are found). Grassland was probably limited in area on the Ethiopian tableland, and the original habitats were heathland and scrub, almost up to the afroalpine zone; most of the open grassy habitats are probably man-made (see Hamilton, 1982). In the Andes, in contrast, grasslands were widespread long before man came into the picture. Hence seedeaters may have had the time to exploit this habitat and its resources longer and over a greater surface area. In the Andes, however, many granivores are also found in bushy habitats in more or less sheltered places, a vegetation type and ecological situation in many ways similar to its Ethiopian equivalent. The question is, then, why a richer avifauna of granivores did not evolve in such apparently favorable habitats in high African mountains.

Guilds from Cliffs and Rocky Slopes (Guilds no. 9, 10, and 11)

A pair of species in guild no. 9 was discussed earlier. Guild no. 10 consists of pigeons and doves, with 1 relatively large species (*Columba*) in Africa, and 3 smaller ones (*Metriopelia*) in the Andes. These two genera are not closely related within the Columbidae. Guild no. 11 contains a variety of birds. In Africa, it includes the Chough and two starlings, in the Andes it includes three tyrant flycatchers. Whether or not these six birds are correctly placed in a guild is debatable without further field work and no further discussion will be given here.

Medium-large Insectivores/Omnivores (Guild no. 12)

This guild contains several species feeding on the ground or in shallow water in wet meadows and marshes. Six species are found in Africa (in six genera), and six species also (but in five genera) in the Andes. Equivalents include the two ibises previously mentioned, two rails (*Rallus*), two lapwings (*Vanellus*), and three snipes (*Gallinago*). Because of the taxonomic similarities, phylogenetic resemblance rather than convergence may influence here the structure of this guild. Note that the very large crane (*Grus*) has no equivalent in the Andes, and that the latter region has two species of ibis (*Plegadis* in addition to *Theristicus* already cited).

As we suggested in our discussion of the guild of large ground seedeaters and/or insectivores (no. 4), some of the species in the present guild could be pooled with those in guild no. 4.

Large Swimming Vegetarians and Piscivores (Guilds no. 15 and 16)

Only two swimming vegetarians are found in Africa, versus 9 in the Andes; no piscivores live in Africa, but 5 do in the Andes, including the flightless endemic grebe *Rollandia microptera*. The discrepancy between Africa and the Andes in water birds of these two guilds clearly reflects the lack of suitable habitat in Africa. Aquatic birds are diverse and reflect a complex history in the Andes (see Chap. 23).

Small Insectivores (Guilds no. 14 and 17)

Birds in the first of these guilds occupy wet meadows and marshes, those in the second the rushes growing around lakes and lagoons. Two species can be placed in the first guild, and three in the second, all in the Andes. The guild's absence from Africa can be attributed to the lack of suitable habitat, especially rushes (*Scirpus*), whereas this belt of lakeshore vegetation is a prominent feature of high Andean lakes and lagoons. In Ethiopia, large lakes with well-differentiated habitats along their shores are found at lower elevations; their rich birdlife clearly cannot find adequate conditions for establishment in the upper altitudinal zones.

In the Andes, rushes are the home of three specialized passerines forming a small community (guild 17) without equivalent in Africa: the Many-colored Rush-tyrant (*Tachuris rubrigastra*; Tyrannidae), the Wren-like Rushbird (*Phleocryptes melanops*; Furnariidae), and the Yellow-winged Blackbird (*Agelaius thilius*; Icteridae). In the Ethiopian highlands, the aquatic vegetation around ponds and marshes simply lacks such specialists.

Summary of African-Andean Comparisons

Several unrelated genera of birds seem to be each other's ecological equivalents in several guilds or trophic groups in Africa and the Andes, respectively. Some of the resemblances are especially close, and deserving of further study, especially among the guild of small ground insectivores (guild no. 5). Some guild-level comparisons have revealed resemblances, some of which could be due to community-level convergence, whereas others are clearly due to phylogenetic resemblances. Among the first, we can cite the small ground insectivores, the small bush insectivores (guild no. 6), the nectarivores-insectivores (no. 7), and the small seedeaters (no. 8). Phylogenetic resemblances include *Buteo* and *Bubo* in the predators, *Anthus* in the small ground insectivores, and the carduelines in the small seedeaters.

Other differences include the usually greater species diversity in the Andean component of the guilds. Exceptions to this rule are the greater number of species of African scavengers and of aerial insectivores. Some guilds are lacking in Africa, for instance the small insectivores of wet meadows (no 14), the large swimming piscivores (no. 16), and the small insectivores of lakeshore rushes (no 17).

INTERTROPICAL COMPARISONS

Using the classification of guilds presented in Table 6–2, we compare in Table 6–4 the composition of guilds on six mountains in the intertropical zone around the world: three sites in the New World (Mexico, Venezuelan Andes, Mt. Itatiaia in southeastern Brazil) and three in the Old World (Mt. Kenya in Africa, Mt. Kinabalu in Borneo, and the Snow Mountains in New Guinea). Although ideally we would have liked to match mountain sites as closely as possible as far as area, altitude, vegetation, and climate are concerned, in practice we were forced to use those mountains for which we had data, available from the literature or from our own field work. We give for each mountain the approximate planar area of páramo vegetation above the continuous timberline, and the altitude of the summit.

For each mountain and for each guild present on that mountain we give in Table 6–4 the number and percent of species. The poorest site is the Mexican one, with only 4 species, the richest is Mt. Kenya with 17. Mt. Kinabalu, with 7 species, is only a little richer than the Mexican site. The other three have 10 or 13 species and are thus more closely comparable. Species richness is clearly influenced by several factors, such as relative area of a given mountain, relative isolation from other high mountains, diversity of vegetation, and climate. We consider the Mexican site to be marginally tropical, but all others to be comparable in their "tropicalness." We are aware of the fact that Kinabalu, unlike all other sites,

TABLE 6–4. Comparisons of guilds in páramos of the New World and the Old World Tropics.

Guild	Mexico Popocatepétl Ixtaccíhuatl (5285 m; a few km^2)		Andes Teta de Niquitao (4000 m; 300 km^2)		Brazil M. Itatiaia (2792 m; a few km^2)		Africa M. Kenya (5195 m; 290 km^2)		Malaysia M. Kinabalu (4094 m; a few km^2)		New Guinea Snow Mountains (4884 m; ?100 km^2)	
	Number of Species	Percent	Number of Species	Percent	Number of Species	Percent	Number of Species	Percent	Number of Species	Percent	Number of Species	Percent
1. Predators					1	10	3	17.6			2	15
2. Scavengers	1	25			1	10	2	11.8				
3. Aerial insectivores			1	8			1	5.9	1	14.3	1	8
4. Large ground seedeaters			1	8	2	20	1	5.9			2	15
5. Small ground insectivores			4	30	1	10	1	5.9	1	14.3	4	31
6. Small bush insectivores	1	25	3	23	3	30	2	11.8	4	57.1	2	15
7. Nectarivores-insectivores			1	8	1	10	1	5.9				
8. Small seedeaters	2	50	2	15	1	10	2	11.8			1	8
Other guilds			1	8			4	23.4	1	14.3	1	8
Totals	4	100	13	100	10	100	17	100	7	100	13	100

Note: Data for Mexico, see Appendix 6–B; data for Teta de Niquitao, see Table 6–9; data for Itatiaia, see Appendix 6–C; data for Mount Kenya, see Table 6–9; data for Kinabalu, see Appendix 6–D; data for Snow Mts., see Appendix 6–E.

lacks grassland (see Chap. 18). It is instructive to note that no two sites are alike in their guild composition.

The majority (2 of 4 species) of birds in Mexico are small seedeaters. No other site has as high a proportion of granivores. The Venezuelan site is dominated by three guilds, the small ground insectivores, the small bush insectivores, and the small seedeaters, which between them account for 68% of the total diversity. Mt. Itatiaia is dominated by the small bush insectivores and the large ground seedeaters (total = 50% of diversity). Mt. Kenya shows no clearcut dominance by any guild, but instead a scatter of species among all guilds. Mt. Kinabalu is dominated by the small bush insectivores (57%), and the Snow Mountains of New Guinea are dominated by the small ground insectivores (31%), although in general guilds there are more "scattered," as on Mt. Kenya.

Absences of guilds are also instructive. Only Mt. Kenya has all the guilds in Table 6–4. The Venezuelan site lacks two (predators and scavengers), and so does the New Guinea site (scavengers and nectarivores-insectivores; although if *Myzomela* had been included it would have lacked only one guild). Mt. Itatiaia only lacks the aerial insectivores. Mt. Kinabalu lacks predators, scavengers, large ground seedeaters, nectarivores-insectivores, and small seedeaters. Finally, the Mexican site lacks predators, scavengers (although it would have one if *Coragyps atratus* were included), aerial insectivores, large ground seedeaters, small ground insectivores, and nectarivores-insectivores.

At present we are not willing to speculate about the possible reasons for absences or dominance of given guilds, but not of other guilds on this or that mountain. We believe that the time is ripe for field work on an intercontinental basis, but in the meantime, taking the data in Table 6–4 at face value, we suggest that intercontinental comparisons are more likely to yield differences in the guild composition than convergences. The question is whether such differences are due to local differences in ecology or history, or both. This is clearly an hypothesis to be tested.

COMPARISONS BETWEEN TIBET, ETHIOPIA, AND THE ANDES

The tropical Andes and the temperate Himalaya, especially the Tibetan plateau, are the most extensive mountain chains in the world. Both consist of high and cold plateaus above which rise even higher mountain peaks, and both are surrounded or flanked by tropical or subtropical habitats. The Ethiopian tableland, although substantial within Africa, is far more modest by comparison. The Tibetan plateau is at an average altitude of about 5000 m, against about 3900 m for the Andean altiplano, and about 3500 m for the Ethiopian tableland. In an earlier publication one of us compared the avifaunas of the Andes and of Tibet (Dorst, 1974). We extend these comparisons here to include the Ethiopian highlands, and we analyze the composition of some guilds in the three regions.

Our comparisons are made between the Outer Plateau of Western Tibet (Vaurie, 1972), an area of about 280,000 km^2 that has about 156 bird species, the puna of the tropical Andes, with an area of about 160,000 km^2 and 148 bird species, and the Ethiopian highlands, with an area of about 4,000 km^2 and 45 species. Because we were not able to make comparisons for all guilds included in Table 6–2, we used only those guilds listed for Tibet in Appendix 6–F: predators, scavengers, large ground seedeaters, small ground insectivores, small bush/ground seedeaters, large swimming vegetarians, and large swimming piscivores.

The avifaunas of these three regions have no close taxonomic affinities, and each has genera restricted to it, but they do share several widespread genera, including *Anas*, *Falco*, *Gallinago*, *Bubo*, *Anthus*, and *Turdus*, represented in each region by different species. More genera are shared pairwise (for instance *Gypaetus* between Ethiopia and Tibet; *Athene* between the Andes and Tibet; or *Vanellus* between Ethiopia and the Andes). In other cases, however, birds from different phylogenetic stocks in each region show what appear to be convergences at the level of pairs or trios of species. Some of these are discussed below.

Among the large ground seedeaters and/or insectivores (guild no. 4), ecological equivalences can be suggested between Tibetan *Tetraogallus* (snowcock), *Perdix* (partridge), and *Syrrhaptes* (sandgrouse); Andean *Nothura* and *Tinamotis* (tinamous), and *Thinocorus* (seedsnipe); and African *Francolinus* (francolins). The resemblance between *Syrrhaptes* and *Thinocorus* would be especially worthy of detailed study.

Among the small ground insectivores (and granivores) (guild no. 5), resemblances exist between Tibetan *Calandrella*, *Eremophila*, *Melanocorypha*, and *Alauda* (larks), Ethiopian *Cercomela* (rock-chat), and Andean *Upucerthia* and *Geositta* (miners), and *Muscisaxicola* (ground tyrant). The similarities between the Tibetan ground jay *Pseudopodoces* and the Andean furnariids *Geositta* and *Upucerthia* appear striking, and would be fascinating to study in the field.

Among the small bush insectivores (guild no. 6), one could suggest that Tibetan *Leptopoecile* (Sylviinae) and Andean *Anairetes* (Tyrannidae) are ecological counterparts.

The small seedeaters guild (no. 8) is represented in Tibet and in the Andes by several genera of finchlike birds that, upon more detailed study, might be found to exhibit convergent resemblances. They include the unrelated Tibetan *Montifringilla* and Andean *Diuca* and *Phrygilus*.

For several Tibetan and Andean birds, ecological resemblances that might be due to convergence include the rupicoline habit, and especially the behavior of nesting underground, sometimes in association with mammals, marmots (*Marmota*), and pikas (*Ochotona*) in Tibet, and tucotucos (*Ctenomys*) in the Andes (see Dorst, 1962b, 1974; see also Meyer de Schauensee, 1984). One of us pointed out in an earlier publication (Dorst, 1962b) that a number of Andean taxa nest underground, and that several of these excavate their own burrows. A resemblance might thus exist between Tibetan *Pseudopodoces* and Andean *Upucerthia*, both of which have strong decurved bills. Among the avian-mammalian nesting associations, we suggest, as possible convergences, those between Tibetan finches *Leucosticte* and marmots, and *Montifringilla* and pikas, and between Andean miners *Geositta* and tucotucos. Note that the finches are placed in the small seedeaters guild (no. 8), whereas the miner is in the small ground insectivores guild (no. 5), but that both of these guilds are basically ground foraging birds. As we have pointed out earlier, although some Ethiopian birds (e.g., *Cercomela*) nest in crevices, the underground nesting habit is either absent or rare in Africa. Thus convergences in this behavioral trait might be restricted to the Andes and Tibet. No breeding association appears to have been reported between mammals (hyrax) and birds in Africa.

We now turn to a numerical analysis of guild composition. Table 6–5 gives the numbers and percentages of species in each of seven guilds from the three regions. The poorest region in terms of numbers of species is Ethiopia (26 species). Note, however, that the area of Ethiopia is very small, and consequently its species diversity cannot really be considered depauperate. The Andean puna and the Tibetan plateau have very similar species numbers, 61 and 58, respectively (but note that Tibet has about twice the area of the Andean puna).

In all three regions the small ground insectivores (no. 5) have the highest proportion of species: 30% in Ethiopia (dominant guild), 30% in the Andes (dominant guild), and 24% in Tibet (co-dominant guild with the small bush/ground seedeaters, guild no. 8, also 24%). The two most important guilds in both the Andes and Tibet are the small ground insectivores and the small bush/ground seedeaters, but they are also important in Ethiopia. When pooled, these two guilds make up 53% of the species numbers in the Andes, 48% in Tibet, and 45% in Ethiopia. But further study of the figures in Table 6–5 reveal differences between Ethiopia on the one hand, and the Andes and Tibet on the other. The next important guild in Ethiopia is the predators (no. 1), with 26%. This guild is much less well represented in Tibet (12%) and in the Andes (8%).

Other differences between Ethiopia and the Andes-Tibet include the 2–3 times greater proportion of scavengers in Ethiopia, the low proportion of large swimming vegetarians in

TABLE 6–5. Comparisons of several guilds in the Ethiopian highlands, the Andean puna, and the Tibetan plateau.

Guild	Ethiopian highlands		Andean puna		Tibetan plateau	
	Number of Species	Percent	Number of Species	Percent	Number of Species	Percent
1. Predators	7	26.0	5	8.0	7	12.1
2. Scavengers (omnivores)	4	15.0	3	5.0	6	10.3
4. Large ground seedeaters/ insectivores	2	7.0	7	11.0	7	12.1
5. Small ground insectivores/ granivores	8	30.0	18	30.0	14	24.1
8. Small bush/ground seedeaters	4	15.0	14	23.0	14	24.1
15. Large swimming vegetarians	2	7.0	9	15.0	7	12.1
16. Large swimming piscivores			5	8.0	3	5.2
Totals	27	100.0	61	100.0	58	100.0

Note: Data for Ethiopia, see Table 6–2; data for Andean puna, see Table 6–2; data for Tibetan plateau, see Appendix 6–F.

TABLE 6–6. Comparison of species richness and planar surface area of African and Andean high-altitude biomes.

Number of bird species		Africa-Andes Ratio × 100	Area (km^2)		Africa-Andes Ratio × 100
Africa	Andes		Africa	Andes	
53 (total)	166 (total)	31.9	6,000 (total)	185,000 (total)	3.2
21 (East Africa)	62 (páramo)	33.9	2,000 (East Africa)	25,000 (páramo)	8.0
45 (Ethiopia)	148 (puna)	30.4	4,000 (Ethiopia)	160,000 (puna)	2.5
53 (total)	62 (páramo)	85.5	6,000 (total)	25,000 (páramo)	24.0

Note: Data for Africa, see Appendix 6–A; data for the Andes, see Chap. 23.

Ethiopia, and the absence of large swimming piscivores in Ethiopia.

The guild composition in these three regions, as we judge from the data presented in Table 6–5, and from more subjective impressions as well, seems to suggest more general convergence at the community level between the Andean puna and the Tibetan plateau than between the Ethiopian highlands and either the Andes or Tibet. Nevertheless, the dominance of small ground insectivores in all three regions is indicative of resemblances, which should certainly be investigated further in terms of possible convergences.

We end up this comparison of the Ethiopian highlands and the Andean puna with the Tibetan plateau by pointing out two other patterns not evident from the date in Table 6–5. Of the guilds listed in Table 6–2, one (aerial insectivores, no. 3) is as poorly represented in Tibet (3–4 species) as in the Andes (3 species), and another (nectarivores-insectivores, no. 7) is totally absent from Tibet. The paucity of aerial insectivores in Tibet can be explained probably by the generally low temperatures and by the ecological importance of strong winds, which must either depress the biomass of aerial plankton, or else must restrict the presence of insects to the lower strata of the vegetation, especially near or on the ground. The absence of nectarivores in Tibet similarly can be assumed to be linked to the lack of adequate resources. Three species of Nectariniidae of the genus *Aethopyga* occur at high altitudes in Tibet (Vaurie, 1972; Meyer de Schauensee, 1984), but they live at the edge of or below the great plateau, in the regions Vaurie designated the Southeastern Plateau and the southern part of the Outer Plateau. Their habitats include montane forests, and rhododendron and juniper scrub up to the timberline. Unlike the Andes and Ethiopia, where some nectarivores adapted to the highest altitudes even with poor nectar resources, the Himalayan Nectariniidae either did not colonize the highest biota or, if they did, became subsequently extinct.

DISCUSSION

In this discussion we will focus especially on the African-Andean comparisons, since these are the two regions we know best and for which we have most data. We start with the observation, made earlier, that the Andean biota are richer in species than the African ones.

Species Richness and Area

In absolute numbers, there are more species in the tropical Andes than in the high mountains of tropical Africa. Table 6–6 gives the relevant figures in the first three columns, where total diversities are compared between Africa and the Andes. The total diversity in Africa is about 32% that in the Andes. African diversity is about 34% that of the Andes when the East African montane archipelago is compared with the Andean páramo archipelago. If the Ethiopian highlands are compared to the Andean puna which include the altiplano, African diversity is about 30% that in the Andes. If all African mountains (Ethiopia and East Africa) are considered as a montane archipelago comparable to the Andean páramo archipelago, then the discrepancy is much less, since Africa has about 86% of the Andes' species diversity.

The last three columns of Table 6–6 compare the respective areas of high-altitude biota in Africa and the Andes. Africa (total, East Africa, or Ethiopia) compared to the total high Andean area, the páramo area, and the puna area, respectively, has between 3% and 8% of the Andean area. When, however, the African area as a whole is compared only to the Andean páramo archipelago, then Africa is seen to have about one fourth the area of the Andes.

Thus, even though Africa has only a very small fraction of the area available for birds in the Andes (2%–24%), it has proportionately many

TABLE 6–7. Comparison of numbers of species per 1000 km^2 in African and Andean high-altitude biomes.

Africa	Number of species per 1000 km^2	Andes	Number of species per 1000 km^2
Total	8.8	Total	0.9
East Africa only	10.5	Páramos only	2.5
Ethiopia only	11.3	Puna only	0.9

more species (30%–86%) than one would expect from a consideration of area alone. One may conclude that there is an area effect, since Africa has fewer species than the Andes because its area is less, but that Africa has proportionately more species than the Andes and is not as depauperate as it appears at first sight.

This idea is further explored in Table 6–7. When the numbers of species per 1000 km^2 (a measure of relative species packing) in Africa and the Andes are compared, it can be seen that Africa has between 9 and 11 species per 1000 km^2 of high altitude biome, whereas the Andes have only between 1 and 3. In other words, Africa has between 3 and 11 times more species per unit area of afroalpine biome than the Andes have of páramo or puna biome.

The figures in Tables 6–6 and 6–7 suggest to us that the high altitude biomes of the Andes and Africa are both probably equally saturated with species, even though they do differ in absolute species densities (more species in the Andes) and relative species densities (more species in Africa). This difference is one of perception, and is related to the fact that the much greater surface area in the Andes is not accompanied by a concomitant increase in habitat diversity. This means that the only proper comparisons are between African and Andean sites that are matched closely for area and species numbers. Table 6–8 summarizes comparisons of species numbers and area for five pairs of mountains in the Andes and in Africa, which have about the same numbers of species and the same surface area. We pursue this analysis below at the guild level, by comparing three of the five pairs of mountains in Table 6–8, and the Ethiopian highlands and the Andean puna.

Intermountain Comparisons of Guilds Between Africa and the Andes

Afroalpine versus Páramo

Tables 6–9, 6–10, and 6–11 list the guilds for the following three intercontinental comparisons: Mt. Kenya (Africa) and Teta de Niquitao (Venezuelan Andes), Mt. Elgon (Africa) and Cendé (Venezuelan Andes), and Crater Highlands (Africa) and Tamá (Venezuelan Andes).

Ethiopian Highlands versus Andean Puna

Table 6–12 lists the guilds in the Ethiopian highlands and at one site, Carpa, in the Peruvian puna. Although the two regions are very different in surface area, they have about the same numbers of species, which is used as the basis for the comparison.

TABLE 6–8. Species numbers and planar surface area of selected pairs of afroalpine and páramo islands.

Pair of montane islands	Number of species	Area (km^2)
Ethiopia (afroalpine)	45	4000
Chimborazo (páramo)	41	3500
Mt. Kenya (afroalpine)	17	290
Teta de Niquitao (páramo)	13	300
Mt. Kilimanjaro (afroalpine)	13	215
Santurbán (páramo)	14	220 (303)
Mt. Elgon (afroalpine)	10	100
Cendé (páramo)	11	100
Crater Highlands (afroalpine)	10	55
Tamá (páramo)	14	100 (67)

Note: Data for afroalpine islands from Appendix 6–A; data for Andean páramo islands, see Vuilleumier and Simberloff (1980); areas in parentheses are from Simpson (1975).

TABLE 6–9. Composition of guilds on Mt. Kenya (afroalpine) and Teta de Niquitao (páramo).

Guild	Mt. Kenya (290 km^2; altitude 5195 m)	Teta de Niquitao (300 km^2; altitude 4000 m)
1. Predators	*Buteo rufofuscus*	
	Aquila verreauxii	
	Bubo capensis	
2. Scavengers (and omnivores)	*Gypaetus barbatus*	
	Corvus albicollis	
3. Aerial insectivores	*Apus melba*	*Notiochelidon murina*
4. Large ground seedeaters/insectivores	*Francolinus psilolaemus*	*Gallinago stricklandii*
5. Small ground insectivores/granivores	*Cercomela sordida*	*Asthenes wyatti*
		Cinclodes fuscus
		Turdus fuscater
		Anthus bogotensis
6. Small bush insectivores	*Bradypterus cinnamomeus*	*Schizoeaca coryi*
	Cisticola hunteri	*Leptasthenura andicola*
		Cistothorus meridae
7. Nectarivores-insectivores	(?*Nectarinia famosa*)	*Oxypogon guerinii*
	Nectarinia johnstoni	
8. Small bush/ground seedeaters	*Serinus canicollis*	*Zonotrichia capensis*
	S. striolatus	*Phrygilus unicolor*
11. Medium-small insectivores/omnivores	*Onychognathus tenuirostris*	
12. Medium-large insectivores/omnivores	*Sarothrura lineata*	
	Gallinago nigripennis	
15. Large swimming vegetarians		*Anas flavirostris*
18. Large insectivores	*Anas sparsa*	
Total number of species (not including "?")	17	13

Note: Data for M. Kenya, see Appendix 6–A; data for Teta de Niquitao, see Vuilleumier and Ewert (1978); guilds as in Table 6–2.

TABLE 6–10. Composition of guilds on Mt. Elgon (afroalpine) and Cendé (páramo).

Guild	Mt. Elgon (160 km^2; altitude 4324 m)	Cendé (100 km^2; altitude about 3550 m)
1. Predators	*Buteo rufofuscus*	*Falco sparverius*
2. Scavengers (and omnivores)	*Corvus albicollis*	
4. Large ground seedeaters/insectivores	*Francolinus psilolaemus*	*Gallinago stricklandii*
5. Small ground insectivores/granivores	*Cercomela sordida*	*Cinclodes fuscus*
		Turdus fuscater
		Sturnella magna
		Anthus bogotensis
6. Small bush insectivores	*Bradypterus cinnamomeus*	*Schizoeaca coryi*
	(?*Cisticola brunnescens*)	*Ochthoeca fumicolor*
	Cisticola hunteri	
8. Small bush/ground seedeaters	*Serinus canicollis*	*Zonotrichia capensis*
	Serinus striolatus	*Phrygilus unicolor*
		Carduelis spinescens
12. Medium-large insectivores/omnivores	*Gallinago nigripennis*	
18. Large insectivores	*Anas sparsa*	
Total number of species (not including "?")	10	11

Note: Data for Mt. Elgon, see Appendix 6–A; data for Cendé, see Vuilleumier and Ewert (1978); guilds as in Table 6–2.

TABLE 6–11. Composition of guilds in the Crater Highlands (afroalpine) and Tamá (páramo).

Guild	Crater Highlands ($55\ km^2$; altitude 3648 m)	Tamá ($100\ km^2$; altitude 3600 m)
1. Predators	*Buteo rufofuscus*	
	(?Bubo capensis)	
2. Scavengers (and omnivores)	*Gypaetus barbatus*	
	Corvus albicollis	
3. Aerial insectivores	*(?Apus niansae)*	*Caprimulgus longirostris*
		Notiochelidon murina
4. Large ground seedeaters/insectivores		*Gallinago stricklandii*
5. Small ground insectivores/granivores	*Cercomela sordida*	*Turdus fuscater*
		Anthus bogotensis
6. Small bush insectivores	*Bradypterus cinnamomeus*	*Schizoeaca fuliginosa*
	Cisticola brunnescens	*Ochthoeca fumicolor*
	C. hunteri	*Cistothorus platensis*
7. Nectarivores-insectivores	*Nectarinia johnstoni*	*Chalcostigma heteropogon*
8. Small bush/ground seedeaters	*Serinus canicollis*	*Zonotrichia capensis*
	S. striolatus	
12. Medium-large insectivores/omnivores		*Gallinago nobilis*
15. Large swimming vegetarians		*Anas flavirostris*
Total number of species (not including "?")	10	14

Note: Data for Crater Highlands, see Appendix 6–A; data for Tamá, see Vuilleumier and Ewert (1978); guilds as in Table 6–2.

TABLE 6–12 Composition of guilds in the Ethiopian highlands and at Carpa in the Peruvian puna.

Guild	Ethiopia (about $4000\ km^2$; altitude about 3600–4200 m)	Carpa (about $4\ km^2$; altitude about 4100 m)
1. Predators	*Buteo rufofuscus*	*Buteo poecilochrous*
	Aquila rapax	
	A. verreauxii	
	Falco tinnunculus	
	F. biarmicus	
	F. peregrinus	
	Bubo capensis	
2. Scavengers (and omnivores)	*Gypaetus barbatus*	*Vultur gryphus*
	Gyps rueppellii	*Polyborus megalopterus*
	Corvus capensis	
	C. crassirostris	
3. Aerial insectivores	*Apus melba*	*Caprimulgus longirostris*
	(*?A. niansae*)	*Petrochelidon andecola*
	Riparia cincta	
	Hirundo fuligula	
	Psalidoprocne pristoptera	
4. Large ground seedeaters/insectivores	*Francolinus castaneicollis*	*Nothura ornata*
	F. psilolaemus	*Thinocorus orbignyianus*
		Colaptes rupicola
5. Small ground insectivores/granivores	*Galerida malabarica*	*Asthenes wyatti*
	Anthus noveaseelandiae	*A. humilis*
	Macronyx flavicollis	*Upucerthia jelskii*
	Lanius collaris	*U. serrana*
	Oenanthe lugens	*Cinclodes fuscus*
	O. bottae	*Geositta tenuirostris*
	Cercomela sordida	*Muscisaxicola juninensis*
	Turdus litsipsirupa	*Turdus chiguanco*
		Anthus correndera
		A. bogotensis

TABLE 6–12 CONTINUED.

Guild	Ethiopia (about 4000 km^2; altitude about 3600–4200 m)	Carpa (about 4 km^2; altitude about 4100 m)
6. Small bush insectivores	*Cisticola brunnescens*	*Leptasthenura andicola*
		L. pileata
		Ochthoeca oenanthoides
		Troglodytes aedon
7. Nectarivores-insectivores	*Nectarinia tacazze*	*Oreotrochilus estella*
	N. famosa	
8. Small bush/ground seedeaters	*Serinus canicollis*	*Zonotrichia capensis*
	S. nigriceps	*Phrygilus punensis*
	S. ankoberensis	*P. fructiceti*
	(?*S. striolatus*)	*P. plebejus*
	Passer griseus	*Sicalis olivascens*
		Diuca speculifera
		Catamenia inornata
		Carduelis atrata
		C. uropygialis
9. Large insectivores/omnivores	*Bostrychia carunculata*	
10. Medium ground seedeaters	*Columbia albitorques*	*Metriopelia melanoptera*
11. Medium-small insectivores/omnivores	*Pyrrhocorax pyrrhocorax*	*Xolmis rufipennis*
	Onychognathus tenuirostris	*Agriornis montana*
	O. albirostris	*A. andicola*
12. Medium-large insectivores/omnivores	*Bostrychia carunculata*	*Plegadis ridgwayi*
	Grus carunculatus	*Vanellus resplendens*
	Rallus rougetii	*Gallinago andina*
	Vanellus melanocephalus	
	Gallinago nigripennis	
13. Large vegetarians	*Cyanochen cyanopterus*	*Chloephaga melanoptera*
14. Small insectivores		*Cinclodes atacamensis*
15. Large swimming vegetarians	*Anas sparsa*	*Anas flavirostris*
	A. undulata	*A. specularioides*
18. Large insectivores	*Anas sparsa*	
Total number of species (not including "?")	44	43
Grand total for each region	45	46

Note: Data for Ethiopia, see Table 6–2 and Appendix 6–A; data for Carpa, see Vuilleumier and Simberloff (1980); species not included for Ethiopia is *Lamprotornis chalybaeus* (feeds on insects, seeds, and fruit and feeds on the ground; not assigned to a guild); species not included for Carpa are *Nycticorax nycticorax*, large piscivore, *Larus serranus*, lake omnivore, and *Grallaria andicola*, ground insectivore (these taxa were not assigned to guilds).

Comments

Table 6–13 summarizes the information in the previous four tables. Several trends appear: there are always more species of predators (guild no. 1) and scavengers (no. 2) in Africa; there are about equal numbers of species of large ground seedeaters/insectivores (no. 4) in each region; there are usually more species of small ground insectivores (guild 5) in the Andes; there are usually more small seedeaters in the Andes (guild 8); and there are usually more medium-large insectivores/omnivores (guild 12) in Africa. Numbers vary for guilds numbers 3 (aerial insectivores), 6 (small bush insectivores), and 7 (nectarivores-insectivores).

In spite of differences, the evidence in these tables suggests to us that there are enough similarities in the guild composition of these four pairs of mountains to hypothesize that convergence at the guild level exists, and that the observed differences could be due to genuine differences in available resources. Such an hypothesis can be tested only by field work on the guild composition in pairs of regions, and on the relative distribution of resources for these guilds in the different regions.

Although the distribution of guilds is not regular, comparisons involving more than one pair of sites (as was done in Table 6–4, with only Kenya and Niquitao) are more significant since variability in guild distribution across pairs of

TABLE 6–13. Summary of results in Tables 6–9 to 6–12.

Guild	Numbers of species in each pair of mountains			
	Kenya : Niquitao	Crater : Tamá	Elgon : Cendé	Ethiopian H.: Carpa
1. Predators	3: 0	2: 0	1: 1	7: 1
2. Scavengers	2: 0	2: 0	1: 0	4: 2
3. Aerial insectivores	1: 1	0: 2	0: 0	4: 2
4. Large ground seedeaters	1: 1	0: 1	1: 1	2: 3
5. Small ground insectivores	1: 4	1: 1	1: 3	8: 10
6. Small bush insectivores	2: 3	3: 3	3: 2	1: 4
7. Nectarivores-insectivores	1: 1	1: 1	0: 0	2: 1
8. Small seedeaters	2: 2	2: 3	2: 3	4: 9
9. Large insectivores/omnivores	—	—	—	1: 0
10. Medium ground seedeaters	—	—	—	1: 1
11. Medium-small insectivores/omnivores	1: 0	—	—	3: 3
12. Medium-large insectivores/omnivores	2: 0	1: 0	0: 1	5: 3
13. Large vegetarians	—	—	—	1: 1
14. Small insectivores	—	—	—	0: 1
15. Large swimming vegetarians	0: 1	—	0: 1	2: 2
18. Large insectivores	1: 0	1: 0	—	1: 0

sites can be assessed. Intercontinental comparisons involving regions other than Africa and the Andes will thus have to take into account several sites in New Guinea, for instance. In the absence of data at this time on resource subdivision, we cannot fruitfully speculate on relative niche breadth. When a given guild is represented by more species in one region or site than another, one could suggest finer resource use and/or greater ecological overlap, but at present such a suggestion is not based on data. We can, however, make some comments on historical factors that might have impinged upon the history of the distribution of species and of guilds, and have thus influenced their compositional similarities and differences in Africa and in the Andes.

Speciation and Colonization

Table 6–2 (Africa and Andes) and Appendix 6–F (Tibetan guilds) show clearly that in some genera more than one species is represented in a given area. This is true, for instance, of the genera *Francolinus*, *Nectarinia*, and *Serinus* in Africa; of *Nothura*, *Muscisaxicola*, and *Phrygilus* in the Andes, and of *Tetraogallus*, *Oenanthe*, and *Montifringilla* in Tibet. In some instances, these congeners are allopatric, and in others they are sympatric. To what extent does the presence of two or more sympatric congeners in a given region reflect past history in terms of speciation in situ and/or colonization from somewhere else. In order to answer these questions, one needs comparable analyses of speciation in Africa, the Andes, and Tibet. We have studied speciation and evolution in Andean birds (Dorst, 1967, 1976; Vuilleumier, 1969, 1970, 1971, 1977; Vuilleumier and Simberloff, 1980; and unpublished data), but not in Tibetan birds, and have only surveyed a few African groups. Speciation has in fact not been studied in Himalayan birds in general. For Africa, we have Moreau's (1966) remarkable overview, as well as the two atlases by Hall and Moreau (1970) and Snow (1978), and the summary by Prigogine (1984).

In Africa, high mountains form ecological islands, each very limited in area (see Appendix 6–A) and separated from other high mountains by stretches of lower country covered with different vegetation. This isolation is geologically recent, and some of the mountains were either united by montane habitats or at least their montane habitats were closer to each other during several pluvial periods of the Pleistocene (Hamilton, 1982; see Chaps. 17 and 22). This situation most likely permitted intercommunication of faunal elements. Later on, during interglacial periods, such as now, greater isolation would have prevailed and have resulted in geographical isolation and possibly speciation on mountaintops. This is particularly true for the very isolated high mountains south of Ethiopia, but it could also have been so for Ethiopia, since the plateau there is far from uniformly distributed from north to south. Pollen analysis clearly shows that ecological conditions were severely affected in Ethiopia, where the vegetation zones fluctuated altitudinally (Bonnefille, 1972).

Although the temperature, humidity, and

vegetation fluctuations were important in East African mountains and in Ethiopia, little or no speciation took place as a consequence of them among the birds of the higher altitudes. Moreau (1966) has suggested that "all but three" species of the East African mountains are derived from lower altitudes, and that in at least the first two of the three species (*Francolinus psilolaemus*, *Nectarinia johnstoni*, and *Cercomela sordida*) "the birds at one stage had more continuous ranges at lower altitudes and . . . their afroalpine association is secondary" (p. 229). In Ethiopia, neither the montane forests nor the afroalpine zone above acted as centers of differentiation (Dorst and Roux, 1973b), but the isolation between the Ethiopian highlands and the high mountains of East Africa has been responsible for some allopatric speciation. Moreau (1966, p. 221), for example, lists the following pairs of species, the first being the Ethiopian one: *Corvus crassirostris*/*C. albicollis*, *Macronyx flavicollis*/*M. sharpei*, and *Onychognathus albirostris*/*O. tenuirostris*. The endemic species of the Ethiopian highlands, the "most remarkable" of which (Moreau, 1966, p. 221) is the goose *Cyanochen cyanopterus*, could either predate Pleistocene events, or else (Moreau, 1966, p. 221) could result from post-glaciation phenomena "due to the [Ethiopian highlands] acting as a more effective faunal refuge, while more extinction took place elsewhere." Moreau (1966, p. 221) adds that the three pairs of allopatric species mentioned above "could conceivably be the result of divergence in the course of the last 18,000 years" (see, however, Chap. 22).

Thus some speciation occurred in Africa, but not very much when compared to the Andes. Indeed, there is ample evidence that speciation has been very active in the latter mountains, since at least one-third of all species are members of superspecies, most if not all of which are the result of Pleistocene speciation events (Vuilleumier, 1969, 1970, 1971; see also Chap. 23). These superspecies are related to the breaking up of formerly more continuous geographic ranges of original stocks by barriers that acted during the Pleistocene history of the biomes. No fewer than five major ecogeographic barriers interrupt the Andes from northern Colombia to central Chile (Vuilleumier, 1969, 1977; Vuilleumier and Simberloff, 1980). The complex relief, deeply cut by inter-Andean and transversal valleys, played a prominent role in such isolation during wide-ranging fluctuations in climate (Dorst, 1967, 1976). Along the cordilleras, successive lowering and raising of vegetation zones greatly modified the distribution of faunal and floral components (see Chaps. 7, 8; Dollfus, 1973; Haffer, 1970; Simpson, 1971, 1975).

Similar opportunities thus existed for speciation in Africa and in the Andes, but little speciation took place in Africa by comparison with the Andes. Furthermore, whereas adaptive radiation can be said to have occurred in several high Andean groups (for example ground tyrants of the genus *Muscisaxicola* [Vuilleumier, 1971], and finches of the genus *Phrygilus*), no such phenomenon took place in the high African biomes. Although the details of allopatric speciation are not known for Tibetan birds, it seems clear that both speciation (see, e.g., Martens, 1975) and some adaptive radiation occurred in the Himalaya, including Tibet (for example among such genera as the redstarts, *Phoenicurus*, the rosefinches, *Carpodacus*, and the snow finches *Montifringilla*).

Another difference between the Andes and African mountains that played a role in speciation and adaptive radiation, and so perhaps ultimately on the development of convergent adaptations among different phylogenetic stocks under similar environmental constraints, is that emigrations to the high-altitude biomes proceeded differently in the two regions. In the Andes several avian groups came from the south of the continent and belong to the Patagonian fauna evolved in southern South America (Chapman, 1917, 1926; Dorst, 1967). Northward colonization was possible because of similarities and relative continuity between habitats at low latitudes in Patagonia and at high altitudes in the Andes. An example of such colonization is the Andean goose *Chloephaga melanoptera*. Other Andean birds, however, have a northern origin and probably emigrated southward along the cordilleras, such as the kestrel *Falco sparverius*.

Unlike the Andes, where southern elements are common, the high East African mountains and Ethiopian highlands have few South African elements (Moreau, 1966, pp. 220–221; see Chap. 17 for a similar phenomenon in the afroalpine flora; see also Chap. 22). Moreau (1966, p. 221) concluded that "on the whole, unless present-day distributions are deceptive (which is quite possible as a result of local extinctions), most montane non-forest birds have originated on either the Abyssinian plateau or the other tropical highlands, but few in South Africa." Of course, South Africa is not located as far south as southern South America, and conditions in South Africa (macchia, savanna, woodlands) are different from those found in southern South America (moorlands, grasslands, steppes). One could probably state that birds living in Patagonia were preadapted to the high montane conditions that developed in the Andes farther north, but that such was not the case for the precursors of today's high African birds.

Moreover, even during the most favourable periods of the Pleistocene, when vegetation zones were depressed altitudinally, montane habitats were not continuous or nearly continuous from south to north in Africa, and several major ecogeographical barriers prevented birds from expanding their range through the whole area. An example of a southern species in the East African mountains is probably *Sarothrura affinis* (see map in Keith et al., 1970).

We wish to point out that very few high-altitude birds came from the north during the Pleistocene. The most conspicuous example is the chough *Pyrrhocorax pyrrhocorax*, which occurs in Africa only in Ethiopia (see also Chap. 22). "It is not difficult to derive this bird from the Palaearctic during the Last Glaciation, by way of the chain of mountains between the Nile and the Red Sea, which can be presumed to have had a climate at that time like that inhabited by this Chough in Morocco today" (Moreau, 1966, p. 221).

Thus, although birds of different geographical origins met on the Ethiopian tableland and on the high East African mountains, for geographic and ecologic reasons the phenomenon is not as widespread as in the Andes.

CONCLUDING COMMENTS

After surveying avifaunas established at high altitudes in Africa and the Andes, and also in other high tropical areas of the Americas and the Old World (Borneo, New Guinea), as well as the temperate fauna of the Tibetan plateau, we would conclude that some pairs of species and some guilds show sufficient resemblances for convergences to be suggested. This is not surprising, for similar resemblances among components of faunas occurring in the same kind of vegetation can be found in other environments. For instance, in lowland tropical rainforests convergences in shape and ecological function are found in birds (Amadon, 1973) and in mammals (Dubost, 1968). It is possible that such convergences are particularly apparent in high mountain habitats, where the smaller number of species and of adaptive types, as well as the relative simplification of the ecosystems make pattern and process easier to detect and to analyze.

Detailed studies of guilds (Root, 1967) and of pairs or trios of species in high tropical mountain biota would, we feel, be highly rewarding. We hope very much that the data we have presented in this chapter, including the species lists in the tables and appendixes, will spur others to pursue this topic. But workers should heed some of the caveats expressed by Blondel et al. (1984) in their critical analysis of community-wide convergence in mediterranean birds. Comparisons will have to be made in regions that have been carefully matched for similarities in climate and vegetation structure to avoid spurious comparisons. Extratropical sites ("controls") will have to be selected to insure that the intratropical comparisons are placed in a broader ecological and evolutionary context. And finally, the respective importance of convergences due to similar selective pressures on distinct genetic stocks, and of similarities due simply to the sharing, at some point in time, of a common ancestor, will have to be carefully evaluated.

APPENDIX 6–A

Bird species living in the afroalpine zone of the Ethiopian highlands and of five high mountains of East Africa.

	Mountain (area in km^2)					
Species	Ethiopia (4000)	Ruwenzori (230)	Elgon (160)	Kenya (290)	Kilimanjaro (215)	Crater Highlands (55)
Threskiornithidae						
1. *Bostrychia carunculata*	+	–	–	–	–	–
Anatidae						
2. *Cyanochen cyanopterus*	+	–	–	–	–	–
3. *Anas sparsa*	+	+	+	+	+	–
4. *A. undulata*	+	–	–	–	–	–
Accipitridae						
5. *Gypaetus barbatus*	+	–	–	+	+	+
6. *Gyps rueppellii*	+	–	–	–	–	–
7. *Buteo rufofuscus*	+	–	+	+	+	+
8. *Aquila rapax*	+	–	–	–	–	–
9. *A. verreauxii*	+	–	–	+	–	–
Falconidae						
10. *Falco tinnunculus*	+	–	–	–	–	–
11. *F. biarmicus*	+	–	–	–	–	–
12. *F. peregrinus*	+	–	–	–	–	–
Phasianidae						
13. *Francolinus castaneicollis*	+	–	–	–	–	–
14. *F. shelleyi*	–	–	–	(?+)	–	–
15. *F. psilolaemus*	+	–	+	+	–	–
Gruidae						
16. *Grus carunculatus*	+	–	–	–	–	–
Rallidae						
17. *Rallus rougetii*	+	–	–	–	–	–
18. *Sarothrura affinis*	–	–	–	+	–	–
Charadriidae						
19. *Vanellus melanocephalus*	+	–	–	–	–	–
Scolopacidae						
20. *Gallinago nigripennis*	+	–	+	+	+	–
Columbidae						
21. *Columba albitorques*	+	–	–	–	–	–
Strigidae						
22. *Bubo capensis*	+	–	–	+	(?+)	(?+)
Apodidae						
23. *Apus melba*	+	+	–	+	+	–
24. *A. niansae*	(?+)	–	–	–	–	(?+)
25. *A. aequatorialis*	–	–	–	–	(?+)	–
Alaudidae						
26. *Galerida malabarica*	+	–	–	–	–	–
Hirundinidae						
27. *Riparia cincta*	+	–	–	–	–	–
28. *Hirundo faligula*	+	–	–	–	–	–
29. *Psalidoprocne pristoptera*	+	–	–	–	–	–
Motacillidae						
30. *Anthus novaeseelandiae*	+	–	–	–	–	–
31. *Macronyx flavicollis*	+	–	–	–	–	–
Laniidae						
32. *Lanius collaris*	+	–	–	–	–	–
Muscicapidae (Turdinae)						
33. *Oenanthe lugens*	+	–	–	–	–	–
34. *O. bottae*	+	–	–	–	–	–
35. *Cercomela sordida*	+	–	+	+	+	+
36. *Turdus litsipsirupa*	+	–	–	–	–	–

APPENDIX 6–A CONTINUED.

Species	Mountain (area in km^2)					
	Ethiopia (4000)	Ruwenzori (230)	Elgon (160)	Kenya (290)	Kilimanjaro (215)	Crater Highlands (55)
Muscicapidae (Sylviinae)						
37. *Phylloscopus umbrovirens*	–	+	–	–	–	–
38. *Bradypterus cinnamomeus*	–	+	+	+	+	+
39. *Cisticola brunnescens*	+	(?+)	(?+)	–	+	+
40. *C. hunteri*	–	+	+	+	+	+
Nectariniidae						
41. *Nectarinia tacazze*	+	–	–	–	–	–
42. *N. famosa*	+	(?+)	–	(?+)	–	–
43. *N. johnstoni*	–	+	–	+	+	+
44. *N. stuhlmanni*	–	+	–	–	–	–
Fringillidae (Carduelinae)						
45. *Serinus canicollis*	+	–	+	+	+	+
46. *S. nigriceps*	+	–	–	–	–	–
47. *S. ankoberensis*	+	–	–	–	–	–
48. *S. striolatus*	(?+)	+	+	+	+	+
Ploceidae						
49. *Passer griseus*	+	–	–	–	–	–
Sturnidae						
50. *Onychognathus tenuirostris*	+	+	–	+	–	–
51. *O. albirostris*	+	–	–	–	–	–
52. *Lamprotornis chalybaeus*	+	–	–	–	–	–
Corvidae						
53. *Pyrrhocorax pyrrhocorax*	+	–	–	–	–	–
54. *Corvus capensis*	+	–	–	–	–	–
55. *C. albicollis*	–	+	+	+	+	+
56. *C. crassirostris*	+	–	–	–	–	–
Totals (not including '?') 53	45	10	10	17	13	10

Notes: Data from Dorst (pers. obs.), Coe (1967), Urban and Brown (1971), Moreau (1966), Chapin (1932), Lamprey (1965), King (1973), Keith at al. (1970), Ash (1979), and Hall (pers. comm.). '+' means presence, and '–' means absence of the species from a given mountain.

APPENDIX 6–B

Birds of the páramo of Popocatépetl and Ixtaccíhuatl, Mexico.

Species	Guild
Troglodytidae	
1. *Troglodytes brunneicollis*	6. Small bush insectivores
Emberizidae	
2. *Oriturus superciliaris*	8. Small bush/ground seedeaters
3. *Junco phaeonotus*	8. Small bush/ground seedeaters
Corvidae	
4. *Corvus corax*	2. Scavengers (and omnivores)

Notes: Data from Paynter (1952) and F. V. (pers. obs.); *Coragyps atratus* (Cathartidae), a scavenger (guild no. 2) might also breed in Mexican páramos.

APPENDIX 6–C
Birds of the páramo of Mount Itatiaia, Brazil.

Species	Guild
Tinamidae	
1. *Rhynchotus rufescens*	4. Large ground seedeaters and/or insectivores
Cathartidae	
2. *Coragyps atratus*	2. Scavengers (and omnivores)
Falconidae	
3. *Falco sparverius*	1. Predators
Trochilidae	
4. *Colibri serrirostris*	7. Nectarivores–insectivores
Picidae	
5. *Colaptes campestris*	4. Large ground seedeaters and/or insectivores
Furnariidae	
6. *Schizoeaca moreirae*	6. Small bush insectivores
7. *Synallaxis spixi*	6. Small bush insectivores
Rhinocryptidae	
8. *Scytalopus speluncae*	6. Small bush insectivores
Motacillidae	
9. *Anthus hellmayri*	5. Small ground insectivores (and granivores)
Emberizidae	
10. *Zonotrichia capensis*	8. Small bush/ground seedeaters

Note: Data from Holt (1928), Pinto (1954), and F. V. (pers. obs.).

APPENDIX 6–D
Birds above timberline on Mt. Kinabalu, Sabah (Borneo).

Species	Guild
Apodidae	
1. *Collocalia esculenta*	3. Aerial insectivores
Pycnonotidae	
2. *Pycnonotus flavescens*	Bush frugivore-insectivore
Muscicapidae (Turdinae)	
3. *Turdus poliocephalus*	5. Small ground insectivores (and granivores)
Muscicapidae (Sylviinae)	
4. *Cettia fortipes*	6. Small bush insectivores
5. *Bradypterus accentor*	6. Small bush insectivores
6. *Phylloscopus trivirgatus*	6. Small bush insectivores
Zosteropidae	
7. *Chlorocharis emiliae*	6. Small bush insectivores (vegetarians)

Note: Data from Smythies (1981) and Gore (1968); *Pycnonotus flavescens* belongs in a guild not included in Table 6–2.

APPENDIX 6–E

Birds of the páramo of the Snow Mountains of New Guinea.

Species	Guild
Anatidae	
1. *Anas waigiuensis*	18. Large insectivores
Falconidae	
2. *Falco cenchroides*	1. Predators
Phasianidae	
3. *Anurophasis monorthonyx*	4. Large ground seedeaters and/or insectivores
Scolopacidae	
4. *Scolopax saturata*	4. Large ground seedeaters and/or insectivores
Tytonidae	
5. *Tyto tenebricosa*	1. Predators
Apodidae	
6. *Collocalia hirundinacea*	3. Aerial insectivores
Motacillidae	
7. *Anthus gutturalis*	5. Small ground insectivores (and granivores)
Muscicapidae (Turdinae)	
8. *Turdus poliocephalus*	5. Small ground insectivores (and granivores)
Muscicapidae (Sylviinae)	
9. *Megalurus timoriensis*	6. Small bush insectivores
Muscicapidae (Muscicapinae)	
10. *Petroica archboldi*	5. Small ground insectivores (and granivores)
Meliphagidae	
11. *Melidectes nouhuysi*	5. Small ground insectivores (and granivores)
12. *Oreornis chrysogenys*	6. Small bush insectivores
Estrildidae	
13. *Lonchura montana*	8. Small bush/ground seedeaters

Notes: Data from Rand (1942), Rand and Gilliard (1968), and Archbold, Rand, and Brass (1942). If *Myzomela rosenbergii* (Meliphagidae) is added to the list, then the guild of nectarivores/insectivores (no. 7) is added to the list of guilds.

APPENDIX 6–F

Composition of selected guilds on the outer plateau of western Tibet.

Guild	Species
1. Predators	*Aquila rapax*
	A. chrysaetos
	Falco cherrug
	F. tinnunculus
	Bubo bubo
	Asio flammeus
	Athene noctua
2. Scavengers (and omnivores)	*Gypaetus barbatus*
	Gyps himalayensis
	Corvus macrorhynchos
	C. corone
	C. corax
	Pica pica
4. Large ground seedeaters and/or insectivores	*Lerwa lerwa*
	Tetraogallus himalayensis
	T. tibetanus
	Alectoris chukar
	Perdix hodgsoniae
	Charadrius mongolus
	Syrrhaptes tibetanus

APPENDIX 6–F CONTINUED.

Guild	Species
5. Small ground insectivores (and granivores)	*Calandrella cinerea*
	C. acutirostris
	Melanocorypha maxima
	Eremophila alpestris
	Alauda gulgula
	Anthus godlewskii
	A. roseatus
	Pseudopodoces humilis
	Saxicola torquata
	Oenanthe pleschanka
	O. picata
	O. deserti
	O. alboniger
	Turdus merula
8. Small bush/ground seedeaters	*Montifringilla adamsi*
	M. taczanowskii
	M. blanfordi
	Serinus pusillus
	Carduelis carduelis
	Acanthis flavirostris
	Leucosticte nemoricola
	L. brandti
	Rhodopechys mongolica
	Carpodacus rhodochlamys
	C. rubicilloides
	C. rubicilla
	C. puniceus
	Emberiza stewarti
15. Large swimming vegetarians	*Anas platyrhynchos*
	A. crecca
	A. strepera
	A. penelope
	A. acuta
	A. clypeata
	Fulica atra
16. Large swimming piscivores	*Podiceps cristatus*
	Phalacrocorax carbo
	Mergus merganser

Note: Data from Vaurie (1972), Meyer de Schauensee (1984) and Schäfer (1938); guilds as in Table 6–2.

REFERENCES

Amadon, D. 1973. Birds of the Congo and Amazon forests: a comparison. In *Tropical forest ecosystems in Africa and South America: a comparative review*, B. J. Meggers, E. S. Ayensu, and W. D. Duckworth, eds., pp. 267–277, Washington D.C.: Smithsonian Institution Press.

Archbold, L., A. L. Rand, and L. J. Brass. 1942. Results of the Archbold expeditions. No. 41. Summary of the 1938–1939 New Guinea expedition. *Bull. Amer. Mus. Nat. Hist*. 79: 197–288.

Ash, J.S. 1979. A new species of serin from Ethiopia. *Ibis* 121: 1–7.

Blondel, J. and J.-P. Choisy. 1983. Biogéographie des peuplements d'oiseaux à différentes échelles de perception: de la théorie à la pratique. *Oecol. Gener*. 4: 89–110.

Blondel, J., F. Vuilleumier, L. F. Marcus, and E. Terouanne. 1984. Is there ecomorphological convergence among mediterranean bird comunities of Chile, California, and France? In *Evolutionary biology*, Vol. 18, M. K. Hecht, B. Wallace, and G. T. Prance, eds. pp. 141–213, New York: Plenum Publishing Co.

Bonnefille, R. 1972. Association polliniques actuelles et quaternaires en Ethiopie (vallées de l'Awash et de l'Omo). Ph.D. Thesis, University of Paris.

Brown, L. H. 1965. Redwinged starlings of Kenya. *J. E. Afr. Nat. Hist. Soc. Nat. Mus*. 25: 41–52.

Brown, L. H. 1969. Observations on the status, habitat and behaviour of the Mountain Nyala *Tragelaphus buxtoni* in Ethiopia. *Mammalia* 33: 546–597.

Bourlière, F. 1957. Un curieux biotope d'altitude: les páramos des Andes de Colombie. *Terre Vie* 104: 297–304.

Cabrera, A.L. 1968. Ecología vegetal de la puna. *Colloq. Geogr. Bonn* 9: 91–116.

Carpenter, F. L. 1976. Ecology and evolution of an Andean hummingbird (*Oreotrochilus estella*). *Univ. Calif. Berkeley. Publ. Zool.* 106: 1–75.

Chapin, J. P. 1932. The birds of the Belgian Congo. Part I. *Bull. Amer. Mus. Nat. Hist.* 75: 1–729.

Chapman, F. M. 1917. The distribution of bird-life in Colombia: a contribution to a biological survey of South America. *Bull. Amer. Mus. Nat. Hist.* 36: 1–729.

Chapman, F. M. 1926. The distribution of bird-life in Ecuador: a contribution to the origin of Andean bird-life. *Bull. Amer. Mus. Nat. Hist.* 55: 1–784.

Cody, M. L. 1970. Chilean bird distribution. *Ecology* 51: 455–463.

Cody, M. L. and H. A. Mooney. 1978. Convergence versus nonconvergence in mediterranean-climate ecosystems. *Ann. Rev. Ecol. Syst.* 9: 265–321.

Coe, M. J. 1967. *The ecology of the alpine zone of Mount Kenya*. Monographiae Biologicae 17. The Hague: Junk.

Corley Smith, G. T. 1969. A high altitude hummingbird on the volcano Cotopaxi. *Ibis* 111: 17–22.

Cuatrecasas, J. 1968. Paramo vegetation and its life forms. *Colloq. Geogr. Bonn* 9: 163–186.

Dollfus, O. 1973. La Cordillère des Andes. Présentation des problèmes géomorphologiques. *Rev. Géogr. Phys. Géol. Dyn.* (2) 15: 157–176.

Dorst, J. 1956. Recherches écologiques sur les oiseaux des hauts plateaux péruviens. *Trav. Inst. Franç. Etudes Andines Paris-Lima* 5: 83–140.

Dorst, J. 1957. The *Puya* stands of the Peruvian high plaeaux as a bird habitat. *Ibis* 99: 594–599.

Dorst, J. 1962a. Nouvelles recherches biologiques sur les Trochilidés des hautes Andes péruviennes (*Oreotrochilus estella*). *Oiseau Rev. Franç. Ornithol.* 32: 95–126.

Dorst, J. 1962b. A propos de la nidification hypogée de quelques oiseaux des hautes Andes péruviennes. *Oiseau Rev Franç. Ornithol.* 32: 5–14.

Dorst, J. 1967. Considérations zoogéographiques et écologiques sur les oiseaux des hautes Andes. In *Biologie de l'Amérique australe*, Vol. 3, C. Delamare-Deboutteville and E. Rapoport, eds., pp. 471–504 Paris: Centre National de la Recherche Scientifique.

Dorst, J. 1974. Adaptations of Andean and Tibetan birds: a brief comparison. *J. Bombay Nat. Hist. Soc.* 71: 506–516.

Dorst, J. 1976. Historical factors influencing the richness and diversity of the South American avifauna. *Proc. 16th Int. Ornithol. Cong.*: 17–35.

Dorst, J. and F. Roux. 1973a. Esquisse écologique sur l'avifaune des Monts Balé, Ethiopie. *Oiseau Rev. Franç. Ornithol.* 42: 203–240.

Dorst, J. and F. Roux. 1973b. L'avifaune des forêts de *Podocarpus* de la Province de l'Arussi, Ethiopie. *Oiseau Rev. Franç. Ornithol.* 43: 269–304.

Dubost, G. 1968. Les niches écologiques des forêts tropicales sudamricaines et africaines, sources de convergences remarquables entre rongeurs et artiodactyles. *Terre Vie* 22: 3–28.

Gore, M. E. J. 1968. A check list of the birds of Sabah, Borneo. *Ibis* 110: 165–196.

Haffer, J. 1970. Entstehung und Ausbreitung nord-Andiner Bergvögel. *Zool. Jb. Syst.* 97: 301–337.

Hall, B. P. and R. E. Moreau. 1970. *An atlas of speciation in African passerine birds*. London: British Museum (Natural History).

Hamilton, A. C. 1982. *Environmental history of East Africa: a study of the Quaternary*. New York: Academic Press.

Hedberg, O. 1964. Features of afroalpine plant ecology. *Acta Phytogeogr. Suecica* 49: 1–144.

Holt, E. G. 1928. An ornithological survey of the Serra do Itatiaya, Brazil. *Bull. Amer. Mus. Nat. Hist.* 57: 251–326.

Jeannel, R. 1950. *Hautes montagnes d'Afrique*. Paris: Publ. Mus. Nat. Hist. Nat. Suppl. No. 1.

Johnsgard, P. A. 1965. *Handbook of waterfowl behaviour*. London: Constable.

Johnson, A. W. 1965. *The birds of Chile and adjacent regions of Argentina, Bolivia and Peru*. Vol. 1. Buenos Aires: Platt Establecimientos Gráficos S. A.

Johnson, A. W. 1967. *The birds of Chile and adjacent regions of Argentina, Bolivia and Peru*. Vol. II. Buenos Aires: Platt Establecimientos Gráficos S. A.

Keast, J. A. 1972. Ecological opportunities and dominant families, as illustrated by the Neotropical Tyrannidae (Aves). In *Evolutionary biology*, Vol. 5, T. Dobzhansky, M. K. Hecht, and W. C. Steere, eds., pp. 229–277, New York: Appleton-Century-Crofts.

Keith, S., C. W. Benson, and M. P. S. Irwin. 1970. The genus *Sarothrura* (Aves, Rallidae). *Bull. Amer. Mus. Nat. Hist.* 143: 1–84.

King, D. G. 1973. The birds of the Shira plateau and west slope of Kibo, Kilimanjaro. *Bull. Brit. Ornithol. Club* 93: 64–70.

Koepcke, M. 1954. Corte ecológico transversal en los Andes del Perú central con especial consideración de las aves. I. Costa, vertientes occidentales y región altoandina. *Mem. Mus. Hist. Nat. "Javier Prado" Lima* 3: 1–119.

Lamprey, H. F. 1974. birds of the forest and alpine zones of Kilimanjaro. *Tanzania Notes Records* No. 64: 69–76.

Langner, S. 1973. Zur Biologie des Hochlandkolibris *Oreotrochilus estella* in den Anden Boliviens. *Bonn. Zool. Beitr.* 24: 24–47.

Lauer, W. 1981. Ecoclimatological conditions of the paramo belt in the tropical high mountains. *Mountain Res. Dev.* 1: 209–221.

Martens, J. 1975. Akustische Differenzierung verwandtschaftlicher Beziehungen in der *Parus* (*Periparus*)-Gruppe nach Untersuchungen im Nepal-Himalaya. *J. f. Ornithol.* 116: 369–433.

Meyer de Schauensee, R. 1982. *A guide to the birds of South America*. 2d rev. ed. Intercollegiate Press (no city given).

Meyer de Schauensee, R. 1984. *The birds of China*. Washington, D.C.: Smithsonian Institution Press.

Monasterio, M. 1980. Las formaciones vegetales de los páramos de Venezuela. In *Estudios ecológicos en los Páramos Andinos*, M. Monasterio, ed., pp. 93–158, Mérida, Venezuela: Universidad de los Andes.

Moreau, R. E. 1966. *The bird faunas of Africa and its islands*. New York: Academic Press.

Nievergelt, B. 1981. *Ibexes in an African environment*. New York: Springer-Verlag.

Orians, G. H. and R. T. Paine. 1983. Convergent evolution at the community level. In *Coevolution*, D. J. Futuyma and M. Slatkin, eds., pp. 431–458, Sunderland, Massachusetts: Sinauer Associates.

Ortiz-Crespo, F. I. and R. Bleiweiss. 1982. The northern limit of the hummingbird genus *Oreotrochilus* in South America. *Auk* 99: 376–378.

Paynter, R. A., Jr. 1952. Birds from Popocatépetl and Ixtaccíhuatl, Mexico. *Auk* 69: 293–301.

Pearson, O. P. 1951. Mammals in the highlands of southern Peru. *Bull. Mus. Comp. Zool.* 106: 117–174.

Prigogine, A. 1984. Speciation problems in birds with special reference to the Afrotropical Region. *Mitt. Zool. Mus. Berlin* 60 Suppl.: 3–27.

Rand, A. L. 1942. Results of the Archbold expeditions. No. 43. Birds of the 1938-1939 New Guinea expedition. *Bull. Amer. Mus. Nat. Hist.* 79: 425–516..

Roe, N. A. and W. E. Rees. 1979. Notes on the puna avifauna of Azángaro Province, Department of Puno, southern Peru. *Auk* 96: 475–482.

Root, R. B. 1967. The niche exploitation pattern of the Blue-gray Gnatcatcher. *Ecol. Monogr.* 37: 317–350.

Schäfer, E. 1938. Ornithologische Ergebnisse zweier Forschungsreisen nach Tibet. *J. f. Ornithol.* 86 Sonderheft: 1–349.

Simpson, B. B. 1971. Pleistocene changes in the fauna and flora of South America. *Science* 173: 771–780.

Simpson, B. B. 1975. Pleistocene changes in the flora of the high tropical Andes. *Paleobiol.* 1: 273–294.

Smythies, B. E. (1981) *The birds of Borneo*, 3rd ed., revised by the Earl of Cranbrook. Kuala Lumpur: The Sabah Society with The Malayan Nature Society.

Snow, D. W., ed. 1978. *An atlas of speciation in African non-passerine birds*. London: British Museum (Natural History).

Snow D. W. 1983. The use of *Espeletia* by paramo hummingbirds in the eastern Andes of Colombia. *Bull. Brit. Ornithol. Club*. 103: 89–94.

Troll, C. 1959. Die tropischen Gebirge. *Bonn. Geogr. Abh.* 25: 1–93.

Urban, E. K. and L. H. Brown. 1971. *A checklist of the birds of Ethiopia*. Addis Ababa: Haile Sellassie I University Press.

Vaurie, C. 1962. A systematic study of the red-backed hawks of South America. *Condor* 64: 277–290.

Vaurie, C. 1972. *Tibet and its birds*. London: Witherby.

Vaurie, C. 1980. Taxonomy and geographical distribution of the Furnariidae (Aves: Passeriformes). *Bull. Amer. Mus. Nat. Hist.* 166: 1–357.

Vuilleumier, F. 1969. Pleistocene speciation in birds living in the high Andes. *Nature* 223: 1179–1180.

Vuilleumier, F. 1970. Generic relations and speciation patterns in the caracaras (Aves: Falconidae). *Breviora* No. 355: 1–29.

Vuilleumier, F. 1971. Generic relationships and speciation patterns in *Ochthoeca*, *Myiotheretes*, *Xolmis*, *Neoxolmis*, *Agriornis*, and *Muscisaxicola*. *Bull. Mus. Comp. Zool.* 141: 181–232.

Vuilleumier, F. 1977. Barrières écogéographiques permettant la spéciation des oiseaux des hautes Andes. In *Biogéographie et évolution en Amérique tropicale*, H. Descimon, ed., pp. 29–51, Paris: Lab. Zool. Ecole Normale Supérieure Publ. No. 9.

Vuilleumier, F. and D. N. Ewert. 1978. The distribution of birds in Venezuelan páramos. *Bull. Amer. Mus. Nat. Hist.* 162: 47–90.

Vuilleumier, F. and D. Simberloff. 1980. Ecology versus history as determinants of patchy and insular distributions in high Andean birds. In *Evolutionary biology*, Vol. 12, M. K. Hecht, W. C. Steere, and B. Wallace, eds., pp. 235–379, New York: Plenum Publishing Corporation.

III

HISTORICAL DEVELOPMENT OF BIOTA

When the high tropical montane environments (páramo and puna in the Andes, afroalpine in East Africa, and tropical-alpine in Malesian mountains) appeared as new habitats toward the end of the Cenozoic, they represented a new zone ready to be colonized by elements from neighboring or distant faunas or floras. Important phases of orogeny, together with substantial local volcanism, permitted the mountain ranges in the intertropical zone to reach great heights where temperature and rainfall were no longer appropriate for the survival of what one normally considers as "tropical" elements. Open and empty landscapes thus became biologically available. We know, of course, that the formation of these new environments was not instantaneous, but instead took place over relatively long periods of time (ecologically speaking). What taxa colonized these new environments? Did they establish themselves successfully, or did they become extinct rather rapidly after their initial implantation? What succession, or turnover, of taxa took place in these environments since their appearance and until the Recent? Can we date the appearance, the disappearance, or the coexistence of various taxa during the history of these environments? In other words, can we trace and describe the historical development of the component elements of the high tropical montane ecosystems? Such knowledge surely would greatly help us understand the present adaptations of the modern taxa.

Once again, as in part II, we deal with hierarchical levels of analysis. We can distinguish the following levels in our search for historical patterns. The first could be called *element*. This term is often left undefined in historical biogeographic studies. We consider an element to be composed of a population of individuals of a given species, which leave the mark of their reproduction. Such elements could have originated in several ways. First, they could have arrived from far away by active or passive dispersal. Second, they could have colonized the new environment by dispersal (either type) from nearby, but different, environments. And finally, they could be the result of local evolution of already present stocks. In practical terms, biologists refer to elements when they state that a given genus appears at a given time in a dated sequence of fossil pollen, for instance. An element could also be viewed in a more abstract way, as for example when one is referring to the appearance of the Gramineae or the Compositae. In this case, the element is composed of a population or populations of several (perhaps many) species. Both kinds of elements are in fact due to two different levels of analysis and perception.

The second hierarchical level can be equated with the concept of *plant formation*. An assemblage of elements makes up a plant formation. Many workers would use here the term "community" instead or as well. Plant formations (and/or communities?) can be arranged into vegetation belts or zones. We have to be careful at this level of analysis not to assume that formations, communities, or vegetation belts are necessarily functional wholes or blocks that evolve as units, with all component elements reacting together. For example, Livingstone (1975, p. 259) wrote: "Biogeographers have frequently assumed that the vegetation belts of mountains represent stable communities, in which the constituent species have an obligatory association with each other. Pollen stratigraphy shows that this has not been the case."

The third level could be considered to be the *biome*, which consists of the integration of plants and animals in ecosystems, including such factors as energy flow between different parts of these systems, and external factors such as climatic variables. The high Andean páramo, for instance, is such a biome, as is the afroalpine of high African mountains, or the montane forest of Africa. In our definition of the term biome we follow authors such as Kendeigh (1961, e.g., pp. 29, 276).

During their history, since their first appearance in the Pliocene or the early Pleistocene, and until the Recent, the high tropical montane environments (the high montane biomes) have been subjected to a series of severe climatic fluctuations, as a result of glacial and interglacial oscillations, as is now well known. One of the most fascinating things about the historical development of these biomes is that although they have remained relatively "stable" as biomes, or physiognomic vegetation units, they have had tremendous vertical and horizontal oscillations in narrow correlation with the climatic cycles. Yet, within each biome, there is ample evidence to show that the component parts (the two previous levels in our hierarchy) have not remained stable, but have greatly varied in their respective composition.

This evidence comes from the specific study of fossil material, especially perhaps fossil pollen and fossil bones of mammals. In this book, we selected examples that show quite clearly some of the possible ways in which elements appeared in the fossil record, and either subsisted or became extinct, and how long-term faunal or floral succession can begin to be understood. Since the Andes have the greatest extension of high tropical montane environments, the examples in the four chapters that follow are all taken from high Andean research. Man, as one of the component elements of the high Andean biomes, appeared relatively recently in their history, but succeeded in modifying them considerably. As we stated previously, we did not include chapters dealing with human adaptations, but we offer here a chapter on one of the multifaceted human influences, that of domestication of some of the native animals, especially mammals.

REFERENCES

Kendeigh, S. C. 1961. *Animal ecology*. Englewood Cliffs, N. J.: Prentice-Hall.

Livingstone, D. A. 1975. Late Quaternary climatic change in Africa. *Ann. Rev. Ecol. Syst.* 6: 249–280.

7

Development of the High Andean Páramo Flora and Vegetation

THOMAS VAN DER HAMMEN AND ANTOINE M. CLEEF

The flora and vegetation now found in the high Andean páramo are the products of evolution in the broadest sense. Environmental changes, evolution of taxa, immigration of taxa, and adaptation are major aspects of the evolution of the páramo ecosystem. Information regarding this process may be obtained not only from historical studies, but also from analysis of the present flora and vegetation. A *conditio sine qua non* of any historical study is adequate knowledge of the recent flora and vegetation. For this reason we begin with a detailed treatment of present-day vegetation and flora, including their origin. We deal next with the historical records from four successive time intervals and conclude with recent and historical data to obtain an overview of the evolution of the páramo ecosystem.

One might question the use of the term "ecosystem" for the entire páramo, because the páramo includes many different types of vegetation and exhibits many local differences in climate, soil, and hydrology. Páramo lakes and peat bogs especially should be regarded as relatively independent systems, as should certain types of páramo scrub or low forest, such as *Polylepis* coppices (Chap. 12), and different types of páramo grassland, such as *Calamagrostis effusa* bunch grasslands. Nevertheless we apply the term here to the entire páramo because of the characteristic climate, flora, fauna, and soil types these ecosystems have in common, as well as the many interrelationships between the different páramo systems. Thus we consider the different components of the páramo biome to be parts of a single major páramo ecosystem.

THE PRESENT VEGETATION BELTS AND THEIR RELATION TO THE CLIMATE

The altitudinal zonation of the páramo vegetation is best known from the Colombian Cordillera Oriental, as a result of extensive surveys begun in 1972 (Cleef, 1978, 1979b, 1981). Especially in the most recent publication a detailed overview is given of the altitudinal zonation and of the syn-systematic vegetation units (associations, communities). Further studies are now being carried out in the Cordillera Central and the Sierra Nevada de Santa Marta. In the Venezuelan Cordillera de Mérida important studies were conducted by Monasterio (1979, 1980) and by Azócar and Monasterio (1979). As most historical data are from the Colombian Cordillera Oriental, an area that may represent the core of the páramo flora and vegetation, this area will be treated in greater detail. The following zonal descriptions coincide partly with Cleef (1981); in this chapter, the names of authors of genera and species have been omitted for the sake of readability. The full names (and the corresponding families) are listed in Cleef (1981).

The altitudinal sequence of zonal páramo vegetation is described separately for the atmospherically dry and the humid sides of the mountains. The subdivisions are based mainly on the differences in floristics and in physiognomy.

Apart from certain edaphic differences, climate is what mostly determines the spatial distribution, physiognomy, and dominant floristic composition in the different altitudinal zones. Accordingly, the slopes facing dry inter-Andean valleys of high plains are atmospherically dry. Slopes exposed to the Amazonian Hylaea, the Orinoco savannas, or the deep Magdalena River valley are atmospherically humid. Marked differences in composition and structure of páramo vegetation correspond to different macroclimates. For example, *pajonales* dominated by tussock grasses (mainly *Calamagrostis effusa*) characterize the arid side of the mountains. *Chuscales*, i.e., open bamboo vegetation consisting mainly of *Swallenochloa* spp., predominate on the atmospherically humid slopes. Thus, a bunch grass páramo occurs on the dry side of the Cordil-

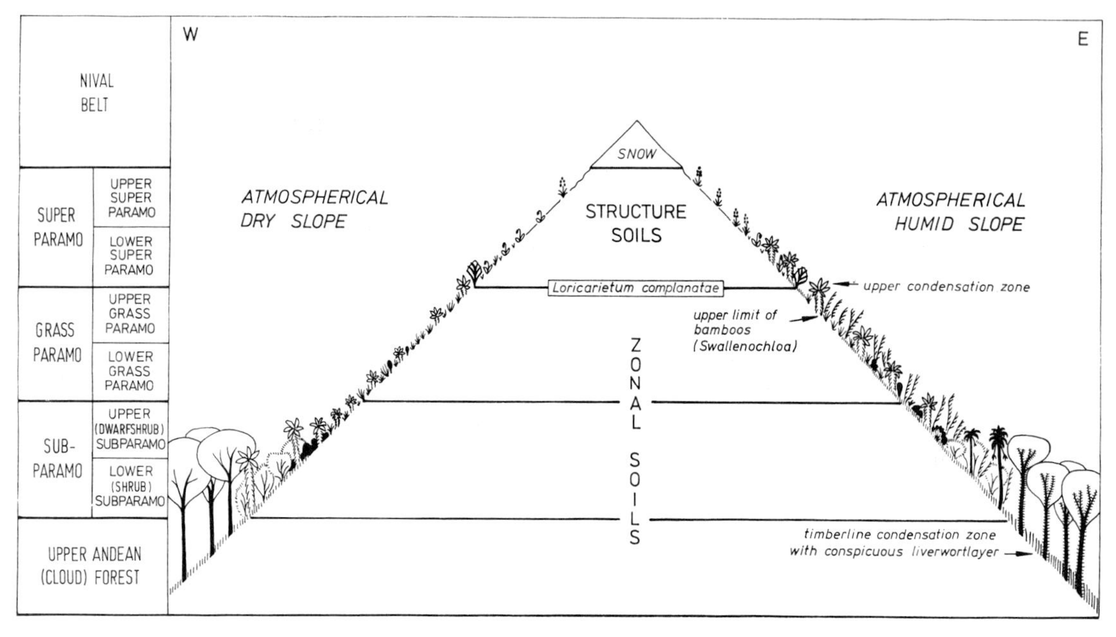

FIG. 7–1. Physiognomic zonation of páramo vegetation in a schematic cross section of the Colombian Cordillera Oriental.

lera, whereas a bamboo páramo is conspicuous on the humid side.

On the basis of purely physiognomic and floristic criteria, Cuatrecasas (1954a, 1958, 1968) subdivided the páramo into three altitudinal belts: (1) the subpáramo, (2) the páramo proper (*páramo propiamente dicho*), and (3) the superpáramo. Based on our studies, and by using similar criteria, we recognize in each altitudinal belt of zonal páramo vegetation a lower and an upper zone, and describe them here for each side of the Cordillera in order of increasing elevation (Fig. 7–1).

Atmospherically Dry Páramo Slopes

In the lower subpáramo or shrub páramo near the upper forestline are bushes of Ericaceae, Compositae, and Melastomataceae. Common and characteristic genera are *Befaria*, *Cavendishia*, and *Macleanea* (Ericaceae); *Ageratina* (= *Eupatorium* p.p.), *Pentacalia* (= *Senecio* p.p.), *Baccharis*, and *Diplostephium* (Compositae); and *Miconia*, *Bucquetia*, and *Brachyotum* (Melastomataceae). Among the species, *Solanum bogotense*, *Sericotheca argentea* (Rosaceae), *Stevia lucida* (Compos.), and *Coriaria ruscifolia* ssp. *microphylla* (Coriar.) are frequently found in this ecotone, as well as scattered small trees of *Buddleia lindenii* (Logan.), *Hesperomeles* (Rosac.), *Gynoxys* (Compos.), *Monnina* (Polygal.), and *Rapanea dependens* (Myrsin.).

At higher elevations in the upper subpáramo or dwarf shrub páramo, the vegetation is dominated mainly by dwarf shrubs of *Arcytophyllum nitidum* (Rubiac.), usually accompanied by *Calamagrostis effusa* and other species of the open grass páramo. *Gaylussacia buxifolia*, a characteristic ericaceous dwarf shrub, and *Chaetolepis microphylla* (Melast.) are frequently encountered. Terrestrial orchids, such as the beautiful, salmon pink flowering *Spiranthes vaginata*, as well as *Altensteinia leucantha*, *A. fimbriata*, and also *Lourteigia* (= *Eupatorium*) *microphylla*, *Miconia parvifolia*, *Paepalanthus paramensis* (Eriocaul.), and *Verbesina baccharidea* are virtually restricted to this zone. The last species always grows in rocky places with a thin soil cover, and is usually associated with small, dense tussocks of *Sporolobus lasiophyllus* (Gramin.)

In the lower zone of the páramo proper (the lower bunch grass páramo), dwarf scrub in general disappears and is replaced by tussocks of *Calamagrostis effusa*, which becomes dominant with increasing altitude. Typical heliophilous plants, which are also present in the upper subpáramo, abound here, for example, *Acaena cylindristachya* (Rosac.), *Azorella* aff. *cuatrecasasii* (Umb.), *Castratella piloselloides* (Melast.), *Lobelia tenera*, *Luzula racemosa* (Junc.), *Oreobolus obtusangulus* ssp. *rubrovaginatus*, and *Rhynchospora paramorum* (Cyperaceae). Endemic species of the arborescent composite genera *Espeletia* and *Espeletiopsis* undoubtedly are the most characteristic feature of the grass páramo.

More or less closed stands of *Calamagrostis effusa* tussock-sward constitute the plant cover in the upper zone of the páramo proper—the upper bunch grass páramo. Higher up the grass cover is sparser and only a few tussocks remain at the superpáramo border. In addition to representatives of the Espeletiinae, taxa such as *Hypericum selaginoides*, *Jamesonia bogotensis* (Polypod.), *Paepalanthus lodiculoides* var. *floccosus* (Eriocaul.), *Stephaniella paraphyllina* (Hepat.), *Diploschistes* sp., and *Lecidea* sp. (Lich.) are conspicuous. *Calamagrostis recta* tussocks, which are dominant in the zonal grass páramos of the Cordillera Central, are, in the Cordillera Oriental, known only at the highest parts covered by grass páramo vegetation. On the Cuchilla Puentepiedra (the nunatak-like watershed between the Páramo Concavo and Alto Valle de Lagunillas on the climatologically dry side of the Sierra Nevada del Cocuy), a continuous (upper) bunch grass páramo, including *Calamagrostis recta* and Espeletiinae, was observed up to at least 4450 m.

In the superpáramo the plant cover is sparse, vegetation occurring only in patches. Low temperatures prevail, and diurnal freezing and thawing cause all kinds of tropical solifluction phenomena, as described by Troll (1958). The number of plant species is smaller than in páramo vegetation at lower elevations, but the percentage of endemics is higher, especially in the vascular flora of the Sierra Nevada del Cocuy.

In the lower superpáramo a narrow belt of the remarkable dwarf shrub *Loricaria complanata* (Compos.) is present, almost continuously fringing the upper limit of the grass páramo. Dense Compositae thickets, consisting mainly of *Pentacalia* (= *Senecio*) *vaccinioides* associated with *P. andicola*, *Diplostephium rhomboidale*, and *D. alveolatum*, occur locally on young terminal moraines along the grass páramo–superpáramo border (4300–4400 m) on the atmospherically drier western slope of the Sierra Nevada del Cocuy. Stems and branches of these interesting bushes are covered with a thick layer of the moss *Zygodon pichinchensis*, especially in sheltered stands. Lower superpáramo vegetation on dry moraines is sparse and consists mainly of vascular plants. A creeping ericaceous dwarf shrub, *Pernettya prostrata* var. *prostrata*, is locally dominant; frequent species are *Agrostis boyacensis*, *A. haenkeana*, *Bartsia* sp., *Diplostephium colombianum*, *Draba litamo* (Crucif.), *Jamesonia goudotii*, *Lachemilla tanacetifolia* (Rosac.) *Luzula racemosa*, *Lycopodium crassum*, *Oritrophium cocuyense*, *Pentacalia* (=*Senecio*) *guicanensis*, *Poa* spp., *Bryum capillare*, *Polytrichum juniperinum* (Musci), and *Cora pavonia* (Lich.). These species are also found on the atmospherically humid slopes of the superpáramo and most of them are found occasionally also in the upper grass páramo.

In the uppermost zone, the upper superpáramo, the biomass and the number of plant species are considerably reduced. Only a few scattered species are left, for instance *Agrostis boyacensis*, *Calamagrostis* sp., *Luzula racemosa*, *Pernettya prostrata* var. *prostrata*, *Poa* sp., *Senecio cocuyanus*, *S. supremus*, *Andreaea rupestris*, *Bryum argenteum*, *Ditrichum gracile*, *Polytrichum juniperinum*, *Racomitrium crispulum*, and *Stereocaulon vesuvianum* var. *nodulosum*. In the Colombian Cordillera Oriental the upper superpáramo is restricted to the Sierra Nevada del Cocuy. At present the permanent snow cap has retired here to about 4800 m, glaciers descending to about 4400 m. The snowline is continuous on the west-facing dip slope of the Cocuy range, but fragmented on the steep east-facing slopes.

Atmospherically Humid Páramo Slopes

Bamboos constitute a very important feature of páramo vegetation on the humid side of the Cordillera, from the forest limit upward almost to the lower limit of the superpáramo. *Swallenochloa* is the most common bambusoid genus, and the species *S. tesselata* is widely distributed in the atmospherically humid páramo of the Cordillera Oriental. Some other bamboo species, such as *Chusquea scandens*, *Neurolepis aristata*, and *Aulonemia triana*, are present at the forest–páramo ecotone and penetrate locally into the subpáramo.

The occurrence of *Swallenochloa* is conceivably attributable to high annual precipitation (Gradstein, Cleef, and Fulford, 1977; Cleef, 1978) and possibly also to the permanently high atmospheric humidity causing a narrow diurnal range of prevailing low temperatures. Further research, especially climatological and ecophysiological, is required before this phenomenon can be explained.

Scrub consisting mainly of Compositae abounds in the lower subpáramo or shrub páramo. Thickets of *Ageratina* (*Eupatorium*) *tinifolia* are most conspicuous (van der Hammen et al., 1980–81). Apart from an abundance of bryophytes, frequently associated vascular taxa are, for example, *Aragoa lycopodioides* (Scroph.), *Baccharis* spp., *Centropogon ferrugineus* (Campan.), *Diplostephium* spp., *Escallonia myrtilloides* var. *myrtilloides* (Escallon.), *Gaiadendron punctatum* (Loranth.), *Gaultheria ramosissima*, *Hypericum* spp., *Oreopanax* sp. (Aral.), *Purpurella grossa* (Melast.), *Miconia* sect. *Cremanium*, *Rapanea dependens*, *Symplocos* spp., and *Ternstroemia meridionalis* (Theac.). The bamboo *Neurolepis aristata* locally forms large and impressive groves.

In the upper subpáramo or dwarf shrub (bamboo) páramo a layer of dwarf scrub and bamboo determines the aspect of the vegetation. As on the dry side, *Arcytophyllum nitidum* is present, but its percentage cover is strongly reduced. It grows on stony slopes associated with *Swallenochloa* and with several species of bryophytes, such as the mosses *Rhacocarpus purpurascens* and *Campylopus cucullatifolius*, and the liverworts *Jamesoniella rubricaulis* and *Lepidozia* sp. Occasionally *Sphagnum magellanicum*, *S. sanctojosephense*, and an unknown species of *S.* section *Malacosphagnum* are present. In general, various ericaceous dwarf shrubs, *Swallenochloa tesselata*, Compositae, and a considerable number of bryophytes are typical of this zone. The ericaceous *Disterigma empetrifolium* plays and important role here. It is accompanied by other dwarf shrubs, such as *Befaria tachirensis*, *Vaccinium floribundum*, *Plutarchia* spp., *Clethra* spp. (Clethrac.), *Ilex* spp. (Aquifol.), *Symplocos* spp., *Ugni myricoides* (Myrtac.), *Ageratina vacciniaefolia*, *Diplostephium huertasii*, *Diplostephium* spp., and tree ferns of the genus *Blechnum* subg. *Lomaria* (Polypod.). *Tofieldia sessiliflora* (= *T. falcata*) with its small white flowers and locally *Spiranthes coccinea* and *Nephopteris maxonii* (Polypod.) are characteristic endemic species of the atmospherically humid zonal subpáramo vegetation.

In the lower bamboo páramo, which is almost entirely dominated by *Swallenochloa tesselata*, some isolated patches of dwarf scrub are still present. In the driest localities *Calamagrostis effusa* is present, but with a reduced percentage cover. Small rosettes of *Castratella piloselloides* (and locally of *C. rosea*) are very common in this belt (but may occur lower down in the dwarf shrub páramo). They are mostly associated with *Rhynchospora paramorum*, *Oreobolus obtusangulus* ssp. *rubrovaginatus*, *Oritrophium peruvianum* ssp. *lineatum*, *Pinguicula elongata*, *Xyris acutifolia* (Xyrid.), and the moss *Rhacocarpus purpurascens*. *Lysipomia muscoides* ssp. *simulans* (Campan.), *Sisyrinchium pusillum*, *Paepalanthus pilosus* and *P. lodiculoides* are regularly encountered. In the lower part of the belt these species are associated with *Arcytophyllum nitidum*. In humid, flat, or gently sloping areas dense *Swallenochloa–Sphagnum* bogs occur, in which the most common *Sphagnum* species are *S. magellanicum*, *S. oxyphyllum*, *S. cuspidatum*, and *S. sancto-josephense*. These bogs have a particularly rich bryophyte flora, the most important elements being the mosses *Breutelia* spp., *Campylopus cavifolius*, *C. cucullatifolius*, *Chorisondontium speciosum*, and *Leptodontium wallisii*; the liverworts *Adelanthus lindenbergianus*, *Anastrophyllum* spp., *Cephalozia dussii*, *Herbertus subdentatus*, *Isotachis multiceps*, *Kurzia verrucosa*, *Lepidozia* spp., *Leptoscyphus cleefii*, *Riccardia* spp., and *Telaranea nematodes* (Gradstein et al., 1977); and the lichens *Cladonia colombiana*, *C. furcata*, and *C. polia* (Sipman and Cleef, 1979).

The upper bamboo bunch grass páramo differs from the previous zone by a general increase of the *Calamagrostis effusa* cover and a decrease of the *Swallenochloa* cover. The upper limit of the *Swallenochloa* dwarf bamboos, at 4100–4200 m and extending along streams up to nearly 4300 m, seems to be determined by nocturnal frost damage and atmospheric drought. In the higher páramos precipitation is apparently reduced with increasing altitude. The floristic composition of the plant cover supports this supposition: the overall vegetation type in the upper part of this zone resembles that of the atmospherically dry slopes and consists of an open sward of bunches of *Calamagrostis effusa* with Espeletiinae, *Cerastium subspicatum*, and *Stephaniella* spp.; *Jamesonia bogotensis* is almost absent.

Along the lower border of the superpáramo the narrow belt of low *Loricaria complanata* bushes is met again, especially on stony and shallow but stable soils. As a consequence of the almost constantly high atmospheric humidity on this side of the Cordillera, *Jamesonia goudotii*, *Lachemilla nivalis*, *Oritrophium peruvianum* ssp. *peruvianum*, and *Valeriana plantaginea* grow here between the clumps of *Loricaria*, apart from many species of bryophytes, including *Rhacocarpus purpurascens*, *Racomitrium crispulum*, *Campylopus pittieri*, *Anastrophyllum nigrescens*, *Gymnomitrion atrofilum*, and *Herbertus subdentatus*. Conspicuous lichens are *Cora pavonia*, *Cladonia* subg. *Cenomyce*, *Peltigera* spp., *Siphula* spp., and *Sphaerophorus melanocarpus*.

On the whole, the more humid climate appears to be responsible for an increase in number of plant species and for a more continuous vegetation cover in areas with more or less permanent fog in the humid lower superpáramo, as com-

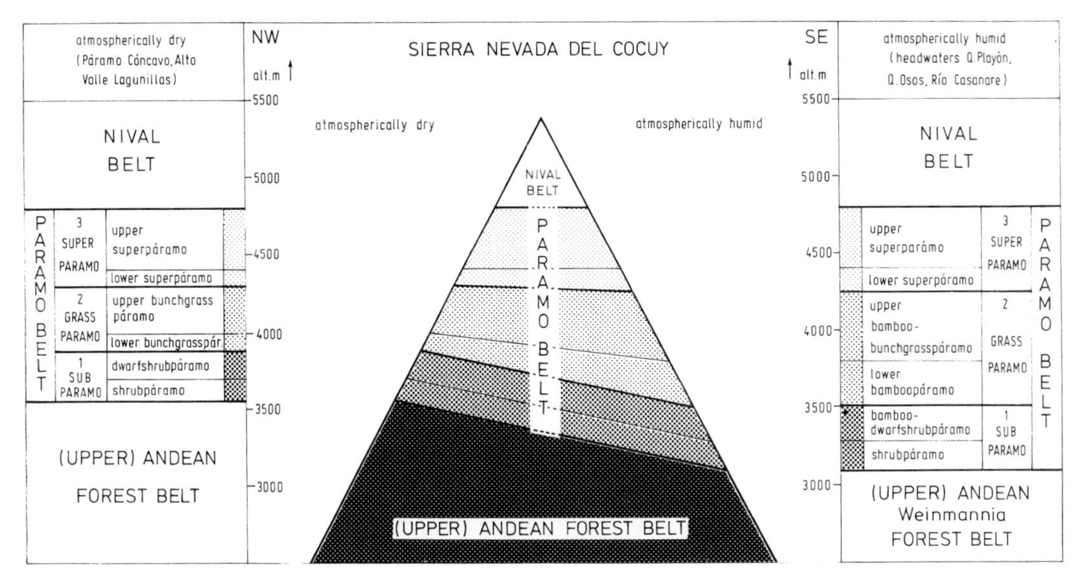

FIG. 7–2. Correlation between altitude and páramo vegetational zonation in a schematic NW–SE cross section of the southern part of the Sierra Nevada del Cocuy (Cordillera Oriental, Colombia). The forest line on the west side lies at a higher elevation, apparently because of ascending warm air currents from the Chicamocha Valley; the forest line on the east side is lower probably because of the extremely wet conditions. Normally the incidence of dry and wet conditions results in higher forest limits on wet slopes and in lower ones on dry slopes.

pared to the opposite dry side of the Cordillera. *Senecio niveo-aureus* is the most conspicuous and characteristic plant in the open zonal communities covering the morainic slopes (with a locally loamy matrix) at this elevation. Other characteristic plants are *Erigeron chionophilus*, *E. ecuadoriensis*, *Senecio* spp., *Diplostephium rupestre* (Compositae); *Valeriana plantaginea*, *Draba* spp. (e.g., section *Chamaegongyle*), *Montia meridensis* (Portul.), *Ourisia muscosa* (Scroph.), *Hymenophyllum trichophyllum* (Hymenoph.), *Arenaria venezuelensis*, and *Cerastium* spp. Some species of the Espeletiinae penetrate into the lower part of this zone. Typical bryophytes are *Breutelia integrifolia*, *Blindia acuta*, *Erythrophyllopsis andina*, *Kingiobryum paramicola*, *Distichium capillaceum*, *Zygodon* spp. (Musci) and *Cheilolejeunea* (subg. *Strepsilejeunea*) and *Anastrophyllum austroamericanum* (Hepaticae). The presence of *Azorella multifida* (Umbellif.), *Bartramia angustifolia*, *Breutelia* sp., *Herbertus subdentatus*, *Plagiochila dependula*, *Rhacocarpus purpurascens*, and *Sphaerophorus melanocarpus* accounts for the persistent atmospheric humidity. Ecological and floristical data about the humid lower superpáramo in the Sierra Nevada del Cocuy were published by Cuatrecasas and Cleef (1978).

Apart from Caryophyllaceae and *Draba* spp. (e.g., *D. hammenii*), the floristic composition and physiognomy of the upper superpáramo seem more or less similar to that of the opposite dry side of the Sierra Nevada del Cocuy.

Páramo Vegetation Belts, Upper Forest Limit, Altitudes, and Climate

The altitude of the páramo vegetation zones of the Colombian Cordillera Oriental was studied in detail in the Páramo Concavo (dry side) and the headwaters of the Río Casanare (humid side), both located in the Sierra Nevada del Cocuy (Fig. 7–2). The vegetational differences between the dry side and the humid side of the mountains are very prominent in the Cocuy region, which can serve as a model area for páramo zonation studies. The local páramo vegetation zones in the Sierra Nevada del Cocuy are apparently arranged asymmetrically (Fig. 7–2). On the atmospherically dry side, warm air currents ascending daily from the very deep and dry inter-Andean Río Chicamocha Valley cause a marked upward shift of the local zones of high-Andean vegetation. A lower bunch grass zone is inconspicuous or nearly absent here. The high-lying watershed in this area almost completely prevents the penetration of humid air from the Casanare slopes toward the dry mountain side.

The upper forest limit in the Colombian Cordillera Oriental in general seems determined mainly by thermal conditions. According to our observations, the *Quercus* cloud forest timberline on the atmospherically humid slopes generally goes up to about 3500 m (Páramo de Guantiva), whereas that of *Weinmannia rollottii* (Cunon.) generally reaches up only to about 3200–3300 m. However, a low forest zone of *Ageratina* (*Eupatorium*) *tinifolia* may be present above it, up to 3500–3600 m (van der Hammen et al., 1980–81).

On the dry side of the Sierra Nevada del Cocuy, *Weinmannia fagaroides* may be present up to about 3500 m. From there to the upper forest limit forest remnants in sheltered places consist mainly of Compositae (e.g., *Diplostephium* spp., *Gynoxys* spp.), Rosaceae (e.g., *Hesperomeles* cf. *lanuginosa*, *Polylepis quadrijuga*), and *Buddleia lindenii* (Logan.), whereas trees of *Weinmannia* are rare or absent. Patches of *Hesperomeles* forest, and of *Polylepis*, or *Gynoxys* forest, may occur up to 3800 m and 4100 m, respectively. Our observations about the timberline here agree largely with those of Cuatrecasas (1958) from other localities on the dry side of the Cocuy range. As regards extreme conditions of humidity, one may wonder whether the presence and extension of *Swallenochloa-Sphagnum* bogs in competition with the forest might have the effect of lowering the upper forest limit. Elsewhere along the northern-Andean region, the upper forest line may fluctuate between the extremes of 3000 m and 4000 m (and mostly between 3200 m and 3800 m). Very dry conditions may lead to páramo elements coming into contact with dry forest or xerophytic vegetation around 3000 m, as observed in some places on the slopes of the Chicamocha Valley. Very wet conditions may result in forest limits of 3900 m and even 4000 m, and occasionally in zones of contact between glaciers and low forest (van der Hammen and Gonzalez, 1960a). This may happen in the Cordillera Central, where *Sphagnum* bogs are scarce or absent.

The length of the total period of atmospheric humidity (which includes periods of both precipitation and fog), together with the total amount of annual precipitation, probably determines the extension of the bamboo cover in the lower páramos. In the upper superpáramo humidity conditions seem to be nearly the same on both sides of the Sierra Nevada del Cocuy.

The special floristic composition and increased density of the vegetation cover of the grass páramo-superpáramo border, especially on the humid side of the Cordillera, probably account for the presence of a marked (third) condensation zone at about the 4000–4200-m level. Gradstein et al. (1977) and Cleef (1978) reported earlier on the botanical richness of the 4000-m condensation zone on the Nevado de Sumapaz. This locality is the best example of this high-altitude condensation zone thus far observed in both the Colombian and Venezuelan Andes. In his climatological study of the Colombian Andes, Weischet (1969) recognized such condensation levels at lower elevations. Guhl (1974) reported the existence of as many as three cloud belts, the uppermost of which is situated in the high páramo. Lauer and Frankenberg (1978) defined the condensation zones from lower levels in the Mexican mountains as altitudinally limited areas of maximum precipitation and longest persistence of fog. They are apparently of great biological and ecological importance.

We conclude that the northern-Andean upper forest limit fluctuates between extremes of 3000 m and 4000 m, but usually mostly between 3200 m and 3800 m. The timberline seems to be determined primarily by thermal conditions. Only in extreme cases of very low and very high rainfall may the position of the treeline be influenced directly by the amount of precipitation (stunted tree growth because of a water and/or nutrient deficit, or the development of *Sphagnum* bogs, and ultimately an excessive development of epiphytes). The thermal conditions that determine the forest limit in all other cases are caused by (a) altitude, (b) ascending warm air currents (when much more low-lying valleys or plains are found downslope), and (c) frequent cloudiness and fog, leading to the suppression of extreme temperatures (such as night frosts).

In most cases three major páramo zones are differentiated (subpáramo, grass páramo, and superpáramo). In some cases the subpáramo zone may be poorly developed or absent, as in the central area of the Cordillera Central, where the high-Andean forest gives way directly to grass páramo at 3800–3900 m. Atmospherically humid páramos are rich in bamboos (*Swallenochloa*) and bogs. Atmospherically dry páramos are more frequently dominated by bunch grasses (*Calamagrostis* spp.). Subpáramo may be found from the forest limit up to 3400–3900 m, grass páramo between the subpáramo and 4200–4500 m, and superpáramo between the grass páramo and about 4800 m (nival belt).

PHYTOGEOGRAPHY OF THE PRESENT FLORA AND VEGETATION

Phytogeography of the Present Flora

Some 300 genera of vascular plants have been recorded up to now from the páramo region, between northern Peru and Venezuela and in

Costa Rica, indicating that it is one of the richest high mountain floras of the world (see Cleef, 1981, 1983, who discusses the composition of the vascular floras of the world's tropical high mountains using figures of Hedberg and others). Cleef's (1979a) list of genera from the Colombian Cordillera Oriental, a focal area of the páramo flora, includes 20 genera (9 families) of Pteridophyta, 62 genera (12 families) of Monocotyledoneae, and 177 genera (52 families) of Dicotyledoneae, in other words a total of 259 genera of vascular plants. A list of the species is not yet available.

It is not yet possible to fully analyze the "genetic" floral elements; hence only "geographic" floral elements have been established. We use here the data of Cleef (1979a), but attempt also to give some indications concerning genetic geographic relations, whenever practicable. In addition, we will discuss certain phytogeographic relationships between the páramo flora and floral areas elsewhere in South America. We will attempt to establish some characteristic distribution patterns and will survey briefly the phytogeographic composition and relationships of the vegetation types. For the phytogeographic records we have used the collections of several herbaria (Bogotá, Utrecht, Washington D.C.), our own collections and field observations, and the relevant literature (Mathias and Constance, 1951, 1962, 1967, 1976; Cuatrecasas, 1954b, 1969, 1979, 1980; Camargo, 1966; Uribe, 1972; Wurdack, 1973; Smith and Downs, 1974; Smith, 1976; Robson, 1977).

Cleef (1979a) distinguished and defined seven geographic floral elements:

1. Páramo element	19 genera	(7.3%)
2. Other Neotropical elements	88 genera	(33.8%)
3. Austral-antarctic element	24 genera	(9.2%)
4. Holarctic element	28 genera	(11.0%)
5. Wide temperate element	51 genera	(19.6%)
6. Wide tropical element	27 genera	(10.4%)
7. Cosmopolitan element	20 genera	(7.7%)
8. Unknown affinity element (probably of wide temperate origin)	2 genera	(1.0%)

We can say that almost 47.5% of the páramo genera are apparently of temperate origin (3 + 4 + 5 + 7), 44.2% are of (Neo) tropical origin (2 + 6), and 7.3% are endemic (1). Approximately half of the páramo genera are thus of (Neo) tropical origin and half are of temperate origin. The contributions of the Holarctic and of the austral-antarctic taxa to the latter half are approximately equal. Table 7–1 presents the genera of the páramos of the Cordillera Oriental. They are given in alphabetical order by family and the "phytogeographic element" or distribution type is indicated for each genus, the numbers corresponding to those of the preceding list.

TABLE 7–1. Checklist of genera of vascular plants of the páramo flora, Cordillera Oriental, Colombia.

Family	Distribution type (floral element)	Genus
Asclep.	2	*Ditassa*
Amaryllid.	2	*Bomarea*
Aquifol.	6	*Ilex*
Araliac.	2	*Oreopanax*
Azollac.	7	*Azolla*
Berberid.	4	*Berberis*
Boraginac	4	*Hackelia*
	4	*Lappula*
	2	*Moritzia*
Bromeliac.	2	*Greigia*
	2	*Puya*
Callitr.	5	*Callitriche*
Campanul.	7	*Lobelia*
	2	*Centropogon*
	2	*Lysipomia*
	2	*Rhizocephalum*
	2	*Siphocampylus*
Caryoph.	5	*Arenaria*
	4	*Cerastium*
	3	*Colobanthus*
	2	*Drymaria*
	5	*Sagina*
	5	*Stellaria*
Clethrac.	6	*Clethra*
Comp./Anthem	3	*Cotula*
Comp./Astereae	2	*Baccharis*
	1	*Laestadia*
	1	*Blakiella*
	6	*Conyza*
	2	*Diplostephium*
	4	*Erigeron*
	1	*Floscaldasia*
	2	*Noticastrum*
	2	*Oritrophium*
	2	*Plagiocheilus*
Comp./Cichor	4	*Hypochoeris*
	5	*Hieracum*
Comp./Eupat.	2	*Ageratina*
	2	*Chromolaena*
	2	*Lourteigia*
	2	*Oxylobus*
	2	*Stevia*
Comp./Inul.	6	*Achyrocline*
	5	*Gnaphalium*
	2	*Loricaria*
	2	*Lucilia*
Comp./Helianth.	1	*Aphanactus*
	7	*Bidens*
	2	*Calea*
	6	*Cosmos*
	1	*Espeletia*
	1	*Espeletiopsis*
	2	*Heliopsis*
	2	*Jaegeria*

TABLE 7–1 CONTINUED.

Family	Distribution type (floral element)	Genus
	1	*Libanothamnus*
	6	*Spilanthes*
	2	*Sabazia*
	6	*Siegesbeckia*
	1	*Tamania*
	2	*Vazquezia*
	2	*Verbesina*
Comp./Mutis	2	*Chaptalia*
Comp./Senecion.	2	*Gynoxys*
	5	*Senecio s.1.* *(incl. Pentacalia)*
	2	*Werneria*
Coriariac.		*Coriaria*
Crassul.	2	*Echeveria*
	5	*Tillaea*
Cruciferae	5	*Cardamine*
	2	*Halimolobus*
	4	*Draba*
Cyperaceae	6	*Bulbostylis*
	5	*Carex*
	6	*Cyperus*
	7	*Eleocharis*
	3	*Oreobolus*
	2	*Phylloscirpus*
	7	*Rhynchospora*
	7	*Scirpus*
	3	*Uncinia*
	1	*Vesicarex*
Elatin.	5	*Elatine*
Equis.	5	*Equisetum*
Ericac.	2	*Befaria*
	2	*Cavendishia*
	2	*Disterigma*
	3	*Gaultheria*
	4	*Gaylussacia*
	2	*Macleanea*
	3	*Pernettya*
	1	*Plutarchia*
	2	*Psammisia*
	4	*Vaccinium*
Eriocaulac.	5	*Eriocaulon*
	6	*Paepalanthus*
Euphorb.	3	*Dysopsis*
	7	*Euphorbia*
Gentiana	5	*Gentiana*
	5	*Gentianella*
	2	*Halenia*
	2	*Macrocarpaea*
Geraniac.	5	*Geranium*
Gramin.	2	*Aciachne*
	5	*Agrostis*
	2	*Aulonemia*
	2	*Axonopus*
	5	*Brachypodium*
	5	*Bromus*
	5	*Calamagrostis*
	4	*Cinna*
	3	*Cortaderia*
	5	*Danthonia*
	2	*Chusquea*
	5	*Brachypodium*
	2	*Aphanelytrum*
	5	*Alopecurus*
	5	*Festuca*
	5	*Hierochloe*
	2	*Lorenzochloa*
	1	*Neurolepis*
	6	*Paspalum*
	5	*Poa*
	6	*Sporobolus*
	1	*Swallenochloa*
	5	*Trisetum*
	4	*Muehlenbergia*
Halorag.	7	*Myriophyllum*
Hymenoph.	6	*Hymenophyllum*
Guttif.	5	*Hypericum*
Iridac.	3	*Orthrosanthus*
	3	*Sisyrinchium*
Isoetac.	5	*Isoëtes*
Juncac.	2	*Distichia*
	5	*Juncus*
	5	*Luzula*
Labiatae	4	*Salvia*
	4	*Satureja*
	4	*Stachys*
Lentibul.	5	*Pinguicula*
	7	*Ultricularia*
Liliac.		*Anthericum*
	1	*Eccremis*
	4	*Tofieldia*
Loganiac.	6	*Buddleia*
	3	*Desfontainea*
Loranth.	2	*Dendrophtera*
	2	*Gaiadendron*
	2	*Phoradendron*
	2	*Tristerix*
Lycopod.	7	*Lycopodium*
Malvac.	2	*Acaulimalva*
Marsil.	5	*Pilularia*
Melastom.	2	*Brachyotum*
	1	*Bucquetia*
	1	*Castratella*
	2	*Chaetolepis*
	2	*Miconia*
	2	*Monochaetum*
	1	*Purpurella* *(=Tibouchina p.p.)*
Myricac.	5	*Myrica*
Myrsin.	2	*Geissanthus*
	2	*Grammadenia*
	6	*Rapanea*
Myrtac.	6	*Eugenia*
	3	*Myrteola*
	3	*Ugni*
Onagrac.	5	*Epilobium*
	3	*Fuchsia*

Family	Distribution type (floral element)	Genus
	4	*Oenothera*
Ophiogloss.	7	*Ophioglossum*
Orchidac.	2	*Altensteinia*
	2	*Elleanthus*
	2	*Epidendrum*
	2	*Gomphichis*
	2	*Masdevallia*
	2	*Maxillaria*
	2	*Myrosmodes*
	2	*Odontoglossum*
	2	*Oncidium*
	2	*Pachyphyllum*
	2	*Pterichis*
	2	*Scaphosepalum*
	4	*Spiranthes*
Oxalid.	7	*Oxalis*
Papilion.	4	*Lathyrus*
	2	*Cologania*
	4	*Lupinus*
	4	*Trifolium*
	4	*Vicia*
Phytolacc.	6	*Phytolacca*
Piperac	6	*Peperomia*
Plantag.	5	*Plantago*
Polygal.	2	*Monnina*
	7	*Polygala*
Polygon.	3	*Muehlenbeckia*
	5	*Rumex*
Polypod. s.l.	7	*Asplenium*
	7	*Blechnum*
	6	*Cheilanthes*
	6	*Ctenopteris* (=*Grammitis*)
	5	*Cystopteris*
	6	*Elaphoglossum*
	2	*Eriosorus*
	2	*Jamesonia*
	1	*Nephopteris*
	7	*Polypodium*
	5	*Polystichum*
	6	*Xiphopteris*
Portul.	3	*Calandrinia*
	5	*Montia*
Potamog.	5	*Potamogeton*
Ranuncul.	5	*Ranunculus*
	4	*Thalictrum*
Rosaceae	3	*Acaena*
	2	*Hesperomeles*
	2	*Lachemilla*
	2	*Polylepis*
	4	*Potentilla*
	5	*Rubus*
	2	*Sericotheca*
Rubiac.	2	*Arcytophyllum*
	5	*Galium*
	3	*Nertera*
	6	*Psychotria*
	2	*Relbunium*
Saxifrag.	3	*Escallonia*
	4	*Ribes*
Scheuchz.	3	*Lilaea*
Scrophul.	2	*Alonsoa*
	1	*Aragoa*
	4	*Bartsia*
	3	*Calceolaria*
	4	*Castilleja*
	5	*Gratiola*
	5	*Limosella*
	5	*Mimulus*
	3	*Ourisia*
	4	*Sibthorpia*
	5	*Veronica*
Selagin.	7	*Selaginella*
Solanac.	2	*Acnistus*
	2	*Cestrum*
	2	*Nierembergia*
	2	*Sessea*
	7	*Solanum*
Symploc.	6	*Symplocos*
Theac.	6	*Ternstroemia*
Umbellif.	2	*Arracacia*
	3	*Azorella*
	7	*Eryngium*
	7	*Hydrocotyle*
	3	*Lilaeopsis*
	1	*Myrrhidendron*
	2	*Niphogeton*
	3	*Oreomyrrhis*
	5	*Daucus*
Urticac.	5	*Urtica*
Valerian.	2	*Phyllactis*
	5	*Valeriana*
Verbenac.	6	*Lippia*
	4	*Vervena*
Violac.	5	*Viola*
Xyridac.	6	*Xyris*

Table 7–2 lists páramo genera occurring outside the Cordillera Oriental.

This list does not include either Costa Rica (with genera like *Westoniella* [Comp.] and the Holarctic *Mahonia* [Berberid.] and *Arctostaphylos* [Ericac.]) or the jalca from Peru south of the Huancabamba depression, with such genera as *Laccopetalum* (Ranuncul.), *Oreithales* (Ranuncul.), *Scleranthus* (Caryoph.), *Ascidiogyne* (Comp.), *Liabum* (Comp.), *Helianthus* (Comp.), and *Anotis* (Rubiac.). The Compositae (41 genera) and the Gramineae (24 genera) are the two families contributing the greatest number of genera (according to the list of the Cordillera Oriental). The following families have between 13 and 7 genera (in descending order): Orchi-

TABLE 7–2. Páramo genera occurring outside the Colombian Cordillera Oriental.

Colombia	Family	Ecuador	Family
Chuquiraga	Comp.	*Aretiastrum*	Valerian.
Gunnera	Halorrhag.	*Astragalus*	Papilion.
Hinterhubera	Comp.	*Bowlesia*	Umbellif.
Neonelsonia	Umbellif.	*Caltha*	Ranuncul.
Ottoa	Umbellif.	*Chrysactinium*	Comp.
Pedicularis	Scrophul.	*Columellia*	Columell.
Perissocoelum	Umbellif.	*Cotopaxia*	Umbellif.
Raouliopsis	Comp.	*Ephedra*	Gymnosp.
Spergularia	Caryoph.	*Eudema*	Crucifer.
		Helichrysum	Comp.
Venezuela		*Hypsela*	Campanul.
Carramboa	Comp.	*Luciliopsis*	Comp.
Chevreulia	Comp.	*Nototriche*	Malvac.
Coespeletia	Comp.	*Perezia*	Comp.
Helleria	Gramin.	*Plagiobothrys*	Boraginac.
Hymenostephium	Comp.	*Rostkovia*	Juncac.
Lagenophora	Comp.	*Rhopalopodium*	Ranuncul.
Ruilopezia	Comp.	*Saxifraga*	Saxifr.
		Stipa	Gramin.
		Tauschia	Umbellif.
		Xylopleurum	Umbellif.

daceae, Polypodiaceae, Scrophulariaceae, Cyperaceae, Ericaceae, Umbelliferae, Melastomataceae, and Rosaceae. Twenty-nine families have 2–5 genera, and 33 families only a single genus.

At present about 35 vascular genera are considered endemic to the páramos of Costa Rica, Venezuela, Colombia, and Ecuador. Few of them (indicated here with an asterisk) extend into the Guatemalan and Mexican high mountains or to Santo Domingo; some also occur in the upper Andean forest. (Those indicated with a dagger are mono- to oligotypic.)

- Compositae (18): *Aphanactus*,* *Blakiella*,† *Carramboa*, *Chrysactinium*, *Coespeletia*, *Espeletia*, *Espeletiopsis*, *Floscaldasia*,† *Hinterhubera*, *Hymenostephium*, *Laestadia*,* *Lasiocephalus*, *Libanothamnus*, *Lourteigia*, *Raouliopsis*,† *Ruilopezia*, *Tamania*,† *Westoniella*
- Cyperaceae (1): *Vesicarex*†
- Ericaceae (2): *Plutarchia*, *Themistoclesia*
- Gramineae (3): *Swallenochloa*, *Neurolepis*, *Helleria* † (status uncertain)
- Melastomataceae (3): *Bucquetia*, *Castratella*,† *Purpurella*† (= *Tibouchina* p.p.)
- Polypodiaceae (1): *Nephopteris*†
- Ranunculaceae (2): *Rhopalopodium*,† *Laccopetalum*†
- Scrophularaiceae (1): *Aragoa*
- Umbelliferae (4): *Myrrhidendron*, *Ottoa*,* *Cotopaxia*,† *Perissicoelum*

These 35 endemic genera out of a total of about 300 constitute about 11% of the flora. In the Cordillera Oriental about 7% of the genera were found to be endemic, but since the lists of genera of areas outside the Cordillera Oriental (and of the Ecuadorian páramos in particular) may be incomplete, generic endemism may be slightly higher.

Many endemic genera occur in the subpáramo and some of them, such as *Libanothamnus*, *Tamania*, *Bucquetia*, *Purpurella*, *Plutarchia*, *Aragoa*, *Neurolepis*, *Myrrhidendron*, and *Swallenochloa*, also may be found in the uppermost Andean forest.

Superpáramo "islands" are much scarcer and smaller than páramo "islands," and their phytogeographic analysis is of interest. Three main superpáramo islands occur in the Cordillera Oriental, a huge one in the Sierra Nevada del Cocuy (fully developed and extending to the nival zone), and two smaller ones (mainly comprising lower superpáramo) in the Páramo de Sumapaz (south of the Cocuy) and in the Páramo del Almorzadero (north of the Cocuy), respectively.

The large Cocuy superpáramo island has about 20 endemic species of higher plants, the Sumapaz 3, and the Almorzadero 2. Endemism is particularly high in the genera *Draba* and *Senecio* s.l.. One hundred and seven species occur in the superpáramo of the Cocuy; endemism at the species level is thus approximately 19%. However, there seems to be only one endemic superpáramo genus, *Floscaldasia* (Comp.), which occurs in the lower superpáramo and uppermost grass páramo.

Seventy-five species of vascular plants were found in the Sumapaz superpáramo island between 4000 m and 4300 m. The following genera are included (those restricted to the superpáramo are indicated with an asterisk, those occurring above 4200 m with a dagger, and those occurring in the superpáramo of the Cocuy with a double dagger).

Lysipomia‡	*Hypericum*‡
Rhizocephalum	*Isoetes**‡
Arenaria†‡	*Luzula*†‡
*Cerastium**†‡	*Satureja*
Baccharis	*Lycopodium**‡
*Diplostephium**‡	*Epilobium*
Espeletia‡	*Ophioglossum*‡
*Erigeron**†‡	*Altensteinia*†‡
*Hypochoeris**†‡	*Lupinus*‡
Laestadia	*Plantago*
*Loricaria**‡	*Muehlenbeckia*‡
Lucilia†‡	*Ctenopteris* (= *Grammitis* p.p.)‡
Gnaphalium‡	*Cystopteris*†‡
*Senecio**†‡	*Jamesonia*†‡
Werneria†‡	*Polystichum*†‡
Carex‡	*Calandrinia*‡
*Draba**†‡	*Montia**†‡
Cardamine	*Lachemilla**†‡
Disterigma	*Arcytophyllum*
Pernettya‡	*Bartsia*†‡
Vaccinium‡	*Castilleja*‡
Gentiana†‡	*Ourisia**†‡
Geranium‡	*Veronica*†
Aciachne†	*Azorella*†
Agrostis†‡	*Niphogeton*‡
*Calamagrostis**†‡	*Oreomyrrhis*†‡
Poa‡	*Valeriana*†‡
Swallenochloa	

The following genera not represented in the superpáramo of the Sumapaz are found in the superpáramo of the Cocuy, in addition to those indicated by a double dagger in the preceding list (those indicated here with a double dagger have not yet been recorded from the Sumapaz, but probably occur there).

Floscaldasia	*Halenia*
Oritrophium	*Bromus*
Aphanactus	*Cortaderia*
Oreobolus	*Hymenophyllum*‡
Phylloscirpus	*Distichia*
Elatine	*Acaulimalva*‡
Paepalanthus	*Asplenium*‡

About two-thirds of the 75 recorded species of the superpáramo of the Sumapaz area also occur in the superpáramo of the Cocuy. Fifty-five genera have been recorded from the Cocuy superpáramo, and an equal number from the Sumapaz above 4000 m. Of the 55 genera recorded in the Sumapaz superpáramo, 26 also occur above 4200 m and 14 represent species restricted to the superpáramo. A phytogeographic analysis of these 55, 55, and 26 genera, respectively, yields the results shown in Table 7–3 (as compared with the total number in the Cordillera Oriental).

Table 7–3 shows that the temperate elements contribute a higher proportion of the genera in the superpáramo, judging by the percentage of Holarctic and especially wide temperate elements, than in the páramo as a whole. Surprisingly, there is also a decrease of the páramo element. Although the local flora is incompletely known, the superpáramos of the Cordillera Central, in a volcanic area, are probably relatively poor in species. The same seems to hold for the grass páramo in the Cordillera Central.

Griffin (1979) made a tentative phytogeo-

TABLE 7–3. Comparison of floral elements (percent) in three superpáramo 'islands' and in the páramo of the Cordillera Oriental (Colombia).

	Páramo	Superpáramo 'islands'		
	Cordillera Oriental	Cocuy (4200 m)	Sumapaz (4000 m)	Sumapaz (4200 m)
1. Páramo element	7.3%	5.5%	5.5%	–
2. Other Neotropical elements	33.8%	23.5%	25.5%	23.0%
3. Austral-antarctic element	9.2%	13.0%	11.0%	12.0%
4. Holarctic element	11.0%	14.5%	16.5%	20.0%
5. Wide temperate element	19.6%	31.0%	36.5%	46.0%
6. Wide tropical element	10.4%	9.0%	2.0%	–
7. Cosmopolitan element	7.7%	3.5%	3.5%	–
Total Neotropical element	51.5%	38.0%	33.0%	23.0%
Total temperate (incl. cosmopolitan) element	48.5%	62.0%	67.0%	77.0%
Total number of genera	259	55	55	26

graphic analysis of the bryophyte flora of the páramos and calculated that 30% of the species are cosmopolitan, 19% range from Bolivia to Mexico, 50% are Andean-austral (South American), while 1% is boreal and 1% is African. There are about 130 genera of Musci in the páramo (Griffin, in prep.). Gradstein et al. (1977) enumerated 62 genera of liverworts. A number of genera are of Holarctic origin (*Blepharostoma*, *Gymnomitrion*, *Lophozia*, *Telaranea*), and many are of austral-antarctic origin (*Clasmatocolea*, *Colura*, *Cryptochila*, *Isotachis*, *Jensenia*, *Lepicola*, *Pseudocephalozia*, *Triandrophyllum*).

There are about 50 genera of macrolichens in the páramos (the crustose genera not included) (H. J. M. Sipman, pers. comm.). Several genera of Stictaceae and Parmeliaceae are (Neo) tropical in distribution, others are of Holarctic (*Peltigera*, *Cetraria*, *Alectoria*) or austral-antarctic origin (*Cladia*, *Neuropogon*).

An analysis of the páramo flora has not yet been made at the species level. However, a preliminary inventory shows that there are some 700 species in the Cordillera Oriental, of which about 250 (over 35%) are endemic. As stated earlier, in the much smaller but more isolated superpáramo of the Cocuy, 19% of species are endemic. In what follows we summarize endemism in a number of páramo genera.

In the Espeletiinae, *Espeletia* has 55 species in the páramos of Colombia, Venezuela, and Ecuador. Of these 13 are found in Venezuela. In Colombia, 36 occur in the Cordillera Oriental, many of them restricted to small areas, but there are only 2 species in the Cordillera Central, 4 in the Troncal Sur and 3 in the Cordillera Occidental. *Espeletiopsis* has 24 species (8 in Venezuela, 17 in the Cordillera Oriental of Colombia). *Ruilopezia* (21 species) and *Libanothamnus* (13) are found in Venezuela only. *Libanothamnus* has arborescent species, one of which (*L. neriifolius*) occurs as low as 1600 m. The Espeletiinae may have originated in the Venezuelan Andean forest zone, where the more primitive genera occur, whereas the more derived genera have their present highest species density in the Cordillera Oriental (*Espeletia*, *Espeletiopsis*) (see Chap. 11).

Diplostephium also has many endemic species. There may be more than 80 species. At least 56 of them occur in Colombia, 31 of which (including 14 endemics) occur in the Cordillera Oriental. The more primitive arborescent species are found in the upper Andean forest and in the subpáramo; the advanced members are more numerous in the high páramos.

Aphanactus has 3 endemic species in the Cordillera Oriental and *Ageratina* (= *Eupatorium* p.p.) several. *Baccharis* has 37 species in Colombia, of which 20 are endemic in the Cordillera Oriental; this genus has representatives as low as 1800 m. Most *Gynoxys* species in the Cordillera Oriental (about 8) are apparently endemic. *Senecio* s.l. has many endemic herbaceous and woody species and generally exhibits a considerable amount of speciation. *Verbesina*, a forest-belt genus, has one subfruticose species in the páramo. *Oritrophium*'s centre of speciation seems to be in Mérida; there are 4 species, 2 of which are endemic, in the Cordillera Oriental.

Aragoa is a most interesting endemic genus, possibly related to the austral-antarctic genus *Hebe* (Scrophulariaceae-Veroniceae). It is a primitive and geographically isolated member of this group (Pennell, 1937). There are nine species, of which 5 are apparently centered in the Cordillera Oriental (*A. abietina*, *A. cupressina*, *A. dugandii*, *A. lycopodioides*, and *A. perez-arbelaeziana*). In the Cordillera de Mérida there are two species (*A. lycopodioides* and *A. lucidula*), in the Sierra Nevada de Santa Marta at least one (*A. kogiorum*), and in the northern Cordillera Occidental one (*A. occidentalis*); see Cleef, (1979a). It is noteworthy that some species are small trees and may occur in the upper mountain forest and in the subpáramo (e.g., *A. perez-arbelaeziana*). *Aragoa* is conspicuously absent from the Cordillera Central.

Berberis has several endemic species in the Cordilleras Oriental, Central and Occidental, in the Cordillera de Mérida, and in the Sierra Nevada de Santa Marta. The northern Andes are apparently a complex, secondary center of speciation for this genus.

Bucquetia probably originated in the Cordillera Oriental, where it has two species; the third species occurs in southern Ecuador.

Draba has a secondary center of speciation in the Cordillera Oriental, and another in the puna.

Gaultheria and *Halenia* are represented by a fair number of endemic species in the Cordillera Oriental, but they are probably not so abundant elsewhere in our area.

Hypericum (sect. *Brathys*) has an important center of speciation in the Cordillera Oriental, most species being endemic (there are about 30 species, at least 20 of which are endemic).

Lachemilla has 19 species in the Cordillera Oriental, 4 of which are endemic (E. G. B. Kieft, pers. comm.)

Lupinus has many endemic species in the páramo, and the Cordillera Oriental is a secondary center.

Miconia sect. *Cremanium* has species in the upper forest and the páramo, especially on the wet side of the mountains, near the boundary between forest and subpáramo. There are 14 species, 7 endemic to the Cordillera Oriental, and 7 elsewhere. *M. chionophila* is found in most of the páramo area from Chachapoyas (Peru) to Tamá (Venezuela).

Niphogeton is a páramo–puna genus with about 18 species, most of which seem to be restricted to the páramo, especially in the Cordillera Oriental. There are 12 species in Colombia, of which 7 are endemic. The genus also developed woody species.

Paepalanthus has approximately as many nonendemic as endemic species in the Cordillera Oriental.

Puya has 18 species in the Cordillera Oriental (secondary center of speciation), two in the Cordillera Central north of the Macizo Colombiano, and 4 in the Macizo.

Valeriana has 6 species in the páramos of the Cordillera Oriental, including herbs, vines, shrubs, and a dwarf tree. Two woody species (*V. triphylla* and *V. arborea*) and one herbaceous species (*V. vegasana*) are endemic. The small cushion plant, *V. stenophylla*, is present in the Cordillera Oriental and in Mérida.

This long list of examples gives an idea of the endemism of vascular plants in the páramo. The Cordillera de Mérida seems to be the cradle of the more primitive Espeletiinae, and the Cordillera Oriental the major center of speciation in *Espeletia* and *Espeletiopsis* (see Chap. 11). A good number of endemic páramo species (or genera) seem to have more primitive arborescent relatives in the upper forest belt and the lower subpáramo.

The Cordillera Oriental seems to be the major area of speciation of a number of genera, perhaps because (1) it contains many isolated páramo islands; (2) it is more stable (versus volcanic and unstable); (3) it is extensive and therefore contains more habitats, created by differences in moisture, elevation, and soil types; and (4) it is possibly somewhat older.

Are there floristic affinities between the páramo and other plant formations in South America? The páramo shares about 12 genera with the open savanna, 15 with the savanna scrub, and 3 with the mountain savanna formations (see the list of savanna species in van Donselaar, 1965). The most important genera shared with savannas are *Bulbostylis*, *Cyperus*, *Eleocharis*, *Rhynchospora*, *Eriocaulon*, *Paepalanthus*, *Axonopus*, *Sporobolus*, *Sisyrinchium*, *Utricularia*, *Xyris*, *Paspalum*, *Tibouchina*, *Polygala*, and *Borreria*. There are even some species in common, for example, *Rhynchospora rugosa* and *Paepalanthus polytrichoides*.

The occurrence of "subpáramo" vegetation on El Avila near Caracas above 2200 m (Steyermark and Huber, 1978) is noteworthy. This subpáramo, at a very low elevation and on relatively low summits (but probably somewhat extended in area owing to anthropogenic influences), has many taxa in common with the subpáramo of higher elevations, but also with the Andean forest. Two genera of the páramo element are *Libanothamnus* and *Excremis*. Some other taxa shared are *Orthrosanthus chimboracensis*, *Arcytophyllum caracasanum*, *Siphocampylus*, *Sporobolus*, *Rapanea dependens*, *Muehlenbeckia thamnifolia*, *Acaena cylindristachya*, and *Castilleja fissifolia*. The mountain flora above Caracas also exhibits a clear relationship to that of the savannas, increasing with decreasing altitude. Vegetation with subpáramo affinities at much lower than normal levels, known as "paramillos", occurs in the Cordillera Oriental of Colombia, especially on mountain tops.

Steyermark (1966, 1979) analyzed the flora of the Guayana highlands, often considered relictual. Most of the taxa represented, however, seem to be derived from those of tropical lowland forests and savannas (see Chap. 13). Of the 459 vascular genera of the summit flora, only 5% are of tropical American-African origin, and 4% are of tropical American-Australasian affinity. The remainder are Neotropical or pantropical. About 11% (50 genera) have a predominantly Andean distribution. Approximately 17% of the genera and more than 90% of the species are endemic. As regards a possible relation with the páramo flora, the shared occurrence of *Oreobolus obtusangulus*, *Swallenochloa weberbaueri*, *Sphagnum oxyphyllum*, *Ugni myricoides*, and other taxa is interesting. The first species occurs on the Cerro Neblina (3014 m), for example, 965 km east of the Nevado de Sumapaz. However, as the main distribution of these taxa is Andean, they are probably of Andean origin. Although there are some floristic relationships between the páramo and the Guayana highlands (see Chap. 13), it seems as if the latter contributed very little if anything to the páramo flora. This is probably attributable to the fact that when the Andes were uplifted, the Guayana area was already quite isolated.

Relationships are much clearer with the puna of Peru, Bolivia, and Argentina. The puna, separated from the Ecuadorian páramo by a lower-lying area, the Huancabamba Depression, is climatologically different from the northern páramo (Chap. 2). A comparison of the páramo flora (Table 7–1) with the check-list of the puna flora of Argentina (Cabrera, 1958) shows that about 30 per cent of the puna genera also occur in the páramo. Of a total of 259 páramo genera (see Table 7–1), about 20 (approximately 7.5 per cent) are found in both puna and páramo. About 80 puna genera are shared with the Cordillera Oriental páramo (and even more with the entire páramo area). Roughly 50 per cent of the genera of the puna flora are of (Neo)tropical and 50 per cent of temperate origin. These figures are similar to those mentioned earlier for the páramo flora. At the species level, finally, 25 taxa are shared between the puna and the páramo.

Although the present record is not complete, the following genera are endemic to both the páramo and the puna: *Lysipomia*, *Rhizocephalum*, *Oritrophium*, *Loricaria*, *Gynoxys*, *Werneria*, *Phylloscirpus*, *Aciachne*, *Lorenzochloa*, *Distichia*, *Acaulimalva*, *Altensteinia*, *Niphogeton*, *Phyllactis*, *Chuquiraga*, *Lucilia*, *Hypsela*, and *Bowlesia*. Some of the genera shared by puna and páramo have reached the puna by means of the páramo (northern origin), but many invaded the páramo from the puna (southern origin). The center of distribution of the genera *Puya*, *Distichia*, *Altensteinia*, *Acaulimalva*, *Loricaria*, and *Werneria* seem to lie in the puna, whereas for *Lachemilla*, *Oritrophium*, *Niphogeton*, and *Baccharis*, it is in the páramo.

Assigning a taxon to the Neotropical element does not always mean that it evolved in situ in the northern Andes. Immigration, presumably from the north, must have accounted for the introduction of genera like *Arcytophyllum*, *Hesperomeles*, *Echeveria*, and *Oxylobus*, which have their closest relatives in the Mexican and Guatemalan mountains. A greater number of genera belonging to the Neotropical element are supposed to have reached the páramos from southern latitudes, e.g., *Acaulimalva*, *Chaptalia*, *Diplostephium*, *Distichia*, *Gaiadendron*, *Loricaria*, *Lysipomia*, *Moritzia*, *Noticastrum*, *Phyllactis*, *Puya*, *Rhizocephalum*, and *Werneria*.

Other, wide temperate taxa are also supposed to have immigrated from the south, for instance *Calamagrostis* sect. *Deyeuxia* (probably), *Geranium* sect. *Andicola*, *Isoëtes* sect. *Laevis*, *Limosella*, *Juncus* subg. *Septati*, and *Plantago*.

Wide temperate taxa migrating by the northern route include *Brachyopodium*, *Eriocaulon*, *Galium*, *Gentiana* (sect. *Chondrophylla*), and *Mimulus*. *Utricularia* (a cosmopolitan element) is represented by a weedy species (*U. obtusa*) apparently introduced by migratory birds from northern latitudes (A. Fernandez P., pers. comm.). *Hypericum* sect. *Brathys* has its nearest relatives in Cuba and Central America, but is also closely related with the most primitive tropical African section *Campylosporus* (Robson, 1977).

In summary, (1) relatively close relationships exist between páramo and puna, between páramo and lower altitude paramillos, and between páramo and upper Andean forest, (2) there seem to be some relationships between the tropical savanna and páramo floras, (3) the open and scrubby vegetation of the Guayana highlands (tepuis) between approximately 500 m and 3000 m elevation has strong relationships with Neotropical savannas, and (4) the floristic relationships with the páramo seems to be from the páramo areas to the Guayana highlands and not vice versa.

Phytogeography of the Present Vegetation Types

We will now survey briefly the phytogeographic relations of the dominant elements of the main páramo vegetation types (Cleef, 1981).

Most zonal open plant communities are dominated by genera belonging to the wide temperate (or cosmopolitan) element.

The bunch grass páramo is dominated by species of the wide temperate genus *Calamagrostis*. These tussock-forming species belong to the section *Deyeuxia*, which is common in tropical high mountains and in cooler parts of the Southern Hemisphere.

The Neotropical, especially tropical-Andean, element is most prominent in the zonal communities of the subpáramo, in particular on the humid side of the mountains, and is, furthermore, strongly represented by the *Loricarietum complanatae*. Bamboo vegetation made up of Chusqueae also belongs here.

In numerous azonal páramo communities (e.g., *Pentacalia* [= *Senecio*] shrub, mire/swamp vegetation, aquatic vegetation), the wide temperate (or cosmopolitan) element is also predominant at the generic level.

The Neotropical element is mainly displayed by the *Oritrophio-Wernerietalia*, which comprise flush vegetation and vascular cushion bogs. Though *Werneria* is truly tropical-Andean in its distribution, *Distichia muscoides* and *Plantago rigida* have related species in the subantarctic region and in the cooler parts of the Southern Hemisphere (Cleef, 1978). *Oreobolus obtusangulus*, a tropical-Andean to Fuegian species, is distributed from Tierra del Fuego to Cerro Neblina and the Colombian páramos. At the generic level the *Oritrophio peruviani-Oreoboletum obtusanguli* displays the austral-antarctic element (Cleef, 1978).

The southern element is also present in the "*Acaenetum cylindristachyae*", the *Philonotido-Isotachidetum serrulatae*, the *Escallonia myrtilloides* dwarf forest, and the *Azorelletum multifidae*. In addition, extensive stands of cushion vegetation of *Azorella aretioides* and mats of *A. pedunculata* are common in the higher reaches of the Ecuadorian páramos (Harling, 1979; Øllgaard and Balslev, 1979).

Other vegetation types constituted by Neotropical (i.e., tropical-Andean) elements are: (1) dwarf forests including *Gynoxys* spp., *Diplostephium*, *Polylepis*, and *Hesperomeles*, and (2) meadows of the *Lorenzochloetum erectifoliae*, *Aciachnetum pulvinati*, *Agrostio-Lachemilletum orbiculatae*, and the *Agrostis breviculmis-Lachemilla pinnata* community types. *Diplostephium* (Cuatrecasas, 1969) and *Polylepis* (Simpson, 1979) are closely related to other austral-antarctic genera.

The *Ditricho-Isoëtion* is made up of a number of associations, each of which is dominated by a different species of *Isoëtes* of the section *Laeves* H. P. Fuchs, which is exclusively tropical-Andean in distribution. This section has its main center of speciation in the páramos and is apparently derived from the amphipolar section *Isoëtes* (Fuchs, 1982). At the generic level *Isoëtes* is ranked as of a wide temperate distribution, although at least one species grows on tidal flats in Guyana. Species of other sections of *Isoëtes* are also found in the páramo, but they do not constitute proper communities. The Holarctic section *Echinatae* Pfeiffer is not represented (Fuchs, 1982).

Other aquatic communities with *Myriophyllum elatinoides* (e.g., the *Hydrocotylo-Myriophylletum elatinoides*) are austral-antarctic at the species level, since the characteristic species, *M. elatinoides*, extends toward Australia and New Zealand.

Running and standing bodies of water in the protopáramo were apparently colonized by species of temperate origin, even when speciation subsequently took place (*Isoëtes karstenii*, *I. cleefii*, *I. palmeri*). Other species are shared with puna lakes and other southern parts of the tropical high Andes (*I. glacialis*, *I. socia*).

The generic páramo element is best exemplified by the *Aragoetum abietinae*, dense *Espeletia* woodland and bamboo vegetation. Espeletiinae-dominated stands from the Colombian and especially the Venezuelan Andes (Vareschi, 1955; Monasterio, 1979) are also considered to belong to this element.

The Holarctic element in páramo communities is rare at the generic level and is represented only by the rare *Myricetum parvifoliae*. *Myrica* is listed as a wide temperate element (Cleef, 1979), but the South American species are derived from Holarctic immigrants.

Xyris bog and *Cyperus* marsh represent the pantropical element and they are present only in the lower part of the páramo belt.

At the specific level most of the 120 recognized páramo plant communities belong to the páramo element, as they are endemic to páramos, a lesser number to the tropical-Andean element (e.g., the *Hyperico-Plantaginetum rigidae* and *Isoëtetum sociae*), and a few (e.g., *Limosella australis* community, *Rhacocarpus-Racomitrium crispulum* vegetation) to the austral-antarctic or (*Dendrocryphaeo-Platyhypnidietum riparioides* and *Eleocharitetum acicularis*) the wide temperate or (*Lemno-Azolletum* [*filiculoides*]) the cosmopolitan elements.

Sphagnum bogs are found mainly in cool temperate regions. The species of *Sphagnum* belong mainly to the pantemperate element (*S. magellanicum*, *S. cuspidatum*); local species are *S. sancto-josephense* (páramo element, but derived from the wide temperate *S. recurvum*) and *S. oxyphyllum*, which is Neotropical and occurs also at lower elevations. Holarctic species are *S. compactum*, *S. cyclophyllum*, and the rare *S. pylaesii*. *S. cyclophyllum* prefers hollows in vascular plant cushion bogs, and *S. pylaesii* is found submerged in páramo lakes (*Ditricho-Isoëtion*).

In summary, the dominant genera (and species) in the vegetation types of the lower part of the páramo (especially the subpáramo) are often of Neotropical (local) origin, and the dominant genera of the vegetation types of the higher parts of the páramo are often of temperate origin. The dominant species of cushion bogs and several kinds of aquatic vegetation are mainly of southern temperate origin; the dominant moss species of *Sphagnum* bogs are mainly of temperate origin.

Some Distribution Patterns of Páramo Taxa

The following four principal distribution patterns of páramo genera and species can be distinguished north of the Huancabamba Depression (Fig. 7–3).

1. Cordillera Oriental (from Sumapaz to Cocuy and/or Almorzadero [Fig. 7–3a]).

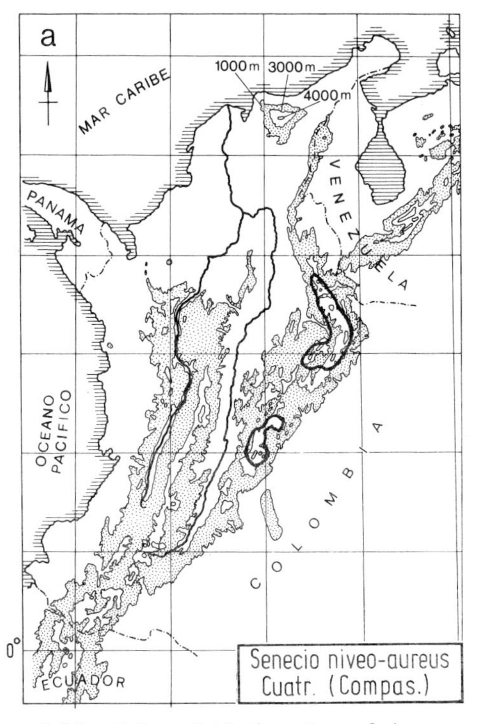

FIG. 7–3 (a–m). Some distribution patterns of páramo taxa.

Examples: *Diplostephium alveolatum* Cuatr., *Espeletia grandiflora* H. and B., *Paepalanthus lodiculoides* Moldenke, *Puya goudotiana* Mez, *P. trianae* Baker, *Pentacalia flos-fragrans* (Cuatr.), *Senecio niveo-aureus* Cuatr., and *Valeriana arborea* Killip and Cuatr.

Within this area the Cocuy (especially the superpáramo) has a number of endemic species, e.g., *Draba hammenii* Cuatr. and Cleef, *Draba litamo* L. Uribe, *Espeletia cleefii* Cuatr., *Oritrophium cocuyensis* Cuatr., *Pentacalia* (= *Senecio*) *cleefii* (Cuatr.) Cuatr., *P. cocuyanus* (Cuatr.) Cuatr., and *P. guicanensis* (Cuatr.) Cuatr.

The area of the Cordillera Oriental from Sumapaz to Almorzadero has 32 species of *Espeletia*, 11 of which occur in the Department of Cundinamarca, 17 in the Department of Boyacá, and 4 in the Sierra Nevada del Cocuy (Cuatrecasas, 1979). Many are endemic in part of the area, constituted by one or by several páramo islands.

2. Cordillera de Mérida (Fig. 7–3b). Examples: *Carramboa*, *Coespeletia*, *Helleria*, *Hymenostephium*, *Ruilopezia*, and *Hinterhubera imbricata*.

3. Cordillera Central and Ecuador. Examples: *Chuquiraga*, *Espeletia hartwegiana* Cuatr., *Gunnera magellanica* Lam., and *Ourisia chamaedrifolia* Benth. Ecuadorian elements such as *Astragalus*, *Bowlesia*, *Columellia*, *Cotopaxia*, *Ephedra*, and *Eudema* increase toward the south.

4. Sierra Nevada de Santa Marta. Examples: *Aragoa kogiorum* Romero, *Diplostephium anactinotum* Wedd., *Hinterhubera harrietae* Cuatr., and *Raouliopsis seifrizii* Cuatr.

Minor areas are:

5. Cordillera de Perijá. Examples: *Espeletia perijaensis* Cuatr. and *Espeletia tillettii* Cuatr.

6. Páramos of the Cordillera Occidental. Example: *Aragoa occidentalis* Pennell.

7. Páramo de Costa Rica. Example: *Westoniella*.

Apart from the distributional patterns restricted to each of the "natural" areas, there are of course species present in all or most of the páramo area, like *Calamagrostis effusa* (Fig. 7–3c). There are also some interesting "disjunct" patterns, which are listed below.

a. Sierra Nevada del Cocuy—Cordillera Central (Fig. 7–3d) Examples: *Floscaldasia hypsophila* Cuatr. and *Nephopteris maxonii* Lellinger.

b. Sierra Nevada de Santa Marta—Cordillera de Perijá (Fig. 7–3e). Examples: *Perissocoelum* Math. and Const., *Symplocos nivalis* Linden.

c. Northern Cordillera Oriental—Cordillera de Mérida (-Sierra Nevada de Santa Marta) (Fig. 7–3f). Examples: *Blakiella bartsiaefolia* (Blake) Cuatr. (not in Santa Marta), *Erigeron paramensis* Arist. and Cuatr., *Lachemilla polylepis* (Wedd.) Rothm., *Libanothamnus* (also Avila), and *Rhynchospora paramorum* Mora (also Costa Rica).

d. Southern Cordillera Oriental (Sumapaz)—

FIG. 7–3. *Continued*

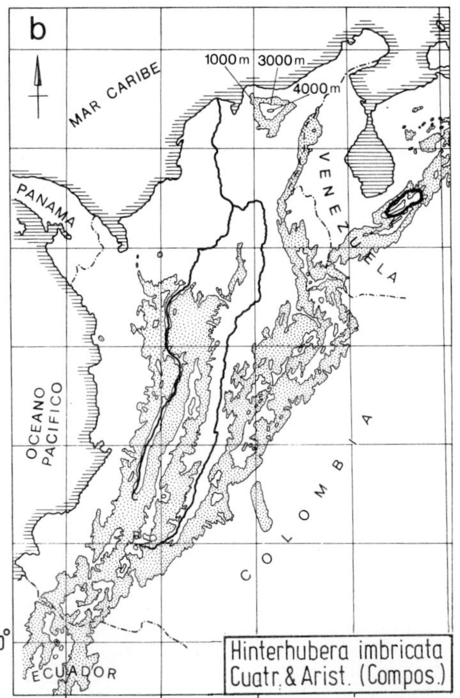

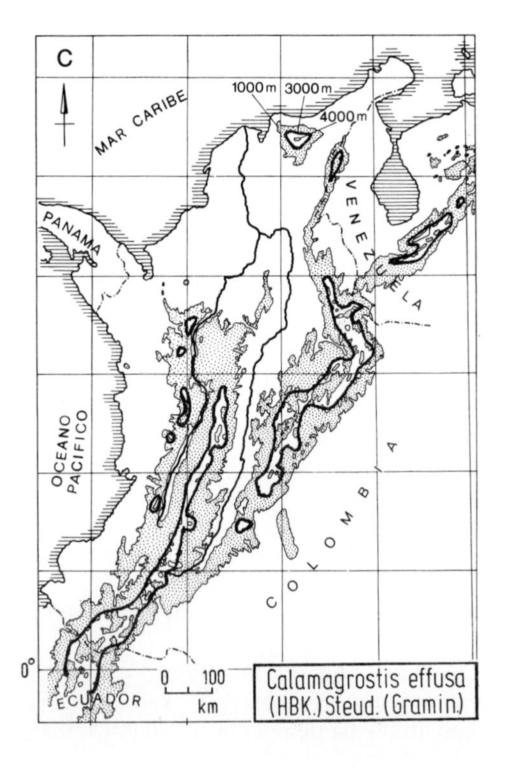

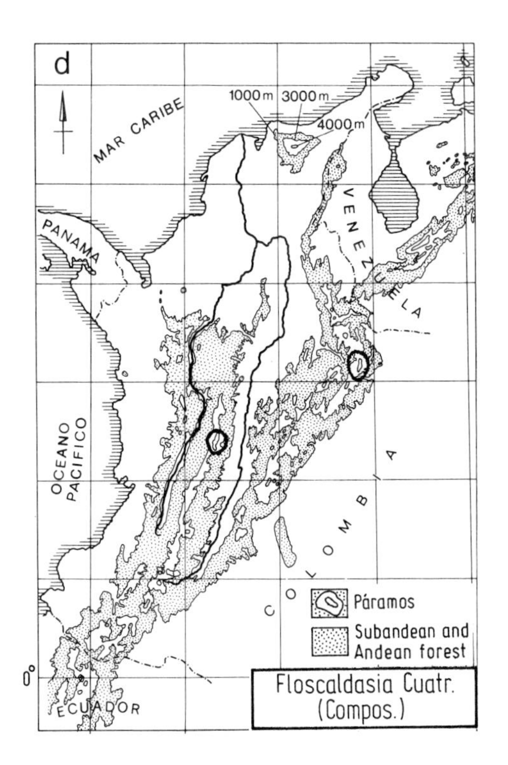
d
MAR CARIBE
1000 m
3000 m
4000 m
PANAMA
VENEZUELA
OCEANO PACIFICO
COLOMBIA
Páramos
Subandean and Andean forest
0°
ECUADOR
Floscaldasia Cuatr. (Compos.)

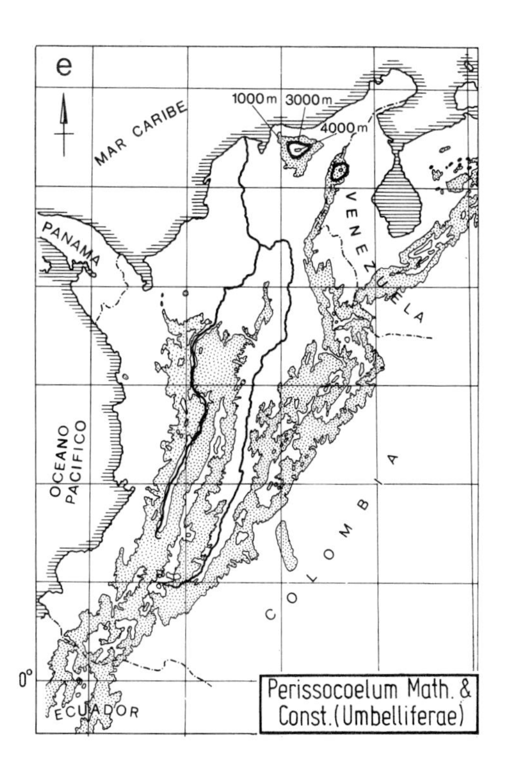
e
MAR CARIBE
1000 m
3000 m
4000 m
PANAMA
VENEZUELA
OCEANO PACIFICO
COLOMBIA
0°
ECUADOR
Perissocoelum Math. & Const. (Umbelliferae)

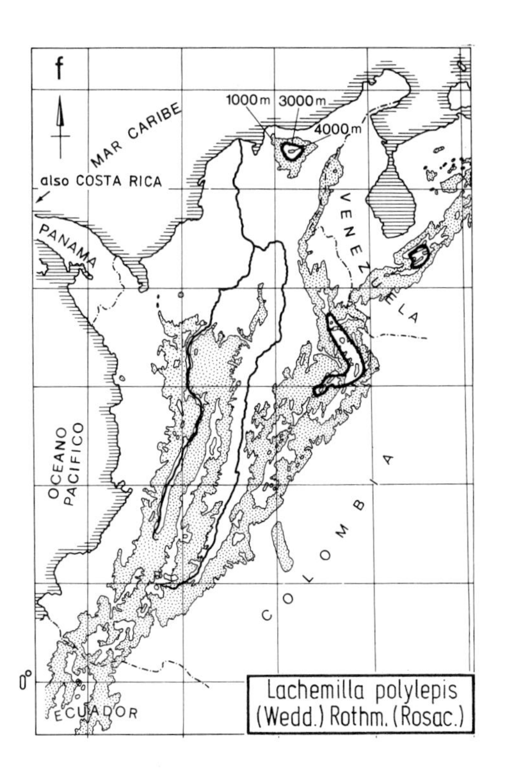
f
MAR CARIBE
1000 m
3000 m
4000 m
also COSTA RICA
PANAMA
VENEZUELA
OCEANO PACIFICO
COLOMBIA
0°
ECUADOR
Lachemilla polylepis (Wedd.) Rothm. (Rosac.)

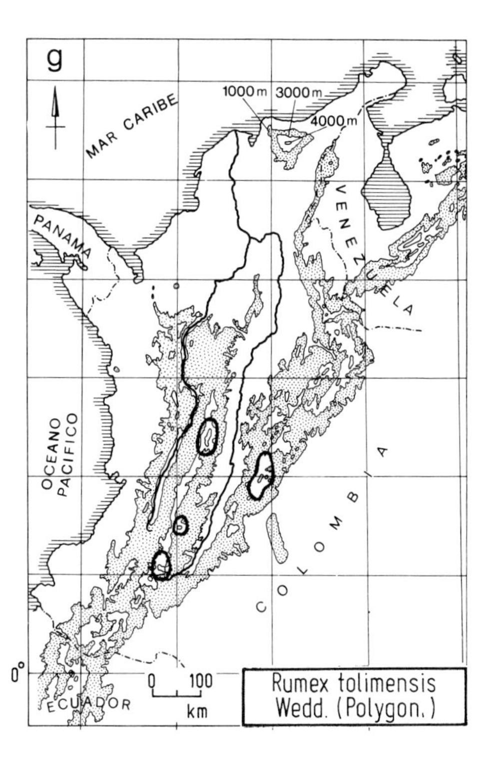
g
MAR CARIBE
1000 m
3000 m
4000 m
PANAMA
VENEZUELA
OCEANO PACIFICO
COLOMBIA
0°
0 100 km
ECUADOR
Rumex tolimensis Wedd. (Polygon.)

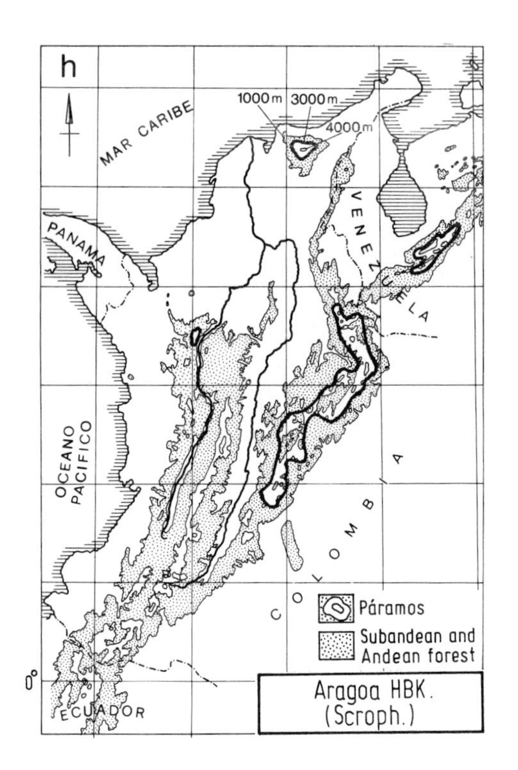
h
MAR CARIBE
1000 m
3000 m
4000 m
PANAMA
VENEZUELA
OCEANO PACIFICO
COLOMBIA
Páramos
Subandean and Andean forest
0°
ECUADOR
Aragoa HBK.
(Scroph.)

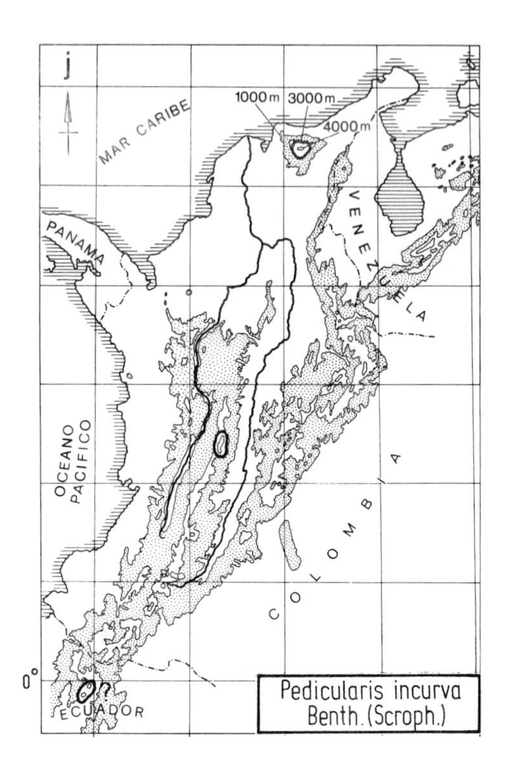
j
MAR CARIBE
1000 m
3000 m
4000 m
PANAMA
VENEZUELA
OCEANO PACIFICO
COLOMBIA
0°
ECUADOR
Pedicularis incurva
Benth. (Scroph.)

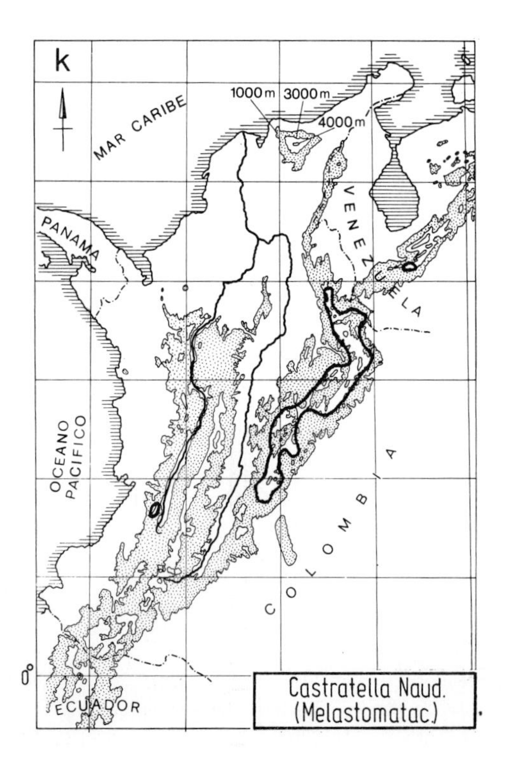
k
MAR CARIBE
1000 m
3000 m
4000 m
PANAMA
VENEZUELA
OCEANO PACIFICO
COLOMBIA
0°
ECUADOR
Castratella Naud.
(Melastomatac.)

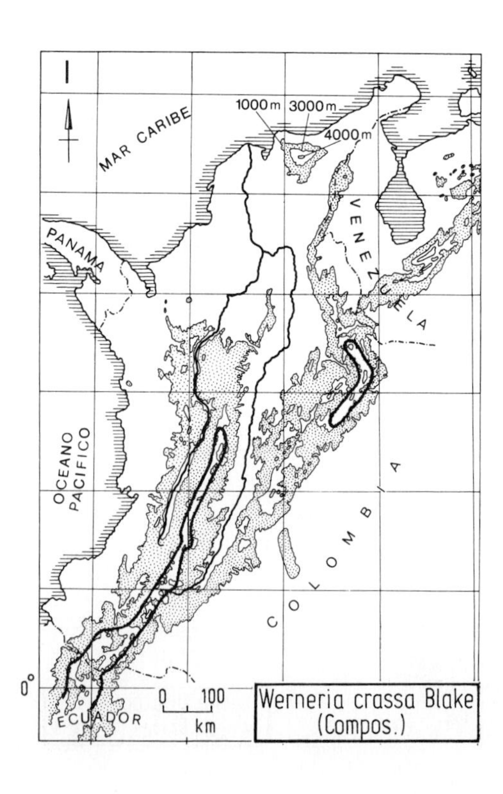
l
MAR CARIBE
1000 m
3000 m
4000 m
PANAMA
VENEZUELA
OCEANO PACIFICO
COLOMBIA
0°
ECUADOR
0 100
km
Werneria crassa Blake
(Compos.)

Cordillera Central—(Ecuador) (Fig. 7–3g). Examples: *Diplostephium rupestre* (H. B. K.) Wedd., *Rumex tolimensis* Wedd., *Senecio summus* Cuatr., *Pentacalia vernicosa* Sch. Bip. ex. Wedd., and *Werneria humilis* H. B. K.

e. Cordillera Oriental—Cordillera de Mérida—Sierra Nevada de Santa Marta—northern Cordillera Occidental (Fig. 7–3h). Example: *Aragoa.*

f. Sierra Nevada de Santa Marta—Cordillera Central—Ecuador (Fig. 7–3j). Examples: *Pedicularis incurva* Benth. and *Diplostephium eriophorum* Wedd.

g. Cordillera Oriental—Cordillera de Mérida—southern Cordillera Occidental (Fig. 7–3k). Example: *Castratella.*

h. Northern Cordillera Oriental (-Cocuy)—Cordillera Central—Ecuador (Fig. 7–3l). Example: *Werneria crassa* Blake, *Distichia muscoides* Nees and Meyen, and *Poa vaginalis* Benth.

i. Cordillera Oriental—Cordillera Occidental (Fig. 7–3m). Example: *Loricaria complanata* (Sch. Bip.) Wedd.

Glacial history played an important role in the establishment of these disjunct distribution patterns, as did the predominant direction of winds (locally in combination with volcanic eruptions), conceivably migratory birds, and the distance between the páramo areas and "islands," at present and in glacial times. In this respect it is interesting to compare the following present distances.

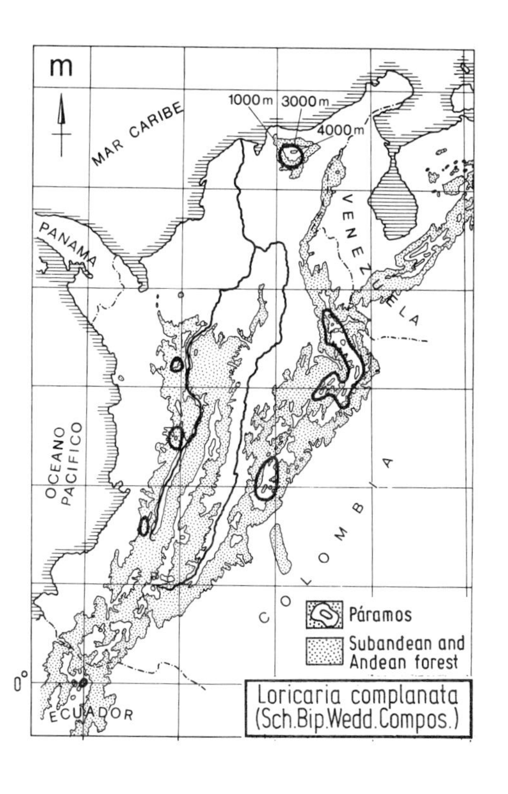

Cordillera de Mérida–Páramo de Tamá (Cord. Oriental)	50km
S.N. de Santa Marta–Sierra de Perijá (northern Cord. Oriental)	75km
Nevado Ruiz (Cord. Central)–Sumapaz (Cord. Oriental)	175km
S.N. del Cocuy–Nevado de Sumapaz (Cord. Oriental)	335km
Nevado Ruiz (Cord. Central)–S.N. del Cocuy (Cord. Oriental)	375km
Pico Neblina (Amazonas)–Sumapaz (Cord. Oriental)	965km
S.N. de Santa Marta (N. Colombia) –Chachapoyas (N. Peru)	1925km
Cord. Talamanca (Costa Rica)/Chiriquí (Panamá)–Cord. Occidental	750km
Chiriquí (Panama)–S.N. de Santa Marta	1000km
Chiriquí (Panama)–Cord. de Mérida	1170km

ORIGINS (PREPÁRAMO VEGETATION)

We call "prepáramo" a hypothetical open type of vegetation that is supposed to have occurred locally on hilltops before the Cordillera rose to its present level (and below the present altitudinal forest line) determined by edaphic and climatic factors (mainly other than temperature) and that may have contained some precursors of later (sub)páramo taxa.

The early (precursor) stage of páramo vegetation and flora (prepáramo) conceivably occurred on the tops of hills or mountains in the area of the present-day northern Andes, which had risen high enough and/or where the environment was extreme enough to support a more or less open type of vegetation instead of forest. In this connection it seems important to discuss briefly the tectonic history of the area. The following summary of the history of the Colombian Andes is based largely on van der Hammen (1961).

During the Upper Cretaceous elements of the Paleozoic basement of the present-day Colombian Cordillera Central apparently formed some kind of north–south separation between the western and eastern geosyncline, thus enabling a different development in the two regions.

Near the end of the Cretaceous, much of the area was near or above sea level. At the beginning of the Paleocene further movements apparently led to a rise of the Paleozoic massifs in the

Cordillera Oriental, dividing the area into several more or less separate basins, and to some upheaval in the area of the actual Cordillera Occidental, separating the area of the actual Cauca Valley from the Pacific region. The separating hills and ridges were probably not very high. During several intervals of the Lower and Middle Tertiary, tectonic-orogenic movements led to the gradual compression of large synclinal areas. The repeated deposition of conglomerates indicates that higher areas were formed. Phases of major importance occurred in the Lower and Middle Eocene; sandstones and conglomerates were deposited in many places, and an uncomformity may be present in different parts of South America. The Middle Eocene is also a major period of bauxite formation. It is highly probable that by that time the important features of the structure of the present Colombian Andes had already been formed. Rather important tectonic-orogenic movements took place in the Upper Oligocene, resulting in an upheaval of the Cordillera Oriental proper, and the termination of geosynclinal sedimentation in the area. Sedimentation continued in the marginal areas (inter-Andean Magdalena Valley and eastern border of the Cordillera Oriental). Embryonic depressions along the flanks of the Cordillera Central subsided considerably and developed into the long, north–south-running grabens of the inter-Andean Cauca and Magdalena Valleys. Nevertheless, there still seem to have been lowland connections between the present Magdalena Valley and the Llanos Orientales during the Miocene. New movements took place at the beginning of the Lower Miocene and sedimentation continued in the inter-Andean valleys during the Miocene. During the interval lasting from the upper Oligocene to approximately the beginning of the Pliocene, the Cretaceous and Tertiary sediments of the Cordillera Oriental were strongly folded (and faulted), according to a system of synclines and anticlines that already partly existed as proto-synclines and proto-anticlines.

By the beginning of the Pliocene the sediments of the inter-Andean Magdalena and Cauca Valleys were folded and the Tertiary geosynclinal sedimentation cycle was practically closed. The whole Andean region, especially the present Cordillera, underwent at that time or somewhat later a strong upheaval, which uplifted the north Andean area to its present altitudes. During the Plio-Pleistocene, volcanic activity was a very important factor in the Colombian Cordillera Central and in Ecuador, and most of the higher elevations (above the present forest line) came into being because of the accumulation of volcanic rocks, including deposits of lava and volcanic ashes.

It is extremely difficult or virtually impossible to ascertain how high the hills were in the northern Andes during the different uplift phases of the Tertiary. One thing is clear: it is inconceivable that areas above 3000 m existed long before the transition from Miocene to Pliocene. It is therefore improbable that areas with a climatic condition closely comparable to that of the actual páramo existed before the Pliocene, or prior to about 5 million years ago. But it is well known that at present open vegetation may occur on isolated hilltops and in boggy depressions at much lower elevations than 3000 m because of local climatic or edaphic conditions. When the general climatic conditions show a more pronounced seasonality, or are marginal for tree growth near climatic savannas or dry forests, this mountain top effect will become much more marked, and open vegetation may be found as low as 1000 m on the summits of otherwise forest-covered hills. A good example of mainly edaphically determined, open vegetation on hilltops is the "savannas" on the 500–3000-m-high sandstone tepuis in the Guayana shield area, which are often completely surrounded by tropical rain forest.

Pollen-analytical data from the Tertiary of the present north Andean area may show the former existence of open areas. The first appreciable representation of grass pollen is found in the Middle Eocene. At that time pollen of *Podocarpus* also appeared, and hills a little over 1000 m may have been present. Grass pollen may well indicate open vegetation related to savannas and/or open swamp conditions. If higher hills really existed, they may have been eroded at a later (Upper Eocene-Oligocene) time, when marine ingressions took place locally.

The proportion of pollen of taxa (Gramineae, Compositae) common in open vegetation types became locally abundant in Miocene sediments. Carbonized (burned) grass cuticles may also be found. It is highly probable that the poaceous pollen is from some type of savanna vegetation. There is no proof of mountains higher than 3000 m, nor is there any indication of the presence of an Andean forest belt. Hills above 1000 m may well have been present again, and the local existence of a savanna climate and/or special edaphic conditions in the area may have been responsible for the occurrence of open areas on hilltops above 1000 m, if they occurred at all.

The Sabana de Bogotá area lay at an elevation of less than 500 m when the sediments of zone I of the Plio-Pleistocene sequence were deposited (Fig. 7–4; see also pollen diagram from Salto

Tequendama, van der Hammen, Werner, and van Dommelen, 1973). Zone I is probably Early Pliocene, i.e., between four and six million years old. The vegetation is of a tropical lowland type. The only indication of somewhat higher hills or mountains in the area is the presence of pollen of *Podocarpus* and *Weinmannia*. Even if the present differences of elevation between the high plain of Bogotá and the mountains of the Eastern Cordillera existed already by that time, the mountains in the area could hardly have been higher than 2000 m, and only the Sierra Nevada del Cocuy would have reached up to about 3000 m. However, the presence of mountains of this elevation in the northern Andes is conjectural.

We thus have no indications that more than about five million years ago any area in the northern Andes had an altitude comparable to the present-day páramos. There might have been open vegetation on hilltops above 1000 m in some parts of the northern Andes, especially from the beginning of the Miocene onward, and also above 2000 m from the Early Pliocene onward.

Theoretically, however, the possibility of open hilltop vegetation at increasingly higher elevations in the course of the Late Tertiary is interesting, especially the example of the tepuis. Moreover, we know that the local paramillos on some mountain tops lower than 3000 m, support a semi-open wet vegetation physiognomically resembling a subpáramo and containing páramo species. We also know that on the slopes of the mountains surrounding the Sabana de Bogotá, where the original forest vegetation has been destroyed by man, a number of páramo species (even *Espeletia argentea* and *Espeletiopsis corymbosa*) grow spontaneously as low as 2600 m. Apparently a factor preventing a number of páramo species from establishing themselves at lower elevations is competiton from woody species, i.e., a light factor, rather than temperature. Páramo species (including *Espeletia* spp.) are also often present in open marsh or bog vegetation in the Andean forest belt, several hundred meters below the altitudinal forest limit. These are windy meadows on hilltops east of Chocontá, between 2800 m and 2900 m, where most of the herbs are páramo species. Thus, if forest vegetation of the Andean forest belt is absent for one reason or another, páramo species and locally even vegetation of the páramo type may replace it down to at least 2500 m on slopes and to at least 2000 m on mountain tops (see the earlier discussion of vegetation of the Avila in Venezuela).

From an examination of the early history it would appear that during the Pliocene and even the Early Pleistocene the altitudinal forest limit was considerably lower (as compared to isotherms) than it is at present. One must keep in mind that during and after the major upheaval of the Cordillera, considerable time must have passed before an Andean and high-Andean forest belt could originate and establish itself. Most Subandean forest taxa evolved from tropical ones, and many may have existed already in the Lower Pliocene. Many Andean forest taxa are of extra-Neotropical origin. These taxa were not readily available, whereas those of Neotropical origin certainly took a long time to evolve and to adapt themselves to the completely changed circumstances. It follows that when the major upheaval of the Cordillera took place during the Pliocene (between pollen zones I and III; Fig. 7–4), the zone between 1000 m and 2400 m was probably easily populated, but it may have taken a long time before the altitudinal forest line shifted from somewhere around 2500 m up to around 3500 m. All this permits the formulation of the following theory.

Possible precursors of páramo vegetation or hilltop "savannas" (prepáramo vegetation) may have been present before the major upheaval of the Cordillera, probably more than four, or possibly five, million years ago. These precursors occurred first on Miocene hilltops above 1000 m in areas with a relatively marked pluvial seasonality or with special edaphic conditions, and later on (Early Pliocene) were found on higher mountains in general. Open vegetation types with herbs and shrubs existed therefore at sites where temperatures were higher than at present in the páramos. During and shortly after the upheaval of the northern Andes, probably about four or perhaps five million years ago, protopáramo vegetation may have developed among other things from floristic elements of the embryonic paramillos above the altitudinal forest limit of that time, then probably situated at approximately 2500 m. Upper Andean forest and páramo belts subsequently evolved more or less simultaneously during the Upper Pliocene and Quaternary, the Andean forest shifting its upper limit upward little by little and replacing the lower páramo zone of that time. The floristic elements of evolving páramo vegetation types came partly from the Mio-Pliocene embryonic paramillos and from Andean forest (the Neotropical elements), partly from the austral-antarctic floral area migrating along the Andes. Later, elements also came from the Holarctic floral region, migrating via stepping stones through the Isthmus of Panama.

If some initial stages of the present páramo flora and vegetation developed in Tertiary hilltop

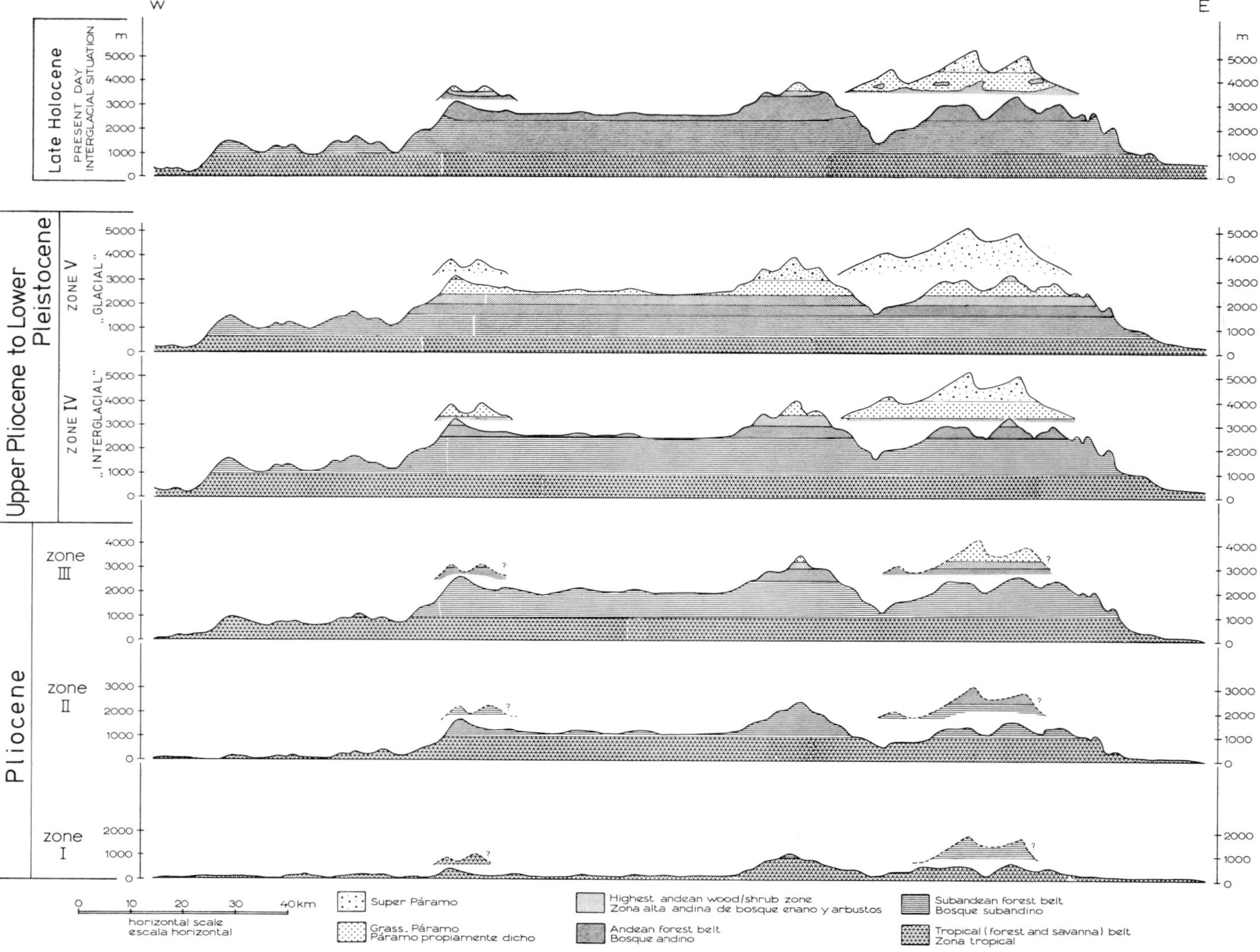

FIG. 7–4. Sections through the Cordillera Oriental (Colombia) showing the tentative reconstruction of vegetation belts dur ing the successive stages of uplift and after its cessation in the Late Pliocene or Early Pleistocene. The uppermost sections show the present situation. The main section is east–west at the latitude of Bogotá, and the small sections shown above each section are from higher areas farther north. After van der Hammen et al. (1973).

"savannas," some floristic relation with present-day savannas may be expected to exist. As mentioned earlier, this seems indeed to be the case, the most interesting examples being provided by *Bulbostylis*, *Eriocaulon*, *Paepalanthus* (even at the species level: *P. polytrichoides*), *Sporobolus*, *Sisyrynchium*, *Xyris*, and *Borreria*.

Another important fact relevant to the early history and evolution of the páramo flora and vegetation is that putative ancestors or near-relatives of a number of Neotropical páramo elements are found in the present Andean forest zone. The most striking example is that the most primitive genus of the Espeletiinae (*Libanothamnus*), with ramified tree forms, has forest species occurring in Venezuela between 2000 and 3000 m. The genera *Espeletia* and *Espeletiopsis*, which largely determine the physiognomy of páramo vegetation today, probably developed from these or from related primitive Espeletiinae. This evolution may have happened in the Venezuelan Andes or in the adjacent northern Cordillera Oriental, the genera later spreading farther south into the Colombian and Ecuadorian Andes. At present the highest species density of both genera is found, however, in the Colombian Cordillera Oriental.

EARLY HISTORY (THE PROTOPÁRAMO)

We call "protopáramo" early páramo vegetation, with typical páramo taxa, but apparently floristically poorer than the present-day páramo. It occurred at elevations above 2000–3000 m and its record was found in sediments of Late Pliocene or Early Pleistocene age.

The major uplift of the Cordillera Oriental started after the deposition of the Tequendama sediments (zone I: Lower Pliocene, a tropical lowland flora). During the interval zone II, sediments were deposited in the same area of the present-day high plain of Bogotá (Fig. 7–4) at an elevation of about 1500 m, and part of the surrounding mountains may have risen to elevations of 2500 m or more. *Weinmannia*, *Clusia*, *Miconia*, *Alchornea*, and Piperaceae were important elements of the local forest flora of the area. *Hieronima*, *Ficus*, Palmae, *Cecropia*, and Apocynaceae were common and Proteaceae and Bombacaceae, possibly not growing in situ but in the vicinity, also occurred. *Hedyosmum* was present but *Myrica* still absent at this stage (van der Hammen et al., 1973; most of this and the following information is borrowed from this publication).

The uplift of the Cordillera continued and considerable quantities of sand, gravel, and, locally, even big blocks of rock were deposited in the area of the high plains. An analysis of a series of lignites from the Tilatá Formation of the Chocontá area (present elevation about 2700 m) yielded a pollen diagram that was assigned to zone III of the Plio-Pleistocene sequence (Fig. 7–4). According to the pollen content, the lignites were deposited at 2300 m, and the local forest type must have resembled that of the higher belt of the recent Subandean forest rather closely. *Alchornea* was the most prolific pollen producer in this forest, other contributors being *Weinmannia*, *Ilex*, *Hedyosmum*, and species of Malpighiaceae. Important is the first and abundant occurrence of *Myrica* pollen (an element of Holarctic origin). The presence and relative abundance of Malpighiaceae in this area at this elevation are striking. Although typical páramo elements had not yet appeared, there are good reasons to suppose that the occurrence of a fair amount of Gramineae, Compositae, and *Hypericum* pollen is indicative of the presence in the surrounding mountains of vegetation developed near and above an altitudinal forest limit (i.e., paramillo or páramo vegetation). The mountain summits may already have attained altitudes of well over 3200 m. As mentioned earlier, the forest limit may well have been lower than it is today, because of the scarcity of woody species adapted to the climatic conditions of the present high-Andean forest zone. The age of zone III may be something like four million years (if the fission-track datings from zone V are correct).

The sedimentation of principally sand, with intercalations of gravel and some clay, continued in the area of the present-day high plain. In the higher part of that series (i.e., the upper part of the Tilatá Formation), pollen-bearing deposits referable to zone IV (Chocontá section) were found (humic clay and peat). The present elevation is 2800 m.

The pollen diagram (Fig. 7–5) represents the oldest páramo vegetation recorded to date. The sediments in question were deposited in an environment where vegetation of the páramo type dominated. The vegetation cover must have been of a very open type and must have been present at least several hundred meters above the altitudinal forest limit of that period. The abundance of Gramineae and the presence of *Valeriana*, *Plantago*, Ranunculaceae, *Aragoa*, and grains of the *Polylepis-Acaena* type, leave no doubt as to the páramo character of the vegetation, although this páramo was still poor in species. A number of taxa regularly represented nowadays in the pollen rain in the páramo are still lacking. The elements belonging to the type of forest growing below the timberline are practically the same as those already encountered in zone III. The presence of some *Alchornea* and of Malpighiaceae, which are totally absent from the diagrams representing zone V, is of special importance in this connection. If we accept the view that the forest limit lay at least 300 m below the site of Chocontá 1, it must have been below 2500 m. If we put the present altitudinal forest limit in the area at 3300 m, this indicates a depression of at least 800 m or possibly even as much as 1000 m compared to its position today. The presence of Subandean elements in the forest apparently only a little below the forest limit, and the very low position of the forest limit at such a relatively early time (still Pliocene, if the datings of zone V are correct), both suggest that the Andean forest belt was not yet fully developed and that the forest limit was lower than it is today under similar climatic conditions. In this respect the relatively abundant representation of elements (*Polygonum*, *Borreria*, *Jussiaea*) no longer found nowadays in comparable páramo vegetation but occurring commonly at lower elevations seems to be significant. On the other hand, a good temperature indicator seems to be the aquatic plant *Myriophyllum* (probably *M. elatinoides*), still mainly restricted in its occurrence to the páramo (van Geel and van der Hammen, 1973), which suggests an altitude comparable to the present altitude of at least 3000 m. It follows that at the time of deposition the temperature was probably at lease somewhat lower than it is today. Van der Hammen et al. (1973) assumed that zone IV represents the first cold phase of the Pleistocene and does not exceed 2.4 million years in age. However, a later fission-track dating of zone V indicating a minimum age of 2.9 million years would render such an age highly improbable. If that date is correct, however, the age of zone IV páramo vegetation would still be Pliocene. This is not so strange in view of accumulating evidence of a period of glaciation in different parts of the world around 3.5 million years ago (see, e.g., Clapperton, 1979). A final dating will have to wait until more reliable dates become available; for the time being we conclude that this early páramo vegetation was present as early as 2 (or possibly even 4) million years ago, in the Late Pliocene or Early Pleistocene.

A more detailed analysis of this early páramo vegetation seems worthwhile. Apart from Gramineae and Cyperaceae, it included Compositae, Ericaceae, the *Polylepis-Acaena* type, *Symplocos*, *Myrica*, *Aragoa*, *Hypericum*, *Miconia*, *Ilex*, Umbelliferae (cf. *Hydrocotyle*), *Borreria*, *Jussiaea* (*Ludwigia*), *Polygonum*, *Valeriana*, *Plantago*, *Ranunculus*, *Myriophyllum*, *Jamesonia*, and *Hymenophyllum*. A phytogeographic analysis of these 17 genera shows that 6 are Neotropical or have wide tropical distribution, 2 are of wide temperate distribution, 3 are austral-antarctic, 2 are Andean, 2 are cosmopolitan, and 2 are of northern origin. The last group includes *Myrica*, which reached South America much earlier in the Pliocene but as an element of the forest zones. Pollen of the *Hypericum* type has been found in Lower Pliocene deposits; the closest and most primitive relatives of *Hypericum* (sect. *Brathys*) are present in tropical Africa (Gondwana element). Most of the elements are therefore of local, Neotropical, and Andean, or austral-antarctic origin. About 50% are of tropical or local origin, and 50% are of temperate zone origin (of which about half is of northern origin).

It seems that the large lake of the Sabana de Bogotá came into existence after the deposition of the Tilatá Formation (inclusive of the sediments corresponding to zone IV), and that sedimentation started during an interval corresponding with what we have called "pollen zone V." These sediments now lie 150–200 m below the surface near Bogotá (Fig. 7–6), but much deeper toward the center of the high plain, and near the surface in the Guasca and Subachoque valleys.

During the formation of zone V, a progressive enrichment of the páramo flora took place. Páramo vegetation seems to have been predominant during most of the long time interval, and to have alternated with a phase in which *Myrica* dominated and with another phase in which *Weinmannia* was dominant. Potassium-argon dates from volcanic ash present in these sediments yielded ages between approximately 1.7 and 3.6 million years (for the plagioclase fraction) and one fission-track date of 3.62 (± 0.67) million years.

The first part of zone V marks the appearance of Caryophyllaceae, *Geranium*, and *Lycopodium* (the two spore types, both those with foveolate and those with reticulate spores), *Sphagnum* and *Isoëtes*. Somewhat later *Gunnera*, *Gentianella corymbosa* type, *Lysipomia*, and Malvaceae (*Acaulimalva*) appeared on the scene. The elements recorded in zone IV remain present.

Of the 26 recorded taxa, 6 are Neotropical or wide tropical, 8 have a wide temperate distribution, 3 are austral-antarctic, 4 Andean, 3 cosmopolitan, and 2 of northern origin, i.e., approximately 40% are of tropical or local origin, and 60% from temperate zones (of which 8% are from northern regions). The immigration of taxa

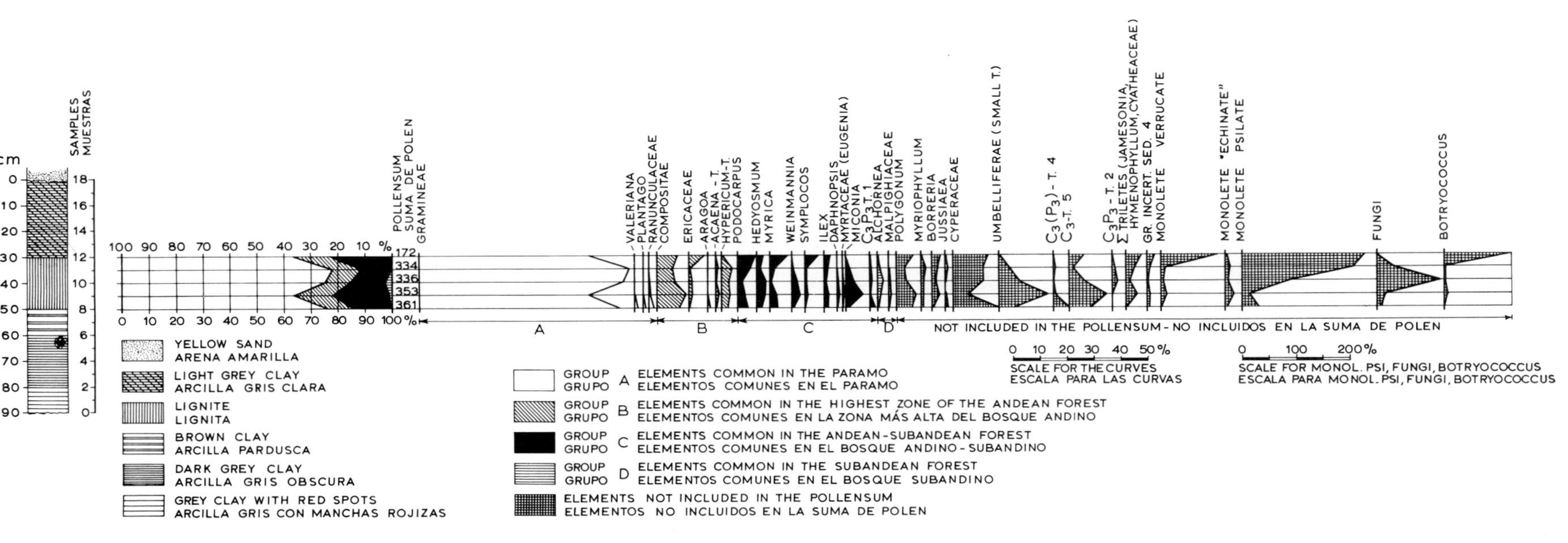

FIG. 7–5. Pollen diagram Chocontá 1, Plio-Pleistocene pollen zone IV.

BOGOTA, CIUDAD UNIVERSITARIA
section C.U.Y. 140 - 195 m

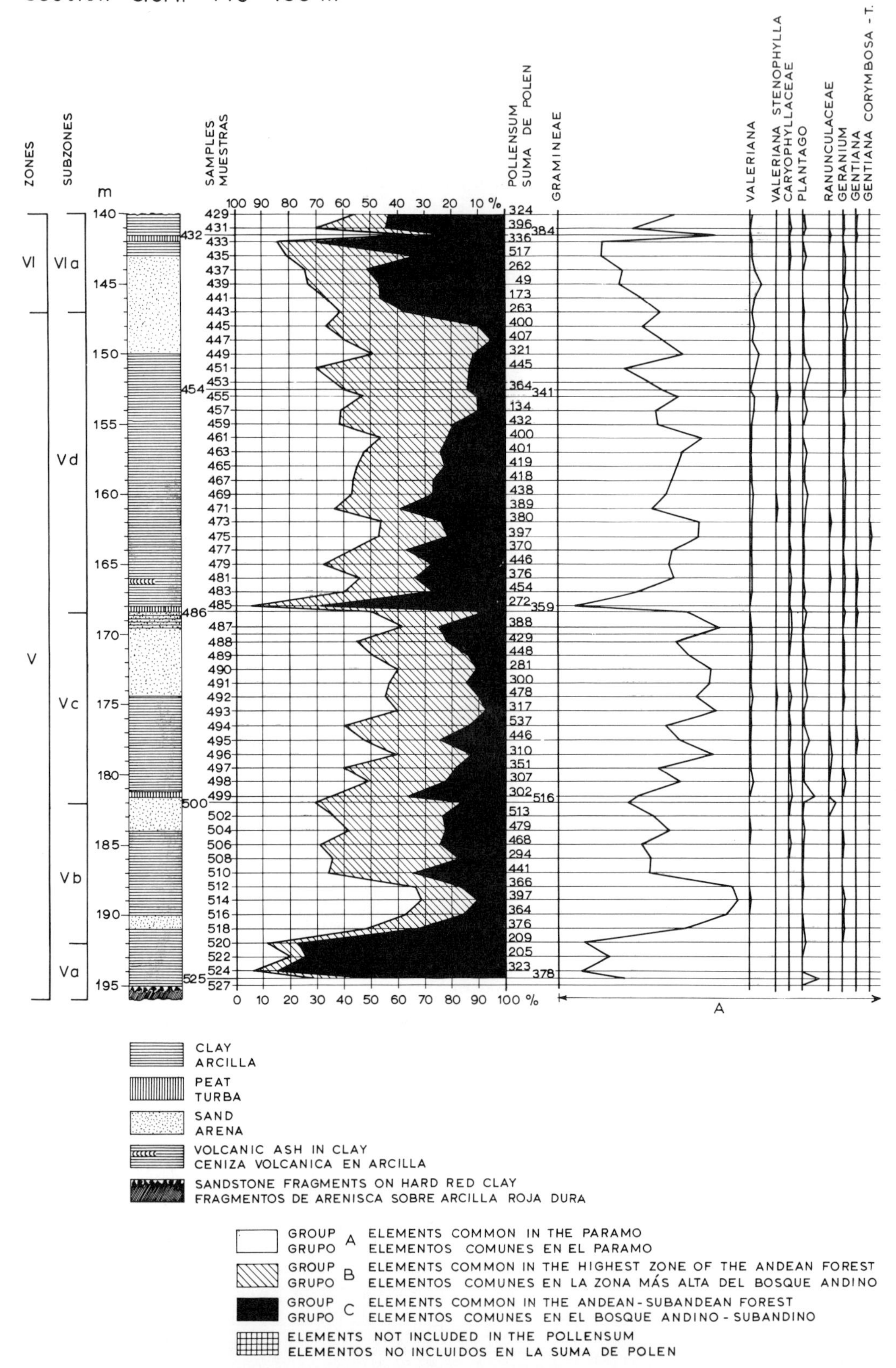

FIG. 7–6. Pollen diagram of the lower part of the 200-m sequence of lake sediments from Bogotá; Plio-Pleistocene pollen zone V and lower part of the pollen zone VI (See also Fig. 7–7, *left*).

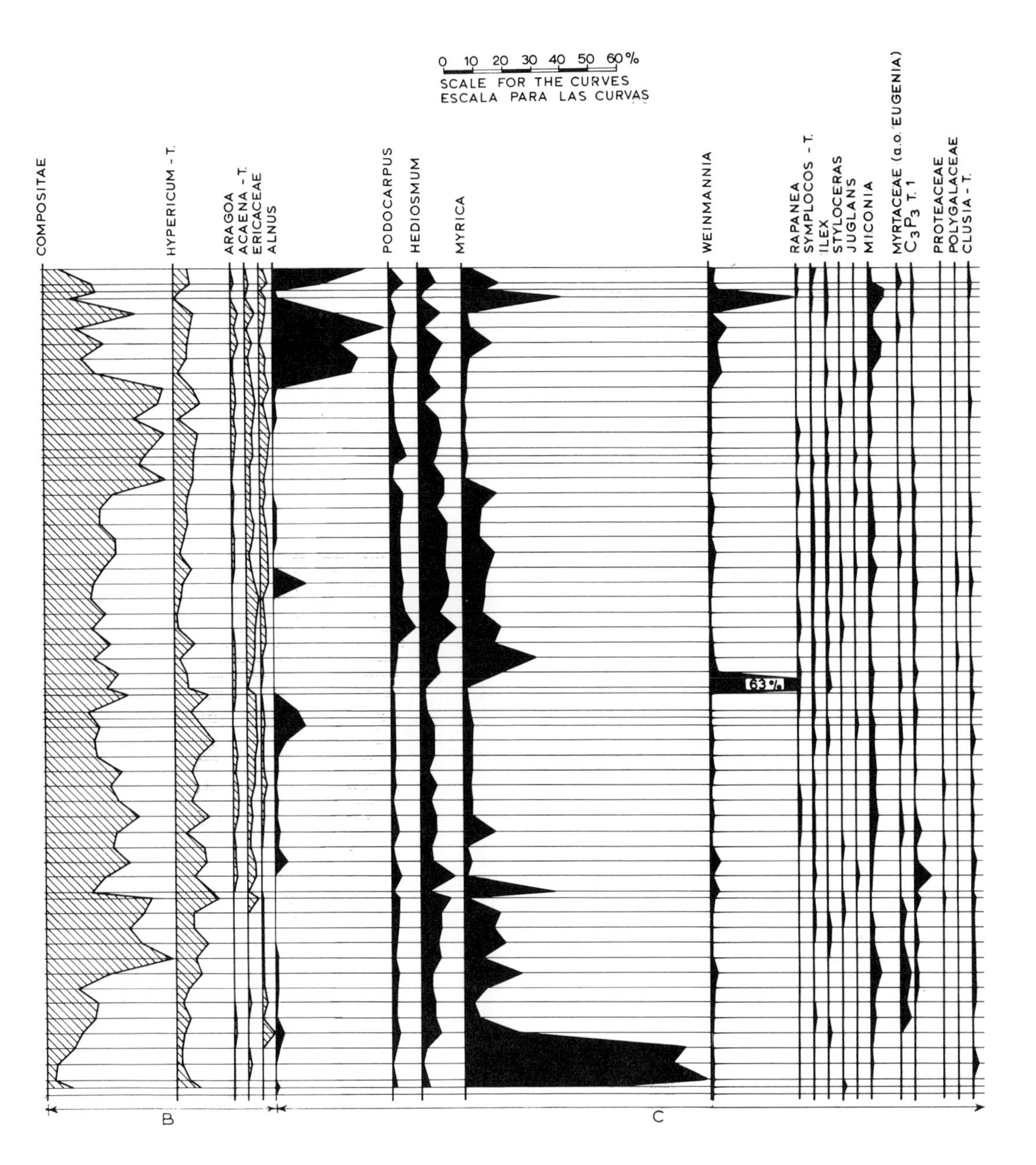
0 10 20 30 40 50 60 %
SCALE FOR THE CURVES
ESCALA PARA LAS CURVAS
COMPOSITAE
HYPERICUM - T.
ARAGOA
ACAENA - T.
ERICACEAE
ALNUS
PODOCARPUS
HEDIOSMUM
MYRICA
WEINMANNIA
RAPANEA
SYMPLOCOS - T.
ILEX
STYLOCERAS
JUGLANS
MICONIA
MYRTACEAE (a.o. EUGENIA)
C_3P_3 T. 1
PROTEACEAE
POLYGALACEAE
CLUSIA - T.
63 %
B
C

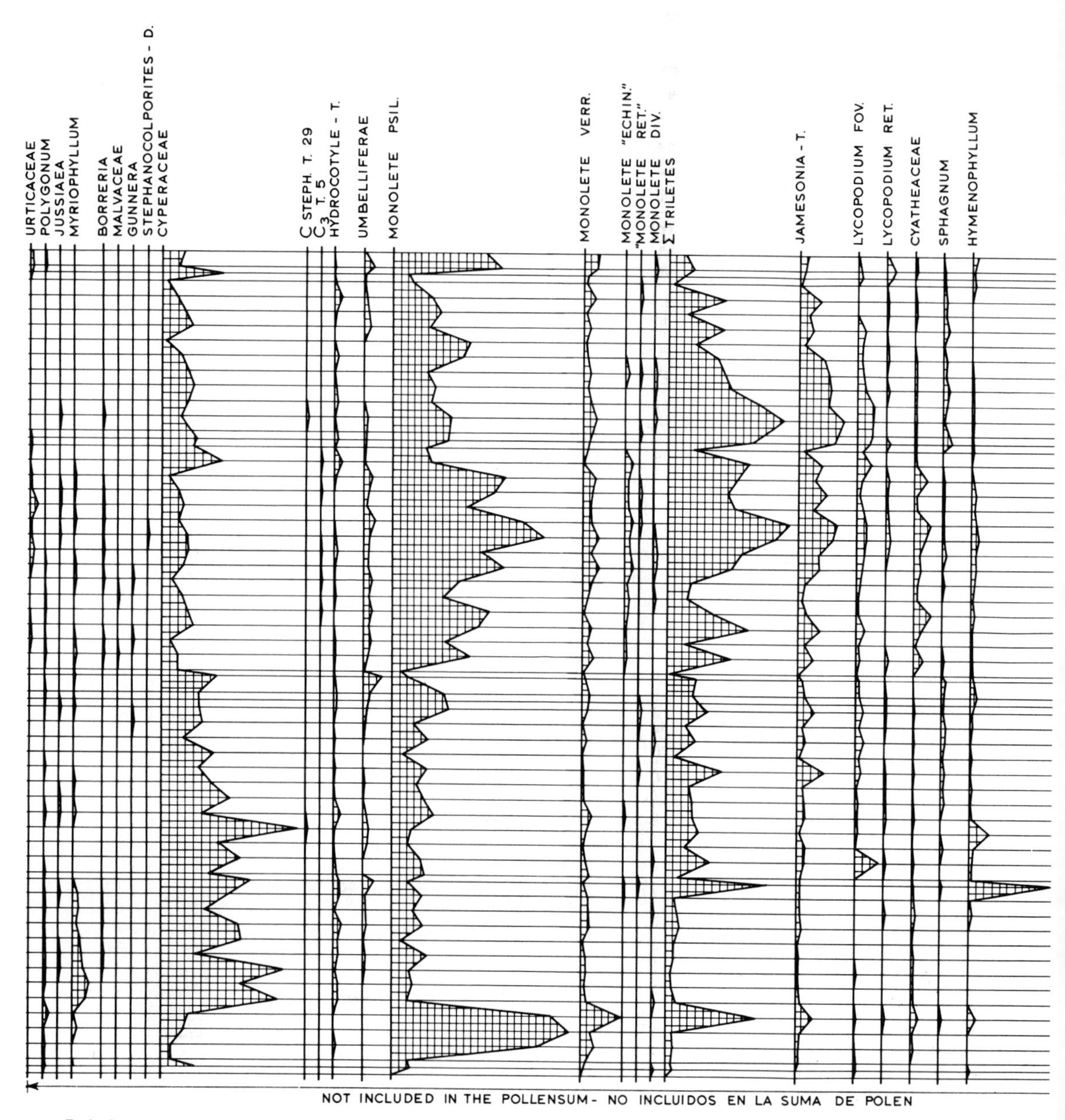

FIG. 7–6. *Continued*

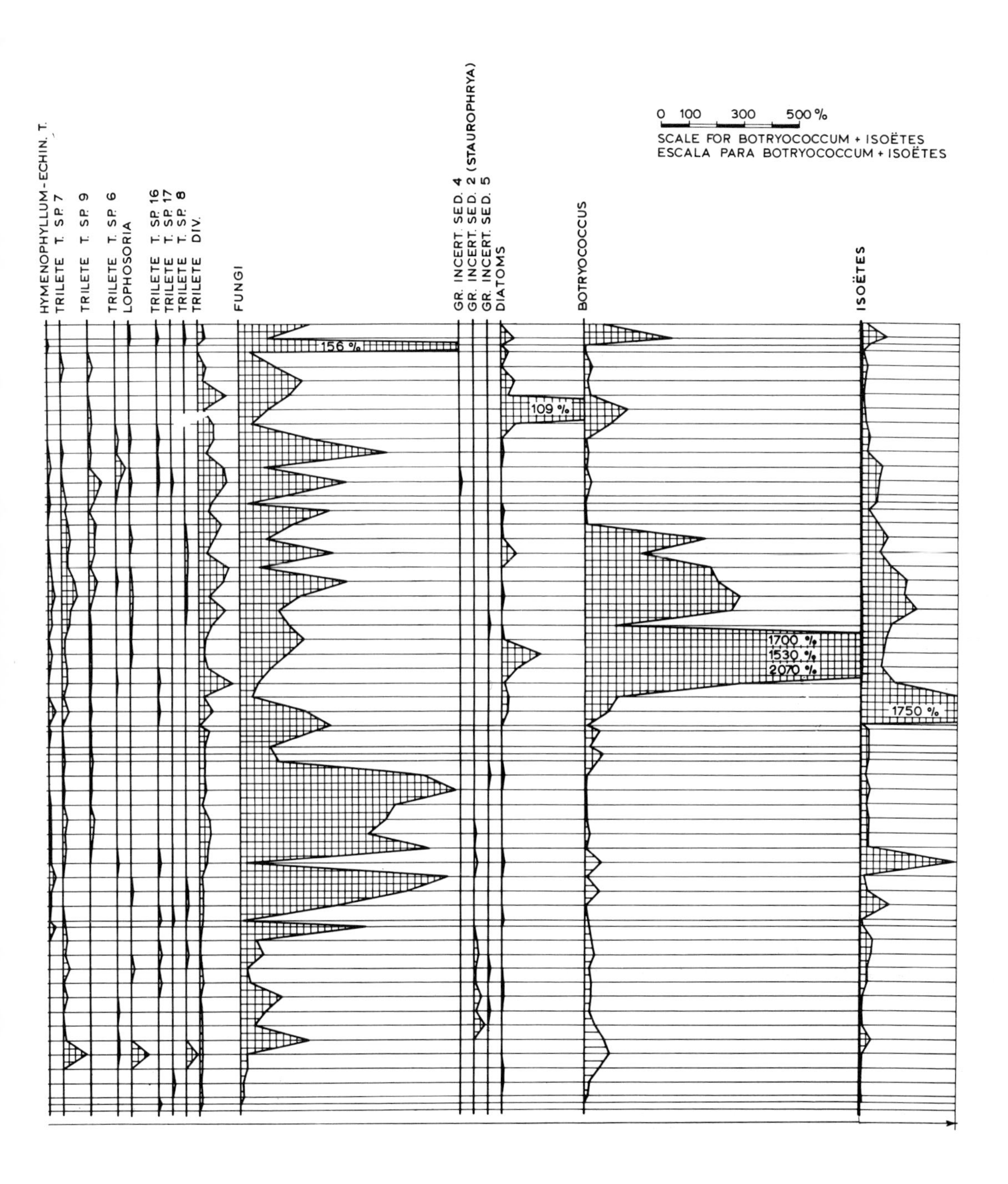

HYMENOPHYLLUM-ECHIN. T.
TRILETE T. SP. 7
TRILETE T. SP. 9
TRILETE T. SP. 6
LOPHOSORIA
TRILETE T. SP. 16
TRILETE T. SP. 17
TRILETE T. SP. 8
TRILETE DIV.
FUNGI
GR. INCERT. SED. 4
GR. INCERT. SED. 2 (STAUROPHRYA)
GR. INCERT. SED. 5
DIATOMS
BOTRYOCOCCUS
ISOËTES
0 100 300 500 %
SCALE FOR BOTRYOCOCCUM + ISOËTES
ESCALA PARA BOTRYOCOCCUM + ISOËTES
156 %
109 %
1700 %
1530 %
2070 %
1750 %

of temperate origin must have been important at that time. The Andean forest zone became enriched by *Styloceras*, *Juglans*, and Urticaceae. *Alnus* apparently migrated a little later into the Andes, but the percentages of representation are low and irregular, and *Alnus* is sometimes entirely absent from the samples.

As stated before, during the arboreal phases the dominant contributors to the pollen spectra were *Weinmannia* or *Myrica*. The lake level may have been lower at that time as can be deduced from the peat layers in the deep section of Bogotá. *Myrica* and perhaps also *Weinmannia* may have been important elements in some swamp forest type on the flat lake bottom, which became dry after a temporary lowering of the water table; they therefore may be locally overrepresented in the peat. However, their contribution during open water phases is so low, and that of open páramo elements and arboreal and fruticose subpáramo elements so high, that they must have been practically absent from the hill slopes around and above the high plain. It seems, therefore, as if lower lake levels coincided with relatively warmer intervals in this case. However, zone V as a whole represents a long lasting period of domination of páramo vegetation. The fact that only one, and probably a relatively short, phase indicates a dominance of Andean forest elements (namely *Weinmannia*) suggests that on the whole the altitudinal forest limit was lower than it is today. This may indicate an extended cold climatic period, or a lower forest limit due to the still incompletely developed higher-Andean forest zone, or both. The presence of *Borreria*, *Polygonum*, and *Ludwigia* suggests somewhat warmer conditions in open stands of vegetation. Today these taxa do not as a rule, or only rarely, grow at altitudes above 3000 m. But the presence of *Myriophyllum*, *Isoëtes*, and *Acaulimalva* possibly indicates a temperature range comparable with that of the present-day páramo. A climate considerably colder than the present one is also suggested by the widespread occurrence of solifluction or of other types of mass movements and by the deposition of a fluvioglacial type of gravel.

PLEISTOCENE HISTORY

The "early history" of páramo vegetation probably also embraces a part of the Lower Pleistocene, but it remains difficult to ascertain where exactly the Pleistocene begins. This difficulty is attributable to some ambiguity of the datings, as well as to the lack of consensus regarding the exact lower boundary of the Quaternary. A date of 1.9 million years may become the official date (corresponding to the stratigraphical limit proposed at the last I.N.Q.U.A. Congress in Paris), but the first clearly glacial phase in northwestern Europe (the Pretiglian) has a probable age of 2.4 million years. The beginning of the apparently first considerable cooling of the oceans was dated at about 3.2 million years, while a major cold phase in South America was dated at around 3.5 million years (Shackleton and Opdyke, 1977; Clapperton, 1979). If the base of the Pleistocene is taken indeed at approximately 2 million years we have to accept the existence of Pliocene glaciations.

Zone V might belong to the oldest Quaternary ("Lower Quaternary"). In the sediments of the Sabana de Bogotá a marked change in the composition of the pollen spectra took place at the base of zone VI (Figs. 7–6 and 7–7). *Alnus* had become firmly established in the area of the high plains and remained one of the most abundant pollen contributors. This influenced appreciably the general makeup of the pollen diagrams, the group of forest elements in the spectra reaching higher values during forest periods than they did before. The age of the base of zone VI ("Middle Quaternary") may be provisionally calculated by extrapolation (or interpolation, respectively) of dates obtained from higher and/or lower levels, and fixed at somewhere between 2.5 and 1.5 million years B.P.

Zone VII comprises the rest of the Quaternary ("Upper Quaternary"). It starts where pollen of *Quercus* (oak, an immigrant from the north), soon to become abundant, appears in the Sabana de Bogotá.

The beginning of zone VII is dated by fission tracks as close to one million years. The pollen diagrams of zones VI and VII (Fig. 7–7) show (for the high plain of Bogotá, at about 2600 m) a repeated alternation of forest and páramo periods, i.e., of interglacials and glacials. At the same time a gradual enrichment of the flora proceeds. This was studied in detail along a 400-m-long section of lake sediments from the center of the high plain of Bogotá (Hooghiemstra, 1984). According to this worker, during the interval zone VI, *Lathyrus*, for example, appears and, by the beginning of the interval zone VII, *Montia* (a wide temperate element).

At the same time, still unidentified pollen types appeared that disappeared from the pollen record somewhat later. The overall picture is that of the combined result of immigration and of adaptation of both endemic and introduced elements, evolution and extinction continually playing a role.

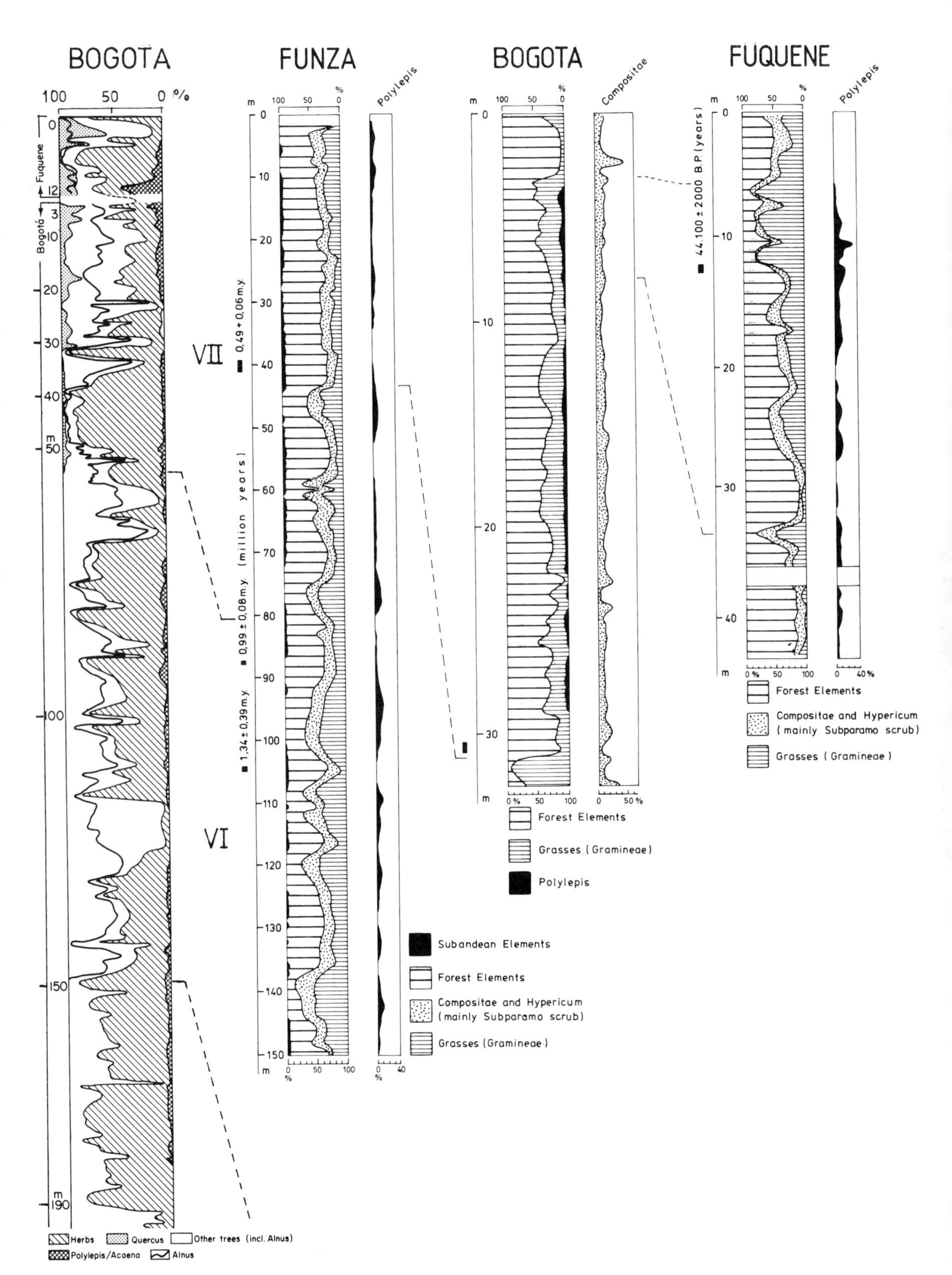

FIG. 7–7. Pollen diagrams from the long sections of the Sabana de Bogotá and Laguna de Fúquene: Bogotá CUY, Funza, Bogotá CUX, and Fúquene 3, respectively. Correlation, zones (VI and VII), depths, and dating are indicated. Altitude of the top of all these sections is about 2580 m.

However, as far as can be deduced from the pollen grains, the addition of new taxa proceeded at a much slower rate in zone VII than in zone V, and possibly the addition of new taxa was at the species rather than at the genus level.

The sequence of intermittent glacials and interglacials is known from several deep sections of the high plain of Bogotá, Bogotá Ciudad Universitaria (van der Hammen and Gonzalez, 1960a, 1964), Funza (van der Hammen, 1974, 1979), and Tarragona (Dueñas, 1979) (see Fig. 7–7). An approximate dating of the sequence has been obtained by potassium-argon and fission-track analyses. During the interval of approximately two million years represented by zones VI and VII, the high plain of Bogotá (between 2550 m and 2600 m above sea level) was, during the warmer phases, situated within the lower part of the Andean forest belt. During most of these "interglacials," there is a slight but clear increase in pollen of Subandean forest elements indicating that this belt was nearby. During the colder phases ("glacials" and "stadials") a marked rise in the percentages of grass pollen and of pollen of typical páramo elements reveals that the high plain then lay within the páramo belt or very close to it.

Polylepis-type pollen in particular attained pronounced maxima during phases with high grass pollen percentages. There were about 15 to 20 cold periods, forming part of an equal number of glacial-interglacial cycles, or stadial-interstadial ones, respectively. A rather detailed pollen diagram of the upper 30 m of the Bogotá section (Fig. 7–7) is available (van der Hammen and Gonzalez, 1960a). When that diagram was published the only other datings available were a few carbon 14 dates from the upper part of the section (representing the last 23,000 years), and the whole section was thought mainly to comprise the last two glacial-interglacial cycles. Volcanic ashes later found in the lower part of the corresponding section (upper 40 meters) of the Funza diagram could be dated at between 400,000 and 500,000 years old. Although these dates seemed too high at first, datings from lower deposits (ca. one million years at a depth of about 87 m) appear to confirm the slow rate of deposition of less than 0.1 mm per year.

Closer to the shore of the great lake of the Sabana de Bogotá (Bogotá section), interglacials are often represented by peat layers indicating that the lake level was lower. This was interpreted to indicate the prevalence of relatively drier conditions (i.e., lower amounts of effective precipitation) during the interglacials and relatively moister conditions during the glacials (van der Hammen and Gonzalez, 1960a). This could in fact be attributable mainly to a higher rate of evaporation during the interglacials. Subsequent and more detailed studies, especially of the last glacial-interglacial cycle, showed that extreme wet phases may also occur close to markedly dry phases in glacial times. In the northern Andes, therefore, interglacials apparently often coincided with lower lake levels (perhaps mainly because of a higher rate of evaporation), whereas glacials often coincided with relatively high lake levels (perhaps partly because of a lower rate of evaporation) but may have included very dry intervals (resulting from a marked fall in annual precipitation).

For a better understanding of the glacial-interglacial cycles in general, and of their effect on the vegetation cover and their importance in connection with such phenomena as plant migration and evolution, more detailed knowledge of the last glacial-interglacial cycle is of considerable importance.

The most complete pollen diagrams of the later part of the Quaternary are from the Laguna de Fúquene, north of the Sabana de Bogotá (Figs. 7–7 and 7–8), but collected at about the same elevation (van Geel and van der Hammen, 1973; van der Hammen, 1979). The first publication covers the most recent period of about 35,000 years; the second presents a generalized diagram of the entire, last interglacial-glacial cycle (probably covering the last 120,000 years). The publication of the complete diagram is in preparation, and there are some carbon 14 dates for the last 45,000 years. Another and probably rather complete sequence is a composite one, from the Sabana de Bogotá (El Abra: Schreve-Brinkman, 1978), and the lower part is well represented in the upper 12 m of sediment in the Sabana de Bogotá (van der Hammen and Gonzalez, 1960a; van der Hammen, 1979).

The last interglacial-glacial cycle in the area of the high plains began with a long and complex period during which forest was present, and was interrupted by some short but pronounced cold periods. This complex history probably corresponds with the Last Interglacial and so-called Early Glacial of the European sequence; lake levels were relatively low. This complex interglacial type of interval is followed by a period with páramo vegetation. From a comparison with sequences from other parts of the world, an age of 70,000 B.P. for that period is acceptable. The lake levels were higher during this cold period, which was followed by several interstadials and stadials, Fig. 7–7 showing a changing proportion of páramo elements and forest elements. A major

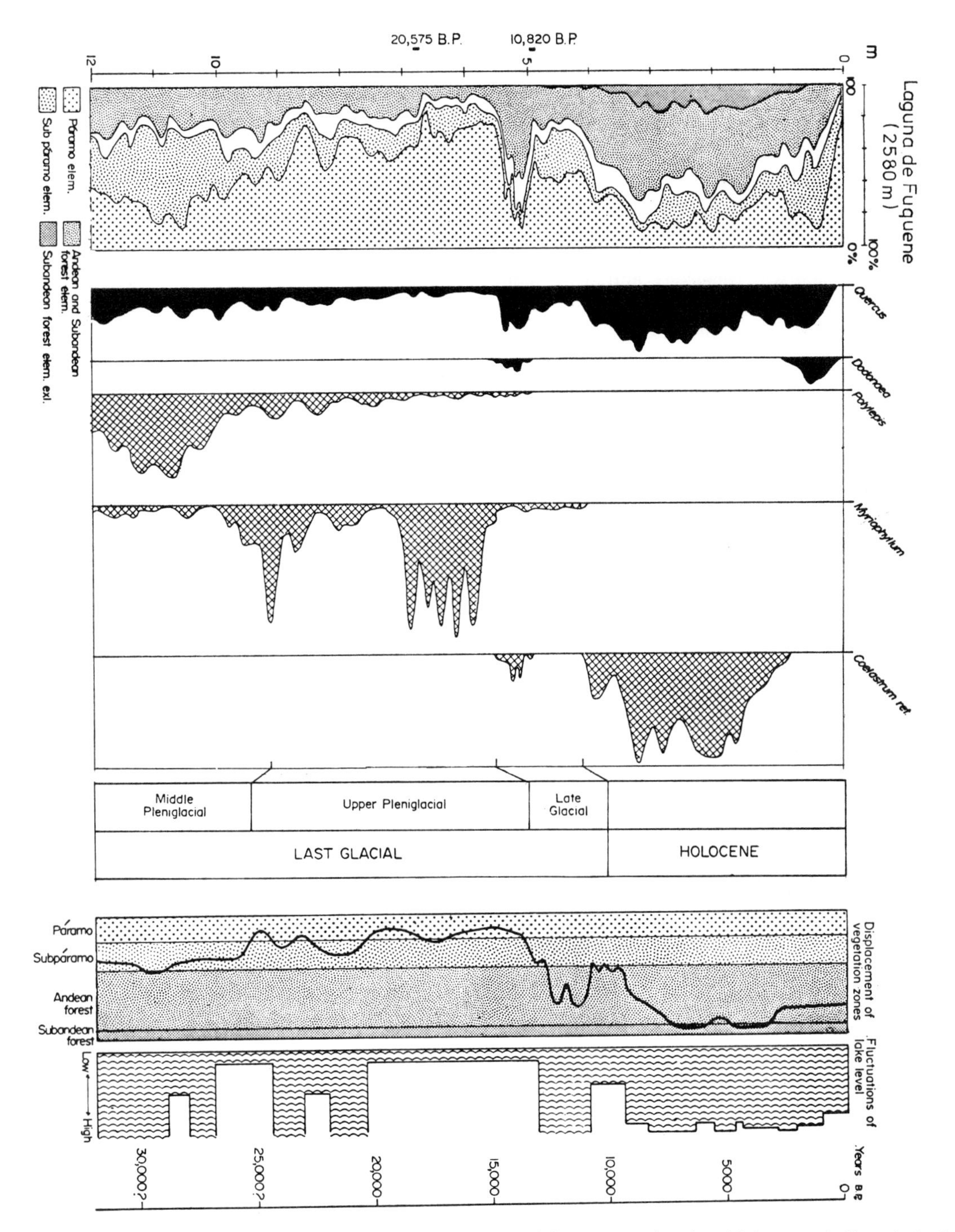

FIG. 7–8. Simplified pollen diagram from Laguna de Fúquene; depth in meters and carbon 14 dates are indicated. At the right is the interpretation (scale in years B.P.): a displacement of vegetation zones and fluctuations of the lake level. The diagram represents approximately the last 32,000 years. Altitude about 2580 m. Adapted from van Geel and van der Hammen (1973).

interstadial seems to have terminated by about 44,000 B.P. (based on a carbon 14 date from Fúquene). The following period until about 21,000 B.P. shows several fluctuations of the lake levels, and also several cold stadials and cool interstadials, but as a whole the páramo vegetation dominated the scene. In this period the *Polylepis* forest apparently formed a broad zone at the altitudinal forest limit; the *Polylepidetum* was certainly much more widespread at that time than it is now. Between 21,000 B.P. and 14,000 B.P. the climate was very cold, and at the same time extremely dry (the lake levels were very low). The rainfall may have been only half the present values, and the mean annual temperature may have been 6–7°C lower than today (or more). A treeless dry páramo covered the high plains during that period, which ended with the beginning of the Late Glacial sequence (at some time between 14,000 and 12,000 B.P.), when lake levels rose again, and the climate was ameliorating. There were several interstadials and stadials in the Late Glacials, which ended with the cooler El Abra stadial (11,000–10,000 B.P.) when the high plains lay within the subpáramo belt.

At the beginning of the Holocene (ca. 10,000 B.P.) the climate ameliorated considerably, forest from then on covering areas above the high plains. The forest limit rose quickly to even higher elevations than at present, and páramo vegetation finally became restricted to its present range, i.e., to above 3300–3500 m.

A dated pollen diagram from a lake at 2000-m altitude, on the western slopes of the Cordillera Oriental (Laguna de Pedro Palo; van der Hammen, 1974), shows that in the Early Late Glacial páramo vegetation was dominant around the lake (Fig. 7–9). This implies a lowering of the forest-páramo limit by more than 1300–1500 m. The conclusion is that during the coldest and driest phase of the Last Glacial (from about 21,000 to about 14,000 B.P.), páramo vegetation covered most of the area above 2000 m. This clearly means that the area covered by páramo vegetation was many times larger than that of the present, and that many of the now isolated páramo "islands" were united and covered fewer but much larger areas (Fig. 7–10).

During the wetter (or very wet) intervals of the Last Glacial (especially between about 44,000 and 21,000 B.P.), the forest limit was at a higher elevation (at about 2500 m ca. 25,000 B.P.), and *Polylepis* forest formed a relatively broad zone near this limit.

So far only the Pleistocene history of the páramo formation outside its present area of occupation has been discussed. Data (pollen diagrams)

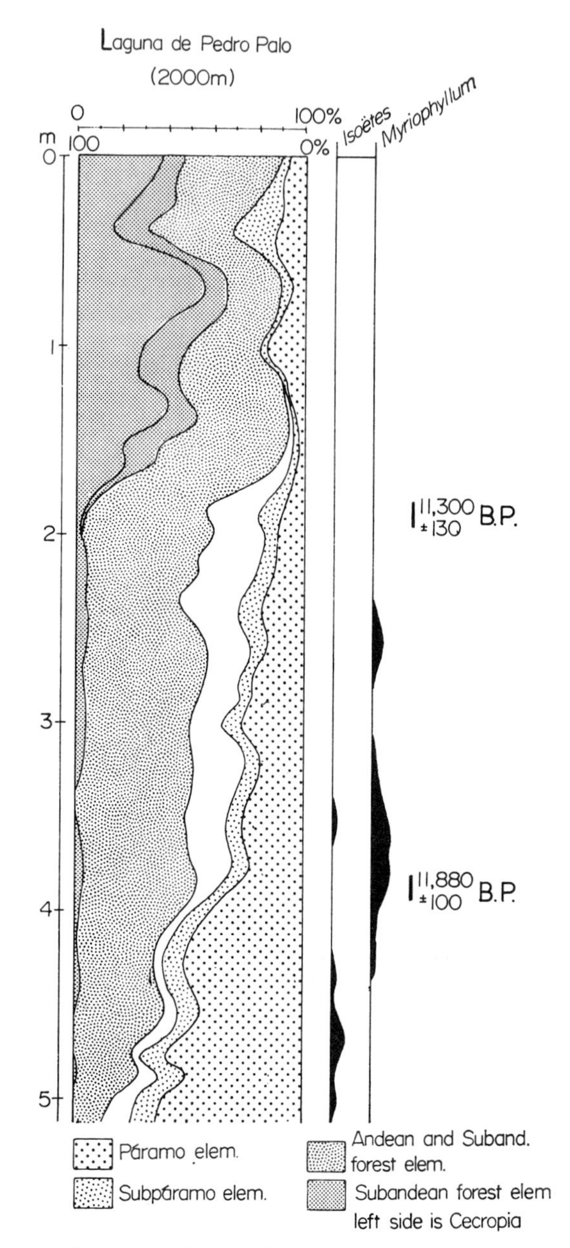

FIG. 7–9. Pollen diagram from the Late Glacial and Holocene of Laguna de Pedro Palo (Cordillera Oriental, Colombia). Altitude about 2000 m. The white area represents the percentage of *Alnus* in the pollen total.

available from within that area go back to approximately 25,000 B.P., but data on the glacial history (e.g., the extension of former glaciers) go back even further. The most relevant data are from the Sierra Nevada del Cocuy (van der Hammen et al., 1980–81); records from other areas cover only the Late Glacial and Holocene (e.g., van der Hammen and Gonzalez, 1960b, 1965). According to the pollen diagrams and the datings from Laguna Ciega, situated at 3500 m on

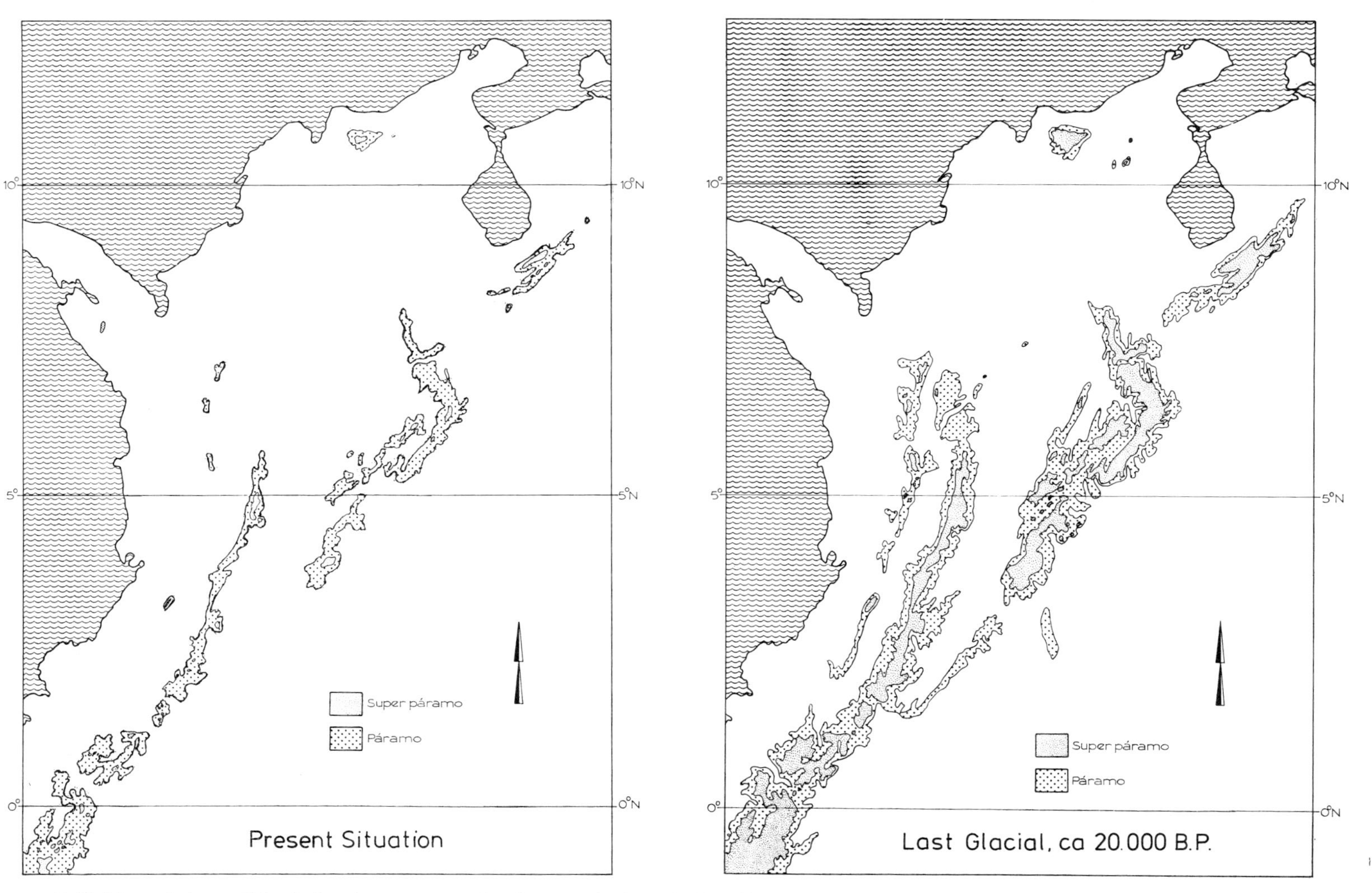

FIG. 7–10. Maps of páramo "islands," as they are now and as they were in the period 21,000–14,000 B.P.

the west flank of the Sierra Nevada del Cocuy (Fig. 7–11), sedimentation started after the disappearance of the glaciers from that site about 23,000 B.P. The diagrams show an alternation of (grass) páramo and subpáramo vegetation. The increasing representation of the subpáramo group of pollen types is often concomitant with a certain increase of the pollen of trees and shrubs, and represents interstadials (the Saravita, Susacá, and Guantiva interstadials, respectively). The period between 20,000 B.P. and 14,000 B.P. shows a dominance of the (upper) grass páramo group. Although from the pollen diagram of Pedro Palo (Fig. 7–9) one could deduce that the lower limit of the páramo lay below 2000 m, at this locality the evidence indicates that the upper limit of the grass páramo was at or above 3500 m. It appears, therefore, that during the period between 21,000 B.P. and 14,000 B.P. the grass páramo zone was found between 2000 m and 3500 m and was apparently broader than it is today.

The glaciers that covered the area of Laguna Ciega came down in places to about 3000 m before they retired, and it is reasonable to suppose that there was ice at this location (Río Concavo stade) in the period between ca. 28,000 B.P. and 24,000 B.P. (van der Hammen et al., 1980). By about 20,000 B.P., the time of maximum glaciation in the northern temperate latitudes, the glaciers already had retired considerably (Early Lagunillas stade, after 21,000 B.P., the ice locally descending to 3300 m). There was, however, a stade earlier than the Río Concavo one, the Río Negro stade, when the ice reached down to 2600 m (and locally even as far down as 2400 or even 2200 m). This stade must be older than 28,000 B.P., and a careful analysis of all available data and datings from the Cocuy and the rest of the Cordillera Oriental makes it probable that it occurred during the wet period(s) of the time between ca. 44,000 B.P. and 34,000 B.P. (van der Hammen et al., 1980). It follows that in the Cordillera Oriental the periods of major glaciation (during the Last Glacial) took place during the wet (and relatively cold) period between 44,000 B.P. and 24,000 B.P. During the very cold but dry period between 21,000 B.P. and 14,000 B.P. the glaciation was far less extensive.

If we combine the vegetational and glacial history data, the following interesting picture emerges (Fig. 7–12). The greatest extension of glaciers probably took place in the period between 45,000 B.P. and 25,000 B.P. The climate was relatively wet at that time and the forest limit lay between 800 and 1000 m below the present one. During the early period of major extension of the ice glaciers and forest may have been

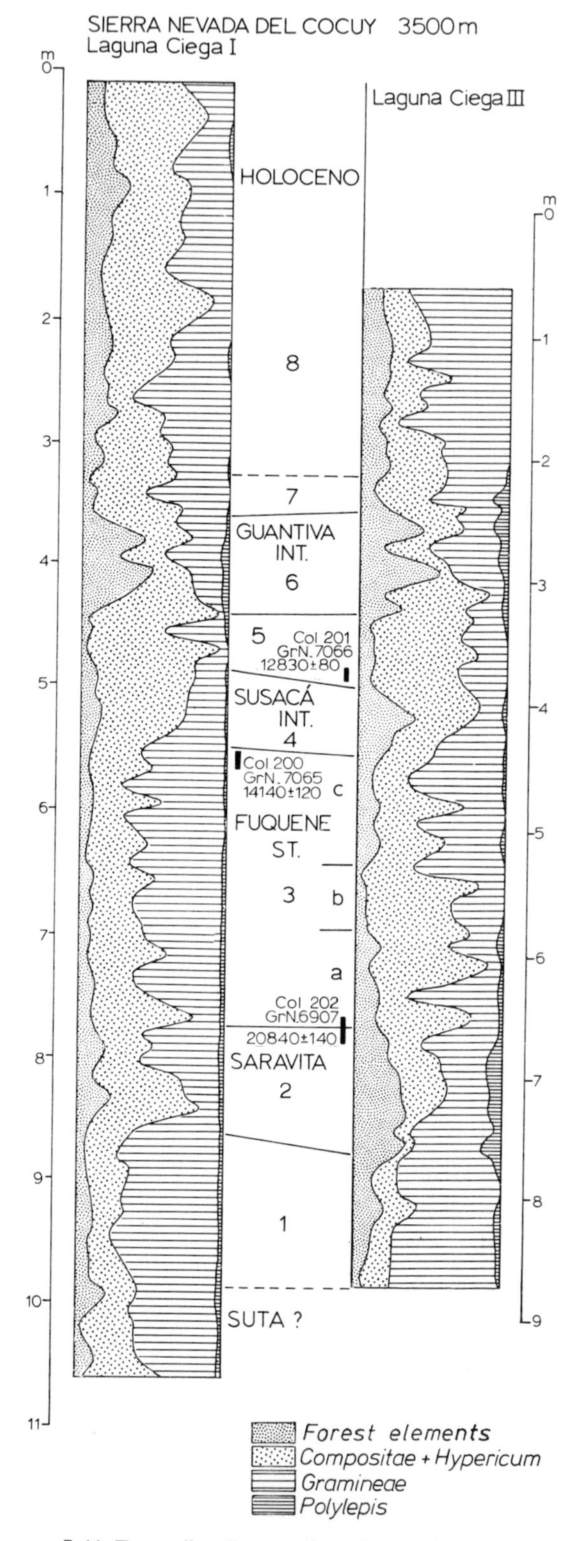

FIG. 7–11. Two pollen diagrams from Laguna Ciega, Sierra Nevada del Cocuy (Cordillera Oriental, Colombia). Altitude about 3500 m. The diagrams represent approximately the last 24,000 years. Simplified from van der Hammen et al. (1980–1981).

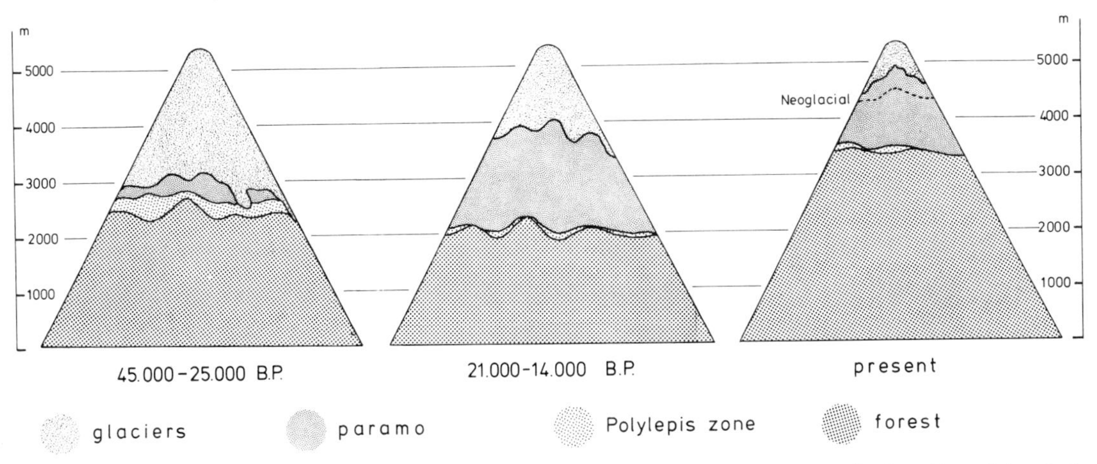

FIG. 7–12. Schematic sections through the Cordillera Oriental, showing the approximate altitudinal position of the glacier ice, the páramo, the *Polylepis* zone and the montane forest during three time intervals of the Upper Quaternary.

locally in contact (at elevations between 2200 m and 2700 m), and the páramo belt must have been relatively narrow and wet, with *Polylepis* abundant in the lower parts. Between 21,000 B.P. and 14,000 B.P., the ice extension was much less, the forest limit lower, and the climate drier, which resulted in a relatively broad and dry páramo belt.

The data available from the Pleistocene suggest that during the many glacials similar circumstances obtained. During the many interglacials, the conditions were similar to those of the Last Interglacial and the Holocene, and the altitudinal forest limit may have risen repeatedly as high as the present one and was perhaps even several hundred meters higher.

During the interglacials the páramo areas shrank and split up into many small islands; atmospheric conditions were probably more or less comparable to present ones. During the glacials, páramo areas extended ultimately to cover a surface many times larger than the present one, especially during the coldest and driest phases. The many minor islands became united to form a few extensive areas, the principal three being the Mérida area, the Cordillera Oriental area, and the Cordillera Central-Ecuador area (in addition there were three minor areas in the Cordillera Occidental and the always isolated Sierra Nevada de Santa Marta, Fig. 7–10). During the wet and cool phases of the glacials, the páramo belt was narrow, and locally even interrupted by glaciers coming into contact with forest, and relatively wet with abundant *Polylepis* forest.

It will be clear from the preceding discussion that the possibilities for immigration of taxa from the northern and southern temperate areas were best during the cold dry phases, as the receiving area was not only larger, especially at times when the temperate sources of origin were also more extensive, but also nearer owing to their displacement toward the equator. The following conclusions, therefore, seem warranted. During glacial phases a maximum immigration of páramo elements from the temperate areas and from the dry puna area took place, with an exchange of species within the separate large páramo islands, and maximum possibilities for interchange of species between these large islands. Interglacial periods were (and are) periods of minimum immigration of páramo elements from the temperate areas, but at the same time periods of maximum isolation of populations in many smaller páramo islands.

HOLOCENE HISTORY

The last major change from glacial to interglacial conditions took place during the Late Glacial. This proceeded with several temporary and colder interruptions. The sequence comprises the Susacá interstadial (ca. 14,000–13,000 B.P.), the Ciega stadial (ca. 13,000–12,400 B.P.), the Guantiva interstadial (ca. 12,400–ca. 11,000 B.P.), and the El Abra stadial (ca. 11,000–10,000 or 9500 B.P.). Before the beginning of the Late Glacial the páramo belt occupied most of the area of the present-day Andean forest belt. When the forest line rose, a succession started in this glacial páramo belt, beginning with the invasion of *Dodonaea viscosa*, *Myrica*, *Rapanea*, and *Miconia*, continuing with *Alnus*, and ending with *Quercus*.

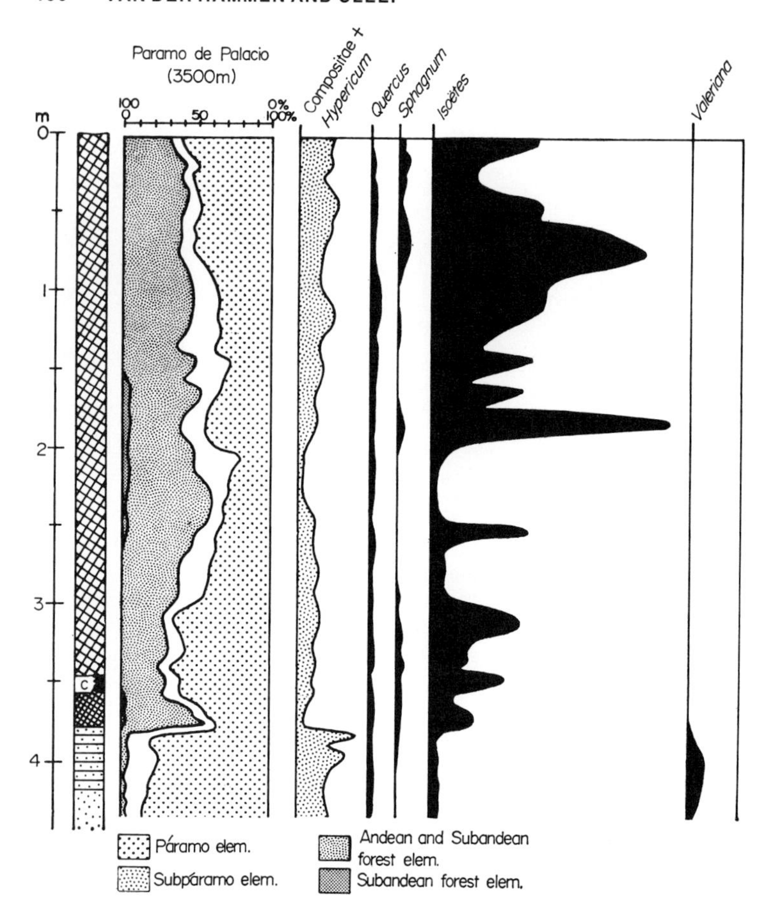

FIG. 7–13. Pollen diagram of the Late Glacial and Holocene (comprising approximately the last 13,000 years) from Páramo de Palacio (Cordillera Oriental). Altitude about 3500 m. Simplified from van der Hammen and Gonzalez (1960b).

In the first part of the Holocene this led to the final reestablishment of *Weinmannia* and *Quercus* forests in the glacial páramo belt, and the forest limit soon reached the position it occupies today.

The Holocene is not, as one might think, a period of complete stability of páramo vegetation. From the pollen diagrams of the present páramo belt (Figs. 7–13 and 7–14), and the corresponding carbon 14 dates, it is clear that relatively important changes occurred. In the first place, the altitudinal forest limit in the period from 7500 to ca. 3000 B.P. was obviously higher than today, and locally may even have extended to altitudes 300–400 m higher. This period is interrupted by a short period with a forest limit near the present one. Two climatic optima can, therefore, be recognized, corresponding to average annual temperatures about 2°C higher than at present. During these two periods, the high Andean forest may have included the present subpáramo belt, at least in several areas. This subpáramo belt in turn may have been displaced to higher altitudes. (However, it is interesting to note that in those areas where the present forest limit is very high—up to 3800–3900 m—as in the Cordillera Central, no clear subpáramo belt can be distinguished.) The area occupied by páramo vegetation was, therefore, notably smaller than it is today, and several minor grass páramo islands below about 3600 m may have disappeared almost completely (van der Hammen, 1979).

The shifting limit between páramo and forest is clearly reflected in the carbon 14 dated pollen diagram from Laguna La Primavera at approximately 3500 m in the Páramo de Sumapaz (Fig. 7–14; Cleef, 1981). The fluctuations registered are principally between subpáramo and grass páramo. The last downward movement of the belts, dated at 2900 B.P., was very sudden, indicating a marked lowering of the temperature. The beginning of an earlier, marked downward shift of the subpáramo belt (and the corresponding cooling of the climate) was carbon 14 dated at 4700 B.P. An earlier, sudden extension of grass páramo vegetation was dated at around 6300 B.P. Most of these kinds of change seem to have affected principally the boundary zones of the páramo belts.

An important aspect of the Holocene evolution of páramo ecosystems is the gradual development of soils and peat bogs. This aspect is most important in the climatically humid páramos because of

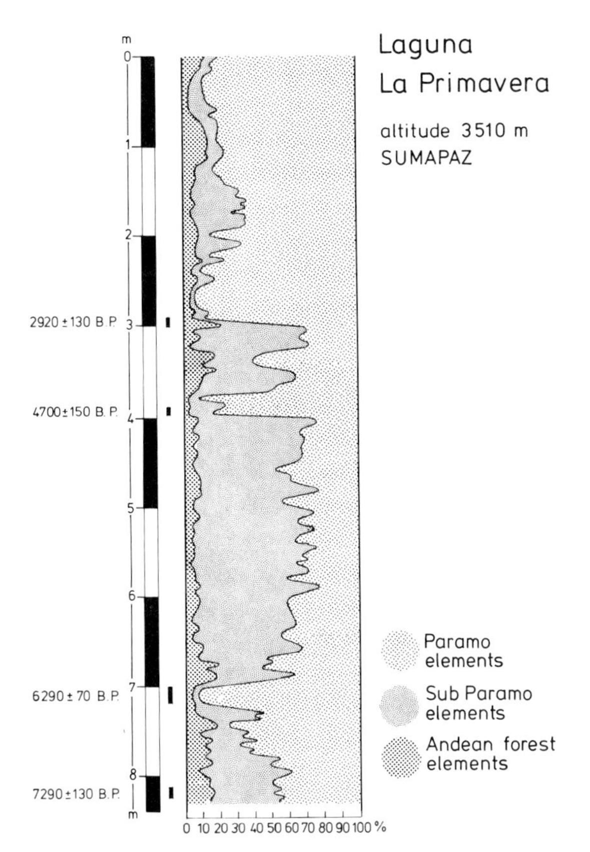

FIG. 7–14. Simplified pollen diagram from Laguna de la Primavera (Sumapaz, Cordillera Oriental) representing the last 7500 years.

the gradual accumulation of thick, black, and peaty soils and peat bogs, in a certain way representing endogenous changes of the ecosystems. *Plantago rigida* bogs may become of considerable importance in the upper grass páramo belt, and *Distichia* bogs in the superpáramo. However, most important, principally in the páramo and subpáramo belts, are the *Sphagnum* bogs, which may have a great extension in humid páramos, and thus replace the grass or bamboo páramo; it is probable that in these very humid areas the open *Sphagnum* bogs in the subpáramo and in the forest páramo ecotone may compete successfully with woody communities and may even influence the altitudinal position of the forest limit.

Carbon 14 dates from the base of peats in the páramo indicate that several bogs started to develop around 5000 B.P. Another date is ca. 7300 B.P. (van der Hammen and Gonzalez, 1960b). Frequently a relatively high percentage of *Plantago* pollen is found at the base, reflecting the vegetation type prevailing before the *Sphagnum* bog developed.

We may conclude that during the Holocene history of the páramo important changes took place in vegetation, climate, and soil. The changing climate, especially the temperature, caused reversible displacements of the boundary zones of vegetation belts of up to several hundred meters, and during the climatic optima the surface area of the páramo as a whole became reduced. Endogenous processes in the ecosystems generate a continuing process of soil development and, in combination with humidity, led to the development of extensive peat bogs and peaty soils, causing irreversible changes in the vegetation. This development is probably typical of most earlier interglacials as well, and one may say that these "irreversible" developments will come to an end only through a major climatic change such as a new glacial period causing a total breakdown of the existing systems and their soils and the creation of new ones at lower elevations.

SUMMARY, DISCUSSION, AND CONCLUSIONS

From the Upper Cretaceous into the Tertiary areas covered with open or half-open vegetation must have existed in northern South America at different elevations, ranging from savannas in the lowlands to mountain savannas and the summit vegetation characteristic of the Guayana highlands and other "tepuis" and mesas. The composition of the tepui flora teaches us that open vegetation types determined by edaphic conditions exist in highlands surrounded by tropical forest vegetation from a few hundred meters to 3000 m. Some of the floral elements of this open vegetation were derived from the surrounding forest and from tropical savannas. The available evidence does not indicate the colonization of the high Andes by pantepui elements, but suggests rather that the older and broken-up tepuis received elements from the younger Andes.

Today similar open areas may occur on low hills and on some high mountains of the Andes below the altitudinal forest line. Their presence may be determined by the local climate (low precipitation and/or marked dry season) and/or by edaphic factors (for instance sandstone hilltops). More extreme ecological conditions prevail on the tops of hills and mountains than lower down, and open or semi-open formations are frequently found there, savanna elements locally occurring at altitudes of up to 1500 m or even 2000 m. On the other hand, certain present-day páramo elements may be found down to 2500 m or even 2000 m, having emigrated to open areas on hilltops (paramillos) or on deforested slopes. Since it is known, moreover, that a good number of (sub) páramo elements have evolved from taxa whose present representatives occur in the Andean forest or even the upper Subandean forest, it is plausible that early open and semi-open vegetation types existed in the Andean area before its final upheaval, especially on hilltops, where plants from lower habitats and forests evolved and became adapted to these higher open environments, at first above 1000 m, later above 1500 m, and finally up to 2500 m. These taxa may have adapted further to become genuine páramo elements. This prepáramo vegetation might have been an important source of the (Neo)tropical and Andean elements in the later páramo vegetation types. We have no direct historical evidence of the existence of such prepáramo vegetation in the Andean area before and during the major upheaval. Their former existence is nevertheless probable. Some lowland Neotropical (or savanna) connections could thereby be explained, including *Bulbostylis*, *Paepalanthus*, *Xyris*, *Tibouchina*, and *Borreria*. The Tequendama pollen diagram pertains to this time and represents the lower part (probably Lower Pliocene) of the Tilatá Formation in the present high plain of Bogotá. It corresponds with zone I of the Plio-Pleistocene sequence and indicates the existence of a tropical lowland forest, of savanna-type vegetation, and, on nearby hills, of *Weinmannia* and *Podocarpus*. Apparently a Subandean forest with several austral-antarctic elements had by then already developed.

More extreme adaptation, especially to night frost, must have occurred as soon as mountain tops and areas above approximately 2300 m had come into being. This probably happened in the Early to Middle Pleistocene, and from that time on one may speak of protopáramo vegetation. The immigration of temperate zone elements became theoretically possible as soon as open vegetation was present above about 2300 m. The first historical indication of the existence of such a protopáramo is provided by a pollen diagram (zone III) from the Middle Pliocene Middle Tilatá Formation. The Cordillera Oriental (or at least the high plain of Bogotá) had by then risen to about 500 m below its present elevation. A Subandean forest type was present at the site, then lying at about 2300 m (now 2700 m). This open vegetation was probably present at a somewhat higher elevation. Only the pollen of abundant pollen producers (Gramineae, shrubby Compositae) could reach the site, but no elements actually restricted in their occurrence to the páramo were recorded.

At that time the Subandean forest zone was certainly well developed, but presumably the (high-)Andean forest zone had not yet fully developed. This is suggested not only by the pollen diagram of zone III, but also by the first pollen diagram of the protopáramo proper (zone IV). This diagram is from the upper part of the Tilatá Formation. The age must be Upper Pliocene; it is older than a layer of volcanic ash dated 3.62(± 0.67) million years B.P. In the diagram of zone III the presence of páramo vegetation not far above the Subandean vegetation zone was suggested, and in the diagram of zone IV (present altitude 2850 m) the presence of Subandean vegetation not too far below the páramo is indicated by the relatively abundant presence of Subandean "background" pollen (of *Alchornea* and of Malpighiaceae). At any rate, the forest line may have been at about 2500 m, i.e. 800 m or more below the present forest limit. This estimate does not take into account the possibility that the site may have been uplifted after the deposition of the material, which would make the relative lowering even greater. It is difficult to explain such displacements by a lowering of the temperature at such an early date, apparently before the beginning of the Pleistocene. This difficulty, combined with the aforementioned indications of a relatively short distance between páramo and Subandean forest, strongly suggests that the Andean forest zone was not yet fully developed and that the forest limit for that reason was lower and corresponded approximately to an average annual temperature of about 13°C (in contrast to today's 9°C). The protopáramo would have had a much larger extension than the páramo has today, would have occupied large areas above 2500 m, and woud have occurred in places with milder temperatures than those prevailing in the present páramos. The upper Andean forest zone and the páramo would have evolved more or less simultaneously. This situation can be attributed perhaps to the relatively fast final upheaval of the Andes, and may be correlated with the difficulty plant taxa of (Neo)tropical origin have in "crossing" the night-frost barrier by adapting to it.

The early protopáramo of zone IV is relatively poor in genera. It was strongly dominated by Gramineae, and apart from Compositae, such typical páramo elements as *Valeriana*, *Plantago*, Ranunculaceae, and the aquatic plant *Myriophyllum* were present. The presence of two woody genera, one endemic in the tropical Andes (*Polylepis*) and one endemic in the páramo (*Aragoa*) is of special interest. *Aragoa* is a relatively isolated taxon with austral-antarctic affinities and must be very old; it may have been an early (Mio-Pliocene) paramillo element. Elements either no longer encountered in the páramo today, or only rarely so, indicate somewhat higher temperatures in the lower páramo than at present; they include *Polygonum*, *Borreria*, and *Ludwigia*. Other early protopáramo taxa are Ericaceae, *Symplocos*, *Myrica*, *Ilex*, cf. *Hydrocotyle*, *Jamesonia*, and *Hymenophyllum*. The two genera that should be of northern and Gondwanian origin (*Myrica* and *Hypericum*, respectively) were already found in the forest zones in much older Pliocene sediments. *Myrica* at least was among the first to cross the Panama landbridge after its formation in the Early Pliocene. This early passage may have been possible through relatively low forest belts. *Jamesonia* evolved locally in the northern Andes, most probably from *Eriosorus*. The precursors of 50% of these early protopáramo taxa are Neotropical (including Andean) and those of the other 50% are temperate. Many were probably upper forest taxa that reached the páramo from below. Possibly none of the elements reached the protopáramo directly from the Holarctic by way of founder species in continental island fashion, but the southern pathway may have been an easy approach because of the low forest line. Indeed the geographic connection and the ecological and botanical relationships between páramo and puna were possibly closer than they are today.

The next phase of protopáramo, zone V of the Cordillera Oriental, lasted for a relatively long time and is much better known. A layer of volcanic ash in this zone has been dated at 3.62 (± 0.67) million years. Another dating from the same zone is on the order of 2 million years. Zone V may possibly turn out to be Early Pleistocene. Most of the new taxa identifiable by their pollen grains enter in this period. First Caryophyllaceae, *Geranium*, *Lycopodium*, *Sphagnum*, and *Isoëtes* appeared on the scene, later followed by *Gunnera*, *Gentianella corymbosa* type, *Lysipomia*, and *Acaulimalva*. The appearance of these taxa means an increase of the temperate element. This long-lasting period with a low forest limit, and accordingly occupying a much larger area than today, clearly favored the fast immigration of elements both from the north and from the south. There is no evidence that in this period the low forest line was formed by a still undeveloped upper Andean forest. Lower temperatures may have been an important factor at that time, possibly in the form of one of several early glacial periods. One possible clue that the forest line in general was even lower than at present is that interglacial periods were not very prominent. It seems, therefore, that the low altitudinal forest limit during this interval is attributable to the combined effects of low temperatures and the presence of a still incompletely developed upper Andean forest. During this period the protopáramo developed rapidly, and toward the end of zone V and the beginning of zone VI, the position of the forest line apparently became more comparable to the present one; the same holds for the páramo flora. Several new introductions reached the páramo from the south. By the end of zone V, many elements now dominant in very typical, azonal vegetation types are present, and probably a number of tropical-Andean communities, some with austral-antarctic affinities, were already well represented. Among them were azonal bog, mire, and lake communities similar to or belonging to the *Hyperico-Plantaginetum rigidae*, *Disticho-Isoëtion*, *Hydrocotylo-Myriophylletum*, and *Sphagneta*. The *Aragoetum*, "*Polylepidetum*" and *Myricetum parvifoliae* or very similar communities must also have existed at that time.

Bunch grass páramo and bamboo páramo may already have been present, and the important genera *Espeletia* and *Espeletiopsis* may already have evolved, probably from the arboreal Andean forest genus *Libanothamnus* although there is no historical proof as yet. This may have happened in the Cordillera de Mérida or in the Tamá area. There is reason to hope that the presence of these genera in the protopáramo will be demonstrated by pollen analysis in the near future.

The base of zone VI may now be provisionally dated between 2.5 and 1.5 million years B.P. It is characterized by the final establishment and sudden abundance of *Alnus*, an element of the Andean forest. The base of zone VII, characterized by the first appearance of *Quercus* in the Andean forest, was dated recently at approximately 1 million years B.P. Zones VI and VII comprise the whole, or at least a considerable part (Middle and Upper), of the Quaternary. Climatic changes in the form of glacials and interglacials (already noticeable in zone V) became obvious in this period; there must have been 15–20 of these cycles. New taxa appeared in this period, e.g., *Lathyrus* and *Montia* arriving from the temperate zones (they are of wide temperate distribution). The successive appearance and disappearance of pollen types that as yet cannot be connected wth any recent páramo taxon presumably reflect the arrival (or, alternatively, the local evolution) and the subsequent decline (perhaps to extinction) of certain species. Although many of the identifiable taxa (mainly genera) still present in the recent páramo pollen rain appeared in the protopáramo phase, evolution at the species level may have been important during zones VI and VII.

During the interglacial times, the páramo was approximately restricted to the "archipelago" it forms today (Fig. 7–10). During glacial times the areal extension of the páramo must have repeatedly approached the situation at the protopáramo stage and may at times even have been larger, the forest limit locally attaining very low altitudes below 2000 m during the extremely cold and dry phases. However, during the cold or cool-wet phases of glacial periods, the forest limit did not reach that extreme, but extended downwards to something like 2500 m; because the extension of the glaciers was considerably greater during the cold-wet phases, the páramo zone was much narrower, and glacier ice may locally have come into contact with stands of forest (Fig. 7–12). If we take as an example the last 100,000 years, for the greater part corresponding with the last interglacial-glacial cycle (and known in greater detail than the earlier cycles), we can say that for more than about 40,000 years interglacial high forest line conditions prevailed (with the forest line at 3200–3700 m); for less than 10,000 years (even possibly no more than 6000 years) extremely dry and cold glacial conditions prevailed (with the forest line locally below 2000 m); and for more than 50,000 years intermediate conditions prevailed (wet-cold, wet-cool, and drier/cool), with forest lines between 2400 and 3000 m.

Although some intervals seem to show that stands of vegetation were more or less stable, we emphasize that there was never complete or virtual stability, and that most of the time the vegetation cover was in a state of flux. It is also important to ascertain whether or not there is such a simple relation as glacial-pluvial. The amount of effective precipitation, at about 2600 m, was high during the greater part of the Last Glacial, and lower during the interglacials preceding and following it; the lowest effective precipitation fell in the last part of the Last Glacial, however. Extrapolating to the entire time interval comprising zones VI and VII (i.e., possibly the last two million years), one may draw the following general conclusions.

During the 15–20 major climatic cycles, the forest line may have been below 2000 m for 5%–10% of the time only; in principle these cold periods created maximum opportunities for the migration of páramo plants, but mainly for those of dry habitats. The páramo zone was consequently broader than it is now and occupied the uppermost part of the present Subandean forest zone, the Andean forest zone, and the páramo zone below glaciers, thus covering a surface area several times larger than that of today. Conditions for the immigration of taxa were optimal.

For about 40% of the time the tree line was more or less at its present position, or a few hundred meters higher; the opportunities for migration of páramo elements were minimal, and so were those for immigration from regions outside the páramo. The rest of the time (50%) the conditions were intermediate, and during the wettest phases the páramo zone may have been a very narrow belt (Fig. 7–12). During the wet-cold phases, the opportunities for the migration of species of wet habitats were better, and so were the conditions for the safe arrival of founder species from far away (but never so favorable as they were for the dry habitat species during the cold-dry times).

In addition, opportunities for migration were always better in the subpáramo than in the grass páramo, and better in the grass páramo than in the superpáramo. Superpáramo islands always had a much smaller surface area and were much more isolated than grass and subpáramos.

Such situations must have had a major influence on the evolution and extinction (as well as on the immigration and migration) of páramo taxa.

Pathways for the overland migration of especially dry habitat páramo plants may have been provided by the 2000-m forest line (5%–10% of the time only; Fig. 7–15). The 2500-m forest line may have been much more favorable, however, for both dry and wet habitat species (50% of the time; Fig. 7–15); whereas the area around 3000 m was available for migrations of páramo plants for at least 60% of the period. Hence, one may expect that páramo islands surrounded by the 3000-m contour line exhibit very close phytogeographic relations, those surrounded by the 2500-m contour relatively close relations (between both dry and wet habitats), and those surrounded by the 2000-m contour more remote connections. The páramo areas not connected by the 2000-m contour line most probably had no Pleistocene overland connections, and their floristic relations were also considerably less; the distribution patterns obtained by connecting areas separated by the 2000-m contour may correspond with those of early prepáramo taxa or with those directly evolved from Andean forest taxa. Or they have been dispersed by land animals inhabiting both páramo and forest, or in the air by air currents, eruptions or by birds. However, several distribution patterns of early taxa may be due to overland migration or to anemochorous distribution during the protopáramo phase, with a forest line remaining between 2000 m and 2500 m during extended intervals.

One must also bear in mind, first, that during

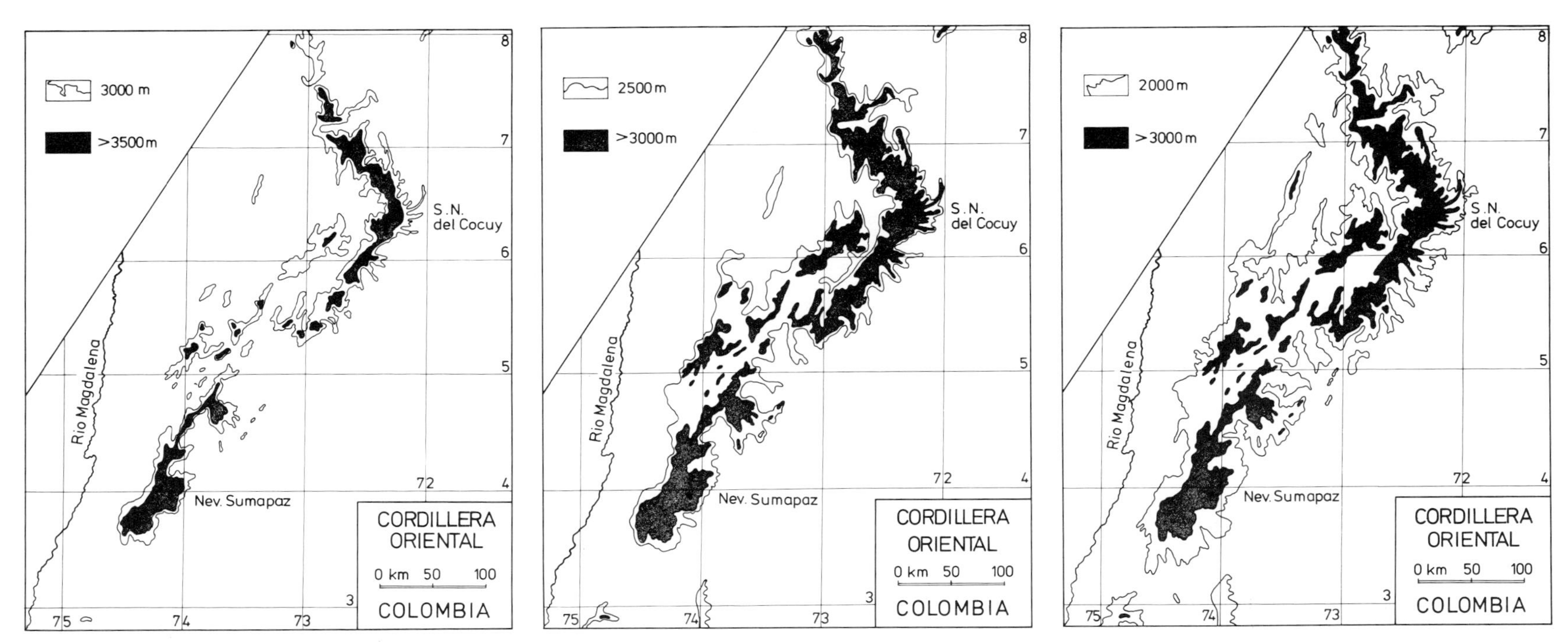

FIG. 7–15. Possible pathways for overland migration of páramo species during the Pleistocene. The present páramo areas are dotted; forest line shown at 3000 m (*left*), 2500 m (*center*), and 2000 m (*right*) respectively.

every glacial period the entire páramo vegetation and flora had to shift to lower altitudinal levels and to areas with different edaphic conditions, and second, that wet periods restricted the opportunities of survival of dry-habitat species, whereas dry periods were unfavorable for wet-habitat species. Moreover, the vegetation cover had to be rebuilt every time. Peat bog ecosystems, evolved during thousands of years, gradually built up thick peat layers, but such a succession had to start all over again in newly invaded territories. Return to previously glaciated and denuded areas necessitated a long-lasting pioneer stage before a sufficient layer of soil was formed to support more stable "climax" communities. Therefore, the story is not simply one of a vertical displacement of vegetation zones without any major change in existing vegetation types, but such shifts also resulted in sometimes considerable differences in the floral composition and distribution of the vegetation types making up the displaced páramo zones. This may have caused the extinction of elements whose ecological niches were temporarily unavailable or severely restricted in their occurrence.

It is also clear that with regard to evolution and speciation, the isolation of populations in different páramo areas became more pronounced during interglacial periods, whereas glacial periods offered far great possibilities for the migration and interchange of floral elements.

From all the evidence accumulated so far an intricate historical picture can be constructed. Obviously the present composition of the páramo flora and the distribution of the individual páramo taxa cannot be explained in any simple way. In view of our knowledge of the history of páramo elements, one can attempt to pinpoint the most probable historical factors responsible for the present composition of the separate páramo islands, for the floristic connections between them, and for the present distribution patterns of páramo taxa.

As most high plains in the Cordillera Oriental lie between 2550 m and 2600 m, a sudden and major change in the surface area of páramo occurred each time the forest limit shifted to below about 2500 m, when most páramo islands became united; a further downward migration to 2000 m made little difference. It seems, therefore, as if most páramo islands in the Cordillera Oriental were more or less effectively connected 60% of the time during the Pleistocene. The superpáramos (Cocuy, Almorzadero, and Sumapaz) never became connected, however, and probably most of the present páramo areas below about 3600 m and both the superpáramos of the Almorzadero and Sumapaz (both below 4374 m), disappeared almost completely during the Middle Holocene warm period (and during at least a part of most interglacials), to become largely repopulated after 3000 B.P. Several species may however have survived in situ under extreme mountain summit conditions.

The Cocuy superpáramo isolation is evident from the fact that almost 20% of the species are endemic to this small area. The fact that not a single genus is endemic to this area, however, and only one to the superpáramo in general, is attributable to the relatively small surface area of the superpáramo island(s), to their relatively young age (certain elevations having been reached only toward the end of the Andean upheaval), and to the probable disappearance of the greater part of the superpáramo habitat during each cold and very wet period of glacial extension.

As regards the grass and subpáramos in general, major connections and separations are successively indicated by the 3000-m (forest) contour (e.g., by several larger areas in the Cordillera Oriental: Sumapaz-Cordillera de Bogotá island, Guantiva island, Boyacá-Cocuy-Almorzadero island; Fig. 7–15). These separations and connections are reflected in the distribution of many species in the genera *Espeletia*, *Niphogeton*, *Puya*, *Diplostephium* and others.

The next level of connection and separation is indicated by the 2500–2000-m (forest) contours (Fig. 7–15). Examples are the Cordillera Oriental from Cáchira to Sumapaz (with more than 7% endemic genera and 35% endemic species), the Cordillera de Mérida, the Cordillera Central, the Cordillera Occidental "islands," the northern Serranía de Perijá, and the Sierra Nevada de Santa Marta. The floristic and vegetational relations between these areas (corresponding with the major Pleistocene páramo, and partly with certain protopáramo islands) cannot be readily explained by simple overland connections and can be understood only from the reconstruction of their early history (i.e., from the early evolution and/or distribution in prepáramo vegetation and upper forest), or by incidental middle- to long-distance (aerial or ornithochorous) dispersal. As mentioned previously, both the possibility and probability of such forms of dispersal were greatly enhanced during times of maximum extension of páramos during the glacial periods. A disjunct area such as that of the apparently young genus *Floscaldasia* (superpáramo, Cocuy, and northern Cordillera Central) can be explained only by long-distance aerial dispersal. The disjunct areas of *Pentacalia vernicosa* and

Rumex tolimensis (Fig. 7–3g; Sumapaz and Cordillera Central) may be the result of long-distance dispersal (volcanic eruptions, favorable winds) combined with overland dispersal, because during the maximum lowering of the forest limit during glacial times páramo may have extended from the Cordillera Central through the southern Cordillera Oriental to very near Sumapaz. A disjunct distribution such as that of *Aragoa* (Fig. 7–3h; Cordillera Oriental, Cordillera de Mérida, Cordillera Occidental, and Sierra Nevada de Santa Marta), may be explained partly by the early history of this genus in prepáramo vegetation, partly by overland dispersal, and partly by aerial dispersal.

Aragoa is absent from the Cordillera Central. In this connection it is important to mention that the páramo flora of the Cordillera Central is relatively poor in species, in comparison to the Cordillera Oriental, for example. This may be explained by accepting a younger age of the higher part of the Cordillera Central and allowing for a shorter time of evolution and immigration. This is not at all unlikely in view of the (Plio-) Pleistocene volcanic origin of this high mountain zone. On the other hand, the continual and often catastrophic volcanic activity may have increased the chances of extinction considerably, so that, as far as the páramos of the Colombian Cordillera Central and Ecuador are concerned, an additional and completely different (namely, geological) factor has to be taken into account to explain the composition of the present-day flora and vegetation cover.

Vuilleumier (1970) compared the bird faunas of different páramos and concluded that they can be explained by a combination of immigration and extinction processes, similar to those of oceanic islands (MacArthur and Wilson, 1967). The comparison of the distribution of páramo areas as isolated "islands" above extensive and almost continuous stands of forest, with a biological archipelago, had already been suggested by Murillo (1951). By assuming that the number of species in isolated páramos is in a state of dynamic equilibrium and that the surface area of the páramos has remained unaltered, Vuilleumier (1970) could predict the number of species by means of an equation in which the distance from the presumed source area and the planar area of the island are the main variables. The resulting high correlation between the predicted and the actual number of species can be taken as a probable indication that the immigration of the birds took place (or reoccurred) in the Holocene and that a state of equilibrium was reached in the last 10,000 years. In a later paper, however, Vuilleumier and Simberloff (1980) emphasized how difficult it is to deduce an equilibrium from the data at hand, since there is no evidence about turnover.

Simpson (1975) made a similar analysis for some plant taxa, mainly of temperate origin. She also found that the principal parameters explaining the number of species were the size of the island and its distance from the source areas. She found a higher correlation coefficient when the size of glacial páramo islands (based on the 2500-m contour) was used. This suggested to her that "insular" immigration (and extinction) took place mainly during glacial times. As pointed out before, páramo "islands" connected by the 2500-m contour line may have existed during a considerable part of the Last Glacial and of the Pleistocene in general (2400–3000-m altitude of the forest line some 50% of the available time), apparently permitting the establishment of some equilibrium. It must be kept in mind that such a situation prevailed also in the protopáramo phase for a considerable length of time. It would also mean that a relatively short interval, such as the Holocene (10,000 years), is not sufficient to enable the attainment of a state of equilibrium among the species of higher plants (this in contrast to birds, which of course can migrate at a much faster rate).

Recently Gilbert (1980) critically examined the studies purporting to validate the equilibrium theory of island biogeography, and concluded that there is but little confirmation of its tenets and that one should certainly not accept it at face value. Nonetheless Vuilleumier's (1970) results are convincing to us, and those of Simpson (1975) at least suggestive. However, as we have learned from our historical studies, there has been a continuous change of flora and vegetation, of vegetation belts, and of the elevation of the forest limit (and accordingly of the size of the páramo islands and of the distribution and extension of minor habitats). This dynamic state of páramo islands and their habitats renders it highly improbable that a veritable equilibrium of the floristic composition is ever reached and that a status quo can only be reached by such widely migrating taxa as birds.

Summarizing the evidence, one may conclude that the evolutionary history of the high Andean páramo flora and vegetation is very complicated and that different islands need not have had the same history. Only very young páramo islands (e.g., those found below 3500 m in the Cordillera Oriental, which are possibly only 3000 years old) may have been populated entirely in the fashion of the oceanic island model, but the available time span may have been too short for the attainment of a state of equilibrium. All other páramo islands were populated in various ways: by evolutionary adaptation of taxa emigrating upward from the upper forest, by dispersal overland and by long- to middle-distance aerial dispersal, and by speciation through isolation. This history began (after a theoretical prepáramo phase) in the Pliocene and Early Pleistocene protopáramo. The forest line at that time seems to have been lower with respect to the isotherms than it is today. Provided the age determinations are correct, the present páramo flora and vegetation must be at least four million years old. During this time span apparently more than 30 endemic genera evolved (out of over 300, i.e., about 10% of the total number), and in the Cordillera Oriental alone some 250 endemic species (of a total of 700, about 35%). A very high percentage of the species growing in the páramo is endemic to the páramo and/or of Neotropical-Andean origin. Of the genera about half are of local and of Neotropical origin, and the other half of temperate origin. Of the temperate genera, about 10% are Holarctic; of the other 40%, 10% are certainly of austral-antarctic origin, but most of the remainder probably invaded the páramo from the south. Immigration from the south must have been much easier than from the north because there were relatively fewer geographic discontinuities. Immigration from the north could start only after the formation of the Panama landbridge. Although the nearest distance between present-day páramo islands on both sides of the landbridge is 750 km, intermediate, stepping stones were probably formed by the hills reaching over 2000 m during glacial times, which reduced the closest distance to about 350 km. The local, Neotropical element is more abundant both in the flora and as dominant species in the vegetation types of the subpáramo. The percentage of the Holarctic element in the flora increases upward and is highest in the superpáramo. Many typical azonal vegetation types in lakes and bogs have austral-antarctic (southern) affinities; some of their principal constituting taxa were already present in the protopáramo, and so were such typical, Andean woody taxa as *Polylepis* and *Aragoa*. The 2500-m contour line indicates the longest-lasting approximate glacial forest line position; this may also have been a long-lasting approximate position during protopáramo times. Interestingly the best results calculated by means of the oceanic island model were obtained by taking this forest line level as defining the "islands." There is apparently no state of equilibrium in the present Holocene situation. Roughly 15–20 glacial cycles can be recognized since protopáramo times in the last 2(± 0.5) million years. During each of these approximately 100,000-year-long cycles páramo zones and their vegetation cover were completely displaced and profoundly changed. There were phases of relatively dry-warm, dry-cold, wet-cold, and wet-cool climates, and a stable situation rarely obtained, the vegetation cover being subjected mostly to slower or faster changes in its floristic composition. The isolation of many páramo islands, in conjunction with the incidental fusion of some of them during glacial times, the great diversity of environments and niches, and the immigration of founder species all contributed to the evolution of a páramo flora and vegetation of great diversity, notwithstanding the probably high rate of extinction caused by repeated and profound changes in the floristic composition, by local catastrophic volcanic eruptions, and by regional glacial events.

REFERENCES

Azócar, A., and M. Monasterio. 1979. Variabilidad ambiental en el Páramo de Mucubají. In *El medio ambiente páramo*, M. L. Salgado-Labouriau, ed., pp. 149–159. Actas del seminario de Mérida, Venezuela, November 5–12, 1978. Caracas: Ediciones Centro de Estudios Avanzados.

Cabrera, A. L. 1958. La vegetación de la puna argentina. *Reg. Invest. Agric.*, Buenos Aires 11: 315–412.

Camargo G., L. A. 1966. Especies nuevas del género *Berberis* de Colombia, Ecuador y Venezuela. *Caldasia* 9: 313–351.

Clapperton, C. M. 1979. Glaciation in Bolivia before 3.27 Myr. *Nature* 277: 375–377.

Cleef, A. M. 1978. Characteristics of neotropical páramo vegetation and its subantarctic relations. In *Geoecological relations between the southern temperate zone and the tropical mountains*, C. Troll and W. Lauer, eds., pp. 365–390. Wiesbaden: Erdwissenschaftliche Forschung, 11.

———. 1979a. The phytogeographical position of the neotropical vascular páramo flora with special reference to the Colombian Cordillera Oriental. In *Tropical botany*, K. Larsen and L. B. Holm-Nielsen, eds., pp. 175–184. London: Academic Press.

———. 1979b. Secuencia altitudinal de la vegetación de los páramos de la Cordillera Oriental, Colombia. *Actas del IV Simposium Internacional de Ecología Tropical*, Vol. 1, pp. 281–297. Panama.

———. 1981. *The vegetation of the páramos of the Colombian Cordillera Oriental*. Dissertationes Botanicae 61. Vaduz: Cramer.

———. (1983). Fitogeografía y composición de la flora vascular de los páramos de la Cordillera Oriental, Colombia. Una comparación con otras montañas tropicales. *Rev. Acad. Col. Cienc. Ex. Fis. Nat.* 15(58): 23–31.

Cuatrecasas, J. 1954a. Outline of vegetation types in Colombia. *Rapports Comm. 8ème Congr. Int. Bot. Sect.* 7: 77–78.

———. 1954b. Synopsis der Gattung *Loricaria* Wedd. *Fedd. Rep.* 56: 149–172.

———. 1958. Aspectos de la vegetación natural de Colombia. *Rev. Acad. Col. Cienc. Ex. Fis. Nat.* 10: 221–269.

———. 1968. Páramo vegetation and its life forms. *Coll. Geogr.* 9: 163–186.

———. 1969. Prima flora Colombiana. 3. Compositae-Asterae. *Webbia* 24: 1–335.

———. 1979. Comparación fitogeográfica de páramos entre varias Cordilleras. In *El medio ambiente páramo*, M. L. Salgado-Labouriau, ed., pp. 89–99. Actas del seminaria de Mérida, Venezuela, November 5–12, 1978. Caracas: Ediciones Centro de Estudios Avanzados.

———. 1980. Miscellaneous notes on neotropical flora, 12. *Phytologia* 47: 1–13.

Cuatrecasas, J., and A. M. Cleef. 1978. Una nueva Crucífera de la Sierra Nevada del Cocuy (Colombia). *Caldasia* 12: 145–158.

Dueñas, H. 1979. Estudio palinológico de los 35 mts. superiores de la sección Tarragona, Sabana de Bogotá. *Caldasia* 12: 539–571.

Fuchs, H. P. (1982): Zur heutigen Kenntnis von Vorkommen und Verbreitung der südamerikanischen Isoëtes Arten. *Proc. Kon. Ned. Akad. Wet., Ser. C* 85: 205–260.

Gilbert, T. S. 1980. The equilibrium theory of island biogeography: Fact or fiction? *J. Biogeogr.* 7: 209–235.

Gradstein, S. R., A. M. Cleef, and M. H. Fulford. 1977. Studies on Colombian cryptogams. II. Hepaticae—oilbody structure and ecological distribution of selected species of tropical Andean Jungermanniales. *Proc. Kon. Ned. Akad. Wet., Ser. C* 80: 377–420.

Griffin, D. 1979. Briófitos y líquenes de los páramos. In *El medio ambiente páramo*, M. L. Salgado-Labouriau, ed., pp. 79–87. Actas del seminario de Mérida, Venezuela, November 5–12, 1978. Caracas: Ediciones Centro de Estudios Avanzados.

Guhl, E. 1974. Las lluvias en el clima de los Andes ecuatoriales húmedos de Colombia. *Cuadernos Geogr.* 1. Bogotá.

Harling, G. 1979. The vegetation types of Ecuador. A brief survey. In *Tropical botany*, K. Larsen and L. B. Holm-Nielsen, eds., pp. 165–174. London: Academic Press.

Hooghiemstra, H. 1984. *Vegetational and climatic history of the highplain of Bogotá, Colombia: A continuous record of the last 3.5 million years.* Dissertationes Botanicae 79. Vaduz: Cramer.

Lauer, W., and P. Frankenberg. 1978. Untersuchungen zur Oekoklimatologie des östlichen Mexico—Erläuterungen zu einer Klimakarte 1:500.000. *Coll. Geogr.* 13: 1–134.

MacArthur, R. A., and E. O. Wilson. 1967. *The theory of island biogeography.* Princeton: Princeton Univ. Press.

Mathias, M. E., and L. Constance. 1951. A revision of the Andean genus *Niphogeton* (Umbelliferae). *Univ. Calif. Publ. Bot.* 23: 405–425.

———. 1962. The Andean genus *Niphogeton* (Umbelliferae) revisited. *Brittonia* 14: 148–155.

———. 1967. Some Umbelliferae of the Andean páramos of South America. *Brittonia* 19: 212–226.

———. 1976. The genus *Niphogeton* (Umbelliferae)—a second encore. *Bot. J. Linn. Soc.* 72: 311–324.

Monasterio, M. 1979. El páramo desértico en el altiandino de Venezuela. In *El medio ambiente páramo*, M. L. Salgado-Labouriau, ed., pp. 117–146. Actas del seminario de Mérida, Venezuela, November 5–12, 1978. Caracas: Ediciones Centro de Estudios Avanzados.

———., ed. 1980. *Estudios ecológicos en los páramos andinos.* Mérida, Venezuela: Universidad de los Andes.

Murillo, L. M. 1951. Colombia, un archipiélago biológico. *Rev. Acad. Col. Cienc. E. F. Nat.* 8: 168–220.

Øllgaard, B., and H. Balslev. 1979. Report on the 3rd Danish expedition to Ecuador. *Rep. Bot. Inst. Univ. Aarhus* 4.

Pennell, F. W. 1937. Taxonomy and distribution of *Aragoa* and its bearing on the geological history of the Northern Andes. *Proc. Acad. Nat. Sciences Philadelphia* 89: 425–432.

Robson, N. K. B. 1977. Studies in the genus *Hypericum* L. (Guttiferae): 1. Infrageneric classification. *Bull. Brit. Mus. (Nat. Hist.)* 5: 291–355.

Schreve-Brinkman, E. J. 1978. A palynological study of the upper Quaternary sequence in the El Abra corridor and rock shelters (Colombia, South America). *Palaeogeogr. Palaeoclimatol. Palaeoecol.* 25: 1–109.

Shackleton, N. J. and N. D. Opdyke. 1977. Oxygen isotope and palaeomagnetic evidence for early Northern Hemisphere glaciation. *Nature* 270: 216–219.

Simpson, B. B. 1975. Pleistocene changes in the flora of the high tropical Andes. *Paleobiology* 1: 273–294.

———. 1979. A revision of the genus *Polylepis* (Rosaceae: Sanguisorbeae). *Smith. Contr. Bot.* 43: 1–62.

Sipman, H. J. M., and A. M. Cleef. 1979. Studies on Colombian cryptogams. V. Taxonomy, distribution and ecology of macrolichens of the Colombian páramos, 1. *Cladonia* subgenus *Cladina. Proc. Kon. Ned. Akad. Wet., Ser. C* 81: 223–241.

Smith, L. B. 1976. Notes on Bromeliaceae 38. *Phytologia* 33: 429–445.

Smith, L. B., and R. J. Downs. 1974. *Pitcairnioidae* (Bromeliaceae). *Flora Neotropica*, monograph 14. New York.

Steyermark, J. A. 1966. Contribuciones a la flora de Venezuela, parte 5. *Acta Bot. Venez.* 7: 9–256.

———. 1979. Flora of the Guayana highland: Endemicity of the generic flora of the summits of the Venezuelan Tepuis. *Taxon* 28: 45–54.

Steyermark, J. A., and O. Huber. 1978. *Flora del Avila.* Caracas: Soc. Ven. Cienc. Nat.

Troll, C. (1958). Structure soils, solifluction, and frost climates of the Earth. U. S. Army Snow, Ice and Permafrost Establ. Translation 143.

Uribe U., L. 1972. Melastomatáceas. In *Catálogo ilustrado de las plantas de Cundinamarca* 5: 61–150.

van der Hammen, T. 1961. Late Cretaceous and Tertiary

stratigraphy and tectogenesis of the Colombian Andes. *Geol. Mijnb.* 40: 181–188.
———. 1974. The Pleistocene changes of vegetation and climate in tropical South America. *J. Biogeog.* 1: 3–26.
———. 1979. Changes in life conditions of earth during the past one million years. A. J. C. Jacobsen Memorial Lecture. *Det Kongelige Danske Videnskabernes Selskab Biologiske Skrifter* 22(6): 1–32.
———. 1979. Historia de la flora y la vegetación en la región montana alta de Colombia. *Actas del IV Simposium Internacional de Ecología Tropical*, Vol. 1, pp. 379–391. Panama.
van der Hammen, T., and E. Gonzalez. 1960a. Upper Pleistocene and Holocene climate and vegetation of the "Sabana de Bogotá" (Colombia, South America). *Leidse Geol. Meded.* 25: 261–315.
———. 1960b. Holocene and Late Glacial climate and vegetation of Páramo de Palacio (Eastern Cordillera, Colombia, South America). *Geol. Mijnb.* 39: 737–746.
———. 1964. A pollen diagram from the Quaternary of the Sabana de Bogotá (Colombia) and its significance for the geology of the Northern Andes. *Geol. Mijnb.* 43: 113–117.
———. 1965. A Late-glacial and Holocene pollen diagram from Cienaga del Visitador (Dept. Boyacá, Colombia). *Leidse Geol. Meded.* 32: 193–201.
van der Hammen, T., J. H. Werner, and H. van Dommelen. 1973. Palynological record of the upheaval of the Northern Andes: A study of the Pliocene and Lower Quaternary of the Colombian Eastern Cordillera and the early evolution of its High-Andean biota. *Palaeogeog. Palaeoclimatol. Palaeoecol.* 16: 1–122.
van der Hammen, T., J. Barelds, H. de Jong, and A. A. de Veer. 1980–81. Glacial sequence and environmental history in the Sierra Nevada del Cocuy (Columbia). *Palaeogeog. Palaeoclimatol. Palaeoecol.* 32.
van Donselaar, J. 1965. An ecological and phytogeographic study of northern Surinam savannas. *Wentia* 14: 1–163.
van Geel, B., and T. van der Hammen 1973. Upper Quaternary vegetational and climatic sequence of the Fúquene area (Eastern Cordillera, Colombia). *Palaeogeog. Palaeoclimatol. Palaeoecol.* 14: 9–92.
Vareschi, V. 1955. Rasgos geobotánicas sobre el Pico de Naiguatá. Monografías geobotánicas de Venezuela I. *Acta Cient. Ven.* 6: 2–23.
Vuilleumier, F. 1970. Insular biogeography in continental regions. I. The northern Andes of South America. *Amer. Nat.* 104: 373–388.
Vuilleumier, F. and D. Simberloff. 1980. Ecology versus history as determinants of patchy and insular distributions in high Andean birds. In M. K. Hecht, W. C. Steere and B. Wallace, ed., *Evolutionary biology*, Vol. 12, pp. 235–379. New York: Plenum Publishing Corporation.
Weischet, W. 1969. Klimatologische Regeln zur Vertikalverteilung der Niederschläge in den Tropengebirgen. *Die Erde* 100 (2–4): 287–306.
Wurdack, J. J. 1973. Melastomataceae. In *Flora de Venezuela*, ed. T. Lasser, Vol. 8. Caracas.

NOTE ADDED IN PROOF

New facts and data have been published since the text of this chapter was written. The relevant additional literature is given below.

Kuhry-Helmens & Kuhry (in press) studied the vegetation and climate of the last half million years in sediments of a small páramo lake in the headwaters of the Subachoque Valley, which is a northern extension of the Bogotá high plain. Kuhry et al. (1983), Melief (1985), and Salomons (in prep.) added substantially to our knowledge of the Late Quaternary of the páramo belt of the Cordillera Central (Parque Los Nevados area) and the Cordillera Oriental (Páramo de Sumapaz). This work is based on palynological and paleoecological studies of lacustrine sediments and soil sections, respectively. Remarkable are the downward shift of the climatic humid superpáramo vegetation dominated by bryophytes in the Nevado de Sumapaz toward about 3000 m in the Last Glacial (diagram Alsacia; Melief, 1985), and the occurrence of páramo-like vegetation at about 1700 m on the western slope of the Central Cordillera near Pereira (Salomons, in prep.).

Exploration of the páramo flora and vegetation continues in Ecuador (Flora of Ecuador project), Venezuela (Mérida Andes, research headed by M. Monasterio; Guayana tepuis, O. Huber) and Costa Rica (Weston; and Chaverri & Cleef, in prep.). In the Colombian Andes exploration was carried out by ECOANDES team members in the summit areas of the Parque Los Nevados, Sumapaz, and Tatamá, western Cordillera, but also in the páramos near Bogotá, the Sierra Nevada de Santa Marta, the Nevado de Huila, Cumbal (Sturm & Rangel, 1984), and elsewhere. Results have been published in the series "Studies on tropical Andean ecosystems" (van der Hammen *et al.*, 1983, 1984, and in prep.) and in "Studies on Colombian cryptogams" (Gradstein, 1982, 1983, and in prep.).

Some new floristic data are interesting in the context of this chapter. In the Páramo de Sumapaz some cushions of *Distichia muscoides* (Juncac.) have been found. *Vesicarex collumanthus* (Cyper.) was assigned to *Carex* by Mora (1983; see also Cleef, 1983). *Senecio niveo-aureus* (Fig. 3a) and *Rumex tolimensis* (Fig. 3g) were found in páramos north of Bogotá (Bekker and Cleef, in press). The number of species of *Espeletia* (Comp.) is expected to increase slightly in the future, since undescribed species were observed during the ECOANDES transects. A second species of *Tamania* (Comp.) apparently turned up in the Cordillera del los Cobardes, Eastern Cordillera (Díaz-Piedrahíta, pers. comm.). The record of *Loricaria complanata* (Comp.) in the Sierra Nevada de Santa Marta (Fig. 3m) is doubtful (Cuatrecasas, pers.

comm.). From the same páramo, however, a second species of *Cotopaxia* (Umbellif.) was described (Constance and Alverson, 1984). *Sericotheca argentea* (Rosac.) seems better placed in *Holodiscus; Holodiscus* has its main distribution center in México.

ADDITIONAL LITERATURE

Bekker, R. and A.M. Cleef. (in press). La vegetación del Páramo de la Laguna Verde (Mnpio. de Tausa, Cundinamarca). Colombia Geográfica. IGAC, Bogotá.

Chaverri, A. and A.M. Cleef. (in prep.). La vegetación de páramos en la Cordillera de Talamanca, Costa Rica. *Brenesia*.

Cleef, A.M. 1982. Distribución y ecología de *Vesicarex collumanthus* Steyermark (Cyperac.). *Acta Biol. Col.* 1:43-49.

Constance, L. and W.S. Alverson. 1984. A second species of *Cotopaxia* (Umbelliferae/Apiaceae), from Colombia. *Caldasia* 14(66): 21-26.

Gradstein, S.R. (ed.). 1982. *Studies on Colombian cryptogams*, 1-10. Vol. I. Utrecht: Institute of Systematic Botany.

Gradstein, S.R. (ed.). 1983. *Studies on Colombian cryptogams*, 11-20. Vol. II. Utrecht: Institute of Systematic Botany.

Kuhry, P., B., Salomons, P.A., Riezebos, and T. van der Hammen. 1983. Paleoecología de los últimos 6000 años en el área de la Laguna de Otún — El Bosque. In T. van der Hammen et al., eds., *La Cordillera Central Colombiana, transecto Parque Los Nevados*. Studies on Tropical Andean Ecosystems 1:227-262.

Kuhry-Helmens, K.F. and P. Kuhry. (in press). Middle and Upper Quaternary vegetational and climatic history of the páramo de Agua Blanca (Eastern Cordillera, Colombia). *Palaeogeogr., Palaeoclimatol., Palaeoecol.*

Melief, A.B.M. 1985. Late Quaternary paleoecology of the Parque Nacional Natural los Nevados (Cordillera Central), and Sumapaz (Cordillera Oriental) areas, Colombia. Thesis, University of Amsterdam.

Mora-Osejo, L.E. 1982. Consideraciones sobre la morfología, antomía y posición sistemática de *Vesicarex* Steyermark (Cyperaceae). *Acta Biol. Col.* 1:31-41.

Salomons, B. (in prep.). Palynology of volcanic soils in Colombia. Late Quaternary climate and vegetation reconstruction of the "Parque Nacional Natural los Nevados," Cordillera Central.

Sturm, H. and O. Rangel. 1985. *Ecología de los páramos andinos: Una visión preliminar integrada*. Bogotá: Instituto de Ciencias Naturales — Museo de Historia Natural, Universidad Nacional de Colombia, Biblioteca José Jeronimo Triana no. 9.

8

Late Quaternary Paleoecology of Venezuelan High Mountains

MARIA LÉA SALGADO-LABOURIAU

QUATERNARY ECOLOGY AS A BACKGROUND FOR PALEOECOLOGY

Three mountain systems are found in Venezuela: the Andes in the west; the Caribbean Mountains along the coast (Cordillera de la Costa); and the Guayana Highlands in the south. Since there are no paleoecological studies of the Guayana Highlands, I will limit myself to the first two mountain systems.

The Mérida Andes

The Andes branch into three roughly parallel cordilleras in Colombia. Near the Venezuelan border, the Eastern Cordillera divides into two ranges (Fig. 8–1) separated by the Maracaibo Basin: the Sierra de Perijá and the Mérida Andes. The Mérida Andes extend northeasterly from 7° to 10° N latitude and are approximately 60–100 km wide and 450 km long. They contain the highest mountains in Venezuela, and consist of two steep ranges separated by a deep valley. Elevations range from about 200 m at both flanks (llanos and Maracaibo basins) up to 5000 m in the Sierra Nevada de Mérida. The Mérida Andes end near Barquisimeto, where they are in contact with the western end of the Caribbean Mountains (see Fig. 8–4).

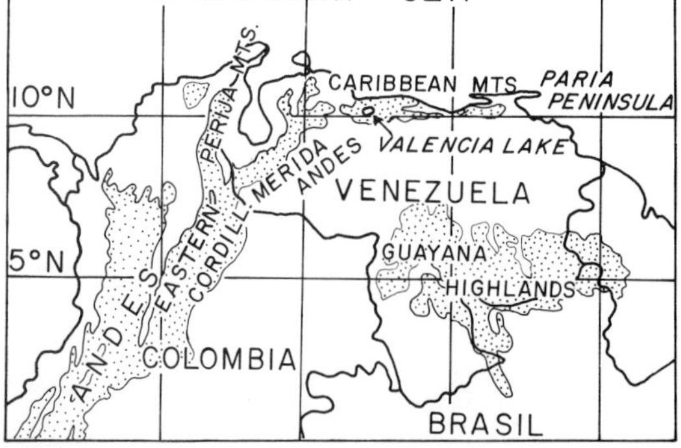

FIG. 8–1. Northern South America, depicting the three mountain systems of Venezuela. Dotted areas: elevations higher than 500 m above sea level.

Evidence from the Eastern Cordillera of Colombia (van der Hammen, 1974, and Chap. 7 this volume) shows that the latest major upheaval took place during the Middle to the Late Pliocene, and that by the end of the Pliocene it had reached its present elevation. Since the Venezuelan Andes are the northeastern continuation of the Eastern Cordillera, it is probable that they too have reached their present elevation before the beginning of the Quaternary (Schubert, 1979a), and the latest period of uplift is said to have begun in the Late Eocene (Shagam, 1972).

Glaciation in the Mérida Andes

At present the snowline is at approximately 4700 m above sea level and only the highest peaks of the Sierra Nevada de Mérida support glaciers. The area covered by glaciers is limited to about 2.7 km^2 (Table 8–1).

Although the glaciers now are very small, glacial action in the past reached much lower elevations. In the present-day periglacial zone

TABLE 8–1. Extent of the Mérida Glaciation in the Venezuelan Andes.

	Glaciation area (km^2)	Periglacial zone (km^2)
Glacial maximum estimation	600	2200
Today's distribution	2.7	1200
Reduction (percent)	99.5	45.5

Source: Recalculated from Salgado-Labouriau and Schubert (1977) and Schubert (1979a).

(3500–4700 m) the landscape is dominated by glacially eroded features: cirques, glacial valleys, moraines, roches moutonnées, arêtes, and horns (Schubert, 1975a). Below it are two major levels of moraines (Schubert, 1974; Schubert and Valastro, 1974).

Between 3000 and 3500 m is a prominent level of well-preserved moraines (Late Stade) now covered by páramo vegetation. At 2600–2700 m there is a level of poorly preserved moraines (Early Stade), which is within the present day forest belt. The morainic levels indicate the existence of a glacial period at the end of the Pleistocene, defined as the Mérida Glaciation (Schubert, 1974) and correlated with similar periods in Colombia and Peru (Schubert, 1979a). No evidence of a glaciation older than this has yet been found in Venezuela, but the possibility of older glaciation is not excluded.

The geologic evidence shows that during the Mérida Glaciation, the snow line was lowered by 1200–1700 m and glaciers covered about 600 km^2 (Table 8–1). Much of the region at present covered by páramo vegetation was then under ice. The altitudes of cirques and moraines suggest the snow line was about 500 m higher in the northwestern part of the Mérida Andes (Schubert, 1975a).

Unfortunately, the lack of direct absolute dating of the Venezuelan moraines does not allow one to date accurately the Mérida Glaciation. Clastic sediments and peat within the morainic loops have yielded minimum radiocarbon ages older than 10,000 years (Schubert and Valastro, 1974; Giegengack and Grauch, 1975; Salgado-Labouriau, Schubert, and Valastro, 1977). A group of fluvioglacial terraces in the Chama River basin (Mesa del Caballo, 3380 m) yielded ages from 16,500 ± 290 to 19,080 ± 820 radiocarbon years B.P. (Schubert, pers. comm). The Mucubají glacial valley, above 3550 m, was already deglaciated at 12,650 ± 130 B.P. and was covered by superpáramo type of vegetation (Salgado-Labouriau et al., 1977). The good preservation of the Late Stade moraines indicate that they are probably of Late Pleistocene age, and that the ice had retreated from the morainic valleys before the beginning of the Holocene.

The Early Stade represented by moraines at lower elevations (2600–2700 m) is older because of its poor preservation and altitudinal position. It probably represents the maximum advance of the Mérida Glaciation (Schubert, 1974).

In the Páramo de Piedras Blancas, above 4000 m, there are numerous small moraines inside large morainic valleys (Schubert, 1975a), which probably resulted from the minor glacial readvances during the general main Late Pleistocene-Holocene glacial retreat. The lack of erosion and weathering of these moraines indicates that these advances occurred during the Holocene (Schubert, 1979a). In the last 100 years glaciers have been retreating (Schubert, 1984).

Fluvial and Fluvioglacial Terraces

During the retreat of the ice of the Mérida Glaciation, fluvioglacial sediments accumulated in the ice-free glacial and morainic valleys. The sediments were subsequently cut into terraces by river erosion. Previously, these terraces were interpreted in terms of a dry climate associated with the glacial conditions (Schubert and Valastro, 1974). However, there is no clear paleoecological evidence of a succession of dry and wet climates in the high Mérida Andes at the beginning of the Holocene. The fluvioglacial terraces dated so far began to be deposited in valleys closed by moraines, in the end of the Pleistocene and during the Holocene (Mucubají Terrace, for example).

There are several levels of fluvial terraces (mesas) below the morainic levels in the Motatán and in the Chama river basins (Fig. 8–2). Several theories attempt to explain the formation of these mesas:(1) each terrace level would correspond to a period of uplift of the Cordillera (Karsten, 1851); (2) the deposition was contemporary with the Pleistocene glaciations (Sievers, 1888); (3) more recently they are thought to have been deposited by torrential streams due to a possible southward displacement of the cyclonic belt during glaciation periods (Tricart and Millies-Lacroix, 1962) or by intermittent, torrential streams in a dry climate (Fairbridge, 1970; Schubert and Valastro, 1980).

It was assumed on stratigraphic grounds that the fluvial terraces of the middle Motatán and Chama rivers were of Late Pliocene and Pleistocene age (Tricart and Millies-Lacroix, 1962; Shagam, 1972). According to these authors the sediments of the Tuñame Terrace (whose pollen analysis is discussed in this chapter) should be of Early or Middle Pleistocene age. Radiocarbon dates from this terrace (Fig. 8–3) yielded ages from 33,700 to 50,600 years (Schubert and Valastro, 1980). Thus, this terrace and the stratigraphically younger terraces in the same formation (Esnujaque Formation) are much younger than previously thought.

Because the Tuñame Terrace contains thick layers of conglomerate with lack of sorting and poor bedding, and is close to the source area, Schubert and Valastro (1980) have suggested that

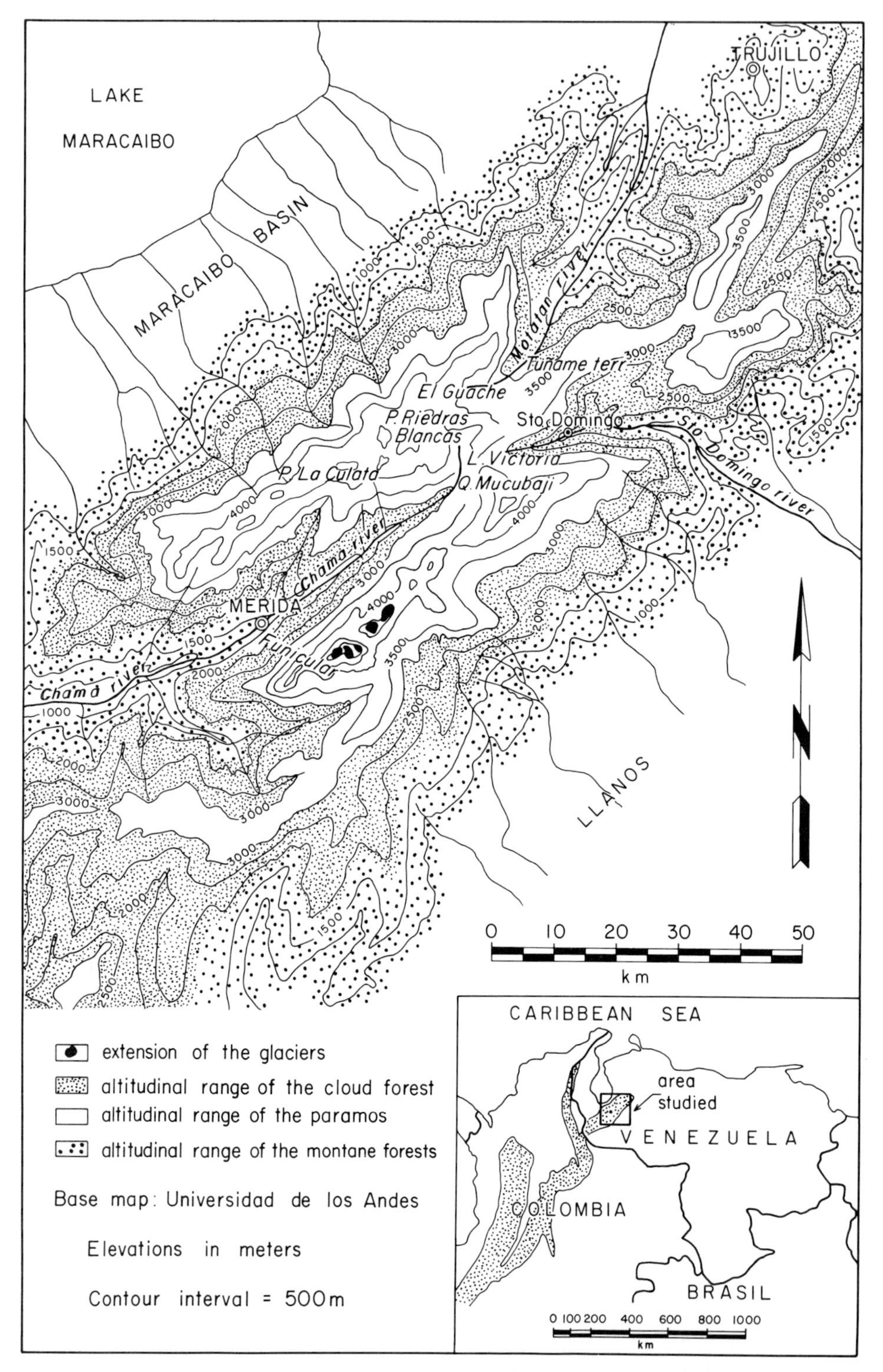

FIG. 8–2. Map of the central Venezuelan Andes, showing the sites of palynological studies. Vegetation belts are represented by their altitudinal limits.

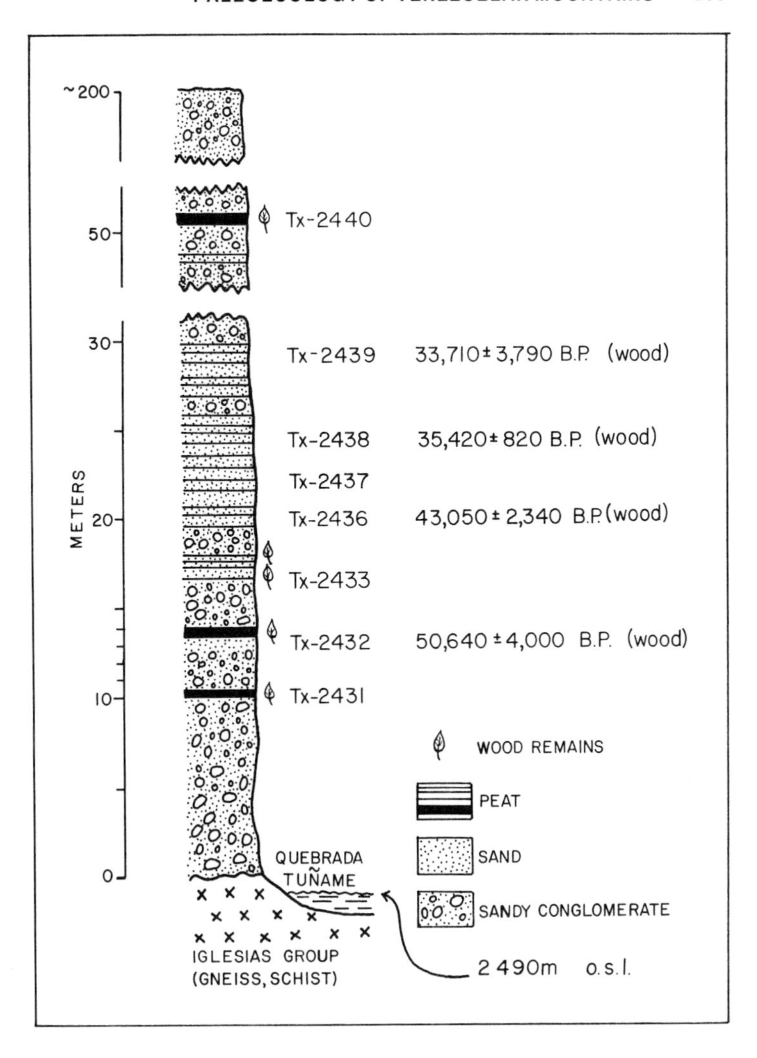

FIG. 8–3. Sketch of the Tuñame Terrace. Adapted from Schubert and Valastro (1980).

the sediments were deposited at the end of the Pleistocene by intermittent, torrential streams under more arid conditions than today.

The Perijá Range

Bordering the Maracaibo Basin to the west is the Perijá Range (Fig. 8–1), which is also a northern extension of the Colombian Eastern Cordillera. Its highest peaks do not reach elevations as high as the Mérida Andes, and no glaciers exist at present. Nevertheless, recent geomorphological studies (Schubert, 1975b, 1979a) revealed evidence of Pleistocene glaciation. Cirques, arêtes, moraines, and so forth are found at elevations above 2700 m. The snow line reached altitudes similar to those of the Early Stade in the Mérida Andes. Although the Perijá range had not been studied in the same detail as the Mérida Andes, glacial retreat was probably synchronous in both mountains.

The Caribbean Mountains

The very abrupt Caribbean Mountains consist of two parallel ranges oriented in an east–west direction along the Caribbean coast of Venezuela. Their western part is in contact with the northern end of the Andes (Figs. 8–1 and 8–4). The mountains extend eastward to the Paria Peninsula and the northern range of Trinidad (Fig. 8–1). Their highest elevations are found north of Caracas (Pico Naiguatá, 2765 m).

The Caribbean Mountains were uplifted during the Tertiary and are considered older than the northern Andes (Menéndez, 1966; Bell, 1971). They are not high enough to support glaciers, but it is possible that during the Mérida Glaciation a

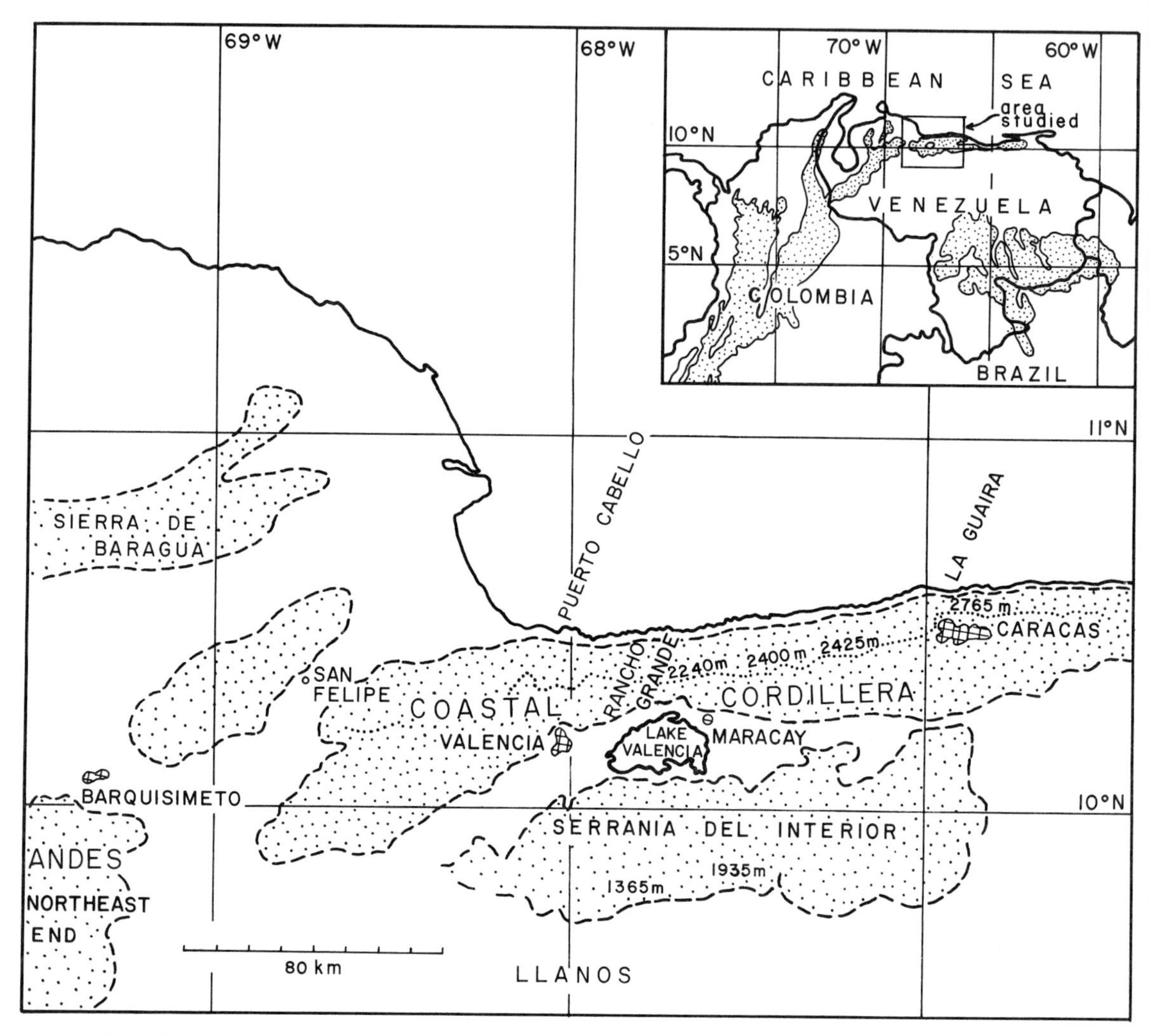

FIG. 8–4. Map of the eastern part of the Caribbean mountains (Cordillera de la Costa).

periglacial zone existed on their higher peaks. A search for glacial and periglacial features should be attempted mainly above the forest belt at the Avila region north of Caracas.

Except for a pollen analysis of a bog at Pico Naiguatá (Vareschi, 1955), there is no paleoecological study from this mountain system. The analysis of the sediments of Lake Valencia given later in this chapter is the only source of paleoecological information on the Late Quaternary of the region.

Lake Valencia (360 km^2) is the largest Venezuelan fresh-water lake. It lies in a tectonic depression between two ranges of the Caribbean Mountains, at 10°20′ N 67°45′ W and 402 m elevation (Fig. 8–4). Geomorphological studies indicate the lake first started in the Late Tertiary or Early Quaternary. Peeters (1973) has suggested that the lake sediments may be 200–300 m thick. A recent seismic investigation showed that lake sediments are present to a depth of at least 100 m (Schubert and Laredo, 1979). Two active faults cross the lake in the east–west direction, of which the northern one (La Cabrera fault zone) shows contemporary displacement.

Detailed studies by Berry (1939), Bonazzi (1958), Peeters (1968, 1970, 1971), and Schubert (1979b) describe several erosional terraces above the present lake level that correspond to the high level of the lake. The maximum level, at 425–427 m above sea level (today the lake is at 402 m), is thought to correspond to the time when the lake drained toward the Orinoco Basin, through the El Paíto gap, south of the city of Valencia. Peeters (1968, 1970, 1971) showed that the lake reached this maximum level several times in its history. He presented evidence, based on the study of the lake terraces and of the deposited clay, of at least three periods of regression between the Late Tertiary and the Late Pleistocene. Based on an assumed sedimentation rate of 1 mm/year, Peeters (1973) suggested that the present lake originated about 15,000 years ago. Recently, radiocarbon dating has shown that this in fact occurred somewhat later. Pollen evidence indicates that the lake was dry about 13,400 B.P. and remained so until about 11,500 B.P. (Salgado-Labouriau, 1980). Evidence from pollen, diatoms, and mollusks shows that the present lake originated at about 10,000 B.P. or shortly before. The history of Lake Valencia in the last 13,000 years is summarized later in this chapter.

PALEOECOLOGY OF THE LATE QUATERNARY

In the Andes as in the coastal mountains, the high elevation of the mountains causes a gradient of climate and, hence, of vegetation. The lowland savannas and dry forests are replaced by rain forest followed by high-altitude grasslands as elevation increases. The presence of these vegetational belts and the good preservation of pollen, spores, and many algae in mountain sediments enables us to reconstruct the succession of past ecosystems.

The Venezuelan Andes

At present paleoecological studies in the Venezuelan Andes are concentrated at the highest elevations of its central part (Fig. 8–2). For a detailed description of the modern vegetation of the region Lamprecht (1954), Vareschi (1970), and Sarmiento et al. (1971), Salgado-Labouriau (1979b), and Monasterio (1980) should be consulted.

The modern pollen deposition at the highest elevations of the Venezuelan Andes, studied by pollen analysis of 18 surface samples collected between 3100 and 4340 m (Salgado-Labouriau, 1979a), shows that the modern palynomorphic assemblages are a good reflection of the vegetation types present there. In all samples the main sources of pollen are Compositae and Gramineae. Other páramo genera have low pollen deposition: *Polylepis*, *Acaena*, *Geranium*. *Montia*, and Caryophyllaceae (Arenaria type). Spores of some páramo pteridophytes are locally abundant, such as *Isoetes* and *Lycopodium*. The genus *Jamesonia*, which characterizes high elevations in tropical South America, has low spore frequency in the different páramos.

Pollen grains and spores from the montane forests are also deposited in the páramos. They are transported by ascending winds and some types are found at the highest elevations studied. The main arboreal pollen deposited in the páramos belongs to the genera *Alnus*, *Podocarpus*, and *Hedyosmum*. Their pollen has high-to-moderate upward dispersion power. The following genera have low deposition in páramo: *Myrica*, *Ilex*, *Juglans*, *Rapanea*, *Alchornea*, and Melastomataceae-Combretaceae. Cyatheaceae spores from the montane forest are relatively abundant in the páramos, indicating a moderate upward dispersion power.

The study of the pollen deposition in relation to the Andean vegetation in the Colombian Eastern Cordillera (Grabandt, 1980) agrees in general with the results from the Venezuelan Andes.

In samples from Venezuelan páramos it is common to find other palynomorphs such as diatoms, *Pediastrum*, *Botryococcus*, and spores of Zygnemataceae and of fungi. Unfortunately, ecological and taxonomic information is scarce, especially on the Chlorococcales, Zygnemataceae and fungi. Therefore, their data cannot yet be included in an ecological analysis. I believe that when this information becomes available, the paleoecological interpretation will be greatly enhanced.

A comparison of the modern assemblages of palynomorphs with those found in the old sediment provides an interpretation of the past environment in the northern Andes.

The Tuñame Terrace

The Tuñame Terrace at 2490 m belongs to a sequence of mesas in the Motatán River basin (discussed earlier). The terrace was deposited during the Late Pleistocene, beginning more than 55,000 years ago. The upper part of the terrace consists of coarse conglomerate. About 50 m of the basal part consists of organic layers separated by sandy conglomerate (Fig. 8–3). Plant remains are found in these layers. An almost complete tree (with roots), embedded in the sediments, was identified as a Myrtaceae, either *Eugenia* or *Myrcia* (Schubert and Valastro, 1980).

The pollen analysis of the organic layers (Fig. 8–5) shows that arboreal pollen dominated over nonarboreal pollen. *Alnus* pollen is the most frequent (always constituting more than 32% of the total pollen) followed by *Podocarpus*. Elements of the Andean forest such as *Hedyosmum* and *Juglans* are present. Myrtaceae are underrepresented in pollen although macrofossils are at the site. Other trees are *Myrica*, *Clarisia*, *Guettarda*, and *Heliocarpus* from the cloud forest; *Escallonia* from the gallery forest; and some species of Melastomataceae-Combretaceae and

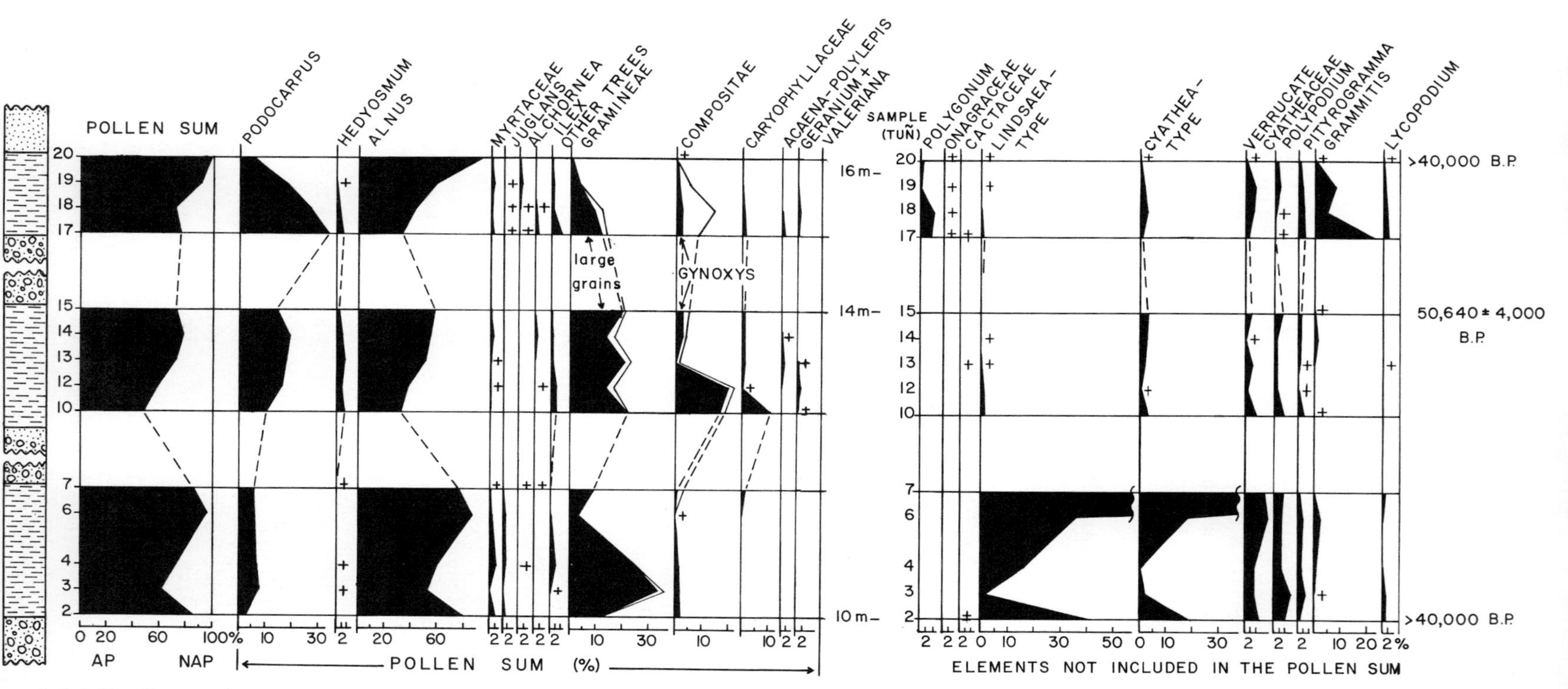

FIG. 8–5. Pollen diagram of the lower part of the Tuñame Terrace, Venezuelan Andes, 2490 m at the river.

Umbelliferae. Most of the Compositae pollen belong to *Gynoxys* (probably *G. meridensis*), which are small trees. *Polylepis* and herb pollen occur in low frequency. This type of assemblage suggests that a montane forest (probably a gallery forest) existed then, an interpretation supported by the presence of spores of tree ferns (Cyatheaceae), *Lindsaea*, and of *Grammitis*, the small fern epiphyte from the cloud forest. The presence at some levels of large pollen grains of grasses (larger than 40 μm in diameter) suggests that they probably belong to a Bambusoideae of the forest (Salgado-Labouriau, 1984).

The presence of a few cactus pollen grains in the sediments suggests that these were carried by the river from dry vegetation above, because the grains are probably too large to be transported by air from below. Unfortunately it is not yet possible to know whether this arid vegetation existed at the same time as the Tuñame forest or if the grains come from the erosion of sediments of a drier climate before 55,000 B.P.

In Europe and North America the interval ca. 100,000 to ca. 10,000 B.P. represents the pleniglacials Würm and Wisconsin respectively. At about the time in which the basal part of the Tuñame Terrace was deposited (55,000–33,710 radiocarbon years B.P.) the northern higher latitudes were probably in an interstadial that was followed by the last glacial advance. Studies of oxygen isotopes in the Caribbean Sea (Emiliani, 1966) also indicate that the temperature was relatively higher at approximately this interval. It is possible that the glaciers in the Venezuelan Andes were retreating then, but absolute dating in most cases is difficult for pleniglacials. Dating is often done stratigraphically or by sedimentation rates based on upper strata. Hence, the comparison between different regions in different continents remains difficult.

Farther south, in the Eastern Cordillera of Colombia (Schreve-Brinkman, 1978; van der Hammen, 1978), the climate was relatively warmer and more humid during the Tuñame forest time and started to be colder and drier only ca. 20,000 years ago (2500 m elevation). The increase in *Polylepis* pollen, which indicates a forest dominated by this genus in the Sabana de Bogotá and at Fúquene, Colombia (van der Hammen, 1974), is coeval with the Tuñame forest. Nevertheless, pollen of *Polylepis-Acaena* is low at Tuñame, suggesting regional differences.

If the Tuñame phase was an interstadial as in the Colombian Andes, the glaciers could have been retreating in the Venezuelan Andes, and torrential streams formed by the melting of the ice could have carried and accumulated the coarse sediments that constituted the terrace. But it is not possible to know yet if the climate in the region around the terrace was drier or wetter than at present.

Postglacial Succession of Vegetation at 3500–4000 m Elevation (Mucubají and La Culata)

During the last Pleistocene glaciation the Páramo de Mucubají (3500–3700 m) (Fig. 8–2) was covered by glaciers. High moraines encased a deep glacial valley within which a sequence of frontal moraines shows that the ice sheet retreated in steps.

The pollen analysis from a fluvioglacial terrace within the Mucubají glacial valley illustrates the sequence of colonization of the region. Gramineae, Compositae, and Montia-type pollen was deposited from 12,650 to 12,250 radiocarbon years B.P. The high frequency of Montia-type pollen (probable *Mona meridensis*; Nilsson, 1966) suggests that the average temperature was at least 2.4°C below the present average of 5.3°C. (Salgado-Labouriau et al., 1977). Mucubají was then a periglacial region. The comparison of the frequency of tree pollen at that time and at present allows one to determine the position of the original tree line. Only long-distance tree pollen was present, and it was very low in frequency. The forest belt was much farther down than now, and a superpáramo (called desert páramo in Chap. 3) existed in the region. This superpáramo assemblage had a much smaller number of species than today's. It probably represents the first plants that occupied the recently deglaciated area (Fig. 8–6).

A short warm interval of ca. 350 years then followed. The cloud forest started moving upward about 12,300 years ago and probably reached its current elevation. The climate would then have been about the same as today's, and the temperature rose very rapidly. This warm interval probably occurred simultaneously at Mucubají and in the Colombian Eastern Cordillera (Guantiva Interstadial; Gonzales, van der Hammen, and Flint, 1965).

Shortly after 11,900 B.P. the average temperature decreased again and the Mucubají region returned to superpáramo conditions. Nervertheless, the pollen assemblage was richer than the previous one and contained the same elements as those of the modern superpáramo assemblages. More species had already established themselves in the region arriving from elevations below the main morainic level. The sequence of their arrival in Mucubají is depicted in Fig. 8–7.

It is easy to assume that while the glaciers kept

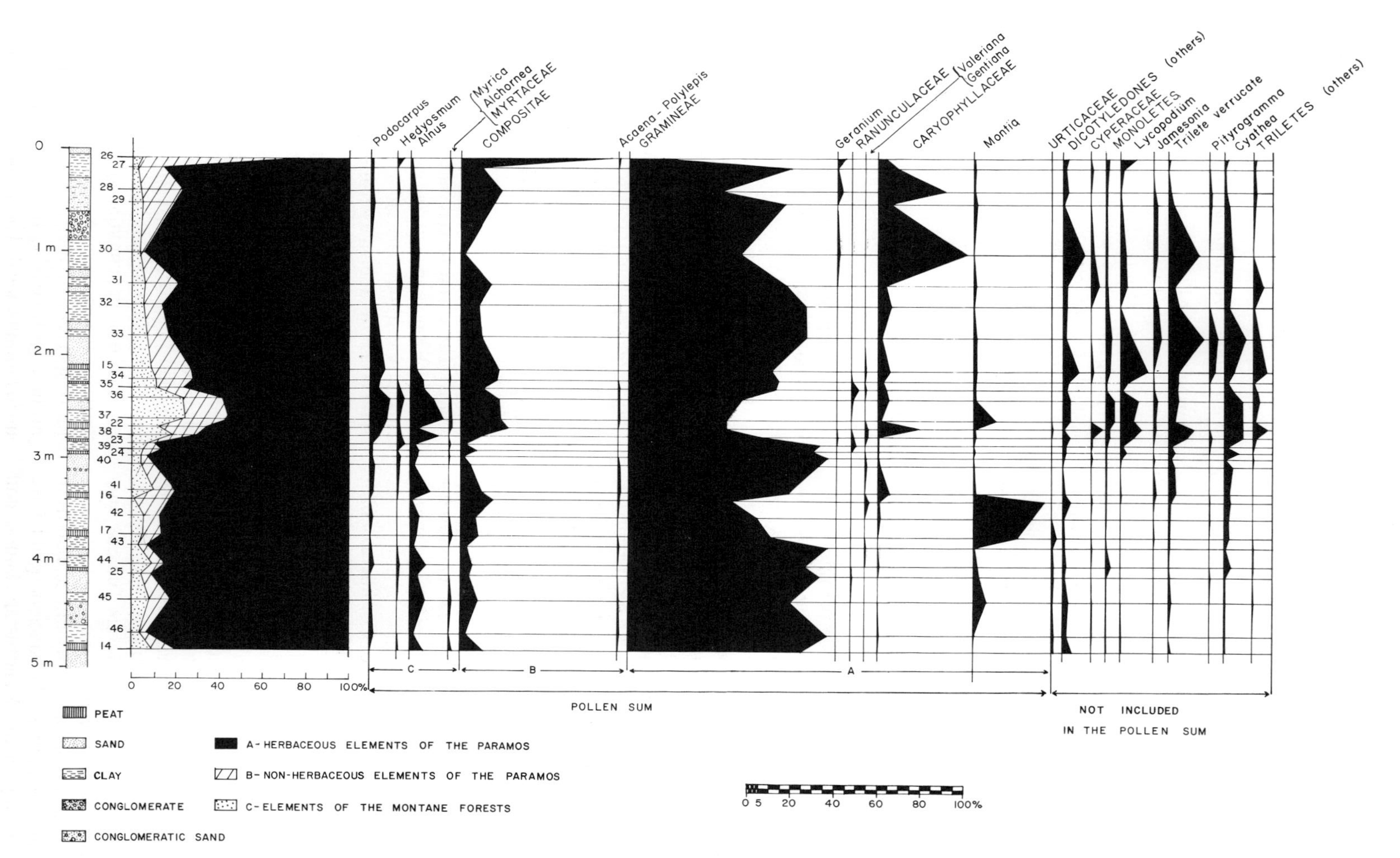

FIG. 8–6. Pollen diagram of the Páramo de Mucubají, 3650 m. After Salgado-Labouriau et al. (1977).

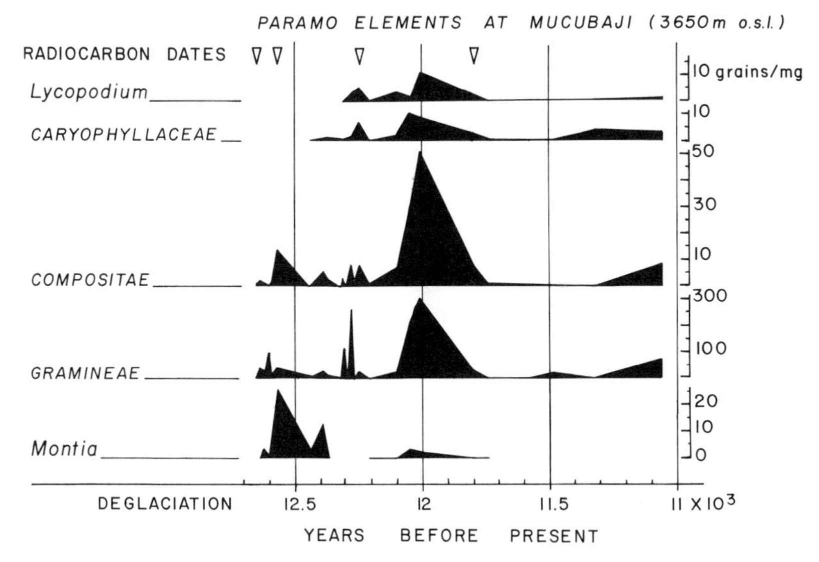

FIG. 8–7. Sequence of arrival of páramo elements after deglaciation in Mucubají. Only the elements reaching concentrations above 5 grains/mg of sediment are depicted. Adapted from Salgado-Labouriau et al. (1977).

retreating, the superpáramo species kept moving up. It is not known when they reached the elevation they occupy today. In Mucubají as temperature rose in the beginning of the Holocene, these species were replaced by the plants from the páramo proper (Andean páramo in Chap. 3).

At about 7500 years B.P. the Páramo de La Culata (approximately 3800 m), which is at a slightly higher elevation than Mucubají, was occupied by a vegetation similar to today's páramo (Salgado-Labouriau and Schubert, 1976), indicating that a cold, humid páramo was already established between 3600 and 4000 m.

At 6150 B.P. the pollen input in La Culata greatly decreased, reflecting poorer local and adjacent vegetation. Although the absolute values of pollen were decreasing, the percentage of tree pollen reached a maximum. The decrease in absolute value of local pollen (grains per milligrams), clearly shows that the increase in relative value of arboreal pollen is an artifact of the method of calculation and not of a rise in the altitude of tree line. Similar conditions were found in the old sediments from the Valle de Lagunillas (3900 m elevation) in the Eastern Cordillera of Colombia (Gonzales et al., 1965) and in the modern pollen deposition in the dry páramos of El Gavilán, Venezuela (Salgado-Labouriau, 1970a). Therefore, at 6150 B.P., La Culata was occupied by a dry páramo vegetation, and the climate was probably drier and colder than it is today. These conditions started to change at 5270 B.P. or shortly after. At 2520 B.P. the humid páramo similar to that encountered today was reestablished.

There is no paleoecological evidence from this time to the present, so it is not known whether other oscillations took place in the last two thousand years at this elevation. A continuous core at lower elevations does show climatic oscillations at the end of the Holocene (Salgado-Labouriau and Schubert, 1977) but since the absolute dating has not been possible, the timing of these oscillations is not known.

The Caribbean Mountains

Vareschi's (1955) analysis of a 0.90-m core in a peat bog from the Naiguatá peak in the Avila region (Fig. 8–4) showed a reduction of *Podocarpus* pollen toward the present in the last 0.30 m, with an increase in *Oyedaea*, grasses, and six páramo species. This was interpreted as clearings in forest.

There is no other pollen diagram from these mountains, but the study of the sediments of Lake Valencia on the southern side of the Cordillera, facing Rancho Grande (Figs. 18–1 8–4), gives further information.

Because this lake is surrounded by steep slopes, mainly on its northern side, vegetational belts in the mountains and surrounding areas provide a source of pollen from different montane plant formations to the lake sediments. Hence it is possible to follow not only the lake's history but also the environmental development of the mountains around it. The Late Quaternary history can be deduced from the analysis of a core in the deepest part of the lake (37 m of water) and

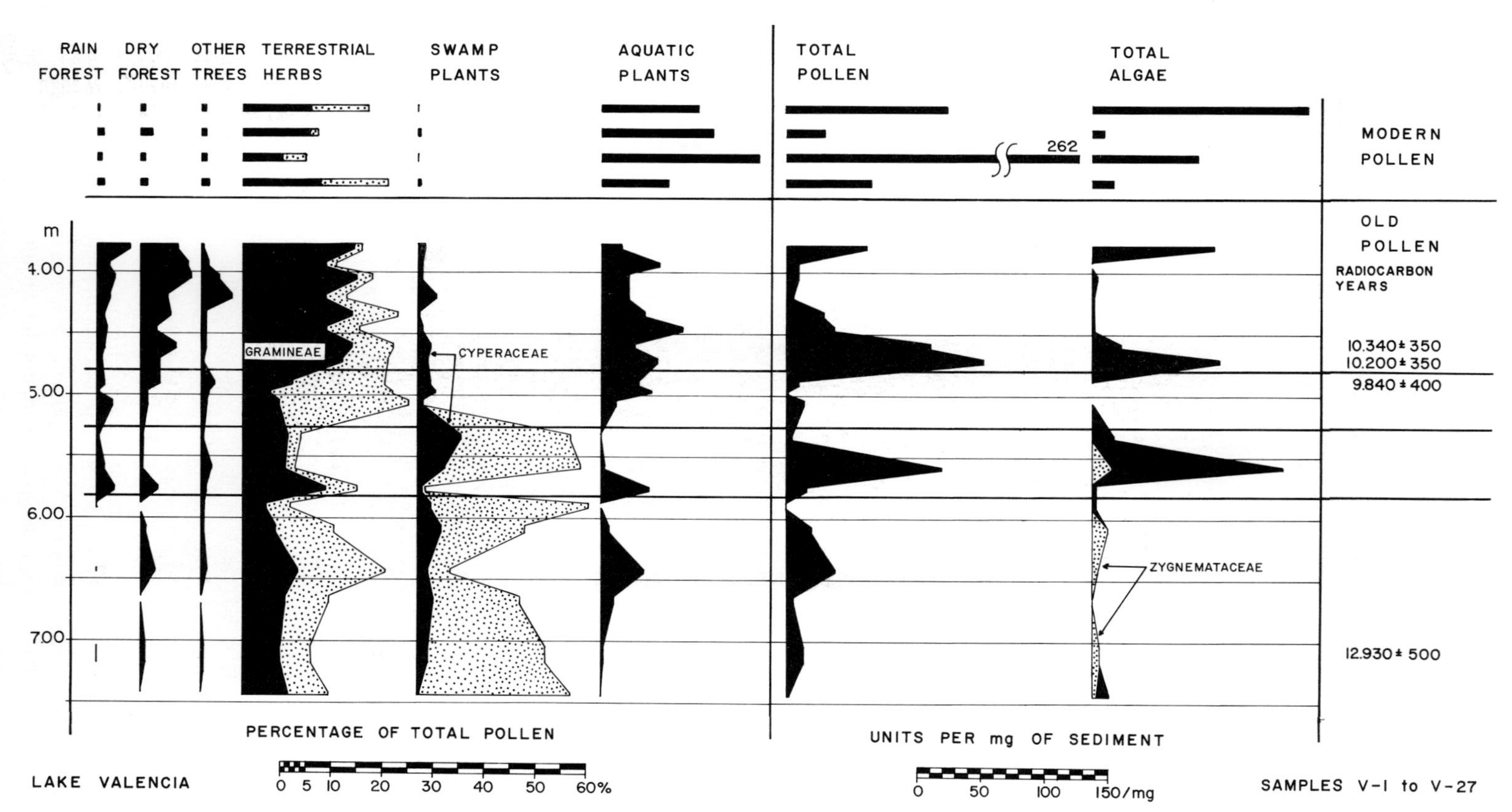

FIG. 8–8. Pollen diagram of the Pleistocene-Holocene boundary of Lake Valencia. After Salgado-Labouriau (1980).

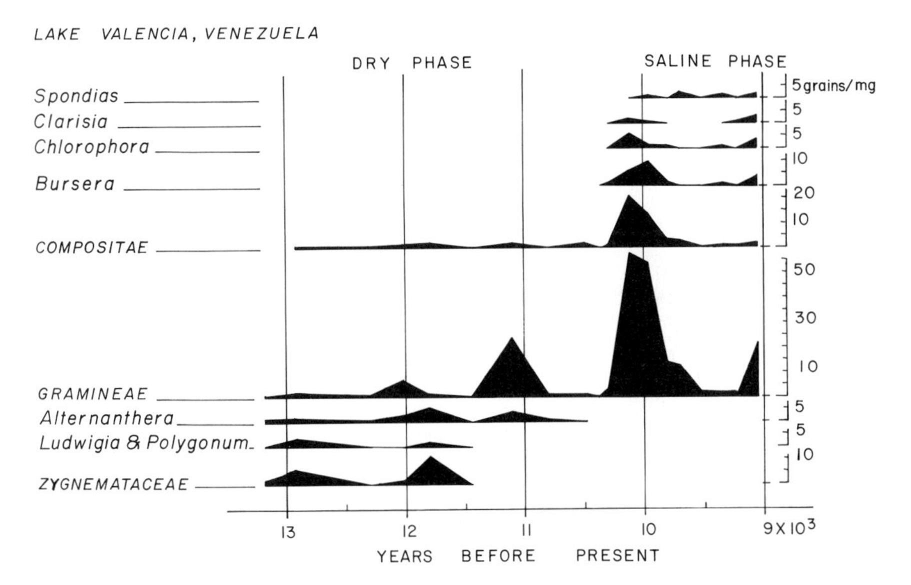

FIG. 8–9. Succession of vegetation in the Lake Valencia region depicting the climatic shift at the Pleistocene-Holocene boundary. Only the elements reaching concentrations above 3 grains/mg of sediment are depicted. Adapted from Salgado-Labouriau (1980).

of two cores near the margin (about 10 m of water). Radiocarbon dates established the chronology of the last 13,000 years and animal and plant microfossils, as well as mineral contents, provided the basis for this reconstruction (Salgado-Labouriau, 1980; Bradbury et al., 1981).

The time interval studied can be divided into three distinctly different environmental phases. (1) Between ca. 13,200 and 10,130 B.P. a dry phase prevailed. The low concentration of pollen, the presence of swamp elements (Salgado-Labouriau, 1980), and the absence of plankton elements (Bradbury et al., 1981) suggest the lake was either drastically reduced or replaced by a swamp in its deeper part. (2) At about 10,130 B.P., at the begining of the Holocene, a richer pollen assemblage began (Salgado-Labouriau, 1980). From then to ca. 9000 B.P. brackish-water diatoms (Bradbury, 1979) and ostracods (Binford, 1979) are found. A saline lake began to form. (3) From 8670 on the lake sediments are rich in microfossils (Bradbury et al., 1981), indicating a true open-water environment.

Succession of Vegetation in the Lake Valencia Region at the Pleistocene-Holocene Boundary

The pollen assemblages found in the sediments of Lake Valencia, besides contributing to the history of the lake, allow one to reconstruct the environment in the mountains around the lake. Pollen can reach the lake by air or by water transport. Therefore, the vegetation covering the highest parts of the mountains (some 20 km in horizontal distance) would contribute to the lake pollen assemblage together with the low-elevation plants from areas around the lake. Since it is a closed basin the rivers carry organic matter from the mountains into the lake, and the sediments deposited remain in the site when the lake dries.

The simple pollen assemblage of the end of the Pleistocene, and its increase in diversity at the beginning of the Holocene, established the chronology of the colonization of the basin by different types of vegetation when the climate shifted from semiarid to relatively wet (Fig. 8–9).

From ca. 13,200 to ca. 10,130 years ago, during the dessiccated phase, the vegetation around the marshes was scarce and most of the soil was bare. The torrential, probably intermittent, rivers and the wind carried very little tree pollen into the swamps or intermittent lake. At 12,930 ± 500 B.P. arboreal pollen was represented only by *Bursera*, *Acacia*, and *Protium*, all in very low frequency (Figs. 8–8, 8–9). At about 12,000 B.P., in addition to these, *Spondias*, *Machaonia*, *Rapanea*, and Myrtaceae are found. The arboreal pollen frequency is so low that trees could not have been in the plains around the marshes. They probably were growing in the surrounding mountains as open dry forest or woodland, probably a thorn forest. The data do not indicate whether these trees also occupied the highest part of the mountains. Nevertheless, it is clear that no rain forest existed in the regions around the lake, otherwise some pollen would have been deposited. This is especially true for *Podocarpus* pollen, which is known to be transported long distances by wind (Hedberg, 1954; Salgado-Labouriau, 1979a).

From ca. 10,000 years on the dry forest elements begin to increase in number and diversity in the mountains around the lake. The first to increase in frequency is *Bursera*, followed by *Spondias*. Other elements, among them *Trema*, *Bravaisia*, *Macrolobium*, *Pera*, *Cecropia*, and other Moraceae, start at this phase, which corresponds to the formation of the saline lake. The diversity and frequency of tree pollen increased in such a way that it reached values higher than those of today. Since the dominant arboreal pollen belongs to the dry forest, a denser, drier formation succeeded the preceding one (Figs. 8–8, 8–9). It cannot have been a humid forest or a semideciduous forest because the spores of pteridophytes had not increased and represented less than 5% of the pollen sum (compare with Tuñame and páramo assemblages). The Late Pleistocene open dry forest (probably a thorn forest) retreated toward the west where it exists today near Barquisimeto, some 200 km west of the lake (Fig. 8–4). It was replaced in the mountains by the dry, deciduous, dense forest that still occurs in the region. It is possible that this dry forest formed a mosaic together with savannas at low elevations. Saline- or subsaline-soil plants such as Chenopodiaceae probably occupied the shores of the saline lake.

The rain forest pollen started in very low frequency at about 11,000 B.P. It increased in frequency approximately 10,000 years ago but it was still low during the saline phase of the lake, suggesting that it was coming from a great distance. The rain forest probably reached the highest parts of the mountains around the lake toward the end of this phase. Where it was before, and how it reached the region afterward, are questions that present information cannot answer.

The succession from a semiarid vegetation to a complex of different types of vegetation covering the lake margins and the surrounding mountains took place only in about 2300 years, when the dry open forest retreated (Fig. 8–9). From then on the types of vegetation that exist today in the Lake Valencia basin were established. Rain forest, dry forest, and savannas occupied the basin during the whole Holocene (Bradbury et al., 1981). It is my interpretation that these three types of vegetation displayed the same spatial pattern as seen today (although their areas of distribution could have differed in size): savannas and dry forest around the lake, followed by dry, semideciduous forest at higher elevations, and by rain forest in the highest parts of the mountains.

Climatic Changes and Oscillations

Paleoecological studies in Venezuela have shown that the climate was not stable during the end of the Quaternary, and geologic evidence has shown that there was at least one major glaciation period at the end of the Pleistocene in the northern Andes (van der Hammen, 1974; Schubert 1979a; and first section of this chapter). The climate became warmer shortly before 13,000 B.P., and subsequent deglaciation was rapid. Table 8–2 summarizes climatic oscillations after deglaciation in the páramo belt of the central Venezuelan Andes.

In the subpáramo the pollen analysis of a 2.5 m continuous core at the margin of a glacial lake also showed climatic oscillations (Salgado-Labouriau and Schubert, 1977). Unfortunately the core could not be dated but since it is continuous it represents the last millenia of the Holocene. At the beginning of the stratigraphic sequence the small glacial Lake Victoria (Fig. 8–2) was larger than at present and was surrounded by grassland páramo. This was followed by an increase in pollen from forest and subpáramo, indicating that the tree line rose in comparison with its present position. A denser vegetation

TABLE 8–2. Late Quaternary climate oscillations in the Mérida Andes.

Radiocarbon years B.P.	Vegetation and climate at 3500–4000 m elevation
0	Humid páramo vegetation, present climate (average temperature 5.3°C)
2520–5270	Vegetation and climate similar to present
6150	Dry páramo vegetation; climate cooler and drier than present
7530–6240	Vegetation and climate similar to present
ca. 11,000–ca.11,700	Humid superpáramo vegetation; climate cooler than present
11,960–12,250	Humid páramo vegetation climate similar to present (Mucubají warm interval)
ca. 12,280–12,650	Superpáramo with scarce vegetation; climate colder and drier than present, average temperature ca. 2.9°C; deglaciation?
Before 12,650	Mérida Glaciation

Source: Based on Salgado-Labouriau and Schubert (1976–1977); Salgado-Labouriau et al. (1977).

replaced the grassland páramo. The lake was still large and the climate humid. Within the last meter of sediment, toward modern times, the lake shrank and the forest retreated. In an aerial photograph from the early 1950s (approximate scale, 1:40,000) the lake is not visible; it either did not exist or was very much reduced.

The dessication of Lake Victoria could have been caused by the erosion of the frontal moraine that dammed it. But the advance of the forest to higher elevations and the change from páramo proper to subpáramo indicate a change from a cooler to a warmer climate, both humid. The retreat of the forest and the reestablishment of páramo conditions in recent times probably were not climatic events and could have been caused by the clearance of vegetation after the Spanish occupation.

While the northern Andes underwent glaciation, the lowlands at Lake Valencia (402 m) and the mountains of the coastal Cordillera around it had a dry climate. At about 11,500 B.P. the climate started to be relatively more humid, and at about 10,000 B.P. there was a dramatic shift from a semiarid to a more humid climate, probably a two-seasonal climate. During this time the lake began to form as a saline lake, which indicates a warm climate in which evaporation was greater than precipitation. At about 8000 B.P. the climate must have been more humid because a fresh-water lake replaced the former saline lake. There are some indications that the lake shrank a little around 6000 B.P., which would suggest a climatic oscillation toward a relatively drier climate. This would occur at the same time as the La Culata dry phase in the Andes. Open-water conditions returned to the lake after this. Finally, the dessication of the lake in the last hundred years is interpreted as having been accelerated by human use of the land and not by climatic events.

The climate during the Late Quaternary at the highest elevations of the coastal Cordillera cannot yet be reconstructed. However, data do show that in the southern part of Rancho Grande (Fig. 8–4) the rain forest elements, absent at 13,000 B.P., started at about 11,000 B.P. This suggests that the upper parts of the mountains were not as humid as they are today and probably were cooler. Temperatures were not very low, however, because the coastal mountains are probably not sufficiently high (mainly at Rancho Grande) to have supported glaciers in the past.

The climatic changes at the Pleistocene-Holocene boundary in the Andes and at Lake Valencia, as well as the climatic oscillations during the Holocene, have been established by microfossils. Therefore, the radiocarbon dates reflect the change in the microfossil assemblage, that is, the response of living beings to climatic events. Some delay in these responses is always possible because some time is necessary for the biota to reach and colonize new environments. This delay is illustrated in Mucubají by the Caryophyllaceae that reached the region some 260 years after the Gramineae and Compositae (Fig. 8–7). The same occurred with *Spondias*, which increased some time after *Bursera* in the Lake Valencia region (Fig. 8–9).

It has been shown that the sea regressed along the coast of Guyana (van der Hammen, 1963) and that dry conditions were found in Colombian savannas that probably correspond to the end of the Pleistocene (van der Hammen, 1974). It is still premature to extend the existence of these dry phases in the lowlands to all of tropical South America. In Zambia, in Africa, the vegetation remained the same during the last 22,000 years (Livingstone, 1971), which is a good reminder that some tropical regions may not have had significant climatic oscillations.

ACKNOWLEDGMENTS

This article was written during the tenure of a John Simon Guggenheim fellowship. The author is grateful to Margaret B. Davis, Carlos Schubert, and Luis F.G. Labouriau for useful suggestions.

REFERENCES

Beebe, W., and J. Crane. 1948. Ecología de Rancho Grande, una selva nublada subtropical en el norte de Venezuela. *Bol. Soc. Ven. Cienc. Nat.* 11: 217–268.

Bell, J. S. 1971. Venezuelan coast ranges. In A. M. Spencer, ed., Mesozoic-Cenozoic orogenic belts. Data for orogenic studies. *Geol. Soc. London*, spec. publ. 4: 683–703.

Berry, E. W. 1939. Geology and paleontology of Lake Tacarigua, Venezuela. *Proc. Am. Philos. Soc.* 81: 547–568.

Binford, M. W. 1979. Holocene paleolimnology of Lake Valencia, Venezuela: Evidence from animal microfossils and some chemical, physical and geological features. Thesis, Indiana University.

Bonazzi, A. 1958. Consideraciones geoquímicas sobre las aguas de la Laguna de Valencia y sobre los suelos del valle de Aragua. *Acta Cient. Ven.* 3: 182–188.

Bradbury, J. P. 1979. Quaternary diatom stratigraphy of Lake Valencia, Venezuela. *Proc. 4th Latin Amer. Geol. Congr.*, 1979, Trinidad and Tobago: 1–52.

Bradbury, J. P., B. Leyden, M. L. Salgado-Labouriau, W. M. Lewis, Jr., C. Schubert, M. W. Binford, D. G.

Frey, D. R. Whitehead, and F. H, Weibezahn. 1981. Late Quaternary environmental history of Lake Valencia, Venezuela. *Science*, 214: 1299–1305.

Emiliani, C. 1966. Paleotemperature analysis of Caribbean cores P6304–8 and P6304–9 and a generalized temperature curve for the past 425,000 years. *J. Geol.* 74: 109–126.

Fairbridge, R. W. 1970. World paleoclimatology of the Quaternary. *Rev. Géogr. Phys. Geol. Dyn.* 12: 97–104.

Giegengack, R., and R. I. Grauch. 1975. Quaternary geology of the Central Andes, Venezuela: A preliminary assessment. *Bol. Geol. Ven.*, publicación especial 7 (Memorias Segundo Congreso Latino Americano de Geología) 1: 241–283.

Gonzales, E., T. van der Hammen, and R. F. Flint. 1965. Late Quaternary glacial and vegetational sequence in the Valle de Lagunillas, Sierra Nevada de Cocuy, Colombia. *Leidse Geol. Meded.* 32: 157–182.

Grabandt, R. A. J. 1980. Pollen rain in relation to arboreal vegetation in Colombian Cordillera Oriental. *Rev. Palaeobot. Palynol.* 29: 65–147.

Hedberg, O. 1954. A pollen-analytical reconnaissance in tropical East Africa. *Oikos* 5: 137–166.

Karsten, H. 1851. Ueber die geognostischen Verhältnisse des nördlichen Venezuela. *Archiv. Min. Geol.* 24: 440–479.

Lamprecht, H. 1954. *Estudios selviculturales en los bosques del valle de la Mucuy, cerca de Mérida.* Facultad de Ingeniería Forestal, ed., Mérida: Universidad de Los Andes.

Livingstone, D. A. 1971. A 22,000 year pollen record from the plateau of Zambia. *Limnol. Oceanogr.* 16: 349–356.

Menéndez, A. 1966. Tectónica de la parte central de las montañas del Caribe, Venezuela. *Bol. Geol.* (Venezuela) 8: 116–339.

Monasterio, M. 1980. Las formaciones vegetales de los páramos de Venezuela. In *Estudios ecológicos en los páramos andinos*, M. Monasterio, ed., pp. 93–158, Mérida: Universidad de los Andes.

Nilsson, 0. 1966. Studies of *Montia* L. and *Claytonia* L. and allied genera. I. Two new genera, *Mona* and *Paxia. Särtryck Bot. Notiser* 119: 265–285.

Peeters, L. 1968. *Origen y evolución de la cuenca del Lago de Valencia, Venezuela.* Instituto para la Conservación del Lago de Valencia, Venezuela.

———. 1970. Les relations entre l'évolution de Lac de Valencia (Venezuela) et les paléoclimats du Quaternaire. *Rev. Géog. Phys. Géol. Dyn.* 12: 157–160.

———. 1971. *Nuevos datos acerca de la evolución de la cuenca del Lago de Valencia (Venezuela) durante el Pleistoceno Superior y el Holoceno.* Instituto para la Conservación del Lago de Valencia, Venezuela.

———. 1973. Evolución de la cuenca del lago de Valencia de acuerdo a resultados de perforaciones. *El Lago* 7: 861–874.

Salgado-Labouriau, M. L. 1979a. Modern pollen deposition in the Venezuelan Andes. *Grana* 18: 53–68.

———, ed. 1979b. *El Medio Ambiente Páramo.* Caracas: Ediciones Centro de Estudios Avanzados.

———. 1980. A pollen diagram of the Pleistocene-Holocene boundary of Lake Valencia, Venezuela. *Rev. Palaeobot. Palynol.* 30: 297–312.

Salgado-Labouriau, M. L. 1984. Late Quaternary palynological studies in the Venezuelan Andes. *Erdwissenschaftliche Forsch.* 18: 279–293.

Salgado-Labouriau, M. L. and C. Schubert. 1976. Palynology of Holocene peat bogs from the Central Venezuelan Andes. *Palaeogeog. Palaeoclimat. Palaeoecol.* 19: 147–156.

———. 1977. Pollen analysis of a peat bog from Laguna Victoria (Venezuelan Andes). *Acta Cient. Ven.* 28: 328–332.

Salgado-Labouriau, M. L., C. Schubert, and S. Valastro, Jr. 1977. Paleoecologic analysis of the Late-Quaternary terrace from Mucubají, Venezuelan Andes. *J. Biogeog.* 4: 313–25.

Sarmiento, G., M. Monasterio, A. Azócar, E. Castellano, and J. Silva. 1971. *Estudio Integral de la Cuenca de los Ríos Chama y Capazón—Vegetación Natural.* Oficina de Publicaciones Geográficas, Instituto de Geografía y Conservación de Recursos Naturales. Mérida: Universidad de los Andes.

Schreve-Brinkman, E. J. 1978. A palynological study of the upper Quaternary sequence in the E1 Abra corridor and rock shelters (Colombia, South America). *Palaeogeog. Palaeoclimatol. Palaeoecol.* 25: 1–109.

Schubert, C. 1974. Late Pleistocene Mérida glaciation, Venezuelan Andes. *Boreas* 3: 147–52.

———. 1975a. Glaciation and periglacial morphology in the northwestern Venezuelan Andes. *Eiszeitalter u. Gegenwart* 26: 196–211.

———. 1975b. Evidencias de una glaciación antigua en la Sierra de Perijá, Estado Zulia. *Bol. Soc. Ven. Espeleol.* 6: 71–75.

———. 1979a. La zona del páramo: Morfología glacial y periglacial de los Andes de Venezuela. In *El Medio Ambiente Páramo*, M. L. Salgado-Labouriau, ed., pp. 11–23. Caracas: Ediciones Centro de Estudios Avanzados.

———. 1979b. Paleolimnología del Lago de Valencia: Recopilación y proyecto. *Bol. Soc. Ven. Cienc. Nat.* 34: 123–55.

Schubert, C. 1984. The Pleistocene and recent extent of the glaciers of the Sierra Nevada de Mérida. *Erdwissenschaftliche Forsch.* 18: 269–278.

Schubert, C., and M. Laredo. 1979. Late Pleistocene and Holocene faulting in Lake Valencia basin, north-central Venezuela. *Geology* 7: 289–92.

Schubert, C., and S. Valastro, Jr. 1974. Late Pleistocene glaciation of Páramo de La Culata, north-central Venezuelan Andes. *Geol. Rundsch.* 63: 516–38.

———. 1980. Quaternary Esnujaque Formation, Venezuelan Andes: Preliminary alluvial chronology in a tropical mountain range. *Zeit. Deut. Geol. Gesellschaft* 131: 927–47.

Shagam, R. 1972. Evolución tectónica de los Andes venezolanos. *Bol. Geol. Ven.*, publicación especial 5 (2): 1201–61.

Sievers, W 1888. Die Cordillere von Mérida nebst Bemerkungen über das Karibische Gebirge. *Geog. Abhandl.* (Penck) 3: 1–238.

Tricart. J., and A. Millies-Lacroix. 1962. Les terraces quaternaires des Andes Vénézuéliennes. *Bull. Soc. Geol. France* 7 (4): 201–18.

Vareschi, V. 1955. Monografías Geobotánicas de Venezuela. I. Rasgos Geobotánicos sobre Pico de Naiguatá. *Acta Cient. Ven.* 6: 2–23.

———. 1970. *Flora de los Páramos de Venezuela*. Mérida: Universidad de Los Andes, Ediciones del Rectorado.

van der Hammen, T. 1963. A palynological study on the Quaternary of British Guiana. *Leidse Geol. Meded.* 29: 125–80.

———. 1974. The Pleistocene changes of vegetation and climate in tropical South America. *J. Biogeog.* 1: 3–26.

———. 1978. Stratigraphy and environments of the Upper Quaternary in the El Abra corridor and rock shelters (Colombia, South America). *Palaeogeog. Palaeoclimatol. Palaeoecol.* 25: 11–162.

9

High Andean Mammalian Faunas During the Plio-Pleistocene

ROBERT HOFFSTETTER

Paleontological collecting during the last few years in the tropical Andes (Bolivia, Peru, Ecuador, and Colombia) has increased considerably our knowledge of their mammalian fossil faunas. Huge gaps nevertheless remain. Ideally one would like to know for each locality the succession of faunas and floras, and the altitudinal variations, all being accurately dated. It would then be possible to reconstruct the local climatic and biotic history, to elucidate ecological roles, to understand dispersal, to estimate substitutions and extinctions, and to distinguish temporary from enduring equilibria. We are far from this possibly unrealizable dream.

A fair number of scattered fossil localities are known in the Andes, but only a few have yielded faunal associations, and their microfauna is almost always lacking. It is relatively easy to determine the general geological age of these localities (Holocene, Pleistocene, Pliocene, or Late Miocene, for instance), but it is usually impossible to be more accurate in dating, for lack of sufficient radiometric or paleomagnetic data. Fossiliferous successions in the same locality can rarely be observed. Paleontological correlations are difficult, because observed differences between two faunas can be explained by geologic age, altitude, general climate, food resources of the environment, or the requirements of each species, whereas the respective roles of these factors cannot be accurately distinguished from each other.

In spite of these caveats, I review in this chapter most of what is known of this topic, a synthesis that has not yet been attempted for the whole of the high tropical Andes. I will take into account as well some unpublished data. First I give a chronostratigraphic reference scale valid for all of South America; then I review data concerning the localities and faunas of each Andean country; thirdly, the dispersal of each of the larger mammal groups that contributed to the faunal occupation of the high Andes is elucidated; and finally I draw general conclusions about the history of these faunas.

STRATIGRAPHIC REFERENCE SCALE

Argentina is the only South American country for which a nearly complete succession of mammal-bearing levels is available for the whole of the Cenozoic. This sequence has allowed paleontologists to establish a biostratigraphic scale, which can be used as a reference for the nonmarine sediments and fossils of the entire continent. We are indebted to the brothers Florentino and Carlos Ameghino for the definition and stratigraphic sequence of most of these "land mammal ages," which are still in use today.

The most delicate problem was the chronological calibration of this succession, and it is mostly on this point that F. Ameghino's propositions have had to be corrected. Paleontological correlation with the other continents is difficult because South America was isolated during most of the Tertiary. As a result its terrestrial faunas are strongly endemic, sometimes up to the order level. Besides the "evolutionary stage" of the local associations, a few paleontologically dated marine intercalations have been used. The result was beyond all expectations and turned out to be one of the major biostratigraphic successes of vertebrate paleontology. The succession established from the Late Paleocene (Riochican) to the Holocene (Recent) has been corroborated remarkably well by the radioisotopic datations recently realized by Marshall et al. (1977, 1979, 1983, 1984). These have been obtained by the potassium-argon method on volcanic samples from 64 to 61 million years B.P. (Salamancan = Danian, below the oldest Tertiary mammal fauna from Patagonia, the Riochican = Late Paleocene) and 35 to 3.5 million years B.P. (Deseadan = Early Oligocene, to Montehermosan = Plio-

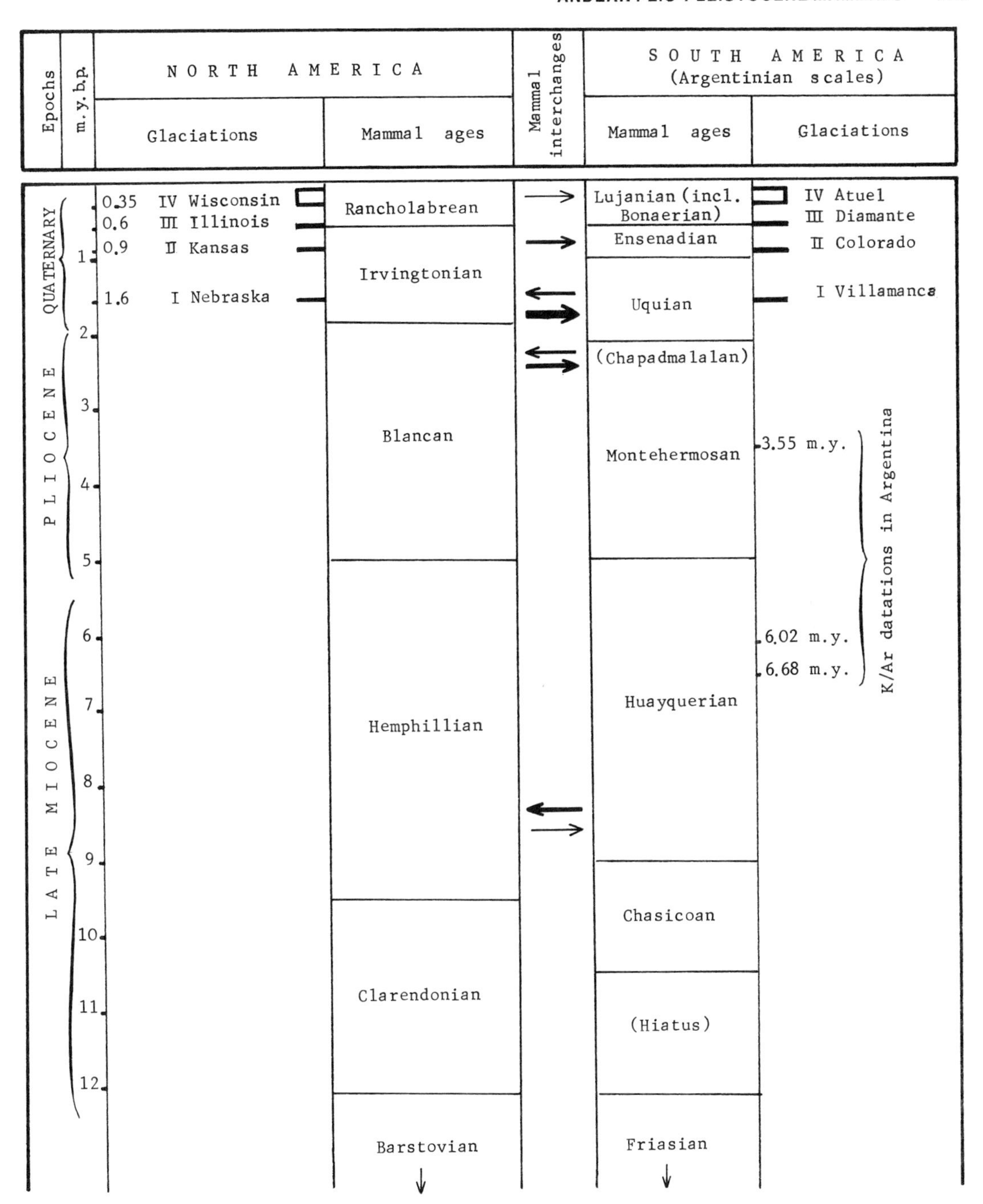

FIG. 9–1. Tentative chronology of mammal ages and glaciations in North and South America. Data from Marshall et al., 1974; Pearson, 1978; Kraglievich, 1952; and Webb, 1976.

cene) intervals. For later levels, the radioisotopic data obtained by the carbon 14 method concern the last 40,000 years only. In the Tarija Pleistocene (Bolivia), datings have been obtained by paleomagnetism and fission-track analysis (MacFadden et al., 1983), and by electron spin resonance (Ikeya in Takai et al. 1982, p. 5).

The following correlation attempts have been made between North and South America for the Late Neogene and the Pleistocene:

(1) Faunal interchanges (mostly mammals) between the Americas. These reveal an approximate synchronism (see Webb, 1976) between: (a) Early Huayquerian and Early Hemphillian (arrival of *Cyonasua* in South America, and of three genera of ground sloths, among which is

Megalonyx, in North America, thanks probably to temporarily favorable conditions for the crossing of the Central American seaway); (b) Chapadmalalan and Late Blancan (important interchanges between the two continents, due probably to the emersion of the Panama isthmus, about 3 million years B.P.); and (c) Uquian + Ensenadan and Irvingtonian (intensification of the interchange, from 2 to 1 million years B.P.).

(2) Glaciations. Four main glaciations have been recognized in the Pleistocene of several parts of the world (northern Europe, the Alps, North America, and the Andes; New Zealand for the last three glaciations). It is likely that they correspond to the same cold phases that affected the entire surface of the globe and that are also identifiable in oceanic deposits. But glacial advances are more numerous, and it is often difficult to distinguish between glacial stadials and glaciations, as well as between interstadials and interglacials. Radioisotopic and mostly paleomagnetic observations have led to a tentative correlation (Fig. 9–1). In the Northern Hemisphere, the four great glacial phases (I to IV) would begin at 1.6, 0.9, 0.6 and 0.35 million years B.P., respectively. At least one older cold phase, in the Pliocene, has also been detected. On this basis, Illinoian is now correlated with Mindel (rather than with the Riss, as was usually supposed before) and Wisconsin with Riss + Würm (Pearson, 1978; Frakes, 1979).

In the Southern Hemisphere, correlations are less certain. As a working hypothesis, however, the four Andean glaciations recognized by Groeber in Argentina (Kraglievich, 1952), and also the four distinguished by Servant in Bolivia (Ballivian, Bles, and Servant, 1978), can be considered as the equivalents of the four glaciations of the Northern Hemisphere, although Servant remains cautious. A paleomagnetic study might allow one to solve the problem, since the beginning of glaciation I (1.6 million years B.P.) corresponds to the Gilsa event, the beginning of glaciation II (0.9 million years B.P.) to the Jaramillo event, and the following glaciations fall within the Brunhes epoch (beginning 0.7 million years B.P.).

Another problem is posed by the limits of the Pliocene. Many geologists, notably in North and South America, have long believed (and some still do), in a "long Pliocene," beginning 12 million years B.P. (including Clarendonian and Chasicoan) or 10 million years B.P. (base of the Hemphillian or of the Huayquerian). Following geologic researches in the Mediterranean, and considering the original definitions, the present trend is to place the Miocene-Pliocene limit between the Tortonian regression and the Plaisancian (Piazencian) transgression, more accurately between the Messinian and Zanclian stages, i.e., around 5 or 5.5 million years B.P. Despite local opposition, still strong in South America, the new concept is becoming increasingly accepted, and general agreement may be expected in a short time (see Hoffstetter, 1980).

The Pliocene-Pleistocene limit, which in any case is conventional, is also the subject of much discussion. An agreement apparently is imminent about the 1.8 million years B.P. date (Olduvai paleomagnetic event), which I adopt here.

MAMMALIAN FAUNAS OF ANDEAN COUNTRIES

Bolivia

Bolivia is the country where a succession of Andean mammal faunas can be illustrated best, because continental formations from the Cretaceous onward show considerable development there. Vast outcrops occur, because of folding followed by erosion (Inca phase toward the end of the Eocene, Quechua phase at the end of the Miocene), and some good paleontological and radioisotopic landmarks are available. Fossil mammal localities include the Latest Cretaceous, the Early Oligocene (Deseadan), the Middle Miocene (Friasian), the Late Miocene (Chasicoan-Huayquerian), the Pliocene sensu stricto (Montehermosan, including Chapadmalalan), and the Pleistocene. Moreover, the geographic configuration of the country is favorable to the settlement of high-altitude faunas and floras. Last, with regard to the Plio-Pleistocene, multidisciplinary studies have been carried out for several years so that important advances are to be expected in the near future.

One of the peculiarities of the Bolivian Andes is the development of a vast altiplano, bordered on the east and west by cordilleras that reach 6000 m or more. The altiplano itself is situated at an altitude of about 4000 m, the lowest points being 3810 m in the north (Lake Titicaca), 3686 m (Lake Poopó) in the center, and 3655 m (great salares) in the south. Mammal-bearing localities are mapped in Fig. 9–2.

Late Miocene

The pre-Pliocene sedimentary series comprises several thousand meters of red beds, folded at the end of the Miocene. I shall briefly consider the

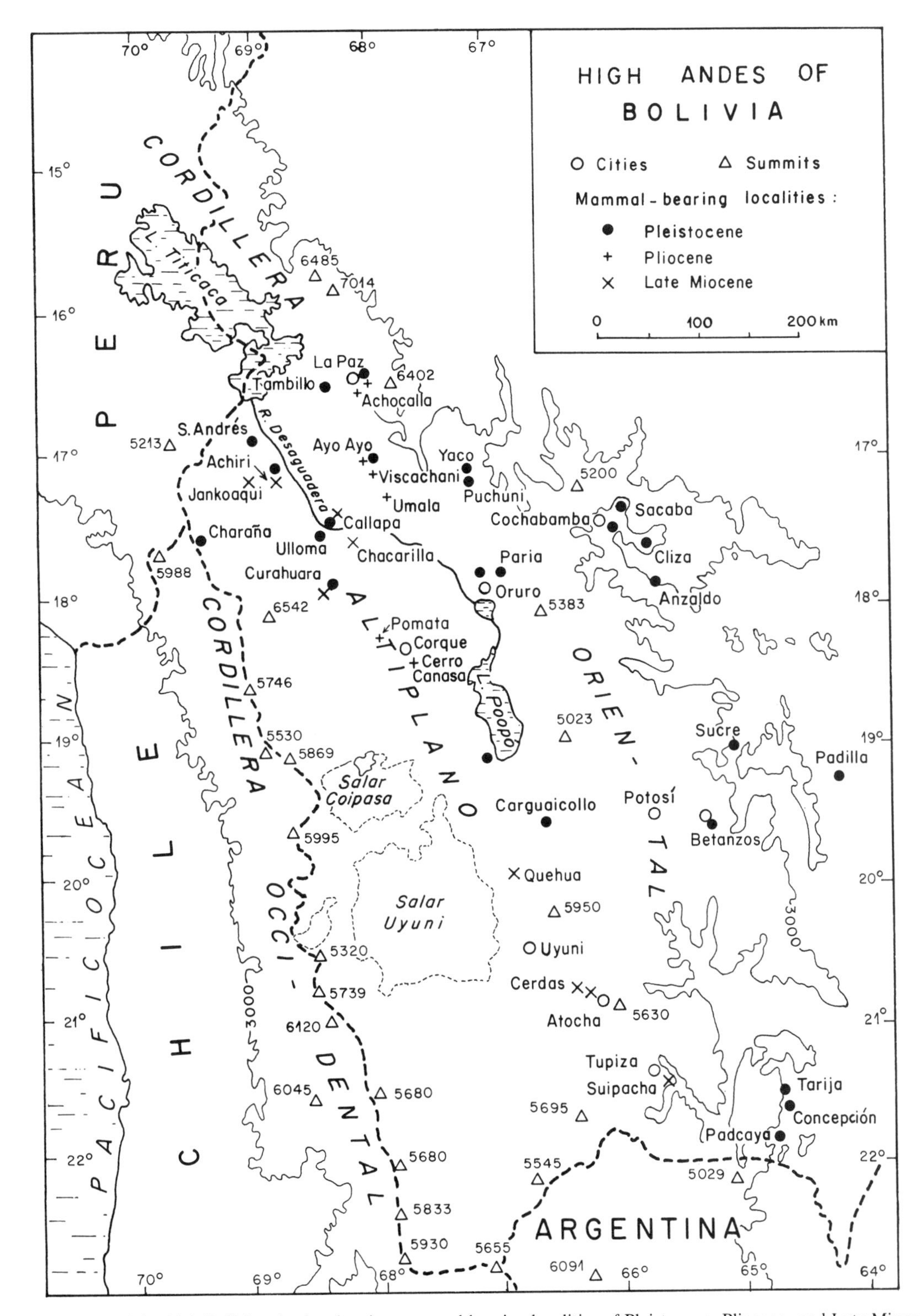

FIG. 9–2. Map of the high Bolivian Andes showing mammal-bearing localities of Pleistocene, Pliocene, and Late Miocene ages. (The contour line corresponds to 3000 m.)

Late Miocene here, in part because many authors still conventionally place it in a "long" Pliocene, but mainly because it allows one to appreciate the sudden faunal turnover that took place about 5.5 million years ago. The Late Miocene is easily identifiable because of its fossil mammals and of "tobas" (cinerites or tuffs) dated by the potassium-argon method. Included in it is notably, at the northwestern end of the altiplano, the Mauri 6 Formation, in which Evernden, Kriz, and Cherrani (1966) have found dates from 5.9 to 10.5 million years B.P. In this formation, some localities that have yielded fossil mammals are Achiri, Jankoaqui, Rosario, and Camacho. Plant remains have also been found: leaves at Jancocata (Berry, 1922) and wood at Jankoaqui (collected by Hoffstetter). Toward the east, this unit passes laterally into the upper part of the Totora Formation, which includes the Ulloma (9.1 million years B.P.) and Callapa (ca. 7.5 million years) cinerites. A few mammal remains have been collected there about 2 km north of Ulloma (Minita Chocopini, near Torini), at Callapa and near Chacarilla, confirming the equivalency with the Mauri 6 Formation. In the southern altiplano, the Quehua Formation (at least the original one, as it outcrops around Quehua) is equivalent paleontologically to the two previously mentioned formations.

The fauna described so far includes an abundant mesotherine, characteristic of the Late Miocene of Bolivia (*Plesiotypotherium* with three species: *P. minus*, *P. achirense*, and *P. majus*, indicating three successive levels: Villarroel [1974, 1978]), Toxodontidae (a medium-sized toxodontine, and a still unnamed large form, with a frontal horn, reminiscent of *Trigodon* from the Pliocene of Argentina, but generically distinct and chronologically older), various ground sloths (Mylodontidae, Megatheriidae, Megalonychidae), Glyptodontidae (Sclerocalyptinae), armadillos (a pampatherine, cf. *Kraglievichia*, at Camacho; a *Chorobates*, rare, at Achiri); and rodents (a dolichotine, *Orthomyctera*, at Quehua).

This fauna evokes a fairly warm climate and a moderate altitude. Note that the occurrence of dolichotines and mesotherids suggests a lowland fauna. Similar indications are given by the flora (leaves) from Jancocata (12 km southeast of Santiago de Machaca, about 4000 m, ca. 10 million years B.P.), described by Berry (1922), who mentioned the genera *Pteris*, *Phragmites*, *Alnus*, *Osteomeles*, *Polylepis*, *Calliandra*, *Cassia*, *Cesalpinia*, and *Melastomites*. However, Berry's conclusions on the humidity of the climate are disputable, because at Jankoaqui, a nearly locality at the same altitude and of the same geologic age, abundant fossil woods have been collected by R. Hoffstetter under the cinerite: all of them (identified by D. Pons) belong to the Mimoseae, and mostly to the genus *Prosopis* ("algarrobo"), which suggests a dry climate, at least locally.

Pliocene (sensu stricto)

The limit between the Miocene sensu lato and the Pliocene sensu stricto is marked in Bolivia by an important tectonic phase (the Quechua phase of Steinmann [1930]), which resulted in folding and uplift, followed by strong erosion, or even peneplaination in the northern part of the altiplano. On this surface rests unconformably the Pliocene (sedimentary or volcanic facies) or directly the Quaternary.

In the altiplano the fossiliferous Pliocene is localized along the Eastern Cordillera, from La Paz to near Patacamayo, and it is found again south of the Rio Desaguadero, notably at Pomata and Cerro Canasa (west of Lake Poopó). Around La Paz, it consists of 500 m of sediments (fine or coarse), the La Paz Formation. A cinerite dated 5.5 million years B.P. by M. Bonhomme (Servant, pers. comm.) has been observed in the lower part at Cota Cota, southeast of La Paz. Fossil localities at La Paz (Gualberto Villarroel, Següencoma) and somewhat farther south (Achocalla) are located below another intraformational cinerite, the Toba Chijini, dated 3.27 million years B.P. (Clapperton, 1979). Farther east, the equivalent Umala Formation also begins with a cinerite (Toba 76) dated 5.5 million years B.P. (Evernden et al., 1966), which also occurs at Pomata (according to the correlation by Martinez). At Cerro Canasa, between Pomata and Lake Poopó, the Remedios Formation is Pliocene in age (stratigraphic studies by Lavenu; fossil mammals identified by Hoffstetter and Villarroel).

The fauna, quite distinct from that of the Late Miocene, indicates a faunal turnover and apparently a higher altitude. It includes a xotodontine toxodontid (*Posnanskytherium*, characteristic of the Bolivian Pliocene, with two species: *P. desaguaderoi* at Següencoma, Ayo Ayo, Viscachani, and Pomata; an undescribed smaller species at Achocalla and Gualberto Villarroel is probably indicative of a distinct level of the Pliocene); a macrauchenid (cf. *Promacrauchenia*); a mylodontine (different from *Glossotheridium*); a megatheriid (related and possibly ancestral to *Megatherium* sensu stricto); a megalonychid (undescribed genus, at Següencoma); a sclerocalyptine glyptodontid (related to *Plohophorops*, at Achocalla); armadillos (at Ayo Ayo: cf. "*Plaina*," common, and *Macroeuphractus*, rare); hydrochoerid (including *Chapalmatherium saa-*

vedra; Hoffstetter, Villarroel and Rodrigo, 1984) and ctenomyid rodents (*Praectenomys*: two species; Villarroel [1975]) and marsupials (*Microtragulus bolivianus* Hoffstetter and Villarroel 1974; a sparassocynid and perhaps a caenolestid (Villarroel, unpublished). The occurrence at Pomata of a large carnivorous phororhacid bird should also be noted.

Noteworthy is the absence of the mesotherids and of the dolichotines. The mesotherids persisted until the Pleistocene, the dolichotines (Mara or Pampas hare) until today in the lowlands of Argentina. The presence of the Ctenomyidae, which are abundant today at high altitudes, should also be noted.

Outside the altiplano, the fossiliferous Pliocene is known only at Anzaldo (55 km southeast of Cochabamba, on the eastern slope of the Eastern Cordillera, about 3000 m), where a Plio-Pleistocene sequence has been studied by M. Montaño (1968). The Pliocene appears at the bottom of the Quebrada Tijraska (3 km southwest of Anzaldo) where a 2-m-thick silstone bed has yielded a femur and a calcaneum of a small macrauchenid (cf. *Promacrauchenia*), Pliocene in age. Above, in a 1-m-thick gravel bed, has been collected a caudal sheath of *Prodaedicurus* cf. *devincenzi* Castellanos 1940 (identified by Hoffstetter), a genus and species described in Uruguay in a local level (Castellanosense) which marks the transition from the Pliocene to the Quaternary (Chapadmalalan or Uquian). At Anzaldo, the overlying beds are typically Pleistocene (with *Megatherium*, *Glyptodon*, *Cuvieronius*, and *Lama*; see Montaño [1968]).

Pleistocene

Since 1973 Servant and his co-workers have carried out multidisciplinary research on the Quaternary of Bolivia, especially on the altiplano and in the Eastern Cordillera, extending the previous contributions of Troll (1927–1928, 1929) and Dobrovolny (1962). It should first be noted that according to Servant and his team, the absolute uplift of the Bolivian Andes seems to be essentially anterior to the Quaternary, although relative vertical motions certainly took place after the Pliocene (Ballivian et al., 1978; Lavenu, 1978). This view is in agreement with B. B. Simpson's ideas (1975, fig. 2, p. 275), but other authors admit the existence of appreciable uplifts of certain sections of the Andes during the Pleistocene.

ALTIPLANO

Near La Paz, the Pleistocene comprises a maximum thickness of 500 m of glacial, fluvioglacial, and lacustrine deposits, with intercalated cinerites, the whole resting on an erosional and alterational surface that truncates the La Paz Formation of Pliocene age. The Pleistocene itself presents phenomena of alteration and pedogenesis at several levels.

The lowest glacial advances are situated at altitudes of 3800–3900 m, in other words 1000 m below present-day glaciers. Servant and Fontes (1978) distinguish four major glaciations: I, Calvario; II, Kaluyo; III, Sorata; and IV, Choqueyapu 1 and 2 (two stadial or glacial advances). They are probably equivalent to the four Pleistocene glaciations recognized by Groeber farther south, in the Chilean-Argentinian Andes. Dobrovolny (1962) admitted an earlier glaciation in Bolivia, called Patapatani, anterior to the Chijini cinerite, and thus within the Pliocene. According to Servant (in Ballivian et al. 1978, p. 105), this glaciation would actually be the Calvario Glaciation, overlain at Patapatani by a Pleistocene cinerite that Dobrovolny mistook for the Chijini cinerite (which is 80 m lower). Even after Clapperton's publication (1979), Servant (pers. comm.) maintains that there is no glaciation below the Toba Chijini. Moreover, Servant distinguishes between Calvario and Choqueyapu, two glaciations that Dobrovolny considered to be a single one (Milluni).

More than moraine remains, the ablation slopes (*glacis d'ablation*) have enabled Servant to distinguish his four glaciations.

Another important aspect of the Pleistocene history of the altiplano is that of lake advances. In the northern part of the altiplano Lake Ballivian extended far beyond Lake Titicaca, northward into Peru, and southward to Ulloma where it was limited by a topographic threshold. Its surface was then about 50% larger than that of the present-day Lake Titicaca. The highest deposits reach the 3880-m mark in the south, but do not exceed 3850 m in the north. The difference may be due to tectonic movements, or to the presence of two distinct bodies of water (Servant and Fontes, 1978, p. 10). To judge from relations with the ablation slopes, the maximum extent of Lake Ballivian would correspond to the retreat phase of glaciation III (Sorata).

At the same time, in the central and southern part of the altiplano, occupied today by Lake Poopó and the salares, a poorly delimited lake advance has left deposits (Escara Formation: Servant and Fontes, 1978, p. 13) as high as 3780 m (i.e., 120 m above the level of the present high salares).

This lacustrine episode was followed by a phase of retreat during which regressive erosion captured a southern tributary of Lake Titicaca, to the profit of the Lake Poopó basin. Hence the individualization of the Rio Desaguadero, which serves as an outlet for Lake Titicaca to the south.

A further lake advance took place in the southern basin, where the huge Lake Minchin, with a length of nearly 500 km (from 17°20′ S to nearly 22° S) and a surface of 60,000 km^2, reached the 3760-m mark (70 m above Lake Poopó, 100 m above the salares). This advance corresponded to the retreat phase of the Choqueyapu 1 Glaciation. It was followed by a retreat, and then by a new lake advance (Lake Tauca, somewhat smaller, 3720-m mark) corresponding to the retreat of the Choqueyapu 2 Glaciation (toward 12,000–10,000 years B.P.).

Servant and Fontes (1978, p. 10) also mentioned a lake advance older than Ballivian, which would have taken place in the Charaña Basin near the Chilean border.

This sequence of geographic events, and the neotectonic studies of Lavenu (1978), provide bases for the establishment of a chronological scale allowing the dating of fossil localities. But this task has hardly begun.

However, the classic locality of Ulloma (southern bank of the Río Desaguadero, 17°30′ S) could already be inserted into this scale. Older works (Sundt, Philippi, Pompeckj, Sefve) and a few recent collections have revealed a fauna of large animals, including a mastodont (*Mastodon bolivianus* Philippi = syn. of *Cuvieronius hyodon* Fisher?); ground sloths (*Megatherium sundti* = synonym of *M. americanum* Cuvier?; *Scelidodon bolivianus* Phil.); a *Glyptodon*; a litoptern (*Macrauchenia ullomensis* Sefve); and an equid (*Parahipparion bolivianum* Phil. = subgenus of *Onohippidium*?). It should be noted that Philippi's species need to be revised. The sediment rests horizontally in marked unconformity on the Totora Formation (Miocene), at an altitude of 3880 m. According to Servant and Fontes (1978, pp. 12–13), who confirmed Troll's (1927–29) opinion, it is a deposit of Lake Ballivian at its maximum extension southward, i.e., during the retreat of glaciation III (Sorata). This is more or less the position admitted by J. L. Kraglievich (1952) for the limit between Bonaerian and Lujanian sensu stricto, or, in the present nomenclature, the top of the early Lujanian sensu lato.

At Ayo Ayo (70 km southeast of La Paz) a fossiliferous Pleistocene (Hoffstetter et al., 1971), with *Macrauchenia* cf. *patachonica* Owen, *Scelidodon* cf. *tarijensis* Amegh., and a cervid, begins with an undated cinerite. The whole rests apparently conformably at 3850–3900 m on the Pliocene (Umala Formation), and Martinez (in Hoffstetter et al., 1971) has concluded that it was a real continuity. It was then tempting to see in the Ayo Ayo cinerite the equivalent of the Toba Chijini and to attribute the fauna collected somewhat above it to the Early Pleistocene. Lavenu (1978), however, found a paleosoil under the cinerite, which indicates a sedimentation gap; moreover, he followed the Ayo Ayo cinerite south of La Paz and considered it to be the probable equivalent of the post-Calvario cinerite, observed at Patapatani and mistaken by Dobrovolny for that of Chijini, which is older. Consequently, the Pleistocene of Ayo Ayo would be later than previously believed, and it could be correlated with Ulloma, which would not be contradicted by the altitudes.

At La Paz itself and farther west, at Tambillo, Pleistocene fossils (mastodonts, glyptodonts) have been collected in the past, but their exact stratigraphic position cannot be determined. The same is true toward the southwest at San Andrés de Machaca and at Achiri, where Ortega (unpublished report to GEOBOL, Servicio Geológico de Bolivia) has mentioned "Pleistocene" fossils, which have not been reexamined and some of which, at least at Achiri, could come from the underlying Miocene.

Still farther west, at Charaña (Chilean border, approximately 4000 m), a collection has recently been made by Marco Blanco in the Charaña Formation, which overlies the Pérez ignimbrite (2.5 million years) and is thus younger than that date. The fossils (determined by Hoffstetter) include *Plaxhaplous* (first discovery of this Pleistocene doedicurine glyptodontid outside Argentina), *Glossotherium*, and *Macrauchenia* cf. *ullomensis* Sefve. A datation is being made on the basis of an intercalated ignimbrite. It may be a fairly old Pleistocene, since Servant and Fontes (1978, p. 10) mentioned the existence of a lake advance older than that of Lake Ballivian (and thus older than the Sorata glaciation) precisely in the Charaña Basin.

In the southern altiplano a few scattered Pleistocene fossils (notably mastodonts) have been recorded from the northwest of Oruro and from Paria, 16 km to the northeast, and on the southwest bank of Lake Poopó, opposite Créqui-Montfort Island (Ahlfeld, 1946, p. 278).

EASTERN CORDILLERA

Outside the altiplano, several intra-Andean basins in the Eastern Cordillera have yielded Pleistocene mammals. The most important localities are located in the basins of Tarija, Concepción (= Uriondo), and Padcaya, in the southern part of the country (21° 20′–21° 55′S; 64° 35′–64° 55′ W). They comprise a considerable accumulation of deposits, dissected by deep *quebradas*, the result being a badland type of landscape. Recently, lithologic and magnetostratigraphic studies have been carried out by MacFadden and Wolff (1981), and MacFadden et al. (1983): these authors located in the Tarija sequence the Brunhes-Matuyana boundary (0.73 million years B.P.) and the Jaramillo subchron (0.90–0.97 million years B.P.); moreover an ash bed from San Blas section, dated 0.7 ± 0.2 million years B.P. (fission tracks in numerous zircons) constitutes another guide mark (perhaps the same proposed by Hoffstetter [1963b], as a boundary between horizons A and B of the Tarija Formation). Fossil bones are observed from the beds corresponding to Jaramillo event, to part of the section corresponding to Brunhes period. But the precise age of the youngest fossils is unknown. According to Ikeya's electron spin resonance dating of some fossil bones, the geologic age of the Tarija Formation would be 0.20–0.25 million years B.P. (Takai et al., 1982, p. 5). It is difficult to reconcile this indication with MacFadden's results.

The altitude is about 2000 m, and so, properly speaking, the Tarija Basin does not belong to the high Andes. However, we do not know what the altitude was at the time of the sedimentation. Moreover, these basins are surrounded by mountains reaching above 4000 m, so that remains of animals living at higher altitudes must have accumulated therein, and some typical mountain forms are indeed found there.

This fauna comprises (Hoffstetter, 1963a, and unpublished identifications of the cricetids by Reig, and later collections):

Marsupials: *Lutreolina* sp.

Xenarthrans: *Nothropus tarijensis* (Burm.); *Megatherium tarijense* Amegh. (cf. *M. americanum* Cuvier); *Glossotherium* (*Pseudolestodon*) *tarijense* (Amegh.); *Lestodon armatus* Gerv.; *Scelidodon tarijensis* (Gerv. and Amegh.); *Scelidodon* sp.; *Chaetophractus tarijensis* (Amegh.); *C. villosus* (Desm.); *Euphractus sexcinctus* (L.) (not much fossilized); *Propraopus* cf. *sulcatus* (Lund) or *grandis* (Amegh.); *Pampatherium* cf. *humboldti* (Lund) or *typus* Amegh.; *Glyptodon reticulatus* Owen; *Chlamydotherium* (? = *Boreostracon*) cf. *sellowi* (Lund); *Neothoracophorus* cf. *elevatus* (Nodot); *Hoplophorus echazui* Hoffst.; *Panochthus* cf. *tuberculatus* (Owen) (scarce).

Rodents: *Coendou magnus* (Lund); *Coendou* sp. (of very large size); *Cavia* sp.; *Galea* cf. *musteloides* Meyen; *Neochoerus* (*Pliohydrochoerus*) *tarijensis* Amegh.; *Hydrochoerus* (?) sp. nov.; *Myocastor* (= *Matyoscor*) *perditus* Amegh. (cf. *M. coypus* [Molina]); *Ctenomys subassentiens* Amegh.; *C. brachyrhinus* Amegh.; *Euryzygomatomys hoffstetteri* Reig; *Kunsia fronto* Winge; *Oxymycterus* cf. *paramensis* Thomas; *Nectomys* cf. *squamipes* Brants; *Phyllotis* cf. *darwini* Waterhouse; *Andinomys* cf. *edea* (Thomas); *Calomys* cf. *lancha* Olfers.

Carnivores: *Canis* (*Aenocyon*) cf. *dirus* (Leidy); *Protocyon* (*Theriodictis*) *tarijensis* (Amegh.); *Chrysocyon brachyurus* Illiger; *Dusicyon proplatensis* (Amegh.) (= *D. gymnocercus tarijensis* [Kragl.]); *Arctodus* (*Arctotherium*) *tarijensis* (Amegh.); *A.* (*Pseudarctotherium*) *wingei* (Amegh.); *Nasua* sp. (not much fossilized); *Conepatus suffocans* Illiger; *Smilodon* sp.; *Leo* (*Jaguarius*) *onca* (*L.*) ssp. cf. *andinus* Hoffst.; *Felis* (*Puma*) *platensis* (Amegh.); *F.* (*Herpailurus*) *yaguaroundi* Geoffroy.

Litopterns: *Macrauchenia patachonica* Owen.

Notoungulates: *Toxodon* cf. *platensis* Owen (scarce).

Mastodonts: *Cuvieronius hyodon* (Fischer) (= *Mastodon andium* auct.); *Notiomastodon* sp.; *Haplomastodon* or *Stegomastodon* sp.

Perissodactyls: *Hippidion principale* (Lund) (= *Equus macrognathus* Weddell = *Stereohippus tarijensis* Amegh.); *H. bonaerense* Amegh.?; *Onohippidium devillei* (Gerv.) (= *Hippidium nanum* Burm. = *Parahipparion meridionalis* Amegh.); *Equus* (*Amerhippus*) *insulatus* Amegh.; *E.* (*A.*) spp.; *Tapirus tarijensis* Amegh.

Artiodactyls: *Platygonus* sp.; *Palaeolama weddelli* (Gerv.); *Palaeolama* sp. (very large size); *Lama* cf. *oweni* (Gerv. and Amegh.); *Lama glama* (*L.*); *L.* (*Vicugna*) *provicugna* (Boule); *Hippocamelus* sp. (? = "*Cervus*" *percultus* Amegh.); *Charitoceros tarijensis* (Hoffst.).

This beautiful fauna (26 families, 52 genera, 63 species) is assuredly the most important one known to this day from the Andean Pleistocene. It has been considered the best reference basis for the Quaternary of the tropical Andes. Most of the genera are also found in Argentina in the Ensenadan–Lujanian complex, and some of them are considered typical of one or the other of these two mammal ages. Does it comprise two successive units? Rovereto (1914, p. 86) made it the type of a particular mammal age, i.e. the Tarijense (Tarijean). And Kraglievich (1934; see *Obras Completas*, III, 1940, p. 421) considered it to be equivalent of the Bonaerian of Argentina, intermediate between the Ensenadan and the Lujanian s.s. (but now united with the latter into a Lujanian s.l.). MacFadden's studies led him to consider the Tarijean as equivalent to the Ensenadan and lower part of the Lujanian s.l. In any case, it is necessary to date the above-mentioned cinerite and to establish the stratigraphic distribution of the faunal components in the three basins. It is likely that appearances, disappearances, and local evolutions of some lineages will be revealed.

Another locality, modest but interesting, is Anzaldo (55 km southeast of Cochabamba), on the eastern slope of the Cordillera at nearly 3000 m (Montaño, 1968). In the Quebrada Tijraska, a Pliocene characterized by a taxon similar to *Promacrauchenia* is exposed. Abov it, a gravel bed has yielded *Prodaedicurus* cf. *devincenzi*, which may belong to the latest Pliocene or to the base of the Pleistocene. In the overlying sediments, Montaño (1968) has collected typically Pleistocene forms: *Megatherium*, *Glyptodon*, mastodonts, and llama.

Findings of mastodonts and glyptodonts have also been reported from the Cochabamba Basin (Sacaba, Cliza, and other localities at 2600 to 2800 m), Sucre Basin (*Glyptodon*; 2800 m), and the Padilla Basin (mastodont; 2120 m).

From Betanzos (50 km northeast of Potosí), at 3390 m, Pick (1944) reported remains of mastodonts, equids, and cervids, to which should be added a camelid (a metapod identifiable in the photograph published by Pick) and *Panochthus* (collection of C. Villarroel, unpublished).

At Carguaicollo (90 km west of Potosí), in the Cordillera Los Frailes, at an altitude of 4100 m, Ahlfeld (1946) mentioned a mastodont tusk and a "*Hippidium*" lower jaw (not illustrated, the genus should be checked).

Lastly, farther north on the western slope of the Eastern Cordillera, 110 km southeast of La Paz, a jaguar lower jaw (*Leo* [*Jaguarius*] *onca*, ssp. cf. *andinus*) has been collected at Yaco (3600 m), and an ulna of *Macrauchenia* at Puchuni (4100 m); the identifications were made by Hoffstetter.

It is remarkable that most of the fossils from these localities also occur (at least at the generic level) in the Tarija fauna. The only exceptions are at Anzaldo (with *Prodaedicurus*) and at Charaña (with *Plaxhaplous*), which could thus represent or include older levels of the Pleistocene (Uquian for the first? Early Ensenadan for the second?) As regards the others, there is nothing to prevent one from referring them to the Tarijean, either intermediate between Ensenadan and Lujanian s.s. of Argentina, or equivalent to late Ensenadan and the lower part of the Lujanian s.l. But one should be careful, because the faunas of the high Bolivian Andes are incompletely known. The absence of some Tarijean genera (*Palaeolama*, *Equus*, *Glossotherium*, *Lestodon*, and mostly *Toxodon*, *Coendou*, and others) is doubtless explainable by the fact that the Bolivian altiplano is above their upper altitudinal limit. But other differences, indicating different geologic ages, may surface.

Peru

Good Tertiary localities are known in coastal areas and also (mostly from the Neogene) in the basin of the Rio Ucayali (Amazonia). Near Lake Titicaca, at Laguna Umayo, 20 km northwest of Puno, has been discovered the first South American Cretaceous mammal locality. But curiously enough the high Peruvian Andes have not yet yielded any remains of Tertiary mammals, with the possible exception of a carapace fragment of a sclerocalyptine glyptodontid, apparently pre-Quaternary (Pliocene?) found by A. Parodi (see Hoffstetter, 1970) on the Hacienda Ymata, west of Puno (approximately 3850 m).

By contrast Quaternary localities are numerous in the Andes, but they seem to represent a fairly late epoch. Some of them are particularly interesting because they illustrate the transition from the Pleistocene to the Holocene, as well as the relationships between man and the fauna (coexistence, hunting, and domestication; see Chap. 10). Mammal-bearing localities are show in Figs. 9–3 and 9–4.

Southern and Central Peruvian Andes

High-altitude localities, most of them located between 3800 and 4500 m, are especially numerous in the central and southern Peruvian Andes south of 9° S. The only exceptions are in the Ayacucho (2800 m) and Cuzco and Huancayo (3300 m) regions. Many sites are cave fillings (the caves served as refuges, habitats, or natural traps), but others are of lacustrine or fluvioglacial origin.

1. Localities: From south to north one can mention the following.

Puno (3830 m) and the surrounding puna; Azángaro (3859 m); Llali (3900 m); Casa del Diablo cave (3819 m), near Tirapata (Nordenskiöld, 1908).

Cuzco Valley (3350 m) (Ramírez Pareja, 1958) and Maras, 30 km to the northwest; to the south, 35 km from Cuzco, Ayusbamba (3800 m) near the Río Apurímac (Eaton and Gregory, 1914).

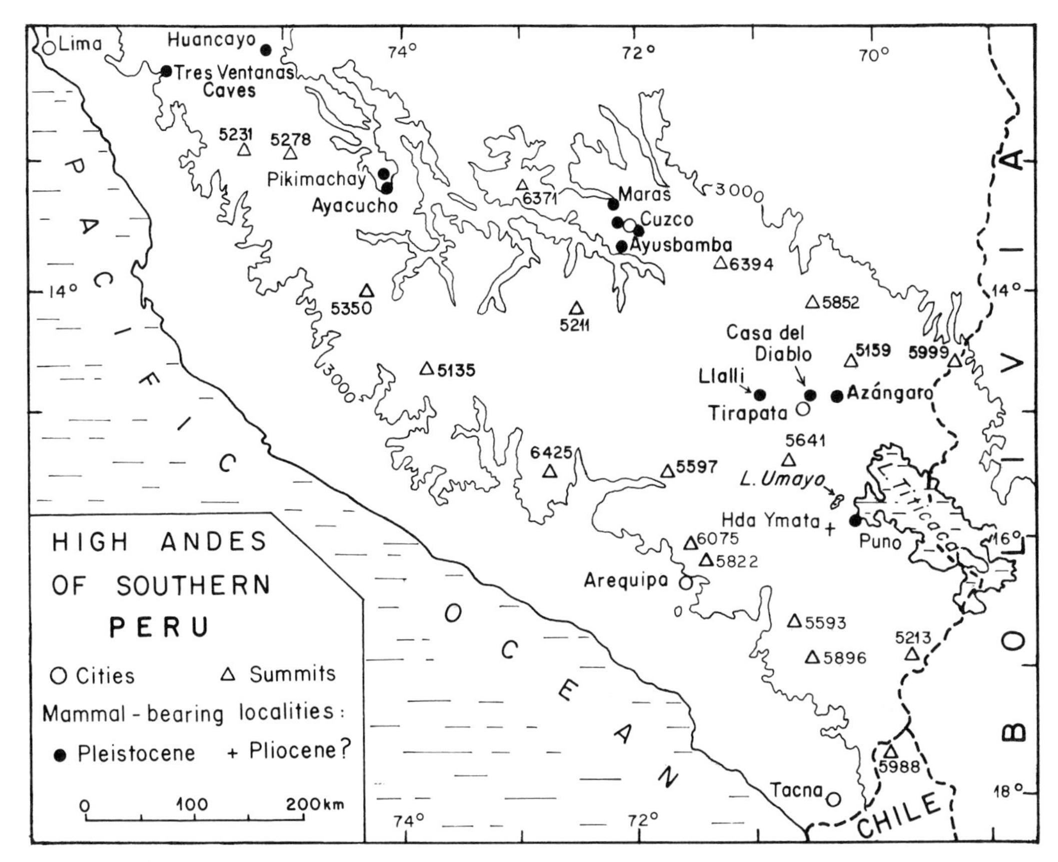

FIG. 9–3. Map of southern Peru showing mammal-bearing localities of Pleistocene and Pliocene (?) ages. (The contour line corresponds to 3000 m.)

Ayacucho area (2800 m) with several caves, among which is Pikimachay (MacNeish, 1971, 1976; MacNeish, Patterson and Browman, 1975).

Tres Ventanas caves (about 4000 m), 100 km east-southeast of Lima, upper valley of the Rio Chilca (F. Engel collect., Lima).

Huancayo area (approximately 3300 m): several poorly recorded localities.

South of Junín: Yantac (4500 m) (megatherid reported by Lisson in 1912 and kept at the Universidad de Ingeniería of Lima); San Pedro de Cajas (ca. 4000 m), Uchcumachay caves (4050 m), Panaulauca, etc. (Wheeler Pires-Ferreira, Pires-Ferreira, and Kaulicke, 1976).

North of Cerro de Pasco: Sanson-Machay cave (4400 m), near Guayllarisquizga (Castelnau collections, published by Gervais in 1855, locality rediscovered by Hoffstetter and Ojeda in 1973); Quishuarcancha cave (approximately 4300 m; A. Pardo, pers. comm.), 23 km northwest of Cerro de Pasco.

Huargo cave (4050 m) near La Unión (Cardish et al., 1973).

2. Faunas: The fossil faunas from the preceding localities include a variety of families.

Gomphotheriidae (mastodonts). *Cuvieronius hyodon* Fischer, in the Cuzco Valley and at Ayusbamba. A few remains near Huancayo, where another mastodont of small size, belonging to an unidentified genus, was also found (Museum of Wariwilca [= Huarivilca]).

Megatheriidae. A species of *Megatherium* s.s. (*M. americanum* or *M. tarijense*), with its characteristic twisted femur has been found in the Cuzco area. It has also been reported (but not illustrated) from Ayacucho, in bed h (MacNeish et al., 1975, identification by T. C. Patterson). The other megatheriid remains belong to a smaller, unnamed genus, with at least two species of different sizes, known from Azángaro (lower jaw), Llali (femur) (University of Arequipa collections), Yantac (subcomplete skeleton, Universidad de Ingeniería, Lima), surroundings of Cerro de Pasco (several bones, "Javier Prado" Museum, Lima), Ayacucho (MacNeish collection). This taxon is differ-

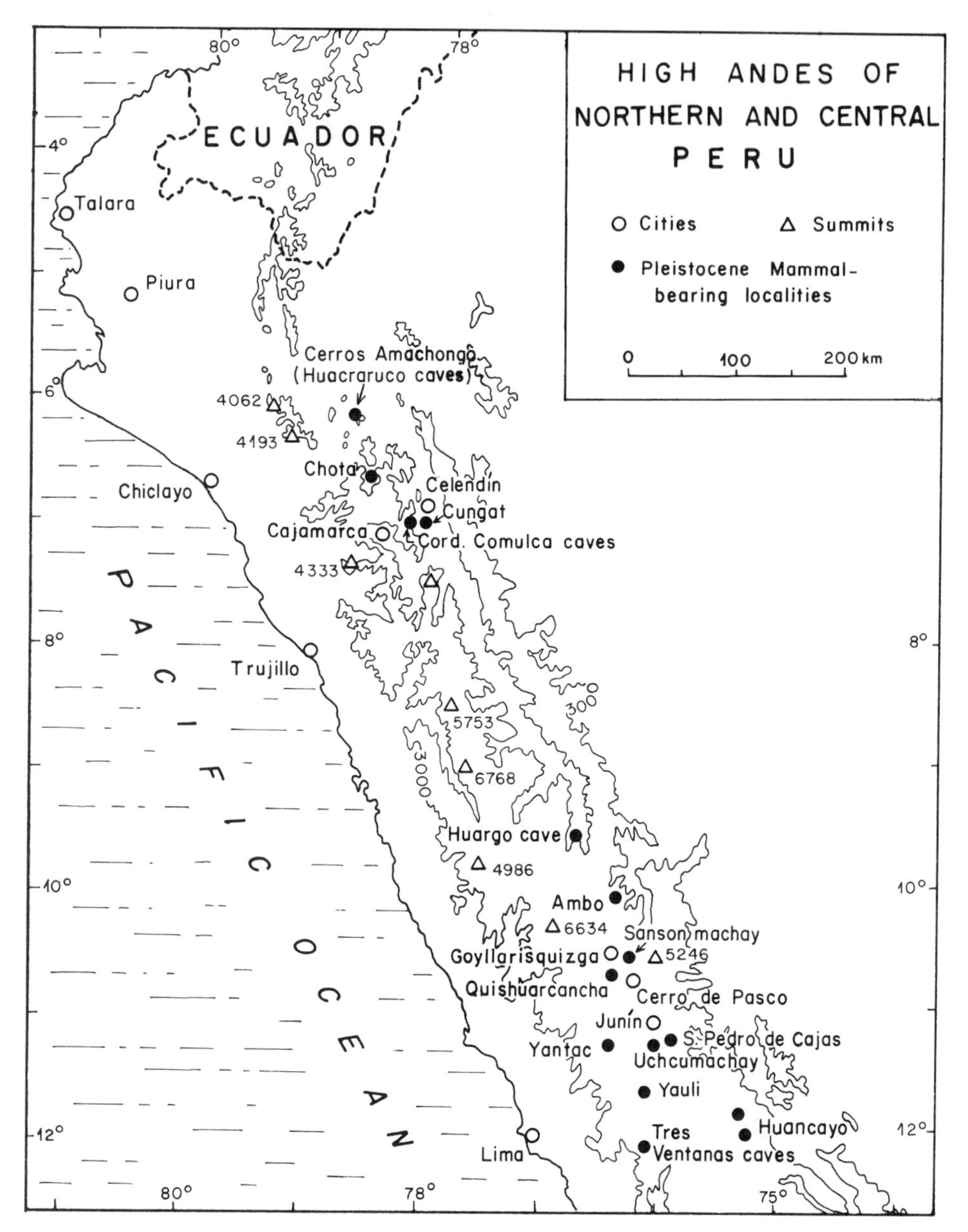

FIG. 9–4. Map of the high Andes of northern and central Peru showing mammal-bearing localities of Pleistocene age. (The contour line corresponds to 3000 m.)

ent from *Megatherium* s.s. and agrees with *Eremotherium* (a tropical genus) by its non-twisted, almost flat femur, and its relatively shallow lower jaw, but it has retained digit II of the manus, whereas *Eremotherium* has lost it. It is reminiscent of "*Megatherium*" (*Pseudomegatherium*) *medinae* from Chile, which also has a flat femur but whose manible is deep, and thus more progressive, like that of "*Megatherium*" s.s. Pending revision, I shall designate the Peruvian taxon as "cf. *Pseudomegatherium*", genus proposed by Kraglievich (1931) for "*M.*" *medinae*.

Scelidotheriinae. This subfamily is well represented in open-air localities (Puno region; Ayusbamba) and also in caves (Casa del Diablo, Pikimachay, Tres Ventanas, Sanson-Machay, Huargo). The fossils are often referred to *Scelidotherium*, but it is actually *Scelidotherium* s.l. (including *Scelidodon*) and it is noteworthy that T. C. Patterson (in MacNeish et al. 1975) reported *Scelidother-*

ium tarijense (i.e., *Scelidodon*) from Pikimachay. Actually, whenever an accurate identification could be made, the taxon was *Scelidodon*. The occurrence of *Scelidotherium* s.s. (*leptocephalum* type) has never been demonstrated outside Argentina; it was apparently a genus restricted to the pampas.

Mylodontinae. A few references to "*Mylodon*" (Huanaro, south of Cuzco; Casa del Diablo; Ayusbamba) could actually pertain to *Glossotherium*, but confirmation is needed, since no significant specimen has been described or illustrated.

Megalonychidae. A 25-cm-long humerus from Casa del Diablo (illustrated by Nordenskiöld, 1908, pl. 2, fig. 2–3) certainly belongs to this family, which is usually rare in the Andean Pleistocene; more accurately, it is referable to the subfamily Nothrotheriinae.

Glyptodontidae. The only remains known from Peru are those of two individuals of the genus *Glyptodon* found near Cuzco by Kalafatovitch in 1955. This locality seems to be the northern limit reached by this genus in the Andes during the Pleistocene (by contrast, east of the Andes the family reached Venezuela and eventually migrated to North America).

Dasypodidae (armadillos). No fossil has been mentioned or observed so far in the Peruvian Andes, perhaps because most localities are at very high altitudes.

Equidae. *Parahipparion* (probably a subgenus of *Onohippidium*) is frequent in most localities. It is known from Ayusbamba, Cuzco, and several caves: Casa del Diablo, Uchcumachay, Quishuarcancha, Huargo (where its remains have been attributed erroneously to *Equus* [*Amerhippus*] sp.). The Peruvian fossils are usually referred to the species *peruanum* (Nordenskiöld, 1908 [type from the Casa del Diablo]), for which Sefve, in 1910, had erected the genus *Hyperhippidium*, which he then brought down to the rank of a subgenus of *Parahipparion* in 1912. A few remains of *Equus* have been reported (Ramírez Pareja [1958] at Cuzco; Lisson [1912] in Yauli Province; MacNeish et al. [1975] at Ayacucho), but these reports have to be confirmed, since there may have been a confusion with the former genus, as happened at Huargo.

Camelidae. *Lama* (including *Vicugna*) is known from the caves of Casa del Diablo, Uchcumachay, Panaulauca, Ayacucho, Sanson-Machay, and Huargo. Some fossil specimens are larger than Recent ones (see chapter 10). But none of them reaches the size of *Palaeolama*, whose presence has never been proven in the Peruvian Andes (although it lived on the coast).

Cervidae. The antlers of a small cervid collected at Cuzco have been illustrated by Ramírez Pareja (1958, fig. 26) under the name cf. *Mazama*; they actually belong to the genus *Charitoceros*, and possibly to the species *C. tarijensis* (Hoffstetter, 1952). This exclusively Andean genus is known from Bolivia (Tarija) and southern Peru. Another cervid from the Peruvian Andes belongs to the genus *Agalmaceros* (founded on *A. blicki*, from Ecuador); an antler fragment collected by Kaulicke at Uchcumachay seemed to be referable to the Ecuadorian species and has been depicted under that name by Wheeler et al. (1976). A superb specimen (complete antlers and brain case) collected by César Yurivilca at San Pedro de Cajas (Quaternary gravels at a depth of 9 m), also referred to *A. blicki* by Wheeler et al. (1976, fig. 4), belongs, however, clearly to the genus *Agalmaceros* (details of the skull, insertion and ornamentation of the antlers), but it is a new species (Hoffstetter, unpublished). It is likely that the specimen from Uchcumachay actually belongs to the same Peruvian species. The genus *Agalmaceros* is strictly Andean, known from central Peru (4000 m) and from Ecuador (2500–3000 m). An incomplete cervid antler from the Casa del Diablo, has been illustrated by Nordenskiöld (1908, pl. 2, fig. 1) under the name of *Furcifer* (?) *wingei* nov. sp. I first interpreted it (1971) as a much larger *Charitoceros* than *C. tarijensis* (which is confirmed by the lower jaw figured by Nordenskiöld). But this antler (the base of which is incomplete) might be the pointed distal part of an antler of the Peruvian species of *Agalmaceros*. If this conspecificity could be demonstrated, then the specimens from San Pedro de Cajas, Casa del Diablo, and Uchcumachay should be named *Agalmaceros wingei* (Nordenskiöld). Other cervids have been reported from Ayusbamba (*Odocoileus*), Uchcumachay (*Odocoileus* and *Hippocamelus*), and Pikimachay (unidentified cervid in the oldest layer of the cave).

Tayassuidae (peccaries). The only record is that of "*Dicotyles*" (a metatarsal and a phalanx) from Casa del Diablo.

Rodentia. Rodents are poorly represented, but

Phyllotis (abundant at Pikimachay), *Lagidium* (Casa del Diablo and Uchcumachay), and Caviidae (abundant at Casa del Diablo) have been reported.

Carnivora. Carnivores are very scarce. Canids (*Dusicyon* [*Pseudalopex*] *peruanus* Nordenskiöld at Casa del Diablo; *D.* [*P.*] *culpaeus* and *Canis familiaris* from the Holocene levels of Uchcumachay); felids ("giant cat" from bed h at Ayacucho and from Casa del Diablo; *Felis* [*Puma*] from bed 7 at Huargo, from bed h-h_1 at Pikimachay, and from the Holocene part of Uchcumachay), and mustelids (*Conepatus* at Ayacucho, bed h-h_1; and at Casa del Diablo, under the name of *Mephitis*) have been reported.

Man. The presence of man is revealed at Ayacucho by artifacts (as early as 25,000 B.P.; according to MacNeish, 1976) and by a few skeletal remains (more than 14,000 B.P.; MacNeish et al., 1975).

3. Dating. A few carbon 14 dates have been obtained from the cave sites:

Tres Ventanas. The deeper layer, with "cf. *Pseudomegatherium*" and *Scelidodon*, is more than 40,000 years old. It is overlain by an archaeological layer that, to judge from the artifacts, is about 10,000 years old.

Pikimachay. Within the Pleistocene beds, MacNeish (1971) distinguishes two fossiliferous complexes: (a) Paccaicasa complex (20,000–16,000 B.C.) with artifacts and fauna including (MacNeish et al., 1975): *Scelidotherium* s.l., *Equus andium*, *Phyllotes*, Cervidae; (b) Ayacucho complex (16,000–13,000 B.C.) with artifacts and an abundant fauna: *Scelidotherium* s.l.; *Megatherium tarijense*, *Equus andium*, *Lama*, large cat, puma, *Dusicyon*, *Conepatus*, *Lagidium peruvianum* and remains of a child's skeleton.

Uchcumachay. The beds are dated on the basis of the artifacts. Level 7 (10,000–7,000 B.C.), with a fauna including *Agalmaceros*, *Parahipparion*, a few rodents; levels 6 and 5 (7,000–2500 B.C. = Holocene) with a fauna comprising only modern genera (*Lama*, *Hippocamelus*, *Odocoileus*, *Dusicyon*, *Canis familiaris*).

Huargo. The deeper beds (10 and 9), undated and without traces of human occupation, contain *Scelidotherium* s.1., *Parahipparion* and *Lama*. Bed 8 (13,400 ± 700 B.P.) indicates temporary occupation and contains the same fauna. The overlying bed 7 is the last to contain extinct genera (*Scelidotherium* s.l., *Parahipparion*), accompanied by *Lama* and *Felis* (*Puma*). Dates of 5520 and 3580 B.P. have been obtained for bed 2.

No date has been obtained for the open-air localities, the only ones in which mastodonts and glyptodonts occur.

Northern Peruvian Andes

Northern Peru is poor in high-altitude faunas. North of the Cordillera Blanca, there is a general lowering of the Peruvian Andes, which connect at a low altitude with those of Ecuador in a deflection zone, the limit coinciding with the Huancabamba Transversal of structural geologists. In the Cajamarca-Celendín region (6–7°S), a few collections made long ago at 2000–2500 m have yielded almost exclusively mastodont remains of the typically tropical genus *Haplomastodon*, which is widely distributed in the northern Andes, as well as in the Pacific littoral strip and eastward in Brazil and Venezuela.

In 1979 (see Sanmartino, Staccioli, and Klein, 1981), the French speleologist Y. Sanmartino and his "Groupe Spéléo de Bagnols-Marcoule" explored the caves of the karstic massifs in the vicinity of the Cordillera de Comulca (20 km southwest of Celendín), and of the Hacienda Huacraruco, Cerros de Amachongo (about 100 km north of Cajamarca, not far from Cutervo). Most of these caves, which open at altitudes between 3350 m and 3550 m, have been animal traps and contain bones, apparently from the Late Pleistocene (with some Holocene components). Unfortunately, only a few specimens could be examined, because most of them were left in situ or deposited at the Cultural Institute in Cajamarca. The first observations (by R. Hoffstetter) reveal the occurrence of *Felis* (*Jaguarius*) *onca andinus* (a lower jaw, Tragadero de la Purla, northeast of the Cordillera de Comulca, 3350 m) and a small mountain horse (probably *Onohippidium* cf. *peruanum*; many bones have been reported, but only a few worn upper molars have been brought back from Gruta del Equus, south of Hacienda Huacraruco, 3528 m), a cervid, *Odocoileus* sp. (identified from a photograph; Tragadero de Lunigan, east of the Cordillera de Comulca, at 3520 m), and a ground sloth brain case (Talahanes cave, 2 km southeast of Huacraruco, at 3650 m). The expedition report suggests a more varied fauna, which will have to be studied in situ. During a new expedition (1982) to the Gruta Blanca, Parque Nacional Cutervo (see Sanmartino et al., 1981, p. 75, spot p. 19), Sanmartino collected fossil bones identified by Hoffstetter (unpublished) as *Tapirus* sp.

(a tibia); cervid cf. *Mazama* (sacrum); and medium-sized megalonychid (ulna, vertebrae, etc.).

Ecuador

Fine lacustrine sedimentary basins developed during the Tertiary in the Ecuadorian Andes, especially in the south around Biblián, Azogues, Cuenca, Nabón, Girón, Loja, and Malacatos. Unfortunately, only two scanty mammal remains have been found there: a rodent skull (*Olenopsis*) at Nabón, and a toxodontid tooth (cf. *Prototoxodon*) at Biblián, both of Miocene age. The Pliocene fauna is completely unknown. However, the Quaternary is well developed and has yielded many fossils (see Fig. 9–5 for localities).

Northern versus Southern Ecuadorian Andes

The northern part of the country (1° N to 2° 30′ S) is the more fossiliferous. The Andes there consist of two parallel cordilleras: the Cordillera Real, with a metamorphic basement, and the essentially Cretaceous Cordillera Occidental; both are crowned with imposing volcanic edifices, several of which nearly reach or are higher than 6000 m; 8 are still active, and about 20 have recently become extinct. Between the two cordilleras, a graben, the interandean corridor, is divided into *hoyas* (at altitudes of 2500–3000 m) separated by thresholds or *nudos*. Intense Quaternary volcanism has produced lavas, lapilli, and cinerites, but mostly "cangahua," which looks like a loess, but is formed by volcanic dust. Transported by wind, cangahua accumulated in considerable amounts, which either cap the reliefs (eolian cangahua), or are deposited in lakes (lacustrine cangahua). Most Pleistocene mammals occur in this cangahua. Sauer (1965) distinguished one fluvioglacial and three glaciations with moraines, separated by three interglacials, and followed by one postglacial. According to him, the cangahua would be an interglacial and postglacial deposit. But according to Bristow (in Bristow and Hoffstetter, 1974), the cangahua is simply a product emitted during a period of intense volcanism and can be glacial, finiglacial, interglacial, or postglacial. Bristow added that it is difficult to distinguish Sauer's second and third interglacials and that he is not sure there were four glacial advances in the Ecuadorian Andes. Radiometric or paleomagnetic dates are badly needed.

The southern part of the country looks quite different. The cordilleras are less individualized, and the reliefs become lower and tend to fan out. The connection with Peru takes place at a low altitude, and according to modern geologists (see Aubouin, 1980) coincides with the Huancabamba Transversal. In this region, the few volcanic edifices have been dissected by erosion and are relatively old (Pliocene-Early Pleistocene?). There is no cangahua. The glaciers have left only moraine fragments and glacial sculpturing, which sometimes occurs at abnormally low altitudes (900 m according to Sauer), thus revealing a recent sinking of the region. As a result, the fossil localities (which are very scanty) are difficult to date.

Puninian Fauna (= Puninense)

The best localities are located north of 2° 30′ S and correspond to Sauer's "cangahua moderna," referred by him to his third (or last) interglacial. More than 30 sites are known at 2500–3300 m, near the following localities from north to south: El Angel, Bolívar, Ibarra (not far from this city, at the foot of the Imbabura Volcano, was found the type tooth of Cuvier's *Mastodonte des Cordilières*); Quito, Alangasí, Latacunga, Ambato, Riobamba, and above all Punín (one of the most famous Andean localities).

The fauna, known through the works of Cuvier, Wolf, Branco, Spillmann, and eventually Hoffstetter (1952 and 1971, with a bibliography), has been designated as Puninian (Puninian age).

Gomphotheriidae. *Haplomastodon chimborazi* (Proaño), with its type coming from Punín, is very widespread. Curiously enough, the occurrence of *Cuvieronius hyodon* (Fisher) = *Mastodon andium* auct. is not demonstrated at this level, although its "type" was collected in a region where all localities are Puninian (see later discussion).

Equidae. *Equus* (*Amerhippus*) *andium* Wagner-Branco is particularly abundant at Punín.

Camelidae. *Palaeolama* (*Protauchenia*) *reissi* (Branco), also frequent at Punín, becomes rarer toward the north.

Cervidae. *Odocoileus peruvianus* (Gray), a possible synonym of *O. virginianus*; *Agalmaceros blicki* (Frick); *Mazama* sp.; *Hippocamelus* (?) sp. (antler fragment at Punín).

Canidae. *Dusicyon culpaeus* (Molina); and possibly *Theriodictis* cf. *platensis* Mercerat (found at Guamote).

Felidae. *Leo* (*Jaguarius*) *onca andinus* Hoffstetter; *Felis* (*Puma*) *platensis* Ameghino; *Smilodon* sp.

Mylodontidae. *Glossotherium* (*Oreomylodon*) *wegneri* (Spillmann).

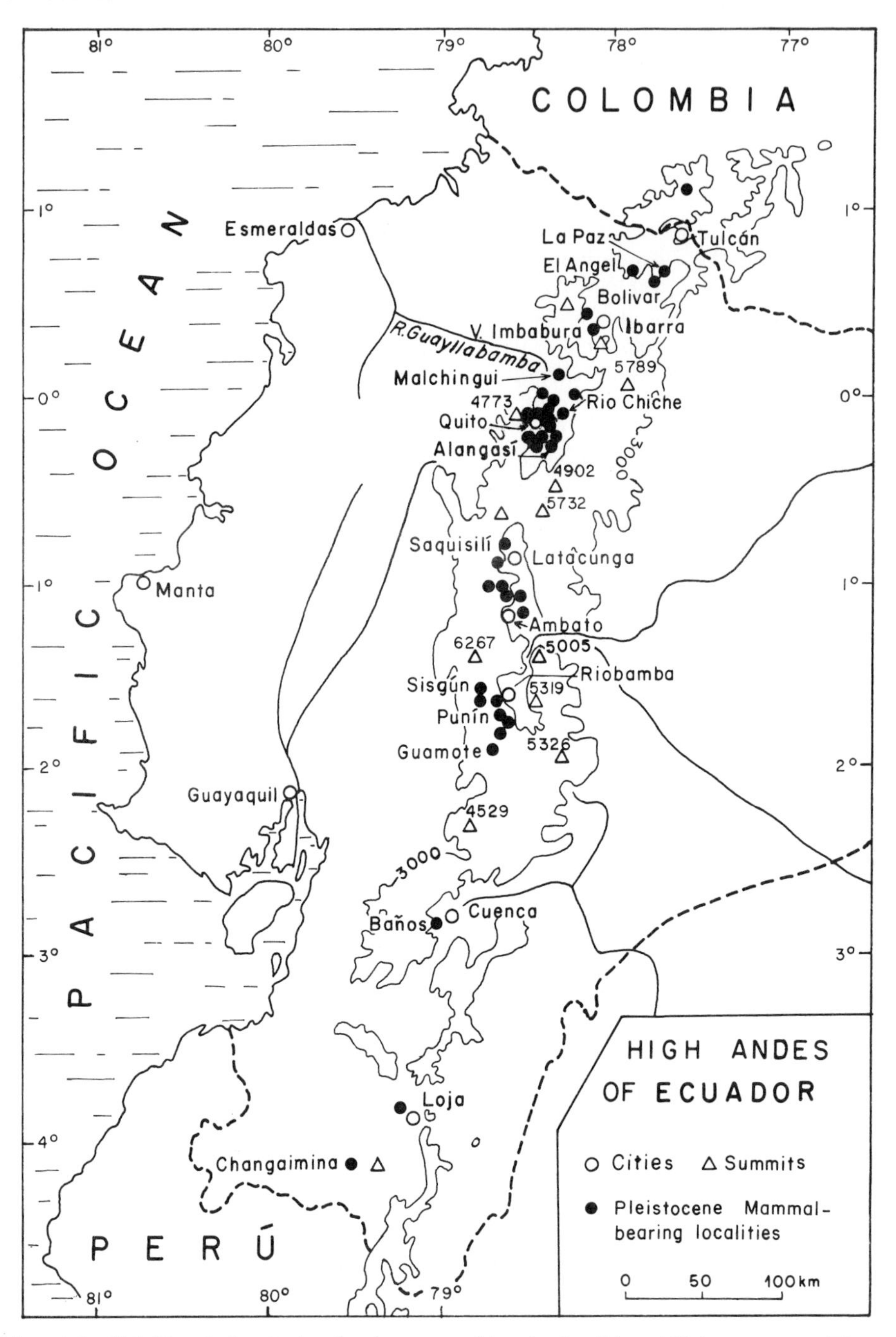

FIG. 9–5. Map of the high Ecuadorian Andes showing mammal-bearing localities of Pleistocene age. (The contour line corresponds to 3000 m.)

Megatheriidae. A few poor fragments of a medium-sized form (perhaps "cf. *Pseudomegatherium*"?), which is rare.

Dasypodidae. *Propraopus magnus* Wolf-Hoffstetter, with its type from Punín; an almost complete armor found at Malchingui (Vasconez collection) has recently been identified (Hoffstetter, unpublished).

This is a well-characterized fauna, recognizable in the whole region where Sauer's cangahua moderna is developed, with nevertheless numerical differences: *Equus* and *Palaeolama* are especially abundant at Punín and its vicinity, whereas *Glossotherium* is much more abundant in the north (region of Quito-Alangasí). It should be noted, however, that only large animals have been collected, probably because of the conditions of fossilization (hence the absence of small carnivores, rodents, lagomorphs, marsupials, etc.). On the other hand, the absence of the Glyptodontidae, the Scelidotherinae, and the genera *Macrauchenia*, *Hippidion*, and *Lama* is worth noting.

Chichean Fauna (= Chichense)

A sample of a slightly older fauna (Chichean of Hoffstetter, 1952) has been collected by Spillman, east of Quito, in the quebrada of the Rio Chiche (= Chichi). As I judge from the local section, the peculiar aspect of the fossils, and the nature of the matrix (adhering volcanic ash, no evidence of cangahua), the site probably corresponds to Sauer's second interglacial. The fauna includes *Equus* (*Amerhippus*) *martinei* (Spillmann) and *Palaeolama crassa* Hoffstetter, two very stoutly built forms. One should add a large undescribed edentate (distal half of a femur very distinct from all classical forms) from the same collection: this is probably the ground sloth mentioned by Spillmann, who also recorded a mastodont and a cervid (these specimens are lost).

The age of both faunas (Chichean and Puninian) remains imprecise. Bristow (in Bristow and Hoffstetter, 1974) tended to unite Sauer's second and third interglacials. He reported that a wood fragment collected at the contact between the cangahua and the underlying Chichi sediments has been given a carbon 14 age of *more than* 48,000 years B.P. This only gives a minimum age for the Chichi sediments, but no indication for the overlying cangahua (the base of which can also lie beyond the possibilities of carbon 14 dating)! In any case, it is certain that the two faunas, Chichean and Puninian, are distinct and follow each other in time. But they may be younger than was formerly believed. Perhaps the Puninian fauna lived from a late interstadial until the end of the Pleistocene, and the Chichean fauna was somewhat older. This would be in better agreement with data from other South American faunas (notably from Argentina, where the genus *Equus* appears only in the Bonaerian (= lower part of Lujanian s.l.), and from Colombia, where a fauna reminiscent of the Puninian seems to correspond to a very late Pleistocene.

Other Faunal Elements

Large canids. A recent collection (made by R. Hoffstetter and E. Torres in 1976) in Quaternary gravels at Guamote (50 km south of Riobamba, at an altitude of 3100 m) includes a lower jaw of *Theriodictis* cf. *platensis* Mercerat (identified by A. Berta), a sacrum and pelvis of *Glossotherium*, and a mandibular ramus of a camelid. Despite the different nature of the sediment, the assemblage could be contemporaneous with the Puninian, the fauna of which would include *Theriodictis* (either a separate genus or a subgenus of *Protocyon*).

Cuvieronius hyodon (Fisher) = *Mastodon andium* auct. Remains of several individuals of this species have been collected at Baños de Cuenca (collection of the Cuenca College), but there is no indication as to stratigraphic position. A much fossilized tusk fragment (with a helicoidal twist and an enamel band) comes from the Guayllabamba Canyon (northeast of Quito); it may be fairly old Pleistocene. A femur of the same genus has been collected at the bottom of the Quebrada Colorada at Punín; it may come from the Puninian or from a lower level. Last, the holotype of the species, a molar with a simple trefoil pattern, poses serious problems; morphologically it is compatible with *Cuvieronius* (in the usual sense, corresponding to the mastodont common at Tarija) and with *Haplomastodon*. However, the locality of this holotype has yielded only the Puninian fauna (in which *Haplomastodon* is the only mastodont known with certainty). *Haplomastodon* might thus be the true *Cuvieronius*. In order not to disturb the taxonomy, the traditional usage of *Cuvieronius* has been retained (Hoffstetter, 1952). The latter genus, in this sense, is certainly present in Ecuador, probably before the Puninian. It is still impossible to decide whether it is absent or rare in the Puninian proper.

Southern *Haplomastodon*. Collections from the southernmost Ecuadorian Andes (vicinity of Loja and Changaimina) include only mastodonts of the genus *Haplomastodon*. Here we find again a zone comparable to that of northern Peru: relatively low altitudes ("saddle" of the Andes at the level of the Huancabamba Transversal) and collections comprising essentially or exclusively *Haplomastodon*.

Parahipparion. In Ecuador, only a proximal phalanx is known from an undetermined locality (level probably older than the Puninian).

Homo. Human remains or artifacts have not yet been observed in situ in Pleistocene beds of Ecuador. The "Punin calvarium" (Sullivan, Hellman, and Anthony, 1925), recently dated by the carbon 14 method (1980, *J. Archeol. Soc.* 7: 97–99), corresponds to 6900 B.P.; it is certainly younger than the Puninian fauna. The age of the Otavalo man has been disputed: the datation of the aragonite matrix gave 28,000–29,000 B.P., but the dating by carbon 14 of the collagen gave

2300 ± 270 and 2670 ± 160 B.P. Apparently the mineral carbonate could be largely of magnetic origin (1973, *Radiocarbon* 15[3]: 467; 1975, *Radiocarbon* 17[1]: 19). More positive data come from the obsidian artifacts from Alangasí, some of which, to judge from the degree of hydration of the surface, would be more than 40,000 years old (Bonifaz, 1977). Unfortunately they are surface findings and the accuracy of the method is disputable. It is certain, however, that as in Peru, man entered the Ecuadorian Andes a little before the end of the Pleistocene.

Colombia and Venezuela

Three cordilleras, which diverge toward the north and are separated by deep valleys, constitute the Colombian Andes. Two of them (Western and Central) are prolongations of the Ecuadorian Andes. In addition, there is the Eastern Cordillera, which divides northward into two branches (Sierra de Perijá and Sierra de Mérida) that surround the Maracaibo Basin in Venezuela. Moreover, an isolated block in the north, the Sierra Nevada of Santa Marta, reaches 5634 m.

The Tertiary mammal localities are located mostly in the valley of the Rio Magdalena, where a fine series, from the Eocene to the Middle Miocene, occurs at altitudes usually lower than 500 m. A few remains from the Caribbean coast, and also an Andean fossil, are referred with some doubts to the Pliocene. The Andean specimen, a fragment of the lower jaw of a peccari, has been collected at Cocha Verde, Nariño, at nearly 3000 m, in a dark green, fine-grained sandstone. Stirton (1946) erected for it a genus and species, *Selenogonus nariñoensis*, and tentatively referred it to the Late Pliocene or Early Pleistocene. It is allied to the Pleistocene genus *Platygonus*, but differs from it by a few specialized characters, which suggest a Quaternary age.

Pleistocene localities (see Fig. 9–6) are fairly numerous and located at various altitudes (see map in Hoffstetter, 1971). But they occur only in very limited areas of the high Andes. A few mastodont teeth have been collected at Túquerres (1° 6′ N, 3200 m) and at Pasto (1° 13′ N, 2600 m) in the southern Department of Nariño. In the Eastern Cordillera, many mammal localities are known in the Department of Cundinamarca: Bogotá, Bosa, Soacha, Mosquera, Sanjón de las Cátedras, Valle de Las Pajas, Madrid, Boyacá, Guasca, Guatavita, and Tocancipá. They correspond to the Sabana Formation (including Stirton's Mondoñedo Formation), a deposit of a Late Pleistocene lake at an altitude of about 2600 m. A few isolated sites are found farther north: Tunja (2820 m), Chiquinquira (2570 m), Leiva (2700 m), Soatá (2045 m), and the Páramo de Cocuy (around 4000 m) in the Department of Boyacá, and Málaga (2040 m) in the Department of Santander. Curiously, no high-altitude locality is known either in the other Colombian cordilleras or in those of Venezuela.

The most important fossiliferous area is that of the Sabana de Bogotá, mentioned in the past (Humboldt and Cuvier) under the name of "Camp des Géants" ("Campo de los Gigantes"). Many collections have been made there, but most of them have disappeared without having been described or illustrated. The following taxa are recognized on the basis of the material available today: *Haplomastodon chimborazi* (Proaño), very abundant; *Equus* (*Amerhippus*) *lasallei* Daniel; *Equus* (*A.*) sp.; *Glossotherium* sp.; *Propraopus* cf. *magnus* (Museo La Salle, from the vicinity of Guasca, determined by Hoffstetter). Papers published locally before 1950 mention many other genera for the Sabana de Bogotá, especially: *Megalonyx*, *Megatherium*, *Mylodon*, *Glyptodon*, *Macrauchenia*, "*Astrapotherium*," "*Meritherium*," "*Palaeomastodon*," "*Tetrabelodon*," *Mastodon*, *Cuvieronius*, "*Dinotherium*," "*Cervus*," *Lama* or *Palaeolama*, *Vicugna*, and *Smilodon* (see the bibliography in Porta, 1961b). Needless to say, these "identifications" are unreliable without figures or descriptions and they cannot be verified since the material has disappeared. Those in quotation marks are completely unlikely, since we are in the South American Quaternary. Nevertheless, camelids, cervids, carnivores, and several families of edentates could occur in this fauna, but proof is lacking. There remains a problem concerning the genus *Cuvieronius*, frequently mentioned under the name *Mastodon andium*. Porta, who revised all the available material, could not confirm its occurrence in Colombia.

Researches by geologists, glaciologists, and palynologists have shed some light on the ancient Bogotá lake (Bürgl, 1980; van der Hammen, 1957, 1965, 1966; van der Hammen and González, 1960, 1964; Julivert, 1961, 1963; Porta, 1961a, b; Porta et al., 1974; see also Chap. 7).

It should first be recalled that in Colombia one observes the traces of a vast glaciation (Aduriameina Stadial), extending down to 2800 m; a later phase (Mamancanaca Stadial) has left moraines down to 3000 or 3500 m. Today, the permanent snow line is at 4600–4900 m. Some authors (van der Hammen, 1957, 1966; Bürgl, 1980) refer these stadials to the Riss and the Würm, respectively. For others (Julivert, 1961, 1963; Porta,

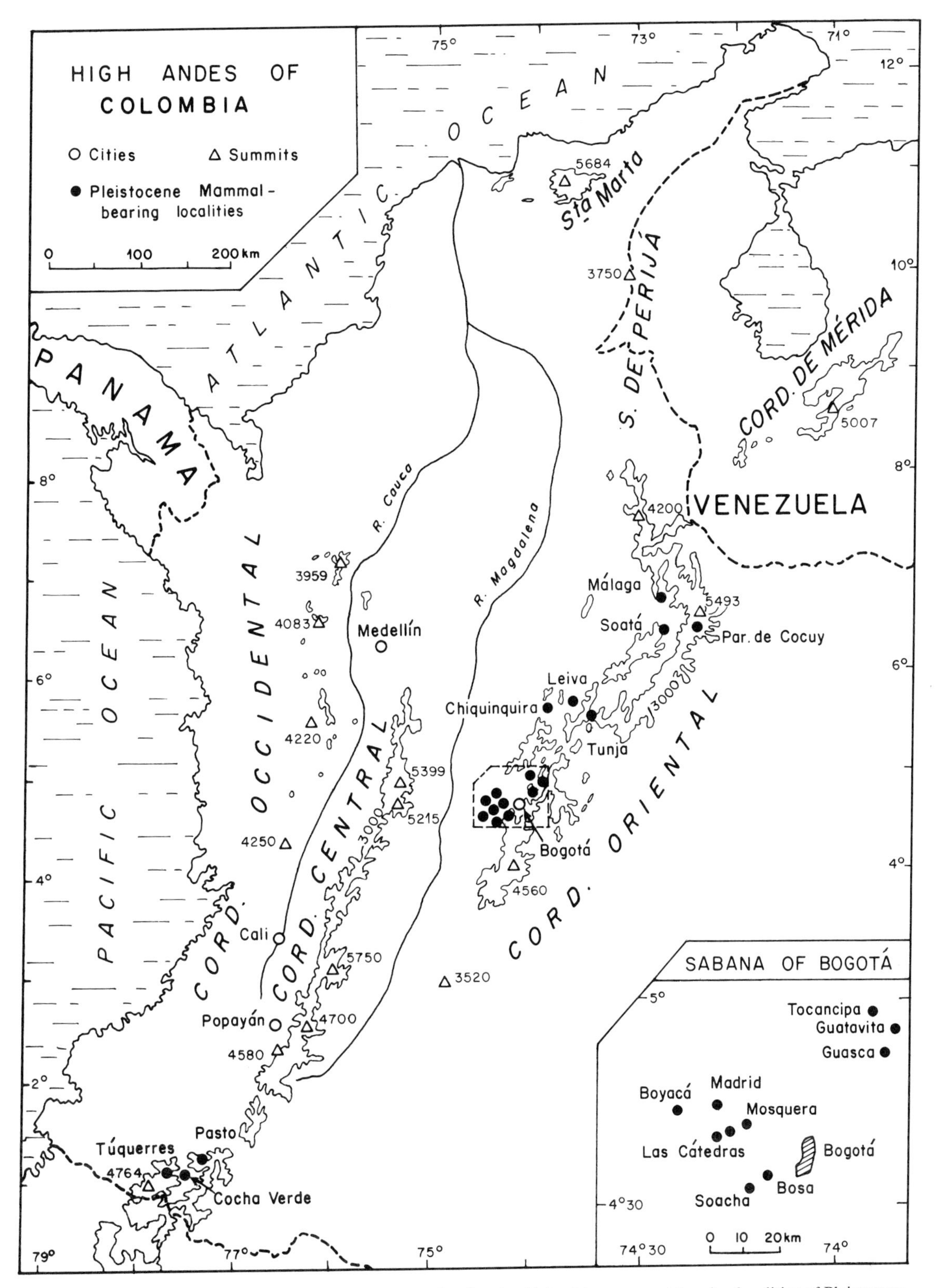

FIG. 9–6. Map of the high Venezuelan and Colombian Andes showing Colombian mammal-bearing localities of Pleistocene age. (The contour line corresponds to 3000 m.)

1961a; Porta et al. 1974), they correspond to the last glaciation and to a retreat phase. According to palynological observations (van der Hammen, 1957; van der Hammen and González, 1960; Chap. 7) the Bogotá lake was surrounded by forests (*Quercus*, *Alnus*, *Podocarpus*) during the interstadial, and by grasses and low shrubs, of páramo type, during the glacial stadials, the upper limit of the forests being located 200 or 900 m lower than at present, at 2400 or 1700 m). It seems that altitudes were then little different from those of today.

Stratigraphic and geomorphological studies (Julivert, 1961, 1963; Porta, 1961a; Porta et al., 1974) show that the Pleistocene includes two local formations; the Tilata Formation related to an upper terrace (5–10 m above the Sabana), and the later Sabana Formation, which is related to a lower terrace (0–5 m). The fossil mammals occupy a well-defined level of the Sabana Formation, between the red silts and the brown silts of Porta (1961a). Considering stratigraphic correlations, this level would be located between two carbon 14 dates: 10,840 and 42,300 B.P. Porta (in Porta et al., 1974, pp. 67–68) even brings forward geomorphological data to locate the faunas between 11,000 and 21,000 years. This is in agreement with G. Correal Urrego (1980) about an archaeological locality (Tibito 1) in the Sabana de Bogotá, where artifacts are associated with remains of mammals (*Amerhippus*, *Haplomastodon*, etc.) dated 11,740 ± 110 B.P. using the carbon 14 method. But according to van der Hammen (1965, p. 384); "A molar of Mastodont was analysed for its fluorine-content by C. J. Overweel; the percentage (0.90%) suggests an approximate age of Early Würm-Glacial or Late Riss-Würm-Interglacial." At any rate, it is Late Pleistocene. However, Porta, who maintained his opinion in 1974, prudently concluded that it is valid only for the fauna of the Sabana and cannot be extended to all Pleistocene localities in Colombia.

From another locality, Soatá, Hernández Camacho and Porta (1960) described a bovid, *Colombibos atactodontus*, but Thenius (1964, p. 275, note 4) and Hoffstetter (1970, p. 948) independently interpreted the specimen as a modern calf, with the dP^4 still in position (this tooth had been interpreted by Soatá et al. as a molariform P^4).

The Páramo de Cocuy is remarkable for its altitude (4000 m). It has yielded a single fossil, a fine mastodont skull (Boule and Thévenin, 1920, fig. 32) dredged from a lake in 1851. It clearly belongs to the *Stegomastodon-Haplomastodon* group, but its molars, with a double-trefoil wear pattern, led me to refer it to the genus *Stegomastodon*, of which it is the only known specimen from the high Andes. A mastodont of the same genus has been found at Pumbeza (15 km from Girardot) at a low altitude (around 350 m) in the Magdalena Valley.

Colombia's contribution to the knowledge of Pleistocene faunas is thus relatively modest. It mainly confirms the distinction between the high-altitude faunas and a warm fauna (with *Eremotherium* and other taxa), which was the source of migrations toward North America.

HISTORY AND DEVELOPMENT OF FAUNAS

Faunal Turnover

It appears that three times during the period under consideration, the fauna has been subjected to profound changes, if not to an almost complete turnover. (This is true mostly in Bolivia, where the evidence is more complete.)

1. The Miocene-Pliocene limit (ca. 6 million years B.P.), which practically coincides with Steinmann's Quechua tectonic phase. This folding episode was apparently accompanied by an uplift, whence higher altitudes and important ecological changes occurred.
2. The Pliocene-Pleistocene limit (ca. 2 million years B.P.), characterized by the massive arrival of North American immigrants (following the emergence of the Panama isthmus) and the extinction of several autochthonous groups; some paleontological data also suggest a rise in altitude at the end of the Pliocene.
3. The Pleistocene-Holocene limit (ca. 10,000 years B.P.), marked by extinction, which mostly affected the large herbivores, whether autochthonous or descendants of Pleistocene immigrants.

These sudden changes are obvious when the compositions of successive large herbivore faunas are compared.

In the Late Miocene (at least in Bolivia), the large herbivores are the abundant mesotherines (with a particular genus, *Plesiotypotherium*), the Toxodontidae (medium-sized Toxodontinae and a large form with a frontal horn), several ground sloths, sclerocalyptine Glyptodontidae (undescribed genus), and the Pampatheriinae (large armadillos).

In the Pliocene, these taxa are replaced by

other large herbivores: macrauchenids of medium size (cf. *Promacrauchenia*), other toxodontids (Xotodontinae, with the particular genus *Posnanskytherium*), different ground sloths from the Miocene ones (mylodontines and megatheriids, the latter apparently ancestral to *Megatherium*), sclerocalyptine glyptodonts (cf. *Plohophorops*) and pampatherine armadillos, and large hydrochoerid rodents (e.g. *Chapalmatherium*).

In the Pleistocene, the large herbivores still comprise a few autochthonous groups, with taxa reaching gigantic sizes. The edentates are especially flourishing (Megatheriidae, Mylodontinae and Scelidotheriinae, various Glyptodontidae), the Macraucheniidae also persist, but the Toxodontidae do not reach high altitudes (they are very rare at Tarija and are absent above 2000 m). In addition, there are immigrants, mastodonts, equids, camelids, cervids, and also peccaries and tapirs, but the latter two are not abundant in the high Andes. Moreover, all carnivores are placentals, which have definitively replaced the borhyaenid marsupials of the Tertiary.

In the Holocene, almost all the large herbivores disappear, with the exception of a few cervids and small camelids (llamas and vicuñas). The date of extinction appears clearly in the section of the Uchcumachay cave in Peru (Wheeler et al., 1976) at 7000 B.C. (i.e., 9000 B.P.), after which date the whole fauna consists of modern genera.

The Pleistocene immigrations and the Holocene extinctions can be observed in all Andean countries. But the crisis at the Miocene-Pliocene transition is illustrated only in Bolivia. It is likely that a similar phenomenon took place elsewhere, at least in Peru where the same Quechua tectonic phase was first recognized. But faunal compositions must have varied somewhat according to the latitude.

Dispersal of South American groups

GROUND SLOTHS (= MEGATHERIOIDEA)

Ground sloths play an important part as plant eaters until the end of the Pleistocene.

The Megatheriidae are represented in the Andes by two lineages. One of them leads to *Megatherium* and is characterized by two apomorphies (derived character, Hennig's terminology): twisting of the femur and strong convexity of the lower profile of the mandible (linked to the increased height of the teeth); this lineage is already present in the Bolivian Pliocene. In the Pleistocene, the much more robust genus *Megatherium* occupies, in the Andes, only southern Peru (north to Cuzco) and Bolivia, where it reaches altitudes greater than 3800 m at Ulloma. The other lineage retains a nearly flat femur and a moderately deep lower jaw; this Pleistocene genus is designated above as "cf. *Pseudomegatherium*." It is known at high altitudes in southern and central Peru, but in the northern Andes it is rare or absent. This genus is closely allied to *Eremotherium*, which differs by its very large size, the loss of the second finger, and the shortening of the third metacarpal; these characters are all apomorphic. It is a tropical Pleistocene genus, very widespread east and west of the Andes, which entered some Andean valleys but always kept to low altitudes. It emigrated to North America in the Pleistocene.

The Mylodontidae comprise two subfamilies. The first, Mylodontinae, are abundantly represented in the Pliocene of Bolivia. In the Pleistocene, they are abundant and varied at Tarija (2000 m) but absent (or rare?) at high altitudes, in Bolivia as well as in Peru. They are numerous but not varied in the Puninian of Ecuador (*Glossotherium* subgenus *Oreomylodon*) (2500–3300 m), but much rarer in Colombia (2600 m). An upper altitudinal limit around 3500 m may explain their absence (or their scarcity) in the higher localities (Bolivia and Peru). They also inhabit the lowlands and have been the source of several emigrations to North America since the Late Miocene. The second subfamily, the Scelidotherinae, has a very different distribution. In the Andean Pleistocene, they seem to be represented only by the genus *Scelidodon*. They are numerous at Tarija (2000 m) and reach high altitudes in Bolivia and Peru, up to 4000 m or more. In northern Peru, Ecuador, and Colombia they no longer occur in the Andes, but they persist on either side of the Andes at low altitudes, west as far as Ecuador and east as far as Venezuela. They did not reach Central or North America.

The Megalonychidae are poorly represented in the Andean localities. It should be noted, however, that they occur in Bolivia in the late Miocene, the Pliocene (an undescribed genus), and the Pleistocene (*Nothropus* at Tarija); a nothrotherine is also known at a high altitude in southern Peru (Casa del Diablo) and a medium-sized megalonychid in northern Peru (Gruta Blanca, Parque Nacional Cutervo).

GLYPTODONTOIDEA

Andean Glyptodontidae are known mostly from Bolivia. The subfamily Sclerocalyptinae is present there in the Late Miocene (undescribed genus), in the Pliocene (cf. *Plohophorops*), and in the Pleistocene at Tarija (*Hoplophorus*,

Panochthus, *Neothoracophorus*—all of them rare); at high altitudes, only some plates of *Panochthus* have been collected near Betanzos (above 3000 m) by C. Villarroel. The Doedicurinae are known only from Anzaldo (approximately 3000 m): *Prodaedicurus* (Pliocene-Pleistocene transition); and from Charaña (4000 m): *Plaxhaplous* (possibly Middle or Early Pleistocene). The Glyptodontinae are by far the most abundant in the Pleistocene of Tarija (*Glyptodon* and allied genera), and are also present, but rarer, at high altitudes (Ulloma-Tambillo).

Outside Bolivia, the only other Andean glyptodontids (with only *Glyptodon*) come from Cuzco (southern Peru, 3350 m), which seems to be their northern limit in the Andes (at least in the Quaternary). But the family is very widespread east of the Cordillera, and the Glyptodontinae entered North America as early as the Late Pliocene.

ARMADILLOS (DASYPODIDAE)

The Pampatheriinae are known from Bolivia in the Late Miocene and in the Pliocene (*Kraglievichia-Plaina* group). In the Pleistocene, they occur at Tarija (*Pampatherium*), but not in the more elevated Bolivian localities; they are also lacking in the northern Andes (Peru, Ecuador, Colombia), but they lived west and east of the Andes and reached North America. The hydrochoerid rodents display the same dispersal pattern and pose the same problems.

The Euphractinae (quirquinchos) have today an essentially austral distribution. A few pre-Quaternary remains are known, and they are common in the Pleistocene of Tarija. They have not been found at higher altitudes, which is puzzling since *Chaetophractus* lives today at high altitudes in Bolivia.

The Dasypodinae (mulitas) have a wider distribution. The Pleistocene genus *Propraopus* (the giant of the subfamily) is known in the Tarijean of Bolivia, the Puninian of Ecuador (Punín, Malchingui), and in Colombia (Sabana de Bogotá). It is absent from Bolivia and Peru, but it is widespread east of the Andes (Argentina, Brazil) and north as far as Florida. *Dasypus* proper is a lowland genus from Argentina to Texas, but unknown in the high Andes.

LITOPTERNA

A moderate-sized representative of the Macraucheniidae (cf. *Promacrauchenia*) is common in the Pliocene of Bolivia (at Anzaldo; and in La Paz and Umala formations). In the Pleistocene, a giant form, *Macrauchenia*, is known at Tarija (2000 m), but also at Ayo Ayo and Ulloma (more than 3800 m); it is also found in southern Peru (a few teeth at Cuzco, 3350 m), but it is unknown farther north in the Andes. On the other hand, it lived east of the Cordillera from Argentina to Brazil and Venezuela.

NOTOUNGULATA

The Toxodontidae are known in Bolivia in the Late Miocene (a medium-sized toxodontine, and a large form with a frontal horn), the Pliocene (a xotodontine, the local genus *Posnanskytherium*), and the Pleistocene (*Toxodon*). The genus *Posnanskytherium*, known only from the Pliocene of the Bolivian altiplano, must have been adapted to high altitudes. But the withdrawal of the Toxodontinae from the high Andes after the Miocene, and then that of the Xotodontinae after the Pliocene, may indicate two uplift phases of the altiplano (at the end of the Miocene and at the end of the Pliocene). The second one is also suggested by the distribution of the pampatheriine armadillos and the hydrochoerid rodents. The genus *Toxodon* is abundant at low altitudes, rare at Tarija, and unknown above 2000 m; it is completely absent farther north in the Andes. It appears that in the Pleistocene the Toxodontinae occupied the whole South American continent east of the Andes, with *Toxodon* in the south and *Mixotoxodon* in the north; they reached Central America (Nicaragua), but could not adapt to high altitudes.

CAVIOMORPH RODENTS

Only a few families have been observed in the Plio-Pleistocene Andean faunas.

Among the Caviidae, the Dolichotinae are worthy of special mention. They are known from the Late Miocene of Bolivia (at Quehua, at about 3800 m), with the genus *Orthomyctera*, but they are lacking in all younger levels and also in other parts of the Andes. But they persist to this day in the southern plains with *Dolichotis*; it is obviously a southern group that could enter the Bolivian Andes only when their altitude was still moderate. With the Mesotheriinae, they indicate that the Quechua tectonic phase led to an important uplift of the Andes. The Caviinae (guinea pigs) have left only a few fossils, because of their small size, but this group likes high-altitude habitats (according to Nordenskiöld, [1980], it has left many remains in the Casa del Diablo; see Chap. 10).

The Ctenomyidae, which appear in Bolivia in the altiplano Pliocene (*Praectenomys*), are common in the Pleistocene of Tarija (*Ctenomys*). They have not been found in the other localities, where only the macrofauna is preserved. But it is an essentially southern group that lives mainly in the mountains.

The Hydrochoeridae also appear in the Pliocene in Bolivia (Ayo Ayo, 3900 m: *Chapalmatherium*). They are also known from the Pleistocene at Tarija (*Pliohydrochoerus* and cf. *Hydrochoerus*), but nowhere else in the Andes. They form an aquatic group living in tropical forests, which ventured only timidly into an inhospitable environment. (The family has emigrated to North America.) It is curious that its representatives could live in the high Andes in the Pliocene. Was there some uplift of the altiplano after the Pliocene? Or had the genera of the two epochs different ecological demands? The pampatheriine armadillos have a similar distribution.

The Myocastoridae are a southern group with an aquatic way of life that only rarely found a favorable habitat in the Andes, where it is known only at Tarija (but where it is common).

The Chinchillidae are represented in the Pliocene of Bolivia (*Lagostomopsis*). The Quaternary forms are poorly known (*Lagidium* in the Holocene of Peru; no *Chinchilla* has been collected), probably because of the conditions of fossilization, since they are essentially Andean genera.

The Erethizontidae occur at Tarija (*Coendou magnus* and another, still larger, species). This is an exceptional occurrence, since this genus mainly lives in the tropical forest.

Other families (Dasyproctidae, Agutidae, Dinomyidae, and even Echimyidae) are essentially tropical groups, which explains their absence among the Andean fossils.

MARSUPIALIA

Marsupials are very poorly represented in Andean localities. However, the occurrence in the Bolivian Pliocene of a microtragulid, a sparassocynid, and perhaps a caenolestid is noteworthy. In any case, it is certain that until and including the Pliocene, the South American flesh-eating mammals were borhyaenid marsupials, and they will certainly turn up in future Andean collections. It should also be noted that, among birds, the Phorusrhacoidea have also played an important part as flesh eaters until the Pliocene (they are known at Pomata, Bolivia). So far, the Andean Pleistocene has yielded only a small didelphid marsupial (*Lutreolina*).

Dispersal of Groups of North American Origin

There are so far no indications of the occurrence of North America immigrants in the Andes before the Pleistocene, unlike Argentina, where they are known in the Late Miocene (*Cyonasua*) and the Pliocene (Mustelidae, Cricetidae, Tayassuidae). Several possible explanations exist, but it is difficult to choose among them. Only Quaternary forms will be discussed here.

MASTODONTS (GOMPHOTHERIIDAE)

Of the four genera to be considered, *Cuvieronius* is the most frequent mastodont at Tarija (2000 m) and the only one known at high altitudes in Bolivia and southern Peru. Its presence has not been demonstrated in northern Peru or in Colombia; in Ecuador it is scarce and of uncertain geologic age. It should be remembered that it is known at low altitudes in Chile, and also in North America (its place of origin). *Haplomastodon* lived throughout the tropical area, including the northern Andes (Colombia, Ecuador, northern Peru), where it always dominates, and sometimes occurs alone. *Stegomastodon* (distinguished from the preceding one by its molar teeth with a double trefoil wear pattern) is frequent in North America and in Argentina; in the Andes, it is known only from Colombia (a skull at Cocuy, 4000 m; a skeleton at Pubenza, 350 m), and perhaps from Tarija (a few tusks). *Notiomastodon* (distinguished from *Stegomastodon* by the enamel band of its tusks) is a poorly known Argentine form; a tusk from Tarija (Bolivia) could belong to it.

This distribution can be explained as follows: a first wave of immigration could have brought *Cuvieronius* (subfamily Cuvieroniinae) all the way to Chile via the Andean route. The other genera belong to the Stegomastodontinae. *Haplomastodon* arrived after *Cuvieronius* and became widespread in the tropical area; it replaced *Cuvieronius* in the northern Andes (where the few finds of the latter may represent an ancient occupation or isolated populations that persisted temporarily). The acquisition of molars with a double trefoil wear pattern could have taken place independently in North America and in Argentina, and the resulting (diphyletic) *Stegomastodon* could then have reached Colombia from the north and Bolivia from the south. Finally, *Notiomastodon*, which retained an enamel band on its tusks (a primitive feature that perhaps still occurs in the first *Haplomastodon*), would also have acquired molars with a double trefoil pattern (which led to a particular branch in a southern habitat).

EQUIDAE

Three genera lived and then became extinct in South America. *Hippidion* arrived very early in the Quaternary; it became widespread in Brazil and Argentina, and from there reached the southern Andes (Tarija), without going higher than 2000 m. *Onohippidium* (including *Parahipparion* and *Hyperhippidium*), closely related to *Hippidion*, is also known from Argentina since the beginning of the Pleistocene; it has also lived in the Andes, mostly in Bolivia and Peru, where it is known from middle altitudes (Tarija), and also at more than 4000 m. It is found at low altitudes as far as the southern tip of Chile (Ultima Esperanza cave); on the other hand, it is very rare in Ecuador and has not yet been recorded from Colombia. It should be noted that this genus survived until 9000 B.P. (Peru, southern Chile) and that it is frequent in caves and is often associated (probably as game) with archaeological sites. *Equus* (subgenus *Amerhippus*) has a later record (from the Bonaerian in Argentina, the Chichean in Ecuador) and has occupied practically all of South America, including the Andes, where species with short, stout legs (mountain habitus) have differentiated; it does not seem to have reached altitudes higher than 3300 m, which would explain its absence in the high Andes of Peru and Bolivia.

TAPIRIDAE

The only known Andean fossils come from Tarija. There is, however, a Recent species (*Tapirus pinchaque*) that lives in the northern high Andes (Peru, Ecuador, Colombia), but there is no paleontological evidence about the date of its arrival at high altitudes. However, a tibia of *Tapirus* (species not identified) has been collected in the northern Peruvian Andes (Gruta Blanca, Parque Nacional Cutervo, altitude not specified, between 2200 and 3500 m).

TAYASSUIDAE (PECCARIES)

The only Andean fossils are *Selenogonus* from southern Colombia (3000 m, probably Early Pleistocene), *Platygonus* at Tarija (Bolivia, 2000 m), and "*Dicotyles*" at the Casa del Diablo (southern Peru, 3800 m). This shows that these animals, which are mainly known from lowland forests, could adapt to high altitudes.

CAMELIDAE

Palaeolama s.l. (including the subgenera *Hemiauchenia*, *Protauchenia*, and *Astylolama*), which had already differentiated in North America, colonized almost all of South America by adapting to various environments. In the Andes, it is particularly abundant at Tarija (Bolivia, 2000 m, with several species) and at Punín (Ecuador, at around 3000 m, one species). But it does not seem to reach higher than 3300 m, hence its absence from the high Andes of Bolivia and Peru. On the other hand, it becomes rarer toward the north, and its (probable) occurrence in Colombia has yet to be confirmed.

Lama (including *Vicugna*). All Recent South American camelids interbreed and produce fertile hybrids; they thus should all be included in the same genus (if not the same superspecies). It seems that the genus *Lama* has differentiated from *Palaeolama* in Argentina, where it is known as early as the beginning of the Pleistocene. In the Andes, it is abundant mostly in Bolivia and Peru. In Ecuador and Colombia, it is not known from the Pleistocene. It is apparently in Peru that it became semidomesticated (5200–4200 B.C.) and then domesticated (4200–2500 B.C.), according to Wheeler et al. (1976) (see Chap. 10). It may have been introduced into Ecuador by man as a domestic animal at a later date.

CERVIDAE

Three genera, two of which are extinct, are exclusively Andean. *Agalmaceros* lived in Ecuador (Puninian, 2500–3000 m), and in central Peru (Late Pleistocene, 4000 m); *Charitoceros* replaces it farther south, at Cuzco, Peru (Late Pleistocene, 3350 m), and in Bolivia (Tarijean, 2000 m). Lastly, *Pudu* (Recent, from Colombia to Chile) is not known as a fossil.

Three other Recent genera are or have been widespread in the lowlands as well as in the Andes. *Hippocamelus* is a southern genus; until recently it occupied southern Patagonia and extended to the Andes, where it reached Ecuador in the north; it is known as a fossil in the Tarijean of Bolivia (2000 m), in the Holocene of central Peru (4000 m), and in the Puninian of Ecuador (rare, at about 3000 m). The other two genera are essentially tropical, with a few Andean representatives: *Odocoileus* is known as a fossil in Ecuador (Puninian) and in the caves of northern (3500 m) and central Peru (Holocene, 4000 m). Only one of the species of *Mazama* is Andean (*M. rufina*); a few fossil remains from the Puninian of Ecuador are referred to it.

There are, moreover, several genera restricted to southern South America, but they apparently did not enter the Andes.

CARNIVORES

Among the Felidae, *Smilodon* has been found in the Andes of Bolivia (Tarijean, 2000 m) and Ecuador (Puninian, approximately 3000 m), but has not been reported above those altitudes. Large jaguars, *Leo* (*Jaguarius*) *onca andinus*, are known from Bolivia (Tarija, 2000 m; Yaco, at about 3600 m), northern Peru (Tragadero de la Purla, 3350 m), and Ecuador (Puninian, around 3000 m). It has been noticed that among Recent and fossil jaguars the largest sizes are found at high altitudes or latitudes. Cougars (*Felis*, subgenus *Puma*) have reached even higher altitudes in the Andes; a particularly powerful species, *F.* (*P.*) *platensis*, lived during the Pleistocene (Tarijean and Puninian, but also the Pampean of Argentina). The Recent species, *F.* (*P.*) *concolor*, is reported from the Holocene and the Late Pleistocene of central Peru at altitudes reaching more than 4000 m. The other subgenera of *Felis*, of small-size, are poorly represented in the fossil record (a yaguarundi from Tarija should be mentioned, however).

The Canidae of the Andean Pleistocene include a few rare representatives of large forms such as *Canis dirus* at Tarija (2000 m) and *Theriodictis* at Tarija and at Guamote (Ecuador, 3100 m). A *Chrysocyon* is also present, but rare, at Tarija. But the most frequent fossils are *Dusicyon*, of medium or small size, very close to the Recent forms.

The other families of carnivores are poorly represented in the Andean Pleistocene. *Conepatus* (Mustelidae) is known from Bolivia (Tarija) and southern Peru (Casa del Diablo). *Arctodus* (Ursidae) is present at Tarija with two well-characterized species; a canine reveals its presence in the Puninian of Ecuador. A little-fossilized *Nasua* (Procyonidae) from Tarija is the only procyonid known from the Andes.

OTHER IMMIGRANT GROUPS

The few finds do not allow one to ascertain their importance and their range. The Cricetidae are relatively well known only at Tarija, where a varied sample has been collected. But for the Sciuridae, the Lagomorpha, and the Insectivora, no fossil is known from the Andean Pleistocene, which can be explained by the late arrival of these immigrants in South America.

CONCLUSIONS

Despite the geographical and temporal gaps in the present paleontological data, the overall characteristics of the Plio-Pleistocene high Andean faunas, and their spatiotemporal successions, are fairly clear.

First of all, the faunas of successive epochs are in striking contrast with each other. Late Miocene, Pliocene s.s., Pleistocene, and Holocene faunas are each distinct. Each epochal limit is marked by a sudden faunal change or turnover, the causes of which are obvious. At the end of the Miocene, folding and uplift strongly modify the environment. These changes resulted in (1) displacements of faunal elements, and (2) local adaptations. In the Late Pliocene, the emergence of the Panama isthmus was followed by immigrations from North America, but these faunal movements did not become clearly visible in the Andes until the Pleistocene. There is evidence of an uplift in Bolivia, which caused the Hydrochoeridae, the Pampatheriinae, and the Toxodontidae to abandon the altiplano. At the beginning of the Holocene, the fossil record shows mass extinctions of the large animals under the joint influence of man and the changing climate.

During each epoch, minor changes probably affected the faunal composition (evolution in situ, local extinctions, and small-scale migrations), but the available data give incomplete indications. The two species of the toxodontid genus *Posnanskytherium*, for example, are never found together. This lack of sympatry (or synchrony) doubtless corresponds to two successive levels of the Bolivian Pliocene. As another example, the Doedicurinae are known in the Andes only toward the Plio-Pleistocene border (*Prodaedicurus* at Anzaldo), and in a relatively old Pleistocene site (*Plaxhaplous* at Charaña). In the Tarija-Padcaya basins, the different species of mastodonts and those of *Palaeolama* and *Lama* have probably different stratigraphic ranges, but these have not yet been established because we lack a good distinction of successive levels. During the Pleistocene, the available data suggest that *Cuvieronius* was replaced by *Haplomastodon* in the northern and central Andes.

During a given Plio-Pleistocene epoch, as today, the Andean faunas vary geographically both with latitude and with altitude, and it is not always possible to distinguish the respective parts played by these two factors. The Late Pleistocene is interesting in this respect, because faunas of this age are known from the various Andean countries, thus allowing one to estimate latitudinal variation, and also because the Andes have probably been subjected only to moderate altitudinal changes during this time, "controlling," as it were, the altitudinal factor.

In the Late Pleistocene, only the following genera or subgenera of large animals are known from high altitudes (3300–4500 m) in Bolivia and Peru: *Cuvieronius*, *Macrauchenia*, *Megatherium*, *Glyptodon*, *Panochthus* (very rare), *Charitoceros*, *Scelidodon*, "cf. *Pseudomegatherium*," *Onohippidium* (*Parahipparion*), *Lama*, *Agalmaceros*,

Odocoileus, *Hippocamelus*, *Jaguarius*(?), and *Puma*. It should be noted that the first six genera of this list do not reach farther north than Cuzco (13° 30′S). Some remains of *Cuvieronius* are known in the Ecuadorian Andes but possibly come from an older Pleistocene level; on the other hand, *Macrauchenia* and *Glyptodon* reach Venezuela in low areas. *Scelidodon* and *Lama*, present in Bolivia and Peru, are unknown in the Pleistocene in the northern Andes. *Agalmaceros* and *Charitoceros*, which are exclusively Andean cervids, replace each other, the first occurring in central Peru and Ecuador, the second in southern Peru and Bolivia. Finally, "cf. *Pseudomegatherium*" is limited at high altitudes to southern and central Peru, but it seems to have close relatives at low altitudes.

The fine faunas from the Puninian of Ecuador and from the Sabana de Bogotá correspond, in the northern Andes, to an area of lower altitudes, from 2400 to 3300 m. They suggest that *Glossotherium*, *Propraopus*, *Haplomastodon*, *Palaeolama*, *Equus* (*Amerhippus*), and *Smilodon* did not reach higher, which would explain their absence from the high Andes of Peru and Bolivia. But it would be useful to know better the middle-altitude faunas of both countries (for instance, Ayacucho and the Huancayo area). This would allow one to check the altitudes reached by the foregoing genera, and also to determine more accurately the southern range limit of the genus *Haplomastodon* (known today from Colombia to northern Peru in the Andes, and as far as southern Peru along the Pacific coastal strip).

The Tarijean is certainly older (Ensenadan and part of Lujanian s.l.), but its localities (approximately 2000 m) do not belong to the high Andes. As a result, its much richer and more varied fauna includes forms that are unusual in the Andes. Most of them are fairly rare (like *Neothoracophorus*, *Hoplophorus*, *Panochthus*, *Toxodon*, *Coendou*, *Canis dirus*, *Chrysocyon*, *Stegomastodon*, and *Notiomastodon*), whereas others are relatively common (*Lestodon*, *Pampatherium*, Hydrochoeridae, and *Myocastor*). But it should be remarked that the Tarijean faunal association includes practically all the genera mentioned earlier from higher-altitude areas, the only exceptions being forms with a more northern distribution, at least in the Andes (*Haplomastodon*, *Agalmaceros*, *Odocoileus*, and "cf. *Pseudomegatherium*"). It is thus chiefly its lower altitude that explains the richness of this fauna but its diversity is also a result of the temporal succession, during a fairly long time span, of animals that lived at various altitudes around the sedimentary basins.

It is interesting to compare the Andean faunas with contemporaneous lowland faunas that occurred west of the Andes (warm faunas of the Ecuadorian and north Peruvian coast, with genera such as *Haplomastodon* and *Eremotherium*), or in Venezuela (warm faunas with *Haplomastodon*, *Eremotherium*, *Mixotoxodon*, *Glyptodon*, etc.), or else in the chaco (more southern faunas with *Stegomastodon*, *Toxodon*, *Mylodon*, *Glyptodon*, and *Panochthus* abundant). Such comparison shows the influence of altitude and latitude and also the role of a barrier played by the Andean Cordillera.

A difficult problem is that of the precise geographic origins of the various Andean animals. Some of them belong to groups that entered the Andean region at an early date, before its major uplift. They may have become adapted to progressively higher-altitude habitats. But on the whole, very few direct filiations can be observed in situ. Thus, for instance, the small Pliocene genus cf. *Promacrauchenia* is replaced without transition by the giant Pleistocene *Macrauchenia*. In most instances these abrupt replacements imply migrations either from within or from outside the Andes. Macroevolution (with punctuated equilibria) seems unlikely, at least in this instance; in any case, it would give only a partial explanation, as each turnover includes the disappearance of some groups and the arrival of others.

Other animals from the tropical Andes were probably derived from the faunas of neighboring warm areas. These taxa could have entered valleys at moderate altitudes (for example, the hydrochoerid rodents, the canid *Chrysocyon*, and the glyptodont *Hoplophorus* from Tarija), and some of them could reach the high Andes (for instance, *Haplomastodon*, the Tayassuidae, and the Tapiridae).

Some groups could have colonized the Andes directly from North America. An instance is that of the mastodont *Cuvieronius*, which apparently did not succeed in colonizing the warm areas of South America (although it had to cross the Panama lowlands), but which occupied the Andes and the Chilean coast. Other North American genera invaded the tropical zone of South America to a greater or lesser extent, but did not succeed in reaching the southern part of the continent. This is true of the cervid *Odocoileus*, whose South American "species" (probably only a subspecies of the ancestral Virginia Deer, *O. virginianus*) are common in the northern part, in low areas as well as in the Andes, but do not extend beyond central Peru toward the south. More limited is the range of the lagomorph *Sylvi-*

lagus and even more that of the northern Andean shrew *Cryptotis* (very close to the North American *Blarina*); both are recent immigrants known only in the northwestern part of South America.

But other Andean animals are derived from stocks that differentiated in the southern temperate zone. They are mostly native South American groups or genera, such as *Megatherium*, the Doedicurinae, *Glyptodon*, the euphractine armadillos, *Macrauchenia*, Ctenomyidae, and also *Toxodon*, *Lestodon*, and *Myocastor* (the last three having a very limited altitudinal and latitudinal range).

It is clear that after they had occupied the warm tropical zone, even North American immigrants could enter the south temperate zone, and there give rise to forms adapted to more severe climates. Some of these taxa could then colonize the Andes from the south, especially *Lama*, *Onohippidium*, *Hippocamelus*, *Hippidion*, *Stegomastodon* (Argentine lineage), and *Notiomastodon* (the last three are known in the Andes only from Bolivia and below 2000 m).

Some cases are more difficult to analyze, for instance, that of the strictly Andean cervids, whose phylogenetic relationships among themselves and with other genera should first be established before their geographic history can be reconstructed.

The Carnivora represent a different case. They are less directly tied to specific vegetation types and are often fairly eclectic in their choice of prey. As a result, their range can extend widely in latitude and in altitude: the subgenus *Puma* is a good example.

ACKNOWLEDGMENTS

I am indebted to Dr. Eric Buffetaut and Dr. François Vuilleumier, who were kind enough to translate and revise this chapter into English. Thanks also are due to Mrs. Elyane Molin, Mrs. Odile Poncy and Mrs. Françoise Pilard for help in the preparation of the manuscript and for the illustrations.

REFERENCES

Ahlfeld, F. 1946. Geología de Bolivia. *Rev. Museo La Plata* (n.s.), secc. Geol. III: 5–570.

Aubouin, J. 1980. Les grandes divisions des Cordillères ouest-américaines. *8è Réunion Annuelle des Sciences de la Terre* 16.

Ballivian, O., J. L. Bles, and M. Servant. 1978. El Plio-Cuaternario de la región de La Paz (Andes Orientales, Bolivia). *Cahiers ORSTOM*, sér. Géol. 10: 101–13.

Berry, E. W. 1922. Late Tertiary plants from Jancocata, Bolivia. *Johns Hopkins Univ. Stud. Geol.* 4: 205–20.

Bonifaz, E. 1977. *Dating of obsidian artifacts of the Ilalo region of Ecuador, according to their hydratation.* Varela 190, Quito, Ecuador: published by the author.

Boule, M., and A. Thévenin 1920. *Mammifères fossiles de Tarija.* Miss. Sci. Créqui-Montfort et Sénéchal de la Grange. Paris: Soudier.

Bristow, C. R., and R. Hoffstetter. 1974. *Ecuador*, 2d ed.: *Lexique stratigraphique international*, vol. V, *Amérique latine*, ed. R. Hoffstetter, fasc. 5a2. Paris: Centre National de la Recherche Scientifique.

Bürgl, H. 1980. Bioestratigrafía de la Sabana de Bogotá y sus alrededores. *Bol. Geol., Serv. Geol. Nac.* (Bogotá) 5: 113–85.

Cardish, A. 1973. Excavaciones en la caverna de Huargo, Perú y 3 Apéndices por L. A. Cardish (dos fechas obtenidas por el método de radiocarbono para el sitio arqueológico de Huargo); Pascual, R., and O. E. Odreman Rivas (Estudio del material osteológico extraido de la caverna de Huargo); Andreis, R. R. and J. Casajus (Sedimentología de los depósitos de la caverna de Huargo). *Rev. Mus. Nac. Lima* 39: 11–47.

Clapperton, C. M. 1979. Glaciation in Bolivia before 3.27 Myr. *Nature* 277: 375–77.

Correal Urrego, G. 1980. Evidencias culturales asociadas a megafauna durante el Pleistoceno tardío en Colombia. *Geología Norandina* (Soc. Geol. Colombia) 1: 29–34.

Dobrovolny, E. 1962. Geología del valle de La Paz. *Minist. Minas Petrol. La Paz, Bol.* 3 (especial): 1–50.

Eaton, G. F. and H. E. Gregory. 1914. Vertebrate fossils from Ayusbamba, Peru. *Amer. J. Sci.* 4th ser., 37: 111–54.

Evernden, J. F., S. J. Kriz, and M. Cherroni. 1966. Correlaciones de las formaciones terciarias de la cuenca altiplánica a base de edades absolutas determinadas por el método potasio-argón. *Hoja informativa* I, *Servicio Geológico de Bolivia*, La Paz.

Frakes, L. A. 1979. *Climates throughout geologic time.* Amsterdam: Elsevier.

Hernández Camacho, J., and J. de Porta. 1960. Un nuevo bóvido pleistocénico de Colombia, *Colombibos atactodontus. Bol. Geol. Univ. Industrial Santander* (Bucaramanga) 5: 41–52.

Hoffstetter, R. 1952. Les mammifères pléistocènes de la République de l'Equateur. *Mém. Soc. Géol. France* (n.s.) 31: 1–391.

——— 1963a. La faune pléistocène de Tarija (Bolivie). *Bull. Mus. Nat. Hist. Nat. Paris* 35: 194–203.

——— 1963b. Les Glyptodontes du Pléistocène de Tarija (Bolivie). I. Genres *Hoplophorus* et *Panochthus. Bull. Soc. Géol. France*, sér. 7, 5: 126–33.

——— 1970. Vertebrados cenozóicos de Colombia, Ecuador, Perú. *Actas IV Congr. Latino-amer. Zool.* (Caracas, 1968) 2: 931–983.

——— 1971. Los vertebrados cenozóicos de Colombia: Yacimientos, faunas, problemas planteados. *Geol. Colomb.* (Univ. Nac. Colombia, Bogotá) 8: 37–62.

——— 1980. Utilización de la escala cronoestratigráfica

internacional en el Terciario mamalífero sudamericano. *Actas II Congr. Argentino Paleont. Bioestratig. y Ier Congr. Latinoamer. Paleont.* (Buenos Aires, 1978) III: 1–8.

Hoffstetter, R., and C. Villarroel. 1974. Découverte d'un marsupial microtragulidé (= argyrolagidé) dans le Pliocène de l'altiplano bolivien. *C. R. Acad. Sci. Paris* (sér. D) 278: 1947–1950.

Hoffstetter, R., C. Martinez, J. Muñoz-Reyes, and P. Tomasi. 1971. Le gisement d'Ayo Ayo (Bolivie). *C. R. Acad. Sci. Paris* (sér. D) 273: 2472–2475.

Hoffstetter, R., C. Villarroel, and G. Rodrigo. 1984. Présence du genre *Chapalmatherium* (Hydrochoeridae, Rodentia) représenté par une espèce nouvelle, dans le Pliocène de l'altiplano bolivien. *Bull. Mus. Nat. Hist. Nat. Paris* (sér. 4) 6: 59–79.

Julivert, M. 1961. Observaciones sobre el Cuaternario de la Sabana de Bogotá. *Bol. Geol., Univ. Industrial Santander* (Bucaramanga) 7: 5–36.

———. 1963. Los rasgos tectónicos de la región de la Sabana de Bogotá y los mecanismos de formación de las estructuras. *Bol. Geol., Univ. Industrial Santander* (Bucaramanga) 13–14: 1–104 and atlas.

Kraglievich, J. 1952. El perfil geológico de Chapadmalal y Miramar, Prov. Buenos Aires. *Rev. Mus. Municipal Cienc. Nat. y Tradicional Mar del Plata* 1: 8–37.

Kraglievich, L. 1931. *Megatherium Lundi Seijoi*, nueva subespecie pleistocena del Uruguay. *Rev. Sociedad "Amigos de la Arqueología,"* Montevideo, 5, 9 pp. (Reprinted in: *Obras completas de L. Kraglievich*, vol. II, 625–632, La Plata [1940]).

———. 1934. La antigüedad pliocena de las faunas de Monte Hermoso y Chapadmalal, deducidas de su comparación con las que le precedieron y sucedieron. Imprenta "El Siglo ilustrado," Montevideo, p. 17–136 (Reprinted in *Obras completas de L. Kraglievich*, Vol. III, 293–433, La Plata [1940].)

Lavenu, A. 1978. Néotectonique des sédiments plio-quaternaires du nord de l'Altiplano bolivien (région de La Paz-Ayo Ayo-Umala). *Cahiers ORSTOM* (sér. Géol.) 10: 115–126.

Lisson, C. S. 1912. Un esqueleto antidiluviano en la Sierra de la Viuda en la provincia de Yauli: El Megaterio de Yantac. *Bol. Soc. Geogr.* (Lima) 1912: 126–129.

MacFadden, B. J., and R. G. Wolff. 1981. Geological investigations of late Cenozoic vertebrate-bearing deposits in southern Bolivia. *Anais II Congr. Latinoamer. Paleont.*, (Porto Alegre, 1981), pp. 765–778.

MacFadden, B. J., O. Siles, P. Zeitler, N. M. Johnson, and K. E. Campbell. 1983. Magnetic polarity stratigraphy of the Middle Pleistocene (Ensenadan) Tarija Formation of southern Bolivia. *Quat. Res.* 19: 172–87.

MacNeish, R. S. 1971. Early man in the Andes. *Scientific. Amer.* 224: 36–46.

———. 1976. Early man in the New World. *Amer. Scientist* 63: 316–327.

MacNeish, R. S., T. C. Patterson, and D. C. Browman. 1975. The Central Peruvian prehistoric interaction sphere. *Papers of the R. S. Peabody Foundation of Archaeology* 7: 1–97.

Marshall, L. G., A. Berta, R. Hoffstetter, R. Pascual, O. A. Reig, M. Bombin, and A. Mones. 1984. Mammals and stratigraphy: Geochronology of the continental mammal-bearing Quaternary in South America. *Palaeovertebrata*, Montpellier, Mém. Extr.: 1–76.

Marshall, L. G., R. F. Butler, R. E. Drake, G. H. Curtis, and R. H. Tedford. 1979. Calibration of the Great American Interchange. *Science* 204: 272–279.

Marshall, L. G., R. Hoffstetter, and R. Pascual. 1983. Mammals and stratigraphy: Geochronology of the continental mammal-bearing Tertiary of South America. *Palaeovertebrata*, Montpellier. Mém. Extr.: 1–93.

Marshall, L. G., R. Pascual, G. H. Curtis and R. E. Drake. 1977. South American geochronology: Radiometric time scale for Middle to Late Tertiary mammal-bearing horizons, Patagonia. *Science* 195: 1325–1328.

Montaño Calderón, M. 1968. Estudio geológico de la región de Anzaldo-Vilaque. Master's Thesis, Universidad Mayor de San Andrés, Fac. Ciencias Geológicas. La Paz, Bolivia.

Nordenskiöld, E. 1908. Ein neuer Fundort für Säugetierfossilien in Peru. *Arkiv Zool.* 4(11): 1–22.

Pearson, R. 1978. *Climate and evolution*. New York: Academic Press.

Pick, J. 1944. La capa fosilífera de Betanzos. *Minería Boliviana* (La Paz) II (13): 36–38.

Porta, J. de 1961a. La posición estratigráfica de la fauna de mamíferos del Pleistoceno de la Sabana de Bogotá *Bol. Geol., Univ. Industrial Santander* (Bucaramanga) 7: 37–54.

———. 1961b. Algunos problemas estratigráfico-faunísticos de los vertebrados en Colombia (con una bibliografía comentada). *Bol. Geol., Univ. Industrial Santander* (Bucaramanga) 7: 83–104.

Porta, J. de., C. Caceres, F. Etayo, R. Hoffstetter, M. Julivert, J. Navas, R. E. Robbins, N. Solé de Porta, B. Taborda, P. Taylor, N. Téllez, and O. Valencia. 1974. *Colombie (Tertiaire et Quaternaire): Lexique stratigraphique international*, Vol. V, *Amérique latine*, R. Hoffstetter, ed., fasc. 4b. Paris: Centre National de la Recherche Scientifique.

Ramírez Pareja, J. A. 1958. Mamíferos fósiles del Departamento del Cuzco. Master's Thesis, Universidad Nacional Cuzco.

Rovereto, G. 1914. Studi di geomorfologia argentina. IV. La Pampa. *Bol. Soc. Geol. Ital.* 33: 75–128.

Sanmartino, Y., G. Staccioli, and J. D. Klein. 1981. *Pérou 79*. Bagnols-sur-Gèze: Publ. Groupe Spéléo Bagnols-Marcoule.

Sauer, W. 1965. *Geología del Ecuador*. Quito: Edit. Minist. Educ.

Servant, M., and J. C. Fontes. 1978. Les lacs quaternaires des hauts plateaux des Andes boliviennes. Premières interprétations paléoclimatiques. *Cahiers ORSTOM*, Sér. Géol. 10(1): 9–23.

Simpson, B. B. 1975. Pleistocene changes in the flora of the high tropical Andes. *Paleobiology* 1: 273–294.

Steinmann, G. 1930. *Geología del Perú*. Heidelberg: Carl Winters Universitätsbuchhandlung.

Stirton, R. A. 1946. A rodent and a peccary from the Cenozoic of Colombia. *Compil. Estudios Geológicos Oficiales Colombia* 7: 317–324.

Sullivan, L. R., M. Hellman, and H. E. Anthony. 1925.

The Punin calvarium. *Anthrop. Pap. Amer. Mus. Nat. Hist.* 23: 309–337.

Takai, F., et al. 1982. Tarija mammal-bearing formation in Bolivia. *Res. Inst. Evol. Biol.* (Tokyo) 3: 1–72.

Takai, F., et. al. 1984. On fossil mammals from the Tarija Department, southern Bolivia. *Res. Inst. Evol. Biol.* (Tokyo) 4: 1–63.

Thenius, E. 1964. Herkunft und Entwicklung der südamerikanischen Säugetierfauna. *Zeitschrift. f. Säugetierkunde* 29: 267–284.

Troll, C. 1927–28. Forschungsreisen in den zentralen Anden von Bolivia und Peru. *Petermann Mitt.* 74: 100–103.

———. 1929. Reisen in den östlichen Anden Boliviens. *Petermann Mitt.* 75: 181–188.

van der Hammen, T. 1957. Estratigrafía palinológica de la Sabana de Bogotá (Cordillera Oriental). *Bol. Geol., Inst. Geol. Nac. Bogotá* 5: 187–303.

———. 1965. The age of the Mondoñedo Formation and the *Mastodon* fauna of Mosquera (Sabana de Bogotá). *Geol. Mijnbouw* 44: 384–390.

———. 1966. The Pliocene and Quaternary of the Sabana de Bogotá (the Tilata and Sabana Formations). *Geol. Mijnbouw.* 45: 102–109.

van der Hammen, T., and E. González. 1960. Upper Pleistocene and Holocene climate and vegetation of the "Sabana de Bogotá" (Colombia, South America). *Leidse Geol. Mededel.* 25: 261–315.

———. 1964. A pollen diagram from the Quaternary of the Sabana de Bogotá (Colombia) and its significance for the geology of the Northern Andes. *Geol. Mijnbouw.* 43: 113–117.

Villarroel, C. 1974. Les Mésothérinés (Notoungulata, Mammalia) du Pliocène de Bolivie. Leurs rapports avec ceux d'Argentine. *Ann. Paléont. (Vert.)* 60: 245–381.

———. 1975. Dos nuevos Ctenomyinae (Caviomorpha, Rodentia) en los estratos de la Formación Umala (Plioceno superior de Viscachani, Prov. Aroma, Depto La Paz, Bolivia). *Act. Ier Congr. Argent. Paleont. Bioestratigr.* Tucumán, vol. II: 495–501.

———. 1978. Edades y correlaciones de algunas unidades litoestratigráficas del altiplano boliviano y estudio de algunos representantes Mesotheriinos. *Rev. Acad. Nac. Cienc. Bolivia* (La Paz) 1: 159–170.

Webb, S. D. 1976. Mammalian faunal dynamics of the great American interchange. *Paleobiology* 2: 220–234.

Wheeler Pires-Ferreira, J., E. Pires-Ferreira, and P. Kaulicke. 1976. Preceramic animal utilization in the Central Peruvian Andes. *Science* 194: 483–490.

10

Domestication of Andean Mammals

ELIZABETH S. WING

In no other place in the Western Hemisphere did the domestication of mammals play as important a role in cultural development as in the Andes. In fact, very few animals were domesticated in the New World outside of the Andes. Guinea pigs, llamas, and alpacas play a very significant role in the ritual and economy of the native people inhabiting the Andes. This complex interaction between domesticated mammals and people is one that developed gradually and had its roots in the early cultural development of this region. Tracing the changing patterns of this use of animals is based on studies of their skeletal remains excavated from archaeological sites in Ecuador, Peru, Bolivia, and Chile.

The evidence to support this discussion comes from the faunal studies of a wide variety of archaeological sites. These sites range geographically from Colombia to Chile and from the puna in the high Andes to the Pacific coast. Because few archaeological excavations have been conducted east of the Andes, and preservation of animal remains is poor in the humid tropics, there are virtually no faunal data from the Amazonian slopes. The chronological range in these data is from approximately 10,000 B.C. to the time of the Spanish conquest.

The archaeological context of the faunal assemblages of course varies greatly, from the refuse of temporary campsites to the refuse from specialized workshops, and to complex urban centers. Added to the biases that result from different uses of animals and disposal of their remains are biases resulting from different recovery and identification methods. Identified animals are tangible evidence of association with human beings. Other animals may have been used but their remains may have been deposited in an unexcavated portion of the site, or they may be lost in any one of the many stages of preservation, recovery, and identification. Two families of mammals that were of prime importance throughout the history of human occupation in the Andes are cervids (including white-tailed deer [*Odocoileus virginianus*] and locally the huemul [*Hippocamelus antisensis*], the brocket [*Mazama gouazoubira*], and pudu [*Pudu mephistopheles*]) and camelids (including the wild guanaco [*Lama guanicoe*] and vicuña [*Lama vicugna* or *Vicugna vicugna*] and the domestic llama [*Lama glama*] and alpaca [*Lama pacos*]). These animals are in the same size range except for the pudu, and preservation and recovery should affect them equally. Guinea pigs (*Cavia porcellus* and other species) differ in behavior, reproductive potential, and human care and use of the domesticated form, and vary in size and delicacy of the bone. These differences may affect the

TABLE 10–1. List of the sites referred to in this chapter, grouped according to elevation: puna above 4000 m and highland valleys between 2000 and 4000 m (these two zones are found at lower elevations in the southern Andes).

Site	Location	Excavator, affiliation, or referenced paper
	Source of data from highland valleys	
Colombia		
Tequendama	Bogotá	Van Gelder-Ottway (1975–77)
Ecuador		
Chobshi Cave	Azuay Prov., Sigsig (2400 m.)	Thomas F. Lynch, Cornell University
La Chimba	Pichincha Prov., 60 km N. of Quito (3160 m)	Alan J. Osborn, Univ. New Mexico
Sacopamba	Northern part of Imbabura Prov. (2200 m)	Alan J. Osborn, Univ. New Mexico
Cochasqui	Pichincha Prov.	Oberem (1975; pers. comm.

Site	Location	Excavator, affilation, or referenced paper
Callejón de Huaylas		
Guitarrero Cave	Callejón de Huaylas (2500 m)	Thomas F. Lynch Cornell University
PAn 3–3, Pl, PIRC	Callejón de Huaylas	late Gary Vescelius Virgin Islands
Ayacucho Valley		
Wari	Ayacucho (2500–3000 m)	Richard MacNeish, R.S. Peabody Foundation
Ayamachay, AC 102		
Wishgana, AS 18		
Puente, AS 158		
Pikimachay, AC 100		
Cuzco to Lake Titicaca		
Marcavalle	Peru, Cuzco (3314 m)	Karen Mohr Chavez, Central Michigan Univ.
Minaspata, PC_z 12–9	Peru, Lucre Valley (3200 m)	Edward Dwyer, RI School of Design
Pikicallepata	Peru, Sicuani (3410 m)	Karen Mohr Chavez, Central Michigan Univ.
Qaluyu	Peru, Pucara (4 km N, 3930 m)	Karen Mohr Chavez, Central Michigan Univ.
Q'Ellokaka	Peru, Pucara (7 km NW, 3930 m)	Sergio Chavez, Central Michigan Univ.
Chile		
Vega Alta	Chile, Atacama Desert	Pollard and Drew (1975)
Loa II	Middle Rio Loa (2500 m)	Pollard and Drew (1975)
	Source of data from the puna	
PAn 12–51	Peru, Upper Callejón du Huaylas	Thomas F. Lynch Cornell University
PAn 12–57	3970–4270 m	
PAn 12–58		
Lauricocha	Peru, Huánuco	Wheeler et al. (1976)
Huanuco Pampa	Peru, Huánuco	Craig Morris American Museum of Natural History
Junín		
Pachamachay	Peru, Junín (4400 m)	Ramiro Matos Universidad de San Marcos, Lima Wheeler et al., 1976 (second sample)
Uchumachay		Wheeler et al. (1976), Lavallée, proyecto Junín-Palcamayo (1975, 1977)
Panalauca		Wheeler et al. (1976)
Acomachay		Wheeler (1975)
Cuchimachay		Wheeler (1975)
Telarmachay		Wheeler et al. (1977)
Chile		
Tulan-52	Chile, Salar de Atacama	Hesse (n.d.)
Puripica		
Tambillo		

deposition, preservation, and recovery of their remains. Thus, in the interpretation of these data one must be aware of their limitations as well as their potential to gain an understanding of the domestication process of mammals and the roles played by domestic animals in human culture.

The faunal samples used to form the data base of this chapter are from studies conducted at the Florida State Museum and others reported in the literature (Table 10–1). The materials from the sites, listed in Table 10–1, were studied at the Florida State Museum unless otherwise cited by reference to a faunal study. The faunal remains from Huanuco Pampa were identified in Peru by Elizabeth Reitz and the author with the help of two Peruvian students, Carmen Rosa Cardoza and Denise Pozzi-Escot. Other students who worked at the Florida State Museum on Andean material were Lynn C. Balck, Kathleen Johnson, Kathleen Byrd, and Gary Shapiro.

CHARACTERISTICS THAT DISTINGUISH DOMESTIC MAMMALS

Changes in Size and Variability

When a live mammal is seen there is rarely any question about whether it is domestic. However, not all of the characteristics of the external features and the behaviors of domesticates are reflected in the fragments of skeletons and dentitions. Changes in color pattern and texture of hair or feathers frequently seen in domestic forms can rarely be documented for prehistoric animals. Changes in size and increase in variability may be reflected in the skeletal remains. These changes, which may be the result of differences in the diet imposed by man, and at least partial release from natural selection, are manifested in the soft parts of the anatomy as well as the skeleton. Data on size changes through time and size ranges or variability are now being amassed, but no conclusive information on size trends is available yet. Progress is being made by a number of workers in at least distinguishing the size of camelids in archaeological samples (Miller, 1979; Hesse, n.d.).

Age Structure

Another anticipated difference is the age structure of the domestic mammals managed by their masters compared with that of the hunted prey animals. Among the herd animals, such as the camelids, or household domesticates, such as the dog and guinea pig, different strategies and circumstances can affect the demography, and thus the age composition, of the faunal assemblage. Animals that are raised for work or products, for example, the llama for burden bearing or alpacas for wool, would most advantageously be maintained for their full productive lives. Natural and cultural factors and events may mitigate this, and animals may be slaughtered or die while in their prime or as juveniles. The high-elevation pasturelands of the puna are rigorous environments, which may be most stressful on diseased and young segments of the population. Relatively large numbers, making up to 60% of the fauna, of newborn camelids have been recovered from three puna sites, Uchcumachay (Wheeler, Pires-Ferriera, and Kaulicke, 1976), Telarmachay (Wheeler, Cardozo, and Pozzi-Escot, 1977), and Pachamachay (Wing, 1975; the camelids were incorrectly aged at 18 months in that publication). This high percentage of newborn individuals has been interpreted as evidence of stress on a vulnerable segment of the population, aggravated by the constraints of domestication. A similar mortality rate is experienced in present-day herds as a result of bacterial infections spread in the confinement of a corral (Fernandez-Baca, 1971; Wheeler, personal communication).

Assigning an age to the skeletal remains of camelids is not simple. The newborn animals are distinguished by the unworn condition of the deciduous premolars and the M_1 lying unerupted in the lower jaw. Subsequently there is great variation in the age at which the deciduous premolars are replaced and in the degree of wear on the molars (Wheeler, personal communication). This variation, which Wheeler documented in alpacas of known age, is yet another aspect of the variability of these domestic mammals. Similar variation is present in the sequence of epiphyseal fusion in camelids. Complete fusion of all epiphyses is very protracted, and some very old animals still have unfused vertebral epiphyses. As a result of this variability, precise aging of the skeletal and dental remains may not be possible, yet assignment to the broader age categories may be sufficiently accurate to detect patterns of use distinct from harvest of wild animals.

Relative Abundance in Faunal Assemblages

In addition to changes in size, increased variability, and modified age structure, clues to control of animal populations through domestication may be manifested in changes in the intensity of its use. Increased intensity of species use may reflect a specialized hunting pattern, but when coupled with other characteristics of domestication, it may indicate use of domesticated animals. This may be further substantiated when the use of an animal is extended into a region where the animal does not characteristically occur. An isolated instance of such an occurrence may not be significant, but with repetition of occurrences a pattern may be revealed.

The natural range of the wild ancestors differs from the range of their domestic forms that are moved around at the will of their masters. The range of wild guinea pigs forms a large arc in the Andes from Venezuela through Colombia, Ecuador, Peru, Bolivia, and north and central Argentina east to Brazil. They live in highland valleys and up into the puna. Guanaco and vicuña have a more restricted range from the Andes of Peru south to Argentina. In the northern parts of this range they inhabit the puna.

INTERPRETATION OF FAUNAL ASSEMBLAGES FROM ARCHAEOLOGICAL SITES

Regional Faunas

The animals that were hunted and fished were chosen from the resources accessible from each site. The natural range of the wild food animals whose remains are found in a site is usually within close proximity to the site. Interesting but rare evidence of trade of animals from distant habitats does occur, for example, the marine fishes from the site of Chavin in the Callejón de Huaylas, many miles from the sea and across a formidable mountain range.

One of the outstanding features of the Andean area is the diversity of habitats and their associated faunas. The steep altitudinal gradient brings different ecological zones into close proximity. Peasants living and farming in such mountainous regions recognize fine distinctions (Brush, 1976; Winterhalder and Thomas, 1978) that are based mainly on the tolerances of different crops to produce optimally under these different climatic conditions. The range of animals is also determined by environmental factors; however, no such fine discriminations are recognized for this analysis. Faunal assemblages from the puna, the highland valleys, and the coast differ and will be analyzed separately. Great differences also exist between the fauna from the northern part of this range, namely Ecuador, and the southern part of the range, including central Peru and south to Lake Titicaca. As it was within these different climates, or resource complexes, that animals became domesticated or were incorporated in the economy, it is appropriate to discuss each of these regional faunas separately.

The Puna

The puna (from about 4000 m elevation) has the fewest different animal species used by prehistoric people. These grazing lands are the prime habitats of the camelids and some cervids, and thus, as one might expect, these two groups form the bulk of the mammals used at this elevation. In fact, in the 13 occupation zones that have been

TABLE 10–2. Puna sites—from the Callejón de Huaylas south to Junín: thirteen occupations inhabited from 7000 B.C.–A.D. 1534.

Abundant species	Percentage of occurrence	Mean percentage of abundance	Range in percentage of abundance
Camelidae	100	85.8	54.9–98.2
Cervidae	100	12.8	1.5–41.7

Scarce species, abundance less than 1%.

Mammals
- Canidae
  - *Canis familiaris*
  - *Dusicyon culpaeus*
- Caviidae
  - *Cavia* sp.
- Chinchillidae
  - *Lagidium peruanum*
- Felidae
  - *Felis* sp.
- Mustelidae
  - *Conepatus rex*

Birds
- Tinamidae

Amphibians
- Leptodactylidae
  - *Batrachophrynus* sp.
- Bufonidae
  - *Bufo* sp.

studied so far, bones of these two animal groups make up 97%–100% of the faunal remains. The small remainder of the fauna is composed of carnivores such as domestic dogs (*Canis familiaris*), fox (*Dusicyon culpaeus*), hognosed skunk (*Conepatus rex*), cats (*Felis* spp.), and large rodents, including guinea pig (*Cavia* sp.) and viscacha (*Lagidium peruanum*) (Table 10–2).

The Highland Valleys

The faunal assemblages from highland valley sites differ from those from puna sites by having a greater variety of animals represented. A measure of this variability is the Shannon diversity index (MacArthur and MacArthur, 1961), which incorporates how many species were used and how much each was relied upon. The average and range in the species diversity in valley sites are 1.0558 (0.2449–1.6829), reflecting a broad-spectrum procurement pattern, compared with the puna diversity of 0.4944 (0.1363–1.2274), reflecting the animal use of a specialized hunter and herder. In valley sites the faunal diversity decreases as domestic animals become the dominant animal used.

As in the puna, deer remains are consistently and abundantly represented in virtually all sites. Camelid remains also become a very prominent feature of these valley site faunas. The general temporal trend in abundance of the remains of these two families is that the Cervidae gradually decrease in importance and reciprocally the Camelidae increase. This shift in relative importance comes around 2000 B.C. in central Peru, but not until later in Ecuador. The earliest faunal sample with the camelid remains from Ecuador is from the site of Pirincay, thought to be from formative times. Guinea pigs (*Cavia* sp.) are the third animal resource that occurs consistently and abundantly in Peruvian sites, at least those south of the Callejón de Huaylas. They are reported to be abundantly represented from two early occupations, Tequendama and El Abra, near Bogotá, Colombia (Van Gelder-Ottway, 1975–1977; Ijzereef, personal communication). In the intervening area of present-day Ecuador, guinea pig remains have not been found in sites dating to times earlier than formative. They have been reported, although the remains are unquantified, from the formative village site of Cotocollao (Peterson, 1977).

The minor elements of the fauna, those species whose remains constitute less than 10% of the total remains, differ as much if not more at the extreme ends of the study area (Ecuador and Peru)—in other words latitudinally, as they do altitudinally. The mammal species identified from the Ayacucho Valley are the same as those from the slightly more northern puna sites, with the addition of two cats (*Felis onca* and *F. wiedii*)

TABLE 10–3. Highland valley sites—Ayacucho Valley: twenty occupations inhabited from 7000 B.C.–A.D. 1425.

Abundant species	Percentage of occurrence	Mean percentage of abundance	Range in percentage of abundance
Camelidae	100	37.0	1–96.6
Caviidae	70	26.4	0–92.4
Cavia sp.			
Cervidae	90	16.1	0–46.7

Scarce species, abundance less than 10 percent

Mammals
- Bovidae
  - *Bos taurus*
  - *Ovis aries*
- Canidae
  - *Canis familiaris*
  - *Dusicyon culpaeus*
- Chinchillidae
  - *Lagidium peruanum*
- Didelphidae
  - *Didelphis sp.*
- Felidae
  - *Felis onca*
  - *Felis wiedii*
  - *Felis* sp.
- Megatheriidae
  - *Eremotherium* sp.
- Mustelidae
  - *Conepatus rex*
- Suidae
  - *Sus scrofa*

Birds
- Accipitridae
- Cathartidae
  - *Cathartes* sp.
- Charadriidae
- Columbidae
- Cracidae
- Falconidae
  - *Falco* sp.
- Picidae
- Psittacidae
- Rallidae
- Tinamidae
- Tytonidae
  - *Tyto alba*

Amphibian
- Bufonidae
  - *Bufo* sp.

Fish
- Catfish

TABLE 10–4. Highland valley sites—Ecuador: three occupations inhabited from 7000–5500 B.C. and AD 650–850.

Abundant species	Percentage of occurrence	Mean percentage of abundance	Range in percentage of abundance
Leporidae *Sylvilagus brasiliensis*	100	41.6	6.7–66.3
Cervidae	100	40.4	15.9–75.8
Camelidae	30	26.9	

Scarce species, abundance less than 10 percent

Mammals
- Canidae
  - *Canis familiaris*
  - *Dusicyon* sp.
- Cebidae
  - *Saimiri* sp.
- Dasyproctidae
  - *Agouti taczanowski*
- Didelphidae
  - *Chironectes* sp.
  - *Didelphis albiventris*
  - *Didelphis* sp.
- Erethizontidae
  - *Coendu bicolor*
- Felidae
  - *Felis concolor*
- Mustelidae
- Procyonidae
- Tapiridae
  - *Tapirus pinchaque*
- Ursidae
  - *Tremarctos ornatus*

Birds
- Columbidae
  - *Columba* sp.
  - *Columbina passerina*
- Phasianidae
  - *Gallus gallus*
- Tinamidae

probably introduced from the moist tropical lands of the Brazilian subregion (Hershkovitz, 1972) and the opossum (*Didelphis* sp.). The big difference between these faunas is the great number and variety of birds from the Ayacucho Valley fauna (Table 10–3). The fauna from highland Ecuador has a distinctive character. The mammals identified include two kinds of opossum (*Didelphis albiventris* and *Chironectes* sp.), a rabbit (*Sylvilagus brasiliensis*), a monkey (*Saimiri* sp.), paca (*Agouti taczanowski*), porcupine (*Coendu bicolor*), puma (*Felis concolor*), spectacled bear (*Tremarctos ornatus*), and tapir (*Tapirus pinchaque*). The birds from these sites, as in the Ayacucho Valley, including one or more doves (Columbidae) and tinamou (Tinamidae) but otherwise do not include the variety of other birds found in the Ayacucho sites (Table 10–4).

The Coast

The coastal sites, close to the rich marine resources, have faunal assemblages most distinct from highland valley and puna faunas. As was seen in the highland faunas, latitude also makes a big difference in the faunal composition. Both guinea pigs (*Cavia porcellus*) and camelids are found in the later occupations where they are introduced into a background of economic dependence on fishing and procurement of marine mammals and birds. In Ecuador guinea pig and camelid remains have been found only in the context of human burials dating from 500 B.C. to A.D. 1155 (Hesse, 1980).

The preceding data are summaries of the economically important animals, based on their abundance in the faunal samples, in the region of the Andes in which domestic animals played a role. The camelids and guinea pigs are commonly represented in the highland valley sites, and camelids are abundant in puna sites. However, in these summaries no distinction was made between domestic, wild, or tame members of these groups. As indicated earlier, no easy way exists for distinguishing the degree to which any individual animal had been subjected to human control. A single herd of animals could include individuals that are domestic as well as others that are wild or tame. Chronicles from Incan times describe roundups of wild animals during which the vicuña were shorn and released, the deer were killed for meat, and some male guanacos were captured and added to the domestic herds (Browman, 1974). Ultimately techniques may be developed to detect such wild or domestic individuals; however, at this stage of knowledge characteristics of the animal remains as well as their context must be relied on to provide evidence for the wild or domestic state of the animals.

CHANGING PATTERNS IN THE USE OF ANIMALS

The greatest impact on the prehistoric economy was the domestication of the herd animals, Camelidae, and the guinea pig. The taming and domestication of these mammals took place in the Andes. Dogs were introduced into South America as domestic animals and occur sporadically throughout the study area. Their relatively rare occurrence does not necessarily indicate, however, that dogs were insignificant in the social cultures of which they were part.

Three other animals are of special interest in regard to human manipulation, but they are not Andean. One of these is the Muscovy Duck (*Cairina moschata*), which was domesticated and reported in association with human burials dating from A.D. 730–1730 at the Ayalan site in the Santa Elena Peninsula of southwestern Ecuador (Hesse, 1980). These remains are not interpreted as evidence of domestic animals, but they do indicate ceremonial use. Another animal also found in ceremonial context of sites in southwestern Ecuador, but several millenia earlier (7500–5500 B.C.) is the native fox (*Dusicyon sechurae*) (Wing, 1980). Although there is no evidence for more than ceremonial use of this animal by Vegan people, convincing arguments have been presented for the domestication of a different species of this genus (*D. australis*) in the Falkland Islands (Clutton-Brock, 1977). The third animal that appeared to have special significance was some species of tinamou (Tinamidae). It is the one bird that was almost universally used during prehistoric times. Remains of the Chilean Tinamou (*Nothoprocta perdicaria*) have been identified from archaeological sites dating from the 14th to the 18th centuries on Easter Island (Carr, 1981). These birds may have been tamed when they were brought from the South American mainland by man.

The principal domesticates, camelids and guinea pigs, profoundly affected South American prehistory and so the remainder of the chapter will be devoted to a discussion of their spread throughout the Andes, and the role they played in the development of a complex society.

DOCUMENTATION OF THE SPREAD OF DOMESTICATES

Sequence of Spread from the Puna to the Highland Valleys and to the Coast

In faunas from puna sites the camelids and cervids predominate, and a general trend evidently existed through time toward the relative increase in camelids at the expense of cervids (Table 10–5). The faunal samples from the earliest time period, 10,000–5500 B.C. show approximately equal numbers of specimens identified as camelid and as deer, which may be interpreted as equal dependence on the two herd groups. In the following time period, 5500–2500 B.C., the remains of camelids increased dramatically relative to the numbers of cervids. From this period to the end of the Incan Empire in the central Peruvian Andes, from the Callejón de Huaylas to Junín, the high level of camelid exploitation is indicated by the presence of over 80% camelids in the faunal assemblages.

The age structure of the camelids in the puna sites varies considerably but tends toward an increase in the number of remains of newborn or fetal animals. In the valley of Junín, newborn animal remains constitute on the average 22.4% (range, 9.4%–45.9%) of the total camelid remains from the 10 occupations that date prior to 5000 B.C., and constitute on the average 51.2% (range, 23.3%–75.1%) of the total camelid remains from the five occupations that date from 1000 to 5000 B.C. (Jane Wheeler, personal communication). Remains of newborn animals are absent or rare in faunal samples from valley sites. One exception to this is the material from Pikimachay Cave in the Ayacucho Valley, which has 39% newborn camelids in the levels that date from 1500 to 4000 B.C.

Several different factors may contribute to the absence or abundance of newborn camelids in faunal assemblages. Newborn animals are exceedingly rare in most archaeological faunal assemblages. This age class is seldom recovered by hunters. Among domestic herds mortality rates are high, ranging from 50% to 60% (Fernandez Baca, 1971). The major contributing cause to this loss is enterotoxemia infection spread among animals confined to crowded corrals (Fernandez Baca, 1971). Animals lost in this way can be retrieved and used by herders as they apparently were in a number of puna sites. Such sites would have been occupied during the rainy season, which extends from December to May and coincides with the period of time when young

TABLE 10–5. Sample size and relative abundance of domestic animals and deer identified from puna sites (data presented are numbers of tooth and bone specimens).

Puna Sites	Total number of identified specimens	Camelidae: Number of specimens	Camelidae: Percent	Cervidae: Number of specimens	Cervidae: Percent	Cavia: Number of specimens	Cavia: Percent	Canis: Number of specimens	Canis: Percent
10,000–5500 B.C.									
PAn 12–58	53	40	75.5	13	24.5	—		—	
Lauricocha[a]	—		59.1		40.9	—		—	
Panalauca[a]	—		26.0		74.0	—		—	
Uchcumachay[a]	175	96	54.9	73	41.7	—		—	
Jaywamachay[e]	9111	2236	24.5	5764	63.3	23	0.3	—	
5500–2500 B.C.									
Lauricocha[a]	—		84.7		13.1	—			2.2
Pachamachay (sample 1)	4618	4471	96.8	90	1.9	24	0.5	—	
Pachamachay (sample 2)[a]	—		97.8		2.0	—			0.2
Panalauca (early)[a]	—		87.6		12.3	—			0.1
Panalauca (late) [a]	—		85.8		13.9	—		—	
Uchcumachay (early)[a]	1194	983	82.3	207	17.3	—		2	0.2
Uchcumachay (late)[a]	881	747	84.8	123	14	—		—	
Tulan-52[b]	14264	12096	84.8	—		—		—	
Puripica[b]	4490	3426	76.3	—		—		—	
Tambillo[b]	2208	1047	47.4	—		—		—	
2500–1750 B.C.									
PAn 12–57	41	39	95.1	2	4.9	—		—	
Acomachay A[c]	68		94.1		5.9	—		—	
Cuchimachay[c]	159	128	80.5	29	18.2	—		—	
Pachamachay[a]	—		96.7		2.7	—			0.1
Telarmachay[a]	162		84.8		13.5	—		—	
1750–450 B.C.									
PAn 12–51	602	533	88.5	69	11.5	—		—	
PAn 12–57	792	761	96.1	30	3.8	—		1	0.1
Pachamachay	3042	2988	98.2	46	1.5	1	0.03	—	
Telarmachay [d] (early)	713	601	84.3	87	12.2	2	0.3	—	
Telarmachay[d] (late)	739	636	86.1	72	9.8	1	0.1	2	0.3
Qaluyu	275	200	72.7	58	21.1	—	—	1	0.4

CONTINUED

TABLE 10–5 CONTINUED.

Puna Sites	Total number of identified specimens	Camelidae		Cervidae		Cavia		Canis	
		Number of specimens	Percent	Number of specimens	Percent	Number of specimens	Percent	Number of specimens	Percent
450 B.C.–A.D. 650									
Qaluyu	94	88	98.6	–		—		1	1.1
A.D. 1425–1534									
Huanuco Pampa	9620	8334	86.6	442	4.6	47	0.5	162	1.7

[a] Wheeler Pires-Ferreira et al. (1976).
[b] Hesse (n.d.).
[c] Wheeler Pires-Ferreira (1975).
[d] Wheeler et al. (1977).
[e] Flannery (n.d.)

are born. Such sites would also be where herd animals were reared rather than locations where the occupants had access to charki (dried meat) and adult animals for their wool and work.

The faunas from the highland valleys present a more complex picture (Table 10–6). The data from faunal samples are available over a vast area extending from Bogotá, Colombia, to the Atacama Desert of Chile. As a result, animal use cannot be expected to be uniform over such a large range. Furthermore, two groups of mammals, the camelids and guinea pigs, played dominant but different roles in these valley economics. The following discussion will deal first with the evidence for use of guinea pigs and second with that for the camelids. This division is made for the sake of simplification and does not imply that the use of these two animals was unrelated; in fact, each is part of an integrated system of animal use.

Guinea pig remains are abundant in sites from the earlier time period. Between 10,000 and 5500 B.C., their remains constitute a substantial portion of the faunas at two sites near Bogotá, Colombia, the site of Tequendama with 37% (Van Gelder-Ottway, 1975–1977) and abundant but unquantified remains at the El Abra rock shelter (Gerald Ijzereef, personal communication) and at two sites in the Ayacucho Valley, Ayamachay with 50% and Puente with 69%. In the intermediate area of Ecuador and the Callejón de Huaylas there are either no or very few remains of guinea pigs. In the following period, 5500 to 2500 B.C., a similar pattern is seen; however, guinea pig remains are relatively more abundant in the two Colombian sites and one of the Ayacucho Valley sites. In subsequent periods guinea pig remains diminish in importance, except at certain sites such as Ayamachay in the Ayacucho Valley and Pikicallepata south of Cuzco where they continue to be abundant. Increased relative abundance is clearly not synonymous with domestication, however. Some initial measurements of the lower jaw have been noted, but more intensive study of skeletal change must be made before the changes related to domestication manifest in the skeleton can be understood (Wing, 1977).

One indication of the special consideration given this animal by the Andean natives is a mummified individual wrapped in woven cloth recovered from the Rosamachay cave in the Ayacucho Valley, and dating from the Chupas period, about 400 B.C. Although a wild animal might be buried wrapped in cloth, such treatment is more likely afforded a domestic one.

The camelid remains, as do those of the guinea pig, show a great variation in relative abundance, even within the sites of one valley during one time period. Admittedly these time periods are long, but the occupations in the Ayacucho Valley probably overlap and the variation may reflect specialization. Even with this variation, a general trend toward increased abundance, and therefore use, of camelids in Peru is evident. During the earliest time period, 10,000–5,000 B.C., camelid remains are absent or scarce. During the following period, 5500–2500 B.C., camelid remains are scarce, although at the Pikimachay site in the Ayacucho Valley camelid remains increased almost fourfold from the earlier to the later half of this time period. Camelid remains increased substantially during the time 2500–1750 B.C. and thereafter are consistently more abundant than deer. The one exception to this is at the Kotosh site in Huánuco where the abundance of camelids lags behind deer during the period 1750–540 B.C.

This time period, 1750–450 B.C., is one during which many social changes were taking place, one of which is the use of animals (MacNeish, Patterson, and Browman, 1975). It marks a time not only when camelids generally become markedly more important in the highlands, but also when dogs, although never very abundant, are frequently represented. After 1750 B.C. dogs occur in over 50% of the sites studied, whereas prior to that time they are seldom reported. An increase in the frequency of occurrence of dogs after 1750 B.C. is also evident in coastal sites.

The frequency of occurrence of the other two domesticates also increases along with the increase of dogs in coastal sites after 1750 B.C. (Table 10–7). Earlier (2500–1750 B.C.) remains of both camelids and guinea pigs have been reported from the coast. These are finds from the sites of Los Gavilanes (PV35–1), Chilca, and Culebras (Lanning, 1967; Bonavia, pers. comm.). In addition to these skeletal materials, hair recovered from Los Gavilanes has been identified as coming from alpaca and from guinea pig (Bonavia, pers. comm.). Camelid remains are abundant in coastal sites only after 450 B.C.

Sequence of Northward Spread of Domestic Animals from the Central Andes

The spread in the use of domestic animals can take place from one altitudinal zone to another, as well as from one region to another at the same elevation. Faunal samples from the puna come from central Peru with the exception of the three samples reported from Chile (Table 10–5). Samples from these two areas do not appear to differ

TABLE 10–6. Sample size and relative abundance of domestic animals and deer identified from highland valley sites (data expressed in numbers of tooth and bone specimens unless otherwise indicated).

Highland valley sites	Total number of identified specimens	Camelidae		Cervidae		Cavia		Canis	
		Number of specimens	Percent	Number of specimens	Percent	Number of specimens	Percent	Number of specimens	Percent
10,000–5500 B.C.									
Tequendama (early)[a]	18(MNI)	—		5	27.5	1	5.5	—	
Tequendama (late)[a]	91(MNI)	—		14	15.4	34	37.4	—	
Chobshi Cave	178	—		135	75.8	—		—	
Guitarrero early (Complex I)	17	—		6	35.3	—		—	
Guitarrero late (Complex II)	221	7	3.2	137	62.0	1	0.5	—	
Ayamachay AC 102	16	1	6.3	5	31.3	8	50	—	
Puente AC 158	1761	47	2.7	376	21.3	1221	69.3	27	1.5
5500–2500 B.C.									
Tequendama[a]	32(MNI)	—		8	25	16	50	—	
Guitarrero Cave (Complex III)	33	4	12.1	19	57.6	—		—	
Ayamachay AC 102	105	1	1.0	6	5.7	97	92.4	—	
Pikimachay AC 100 (early)	50	5	10	17	34	22	44	—	
Pikimachay AC 100 (late)	141	54	38.3	1	0.7	44	31.2	—	
Puente AC 158	1239	41	3.3	759	61.3	243	19.6	23	1.9
2500–1750 B.C.									
Kotosh-Mito	543	67	12.3	309	56.9	166	30.6	—	
Pikimachay AC 100	118	51	43.2	20	16.9	15	12.7	—	
Puente AC 158	15	4	26.7	7	46.7	3	20	—	
1750–450 BC									
Pirincay		√[d]		√		—		—	
PAn 3	397	353	88.9	16	4.0	7	1.8		
Pi	44	24	54.6	2	4.6	8	18.2	4	9.1
PIRC	225	203	90.2	22	9.8	—		—	
Kotosh (early)	200	40	20	99	49.5	50	25	—	
Kotosh (late)	663	322	48.6	329	49.6	10	1.5	1	1.2
Pikimachay AC 100	73	30	41.1	7	9.6	12	16.4	—	

Wishgana AS 18 (early)	43	43	100	—		—		—	
Wishgana AS 18 (late)	670	531	79.3	—		4	0.6	128	19.1
Marcavalle	3304	3111	94.2	145	4.4	30	0.9	3	0.1
Minaspata	233	157	67.4	51	21.9	17	7.3	—	
Pikicallepata (early)	64	37	57.8	7	10.9	14	21.9	—	
Pikicallepata (late)	447	343	76.7	19	4.3	20	4.5	4	0.9
Q'Ellokaka	112	83	75.9	6	5.4	9	8.0	1	0.9
450 B.C.–A.D. 650									
Guitarrero (Complex IV)	456	36	7.9	54	11.8	4	0.9	2	0.4
Kotosh	387	263	69.6	69	17.8	47	12.2	—	
Ayamachay AC 102	24	6	25.0	4	16.7	13	54.2	—	
Pikimachay AC 100	50	33	66.0	7	14.0	-		1	2.0
Wishgana AS 18	840	811	96.6	6	0.7	10	1.2	1	.1
Marcavalle	74	48	64.9	2	2.7	6	8.1	1	1.4
Minaspata	417	272	65.2	60	14.4	63	15.1	3	.7
Pikicallepata	262	85	42.1	9	4.5	97	48.0	1	.5
Vega Alta II[b]		√							
Loa II[b]		√							
A.D. 650–850									
Sacopampa	145	39	26.9	23	15.9	2	1.4	3	2.1
La Chimba	10039	-		2968	29.6	—		6	.06
Ayamachay AC 102	17	14	82.4	1	5.9	—		—	
Pikimachay AC 100	13	7	53.9	1	7.7	—		—	
A.D. 850–1425									
Cochasqui[c]		√		√		√		√	
Ayamachay AC 102	31	11	35.5	2	6.5	18	58.1	—	
Pikimachay AC 100	17	9	52.9	2	11.8	—		—	
A.D. 1425–1534									
Tarma	308	266	86.4	40	13.0	1	.3	1	.3
Minaspata	130	106	81.5	3	2.3	15	11.5	—	

[a] Van Gelder-Ottway (1975–77).
[b] Pollard and Drew (1975).
[c] Oberem (1976).
[d] √ indicates presence, but unquantified

TABLE 10–7. Sample size and relative abundance of domestic animals and deer identified from coastal sites (data presented are numbers of tooth and bone specimens; check mark indicates presence but not quantified).

Coastal sites	Total number of identified specimens	Camelidae		Cervidae		Cavia		Canis	
		Number of specimens	Percent	Number of specimens	Percent	Number of specimens	Percent	Number of specimens	Percent
10,000–5500 B.C.									
OGSE 38, Ecuador [a]	17	—		—		—		—	
OGSE 80, Ecuador[a]	302	—		17	5.6	—		—	
PV7–19, N. Peru	71	—		—		—		—	
Cupisnique sites (5), N. Peru	5990	—		—		—		—	
Quebrada las Conchas[b] 9410 ± 160 B.P. 9680 ± 160 B.P.		√		—		—		—	
5500–2500 B.C.									
OGSE 63, Ecuador[a]	79	—		59	74.7	—		—	
PV7–16, N. Peru	52	—		—		—		—	
Los Gavilanes PV35–1 north-central Peru	246	—		—		√[c]		—	
Abtao I 1st occupation[b]		—		—		—		—	
2500–1750 B.C.									
Real Alta, Ecuador[a]	806	—		121	15	—		—	
Loma Alta, Ecuador[a]	628	—		272	43.3	—		28	4.5
Valdivia, Ecuador[a]	548	—		286	52.2	—		—	
OGSE 62, Equador[a]	290	—		—		—		—	
OGSE 62C, Ecuador[a]	224	—		—		—		—	
Padre Alban[d]		—		—		—		—	
Alto Salaverry[d]		—		—		—		—	
Las Aldas, north-central Peru	213	—		—		—		—	
Los Gavilanes PV=35–1, north-central Peru	116	1	0.9	2	1.7	√[c]		—	
Chilca central Peru	23	1	4.4	—		—		—	
Culebras[e]		?		?		√		?	
Abtao, 2d occupation[b]		—		—		—		—	

1750–450 B.C.									
OGCH 20, Ecuador[a]	1906	—		265	13.9	—		—	
Gramalote[d]		—		—		—		—	
Caballo Muerto[d]		√		√		—		√	
Las Aldas, north-central Peru	566	—		—		—		2	0.4
Curayacu, central Peru	918	4	0.4	5	0.6	1	0.1	2	0.1
Abtao I, 3rd occupation[b]		—		—		—		—	
450 B.C.–A.D. 650									
OGSE 46U, Ecuador[a]	459	—	—	—		—	—		
OGSE 46D, Ecuador[a]	1038	—		3	0.3	—		2	0.2
Ayalan[f] (500 B.C.–A.D. 1155)		√		√		√		√	
PV7–18, N. Peru	26	12	46.2	—		—		—	
Cerro Arena[d]		√		—		√		—	
Moche Huacas[d]		√		—		√		√	
Viru 434[g]	362	181	50.0	7	1.9	—		13	3.6
Viru 604, N. Peru	519	4	0.8	—		12	2.3	—	
Viru 368, 633, 636, N. Peru	195	4	2.1	—		4	2.1	—	
Viru 632, N. Peru	229	8	3.5	—		—		—	
Bermejo, north-central Peru	74	1	1.4	—		—		1	1.4
Chilca, Central Peru	1942	—		—		—		—	
A.D. 650–850									
Pampa Grande[h]	828	581	70.2	1	0.2	25	3	75	9
Galindo[d]		√		—		√		√	
Viru 631, N. Peru	180	3	1.7	—		1	0.6	1	0.6
A.D. 850–1425									
Ayalan (A.D. 750–1730)[f]		√		√		√		√	
OGSE 41E, Ecuador	11	—		—		—		—	
Chan Chan[d]		√		—		√		√	
Caracoles[d]		√		—		—		√	
Cerro la Virgen[d]		√		—		√		—	
Choroval[d]		√		—		—		—	
Chilca, central Peru	3068	7	0.2	—		—		—	0.2

[a] Byrd (1976); [b] Llagostera Martinez (1979); [c] Bonavia (pers. comm.); [d] Pozorski (1979); [e] Lanning (1967); [f] Hesse (1980); [g] Reitz (1979); [h] Shimada (1979)

greatly in respect to the relative abundance of camelids.

The highland valley sites are distributed over a greater geographic range and show regional differences (Table 10–6). As mentioned earlier, even in one region such as in the Ayacucho Valley, the relative abundance of animals shows great variation within one period. Full understanding of these variations awaits more complete study of the archaeological contexts of these faunal assemblages. In the earlier time periods great relative abundance of guinea pigs is seen in Colombia and Peru but not in the intermediate region of Ecuador. The trend of increasing abundance of camelids relative to cervids is observed in central and southern Peru. It is only around A.D. 700 that camelids appear in the archaeological record of Ecuador but they have not been reported from Colombia. Prior to Spanish conquest of the New World guinea pigs had spread to Venezuela and the West Indies (Antigua and Dominican Republic; Wing, Hoffman, and Ray, 1968). Archaeological faunas south of Peru have been insufficiently studied to reveal the patterns of camelid and guinea pig use there.

At about the same time camelid remains appeared in the highland sites of Ecuador, camelid and guinea pig remains occurred in ceremonial contexts on the coast.

In the following summary of these diverse lines of evidence pertaining to domestic animals, it is good to remember that the abundance of the remains identified is proportional to the intensity of prehistoric use of the animal.

1. Hunting of wild camelids in the puna and occasionally in the highland valleys took place from 10,000 to 5500 B.C. At the same time intensive use of guinea pigs was made in the highland valleys of Bogotá, Colombia, and Ayacucho, Peru (Figs. 10–1 and 10–2).
2. Intense use and beginnings of control of camelid herds occurred in the puna from 5500 to 2500 B.C. Domestic guinea pigs existed in the highland valleys (Figs. 10–1 and 10–3).
3. Domestication of camelids is indicated by continued intensive use in the puna and increased use in the highland valleys. Both domestic camelids and domestic guinea pigs were introduced to the coast during the

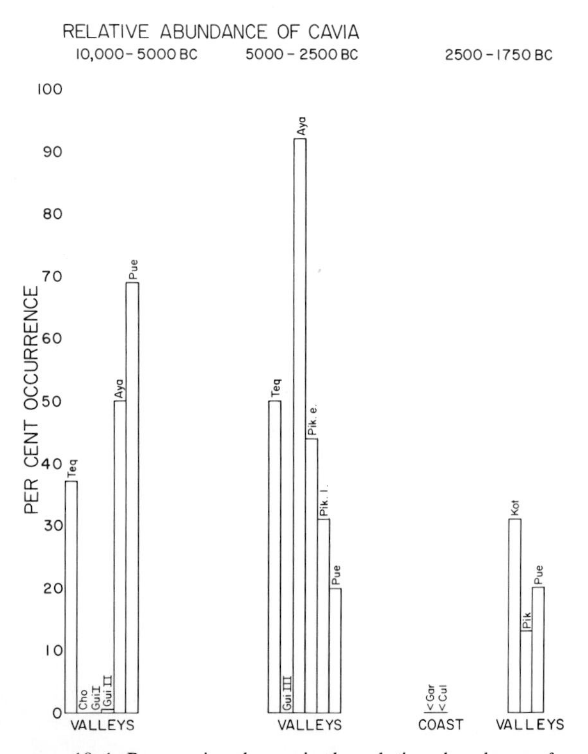

FIG. 10–1. Progressive change in the relative abundance of *Cavia* through time.

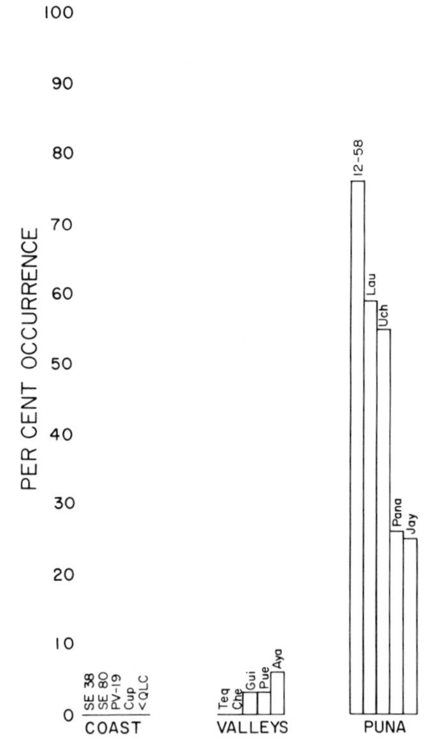

FIG. 10–2.

FIGS. 10–2 to 10–6. Progressive change in the relative abundance of camelids through time. The percentage is based on the numbers of specimens identified as camelid in the major habitat groups of coastal, highland valleys, and puna. Within these groups they are arranged latitudinally from north (*left*) and south (*right*).

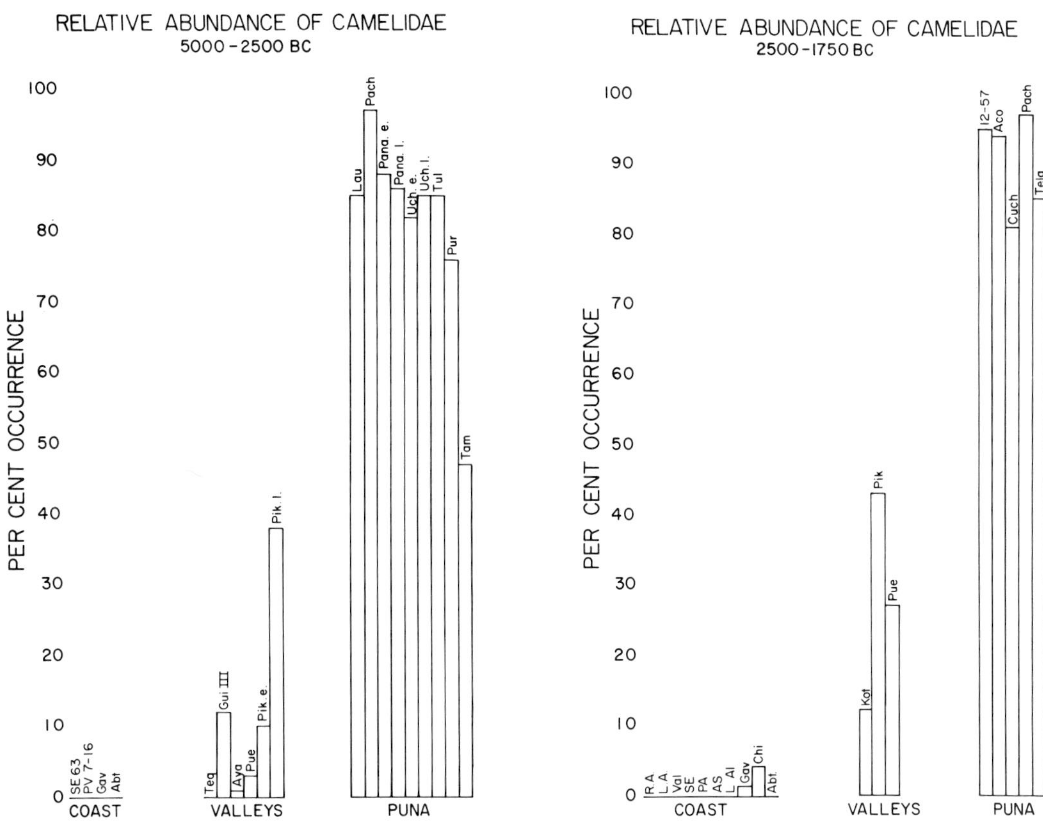

FIG. 10–3.

FIG. 10–4.

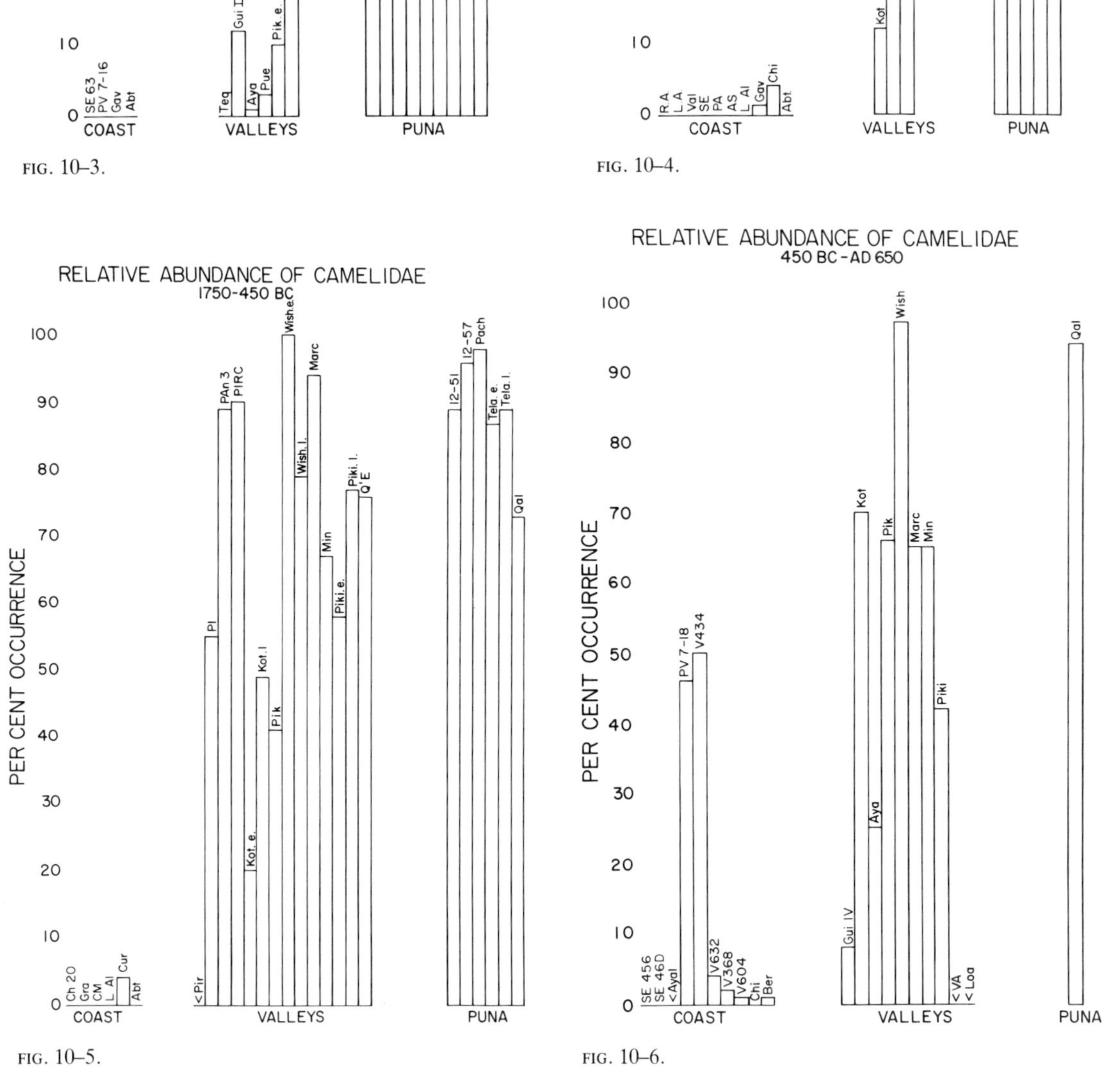

FIG. 10–5.

FIG. 10–6.

period from 2500 to 1750 B.C. (Figs. 10–1 and 10–4).
4. Intense use of camelids and more widespread use of dogs took place in highland valleys from 1750 B.C. on through prehistoric times (Fig. 10–5).
5. Evidence exists in the coastal textiles for the development of improved wool-producing animals by 500 B.C. (Fig. 10–6).

ROLE OF DOMESTIC ANIMALS IN PREHISTORIC CULTURES

Domestic animals fill many roles in the lives of Andean people today as they did in the historic past (Gade, 1967, 1969, 1975; Flores Ochoa, 1968). Possession of animals, particularly the large ones, is a measure of wealth. This wealth can accumulate and convey the status of the owner or be used as a medium of exchange or influence. The meat of these animals is eaten. The bone, hide, and dung of the larger animals are also used. These animals are used for sacrifice and divination. Both llama and alpaca have wool that is woven, the llama wool being used for coarser utilitarian fabrics and ropes and the alpaca wool for fine cloth. In addition, the llama is important as a beast of burden.

Undoubtedly, many of these uses have great antiquity, but the fragmentary skeletal remains in archaeological sites do not reveal the full impact of an animal on a prehistoric people. The clearest indication of special treatment, and thus presumed regard, is an animal burial. A number of burials have been found, which appear in the form of a simple burial or of an offering. Dogs are perhaps the most frequently found animal in burials. A number of dog burials were excavated from the Real Alto site occupied as early as 2500–1750 B.C. on the Santa Elena Peninsula. A dog burial was also found at Rosamachay in the Ayacucho Valley. This is the same site and time, 400 B.C.–A.D. 300, from which the fabric-wrapped and mummified guinea pig cited earlier was recovered. Camelid feet have been incorporated in burials, probably of highland people, in coastal sites of southern Peru and northern Chile. A young camelid was excavated from under the temple entrance at Pampa Grande (Melody Shimada, pers. comm.). It is possible that this animal was sacrificed at this temple. Llama, dog, guinea pig, and muscovy duck were incorporated in urn burials at Ayalan on the Santa Elena Peninsula. Again, these may have been sacrificial animals (Hesse, 1980). These are quite varied examples occurring widely in the study area and extending quite far back in time.

Further insight is provided by artistic representations of animals, particularly when the artwork is as representational as the Moche art. The Moche people flourished along the northern coast of Peru about 1000 years before the ascendancy of the Incan Empire. Modeled and drawn llamas are shown with mottled coats, notched ears, and bearing patterned, woven bags (Donnan, 1976). Guinea pigs are also portrayed (Gade, 1967). These explicit and detailed figures show how the animals functioned in a prehistoric society. This sort of information cannot be revealed by remains, although we do know from the skeletal remains that the flesh of these animals was eaten and the bone was used to make tools.

WHY THE ANDES BECAME A CENTER OF DOMESTICATION

Why was the Andes the only real center of animal domestication in the New World? In other words, what is it about the particular conformation of land forms, the native fauna, or the early inhabitants of the Andes that made this area a center of animal domestication?

One can assume that hunters and fishermen know a great deal about the habits of at least their predominant prey animals. People who depend on wild animals for meat know not only where these animals can be found and at what times of year, but also their daily activity patterns, their breeding schedule, and the number of young they produce each year. Generally this knowledge was enough for people to control certain animals, if they chose to do so. With control comes a responsibility to maintain the investment.

Gade (1967) has proposed that guinea pigs may have virtually "invited" human control or at least association. Guinea pigs may have been attracted to the warmth and protection of human shelter, food refuse being an added enticement. It may have required little effort for people to foster this attraction, tame the animals, and ultimately control them as domestic house animals.

The large herbivores clearly would present different problems and potentials to the puna hunters. Deer (members of the family Cervidae) are one of the prime game animals all over the prehistoric world. They have never been domesticated and have rarely been tamed. Herd animals belonging to the family Camelidae, the camel and dromedary, have been domesticated in Asia, where they are used as beasts of burden and for their hair. There may be some intrinsic characteristic of camelid behavior that allows them to tolerate association with humans. In the Andes the motivation for taming and controlling the camelids may never be documented. Whether it was primarily to have a ready supply of meat, wool, or dung, or whether the larger animals were selected to share the work of burden bearing as a secondary consideration, may never be known. It seems likely that the great altitudinal relief seen in the Andes and the isolation of different commodities at different elevations may have been a stimulus to the exchange of goods, and llamas may have taken part early in this exchange. If so, a primary motivation for domestication of llamas may have been economics.

Whatever the actual sequence of events in these domestications, it is clear that many factors of animal, culture, and environment had to come together in the right combination to result in this center of animal domestication. The domestication of plants accompanied that of animals, and together they played a profoundly important part in the development of the complex cultures of the Andes (Pickersgill and Heizer, 1977).

ACKNOWLEDGMENTS

This research could not have been done without the generous support of the National Science Foundation (grants GS 1954, GS 3021, and Soc 74–20634), the many archaeologists who have entrusted their excavated faunal samples to this study, and fellow zooarchaeologists who have unstintingly shared their ideas and data. The students who have painstakingly helped in the identification of the faunal remains include Lynn Cunningham Balck, Kathleen Johnson, Kathleen Byrd, Gary Shapiro, and Elizabeth Reitz. My sincerest thanks to all.

REFERENCES

Browman, D. L. 1974. Pastoral nomadism in the Andes. *Curr. Anthrop.* 15: 188–96.

Brush, S. B. 1976. Man's use of an Andean ecosystem. *Human Ecol.* 4: 147–66.

Byrd, K. M. 1976. Changing animal utilization patterns and their implications: Southwest Ecuador (6,500 BC–AD 1,400). Ph. D. diss., University of Florida.

Carr, G. S. 1981. Historic and prehistoric avian records from Easter Island. *Pacific Sci.* 34: 19–20.

Clutton-Brock, J. 1977. Man-made dogs. *Science* 197: 1340–42.

Donnan, C. B. 1976. *Moche art and iconography.* UCLA Latin American Studies, Vol. 33. Los Angeles: University of California Press.

Fernandez Baca, S. 1971. *La Alpaca: Reproducción y crianza.* Centro de Investigación, Instituto Veterinario de Investigaciones Tropicales y de Altura, Lima, Peru, Bol. No. 7: 7–43.

Flannery, K. V., C. S. Jacobson, and E. Baker. n.d. The animal bones from caves above 3000 meters. Unpublished manuscript.

Flores Ochoa, J. A. 1968. Los Pastores de Paratia: Una introducción a su estudio. Instituto Indigenista Interamericano. *Anthropología Social* (Mexico) 10: 1–159.

Gade, D. W. 1967. The guinea pig in Andean folk culture. *Geog. Rev.* 57: 213–24.

———. 1969. The llama, alpaca, and vicuña: Fact vs. fiction. *J. Geog.* 68: 339–43.

———. 1975. *Plants, man, and the land in the Vilcanota Valley of Peru.* Biogeographica, vol. 6. The Hague: Junk.

Hershkovitz, P. 1972. The recent mammals of the neotropical region: A zoogeographic and ecological review. In *Evolution, mammals, and southern continents.* A. Keast, F. C. Erk, and B. Glass, eds., pp. 311–431. Albany: State University of New York Press.

Hesse, B. 1980. Archaeological evidence for Muscovy duck in Ecuador. *Curr. Anthrop.* 21: 139–140.

———. n.d. Harvest profiles for small and large camelids from archaic sites in Northern Chile. Unpublished manuscript.

Lanning, E. P. 1967. *Peru before the Incas.* Englewood Cliffs, N.J.: Prentice-Hall.

Llagostera Marginez, A. 1979. 9,700 years of maritime subsistence on the Pacific: An analysis by means of bioindicators in the North of Chile. *Amer. Antiquity* 44: 309–324.

MacArthur, R. H., and J. W. MacArthur. 1961. On bird species diversity. *Ecology* 42: 594–598.

MacNeish, R. S., T. C. Patterson, and D. L. Browman. 1975. The Central Peruvian prehistoric interaction sphere. *Papers Robert S. Peabody Found. Archaeol.* 7: 3–97.

Miller, G. R. 1979. An introduction to the ethnoarchaeology of the Andean Camelids. Ph.D. diss. University of California, Berkeley.

Oberem, U. 1975. Informe de trabajo sobre las excavaciones de 1964/1965 in Cochasqui, Ecuador. *Bonner Amer. Stud.* 3: 71–79.

Peterson, E. 1977. Cotocollao, a formative period village in the Northern highlands of Ecuador. Paper presented at the 42nd annual meeting Society for American Archaeology, New Orleans.

Pickersgill, B., and C. B. Heiser, Jr. 1977. Origins and distribution of plants domesticated in the New World tropics. In *Origins of agriculture*, Charles A. Reed, ed., pp. 803–835. The Hague: Mouton.

Pollard, G. C., and I. M. Drew. 1975. Llama herding and settlement in Prehispanic northern Chile: Application of an analysis for determining for domestication. *Am. Antiquity* 40: 296–305.

Pozorski, S. G. 1979. Prehistoric diet and subsistence of the Moche Valley, Peru. *World Archaeol.* 11: 163–84.

Reitz, E. J. 1979. Faunal materials from Viru 434: An Early Intermediate Period site from Coastal Peru. *Florida J. Anthrop.* 4: 76–92.

Shimada, M. 1979. Paleoethnozoological/Botanical analysis of Moche V economy at Pampa Grande, Peru. Unpublished manuscript.

Van Gelder-Ottway, S. M. 1975–1977. Mammal remains from Tequendama. Unpublished manuscript.

Wheeler Pires-Ferreira, J. 1975. La fauna de Cuchimachay, Acomachay A, Acomachay B, Tellarmachay y Uto I. *Rev. Museo Nac. Lima* 41: 120–127.

Wheeler Pires-Ferreira, J., E. Pires-Ferreira, and P. Kaulicke. 1976. Preceramic animal utilization in the Central Peruvian Andes. *Science* 194: 483–490.

Wheeler, J., C. R. Cardoza, and D. Pozzi-Escot. 1977. Estudio provisional de la fauna de las capas II y III de Telarmachay. Apendice I. *Rev. Museo Nac. Lima* 43: 97–102.

Wing, E. S. 1975. Informe preliminar acerca de los restos de fauna de la Cueva de Pachamachay, en Junín, Perú. Apendice III. *Rev. Museo Nac. Lima* 41: 79–80.

———. 1977. Animal domestication in the Andes. In *Origins of agriculture*, Charles A. Reed, ed., pp. 837–859. Hague: Mouton.

———. 1980. The desert fox. Paper presented at the 45th annual meeting of the Society for American Archaeology, Philadelphia.

Wing, E. S., C. A. Hoffman, Jr., and C. E. Ray. 1968. Vertebrate remains from Indian sites on Antigua, West Indies. *Carib. J. Sci.* 8: 123–139.

Winterhalder, B. P., and R. B. Thomas. 1978. Geoecology of southern highland Peru: A human adaptation perspective. *Inst. Arctic Alpine Res. Occas. Paper* No. 27: 1–91.

IV

DIVERSIFICATION AND ADAPTIVE RADIATION

The biota of the high tropical mountains contain excellent examples of diversification among taxa and of adaptive radiation. What ecological roles do these diversified taxa play within some of the high tropical biomes? How did these taxa differentiate? What specific ecological niches are filled during the process of adaptive radiation? What kinds of morphological diversification accompany the ecological diversification?

In part IV we explore such questions with concrete examples taken from two rather distinct high tropical montane environments. The first, the tepuis or table mountains of the Guayana Highlands, is composed of habitats found on isolated mountaintops that owe their dissection to the erosion of a very old geologic formation (see, e.g., Haffer, 1974; B.B. Simpson, 1979). These habitats are not necessarily, of course, as old as the geologic substrata that support them. Nevertheless the tepui plant formations of the summits, the most interesting ones in this connection, are undoubtedly older than the páramo or puna biomes of the high tropical Andes. The latter, which form the second environment used here as background for an examination of diversification and adaptive radiation, are, in G. G. Simpson's (1965, p. 216) words, "one of the most recent habitats in South America." The tepui environments, as a theater for diversification and adaptive radiation, are thus old, whereas the páramo and the puna are not only recent but also new, since they occur in a cold, glacial or periglacial climatic regime that appeared in the tropics only in the last two million years or so.

We quote here from G. G. Simpson (1953, p. 223) two citations that are especially pertinent to our concerns. The first relates diversification to the appearance of a new environment:

> . . . diversification follows, and . . . begins almost immediately, when a group spreads to a new and, for it, ecologically open territory. The extent of adaptive diversity eventually reached tends, although rather roughly, in the first case to be proportional to the distinctiveness of the new adaptive type and in the second to the extent and diversity of the new territory.

In the case of the recent and new environments (and "desert" in the sense of both empty when first open to colonization, and barren because of climatic conditions, as in high montane desert), we should be able to analyze very clearly the evolution of adaptive diversity along the lines suggested by G. G. Simpson. We offer *Espeletia* and *Polylepis*, among plants, and *Telmatobius* and *Orestias*, among animals, as examples.

The second quote makes a distinction between adaptive radiation and progressive modification:

> So far as adaptive radiation can be distinguished from progressive occupation of numerous zones, a phenomenon with which it intergrades, the distinction is that adaptive radiation strictly speaking refers to more or less simultaneous divergence of numerous lines all from much the same ancestral adaptive type into different, also diverging adaptive zones.

After studying the chapters on diversification and radiation in the tepui flora of the summits, and in the páramo and puna rodent fauna, readers will be able to see clearly which elements or components have evolved through sequential occupation of adaptive zones and which others have diversified through adaptive radiation into different phyletic and ecological lines. They also will be able to compare the effects of a relatively old (tepui summit habitats) and a relatively recent and new (páramo and puna habitats) environment on the divergence and diversification of the various taxa that first succeeded in occupying these habitats.

As in earlier parts of the book, the material is organized hierarchically. Two more or less parallel hierarchies of levels can be found in part IV. One is essentially taxonomically based. Diversification can thus be, and is, studied at the level of the genus, or of a higher taxonomic grouping (family, for instance), or else at the level of the aggregate of taxa (of whatever rank) that form all or part of a flora or fauna. The second hierarchy is fundamentally ecological, since the diversification and radiation are viewed in the context of a series of units, from the habitat on up to the biome.

In some ways, *Espeletia* (or rather the Espeletiinae in the new classification of Cuatrecasas) represents perhaps the best example we have of adaptive radiation in the new and recent tropical high montane environments. Indeed, the processes of fragmentation of species into populations and subsequent speciation, of diversification of new taxa into several divergent life forms, of refinement of reproductive patterns in subtle adaptive terms, and of relative duration of life cycles are all part of one of the most intriguing radiations in the high tropical biota. Studies of various aspects of adaptive strategies in part II and of diversification in part IV thus tie together these two sections of the book.

REFERENCES

Haffer, J. 1974. *Avian speciation in tropical America*. Publ. Nuttall Ornithol. Club, No. 14. Cambridge, Mass.

Simpson, B. B. 1979. Quaternary biogeograpy of the high montane regions of South America. In *The South American herpetofauna: Its origin, evolution, and dispersal,* W. E. Duelmann, ed., pp. 157-188, Museum of Natural History, Univ. of Kansas, Monograph 7.

Simpson, G. G. 1953. *The major features of evolution*. New York: Columbia Univ. Press.

———. 1965. *The geography of evolution. Collected essays*. New York: Capricorn Books.

11

Speciation and Radiation of the Espeletiinae in the Andes

JOSÉ CUATRECASAS

CHARACTERIZATION OF THE ESPELETIINAE

The Espeletiinae (Asteraceae, tribe Heliantheae) comprise a group of about 130 species of highly evolved flowering plants. They are geographically and ecologically highly specialized to the peculiar life conditions of the northern section of the tropical Andes.

Detailed characterizations are available in the original description of the subtribe (Cuatrecasas, 1976), and in the author's monograph, which is in final preparation. At this time it may suffice to point out the most outstanding differential features of the subtribe Espeletiinae within the Heliantheae: (1) spiral phyllotaxis; (2) obpyramidal shape of the achenes, which are, in addition, smooth, without striations, glabrous, and epappose (with a minimal exception); (3) xeromorphic structure and organization of the biotypes, including heads and leaves; (4) fertile female ray flowers and functionally male disc flowers; (5) basic chromosome number $N = 19$. These characters separate the Espeletiinae from other groups of the tribe such as Melampodiinae and Polymniinae, as described in other works (Bentham, 1873; Hoffmann, 1890–1894; Stuessy, 1973, 1977; Robinson, 1981).

CHARACTERIZATION OF INFRA SUBTRIBAL TAXA

Variations in number, shape, size, color, and kind or density of trichomes for every given part provide floral or vegetative characteristics of taxonomic value at the specific level and to some extent at the generic level. Variations in leaf outline and type of indumentum are basic considerations for specific distinctions. Some aspects of the leaf structure and architecture are important in generic differentiation, and phylogenetic trends in the Espeletiinae may be traced on that basis. Other important taxonomic features are found in the inflorescences and growth forms. Some of these structural variations undoubtedly represent evolutionary trends that are influenced by the extreme environmental factors prevailing in the northern tropical Andes. Three morphological features seem most significant in the differentiation and evolution of the Espeletiinae: leaves, inflorescences, and biotypes.

Leaves

The leaves are usually large, up to 50 × 12 (70 × 25) cm, the blades variously oblong, elliptic to linear, only slightly attenuate at the base, therefore sessile, or attenuate and proximally contracted into a more or less distinct petiole. The margin is revolute, entire or minutely remote toothed; the venation is pinnate, usually prominent with a robust middle nerve. The sheaths are usually enlarged, amplectant or semiamplectant to the stem, mostly imbricated on it; they are either open and flat, or curving only to embrace the stem with free margins, or encircling the stem and tubular. The blades may be moderately pubescent or more usually densely vested; at least on the lower side, the vesture is frequently very thick, the blades lanate or sericeous. The sheaths are usually sericeous on the abaxial side.

The great range of variation indicated in the shape and size of the blades, venation, and quality and density of indumentum is important when considering speciation. The structure of the leaf base has importance at the generic level; e.g., tubular sheaths characterize *Libanothamnus* and *Carramboa*, open sheaths the other genera (Fig. 11–1).

Important trends of evolution in the Espeletiinae are on one hand the trend that has favored the tubular type of sheaths and, on the other, a

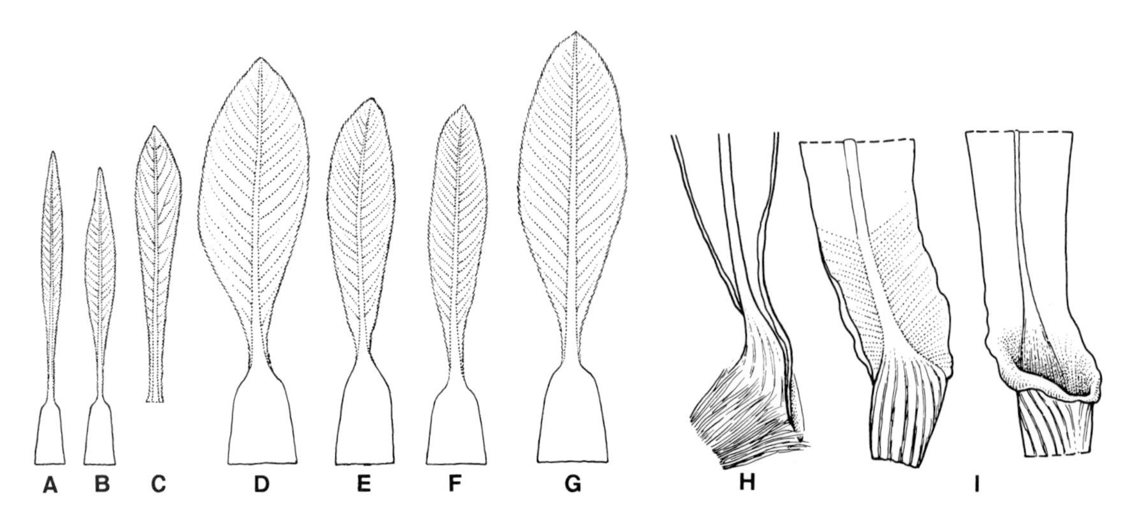

FIG. 11–1. A–G. Typical shape of leaves in *Espeletia*, always with large, flat sheaths. *H,I*, Tubular leaf sheaths in *Libanothamnus*.

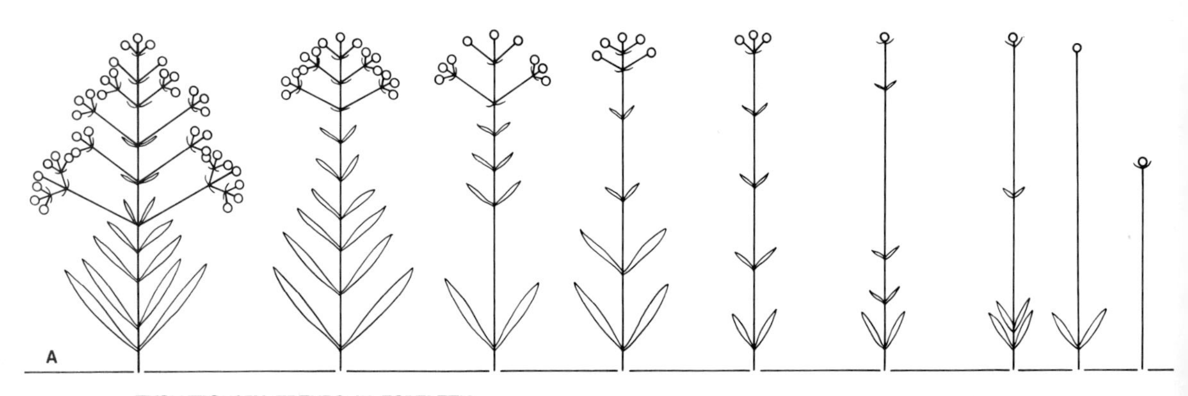

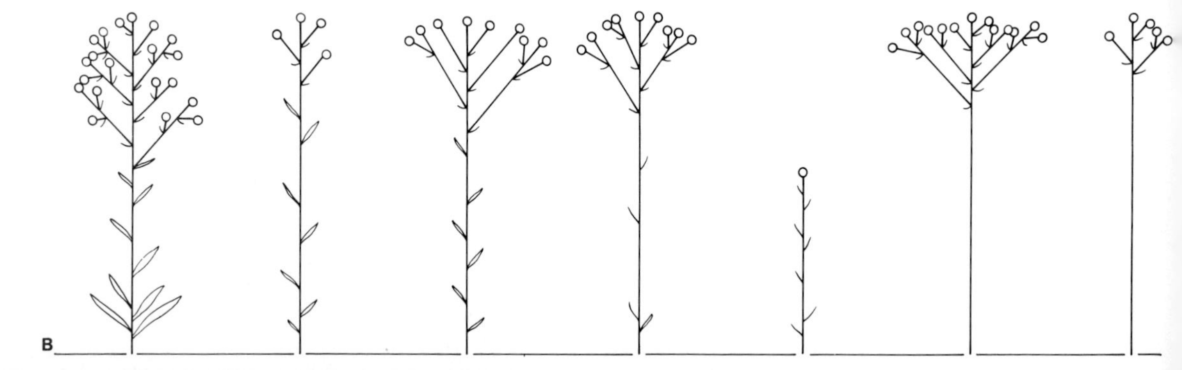

FIG. 11–2. Evolutionary trends in *Espeletia* (A) and *Espeletiopsis* (B) inflorescences.

trend to enlarge the flat leaf sheath favoring the imbricate kind of cover that protects the body of the caulirosulas. Another trend is the progressive reduction of the pseudopetiole and consequent broadening of the lamina base, becoming sessile.

Inflorescences

The inflorescence structural system in the Espeletiinae is monotelic, usually compound or, as called by W. Troll, a synflorescence. The term *monotelic* was established by W. Troll for the kind of inflorescences characterized by the limited growth of the axes, which are all terminated or closed by an apical flower. These are the inflorescences usually termed definite, determinate, closed, and cymose (W. Troll, 1957, 1964–1969; Cuatrecasas, 1980).

In a synflorescence two zones can be considered: (1) the lower, vegetative, leafy part, and (2) the upper, fertile, floriferous part. Within the monotelic synflorescences in the Espeletiinae I distinguish two types: (1) dichasial synflorescences, with opposite branches, bracts, and leaves. They are well developed in *Espeletia* (Fig. 11–2A). (2) monochasial synflorescences, with alternate branches, bracts, and leaves. They are found in other genera with many variations, e.g., *Espeletiopsis* and *Libanothamnus*. This type is considered derived from the former by the alternate development of only one of the two lateral opposite elements (Fig 11–2B).

These two kinds of synflorescences are fundamentally distinct in the Espeletiinae and have ecological and taxonomic implications.

The perfect theoretical dichasial model can be found in many species of *Espeletia*, but the individuals in fact often fail to develop some parts of the synflorescence. Variations that do not affect

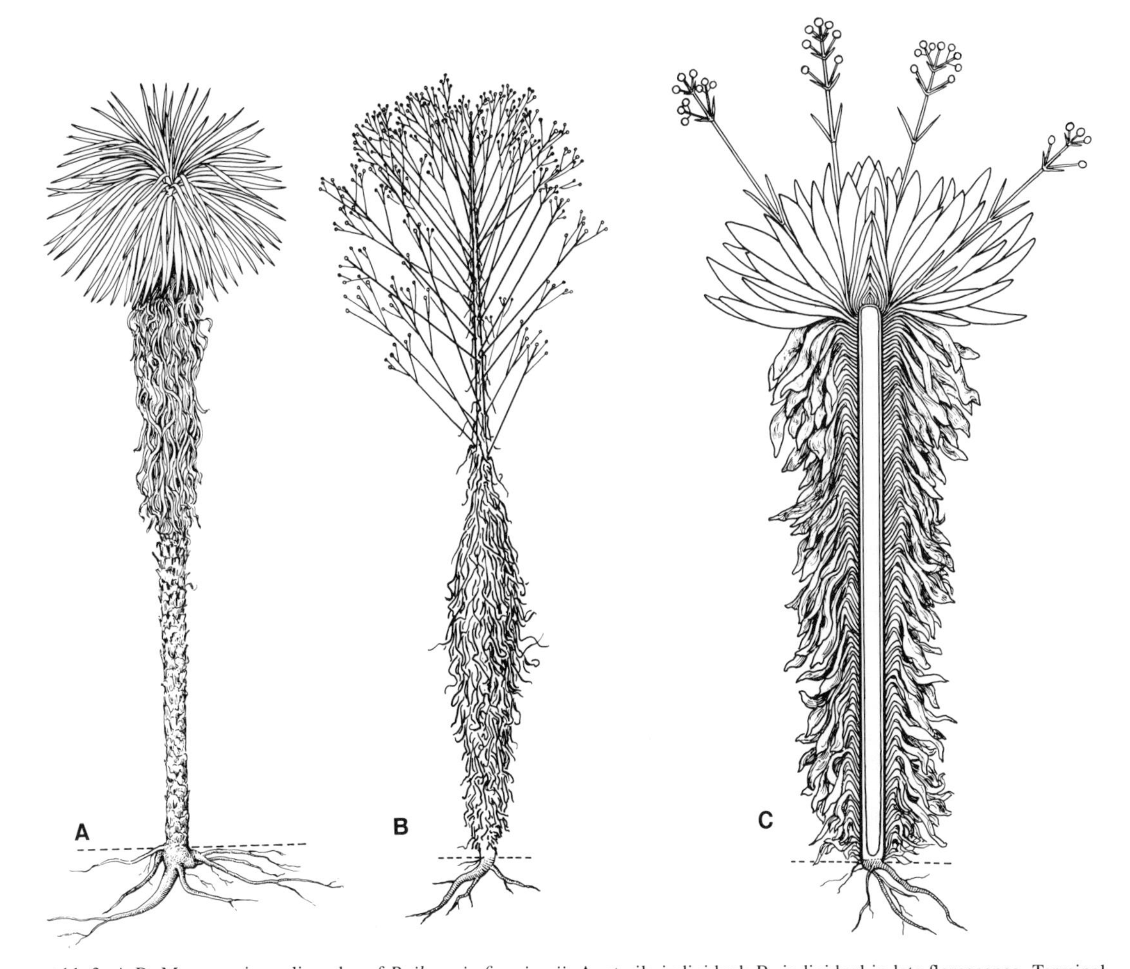

FIG. 11–3. A,B. Monocarpic caulirosulas of *Ruilopezia figueirasii:* A, sterile individual; B, individual in late florescence. Terminal synflorescence. C, Polycarpic caulirosula of *Espeletia grandiflora*, diagrammatic in longtitudinal section. Axillary (lateral) synflorescences.

its basic structure and symmetry are those related to the number of pairs of leaves or pairs of branches or peduncles. Other types of variations are asymmetrical reductions, which take place by suppression or displacement of one or the two leaves or one of the branches or peduncles in a given pair, which then become alternate. Significant evolutionary trends in dichasial synflorescences, exemplified by *Espeletia*, are (1) reduction of branching and hence of the number of capitula to a small number, down to the single head; (2) reduction of the length of the internodes, peduncles, and pedicels leading to the concentration of a number of capitula into a more or less compact glomerule; (3) reduction of the number of sterile leaves or bracts of the vegetative section of the synflorescences (the final step being a naked sterile part of the axis); and (4) asymmetrical reduction of branches, peduncles, bracts and sterile leaves, with partial effect of alternate branching. This kind of variation is present in some species (e.g., *E. schultzii*), but never affects the whole synflorescence.

Lateral and terminal synflorescences

The monochasial synflorescences also vary with respect to the extent of branching and hence to the number of capitula, which in rare cases are reduced to a single one. The amount of foliage of the vegetative portion also varies and may show an evolutionary tendency toward reduction to a naked axis (Fig. 11–2).

In many cases the synflorescences are flowering branches axillary to the median leaves of the rosettes. The central or apical bud (meristem) of the rosette is continuously active for many years, the single stem growing vertically without interruption. The axillary (lateral) synflorescences are produced periodically; in the Espeletiinae they can be of the dichasial type (*Espeletia, Carramboa*) (Fig. 11–3C) or monochasial (*Espeletiopsis, Coespeletia*).

Terminal synflorescences are produced directly by the apical meristem of a stem or branch; this implies the end of the life of the stem or branch (Fig. 11–3B). In the Espeletiinae, the terminal synflorescences are always of the monochasial type, usually corymbiform and abundantly floriferous.

The ultimate terminal inflorescence in fact is the capitulum, which in Compositae plays the role the flower has in other families. Variations in size and shape of capitula partially reflect trends of evolution and often are coordinated with other features.

Life Forms

The Espeletiinae have diversified growth forms adapted to various extreme environmental conditions of the high tropical Andes. The xeromorphic characteristics of most of the páramo plants are generally recognized and have been described elsewhere (Cuatrecasas, 1934–1979; C. Troll, 1958a; Vareschi, 1970; Cleef, 1978; Sturm, 1978). The Espeletiinae, in addition to many of those general vegetative anatomic features, have developed a few exceptional and distinctive growth forms (with coordinated structural features that protect their biological cycle against cold and physiological dryness), which have been extraordinarily successful. Some, such as the caulirosulate *Espeletia* and *Coespeletia*, dominate extensive stressful páramo regions. This type of adaptation is exceptional among dicotyledons, although convergent forms are well known from the high mountains of tropical East Africa with *Dendrosenecio* and *Lobelia* (Hedberg, 1964; Mabberley, 1973; Hedberg and Hedberg, 1979; see Chap. 4) and from Hawaii with *Wilkesia*, *Argyroxiphium*, and *Dubautia* (Keck, 1936; Carlquist, 1974).

In the Espeletiinae we find growth forms from true trees (woody plants with a main massively woody trunk undivided in the lower half to one fourth, and profusely branched upward), reaching 20 m high, to small, sessile rosette plants no more than 20 cm broad. They can be classified as follows:

1. *Trees*. Woody plant with an erect distinct trunk branched profusely at a considerable distance from the ground (Fig. 11–4).

a. Monopodial trees, with lateral buds and axillary dichasial synflorescences (model: *Carramboa trujillensis*). The main trunk and branches exhibit monopodial, indefinite, apical growth; the leaves and branches are spirally alternate. The leaves are usually large and broad, and green with tubular sheaths, mostly crowded at the end of the branches (Fig. 11–4B).

b. Basically monopodial trees, with lateral buds and branches and terminal monochasial synflorescences (model: *Libanothamnus neriifolius*). Leaves are spirally alternate, with tubular sheaths, the laminas oblong, coriaceous, and xeromorphic; they are usually clustered at the end of the branchlets. The main trunk is monopodial. The main branches are spirally alternate by often pseudoverticils are produced by two to five branches arising very closely to each other; these pseudoverticils may appear only once or several times over the same trunk (fig. 5–1 in Cuatreca-

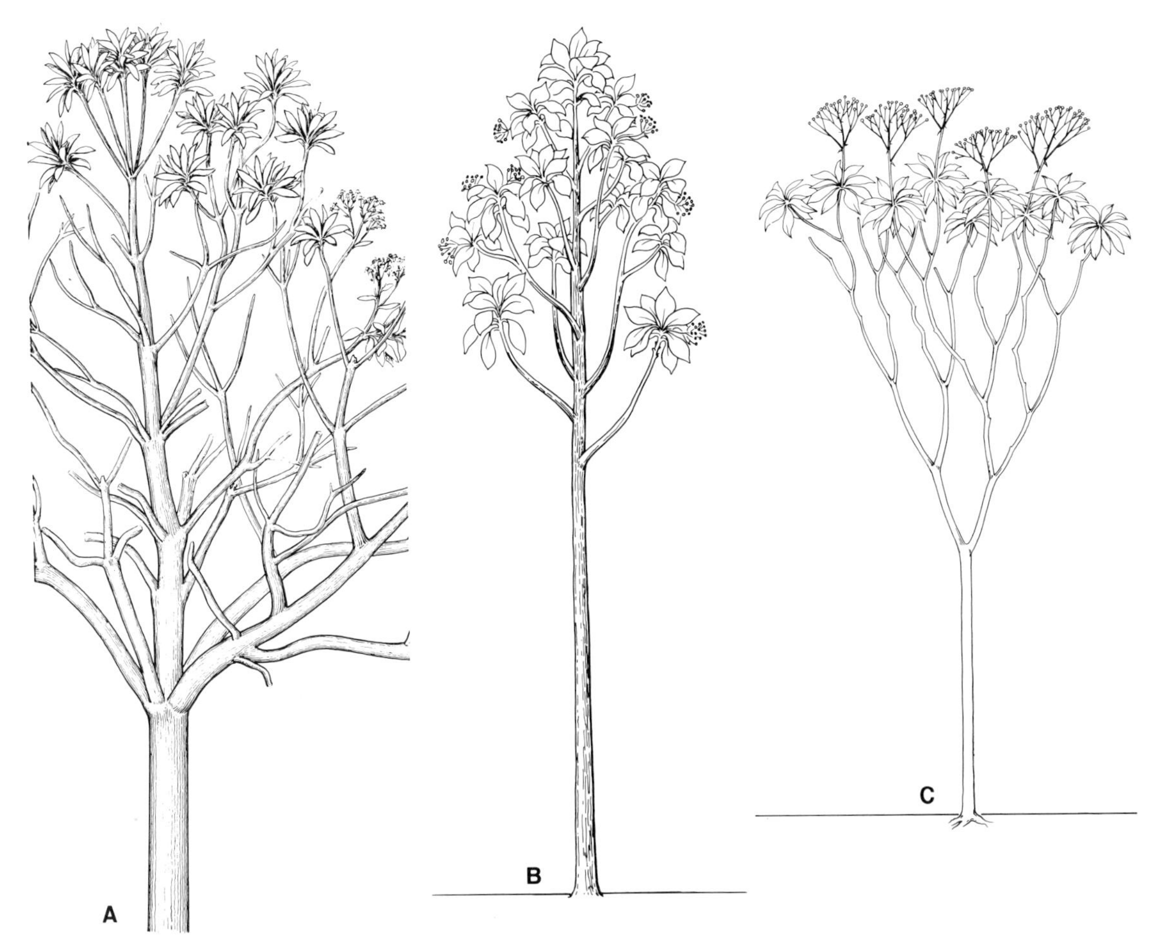

FIG. 11–4. A. Branching in a tree of *Libanothamnus neriifolius* var. *boconensis* (partial). B, Tree of *Tamania chardonii*. Pseudodichotomic branching, sympodial. C, Tree of *Carramboa trujillensis*. Monopodial alternate branching.

sas, 1979a); main branches and branchlets are also monopodial. In regularly developing trees, the distal branches with terminal synflorescences have further sympodial growth by way of subapical axillary shoots. In short, the basic central architecture of such trees is monopodial, but the peripheral architecture of the crown is sympodial (Fig. 11–4A).

When trying to describe a structural type of a particular genus or species, it must be realized that the trees are not always allowed to grow and develop freely in nature. There are irregular environmental factors in a particular locality, such as the lack of necessary shade for young plants, excessive vegetation and competition, and mutilations, often produced by animals incidentally or by their eating the tender shoots. Examination of many trees of *Libanothamnus* indicates considerable variation in the number of single branches and pseudoverticillate branches on the same species and in a single tree. The amount of sympodial branching in a single tree in proportion to the monopodial central branching can vary considerably due to changes of environmental factors that affect its apparent architecture. This has been especially noticed in populations of *L. occultus*, in which the peripheral sympodial branching may predominate over a greatly reduced central monopodial branching.

c. Trees have monopodial trunks, with branching pseudodichotomous and sympodial, and with synflorescences terminal and monochasial (Fig. 11–4C) (model: *Tamania chardonii*; fig. 5–2 in Cuatrecasas, 1979a). Leaves are spirally alternate, with open sheaths, and the laminae are coriaceous and densely vested beneath; leaves are more or less crowded in terminal clusters. The mature trunk is massively woody. The architecture of such trees agrees with the "Leuwenberg model" of Hallé and Oldeman (1970). *Tamania* develops an initial, robust, erect stem with sparse, spirally alternate leaves, forming a large cluster or rosette terminally. After an indefinite time, probably several years, the terminal meristem goes into the reproductive phase and develops a large robust synflorescence whose axis is the continuation of the main stem of the tree. At the same time, one or more dormant axillary buds, located immediately below the inflorescence, begin to develop into shoots. These shoots become the branches that continue the growth of the tree, usually forming a false dichotomy. Occasionally more than two buds may develop at the top of the main stem, producing a pseudopleiochasial branching.

2. *Caulirosulas*. Here we see multifoliate rosettes whose leaves have enlarged imbricate sheaths arranged in multispiral circles at the end of a long or short pachycaul stem. The terminal, apical, vegetative bud continuously produces new leaves, while the older proximal leaves are successively drying, withering, and remaining attached to the stem, embracing it with the closely imbricate sheaths. Generalities about caulirosula growth form have been given in earlier publications (Cuatrecasas, 1934, 1949b, 1968, 1976) and an abridged account of its architecture and of the different types, with bibliography, was presented by the author at the 1978 Aarhus Symposium (Cuatrecasas, 1979a). The anatomy of the apical meristem and its function were studied in detail by Weber (1956) and Rock (1972, 1981). The following is a brief review of the various kinds of caulirosulas found in the Espeletiinae (Fig. 11–3).

a. *Polycarpic caulirosula*. The usually large rosettes terminate an erect, simple, straight stem or trunk 10 cm to 15 m high. Inflorescences arise from the axils of the rosette leaves. The leaf blades are coriaceous, flexible or rigid, rather oblong, xeromorphic, and covered at least on the undersurface by a thick white or yellowish indumentum. The large sheaths, as wide as or wider than the lamina, are closely imbricated and appressed, tightly surrounding the stem. The older, hanging, marcescent leaves can persist indefinitely, forming a bulky covering on the stem. The trunk appears to be several times thicker than the real stem and had a columnar aspect. The stem, usually with a diameter of 6–12 (sometimes 15) cm, has a very thick central soft pith surrounded by a hard, rigid, woody vascular cylinder and a tight bark.

In this kind of caulirosula the apical meristem remains indefinitely active, promoting the constant upward growth of the stem and renewal of the rosette leaves. Florescence occurs by way of lateral synflorescence branches from the axils of the median adult leaves of the rosette. A rosette may produce many synflorescences at one time, subverticillately, spirally arranged like the leaves. A plant may continuously produce inflorescences, although usually the florescence is intermittent. There can be many cycles in a year and the timing is not yet well known. (Studies on floration cycles have been carried out in Venezuela by Monasterio [1979]; see also A. P. Smith [1975]; Chap. 3) It has been observed that in dry periods inflorescences are rare and in rainy periods they are profusely (often continuously) produced; after fructification the inflorescences die out but usually remain attached to the trunk, hanging among the marcescent leaves (model: *Espeletia grandiflora* H. and B.) (Fig. 11–3C). Polycarpic caulirosulas are essentially unbranched (monocaulous), but due to accidental injury of the apical bud, few branches may replace the otherwise single axis. Rarely, true spontaneous branching is produced. With respect to the branching architecture, its monocaul form has been superficially placed into the "sensu amplo" Corner model group of Hallé and Oldeman (1970) classification. However, *Espeletia* caulirosula has differentiated reproductive branches; the main trunk has spiral phyllotaxis whereas the axillary, lateral inflorescences have dichasial branching with decussate phyllotaxis; this places *Espeletia* into another architectural type apart from the Corner model.

The following four variations of polycarpic caulirosulas can be considered:

(i) *Polycarpic caulirosula proper*. These are tall caulirosulas with an elevated monopodial stem, typified by *Espeletia*. It is represented by most species of *Espeletia*, *Coespeletia*, and *Espeletiopsis*; it is the component of the "caulirosuleta" formations that dominate large Andean páramo areas, characterized by their impressive peculiar physiognomy.

(ii) *Polycarpic caulirosula, sessile*. These are the caulirosulas with short stems (not growing higher than 20 cm), enclosed by their own rosette and the compact pack of marcescent leaves. Although they do not grow in height, they can become extremely old, increasing considerably the thickness of the wood cylinder. Due to the relative shortness of the stem and the bulk of the rosettes and the marcescent leaf layers, the rosette seems to be sessile. They are often described as acaulirosula, or acaulescent, which are not appropriate terms. Among the *sessile* caulirosulas can be mentioned *Espeletia argentea*, *E. boyacensis*, *Espeletiopsis muiska*, *Espeletiopsis pannosa* (fig. 5–11 in Cuatrecasas, 1979a). Caulirosulan *Espeletia* of slow growth can be called sessile in their earlier stages.

(iii) *Polycarpic caulirosula, rhizomatic*. The real stem is subterranean and the rosette sessile. The rhizome can be unbranched or sparingly branched, each branch ending with a rosette. These are small plants, like *Espeletia marthae* and *Espeletiopsis caldasii* (model) (Cuatrecasas, 1979a, fig. 5–14.)

(iv) *Polycarpic caulirosula, tuberose.* The stem is a subterranean tuber, usually ellipsoid or irregularly cylindrical and thick. The rosette is sessile. These are usually small plants, such as *Espeletia batata* Cuatr. (model), *Espeletia weddellii* Sch. Bip., and *E. nana* Cuatr. The common name of this biotype is *frailejón batato* (Cuatrecasas, 1979a, figs. 5–2, 5–12).

b. *Monocarpic caulirosula.* This type of caulirosula has a vegetative structure similar to the polycarpic caulirosula typified by the *Espeletia* model. The biological cycle is different. After a time, usually many years of vegetative growth, the apical meristem abruptly changes from a vegetative to a sexual phase, bringing forth a large terminal synflorescence, the whole plant dying after fructification; it is monocarpic, hapaxantic (Fig. 11–3 A, B). The model is *Ruilopezia figueirasii* (Cuatrecasas, 1976, fig. 2; 1979a, figs. 5–3, 5–14, p. 402). The following four variants can be considered:

(i) *Monocarpic caulirosula proper.* These are tall caulirosulas, monopodial, agreeing with the model described earlier. Limited branching is produced only by accident. Other species, besides the model, are *Ruilopezia ruizii*, *R. marcescens*, *R. lopez-palacii*, and *R. paltonioides*. This caulirosula biotype can be placed architecturally under the Holttum model of Hallé and Oldeman (1970).

(ii) *Monocarpic caulirosula, rhizomatic.* The rosette is sessile; the stem for the most part is subterranean, except for the distal portion that bears the terminal rosette. It can produce subaxillary basal shoots, which may develop into branches topped by large terminal rosettes. In such cases each branch and its rosette are also monocarpic. *Ruilopezia atropurpurea* is the model of this biotype.

(iii) *Monocarpic caulirosula, tuberose.* The rosette is sessile; the caudex is subterranean, thick, and tuberose, with a long vegetative life. At the time of florescence it produces a large and showy terminal synflorescence that exhausts all reserves of the tuberous stem, and the whole plant dies. Examples are *Ruilopezia jabonensis* (model), *R. margarita*, and *R. floccosa*.

(iv) *Monocarpic caulirosula, low and more or less branched.* The rosettes are short stemmed and sessile or subsessile. The plant can be shortly monocaul or branched; the branches are alternate, often short and infrequent or more ramified with a pseudodichotomic or ternate manner of growth; the branches are thickly covered by the marcescent leaves or sheaths, and procumbent, each with a large and heavy terminal rosette resting on the ground. The terminal, richly flowered synflorescence stands erect with a thick axis that is a continuation of the stem. At fructification, the whole rosette dies. The bushy character is not usually apparent, because each of the large rosettes appears to be on the ground, usually crowded in a gregarious way; large colonies in a population are made by many individuals supporting from one to several rosettes. The model of this growth form is *Ruilopezia jahnii* (Cuatrecasas, 1979a, fig. 5–5).

THE ORIGIN OF THE COMPOSITAE (HELIANTHEAE)

There is much speculation about the time and place of origin of the Compositae. It had been generally assumed that they originated in the Lower Oligocene or the Late Cretaceous (Turner, 1977). The earliest unquestionable fossil record, however, dates only from the Miocene, and recent palynological data show that from the Middle Miocene to the present, the amount of Compositae pollen in the sediments increases dramatically (Germeraad, Hopping, and Muller, 1968; Muller, 1970; van der Hammen, 1974, 1979; Stuessy, 1977; see Chap. 7). Unfortunately, palynology still does not allow much distinction of genera or groups of genera in the Asteraceae, the pollen morphology being rather uniform for the largest part of the family. Actually such uniformity may be another indication of the youth of the Asteraceae.

The following statement by Germeraad et al. (1968, p. 283) is considered valid still: "Earlier views (Kuyl et al., 1955) that, on the basis of the palynological record, Asteraceae emerge as one of the youngest developments within the angiosperms are thus confirmed. The place of origin is still obscure, however, and may very well have been situated outside the tropical belt, the time of origin presumably being mid-Tertiary."

It is not possible to determine or even discuss at this time which tribe might be the original group or the more primitive representative of the Asteraceae. Cronquist (1955) postulated the Heliantheae to be the most primitive group of the family. This theory, as well as the place of origin, has been discussed by several workers (Carlquist, 1966, 1976; Stuessy, 1977; Turner, 1977; Robinson, 1981).

Observing the variety of Compositae currently

distributed throughout tropical America reveals the amazing capability of the earlier Asteraceae to differentiate into many high category groups preserving the fundamental structure and organization of the family embodied in the capitulum. They have the flexibility to adapt to the most diverse kinds of habitats and the ability to transform their vegetative structures as protective devices against stressful environmental conditions. Everything proves the genetic capacity of the Asteraceae for virtually unlimited kinds of mutations that provide opportunities for overcoming most obstacles to survival posed by continuous ecological changes.

Such a strong biological "drive" of the Asteraceae has produced spectacular growth in a number of taxa and individuals, hence of the populations of this family, during a relatively short period of geologic time. The tribe Heliantheae, with 265 genera and about 3000 species according to recent revision (Robinson, 1981), and with the greatest variety of taxa and life forms, are heavily concentrated in the Neotropics and are clearly Neotropical in origin and development. The tribe exceeds other major tribes in the area (e.g., Senecioneae, Astereae, Eupatorieae, and Mutisieae) in numbers and diversity. The subtribe Espeletiinae originated in this evolutionary context, with the ability to adopt a variety of morphological and reproductive strategies that improved their ability to survive and permitted them to play a predominant role among the overall vegetation of large areas in the high Andes.

PLACE OF ORIGIN OF THE ESPELETIINAE

The subtribe has about 130 species in 7 genera distributed in an area approximately 1700 km long, from northern Ecuador (Llanganatis, Tungurahua) to the littoral mountain chain of El Avila (Venezuela). It is obvious, looking at the distribution (Fig. 11–5), that the major concentration of taxa occurs in the orographic massif of the Sierra Nevada de Mérida with its various extensions northeastward and southwestward. This Venezuelan branch of the Northern Andes is separated by the Maracaibo Basin and the Cúcuta Depression from the Eastern Colombian Cordillera ending in the Páramo de Tamá-El Cobre Massif; from this point to its extreme north, at Páramos de Nepes and de Barbacoas (Trujillo-Barinas), this orographic unit is about 300 km long; the maximum width of the largest central section at 3000 m is 50 km; its base at 1000 m is about 77 km broad.

In this Venezuelan segment of the Andes are found 6 genera and 63 endemic species of Espeletiinae. All 6 genera are widespread in this Andean range, 4 of them having their maximum development in the area. Two genera (*Carramboa* and *Coespeletia*) are endemic, *Ruilopezia* is nearly endemic, for of its 21 species only 1 is found outside the area (although very close to it). Of the two other genera, *Espeletiopsis* is widely distributed in the Colombian Oriental chain and *Libanothamnus*, concentrated in the Venezuelan

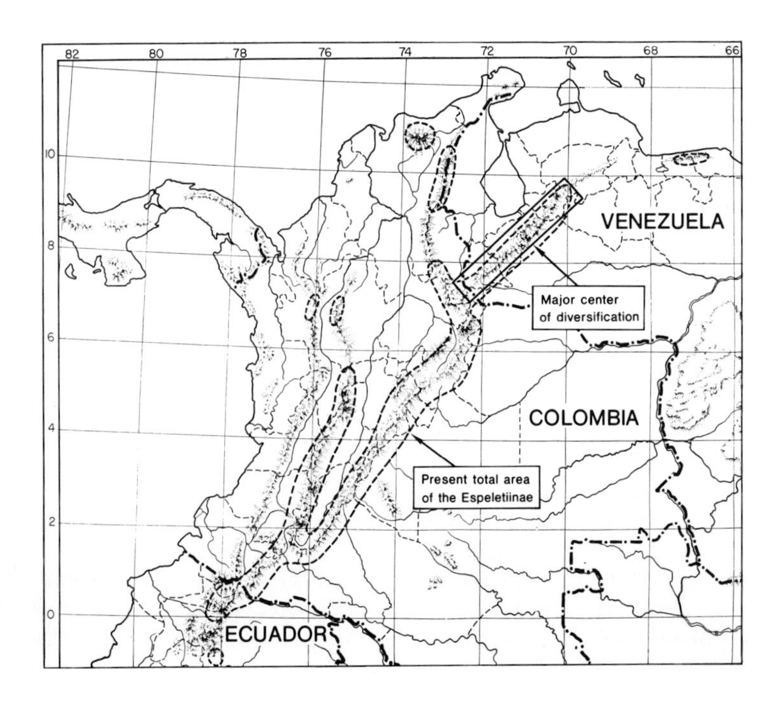

FIG. 11–5 Original center of the Espeletiinae. Venezuelan Andean branch with 65 species and 7 genera, 4 of them endemic. The species are all endemic.

Andes, radiates to neighboring mountains of Colombia. Only one genus, *Espeletia*, is spread throughout Colombia along its two principal Cordilleras southward to Ecuador, to stop at 1°S.

If we include the Venezuelan border of the Colombian Eastern Cordillera (Páramo de Tamá and Sierra de Perijá), a total of 71 species and all 7 genera of the Espeletiinae are found in this area. The monotypic genus *Tamania* is restricted to the mountains around the Páramo de Tamá-El Cobre massif.

This unbalanced distribution of taxa with major concentration in the Venezuelan Andes suggests this region as the original place of formation and diversification of the Espeletiinae.

THE ANCESTOR OF THE ESPELETIINAE, THE PROTO-ESPELETIINAE

The Heliantheae, predominantly opposite leaved, probably were already established in northern South America in Miocene time (e.g., Turner, 1977). During the Pliocene, a common ancestor of the present subtribe Espeletiinae probably developed along the Venezuelan Andes during the last phases of their uplift. This ancestor had already attained a stable type of floral structure and capitulum organization. It was an heterogamous capitulum with ray female flowers and functionally male disk flowers, the primitive pappose achene would evolve into an epappose one in response to the inefficiency of a pappus in the non windy environment of the tropical mountain forests. The achene was prismatic or pyramidal 3–4-angular with a smooth surface. The involucral phyllaries would keep the fruits confined in the heads for some time, and the receptacle was kept chaffy, the paleas providing protection for the flowers against dessication and cold.

The basic chromosome number was already 19, which indicates a primitive constituency for the subtribe. The woody basic habit of the Espeletiinae coincides with the general acceptance of the idea that woodiness is an ancestral feature of the Asteraceae, especially of the Heliantheae (Carlquist, 1976; Cronquist, 1977; Robinson, 1981). I consider the woody structure to be a common feature of the dicots growing in humid or mesic tropics.

Derived from some older ancestor with decussate leaves, the immediate ancestor of the Espeletinae had evolved into a form with alternate spiral phyllotaxy. This leaf disposition probably arose in response to strong local climatic changes and diverse conditions encountered in its emigrations into more open areas, usually much colder or drier, and with less competition from other groups.

This ancestor, or "proto-Espeletiinae", was a perennial, evergreen, woody plant occurring in a tropical, rather mesic, climate favoring uninterrupted growth during a long life span. It was shrubby and branched with xeromorphic foliage, capable of evolving into diversified growth forms. These initial shrubs were distributed in an early, probably semiopen, montane forest alternating in a mosaic with patches of scrubby vegetation, and associated with trees of *Podocarpus*, *Weinmannia*, *Myrica*, Lauraceae, *Hypericum*, etc., known to be inhabitants of the region since the Late Pliocene (van der Hammen, 1973, 1974, 1979a,b). The area was at an altitude similar to that of the present Andean forest, greatly extended horizontally and therefore easily exposed to climatic variables caused by the topographic limits on the one hand and glacial fluctuations on the other. This Andean zone was ecologically intermediate between humid and xeric and between temperate-cold and cold habitats, conditions modified by intermittent local changes of climate reflected in glaciations. Stebbins (1977) wrote that, "evolution in the Compositae . . . has been a succession of adaptive radiations rather than a series of trends in any paricular direction" and also that the adaptive radiation goes from semixeric habitats in one direction to specialized mesic or even hydric conditions. Radiation and evolution of the Espeletiinae have certainly followed such a spectrum.

The hypothetical pre-Espeletiinae shrub would diversify in different ways and adapt to mesic or humid habitats on the one hand, or to xeric and cold or very cold (páramos or superpáramos) on the other hand. In this process, shrubs would evolve into mesophytic or xerophytic trees, unbranched caulirosulas, or sessile caulirosulas; they would also diversify from polycarpic to monocarpic growth forms.

LIVING TAXA OF THE ESPELETIINAE

The following section describes the possible course of adaptive differentiation and evolution that had given way to the present living taxa of the Espeletiinae (Fig. 11–6).

Tamania

Tamanias are trees of medium size with flexible coriaceous leaves and terminal inflorescences; the flowers have a caducous pappus. The growth

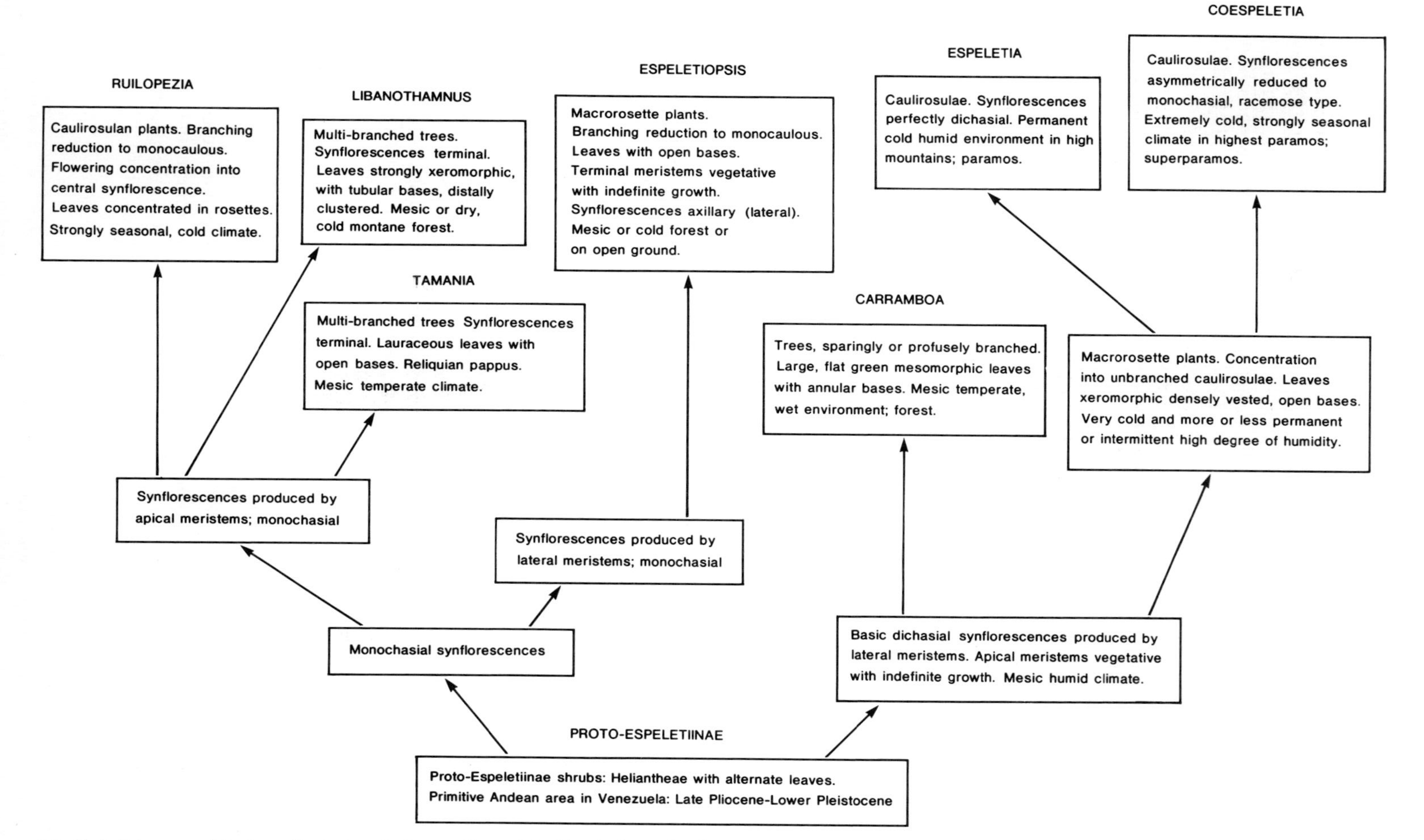

FIG. 11–6. Diagram of the hypothetical evolutionary lines of the Espeletiinae, leading to seven contemporary genera.

form is perfectly fitted to thrive in the mesic or humid, temperate-cold environment of Andean forests. The genus is represented by a single species, which could have been an active element of the earlier humid Andean forest probably stretching along the western section of the Venezuelan Andes to the eastern part of the Eastern Cordillera of Colombia, during a humid period of the earlier Pleistocene. At present, *Tamania* is found only in the Andean forests of the eastern slopes of the Eastern Cordillera, in the El Cobre-Páramo de Tamá region, and in the upper Margua Valley (Norte de Santander). It thrives between 2200 and 2900 m, in forests including *Weinmannia tamana*, *Myrica parviflora*, *Ocotea calophylla*, *Libanothamnus neriifolius*, *Ladenbergia macrocarpa*, *Clusia androphora*, *Brunellia trigyna*, *Hedyosmum colombianum*, *Ceroxylon sclerophyllum*, *Schefflera samariana*, and *Befaria glauca* (Páramo de Tamá, Colombia). It was found also in association with *Weinmannia tomentosa*, *Brunellia trigyna*, *Podocarpus rospigliosii*, *Clusia* sp., *Diplostephium rosmarimifolium*, and others, in a damaged forest at 2700 m (Venezuela). This present limited area should be considered a refuge for *Tamania*, an ancestral element. The presence of a residual kind of "caducous" pappus and the high percentage of sterile vacuous pollen grains (Marticorena, unpublished) indicate the ancestral relict character of this genus. Its absence from the main Venezuelan Cordillera where it could have originated can be explained by a very cold and possibly dry phase, supposed to have existed at the beginning of the Pliocene, which did not seriously affect the moister Colombian Eastern Cordillera.

Libanothamnus

Another direct successor of the theoretical pre-Espeletiinae element is certainly the present genus *Libanothamnus*. This genus probably developed under the stress of more severe, drier conditions in a period of the Late Pliocene or Lower Pleistocene along the Venezuelan montane area under the influence of a longer dry season or a dry glacial period. This condition forced the trees to adopt a marked xeromorphic structure by developing a thicker sclerophyllous type of leaves, dense abaxial lanate indument, and amplexicaul tubular leaf bases. These characters, in addition, become a better protective device of the terminal meristems or buds and the young shoots by being distally imbricate.

A prototype species of *Libanothamnus* with these features was surely capable of spreading through the upper belt of the Andean forest. It was well prepared for emigration to more open, drier areas, or to enter colder, probably more open areas, in the upper reaches of the mountains at a time of recession of glaciers in interglacial periods. Migration and adaptation to higher altitudes have led to the origin of new species of *Libanothamnus*. Two initial types can be considered in this genus: one of them with rather separated leaf veins is represented by *L. neriifolius*, the most widespread species both in latitude and altitude, with a broader range of adaptive flexibiliy. It extends from the Colombian Páramo de Tamá along the entire Venezuelan Cordillera to its northern end; it reappears in the Coastal Cordillera (El Avila). Its optimum environment is the mesic montane forest between 2000 and 3000 m, but it is also found from 1600–1800 to 3200 m. It thrives in cloudy, humid forests as in Norte de Santander-Tamá and Páramo del Batallón, as well as in apparently semidry forests and open deforested mountain slopes of Lara and Trujillo States. It is known for its rapid spread in new, open montane areas where it may dominate the landscape. Having been found as low as 1600 m, it is easy to believe that during the period of the Last Glacial with the considerable lowering of the forest belt *L. neriifolius* was able to cross the low hilly land on the way from Lara State to the Coastal Cordillera, where a now prosperous new population was founded. This population was and remained definitely isolated when, following the last glaciation, a warmer and arid climate prevailed in the area, leaving the lower lands lacking *Libanothamnus* and mesic forest (Fig. 11–7).

Inhabiting a lower, larger, and more continuous area than any other species of the genus, with demonstrated ability to colonize new areas and having a broad range of altitudinal tolerance, *L. neriifolius* has fairly constant features with only a few local and regional variations. These variations are not always correlated and are taxonomically at a lower level than the subspecies. They show, nevertheless, trends of differentiation that are general in the genus. *L. neriifolius* might be either the most primitive species of the genus or very close to the extinct ancestor. A thoroughly distinctive species, *L. liscanoanus*, presumably is a microthermic species derived from it. The other type of *Libanothamnus*, characterized by its thicker, more rigid leaves, with an increased number of close, thicker secondary veins and more compact indument, emerged at the uppermost area of the forest. This early species, probably very close to (if not the same) as *L. occultus*, spread along the Venezuelan Andes from the eastern front of the Colombian Cordillera to the

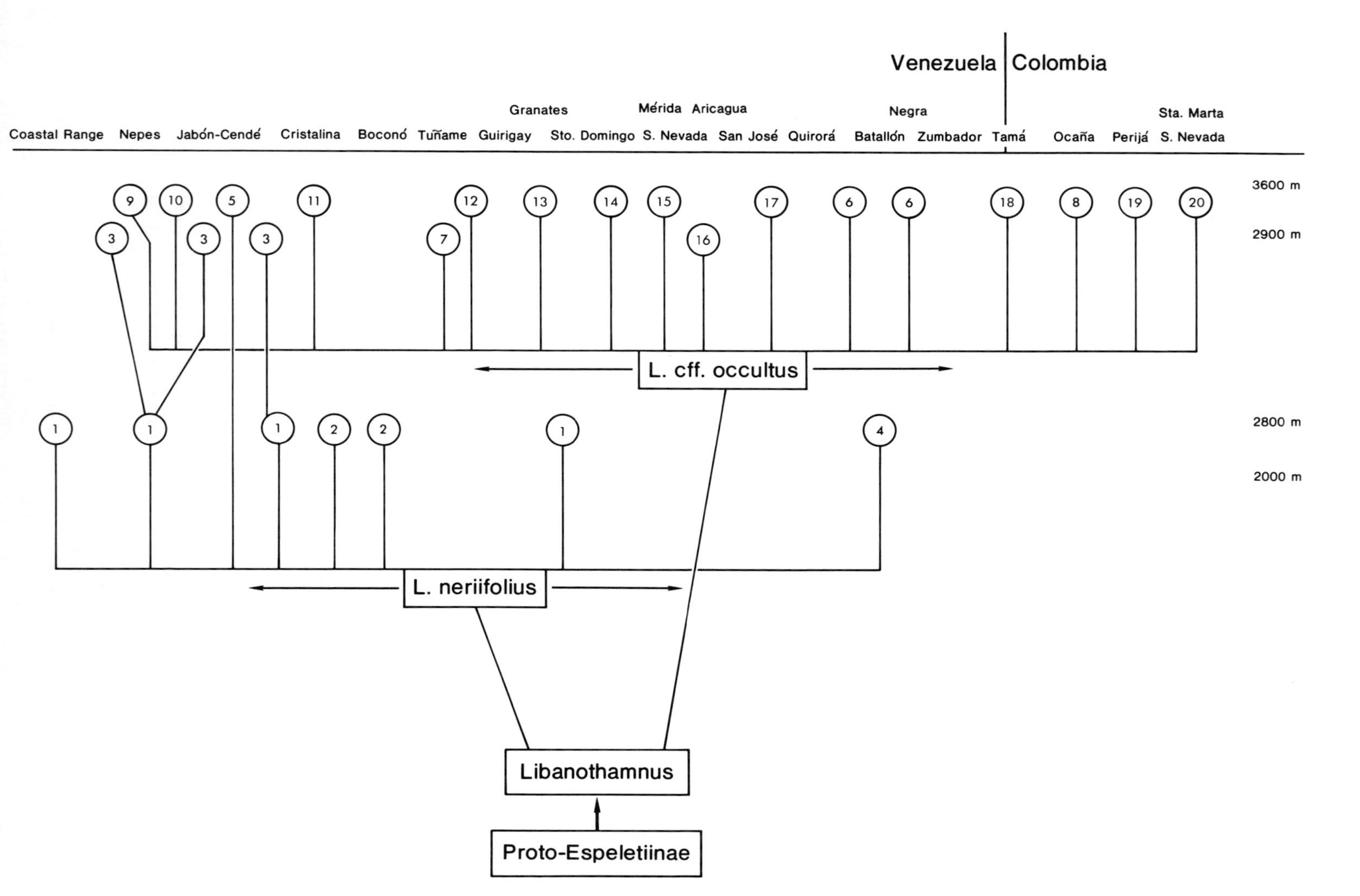

FIG. 11–7. Phylogenetic tree of *Libanothamnus* Ernst.

northern end of the Venezuelan Cordillera during a cold Pleistocene period which facilitated migration across some depressions in the mountain ranges. This species also passed across the Táchira Depression to the Colombian Cordillera Oriental and from there to the end of the Sierra de Valledupar and Sierra Nevada de Santa Marta. Likewise, it could have crossed the Chama River to the northern end of the Mérida Cordillera and to the cerros of Trujillo State, etc. Successive glacial recession, followed by ascension of this *Libanothamnus* to higher parts of the mountains and valleys, led to new populations isolated from each other by the rough mountain-

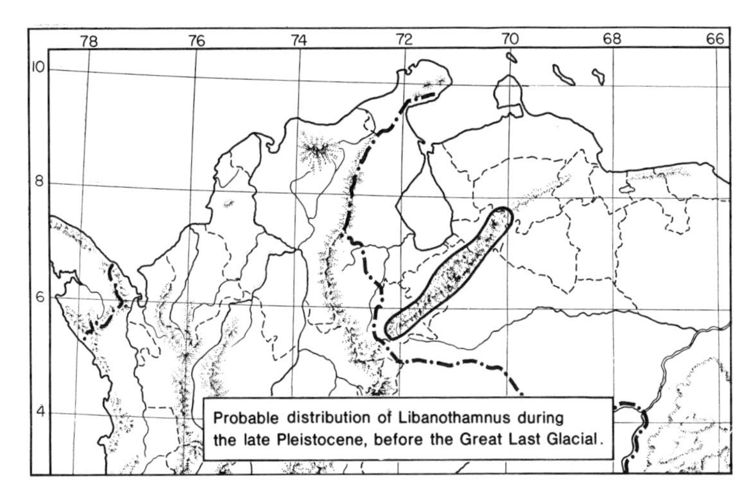

FIG. 11–8. Probable distribution of *Libanothamnus* in the Late Pleistocene, before the Last Glacial.

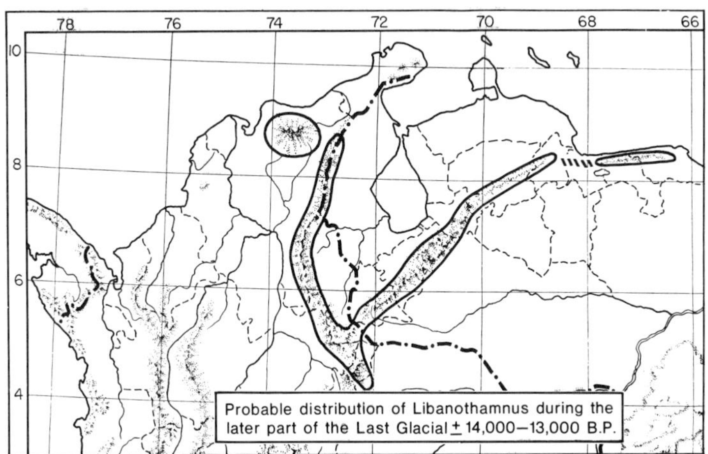

FIG. 11–9. Probable distribution of *Libanothamnus* in the later part of the Last Glacial, ± 14,000–13,000 B.P. (Susacá Interstadial).

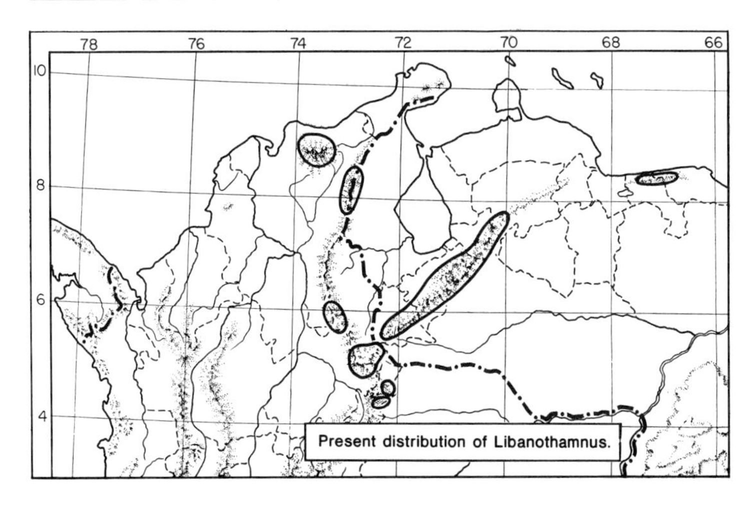

FIG. 11–10. Present distribution of *Libanothamnus*.

ous topography characteristic of the region. This isolation, in time, allowed differentiation leading to many endemic species. In this way, *L. occultus* gave rise more or less directly to *L. humbertii*, *L. banksiaefolius*, *L. granatesianus*, *L. lucidus* in the main Mérida massif, *L. parvulus* and *L. cristamontis* in the Cendé-Barbacoas range, *L. arboreus* and *L. griffinus* in the Guirigay-Boconó region, *L. divisoriensis* in the Serranía Valledupar, *L. tamanus* at the Colombian-Venezuelan border of Tamá, and *L. glossophyllus* in the Sierra Nevada de Santa Marta. In short, 10 or 12 species were differentiated along several sections of the cordillera (see Fig. 11–10). Besides, few varieties or subspecies have been detected.

Summarizing, *Libanothamnus* probably originated in the main Venezuelan Andes, probably not far south of the present Mérida location, during Late Pliocene or Early Pleistocene time, when the Cordillera had not yet reached its present elevation. The initial species, *L. neriifolius*, or another very close to it, spread very quickly, mostly horizontally, along the whole Cordillera at a variable altitude in a primitive mesic, montane forest; its ability to thrive under relatively broad extremes of temperature and dryness allowed the species to cover a great area. From *L. neriifolius* stock, evolving into a type fitted for colder climate, emerged a second species, like *L. occultus*, more markedly psychromorphic. This new species established itself parallel to the first, forming a new population belt of *Libanothamnus* at an irregular colder climatic zone at a higher altitude caused by the elevation of the Cordillera. Both species extended as far as possible and both reached the eastern front of the Colombian Cordillera Oriental, where they persisted. The second species, or *L. occultus*, split into local populations along the Cordillera, separated by deep valleys and steep rocky ridges; the result was isolation and differentiation of new species and subspecies. It was probably during the northern Andean Last Glacial that lowering of vegetation belts facilitated the emigration of *L. occultus* northward to the Sierra de Perijá, and across the deep César Depression to the Sierra Nevada de Santa Marta, where the new population, in isolation, evolved into the polymorphous and more markedly xeromophous species, *L. glossophyllus* (Figs. 11–7, 11–8, 11–9, 11–10).

Ruilopezia

The genus *Ruilopezia* probably originated from the hypothetical primitive shrubby ancestor involving an evolutionary trend toward concentration of both the vegetative and the reproductive parts. This trend is accomplished by reduction of the vegetative branching to a few branches and concentration of the leaves into large terminal rosettes; in addition, the florescences, always terminal, are reduced in number but increased in size. The most primitive form in this series is the branched caulirosula usually with few prostrate branches, each terminated by a monocarpic rosette, e.g., *R. jahnii*, *R. bromelioides* (Cuatrecasas, 1979a, Figs. 5–3, 5–4). The trend toward simplification in this line of evolution is toward the single monocarpic caulirosula with a single central inflorescence; it is achieved by the monocaul sessile caulirosula (whether rhizomatic, e.g., *R. atropurpurea*, or tuberose, e.g., *R. jabonensis*) and by the tall monocarpic caulirosula (e.g., *R. paltonioides*, *R. ruizii*, *R. figueirasii*) (Cuatrecasas, 1976, fig. 2; 1979a, figs. 5–3, 5–4, 5–5) (Fig. 11–4A, B.).

The monocarpic caulirosula in *Ruilopezia*, whether sessile or tall, combined with a broad range of variations in the xeromorphic structure of the leaves, was an adequate growth form to adapt to difficult conditions such as long periods of seasonal dryness and low temperature. The harshness of the climate, and in addition the stress inflicted on the plants by continuous migratory oscillations caused by alternating glaciations, generated serious difficulties for the survival of the species during the Pleistocene, especially in the northeastern part of the Venezuelan Andes. *Ruilopezia* evolved a series of strategies to overcome these difficulties. The adoption of a monocarpic biological cycle was an adaptation to the extreme seasonal macroclimate often prevailing in the area. Here some of the sessile caulirosulas grow in semiopen locations in the upper montane forest and in the open bushy subpáramo zone; others with woolly leaf structure similar to that of some *Espeletia* became perfectly adapted to open grass páramos (e.g., *R. floccosa*, *R. jabonensis*). The derivatives with tall caulirosulas were perfectly fit to grow among the trees in the Andean forest, especially at its upper level. Some of the newly originated species had the flexibility to take advantage of the new open areas, whether at high elevations (e.g., *R. jabonensis*) or lower, where the trees for some reason receded or were destroyed (e.g., *R. paltonioides*, *R. figueirasii*). However, the opposite also took place; during the periods of recession of the glaciers, with elevation of the thermal belts, the montane or Andean forest spread upward, forcing the subpáramo and páramo species up to the summits with subsequent reduction of their area. This reduction sometimes (especially in the period

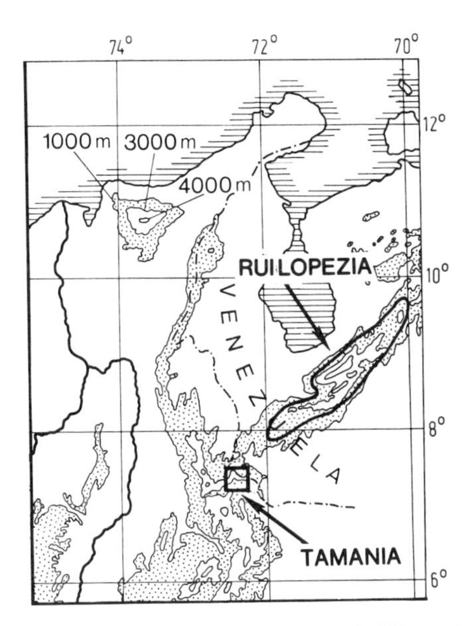

FIG. 11–11. Distribution of *Ruilopezia* and of *Tamania*.

consecutive to the Last Glacial) was considerable, fragmenting the continuity of populations, creating biological islands, some of them very small, which in fact were refuges in which new trends of differentiation and speciation might have taken place. The extreme reduction of the ecological areas (islands) also caused extinction of species populations and even of whole species.

In any case, the evolutionary line of *Ruilopezia*, with its plasticity to evolve a broad range of morphological variations, enabled the group to prevail in a variety of habitats along the whole Venezuelan Cordillera in spite of the rough conditions imposed on it during the Pleistocene (Fig. 11–11).

Speciation in *Ruilopezia*, as it is seen today, took place in a relatively narrow altitudinal zone, the ecotone between montane forest and páramo. This zone probably was the most affected by the glacial fluctuations because of its broken physiography with many partially disconnected hills, mountain peaks, and ridges of median altitude, which have alternated between periods of biological isolation and communication. These repeated environmental changes probably have been responsible for a great amount of selection followed by successful differentiation in *Ruilopezia*. From the shrubby ancestor of the Espeletiinae there developed rather low plants with few branches, usually prostrate with large and dense rosettes adapted to grow in open páramo with rather rocky ground. Specialized end products of this line of evolution are, e.g., *R. jahnii* and *R. bromelioides*. Most of the species developed with basically a single rosette, either sessile (e.g., *R. atropurpurea*, *R. bracteosa*) or stipitate. The majority of the living species of the genus are tall caulirosulas growing in the upper zone of the forest at timberline or in the subpáramo zone. Altitudinal lines for the total geographic area are difficult to trace because of the irregularity of the region, and because the zonation lines vary from one mountain to the other.

Three of the eight tall caulirosulan *Ruilopezia* grow in the hills of the northeastern part of the Cordillera; the other five grow along the southwestern half. The two or three sessile species mentioned earlier belong also to the northeastern section (Trujillo-Lara), whereas the two branched ones grow in the southwestern ranges. Another group of species has specialized by evolving a sessile, more or less tuberous, cormus, a dense mass of linear, white, woolly leaves, and a central white inflorescence with bright or pale yellow long rays. This group has evolved at the upper zone of high páramo and at present it is represented by four known species found on open summits (*R. jabonensis*, *R. margarita* [northeast], *R. leucactina* [southwest], and *R. floccosa* [central area, Santo Domingo]) (Fig. 11–12).

Summarizing, *Ruilopezia* probably arose during the Early Pleistocene in the Venezuelan Cordillera, an initial form spreading from Táchira to Trujillo-Lara, and then differentiating regionally into the subpáramo zone. This zone produced 17 species, 5 in the northeastern subcenter of speciation and 12 in the southwestern subcenter of speciation. Also in the páramo zone, four additional species were produced, three of them in the northeastern section (Lara-Trujillo-Mérida), the other in the southwestern section (Mérida-Táchira). Probably several more species existed in the past and more might be discovered in the future. Some of the species live today in very restricted areas of the broken mountains of the Sierra Nevada and its southern slopes and branches (e.g., *R. josephensis*, *R. ruizii*, *R. bromelioides*, *R. cuatrecasasii*, *R. coloradarum*, *R. hanburiana*), an indication that other species may be found in some of the many *quebradas* or hills still unexplored. Some of the local endemics are relicts from larger populations occupying more extensive páramo-subpáramo surfaces that existed in lower glacial and dry periods. Another example of a relict species is *R. lopez-palacii*, which is found only at the Páramo del Guaramacal. This "páramo" is in fact a subpáramo formation of limited size, at the top of Guaramacal hill, at 3100 m or 3200 m. In this locality it is associated with *R. viridis*, and in a more open spot there it is flanked by a formation of *R. jabonensis*. But these páramo-subpáramo species are found

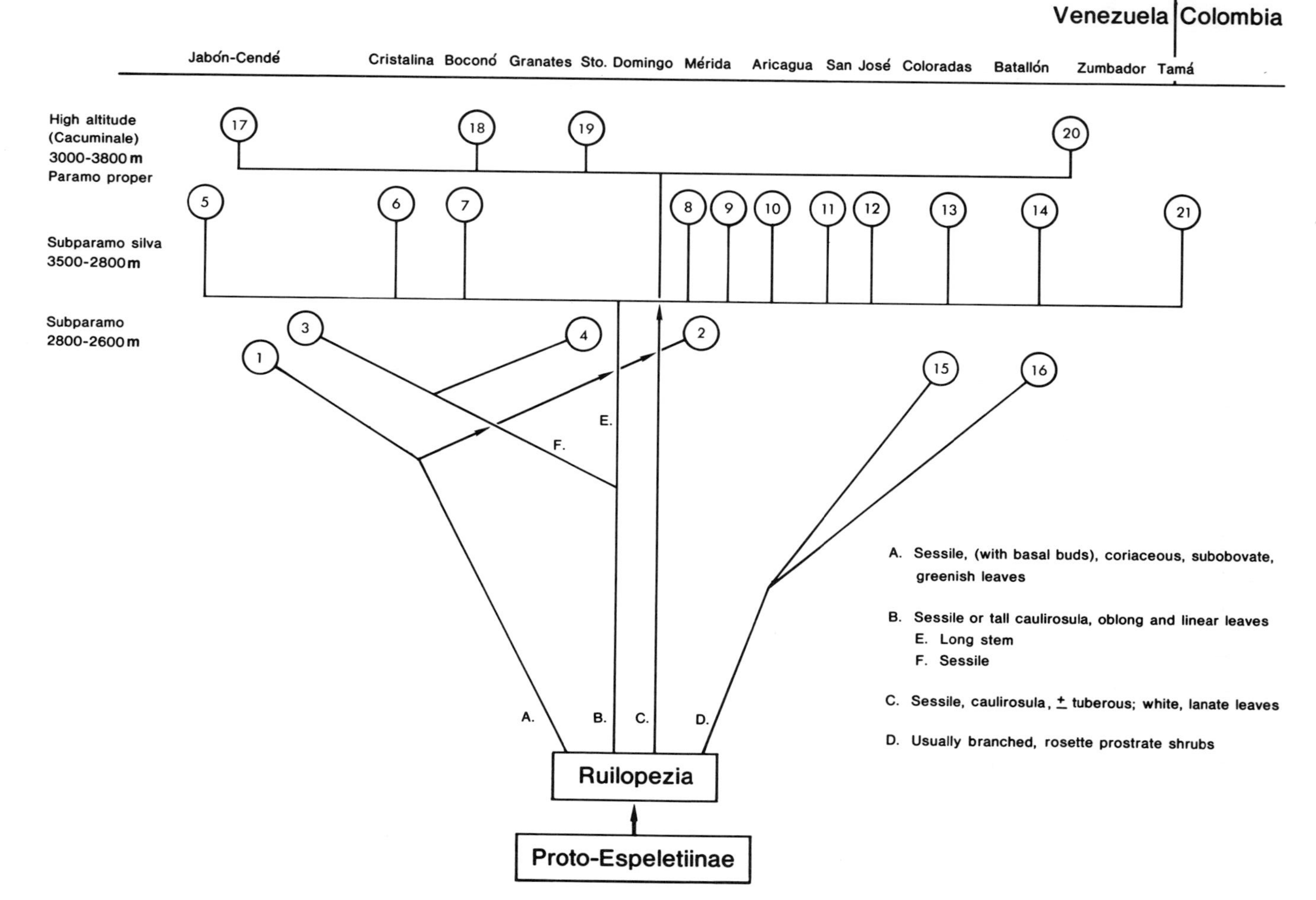

FIG. 11–12. Phylogenetic tree of *Ruilopezia* Cuatr.

only at the top of the mountain. Heavy montane (Andean) forest surrounds the "páramo" and covers all the slopes of the mountain except for some sections that have been opened recently and in which the forests are being progressively destroyed. This "páramo" is, no doubt, a relict and this summit of the Guaramacal hill a refuge for the species that in former times occupied a larger part of the mountain, which later was invaded by the forest after the recession of the glaciers.

The extreme diversification represented by the more than 20 species of *Ruilopezia* reflects its relatively long evolutionary history.

Espeletiopsis

The prototype of *Espeletiopsis* seems to have been a sparsely branched shrub or tree with axillary inflorescences, growing in a mesophytic Andean forest in the Venezuelan Andes and Colombian Eastern Cordillera. The climate probably was like the present one, rather humid with a short seasonal subdry period, favoring continuing growth and polycarpism. As in other groups, *Espeletiopsis* shows a strong tendency toward simplification, adopting the caulirosula growth form. *E. purpurascens*, growing in the Páramo de Tamá area of Colombia and Venezuela, and *E. insignis* from Norte de Santander are good examples of tall caulirosulas with naked stems up to 12 m long, thriving in the Andean montane forest and raising their large sclerophyllous rosette to the top of the canopy. In undisturbed climax forests of the high Chitagá Valley, in 1940, these large palmlike rosulas of *E. insignis* could be seen from a distance and from above, scattered in the forest (Figs. 11–13, 11–14).

Correlated with their ascent to the subpáramo

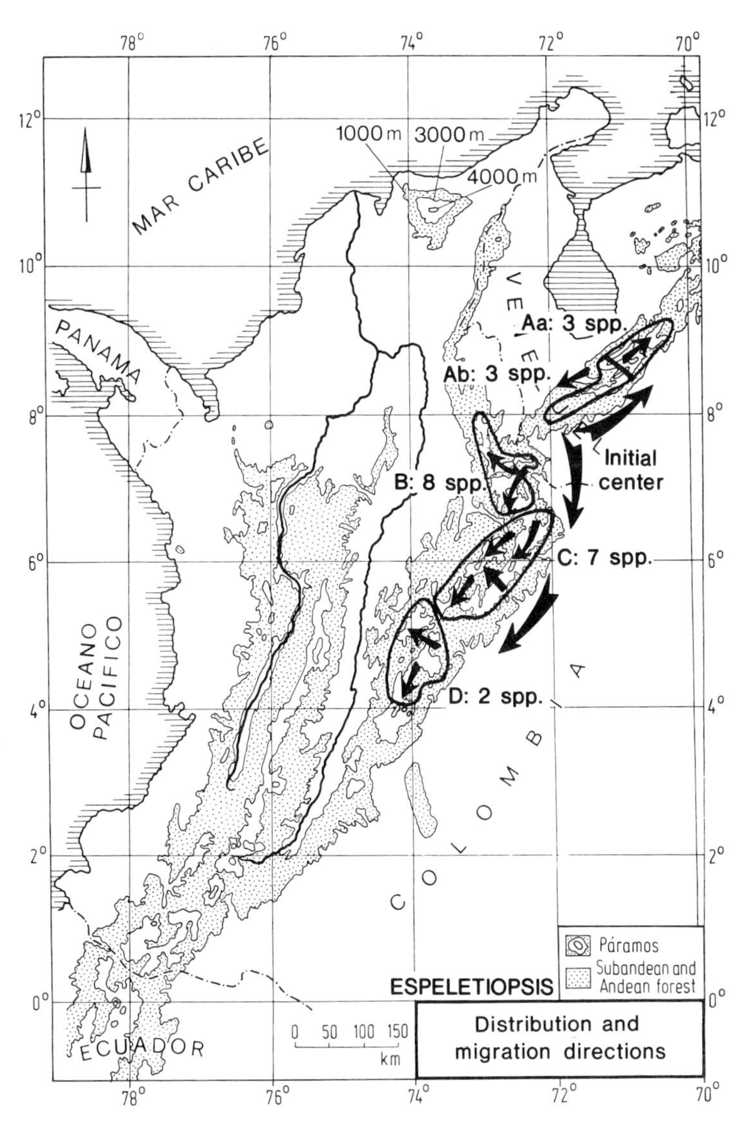

FIG. 11–13. Present distribution of *Espeletiopsis*. Indication of subareas and directions of radiation. Subareas or páramo blocks: Aa, Trujillo; Ab, Mérida; B, Santander; C, Boyacá; D, Cundinamarca.

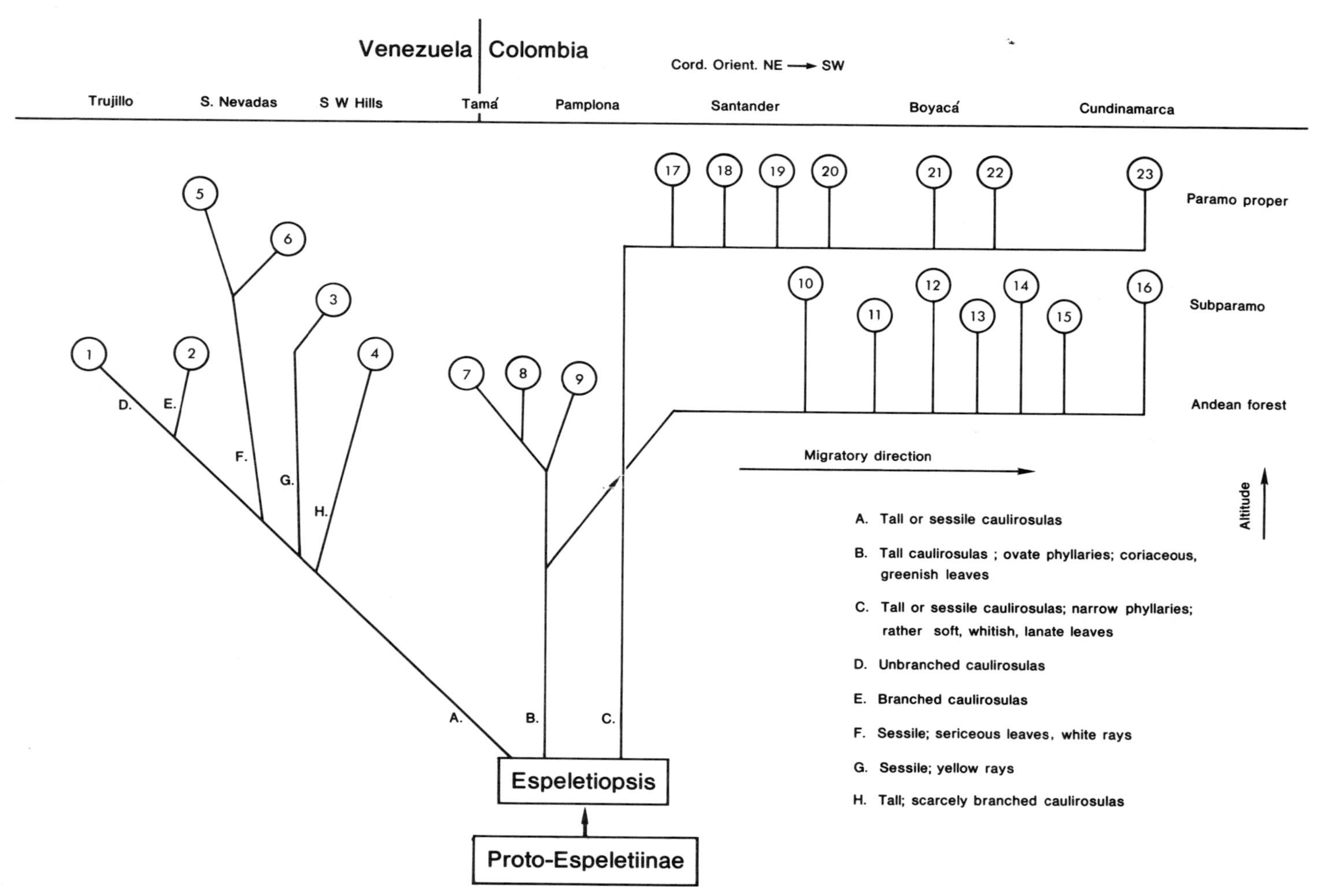

FIG. 11–14. Phylogenetic tree of *Espeletiopsis* Cuatr.

TABLE 11–1. Distribution patterns of *Espeletia*, *Espeletiopsis*, *Diplostephium*, and *Puya* in the Colombian Cordilleras.

Cordillera Oriental							
Espeletiopsis		*Espeletia*		*Puya*		*Diplostephium*	
total	endem	total	endem	total	endem	total	endem
17	17	38	38	18	15	31	25
Sierra de Perijá							
0	0	2	2	2	1	3	2
Santander-Norte de Santander-Tamá							
8	8	7	7	6	4	16	9
Páramos de Boyacá							
7	7	18	17	10	3	12	4
Páramos de Cundinamarca							
2	2	11	10	7	1	12	3
Sierra Nevada de Santa Marta							
0	0	0	0	4	3	12	11

Cordillera Central					
Espeletia		*Puya*		*Diplostephium*	
total	endem	total	endem	total	endem
4	3	5	2	17	9
Cordillera Central proper					
2	2	2	1	15	5
Troncal Sur					
2	1	4	1	8	1
Cordillera Occidental					
3	2	2	1	6	2
Páramo del Chaquiro					
1	0	?		1	0
Páramos de Frontino					
2	2	?		?	
Páramo de Tatamá					
0	0	?		2	2
Páramo de Los Farallones de Cali					
0	0	1	1	3	1

Note: Number of species for each genus and corresponding number of endemics are tabulated for each of the three Cordilleras (Oriental, Central, Occidental) and the Sierra Nevada de Santa Marta, and for their significant páramo sections or páramo blocks; note that *Espeletiopsis* is absent from the Cordilleras Central and Occidental.

and the páramo zones, the caulirosulas evolved, keeping their marcescent leaves, a condition found in most of the species. A final growth form in the series of *Espeletiopsis* species at high altitude is the subsessile and sessile rhizomatous caulirosulas. Undoubtedly, some branched species did spread initially throughout the total area, and subsequently diversified, but at present there are few branched species. Two are known in Venezuela, *E. cristalinensis* and *E. meridensis*, which are subpáramo caulirosulas with usually few branches and thickly whitish lanate leaves. The first species extends from Trujillo to the middle Sierra Nevada de Mérida (south), whereas *E. meridensis* is found in the extreme southwest, in Páramos La Negra and Batallón. Two other often branched species are found in Colombia, which belong to another group, notable for their rather hard sclerophyllous leaves. These two species (*E. garciae* and *E. pleiochasia*) belong to the forest zone and occasionally to the subpáramo; they are only sparsely branched or monocaul. This branching, being a primitive character, contrasts with the advanced nature of the vegetative part of the inflorescences of these two species, which lack bracts. This absence of bracts is found only in a few other species that have evolved independently (e.g., *E. corymbosa* and *E. glandulosa*).

Two major groups of species can be recognized in *Espeletiopsis*: the first can be called sclerophyllous because their leaves are mostly hard coriaceous and usually green or greenish. This group is centered in Colombia, having at present about 10 species distributed regionally throughout the Eastern Cordillera. Only one reaches Venezuela (*E. purpurascens*) on the Páramo de Tamá, at the eastern end of the Colombian Cordillera Oriental. Subcenters of speciation are the páramo islands of Norte de Santander (4 spp.), Boyacá (7 spp.), and Cundinamarca (2 spp.). One species (*E. glandulosa*) departs considerably from the others in the group by having linear revolute leaves, a high degree of glandulosity, and herbaceous outer involucral bracts.

The second group consists of subpáramo and páramo species, all caulirosulas with white lanate or sericeous leaves, and a short or up to 1 m high trunk. In this group, one center of diversity is in the Venezuelan Andes, where there are four subpáramo species with yellow rays (*E. jajoensis*, *E. cristalinensis*, *E. pozoensis*, and *E. meridensis*) and two high páramo species with white rays (*E. pannosa* and *E. angustifolia*). Such white-rayed species are exceptional in the genus, the first from the Venezuelan central massif and North Cordillera, and *E. angustifolia* being mostly from the southern hills. The second center is located in Colombia, with the subcenter islands producing four species in Norte de Santander-Santander (Romeral-Almorzadero), two in Boyacá, and one in Cundinamarca (Fig. 11–14; Table 11–1).

It is important to note that *Espeletiopsis* does not occur south of Bogotá and does not enter the páramo zone of the large, high Páramo de Sumapaz. In fact, the nemoral species can live well in atmospherically humid areas like the mesophytic Andean forest and are found from the forested slopes of Páramo de Tamá southwestward to the hills east of Bogotá. But the páramo or subpáramo species of *Espeletiopsis* are missing in the true páramo zone of Páramo de Tamá because they may need some peculiar environmental conditions probably lacking in that area, e.g., higher temperature, different soil, rocky ground, or better soil drainage. The absence of some of the necessary conditions also probably prevents these species from migrating farther south of Cundinamarca. The lowering of the climatic zones during glaciation brought the subpáramo belt to a 1000-1200-m lower level and allowed the Espeletiae, but not *Espeletiopsis*, to migrate southward along the Eastern Cordillera. Nevertheless, the broad range of variation and the high degree of specialization in *Espeletiopsis* indicate its considerable antiquity. Its origin probably dates from the Lower Pleistocene, and its development in Colombia from the Middle Pleistocene, long before the Last Glacial.

Carramboa

Carramboa has about seven arboreal species found only in the Mérida Andes, from the northern limits of the states of Trujillo and Lara to the southern end at Páramo del Zumbador in Táchira. The area is split into two sections. The northern section, in Trujillo State and neighboring Lara, has two species growing in the montane forest in rather protected and humid parts. One of the species (*C. trujillensis*), which has been found on both sides of the Serrania del Jabón-Cendé, the divide between Trujillo and Lara (Humocaro Bajo-Agua de Obispo), also grows on the Páramo de La Cristalina (Trujillo-Boconó) and on the slopes of Páramo del Guaramacal (close to Boconó), between 2400 and 2800 m. The other species, *C. wurdackii*, grows in a limited locality of Quebrada del Turmal, 2400 m altitude, in the western drainage of Páramo del Jabón.

The second section of the area includes both lower slopes of the Chama Valley (Mérida) and the mountain ranges between Sierra Nevada de

Mérida and Páramo del Zumbador. *C. badilloi* grows along forested banks of creeks of El Morro and Aricagua as well as in the region of the Páramo de San José, at 2400–3000 m altitude. *C. pittieri*, the most widespread species, is found in forests, especially in deep ravines of the slopes of the Páramo de Las Coloradas, Páramos de Mijará and de Quirorá, Mesa de Quintero, Pico Horma, Páramo del Batallón, Páramo del Zumbador (near Queniquea) between 2500 and 2800 (sometimes 3000) m. Other species of limited distribution are *C. rodriguezii*, growing in the forested area of Páramo de Las Coloradas towards El Molino, 2600–2800 m, and *C. tachirensis* of Portachuelo de La Grita toward Pregonero, 2800–2950 m altitude. *C. littleii*, most closely related to *C. badilloi*, grows in the forested area of San Eusebio in the Cordillera del Norte, at 2400–2600 m.

The genus *Carramboa* seems to have developed directly from the common Espeletiinae ancestor, in tree form. It retains the basic spirally alternate phyllotaxis with usually large or very large, green, sparsely or moderately pubescent leaves. The inflorescences are axillary, with the main axis and branches having indefinite terminal growth. The branching of the inflorescences is primarily dichasial, having at least the lower branches and bracts opposite, whereas the upper ones may be opposite or alternate. It seems that *Carramboa* was originally adapted to a humid montane forest and originally spread horizontally around the Sierra Nevada de Mérida and into valleys north and southward at somewhat lower elevations. The isothermic and permanently humid condition of the cloud forest, where the genus originated and developed, probably favored the retention of the dichasial inflorescences, which contrasts with the marked spirally alternate vegetative leaves and branches characteristic of the subtribe. In the humid periods of the Pleistocene, the genus probably spread throughout the area, but the vertical oscillations of the vegetation caused by the alternating periods of glaciation eliminated it from large parts of the area. The currently known localities represent, in great part, refuge habitats, especially for Trujillo State and other deforested areas. But *Carramboa* persists in deep, less accessible valleys and along small, still-forested creekbanks. I have no doubt that more localities, and maybe undescribed species, will be found with future exploration of the rugged Venezuelan Andes (Figs. 11–6, 11–15).

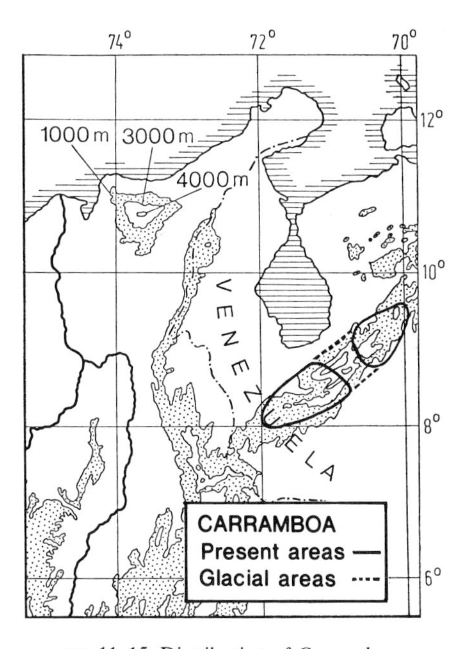

FIG. 11–15. Distribution of *Carramboa*.

Espeletia

This genus probably originated in an earlier glacial period of the Lower Pleistocene or perhaps of the Upper Pliocene when cold, wet páramo conditions prevailed in a zone between the Western Venezuelan and Eastern Colombian Cordilleras (Fig. 11–16). The tall caulirosula emerged first as the biotype able to thrive in the wet subpáramo scrub zone. The first stock of this series established itself in that environment in the form of polycarpic caulirosulas with indefinite terminal growth and axillary inflorescences, well equipped to survive in the perennially cold climate, humid atmosphere, and wet ground.

The primary evolutionary process that caused the ancestor of the Espeletiinae to develop its basic spiral phyllotaxis changed direction somewhat at a certain point in its phylogenetic development. In the preceding genera, except for *Carramboa*, spiral phyllotaxis and alternate branching were general in both the vegetative and reproductive parts. However, in one of the phyletic lines the inflorescences retained the primitive dichasial structure along with decussate phyllotaxis. Presumably, relatively permanent humid atmospheric conditions have favored the persistence of the dichasial condition in the ephemeral reproductive branching. This phyletic line produced *Carramboa* in the cool-temperate nemoral environment, and led to *Espeletia*, a stock potentially best fit to compete in a continually cold, wet, open habitat.

Espeletia is fundamentally gregarious and specialized for cold, humid habitats with humifer-

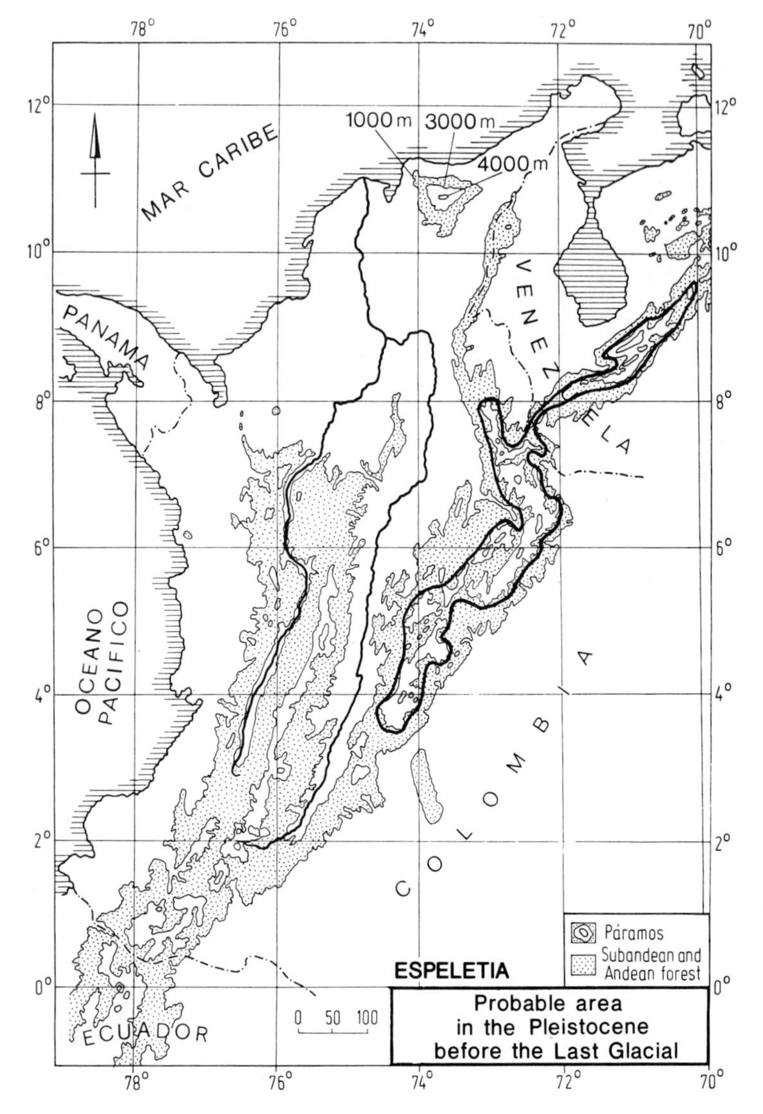

FIG. 11–16. *Espeletia.* Probable area occupied by the genus in the Pleistocene, before the Last Glacial.

ous soil, preferably in wet or marshy páramos, and sometimes rocky, very humid places. Its rather heavy seeds with no dispersal mechanisms are profusely produced several times in a year, but the local reproductive rate is low. The gregarious populations may cover several square kilometers of páramo but they are often less extensive because of topographic or ecological obstacles to continued propagation. *Espeletia* species may form dominant communities around the upper part or peaks of mountains and are usually limited at their base by Andean forests. In this way populations on emerging mountains, hills, or peaks become isolated. Other preferred habitats are also limited by forests. The bottoms of small-to-large, wet valleys, and flat or moderately inclined grasslands become quickly covered by *Espeletia* communities, like islands completely surrounded by heavy Andean forest.

These isolated communities may develop some degree of differentiation and in the long run evolve into new subspecies or species. In the process of slow, long-distance migration, adaptation to various degrees of humidity, differences in drainage, or chemical and physical soil conditions, apparently small variations in local habitats caused or facilitated evolutionary changes and speciation. New taxa were produced and distributed according to ecological preferences.

Espeletia is the most representative genus of the ecological concept of a páramo. Most of its species develop best in the páramo proper zone. They are found also in the subpáramo zone, especially in mosaic distribution or in ecotonal communities. *Espeletia* is rather rare in the superpáramo zone, approaching only its lower border. Very few species, e.g., *E. cleefii* and *E. lopezii* f. *alticola*, may thrive abundantly in rocky habitats

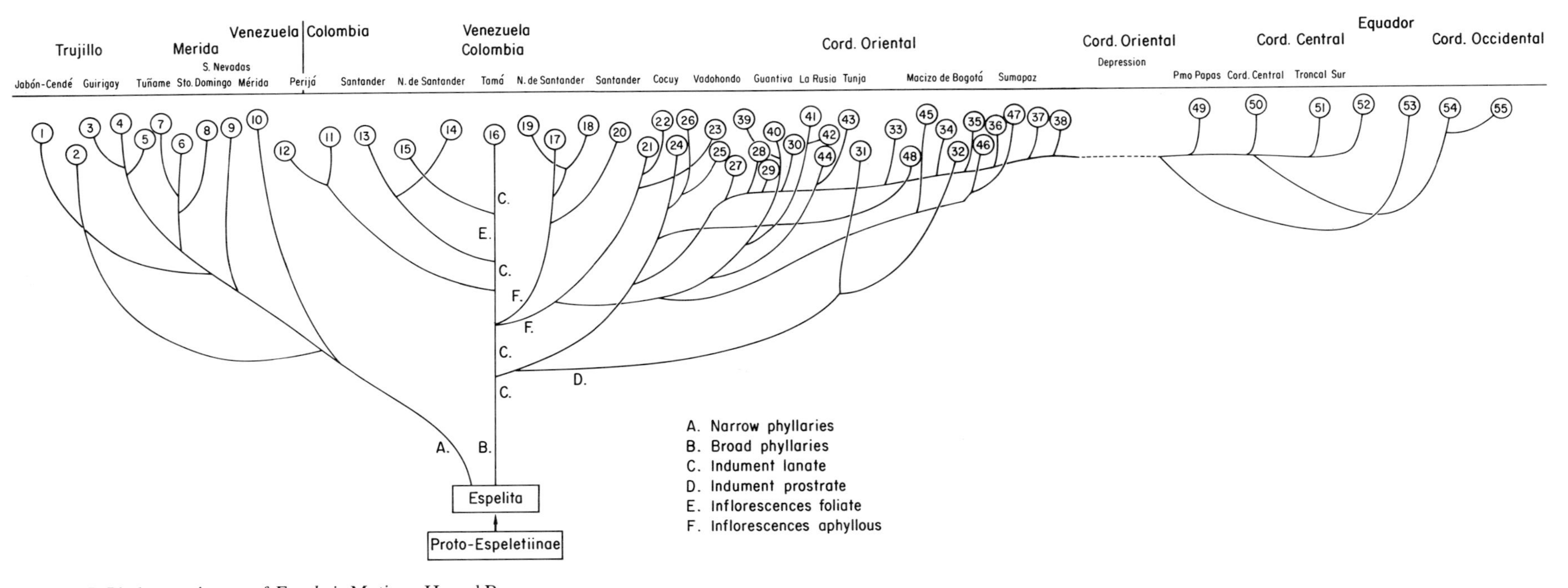

FIG. 11–17. Phylogenetic tree of *Espeletia* Mutis ex H. and B.

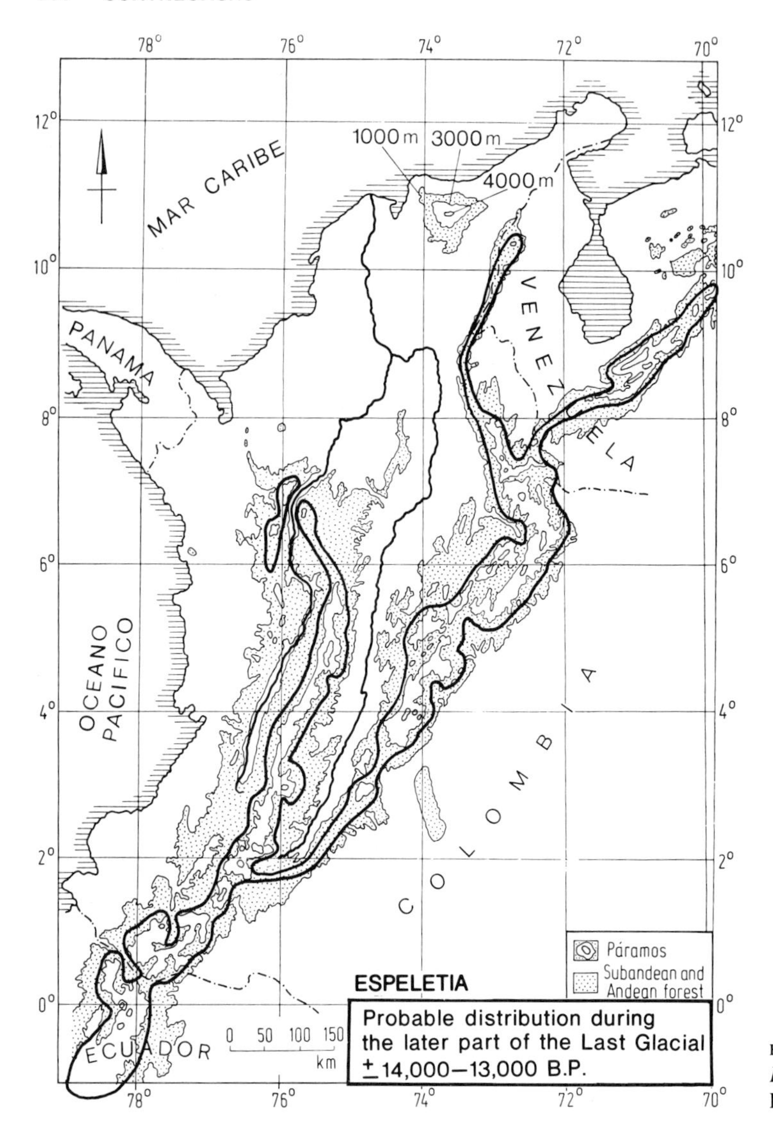

FIG. 11–18. Probable distribution of *Espeletia* during the later part of the Last Glacial, 14,000–13,000 B.P.

in the superpáramo slopes of the Sierra Nevada del Cocuy (Colombia) and, sporadically, *E. semiglobulata* and *E. schultzii* in the Sierra Nevada de Mérida, defying the characteristic frost and solifluction common in that periglacial area (see Chap. 2).

Main Trends in Specific Differentiation

The most noticeable trends are related to the size and kind of inflorescences, the shape of the leaves, and type of indumentum. The large, tall, multicapitulate inflorescences are primitive; reduction in total length, reduction and suppression of bracts of the proximal sterile portion, and reduction in number and increase in size of capitula are derived phyletic trends (Fig. 11–2A). Pseudopetiolate leaves are primitive and evolutionarily the general trend is toward reduction and elimination of the pseudopetiolar section, the derived form of lamina being sessile. Lanate indumentum is probably primitive, becoming more or less dense or thick according to environmental situations; divergent phyletic lines are characterized by prostrate sericeous indumentum or by a straight patulous, velutinous type. In several groups, there is a minor reduction in the ray corollas, approaching obsolescence. The length of the tube of the ray corollas initially increases to some extent; however, a major trend toward reduction of the tube is obvious along the migration route of *Espeletia* from Boyacá to the

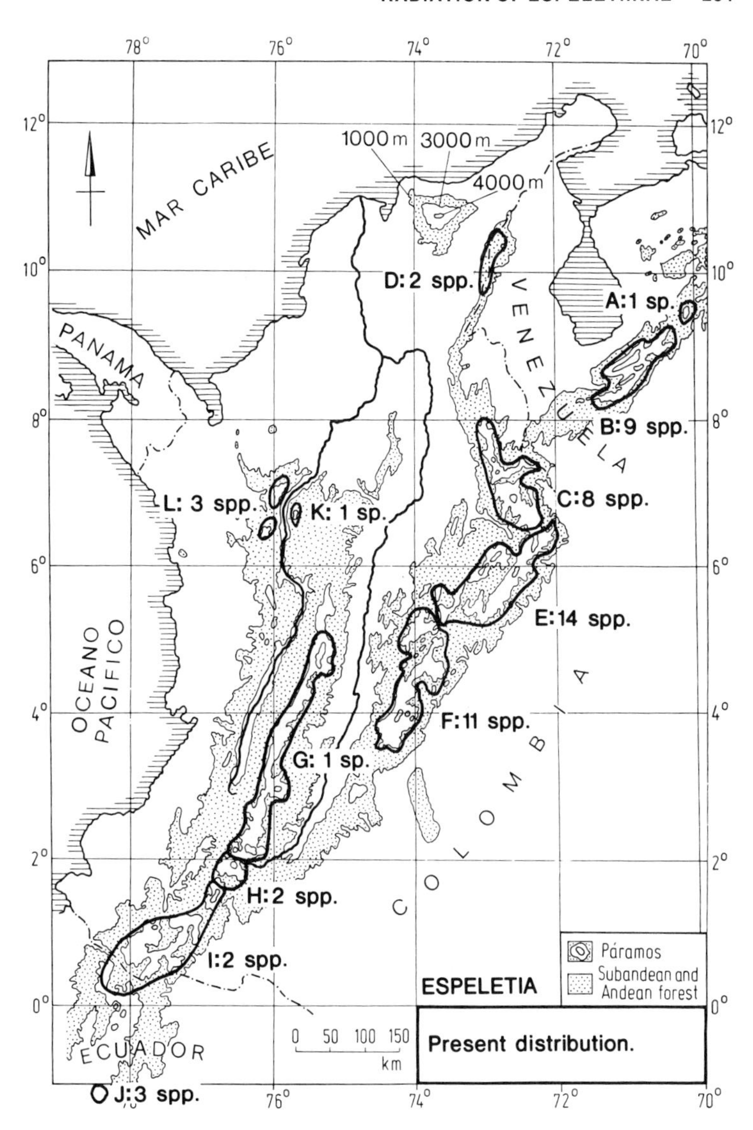

FIG. 11–19. Present areas of distribution of *Espeletia*; areas and subareas: A, Cendé-Jabón; B, Sierras de Mérida-Trujillo; C, Tamá-Santander and Norte de Santander; D, Sierra de Perijá-Valledupar; E, páramos of Boyacá; F, páramos of Cundinamarca; G, main Central Cordillera of Colombia; H, Macizo Colombiano nucleus; I, Troncal Sur of the Central Cordillera; J, Sierra de Los Llanganatis, Ecuador; K,L, northern parts of Western and Central Cordillera Chaquiro-Frontino páramos.

extreme southern limit of the genus (Figs. 11–17, 11–18, 11–19, 11–20).

Two Main Directions of Radiation

COLOMBIAN CORDILLERAS

Eastern Cordillera: *Espeletia* first emigrated horizontally in two opposite directions, one along the Venezuelan Andes, the other from the Venezuelan border to the Eastern Cordillera of Colombia. This latter stock was extremely successful in establishing population in wet places in the subpáramo zone and eventually in the upper level of forested ecotonal zones, migrating from the present Cobre-Tamá massif southward along the Eastern Cordillera of Colombia through a semiopen or open upper belt (subpáramos or páramos), preferably in the more humid eastern drainage. Subsequently, colonies spread aggressively across the heights of the Cordillera and transversally to the western upper slopes and at the same time radiated southward (Fig. 11–20).

The Colombian Cordillera Oriental, the main Andean range of northern South America, is extraordinarily large and complicated, having extensive areas above 3000 m up to 4200 m altitude, not to mention the much higher altitudes of particular hills or high individual massifs such as the Sierra Nevada del Cocuy (5493 m) and the Nevado de Sumapaz (4250 m). Higher massifs and deep valleys are well-known barriers to the dissemination of páramo plants but in addition,

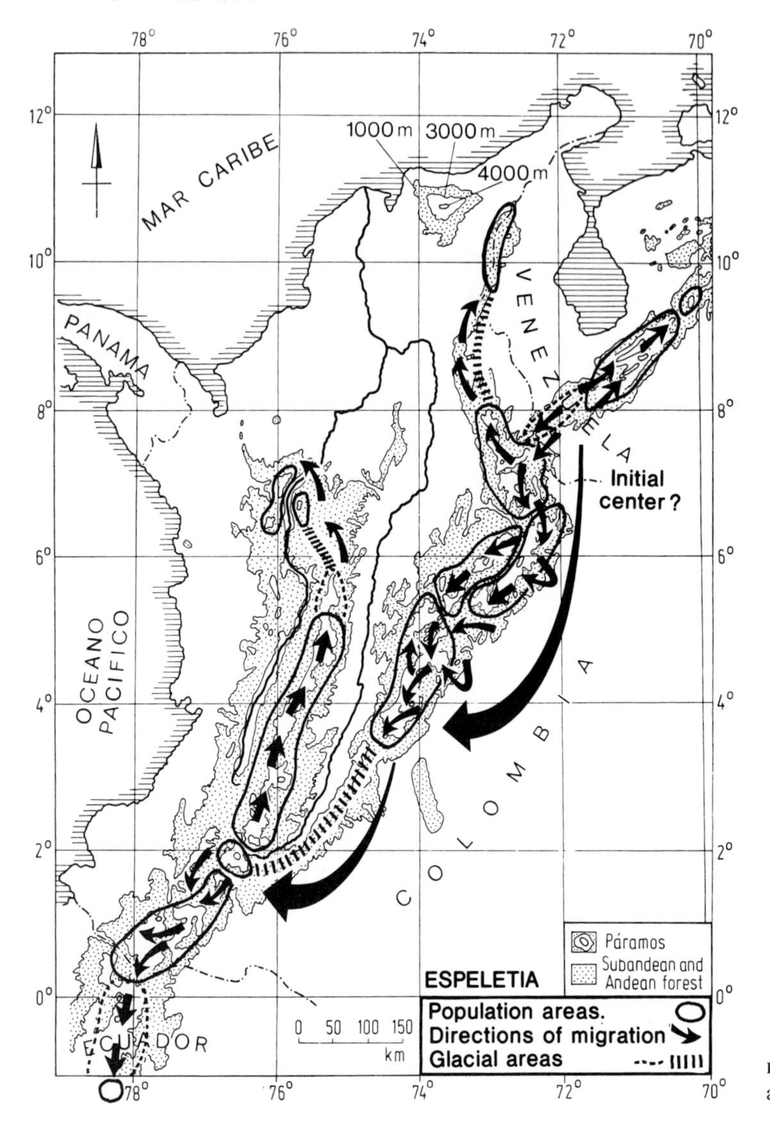

FIG. 11–20. Directions of migration and radiation in *Espeletia*.

innumerable smaller valleys or breaks, peaks and ridges on uplands constitute physiographic barriers to the migration of páramo species and cause isolation with consequent speciation.

Espeletia presumably migrated across and along the Eastern Cordillera southward in an early or middle period of the Pleistocene. Long before the Last Glacial its populations had reached and established themselves in Cundinamarca and had succeeded in dominating the páramo landscape of the region. But at the southern border of the Páramo de Sumapaz the advance of *Espeletia* was stopped by a depression of the Cordillera and by the dense Andean forest that covered the steep slopes and ridges of that sector of the range; moreover, there were no páramos from about 3.7° N south to 2° N where the Eastern and Central Cordilleras join. In the lower and narrower northern extremes of the Cordillera north of Ocaña, only small scattered páramos or subṕramos are found in the Sierra de Valledupar and Sierra de Perijá, which apparently escaped early colonization by *Espeletia* (Fig. 11–19,D).

The major part of the Cordillera Oriental, from Ocaña to Páramo de Tamá and from there to Páramo de Sumapaz, an area roughly 570 km long, supports 36 recognized species of *Espeletia*, besides two more at the northern end of Sierra de Perijá. These numbers indicate a considerable degree of speciation for the region. The study of species distribution reveals the existence of three main areas or centers of speciation on the Cordillera Oriental that are coincident with major orographic nuclei (Fig. 11–19, C, E, F). The first area includes the páramos of Santander and Norte de Santander-Táchira; the second area, páramos of Boyacá; and the third area, páramos of Cundinamarca. Of the species growing in each major area, some may occur more or less throughout the area, but most are restricted to a smaller ecological or topographic range inside each area. The same centers of speciation have already been considered for other groups (e.g., *Espeletiopsis*).

The first center of radiation is Santander and Norte de Santander-Táchira. This comprises the páramos of Tamá and Cobre, Fontibón-Pamplona, Santurbán, Romeral, Berlin, Ocaña, and Almorzadero in a broad sense. These are more or less limited páramos, which were to varying degrees connected, at least in several past periods of time. They support a total of eight species. The primitive type of the genus (with large paniculate, tall inflorescences, and pseudopetiolate, sublanceolate leaf laminas) is represented by *E. steyermarkii* at the Venezuelan side of the Cordillera and *E. standleyana* in the lower part of Páramo de Santurbán-Berlin and eastern slopes of Páramo del Almorzadero. Reduction of inflorescences with suppression of sterile leaves was experienced by *E. brassicoidea* in formerly luxuriant communities of the Páramo de Fontibón (Pamplona), now near extinction. Further reduction of inflorescences, in total length and number of capitula, leads to *E. dugandii* (limited to a section of Páramo del Almorzadero), *E. canascens* (top, 4200 m, of Páramo del Romeral), and *E. conglomerata* extending from Almorzadero to the páramos of Santurbán, Vegas, and Romeral, and to the heights of Páramo de Tamá. Extreme specialization was reached in this area by *E. estanislana*, an hygrophilous species with single large heads, found in small reduced areas in Páramos del Almorzadero and Berlin. Related to this is *E. roberti*, an isolated relict at the cacuminal páramo of Jurisdicciones (Ocaña). A subcenter of speciation can be considered the Sierra de Perijá. at the northern end of the Cordillera with two related species, *E. tilletii* and *E. perijaensis*.

The second center of radiation is Boyacá. Three subcenters of speciation may be considered:

1. Sierra Nevada del Cocuy or Guicán: In this area there are four species, all tall caulirosulas, two belonging to the primitive mainstream of the genus, *E. curialensis* and *E. cleefii*, and the other two, *E. annemariana* and *E. lopezii*, representing a parallel line of evolution with aphyllous inflorescences. *E. lopezii* forms the largest communities in the area with a broad range of altitudes. *E. curialensis* and *E. annemariana* grow in median and low páramo and subpáramo zones, spreading along the eastern front of the Cordillera; *E. lopezii* and *E. cleefii* dominate the upper parts of the Cocuy Massif ascending well to the rocky superpáramos. *E. lopezii*, with short naked inflorescences of three large heads, forms extensive communities in the páramo proper of the western drainage, in addition to its aggressive peregrination to the south along the páramos of the eastern slopes, e.g., Páramos de Socha and Pisba (Fig. 11–20).
2. Páramos of Guantiva-La Rusia, Alto Las Cruces (between Duitama and Encino and Belén-Susacón-Gonzaga, with a maximum altitude of 4200 m), a great massif with extensive páramos or valleys between 3000 and 4000 m: In this region a large variety of caulirosuletas cover large flat areas or steep slopes. In these landscapes *Espeletia* species form monospecific consociations. It is the richest existing center of *Espeletia*, with 13 living species (besides varieties and subspecies). The mainstream of the genus is represented here by a variant of *E. grandiflora* (ssp. *boyacana*) in small refuges at the heights of Páramos de La Rusia and Las Cruces. Other species with leafy inflorescences are *E. azucarina* and *E. congestiflora*; this last, with extremely thick, compact, curled lanate, yellowish indumentum, and few capitula compactly glomerate in a compound head, is a very specialized form suited to less humid habitats of well-drained slopes. Six other species belong to the parallel line of evolution with aphyllous inflorescences, with decreased numbers of capitula (*E. murilloi*, *E. nemekenei*, *E. incana*, *E. brachyaxiantha*, *E. arbelaezi*, and *E. discoidea*). *E. nemekenei* and *E. murilloi* have 5–11(–17) capitula on very exserted inflorescences; the other species, living at higher altitudes, have 3–5 capitula, mostly 3, the inflorescence shorter than or equaling the leaves. In addition, *E. rositae* usually has 5 large capitula, but a variety has developed with a single head. Another species with well-developed inflorescences, *E. boyacensis*, represents a distinct line of evolution characterized by prostrate sericeous indument (sect. *Bonplandia*), in addition to very large, long inflorescences with smaller heads.
3. An irregular series of páramos at the eastern side of the Cordillera separated from the Guantiva region by the Chicamocha Valley: This is the region of Páramos de Socha and Pisba, de Vadohondo in the valley of the Rio

Cusiana, Páramo del Arnical, Páramo de la Sarna near Laguna de Tota, and hills around Sogamoso. In this region are three endemic species of the primitive type, *E. jaramilloi*, *E. tunjana*, and *E. oswaldiana*, and varieties of other species, such as *E. annemariana* var. *rupestris*. In addition, *E. lopezii*, *E. boyacensis*, and *E. murilloi* are present in abundance. Also present is *E. congestiflora*. *E. tunjana* is a remarkable species because of its large, branched inflorescences.

In short, 17 species (plus a subspecies from the next area) are known from the Boyacá center.

The third center of radiation is Cundinamarca. The main subcenter is the Bogotá-Sumapaz Massif in a series of páramos in a large, broken area. A secondary subcenter is the Zipaquirá-Neusa subcordillera, west of the Sabana de Bogotá.

1. The Zipaquirá-Neusa area: Here three endemic species are considered: *E. chocontana* and *E. cayetana*, belonging to the foliate group in the same phyletic line as *E. grandiflora*, and *E. barclayana* of the aphyllous inflorescence series. *E. grandiflora* is also present as it is all around the Sabana de Bogotá as well as *E. argentea*; but these two species have their origin in the large Bogotá Massif at the eastern side of the Sabana.
2. Bogotá Massif-Páramo de Sumapaz: This is a large massif reaching altitudes above 3800 m at many points and also with many flat or moderately inclined surfaces, various degrees of humidity from deep swampy to drained areas, alternating with rocky spots, crested ridges, and abrupt broken cliffs with green grassy and bushy areas. Forests, especially on the east-facing slopes, may cover the terrain up to a very high altitude, 3500 m and more, where it has not been destroyed as happens on steep rocky ridges.

Eight species are found in the area, the most dominant of which, *E. grandiflora*, is found throughout the whole area from the top of some hills and grasslands of Páramo de Sumapaz at 4000 m altitude, down to near the Sabana de Bogotá sporadically at 2700 m in open spots. The species develops best above 3000 m, forming extensive gregarious caulirosulan communities in grasslands. This species is aggressive in occupying new open areas and can be used as an example of how *Espeletia* has managed to dominate the páramos of the whole Cordillera. Its local variations and several local varieties or subspecies also indicate the plasticity of the genus. In this context, it is worth mentioning that *E. grandiflora* retains the primitive characteristics of pseudopetiolate lanceolate leaves and large paniculate inflorescences with leafy vegetative portions. Other species of the same phyletic line are *E. miradorensis* and *E. tapirophila* forming communities in restricted areas on the upper Páramo de Sumapaz. *E. uribei* is an important species growing in the upper zone of the Andean forest and in subpáramo bushes. It is found in some open areas as a remnant of damaged forests. The species has the primitive characteristics given for the genus and in addition can develop a high stem up to 14 m tall under forest protection, and a trunk of 16-cm diameter with 10-mm-thick corky bark and a thick cylinder of wood. All this indicates that this biotype may represent one of the most primitive type of *Espeletia*. Another species close to it with similar morphological characteristics and ecological preferences is *E. curialensis* of the eastern slopes of the Cordillera in Boyacá.

This study would support the idea that the genus originated in the more humid, cloudy environment of the upper woody belt of the eastern slopes of the Cordillera. Reduction has occurred with the migration to the open páramos. Two other important species of the region are of the aphyllous-inflorescence stock: *E. killipii*, abundant in the Guasca-Bogotá area, preferably in eastern-drainage, grassy, wet páramos, and its close relative, *E. summapacis*, a showy species with snow-white inflorescences consistently with three large heads and an endemic of high areas (3900–4000 m) in Páramo de Sumapaz. *E. cabrerensis* from the southwestern slopes of Páramo de Sumapaz is a multicapitulate species with aphyllous inflorescences, probably independently derived from a foliate-type species in the same Páramo de Sumapaz. A distinctive species of the Bogotá area in Cundinamarca, *E. argentea* (with an eradiate form), is widespread in páramos below 3500 m altitude. It has flattened, silvery, sericeous indumentum and belongs to an independent stock (sect. *Bonplandia*) that produced the already mentioned *E. boyacensis* in the Boyacá region.

Central Cordillera: The Central Cordillera of Colombia that departs from the so-called Troncal Sur mountain range, which continues southward into Ecuador, diverges from the Cordillera Oriental in the region of Páramo de las Papas or "Macizo Colombiano," somewhat below 2° N. It is a huge Cordillera, 420 km long and above 3000 m in altitude which includes several of the highest peaks of Colombia (the Quindío group of volcanoes: Tolima, 5215m; Ruiz, 5400 m; Santa Isabel, 5100 m; Quindío, 5150 m; Nevado del Huíla, 5750 m; Puracé, 4646 m). The Cordillera formerly was covered by dense forests on both slopes and a series of almost uninterrupted páramos above 3500 m from the southern Páramo de las Papas to the Quindío Massif at about 5° 10′ N. Smaller páramos are also scattered at the northern lower extension of the Cordillera up to near 7° N. From the Páramo de las Papas southward, the páramos continue along the Troncal Cordillera and its lateral extensions and across the volcanic region of Pasto, Cumbal, Tulcán, and Chiles of southern Colombia and northern Ecuador south to the Páramos del Angel to about 0°30′ N (Fig. 11–19).

Only one species, *E. hartwegiana*, is found in the great extension of páramo surfaces along the Cordillera Central from Quindío to the Macizo Colombiano (Fig. 11–19,G). In the Páramo de las Papas, another species, *E. idroboi*, forms small colonies, whereas *E. hartwegiana* is also found southward sporadically in Nariño (Fig. 11–19, H). Southward of the Páramo de las Papas, two other species are present: *E. schultesiana* around the Laguna de La Cocha, Cerro de Patascoy (Putumayo, Nariño), and *E. pycnophylla* (Fig. 11–19, I). *E. hartwegiana* and *E. schultesiana* both have inflorescences of the leafy type and rather sessile leaves; *E. idroboi* had pseudopetiolate sublanceolate leaves; *E. pycnophylla* has naked inflorescences and broad obovate-oblong sessile leaves. The dominant species are *E. hartwegiana* along the whole Cordillera Central and *E. pycnophylla* in the troncal series of páramos extending into Ecuador down to Páramos del Angel, and farther south still in the Llanganatis (or Llanganates). Another species occurs in páramo-subpáramo at the northern end of the Cordillera Central in Antioquia: *E. occidentalis* ssp. *antioquensis* (Fig. 11–19, K).

There is a marked contrast between the Colombian Cordilleras with respect to speciation in *Espeletia*: the Cordillera Oriental has 38 species, whereas the Cordillera Central has only 5 although its whole extension is similar (about 400 km). The contrast in species diversity is particularly noticeable when individual limited massifs are compared, for example, the Nevado del Cocuy-Chita, which has 4 species of *Espeletia*, whereas the Tolima-Ruiz Massif with as large or much larger páramo surfaces covered by luxuriant caulirosuleta of *Espeletia*, had only 1 species. Páramo de la Rusia-Guantiva has 10 species and Páramo de Sumapaz 7 species; in contrast, the Huila Massif in the Cordillera Central, as well as Puracé, has only 1 species. Thus, the fact that the Cordillera Oriental is wider with a larger surface is insufficient to explain the great contrast. What counts is the age of the area. The Cordillera Central is very old and the páramo environment and its vegetation are known to have been well developed and capable of supporting *Espeletia* since the Early and Middle Pliocene. But the Colombian *Espeletia* stock came from the border of the Venezuelan and Colombian Andes, as already explained, and in its migration along the Colombian Cordillera Oriental was halted at the depression south of Páramo de Sumapaz where the mountain range drops considerably with no páramos for 270 km until the Macizo Colombiano, where both Cordilleras meet (Fig. 11–16). It was only in one of the last glacial periods, certainly during the wet Late Glacial, ca. 14,000/12,000 B.P.–ca. 10,000 B.P. (van der Hammen, 1979a,b; van der Hammen et al., 1980) that the páramo belt was lowered to an extent that it covered the southern section of the Eastern Cordillera (Fig. 11–18). It was at this time that *Espeletia* was able to spread farther south than the Páramo de Sumapaz and reach the Cordillera Central at Páramo de las Papas in the Macizo Colombiano. One species of *Espeletia* close to *E. grandiflora* did reach that far. *E. idroboi* is found in limited colonies at the Macizo Colombiano, but two other related species having broader, sessile leaves with shaggier, thicker lanate indumentum developed: (1) *E. hartwegiana*, which moved mainly to the north, and with tremendous aggressiveness spread along the whole Central Cordillera, succeeding in dominating its páramos. The humid habitat and humiferous soil in the region facilitated rapid settlement of the invader; and (2) *E. schultesiana*, which spread southward to the Sibundoy region, Cerro Patascoy, Laguna de La Cocha, the upper headwaters of the Rio Guamues. The most important species in this southern region is, however, *E. pycnophylla*, which can be considered a descendant of one of the former species or of their ancestor, mainly by losing the leaves of the proximal part of the inflorescences. *E. pycnophylla* may develop very tall caulirosulas (up to 10 m) and build widespread communities extending from the Páramos de Quilinsayaco, del Bordoncillo, del Galeras, etc., southward in Troncal Sur, to the large volcanic region at the Colombia and Ecuador border, on widespread grasslands around the volcanoes of Cumbal, Chiles, and Tulcán. From here the páramos with *E. pycnophylla* go farther to the Páramos del Angel and southward to about 0°30′ N. At this point the *Espeletia* population stops (Fig. 11–19). Nevertheless, a long distance from here suddenly reappears a large community of *Espeletia* in the isolated Sacha-Llanganatis mountains between 2600 and 3600 m altitude, at 1° S in Ecuador. This range is the easternmost of the complex of three parallel mountain ranges called the Cordillera de los Llanganatis. The presence of *Espeletia* in the Sacha-Llanganatis, so far away from the aforementioned populations of the genus, and its apparent absence in the two other Llanganatis Cordilleras (Andrade Marín, 1940) are difficult to explain. It must be said that according to Andrade Marín, the easternmost of the three Llanganatis, in contrast to the other two, is extremely wet and is permanently cloudy or rainy or foggy; the ground is swampy with many lakes, the terrain broken and rocky, and the luxuriant vegetation covered with a great mass of mosses, liverworts, and ferns. An explanation that could be given for the existence of this isolated *Espeletia* population is some sort of long-distance dispersal. However, probably most *Espeletia* at the end of the Last Glacial spread much farther south of the present limits of the genus in Carchi Province. Moving through the continuous series of wet páramos of that time, *Espeletia* was able to reach the Llanganatis. Finding in those mountains its perfect habitat, *Espeletia* has remained firmly established up to the present. After glaciation, with the general lifting of the páramo belt, the Llanganatis páramos were isloated from those of the main Ecuadorian Cordillera. Furthermore, a period of intense dryness in northern Ecuador probably caused the elimination of former *Espeletia* populations situated farther south of the present limits of the genus in Carchi Province.

E. hartwegiana is the only species of the main bulk of the Cordillera Central. It seems that *Espeletia*, having invaded the Cordillera from the south and spreading toward the north, has not yet had the time for further speciation; also the Cordillera Central had a relatively uniform environment, particularly during the Holocene. Nevertheless, variations can be seen in populations in various parts of the area, variations that seem to be genetically fixed but not differentiated enough to warrant specific status. The population of the Ruiz-Tolima Massif has features consistent enough to be considered a subspecies (ssp. *centroandina*) as do the plants of Páramos de Barragán and Hermosas (ssp. *barragensis*); other variants are found in Páramo de Miraflores (var. *vegasana*) and in the Páramos de Santo Domingo, de Moras, and de Las Papas (var. *morarum*), the latter with the tendency to reduce the lower pair of leaves of the inflorescences. This trend terminates in another species (*E. pycnophylla*). These variations are signs of incipient speciation.

A similar situation occurs with the widespread population of *E. pycnophylla*, probably the most recent species of *Espeletia*. It shows some regional differentiations of infraspecific importance, such as var. *galerana* of the Volcano Galeras (Pasto), ssp. *angelensis*, spread mostly in northern Ecuador in Carchi Province at Páramos del Angel and bordering Colombia, in the volcanic region with volcanoes Cumbal, Chiles, Tulcán, Túquerres, and so on. Although it is only known from one collection, the Llanganatis population shows small differences but they are apparently sufficient to warrant subspecific recognition (ssp. *llanganatensis*) (Fig. 11–19).

The northern extreme of the Cordillera Central has rather poor representation of *Espeletia* populations. The species of the area, *E. occidentalis* ssp. *antioquensis*, is very different from *E. hartwegiana* and is limited to the lower páramos or subpáramos of that region of Antioquia. The origin of this population might be different from the main population of the Cordillera.

Western Cordillera. Only in the northern extremity of the Western Cordillera are there páramos with *Espeletia*. Three species have been found in the region: *E. occidentalis* ssp. *occidentalis*, *E. praefrontina*, and *E. frontinoensis*. One hypothesis about their origin is that they may have crossed over the Rio Cauca in Antioquia through a narrow part of the valley separating the Cordilleras. Here, past volcanic activities have formed bridges between the Cordilleras across the primitive river basin. The ancestor could have been some form of *E. hartwegiana*, giving rise in the new habitat to *E. occidentalis* and *E. praefrontina*; from this was derived *E. frontinoensis* (Fig. 11–19, L).

VENEZUELAN CORDILLERA

The other main branch of the first *Espeletia* stock originated in the lower páramo zone in the humid western section of the Venezuelan Cordillera. It took a northwestern direction following the main mountain range on both sides of the Chama Valley, reaching the upper parts of the Cordilleras of Mérida, La Culata, and Santo Domingo and continued northward across the "Trujillo Cordillera." The primary populations probably suffered from drastic climatic changes in the region and from too many oscillations of the altitudinal thermal belts. It seems clear that the periods of higher temperature eliminated *Espeletia* from lower mountain ranges as, for instance, the southern section of the Cordillera (Páramos del Batallón, La Negra, and Zumbador, etc.). The same forces reduced some species to the status of relicts facing extinction. In any case, it is certain that the continual environment changes and strong seasonal characteristics of the climate in most parts of the region favored marked taxonomic differenation. This is evident in the living Venezuelan species of *Espeletia*. The species closer to and more representative of the ancestor of the group is *E. schultzii*, ordinarily with large inflorescences, thickly white lanate leaves and a trunk usually no more than 1 m tall; it thrives best in a wet grassland environment at high altitudes (e.g., 3600 m). *E. schultzii* has great vitality with a tremendous capacity to accommodate to less than optimal conditions. As a result, its populations cover large expanses of páramo at high and middle, and even low (2600 m), altitudes, and readily colonize newly open territories such as abandoned fields and cleared forestland. The area covered by *E. schultzii* today is much larger than it would be without human interference (Fig. 11–19, B).

Of the other nine species growing in this Cordillera, two strongly diverged from the primitive genetic type by adopting a distinctive kind of indumentum. In *E. semiglobulata* the indumentum is white and superficially appressed. In addition, its heads are smaller, and the involucres and flowers also show marked differences. *E. aristeguietana* has an indumentum unique in the genus, tomentose-velutinous, surely marking a distinctive genetic line. *E. semiglobulata* grows in strictly wet edaphic conditions at high altitudes (Sierra Nevada-Culata Massif); *E. aristeguietana* is only known from Trujillo in Cerros de La Cristalina (2600–2800 m), possibly a relict of a former larger area. The other seven species are closely related to *E. schultzii*, sharing with it the same kind of involucre made up of narrow, oblong, acute phyllaries. They must have been derived from this species or from an ancestor having an hirsute receptacle. These seven species all grow in humid true páramos or in superpáramos, most of them (five) in Páramos de Apartaderos-Piñango-Timotes-Santo Domingo, and in the neighboring region of Trujillo. *E. cuniculorum*, only in Páramo de Los Conejos, has, like the original type of *E. schultzii*, rather perfect dichasial inflorescences but their foliar structure and the consistently hirsute receptacle make a divergent endemic type. The tendency for reduction of the number of elements, and for asymmetry of the inflorescence, already strongly developed in *E. schultzii*, has led to a group of species of smaller size, with a subterranean, often tuberose, caudex, with very few or single heads in the inflorescences. The center of radiation of this group is probably the páramo regions of Tuñame-Guirigay-Niquitao. The species spread to Páramo de Los Granates-Santo Domingo and to Apartaderos, and also to the Cordillera de Trujillo northward to Páramos de Cendé-Jabón. Today *E. ulotricha* is a relict of once more widespread populations, now restricted to the summits of Páramos de Cendé-Jabón-Las Rosas (Fig. 11–19, A). The group of smaller plants with tuberose corms is concentrated in páramos of southern Trujillo and Mérida State, being endemic to small páramo areas and adapted to very wet ground (e.g., *E. tenore*, *E. marthae*); some may be more broadly distributed but are ecologically limited to marshy or less wet areas, e.g., *E. batata*, *E. nana*, and *E. weddellii*. The center of speciation of *Espeletia* in the Venezuelan Cordillera probably was the region of Apartaderos, which is continuous with the páramos of Piñango and Timotes. A subcenter was the southern Trujillo region of Páramos del Guirigay-Llano Corredor, extending to the border of Trujillo-Mérida.

Coespeletia

The genus *Coespeletia* is an earlier stock close to *Espeletia*, also with polycarpic caulirosulan biotype, but diverging by a monochasial type of inflorescence. It spread throughout the main Mérida Andes and reached their summits, adapting to the extreme minimal thermal conditions combined with the marked seasonal climate of these páramo and superpáramo areas. The primitive species of *Coespeletia*, probably growing at median altitudes (3100–3500 m), had paniculate-racemiform inflorescences, that is, determinate racemes with shortly branched peduncles (e.g., *C. thyrsiformis*). Through emigration and settlement at high altitudes (4000–45000 m), a strict, simple-racemose type of inflorescence was, however, firmly established. Further evolutionary steps involved the reduction of the number of heads in each inflorescence combined with head-size increase (*C. spicata* to *C. timotensis*). A final step resulted in the monocephalous type of *C. moritziana*, which has the largest capitula in the genus. This is also the species best equipped to thrive in the highest altitudes of the Venezuelan superpáramo (desert páramo in Chap. 3), which, besides being extremely cold, is a rocky, sandy, or gravelly, arid, and exposed habitat, with frequent or regular diurnal freeze-thaw cycles of the soil, a characteristic of the periglacial zone (Fig. 11–21).

The heights of the orographic nucleus of the Sierra Nevada de Mérida sensu lato, and the intermediate zone of Apartaderos between the Sierra Nevada of Mérida, the Sierra of Santo Domingo and the Sierra de La Culata (del Norte), were the center of speciation of *Coespeletia*, One of the more primitive species, *C. elongata*, spread along the Sierra del Norte, and later

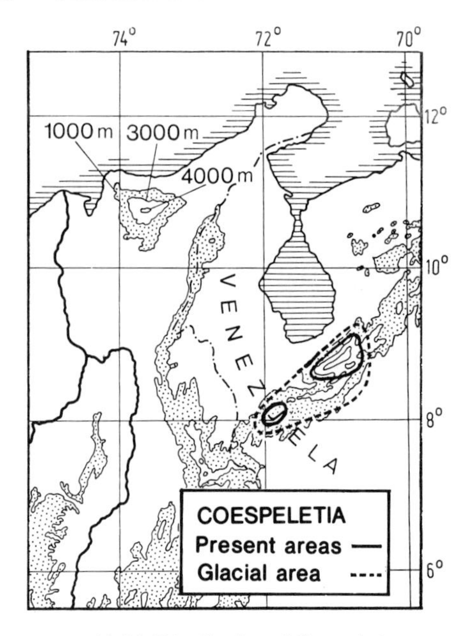

FIG. 11–21. Distribution of *Coespeletia*.

gave rise to *C. spicata*, *C. timotensis* (with forma *lutescens*), and *C. moritziana*, which migrated in all directions horizontally and upward to the mountain slopes, settling locally according to ecological preferences. Today, they dominate the plant communities and the landscape of the largest part of the superpáramo zone (or desert páramo) of the central massif of the Mérida Cordilleras. This zone in great part coincides with the periglacial region (Schubert, 1975, 1979). The area was described by Monasterio (1979, Chap. 3), as "Páramo Desértico," because of the scarcity of other plants in association with showy *frailejones*. Nevertheless, throughout the periglacial zone a natural landscape of caulirosuletum of variable density is widespread, consisting of a few species of *Coespeletia*, mainly *C. timotensis*, which is extremely impressive and unique. In spite of the great difficulties solifluction poses for seedling survival (the mortality rate may reach 100%), the species have managed not only to survive but to dominate large areas of this hostile habitat. Well-based estimates (A. P. Smith, 1978) indicate that the adult individuals of *C. timotensis* usually surpass 100 years of age. The caulirosulas of *Coespeletia* are best equipped morphologically to grow in this environment. The massive, columnar trunk has a cover of marcescent, exceedingly tight leaves, incomparably thicker, more compact, and insulating than that of any Espeletiinae of the wetter Colombian páramos. The considerable thickness of these giants of the páramo flora gives the impression of stability and majesty. Considering the complexity of its morphology, one may conclude that the reproductive and morphological strategies developed by *Coespeletia* represent the major success in the evolution of the Espeletiinae (see Chap. 3).

COMPARATIVE DISTRIBUTION OF *ESPELETIA* WITH OTHER TAXONOMIC GROUPS

The pattern of distribution of *Espeletia*, showing an overwhelming number of 38 species in the Eastern Cordillera, against only 4 spp. in the Central and 3 in the northern end of the Western Cordillera (see Table 11–1), compares well with a few other taxonomic groups, e.g., *Puya* (Bromeliaceae) (Fig. 11–22). This genus has 8 species in the Eastern Cordillera páramos, only 2 species in the Central Cordillera, and 1 species in the southern section of the Western Cordillera. Although the total number of species is lower than that of *Espeletia*, the pattern of distribution is parallel. But *Puya* has in addition 4 species in the Sierra Nevada de Santa Marta, where *Espeletia* is lacking, and 1 species (*P. occidentalis*) in the Farallones de Cali, a highland páramo without *Espeletia*. Besides, species of *Puya* are also present in the páramos of Costa Rica and in those of Ecuador and northern Peru, where *Espeletia* is lacking (Cuatrecasas, 1979b; Smith and Downs, 1974). In predicting the number of species in páramo islands and in general in high Andean "islands," some workers consider the size of the islands and the distances separating them to be most important variables (Simpson Vuilleumier, 1971; Simpson, 1975; Vuilleumier and Ewert, 1978). In order to understand patterns of speciation, radiation, and distribution, it is important also to consider the kind of morphological devices the propagules have to facilitate their distribution. The absolute lack of any such devices in *Espeletia* restricts its distribution to the slow path of progressive local expansion of each gregarious population. The control on means of propagation limits the area of interbreeding populations, causing more frequent speciation. *Espeletia* populations, separated by ecological or physical barriers, even small ones such as a band or girdle of forest, or a rocky crest, may remain forever isolated without opportunities for interbreeding; the adjacent populations will continue to differentiate indefinitely. Conversely, the seeds of *Puya* are light and have a membranous marginal wing that facilitates dispersal; some species, furthermore, e.g., *P. goudotiana*, may have a long, thick, spiciform, multiflorous panicle topping a robust axis 2–6 m high, which releases thousands

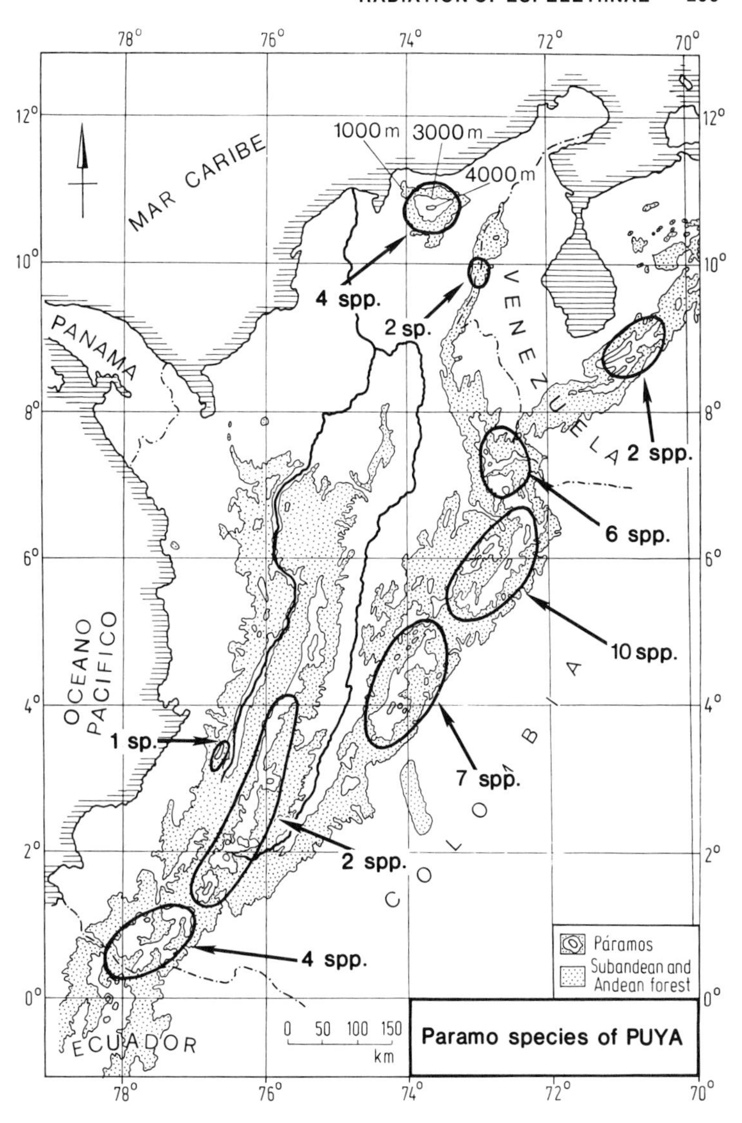

FIG. 11–22. Distribution of páramo species of *Puya* with number of species per páramo block. Compare with Table 11–1 for figures pertaining to *Espeletia*.

of winged seeds that are easily carried by the winds from one páramo area to another. Possibly for this reason, *P. goudotiana*, with its tall inflorescences, is the most widespread species in the páramos of the Colombian Eastern Cordillera, from Bogotá to Norte de Santander. The tall *P. aristeguietae* also has a wide range, from southern Trujillo to Mérida and Táchira, to the Sierra de Perijá and the Nevado del Cocuy. Similarly *P. hamata*, 2–4 m tall, has spread from the Central Cordillera of Colombia, southward through Ecuador as far as Cajamarca, Peru. The Cordillera Oriental is obviously a center of speciation for *Puya*, although most of the species have a larger range than do individual species of *Espeletia*. For example, the Cundinamarca area has 11 species of *Espeletia*, most of them with a restricted subarea, whereas its 7 species of *Puya* are more widespread and at least 5 of them extend farther through the Boyacá area and 1 to Norte de Santander. Probably from this center of speciation (Cordillera Oriental), *Puya* radiated to the Mérida Cordillera, to the Serranía de Valledupar-Perijá, to the Sierra Nevada de Santa Marta, to the Western Cordillera of Colombia and to the páramos of Costa Rica, either by progressive terrestrial propagation or by long-distance transportation, or both.

An interesting comparison can be made between the distribution of *Espeletia* and *Puya*, and that of *Diplostephium* species in Colombia. *Displostephium* is a genuine Neotropical Andean genus of Compositae whose maximum development has been in Colombia (57 species) and Ecuador-Peru (50 species). The adaptation to wind dispersal of the *Diplostephium* propagules, which are light achenes with a pappus, has facilitated interpáramo diffusion and intercordilleran transportation. *Diplostephium* has thus not been exclusively dependent on the lowering of temperature and vegetation belts to expand its populations. Long-distance wind transport is a reasonable explanation for the presence of *Diplostephium* in the Sierra Nevada de Santa Marta, Cerros de Tatamá, and Farallones de Cali (Cordillera Occidental) and the Costa Rican páramos, all lacking *Espeletia*. It also explains the great profusion of species of *Diplostephium* in the Central Cordillera, in contrast to the limited species of the non-wind-distributed *Espeletia* in this area.

The main center of *Diplostephium* speciation in the northern Andes is also the Colombian Eastern Cordillera with 31 species, 25 of them endemic to the area. The Sierra Nevada de Santa Marta follows as a subcenter of speciation with 12 species, of which 11 are endemic to the "island." The Central Cordillera has 17 species with 9 endemic to it or extended to its continuation, the Troncal Sur (8 species, including 1 additional endemic). The Western Cordillera has 6 species with 3 endemics. Only 2 species are found in the main Venezuelan Cordilleras (Mérida and Trujillo); the other 6 cited from Venezuela (including 3 endemics), orographically correspond to the Venezuelan eastern side of the main Colombian Eastern Cordillera. It indicates that *Diplostephium* radiated from the Colombian center of speciation, from the southwest to the northeast,

TABLE 11–2. Comparison of the number of species and endemics (in brackets) of *Espeletia* and *Diplostephium* found in comparable páramo sections of the two main Colombian Cordilleras.

Cordillera Oriental		Cordillera Central	
Espeletia	*Diplostephium*	*Espeletia*	*Diplostephium*
Páramos Almorzadero-Romeral		Macizo Ruiz-Tolima	
6(6)	11(5)	1(1)	6(1)
estanislana	*ellipticum*	*hartwegiana*	*eriophorum*
standleyana	*dentatum*	ssp. *centroandina*	*rupestre*
canescans	*oblongifolium*		*schultzii*
brassicoidea	*mutiscuanum*		*violaceum*
conglomerata	*tolimense*		*revolutum*
dugandii	*juajibioi*		*tolimense*
	lacunosum		
	revolutum		
	rosmarinifolium		
	glutinosum		
	apiculatum		
Nevado del Cocuy		Nevado del Huíla	
4(2)	7(0)	1(1)	7(2)
lopezii	*colombianum*	*hartwegiana*	*glandulosum*
cleefii	*lacunosum*	var. *morarum*	*ritterbushii*
curialensis	*revolutum*		*schultzii* v. *lehmanianum*
annemariana	*alveolatum*		*cinerascens* v. *puracense*
	rhomboidale		*pittieri*
	tenuifolium		*floribundum*
	huertasii		*bicolor*
Macizo Bogotá-Sumapaz		Páramos del Puracé	
11(10)	10(3)	1(1)	9(0)
grandiflora	*ochraceum*	*hartwegiana*	*rupestre*
miradorensis	*rupestre*	ssp. *hartwegiana*	*hartwegii*
tapirophila	*rhomboidale* f.		*spinulosum*
chocontana	*revolutum*		*glandulosum*
cayetana	*alveolatum*		*schultzii* v. *lehmanianum*
barclayana	*phylicoides*		*cinerascens* v. *puracense*
killipii	*fosbergii*		*pittieri*
summapacis	*huertasii*		*floribundum* v. *putumayense*
cabrerensis	*juajibioi* ssp.		*bicolor*
uribei	*heterophyllum*		
argentea			

but stopped abruptly at the eastern border of the Cordillera Oriental. This suggests that the strong northeastern winds blowing over the Venezuelan llanos and against the Andes are mainly responsible for the limitation of the *Diplostephium* migration into Venezuela. Conversely, the Espeletiinae with their heavy, bare achenes, and their slow progression on the ground, would have no difficulty in migrating in any direction with respect to winds.

The capability for plurispecific populations of *Diplostephium* in the Colombian Cordillera Central is shown in a comparison of single páramo areas in both Cordilleras (Table 11–2).

As Table 11–2 indicates, single páramo nuclei in the Cordillera Central, populated by a single species of *Espeletia*, may have as many species of *Diplostephium* as comparable páramos of the Cordillera Oriental. The Cordillera Central páramo groups of Ruiz-Tolima, Huila and Puracé (all with only *E. hartwegiana*) have, respectively, 6, 7, and 9 species of *Diplostephium*. This compares with the páramo groups of Almorzadero-Romeral, Cocuy and Bogotá-Sumapaz, having, respectively, 6, 4, and 11 *Espeletia* species and 11, 7, and 10 *Diplostephium* species. Note the higher number of local endemic species of *Espeletia* in the Eastern Cordillera and the lower number of local *Diplostephium* endemics in both Cordilleras; the easier diffusion ability of the pappus-bearing *Diplostephium* propagules explains the difference.

The effects of an insular situation on differentiation and speciation of *Diplostephium* are extremely visible in the Sierra Nevada de Santa Marta (Table 11–1) where the genus is represented by 11 endemic, much-diversified species, in addition to one (*D. rosmarinifolium*) rather common in the Eastern Cordillera. The genus must be very ancient on that mountain, but the date of origin cannot be obtained because of the difficulties in identifying Compositae genera from pollen depositions.

ACKNOWLEDGMENTS

The continued support of Dr. Dieter Wasshausen, Chairman of the Department of Botany, is much appreciated, as is the collaboration of Miss Alice Tangerini, who prepared most of the illustrations. I am grateful to Drs. Harold Robinson and Richard Cowan for their assistance in the final editing of the manuscript. Appreciation is also expressed to Drs. T. van der Hammen, Leo J. Hickey, and A. M. Cleef for valuable information on the geologic history and ecology related to the subject.

REFERENCES

Aristeguieta, L. 1964. Compositae. In *Flora de Venezuela*, Vol. 10, pt. 1, pp. 407–461. Caracas: Instituto Botánico.

Azocar, A., and M. Monasterio. 1979. Variabilidad ambiental en el Páramo de Mucubají. In *El Medio Ambiente Páramo*, M. L. Salgado-Labouriau, ed., pp. 119–159. Caracas: Ediciones Centro Estudios Avanzados.

Bentham, G. 1873. Compositae. In *Genera Plantarum*, G. Bentham and J. Hooker, eds., Vol. 2, no. 1, pp. 163–533. London: Lovell Reeve, Williams and Norgate.

Carlquist, S. 1966. Wood anatomy of Compositae: A summary, with comments on factors controlling wood evolution. *Aliso* 6: 25–44.

———. 1974. *Island biology*. New York: Columbia University. Press.

———. 1976. Tribal interrelationships and phylogeny of the Asteraceae. *Aliso* 8: 465–492.

Cleef, A. N. 1978. Characteristics of neotropical páramo vegetation and its subantarctic relations. In *Geo-ecological relations between the southern temperate zone and the tropical mountains*, C. Troll and W. Lauer, eds., Erdwissenschaftliche Forschung, Vol. 11, pp. 365–390. Wiesbaden: Franz Steiner Verlag.

———. 1979. The phytogeographical position of the neotropical vascular paramo flora—with special reference to the Colombian Cordillera Oriental. In *Tropical botany*, eds., K. Larsen and L. Holm-Nielsen, pp. 175–184. London: Academic Press.

———. 1981. *The vegetation of the páramos of the Colombian Cordillera Oriental*. Dissertationes Botanicae 61. Vaduz: Cramer.

Cronquist, A. 1955. Phylogeny and taxonomy of the Compositae. *Amer. Midl. Nat.* 53: 478–511.

———. 1977. The Compositae revisited. *Brittonia* 29: 137–153.

Cuatrecasas, J. 1934. Observaciones geobotánicas en Colombia. *Trab. Mus. Nac. Cienc. Nat., Ser. Bot.* 27: 1–144.

———. 1948. New mural shows plant life of Colombia's high Andes. *Bull. Chicago Nat. Hist. Mus.* 19: 1–3.

———. 1949a. Les especies del género *Espeletia*. *Bull. Inst. Catal. Hist. Nat.* 37: 30–41.

———. 1949b. Rosette trees, a tropical growth form. *Bull. Chicago Nat. Hist. Mus.* 20: 6–7.

———. 1954a. Outline of vegetation types in Colombia. *Rapports Comm. 8ème Congr. Int. Bot. Sect.* VII: 77–78.

———. 1954b. Distribution of the genus *Espeletia*. *Rapports Comm. 8ème Congr. Internat. Bot. Sect.* IV: 131–132.

———. 1957. A sketch of the vegetation of the North Andean Province. *Proc. 8th Pacific Sci. Congress, Botany* 4: 167–173.

———. 1958. Aspectos de la vegetación natural de Colombia. *Rev. Acad. Colomb. Cienc.* 10: 221–268.

———. 1968. Páramo vegetation and its life forms. *Colloquium Geographicum* 9: 163–186.

———. 1976. A new subtribe in the Heliantheae (Compositae): Espeletiinae. *Phytologia* 35: 43–61.

———. 1979a. Growth forms of the Espeletiinae and their correlation to vegetation types. In *Tropical botany*, K. Larsen and L. Holm-Nielsen, eds. pp. 397–410. London: Academic Press.

———. 1979b. Comparación fitogeográfica de páramos entre varias cordilleras. In *El Medio Ambiente Páramo*, M. Salgado-Labouriau, ed., pp. 89–99. Caracas: Ediciones Centro de Estudios Avanzados.

———. 1980. La inflorescencia en la taxonomía de las Espeletiinae (Heliantheae, Compositae). *Memorias del 6° Congreso Venezolano de Botánica*, pp. 191–194. Maracay: Universidad Central de Venezuela, Facultad de Agronomía.

Germeraad, J. H., C. A. Hopping, and J. Muller. 1968. Palynology of Tertiary sediments from tropical areas. *Rev. Palaebot. Palynol.* 6: 189–348.

Hallé, F., and R. A. A. Oldeman. 1970. *Essai sur l'architecture et la dynamique de croissance des arbres tropicaux.* Paris: Masson.

Hedberg, O. 1964. Features of afroalpine plant ecology. *Acta Phytogeog. Suecica* 49: 1–144.

———. 1969. Evolution and speciation in a tropical high mountain flora. *Biol. J. Linn. Soc.* 1: 135–148.

———. 1973. Adaptive evolution in a tropical-alpine environment. In *Taxonomy and ecology*, V. N. Heywood, ed., pp. 71–92. London: Academic Press.

Hedberg, O., and I. Hedberg. 1979. Tropical-alpine lifeforms of vascular plants. *Oikos* 33: 297–307.

Hoffmann, O. 1890–1894. Compositae. In *Die natürlichen Pflanzenfamilien*, ed. A. Engler and K. Prantl, eds., Vol. 4, pt. 5, pp. 87–394. Leipzig: Engelman.

Keck, D. D. 1936. The Hawaiian silverswords. Systematics, affinities and phytogeographic problems of the genus *Argyroxiphium*. *Bernice P. Bishop Mus., Occas. Papers* 11: 1–38.

Kempt, E. M., and K. H. Wayne. 1975. The vegetation of Tertiary islands on the Ninetyeast Ridge. *Nature* 258: 303–307.

Lauer, W. 1979a. La posición de los páramos en la estructura del paisaje de los Andes tropicales. In *El Medio Ambiente Páramo*, M. L. Salgado-Labouriau, ed., pp. 29–45. Caracas: Ediciones Centro Estudios Avanzados.

———. 1979b. Die hypsometrische Asymmetrie der Páramo—Höhenstufe in der nördlichen Anden. *Innsbrucken Geog. Stud.* 5: 115–130.

Mabberley, D. J. 1973. Evolution in the giant groundsels. *Kew Bull.* 28: 61–96.

Monasterio, M. 1979. El Páramo Desértico en el altiandino de Venezuela. In *El Medio Ambiente Páramo*, M. L. Salgado-Labouriau, ed., pp. 117–146. Caracas: Ediciones Centro Estudios Avanzados.

———. 1980. (ed.). *Estudios ecológicos en los Páramos Andinos*. Mérida: Ediciones Universidad de los Andes.

Muller, J. 1970. Palynological evidence on early differentiation of angiosperms. *Biol. Rev.* 45: 417–450.

Raven, P. H., and D. N. Axelrod. 1974. Angiosperm biogeography and past continental movements. *Ann. Miss. Bot. Gard.* 61: 539–673.

Robinson, H. 1981. Revision of the Tribal and Subtribal limits of the Heliantheae (Asteraceae). *Smithsonian Contr. Bot.* No. 51.

Rock, B. N. 1972. Vegetative anatomy of *Espeletia* (Compositae). Ph.D. thesis, University of Maryland.

———. 1981. Apical stem growth in arborescent members of the Espeletiinae (Compositae). Progress Report.

Salgado-Labouriau, M. L. 1979. Modern pollen deposition in the Venezuelan Andes. *Grana* 18: 53–68.

Salgado-Labouriau, M. L., C. Schubert, and S. Valastro, Jr. 1977. Paleoecologic analysis of a Late-Quaternary terrace from Mucubají, Venezuelan Andes. *J. Biogeog.* 4: 313–325.

Schubert, C. 1974. Late Pleistocene Mérida Glaciation Venezuelan Andes. *Boreas* 3: 147–152.

———. 1975. Glaciation and periglacial morphology in the northwestern Venezuelan Andes. *Eiszeitalter u. Gegenwart* 26: 196–211.

———. 1979. La zona del Páramo: Morfología glacial y periglacial de los Andes de Venezuela. In *El Medio Ambiente Páramo*, M. L. Salgado-Labouriau, ed., pp. 11–27. Caracas: Ediciones Centro Estudios Avanzados.

Simpson Vuilleumier, B. 1971. Pleistocene changes in the fauna and flora of South America. *Science* 173: 771–780.

Simpson, B. 1975. Pleistocene changes in the flora of the tropical Andes. *Paleobiology* 1: 273–294.

Smith, A. C., and M. Koch. The genus *Espeletia*: A study in phylogenetic taxonomy. *Brittonia* 1: 479–530.

Smith, A. P. 1974. Bud temperature in relation to nyctinastic leaf movement in an Andean giant rosette plant. *Biotropica* 6: 263–266.

———. 1975. Population dynamics and life form of *Espeletia* in the Venezuelan Andes. Ph.D. Thesis, Duke University.

Smith, L. B., and R. J. Downs. 1974. Pitcairnoideae (Bromeliaceae). In *Flora Neotropica*, ed. Org. f. Flora Neotropica, Monograph 14. New York: Hafner Press.

Stebbins, G. L. 1977. Development of comparative anatomy of the Compositae. In *The biology and chemistry of the Compositae*, V. H. Heywood, J. B. Harborne and B. L. Turner, eds., Vol. 1. pp. 91–109. London: Academic Press.

Stuessy, T. F. 1973. A systematic review of the subtribe Melampodiinae (Compositae, Heliantheae). *Contr. Gray Herb.* 203: 65–80.

———. 1977. Heliantheae-systematic review. In *The biology and chemistry of the Compositae*, V. H. Heywood, J. B. Harborne, and B. L. Turner, eds., Vol. 2, pp. 621–671. London: Academic Press.

Sturm, H. 1978. *Zur Oekologie der Andinen Paramoregion*. Biogeographica No. 14. The Hague: Junk.

Troll, C. 1958a. *Zur Physiognomik der Tropengewächse.* Jahresbericht der Gesellschaft von Freunden und Förderer der Rheinischen Friedrich-Wilhelms-Universität Bonn. Bonn.

———. 1958b. *Structure soils, solifluction and frost climates of the earth.* U.S. Army Snow Ice and Permafrost Establ. Translation 143. (Translation of Troll, C. 1944. Strukturböden, Solifluktion und Frostklimate der Erde. *Geol. Rundschau* 34: 546–694.)

Troll, W. 1957. *Praktische Einführung in die Pflanzenmorphologie*, Vol. 2. Jena: Fischer Verlag.

———. 1964–69. *Die Infloreszenzen*. Jena: Fischer Verlag. Vol. 1, 1964; Vol. II part 1, 1969.

Turner, B. L. 1977. Fossil history and geography. In *The biology and chemistry of the Compositae*, V. H. Heywood, J. B. Harborne, and B. L. Turner, eds., Vol. 1, pp. 21–39. London: Academic Press.

van der Hammen, T. 1974. The Pleistocene changes of vegetation and climate in tropical South America. *J. Biogeog.* 1: 3–26.

———. 1979a. History of flora, vegetation and climate in the Colombian Cordillera Oriental during the last five million years. In *Tropical botany*, K. Larsen and L. B. Holm-Nielsen, eds., pp. 25–32. London: Academic Press.

———. 1979b. Historia y tolerancia de ecosistemas parameros. In *El Medio Ambiente Páramo*, L. M. Salgado-Labouriau, ed., pp. 55–66. Caracas: Ediciones Centro Estudios Avanzados.

van der Hammen, T., J. Barelds, H. De Jong, and A. A. De Veer. 1980. Glacial sequence and environmental history in the Sierra Nevada del Cocuy (Colombia). *Palaeogeog. Palaeoclimat. Palaeoecol.* 32:

van der Hammen, T., J. H. Werner, and H. van Dommelen. 1973. Palynological record of the upheaval of the northern Andes. *Rev. Paleobot. Palynol.* 16: 1–122.

Vareschi, V. 1970. *Flora de los Páramos de Venezuela*. Mérida, Venezuela: Universidad de los Andes.

Vuilleumier, F., and D. N. Ewert. 1978. The distribution of birds in Venezuelan páramos. *Bull. Am. Mus. Nat. Hist.* 162; 47–90.

Weber, H. 1956. Histogenetische Untersuchungen am Sprossscheitel von Espeletia mit einem Überblick über das Scheitelwachstum überhaupt. *Abb. Akad. Wiss. u. Lit. Mainz. Mathem. naturm. K. 1956*: 566–618.

12

Speciation and Specialization of Polylepis *in the Andes*

BERYL B. SIMPSON

Although the Rosaceae is predominantly a family of the north temperate zone that is primarily known for the numerous economically important species it contains, it also includes several taxa that have austral centers of development. In particular, the Sanguisorbeae tribe of the Rosoideae contains several genera that have their greatest development in the Southern Hemisphere. Included in this group are the closely related genera, *Acaena*, *Margyricarpus*, and *Polylepis*. The first of these genera has species in South America, Africa, Tristan da Cunha, New Zealand, Tasmania, and Australia as well as a few that range into North America. This wide-ranging distributional pattern seems to represent the remnants of a former, more continuous range in the Southern Hemisphere. The other two genera are restricted to the montane areas of western South America. Of the two, *Polylepis* species are particularly conspicuous elements of some high tropical elevation habitats because they can be the only trees in areas usually dominated by low grasses, herbs, and shrubs (Fig. 12–1D). Particularly noticeable in this regard are *P. besseri* in southern Peru at 3800 m, and *P. tarapacana* along the border of Chile and Bolivia (between 4200 and 5200 m). Other arborescent taxa such as some *Espeletia* species (Asteraceae) (or Espeletiinae, see Chap. 11) in páramos and *Puya raymondii* (Bromeliaceae) in the puna also occur at elevations up to 4200 m, but compared to *Polylepis* they have very restricted distributions. *Polylepis* species are rarely dominant over broad areas of terrain but they can be locally abundant in small copses (Fig. 12–1C) or in bands limited to particular altitudinal zones. The restriction of *Polylepis* woodlands to specific sites can be attributed in part to the sporadic occurrence of suitable microhabitats and in part to human influences. Most of the 16 species of *Polylepis* occur in parts of the Andes that reached their present elevation in the late Tertiary or Pleistocene. Throughout this period, the high Andes were subjected to a series of climatic fluctuations caused by worldwide glacial cycles. Consequently, much of the evolution of *Polylepis* at the species level must have occurred during the last few million years and under highly unstable climatic conditions. Palynological work (van Geel and van der Hammen, 1973; van der Hammen, Werner, and van Dommelen, 1973; Salgado-Labouriau, Schubert, and Valastro, 1977; Graf, 1979; Flenley, 1979, and references therein; Ochsenius, 1980) has amply documented fluctuations in population sizes of *Polylepis* species during the last million years. Superimposed on these natural geologic and climatologic events have been human actions, altering the geography and biology of *Polylepis* species by both destruction and cultivation. Because of these human influences, it is unfortunately impossible precisely to reconstruct the evolutionary history of the genus, but an analysis of the modifications of several morphological characters for particular ecological conditions and the relationships of the species of the genus allows one to make some inferences about specializations and speciation patterns within the genus.

THE GENUS *POLYLEPIS*

Systematic Position within the Rosaceae

The approximately 3000 species in the Rosaceae are dispersed, according to most botanists, among four subfamilies, the Rosoideae, Amygdaloideae (Prunoideae), Maloideae (Pomoideae), and Spiraeoideae. The subfamilies are distinguished primarily on the basis of floral morphology and fruit characters. The Rosoideae differs from the other subfamilies in having dry, dehiscent fruits, usually achenes, or aggregate fruits with druplets or achenes. The subfamily has

FIG. 12–1. Morphology and habitats of *Polylepis* species. A. A branch of *P. weberbaueri* from Azuay, Ecuador, showing the dimorphic areas of the leafing branches. B. An inflorescence of *P. racemosa* from central Peru showing the numerous, exposed, pubescent anthers and small, inconspicuous flowers. C. A woodland of *P. racemosa* near La Quiñua, Peru. D. A windswept tree of *Polylepis tarapacana* near Sajama, western Bolivia, close to the Chilean border.

eight tribes (Robertson, 1974a–c), one of which, the Sanguisorbeae, is peculiar within the entire family because of the tendency of genera within it to have species with inconspicuous, wind-pollinated flowers. Most of the species of the family have rather generalized flowers that are insect pollinated (Müller, 1883; Free, 1970; Robertson, 1974a; McGregor, 1976). There is relatively little diversity in floral structure correlated with divergent pollination syndromes except in this small subtribe, which contains perhaps 300 species (the presence of apomixis in *Alchemilla* distorts the figures somewhat). A few genera of the Sanguisorbeae have flowers that have reached an end point in specialization for wind pollination. They are unisexual and apetalous. *Polylepis* has apetalous but perfect flowers (except in perhaps one population of *P. serrata*).

Clustered in the Sanguisorbeae is a group of genera conservatively treated as *Sanguisorba*, *Sarcopoterium*, *Bencomia*, *Cliffortia*, *Acaena*, *Margyricarpus*, *Tetraglochin*, and *Polylepis*. Within this group, the first three genera are more closely related among themselves than to the other five genera (Nordborg, 1966). *Sanguisorba* is a widespread genus in the Northern Hemisphere, *Bencomia* consists of monoecious or dioecious shrubs and trees found in the Canary and Madeira Islands, and *Sarcopoterium* is a monotypic genus native to Italy. *Cliffortia* also contains monoecious and dioecious trees and shrubs, but occurs in Africa. *Acaena*, *Margyricar-*

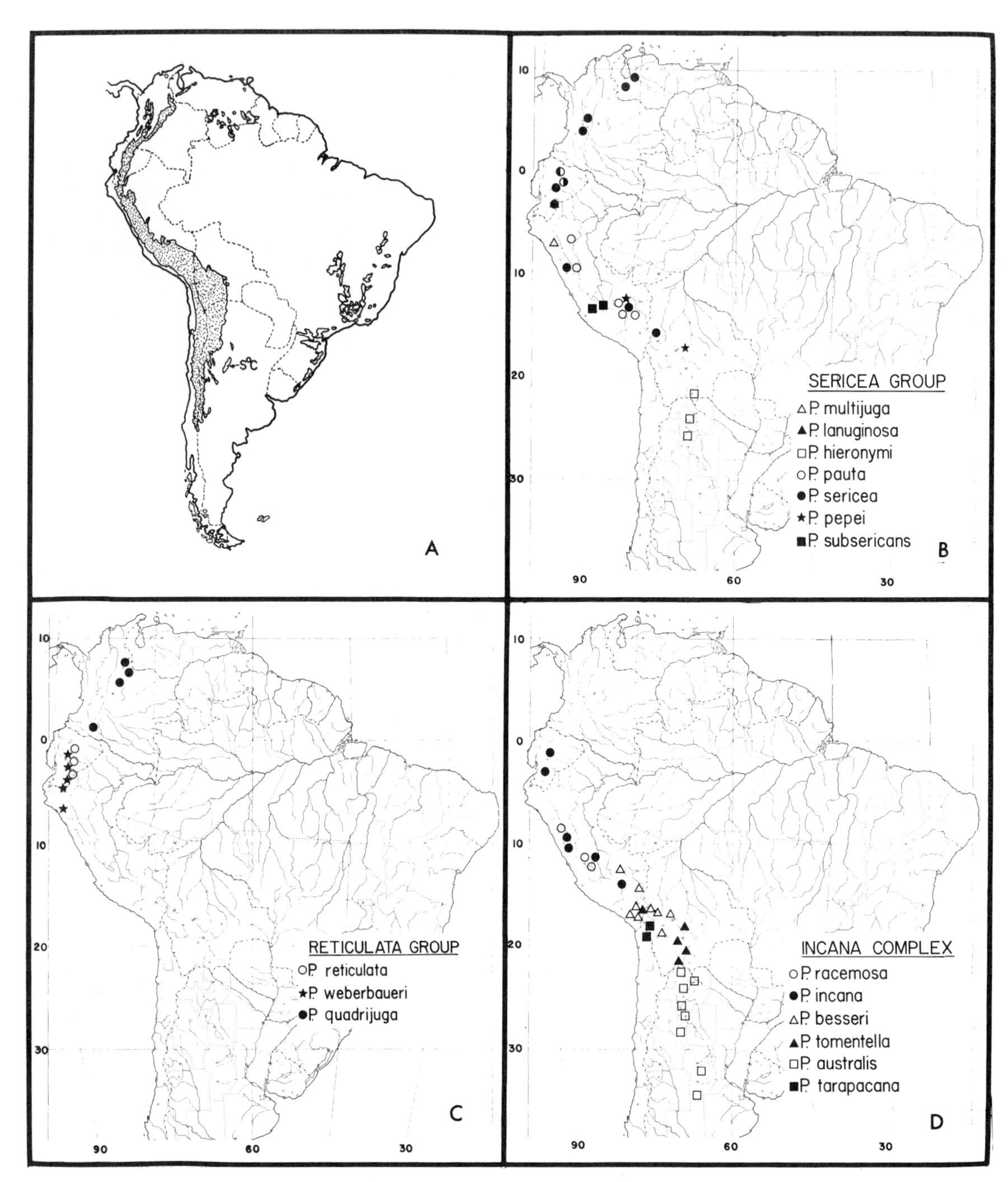

FIG. 12–2. High montane areas of South America and distributions of species of *Polylepis*. A. Relief of South America. Areas about 2000 m in western South America stippled. In eastern South America, the open outlines indicate the 1000-m contour line. S°C = Sierra de Córdoba. B. Distributions of members of the *sericea* group. C. Distributions of species included in the *reticulata* group. D. Distributions of the members of the *incana* complex. In B through D, symbols that are half dark and half light indicate sympatry of the two species with the respective completely dark and open symbols.

pus (including *Tetraglochin*), and *Polylepis* constitute an alliance of closely related genera. Species of *Acaena* are herbaceous, whereas both *Margyricarpus* and *Polylepis* contain only woody species. However, essentially all species of *Polylepis* are tropical (although *P. australis* grows primarily in subtropical regions; Fig. 12–2D), whereas most *Margyricarpus* species are temperate.

Distinctive Characters of *Polylepis*

As in the case of all members of the Sanguisorbeae, *Polylepis* species have inconspicuous (about 4–10-mm diameter) flowers borne in inflorescences. However, flowers of *Polylepis* are apetalous and have numerous stamens (more than 10), with trichomes covering, or at the apex of, the anther sacs. The single pistil is terminated by a flat, expanded, fimbrillate stigma. These floral characters are similar to those of wind-pollinated flowers throughout the angiosperms (Whitehead, 1969). The fruits are small achenes (2–12 mm in length), variously adorned with spines, tubercles, or wings (Fig. 12–4). All of the members of the genus are trees or shrubs and all have a characteristic, exfoliating red bark composed of numerous, very thin, brittle layers of periderm. The leaves are pinnately compound with a terminal leaflet. As in most Rosaceae, there is a pair of stipules at the base of each leaf, but in *Polylepis* these are fused around the branch bearing the leaf. The upper part of the fused stipules is expanded so that they form a cone encircling the branch with the leaf petiole emerging from one side. Since most of the species of the genus have their leaves congested on portions of the branches, the stipule sheaths overlap (Figs. 12–1A, 12–3), resulting in a stack of inverted cones with the leaves emerging from among them. Although similar in many respects to members of *Acaena* section *Elongata* (Bitter, 1911b), species of *Polylepis* differ from them in habit and in the presence of pubescence on the anther sacs.

FIG. 12–3. Branching patterns and leaf morphologies of members of the three species groups of *Polylepis*. All figures × 1/4. A. *P. multijuga* in the *sericea* group. B. *P. reticulata* of the *reticulata* group. C. *P. incana* of the *incana* complex.

Species Groups within *Polylepis*

Bitter (1911a) arranged the 33 species of *Polylepis* he recognized into two sections and eleven groups of lower rank, but my recent treatment (Simpson, 1979a) reduced the number of species to 13 and added two taxa described after 1911. These 15 species were clustered into three, apparently natural, species groups. The first of these, the *sericea* species group, contains seven species that tend to have many-flowered inflorescences, leaves with numerous, thin leaflets (Fig. 12–3A) bearing long and/or silky trichomes on the under leaflet surfaces. The *reticulata* species group encompasses three species with relatively few leaflets per leaf, rugose or shiny upper leaflet surfaces, notched leaflet apices (Fig. 12–3B), and feltlike coverings on the under leaflet surfaces. The last group is an informal species group designated the *incana* complex. It is a confusing assemblage of very similar species with variable numbers of leaflets (Fig. 12–3C) and glabrous, glandularly pubescent, or sporadically villous under leaflet surfaces. In Simpson's treatment (1979a), five species were included in the complex. It now appears that six are actually involved (see discussion later in the chapter). Within each of these species groups there has been a radiation leading to the present number of species. In some cases, hybridization between species (e.g., betwen *P. incana* and *P. racemosa* in central Peru) occurs sporadically today (usually due to human influences) and may have occurred in the past when climatic changes allowed species ranges to come into contact.

Morphological changes have proceeded in parallel fashion within the different groups. That such parallel evolution should have occurred in the different groups is understandable when the causes of the changes in characters are analyzed. Almost all of the specializations in diverse characters appear to be adaptations to high-elevation and/or dry habitats. Radiation into such habitats from more mesic, lower elevational areas has occurred independently in each species group. Nevertheless, although directions of morphological change have proceeded similarly in the three groups, the historical factors that have led to speciation have not been the same in each. The differences in speciation patterns can be attributed in part to the fact that the groups are centered in different regions of the Andes, and in part to the lack of uniformity of Andean historical geology and climatology. The *sericea* group extends from Venezuela through Colombia southward to Bolivia with species in the southernmost part of its range located primarily on the eastern slope of the Andes (Fig. 12–2B). The *reticulata* group is found in Colombia, Ecuador, and extreme northern Peru (Fig. 12–2C). Members of the *incana* complex occur from Ecuador to north-central Argentina with a concentration of species in southern Peru and Bolivia (Fig. 12–2D).

METHODOLOGY

Sources of Data

Eleven of the sixteen species of *Polylepis* were collected in the field and habitat characteristics noted. My own ecological data were supplemented by reports of *Polylepis* habitats in the literature. Geographic information was compiled using personal field data, herbarium label data, and published accounts of distributions. Numerous morphological characters involving aspects of branching pattern, leaf and leaflet morphology, stipule and stipule sheath patterns, vestiture, inflorescence size and vestiture, and flower and fruit parameters were measured or scored for over 650 specimens.

Geographic Data

It is impossible to use modern distributional herbarium data to reconstruct adequately the prehuman distributions of *Polylepis*. It is impossible also to rely on descriptions of early explorers since much of the disturbance of the woodlands was accomplished by native people before the arrival of the Spanish. Also, since the taxonomy of *Polylepis* was not stabilized until recently, the use of a name in a historical text might, or might not, actually refer to the species mentioned. Even trained botanists such as Weberbauer (1945) usually referred simply to *Polylepis* without any species designation. Voucher specimens were rarely collected by early explorers. In particular, writings between 1700 and 1900 tend to refer to almost any species from Ecuador, Peru, or Bolivia as *P. incana*. Consequently, the writings of explorers and botanists prior to 1900 were used primarily as a guide to the types of habitats and localities of the genus before the spate of recent cutting.

I do believe, however, that the range limits of the various species still reflect the natural boundaries (within a few hundred kilometers) of the species distributions and that an analysis of these limits can give indications about barriers that have acted in the past to shape the speciation patterns we now see. In a few cases, planting by man has artificially increased the geographic distributions of some species and produced artificial hybrid zones. This phenomenon is particularly apparent in the case of *P. racemosa*, but has probably occurred elsewhere as well.

Morphological Data

For many of the characters of *Polylepis* studied, I have hypothesized changes in morphology. Obviously, it is impossible to know with certainty the actual sequences of changes that have occurred in a given character for a given taxon. However, when a morphological character has several different character states that appear to represent selected adaptations for different environmental conditions, and the environmental conditions can be ranked from least to most severe and/or recent, we can infer sequential modifications of the character. Since the very high elevation habitats (3800 m and above) of the Andes are quite young (in most cases younger than five million years; Fernandez, 1970; Simpson, 1979b), and the arid habitats are younger than the mesic ones (Simpson, 1975, 1979b), species adapted to these habitats should be derived from ancestors growing at older, lower, and more mesic habitats. Despite these general trends, however, each character was assessed in the context of each species and species group. The morphological measurements made were not employed to produce a phenogram, or diagram that shows species connected by distances reflecting their overall similarity, for several reasons. Phenograms can be constructed by calculating a coefficient of similarity or Manhattan distance between each item or specimen (often called an OTU, or operational taxonomic unit; Sokal and Sneath, 1963) or by computing a measure of the Euclidean distance between items. The calculated distances or similarities are then used to draw a nested diagram (the phenogram). In the calculation of the distance or similarity measures, missing data can seriously distort the final results. For Euclidean distance measures, OTUs must be grouped into populations so that each character will have a mean and a variance. In this study, few of the 650 specimens used for measurements and scoring yielded a complete set of characters. In addition, it was impossible to cluster specimens into meaningful populations because I could not tell if the same collection number of a particular collector represented branch segments of a single individual or of several individuals, or a combination of both. Locality data on specimens were rarely precise enough to permit the specimens of two or more collectors to be lumped together as a population sample. In other words, the number of measurements per specimen was too incomplete to allow me to use specimens as OTUs, and collection records were too poor to permit either the combining of data from specimens into one "complete" specimen or the clustering of specimens into meaningful populations. The measurements did, however, provide a basis for the assessment of variation within a taxon.

PATTERNS OF MORPHOLOGICAL SPECIALIZATION IN *POLYLEPIS*

General Patterns

Starting with the most basic of morphological characteristics, plant growth form, there is a change in *Polylepis* species from tall trees (up to 27 m) to small, branched trees to shrubs about 1.5 m tall. Ancestrally, *Polylepis* probably had a tree growth form. Evidence for this supposition is derived from the fact that the high-elevation habitats in which small tree-forming or shrubby species of *Polylepis* occur are much younger than the upper montane habitats in which tall arborescent species of the genus are found. Plants of several species such as *P. multijuga*, *P. pauta*, and *P. hieronymi* are large trees that today inhabit the upper portions of the montane forests of northern Peru, the Cordillera Blanca of Peru, and the eastern slopes of the Cordillera Oriental in northern Argentina, respectively (Fig. 12–2B). It is possible that the tree growth form could have been derived from a shrubby growth form, but such a change is rarer than the converse (Carlquist and Cole, 1970). In *Polylepis*, no anomalies in wood formation appear to be present (Metcalfe and Chalk, 1950; personal observations). Moreover, shrubby species such as *P. weberbaueri* and *P. pepei* grow in relatively arid areas of the inner slopes of the Ecuadorian Andes and the high Eastern Cordillera of Bolivia (Figs. 12–2B, C). This is not to say that all species of *Polylepis* at high elevations and/or in arid habitats are shrubs. To the contrary, *P. tarapacana*, which has been recorded at higher elevations than any other species (over 5000 m) and which occurs in very dry areas of northern Chile and adjacent Bolivia, is a small tree (Fig. 12–1D). Nevertheless, within the different species groups, where both trees and shrubs are found, the frutescent habit appears to be the derived one.

The form of the branches of many species of *Polylepis* is so peculiar that it immediately attracts attention. Instead of bearing leaves evenly along the young branches, species of the genus usually exhibit a pattern of a long, bare branch segment followed by an area in which the leaves are borne congested together (Fig. 12–1A, 12–3A, B, C). Bitter (1911a) called this a long shoot–short shoot organization, but such terminology is incorrect according to modern usage. The term *long shoot* should refer to a branch that is elongate and *short shoot* to an entire, reduced branch. Actually the best way to describe the branches of *Polylepis* is to say that each leafing shoot is dimorphic with some portions of the shoot bearing leaves, and the other portions bare (Fig. 12–1A, 12–3A, B, C). That the pattern of bearing leaves on restricted portions of a branch is genetically, rather than environmentally, controlled has been shown by Olsen (1976), who found the dimorphic pattern of leaf production was retained in *P. australis* growing in a Danish garden where climatic conditions were quite different from those of its native home in northern Argentina. Likewise, plants of *P. sericea* from the Mérida Andes growing in the phytotron at Duke University retained the pattern of leaf production found in nature (personal observation).

The pattern of variable leaf production occurs in all species of *Polylepis*, but is most pronounced in species of very high elevation or very dry habitats. Species such as *P. multijuga* (Fig. 12–3A) and *P. sericea* show the fewest differences in the parts of a leafing shoot. The most extreme examples of intrabranch dimorphism are found in *P. pepei*, a shrubby species of very high elevations in the Bolivian Andes, and *P. weberbaueri* (Fig. 12–1A), a shrubby species of the arid inter-Andean valleys of Ecuador. This modification of a typical angiosperm pattern of leaf production (in which leaves are borne evenly along a branch), is not, however, an adaptation to elevation or aridity per se, but probably to seasonality of the high tropical environment. Except for *P. hieronymi* and *P. australis*, both native to northern Argentina and adjacent Bolivia, all species of the genus appear to be evergreen in their native habitats. Photosynthesis and growth, therefore, usually continue all year, but flushes of growth and leaf production tend to occur during the wettest parts of the year. Jäger (1962) has documented the presence of growth rings in trunks of *P. sericea* from about 4000 m in the tropical Andes of Venezuela that appear to reflect seasonal changes in soil moisture availability. Strong seasonality of climate in the Andes is related to both height and latitude and would have increased as the cordillera was uplifted. There is a pronounced diurnal variation in temperature in high tropical mountains (over 3500 m; Guhl, 1968; Troll, 1968), and seasonal variations in rainfall in both the páramo (defined here as the habitats above tree line in Venezuela and Colombia and most of Ecuador) and the puna (considered here to be the supraforest habitats of southern Ecuador, Peru, Bolivia, and northern Argentina and Chile) (Walter, Harnickell, and Mueller-Dombois, 1975; Johnson, 1976; Snow, 1976; Simpson, 1979a). However, the amount of total precipitation decreases from Venezuela to northern Argentina and adjacent Chile (from 1779 mm in Mérida, Venezuela, to 43 mm at Tacna, Peru) and the severity of the dry season increases (see Walter et al., 1975 for a complete set of climate diagrams). More details about climatic fluctuations are given in Chapter 2. Because of the relationship between flushing of leaf production and strong seasonality, which is, in turn, a relatively recent syndrome in the Andes, extremes in the intrabranch pattern of leaf production are considered derived conditions.

The leaves of various species of *Polylepis* often differ so much from one another that leaf morphology provides one of the best characters for distinguishing species (Simpson, 1979a). Leaves of species differ in size, the pubescence of the rachises, the number of leaflets per leaf, the thickness of the leaflets, the location of the stomates on the leaflets, and the kind of pubescence on the under leaflet surfaces. It has been shown from a number of studies (Hedberg, 1964; Parkhurst and Loucks, 1972; Boyer, 1973) that features such as thick cell walls (and thick cuticles), sunken stomates, coverings of trichomes or scales, and the reduction of leaf surface area by the production of compound leaves with small leaflets are correlated with conditions of physiological xerophytism produced by either the actual lack of soil moisture (arid climates) or semifrozen soil (as occurs in high tropical habitats). Species of *Polylepis* that occur at relatively low elevations (1800–3000 m) on the eastern side of the Andes or in northern Peru tend to have relatively large leaves with numerous thin leaflets, thin epidermal cell walls and cuticles, and stomates that are not deeply sunken (Figs. 3–5 in Simpson, 1979a). Species such as *P. tarapacana* that occur at extremely high elevations (over 4000 m) in arid habitats, which are among the youngest in South America (Simpson, 1979b), have small leaves with few, thick leaflets with a pronounced thickening of the outer cell walls of the epidermis, a thick cuticle overlain by a waxy coating, and deeply sunken stomates. Specializations in leaf morphology in the genus thus appear to include reductions in overall leaf size, a diminution of the number of leaflets, an increase in leaflet thickness and in the thickness of the cell walls of the epidermal cells, and a restriction in the number and depth of the stomates.

Although all species of *Polylepis* are wind pollinated, there is considerable variation in the lengths of the inflorescences and in the number of flowers they contain. In general, the morphology of individual flowers is consistent throughout the genus. They are apetalous with four or five, often red-tinged, green sepals. The number of stamens in a flower varies considerably, even within a species. Although the color of the anther sacs has been used as a specific character (Bitter, 1911a), it is probably also variable within a species. The stigma is uniformly expanded and fimbrillate. The flowers are borne alternately on pendent racemes (Fig. 12–1B). Oddly enough, despite the fact that species of *Polylepis* have flowers with numerous anthers abundantly filled with pollen (personal observation) and pendent inflorescences easily shaken by the wind, there appears to be little pollen dispersal (Salgado-Labouriau, 1979), at least in Venezuela (see Chap. 8). Moreover, in the one species studied, *P. australis*, geitonogamy leading to self-fertilization apparently occurs regularly (Olsen, 1976). Geitonogamy, the pollination of a flower by pollen from a different flower on the same plant, is opposed to autogamy, or self-pollination by pollen from within the same flower. Temporal separation of stigma receptivity and anthesis (*Polylepis* flowers are strongly protogynous) within a flower precludes autogamy. Lack of gene flow because of restricted movement of pollen, combined with the potential for inbreeding, should have pronounced effects on the genetic structure of populations and the rapidity with which differentiation between populations can occur. Numerous studies have shown that inbred plants, or populations of plants that usually inbreed, contain less genetic variability than those that normally outcross (Crumpacker, 1967; Allard et al., 1968, 1972; Solbrig, 1972; Selander and Kaufman, 1973; Sanders and Hamrick, 1980) and that selection may therefore act much more quickly than in populations of obligate outcrossers (Wright, 1940; Stebbins, 1957; Rehfeldt and Lester, 1969; Allard, 1975; Jain, 1976).

Differences in the inflorescence structure between species of *Polylepis* involve primarily a reduction in the number of flowers per inflorescence and in the type of vestiture on the rachis. In extreme cases, the inflorescences are so reduced, bearing only one or two flowers, that the inflores-

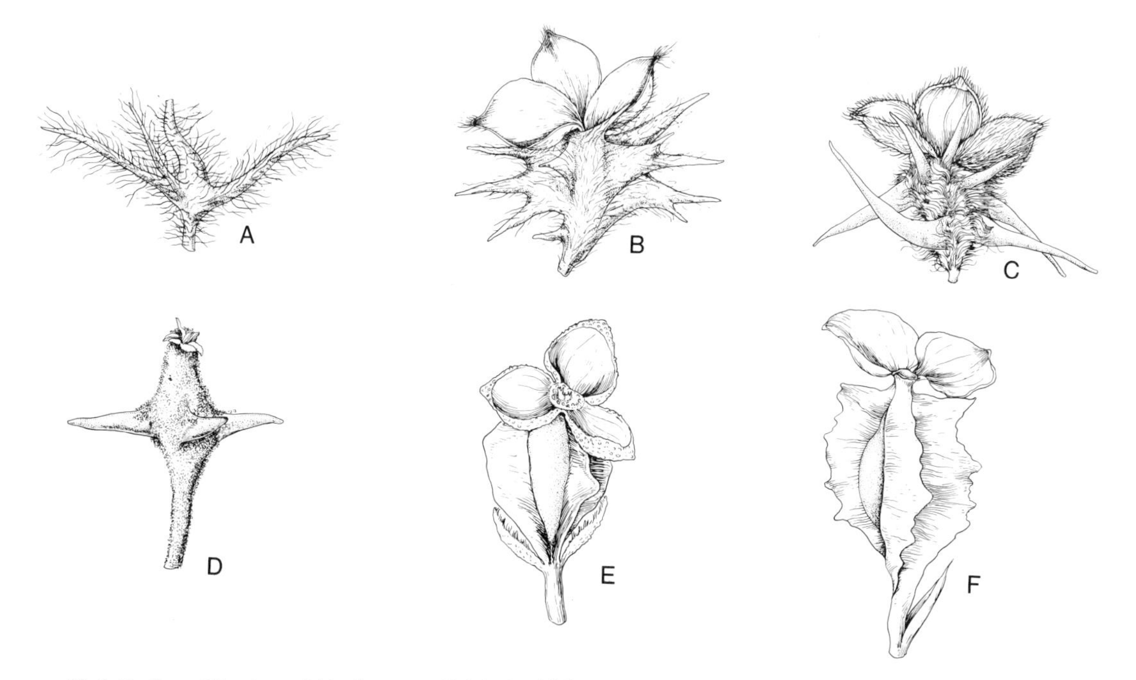

FIG. 12–4. Fruit modifications within the genus *Polylepis*. All figures × 4. A. *P. multijuga*. B. *P. pauta*. Both members of the *sericea* group. C. *P. quadrijuga*. D. *P. reticulata*. Both members of the *reticulata group*. E. *P. incana*. F. *P. australis*. Both members of the *incana* complex.

of the Peruvian and Bolivian Andes (Fig. 12–2B, D), and their fruits may, in fact, be passively dispersed. A final fruit type (Fig. 12–4F) in the genus, most pronounced in *P. australis*, has expanded wings that appear to aid in dispersal by wind. There are no directional trends in the genus as a whole. Changes from one type of fruit adorncence is no longer pendent. Because this reduction occurs in the very high elevation species *P. tarapacana*, it, like many other reductions, is considered to be a specialization for the extremely high and arid habitats in which this taxon occurs. At high tropical elevations, gametophytic tissue (in particular) must be protected from nightly frosts (Hedberg, 1964), but selection in wind-pollinated species is toward increased exposure of the sexual organs resulting in dangling anthers, large stigmas, and a loss of floral parts that might obstruct pollen movement. I have argued elsewhere (Simpson, 1979a) that at elevations this high and with such xeric conditions, there are consequently opposing selective forces that have acted to protect the gametophytic tissue on the one hand and to expose the anthers and stigma so as to effect pollination on the other.

The fruits of all species of *Polylepis* are relatively small, indehiscent, and green (sometimes tinged with red). In all cases they are decorated with some sort of sculpturing (Fig. 12–4). Bitter (1911a) stated that the fruits could not be animal dispersed since there were no animals that could serve as dispersal agents at the elevations where *Polylepis* species occur. However, more recent evidence suggests that animals are the major dispersers of the fruits of most species. Some species of the genus (*P. multijuga*, *P. sericea*, *P. pauta*, *P. reticulata*, *P. weberbaueri*) have spines, or even recurved spines (*P. hieronymi*), that anchor the fruits to the feathers or fur of animals moving in or under plants (Fig. 12–4A, B, C, D). Although there has been little documentation of mammals using trees of *Polylepis* for foraging, there has been ample evidence that birds (particularly *Oreomanes fraseri*, Coerebidae or Thraupidae, the Giant Conebill) spend an appreciable amount of time in the trees (M. Koepcke, 1954; H.-W. Koepcke, 1961; see Chap. 23 for a discussion of birds in *Polylepis* woodlands). Other species of the genus (e.g., *P. pepei*, *P. besseri*, *P. incana*, *P. tomentella*) have irregular knobs or ridges on the achene surface that do not seem particularly suited for animal or wind dispersal (Fig. 12–4E). Most of these species occur in relatively dry, high (over 3800 m) areas ment to another have occurred within each of the different species groups, sometimes more than once.

The *sericea* Group

The *sericea* species group contains seven species (*P. multijuga*, *P. lanuginosa*, *P. hieronymi*, *P. pauta*, *P. sericea*, *P. pepei*, and *P. subsericans*). In the monograph of the genus (Simpson, 1979a), *P. subsericans* was discussed with this group but was not listed as a member because its leaf morphology shows characters of both this group and the *incana* complex. Specimens of *P. subsericans* do not include any flowering or fruiting material, so its assignment to the group must still be considered tentative. Eventually, *P. subsericans* might be shown to be an aberrant member of the *incana* complex that resembles a species of the *sericea* group in vegetative morphology.

There is, however, enough information about the other members of the *sericea* group to allow some comments to be made about their relationships to one another and possible times and directions of evolution. As previously described by Bitter (1911a) and emphasized by Simpson (1979a), *P. multijuga* exhibits many features that are associated with unspecialized habitat conditions and that show similarities to members of section *Elongata* of *Acaena*. *Acaena* was considered by both Bitter (1911a) and Simpson (1979a) to be more closely related to *Polylepis* than to any other taxon. The morphological characters of *P. multijuga* that appear to resemble most closely the postulated ancestral traits in the genus are large leaves with numerous, thin leaflets, long, multiflowered inflorescences, and fruits adapted for animal dispersal. It is noteworthy that this species is restricted to the low montane areas of northern Peru, an area known for high endemism and relict species (Simpson, 1975). Presumably, for plants at least, this area has had a high stability of climate relative to other areas of the Andes since about the middle of the Tertiary.

North of the populations of *P. multijuga*, across the low area separating Peru from Ecuador, are populations of *P. lanuginosa*, undoubtedly the closest relative of *P. multijuga*. Although similar in many respects to *P. multijuga*, *P. lanuginosa* has adaptations such as thick leaves and a small tree growth form that indicate responses to selection in a relatively arid environment. *Polylepis lanuginosa* may either represent a northern isolate of a former, more continuous distribution of *P. multijuga* to the north (in other words, a disruption of an old continuous range), or more probably, a species that resulted from differentiation following establishment of a propagule (long-distance dispersal) from the Peruvian populations of *P. multijuga*. Like that of *P. multijuga*, the distribution of *P. lanuginosa* is geographically restricted.

Two other species derived from the *P. multijuga–P. lanuginosa* lineage are *P. pauta* and *P. sericea*. Of the two, *P. pauta* is more similar to *P. multijuga* and exhibits more unspecialized characters. Its distribution is broad, ranging from Ecuador to southeastern Peru (Fig. 12–2B). In many cases, particular populations of the species are recognizable because they differ from one another in morphology.

The second species, *P. sericea*, has the most extensive latitudinal range of any species of *Polylepis*, extending from Venezuela to northern Bolivia. However, in contrast to *P. pauta*, *P. sericea* shows little geographic variation across this broad range. In some areas (central Ecuador, Fig. 12–2B), it is apparently sympatric with *P. pauta*, but no evidence of hybridization has been found. The pattern exhibited by *P. sericea* thus seems to be one of a species that rapidly spread along the Andes in the last million years as a result of climatic changes. Inferential evidence for a recent expansion comes from paleopalynological data. Although it is impossible to identify individual species of *Polylepis* from their pollen, we might suppose that the fossil record of *Polylepis* from the Sabana de Bogotá (van der Hammen, 1974, and references therein; see Chap. 7) was left by this species since it is today the only member of the genus in this part of Colombia or neighboring Venezuela. The palynological record of the Sabana de Bogotá dates back about two million years, but van der Hammen (1974) and van der Hammen et al. (1973) have found evidence of *Polylepis* pollen for only the last 600,000 years.

Presumably the same type of climatic conditions that permitted the expansion of *Polylepis* into the Central (and Eastern ?) Cordilleras of Colombia about 600,000 years ago would have led also to range extensions southward along the Andes. Shortly after the initial 600,000 years B.P. appearance of *Polylepis* pollen grains in the sediments, they begin to decrease in frequency. Later, at about 30,000, a second *Polylepis* population increase is recorded in the Sabana cores. Although neither the 600,000 nor the 30,000 years B.P. dates can be assumed with certainty to be times of expansion into other areas of the Andes, they do support the idea of recent range extensions of *Polylepis* (see Chap. 7). *P. sericea* (as far as could be ascertained from specimens), does not extend into the Eastern Cordillera of Colombia even though it occurs in the Mérida Andes of Venezuela. Its migration into Venezuela from Colombia could have been by long-distance dispersal or the species formerly could have been distributed across the Eastern Cordillera and suffered extinction there because of habitat changes or displacement by *P. quadrijuga*, the only species now present in the eastern ranges of Colombia.

Finally, *P. pepei*, a recently described member of this group, has several morphological characters that are similar to those of *P. sericea*. Presumably, both *P. pepei* and *P. sericea* were derived from some common ancestor. Collection records show that *P. pepei* occurs in the highest ceja de la montaña of the Eastern Cordillera of Peru and northern Bolivia. However, specimens of the species are so few that it is impossible to determine if the apparent large disjunctions between populations are real or artifacts.

The *reticulata* Group

The *reticulata* group comprises three taxa, *P. reticulata*, *P. weberbaueri*, and *P. quadrijuga*. The first two species, both of which occur in Ecuador (Fig. 12–2C), morphologically are very similar. The third species, which is restricted to the Eastern Cordillera of Colombia, is similar in some ways to *P. lanuginosa* of the *sericea* group. In contrast to *P. sericea*, *P. quadrijuga* does not reach Venezuela, nor does it occur in any of the other ranges of the Colombian Andes. Its similarity to members of the *sericea* group and presence on the oldest high chain of the Colombian Andes may indicate its early northward spread into Colombia. The two more southerly distributed species occur sporadically on the slopes of the Ecuadorian volcanoes with *P. weberbaueri* reaching into northern parts of Peru. As in many other species of *Polylepis*, therefore, the northern Peru low area does not act as a barrier or constitute a distributional boundary. The relatively minor influence of the Huancabamba Depression as a barrier to the migration of some species can be ascribed to the fact that they grow at midaltitude elevations (commonly to 2800 m). Evidently, the depression does not pose the same kind of ecological barrier that it does to taxa restricted to very high elevational habitats (see F. Vuilleumier, 1984, p. 720, for similar conclusions about avian taxa).

Although collection records would indicate that *P. weberbaueri* and *P. reticulata* can occur together (e.g., on Chimborazo volcano, Ecuador), I have not seen them growing within a hundred miles of one another. Their spotty distributional patterns in Ecuador reflect the patchy nature of areas of moist microhabitat. Areas of woodland coincide, for example, with a zone of humidity north of Cañar caused by a break in the Western Cordillera that allows moist air to enter the inter-Andean valley. Likewise, *P. weberbaueri* occurs in northern Peru sporadically on mountains in areas wetter than most of the surrounding terrain. Habitat data suggest that within these microhabitats, *P. weberbaueri* is generally restricted to drier microhabitat sites than *P. reticulata*. Undoubtedly, the two are closely related and derived from a common ancestor.

The *incana* Complex

The *P. incana* complex was designated as a complex rather than a species group (Simpson, 1979a), because the taxa it comprises are all very similar morphologically, exhibit an array of confusing characters that may indicate hybridization, and have been more affected by human influences than the species of the other groups (Pulgar, 1967; Winterhalder and Thomas, 1978). In one case, the circumscription of a taxon of this group, *P. tomentella* (Simpson, 1979a), is now known to be incorrect. From specimen data available at the time of the monograph, it appeared that there was a pattern of clinal variation from eastern to western Bolivia, a pattern found in some taxa ranging across the altiplano (B. Vuilleumier, 1970). The eastern populations corresponded to the type of *P. tomentella*, and the western populations to the type of *P. tarapacana*. Recent field observations by Chilean botanists (Marticorena, in correspondence) have now found both occurring within 25 km of one another in the Cordillera of Arica in northern Chile without any evidence of intermediates. The two seem to be separated altitudinally with *P. tomentella* in this area growing between 3200 and 3400 m and *P. tarapacana* occurring between 4500 and 4900 m. Consequently, *P. tarapacana* should be considered a separate species distinct from *P. tomentella*. Another species in this complex, treated as *P. besseri* (Simpson, 1979a), will probably also be shown in the future to contain more than one biological species. The other three species in the complex, *P. incana* (Fig. 12–3C), *P. racemosa*, and *P. australis* pose fewer taxonomic problems and have more clearly defined specific limits than those mentioned for the other species.

Four of the species of this complex, *P. besseri*, *P. tomentella*, *P. tarapacana*, and *P. australis*, occur on the altiplano of southern Peru, Bolivia, and northern Argentina and Chile. In numerous other groups of flowering plants, imperfect or incomplete differentiation of species across this plateau has been shown. Hybrid zones are common (Simpson, 1975, 1979b) and/or species often display stepped clines that are hard to interpret. *Polylepis* is no exception. Besides its complex geologic and climatic history, the region from central Peru to northern Argentina was the home of the great Incan Empire and has been heavily used by man for millenia. Although man no doubt has influenced the distributions of many taxa, it is probably species of genera like *Polylepis* that provided a commodity such as firewood that have suffered most from human influence (Winterhalder and Thomas, 1978).

P. incana and *P. racemosa* are both found in Peru and now co-occur in several areas, particularly throughout the broad, central Mantaro River valley from Cerro de Pasco to Huancavelica. In areas where the two grow together, there is morphological evidence of hybridization. An assessment of the distributions and habitats of the two species elsewhere indicates that under natural conditions, they would probably rarely occur together. *P. incana* is generally found in more arid situations than *P. racemosa*. According to Pulgar (1967), the latter is naturally confined to the slopes flanking the broad valley. Planting by man has led to the extensive mixed populations along this densely inhabited valley.

In southern Peru, *P. incana* is replaced by *P. tomentella* (similar to it in morphology), *P. tarapacana*, and *P. besseri*. These last three species extend into Bolivia (Fig. 12–2D). *P. besseri* occurs to the north of Lake Titicaca and around the northeastern curve of the altiplano near the border between Bolivia and Peru. *P. tomentella* occurs at the southern end of Lake Titicaca and in the central and western altiplano. *P. tarapacana* is restricted to the westernmost area of the altiplano along the Bolivian-Chilean border and in northern Chile. Similar distributional patterns among the related species of a genus have been illustrated in Simpson (1975). *P. australis*, the last species in this complex, is located farther south in tropical and extratropical montane areas of Argentina, including the Sierra de Córdoba in the province of Córdoba (Fig. 12–2A, D). This southern species, one of the most specialized in the genus, has adaptations, not for high-elevation habitats, but for temperate (to subtropical) conditions. Of all the species of the genus, this is the only one with fruits clearly adapted for wind dispersal (Fig. 12–4F). In this case, there has been a reinvasion of a moderate-elevation habitat and a secondary evolution of leaves with numerous, thin leaflets.

PATTERNS OF SPECIATION

It is obviously very difficult to reconstruct with any precision the barriers that have led to isolation and eventual speciation within *Polylepis*. The natural restriction of populations to areas that are not only ecologically unknown but often unmapped makes interpretations of distributions tentative. Both human influence in the destruction or artificial construction of populations and spotty collections further complicate the problem of deciphering biogeographic patterns. A few things about the patterns of speciation based on an assessment of morphological variation and general distributional patterns can nevertheless be proposed.

First, *Polylepis* species, with the possible exception of *P. australis*, appear to have special microhabitat requirements for establishment, growth, and survival. It is possible that species like *P. multijuga*, *P. pauta*, and *P. hieronymi*, which grow in areas contiguous with the upper montane forest, have fewer ecological specializations than other members of the genus, but they apparently cannot compete successfully with arborescent species (e.g., species of the Leguminosae, Moraceae, Myristicaceae, and Bignoniaceae) of lower elevations. Other *Polylepis* species, many of which occur in interior Andean valleys, form small populations in areas where microhabitat conditions permit their survival. Although the physiological requirements of seedling establishment have not been definitely proved, Smith's work (1977) indicates that protection of the seedling from frost heaving or freezing of the roots is necessary for successful establishment and may help to explain the usual occurrence of high-elevation *Polylepis* trees or shrubs in crevices, rocky soil, or on hillsides.

Second, the genus appears to date from the middle of the Tertiary. This estimate probably represents the maximum age since only at about this time were there any mountains high enough to support species growing at the lowest elevation recorded for modern species. As the Andes rose, species of *Polylepis* apparently spread along the eastern slopes of the Andes and across the mountains into areas moist enough to support plant growth. Drying of the continent and isolation of populations in pockets on mountain slopes combined with an apparent potential for inbreeding would have fostered differentiation. Continued drying of the inter-Andean regions and the altiplano during the final phases of uplift would have led to increased specializations within various populations.

Finally, during the Pleistocene, there were undoubtedly many episodes of range expansions and times of local population retraction or extinction. Such sequences have been documented for central Colombia and Venezuela (see Chaps. 7 and 8). Among the modern speciation patterns that appear to be attributable to Pleistocene events are the long latitudinal distribution of *P. sericea* with little population differentiation; the splitting of an ancestral stock into *P. weberbaueri* and *P. reticulata* on the volcanoes of Ecuador; the fragmentation of the members of the *incana* complex across the altiplano into *P. besseri*, *P. tomentella*, and *P. tarapacana*; and the restriction of the progenitor of *P. australis* to northern Argentina.

REFERENCES

Allard, R. W. 1975. The mating system and microevolution. *Genetics* 79: 115–126.

Allard, R. W., C. R. Babbel, M. T. Clegg, and A. L. Kahler. 1972. Evidence for coadaptation in *Avena barbata*. *Proc. Natl. Acad. Sci.* (U.S.A.) 69: 3043–3048.

Allard, R. W., S. K. Jain, and P. L. Workman. 1968. The genetics of inbreeding populations. *Adv. Genetics* 14: 55–131.

Bitter, G. 1911a. Revision der Gattung *Polylepis*. *Bot. Jahrb. Syst.* 45: 564–656.

———. 1911b. Die Gattung *Acaena*. *Biblioth. Bot.* 74: 1–80.

Boyer, J. S. 1973. Response of metabolism to low water potentials in plants. *Phytopathology* 63: 466–472.

Carlquist, S., and M. J. Cole. 1970. *Island biogeography*. New York: Columbia Univ. Press.

Crumpacker, D. W. 1967. Genetic loads in maize (*Zea mays* L.) and other cross-fertilized plants and animals. *Evolutionary Biol.* 1: 306–424.

Fernandez, J. 1970. *Polylepis tomentella* y orogenía reciente. *Bol. Soc. Argent. Bot.* 13: 14–30.

Flenley, J. R. 1979. *The equatorial rain forest.* Boston: Butterworths.

Free, J. B. 1970. *Insect pollination of crops*. New York: Academic Press.

Graf, K. J. 1979. Untersuchungen zur rezenten Pollen- und Sporenflora in der nördlichen Zentralkordillere Boliviens und Versuch einer Auswertung von Profilen aus postglazialen Torfmooren. Ph.D. thesis. Univ. of Zurich. Zurich: Juris.

Guhl, E. 1968. Los páramos circundantes de la sabana de Bogotá, su ecología y su importancia para el regimen hidrológico de la misma. *Colloq. Geog. Bonn* 9: 195–212.

Hedberg, O. 1964. Features of afroalpine plant ecology. *Acta Phytogeog. Suecica* 49: 1–144.

Jäger, E. T. 1962. La formación de zonas generatrices en plantas leñosas del límite selvático andino. *Acta Cient. Venezolana* 13: 126–134.

Jain, S. K. 1976. The evolution of inbreeding in plants. *Ann. Rev. Ecol. Syst.* 7: 469–495.

Johnson, A. M. 1976. The climate of Peru, Bolivia and Ecuador. In *Climates of Central and South America*, W. Schwerdtfeger, ed., pp. 147–218. New York: Elsevier.

Koepcke, H.-W. 1961. Synökologische Studien an der Westseite der peruanischen Anden. *Bonner Geog. Abhand.* 29: 1–318.

Koepcke, M. 1954. Corte ecológico transversal en los Andes del Perú central con especial consideración de las aves. *Mem. Mus. Hist. Nat. "Javier Prado"* Lima 3: 1–119.

McGregor, S. E., ed. 1976. *Insect pollination of cultivated crop plants*. U.S.D.A. Agric. Res. Service. Agric. Handbook 496. Washington, D.C.: Government Printing Office.

Metcalfe, C. R., and L. Chalk. 1950. *Anatomy of the dicotyledons*. Vol. 1. Oxford: Clarendon Press.

Müller, H. 1883. *The fertilization of flowers*. Translated and edited by D'Arcy W. Thompson. London: Macmillan.

Nordborg, G. 1966. *Sanguisorba* L., *Sarcopoterium* Spach, and *Bencomia* Webb et Benth. Delimitation and subdivision of the genera. *Opera Bot.* 11: 1–103.

Ochsenius, C. 1980. *Cuaternario en Venezuela*. Coro: Univ. Francisco de Miranda.

Olsen, O. 1976. *Polylepis australis* Bitter—en ny hårdfor vedplante for Denmark. *Dansk Dendrol. Arsskr.* 5: 21–31.

Parkhurst, D. F., and O. L. Loucks. 1972. Optimal leaf size in relation to environment. *J. Ecol.* 60: 505–537.

Pulgar, V. J. 1967. *Geografía del Perú*. Lima: Editorial Universo.

Rehfeldt, G. E., and D. T. Lester. 1969. Specialization and flexibility in genetic systems of forest trees. *Silvae Genet.* 18: 97–144.

Robertson, K. R. 1974a. The genera of the Rosaceae in the southeastern United States. *J. Arnold Arbor.* 55: 303–332.

———. 1974b. The genera of the Rosaceae in the southeastern United States. *J. Arnold Arbor.* 55: 344–401.

———. 1974c. The genera of the Rosaceae in the southeastern United States. *J. Arnold Arbor.* 55: 611–662.

Salgado-Labouriau, M. 1979. Modern pollen deposition in the Venezuelan Andes. *Grana* 18: 53–59.

Salgado-Labouriau, M., C. Schubert, and S. Valastro, Jr. 1977. Paleoecologic analysis of a late Quaternary terrace from Mucubaji, Venezuelan Andes. *J. Biogeog.* 4: 313–325.

Sanders, T. B., and J. L. Hamrick. 1980. Variation in the breeding system of *Elymus canadensis*. *Evolution* 34: 117–122.

Selander, R. K., and D. W. Kaufman. 1973. Self-fertilization and genetic population structure in a colonizing land snail. *Proc. Natl. Acad. Sci.* (U.S.A.) 70: 1186–1190.

Simpson, B. B. 1975. Pleistocene changes in the flora of the high tropical Andes. *Paleobiology* 1: 273–294.

———. 1979a. A revision of the genus *Polylepis* (Rosaceae: Sanguisorbeae). *Smithsonian Contr. Bot.* 43: 1–62.

———. 1979b. Quaternary biogeography of the high montane regions of South America. In *The South American herpetofauna: Its origin, evolution and dispersal*, W. E. Duellman, ed., pp. 157–188. Lawrence: University of Kansas Press.

Smith, A. 1977. Establishment of seedlings of *Polylepis sericea* in the paramo (alpine) zone of the Venezuelan Andes. *Bartonia* 45: 11–14.

Snow, J. W. 1976. The climates of northern South America. In *Climates of Central and South America*, W. Schwerdtfeger, ed., pp. 295–379. New York: Elsevier.

Sokal, R. R., and P. H. A. Sneath, 1963. *Principles of numerical taxonomy*. New York: Freeman.

Solbrig, O. T. 1972. Breeding system and genetic variation in *Leavenworthia*. *Evolution* 26: 155–160.

Stebbins, G. L. 1957. Self-fertilization and population variability in the higher plants. *Amer. Nat.* 41: 337–354.

Troll, C. 1968. The cordilleras of the tropical Americas. Aspects of climate, phytogeography and agrarian ecology. *Colloq. Geog. Bonn* 9: 15–56.

van Geel, B., van der Hammen, T. 1973. Upper Quaternary vegetational and climatic sequence of the Fuquene area (Eastern Cordillera, Colombia). *Paleogeog. Palaeoclimat. Palaeoecol.* 14: 9–92.

van der Hammen, T. 1974. The Pleistocene changes of vegetation and climate in tropical South America . *J. Biogeog.* 1: 3–26.

van der Hammen, T., J. H. Werner, H. van Dommelen. 1973. Palynological record of the upheaval of the northern Andes: A study of the Pliocene and lower Quaternary of the Colombian Eastern Cordillera and the early evolution of its high Andean biota. *Rev. Palaeobot. Palynol.* 16: 1–122.

Vuilleumier, B. S. 1970. The systematics and evolution of *Perezia* sect. *Perezia* (Compositae). *Contr. Gray Herb.* 99: 1–163.

Vuilleumier, F. 1984. Zoogeography of Andean birds: Two major barriers; and speciation and taxonomy of the *Diglossa carbonaria* superspecies. *Nat. Geog Soc. Res. Reports* 16: 713–731.

Walter, H., E. Harnickell, and D. Mueller-Dombois. 1975. *Climate-diagram maps*. New York: Springer.

Weberbauer, A. 1945. *El Mundo Vegetal de los Andes Peruanos*. Lima: Ministerio de Agricultura.

Whitehead, D. R. 1969. Wind pollination in the angiosperms: Evolutionary and environmental considerations. *Evolution* 23: 28–35.

Winterhalder, B. P., and R. B. Thomas. 1978. Geoecology of southern highland Peru: A human adaptive perspective. *Inst. Arctic. Alpine Res. Occas. Paper* 27: 1–91.

Wright, S. 1940. Breeding structure of populations in relation to speciation. *Amer. Nat.* 74: 232–248.

13

Speciation and Endemism in the Flora of the Venezuelan Tepuis

JULIAN A. STEYERMARK

Geographically, the subject of this chapter is an area of approximately 1,200,000 km^2 (Priem et al. 1973), which stretches to a large extent in Venezuela south of the Orinoco river, but which also includes peripheral areas in western Guyana, southeastern Colombia, northern Brazil, and an isolated outlier in Surinam. I consider the Venezuelan Guayana to include both the Territorio Federal Amazonas and the Estado Bolívar.

The Guayana region as defined here is covered by hundreds of thousands of square kilometers of tall primary forest, which forms in many sectors an unbroken green mass, whereas elsewhere are open, treeless or nearly treeless savannas of straw-colored hues interrupted by gallery forests or moriche palms along water courses. In the eastern sector of Venezuela and adjacent Guyana the savannas are more common. Irregularly interspersed within this vast area are elevated and conspicuous mountains with usually high, perpendicular-walled, sandstone bluffs of white, rose, pink, and buff, towering to relatively flat or tablelike summits that stand out prominently above and in contrast to the level or slightly rolling landscape. The whole area is part of the ancient Guayana Shield of northeastern South America, which, together with the Brazilian Shield, forms the old nuclear core of the South American continent.

These dramatically isolated mountains are known in the state of Bolívar in eastern Venezuela as tepui (plural, tepuis); in Spanish, usually *Tepuy* (plural, *Tepuyes*). The term tepui comes from the Taurepán, Arekuna, and Kamarakota Indian tribes who speak the Pemón language, belong to the Carib linguistic stock, and live in eastern Bolívar, especially the sector known as the Gran Sabana (Armellada, 1944). To these Indians tepui signifies a mountain, especially one with the truncate summit of the table mountains. Farther westward in Venezuela, and especially in the Territorio Federal Amazonas, the same tabular sandstone formations are referred to generally as *cerro* (plural, *cerros*) and in the Yekuana dialect, spoken by the Maquiritare (Makiritare) Indians of the forested area encompassing the Caura, Ventuari, Cunucunuma, and Padamo river sectors, such cerros are called jidi, such as Jáua-jidi, Guanacoco-jidi, and Marahuaca-jidi. In northern Brazil such mountains are termed *serras* (singular *serra*); in Guyana they are referred to as "mount," "mountain," or as "*tipu*" in the language of the Akawaio Indians.

DISTRIBUTION AND TOPOGRAPHY OF THE TEPUIS

Although the great majority of these sandstone mountains attain their maximum development in Venezuela, extensions of the same formations are conspicuous but less abundant in adjacent countries. To the east in Guyana they are represented by the Pakaraima Mountains, Merume Mountains, Ayanganna, Mt. Twek-quay, Holitipu, Cerro Venamo, Mt. Tulameng, Haiamatipu, Caburai-tepe, Maccari (Makari) Mountains, Parish's Peak, and others. The easternmost limit of the sandstone table mountains occurs in Tafelberg, Surinam, and in Colombia the westernmost extension is the Macarena Massif, which reaches a height of approximately 1500 m. The smaller of these sandstone or quartzitic mountains are generally only 240–360 m above sea level in the Vaupés and Caquetá river basins of southeastern Colombia (Cerro Macuye, Camarote, Cerro de la Iglesia, Cerro del Gigante, Cerro del Castillo, Cerro Yapobodá, Cerro Circasia, Cerro Cuduyarí, Cerro de la Campana, and Cerro Chiribiquete). Some of the larger of these Colombian mountains attain 700 m (Schultes, 1944, 1945). Along the Venezuelan-Brazilian border, the Pakaraima Mountains extend the Roraima Formation southward and are prominent as the Serra do Sol, Serra Tepequem, and Serra Marutani (Uru-

FIG. 13–1. Location of tepuis in Venezuela.

tany). Westward the Brazilian sandstone mountains extend to the Serras Tapirapeco and Imeri, where they attain 3014 m at Pico da Neblina, the highest altitude of the sandstone mountains of the Roraima Formation. Other sandstone mountains are found in northernmost Brazil in Amazonas State and Roraima Territory (Serra Pirapucu and Serra dos Surucucus, respectively).

In Venezuela there are hundreds of sandstone elevations spread over the southern portion of the country. If we limited ourselves to the larger or more prominent ones, then somewhat over 50 are evident (Fig. 13–1).

The spatial distribution of the Venezuelan table mountains is quite irregular. Proceeding from east to west in the Estado Bolívar, tepuis are concentrated along the Venezuelan-Guyanan border, comprising the Roraima chain, with Ror-

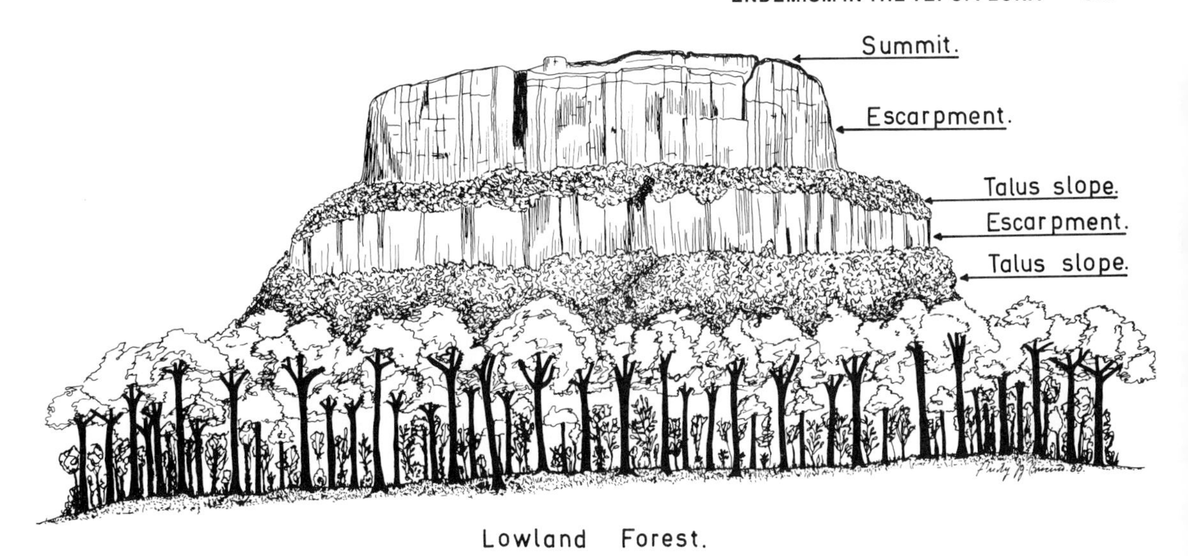

FIG. 13–2. Aspect of a typical tepui.

aima, Kukenán, and Ilú-tepui as dominant topographic features. Ninety to 150 kilometers to the northwest, another group of tepuis is conspicuous, among which are Sororopán-tepui, Ptari-tepui, Camarcaibara-tepui, and Aparamán-tepui, while slightly westward and southwestward toward the Caroni River appear the huge massifs of Auyán-tepui and Chimantá-tepui, together with the adjacent mountains of Aprada-tepui, Upuigma-tepui, and others. West of the Caroni River, another 100 km, the large Cerro Guaiquinima is encountered east as far as the Paragua River. Westward from Cerro Guaiquinima no major mountains appear for another 100 km to the southwest, until the gigantic group of Jáua, Sarisariñama, and Guanacoco is encountered, together with smaller and less conspicuous adjacent elevations.

Another 150 km westward a group of mountains begins in the Territorio Federal Amazonas. The first group to be noted is the chain of prominences known as Parú or Asisa. About 130 km to the north of Parú is Yaví, while approximately another 100–120 km south of Parú three prominent mountains are concentrated: Duida, Huachamacari, and Marahuaca. Westward and southwestward from Cerro Yaví, within a radius of 25–60 km, lies another cluster of sandstone mountains, which are, from north to south, Guanay, Corocoro, Camani, Yutaje, and Morrocoy. Between the latter group of mountains and the Orinoco boundary separating Venezuela and Colombia, is a hiatus of nearly 100 km before the appearance of the large sandstone mountain mass of Sipapo and the nearby towerlike Autana. Southeastward from here at a distance of another 130 km, on the other side of the Ventuari River, appears the isolated Yapacana mountain. Finally, after a stretch of 250–320 km southeastward toward the southern border of Venezuela three prominent mountains come into evidence: Aracamuni, Avispa, and finally, Neblina.

The irregular distribution of the tepuis, together with their different sizes, shapes, and heights, accounts for a highly diversified topographic pattern in the landscape. While the summits generally range between 1000–1300 and 2500 m, their highest elevations are Pico da Neblina, 3014 m (Brazilian side of Serra da Neblina, Venezuelan-Brazilian frontier in northwestern Amazonas, Brazil), and Pico 31 de Marzo, 2992 m (Venezuelan side of Cerro de Neblina). The other principal sandstone mountains in Venezuela that attain 2500 m or more are Marahuaca, Ilú-tepui, Roraima, Kukenán, Auyán-tepui, and Ptari-tepui. The lowest summits are on Cerro Perro at 700 m along the Paragua River of the Estado Bolívar, and on the quartzitic mountains of southeastern Colombia at 200–400 m. On the summit of Cerro Guaiquinima, along the Paragua River, a depressed portion of the summit is only at 700–750 m, although other parts of the summit range from 1300 to 1700 m.

The size of the sandstone mountains varies from small, towerlike eminences 1 km or less long by 200 m or less wide, such as Autana, Upuigma-tepui, Guadacapiaque-tepui, and Tremón-tepui, to such gigantic massifs as Chimantá, Jáua-Sarisariñama, Auyán, Guaiquinima, Duida, and Neblina, whose summits may cover from 600 to 2500 km^2 or more, and be from 25 to 80 km long.

Characteristically, the sandstone mountains

have more or less flat summits and are flanked on all or most sides by one to three tiers of sheer vertical cliffs or escarpments (themselves 100–500 m or more high), which are broken up usually by forested, terraced talus slopes (Fig. 13–2). This topography is typical of most of the tepuis of eastern Venezuela, especially those between Roraima and Auyán-tepui, or west to the Caroni River. But proceeding westward into the Territorio Federal Amazonas, the larger sandstone mountains of Duida, Sipapo, Guanay, and Neblina present more undulating and irregular summit contours as a result of past geologic faulting and diastrophism. Some of the mountains appear flat-topped, but their summits often possess depressed portions in the center.

GEOLOGIC HISTORY

The area under discussion is included within the Guayana Shield, one of the two most ancient landmasses of South America. The Guayana Shield is the northern part and the Brazilian Shield is the southern member of this old sector of South America, encompassing a large territory of central and southern Brazil. The two shields together comprise part of the Gondwana continent. The oldest portion of the Guayana Shield consists of igneous and metamorphic rocks (highly crystalline granites, porphyries, gneisses, and schists of Proterozoic age). Over this crystalline basement peneplain of the Guayana Shield a layer of sediments, the Roraima Formation, was deposited, totaling over 1,000,000 km^2 of which 800,000 km^2 has been removed by erosion, leaving 200,000 km^2 exposed in the present topography as sandstone outcrops (Gansser, 1974). These sediments, with an accumulated depth of as much as 3000 m, range over a very extensive region of about 1900 km from Tafelberg in Surinam in the east to extreme western Venezuela and southeastern Colombia in the west, with a minimal surface embracing 250,000 km^2 of sediments (Ghosh, 1977). Studies by Colvée (1971, 1972), Gansser (1974), Talukdar and Colvée (1974), and Ghosh (1977) indicate that the Roraima Formation was a continental sedimentation that possibly occurred at various ages in separate and different small basins, instead of only one extensive, contemporaneous, and uniform continental sedimentation, as previously believed. The differences in ages of the basins in which the sediments were deposited are thought to be due to different periods of tectonic movements. These different sedimentations are believed to be between 1500 and 1800 million years old, between the close of the Lower Precambrian or Middle Proterozoic and the beginning of the Precambrian (McConnell, 1959, 1968; McDougall, Compston and Hawkes 1963; Snelling, 1963; Snelling and McConnell, 1969; McConnell and Williams, 1970; Colvée, 1972; Ghosh, 1977). Recent studies suggest that the sedimentation process evolved toward the west, and is, therefore, younger than the sediments in the eastern part of the Roraima Formation. Subsequent to these initial ancient depositions, later sedimentations may have occurred in the western sector of the Roraima Formation east of the Cordillera Macarena of Colombia in the Upper Eocene and Lower Oligocene, 80 million years B. P. (Paba Silva and van der Hammen, 1960; Gansser, 1974; Maguire, 1979).

Conclusive fossil evidence to support the age of the Roraima Formation is still lacking. However, some samples of microfossils of sponge spicules from Brazil and Guyana and of micro plant pollen and spores from Guyana and southeastern Venezuela have suggested ages ranging from Cretaceous to Miocene (Gansser, 1974). However, these samples have not been accepted as positive evidence due to the possibility of contamination or to unconvincing data.

The greatest concentration of the Roraima Formation is in eastern Venezuela, and other concentrations are centered in central-western Venezuela and northern Brazil. The sandstone mountains, which represent the eroded remnants of a once greater sedimentary mass, rest at different altitudes above sea level on the older crystalline Precambrian basement. In the western portion of the Venezuelan Guayana of Territorio Federal Amazonas, the base of the mountains begins as low as 100–150 m above sea level in lowlying savannas and rain forest adjacent to streams draining into the Orinoco or Ventuari rivers. Eastward in Venezuela the mountains rise from plateau levels in the Estado Bolívar from slightly higher levels of 250–300 m above sea level, and in the Caroni River area from still higher levels of 300–500 m. In the extreme eastern portion, the plateaus on which the mountains rest reach 1000–1300 m.

The question of the actual time of erosion of the sandstones to produce the present tepui landscape is still a matter of disagreement. Some authors have suggested that erosion began in the Late Cretaceous and the Early Tertiary, and was followed by another period of erosion in the Late

Tertiary (McConnell, 1959a; Gansser, 1974) that resulted in the present topography of the table mountains.

ASPECTS OF THE VEGETATION AT DIFFERENT ALTITUDES BELOW THE SUMMITS

Vegetation at the Base of the Mountains

As already indicated, the mountains in the western portion of the Guayana Shield start at lower altitudes than those at the eastern extreme, with a difference of as much as 1000 m between the two sectors. The western mountains rise either from lowland savanna areas situated on white sand or igneous basement rock, or from rain forest terrain. The mountains in the eastern sector rise from savannas or forests occurring on higher plateaus covered with sandstone, lateritic, or igneous-based rock.

The greater percentage of forests in the Guayana Shield are situated in low erosional valleys and river systems lying between the elevated sandstone mountains and plateaus, or form a green cover the associated igneous-based mountains. The taxa of such forests, which comprise a rich and diversified flora, have been derived either mainly from the south, dominated by lowland Amazonian elements, or else from the east with a dominance of the lowland Guianan element. Few, if any, elements appear to have been derived from the flora of the Guayana Shield itself.

These forests consist primarily of evergreen rain forests between 25 and 45 m tall, at altitudes generally less than 500 m, often forming an interminable mass of green cover between the mountains. In the Estado Bolívar of Venezuela they appear along the middle and lower Rio Caura between 100 and 200 m, along the upper Caura and Rio Erebato at 250–350 m, along the Rio Paragua at 250–350 m, along the Paramichi and headwaters of the Paragua River at 400–500 m, in the lower Rio Caroni at 100–250 m, and in the middle and upper Caroni at 250–500 m. On the other hand, in the Territorio Federal Amazonas the greater portion of the forested areas is found at lower altitudes of 100–200 m along the Orinoco and Ventuari rivers and their tributaries. This is due to the removal of intervening strata of the Roraima Formation by periods of erosion leaving only the basal, largely igneous, basement complex (Cuchivero-Pastora) intact (Lexico Estratigráfico de Venezuela, 1970; Ghosh, 1977; Maguire, 1979).

The riverine forest species of the Estado Bolívar and Territorio Federal Amazonas frequently occupy annually flooded terrain and often show a broad distributional range in the Guianas and Amazonia. This type of distribution is exemplified by *Caryocar microcarpum* (Caryocaraceae), *Caraipa densifolia* subsp. *densifolia* (Guttiferae), *Macrolobium acaciaefolium* (Leguminosae), *Abuta grandifolia* (Menispermaceae), and *Panopsis rubescens* (Proteaceae). Such taxa are generally well distributed in Estado Bolívar and in Territorio Federal Amazonas.

On the other hand, lowland forest species inhabiting nonflooded forest terrain may exhibit an entirely different type of distributional pattern. For example, both *Sciadotenia cayennensis* (Menispermaceae) and *Strychnos jobertiana* (Loganiaceae) are confined in Venezuela to only one or two localities in the Territorio Federal Amazonas, whereas *Strychnos mitscherlichii* (including varieties), occurs only in the eastern part of the Venezuelan Guayana in the Estado Bolívar, but elsewhere is widely distributed throughout Amazonas and the Guianas. Other lowland forest species, such as *Sacoglottis cydonioides* and *Panopsis sessilifolia*, limited to the eastern part of the Venezuelan Guayana of Estado Bolívar, are more restricted geographically, being found only in eastern Amazonas and the Guianas. The distributional pattern of various Venezuelan forest taxa occurring in the Guayana Shield supports the data given by Prance (1973, 1978) for distributional patterns of species inhabiting flooded and nonflooded forest terrain.

Vegetation on Talus Slopes

The evergreen rain forest generally found up to 500 m gives way to a humid montane forest as the slopes of the table mountains are ascended. The physiognomy of the forest continues to consist of trees of generally tall stature. But the differentiation of the predominantly lowland forest species becomes more marked as contact is made with the sandstone talus slope (Fig. 13–2). As an example the genus *Kotchubaea* (Rubiaceae) (Fig. 13–3) may be cited. Most of the species of this genus are distributed in lowland elevations of Amazonian Brazil and Peru as well as the Guianas, but on the lower sandstone slopes of the Chimantá Massif and Cerro de la Neblina are found *K. longiloba* and *K. neblinensis*, respectively.

Higher on the talus slopes and in closer proxi-

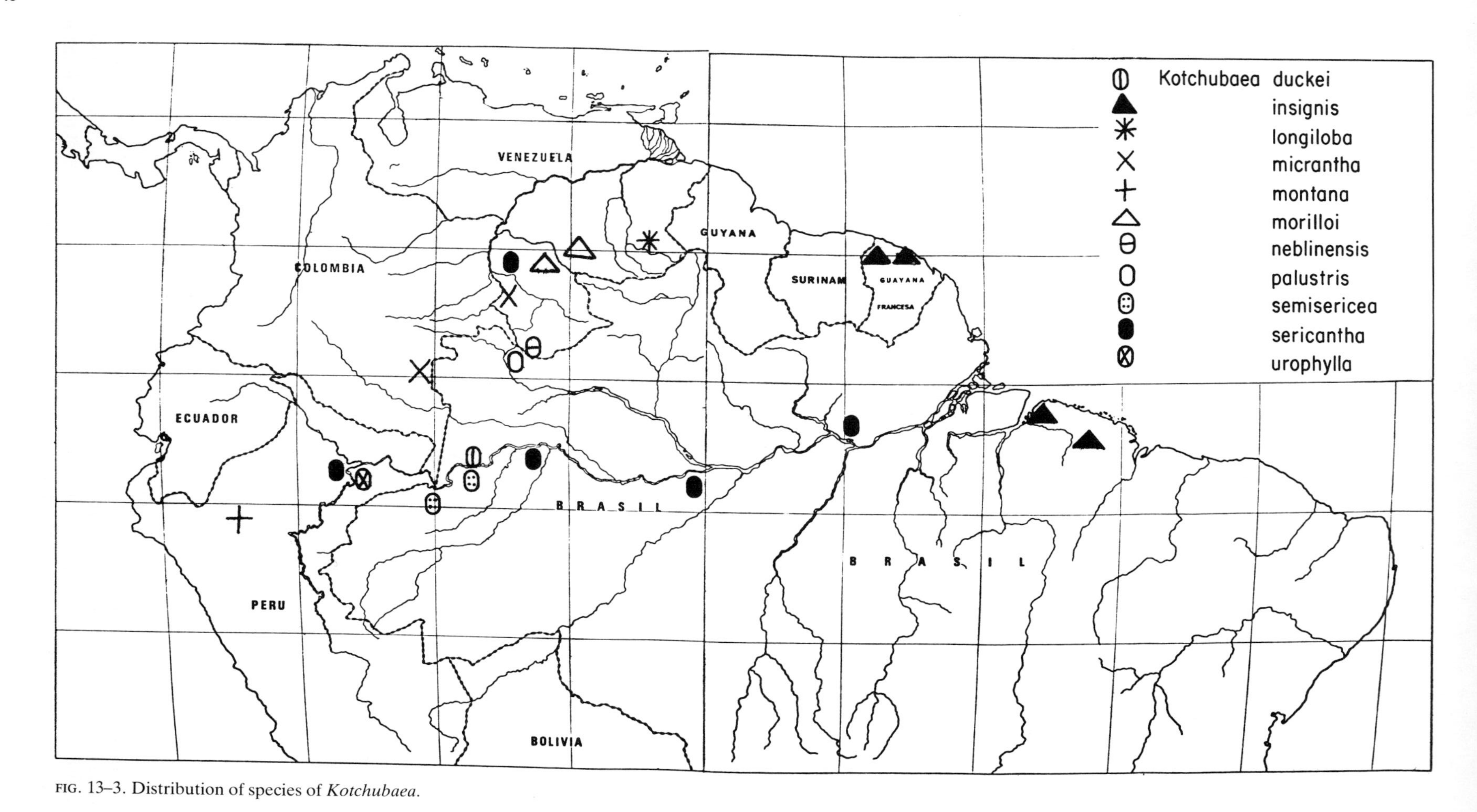

FIG. 13–3. Distribution of species of *Kotchubaea*.

mity to the base of the sandstone escarpment the change in forest flora becomes more marked. Many genera drop out and are replaced by others such as *Clusia* and *Moronobea* (Guttiferae), *Miconia* and *Graffenrieda* (Melastomataceae), *Magnolia* (Magnoliaceae), *Myrcia* (Myrtaceae), *Schefflera* (Araliaceae), *Drimys* (Winteraceae), *Viburnum* (Caprifoliaceae), *Ocotea* and *Nectandra* (Lauraceae), *Brunellia* (Brunelliaceae), *Weinmannia* (Cunoniaceae), *Vochysia* (Vochysiaceae), *Hieronyma* (Euphorbiaceae), *Ficus* (Moraceae), and others. Ericaceae and ferns are especially common as the base of the sandstone escarpment is approached.

Additionally, many species of Andean ancestry have evolved in the cooler, humid forests of the upper talus slopes of the tepuis of Estado Bolívar and cerros of Territorio Federal Amazonas. This is evidenced by the numerous instances in which most of the members of a genus have the greater part of their geographic range centered in various portions of the Andes, but have one or more outlying disjunct species or subspecies isolated in some part of the Guayana Shield, and more particularly on the elevated sandstone mountains. The following endemic species, belonging to genera of predominantly Andean distribution, have evolved on the middle or upper forested talus slopes overlying the sandstone of the Roraima Formation: *Mandevilla pachyphylla* (Apocynaceae); *Schefflera duidae* (Araliaceae); *Weinmannia roraimensis* (Cunoniaceae); *Sphaeradenia rubescens* (Cyclanthaceae); *Carex roraimensis* (Cyperaceae); *Vaccinium euryanthum* (Ericaceae); *Aulonemia*, *Chusquea*, *Cortaderia roraimensis*, and *Neurolepis* spp. (Gramineae); *Hypericum roraimense* (Guttiferae); *Nectandra* spp. (Lauraceae); *Tofieldia schomburgkiana* (Liliaceae); *Chaetolepis anisandra*, *Bertolonia venezuelensis*, *Macrocentrum minus*, and *Salpinga maguirei* (Melastomataceae); *Grammadenia ptariensis* (Myrsinaceae); *Acineta alticola*, *Brachionidium brevicaudatum*, *Lepanthes duidensis*, *Stelis grossilabris*, and *S. obovata* (Orchidaceae); *Podocarpus roraimae*, *P. steyermarkii*, and *P. tepuiensis* (Podocarpaceae); *Palicourea obtusata* (Rubiaceae); and *Ternstroemia* spp. (Theaceae).

Vegetation at the Base of Escarpments

The high vertical sandstone bluffs separating the forested talus slopes from the summit of the table mountains afford a distinct habitat for those species that have become adapted to a lithological medium (Fig. 13–2). Most of the plants that grow on the bare sandstone strata or are lodged in its crevices belong to endemic species closely associated with this particular type of habitat. Apparently, few taxa have successfully adapted to this restricted ambience with its scarcity of soil or support for root growth. This ability to lodge on dry or moist, shaded ledges or in crevices of high bluffs gives these plants a certain advantage in competition with other plants that may require more soil, root space, or a forest medium.

The family Bromeliaceae has been especially successful in this type of habitat, with the genera *Cottendorfia*, *Navia* and *Brocchinia* conspicuously prominent in the evolution of endemic taxa originating on different table mountains. The following species of Bromeliaceae are among some of these cliff dwellers: *Cottendorfia serrulata*, *C. longipes*, *Brocchinia secunda*, *Navia lindmanioides*, *N. splendens*, *N. steyermarkii*, *N. pulvinata*, *N. barbellata*, *N. latifolia*, *N. cretacea*, *N. ocellata*, *N. immersa*, and *Connellia nutans*.

In addition, many other genera belonging to diverse families have evolved species restricted to the bluff habitat. Among these may be mentioned *Ichnanthus duidensis*, *I. longifolius* and *Axonopus steyermarkii* (Gramineae); *Rhynchospora ptaritepuiensis* (Cyperaceae); *Xyris ptariana* (Xyridaceae); *Paepalanthus scopulorum* (Eriocaulaceae); *Stegolepis gleasoniana*, *S. perligulata*, and *S. breweri* (Rapateaceae); *Sauvagesia duidae* and *S. longipes* (Ochnaceae); *Miconia curta* subsp. *curta*, *Macrocentrum longidens*, and *M. steyermarkii* (Melastomataceae); *Nautilocalyx resioides* and *Tylopsacas cuneatus* (Gesneriaceae); *Utricularia aureomaculata* and *U. heterochroma* (Lentibulariaceae); and *Alomia ballotaefolia* (Compositae).

Some of these bluff dwellers, such as *Paepalanthus*, *Xyris*, *Utricularia*, and *Stegolepis*, have adapted to an abundance of moisture on the bluff ledges. Others, such as *Navia* and *Macrocentrum* seem to require a drier habitat under overhangs and away from the spray of waterfalls or torrential rains. Although many of these species are restricted to the base of the sandstone escarpments, others occupy equally diversified habitats on the sandstone summit of the tepuis.

SUMMIT VEGETATION, SPECIATION AND ENDEMISM

On the summits of the sandstone table mountains are found (1) forests of tall or dwarfed trees, including riverine forests; (2) epiphytes in forest associations; (3) shaded crevices of rocks, bluffs, and ledges; (4) wet or dry open savannas without rock outcrops; and (5) exposed rock outcrops, open sandy or rocky areas. Of these environments, forest associations are occupied by the majority of the genera present (amounting to 250 angiospermous and 33 pteridophyte genera of a total of 460 genera listed in the Appendix). Savannas and rocky open areas, sand, and exposed rocks are the next most commonly occupied (117 genera appearing on savannas, and 118 on sand, rocks, open areas, or exposed rocks; Steyermark [1979]). Note that 71 genera ordinarily in savanna habitats may inhabit other open sites and that 80 genera usually found on rocky outcrops, sandy, or rocky open areas also may occur elsewhere.

The percentage of endemic genera strictly limited to the summits amounts to only 8.5%, or 39 genera of a total of 460 known genera. These genera are as follows: *Hymenophyllopsis*, *Salpinctes*, *Crepinella* (now relegated to *Schefflera*), *Cardonaea*, *Chimantaea*, *Duidaea*, *Eurydochus*, *Glossarion*, *Tyleropappus*, *Guaicaia*, *Neblinaea*, *Quelchia*, *Mycerinus*, *Tepuia*, *Roraimanthus* (of doubtful status), *Wurdackia* (also doubtfully distinct), *Celianella*, *Myriocladus*, *Pyrrorhiza*, *Mallophyton*, *Neblinanthera*, *Adenanthe*, *Adenarake*, *Tyleria*, *Phelpsiella*, *Saccifolium*, *Cephalodendron*, *Chondrococcus*, *Coryphothamnus*, *Maguireocharis*, *Duidania*, *Pagameopsis*, *Achnopogon*, *Tateanthus*, *Neotatea*, *Neogleasonia*, *Neblinaria*, *Ayensua*, and *Achlyphila*.

Some of these genera are restricted to only one or two mountain summits, but others are more generally distributed. Cerro de la Neblina, on the southwestern border of Venezuela, possesses the greatest number (12) of endemic genera restricted to the summit of a single mountain: *Eurydochus*, *Glossarion*, *Guaicaia*, and *Neblinaea* (tribe Mutisieae, Compositae); *Pyrrorhiza* (Haemodoraceae), *Neblinanthera* (Melastomataceae); *Adenarake* (Ochnaceae); *Neblinaria* (Theaceae [Bonnetiaceae]); *Cephalodendron* and *Maguireocharis* (Rubiaceae); *Saccifolium* (Saccifoliaceae); and *Achlyphila* (Xyridaceae).

The summits of other table mountains harbor smaller numbers of genera so restricted to their summits, as may be noted in Table 13–1.

Of the 39 strictly endemic summit genera, only 7 (17.95%) inhabit forest environments (*Tyleria*, *Neblinanthera*, *Tateanthus*, *Pagameopsis*, *Cephalodendron*, *Duidania*, and *Eurydochus*), but even some of these may inhabit more open dwarf forest. No endemic summit genera are epiphytic. The three endemic summit genera (7.7%) principally inhabiting shaded bluffs, ledges, and rock crevices are *Wurdackia* (Eriocaulaceae,) *Saccifolium* (Saccifoliaceae), and *Hymenophyllopsis*

TABLE 13–1. Relative occurrence of 18 endemic genera on table mountain summits, excluding Cerro de la Neblina.

Genus (Family)	Mountain(s)
Restricted to the summit of one mountain	
Wurdackia (Eriocaulaceae)	Chimantá
Mallophyton (Melastomataceae)	Chimantá
Adenanthe (Ochnaceae)	Chimantá
Phelpsiella (Rapateaceae)	Parú
Chondrococcus (Rubiaceae)	Parú
Coryphothamnus (Rubiaceae)	Auyán-tepui
Salpinctes (Apocynaceae)	Duida
Restricted to the summits of two mountains	
Cardonaea (Compositae)	Sarisariñama and Marahuaca
Duidaea (Compositae)	Duida and Marahuaca
Tyleropappus (Compositae)	Duida and Marahuaca
Achnopogon (Compositae)	Auyán-tepui and Chimantá
Tepuia (Ericaceae)	Auyán-tepui and Chimantá
Ayensua (Bromeliaceae)	Auyán-tepui and Uaipán-tepui
Found on the summits of three or more mountains	
Tyleria (Ochnaceae)	Known from five different summits
Tateanthus (Melastomataceae)	Duida, Marutani, and Sarisariñama
Chimantaea (Compositae)	Chimantá, Auyán-tepui, and Aprada-tepui
Pagameopsis (Rubiaceae)	Known from five different summits
Hymenophyllopsis (fern)	Known from seven different summits

(Pteridophyta). Seven endemic summit genera (17.95%) are restricted to wet or dry savannas (*Salpinctes*, *Achnopogon*, *Mycerinus*, *Mallophyton*, *Adenarake*, *Neblinaria*, and *Neotatea*), and 10 (25.6%) to open scrub, bush, forest margins, or even savanna-like habitats (*Guaicaia*, *Sipapoanthus*, *Myriocladus*, *Adenanthe*, *Tyleria*, *Phelpsiella*, *Aphanocarpus*, *Pagameopsis*, *Neogleasonia*, and *Achlyphila*). Twelve (30.8%) of the endemic summit genera favor rock outcrops, or open sandy and rocky areas (*Cardonaea*, *Duidaea*, *Glossarion*, *Neblinaea*, *Tyleropappus*, *Tepuia*, *Roraimanthus*, *Celianella*, *Myrtus alternifolia*, *Chondrococcus*, *Coryphothamnus*, and *Maguireocharis*). An additional 13 other genera are usually found on rock outcrops or sandy open places, but may also frequent scrub forest or forest margins (*Ayensua*, *Chimantaea*, *Guaicaia*, *Quelchia*, *Mycerinus*, *Sipapoanthus*, *Myriocladus*, *Adenanthe*, *Tyleria*, *Phelpsiella*, *Cephalodendron*, *Pagameopsis*, and *Neogleasonia*.

The majority (27 or 69%) of the endemic genera restricted to the summits are monotypic, while three (7.7%) genera (*Achnopogon*, *Neogleasonia*, and *Neotatea*) consist of three species, and only five genera (*Hymenophyllopsis*, *Chimantaea*, *Quelchia*, *Tyleria*, and *Myriocladus*) comprise four or more species. The last-mentioned genus, *Myriocladus*, has actually speciated into 20 taxa distributed on the summits of various tepuis.

Although nearly two thirds of the summit genera with the greatest diversity of species occupy forest habitats (dwarf, tall, or riverine), only eight of the strictly endemic summit genera (20.5%) are usually associated with the forest habitat. On the other hand, the more impoverished habitats of open savannas are the sites where the greater number of strictly endemic summit genera are found. It is thus likely that speciation took place in such open areas.

In the summit flora of the Jáua Massif (Steyermark and Brewer-Carias, 1976), 32 families, 78 genera, and 120 species are enumerated from the savannas and open rocky and sandy habitats, whereas much larger totals of 75 families, 207 genera, and 402 species are recorded from the forested habitats. On the summits of the larger table mountains (Chimantá Massif, Jáua, Duida, Guaiquinima, and Neblina), where forest habitats are common, total plant collections indicate a great difference in the numbers of species, the forested habitats containing by far the greater number of genera and species. The family Orchidaceae, the largest family occurring on the table mountains, has the greatest number of species in the forested habitats, thus augmenting the total numbers of genera and species. By contrast, some of the smaller tepuis, such as Roraima and Ptari-tepui, whose summits consist largely of bare rock outcrops and some savanna-like areas, have a greatly reduced and impoverished flora. Upon such mountains it is the forest habitats that possess the higher number of families, genera, and species (Steyermark, 1974b).

Speciation and Adaptations

Following periods of uplift, subsequent erosion resulted in separation of the sandstone table mountains, and in isolation of distinct summits. Many ancestral plant genera related to lowland, Andean, and pantropical floral elements have undergone speciation on the summits. As a result, a far larger proportion of endemic species occur on the summits of the sandstone mountains than on their forested talus slopes or on their high vertical escarpments. The number of endemic species on a given summit of a table mountain appears to be correlated with size of mountain, altitude, and diversity of habitat. In general, the greater the surface area available, the greater the number of endemic species. Mountains with summits between 1500 and 3000 m appear to have more endemic species than those below 1500 m. The greater the diversity of habitats on any summit (rock outcrops, swampy, open, savanna-like areas, streams, dwarf and tall forests), the greater the total numbers of species and endemic taxa present. Thus, each of the larger mountains (Chimantá Massif, Auyán-tepui, Duida, Sipapo, Neblina, Meseta de Jáua-Sarisariñama, and Guaiquinima) has varied habitats and a high proportion of endemic species. On the other hand, the summits of Roraima and Ptari-tepui, although attaining 2610 and 2620 m, respectively, have more uniform, largely bare rocky surfaces with little variation in habitats and relatively small surface areas. They have fewer endemic species. Mount Roraima was the first to have its summit explored botanically and its endemic flora was reckoned at nearly 90% of its species. With subsequent exploration of other Venezuelan mountains on whose summits the same species were found, the number of endemic species on Roraima fell to 54% (including the upper slopes). On the summit of Ptari-tepui, likewise, only 10% of the summit flora is endemic, and on the very reduced area of the summit of Cerro Autana only 5.3% is endemic.

Especially harsh environmental factors prevail

on the rockiest and most exposed sandy portions of the summits. As a result of a combination of high acidity with low nutrient content of the sandstone substrate, rapid runoff following strong precipitations, winds and subsequently high evaporation rates, reduced temperatures, and intense sunlight, the flora is largely xeromorphic, despite heavy rainfall most of the year. This is attested to by the evolution of bizarre forms with thick, sclerophyllous, highly reduced, glossy, waxy, or revolute leaves, often in tufts or rosettes, or covered with a sericeous gray, white, or brown tomentum (Schultes, 1944, 1945; Steyermark, 1966, 1967a; Steyermark and Brewer-Carias, 1976). Frequently, the stem is greatly shortened or elongated, simple, and virgate, giving a weird appearance to the landscape. The unique summit genera showing this highly modified growth related to a severe habitat are exemplified by *Aphanocarpus*, *Adenanthe*, *Ayensua*, *Bonnetia*, *Chimantaea*, *Coryphothamnus*, *Duidaea*, *Orectanthe*, *Neblinaria*, *Neotatea*, *Neogleasonia*, *Maguireothamnus*, *Tepuia*, *Tyleria*, and *Stomatochaeta*.

A close correlation exists between the occurrence of these specialized genera and their adaptation to open rocky sandstone or sandy terrain. The various changes and disturbances, concomitant with the erosional history of the mountains, would have produced, at different times, freshly exposed rocky or sandy habitats. These newly available habitats would be occupied by those pioneering taxa, which, during the stages of evolutionary development, could adapt to this xerophytic environment. Such a view agrees, at least in part, with that of Raven and Axelrod (1974), who suggested that many taxa of endemic families may have survived or evolved under edaphic desert or xerophytic conditions.

The great cycles of speciation in the tepuis appear to have been correlated with changes induced by past geologic and paleoclimatologic events that have produced a particular combination of edaphic conditions characteristic of the talus slopes, sandstone escarpments, and exposed summits. The present percentage of speciation of endemic taxa would appear to be associated with, and to increase in direct proportion to, the presence and dominance of the sandstone substrate. Apparently, speciation has proceeded at an unequal rate on the tepuis. This nonuniformity of speciation is well shown by the differences expressed in the geographic distribution of the taxa encountered on the summits, bluff escarpments, and talus slopes. These different patterns of distribution are now discussed in sequence, beginning with the generally distributed taxa not showing speciation and ending with those that are highly localized.

Generally Distributed Genera and Species Not Showing Speciation

Although the actual summits of the sandstone table mountains demonstrate the greatest degree of speciation in the flora found thereon, it is important to draw attention to the fact that a considerable number, although small percentage, of summit species are generally distributed elsewhere. These may occur not only on the slopes and summits or other sandstone mountains, but are found in other portions of the Guayana Shield at lower elevations, as well as in different sectors of South or even of Central America.

Many of the Orchidaceae have a wide geographic range outside the tepui summits. Since the Venezuelan tepuis are located within the states of Bolívar and Territorio Federal Amazonas, it is noteworthy that 46% of the approximately 1200 species of Orchidaceae recorded for Venezuela (Dunsterville and Garay, 1959–76; Foldats, 1969–70) occur in the Guayana Shield of the combined states of Bolívar and Territorio Federal Amazonas (Dunsterville and Garay, 1959–1976, 1979). Moreover, 76% of all Venezuelan species are shared by the state of Bolívar and the Andes, and 63% of all Venezuelan species occur in the combined Territorio Federal Amazonas and the Andes. On the summits of the sandstone mountains occur many species of orchids whose range extends widely outside of Venezuela: *Elleanthus kermesianus*, *E. linifolius*, *Epidendrum coriifolium*, *E. fragrans*, *E. nocturnum*, *E. ramosum*, *E. repens*, *E. secundum*, *E. teretifolium*, *Habenaria caldensis*, *Lepanthes lindleyana*, *Lepanthopsis floripecten*, *Maxillaria alpestris*, *M. aurea*, *M. brunnea*, *M. violaceopunctata*, *Pleurothallis ciliaris*, *P. foliata*, *P. imraei*, *P. sclerophylla*, *Pogonia rosea*, and *Stelis cucullata*.

The preceding examples represent a small percentage of the species comprising the flora that show no geographic speciation, but that nevertheless occur on the summits of the table mountains. Although some of them, such as ferns, bromeliads, and orchids, belong to groups that possess lightweight, wind-dispersed seeds or spores, others have fleshy fruits that are not wind-distributed (e.g. *Gnetum*, *Calyptranthes*, *Miconia*, and *Sphyrospermum*).

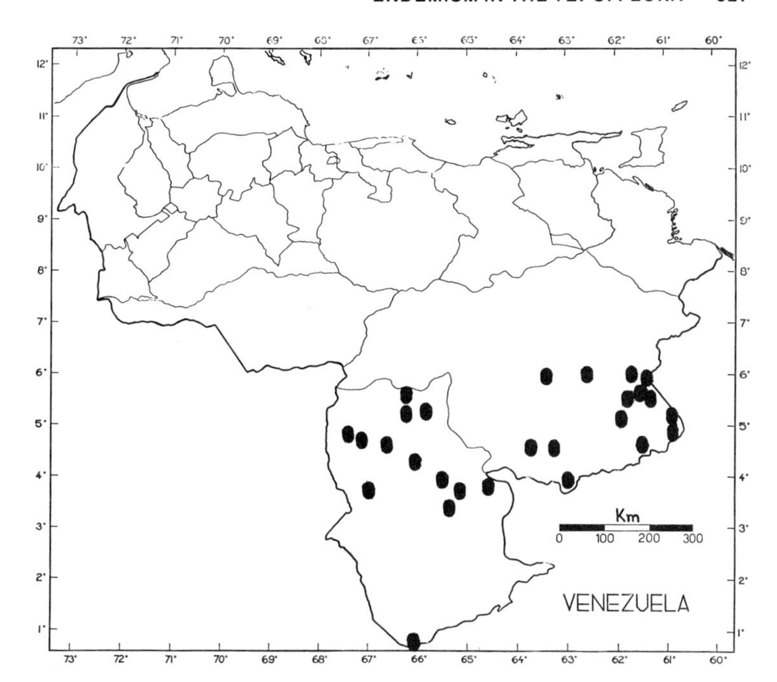

FIG. 13–4. Distribution of *Ilex retusa,* an example of a species autochthonous to the tepuis but of general distribution in the Guayana Highland.

Generally Distributed Genera and Species Showing Speciation

Perhaps the highest percentage of the summit flora comprises species whose range extends across many of the larger table mountain summits. Indeed, some of these summit species, such as *Drosera roraimae*, *Utricularia humboldtii*, and *Ilex retusa* (Fig. 13–4), occur on most or all of the sandstone summits that have been explored thus far. Other species, however, are less widespread. It can only be conjectured whether such generally distributed taxa owe their widespread range to inherent biological factors that have enabled them to adapt better to their environment following a succession of past geologic events, to their pioneering ability to occupy available summit habitats, or to an original widespread distribution related to a period when (and if) the Guayana Highland summit area was more continuous. I can only state at present that such summit taxa have successfully maintained a broad range throughout the table mountain area, and at the same time have not speciated.

Generic Segregation into Eastern and Western Zones

An analysis of the summit flora shows that in addition to the generally distributed species found in numerous genera, there is also a geographic restriction of many genera to eastern or western sectors of the Guayana Highland, respectively, as indicated in Table 13–2, and as shown in Figs. 13–5 and 13–6. This generic partitioning and speciation probably originated at some remote period during the past geologic history of the Guayana Highland area, following periods of erosion or uplift when more extensive or continuous mountains of sandstone became separated from one another. These periods of uplift and/or erosion were probably uneven, both in time and space, and may have altered various sectors of the Guayana Highland at different time intervals. The western mountains, such as Neblina, Sipapo, Parú, Guanay, and Duida, have a much more irregular physiographic summit contour, owing to more block faulting and/or other metamorphic and diastrophic events, than the remarkably flat summits of Roraima, Ptari-tepui, Chimantá, and Auyán-tepui of southeastern Venezuela, which by comparison suffered fewer geologic disturbances or less metamorphism. Cerro de la Neblina, with its peaks projecting above the level of the rest of the summit, together with its sharp ridges, steep valleys, and deep canyons, is strikingly different from the other sandstone mountains.

The uneven series of geologic events, with sedimentation processes occurring at different times and in different places (basins) already referred to (Gansser, 1974; Ghosh, 1977), may have been primary factors affecting the different types of

TABLE 13–2. Summit genera with marked geographic segregation.

Eastern (Estado Bolívar)	Western (chiefly Terr. Fed. Amazonas)
Eriocaulaceae	Xyridaceae
Wurdackia	*Achlyphila*
Bromeliaceae	Rapateaceae
Ayensua	*Phelpsiella*
Connellia	
Ochnaceae	Ochnaceae
Adenanthe	*Adenarake*
	Tyleria
Melastomataceae	Melastomataceae
Mallophyton	*Neblinanthera*
	Tateanthus (east to Marutaní, Estado Bolivar)
Araliaceae	Haemodoraceae
Crepinella	*Pyrrorhiza*
Ericaceae	Apocynaceae
Tepuia	*Salpinctes*
Rubiaceae	Rubiaceae
Aphanocarpus	*Cephalodendron*
Coryphothamnus	*Chondrococcus*
	Duidania
	Maguireocharis
	Neblinathamnus
Compositae	Compositae
Achnopogon	*Duidaea*
Chimantaea	*Eurydochus*
Quelchia	*Glossarion*
	Guaicaia
	Neblinaea
	Tyleropappus
	Theaceae (Bonnetiaceae)
	Neblinaria
	Neogleasonia
	Neotatea
	Saccifoliaceae
	Saccifolium

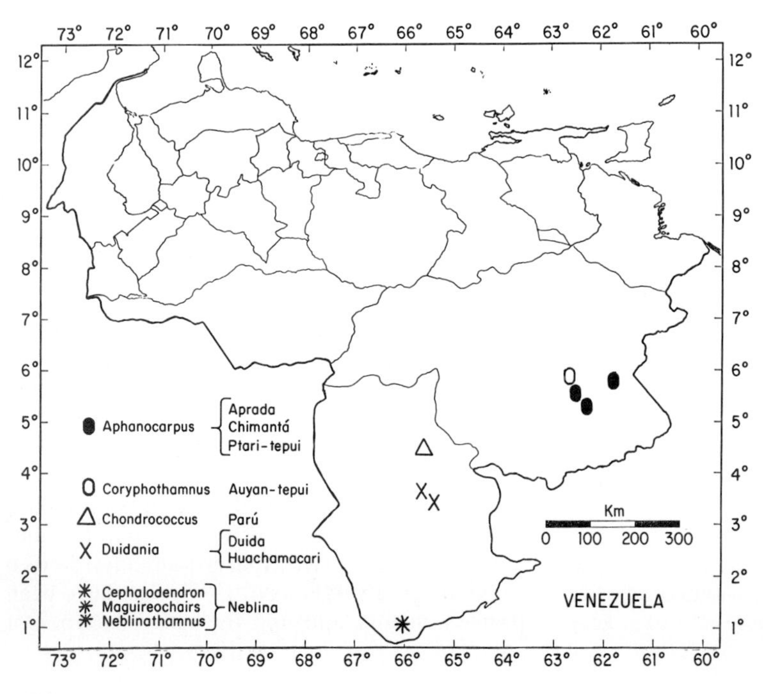

FIG. 13–5. Summit genera of Rubiaceae segregated into eastern and western zones.

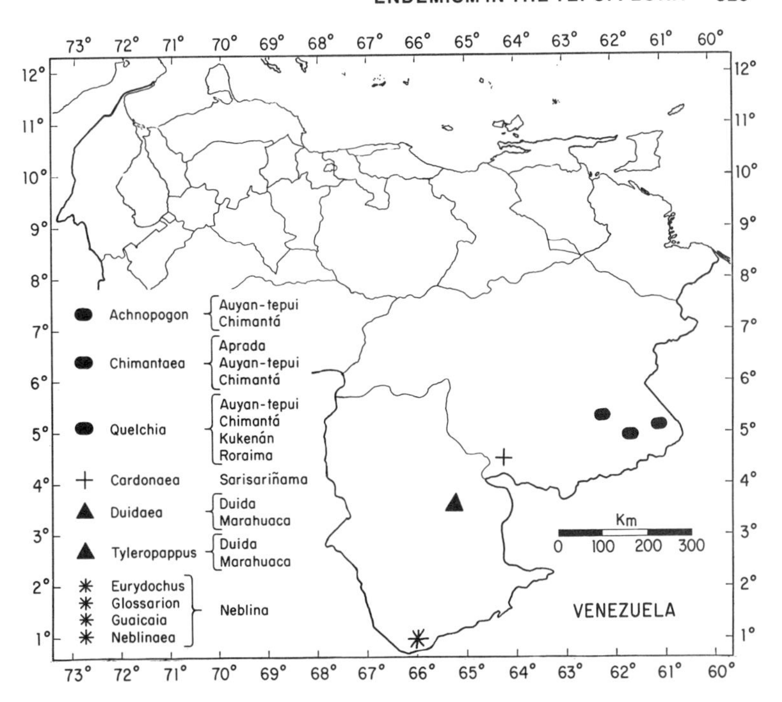

FIG. 13–6. Summit genera of Compositae segregated into eastern and western zones.

contour and physiography of the tepuis, ultimately related to the generic segregation of a number of genera of the Guayana Highland flora, in which lack of uniformity of geographic distribution is the usual feature rather than the exception, among the genera. On the other hand, some of the endemic summit genera have a more widespread distribution in the Guayana Highland area, where they are represented by one (*Didymiandrium*, *Vellozia*, *Celianella*), two (*Nietneria*, *Mycerinus*, *Pagameopsis*), or more species.

Speciation in Various Eastern and Western Tepuis

Geographic segregation of the taxa including the Guayana Highland, and especially of those taxa found on the summits or upper escarpments of the tepuis, is most clearly evidenced in the speciation of thousands of specific and subspecific taxa of limited occurrence, or of those confined to only certain tepuis. Maguire (1970) has indicated a total of some 4000 endemic species already described from the Guayana, with an estimated total flora of about 8000 species, "of which more than 75 percent would be confined to the Guayana region." This estimate more recently has been raised to "in excess of 10,000 species" by Maguire (1979). This wealth of flora, much of it of unique character, is due to a combination of factors: (1) the long geologic history of the terrain together with the probable availability for plant occupation possibly since the Late Cretaceous; (2) isolation of tepuis into islands with resultant specific segregation following or during uplift and erosion of the mountains at different time intervals; (3) combinations of particular edaphic conditions peculiar to the Guayana Highland, such as soils of high pH, with high acidity, high rainfall, high wind, and peculiar sunlight and temperature parameters; and (4) former connections with floral elements derived from forest refugia originating from the Amazonian Hylaea and the Andes preceding or during Pleistocene climatic changes, as well as ancient relict relationships with western Gondwana or Malaysian-Australasian floral elements (Simpson and Haffer, 1978; Maguire, 1979; Steyermark, 1979a,b,c). Thus a whole set of conditions has contributed to the present flora and helps to account for the apparent lack of uniformity evidenced by the geographic distribution of many of the specific taxa. The geologic events of the periods of uplifts and erosion, and the differences of deposition of sediments at various time intervals and in different basins would also help explain the lack of uniformity throughout the Venezuelan Guayana.

Examples are legion in the flora of the Guayana Highland of specific and subspecific taxa that follow various patterns of geographic isolation and segregation. The genus *Navia* of the Bromeliaceae, mainly confined to sandstone crevices and outcrops on the summits, or to sandy savannas, rarely on igneous outcrops, is distributed throughout the limits of the Guayana High-

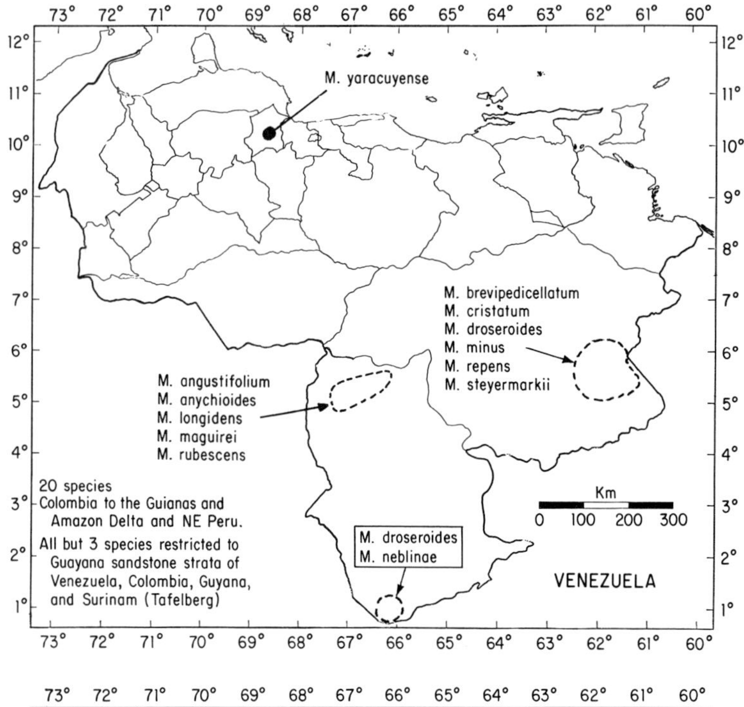

FIG. 13–7. Geographic segregation of the genus *Macrocentrum* into eastern and western zones.

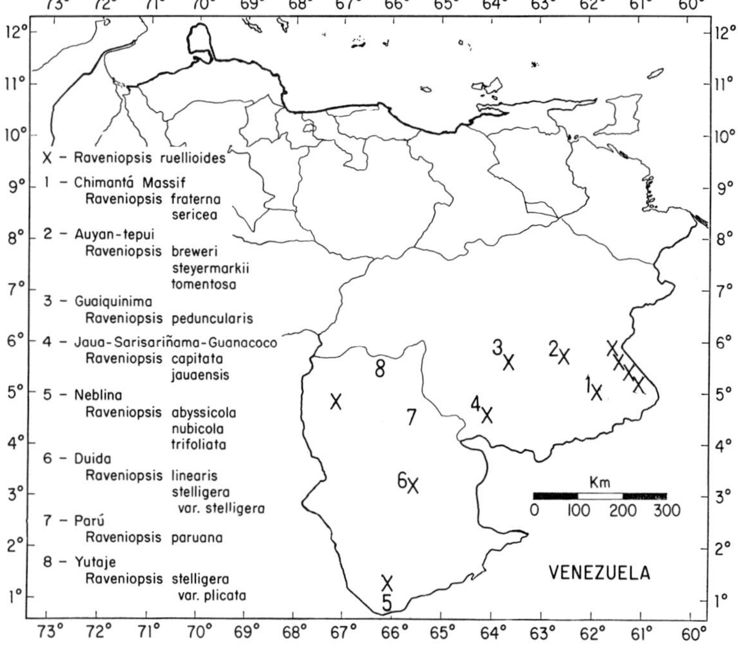

FIG. 13–8. Geographic distribution of *Raveniopsis*, a genus autochthonous to the Guayana Highland, found mostly on the summits of the tepuis.

land from Tafelberg in Surinam, Guyana, Venezuelan Guayana, southeastern Colombia, and northern Brazil. In this stretch of territory it has speciated into nearly 80 taxa. However, most of these taxa are limited to only one or two mountains each. Similar limitations are noted in the speciation of other genera characteristic of the Guayana Shield, exemplified by *Bonnetia* (Theaceae [Bonnetiaceae]); *Gongylolepis* and *Stenopadus* (Compositae); *Salpinga*, *Macrocentrum* (Fig. 13–7), *Phainantha*, and *Graffenrieda* (Melastomataceae); *Raveniopsis* (Fig. 13–8) and *Spathelia* (Rutaceae); *Stegolepis* and *Saxofridericia* (Rapateaceae); *Paepalanthus*, *Syngonanthus*, and *Leiothrix* (Eriocaulaceae); *Xyris* (Xyridaceae); *Ilex* (Aquifoliaceae); *Diacidia* (including *Sipapoa*) (Malpighiaceae); *Abolboda* (Abolbodaceae); and various other genera.

Although speciation is well marked in the Guayana on the sandstone summits, talus slopes,

TABLE 13–3. Examples of geographic segregation within genera and species of Guayana.

	Eastern taxon	Western taxon
Abolbodaceae		
Orectanthe	*sceptrum* subsp. *sceptrum*	*sceptrum* subsp. *occidentalis*
Rapateaceae		
Saxofridericia	*regalis*	*compressa*, *duidae*, *grandis*, *spongiosa*, and others
Stegolepis	*angustata*, *guianensis*, *parvipetala*, *steyermarkii*, *vivipara*	*celiae*, *grandis*, *hitchcockii*, *pauciflora*, *pungens*
Stegolepis	*ptaritepuiensis* closely related to	*neblinensis*
Rutaceae		
Raveniopsis	*breweri*, *capitata*, *fraterna*, *peduncularis*, *sericea*, *tomentosa*	*abyssicola*, *linearis*, *nubicola*, *paruana*, *stelligera*, *trifoliata*
Theaceae		
Bonnetia	*sessilis* related to	*crassa*
	steyermarkii related to	*tristyla* (east to Jáua)
Melastomataceae		
Macairea	*aspera*, *cardonae*, *chimantensis*, *parvifolia*	*duidae*, *linearis*, *neblinae*, *rigida*
Macrocentrum	*brevipedicellatum*, *cristatum*, *minus*, *repens*, *steyermarkii*	*angustifolium*, *anychioides*, *longidens*, *maguirei*, *neblinae*, *rubescens*
Graffenrieda	*jauana*, *obliqua*, *sessilifolia*, *steyermarkii*	*cinnoides*, *fantastica*, *fruticosa*, *hitchcockii*, *lanceolata*, *pedunculata*, *reticulata*, *rufa*, *sipapoana*, *tricalcarata*
Apocynaceae		
Aspidosperma	*steyermarkii* related to	*neblinae*
Galactophora	*schomburgkii*	*crassifolia*
Rubiaceae		
Maguireothamnus	*speciosus* related to	*tatei*
Retiniphyllum	*schomburgkii* subsp. *schomburgkii*	*schomburgkii* subsp. *occidentale*
Retiniphyllum	*scabrum* subsp. *scabrum*	*scabrum* subsp. *erythranthum*
Compositae		
Gongylolepis	*benthamiana*	*paniculata* and other species
Stenopadus	*cardonae*, *chimantensis*, *connellii*, *sericeus*, *talaumifolius*	*cucullatus*, *eurylepis*, *huachamacari*, *kunhardtii*, *neblinensis*, and others

and escarpments of the table mountains, it is also conspicuous in the sandy soils developed from the white sands and other types of Guayanan savannas, especially in the Territorio Federal Amazonas, but also in the Gran Sabana of southeastern Estado Bolívar. The speciation in savanna habitats at the lower altitudes is especially outstanding in *Xyris* (Xyridaceae); *Paepalanthus* and *Syngonanthus* (Eriocaulaceae); *Rapatea*, *Duckea*, *Monotrema* and *Schoenocephalium* (Rapateaceae); *Abolboda* (Abolbodaceae); *Sipaneopsis* (Rubiaceae); and *Pachyloma* (Melastomataceae).

Geographic segregation of species in the Venezuelan Guayana, and especially in the Guayana Highland flora, occurs throughout various families of plants. Some of these examples are shown in Table 13–3.

Highly Localized Endemic Genera and Species

Perhaps the plants of the table mountains upon which most attention has been focused are those belonging to highly localized endemic genera and species. These are the taxa found nowhere else in the world, which are either autochthonous to the sandstone summits, escarpments, or talus slopes of the tepuis, or whose ancestors have migrated from some other area and speciated upon contact

TABLE 13–4. The most highly localized monotypic genera on the summits of Venezuelan table mountains.

Xyridaceae *Achlyphila disticha* (Neblina) Rapateaceae *Phelpsiella ptericaulis* (Parú) Haemodoraceae *Pyrrorhiza neblinae* (Neblina)	Melastomataceae *Mallophyton chimantense* (Chimantá) *Neblinanthera cumbrensis* (Neblina) Saccifoliaceae *Saccifolium bandeirae (Neblina)*
Eriocaulaceae *Wurdackia flabelliformis* (Chimantá) Bromeliaceae *Ayensua uaipanensis* (Uaipán-tepui and adjacent Auyán-tepui)	Rubiaceae *Coryphothamnus auyantepuiensis* (Auyántepui) *Cephalodendron globosum* (Neblina) *Chondrococcus laevis* (Parú)
Ochnaceae *Adenarake muriculata* (Neblina) *Adenanthe bicarpellata* (Chimantá)	Compositae *Eurydochus bracteatus* (Neblina) *Glossarion rhodanthum* (Neblina) *Guaicaia bilabiata* (Neblina) *Neblinaea promontorium* (Neblina)
Theaceae (Bonnetiaceae) *Neblinaria celiae* (Neblina)	

Note: Of the 18 taxa listed, 11 are restricted to Neblina, 3 to Chimantá, 2 to Uaipán-Auyán and 2 to Parú.

with the table mountains, or portions of the Guayana Shield.

The list of 18 taxa (Table 13–4) includes the most highly localized monotypic genera, so far as known confined to the summit of only one Venezuelan tepui. In each case the tepui on which the genus is encountered is given in parentheses. In addition, there are other highly localized monotypic genera found elsewhere in the area delimited by the Roraima Formation within and outside Venezuela, which are known at present from only one locality. These include the orchid, *Dunstervillea mirabilis* of southeastern Venezuelan Guayana, various monotypic Melastomataceae of the Guyana sandstone mountains (*Maguireanthus ayangannae*, *Boyania ayangannae*, *Ochthephilus repentinus*, *Tryssophyton merumense*), two Rapateaceae of Guyana (*Potarophytum riparium* and *Windsorina guianensis*), and *Chronocentrum cyathophorum* of the Euphorbiaceae. *Jasarum steyermarkii*, a monotypic aquatic Araceae, is now known from a few local streams of the Gran Sabana of eastern Venezuela and adjacent Guyana.

ANALYSIS OF THE PERCENTAGE OF GENERIC ENDEMISM

In addition to the 39 strictly endemic genera, another 38 genera are found mainly, but not exclusively, on the summits, and are encountered either on some portions of the lower talus slopes, or at the lower altitudes of the surrounding Gran Sabana or lowland forests of Estado Bolívar, or adjacent Guyana, or at still lower elevations of the savannas or forests of Territorio Federal Amazonas (Venezuela), or in adjacent Amazonian Colombia, Brazil, or the Guianas. If these additional quasi-endemic genera are included, the total number of endemic genera is elevated to 77, or 16.7% (Fig. 13–9) of the summit flora (Steyermark, 1979a). These additional 38 genera are the following: *Pterozonium* (ferns); *Orectanthe* (Abolbodaceae); *Cephalocarpus*, *Everardia*, *Didymiandrium*, and *Rhynchocladium* (Cyperaceae); *Amphiphyllum*, *Kunhardtia*, *Saxofridericia*, and *Stegolepis* (Rapateaceae); *Brocchinia*, *Connellia*, *Cottendorfia*, and *Navia* (Bromeliaceae); *Hexapterella* (Burmanniaceae); *Carptotepala* (Eriocaulaceae); *Apocaulon*, *Raveniopsis*, and *Spathelia* (Rutaceae); *Blepharandra*, and *Diacidia* (Malpighiaceae); *Philacra* and *Poecilandra* (Ochnaceae); *Bonnetia* (Theaceae [Bonnetiaceae]); *Macrocentrum* (Melastomataceae); *Elaeoluma* (Sapotaceae); *Chorisepalum* (Gentianaceae); *Ledothamnus* and *Notopora* (Ericaceae); *Velloziella* (Scrophulariaceae); *Tylopsacas* (Gesneriaceae); *Nietneria* (Liliceae); and *Gongylolepis*, *Maguireothamnus*, *Merumea*, *Neblinathamnus*, *Stenopadus*, and *Stomatochaeta* (Compositae).

Tables 13–5 and 13–6 yield a total of 119 genera autochthonous to some part of the Guayana Shield. Thus, the proportion of endemic genera within the Guayana Shield amounts to about 24% of the total generic flora (90 genera from Table

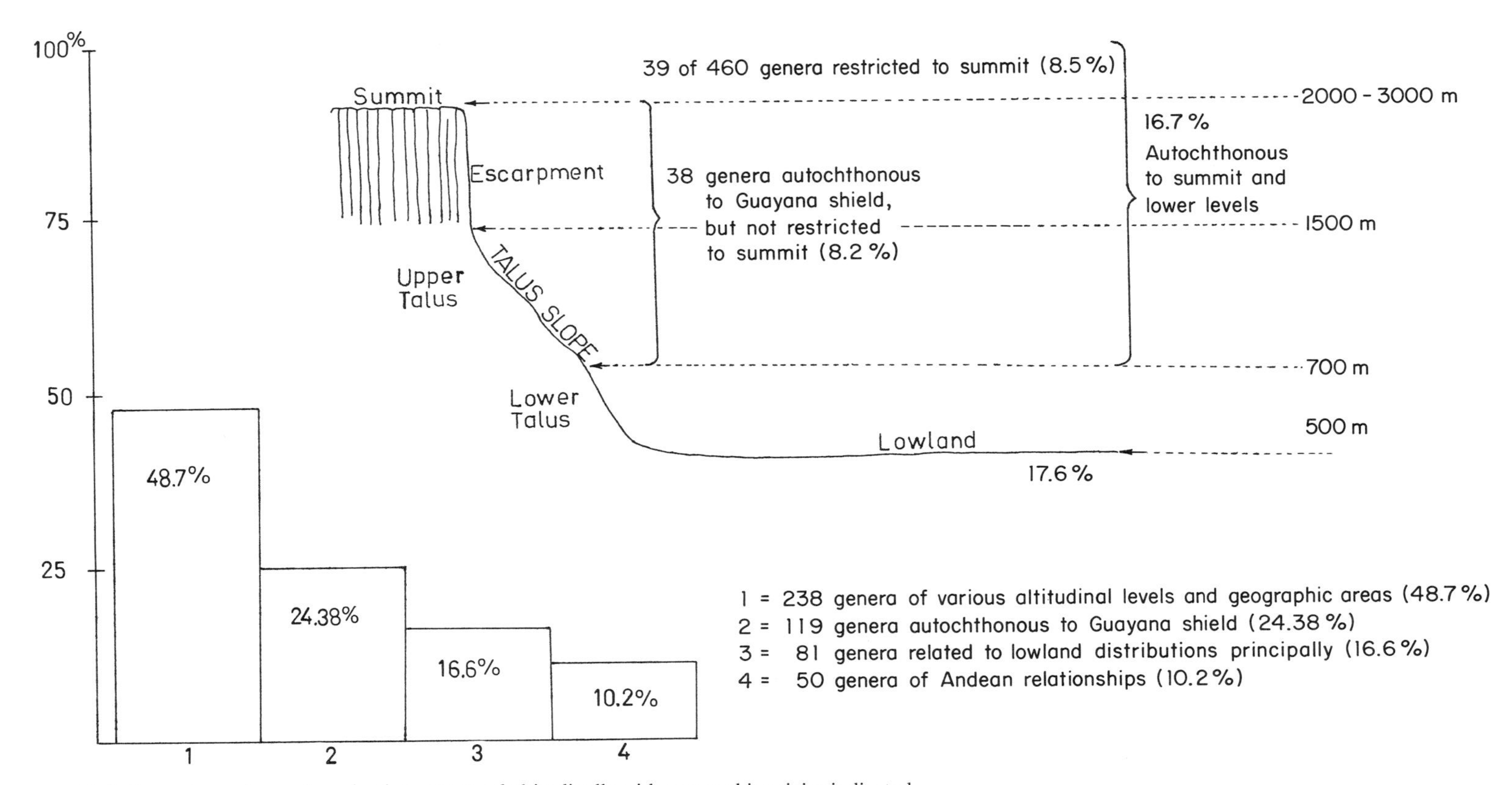

FIG. 13–9. Percentage of generic endemism, arranged altitudinally with geographic origins indicated.

TABLE 13–5. Examples of autochthonous genera and/or species on the Venezuelan table mountains.

Ferns
- *Hymenophyllopsis*
- *Pterozonium*

Gramineae
- *Myriocladus*

Cyperaceae
- *Didymiandrium*
- *Everardia*
- *Rhynchocladium*

Xyridaceae
- *Achlyphila*

Abolbodaceae
- *Orectanthe*

Rapateaceae
- *Amphiphyllum*
- *Guacamaya*
- *Kunhardtia*
- *Phelpsiella*
- *Saxofriedericia*
- *Stegolepis*

Haemodoraceae
- *Pyrrorhiza*

Eriocaulaceae
- *Carptotepala*
- *Roraimanthus*
- *Wurdackia*

Bromeliaceae
- *Ayensua*
- *Brocchinia*
- *Connellia*
- *Cottendorfia*
- *Navia*

Liliaceae
- *Nietneria*

Velloziaceae
- *Vellozia (tubiflora)*
- *Barbacenia (celiae)*

Burmanniaceae
- *Hexapterella*

Orchidaceae
- *Dunstervillea*

Santalaceae
- *Austroamericium (tepuiense)*

Loranthaceae
- *Ixidium (antidaphneoides)*

Sarraceniaceae
- *Heliamphora*

Rutaceae
- *Apocaulon*
- *Raveniopsis*
- *Spathelia*

Tepuianthaceae
- *Tepuianthus*

Tetrameristaceae
- *Pentamerista*

Malpighiaceae
- *Blepharandra*

Euphorbiaceae
- *Celianella*
- *Senefelderopsis*

Ochnaceae
- *Adenarake*
- *Adenanthe*
- *Philacra*
- *Poecilandra*
- *Tyleria*

Dipterocarpaceae
- *Pakaraimaea*

Theaceae (Bonnetiaceae)
- *Bonnetia*
- *Neblinaria*
- *Neogleasonia*
- *Neotatea*

Flacourtiaceae
- *Euceraea (sleumeriana)*

Myrtaceae
- *Myrtus (alternifolia)*

Melastomataceae
- *Farringtonia*
- *Macrocentrum*
- *Mallophyton*
- *Neblinanthera*
- *Phainantha*
- *Tateanthus*

Araliaceae
- *Crepinella*

Ericaceae
- *Ledothamnus*
- *Mycerinus*
- *Notopora*
- *Tepuia*

Gentianaceae
- *Chorisepalum*

Saccifoliaceae
- *Saccifolium*

Apocynaceae
- *Salpinctes*

Scrophulariaceae
- *Velloziella*

Gesneriaceae
- *Tylopsacas*

Rubiaceae
- *Aphanocarpus*
- *Cephalodendron*
- *Chondrococcus*
- *Duidania*
- *Maguireocharis*
- *Maguireothamnus*
- *Merumea*
- *Neblinathamnus*
- *Pagameopsis*

Compositae
- *Achnopogon*
- *Cardonaea*
- *Chimantaea*
- *Duidaea*
- *Eurydochus*
- *Glossarion*
- *Gongylolepis*
- *Guaicaia*
- *Neblinaea*
- *Quelchia*
- *Stenopadus*
- *Stomatochaeta*
- *Tyleropappus*

TABLE 13–6. Examples of additional autochthonous genera in other areas of the Guayana Highland.

Araceae *Jasarum* Cyperaceae *Cephalocarpus* *Exochogyne* Rapateaceae *Potarophytum* *Windsorina* Annonaceae *Pseudephedranthus* Rutaceae *Decagonocarpus* *Myllanthus* Malpighiaceae *Diacidia* *Pterandra*	Euphorbiaceae *Chonocentrum* *Dodecastigma* *Haematostemon* *Hevea* Icacinaceae *Emmotum* Guttiferae *Thysanostemon* Melastomataceae *Acanthella* *Boyania* *Maguireanthus* *Ochthephilus* *Salpinga* *Tryssophyton*	Sapotaceae *Elaeoluma* Apocynaceae *Galactophora* Bignoniaceae *Digomphia* Rubiaceae *Chalepophyllum* *Dendrosipanea* *Gleasonia* *Sipaneopsis*

13–5 plus 29 genera from Table 13–6, giving a total of 119 genera out of 489 [29 genera from Table 13–6 plus 460 genera mentioned earlier]).

ANALYSIS OF THE PERCENTAGE OF SPECIFIC ENDEMISM

The percentage of specific endemicity of the total tepui flora is much greater than that of generic endemicity. Maguire (1970) stated that "some 4000 endemic species . . . have been described from Guayana." He also stated that "the total floras of Guayana would approach an estimated magnitude of 8000 species of which greatly more than 75 percent would be confined to the Guayana region. The summit area, it is believed, will have been found to contain more than one-third of the species of Guayana, nearly three-quarters of the total endemic species and the preponderance of endemic genera of the entire area." Maguire concluded that of the estimated 8000 species, 2000 of these may be encountered on the summits of the Guyana Highland, of which 90%–95% would be endemic. Of the remaining 6000 species, Maguire calculated that approximately 3000 occur at low altitudes below 1000 m, whereas the other 3000 found at middle altitudes above 1000 m below the summit, but on the talus slopes and upland plateaus, would reach an endemicity of 70% or more.

A detailed analysis of the species found on the summits of the table mountains reveals, however, a much lower average percentage of endemic elements confined to the summit area than the 90%–95% estimated by Maguire, and a percentage much lower than 70% for those species occurring at the middle altitude above 1000 m and below the summits. When the floras of all the sandstone mountains of the Venezuelan Guayana are contrasted, the summit flora of Cerro de la Neblina shows the highest percentage of endemic specificity, reaching over 60%.

The percentage of specific endemicity of the summit floras varies considerably from one tepui to another. A few examples will illustrate this point. On Auyán-tepui, 78 species of a total of 826 vascular plants, or 9.4% of the entire vascular flora, are endemic. Of these 78 species, 67 are endemic to the summit and represent 8.1% of the vascular flora. Only 10 species are endemic to the lower slopes of the mountain between 700 and 1500 m. Of the 116 species, 1 subspecies, and 6 varieties described from Auyán-tepui as new to science, 35 (28%) are now known from other table mountains, and 10 species have fallen into synonymy.

On the Meseta de Jáua, including Cerro Jáua and Cerro Sarisariñama, 526 species of vascular plants are known from the summit, of which 76 (14.4% of the total vascular flora) are endemic. On the upper slopes and summit of Mount Roraima, the percentage of endemic species has declined from an original high of 90% estimated by Gleason (1929), to the present 54% (Steyermark, 1966). On the summit of Cerro Autana only 7 out of 132 vascular plant taxa, or 5.3% of the total vascular flora, are endemic (Steyermark, 1975). On a recent expedition to the previously unexplored summit of Ptari-tepui, only 10% of the vascular flora was found to be endemic, the remaining 90% consisting of taxa already previously known from the lower and upper talus slopes of that table mountain (Steyermark, 1957, 1966). On another recently explored summit of Cerro Kukenán, only 31.3% of the flora proved to be endemic, the remainder already known from the lower talus slopes of other table mountains.

The percentages of specific endemicity on the summits of Cerro Duida and the Chimantá Massif are relatively high and approach 40%. At the present writing, no figures for endemicity are available for many of the other table mountains, but judging from an examination of collections taken from the other table mountains, there would appear to be considerable repetition of species on the various summits, since numerous taxa common to the summits of the various table mountains have already been collected. As collecting on the summits of the numerous unexplored tepuis intensifies, the percentage of the specific endemicity of the summit vegetation may decrease as the same species is encountered on more than one mountain, or because of its localization at lower altitudes.

RELATIONSHIPS OF THE TEPUI FLORA

We turn now to the relationship this flora has with other parts of the world. Since the Guayana Shield may have been available for plant occupation since the Late Cretaceous (Maguire, 1970), it is likely that it could have received floral elements from a number of geographic sources at various intervals from that time on to the Late Tertiary and Pleistocene. An inventory of the flora reveals that speciation has occurred at different altitudinal levels, not only on the higher isolated summits of the tepuis, talus slopes, and bluffs, but equally prominently throughout the lowlands and plateau areas at the base of the tepuis. Thus, the plateaus of the Gran Sabana and adjacent Kaieteur Plateau in the eastern sector of the Roraima Formation, the lower white-sand savannas, and areas of igneous outcrops of the Rio Negro-Guainía-Orinoco drainage of the western sector all demonstrate the phenomenon of speciation.

Summit Relationships

Of the 460 genera known from the summits of the Venezuelan tepuis (see Appendix 13–A), about 305 (64%) are confined to the Neotropics, about 31 (6.5%) are mainly Neotropical, and another 81 (17%) are Pantropical, about 28 (6%) are mostly Paleotropical, about 23 (5%) are associated with Africa, Madagascar, or the Mascarenes, and about 7 (1.5%) are related to genera distributed in Malaysia and Australasia (Fig. 13–9). Although about two thirds of the summit flora consists of Neotropical (including endemic) elements, it is noteworthy that about 31.5% includes elements of floras related to the Old World tropics. Thus, alongside such endemic genera as *Spathelia* or *Raveniopsis* others such as *Sloanea*, *Phoebe*, and *Laplacea* have representative species in Malaysia, Indonesia, or adjacent areas, and others still, such as *Amanoa*, *Ocotea*, *Swartzia*, *Ternstroemia*, and *Vellozia* have African and Malagasy representatives. Of a similar nature is the mixture of many species endemic to the tepuis with taxa of a more general or widespread distribution, such as *Cyrilla racemiflora* of the West Indies, Central, and South America, or *Utricularia alpina* of the West Indies, Central, and South America, or the widely distributed South American *Befaria glauca* and *Psila brachylaenoides*.

Additionally, an analysis of the summit genera reveals a clearcut relationship with the Andean flora or with the floras of lowland areas peripheral to the Venezuelan Guayana, especially the floras originating from the Amazon Basin. Andean elements comprise 50 genera, or about 11% of the summit flora, and include preponderantly Andean genera or genera with Andean centers of distribution, such as *Stenospermation*, *Schefflera*, *Viburnum*, *Hedyosmum*, *Clethra*, *Weinmannia*, *Oreobolus*, *Uncinia*, *Disterigma*, *Pernettya*, *Thibaudia*, *Hieronyma*, *Aulonemia*, *Chusquea*, *Cortaderia*, *Neurolepis*, *Hypericum*, *Oedomatopus*, *Nectandra*, *Excremis*, *Chaetolepis*, *Monochaetum*, *Grammadenia*, *Lepanthes*, *Pachyphyllum*, *Stelis*, *Podocarpus*, *Monnina*, *Palicourea*, and *Drimys*. In a study of pteridophytes collected on the talus slopes of Ptari-tepui below the summit from 1200 to 2400 m, Morton (1957) noted that the "primary relationship of the fern flora is perhaps with that of the eastern Andes of the region of Mérida." Examples of Andean relationships include such taxa as *Paesia acclivis* var. *polystichoides*, *Sticherus rubiginosus*, *Cyathea meridensis* and *C. petiolulata*, *Oleandra lehmannii*, *Eschatogramme furcata* var. *bicolor*, and *Selaginella rigidula*. The occurrence of the Andean *Oreobolus* and *Uncinia* in the summit flora is further evidence of relationships with the Andean flora, even though these two genera are also represented in parts of Malaysia, Australasia, and some Pacific islands.

Moreover, frequently the endemic species found on the talus slopes and quebradas of the tepuis have a closely related, allopatric species in

TABLE 13–7. Examples of genera with a Venezuelan tepui endemic and a related Andean species.

Genus	Venezuelan endemic	Related Andean species
Paepalanthus	*P. scopulorum*	*P. killipii*
Anthurium	*A. ptarianum*	*A. conjunctum* and *A. lechlerianum*
Tillandsia	*T. turneri* var. *orientalis*	*T. turneri* var. *turneri*
Octomeria	*O. rhizomatosa*	*O. longifolia*
Brunellia	*B. comocladifolia* subsp. *ptariana*	*B. comocladifolia* subsp. *comocladifolia*
Weinmannia	*W. balbisiana* var. *ptariana*	*W. balbisiana* var. *balbisiana*
Weinmannia	*W. pinnata* var. *ptaritepuiana*	*W. pinnata* var. *pinnata*
Tovomita	*T. angustata*	*T. weddelliana* and *T. longicuneata*
Gaultheria	*G. lepida*	*G. odorata*
Grammadenia	*G. ptariensis*	*G. lehmannii*
Psychotria	*P. speluncae*	*P. hazenii*
Viburnum	*V. tinoides* var. *roraimensis*	*V. tinoides* var. *tinoides*
Paesia	*P. acclivis* var. *polystichoides*	*P. acclivis* var. *acclivis*

the Andes. Often these relationships bridge a broad geographic gap between the occurrence of the genus on the slopes and that of its closest relative in the Andes. For example, *Vaccinium euryanthum* from the talus slopes of Ptari-tepui (A. C. Smith, 1953) has as its closest relatives the Andean *V. dependens* and *V. sphyrospermoides* of Peru. Similar tepui-Andean relationships at the intrageneric level are listed on Table 13–7.

Table 13–8 lists examples of Andean relationships at the intraspecific level (sometimes also with the Venezuelan Coastal Cordillera). This list includes primarily Andean species that reappear on the quebradas, talus slopes, or summits of the tepuis.

Affinities and relationships are also shown between summit taxa and taxa from the floras of the Antilles, Mexico, Central America, the Guianas, Amazonian Brazil, and the Brazilian Shield (Steyermark, 1966). Such affinities are manifest also among those of the surrounding lower plateaus, savannas, and lowland forests. About 18% of the generic floral element occurring on the summits of the Venezuelan tepuis is related mainly to a lowland flora. Thus, the endemic apocynaceous *Aspidosperma decussata*, *A. neblinae*, and *A. steyermarkii* of the tepui summits and upper talus slopes belong to a genus most of whose species are distributed in southern and southeastern Brazil, Paraguay, Argentina, Bolivia, and Amazonia. Other predominantly lowland genera of wide distribution that appear on the summits of one or more of the Venezuelan tepuis are *Mauritia*, *Schiekia*, *Paepalanthus*, *Syngonanthus*, *Calyptrocarya*, *Ficus*, *Coussapoa*, *Cecropia*, *Coccoloba*, *Marathrum*, *Biophytum*, *Urospatha*, *Licania*, *Rourea*, *Calliandra*, *Macrolobium*, *Pithecellobium*, *Swartzia*, *Doliocarpus*, *Ruizterania*, *Dulacia*, *Conceveiba*, *Tabebuia*, *Codonanthe*, *Couma*, *Capirona*, *Duroia*, *Geophila*, *Kotchubaea*, *Pagamea*, *Platycarpum*, and *Vernonia*.

Relationships at Levels below the Summit

A consideration of the geographic relationships of the floras surrounding the base of the table mountains, at lower elevations such as Gran Sabana of Estado Bolívar, or the lowland Rio

TABLE 13–8. Examples of species occurring in the Venezuelan table mountains and in the Andes.

Guzmania retusa	*Sphyrospermum buxifolium*
Guzmania squarrosa	*Emmeorrhiza umbellata* var. *septentrionalis*
Isachne ligulata	*Psila brachylaenoides* var. *brachylaenoides*
Carex polystachya	*Podocarpus magnifolius*
Oreobolus obtusangulus	*Isoetes killipii*
Maxillaria meridensis	*Cyathea meridensis*
Maxillaria aurea var. *gigantea*	*Cyathea petiolulata*
Peperomia omnicola	*Oleandra lehmannii*
Roucheria laxiflora	*Elaphoglossum stenopteris*
Zinowiewia australis	*Hymenophyllum apiculatum*
Miconia tinifolia	*Selaginella rigidula*
Monochaetum bonplandii	
Ternstroemia camelliaefolia	
Gaultheria alnifolia	

FIG. 13–10. Guayana Shield–Brazilian Shield disjunction, represented by the species *Austroamericium* (*Thesium*) *tepuiense* in Venezuelan Guayana.

Negro white-sand savannas and other lowland savannas of the Territorio Federal Amazonas, and the intervening forests, provides us with additional data with respect to relationships of the tepui flora. The bromeliaceous genera *Navia*, *Brocchinia*, and *Cottendorfia*, whose evolution is centered in the Guayana Highland, with preponderance of speciation on the summits and escarpments of the tepuis themselves, have also speciated widely at lower altitudes in the savannas and plateaus surrounding the base of the tepuis. Of the 18 described species of *Brocchinia*, four (*B. serrata*, *B. bernardii*, *B. prismatica*, and *B. steyermarkii*) have speciated between 100 and 700 m. *Navia*, with nearly 80 species, mostly evolved on the higher altitudes of the Guayana Highland, has no fewer than 22 taxa (*N. crispa*, *N. cataractarum*, *N. hohenbergioïdes*, *N. gracilis*, *N. brocchinioides*, *N. reflexa*, *N. ramosa*, *N. caulescens*, *N. lopezii*, *N. myriantha*, *N. navicularis*, *N. breweri*, *N. octopoides*, *N. bicolor*, *N. heliophila*, *N. acaulis*, *N. connata*, *N. schultesiana*, *N. graminifolia*, *N. sandwithii*, *N. arida*, and *N. fontoides*), which speciated at lower altitudes, mostly from 60 to 700 m. Most of these species have evolved in areas associated with the Roraima formation, either on the sandstone outcrops themselves, or in sandy savannas or forests in soils derived from the formation.

On the terraces and plateaus, and in gallery forests of the Gran Sabana of southeastern Venezuela between 350 and 1300 m, occur a large number of endemic species. Among these may be mentioned *Eriocaulon dimorphopetalum*, *Paepalanthus steyermarkii*, *Syngonanthus venezuelensis*, *Trimezia fosteriana*, *Cecropia kavanayensis*, *Euplassa venezuelana*, *Roupala minima*, *Licania lasseri*, *Spathelia fruticosa*, *Tetrapteris pusilla*, *Qualea ferruginea*, *Vochysia ferruginea* and *V. rubiginosa*, *Phoradendron venulosum*, *Poecilandra pumila*, *Caraipa longipedicellata*, *Bonyunia minor*, *Clusia pusilla*, and *Orthaea crinita*. Most of these taxa belong to genera of broad distribution outside the Guayana Shield area, often with

FIG. 13–11. Guayana Shield–Brazilian Shield disjunction, exemplified by Velloziaceae. Modified from L. B. Smith, Fig. 28 (1962), with additional data from Venezuela.

centers of evolution in the Amazonian Hylaea. Such species are judged to have evolved under the special edaphic conditions peculiar to the Gran Sabana area.

Many of the arboreal endemic species (and other species) of the gallery forests and quebradas of the lower slopes and savannas show close affinities with species to the south in Brazil. *Macropharynx strigillosa* (Apocynaceae), for example, is most closely related to *M. spectabilis* of the lower Amazonian flora (Woodson, 1953). Some of the endemic species of the Gran Sabana may have speciated from taxa originating from the Brazilian Shield. An example of the latter relationship is *Thesium tepuiense* (separated by Hendryck [1963] as a distinct genus, *Austroamericium*). This remarkable disjunct endemic of the Gran Sabana and sandstone table mountains of Venezuela has become differentiated over a period of time from the southeastern Brazilian *T. brasiliense* and *T. aphyllum* (Fig. 13–10). Other obvious relationships with the Brazilian Shield are manifest in the melastomataceous genera *Marcetia*, *Microlicia*, and *Tibouchina* (Wurdack, 1973), each of which has a large concentration of taxa in the Brazilian planalto and in southeast Brazil. In the case of *Marcetia*, only one species is found in the Venezuelan Guayana, whereas the other 14 species are known only from the Brazilian planalto and Uruguay. Of the 130 species known in *Microlicia* most are in the Brazilian planalto, whereas only two are found in the Venezuelan Guayana. Likewise, the main center of evolution for the genus *Vellozia* of the unique family Velloziaceae is in the Brazilian planalto with nearly 75 species (L. B. Smith, 1962), whereas only one species, *V. tubiflora* (Fig. 13–11), is known in the Venezuelan Guayana. Other genera whose centers of evolution are in the Brazilian Shield are the rubiaceous *Borreria*, *Declieuxia*, and *Psyllocarpus* (Kirkbride, 1976, 1979). In such cases, where the number of species is overwhelmingly in the Brazilian Shield, the immigration would appear to have taken place into the Guayana Shield during one of the arid phases of Pleistocene or post-Pleistocene times. Genera inhabiting the forested areas of the Guayana Shield related to probable origins and centers of distribution in southern and central Brazil on the Brazilian Shield are exemplified by *Aioua* (Lauraceae), *Eugenia* (Myrtaceae), *Euplassa* (Proteaceae), and *Amaioua* (Rubiaceae). *Tapirira guianensis* of the Anacardiaceae is an example of a rain-forest species that inhabits mainly Amazonian Colombia, Ecuador, Peru, Guyana, northern Brazil, and mainly southern Venezuela, but which reappears in the Brazilian Shield portion of southern Brazil. It has been postulated (Smith, 1962) that the distribution of this and other taxa with similar geographic ranges shows evidence of having bridged the barrier of the Brazilian planalto. Their migrations between the Brazilian and Guayanan Shields may have been effected by means of connecting gallery forests during arid phases or even continuous broad rain forest that may have existed during one of the humid phases of the Pleistocene. The subsequent occurrence of a drier climatic phase disrupted the pattern of continuous forest distribution, leaving the extensive Brazilian planalto as an effective barrier to separate the northern from the southern portions of the geographic ranges.

Among wide-ranging species of Gramineae and Cyperaceae encountered on the Gran Sabana are *Andropogon leucostachys*, *Elionurus adustus* and *Thrasya petrosa* (Gramineae), *Fimbristylis dichotoma* and *Fuirena umbellata* (Cyperaceae), whereas associated members of the same families occurring also in the Gran Sabana are restricted in their geographic distribution, for example *Panicum kappleri* and *P. tatei*, *Axonopus kaieteurensis*, and *Paspalum gossipinum*. The same situation applies to the Polygalaceae, where such species as *P. longicaulis* and *P. paniculata* have a relatively broad geographic range, whereas *P. appressa* is very restricted.

In the Amazonian savannas at still lower elevations of mainly 100–200 m, especially well developed in the drainage of the Rio Guainía, the lower Ventuari, Pacimoni, Pimichín, and Atabapo rivers, special edaphic soils of high acidity and low water-holding have given rise to unusual types of vegetation containing many endemic plant taxa. A few of these taxa, such as *Sauvagesia nudicaulis*, *Rapatea yapacana*, *Licania lanceolata*, *L. savannarum*, and *Macrolobium savannarum*, belong to genera of wide distribution; others, such as *Duckea junciformis*, *Guacamaya superba*, *Sipaneopsis maguirei*, and *Dendrosipanea wurdackii*, pertain to genera of more restricted geographic range, principally of the upper Orinoco and Rio Negro. Especially noteworthy is the localization of *Pentamerista neotropica*. This species, restricted to a few savannas bordering the upper Orinoco, Pacimoni, Atabapo, and adjacent portions of the Territorio Federal Amazonas, belongs to the monotypic genus *Pentamerista* of the Tetrameristaceae, a family known elsewhere only in Malaysia and represented there by the genus *Tetramerista*. This, and other similar disjunct distributions of genera (*Pakaraimaea*, for example, the only known neotropical genus of the Old World family Dipterocarpaceae) and families (Rapateaceae), suggest an old historical connection between the Guayana Shield flora and floras elsewhere, such as Malaysia or Gondwanaland.

Igneous outcroppings in the area of the Orinoco and Yatua rivers, sometimes adjacent to the sandy savannas of the Territorio Federal Amazonas harbor a flora frequently inhabited by endemic species. *Borreria pygmaea* (Fig. 13–12), *Rudgea maypurensis* (Fig 13–12), and *Tocoyena brevifolia* of the Rubiaceae are noteworthy endemic taxa speciated on the igneous "lajas" in the vicinity of Puerto Ayacucho, whereas *Saxofridericia spongiosa*, *Pitcairnia wurdackii*, *Paradrymonia yatuana*, and *Decagonocarpus oppositifolius* are local endemics on the igneous outcrops along the Rio Yatua.

I have noted earlier that a number of genera occurring in the Venezuelan Guayana with a preponderance of lowland species, such as *Kotchubaea*, and distributed mainly in the Guianas and Amazonia, have speciated into endemic taxa upon coming into contact with the talus slopes and sandstone escarpments or summits of the tepuis. This type of speciation, correlated with the highly acid soils derived from sandstone and igneous parental material of the Guayana and resultant leaching of humic acid in the associated black-water streams (Janzen, 1974; Cowan, 1975), is reflected apparently in the distribution of many species of the forest flora confined mainly to the Guayana, such as *Anthrodiscus*

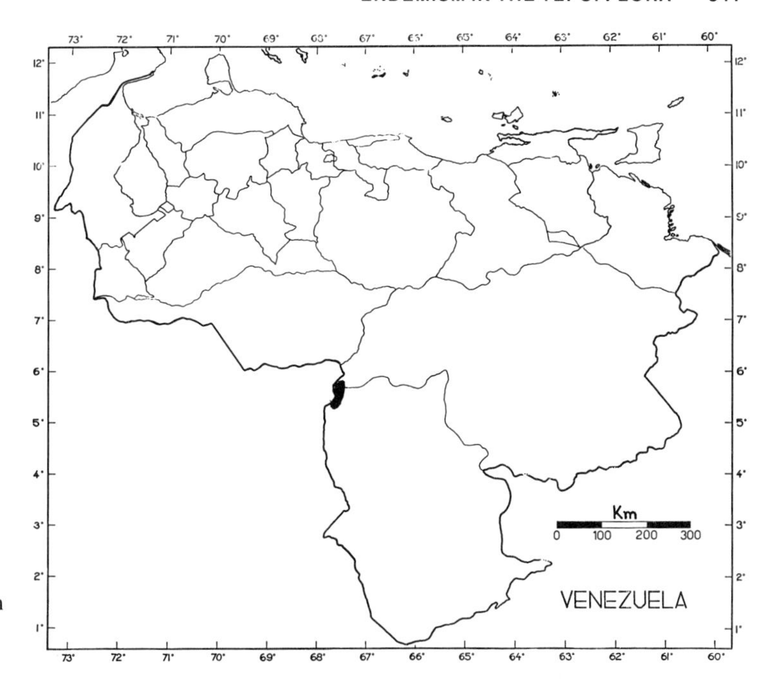

FIG. 13–12. Distribution of *Borreria pygmaea* and *Rudgea maypurensis* on igneous outcrops in the Guayana Shield.

mazarunensis, *Gleasonia*, *Pagamea*, and *Retiniphyllum* spp. The centers of distribution of many of the Guayanan genera with tropical lowland affinities in Amazonia or Amazonian Brazil are demonstrated by the Caryocaraceae, Chrysobalanaceae (*Hirtella*, *Licania*), Hippocrateaceae, Leguminosae (*Heterostemon*, *Sclerolobium*, *Tachigalia*), Loganiaceae (*Strychnos*), Menispermaceae (Krukoff and Barneby, 1970), Moraceae (*Sorocea*), Ochnaceae (*Ouratea*), Rubiaceae (*Gleasonia*), and Vochysiaceae, as well as many other genera and families. Hence, it is postulated that the preceding genera and families with similar distributional patterns have entered the periphery of the Guayana Highland areas from Amazonian centers to the south, such as the Manaus, Olivença, Tefé, and Belém-Xingú forest refuges (Prance, 1973).

MIGRATIONS OF FLORAL ELEMENTS.

Vertical Migrations

The impression has often been given that the summits of the sandstone table mountains, which protrude above the general landscape as isolated, rock-bound, vertical-walled fortresses, have prohibited vertical movements of the floras downward or upward. This point of view has been published by Maguire (1970). He contends that the species of the summits of the sandstone mountains "do not establish themselves at lower altitudes," that conversely, taxa of lower altitudes would have equal difficulty moving upward, and that the absence of species common to both the summits and "contiguous low altitude terrain attests to nonsuccessful dispersal."

Contrary to this point of view, however, is a mass of evidence that has accumulated during recent years of exploration on the tepuis. These data indicate positively that a significant lowland floral element has established itself on the summits of at least some of the sandstone mountains. The most noteworthy example of the successful dispersal and establishment of the lowland flora on the summit of a tepui can be seen in the flora of Cerro Guaiquinima, situated by the Paragua River in Estado Bolívar (Steyermark and Dunsterville, 1980) Here, out of a total of 237 genera of flowering plants, 61 (25.7%) are those with the predominant portion of the species found mainly at low altitudes of 50–500 m. Some of the genera, such as *Odontadenia* (Apocynaceae); *Urospatha* (Araceae); *Nephradenia* (Asclepiadaceae); *Exochogyne* (Cyperaceae); *Doliocarpus* (Dilleniaceae); *Codonanthe* (Gesneriaceae); *Schiekia* (Haemodoraceae); *Dimorphandra*, *Dipteryx*, and *Macrolobium* (Leguminosae); *Jessenia* and *Mauritia* (Palmae); *Rhyncholacis* (Podostemonaceae); *Duroia* and *Geophila* (Rubiaceae); and *Trigonia* (Trigoniaceae), are nearly exclusively characteristic of the *tierra caliente* zone. Most of the species of other genera, such as *Aspidosperma* (Apocynaceae), *Billbergia* (Bromeliaceae), *Myrmidone* (Melastomataceae), *Ouratea*

(Ochnaceae), and *Roupala* (Proteaceae), are usually found at lowland altitudes.

Of the total of 449 species of flowering plants at present recorded from the summit of Cerro Guaiquinima, 97 (21.6%) are distributed chiefly at low elevations between 50 and 500 m. Thus one encounters such lowland species as *Urospatha sagittifolia*, *Nephradenia linearis*, *Billbergia macrolepis*, *Tillandsia paraensis*, *Exochogyne amazonica*, *Rhynchospora candida*, *Curtia tenella*, *Codonanthe calcarata*, *Caraipa grandifolia* subsp. *grandifolia*, *Schiekia orinocensis* subsp. *orinocensis*, *Dalbergia monetaria*, *Myrmidone macrosperma*, *Jessenia bataua*, *Mauritia martiana*, *Geophila orbicularis*, *Retiniphyllum concolor*, *Abolboda americana*, *Xryis involucrata*, *X. subuniflora*, and *X. uleana*, all successfully established on the summit.

All these lowland species occur on the summit at one of the lowest and most depressed sites of Cerro Guaiquinima, at altitudes of 700–750 m. Such lowland taxa have apparently been able to migrate upward, following the largest reentrant valleys or passes located on the south flank of the mountain. At such places, no physical barriers, such as high escarpments, impede the dispersal of diaspores along the dissected portions of the mountain. At these sites the lowland taxa occur alongside of and are associated with some of the typical summit floral elements with such endemic species as *Stegolepis squarrosa*, *Biophytum cardonae*, *Raveniopsis peduncularis*, *Blepharandra fimbriata*, *Stomatochaeta cylindrica*, and *Bonnetia lancifolia*. On the other hand, the greater percentage of species of the summit flora of the higher altitudinal levels of Cerro Guaiquinima have not descended to the lowest depressed parts of the summit. Apparently the species of the higher levels, 1500–1650 m, have been unable to survive at the warmer temperatures of the lower levels, because of previous long-adjusted adaptation to cooler temperatures.

The summit floras of the other tepuis, such as Autana, Jáua, Sarisariñama, and Duida, provide further examples of the upward migration of the lowland floral element (Steyermark, 1974a, 1975) and evidence of the vertical movements between the floras. Cerro Autana, 1300 m, is one of the smallest of the sandstone mountains, with a summit only 1 km long. It stands isolated with its vertical-walled escarpments in the western portion of Territorio Federal Amazonas. Despite its formidable appearance as a barrier to the vertical dispersal of plants, its summit flora harbors a surprising number of lowland species, which are evidence of interchange and vertical movement of the flora at some time in the past. For instance, *Acanthella sprucei* and *Myrmidone macrosperma* (Melastomataceae) are two characteristic lowland species represented on its summit. Similarly, we find on its summit *Capirona decorticans* (Rubiaceae), a tree occurring elsewhere only at low elevations of 100–125 m in Amazonian Brazil and southwestern Venezuela, *Epistephium hernandii* of lowland Amazonian Colombia, and a form of *Utricularia amethystina*, which, according to Taylor (1967), the leading authority of the family Lentibulariaceae, is the "typical low altitude form of the species." Another species, *U. longicaulis*, is found in Venezuela chiefly in the lowlands of Estado Bolívar and Territorio Federal Amazonas but occasionally occurs isolated on the summits of various tepuis, including Cerro Autana.

Other examples exist on the Meseta de Jáua (Steyermark and Brewer-Carias, 1976) where at altitudes of 1320–1450 m trees and lianas, which ordinarily are associated with the surrounding lowlands at the base of the mountains, are isolated on parts of the summit. These include *Licania discolor* and *L. incana* (Chrysobalanaceae), *Rourea sprucei* (Connaraceae), *Gnetum nodiflorum* (Gnetaceae), and *Viroli pavonis* (Myristicaceae). At 1800 m is the noteworthy occurrence of *Jenmaniella ceratophylla* (Podostemonaceae), a rare aquatic plant elsewhere known only at low altitudes.

Such successful interchanges of floral elements from different altitudinal levels, especially perhaps movements from the lowlands upward to the summits, support the view that there have been vertical movements and dispersals of the same species present on the tepuis, and may help explain some of the speciation events among the summit endemics.

Centrifugal versus Centripetal Migration

I have noted earlier that only 39 (8.5%) of the genera inhabiting the summits of the Venezuelan sandstone tabular mountains are restricted to the summit, and that this total increases to 77 (16.7%) if one includes genera that occur on the vertical escarpments, some portion of the talus slopes, or the surrounding lower altitudes of the plateaus and quebradas of the Gran Sabana and contiguous areas (Fig. 13–9). These 77 genera, of undoubted autochthonous origin in the Venezuelan Guayana Shield in some instances, have dispersed centrifugally to contiguous outlying areas. For example, the genera of Mutisieae of the tepuis, which are endemic to the Guayana Highland and mainly endemic to their summits (*Ach-*

nopogon, *Chimantaea*, *Duidaea*, *Stomatochaeta*, *Cardonaea*, *Guaicaia*, *Eurydochus*, *Neblinaea*, *Stenopadus*, *Gongylolepis*, *Glossarion*, and *Quelchia*), have evolved relatively few species that have migrated beyond the principal area of the Guayana Shield. This kind of autochthonous origin with centrifugal dispersal is evident in the 77 endemic genera. Among these 77 autochthonous genera *Navia* (Bromeliaceae) is the most prolifically speciating genus, with nearly 80 species spread throughout the tepuis on usually sandstone outcrops and substrate, its optimum environment for adaptive radiation. The tepuis have served as a rich center of evolution for other Bromeliaceae, such as *Brocchinia* and *Cottendorfia*. An analysis of floral elements reveals that only 16.7% of the summit flora is autochthonous, whereas 17.7% of the flora is related to a tropical lowland element which dominates the Amazonian Hylaea, or which has originated outside the Guayana Highland (Fig. 13–9).

But what of the remaining 76.8% of the nonendemic genera found on the summits that are not restricted to the summits? This percentage would indicate that the greater part of the generic flora of the summits have migrated to the mountains from other areas during past geologic time and are, therefore, centripetally distributed. This centripetal pattern of immigration is manifested (1) by the 17.6% (Fig. 13–9) of predominantly lowland genera found on the tepuis and strongly indicative of an Amazonian Basin origin, and (2) by the 48.4% that apparently have speciated from a more widespread neotropical distribution (derivatives of west Gondwanaland-tropical American, Australasian-Malaysian, or Antarctic-tropical American elements) and/or a pantropical distribution.

If the distribution of the total number of species comprising a genus is analyzed, the significance of the centripetal pattern of distribution becomes more evident. Maguire (1956) suggested that *Eupatorium* and *Calea* (Compositae) manifested a centripetal dispersal, since most of their species were developed outside the Guayana Highland. A few cases from numerous genera may be added to exemplify this centripetal pattern of dispersal common to most of the flora. *Marcgravia*, with 40 or more neotropical species, has only 3 taxa known from the Guayana Highland. *Syngonanthus* and *Paepalanthus*, common summit genera of the tepuis, are chiefly neotropical with 195 taxa in *Syngonanthus* and 484 taxa in *Paepalanthus*, but both genera also have taxa in Africa and Madagascar. *Syngonanthus* and *Paepalanthus* have numerous endemic species autochthonous to the summit and upper slopes of the tepuis, yet speciation has been predominantly in the lowlands with numerous species to the south of the Guayana Shield in Brazil and within the Brazilian Shield area. Also, some of the species of both genera are found both in the lowlands and on the summits. The genus *Drosera*, with approximately 100 species, ranges in North and tropical America, Africa, Madagascar, Hawaii, China, and Australasia, but only 12 species are scattered in Venezuela and in the Guayana Highlands. Of these 12, *D. roraimae* ranges widely over the summits of the tepuis, but the genus as a whole is clearly centripetal so far as the Guayana area is concerned. The Podocarpaceae are represented in the Guayana Highland by 6 species, mainly on the summits and upper slopes. However, the family has nearly 100 species, of which 30 are neotropical, and the remainder pantropical. The Proteaceae are represented in the Guayana Highland by the genera *Euplassa*, *Panopsis*, and *Roupala*. A few species have evolved by speciation in each of these three genera, but *Euplassa* with 25 species and *Roupala* with 51 species are best represented in Brazil, and have evidently dispersed centripetally to the Guayana Highland. The genus *Symplocos*, with 282 recorded species, is represented outside of tropical America with approximately 167 taxa in Malaysia and Australasia, but in the Guayana Highland has speciated only 8 taxa (Maguire, 1978). Similarly, *Styrax*, with approximately 100 species in the New World, is represented in the Guayana Shield area by only 8 endemic taxa, with a ninth (*S. guianensis*) of wider distribution outside the Guayana area in the Amazonian Basin (Maguire, 1978). *Styrax* is represented in the Old World by more taxa than in the New.

The greater part of the generic flora present in the tepuis has been centripetally derived from other sources, and past immigrations and dispersals from other areas can explain the present percentage of taxa now in existence. This diversified geographic origin would help to explain some of the more distant immigrations of the centripetal dispersal type apparently originating from the Amazonian Hylaea, as well as from widespread neotropical distribution patterns that have entered the Guayana area in past epochs from West Gondwanaland, Australasia, Malaysia, and/or pantropical-cosmopolitan sources.

Ornithological data from the tepui area would tend to support the botanical data, as it has been suggested (Mayr and Phelps, 1967, p. 304) that "it is quite conceivable that Pantepui [the table mountain area] has been a minor center of differentiation, which has contributed elements to other parts of South America, yet many elements of Pantepui give the impression of being relicts of formerly more widely distributed types." Haffer (1974) has reviewed the various theories pertaining to avian evolution in the tepuis.

During the long history of the Guayana Shield, there has been ample opportunity for expansion and contraction of the floras, ample contact with other floras, and modifications and adaptations of the flora to new ecological conditions and habitats, as a result of altitudinal changes through uplifts, or due to changes in erosion, extent of rock exposures available, and varying patterns of river systems and their orientation. Evolutionary development during this long interval has proceeded with speciation of thousands of taxa.

Concomitant with or following these changes, the alternating cooler and warmer, as well as drier and wetter, phases of the glacial and interglacial stages of the Pleistocene and post-Pleistocene periods, must have greatly influenced not only the forested and nonforested portions of the Guayana Shield area, but also the distribution and migrations of many taxa therein. The phenomenon of lowland species isolated on the summits of the tepuis may be explained, in some cases perhaps, by present-day dispersal following available migration routes over low passes into the mountains, or by upward migrations during one of the warmer interglacial stages of the Pleistocene, when warmer and/or more humid climatic conditions prevailed. Doubtless, evolution of species within the Guayana Shield has been influenced by numerous factors, including paleoclimatology, isolation, and edaphic conditions.

APPENDIX 13–A.

Geographical affinities of 460 genera that occur on the summits of the Venezuelan table mountains (total number of species based on latest estimates).

Genus	Total species	All neotropic	Mainly neotropic	Africa, Madagascar, Mascarenes	Malaysia, Australia, and others	Cosmopolitan or pantropic	Mainly Old World	Venezuelan distribution or comments on tepuis
PTERIDOPHYTA								
Adiantum	200		X					
Arachniodes						X		
Asplenium	650					X		20 Guayana Highland
Blechnum	220					X (mainly S. Hemisphere)		
Blotiella	15			X				
Cheiroglossa vulgatum	1	X						
Cochlidium	7	X						1 on tepuis, down to 600 m
Culcita	5				X			
Cyathea	40	X						Andes center of diversity
Danaea	30	X						2 of 5 Ven. spp. on tepuis
Dennstaedtia	70					X (pantropic)		
Dicranopteris	10					X (pantropic)		2 of 5 Ven. spp. from Andes and tepuis
Diplazium	400					X (pantropic)		
Doryopteris	26		21 spp. (mainly (Brazil)		X	X		*D. lomariacea* of tepuis; also Peru, Brazil, and Paraguay
Dryopteris	150					X		
Elaphoglossum	400					X		Andes center of origin (70–100 Ven. spp.)
Enterosora	4	X						2 tepui summits
Eriosorus	30	X						2 tepui spp. and 2 endemic tepui varieties

APPENDIX 13–A CONTINUED.

Genus	Total species	All neotropic	Mainly neotropic	Africa, Madagascar, Mascarenes	Malaysia, Australia, and others	Cosmopolitan or pantropic	Mainly Old World	Venezuelan distribution or comments on tepuis
Grammitis	150 (50–Ecuador)				X	X (pantropic)		
Hymenophyllopsis	5	X						+ endemic tepui summits
Hymenophyllum	60					X		4 Ven. spp.
Isoetes	75					X		2 on tepuis
Lindsaea	200		X			X (mainly Asiatic)		Guayana one of centers
Lonchites	2			X				
Lycopodium	450					X		31 Ven. spp.
Nephrolepis	30					X (pantropic)		
Oleandra	40					X		3 Ven. spp. of wide range and tepuis
Osmunda	7					X (cosmopolitan)		
Peltapteris	4	X						Andes, Coastal Cord., tepuis
Pityrogramma	11		X (2 in Old World)	X				*P. tartarea* on tepui of wide range
Polypodium	75					X		some tepui endemics, but mainly Andes and Coastal Cordillera
Pterozonium	12	X						mainly Guayana Highland
Selaginella	750					X (pantropic)		
Schizaea	20				X	X		
Sphaeropteris	120					X	X	
Sticherus	100					X		9 spp. Ven., 3 on tepuis at lower alt. than other genera
Trichipteris	90	X (51 S. Am. 12 C. Am.)						

Trichomanes	31	X			
Vittaria	72			X (60 spp.)	
ACANTHACEAE					
Justicia	300		X		
ANNONACEAE					
Rollinia	65	X (mostly S. Am.)			
APOCYNACEAE					
Aspidosperma	80	X			mainly lowland spp.
Couma	6	X (5 in S. Am.)			mainly lowland spp.
Galactophora	6	X (all S. Am.)			lowland spp. frequent
Mandevilla	108	X			
Odontadenia	30	X			mainly low altitudes + tepui endemic summit
Salpinctes	2	X			
AQUIFOLIACEAE					
Ilex	400		X		55 Guayana High. spp.
ARACEAE					
Anthurium	700	X			
Philodendron	350	X			
Spathiphyllum	45		X		most spp. from 100–1100 m.
Stenospermation	21	X (mainly Andean)			
Urospatha	20	X			lowlands
ARALIACEAE					
Crepinella	1	X			+ summit Roraima endemic
Didymopanax	40	X			endemic tepui summit spp.
Oreopanax	120	X			
Schefflera	200		X (New World spp. mainly in S. Am. Andes)		endemic tepui summit spp.

APPENDIX 13–A CONTINUED.

Genus	Total species	All neotropic	Mainly neotropic	Africa, Madagascar, Mascarenes	Malaysia, Australia, and others	Cosmopolitan or pantropic	Mainly Old World	Venezuelan distribution or comments on tepuis
ASCLEPIADACEAE								
Blepharodon	30	X				X		
Cynanchum	150					X		
Ditassa	75	X (S. Am.)						
Metastelma	100	X						
Nephradenia	10	X						
BIGNONIACEAE								
Anemopaegma	30	X						mainly low alt. in Ven.
Digomphia	3	X (S. Am.)						2 spp. tepui; 1 sp. savanna at low alt.
Distictella	10	X						Guayana is center, but low alt.
Tabebuia	100	X						mainly low alt.
BOMBACACEAE								
Rhodognaphalopsis	13			X	X	X (pantropic)		
BROMELIACEAE								
Ayensua	1	X (Ven.)						lowland in Ven. + endemic summit
Billbergia	50	X						mainly low alt.
Brocchinia	18	X (Ven. Guayana)						mainly Ven. Guayana; also Colombia and Guyana
Connellia	4	X (Ven.)						mainly on tepui summit
Cottendorfia	24	X (Ven., Brazil, Guyana)						mainly Ven., Guay. High., and Guyana and Brazil
Guzmania	126	X						

Navia	80	X					mainly Ven. tepuis; also Guianas and Brazil
Pitcairnia	260		X	X (1 Africa same sp. as Am.)			
BURMANNIACEAE							
Hexapterella	2	X					
Burmannia	57					X	Ven. spp. in Guayana; others in Brazil
Dictyostega	2		X		X		
BURSERACEAE							
Protium	90 (mainly Brazil)		X	X (3 Madag.)	X	X	
CAMPANULACEAE							
Centropogon	230	X					
Siphocampylus	215	X					
CAPRIFOLIACEAE							
Viburnum	200					X (Asia and N. Am. especially)	mostly Andes and temperate
CELASTRACEAE							
Maytenus	70	X					
CHLORANTHACEAE							
Hedyosmum	40		X (1 sp. China)				mainly Andes of Colombia to Peru
CHRYSOBALANACAE							
Licania	152		X (1 sp. Malaysia)				mainly low alt. in Amazon and Guiana
CLETHRACEAE							
Clethra	65				X	X (Canary Is.)	4–5 Ven. spp. mainly Andes and Coast. Cordillera

APPENDIX 13–A CONTINUED.

Genus	Total species	All neotropic	Mainly neotropic	Africa, Madagascar, Mascarenes	Malaysia, Australia, and others	Cosmopolitan or pantropic	Mainly Old World	Venezuelan distribution or comments on tepuis
COMBRETACEAE								
Terminalia	250					X		5 endemic Guay. High.
COMPOSITAE								
Achnopogon	3	X						+ endemic tepui summit
Baccharis	400	X (mainly S. Am.)						
Calea	90	X						many spp. Ven. on tepuis
Cardonaea	1	X						+ endemic summit tepui
Chimantaea	8	X						+ endemic summit tepui
Duidaea	4	X						+ endemic summit tepui
Erechtites	5	X (3 S. Am.)						
Eupatorium	600		X			X		
Eurydochus	1	X						+ endemic tepui summit
Glossarion	1	X						+ endemic tepui summit
Gongylolepis	13	X						all Guayana, except 1 Andes and 1 lowland
Guaicaia	1	X						+ endemic tepui summit
Mikania	275		X (1 sp. Old World)					
Neblinaea	1	X						+ endemic tepui summit
Oyedaea	30	X						
Piptocarpha	50	X						
Quelchia	5	X						+ endemic tepui summit

Senecio	1500				X	Ven. spp. mainly in Andes and Coast. Cord.
Stenopadus	15	X				mainly tepuis also Colombia and n. Brazil
Stomatochaeta	4	X				tepuis and Gran Sabana, Guyana
Tyleropappus	1	X				+ endemic tepui summit
Verbesina	250	X				
Vernonia	600		X (few spp. N. Am.)			mainly low altitudes
CONNARACEAE						
Rourea	90			X		
CUCURBITACEAE						
Gurania	35	X				10 spp. Ven.
CUNONIACEAE						
Weinmannia	170	X	X	X	X	common Andes
CYCLANTHA-CEAE						
Sphaeradenia	38	X (mainly Andes)				common Andes; Guayana High. one of centers
CYPERACEAE						
Bulbostylis	100				X	
Calyptrocarya	5	X				mainly lowland
Carex	2000				X (especially temperate)	
Cephalocarpus	3	X				
Cladium	3	X (and N.Am.)				1 tepui endemic
Dichromena	60	X (and N. Am.)				
Didymiandrium	1	X				tepuis, Gran Sabana, and Guyana
Eleocharis	200				X	
Everardia	13	X				tepuis, Guyana, Surinam, Colombia

APPENDIX 13–A CONTINUED.

Genus	Total species	All neotropic	Mainly neotropic	Africa, Madagascar, Mascarenes	Malaysia, Australia, and others	Cosmopolitan or pantropic	Mainly Old World	Venezuelan distribution or comments on tepuis
Exochogyne	1	X (S. Am.)						low altitudes
Hypolytrum	50		X (mostly Brazil)			X		mainly low alt.
Lagenocarpus	12	X						mainly low alt. savannas, also tepuis
Mapania	45				X (center in Malaysia and Indo-China)			tepui slopes, Imataca, Colombia, low alt. Panama
Oreobolus	10				X			Andes and summit of Neblina
Rhynchocladium	1	X						endemic slopes and tepui summit
Rhynchospora	250					X (tropics to subarctic)		50 spp. Guayana Highland; many Andean affinity
Scleria	31					X (subtropic to temperate)		tepui spp. widely distrib.
Uncinia	35				X			Andes and Neblina
CYRILLACEAE								
Cyrilla	1	X (SE U.S.)						widely distributed
Purdiaea	11	X (Cuba, Colombia, Peru, Venezuela)						Ven. sp. close to Cuban sp.
DILLENIACEAE								
Doliocarpus	26	X						principally lowland
DIOSCOREACEAE								
Dioscorea	600					X		
DIPTEROCARPACEAE								
Pakaraimea	2	X						New World representative of Paleotropic family

DROSERACEAE					
Drosera	100		X	X	12 Ven. spp. mainly in Guay. High.
ELAEOCARPACEAE					
Sloanea	120 (65 neo-tropics)		X		largest no. Am. spp. in Guianas and Amazon
ERICACEAE					
Befaria	25	X (NW. S. Am. center)			6 spp. Guay. High
Cavendishia	140	X (70% spp. Colombian)			4 spp. Guayana High.
Disterigma	22	X (mainly Andean)			2 spp. Guayana
Gaultheria	147 (85 spp. C. Am. and S. Am.)		X		6 spp. Guayana High.
Gaylussacia	49 (40 spp. S. Am. 35 spp. SE Brazil)	X			1 spp. tepui throughout Ven. and in Colombia
Ledothamnus	9	X			endemic to Gran Sabana and tepui area
Macleania	45	X			+ endemic tepui summit
Mycerinus	3	X			
Notopora	5	X			endemic to Gran Sabana and tepuis
Orthaea	22	X			9 spp. Guayana Highland and Gran Sabana
Pernettya	19 (Andes center of distrib.)		X		1 sp. endemic to tepui summit
Psammisia	50 (34 spp. Colomb.)	X			
Sphyrospermum	18 (10 spp. Ecuador center of distrib.)	X			4 spp. on tepuis wide ranging
Tepuia	8	X			+ endemic tepui summits

APPENDIX 13–A CONTINUED.

Genus	Total species	All neotropic	Mainly neotropic	Africa, Madagascar, Mascarenes	Malaysia, Australia, and others	Cosmopolitan or pantropic	Mainly Old World	Venezuelan distribution or comments on tepuis
Thibaudia	60 (Andes center)	X						10 spp. Guay. High.
Vaccinium	450 (240 spp. Malesia)					X		4 spp. Guay. High.
ERIOCAULACEAE								
Carptotepala	1	X						Ven., Guyana
Eriocaulon	400					X		
Leiothrix	65	X						
Paepalanthus	485		X (484)	X (1)				mainly lowland + endemic tepui summit
Roraimanthus	1	X						
Syngonanthus	196		X (195)	X (1)				mainly lowland + endemic tepui summit
Wurdackia	1	X						
ERYTHROXYLACEAE								
Erythroxylum	200 (77 Brazil)				X (chiefly Am. and Madag.)			21 spp. Ven., Guyana, Brazil
EUPHORBIACEAE								
Amanoa	9 (Guiana and Brazil center)		X					
Celianella	1	X						+ endemic summit tepuis
Chaetocarpus	10 (3 centers of distrib.)				X (India, E. Indies W. Africa			5 spp. S. Am.
Conceveiba	6	X (S. Am.)						mainly lowland
Croton	1000 (750 spp. neotropics		X					40 spp. Guay. High., mostly low to middle alt.
Hieronyma	31	X (7 Andes, 11 W. Ind. 55 Brazil						E part of Guay. High., Andes, Coast. Cord.
Mabea	51	X						18 spp. Guay. High.

Phyllanthus	700		X	X	X (Java, Cuba, New Guin.)	41 spp. Guayana High.
Sapium	95				X	E sector of Guayana
FLACOURTIACEAE						
Euceraea	2	X				Gran Sabana, tepui summit, and Brazil
GENTIANACEAE						
Chelonanthus	20	X (S. Am.)				
Chorisepalum	6	X (S. Am.)				summits and slopes, tepuis
Curtia	10	X (S. Am.)				
Irlbachia	1	X (S. Am.)				
Lisianthus	50	X				
Macrocarpaea	35	X				
Symbolanthus	25	X				
Tachia	9	X (S. Am.)				lowland
Tapeinostemon	6	X (S. Am.)				5 spp. in Ven.; 4 on tepuis
Voyria	40		X			
GESNERIACEAE						
Codonanthe	15	X				mainly lowland
Dalbergaria	66	X				
Episcia	8	X				
Nautilocalyx	45	X (S. Am.)				
Tylopsacas	1	X				Ven. and Guyana
GNETACEAE						
Gnetum	30		X	X (Indo-Malesia to Fiji)	X	
GRAMINEAE						
Andropogon	200				X	
Arthrostylidium	20	X				4 Ven. spp., 1 on tepui summit
Aulonemia	24	X				3 Ven. spp.
Axonopus	80		X	X	X	

APPENDIX 13–A CONTINUED.

Genus	Total species	All neotropic	Mainly neotropic	Africa, Madagascar, Mascarenes	Malaysia, Australia, and others	Cosmopolitan or pantropic	Mainly Old World	Venezuelan distribution or comments on tepuis
Chusquea	96 (mainly Andean)	X						1 spp. tepui summit
Cortaderia	15	X (S. Am.)						
Echinolaena	2				X			
Ichnanthus	52		X (mainly S. Am)		X			
Isachne	25					X (mostly trop, Asia)		
Ischaemum	50					X	X	
Myriocladus	21	X						+ endemic tepui summits
Neurolepis	9 (mainly Andean)	X (S. Am., Trinidad)						
Olyra	25		X	X (1)				
Panicum	600					X		
Paspalum	400		X (Brazil a center)			X		
Raddiella	6	X						
Trachypogon	15		X (S. Am.)	X (4)				
GUTTIFERAE (CLUSIACEAE)								
Caraipa	21	X (S. Am.)						7 spp. Guay. High., 4 endemic
Clusia	145			X	X			
Hypericum	400					X		
Mahurea	2	X (S. Am.)						Guay. High. and Amazonia
Moronobea	7	X (S. Am.)						4 spp. Guay. High., mainly lowland
Oedematopus	10	X (S. Am.)						
Vismia	52		X (45 Am.)	X (7)				
HAEMODORACEAE								
Pyrrorhiza	1	X						+ endemic tepui summit
Schiekia	1	X (S. Am.)						lowland

HALORAGIDACEAE							
Laurembergia	22			X	X		
HUMIRIACEAE							
Humiria	3	X (S. Am.)					
ICACINACEAE							
Emmotum	13	X (S. Am.)					7 spp. Guay. High.
IRIDACEAE							
Trimezia	6	X					1 endemic of Guay. High.
LAURACEAE							
Aiouea	30	X (20 spp. Brazil)					
Nectandra	100	X (most Andes)					Guianas, Antilles, and Amaz. Hylaea centers
Ocotea	400		X	X (3)			
Persea	102 (58 S. Am.)		X				
Phoebe	80 (48 Mex. and C. Am., 30 S. Am.)		X		X		
LEGUMINOSAE							
Calliandra	100			X	X		mainly lowland
Cassia	500					X	mainly lowland in Ven. Guayana
Clitoria	35					X	
Dalbergia	100					X	
Dimorphandra	25	X					Gran Sabana and lowland
Dipteryx	13	X					mainly lowland
Machaerium	140		X	X			
Macrolobium	50	X					mainly lowland
Pithecellobium	200					X	lowland
Swartzia	125			X			
LILIACEAE							
Excremis	1	X (S. Am.) Andean					Andean

APPENDIX 13–A CONTINUED.

Genus	Total species	All neotropic	Mainly neotropic	Africa, Madagascar, Mascarenes	Malaysia, Australia, and others	Cosmopolitan or pantropic	Mainly Old World	Venezuelan distribution or comments on tepuis
Nietneria	2	X (S. Am.)						Gran Sabana and tepui summits of Ven., Guiana
Smilax	350			X	X	X		
Tofieldia	20	X (S. Am.)						Andes mainly
LENTIBULARIACEAE								
Genlisea	15 (11 American)			X (3)				7 spp. Guay. High.
Utricularia	120					X		41 spp. Guay. High., of which 16 are restricted
LINACEAE								
Ochthocosmus	6	X (S. Am.)						4 spp. Guay. High., 1 sp. Ven. lowland
Roucheria	8	X (S. Am.)						
LOGANIACEAE								
Bonyunia	4							2 spp. Guayana High.
LORANTHACEAE								
Dendrophthora	60	X (mostly W. Ind.)						
Eubrachion	2	X (S. Am.)						
Gaiadendron	7	X (mainly Andes)						
Ixidium	4	X (Cuba, Ven., Brazil)						1 sp. tepui
Oryctanthus	10	X						
Phoradendron	300	X (mostly S. Am.)						
Phrygilanthus	30		X		X			1 sp. on tepui with wide range

Phthirusa	60	X			15 spp. Ven.
Psittacanthus	80	X			lowland spp. in Amazonia
Struthanthus	70	X (S. Am. mainly)			13 spp. in Ven. some tepui spp.
LYTHRACEAE					
Cuphea	200 (136 S. Am.)	X			
MALPIGHIACEAE					
Blepharandra	4	X			3 spp. Guay. High.
Byrsonima	105	X			
Diacidia	12	X (S. Am.)			9 spp. Guay. High.
Heteropterys	100		X	X (1)	
Tetrapterys	70	X			
MARCGRAVIACEAE					
Marcgravia	40	X			
Norantea	35	X			
MELASTOMATACEAE					
Acanthella	2	X (S. Am.)			
Adelobotrys	20	X			usually low elevations
Chaetolepis	11	X (mainly páramo and subpáramo)			1 sp. on escarpments
Clidemia	160	X			51 spp. Ven.
Comolia	30	X (S. Am.)			many lowland spp. in Ven. few tepui spp.
Ernestia	12	X (S. Am.)			in Ven. confined to Guay. and with 2 tepui spp.
Graffenrieda	40	X			28 spp. Ven., some on tepuis
Leandra	200	X			20 spp. Ven.
Macairea	35	X (centers in Guayana and Braz. Planalto)			

APPENDIX 13–A CONTINUED.

Genus	Total species	All neotropic	Mainly neotropic	Africa, Madagascar, Mascarenes	Malaysia, Australia, and others	Cosmopolitan or pantropic	Mainly Old World	Venezuelan distribution or comments on tepuis
Macrocentrum	20	X						17 spp. on Guayana tepuis
Mallophyton	1	X						+ endemic tepui summit
Marcetia	15	X (all Brazil and Uruguay except one species)						
Meriania	50	X						
Miconia	1000	X						
Microlicia	130	X (S. Am., mainly Braz. Planalto)						2 Ven. spp. of tepuis and Gran Sabana
Monochaetum	35	X						mainly Andes of Ven.
Myrmidone	2	X (S. Am.)						lowland
Neblinanthera	1	X						+ endemic tepui summit
Phainantha	4	X (S. Am.)						4 spp. Ven. Guayana
Salpinga	7	X (S. Am.)						4 spp. Ven. Guayana
Siphanthera	15	X (S. Am.)						Ven. spp. on savannas and tepuis
Tateanthus	1	X (Ven., Brazil)						+ endemic tepui summit
Tibouchina	350	X (esp. SE Brazil)						many spp. restricted to Ven. Guayana
Tococa	50	X						
Topobea	25	X						3 Ven. spp. 1 on tepui

MORACEAE				
Cecropia	80	X		2 on tepui summits, mainly lowland spp.
Coussapoa	30	X (chiefly S. Am.)		mainly lowland
Ficus	600		X	mainly lowland
MYRISTICACEAE				
Virola	38	X		mainly lowland
MYRSINACEAE				
Cybianthus (including *Conomorpha)*	61	X		
Grammadenia	10	X		1 on tepuis
Rapanea	136		X	1 on tepuis
MYRTACEAE				
Calycolpus	10	X		2 spp. tepuis
Calyptranthes	100	X (many in S. Brazil)		16 spp. Guayana mainly lowland
Eugenia	275	X (mainly Brazil)		
Gomidesia	40	X		1 Guayana sp.
Marlierea	95	X		25 spp. in Guayana, one of centers
Myrcia	300	X		50 spp. in Guayana
Myrtus		X (mostly W. Ind.)		1 sp. unique on tepui summits
Plinia	31	X (25 W. Ind.)		1 sp. on tepui
Siphoneugena	4	X		
Ugni	10	X (Andean)		1 sp. on tepuis, Andean
NYCTAGINACEAE				
Neea	72	X (mainly S. Am. but Mex., C. Am., W. Ind.)		
OCHNACEAE				
Adenanthe	1	X		+ endemic summit tepui

APPENDIX 13–A CONTINUED.

Genus	Total species	All neotropic	Mainly neotropic	Africa, Madagascar, Mascarenes	Malaysia, Australia, and others	Cosmopolitan or pantropic	Mainly Old World	Venezuelan distribution or comments on tepuis
Adenarake	1	X						+ endemic summit tepui
Elvasia	11	X						lowland
Ouratea	100		X	X	X			
Philacra	4	X (S.Am.)						tepui summit and slopes in Ven. and Brazil
Poecilandra	3	X (S. Am.)						Ven. tepui summits, slopes and Gran Sabana
Sauvagesia	25		X	X (1)				
Tyleriia	9	X						+ endemic tepui summit
OLACACEAE								
Dulacia	14	X (S. Am.)						mainly lowland
Schoepfia	34		X		X	X		
ORCHIDACEAE								
Achineta	10	X (Colombia best center)						
Bifrenaria	30	X (Brazil best center)						
Bletia	30	X						
Brachionidium	18	X						
Bulbophyllum	500					X	X	
Catasetum	75	X						
Cattleya	20	X						some lowland species
Chamelophyton	1	X						
Cyrtopodium	20	X (Brazil best center)						
Dichaea	35	X						
Dryadella	15	X (mainly Brazil and Ecuador)						1 sp. on Ven. tepui
Duckeella	3	X (S. Am.)						2 of 3 Ven. spp. are lowland

Elleanthus	70	X (most numerous in Costa Rica and Peru)				
Epidendrum	1000	X				
Epistephium	25	X (S. Am. and Trinidad)				N. Brazil is center
Eriopsis	8	X				
Galeandra	24	X				4 of 8 Ven. spp. are lowland
Gomphicis	20	X (mainly Andean)				
Habenaria	500			X		
Hexadesmia	15	X				
Houlletia	5	X				
Lepanthes	160	X (best developed in Andes)				
Lepanthopsis	12	X				6 spp. Ven. 6 spp. W. Indies, 1 Guayana High.
Manniella	2		X (1)			
Masdevallia	300	X				
Maxillaria	300	X (Brazil and Colombia centers)				
Mendoncella	3	X				
Octomeria	100	X (Brazil a center)				
Oncidium	400	X				
Otoglossum	7	X				1 sp. on tepui summit
Pachyphyllum	24	X (mostly Andean)				
Pinelia	5	X (S. Am.)				1 sp. on tepui summit
Pleurothallis	700	X				
Pogonia (Cleistes)	40				X	
Polystachya	200		X (S. Africa)		X	
Prescottia	40	X (Brazil has 1/2 number of species)				
Reichenbachanthus (Hexisea)	6	X (Costa Rica is center)				2 spp. at low alt. in Ven.
Scaphosepalum	25	X (Colombia and Costa Rica best centers				1 sp. on tepui some on Coastal Cord.

APPENDIX 13–A CONTINUED.

Genus	Total species	All neotropic	Mainly neotropic	Africa, Madagascar, Mascarenes	Malaysia, Australia, and others	Cosmopolitan or pantropic	Mainly Old World	Venezuelan distribution or comments on tepuis
Scaphyglottis	50	X (Costa Rica has 1/2 number of species)						
Scuticaria	4	X (S. Am.)						1 Ven. sp. mainly lowland
Sobralia	80	X (Colombia and Costa Rica are centers)						some endemic Guyana spp.
Spiranthes	200						X	Brazil is dispersal center in S. Am.
Stelis	200	X (Andes center)						
Trigonidium	12	X						mainly lowland
Vanilla	50						X	
Vargasiella	2	X (Peru and Ven.)						1 sp. on tepui slope
Xerorchis	2	X						2 spp. Ven. in lowland
Zygosepalum	20	X						
Barbosella	15	X (mainly S. Brazil)						
Jacquiniella	5	X						
Notylia	40	X (Brazil is center)						
Brassia	50	X (Andean countries, Peru especially)						
OXALIDACEAE								
Biophytum	50			X	X			lowland
PALMAE								
Bactris	200	X						usually lowland, several spp. in Guayana High.
Dictyocaryum	2	X						
Euterpe	50	X						several endemic spp. Guay. High.
Geonoma	75	X						2 endemic spp. Guay. High.

Jessenia	10	X					lowland
Mauritia	6	X					lowland
PASSIFLORACEAE							
Passiflora	400		X	X (1 Madagas.)	X (40 Asia)		
PIPERACEAE							
Peperomia	1000					X	
Piper	2000					X	
PODOSTEMONACEAE							
Jenmaniella	7	X (S. Am.)					lowland, 1 sp. on tepui summit
Marathrum	25	X					lowland, 1 sp. on tepui summit
Rhyncholacis	25	X (S. Am.)					lowland, 1 sp. on tepui summit
PODOCARPACEAE							
Podocarpus	100				X		6 spp. tepui summits and slopes
POLYGALACEAE							
Bredemeyera	60				X		
Monnina	85	X (Andes center)					
Polygala	500 (180 N. Am., 35 Mex. & C. Am.)						
Securidaca	30			X	X		
POLYGONACEAE							
Coccoloba	263 (162 S. Am.)	X					
Euplassa	25	X (Brazil center)					2 spp. tepuis 4 spp. tepuis
Panopsis	20	X					
Roupala	51	X (Brazil center)					5 spp. tepuis
RAPATEACEAE							
Amphiphyllum	2	X (Ven.)					2 endemic tepui spp. Ven.
Kunhardtia	2	X (Ven.)					1 summit endemic sp.,

APPENDIX 13–A CONTINUED.

Genus	Total species	All neotropic	Mainly neotropic	Africa, Madagascar, Mascarenes	Malaysia, Australia, and others	Cosmopolitan or pantropic	Mainly Old World	Venezuelan distribution or comments on tepuis
Phelpsiella	1	X (Ven.)						1 lowland sp. + endemic summit
Saxofridericia	9	X (S. Am.)						mainly tepuis of Ven., also lowland
Stegolepis	27	X (Ven., Guyana)						mainly tepuis summits, Gran Sabana
RHAMNACEAE								
Rhamnus	75				X			few spp. S. Am.
ROSACEAE								
Prunus	200				X (Europe, Asia, Am.)			
RUBIACEAE								
Aphanocarpus	1	X (Ven.)						+ endemic tepui summit
Capirona	6	X (S. Am.)						lowland
Cephalodendron	1	X (Ven.)						+ endemic tepui summit
Chiococca	20	X						1 summit sp. on tepui
Chondrococcus	1	X (Ven.)						+ endemic tepui summit
Coryphothamnus	1	X (Ven.)						+ endemic tepui summit
Duidania	1	X (Ven.)						+ endemic tepui summit
Duroia	28	X (S. Am.)						mainly lowland; 13 Ven. spp., 4 on tepuis
Elaeagia	15	X (S. Am.)						6 Ven. spp., 1 on tepuis
Geophila	30		X	X		X		mainly lowland
Gleasonia	4	X (S. Am.)						3 low alt., 1 on tepui

Hillia	20	X			10 spp. Guay. high., 4 on tepuis
Kotchubaea	9	X (S. Am.)			lowland 1 sp. summit, 3 spp. slopes
Ladenbergia	30	X (S. Am.)			lowland 3 spp. slopes and summit
Maguireocharis	1	X (Ven.)			+ endemic tepui summit
Maguireothamnus	3	X (Ven.)			tepui slope and summit
Malanea	26	X			10 spp. tepui slope and summit
Merumea	2	X (Ven., Guyana)			1 sp. tepui summit
Neblinathamnus	3	X (Ven., N. Brazil)			tepui summit and slopes
Pagamea	23	X (S. Am.)			mostly lowland, 10 spp. on tenuis
Pagameopsis	2	X (Ven.)			+ endemic tepui summits
Palicourea	200	X			1 sp. tepui summit
Perama	12	X (mainly S. Am.)			tepui spp. widely distributed
Platycarpum	10	X (S. Am.)			mainly lowland
Psychotria	700			X	lowland and summits
Relbunium	30	X			1 tepui summit sp. widely distributed
Remijia	35	X (S. Am.)			mainly lowland 6 spp. tepui slopes or summit
Retiniphyllum	21	X (S. Am.)			mainly lowland 3 spp. tepuis
Rudgea	150	X			1 endemic tepui summit sp.

APPENDIX 13–A CONTINUED.

Genus	Total species	All neotropic	Mainly neotropic	Africa, Madagascar, Mascarenes	Malaysia, Australia, and others	Cosmopolitan or pantropic	Mainly Old World	Venezuelan distribution or comments on tepuis
Schradera	35	X						1 sp. summit
RUTACEAE								
Apocaulon	1	X						tepui slopes and summit
Myllanthus	3	X (Ven., Brazil)						1 sp. lowland Brazil, 2 spp. Ven. summit and lowland
Raveniopsis	14	X (Ven.)						mainly tepui summits
Spathelia	20	X (14 W. Ind., 5 Guayana, 1 Brazil)						5 Ven. spp. on summit and slopes, Gran Sabana
SACCIFOLIACEAE								
Saccifolium	1	X (Ven.)						+ endemic summit tepui
SANTALACEAE								
Thesium	245	X (*Austro americium*)				X (*Thesium*)	X (*Thesium*)	1 sp. tepui summit and Gran Sabana
(*Austroamericium*)	(3)	(Ven., Brazil)						
SAPINDACEAE								
Matayba	46	X						
SAPOTACEAE								
Chrysophyllum	150					X		
Ecclinusa	21		X (20)	X (1)				
Elaeoluma	2	X						
Glycoxylon	5	X (Ven., Brazil)						
Micropholis	70	X						
Neoxythece	3	X						

SARRACENIACEAE							
Heliamphora	6	X (Ven.)					summit, slope, and Gran Sabana
SCROPHULARIACEAE							
Velloziella	2	X					
SIMARUBACEAE							
Picramnia	40 (23 S. Am.)	X					
Simaruba	9	X					
SOLANACEAE							
Cestrum	250	X					
Markea	20	X		X	X		
STYRACACEAE							
Styrax	200					X	10 spp. Guayana
SYMPLOCACEAE							
Symplocos	282				X (167 Australia)		8 spp. Guayana
THEACEAE							
Archytaea	3	X (S. Am.)					2 of 3 spp. lowland
Bonnetia	21	X					most Ven. spp. on tepuis at low or high alt.
Freziera	34	X					
Laplacea	18				X (10)		
Neblinaria	1	X (Ven., Brazil)					+ endemic tepui summit
Neogleasonia	3	X (Ven.)					+ endemic tepui summit
Neotatea	3	X (Ven., Colombia)					+ endemic tepui summit
Ternstroemia	50		X (Andes of N. S. Am. center)	X (2)			
TEPUIANTHACEAE							
Tepuianthus	6	X (Ven., Colombia)					tepui summits and savanna of Terr. Fed. Amaz.

APPENDIX 13–A CONTINUED.

Genus	Total species	All neotropic	Mainly neotropic	Africa, Madagascar, Mascarenes	Malaysia, Australia, and others	Cosmopolitan or pantropic	Mainly Old World	Venezuelan distribution or comments on tepuis
THYMELEACEAE								
Daphnopsis	46	X						2 spp. tepui, summit, slope
TRIGONIACEAE								
Euphronia (*Lightia*)	3	X (S. Am.)						Gran Sabana, tepuis
Trigonia	30	X						
VELLOZIACEAE								
Vellozia	124		X (Brazil Planalto center)	X				1 Ven. spp. tepui summit and lowland
VERBENACEAE								
Aegiphila	150	X						few on tepuis
VOCHYSIACEAE								
Ruizterania	18	X (S. Am.)						mainly lowland
Vochysia	80	X			X			
WINTERACEAE								
Drimys	70				X			4 spp. New World
XYRIDACEAE								
Abolboda	17	X (S. Am.)						mainly lowland
Achlyphila	1	X (Ven.)						+ endemic summit tepui
Orectanthe	2	X (Ven.)						tepui summits, Gran Sabana
Xyris	250		X (160)	X	X (20)	X		72 spp. Guayana, 90 Brazil High.

REFERENCES

Aristeguieta, L. 1964. Compositae. In T. Lasser, ed., *Flora de Venezuela*. 10 (1–2): 1–941. Caracas: Inst. Bot. Minist. Agric. y Cría.

Armellada, C. de. 1944. *Gramática y diccionario de la lengua Pemón (Arekuna, Taurepan, Kamarakoto). 2. Diccionario*. Caracas: C. D. Artes Graficas.

Ayensu, E. S. 1973. Phytogeography and evolution of the Velloziaceae. In *Tropical forest ecosystems in Africa and South America: A comparative review*, B. J. Meggers, E. S. Ayensu, and W. D. Duckworth, eds., pp. 105–119. Washington, D.C.: Smithsonian Institution Press.

Barneby, R. C., and B. A. Krukoff. 1971. Supplementary notes on American Menispermaceae, VIII. *Mem. N.Y. Bot. Gard.* 22: 1–89.

Bernardi, L. 1961. Revisio generis *Weinmanniae*, I. *Candollea* 17: 123–189.

——— 1963. Revisio generis *Weinmanniae*, II. *Candollea* 18: 285–334.

——— 1964. Revisio generis *Weinmanniae*, III. *Bot. Jahrb.* 83: 126–221.

Buchholz, J. T., and N. E. Gray. 1948. A taxonomic revision of *Podocarpus*, II and IV. *J. Arnold Arb.* 29: 64–76.

Camp, W. H. 1947. Distribution patterns in modern plants and the problems of ancient dispersals. *Ecol. Monogr.* 17: 159–183.

——— 1952. Phytophyletic patterns on land bordering the South Atlantic basin. *Bull. Amer. Mus. Nat. Hist.* 99: 205–216.

Colvée, P. 1971. *Inf. Codesur 5-II-A*. Ministerio de Obras Publicas, Comisión para el Desarrollo del Sur. Caracas.

——— 1972. *Consideraciones geológicas sobre el Cerro Autana*. Ministerio Obras Publicas, Comisión para el Desarrollo del Sur. Caracas.

Cowan, R. S. 1953. A taxonomic revision of the genus *Macrolobium* (Leguminoseae-Caesalpinioideae). *Mem. N.Y. Bot. Gard.* 8: 257–342.

——— 1975. A monograph of the genus *Eperua* (Leguminosae: Caesalpinioideae). *Smithson. Contrib. Bot.* 28: 15–17.

Dunsterville, G. C. K. 1973. Orchids of Cerro Autana. *Bull. Am. Orchid Soc.* 42: 388–401.

Dunsterville, G. C. K., and L. A. Garay. 1959–1976. *Venezuelan Orchids Illustrated*. 6 Vols. London: Deutsch.

——— 1979. *Orchids of Venezuela, an illustrated field guide*. 3 Vols. Cambridge, Mass.: Harvard Univ. Press.

Ewan, J. 1947. A revision of *Chorisepalum*, an endemic genus of Venezuelan Gentianaceae. *J. Wash. Acad. Sci.* 37: 392–396.

Foldats, E. 1969–1970. Orchidaceae. In T. Lasser, *Flora de Venezuela*. 15 (1–5). Caracas: Inst. Bot. Minist. Agric. y Cría.

Gansser, A. 1974. The Roraima problem. *Verhandl. Naturf. Ges. Basel.* 84: 80–100.

Ghosh, S. 1977. Geología del Grupo Roraima en Territorio Federal Amazonas, Venezuela. *Mem. 5o Congreso Geol. Ven.* pp. 167–173.

Gleason, H. A. 1929. The Tate collection from Mt. Roraima and vicinity. *Bull. Torr. Bot. Club.* 56: 391–408.

——— 1931. Botanical results of the Tyler-Duida expedition. *Bull. Torr. Bot. Club.* 58: 277–506.

Gleason, H. A., and E. Killip. 1939. The flora of Mount Auyan-tepui, Venezuela. *Brittonia* 3: 141–204.

Good, R. 1974. *The geography of the flowering plants*. 4th ed. London: Longman.

Haffer, J. 1974. Avian speciation in tropical South America. *Publ. Nuttall Ornithol. Club*, no. 14: 1–390.

Hendryck, R. 1963. *Austroamericium*, género nuevo. *Bol. Soc. Arg. Bot.* 10: 120–128.

Jablonski, E. 1967. Euphorbiaceae. In B. Maguire and collaborators, The botany of the Guayana Highland, pt. VII. *Mem. N.Y. Bot. Gard.* 17: 80–190.

Janzen, D. H. 1974. Tropical blackwater rivers, animals, and mast fruiting by the Dipterocarpaceae. *Biotropica* 6: 69–103.

Kirkbride, J. H., Jr. 1976. A revision of the genus *Declieuxia* (Rubiaceae). *Mem. N.Y. Bot. Gard.* 28: 1–87.

——— 1979. Revision of the genus *Psyllocarpus*. *Smithson. Contrib. Bot.* 41: 1–32.

Koyama, T. 1972. Cyperaceae. In B. Maguire and collaborators, The botany of the Guayana Highland, pt. IX. *Mem. N.Y. Bot. Gard.* 23: 23–89.

Krukoff, B. A. 1972. American species of *Strychnos*. *Lloydia* 35: 193–271.

Krukoff, B. A., and R. C. Barneby. 1969. Supplementary notes on the American species of *Strychnos*, VIII and IX. *Mem. N.Y. Bot. Gard.* 20: 1–99.

——— 1970. Supplementary notes on the American Menispermaceae, VI. *Mem. N.Y. Bot. Gard.* 20: 1–70.

Kubitzki, K. 1978. *Caraipa*. In B. Maguire and collaborators, The botany of the Guayana Highland, pt. X. *Mem. N.Y. Bot. Gard.* 29: 82–138.

Langenheim, J., Y. Lee, and S. Martin. 1973. An evolutionary and ecological perspective of Amazonian Hylaea species of *Hymenaea* (Leguminosae-Caesalpinioideae). *Acta Amazonica*. 3: 5–38.

Leeuwenberg, A. J. M. 1969. Notes on American Loganiaceae, III. Revision of *Bonyunia* Rich. Schomb. *Acta Bot. Neérl.* 18: 152–158.

Lellinger, D. B. 1967. *Pterozonium*. In B. Maguire and collaborators, The botany of the Guayana Highland, pt. VII. *Mem. N.Y. Bot. Gard.* 17: 2–23.

Léxico Estratigráfico de Venezuela. 2d ed. 1970. Bol. Geol. Publ. Esp. No. 4. Minist. Minas e Hidroc. Caracas: Editorial Sucre.

Maguire, B. 1956. Distribution, endemicity, and evolution patterns among Compositae of the Guayana Highland of Venezuela. *Proc. Am. Philos. Soc.* 100: 467–475.

——— 1958a. Xyridaceae. In B. Maguire and collaborators, The botany of the Guayana Highland, pt. III. *Mem. N.Y. Bot. Gard.* 10: 1–19.

——— 1958b. Rapateaceae. In B. Maguire and collaborators, The botany of the Guayana Highland, pt. III. *Mem. N.Y. Bot. Gard.* 10: 19–49.

——— 1965a. In B. Maguire and collaborators, The botany of the Guayana Highland, pt. VI. *Mem. N.Y. Bot. Gard.* 12: 1.

——— 1965b. Rapateaceae. In B. Maguire and collaborators, The botany of the Guayana Highland, pt. VI. *Mem. N.Y. Bot. Gard.* 12: 69–102.

——— 1970. On the flora of the Guayana Highland. *Biotropica* 2: 85–100.

——— 1972. Bonnetiaceae. In B. Maguire and collaborators, The botany of the Guayana Highland, pt. IX. *Mem. N.Y. Bot. Gard.* 23: 131–165.

——— 1978. Symplocaceae and Styracaceae. In B. Maguire and collaborators, The botany of the Guayana Highland, pt. X. *Mem. N.Y. Bot. Gard.* 29: 36–61, 204–229.

——— 1979. Guayana, region of the Roraima sandstone formation. In *Tropical Botany*, K. Larsen and L. Holm-Nielson, eds., pp. 223–238. New York: Academic Press.

Maguire, B., and J. M. Pires. 1978. Saccifoliaceae. In B. Maguire, The botany of the Guayana Highland, pt. X. *Mem. N.Y. Bot. Gard.* 29: 230–244.

Maguire, B., and L. B. Smith. 1964. Xyridales. In B. Maguire, The botany of the Guayana Highland, pt. V. *Mem. N.Y. Bot. Gard.* 10: 7–37.

Maguire, B., and J. J. Wurdack. 1957a. Droseraceae. In B. Maguire and J. J. Wurdack and collaborators, The botany of the Guayana Highland, pt. II. *Mem. N.Y. Bot. Gard.* 9: 331–336.

——— 1957b. Compositae. In B. Maguire and J. J. Wurdack and collaborators, The botany of the Guayana Highland, pt. II. *Mem. N.Y. Bot. Gard.* 9: 366–392.

——— 1961. Ochnaceae. In B. Maguire and J. J. Wurdack and collaborators, The botany of the Guayana Highland, pt. IV (2). *Mem. N.Y. Bot. Gard.* 10: 6–21.

Maguire, B., J. A. Steyermark, and J. J. Wurdack. 1957. Botany of the Chimantá Massif, I. *Mem. N.Y. Bot. Gard.* 9: 393–439.

Markgraf, F. 1930. Monographie der gattung *Gnetum*. *Bull. Jard. Bot. Buitenzorg*, ser. III, 10: 407–499.

Mayr, E., and W. H. Phelps, Jr. 1967. The origin of the bird fauna of the South Venezuelan highlands. *Bull. Amer. Mus. Nat. Hist.* 136: 269–328.

McConnell, R. B. 1959. The Takutu formation in British Guiana and the probable age of the Roraima Formation. *Trans. Second Caribbean Geol. Conf.*, Puerto Rico, 1959. pp. 163–170.

——— 1968. Planation surfaces in Guyana. *Geog. Jour.* 134: 507–520.

McConnell, R. B., and E. Williams. 1970. Distribution and provisional correlation of the Precambrian of the Guyana Shield. *Proc. 8th Guiana Geol. Conf.* I, 1–20. Georgetown, Guyana: Dept. Geol. Mines.

McDougall, I., W. Compston, and D. D. Hawkes. 1963. Leakage of radiogenic argon and strontium from minerals present in Proterozoic dolerites from British Guiana. *Nature* 193: 564–567.

Ministerio de Obras Publicas, Codesur, Territorio Federal Amazonas y Distrito Cedeño del Estado Bolívar. Mapa. 1975. Edición Preliminar.

Morton, C. V. 1957. Pteridophyta (Ptari-tepui). In J. A. Steyermark, ed., Contributions to the flora of Venezuela. *Fieldiana, Bot.* 28: 729–741.

Müller, P. 1973. *The dispersal centres of terrestrial vertebrates in the Neotropical Realm. Biogeographica* 2. The Hague: Junk.

Paba Silva, F., and T. van der Hammen. 1960. Sobre la geología de la parte sur de la Macarena. *Bol. Geol. (Colombia)* 6: 7–30.

Prance, G. T. 1972. Chrysobalanaceae. In *Flora Neotropica*. Monograph 9. New York: Hafner.

——— 1973. Phytogeographic support for the theory of Pleistocene forest refuges in the Amazon Basin, based on evidence from distribution patterns in Caryocaraceae, Chrysobalanaceae, Dichapetalaceae, and Lecythidaceae. *Acta Amazonica* 3: 5–28.

——— 1978. The origin and evolution of the Amazon flora. *Interciencia* 3: 207–223.

Priem, H. N. A., N. Boelrijk, E. Hebeda, E. Verdurmen, and R. Verschure. 1973. Age of the Precambrian Roraima Formation in northeastern South America: Evidence from isotopic dating of Roraima pyroclastic rocks in Surinam. *Geol. Soc. Amer. Bull.* 84: 1677–1684.

Raven, P. H., and D. I. Axelrod. 1974. Angiosperm biogeography and past continental movements. *Ann. Mo. Bot. Gard.* 61: 539–673.

Schultes, R. E. 1944. Notes on the ecology of some isolated sandstone hills of the Uaupes region. *Caldasia* 3: 124–130.

——— 1945. Glimpses of the little known Apoporis river. *Chron. Bot.* 9: 123–127.

Simpson, B. B. 1979. Quaternary biogeography of the high montane regions of South America. In *The South American herpetofauna: Its origin, evolution, and dispersal*, W. D. Duellman, ed., Museum of Natural History, Monograph 7: 155–188. Lawrence: Univ. of Kansas Press.

Simpson, B. B., and J. Haffer. 1978. Speciation patterns in the Amazonian forest biota. *Ann. Rev. Ecol. Syst.* 9: 497–518.

Sleumer, H. 1954. Proteaceae americanae. *Bot. Jahrb.* 76: 139–211.

Smith, A. C. 1953. Ericaceae, In J. A. Steyermark, ed., Contributions to the flora of Venezuela. *Fieldiana, Bot.* 28: 453–454.

Smith, L. B. 1962. Origins of the flora of southern Brazil. *Contrib. U.S. Nat. Herb.* 35: 215–249.

Smith, L. B., and E. S. Ayensu. 1976. A revision of American Velloziaceae. *Smithson. Contrib. Bot.* 30: I–VII, 1–172.

Smith, L. B., and R. J. Downs. 1974. Bromeliaceae: Pitcairnioideae. In *Flora Neotropica*. Monograph 14 (1): 212–231; 437–500. New York: Hafner.

——— 1977. Bromeliaceae: Tillandsioideae. In *Flora Neotropica*. Monograph 14 (2): 663–1492. New York: Hafner.

Snelling, N. J. 1963. Age of the Roraima Foundation, British Guiana. *Nature* 198: 1079–1080.

Snelling, N. J., and R. B. McConnell. 1969. The geochronology of Guyana. *Geol. Mijnb.* 48: 201–213.

Steyermark, J. A. 1949. The genus Oreobolus. *Bol. Soc. Ven. Cienc. Nat.* 11: 306–311.

——— 1957. Contributions to the flora of Venezuela, part 4. *Fieldiana, Bot.* 28: 679–1190.

——— 1966. Contribuciones a la flora de Venezuela, parte 5. *Acta. Bot. Venez.* 1: 129–168.

——— 1967a. Flora del Auyán-tepui. *Acta. Bot. Venez.* 2: 44–47; 64.

——— 1967b. Rubiaceae. In B. Maguire and collaborators, The botany of the Guayana Highland, pt. VII. *Mem. N.Y. Bot. Gard.* 17: 369–435.

——— 1972. Rubiaceae. In B. Maguire and collaborators, The botany of the Guayana Highland, pt. IX. *Mem. N.Y. Bot. Gard.* 23: 227–832.

——— 1974a. The summit vegetation of Cerro Autana. *Biotropica* 6: 7–13.

——— 1974b. Rubiaceae. In T. Lasser, ed., *Flora de Venezuela.* 9: 1–2070. Caracas: Inst. Bot., Minist. Agric. y Cría.

——— 1975. Informe sobre la flora del Cerro Autana. *Acta Bot. Venez.* 10: 219–233.

——— 1979a. Flora of the Guayana Highland: Endemicity of the generic flora of the summits of the Venezuelan tepuis. *Taxon* 28: 45–54.

——— 1979b. Plant refuges and dispersal centres in Venezuela: Their relict and endemic element. In *Tropical botany*, K. Larsen and L. Holm-Nielson, eds., pp. 185–221. New York: Academic Press.

——— 1980. Relationships of some Venezuelan forest refuges with lowland tropical floras. In *The model of biological diversification in the humid tropics*, G. T. Prance, ed., pp. 182–220. New York: Columbia Univ. Press.

Steyermark, J. A., and C. Brewer-Carias. 1976. La vegetación de la cima del Macizo de Jaua. *Bol. Soc. Ven. Cienc. Nat.* 32: 179–405.

Steyermark, J. A., and G. C. K. Dunsterville. 1980. The lowland floral element on the summit of Cerro Guaiquinima and other cerros of the Guayana highland of Venezuela. *J. Biogeog.* 7: 285–303.

Steyermark, J. A., and B. Maguire. 1967. Botany of the Chimantá Massif, II. *Mem. N.Y. Bot. Gard.* 17: 440–464.

——— 1972. The flora of the Meseta del Cerro Jaua. *Mem. N.Y. Bot. Gard.* 23: 833–892.

Talukdar, S. C., and P. Colvée. 1974. *Geología y estratigrafía del área Meseta del Viejo-Cerro Danto, Territorio Federal Amazonas, Venezuela.* Ministerio de Obras Públicas, Comisión para el Desarrollo del Sur. Caracas.

Taylor, P. 1967. Lentibulariaceae. In B. Maguire and collaborators, The botany of the Guayana Highland, pt. VII. *Mem. N.Y. Bot. Gard.* 17: 201–227.

Thomas, J. L. 1960. A monographic study of the Cyrillaceae. *Contrib. Gray Herb.* 186: 1–114.

Tryon, R. 1972. Endemic areas and geographic speciation in tropical American ferns. *Biotropica* 4: 121–131.

——— 1975. The biogeography of endemism in the Cyatheaceae. *Fern Gaz.* 1: 73–79.

Van Royen, P. 1951. The Podostemaceae of the New World, I. *Med. Bot. Mus. Herb. Rijksun.* (Utrecht) 107: 1–150.

——— 1953. The Podostemaceae of the New World, II. *Act. Bot. Néerl.* 2: 1–21.

Vareschi, V. 1969. Helechos. In T. Lasser, ed., *Flora de Venezuela.* 1: 1–1033. Caracas: Inst. Bot., Minist. Agric. y Cría.

Woodson, R. E. 1953. In J. Steyermark, ed., Contributions to the flora of Venezuela, part 3. *Fieldiana, Bot.* 28: 499–500.

Wurdack, J. J. 1973. Melastomataceae. In T. Lasser, ed., *Flora de Venezuela.* 8: 1–819. Caracas: Inst. Bot., Minist. Agric. y Cría.

Note Added in Proof

Since the Appendix was prepared, additional genera were found on the summits of the Venezuelan table mountains, which bring the total number to 471.

Bromeliaceae
- *Brewcaria*—endemic to tepuis of Terr. Fed. Amazonas, Venezuela
- *Steyerbromelia*—endemic to Cerro Marahuaca, Terr. Fed. Amazonas, Venezuela

Compositae
- *Chionolaena*—Tepuis (Cerro Marahuaca, Terr. Fed. Amazonas) and high mountains of southern Brazil
- *Oritrophium*—Tepuis (Cerro Marahuaca, Terr. Fed. Amazonas) and páramos of the Andes of South America

Gentianaceae
- *Celiantha*—endemic to tepuis (Cerro de la Neblina), Terr. Fed. Amazonas, Venezuela
- *Neblinantha*—endemic to tepuis (Cerro de la Neblina), Brazil side
- *Wurdackanthus*—endemic to tepuis (Cerro de la Neblina), Brazil side

Ochnaceae
- *Perissocarpa*—Andes and Coastal Cordillera of Venezuela and tepuis of Venezuela and northern Brazil (Aracá)

Rapateaceae
- *Marahuacaea*—endemic to tepuis (Cerro Marahuaca), Terr. Fed. Amazonas, Venezuela

Rutaceae
- *Rutaneblina*—endemic to tepuis (Cerro de la Neblina), Terr. Fed. Amazonas, Venezuela

Theaceae
- *Acopanea*—endemic to tepuis (Chimantá Massif), Edo. Bolívar, Venezuela

14

Speciation and Adaptive Radiation in Andean Telmatobius *Frogs*

JOSÉ M. CEI

Although several groups of anurans probably started to spread on the rising cordilleras through the Early and Middle Tertiary, herpetofaunal elements at altitudes above 2000 m could not have emigrated into high-elevation habitats until the Late Pliocene or Pleistocene (Simpson, 1979). The potential adaptive trends of early invaders enabled them to colonize the available niches in the recently uplifted environments, where they began to differentiate. Most of the first Andean anurans belong to austral or Gondwanian stocks such as leptodactylid or hylid frogs, which evolved in the geographically isolated Neotropical continent and underwent separate radiations from the Eocene through the Miocene. Several anuran groups have interesting distribution and speciation patterns in the high Andes. For example, a narrow-skull lineage of South American toads, the *Bufo spinulosus* group, is divided into several forms distributed from southern Ecuador to about 46° S in Patagonia. Among the hylids, some species of *Hyla* (*H. pulchella andina* Muller, *H. charazani* Vellard) extend through the central plateau, and marsupial frogs of the genus *Gastrotheca* are found from Colombia to Argentina. Members of the Leptodactylinae and Eleutherodactylinae also occur at remarkable altitudes, such as *Pleurodema marmorata* (Duméril and Bibron) and *P. cinerea* Cope from the Bolivian highlands, or some Ecuadorian species of *Eleutherodactylus*. However, no other Andean anuran stock seems to exhibit such an outstanding radiation and such a broad distribution in the high Andes as the primitive leptodactylid branch Telmatobiinae, especially the genera *Telmatobius* and *Batrachophrynus*.

According to its etymological roots the Greek name *Telmatobius* means "life in the summits". The genus comprises about 30 species, scattered from the equator (*T. niger* Barbour and Noble) to 29° S on the eastern slopes of the Andes (*T. contrerasi* Cei), and to 33° on the western slopes (*T. montanus* Lataste) (Fig. 14–1). The closely allied genus *Batrachophrynus* is probably a specialized ancestral branch of the main *Telmatobius* lineage. But other authors have reported it as an ancient relict of a group having apparently a widespread distribution southward, prior to the uplift of the Andes (Lynch, 1978; Duellman, 1979). Only two localized species are known from the central Peruvian Andes, the giant frog *Batrachophrynus macrostomus* Peters from the shallow rush-covered lagoons of Junín, at about 4000–4200 m, and the moderate sized *B. brachydactylus* Peters living in the small creeks near these lagoons. This second species has been recently placed in the monotypic genus *Lynchophrys* (Laurent, 1983).

Morphological and ecological evidence indicate that close evolutionary relationships exist between *Telmatobius* and other Andean and extra-Andean telmatobiines such as *Alsodes*, *Atelognathus*, *Eupsophus*, and *Somuncuria*. This chapter incorporates new data now available for an evaluation of the evolutionary history and adaptive features of *Telmatobius*. These data include both the latest reassessment of the formerly described "Patagonian" *Telmatobius* south of 38° S (Lynch, 1978) and the results of Laurent's research on species of forest-inhabiting *Telmatobius* found at relatively low altitudes in subtropical northwestern Argentina (Laurent, 1970a, 1973). Lest the reader believe that the evidence at hand is sufficient for a conclusive analysis of genetic and evolutionary processes in such a widespread and confusing genus, however, I shall quote Duellman's statement (1979) that "the systematic relationship of the species of *Telmatobius* presently are too poorly known to assess fully the historical biogeography of the group."

ORIGIN OF THE MAIN *TELMATOBIUS* LINEAGE

The great antiquity and early distribution of the ancestors of telmatobiine genera are suggested by

FIG. 14–1. Geographic distribution of the genera *Telmatobius* (*stippled areas*) and *Batrachophrynus* (*dotted circle*) in the Andes. Black areas indicate scattered southern relicts of the primitive atelognathid stock.

Early Tertiary leptodactylid frog communities (Schaeffer, 1949; Heyer, 1975; Lynch, 1978). Lynch advocated relationships between modern but primitive central and southern Chilean faunal elements and the several genera radiating out from austral Cenozoic anurans. The formerly named "Patagonian" *Telmatobius*, or atelognathid frogs (*Atelognathus*, *Somuncuria*) living as relicts on isolated basaltic tablelands (Cei, 1969, 1972; Lynch, 1978), may also be related to these ancestral genera, which were probably distributed in areas of high environmental equability (Axelrod and Bailey, 1969).

As in the most primitive living anurans, leiopelmatid features of the atelognathids are suggested by the apparent stegochordal centra and the presence of nine presacral vertebrae, the latter condition being reported also in *Batrachophrynus*. Phyletic relationships with Jurassic leiopelmatid frogs of Patagonia have been claimed and hypothetical intermediate ancestors postulated (Lynch, 1978). However, several morphological and physiological traits link *Telmatobius* with *Atelognathus*, in spite of some very important diverging character states such as the lack of a quadratojugal in the latter. This taxonomic link is clearly shown by the most parsimonious but consistent cladograms in Lynch's (1978) cladistic analysis; it is also supported by a general comparison of intergeneric serological relationships (Cei, 1970). In precipitine reactions across taxa, I found that *Telmatobius hauthali* Koslowsky is only moderately distant serologically from *Atelognathus reverberi* (Cei), *Somuncuria somuncurensis* (Cei), and *Telmatobufo*, but more distant from *Alsodes*, *Caudiverbera*, and the Leptodactylinae.

A Cenozic "lower telmatobiine" stock cannot be questioned, probably in southern continental areas of high climatic equability. Moreover, patterns of distribution and ecological features both of the genus *Telmatobius* and of atelognathid relicts suggest a probably independent beginning of the speciation and adaptive radiation of *Telmatobius*, prior to the Late Pliocene crisis and intercontinental reconnection. The major geological events taking place at that time were the uplift of the precordilleras of western Argentina, the newly folded Subandean range of northern Argentina and Bolivia, and the rising eastern border of the Cordillera Real of Ecuador (Harrington, 1962). All of these tectonic events played an important geomorphological role and contributed to the geographic isolation of the just established high Andean fauna by increasing or modifying topographic and climatic barriers.

Morphological and biogeographic kinds of evidence indicate that extant atelognathids should be considered modified relicts of the basic evolutionary step in telmatobiine radiation. The peculiar position of *Somuncuria* between lower telmatobiine frogs and leptodactyline *Pleurodema* (Lynch, 1978) seems to strengthen this idea. During climatic and vegetation fluctuations of the Pleistocene, a wide elimination of Tertiary anuran communities took place in Patagonia, but few atelognathids could survive in the dry northern plateaus of Neuquén. The uncommon southernmost species of *Atelognathus*, *A. grandisonae* Lynch, lives in the moorlands of Puerto Eden, southern Chile, at about 50° S, and the recently discovered *A. salai* (Cei, 1984) inhabits small basaltic lagoons north of Lago Buenos Aires in Santa Cruz Province, southern Argentina.

The late uplift phases of the Cordilleras and the Pleistocene glaciations resulted in drastic changes in climate and morphology of the young southern Andean landscapes south of 35° S. Thus, their periglacial environments became a rather unsuitable habitat for any further southward expansion of the northernmost *Telmatobius* communities.

FIRST ANDEAN *TELMATOBIUS* DISPERSAL

Vellard's statement (1951) that the lower altitudinal boundaries of the distribution of *Telmatobius* in Argentina lie at 2000 m, at about 25° S, was modified, surprisingly, by the discovery of these anurans in wet subtropical forests between 23° and 28° S. *Telmatobius barrioi* Laurent has been reported at about 1,600 m from the streams of the warm Calilegua forest, Jujuy Province (Laurent, 1970a). *Telmatobius ceiorum* Laurent from Banderita, Catamarca Province, lives in forest streams from 1300 to 1900 m, among submerged crags forming shallow ponds and cascades surrounded by tropical vegetation. This habitat clearly recalls, for example, the eastern subtropical mesic environments of Brazil, inhabited by a number of primitive leptodactylid forms such as Elosiinae, Gripiscini, *Thoropa*, and *Limnomedusa*. These taxa are primarily distributed in areas of equable climates, such as humid northeastern Argentina, or the "serras" of the Brazilian littoral.

I would like to emphasize the remarkable biogeographic and evolutionary interest exhibited by the occurrence of *Telmatobius* species in discontinuous lower Andean forests. Wet forest remnants of western Argentina were maintained from the Pliocene through the Pleistocene by the rainshadow effect of the high mountains. Their present anuran relicts originally belonged to a broader range of lower telmatobiine stocks, probably prior to the Miocene Andean uplift and the subsequent progressive dessication of central regions of South America, giving rise to today's arid belt from the chaco to the caatinga. The presumed climatic and ecological stability of Subandean forested areas later allowed the survival of several faunal elements through humid–arid cycles of the Pleistocene. Such forests acted as refuges on Western Cordilleran slopes at this latitude (Simpson Vuilleumier, 1971).

Laurent (1977) pointed out the probable conspecificity of *Telmatobius barrioi* and *T.oxycephalus* Vellard from Cerro Escalera, at 3800 m, Salta Province, and other montane localities such as Rio Yala, Jujuy Province, a transition zone on the eastern slope of the Subandean relief (Fig. 14–2). Laurent also suggested a similar pattern of distribution for the highland populations of *T. ceiorum*. Barriers to gene flow are apparently unreported, but because we know very little about the biological relationships between highland and lower subtropical populations of both the *oxycephalus-barrioi* group and of the *ceiorum* group, no further statement can be made here about their probable potential adaptedness, systematic status, or incipient speciation.

Remnants of past steps in the northward dispersal of *Telmatobius* are likely to be still represented by some other forms from streams or pools in the Subandean transition zone of mountain forests. Trueb's (1979) careful revision of the Ecuadorian species provides information on ecological and behavioral characteristics both of the páramo inhabitant *T. niger* Barbour and Noble and of the subpáramo or upper humid forest inhabitants *T. cirrhacelis* Trueb and *T. vellardi* Musterman and Leviton. Trueb pointed out that *T. cirrhacelis* lives in wet forest. At about 4° S and 79° W the cloud forest, at 2700 m, is characterized by thick layers of moss, dense bushes, bromeliads, and bamboos. The observed daily temperature range was moderate (between 8° and 17° C), and rather different from the uniformly low annual páramo temperature, or the strong daily variation in the puna. It should be emphasized that the geographic location of the findings—the

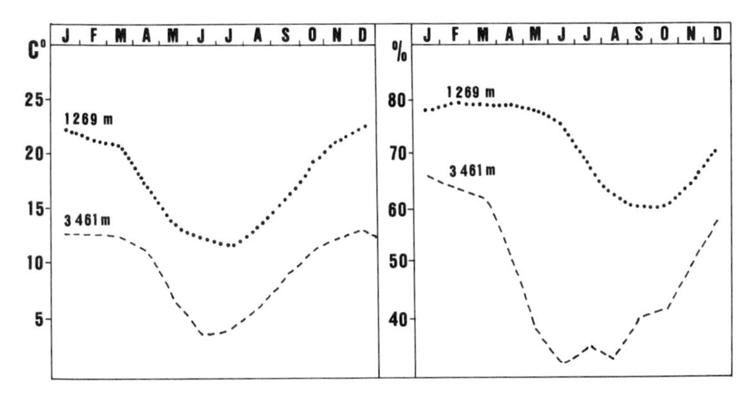

FIG. 14–2. Climate of different localities within the altitudinal range of populations of the *Telmatobius oxycephalus* group. *Left*: Means of monthly temperatures at Jujuy (1269 m) and La Quiaca (3461 m). *Right*: Means of monthly relative humidity (%) at the same localities. Drawn in accordance with data of the Estadísticas Climatológicas, Dirección Meteorología, Geofísica e Hidrología, Rep. Argentina. Buenos Aires (1944).

Loja-Zamora ridge and Loja-Saraguro basins—lies in the presumed glacial refuge area of the lower eastern Andean slopes (Simpson Vuilleumier, 1971).

In summary, all available data suggest that the early origin and distribution of *Telmatobius* ancestors were in peripheral, subtropical, and more equable environments, prior to the final uplift of the Andes. Subsequently, the colonizers of different levels and vegetation zones expanded or reduced their ranges (see Fig. 14–2). Prolonged isolation took place during the ecological fluctuations and climatic shifts resulting from the glaciations. Extremely fragmented or reproductively isolated populations occurred whenever Pleistocene humid–arid cycles repeatedly altered the Cordilleran landscapes and environmental conditions. Finally, together with the numerous actively speciating forms in the páramo-puna habitats, relicts of lower populations remain, such as the forest-inhabiting species from western Argentina or southeastern Ecuador. Presumably they represent living relicts of early steps in the altitudinal colonization of the high Andes by anurans.

SPECIATION AND ADAPTIVE RADIATION IN THE HIGHLANDS

Evolutionary trends in the genus *Telmatobius* are reflected by the present patchy distribution and still uncertain taxonomic status of many species groups. Taxonomic assignments of geographic isolates to species or subspecies often poses difficult problems. Altitudinal limits vary in the central Andes from 5000 m, reported for *Telmatobius marmoratus marmoratus* (Duméril and Bibron) in Bolivia, down to 1300 m, reported for the subtropical forest species in western Argentina (*T. barrioi*, *T. ceiorum*). Among the described species and subspecies of *Telmatobius*, about 70% are represented by páramo-puna inhabitants, found chiefly at altitudes upwards of 3000–3500 m. Mechanisms of allopatric or geographic speciation (Mayr, 1963) provide a basis to attempt a general understanding of complicated distributional shifts and transient genetic arrangements of *Telmatobius* population units. The initial speciation of a presumed genetically predisposed anuran invader such as *Telmatobius* was probably explosive. Rapid adaptive radiation may have been facilitated by the several unoccupied niches that resulted from the progressive uplift and final geomorphological modeling of mountain landscapes. The rate of speciation was later affected by modifications in the physical extent and frequency of environmental barriers. Major glacial lakes, fluctuations in vegetation belts followed by disjunctions in biome distribution, and local ice expansion or retreat, acted as predominant ecological and climatic factors that influenced the rate of speciation of widespread Andean populations. The increased frequency of geographic isolation of population complexes led to repeated species splittings. The degree of morphophysiological diversity so achieved was in agreement with the different degree of selection pressure locally required for the adaption to changing habitat and altered ecological factors.

The *culeus* and *albiventris* Groups

If the taxa reported to date are analyzed, some adaptive lineages or species groups become evident. According to Vellard's studies (1951, 1953, 1960) two well-differentiated species groups—the *culeus* and *albiventris* groups—spread in the Lake

Titicaca basin. Lake Titicaca is only a remnant of the much greater Pleistocene Lake Ballivian (see Chap. 9).

Besides the nominate form described by Garman, several subspecific taxa belong to the *culeus* group, such as the large *culeus escomeli* Angel from the nearby Lake Lagunillas, *culeus dispar* Vellard and *culeus fluviatilis* Vellard from the neighboring rivers flowing into the lake, and *culeus lacustris* Vellard from the small Lake Umayo connected to Titicaca by the Vilique River. Some still unanalyzed ancient climatic fluctuation may explain, for example, the strange location of the isolated *T. culeus exsul* Vellard from the Yura River, at 2500 m in the Arequipa region.

The *albiventris* complex includes two inhabitants of the southern embayment of Lake Titicaca (*albiventris albiventris* Parker and *albiventris globulosus* Vellard) in the large Paucarcolla or Puno bay, and a more localized form in the Arapa lagoon, close to the northern border of Lake Titicaca (*albiventris parkeri* Vellard). Unquestionable sympatry suggests specific rank for the representatives of the *culeus* and *albiventris* groups in the Lake Titicaca biotopes (Fig. 14–3). However, two isolated forms of the *crawfordi* group, reported later, *crawfordi crawfordi* and *crawfordi semipalmatus*, live in the nearby small Lake Saracocha, south of the Rio Cabanillas. Vellard (1953, 1954) stressed the lack of significant taxonomic boundaries between the widely intergrading *Telmatobius* populations from the plateau. Vellard's phylogenetic arrangement consists of a single polymorphic evolutionary lineage assembling any intergrading population of the widespread *albiventris*, *culeus*, *crawfordi*, and *marmoratus* group, probably related to a somewhat undifferentiated central form such as *T. marmoratus hintoni* Parker from lower montane environments in Cochabamba, Bolivia. At the same time Vellard proposed a pragmatic taxonomic assessment in which he grouped into the aforementioned specific taxa the morphologically and ecologically recognizable subspecific units. Earlier, Parker (1940) had not reached satisfactory conclusions in his discussion of the status of Titicaca frogs. The assumption that *culeus* and *marmoratus* were only racially distinct was rejected by Parker because of its "practical disadvantages."

The difficulties facing systematists trying to arrange these taxa into a biologically valid scheme are not surprising in view of the ill-defined and poorly known genetic features of these Andean anurans. The taxonomy of these groups has hardly been improved upon since Parker's or Vellard's time. Intraspecific races correspond to most of the pragmatically proposed subspecies in the present arrangement of Lake Titacaca dwellers. Gene flow between topographically isolated populations depends on occasional alternating restrictions or expansions of their ecological niches. It may explain the intriguing intermediate individuals between the races *culeus*, *dispar*, or *fluviatilis* commented on in Vel-

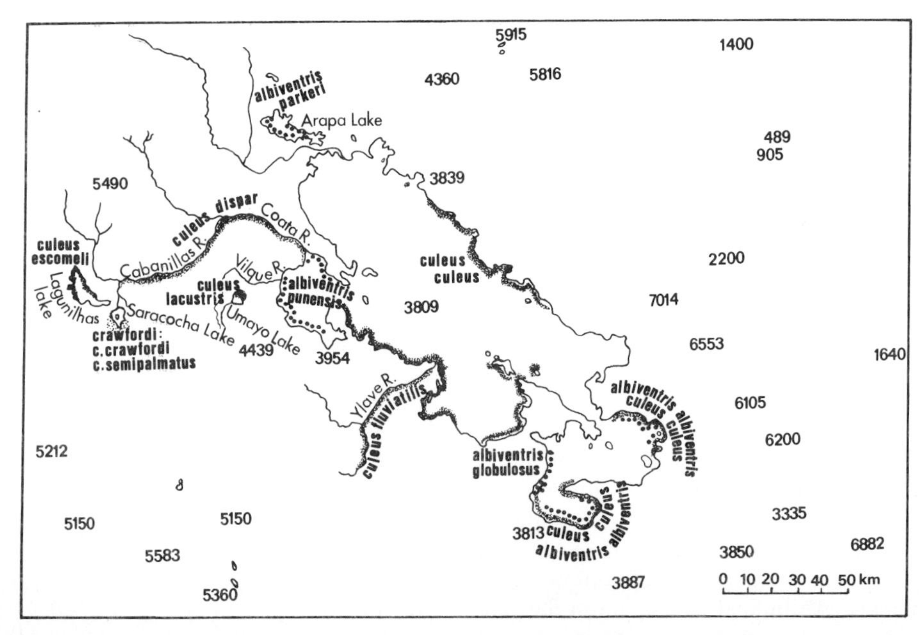

FIG. 14–3. Allo- and sympatric subspecific forms of the *Telmatobius culeus*, *T. albiventris*, and *T. crawfordi* groups around Lake Titicaca. Scattered numbers indicate some of the major altitudes of this region.

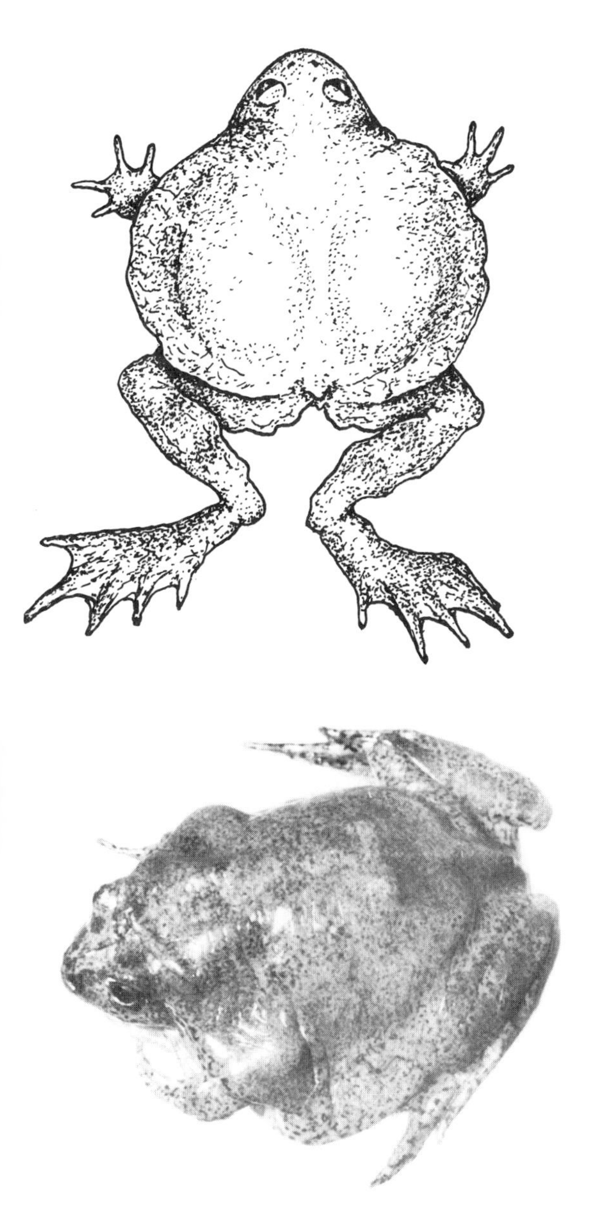

FIG. 14–4. Bagginess in *Telmatobius* and *Atelognathus*. *Top*: *T. culeus* from Lake Titicaca. *Bottom: A. patagonicus* from Lake Laguna Blanca, Neuquén, Argentina.

FIG. 14–5. Absence of bagginess in *Telmatobius* and *Atelognathus*. *Top*: *T. marmoratus* from near Lake Titicaca. *Bottom*: *A. praebasalticus* from near Lake Laguna Blanca, Neuquén, Argentina.

lard's reports. They probably attest to some former and/or accidental gene exchanges in currently separated, no longer interbreeding populations.

Characters of Lacustrine and Nonlacustrine *Telmatobius*

Let me briefly state the morphological traits used in taxonomic screening of the largest lacustrine *Telmatobius* representatives. Globose body shape, narrow or enlarged head, relative length of palmate hind legs, development of vomerine teeth, smooth skin surface, soft lateral skin folds or "bagginess," a peculiar thick dorsal disk under dermal layers, and different color patterns are the phenotypic characters on which systematic diagnosis or descriptions commonly depend. Most of these characters suggest specialized swimmers living in deep waters or coastal bogs. Some of these characters can be considered to represent ancestral diverging phyletic trends of a general telmatobiine stock. The bagginess, for example, is also exhibited by *Alsodes* or *Atelognathus*, as in *Atelognathus patagonicus* (Gallardo) and *A. nitoi* (Barrio) (Figs. 14–4 and 14–5).

All of these characters can be used to distinguish the *culeus* or *albiventris* groups from the smaller, warty, and short-legged *marmoratus* forms, emphasized as a central evolutionary stock in Vellard's assumptions. The *marmoratus* forms lack the dorsal disk and skin bagginess. They are found under stones, in shallow pools and streamlets of páramos, or in running rivers along

TABLE 14–1. Taxa of the *Telmatobius marmoratus* group.

Taxon	Distribution and remarks
m. rugosus Vellard	Eastern borders of Lake Titicaca basin and Bolivian plateau (Carangas Mts.)
m. riparius Vellard	Western stony shores of Juli and Pomata bays, near Juliaca
m. angustipes (Cope)	Lake Umayo: on southwestern borders of Lake Titicaca basin
m. microcephalus Vellard	Cuzco Mountains
m. marmoratus (Duméril and Bibron)	Widely distributed in Peruvian and Bolivian highlands
m. gigas Vellard	Huallamarca, 3800 m: a giant form, converging to the *culeus* forms from Lake Titicaca
m. verrucosus Werner	Eastern Andean slopes; lives sympatrically with *Hyla charazani* in running waters of the Hachuri River, east of Cordillera Real, 2800–3200 m
m. bolivianus Parker	Yungas de Unduavi (2800 m): humid, montane tropical environment
m. simonsi Parker	Sucre: lower eastern slope of the Sucre Mountains.

Andean slopes, from Cochabamba (*marmoratus hintoni*) northward to the Peruvian Vilcanota River (*marmoratus pseudojelskii* Vellard, *marmoratus pustolosus* [Cope]). Besides the peripheral forms (*pseudojelskii*, *pustolosus*, *hintoni*) reaching lower and more temperate altitudes in the Cuzco and Cochabamba valleys, many other subspecific taxa have been described. These taxa are tentatively listed in Table 14–1, together with their general distribution and/or ecological preferences.

The brackish or salty, dry plateau south of the flooded Lake Poopó depression is apparently unfavorable to anuran distribution, but several poorly known populations of the *marmoratus* group have been reported from Villazón, on the Bolivian-Argentine border, and from the highlands of Jujuy, in northern Argentina. The range of the so-called subspecific *marmoratus* complex would thus extend through no less than 1500 km, in some of the most broken-up and rough geomorphological regions of the high Andes.

The *crawfordi* Group

Telmatobius crawfordi crawfordi (Parker) and *T. crawfordi semipalmatus* Vellard from the Peruvian plateau constitute a fourth morphological group, closely related to *marmoratus* in spite of its reduced size, shorter head, smoother skin, and shorter, slightly palmate, hind legs and feet. Both subspecies of *crawfordi* are found at a high altitude (4250 m) in the small Lakes Saracocha and Chojcora, near Lagunillas, and are probably separated by topographical barriers from the large frogs of the *culeus* group.

I must emphasize again that such patchy distribution and the genetic relationships of these taxa should be easier to understand if Recent or Pleistocene sequences of population disjunctions and secondary contacts could be shown. However, much information is required to make clear the effective biological gaps between related forms of the *marmoratus*, *culeus*, *albiventris*, and *crawfordi* groups. It is at present virtually impossible to offer a suitable evaluation of the genetic discontinuities and taxonomic status of the innumerable intermediate samples reported by Vellard (1951, 1953, 1954).

Telmatobius in Northwestern Argentina

Interesting data have been gathered by Laurent (1970a, b, 1973, 1977) on inter- and intraspecific relationships of *Telmatobius* in northwestern Argentina. His studies in the Tucumán region drew attention to the present imbalance of genes related with size regulation in two geographically localized forms, *T. laticeps* Laurent and *T. hauthali pisanoi* Laurent, south of and on the western side of the Cumbres Calchaquies, respectively (Laurent and Terán, 1981). Formerly confused under the erroneous name *T. schreiteri*, the large and warty *T. laticeps* populations live in the Tafí Valley, at about 3000 m, while the smaller and only slightly warty *T. hauthali pisanoi* occurs northward on the surrounding high mountains. Laurent postulated a former major lake in the Tafí Valley, positively acting as "a selection factor on growth regulating alleles of large *laticeps* phenotypes." Similar effects have been invoked for the large *culeus* frogs inhabiting Lakes Titicaca and Lagunillas. Such a generalization may be weakened by the reported *T. marmoratus gigas* (190 mm, snout–vent length) from the Carangas Mountains, far away from lacustrine habitats. However, the tremendous Pleistocene Lake Minchin, some 400 km northwest of Uyuni, could have harbored large forms, some of which now survive, away from water, as gigantic peripheral relicts. Since the drying out of the putative lake in the Tafí Valley, zones of introgression occurred

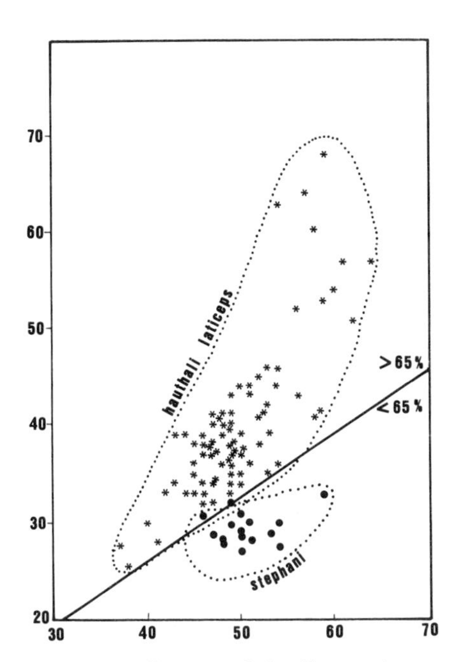

FIG. 14–6. Scatter diagrams of the distance from eye to mouth (*ordinate*) plotted against diameter of the eye (*abscissa*) in populations of *Telmatobius laticeps* and *T. stephani* from the Tucumán mountain region. Scales are in 1/10 mm. After Laurent (1973).

in neighboring *laticeps* and *pisanoi* populations, as supported by the biometrical analysis of several morphological characters (Laurent, 1977). Intermediate individuals are often observed and the "lacustrine" *laticeps* should perhaps gradually decline under the genetic pressure of dominant *pisanoi* characters.

Introgression zones also have been reported in the peripheral ranges of *Telmatobius laticeps*, *T. ceiorum* and *T. stephani* Laurent. However, a detailed analysis of the scatter diagrams, reflecting their negative or positive allometric characters, seems to support the specific status of these forms (Fig. 14–6). *T. hauthali*, whose type was discovered at 4000 m in the "Catamarca Cordilleras" (Koslowsky, 1895), where it has recently been studied again by Lavilla (1984), seems to have evolved as the main southernmost lineage from the parental stock of the genus. Among its characteristics one may cite a strongly cornified epidermis, sexual dimorphism, and impressive skin excrescences on chest and thumb. *T. hauthali* occurs in a variety of biotopes, from running streams to shallow montane pools or springs. The brightly spotted *T. stephani* from Sierra del Manchao, 2000 m, Catamarca, *T. contrerasi* from high in the San Juan Valley, and *T. schreiteri* from the Sierra de la Rioja belong to the same complex. The last species is not a true Andean taxon. Its disjunct populations live at about 3000 m in the Sierra de Velasco (Laurent, 1977) and at about the same altitude in running waters of the parallel Sierra de Famatina (6250 m at the summit). Adults and larvae of the Famatina populations are rheophilous amphibians, swimming in small, cold streams flowing into a major stream, the Amarillo River (Cei, per. obs.). This river flows down the upper levels of this ancient Ordovician massif, which was independently uplifted during the Andean orogenesis (De Alba, 1979). The yellow waters of the Amarillo River run across thick cupriferous pyrite and limonite beds, soon reaching a very unusual degree of acidity (pH 2.5). Any frog falling in the sulfuric waters of the Amarillo River dies; a curious but insurmountable chemical barrier is thus established and isolates *T. schreiteri*. As in other Famatina endemics, patterns of distribution of this species may be the result of alternating faunal interchanges between the Famatina highlands and the Andes, through glacial and interglacial stages of the Pleistocene (Cei, 1980).

Pacific Slope *Telmatobius*

Stocks of *Telmatobius* have undergone a more restricted and irregular dispersal along the Pacific Andean slopes than along the eastern ones. Geologic features of the recently uplifted Cordilleran borders produced the geomorphological factors that resulted in almost incomprehensible distribution patterns. Chilean *Telmatobius montanus* and *T. laevis* Philippi are reported at a slightly higher latitude than the western biotopes where Argentine forms occur. Long doubtful and neglected, these taxa have been revised recently. Physiological and ecological traits of *T. montanus* from Cordilleran environments near Santiago de Chile were pointed out by recent research (Busse, 1980). The somewhat generalized character of this frog suggest rheophilous trends in both adult and larval stages. Some 15 degrees of latitude separate southern Chilean *Telmatobius* from other and more specialized aquatic forms found in the arid plateaus of Atacama and Tarapacá. These taxa are scattered along the boundaries of the almost desert high puna, characterized by inhospitable salt flats or *salares*. As far as is known the only *Telmatobius* described from the puna is *atacamensis* Gallardo, a sluggish, flat-headed frog living in the scanty pools and creeks of such a dry habitat. A significant member of the Atacama fauna is *T. halli* Noble, living in hollowed edges of the San Pedro de Atacama River (2500 m), or in isolated warm springs such as Ollague, close to the dry Salar de Carcote, some

80 km northward. It is a completely aquatic anuran whose morphological peculiarities are its especially weak or absent teeth, and very flattened snout. *Telmatobius pefauri* Veloso and Trueb (1976) from the streams of Murmuntani (3200 m) in the mountains of Tarapacá, is a very different frog, long-legged and tooth-bearing, showing a concealed tympanum, incompletely developed under the skin. Remnants of the tympanic annulus beneath the skin are also present in *T. halli* and *T. peruvianus* Wiegmann, a third and easily distinguishable northernmost form from the Huatilla Mountains.

It is easy to imagine that the almost clearcut systematic status and the scattered distributions of western subtropical *Telmatobius* species are consistent with marked geographic isolation and complete climatic and/or topographic barriers to gene flow. Ancestors of the present taxa probably became disjunct early and irreversibly during Quaternary periods of increased environmental dryness in the puna and neighboring regions.

Telmatobius of the Peruvian Plateau

The broad structural uplift of the Peruvian plateau trends northwestward into the steep ridges of the Cordillera Blanca and the central Cordillera, encasing the long valley of the Marañon River. This region is bordered on the west and east by the high crests of the Andes, on the south by the broad highlands of the Apurimac Basin, and is contiguous to the volcanic peaks of the Arequipa Mountains. In such an extensive and geomorphologically complicated Late Pleistocene landscape, initial colonization of *Telmatobius* resulted in the extraordinary multiplication of geographic isolates, with subsequent genetic introgressions or intergrading of formerly discontinuous population units. One can thus see a range of speciation levels, from incomplete to complete. As a consequence of this history, complexes of taxa even more puzzling than the *Telmatobius* of the Titicaca Basin evolved. Vellard's (1955) pioneer work is here again the only major source of available information. As in the case of the main evolutionary lineages of the *marmoratus* group, Vellard's dialectic approach resulted in his opposing a fundamental, widespread stock—or *jelskii* lineage—to a number of "operational" specific and subspecific units, corresponding nearly to physiological and biogeographic features of the region. Trueb (1979) wisely recommended that it would be inappropriate to suggest intrageneric relationships "until such time as a great deal more is known about the species that comprise *Telmatobius*."

In order to point out the general biogeographic and evolutionary interest of the adaptive radiation of *Telmatobius* in the Andes, Vellard's (1955) provisional arrangement of the so-called *jelskii* group shall be used. This arrangement consists of four morphologically distinguished population assemblages: the *arequipensis* forms in the south, the central or "*eujelskii*" complex, the *brevirostris* complex from the Huánuco and Ambo region, and the *rimac* forms of the high coast valleys between Lima and Casma, along the Chillón and Pativilca rivers (Fig. 14–7). *Telmatobius intermedius* Vellard, a poorly defined taxon, could be added also. It was found at 3300 m near Puquio, on the Pacific slope, and it appears related to *T. peruvianus* from Tacna, about 500 km to the south.

According to Vellard's (1955) descriptions, moderate-sized members of the *jelskii* group (sensu lato) can be characterized by having the head almost as long as wide (significantly wider in the *culeus-albiventris-crawfordi* complex; slightly wider in *marmoratus*), the tympanum often visible, large hind legs, soft or moderately tuberculate skin without dorsal disk or bagginess, well-developed secondary sex characters on thumbs and chest, and brilliant orange-yellow ventral colorations. Four subspecies of the central or "*eujelskii*" complex are reported from the highlands on the western side of the Apurimac Basin: *jelskii jelskii* Peters, *jelskii walkeri* Shreve, *jelskii longitarsis* Vellard, and *jelskii bufo* Vellard. The *brevirostris* complex is distributed through the mountains of Huánuco (2000–3000 m) on the upper basin of the Huallaga River, running almost 200 km westward of the parallel Apurimac-Ucayali system. Morphological traits differentiating Vellard's taxa (*brevirostris brevirostris*, *brevirostris parvulus*, *brevirostris punctatus*) from the central or "*eujelskii*" complex are too weak or subjective to provide for an easy diagnosis or identification. Nevertheless these easternmost stream dwellers are geographically well separated from the typical *jelskii* frogs of the Tarma or Palca rivers. The interposed Junín Plateau, home of the endemics *Batrachophrynus* and *Lynchophrys*, apparently lacks any representative of the genus *Telmatobius*, except the small, webless, and terrestrial *T. juninensis* (Shreve), a species transferred to *Telmatobius* by Lynch (1978), mostly on osteological grounds. Geographic barriers must have prevented interbreeding between *jelskii* and *brevirostris* populations and *rimac* (*rimac rimac* Schmidt, *rimac meridionalis* Vellard), found in montane stony creeks of the Ancash and Lima regions. Among the most distinctive characters of *rimac*, the evident thumb pad reduction in males was pointed out by Vel-

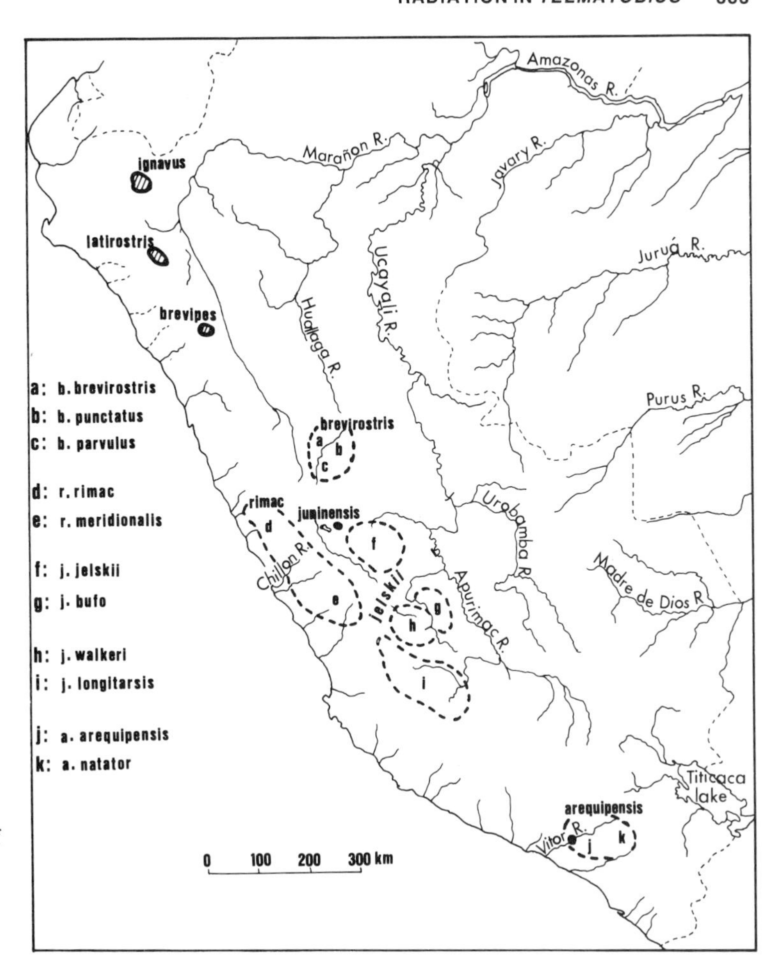

FIG. 14–7. Geographic distribution of Peruvian species and subspecific complexes of *Telmatobius* from the Huancabamba Depression to Arequipa.

lard (1955). This condition is quite distinct from the acuminate and cornified black excrescences on the thumb and chest of *T. arequipensis* (*arequipensis arequipensis* Vellard, *arequipensis natator* Vellard), reported from shallow ponds and flooded fields in different localities of Arequipa, at about 2500 m. The present isolation of *arequipensis* may have been less strong in earlier times, and the steep cliffs surrounding at present the Arequipa region may not have been such efficient barriers in the past as to lead to complete population disjunction. However, no morphological intermediates have been observed yet, either with the nearest *Telmatobius jelskii* (*longitarsis* from Puquio, *bufo* from the eastern Ayacucho Mountains), or the sympatric *T. culeus exsul* of the springs at Yura, only 30 km from the town of Arequipa.

A few scattered species of *Telmatobius* have been reported from 8° S to the Ecuadorian border. They are *T. brevipes* Vellard from Huamachuco, near the Criqueguas River in the Marañon Basin; *T. latirostris* Vellard from Cutervo, Cajamarca, at about 2300 m; and *T. ignavus* Barbour and Noble from the Huancabamba Depression (2000 m), a physiographic boundary separating the Peruvian from the Ecuadorian forms of the genus. Present evidence does not enable one to establish any intrageneric relationship involving these scarely known, almost terrestrial frogs, found along the edge of dry subtropical woods and the more humid environments of the "ceja de montaña" or cloud forest. Until adequate taxonomic information is available, any inference on the evolution of species of *Telmatobius* from northern Peru will remain speculative.

CONCLUSIONS

Despite inadequate knowledge of taxonomic relationships at the species and subspecies levels, there is little doubt that the genus *Telmatobius* has undergone an outstanding adaptive radiation. It can be postulated that during the Middle and Late Tertiary a parental stock of these frogs colonized the just uplifted Cordilleran environments south of the equator. These early invaders were probably related to some ancient telmatobiine branch as well as to the southern relicts grouped by Lynch (1978) in the specialized atelognathid assemblage from Patagonia. The most primitive step in Andean amphibian dispersal can be inferred from the present forest-inhabiting species of *Telmatobius* found along eastern Andean slopes, of which the *oxycephalus-barrioi* complex and the *ceiorum* complex provide good examples. Moreover, other significant forms are known, or shall surely be discovered, at relatively low altitudes in the subtropical forest belt along the eastern slopes from Bolivia northward to Ecuador. Haffer (1969) suggested a lowering of vegetation zones and an increase in the extension of montane forest along the Andes during glacial maxima. Populations of *oxycephalus* are reported at present from localities that range from 1000 m to 4000 m.

The presumed diversity of unoccupied niches in early montane biotopes and the corresponding lack of ecological competition may have facilitated the multiplication of diverging populations; therefore the rate of speciation was probably related to the increased frequency of geographic isolates. In other words, speciation patterns of ancestral telmatobiine invaders may have been explosive. Geomorphological changes and environmental crises following climatic Pleistocene events during glacial–interglacial cycles resulted alternately in prolonged geographic and genetic isolation, and in gene flow or introgression. Repeated changes of topographic and climatic barriers are hypothesized to have widely affected the evolution of Andean *Telmatobius* populations, allowing "pulverization" of their widely intergrading systematic characters, as already emphasized in Vellard's theoretical and pragmatic scheme. Preliminary examinations of the herpetofauna of the northern Andes have shown that specific differentiation and dispersal of *Telmatobius* mainly developed above 3000 m in páramo environments around large lacustrine basins and their drainage. Vellard (1953) reported four widely sympatric specific assemblages of species, namely, the *culeus*, *albiventris*, *crawfordi*, and *marmoratus* complexes near Lakes Titicaca and Poopó. They have been interpreted as a single polymorphic evolutionary line, all populations of these widespread groups probably being related to a rather undifferentiated central form such as *Telmatobius marmoratus hintoni* from the Cochabamba region of Bolivia. Other morphologically defined *Telmatobius* lines exist in the Peruvian highlands, eastward and northward of the preceding four major groups centered on Lakes Titicaca and Poopó, the remnants of the much larger Ballivián and Minchín Pleistocene basins. They are the *jelskii*, *brevirostris*, *rimac*, and *arequipensis* lines occurring along the western Andean slopes, through the Apurimac Basin, the coastal, central, and southern mountain range, and the inland Huánuco region near the Marañon Valley. Conclusions similar to those given for Titicaca frogs also emerge from Vellard's studies of these complexes. There is unfortunately little genetic and taxonomic research to permit one to discuss his statements further.

Peculiar patterns of adaptive radiation of *Telmatobius* and its puzzling taxogenetic differentiation have been also stressed by Laurent (1977) during his research on the *hauthali* group and allies from northwestern Argentina. An analysis of the factors of speciation was completed for specific and subspecific units such as *Telmatobius laticeps* and *T. hauthali pisanoi*, living in the Tafí Valley and on the western slopes of the Calchaquies Mountains. Geographic isolation and genetic introgression each played a prolonged role leading to the present imbalance of genes dealing with size regulation in both forms. A Quaternary Tafí high mountain lake was postulated to have acted as a selection factor increasing the frequency of growth genes sufficiently to allow significant differentiation of isolated populations. Besides larger *laticeps* populations, this adaptive trend could explain giant *culeus* frogs from Lake Titicaca, or rare relicts of giant forms from Oruro. The Pleistocene system of large glacial bodies of water in the altiplano effectively acted as physiographic and ecological barriers for the many isolates, and caused divergences of population on either side of the lakes.

Repeated episodes of expansion and isolation in accordance with glacial–interglacial cycles may likewise explain disjunct distribution patterns as well as the disjunct populations of *Telmatobius schreiteri* from the Famatina and Velasco mountains facing the Cordilleras at about 29° S. Strong environmental barriers provided by the arid salt flats of the puna during the Pleistocene can explain the clear-cut genetic and morphological differentiation of isolated species such as *T. halli* or *T. atacamensis*, and probably also of the Chilean *T. montanus*, assigned to *Alsodes* by Diaz and Valencia (1985), or *T. pefauri*. Geomorphological and/or climatic barriers can be correlated with the speciation of peripheral isolates that achieved genetic divergence from their hypothetical parental stocks or precursors.

I must finally mention some morphophysiological traits that characterize the adaptive radiation and evolution of telmatobiine frogs. These adaptations are consistent with the physical and biotic features of the environments in which *Telmatobius* frogs live, keeping in mind the assumption that ancestral organisms "lived under ecological conditions similar to those under which their closest modern relatives now live" (Simpson Vuilleumier, 1971). As reported first in the analysis by Vellard (1951), and later in the work of de Macedo (1960), adaptive alterations especially concern tegumentary, respiratory, and reproductive systems. These alterations result in an increase in oxygen uptake in the poorly oxygenated waters of high mountains, and in a more efficient reproductive strategy through much of the year and under adverse weather conditions. Skin bagginess and vascularization, dermal disks, keratinous dermal layers and warts, nuptial thumb pads and horn points on chest, and glutinous gland secretions are the most remarkable of these characters, some of which are often very significant at the level of specific differentiation. Only scarce or fragmentary data are so far available on annual reproductive cycle and endocrine regulation in anurans from the highest altitude. Continuous gonadal activity and exceptional cold-resistant germ cells have been observed in *Telmatobius laticeps* (Cei, 1949), or in *T. halli* (Cei, unpublished data).

Telmatobius frogs are scarce north of Lake Junín and the Junín plateau, which are the home of the closely related telmatobiine *Batrachophrynus* species, but form a rather impressive gap in the distribution area of the genus *Telmatobius*. The currently recognized taxa *brevipes* from Huamachuco, *latirostris* from Cajamarca, and *ignavus* from the Huancabamba Depression probably do not represent the real geographic distribution in northern Peru of such an invasive anuran group, for lack of adequate information. Distribution pattern, taxonomic status, and ecology are better known for Ecuadorian representatives, which are mostly concentrated in lower southern ranges of Azuay, Loja, and Zamora Provinces. Trueb's (1979) study provides information on these well-characterized peripheral species: *T. niger*, *T. vellardi*, and the bamboo forest inhabitant *T. cirrhacelis*. North of the so-called nudo del Azuay, only the páramo dwelling *T. niger* is found, apparently the northernmost environment reached by these Andean frogs during past migrations along the whole Cordilleran range. The Late Pleistocene volcanic activity north of the nudo del Azuay may have been a factor limiting some of the dispersal routes of these peculiar amphibians.

ACKNOWLEDGMENTS

I am grateful to Esteban Lavilla, Instituto M. Lillo, Universidad Nacional de Tucumán, Argentina, for his kind assistance during my fieldwork in the Tucumán region. I thank Raymond F. Laurent of the same institution for his kind permission to use original diagrams of his papers. Finally, I thank Mrs. Maria T. Bernardo Lopes of the Museu Zoologico, Universidade de Lisboa, Portugal, for preparing several of the diagrams and figures.

REFERENCES

Axelrod, D. I., and H. P. Bailey. 1969. Paleotemperature analysis of Tertiary flora. *Paleogeogr., Paleoclimatol., Paleoecol.* 6: 163–195.

Busse, K. 1980. Zur Morphologie und Biologie von *Telmatobius montanus* Lataste 1904, nebst Beschreibung seiner Larve (Amphibia, Leptodactylidae). *Amphibia-Reptilia*. 1: 113–125.

Cei, J. M. 1949. Sobre la biología sexual de un batracio de gran altura de la región andina (*Telmatobius schreiteri* Vellard). *Acta Zool. Lilloana* 7: 467–488.

———. 1969. The Patagonian telmatobiid fauna of the volcanic Somuncura plateau. *J. Herpetol.* 3: 1–18.

———. 1970. La posición filética de Telmatobiinae, su discusión reciente y significado crítico de algunos immunotests. *Acta Zool. Lilloana*, 27: 181–192.

———. 1972. Herpetología Patagónica. V. Las especies extra-cordilleranas alto-patagónicas del genero *Telmatobius*. *Physis* (Buenos Aires) 31: 431–449.

———. 1980. New endemic lizards from the Famatina mountains of western Argentina. *J.Herpetol.* 14: 57–64.

———. 1984. A new leptodactylid frog, genus *Atelognathus*, from southern Patagonia, Argentina. *Herpetologica* 40: 47–51.

De Alba, E. 1979. Sistema del Famatina. In *Segundo Simposio de Geología Regional Argentina*, Vol I. Córdoba: Edit. Acad. Nac. Ciencias. pp. 349–395.

Diaz, N. F., and J. Valencia. 1985. Larval morphology and phenetic relationships of the Chilean *Alsodes*, *Telmatobius*, *Caudiverbera* and *Insuetophrynus* (Anura: Leptodactylidae). *Copeia*: 175–181.

Duellman, W. E. 1979. The herpetofauna of the Andes: Patterns of distribution, origin, differentiation, and present communities. In *The South American herpetofauna: Its origin, evolution and dispersal*, W. E. Duellman, ed., Museum of Natural History, Monograph. 7: 371–459. Lawrence: Univ. of Kansas Press.

Haffer, J. 1969. Speciation in Amazonian forest birds. *Science* 165: 131–137.

Harrington, H. J. 1962. Paleogeographic development of South America. *Bull. Amer. Ass. Petrol. Geol.* 46: 1773–1814.

Heyer, W. R. 1975. A preliminary analysis of the intergeneric relationships of the frog family Leptodactylidae. *Smithsonian Contrib. Zool.* 199: 1–55.

Koslowsky, J. 1895. Batracios y reptiles de La Rioja y Catamarca. *Rev. Mus. La Plata* 6: 360–371.

Laurent, R. F. 1970a. Dos nuevas especies argentinas del género *Telmatobius* (Amphibia, Leptodactylidae). *Acta Zool. Lilloana* 25: 207–226.

———. 1970b. Contribución a la biometría de algunas especies argentinas del género *Telmatobius*. *Acta Zool. Lilloana* 25: 279–302.

———. 1973. Nuevos datos sobre el género *Telmatobius* en el noroeste argentino con descripción de una nueva especie de la Sierra del Manchao. *Acta Zool. Lilloana* 30: 163–187.

———. 1977. Contribución al conocimiento del género *Telmatobius* Wiegmann (4ª Nota). *Acta Zool. Lilloana* 22: 189–206.

———. 1983. Heterogeneidad del género *Batrachophrynus* Peters (Leptodactylidae). *Acta Zool. Lilloana* 37: 107–113.

Laurent, R. F., and E. Terán. 1981. Lista de los anfibios y reptiles de la Provincia de Tucumán. *Misc. Fund. M. Lillo* 171: 5–15.

Lavilla, E. 1984. Redescubrimiento de *Telmatobius hauthali* Koslowsky (1895) y descripción de su larva. *Acta Zool. Lilloana* 38: 57–58.

Lynch, J. D. 1978. A re-assessment of the telmatobiine leptodactylid frogs of Patagonia. *Occas. Pap. Mus. Nat. Hist. Univ. Kansas* 72: 1–57.

Macedo, H. de. 1960. Vergleichende Untersuchungen an Arten der Gattung *Telmatobius* (Amphibia, Anura). *Z. Wiss. Zool. Abt. A* 163: 355–396.

Mayr, E. 1963. *Animal species and evolution*. Cambridge, Mass.: Harvard Univ. Press.

Parker, H. W. 1940. The Percy Sladen Trust expedition to Lake Titicaca in 1937. Amphibia. *Trans. Linn. Soc. London* 1: 203–216.

Schaeffer, B. 1949. Anurans from the Early Tertiary of Patagonia. *Bull. Amer. Mus. Nat. Hist.* 93: 41–68.

Simpson, B. B. 1979. Quaternary biogeography of the high montane regions of South America. In *The South American herpetofauna: Its origin, evolution, and dispersal*, W. E. Duellman, ed., Museum of Natural History, Monograph 7: 157–188. Lawrence: Univ. of Kansas Press.

Simpson Vuilleumier, B. B. 1971. Pleistocene changes in the fauna and flora of South America. *Science* 173: 771–780.

Trueb, L. 1979. Leptodactylid frogs of the genus *Telmatobius* in Ecuador with the description of a new species. *Copeia* 4: 714–733.

Vellard, J. 1951. Estudios sobre batracios andinos. I. El grupo *Telmatobius* y formas afines. *Mem. Mus. Hist. Nat. "Javier Prado," Lima* 1: 3–89.

———. 1953. Estudios sobre batracios andinos. II. El grupo *marmoratus* y formas afines. *Mem. Mus. Hist. Nat. "Javier Prado," Lima* 2: 1–53.

———. 1954. Études sur le lac Titicaca. V. Les *Telmatobius* du haut plateau interandin. *Travaux Inst. Français Etud. Andines Lima* 4: 1–57.

———. 1955. Estudios sobre batracios andinos. III. Los *Telmatobius* del grupo *jelskii Mem. Mus. Hist. Nat. "Javier Prado," Lima* 4: 1–28.

———. 1960. Estudios sobre batracios andinos. VI. Notas complementarias sobre *Telmatobius. Mem. Mus. Hist. Nat. "Javier Prado," Lima* 10: 1–20.

———. 1970. Contribución al estudio de los batracios andinos. *Rev. Mus. Arg. Cienc. Nat. Zool.* 10: 1–25.

Veloso, M. A., and L. Trueb. 1976. Description of a new species of telmatobiine frog, *Telmatobius* (Amphibia, Leptodactylidae) from the Andes of northern Chile. *Occas. Pap. Mus. Nat. Hist. Univ. Kansas* 62: 1–10.

15

Speciation and Adaptive Radiation in Andean Orestias *Fishes*

WOLFGANG VILLWOCK

According to Berg (1958) the order Cyprinodontiformes (Microcyprini or Cyprinodontes) includes seven families, six of which belong to the suborder Cyprinodontoidei. This suborder is divided into two family groups, which are based on oviparous forms and on viviparous ones, respectively. The oviparous cyprinodonts, which are the ones that interest us here, are divided into two families, the Cyprinodontidae and the Adrianichthyidae. The subfamilies Lamprichthyinae and Orestiinae, belonging to the Cyprinodontidae, occupy an exceptional position in that both represent geographically restricted species, *Lamprichthys tanganjicanus* from Lake Tanganyika (now Lake Tanzania) in East Africa, and *Orestias* spp. from the inter-Andean highlands of Peru, Bolivia, and northeastern Chile.

HISTORY, GEOGRAPHY, AND GENERAL HYDROGRAPHY

The main area of distribution of the genus *Orestias* and its recent species flock is the southern part of the inter-Andean highland. This region, the altiplano, extends from southern Peru for about 900 km south as far as Bolivia and beyond to northwestern Argentina and northeastern Chile. An inland depression with no drainage to the sea (Barth, 1972) between the western and the eastern Cordilleras, the altiplano is divided into smaller basins. The northernmost of these basins is covered in large part by Lake Titicaca (3811 m above sea level), followed southward by Lake Poopó (3686 m), Salar de Coipasá (3657

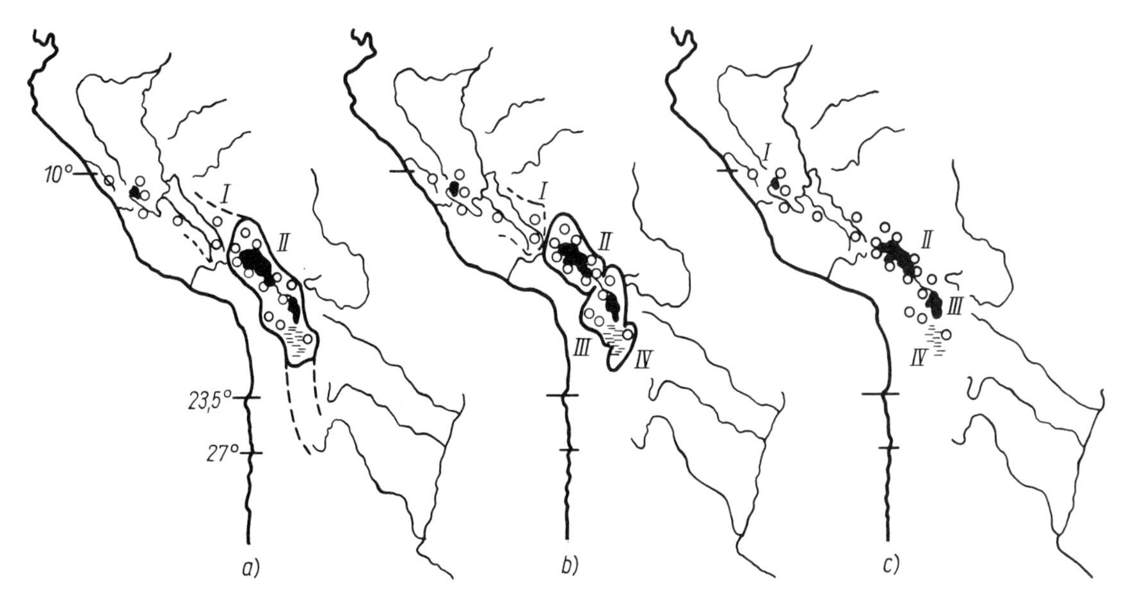

FIG. 15–1. Historical development of the inter-Andean basin (modified from Moon [1939]). (a) Pliocene – Pleistocene. I: Isolation of the northern part of the inter-Andean basin (more or less completed); II: Lake Ballivian. (b) Pleistocene. I: Isolated northern part; II: ancient Lake Titicaca; III: Lake Minchin; IV: Lake Reck. (c) Present. I: Junín Basin; II: Lake Titicaca; III: Lake Poopó; IV: Salar de Uyuni. o, recent localities of *Orestias*; – – – –, doubtful border of ancient inter-Andean basin; —, more or less certain borders of ancient lakes.

m), and Salar de Uyuni (3653 m). Today the southernmost part of the altiplano is characterized by some smaller lakes in the high Andes between altitudes of about 4300 m and 5000 m.

The inter-Andean basin is, according to Moon (1939) and other authors (e.g. Newell, 1949), a rather young formation, the origin of which can be dated to the end of the Miocene. According to Newell (1949, p. 10), "There is abundant evidence that Miocene or younger mountains were eroded to a plain of slight relief before uplift to the present Andean arch." Even in the Pliocene the forerunner of the modern altiplano was situated only a few hundred meters above sea level and, because of its closeness to the equator, was occupied by a semitropical flora and fauna (Moon, 1939). Further, Moon considers the elevation of the Andean mountains and, in connection with this, the rise of the inter-Andean highlands to have started at the end of the Pliocene or beginning of the Pleistocene. "Evidence suggesting rapid uplift is found along the coast of

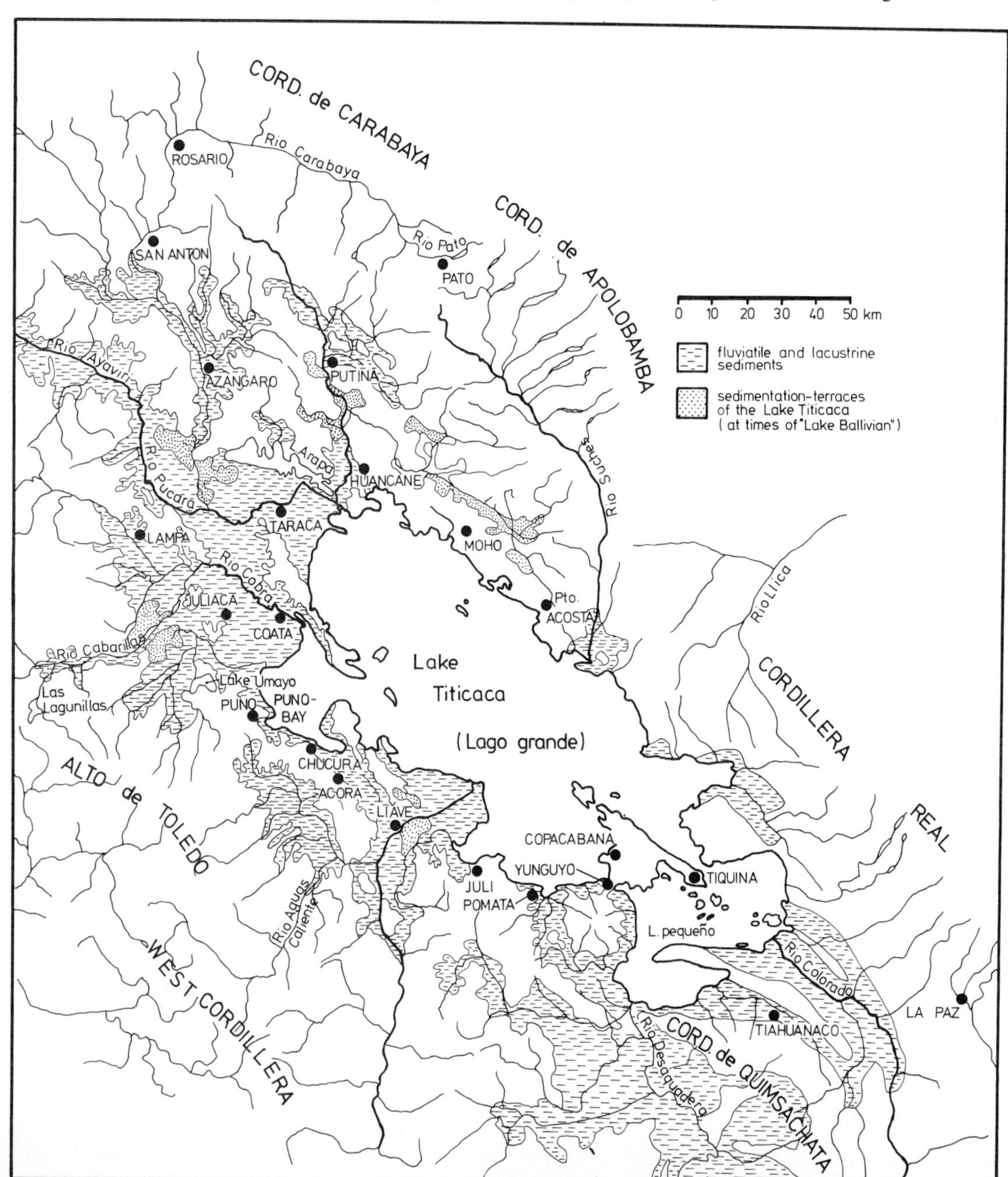

FIG. 15–2. Lake Titicaca region, its recent tributaries, and ancient extension. Modified from Monheim (1956).

the Peruvian Gulf in Pliocene or Pleistocene marine deposits, high above sea-level" (Newell, 1949, p. 10). One consequence of the change in elevation above sea level was the extinction of the warm-adapted elements of the ancient flora and fauna (see also Chap. 9). Only those ancestors of recent species survived that possessed the ability, or in terms of evolutionary biology, the preadaptation for further changing conditions. The unknown ancestor of *Orestias* is assumed to have been among those, because it is very hard to imagine that the ancestor should have migrated into the area after the final uplift had ended. It might be helpful for a better understanding of the microevolutionary processes, i.e., the modes of speciation, that took place in the ancient *Orestias* species swarm to review Moon's (1939) conclusions. He postulated that an old, already elevated body of water became divided into at least two or three smaller ones because of decreasing rainfall and cooling during the Early Pleistocene. The oldest, as well as the largest, of these old lakes has been called Lake Ballivian (see Fig. 15–1a) and probably covered the entire inter-Andean high plateau, including the altiplano with the present-day Lakes Titicaca and Poopó as well as the isolated salares in the south. With further hydrographical changes Lake Ballivian shrank and became divided into different, more or less separated parts (see Fig. 15–1b). The present situation (see Fig. 15–1c) was reached during glacial and postglacial times, when the waters of the early altiplano were subdivided into smaller ones again, and became brackish (Lake Poopó) or even salt pans (Salares de Coipasá and Uyuni). The same fate of increasing salinity is predicted for Lake Titicaca itself, but has not yet happened, however, because of its extent and depth (Figs. 15–2 and 15–3). For more details on the geology of the altiplano see Chapter 9.

Modern Lake Titicaca is located beween 15°14′ and 16°35′ S and 68°37′ and 70°02′ W at an altitude of 3811m, is about 176 km long (NW to SE), and has a maximum width of 75 km. The surface area is about 8100 km^2. Differences in measurements of surface area from 6900 km^2 to 8300 km^2

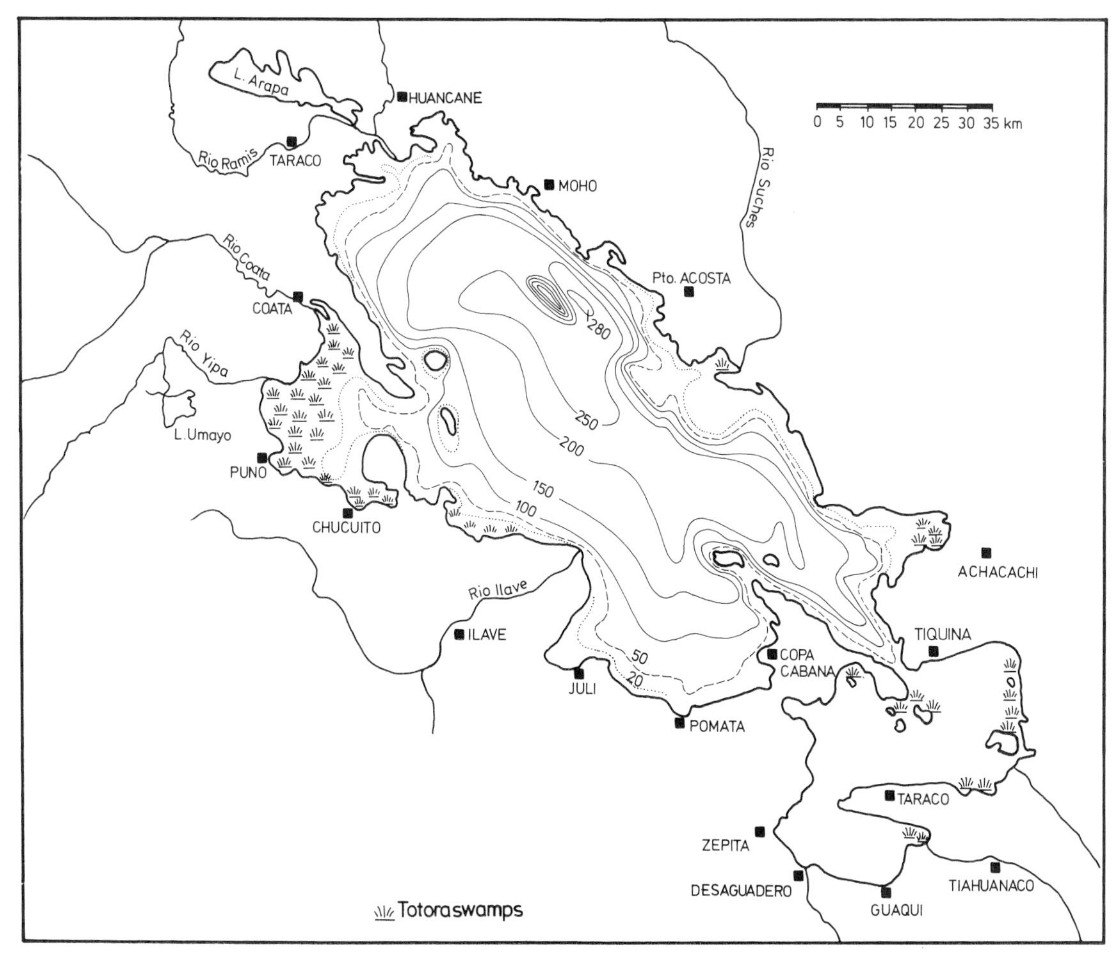

FIG. 15–3. Lake Titicaca and its hydrographic structure (isobatic depthlines according to Monheim [1956]).

appear to depend on changes in its water level, which are influenced by climatic factors such as the amount of rainfall, the melting of snow, and temperature. The amount of variation may be exemplified by the fact that in 10 years (1933–43) the sea level dropped by about 5 m (Monheim, 1956), which means dramatic alterations in the distribution of aquatic organisms within the extended shallow bays and lagoons of Lake Titicaca (see Fig. 15–3). Three main parts of Lake Titicaca may be distinguished, Lago Grande, Puno Bay, and Lago Pequeño, which is connected with Lago Grande through the Strait of Tiquina (see Figs. 15–2 and 15–3). The average depth of Lago Grande is more than 100 m (maximum 281 m), whereas the other two areas, Puno Bay and Lago Pequeño, show an average depth of only 10 m or less (Monheim, 1956).

ECOLOGICAL ASPECTS

The present Lake Titicaca shows a remarkable diversity in ecological conditions because of its size and geomorphological structures. There are not only extensive shallow bays, such as Puno Bay, Lago Pequeño, and smaller ones (see Fig. 15–3), but also coastal sections with rocky slopes and drop-offs with depths of more than 20 m near the banks (e.g., near Moho, northeastern part of Lago Grande).

The bottom of shallow bays and lagoons is composed predominantly of mud or fine-grained sand, and the substrate off the rocky coastline consists mainly of larger stones or at least pebbles.

Shallow areas are mostly covered with immense fields of totora (Fig. 15–3), an endemic *Scirpus* (*S. tatora* according to Löffler, 1968), submerged algae, and other plants such as *Fontinalis* sp., *Myriophyllum* sp., and *Potamogeton* sp., which fill large sections of the low-lying coast. The deeper rocky or pebbly parts lack any noticeable dense vegetation, except for some isolated *Potamogeton* plants.

Only two relatively large lakes in the vicinity of Lake Titicaca reflect ecological conditions similar to those of the main lake itself, namely Lake Arapa to the north and Lake Umayo (to the west of Lago Grande), to which they are connected by small rivers (Fig. 15–2). All the other bodies of water occupied by *Orestias*, not only within the Lake Titicaca region, but also north of the watershed of the La Raya Mountain chain in the Urubamba Valley and still farther north in the Junín Plain, show remarkable similarities to the large bays of Lake Titicaca (such as totora and submerged plants).

TABLE 15–1. Temperatures in Lake Titicaca at different depths.

Depth	Summer (Jan.–March)	Depth	Winter (May-July)
surface (average)	13.5°C	1–50 m	11.4°C
		50–100 m	11.2°C
−135 m	12.5°C	100–150 m	11.1°C
		150–200 m	11.1°C
		200–250 m	11.0°C
−276 m	12.8°C	>250 m	10.9°C

The data in Table 15–1 indicate water temperatures in Lago Grande at different times of the year (modified from Monheim, 1956).

According to Monheim (1956), the average temperature of Lago Pequeño during the summer months is 13.5°C at the surface and 14.5–18.6°C at depths of between 5 m and 13 m among submerged vegetation (for more detailed information, see also Schindler, 1955).

The chemistry of different waters in the Titicaca Basin is virtually unknown, only some general aspects having been published for Lago Grande. One sample taken by Löffler in April 1954 (cited by Monheim, 1956) showed the composition indicated in Table 15–2.

This composition allows one to characterize Lago Grande as a freshwater lake. Salt composition as well as concentration should be distributed more or less evenly within Lago Grande, because the limnological status as a whole was indicated by Löffler (1968, p. 59) for "the world's biggest tropical high-mountain lake" as being "cold polymictic."

No precise information is available on the chemistry of Lake Poopó, the southern salares or any of the other isolated lagoons, ponds, and running waters of the inter-Andean area. Only some scattered data are found in the literature. However, this justifies the statement that all *Orestias*-bearing waters with the exception of Lake Poopó

TABLE 15–2. Composition of several chemical salt components of water in Lake Titicaca.

pH 7.5	Total dh 17.3[a]
Ca	0.054 g/l
Mg	0.041 g/l
Na	0.176 g/l
K	0.014 g/l
Cl	0.244 g/l
SO_4	0.251 g/l
	0.780 g/l

[a] dh = degree of hardness (according to German standards)

and the salares seem to be freshwater. The vegetation and fauna associated with *Orestias* in Lake Titicaca and its tributaries as well as in the Junín Plain supports this conclusion. The only other native vertebrates in the Titicaca Basin and its different waters are the siluroid fish genus *Trichomycterus* (Tchernavin 1944b) and the endemic frog genus *Telmatobius* (see Chap. 14). The introduction of different salmonids since 1937 (according to Urquidi, 1969) or 1942 (according to Everett, 1971) in Lake Titicaca has led to the extinction of at least one of the largest species, *O. cuvieri* (Villwock, 1963a, 1972).

THE GENUS *ORESTIAS*

Several authors have considered this group of cyprinodonts, endemic to the inter-Andean basin of Peru and Bolivia, to have the rank of a family or at least of a subfamily. According to Tchernavin (1944a) the different forms of *Orestias* might best be described as a subfamily of Cyprinodontidae (see introductory discussion in this chapter). Tchernavin (1944a, p. 149) suggested that "there is good reason to consider *Orestias* as a group of more than generic significance, which could be divided into several genera." This will be briefly discussed at the end of this chapter.

The main characters of the genus *Orestias* are as follows: Ventral fins are absent in all known species. Scaling is irregular and is reduced in most of the species, especially on the chest, belly, and neck, which are partly or totally naked (Fig. 15–4). The lateral line, which is always distinct and rather uniform in all species, consists of a more or less regular row of perforated and grooved scales along the sensory canal except where scales cover the basis of fin rays of the caudal fin. The teeth are conical or blunt, forming usually more than one irregular row. Sexual dimorphism is marked. Spines and hooks, which are present on the scales and fin rays of sexually mature males, are fewer in number and in degree of development in ripe females. "In some species, small local forms are found and some specimens ripen long before reaching the limit of growth. Such small specimens have imperfectly developed scales, or have hardly any scales at all" (Tchernavin, 1944a, p. 150). More detailed information on osteological and other characters of *Orestias* is given by Tchernavin (1944a).

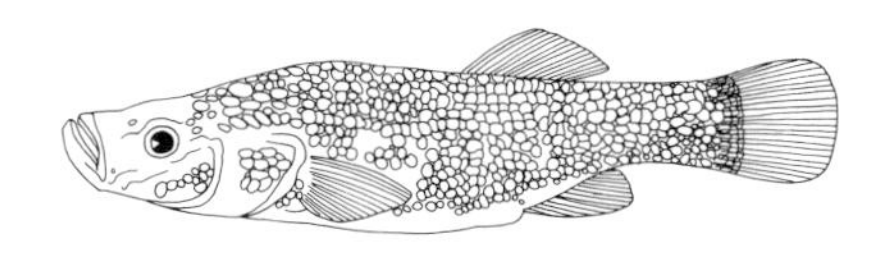

FIG. 15–4. Schematic draft of scale reductions (scaliness) most common in *Orestias* (according to a photo of *O. cuvieri*; see also FIG. 15–12).

SPECIES OF THE GENUS *ORESTIAS*

The first collection of *Orestias* was that of Pentland, brought to France in the 1830s. Valenciennes (1839) established the new genus *Orestias* and described 10 species in 1846 (Cuvier and Valenciennes, 1846). A second, larger collection (apart from several smaller ones) was that of Agassiz and Garman, first described by Garman in 1876 and then revised by himself in 1895. The British Titicaca expedition brought back the third large collection in 1937. The revision by Tchernavin was based mainly on this material. A further collection of these cyprinodonts was made by the author (under the direction of Professor Dr. C. Kosswig) and brought back to the Zoological Institute and Zoological Museum, University of Hamburg, in 1960, a collection that contains about 3620 specimens, of which 1860 were sexually undifferentiated juveniles. This material was obtained at 21 different localities, from the known distribution area of *Orestias*, except from Lake Poopó and the salares. In addition to my own material I have examined *Orestias* specimens available in museums in Washington, London, Paris, Munich, and Hamburg (Fig. 15–5). The present chapter is based on nearly 5000 specimens, which, according to Tchernavin (1944a), belong to 21 species and subspecies.

BRIEF DESCRIPTION OF THE MAIN DISTRIBUTION PATTERNS, WITH REMARKS ON ECOLOGY AND FEEDING HABITS

The species of the genus *Orestias* may be reduced to four main groups (groups I–IV sensu Tchernavin, 1944a). Tchernavin's group IV (Fig. 15–6) comprises four species:

O. mülleri Cuv. and Val. 1846
O. incae Garman 1895
O. crawfordi Tchernavin 1944
O. mooni Tchernavin 1944

These species may be easily distinguished from those of all other groups, not only by their distribution patterns, but also by taxonomically valuable characters that are discussed further on. These species were caught in the northeastern

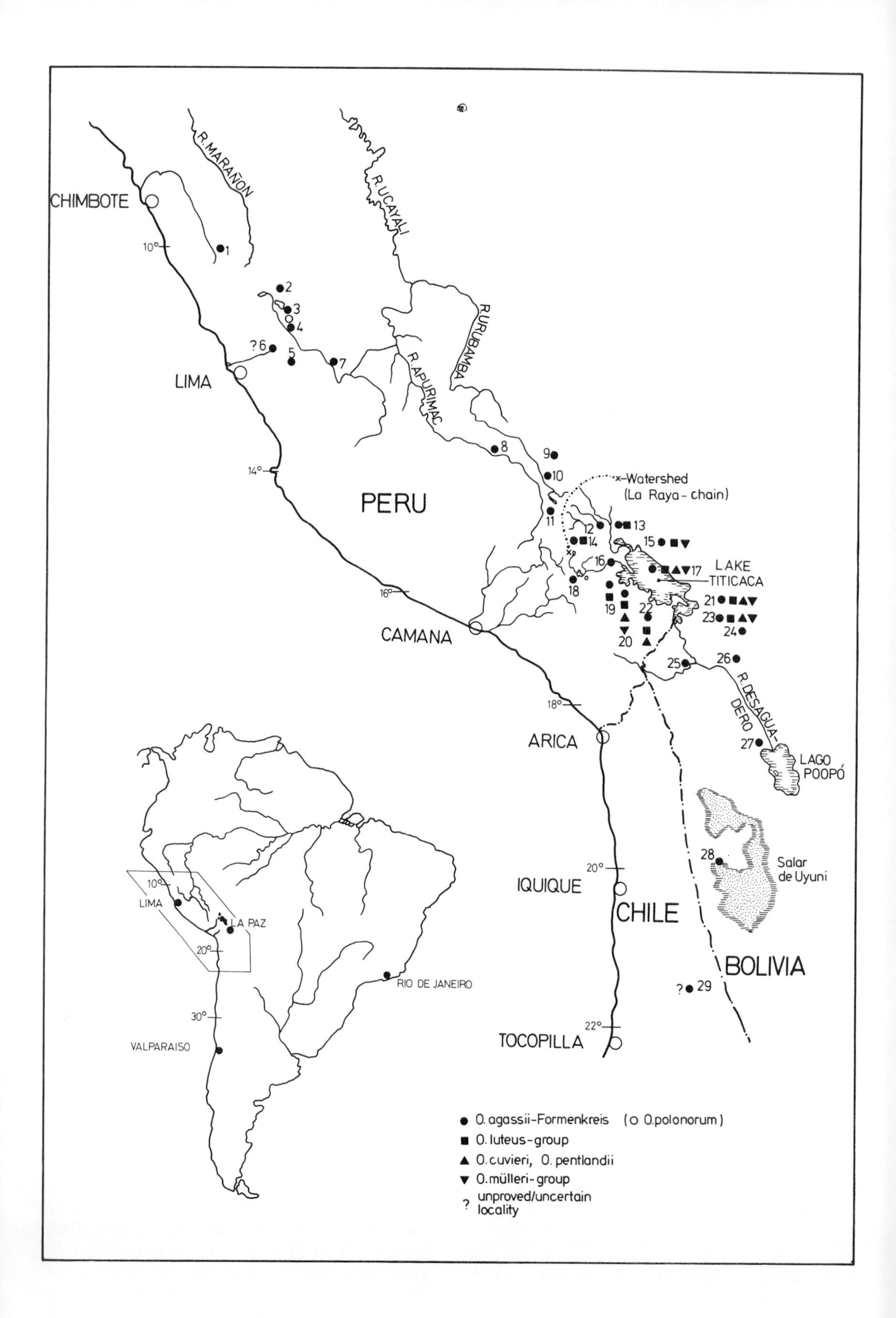
CHIMBOTE
R. MARAÑON
R. UCAYALI
R. URUBAMBA
R. APURIMAC
LIMA
PERU
Watershed
(La Raya - chain)
LAKE
TITICACA
CAMANA
ARICA
R. DESAGUA-
DERO
LAGO
POOPÓ
Salar
de Uyuni
IQUIQUE
CHILE
BOLIVIA
TOCOPILLA
LA PAZ
RIO DE JANEIRO
VALPARAISO
O. agassii-Formenkreis (o O. polonorum)
O. luteus-group
O. cuvieri, O. pentlandii
O. mülleri-group
unproved/uncertain locality

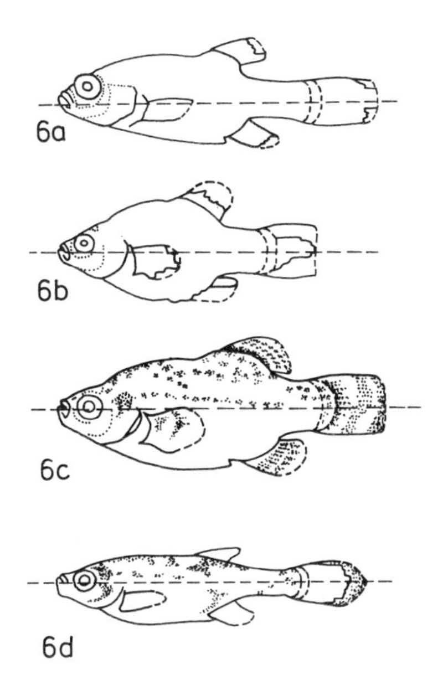

FIG. 15–6. Group IV species (modified from Tchernavin [1944a]). (6a) *O. mülleri*, (6b) *O. incae*, (6c) *O. crawfordi*, (6d) *O. mooni* (scale about 2/3, except O. *mooni*, about $1\frac{1}{2}$).

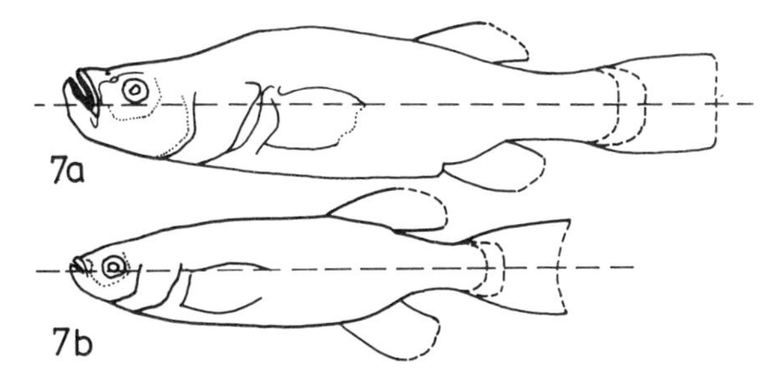

FIG. 15–7. Group I species (modified from Tchernavin 1944a). (7a) *O. cuvieri*, (7b) *O. pentlandii* (scale about $\frac{2}{3}$).

part of Lago Grande and Puno Bay (Fig. 15–5), all at a considerable depth (maximum, 28 m). As an adaptation to their preference for inhabiting deeper water all specimens investigated (by Tchernavin, no material collected by the author) showed an unpigmented (white) peritoneum.

The so-called first group (group I *sensu* Tchernavin) includes the largest of all species of the genus, *O. cuvieri* Val. 1839 (Indian name, "Umanto"), and *O. pentlandii* Val. 1839 (Indian name, "Boga") (Fig. 15–7). Both might be defined as "swarm fish" in that they form small shoals that are found pelagically in the open parts of Lago Grande and Puno Bay (according to Tchernavin, 1944a). Tchernavin considered them to be close relatives and to be able to hybridize. Histological investigations carried out on the "hybrid" gonads (two sexually mature males; Tchernavin, 1944a, p. 161) showed normal meiotic segregation and completely normal sperm formation (Villwock, 1964a). Tchernavin's statement (p. 161) that "both were taken among a school of ripe *Orestias cuvieri*" and his own data on the taxonomic characters of the assumed parental species prove that the two male specimens in question are members of *O. cuvieri*. (More than doubtful are the "probable hybrids" [Tchernavin, 1944a, p. 200] between *O. a. owenii* × *O. olivaceus*, because the specimens in question proved to be undifferentiable juveniles.)

To decide whether *O. cuvieri* and *O. pentlandii* might be closely related, or only shoaling together, a representative number of individuals of each taxon were examined for gut content. The results of these studies showed that the species are valid, definable as far as their food is concerned. Thus the intestines of *O. pentlandii* contained the remains of *Cladocera*, and larvae and/or pupae of mosquitoes, and the digestive tracts of *O. cuvieri*, when filled, contained scales (Villwock, 1965). The statement by Zuñiga (1941, p. 83) that *O. pentlandii* is "una especie fitófaga" (i.e., plant-eating) was not confirmed, as the remains of lower plants were not dominant in the counts but formed only a very small part of the whole gut content, arthropod remains forming the main part. *O. pentlandii* represents a more "herringlike" type of swarm fish that feeds mainly on zooplankton and insects, whereas *O. cuvieri* apparently is a "pikelike" species that preys on other fish (see Villwock, 1965).

It is worthwhile to mention at this point that no *O. cuvieri* specimens were caught during the author's 1960 trip to the area. The species has apparently become extinct after the successful

FIG. 15–5. Distribution of *Orestias* (modified from Villwock [1966]). (1) Laguna de Conococha, (2) Llenamat, (3) Lago Junín, (4) Rio Mantaro, (5) small running waters along road La Oroya-Morococha, (6) Rio Rimac (doubtful locality), (7) Huancayo, (8) Rio Apurímac-Cuzco region, (9) Rio Urubamba, (10) Laguna Pomacanchi, (11) Laguna de Langui, (12) Rio Tirapata, (13) Lagunas de Checayani-Azángaro, (14) northern headwaters of Lago Arapa, (15) several localities at Huancané, Vilque Chico and Moho, (16) Juliaca, (17) Capachica Peninsula, (18) Las Lagunillas, (19) Lago Umayo, (20) Puno Bay, (21) Huarina-Lago Pequeño, (22) Desaguadero-L. Pequēno, (23) Guaqui-L. Pequeño, (24) running waters in the surroundings of La Paz, (25) Rio Mauri, (26) Rio Desaguadero (Corocoro), (27) Yarona Fountain-Lago Poopó, (28) Salar de Uyuni (near Llica), (29) Lago Ascotan-Chile (uncertain locality).

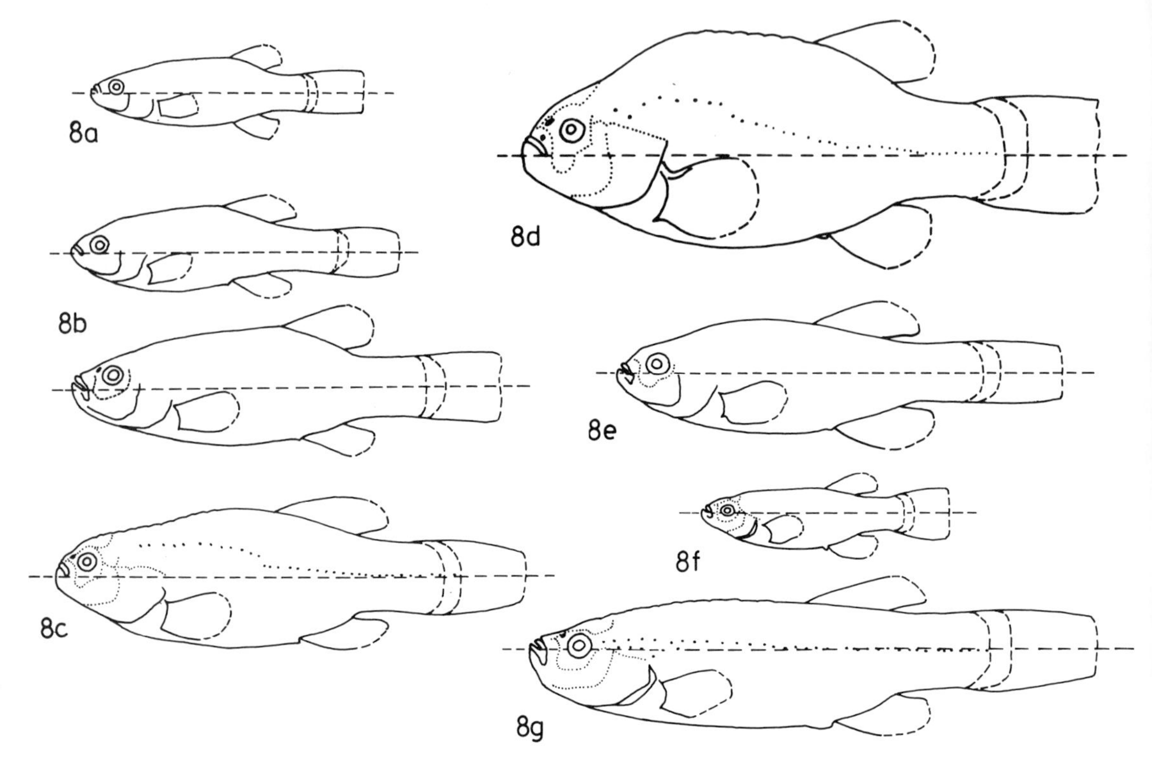

FIG. 15–8. Group II (modified from Tchernavin [1944a], in part). (8a) *O. agassii* forma typica, (8b) *O. a.owenii* (two different adult specimens), (8c) *O. a. tschudii*, (8d) *O. jussiei puni*, (8e) *O. a. pequeni*, (8f) *O. langui*, (8g) *O. polonorum* (scale about $\frac{2}{3}$).

introduction of salmonids into its habitat (see Villwock, 1972).

The two species of the group discussed here are restricted to Lake Titicaca, but *O. pentlandii* is possibly also found in nearby Lake Arapa (Figs. 15–2, 15–5). Reports of *O. pentlandii* in the Cuzco Valley (Urubamba Valley, north of the La Raya mountain chain) have proved to be unfounded.

The species of group II, sensu Tchernavin (1944a), are as follows (see Figs. 15–8 and 15–9):

- *O. agassii* Val. 1839 (typical form)
- *O. a. owenii* Val., infraspecies sensu Tchernavin (=*O. owenii* Cuv. and Val. 1846)
- *O. a. tschudii* Castelnau, infraspecies sensu Tchernavin (=*O. tschudii* Castelnau 1855)
- *O. a. pequeni* Tchernavin 1944, infraspecies
- *O. a. elegans* Garman, subspecies sensu Tchernavin 1944 (=*O. elegans* Garman 1895)
- *O. uyunius* Fowler 1940
- *O. jussiei* Val. 1839
- *O. j. puni* Tchernavin 1944
- *O. polonorum* Tchernavin 1944
- *O. langui* Tchernavin 1944
- *O. olivaceus* Garman 1895
- *O. luteus* Val. 1839
- *O. albus* Val. 1839

This group probably includes at least two valid species with a certain number of "infraspecies" sensu Berg (1958, according to Tchernavin, 1944a). These two species, easily distinguishable from each other (and from the ones discussed earlier) by their local distribution as well as by their environmental demands, are *O. agassii* Cuv. and Val. 1846 (Valenciennes, 1839: nomen nudum) and *O. luteus* Cuv. and Val. 1846 (Valenciennes, 1839: nomen nudum).

The smaller members of this species flock are common in rivers, ponds, and lagoons as well as in the shallow bays of Lake Titicaca and its adjoining larger lakes, where they are found in submerged vegetation. This generalized form might best be interpreted as *O. agassii* forma typica (see also Tchernavin, 1944a, p. 174). Tchernavin (1944a) reported one specimen of Garman's collection from "River Rimac" (p. 163). This informtion is incorrect, because this single individual would be the only one that would have existed west of the Cordilleras (see Fig. 15–5). This mistake could have been due to a confusion between the names River Rimac and River Apurímac, the latter belonging to the Urubamba system. The larger specimens of *O. agassii* are only found in Lake Titicaca and larger (i.e., deeper) lagoons, where they occur mainly

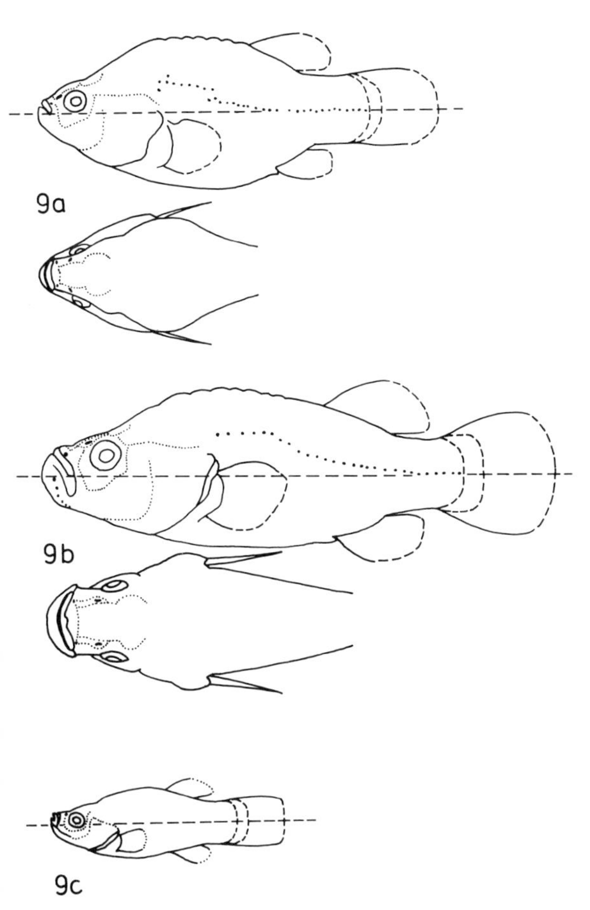

FIG. 15–9. Group II species (modified from Tchernavin [1944a], in part). (9a) *O. luteus* (lateral and dorsal view), (9b) *O. albus* (lateral and dorsal view), (9c) *O. clivaceus* (scale about $\frac{2}{3}$).

beyond the totora girdle at the edge to deeper water. Investigations of the gut content of 287 individuals from 16 different localities, undertaken because the different forms of *O. agassii* were not clearly definable by means of classic taxonomic characters, showed that the seemingly typical form of *O. agassii* has no food preference, whereas larger individuals attributable to *O. a. tschudii* Tchernavin 1944 were found to be plant feeders (see also Zuñiga, 1941). Both forms also include ostracods in their diet, although ostracods make up different proportions of their food, being found only occasionally in the gut of *O. a. tschudii*, but with regular frequency in *O. agassii* forma typica. Furthermore, when closely examined it was found that the species of ostracods taken by the two forms of *Orestias* differed markedly. Thus only two species (an unknown species of *Paracythereis* and *Clamydotheca incisa*) were found in *O. a. tschudii*, but both were absent, as far as could be established, in the diet of *O. agassii* forma typica. We must conclude from this that the different forms of *O. agassii* occupy different ecological niches, at least as adults (Villwock, 1965).

The second of the valid species belonging to Tchernavin's group II is *O. luteus* Cuv. and Val. 1846. This species and its relatives appear to be restricted to suitable deeper waters on the altiplano. They have never been reported or recorded with any certainty from the northern part of the range of *Orestias*. The species represents an altogether different ecological type when compared to all the other species mentioned thus far. They rarely feed on plants or parts of plants, but almost exclusively take mollusks and arthropods from the bottom. This knowledge, obtained after investigations of the gut contents of existing material, led me to collect most of the adults of *O. luteus* during the 1960 expedition with an eel-

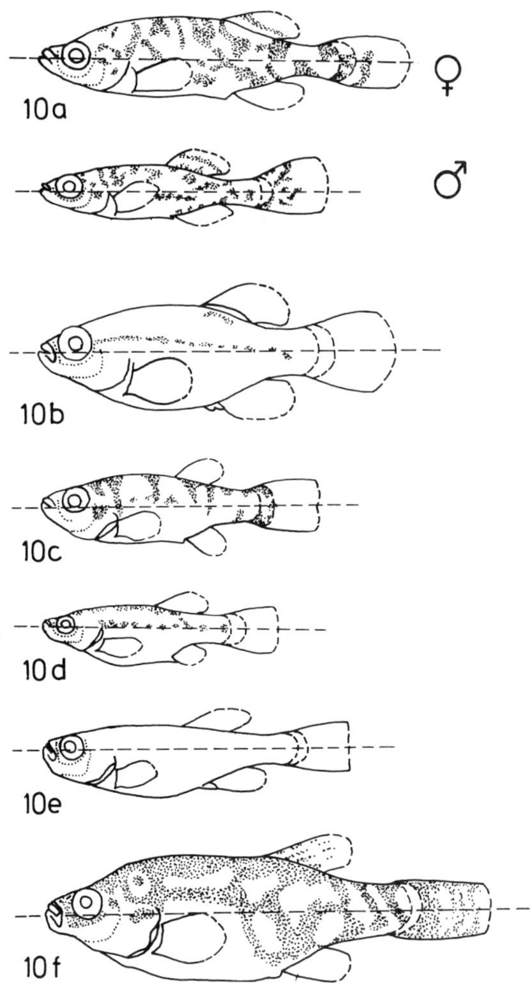

FIG. 15–10. Group III species (modified from Tchernavin [1944a]). (10a) *O. gilsoni* (female above, male), (10b) *O. tutini*, (10c) *O. taquiri*, (10d) *O. minimus*, (10e) *O. minutus*, (10f) *O. uruni* (scale about $1\frac{1}{2}$).

basket-like net or gill net placed on the lake bed beyond the totora girdle. Juvenile individuals on the other hand were caught mostly in shallower water, very often shoaling with the juveniles of other species.

The species of group III (Tchernavin, 1944a, p. 212) comprise "apparently a rather artificial group" of smaller species, all less than 60 mm long. According to Tchernavin all these *Orestias* species are ecologically restricted, i.e., found only in a single habitat. The first four species in the following list were recorded from depths greater than normal; the other two correspond in their habitats completely to *O. agassii* elsewhere in the Lake Titicaca region (Fig. 15–10):

Species	Locality	Depth
O. gilsoni Tchernavin 1944 *O. tutini* Tchernavin 1944 *O. taquiri* Tchernavin 1944 *O. minimus* Tchernavin 1944	Taquiri Island, Lago Pequeño	2.7–2.8 m
O. minutus Tchernavin 1944 *O. uruni* Tchernavin 1944	Juli, Lago Grande Uruni Bay, Lago Grande	10–11 m

SOME CRITICAL REMARKS ON THE TAXONOMY AND SYSTEMATICS OF *ORESTIAS*

It is impossible to define the systematic position of the different "species" of the genus *Orestias* by means of classic taxonomic characters only as most of these characters are either too similar to each other (e.g., lateral line system, tooth structure, and tooth arrangement on the jaws) or are highly variable and only of little value when small numbers of individuals are examined. Characters in regression (scaliness) normally show a large amount of variability (Aksiray and Villwock, 1962; Villwock, 1963b, 1964b), and should not be used for taxonomic statements without a knowledge of the genetic constitution of the characters concerned. The best way to obtain sufficient information is to carry out crossbreeding experiments, or at the very least by statistical evaluation of classic characters. Neither possibility exists—or at least not yet. Some initial calculations on scaliness showed this character to be unsuitable as a tool for making helpful decisions. The same is also true of body indexes, especially those obtained from specimens apparently allowed to dry out several times and rehydrated with overconcentrated preservation fluid (i.e., formalin) after they were collected, a practice that resulted in many specimens becoming soft and flaccid. This fact leads to the difficulty of getting reliable measurements, because the same individual can be deformed to become "high" and "short" or "low" and "long."

The investigations of Tchernavin (1944a) represent the most valuable attempt at a revision of the Orestiinae. It is no doubt correct to separate the species of group IV from the other species of *Orestias*, because they are the only ones to be totally scaled. The problem that remains, however, is how to distinguish their juveniles from those of the other species, since all *Orestias* in their juvenile stage show scale reductions. Second, group IV species occur in rather deep waters and consistently have an unpigmented white peritoneum.

Taking these critical remarks into consideration, the systematic status of *O. mülleri* Val. 1839 seems to be the only one valid within the species of group IV. *O. incae* and *O. crawfordi* are too similar to each other to be recognized as two separate species. This is especially so because of the very small number of individuals available for detailed investigations and their poor state of preservation (Tchernavin, 1944a, p. 222: "All specimens soft after over a hundred years of preservation"). The number of available specimens for the preceding three species was four, one, and two, respectively. The same reservation has to be made for *O. mooni*, based on 37 specimens, considering Tchernavin's own remarks (p. 238) that "all the specimens were dissected when collected, and their intestines extracted; fins in almost all specimens are broken," and that "their state of preservation [was] not perfect." According to present knowledge, group IV species should be described as *O. mülleri* or *O.* cf. *mülleri* (see Fig. 15–11).

Group I species, *O. cuvieri* and *O. pentlandii*,

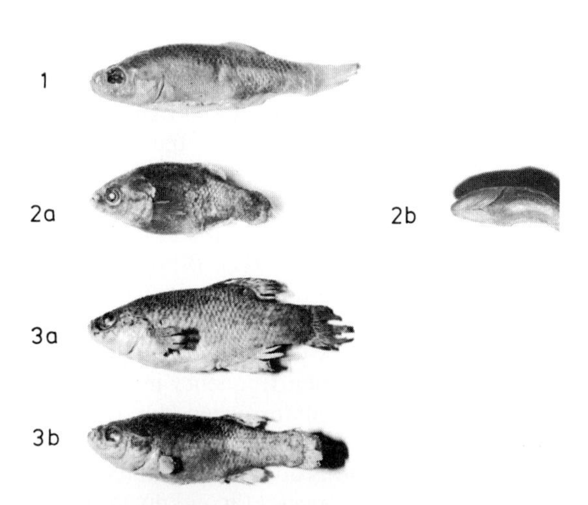

FIG. 15–11. (1) *O. mülleri*, (2) *O. incae* (a, lateral view; b, ventral view to complete scaliness), (3) *O. crawfordi* (a, b two different specimens from the same population; note preservation condition). All photos by the author (scale about ⅔).

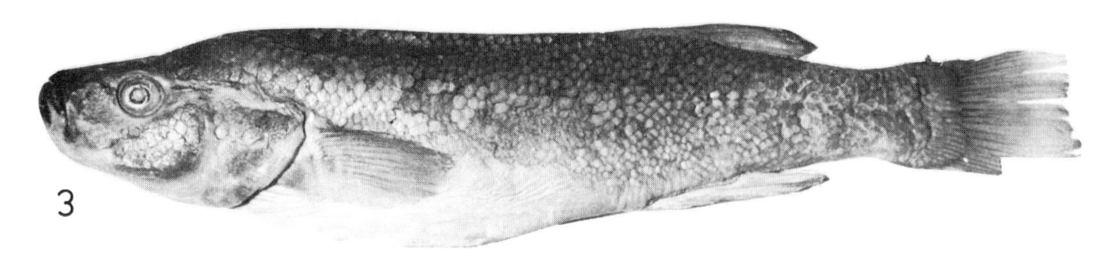

FIG. 15–12. (1) *O. cuvieri*, (2) *O. pentlandii*, (3) *O. cuvieri* × *O. pentlandii* (according to Tchernavin [1944a], see also text, section on taxonomy and systematics). All photos by the author (scale about $\frac{2}{3}$).

have to be accepted as valid for the aforementioned reasons (see Fig. 15–12).

The so-called species of group II have to be reduced to two, including all the others called "infraspecies" of one or the other of the two basic species. One of the two species is *O. agassii* with the widely distributed forma typica. According to Tchernavin the following infraspecies can be recognized (see Fig. 15–13, 1–5):

O. a. owenii Tchernavin 1944
O. a. tschudii Tchernavin 1944
O. a. pequeni Tchernavin 1944
O. a. elegans Tchernavin 1944 (not subspecies sensu Tchernavin.)

In the course of a further systematic revision the description of infraspecies should be avoided. Instead the term *population* and additional application of the locality of capture, for instance, *O. agassii* (pop. Chucuito), should be used. This description would be much more correct from the standpoint of an evolutionary biologist than the use of *infraspecies*. Taking these critical remarks into consideration, the following taxa also have to be understood as members of the *O. agassii Formenkreis* (or superspecies) (see Fig. 15–13, 6–9):

O. a. minutus
O. a. uruni
O. a. gilsoni
O. a. taquiri

Very probably, *O. minimus* Tchernavin 1944 has to be added to the long list of *O. agassii*-like forms since the greater depth at which these specimens were caught is not enough to erect a separate systematic unit. The idea that this form might be close to *O. mülleri*, because of the proximity of its distribution, has to be rejected because of its incomplete scaling and pigmented peritoneum, two characters typical of *O. mülleri* and its relatives. *O. tutini* Tchernavin 1944 resembles in general more *O. luteus* than *O. agassii*. My reexamination, however, in connection with the identical type locality convinced me that

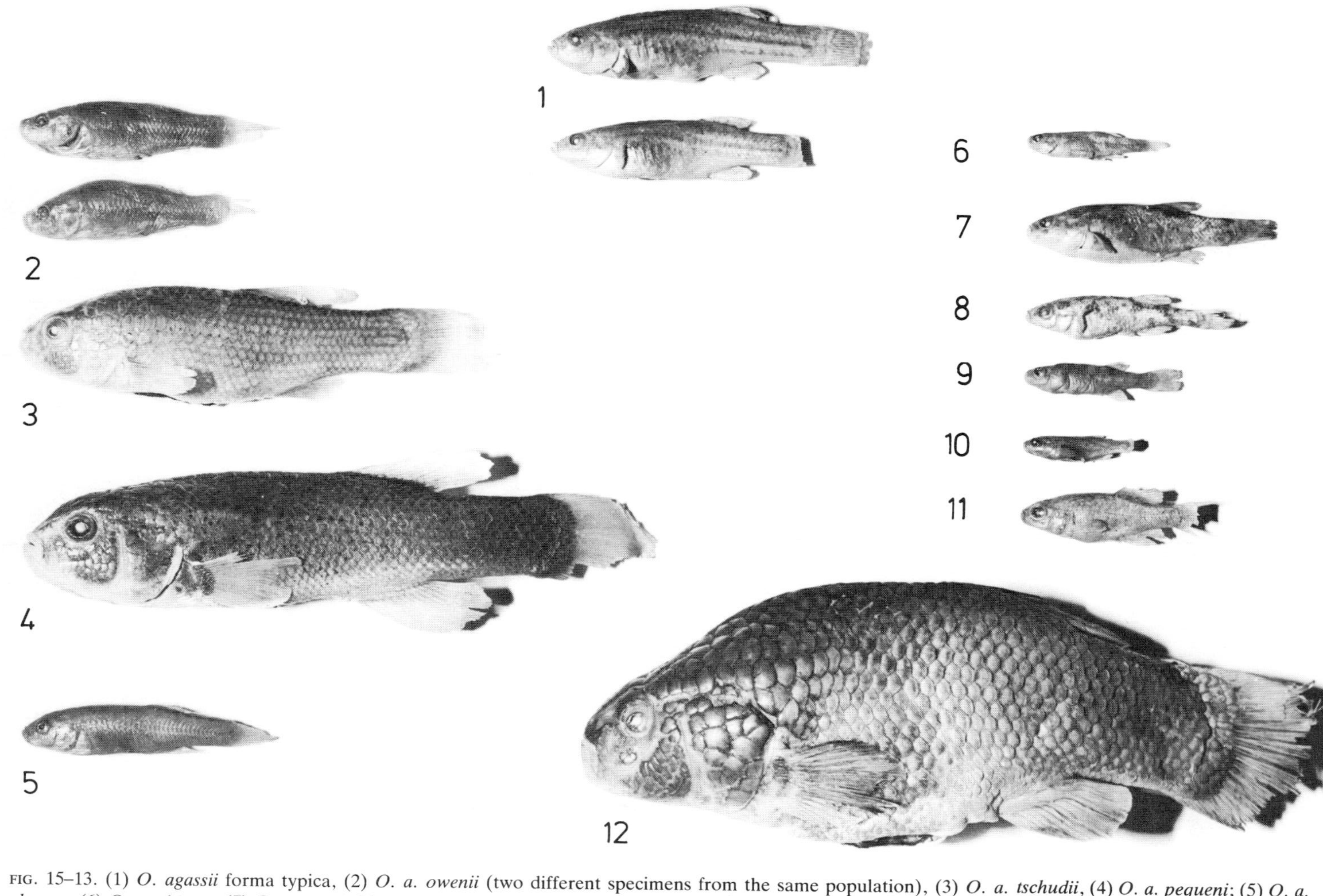

FIG. 15–13. (1) *O. agassii* forma typica, (2) *O. a. owenii* (two different specimens from the same population), (3) *O. a. tschudii*, (4) *O. a. pequeni*; (5) *O. a. elegans*, (6) *O.a. minutus*, (7) *O. a. uruni*, (8) *O. a. gilsoni*, (9) *O. a. taquiri*, (10) *O. a. minimus*, (11) *O. a. tutini*, (12) *O. a. jussiei* (pop. Puno = *O. jussiei puni* Tchernavin [1944a]). All photos by the author (scale about $\frac{2}{3}$).

O. tutini must also be united with *O. agassii*. Moreover, there is no good reason to accept more than one "infraspecies" name for all three forms, all caught together at Taquiri Island, Lago Pequeño, at the same depth.

Finally, the systematic status of two more species (with one subspecies) remains unproved. *O. uyunius* Fowler 1940, *O. jussiei* Cuv. and Val. 1846, and *O. j. puni* Tchernavin 1944, all of which live in Lake Titicaca or the southern altiplano. Concerning the status of *O. uyunius* Tchernavin (1944a, p. 189) himself wrote: "There is no reason in hesitating to include this 'species' into the synonymy of *O. agassii* as southernmost representative of this all-round species." The same is proposed for *O. jussiei* and *O. j. puni*, which are in reality very close to *O. a. tschudii*. In view of the present thinking, the list of "infraspecies" of *O. agassii* has to be completed as follows (see Fig. 15–13, 10–12):

O. a. minimus
O. a. tutini
O. a. uyunius
O. a. jussiei (with two different populations)

The question of the systematic status of *Orestias* species found north of the watershed formed by the La Raya mountain chain (Fig. 15–5) is still open. Apart from *O. agassii*, two other species, *O. langui* Tchernavin 1944 and *O. polonorum* Tchernavin 1944, have been recorded there.

After having collected many specimens from the Lake Junín basin I contend that the smaller *Orestias* specimens caught in the northern part of the old inter-Andean basin have to be considered to be *O. agassii* (see extension of Lake Ballivian in Fig. 15–1). There is really no taxonomic evidence for the separation of any *Orestias* fitting the general features of *O. agassii*. Consequently, Tchernavin's species *O. langui* has also to be added to the long list of *O. agassii* forms.

O. polonorum Tchernavin 1944, type locality Lake Junín, might well be a valid species. New material collected from the Llenamat, an isolated deep lagoon in the Junín Basin, shows some characters that might be of specific value, e.g. tooth formation and coloration in life. On the other hand, however, *O. polonorum* resembles in many aspects the ecologically specialized *O. a. tschudii* from Lake Titicaca, and it may be suspected that *O. polonorum* is the corresponding deeper water form in the northern distribution of *Orestias*. This is the idea I favor.

The total list of "infraspecies" therefore has to be extended by: *O. a. langui* and probably *O. a. polonorum* (investigations of the gut contents remain to be made). (See Fig. 15–14, 1 and 2.)

The second species of group II, *O. luteus*, has to be recognised as valid with respect not only to its ecological status, coloration in life (see species name), and body indexes from freshly killed specimens, but also other characters, e.g. scale structures, especially at the neck of adult individuals, which show that this species is distinct in nearly every regard.

Critical studies of *O. olivaceus* Garman 1895, another member of group II sensu Tchernavin, gave rise to the assumption that it is nothing but a

FIG. 15–14. (1) *O. a. langui* (female above, male), (2) *O. a. polonorum* (? *O. polonorum* Tchernavin [1944a]; see also text, section on taxonomy and systematics). All photos by the author, natural scale.

FIG. 15–15. (1) *O. luteus* (a, lateral view; b, dorsal view, not the same specimen as a), (2) *O. albus* (2a, lateral view; b, dorsal view, same specimen), (3) *O. luteus* (undifferentiated juvenile = *O. olivaceus* Tchernavin [1944a].) All photos by the author (scale about $\frac{2}{3}$). Except for *O. luteus* (freshly killed), all photos were taken from the preserved material investigated by Tchernavin (1944a) (Rolleiflex 4 × 4 cm, flash photos).

juvenile stage of *O. luteus*. This was confirmed by a complete series of individuals I collected in 1960 at Lake Umayo, the specimens ranging from juveniles to ripe males and females. There is no doubt in my mind that *O. olivaceus* is a synonym of *O. luteus*.

A situation similar to *O. agassii* is found between *O. luteus* and the so-called *O. albus* Val. 1839. *O. luteus* and *O. albus* were named by Valenciennes (1839), but only as nomina nuda (p. 118). Cuvier and Valenciennes (1846) described both species, with page priority for *O. albus* (p. 242) against *O. luteus* (p. 243). However, the name *O. luteus* is preferred since only light or dark yellow specimens were found in nature, never "white" ones. Because there is no reliable taxonomic reason for maintaining *O. albus* as a species, I propose combining both in a single species, namely, *O. luteus* (note also the identical Indian name for *O. luteus* as well as "*O. albus*": "Carache"). The dorsal view shown in Tchernavin's fig. 16 (p. 210) is not correct with respect to the broad mouth. My own photographs of the very specimen drawn and published by Tchernavin (1944a) show much smaller differences in this character when compared with *O. luteus* than does Tchernavin's figure (see Fig. 15–15, 1–3).

After a critical analysis I would like to state that the diversity of species in the genus *Orestias* should be reduced to the following:

1. *O. agassii* Cuv. and Val. 1846 (highly variable species, most common)
2. *O. luteus* Cuv. and Val. 1846
3. *O. pentlandii* Val. 1839
4. *O. cuvieri* Val. 1839
5. *O. mülleri* Cuv. and Val. 1846

Final conclusions cannot be drawn before additional investigations have been carried out, such as statistical evaluations of some meristic characters, detailed studies on gill rakers in connection to feeding habits, scanning microscope studies of ultrastructures of teeth and scales, the last carried out at present by the author.

It is unfortunate that only very few cytogenetic

studies have been done (Lueken, 1962) and that, for instance, absolutely no studies have been carried out on enzyme patterns, both modern techniques that demand new investigations in situ with living fresh material.

EVOLUTIONARY ASPECTS

Concrete information on the early development of the inter-Andean basin is regrettably still fragmentary. Assuming that Moon's (1939) and Newell's (1949) considerations are true, then the inter-Andean region should have been covered about 10 million years ago by a more or less continuous body of water. According to its geographic position the climatic conditions must then have been semi-tropical montane. The elevation to its present great height, together with climatic changes in general, would have resulted in a cooling of the whole region. Species that were not "preadapted" became extinct, and only a few survivors, the ancestors of *Orestias*, were able to adapt to changing conditions.

The last changes toward the present situation should have taken place within the last two million years. Speciation of the old macro-population of proto-*Orestias* might have started enclosed in the old basin, because it was divided by mountain building into a northern and a southern part (the La Raya Mountains). The still widespread occurrence of the common "standard" form *O. agassii* in the northern and the southern part of the distribution area of the genus suggests that the assumption of a relatively late beginning of diversificaton is correct, and that this diversification has not yet come to an end. Some information on the karyotypes of different populations ("infraspecies" sensu Tchernavin, 1944a) of *O. agassii* and *O. luteus* may underline this assumption: the number of chromosomes, for instance, is N = 24 in all specimens tested by Lueken (1962).

Speciation takes place most commonly by means of geographic isolation of populations. At least two or more new species may be derived from one ancestor, depending on the number of groups of individuals which became isolated (Mayr, 1948; Kosswig, 1963; Lowe-McConnell, 1969; Banarescu, 1978; White, 1978). This mode of speciation does not apply to *Orestias* because the main diversification is not found in isolated waters but within one body of water, namely, Lake Titicaca. Even the assumption that the different *Orestias* might have developed separately and then became reunited later within the lake is not very good, as at least some of those species that evolved allopatrically should have survived in their original habitats.

That the contrary is true may easily be proved, however. Totally isolated lagoons or lakes have only a single, more or less generalized species, *O. agassii*. The modified assumption that changes in the level of Lake Titicaca might have caused the diversification, similar to the surprising course of evolution among East African cichlids (Fryer, 1959; Kosswig, 1960, 1963; Greenwood, 1964, 1965, 1974) is also of little value here. Certainly, evidence exists that the level of the ancient as well as of the modern Lake Titicaca changed repeatedly (Monheim, 1956; Dolfus, Taltasse, and Tricart, 1966). However, it is difficult to believe that the time elapsed since then might be sufficient for species formation to take place in *Orestias*. There is no evidence of reproductive isolation such as exists in *Salmo letnica* of Lake Ohrid, Yugoslavia (Brooks, 1950), or ethological isolation as is shown in the "preferential mating" in some of the East African cichlids (Kosswig, 1963). Both these mechanisms may indeed lead to quick specific divergence as has been proved in the preceding examples.

Thus the only explanation left is that of (primarily) ecological isolation, perhaps secondarily enhanced by means of microgeographic isolation, i.e., separation of populations along the borders of different ecological niches, according to the hypothesis of "non-competition" and its genetic implications. In a second (or third) step, competition by differentially adapted populations might have helped to prevent potential interbreeding between members of different populations. Only if evolution is understood as a statistical phenomenon according to which individuals with the same ecological preference breed more often with their own kind than with individuals living in different ecological niches can this type of intralacustrine speciation be accepted. The reality of adaptive radiation in closed bodies of water is indisputable. How else should we interpret either the famous Lake Baikal fauna (Kozhow, 1963) or the parallelism in species diversification of other Lake Titicaca organisms, such as the siluroid fish *Trichomycterus* (Tchernavin, 1944b), the frog genus *Telmatobius* (see chap. 14), the amphipod *Hyalella* (Brooks, 1950), and mollusks? The number of similar phenomena, convergent ecological adaptations in Lakes Baikal, Lanao, and Titicaca, also support the idea of intralacustrine speciation, with its different modes. Note that "intralacustrine" does not mean "sympatric," at least in a restricted sense, as White (1978) has already pointed out (see also Tauber and Tauber, 1977). According to this concept, one can expect as many species as there are different ecological niches. As far as Lake Titicaca, then, is concerned, the unanswered question that remains is, to what degree have the cyprinodontids which are endemic to this lake progressed toward becoming biological species. In other words, to what extent do they show reproductive isolation? The answer might be that the number of recent species of *Orestias* has to be reduced further. Even such a reduction in the number of valid species, however, will not contradict the principle of adaptive radiation.

The relationship of *Orestias* within the family of oviparous Cyprinodontidae is completely open. Nothing is known about their nearest relatives or about their origin. Thus the fact that the recent representatives of the genus *Orestias* are widespread within the ancient inter-Andean basin of the modern altiplano allows me to state that they all originated from a single ancestor. It is highly improbable that the regressive character of the lack of ventral fins might have evolved more than once.

EDITORS' POSTSCRIPT

While the book was in press Parenti's revision (1984a) and summary of the distribution (1984b) of *Orestias* were received. After a cladistic analysis (Hennig, 1966), Parenti (1984a) recognized 43 species which she (1984b, p. 85) placed into four "monophyletic species complexes." Although Parenti (1984b, p. 85) considers *Orestias* "a well-defined monophyletic genus," she does not view the genus as a single species flock (but see Mayr, 1984 and Greenwood, 1984). Parenti (1984a, b) believes that Andean *Orestias* and Anatolian *Kosswigichthys* are sister groups that comprise the tribe Orestiini, a view Mayr (1984) thinks is unlikely. Intralacustrine speciation and adaptive radiation have clearly taken place in *Orestias*, as well as interbasin speciation.

REFERENCES

Greenwood, P. H. 1984. What *is* a species flock? In *Evolution of fish species flocks*, A. A. Echelle and I. Kornfield, eds., pp. 13–19. Orono: Univ. Maine at Orono Press.

Hennig, W. 1966. *Phylogenetic systematics*. Urbana: Univ. of Illinois Press.

Mayr, E. 1984. Evolution of fish species flocks: a commentary. In *Evolution of fish species flocks*, A. A. Echelle and I. Kornfield, eds., pp. 3–11. Orono: Univ. of Maine at Orono Press.

Parenti, L. R. 1984a. A taxonomic revision of the Andean killifish genus *Orestias* (Cyprinodontiformes, Cyprinodontidae). *Bull. Amer. Mus. Nat. Hist.* 178: 107–214.

Parenti, L. R. 1984b. Biogeography of the Andean killifish genus *Orestias* with comments on the species flock concept. In *Evolution of fish species flocks*, A. A. Echelle and I. Kornfield, eds., pp. 85–92. Orono: Univ. of Maine at Orono Press.

ACKNOWLEDGMENTS

The author wishes to express his sincere thanks to Prof. Kosswig, who enabled him to carry out all the investigations in *Orestias*. The author also thanks the curators of the ichthyological collections of the following museums: the Smithsonian Institution, Washington, D.C.; the British Museum (Natural History), London; the Museum d'Histoire Naturelle, Paris; and the Munich Museum and Zoological Museum, Hamburg. The author also extends thanks to Mrs. Hänel (for the drawings), Prof. Hartmann (for determining the ostracods), and last but not least to Dr. Bürkel for his kind efforts in correcting the English text of this chapter. Mrs. Hänel, Prof. Hartmann, and Dr. Bürkel all belong to the staff of the Zoological Institute and Zoological Museum, University of Hamburg. The research on which this chapter is based was supported by a grant from the Deutsche Forschungsgemeinschaft.

REFERENCES

Aksiray, F., and W. Villwock. 1962. Populationsdynamische Betrachtungen an Zahnkarpfen des südwest-anatolischen Aci-(Tuz-) Gölü. *Zool. Anz.* 168: 87–101.

Banarescu, P. 1978. Some critical reflexions on Hennig's phyletical concepts. *Z. zool. Syst. Evolut.-forsch.* 16: 91–101.

Barth, W. 1972. Die geowissenschaftliche Literatur Boliviens in den Jahren 1960–1971: Ein Überblick. *Zbl. Geol. Paläont.* Teil I (1–2): 100–130.

Berg, L. S. 1958. *System der rezenten und fossilen Fischartigen und Fische*. Berlin: VEB Deutscher Verlag der Wissenschaften.

Brooks, J. L. 1950. Speciation in ancient Lakes. *Quart. Rev. Biol.* 25: 30–60; 479–526.

Cuvier, G. L. and A. Valenciennes. 1846. *Histoire naturelle des poissons*. Vol. *18*. 225–230. Paris.

Dollfus, O., P. Taltasse, and J. Tricart. 1966. Étude des

formations alluviales et lacustres de la region d'Ilave (Dépt. de Puno, Pérou). *Bull. Assoc. franç. pour l'étude du Quarternaire* 2: 152–162.

Everett, G. V. 1971. *A study of the biology, population dynamics, and commercial fishery of the rainbow trout (Salmo gairdneri Richardson) in Lake Titicaca.* FAO Report 1–24.

Fowler, H. W. 1916. Notes on fishes of the orders Haplomi and Microcyprini. *Proc. Acad. Nat. Sci. Philad.* 68: 425–439.

Fryer, G. 1959. The trophic interrelationships and ecology of some littoral communities of Lake Nyassa with especial reference to the fishes, and a discussion of the evolution of a group of rock frequenting Cichlidae. *Proc. Zool. Soc. London* 32: 153–281.

Garman, S. 1876. Exploration of Lake Titicaca by A. Agassiz and S. W. Garman. *Bull. Mus. Comp. Zool. Harv.* 3: 274–278.

———. 1895. The cyprinodonts. *Mem. Mus. Comp. Zool. Harv.* 19: 1–179.

Greenwood, P. H. 1964. Explosive speciation in African Lakes. *Proc. Roy. Inst.* 40: 256–269.

———. 1965. The cichlid fishes of Lake Nabugabo, Uganda. *Bull. Brit. Mus. (Nat. Hist.), Zool.* 12: 315–357.

———. 1974. Cichlid fishes of Lake Victoria, East Africa: The biology and evolution of a species flock. *Bull. Brit. Mus. (Nat. Hist.), Zool.*, Suppl. 6: 1–134.

Hubbs, C. L. 1940. Speciation of fishes. *Am. Nat.* 74: 198–211.

Kohzov, M. 1963. *Lake Baikal and its life.* The Hague: Junk.

Kosswig, C. 1960. Bemerkungen zum Phänomen der intralakustrischen Speziation. *Zool. Beitr.*, n.f. 5: 497–512.

———. 1963. Ways of speciation in fishes. *Copeia* 2: 238–244.

Kosswig, C., and W. Villwock. 1964. Das Problem der intralakustrischen Speziation im Titicaca- und im Lanaosee. *Verh. Dtsch. Zool. Ges. Kiel: 95–102.*

———. 1974. Remarks concerning the phylogenetic status of the cyprinids of Lake Lanao (Mindanao). *Philippine Scientist* 11: 21–25.

Loffler, H. 1968. Tropical high mountain lakes. Their distribution, ecology and zoogeographical importance. *Colloq. Geogr.* 9: 57–76.

Lowe-McConnell, R. H. 1969. Speciation in tropical freshwater fishes. *Biol. J. Linn. Soc.* (London) 1: 51–75.

Lueken, W. 1962. Chromosomenzahl bei *Orestias* (Pisces, Cyprinodontidae). *Mitt. Hamburg. Zool. Mus. Inst.* 60: 195–198.

Mayr, E. 1947. Ecological factors in speciation. *Evolution* 1: 263–288.

———. 1948. The bearing of the New Systematics on genetical problems. *Adv. in Genetics* 2: 205–237.

———. 1962. Zufall oder Plan, das Paradox der Evolution. In *Evolution und Hominisation.* Stuttgart: Fischer-Verlag. 21–35.

Monheim, F. 1956. Beiträge zur Klimatologie und Hydrologie des Titicacabeckens. *Heidelb. Geogr. Arb. H.* 1: 1–152.

Moon, H. P. 1939. The geology and physiography of the altiplano of Peru and Bolivia. *Trans. Linn. Soc.* (London) 1: 27–43.

Newell, N. 1949. Geology of the Lake Titicaca region, Peru and Bolivia. *Geol. Soc. Am., Memoir* 36: 1–111.

Schindler, O. 1955. Limnologische Studien am Titicaca-See. *Archiv f. Hydrobiol.* 51: 118–124.

Tauber, C. A., and M. J. Tauber. 1977. A genetic model for sympatric speciation through habitat diversification and seasonal isolation. *Nature* 268: 702–705.

Tchernavin, V. V. 1944a. A revision of the subfamily *Orestiinae. Proc. Zool. Soc. London*, 114: 140–223.

———. 1944b. A revision of some *Trichomycterinae* based on material preserved in the British Museum (Natural History). *Proc. Zool. Soc. London.* 114: 234–275.

Urquidi, W. Terrazas. 1969. Problemas de conservación de los recursos pesqueros de Bolivia. *Bol. Experimental* 40: 1–16 (Ministero de Agricultura, La Paz, Bolivia).

Valenciennes, S. A. 1839. Rapport sur quelques poissons d'Amérique, rapportés par M. Pentland. *L'Institut* 7: 118.

Villwock, W. 1963a Die Gattung *Orestias* (Pisces: Microcyprini) und die Frage der intralakustrischen Speziation im Titicaca-Seengebiet. *Verh. Dtsch. Zool. Ges. Wien 1962*: 610–624.

———. 1963b. Genetische Analyse des Merkmals "Beschuppung" bei anatolischen Zahnkarpfen (Pisces, Cyprinodontidae) im Auflöserversuch. *Zool. Anz.* 170: 23–45.

———. 1964a. Vermeintliche Artbastarde in der Gattung *Orestias* (Pisces, Cyprinodontidae). *Mitt. Hamburg. Zool. Mus. Inst.*: 285–291.

———. 1964b. Genetische Untersuchungen an altweltlichen Zahnkarpfen der Tribus *Aphaniini* (Pisces, Cyprinodontidae) nach Gesichtspunkten der Neuen Systematik. *Z. zool. Syst. Evolut.-forsch.* 2: 267–382.

———. 1965. Zur Biologie der *Orestiinae* (Pisces, Cyprinodontidae), unter besonderer Berücksichtigung von Darminhaltsuntersuchungen. Ein Beitrag zum Problem der Speziation in der Gattung *Orestias. Abhandl. u. Verh. Naturwiss. Ver. Hamburg,* n.f. 10: 153–166.

———. 1972. Gefahren für die endemische Fischfauna durch Einbürgerungstversuche und Akklimatisation von Fremdfischen am Beispiel des Titicaca-Sees (Peru/Bolivien) und des Lanao-Sees (Mindanao/Philippinen). *Verh. Internat. Ver. Limnol. 18: 1227–1234.*

Wahl. E. 1976. *Morphologisch-Ökologische Untersuchungen zur Systematik der Lanao-Cypriniden.* Ph.D. diss., Univ. Hamburg.

White, M. J. D. *1978. Modes of speciation.* San Francisco: Freeman.

Wilkens, H., N. Peters, and C. Schemmel. 1979. Gesetzmässigkeiten der regressiven Evolution. *Verh. Dtsch. Zool. Ges. Regensburg* 1979: 123–140.

Zuñiga, E. 1941. Regimen alimenticio y longitud del tubo digestivo en los peces del género *Orestias. Bol. Mus. Hist. Nat. "Javier Prado", Lima* 5: 79–86.

16

Diversity Patterns and Differentiation of High Andean Rodents

OSVALDO A. REIG

Prudent specialists may think that to try to assess the evolutionary history of high Andean rodents is an audacious undertaking and that an author who contributes to the topic must be bold enough to want to tread on risky ground. These persons are not entirely wrong. Being in the shoes of the author who was asked to make that contribution is thus not easy. The reasons for this insecurity are several.

1. The organisms for which one attempts to explain the history of the distribution patterns are still in an unstable taxonomic situation. Forty years ago, referring to the South American cricetid rodents (which make up the majority of the rodents on this continent; see Fig. 16–1), J. R. Ellerman (1941, p. 327) stated: "Directly Panamá is passed, an enormous list of names described for the most part binominally, and in appalling chaos, is reached." That chaotic situation has been partially overcome since Ellerman wrote those words, however, thanks to some valuable revisionary attempts (see especially Hershkovitz, 1944, 1955, 1960, for *Nectomys*, *Holochilus*, and *Oecomys*, respectively; Pearson, 1958, and Hershkovitz, 1962, for the phyllotines; and Hershkovitz, 1966a, for scapteromyines). But some of the results of these revisions have been challenged by recent chromosomal investigations (see especially Gardner and Patton, 1976; Pearson and Patton, 1976), which demonstrated that species recognition among South American cricetid mice may be misleading without the help of cytotaxonomic evidence. In fact, specialists in the field of systematics of South American cricetid rodents agree that a stable classification of these rodents still does not exist. Even when the

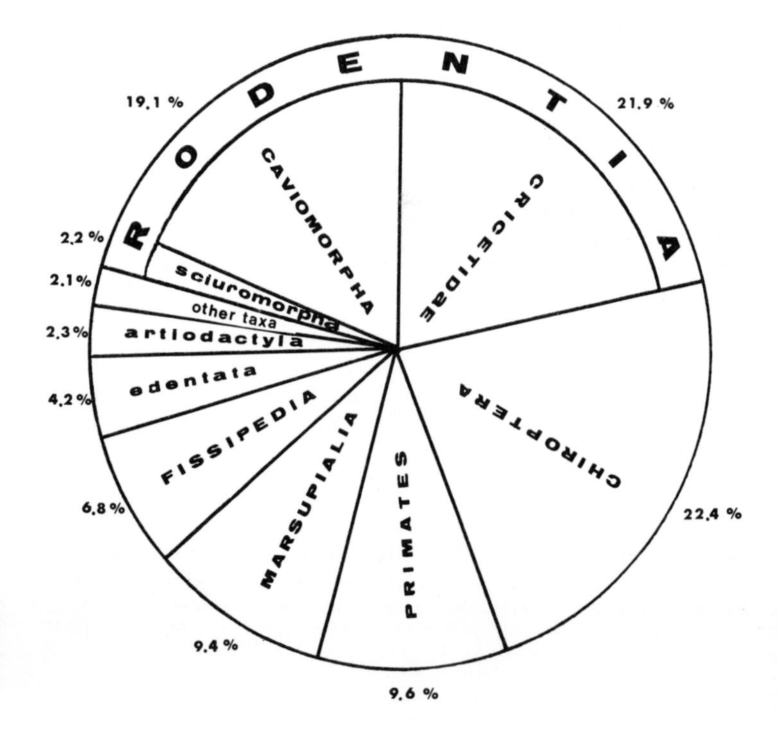

FIG. 16–1. Relative species diversity in the different taxa of South American living mammals, expressed as a percentage of the total number of species of mammals in the present South American fauna. Original data from Cabrera (1961) modified according to various sources, including partial revisions by the author.

classification is rather firm at the generic and suprageneric levels, everybody recognizes that much work remains to be done to clarify the status of the species. The situation is probably a bit better as regards the caviomorphs (which make up the other important component of South American living rodents; see Fig. 16–1), but we can hardly conclude that their systematics is firm enough at the species level.

2. The origin and evolutionary history of South American rodents as a whole is a topic of extensive disagreement. Notwithstanding the wealth of fossil caviomorphs in deposits of the Tertiary and Quaternary of South America, doubts are still extensive as regards the relationships between the different suprageneric taxa of the Caviomorpha, and the geographic origins of the whole group, which is the subject of a lasting and as yet unsettled controversy (see Hoffstetter, 1975; Lavocat, 1971, 1974; Patterson and Pascual, 1972; Patterson and Wood, 1982; Reig, 1981c; G. G. Simpson, 1980; Wood, 1974, 1975). Additionally, the effect of the unwarranted maintenance of a traditional dichotomy between the study of living and fossil rodents in South America is that we are still far from having a good understanding of the relationships between the fossil and living representatives of the Caviomorpha. Controversies are even hotter concerning the currently more abundant group of the South American cricetids. In their case, the fossil record is less significant than for the caviomorphs, and few specialists have been inclined to try to read the lost pages of the history of these mice from the wealth of data provided by the living representatives of the group. This is one of the bases (another being perhaps strong prejudices of opposing biogeographic schools) of widespread disagreement about the origination time, the place of major diversification, and the overall evolutionary history of South American cricetid rodents (see Hershkovitz, 1966b, 1972; Patterson and Pascual, 1972; Baskin, 1978; Reig, 1978, 1980; Marshall, 1979; G. G. Simpson, 1980).

3. With a few exceptions (Pearson, 1951, 1982; Dorst, 1958) regional studies of the Andean rodent fauna are nonexistent, which obliges one to reconstruct diversity patterns for given Andean biomes from an overwhelming amount of disparate data scattered in an extensive, mostly descriptive and taxonomic literature. Such reconstructions are not always accurate as regards habitats and other ecological characteristics of the species.

These difficulties are certainly cumbersome, but at the same time they are challenging. Notwithstanding the inherent risks of the enterprise I decided to dig as deep as possible in the topic. After a painstaking compilation of the available evidence and critical evaluation of the existing theories, I concluded that the results might be of interest to my colleagues. Therefore I will attempt here to describe the diversity patterns and to reconstruct a scenario of the probable evolutionary history of the rodent fauna of the high Andes. I am convinced that the former can be based on enough reliable information, and that the latter, even when tentative, is the more reasonable inference to be drawn from the best body of evidence and theory so far available.

DEFINITIONS

What kinds of rodents and which biotic and ecological units must one deal with in a study of high Andean rodents? The taxonomic groups of rodents relevant to this study have been for a long time in a state of nomenclatural confusion. The biotic and ecological units have been defined differently by various authors. It is necessary, therefore, to define the faunistic, taxonomic, and biotic names I will be using in this chapter.

High Andean rodent fauna: the assemblage of rodent species inhabiting the present puna and páramo biomes.

Puna biome: high Andean biotic community characterized by treeless alpine-type arid plant formation of grass tussocks (pajonales) and sagelike bushes (tolar), extending continuously in the form of a dry, barren steppe from central Peru (south of Cajamarca) southward to northwestern Argentina (9° S to 27° S) above about 3700 m. This concept agrees with Weberbauer's (1936) and Vuilleumier's (1979) concept of puna and with Pearson's (1951) and B. B. Simpson's (1975) concept of altiplano, but is more restricted than Chapman's (1921) and Osgood's (1943b) concept of puna.

Páramo biome: high Andean biotic community of cold, humid, sometimes boggy meadows characterized by open vegetation of tussocks of grasses, composites, and flowering herbs and bushes, but also including islands of woodland vegetation. The páramo biome is discontinuously distributed from northern Colombia and western Venezuela through most of Colombia and Ecuador up to northern Peru (11° N to 8° S) above the timberline (usually above 3000 m). This concept is in agreement with Cuatrecasas's (1957) and

TABLE 16–1. Tribal and generic classification and number of recognized species of South American sigmodontine rodents.

Family Cricetidae, subfamily Sigmodontinae

Tribe Oryzomyini (13, 110)
- Genus *Oryzomys* (45) [a]
- Genus *Thomasomys* (25)
- Genus *Oecomys* (10) [b]
- Genus *Rhipidomys* (9)
- Genus *Nectomys* (5) [c]
- Genus *Neacomys* (4)
- Genus *Delomys* (3)
- Genus *Aepeomys* (3)
- Genus *Nesoryzomys* (3)
- Genus *Phaenomys* (1)
- Genus *Chilomys* (1)
- Genus *Scholomys* (1)
- Genus *Wilfredomys* (1)

Tribe Ichthyomyini (4, 8)
- Genus *Ichthyomys* (3)
- Genus *Anotomys* (2) [d]
- Genus *Daptomys* (2) [e]
- Genus *Neusticomys* (1)

Tribe Akodontini (10, 62)
- Genus *Akodon* (34) [f]
- Genus *Oxymycterus* (9)
- Genus *Bolomys* (6) [g]
- Genus *Chelemys* (4) [h]
- Genus *Microxus* (3) [i]
- Genus *Notiomys* (2)
- Genus *Blarinomys* (1)
- Genus *Podoxymys* (1)
- Genus *Lenoxus* (1)
- Genus *Juscelinomys* (1)

Note: Figures in parentheses after each genus represent the number of recognized species. Figures in parentheses after each tribe represent, first, the number of genera, and second, the number of species in the given tribe. After O. A. Reig's unpublished provisional revision of the South American Sigmodontinae.

[a] I include in *Oryzomys* the following subgenera, with the number of their recognized species in parenthesis: *Oryzomys (28)*, *Microryzomys (1)*, *Melanomys (3)*, *Macruroryzomys (1)*, *Sigmodontomys (1)*, *Oligoryzomys (11)*. I include here only the South American forms. The number of recognized species either in *Oryzomys* s.s. or in *Oligoryzomys* has been reassessed mostly from the chromosomal work of Gardner and Patton (1976) and of others, including unpublished results by the author. *Oligoryzomys* is here considered as a subgenus or *Oryzomys* only in a provisional way. I am increasingly inclined to think of it as a full genus.

[b] Recent chromosomal results by Gardner and Patton (1976) and unpublished chromosomal work by O. A. Reig and collaborators clearly demonstrate that *Oecomys bicolor* and *Oecomys concolor* are composite. For the recognition of *Oecomys* as a full genus, see Gardner and Patton (1976).

[c] Chromosomal information also indicates that *Nectomys squamipes* is composite. From my own results and from the karyological data of several authors, including Gardner and Patton (1976) I recognize the following species in *Nectomys*: *N. squamipes*, *N. melanius*, *N. apicalis*, *N. palmipes* and *N. garleppii*.

[d] I include in *Anotomys* not only the type species, *A. leander* Thomas 1906, but also the species *trichotis* Thomas 1897, which Thomas wrongly referred to the Central American genus *Rheomys*.

[e] Musser and Gardner (1974) recently described *Daptomys peruviensis*, which adds to the formerly single species of the genus, *D. venezuelae*.

[f] I include in *Akodon* the following subgenera, with the number of their recognized species in parentheses: *Akodon (25)*, *Abrothrix (6)*, *Deltamys (1)*, *Hypsimys (1)*, and *Chroeomys (1)*. Recent chromosomal information clearly indicates that *Abrothrix* is very close to *Akodon s.s.*, and not necessarily closer to *Oxymycterus*.

[g] *Bolomys*, as typically represented by *B. amoenus*, is as different from *Akodon* as are widely recognized genera like *Oxymycterus* or *Notiomys*. Its recognized species, *B. amoenus*, *B. lactens*, *B. lenguarum*, *B. obscurus*, *B. lasiurus* and a new undescribed species, are uniform in showing a $2n = 34$ karyotype distinctive as regards the remaining Akodontini. *Cabreramys* is a junior synonym of *Bolomys*, which also comprises the 'southern forms' of *Zygodontomys* in Hershkovitz (1962), and '*Akodon*' *arviculoides*, which I tentatively only recognize as a subspecies of *Bolomys lasiurus*.

[h] *Chelemys* and *Notiomys* greatly differ from each other in molar morphology, as well as in the structure of the palate and in several other skull characters. O. P. Pearson (pers. comm.) found that they also differ strikingly in the morphology of the digestive system. Therefore, I revalidate *Chelemys* as a full genus of the Akodontini, with the species *C. megalonyx*, *C. augustus*, *C. macronyx* and, probably, also *C. delfini*.

[i] There is no doubt that *Microxus*, as typically represented by *M. mimus*, and including *M. bogotensis* and *M. latebricola*, is quite different from *Abrothrix* in skull and dental morphology, in spite of the contrary opinion of Hershkovitz (1966a). Recent chromosomal results on a Venezuelan population referred to *Microxus bogotensis* (Barros and Reig, 1979) clearly confirm the full generic status of this taxon.

Tribe *Scapteromyini* (3, 6)
Genus *Bibimys* (2) [j]
Genus *Scapteromys* (2) [k]
Genus *Kunsia* (2)

Tribe *Wiedomyini* (1, 1)
Genus *Wiedomys* (1)

Tribe Phyllotini (14, 45)
Genus *Phyllotis* (13)
Genus *Calomys* (9) [l]
Genus *Euneomys* (5)
Genus *Auliscomys* (4) [m]
Genus *Graomys* (3)
Genus *Andalgalomys* (2) [n]
Genus *Eligmodontia* (3)
Genus *Pseudoryzomys* (1)
Genus *Galenomys* (1)
Genus *Reithrodon* (1)
Genus *Chinchillula* (1)
Genus *Neotomys* (1)
Genus *Andinomys* (1)
Genus *Irenomys* (1)

Tribe Sigmodontini (2, 10)
Genus *Sigmodon* (4) [o]
Genus *Holochilus* (6) [p]

Sigmodontinae *incertae sedis* (4, 7)
Genus *Zygodontomys* (4) [q]
Genus *Rhagomys* (1) [r]
Genus *Punomys* (1) [s]
Genus *Abrawayaomys* (1)

---

[j] *Bibimys* has been recently proposed as a genus of the Scapteromyini (Massoia, 1979). Besides its type species, *Bibimys torresi*, from the Paraná Delta, Massoia (1980) just referred to *Bibimys* the chacoan species *chacoensis*, formerly described as a species of *Akodon* (Shamel, 1931).

[k] Recent chromosomal work demonstrated that *Scapteromys tumidus* ($2n = 24$) and *S. aquaticus* ($2n = 32$) must be treated as full specied (Gentile de Fronza, 1970; Brun-Zorilla, Lafuente, and Kiblisky, 1972).

[l] The status of the species of *Calomys* was confusing after the excess of lumping in the revision of Hershkovitz (1962). Chromosomal work by Pearson and Patton (1976) and others, together with my own revision of types in the British Museum, allowed the recognition of the following species. *C. callosus* ($2n = 36$, $FN = 48$), *C. boliviae* (including *fecundus*) ($2n = 50$, $FN = 66$), *C. sorellus* ($2n = 62$–64, $FN = 68$), *C. lepidus* ($2n = 36$, $FN = 68$, *C. musculinus* (including *cortensis*) (chromosomes unknown, but morphologically different from the next), *C. murillus* ($2n = 38$), $FN = 48$), *C. laucha* ($2n = 62$, $FN = 82$), *C. tener* (I provisionally assign to this species the $2n = 66$, $FN = 68$ form recently described by Yonenaga, 1975), and *C. humelincki* (chromosomes unknown but different from *C. laucha* in morphology). It is also possible that *C. bimaculatus* should be validated, if the $2n = 56$ karyotype from Uruguay reported by Brum (1965) is confirmed.

[m] Following a suggestion of Pearson and Patton (1976) I include in *Auliscomys* not only the species *pictus*, *boliviensis*, and *sublimis*, but also *micropus*, usually assigned to the genus *Phyllotis* as a subgenus *Loxodontomys*. The chromosomes of *micropus* are quite similar to those of typical species of *Auliscomys* (Venegas, 1974).

[n] For this genus of phyllotines recently proposed for the species *olrogi* and *pearsoni*, see Williams and Mares (1978).

[o] After a preliminary revision based mostly on morphological characters but also including some chromosomal information, I recognize the following living species of *Sigmodon* in South America: *S. alstoni* (including *alstoni*, *savanarum*, and *venester* as subspecies), *S. hispidus* (including *hirsutus* and *bogotensis* as subspecies), *S. peruanus* (including *peruanus*, *simonsi*, *chonensis* and *puna* as subspecies), and *S. inopinatus*.

[p] *Holochilus* is badly in need of revision. It is now certain that *Holochilus brasiliensis* is composite. At least *brasiliensis*, *magnus*, *balnearum*, and *amazonicus* seem to deserve full species status. Other forms, such as *guianae*, *venezuelae*, and *chacarius*, may also deserve species recognition, but their status is not yet clear. The number of six recognised species is therefore a conventional approximation.

[q] *Zygodontomys* is here restricted to the 'northern forms' of Hershkovitz's (1962) provisional revision. After revision of most of the type and original material, I recognize the following species and subspecies in South America: *Z. microtinus microtinus*, *Z. microtinus thomasi*, *Z. microtinus stellae*, *Z. punctulatus punctulatus*, *Z. punctulatus griseus*, *Z. punctulatus fraterculus*; *Z. brevicauda brevicauda*, *Z. brevicauda tobagi*, *Z. brunneus brunneus*, and *Z. brunneus sanctaemartae*. *Z. reigi*, recently proposed as a new species by Tranier (1976) is almost a nomen nadum. I consulted the specimens upon which the new species was based in Paris, and I could not arrive at a definite decision about its status.

[r] *Rhagomys* is a very primitive genus whose relationships are uncertain (See Reig, 1980)

[s] *Punomys* may be a well-differentiated offshoot of the Phyllotini, but its relationships are still uncertain (Osgood, 1943a).

TABLE 16–2. Classification of the genera of South American caviomorph rodents and number of their recognized living species.

Infraorder Caviomorpha

Superfamily Octodontoidea
- Family Octodontidae (6, 52)
  - Subfamily Octodontinae
    - Genus *Octodon* (3)
    - Genus *Octomys* (2) [a]
    - Genus *Octodontomys* (1)
    - Genus *Pithanotomys* (1) [b]
    - Genus *Spalacopus* (1)
  - Subfamily Ctenomyinae
    - Genus *Ctenomys* (44) [c]
- Family Abrocomidae (1, 2)
  - Genus *Abrocoma* (2)
- Family Echimyidae (15, 65)
  - Subfamily Echimyinae
    - Genus *Makalata* (1) [d]
    - Genus *Echimys* (10)
    - Genus *Isothrix* (3)
    - Genus *Mesomys* (4)
    - Genus *Lonchothrix* (1)
    - Genus *Diplomys* (3)
    - Genus *Proechimys* (30) [e]
    - Genus *Euryzygomatomys* (2) [f]
    - Genus *Carterodon* (1)
    - Genus *Hoplomys* (1)
    - Genus *Trichomys* (1) [g]
  - Subfamily Dactylomyinae [h]
    - Genus *Dactylomys* (3)
    - Genus *Thrinacodus* (3)
    - Genus *Kannabateomys* (1)
  - Subfamily Chaetomyinae [i]
    - Genus *Chaetomys* (1)
- Family Myocastoridae (1, 1) [j]
  - Genus *Myocastor* (1)

Note: Figures in parentheses after each genus represent the number of species recognized by the author. Figures in parentheses after each family represent, first, the number of genera, and second, the number of species in the given family. Based mostly on Cabrera (1961) and Honacki et al. (1982), with modifications from several sources, as indicated in the notes.

[a] I include here *Tympanoctomys barrerae*, obviously a distinctive species, but hardly separable from *Octomys mimax* at the generic level.

[b] *Aconaemys* Ameghino 1891 is not separable at the generic level from *Pithanotomys* Ameghino 1887, a generic name that has been applied so far only to fossil species. Differences between *Aconaemys fuscus* and the fossil *P. columnaris* are not greater, for instance, than differences between *Ctenomys australis* and *Ctenomys latro*, which no one attempted to separate in different genera. *Pithanotomys* has priority over *Aconaemys*, and therefore the former would be the generic name to use for the living species *fuscus*. This conclusion is based on the study of complete skulls and postcranial bones in the collection of the Museum of Mar del Plata, from Pliocene and Pleistocene deposits of Buenos Aires Province, which I plan to present in a forthcoming paper.

[c] The number of recognized species of *Ctenomys* that I report here is a tentative approximation, based on my own revisions of the type specimens in the British Museum and on the results of chromosomal investigations I undertook with P. Kiblisky (Reig and Kiblisky, 1969) and that have been continued by Gallardo (1979) and others.

[d] For the genus *Makalata* and the validation of *Sphiggurus*, see Husson (1978).

Superfamily Erethizodontoidea
Family Erethizodontidae (3, 7)
Subfamily Erethizondontinae
Genus *Coendu* (2)
Genus *Echinoprocta* (1)
Genus *Sphiggurus* (4) [d]

Superfamily Cavioidea
Family Caviidae (5, 15)
Subfamily Caviinae
Genus *Cavia* (6)
Genus *Microcavia* (3)
Genus *Galea* (3)
Genus *Kerodon* (1)
Subfamily Dolichotinae
Genus *Dolichotis* (2)
Family Hydrochoeridae (1, 1)
Genus *Hydrochoeris* (1)

Superfamily Chinchilloidea
Family Chinchillidae (3, 6)
Subfamily Chinchillinae
Genus *Chinchilla* (2)
Genus *Lagidium* (3)
Subfamily Lagostominae
Genus *Lagostomus* (1)
Family Dinomyidae (1, 1)
Genus *Dinomys* (1)

Caviomorpha *incertae sedis*
Family Dasyproctidae (2, 9)
Genus *Dasyprocta* (7)
Genus *Myoprocta* (2)
Family Agoutidae (2, 2)
Genus *Agouti* (1)
Genus *Stictomys* (1)

---

[e] *Proechimys* is a genus whose species are difficult to recognize at the morphological level, hence the lumping tendencies in most of the revision of species of this genus (Hershkovitz, 1948; Moojen, 1948). The present assessment of its number of species is approximate and is based mostly on the results of modern chromosomal work (Gardner and Patton, 1976; Reig et al., 1979a, 1980).

[f] *Cliomys laticeps* is indeed a distinctive species, but after revizing the types and additional material of this species and of *Euryzygomatomys spinosus* (including *catellus*) in the British Museum, I could not find reasons to recognize more than two species of a single genus, i.e. *Euryzygomatomys*, in that material. The status of *Hypudaeus guiara* Brandt, referred to *Euryzygomatomys* by most modern authors, is dubious. It is probably a junior synonym of *spinosus* (G. Fischer).

[g] For the priority of *Trichomys* over *Cercomys*, see Petter (1973).

[h] I follow the traditional view by placing the dactylomyines as a subfamily of the Echimyidae. However I have serious doubts that this is the best course of action. I believe that the alternative view that the Dactylomyinae are better placed as a subfamily of the family Capromyidae must be taken seriously into account.

[i] In the assignment of the Chaetomyinae to the Echimyidae, I follow Woods (1982).

[j] Needless to say, I do not consider here the West Indian capromyids. For the validation of the Myocastoridae as a full family different from Capromyidae, I follow the recent anatomical results of Woods and Howland (1979). I recognize in the Capromyidae three subfamilies: Capromyinae (*Capromys, Mysateles, Mesocapromys, Geocapromys;* see Kratochvíl, Rodríguez, and Barus, 1978), Plagiodontinae (*Plagiodontia, Isobolodon*), and Heteropsomyinae (*Heteropsomys, Boromys, Brotomys*, etc.) (see Woods, 1982).

Vuilleumier's (see Vuilleumier, 1979; Vuilleumier and Simberloff, 1980) concept of páramo.

Puna rodent fauna: the 45 species of sigmodontine cricetid and caviomorph rodents reported by different authors to occur in localities belonging to the puna biome. Where explicit reference that a given species belongs to the puna biome is missing, altitudinal and geographic location data in the area of the puna from 3700 m and above were taken into account to decide whether that species belonged to the puna biome.

Páramo rodent fauna: the 27 species of sigmodontine cricetid, caviomorph, or sciurid rodents that have been reported by different authors to occur in one or more of the different recognized páramos (for an up-to-date account of the location of 23 ornithologically explored páramos, see Vuilleumier and Simberloff, 1980). Where explicit reference for a given species to belong to one or more páramos is missing, geographic location and an altitude from 3000 m and above were used to decide whether the species belonged to one or more páramos. No interpáramo distributional differences have been considered in this study. A few of the species reported here as páramo species may well be also members of the ecotone páramo-montane forest, or species preferentially living in the high-altitude montane forest below the páramos properly, but showing marginal distribution in the latter.

Sigmodontine cricetid rodents: any rodent species, genus, or tribe belonging to the subfamily Sigmodontinae of the family Cricetidae as defined by Reig (1980). An account of the tribal and generic classification of the South American living Sigmodontinae is given in Table 16–1.

Neotomine cricetid rodents: any rodent species, genus, or tribe belonging to the mostly North American subfamily Neotominae of the family Cricetidae, as defined by Reig (1980).

Caviomorph rodents: any rodent species, genus, or taxon of familial level belonging to the infraorder Caviomorpha. The classification of the South American living representatives of the caviomorph rodents is given in Table 16–2.

Sciurid rodents: any rodent species or genus belonging to the squirrel family Sciuridae. Reference to sciurid rodents in this chapter covers only members of the arboreal tribe Sciurini, as defined by Moore (1959).

Protopuna: region extending approximately in the area of the present puna biome in Miocene-Pliocene times, before the Plio-Pleistocene uplift, which raised the puna surface to its present level.

MATERIAL AND METHODS

The evidence on which this chapter is based comes mainly from the distributional data of the relevant species recorded in the literature. Although the systematic lists of South American rodents such as those of Cabrera (1961), Ellerman (1941), Gyldenstolpe (1932), and Tate (1932a–h, 1935) have been most useful in compiling preliminary systematic and distributional data, all the published revisionary monographs for the different taxa have been consulted (Oliveira Pinto, 1931; Hershkovitz, 1944–1966b; Moojen, 1948, Cabrera, 1953; Moore, 1959; Ojasti, 1972; etc.), as well as most of the original publications of authors such as J. A. Allen, H. E. Anthony, W. H. Osgood, and O. Thomas. Additionally, faunal revisions for different South American countries such as those of Moojen (1952), Osgood (1943b), Pine, Miller, and Schamberger (1979), Soukup (1961), and many others have been consulted to bring up to date the distribution of the species.

In addition, these distributional data have been evaluated, particularly in the case of the sigmodontine cricetids, octodontids, and echimyids, through the author's personal examination of specimens (including the majority of the type specimens) in the major European, North American, and South American museums. An important additional source of information for this evaluation was the recent karyological data, either reported by several authors (summary in Gardner and Patton, 1976) or obtained through the research of the author and his collaborators, part of which is still unpublished. The chromosomal information was critical in the assessment of the species I recognized in many of the most problematic genera.

These evaluations resulted in a provisional reassessment of the species in the different genera, the full report of which will be published by the author in the near future. A résumé of the living genera recognized, and of the number of their recognized species, is given in Table 16–1 for the Sigmodontinae, and in Table 16–2 for the Caviomorpha.

After the lists of the recognized taxa were compiled, all localities (including geographic location and altitude) where each of the species was reported to occur were recorded. These locality records allowed me to estimate N_g, the total number of known geographic occurrences of each analyzed taxon (species, genus, tribe, and so on). Obviously, the N_g of a given species may cover various geographic regions and altitudinal locations. Consequently, different components of the N_g of each species were scored in terms of six geographic distributional units and altitudinal classes.

The six geographic units (sections of the Andes defined as in B. B. Simpson, 1975) are as follows: (1) Northern Andean (NA): all localities, whatever their altitude, north of the Huancabamba deflection, including the Western Cordillera of Colombia and Ecuador, the Central Andes of Colombia and the Eastern Andes of Ecuador, the Perijá Mountains, the Sierra Nevada de Santa Marta, and the Mérida Andes; (2) North-Central Andean (NCA): all localities, whatever their altitude, situated in the Cordillera Occidental of Peru, the puna surface and the Cordillera Oriental, between the Huancabamba deflection in the north and 15° S in the south; (3) South-Central Andean (SCA): all localities, whatever their altitude, located in the Principal Cordillera, the puna surface and the Cordillera Oriental, in Peru, Bolivia, northern Chile, and northwest Argentina, between 15° S and 25° S; (4) North-Southern Andean (NSA): all sites, whatever their altitude, situated in the Principal Cordillera, the southern portion of the puna to which I add the Pampean Ranges (excluding the Sierras de Balcarce, del Tandil, and de La Ventana), in northern and central Argentina and in central Chile, between 25° S and 35° S; (5) South-Southern Andean (SSA): all sites, whatever their altitude, in the Coastal Cordillera of Chile, the southern portion of the principal Cordillera, and the Patagonian Cordillera, between 35° S and southern Tierra del Fuego. (6) Non-Andean (NOA): all localities, whatever their altitude, not belonging to any of the preceding five Andean units. The five altitudinal classes are below 1000 m, 1000–2000 m, 2000–3000 m, 3000–4000 m, and above 4000 m.

The data of the scoring of the N_g components of each species (i) for the six different geographic units, as well as for the five different altitudinal classes, were the basis for estimating the frequency (f) of the distribution of each corresponding taxon of supraspecific rank (T) in each of the units and classes (U), from the expression:

$$f_T \text{ in } U = \frac{\sum f_i \text{ in } U}{\sum N_{g_i}}$$

The f_T distributions allow us to estimate the placement of the areas of original differentiation (AOD), a concept introduced here to replace the rather vague concepts of "center of origin" and "dispersal center." An AOD is defined as "the geographic space within which a given taxon experienced the main differentiation (cladogenesis) of its component taxa of subordinate rank." An AOD is not precisely a center of origin in the sense that the original stock of a taxon that shows a given AOD may not have, and usually has not, originated in the same area where the taxon underwent its main differentiation. At the same time an AOD of a given taxon is here considered to be the area where a new taxon derived from the former originated. In this sense, in defining an AOD, I have adopted a modified version of the Henning-Brundin concept of center of origin (see Pielou, 1979). According to this modified version, the AOD of a given taxon has the followed characteristics: (1) it is the area reached by the ancestral stock of the taxon after its dispersal from the AOD of its parental taxon; (2) it is the area where the taxon split most actively into subordinate taxa, acquiring through this process its defining properties (i.e., the set of attributes that defines it as a taxon concept of its rank; see Reig, 1979); and (3) it is the area in which the taxon split into an apomorphic (derived) species, which dispersed into a new AOD to generate there a new sister taxon of the same rank.

Following a well-established biogeographic practice, I identified the AOD of a given taxon by plotting the ranges of its component species on a map, thus delimiting the area of major overlap in species distribution. These areas of congruence (*Arealkerne*; Reinig, 1950) were chosen as a first approximation to represent the probable AOD of the taxon under consideration (usually a taxon of family level for caviomorphs, but of tribal level for sigmodontine rodents). I decided on the location of the AOD of the given family or tribe after considering the range overlap of genera in the area, and the percentage of endemic genera. Thus, the procedure I used to define the AODs is rather similar to the way in which P. Müller established his "dispersal centers" (Müller, 1973), though my procedure is based on more quantitative evidence.

In assessing the criteria to define the AODs I endorse implicitly the traditional Wallacean view (see Vuilleumier, 1978) that local origination, dispersal, and local differentiation are the processes responsible, at least together with other processes, for the patterns of geographic distribution exhibited by living organisms. This endorsement does not necessarily mean a rejection of "vicarianism" (Croizat, Nelson, and Rosen, 1974; Rosen, 1975), which only in the extreme form of an exclusive school might be considered as an alternative opposed to the "dispersalist" view of those who advocate centers of origin. On the contrary, I am convinced that vicariance (in other words the splitting of taxa or biota by extrinsic barriers) is a compelling explanation for many geographic patterns of distribution. But I am also convinced, as is Pielou (1979), that the reality of species-rich centers supporting the idea of special areas of local differentiation is pervasive and that the observed "centers of origin" remain unexplained when one adopts an exclusive version of the vicariance school. I strongly believe that progress in building a cogent biogeographic theory demands the rejection both of exclusive school membership and of that odd sort of militant fundamentalism that unfortunately is hampering the necessary dialogue between members of different contemporary schools of biogeography.

THE MAJOR DIVERSITY PATTERN OF HIGH ANDEAN RODENTS

According to my analysis, the puna rodent fauna comprises 47 species belonging to 24 taxa of generic level (genera and subgenera) (Tables 16–3 and 16–5), whereas the páramo fauna consists of 27 species belonging to 15 different genera or subgenera (Tables 16–4 and 16–5).

I include in these totals every species whose occurrence has been demonstrated for either the puna or at least one of the páramos. However, several of the species considered are not restricted to puna or páramo habitats, either showing a broader ecological range or being more typical of montane forest habitats below the puna or the páramos. Therefore, I also estimated, both from the data in the literature and from personal experience, the number of species that occur only in the puna or páramo habitats, thus defined as the number of endemic species in each of the two biomes. After culling out the nonendemic species the number of species belonging strictly to the puna and páramo were 32 and 13, respectively.

In Table 16–3 and Table 16–4 I list species and in Table 16–5 I summarize the information for species diversity in each biome. It is evident from these data (a) that there are about twice as many species of rodents in the puna as in the páramos, (b) that there is more endemism at the species level in the puna (68%) than in the páramo (48%) rodent fauna, and (c) that the two faunas are quite different in species composition, Simpson's resemblance index $100(C/N_1)$ giving a value of 3.8 (i.e., only about 4% of páramo species [the smaller fauna] are present in the puna biome). If we take into account the fact that this figure is obtained by the presence of a single shared species, *Oryzomys* (*Microryzomys*) *minimus*, and that this species is restricted to northern puna localities, the lack of resemblance in species composition between the puna and páramo rodent faunas is even greater. In fact, the resemblance becomes nil if we compare the rodent fauna of the central and southern puna with the páramo rodent fauna.

As is to be expected, the resemblance between the puna and the páramo rodent faunas is greater if we compare the two faunas at the generic (genus and subgenus) level. But even so, Simpson's index of faunal resemblance is low (33.3, see Table 16–6), which means that 66.7% of the genera of páramo rodents are not found in the puna. The analysis at the generic level also shows a much greater richness in diversity in the puna (24 genera or subgenera) than in the páramo (15 genera or subgenera). It finally demonstrates a striking difference between the two faunas in percentage of endemism, as the páramo fauna shows no endemic genera or subgenera, whereas 6 (25%) of the 24 genera or subgenera of rodents of the puna are endemic, the endemic taxa being *Andinomys*, *Chroeomys*, *Chinchillula*, *Galenomys*, *Neotomys*, and *Punomys*.

The differences between the rodent faunas of the two biomes are even more evident if we compare the suprageneric classification of the species and genera of each fauna (Table 16–5, Fig. 16–2). The puna rodent fauna may be characterized as a phyllotine-caviomorph-akodontine assemblage, as 95.7% of the puna species belong to one of those taxa, the respective proportions being 44.7% phyllotines, 31.9% caviomorphs, and 19.1% akodontines. The remaining 4.3% of the puna rodent species belong to the Oryzomyini. By contrast, the páramo rodent fauna is mostly an oryzomyine fauna, with members of this tribe of sigmodontine cricetids amounting to 59.2% of the total species number. The remaining species are scattered among a greater number of suprageneric taxa, akodontines coming next with 14.8% of the species, followed by the caviomorphs and

TABLE 16–3. Rodent species living in puna sites in the Andes of South America.

Myomorpha	Caviomorpha
Cricetidae	Caviidae
Sigmodontinae	*Galea musteloides* [c]
Oryzomyini	*Cavia tschudii*
Oryzomys (Microryzomys) minutus [a]	*Microcavia niata*
Thomasomys gracilis [b, c]	*Microcavia shiptoni*
Akodontini	Octodontidae
Akodon (Akodon) albiventer	Octodontinae
Akodon (Akodon) andinus	*Octodontomys gliroides* [c]
Akodon (Akodon) boliviensis [c]	Ctenomyinae
Akodon (Akodon) pacificus	*Ctenomys budini*
Akodon (Akodon) puer	*Ctenomys frater*
Akodon (Chroeomys) jelskii	*Ctenomys leucodon*
Bolomys amoenus [d]	*Ctenomys lewisi*
Bolomys lactens	*Ctenomys opimus*
Oxymycterus paramensis [c]	*Ctenomys peruanus*
Phyllotini	Abrocomidae
Auliscomys boliviensis	*Abrocoma cinerea*
Auliscomys pictus	Chinchillidae
Auliscomys sublimis	*Chinchilla brevicaudata*
Andinomys edax	*Lagidium viscaccia*
Calomys lepidus	*Lagidium peruanum*
Calomys musculinus [c, e]	
Calomys sorellus	
Chinchillula sahamae	
Eligmodontia puerulus	
Galenomys garleppi	
Graomys domorum [c]	
Neotomys ebriosus	
Phyllotis andium [c]	
Phyllotis caprinus [c]	
Phyllotis darwini [c]	
Phyllotis definitus [c]	
Phyllotis magister [c]	
Phyllotis osilae [c]	
Phyllotis osgoodi [f]	
Phyllotis wolffsohni [c]	
Punomys lemminus [g]	

[a] Based on *Microryzomys minutus aurillus* Thomas, found at 4200 m in Department Cuzco (Thomas, 1926). *M. minutus altissimus*, found in Cerro de Pasco (Dept. Pasco, Osgood, 1943a) and in Dept. Ancash (Gardner and Patton, 1976) does not seem to reach above 3500 m.
[b] *Thomasomys gracilis* was originally described by Thomas (1917) from Machu Picchu, Cuzco, at 3600 m. It seems to be mostly an inhabitant of the montane forests of the eastern slopes of the puna that extends only secondarily to the puna proper.

[c] Species found in puna sites, but also, in some cases more typically, at lower altitudes.

[d] For the validity of *Bolomys* as a full genus, and for its species, see Table 16–1, fn. *g*.

[e] I restrict *Calomys musculinus* (including *cortensis*, which does not seem to deserve subspecies status) to populations of northwestern Argentina. The types and original series of Thomas of *musculinus* and *cortenis* in the British Museum are larger and distinctive in skull characters when compared to *murillus*, which is currently considered a subspecies of *musculinus*. Therefore, and pending a confirmation from the knowledge of the chromosomes of *musculinus* proper. I think the best course of action, given the present state of knowledge, is to place *musculinus* and *murillus* in two different species. *Calomys musculinus* has been reported from the type locality (Maimará, Jujuy) at 2300 m, and from lower altitudes in San Salvador de Jujuy and La Puntilla, near Tinogasta, Catamarca. Its occurrence in a puna site comes from another record in Santa Catalina (Jujuy), at 3802 or 4500 m (Thomas, 1920; Hershkovitz, 1962).

[f] For a recent reevaluation of *Phyllotis osgoodi* Mann 1950, see Spotorno (1976) and Walker et al. (1979).

[g] *Punomys lemminus* is hard to place in any of the recognized tribes of the Sigmodontinae (see Reig, 1980). I consider it certainly as a Sigmodontinae *incertae sedis* and place it here in the list of phyllotines for the sake of simplicity.

TABLE 16–4. Rodent species living in one or more of 23 páramos in the northern Andes of South America.

Myomorpha

Cricetidae

Sigmodontinae

Oryzomyini

Aepeomys vulcani [a]
Oryzomys (Oryzomys) pectoralis [b]
O. (Oryzomys) munchiquensis [c]
O. (Microryzomes) minutus
O. (Oligoryzomys) novus
Rhipidomys venustus [c]
Thomasomys aureus [c]
Thomasomys boeops
Thomasomys bombycinus
Thomasomys cineriventer [c]
Thomasomys laniger [c]
Thomasomys ladewi
Thomasomys melanochronos
Thomasomys paramorum
Thomasomys pyrronotus
Thomasomys rhoadsi [c]

Ichthyomyini

Neusticomys monticolis
Anotomys leander

Akodontini

Akodon mollis [c]
Akodon tolimae [c]
Akodon urichi [c]
Microxus bogotensis

Phyllotini

Phyllotis haggardi [c]

Sigmodontini

Sigmodon inopinatus

Sciuromorpha

Sciuridae

Sciurinae

Syntheosciurus granatensis [c, d]

Caviomorpha

Caviidae

Cavia porcellus [c, e]

Agoutidae

Stictomys taczanowski [c]

---

[a] This form, given only subspecies status in some papers under the species *Aepeomys luggens*, is here provisionally recognized as a full species on the basis of its geographic differentiation.

[b] This is quite probably the name to apply to the $2n = 66$, $FN = 94$ species of the *Oryzomys albigularis* complex, described by Gardner and Patton (1976).

[c] Species found in páramo localities, but also, in some cases more typically, at lower altitudes.

[d] For the inclusion of *granatensis* in *Syntheosciurus*, see Moore (1959). The occurrence of this species in páramo localities is based on reported occurrences of *S. g. quindianus*, *S. g. söderströmi*, and, more doubtfully, *S. g. meridensis*.

[e] *Cavia porcellus anolaimae*, from the region of Bogotá (Allen, 1916).

TABLE 16–5. Total number of species and number of endemic species distributed among suprageneric taxa of high Andean rodents.

Suprageneric taxon	Puna			Páramo		
	Species (*N*)	Endemic species (*N*)	Endemism (%)	Species (*N*)	Endemic species (*N*)	Endemism (%)
Sigmodontinae						
Oryzomyini	2	–	–	16	9	56.2
Ichthyomyini	–	–	–	2	2	100.0
Akodontini	9	7	77.8	4	1	25.0
Phyllotini	21	12	57.1	1	–	–
Sigmodontini	–	–	–	1	1	100.0
Total Sigmodontinae	32	19	59.4	24	13	54.2
Caviomorpha						
Octodontidae	7	6	85.7	–	–	–
Caviidae	4	3	75.0	1	–	–
Abrocomidae	1	1	100.0	–	–	–
Chinchillidae	3	3	100.0	–	–	–
Agoutidae	–	–	–	1	–	–
Total Caviomorpha	15	13	86.7	2	–	–
Sciuromorpha						–
Sciuridae	–	–	–	1	–	
Total rodent species	47	32	68.1	27	13	48.1

TABLE 16–6. Number of living rodent genera in the puna and páramo biomes of the high Andes of South America.

	In puna and páramo N	In páramo N_1	In puna N_2	Common to both C	Index of faunal resemblance $100(C/N_1)$
Total number of genera	39	15	24	5	33.3
Number of endemic genera	6	–	6	–	–

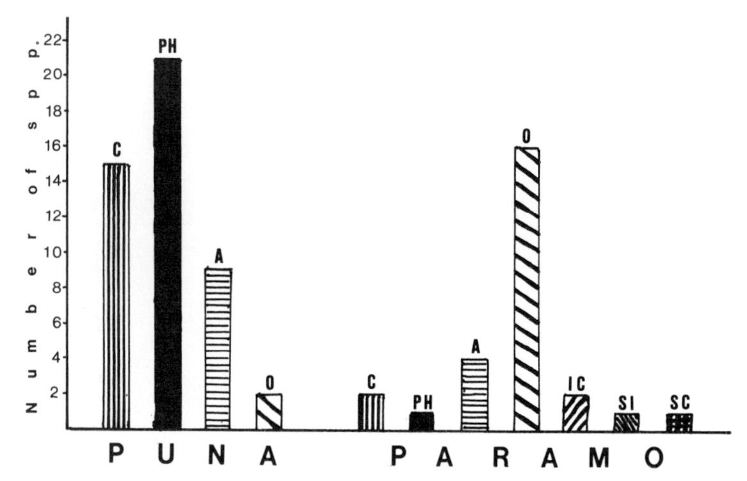

FIG. 16–2. Bar diagram of species numbers in the different taxa of puna and páramo rodent fauna. C, Caviomorpha; PH, Phyllotini; A, Akodontini; O, Oryzomyini; IC, Ichthyomyini; SI, Sigmodontini; SC, Sciuridae.

ichthyomyines, each with 7.4%, and next by the phyllotines, Sigmodontini, and Sciuridae, each with 3.7% of the species. The striking difference between puna and páramo faunas lies in the inverted proportion of phyllotines and oryzomyines in each of the faunas, in the absolute lack of ichthyomyines, Sigmodontini, sciurids, and agoutids in the puna, and of octodontids and chinchillids in the páramos. It must also be mentioned that the single representative of the phyllotines in the páramos is restricted to the southern páramos of Ecuador. Finally, only one taxon of the Caviomorpha, the Caviidae, is shared by the puna and páramos faunas, but the single representative of this family in the páramos is not a typical páramo species.

The striking difference in the diversity of patterns of the páramo and puna rodent faunas suggests that they are the result of different historical processes, with in situ differentiation playing an important role in the shaping of their present features. In the following sections I shall attempt to reconstruct the evolutionary and biogeographic histories of the involved taxa, as a necessary foundation for a description of a probable scenario of the history of the páramo and puna rodent faunas as a whole.

THE PROBABLE HISTORY OF ANDEAN CAVIOMORPH RODENTS

Table 16–7 shows that most (68.1%) of the extant South American caviomorphs are non-Andean in distribution, and that only 31.9% now live in the Andes. A reverse situation holds for the living representatives of the Sigmodontinae, 71.6% of which are distributed in Andean localities and only 28.4% in non-Andean regions.

I contend that the striking differences in the present distribution of South American caviomorph and sigmodontine rodents reflect different biogeographic histories. I suggest that the caviomorphs originally evolved in lowland and forest biomes, and secondarily invaded Andean zones. In the case of the Sigmodontinae, on the other hand, the data suggest that the Andes were their main region of differentiation and that these rodents reached eastern lowland life zones from their Andean areas of original differentiation during different episodes. This interpretation is corroborated by an intrataxon analysis of geographic ranges of the species, and the study of endemism of species and genera.

The Overall Pattern of Caviomorph Origins and Their Colonization of the High Andes

Caviomorph rodents are an ancient immigrant stock in the South American fauna. The evidence is conclusive that they had no possible ancestors among members of the ancient autochthonous South American mammalian stock, which consisted of marsupials, xenarthrans, and meridiungulates plus three families of dubious classification, the Necrolestidae, Groeberiidae, and Microtragulidae (Reig, 1981c). Caviomorphs are, in fact, typical members of the source faunal unit that I have called the South American allochthonous stock (Reig, 1981c), consisting of an array of mammalian taxa that colonized South America from elsewhere during times when the continent was geographically isolated.

The first representatives of the Caviomorpha occur in strata of the Deseadan (i.e., Lower Oligocene) of Patagonia and Bolivia, where they suddenly appeared, and differentiated in at least four families. Recently (Reig, 1981c), I con-

TABLE 16–7 Partitioning of the distribution of South American representatives of sigmodontine tribes and caviomorph families in Andean and non-Andean localities.

Taxon	Number of genera	Number of species	Andean		Non-Andean	
			N_g	f_T	N_g	f_T
Caviomorpha						
Octodontidae	6	52	130	0.647	71	0.353
Abrocomidae	1	2	15	1.000	—	—
Echimyidae	15	65	10	0.039	244	0.961
Chinchillidae	3	6	30	0.600	20	0.400
Myocastoridae	1	1	—	—	35	1.000
Caviidae	5	15	87	0.439	111	0.561
Hydrochoeridae	1	1	—	—	38	1.000
Dinomyidae	1	1	4	0.800	1	0.200
Dasyprocticae	2	10	16	0.145	94	0.856
Agoutidae	2	2	10	0.263	28	0.737
Erethizodontidae	3	7	12	0.308	27	0.692
Total	40	161	314	0.319	669	0.681
Sigmodontinae						
Oryzomyini	13	110	298	0.685	137	0.315
Ichthyomyini	4	8	9	0.692	4	0.308
Akodontini	10	62	302	0.712	128	0.298
Scapteromyini	3	6	—	—	45	1.000
Wiedomyini	1	1	—	—	4	1.000
Phyllotini	14	45	724	0.809	171	0.191
Sigmodontini	2	10	7	0.205	27	0.794
Genera *incl. sedis*						
Zygodontomys	1	4	5	0.250	15	0.750
Punomys	1	1	2	1.000	—	—
Rhagomys	1	1	—	—	1	1.000
Abrawayaomys	1	1	—	—	—	—
Total	51	249	1347	0.716	533	0.284

Note: N_g is the number of geographic occurrences (localities) for a given taxon in either Andean or non-Andean partitions. f_T is the frequency of the given taxon in any of the two partitions, as calculated from the former absolute numbers (see Materials and Methods section).

cluded that the more reasonable hypothesis about their origins is that they arrived in South America from Africa by transatlantic rafting, as advocated, among others, by Hoffstetter (1975, 1976), Lavocat (1971, 1973), and Raven and Axelrod (1975). This hypothesis is supported by the fact that the closest living and fossil relatives of caviomorphs are the African rodents of the infraclass Phiomorpha. The relationships between caviomorphs and phiomorphs are so close and multiple, that in spite of opposing views of some still active dissidents (see, for instance, Wood, 1974, 1975, and, for a recent version of a similar view, Tasser Hussain, de Bruijn, and Leinders, 1978), little doubt remains that the two groups represent sister taxa of a common origin and that the caviomorphs were directly derived from the phiomorphs.

The fact that caviomorphs are already differentiated to a certain extent in the Early Oligocene obliges one to date their arrival in South America further back in time, Middle or Upper Eocene being a reasonable estimate. Where did the original implantation of the immigrant ancestral caviomorph stock take place? In the Early Tertiary, the shortest distance between South America and Africa was between northwestern Africa and northeastern Brazil (see, for instance, Cox, 1974, fig. 9). This location is a plausible one for a successful implantation of transatlantic rafters coming from Africa along the east–west Benguela surface water circulation, which dominated the equatorial Atlantic ocean since the Upper Cretaceous (Berggren and Hollister, 1974).

Seemingly, the early caviomorphs came from lowland tropical phiomorphs, and there are reasons to believe that the former first expanded and diversified in tropical humid lowland forest biomes, which were much more extensive in South America in Early and Middle Tertiary times (Sarmiento, 1976; Solbrig, 1976). A retraction of the forest biomes northward occurred in correlation with an important episode of Andean uplift which began in Early Miocene times (Groeber, 1951; Harrington, 1956; Jenks, 1956; Lohmann, 1970). This phase of Andean orogeny was the more probable cause of the outcome of drier and cooler climates south of the region of the present Amazon Basin that resulted in regions of mesic and arid open vegetation to which caviomorphs gradually adapted. Adaptations to aridity in the monte biome during Miocene and Pliocene times (Mares, 1975) may have functioned as preadaptations for later colonizing success of the puna when the altiplano started to uplift, reaching its present height after the Middle Pliocene (Ahlfeld, 1970). In the region of the Amazon Basin and northward of it, where the conditions of tropical humidity were not so drastically affected, forest caviomorphs gradually invaded mountain forests of the uprising Andean slopes. A few of them became tolerant of higher altitudes and eventually adjusted to páramo conditions.

The Echimyids as Ancestral Octodontoids

The early caviomorphs that dispersed from northeastern Brazil and radiated in the extensive humid lowland forest biomes of Eocene and Oligocene times are likely to have been the Echimyidae. This family is represented in the Early Oligocene of Patagonia by the extinct subfamily Adelphomyinae (Patterson and Pascual, 1968), and in Bolivia by *Sallamys*, closely related to the living taxa (Patterson and Wood, 1982). I believe that the Patagonian adelphomyines reached the humid forested regions of what is now Patagonia from a Brazilian area of original differentiation of the Echimyidae. The echimyids are, I contend, the most primitive of the living and fossil caviomorphs. The octodontids, currently considered the most primitive group of the Octodontoidea, are in my opinion a derived sister taxon of the Echimyidae. That the latter first differentiated in the Amazonian region is indicated by the overwhelmingly greater number of species and genera of this family now living in the fauna of the Amazon Basin. That the Echimyidae are the most primitive octodontoids is supported by the fact that living genera of this family now belonging to the Amazonian fauna, for instance, *Mesomys* and *Lonchothrix*, show brachyodont and pentalophodont molar teeth of the type one would expect in an ancestral caviomorph. The greater differentiation of echimyids in the living fauna (15 genera and 65 species; see Table 16–7), further supports the idea that they are the older South American caviomorphs.

Most of the echimyids inhabit tropical and subtropical forests, and many of them show arboreal adaptations. A few of them are also adapted to open mesic environments, both as terrestrial forms (*Trichomys*) or as quasi-subterranean forms (*Carterodon*, *Euryzygomatomys*). They are probably the survivors of a group of fossils that populated steppe and semiarid biomes in the Upper Miocene and Pliocene (*Pattersonomys*, *Trichomys*, *Eumysops*) and that were displaced to the north by the increasing expansion of the octodontids. None of the echimyids, however, seems to have become adapted to strictly arid biomes. They did reach Andean forest biomes, but only 4% of their living representatives did so (Table 16–7), and they never attain high altitudes.

Octodontids and Abrocomids

The more important caviomorph group in the present puna biome are the octodontids. Apart from their seven (13.5%) high Andean representative species, the remaining octodontid species (86.5%) inhabit lower-altitude Andean sites and lowland mesic and arid habitats (Table 16–7). The living octodontids are usually divided into two subfamilies, the more generalized Octodontinae and the more specialized Ctenomyinae. The Octodontinae include mostly terrestrial, surface-dwelling genera (*Octodon*, *Octomys*, *Octodontomys*), but also quasi-subterranean (*Pithanotomys*; see footnote *b* in Table 16–2) and fully subterranean (*Spalacopus*) genera. The living ctenomyines comprise only species fully adapted to the subterranean habitat, all of them in the genus *Ctenomys*.

The octodontids are closely related to the echimyids. The former probably represent an apomorphic sister group of the latter that adapted to the more arid environments of southern South America, which began extending in the Miocene. It is now widely accepted (Pascual, Pisano, and Ortega, 1965; Pascual, 1967) that octodontids orginated in the austral regions of South America. Although putative early octodontids have been reported since the Deseadan (Wood and Patterson, 1959; Patterson and Wood, 1982), unquestionable members of this family are known since the Early Miocene. *Chasicomys* (Pascual, 1967) may be considered an early representative of the Octodontinae, whereas the Ctenomyinae appear to be first represented in the genus *Proctenomys* (Pascual et al., 1965) of the Earliest Pliocene or Late Miocene of southern Buenos Aires Province. In the Upper Pliocene of Bolivia ctenomyines were represented by two species referred to *Praectenomys* (Villarroel, 1975). About 65% of the living octodontids are now distributed in southern or south-central Andean localities (Table 16–7). This pattern suggests that octodontids, although probably first evolving from, and in, lowland mesic to semiarid biomes of southern South America, became increasingly tied to proto-Andean arid zones where they experienced their main differentiation. Here they probably acquired preadaptations that eventually permitted them to thrive successfully in the puna habitats, once the altiplano reached its present height.

Abrocomids are also members of the puna rodent fauna, though they are also represented in lower south-central and southern Andean locations. Taking together their present ranges, 100% of their living representatives are Andean in distribution (Table 16–7). However, in the Pliocene (Rovereto, 1914) and in the Pleistocene (unpublished remains from the Lower Pleistocene of the Chapadmalal region in the Museum of Mar del Plata), they were represented both in deposits close to the Andes and in the pampa region. Abrocomids are so closely related to octodontids that it has been suggested that they may represent merely a subfamily of the latter (Pascual, 1967, p. 269 n.). In any case, they are a distinct offshoot of the octodontoids that evolved during Upper Miocene or Early Pliocene times. As the octodontids have, they may also have evolved in mesic to arid lowland biomes, to become actually further restricted to Andean semiarid to arid zones, where they may have acquired adaptations for successful life in the puna.

Caviids and Chinchillids

Caviids and chinchillids are two other important groups of puna rodents. They are present in the fossil record since the Oligocene, and they may have had an origin independent of that of the octodontoid complex. Recent macromolecular investigations indicate that caviids, chinchillids, dasyproctids, and erethizodontids compose a clade based on protein affinities that is clearly separated from the Octodontoidea (Sarich and Cronin, 1980). Chinchillids predominate (60%) in the living fauna of Andean localities, whereas caviids are more frequent in lowland, mostly mesic or semiarid environments (56%), being nevertheless well represented (44%) in Andean sites.

Ancestors of the cavioid-chinchilloid complex probably adapted to unforested habitats since the beginning of their origination from their hypothetical northeastern Brazilian place of implantation, later differentiated in more southern mesic and arid biomes, and eventually became increasingly connected with the Andes. By Miocene times some of them adapted to arid Andean valleys from where they were uplifted to the present puna. This hypothesis may well fit the case of the chinchillids *Lagidium* and *Chinchilla*, which are Andean endemics. This history of caviids is less clear because the three genera found in the puna, *Cavia*, *Galea*, and *Microcavia*, are widespread in distribution, being now also common in lowland mesic biomes, and in the case of *Cavia* also occurring in forested habitats. *Galea* and *Microcavia* evolved endemic species in the puna, but the species of *Cavia* found there is not endemic, being also common at lower altitudes. It is probable that *Cavia*, a domestic animal in most Andean cultures, was brought by early Amerindians to some of its present Andean habitats (see Chap. 10). The euryoecious genus *Cavia* also expanded to the north, becoming adapted, probably secondarily, to montane northern Andean forests to occasionally reach some of the páramos.

Agoutids

Agoutids are totally absent from the puna, but one species is well adapted to some of the páramos. They are unknown in the Tertiary fossil record, their evolution and phylogeny being rather obscure. They are considered to be relatives of dasyproctids, a family that is predominantly non-Andean (86%) in distribution and that typically lives in forested habitats. Living agoutids are also mostly distributed in forested lowland habitats (73.7%). The monotypic genus *Stictomys* is typical of the northern Andes, where it inhabits humid montane forests in Ecuador, Colombia, and Venezuela, reaching up to the páramo of Pichincha (*S. taczanowski andina*) and also probably to the páramo of Mérida (*S. taczanowski sierrae*). *Stictomys* probably differentiated in a northern Andean area in adaptation to montane forests and spread secondarily into some of the páramos.

Overview of the History of High Andean Caviomorphs

The emerging picture of the history of caviomorphs from the puna is clearly different from that of the páramo caviomorphs.

Puna caviomorphs comprise taxa that originally differentiated in mesic and semiarid lowlands of southern South America east of the Andes. The more proximate puna caviomorph ancestors probably evolved adaptations to aridity in the monte biome and in the arid valleys of northern regions of the southern Andes and in the southern regions of the central Andes, during Miocene-Pliocene times. They reached present heights of the puna with the uplift of the altiplano that occurred since the Middle Pliocene.

The few caviomorphs now represented in the páramos have evolved most likely from humid lowland forest dwellers from northern South America. Their immediate ancestors became adapted to high montane northern Andean forests, from where they eventually reached the páramos in Plio-Pleistocene times, when the main Andean uplift occurred.

THE PROBABLE HISTORY OF ANDEAN CRICETID RODENTS

Contrasting Views on the Antiquity and Place of Differentiation of the South American Cricetid Rodents

Most of the South American cricetid rodents belong to an assemblage of genera diversely grouped under the names of "hesperomyines," "complex penis cricetines," "South American cricetines," or "Sigmodontini." In a recent paper (Reig, 1980), I gave arguments that support the view that this group of genera represents a subfamily of its own, for which the more appropriate name is Sigmodontinae. At the same time, I attempted to demonstrate that the genera referred to this subfamily make up a cohesive natural taxon distinct from other cricetid groups in several character states, not only the presence of a complex penis. In fact, the genera of the Sigmodontinae are closely linked to each other in the morphology of several organ systems, including penis morphology, myology, digestive system, male accessory sexual glands, hair structure, and tooth and skull anatomy (Reig, 1980, 1981c; see also new evidence supporting this view in Carleton, 1980; Voss and Linzey, 1981). Their close relationship is also supported by chromosomal (Gardner and Patton, 1976; Pearson and Patton, 1976) and by ectoparasite data (Wenzel and Tipton, 1966).

As regards the antiquity of the Sigmodontinae in South America, and the place of their original differentiation, three main views are now under discussion. The classical theory, as typically advocated by Baskin (1978), Patterson and Pascual (1972 and earlier), and G. G. Simpson (1950, 1969, 1980), holds that these mice are recent invaders in South America. They differentiated in tropical North America into most of their living diversity before spreading to South America by overland dispersal after the Panamanian land bridge was established by late Pliocene times (i.e., according to modern age determinations, around three million years B.P.; see Marshall et al., 1979). The main arguments in support of this view are, first, that fossil cricetid rodents are known in South America only since the Chapadmalalan (Upper Pliocene), and second, that phyllotines and sigmodontines are allegedly present in the Miocene and Early Pliocene, respectively, of North America (Baskin, 1978; Jacobs and Lindsay, 1981).

Disregarding some earlier occasional advocaters, the second view was originally proposed by Hershkovitz (1966a, 1972), later supported by Savage (1974), and developed more recently by the author (Reig, 1975, 1978, 1980, 1981c). All three authors hold the view that the pattern of extensive endemic differentiation of the South American cricetids obliges one to assume that they diversified locally from an invading ancestor that entered South America by Miocene times. My own version is that the Sigmodontinae are members of an ancient source unit of the present South American fauna: the South American allochthonous stock (Reig, 1981c). They evolved most of their genera and species on this continent, during different episodes of local differentiation. The original North American ancestor entered South America through its northwestern corner by overwater dispersal prior to the establishment of the Panamanian bridge, most probably in the Early Miocene. I also argued that the primitiveness of the more generalized living Sigmodontinae, namely the Oryzomyini, makes it necessary to postulate that they are the direct descendants of the Oligocene North American Eucricetodontinae (Martin, 1980), whereas the Neotomyinae probably originated from a different cricetid fossil stock and differentiated in the northern continent. Additionally, I demonstrated (Reig, 1978) that fossil cricetids occurred in South America earlier than the Chapadmalalan, in strata of Montehermosan age dated some four million years B.P. (Marshall et al., 1979), and that the Montehermosan and Chapadmalalan fossil mice belonged to highly evolved taxa, suggesting a much earlier presence of the group in South America. I finally claimed that the presence of phyllotines in the Miocene of North America has not been demonstrated clearly, because the alleged North American representative of *Calomys* reported by Baskin (1978) most probably does not belong to this genus, but to a typical neotomine (*Bensonomys*) showing some features in the molar teeth that converge with *Calomys*.

Recently, Marshall (1979) presented a third theory on the antiquity and origins of the South American cricetids, which can in some ways be regarded as a compromise between the other two. According to Marshall (1979), the Sigmodontinae (called Sigmodontini by him) evolved in North America before seven million years B.P. (i.e., before the Late Hemphillian, considered now to be Upper Miocene) and reached South America by waif dispersal from Central America about six million years ago, i.e., in the Upper Miocene. This date was inferred from the record of a lowering of the sea level at those times, which reduced the width of the marine barrier in the Bolívar Trough, thus facilitating overwater dispersal. This date is also indirectly indicated by the first occurrence of *Cyonasua* in South America, which may have entered at the same time. The first invading sigmodontines were sylvan, and they differentiated into pastoral forms in the savannas of northern South America. These pastoral cricetids later spead southward along the eastern foothills of the Andes when savanna biomes expanded in Lower Pliocene times, to eventually reach the Argentine pampas 3.5–4 million years ago, when their fossils are preserved in the Monte Hermoso formation.

I believe that the first theory has severe shortcomings because it cannot convincingly explain the hierarchy and extent of diversification and the high degree of endemism of the South American sigmodontines. Significantly, its proponents have not had firsthand experience with living cricetid rodents; they are mostly paleontologists who are too exclusively attached to inductive reasoning based solely on the fossil record. Curiously enough, one can argue that the fossil record does not provide firm support for their views, because it lacks convincing evidence of the past diversification of sigmodontines in tropical North America, and because the Pliocene akodontines and phyllotines found in Montehermosan and Chapadmalalan deposits are of advanced nature.

Marshall's theory offers evolutionists a more reasonable explanation, because it gives room to account for the endemism and progressive diversification of these rodents and it is more in tune with the fossil and geologic data. However, it can be refuted on two important grounds: first, the evidence suggesting that the major episodes of sigmodontine diversification occurred in different regions of the Andean chain, and second, the evidence suggesting that akodontines and phyllotines dispersed to the pampean region from central and southern areas of the Andes, which contradicts the view that pastoral sigmodontines (i.e., various akodontines and most of the phyllotines) evolved in the savannas of northern South America to disperse southward through a savanna-grassland corridor along the eastern foothills of the Andes. Additionally, it does not properly account for the geologic time necessary for the extensive and graded diversification of the different groups of the Sigmodontinae to have occurred.

In the next sections I shall try to further support the second theory, providing new evidence from the analysis of the range and areas of probable differentiation of the main sigmodontine tribes.

Diversity and Major Distributional Patterns of South American Sigmodontinae

The Sigmodontinae are extensively and complexly diversified in South America. The 51 genera and 249 species (Table 16–1) recognized for this continent can be grouped in seven different tribes (Reig, 1981c): Oryzomyini, Ichthyomyini, Akodontini, Scapteromyini, Wiedomyini, Phyllotini, and Sigmodontini. These tribes differ greatly in diversity, as 87.1% of the living species belong to one of three tribes: the Oryzomyini (44.2%), the Akodontini (24.9%), or the Phyllotini (18.1%). This pattern strongly suggests that the differentiation of the Sigmodontinae took place in South America through three main cladogenetic episodes.

Of the three numerically dominant tribes, the Oryzomyini are the most primitive in dental and skull characteristics, being closely allied to the Oligocene Eucricetodontinae (Reig, 1977, 1980, 1981c); they include mainly sylvan omnivorous to insectivorous life forms. The Akodontini are probably oryzomyine derivatives adapted to more open lands. Though many akodontines retained a mainly omnivorous-insectivorous diet, a significant portion of their genera (*Notiomys*, *Oxymycterus*, *Lenoxus*) specialized toward a predominantly animal diet. The phyllotines comprise mostly herbivorous forms inhabiting unforested lands, including a considerable number of typically grassland-dwelling genera and species. They are more advanced than the akodontines in possessing several derived character states in their dentition, skull morphology, and digestive system (Vorontzov, 1967). They may be derived from the akodontines or directly from an oryzomyine stock. Whatever the case, the morphological, chromosomal (Gardner and Patton, 1976), and distributional evidence clearly indicates that these tribes are closely related to each other and that they form three well-defined clades. The oryzomyines represent certainly the ancestral group, whereas the akodontines and phyllotines are apomorphic, derived taxa.

All three major tribes of the Sigmodontinae clearly have a predominantly Andean distribution in the present fauna. The f_T analysis summarized in Table 16–7 shows that 68.5% of the known occurrences of oryzomyines are distributed in Andean sites; the percentage for the akodontines is 71.2% and that for the phyllotines 80.9%. This strongly suggests that the cladogenetic episodes responsible for the differentiation of these three tribes occurred in the Andes.

There is also strong evidence supporting the view that each of these three cladogenetic episodes occurred in a different Andean region. The determination of these regions (the AODs; see Materials and Methods section) is based on the analysis of the f_T distributional partitioning in each of the Andean geographic units defined earlier, and in the location of the areas of major generic richness and of major endemism for each tribe. These analyses are made in detail in the next sections.

TABLE 16–8. Partitioning of the distribution of the mainly Andean tribes of South American Sigmodontinae into different Andean units and non-Andean localities.

	Total	Non-Andean		NA		NCA		SCA		NSA		SSA	
Tribe	N_g	N_g	f_T	N_g	f_T	N_g	f_T	N_g	f_T	N_g	f_T	N_g	f_T
Oryzomyini	435	137	0.315	185	0.425	62	0.143	39	0.090	4	0.009	8	0.018
Phyllotini	895	171	0.191	28	0.031	147	0.194	346	0.387	103	0.115	73	0.081
Akodontini	430	128	0.298	41	0.095	23	0.053	63	0.145	89	0.207	86	0.200
Ichthyomyini	13	4	0.307	8	0.651	1	0.077	–	–	–	–	–	–

Note: See Table 16–7 and Materials and Methods section for explanation of abbreviations.

TABLE 16–9. Distribution of the oryzomyine genera and subgenera in Andean units and in non-Andean localities.

Genera and subgenera	Non-Andean	NA	NCA	SCA	NSA	SSA
Oryzomys s.s.	+	+	+	+	—	—
Oligoryzomys	+	+	+	+	+	+
Microryzomys	—	+	+	—	—	—
Melanomys	+	+	—	+	—	—
Macrororyzomys	—	+	—	—	—	—
Sigmodontomys	—	+	—	—	—	—
Thomasomys	+	+	+	+	—	—
Oecomys	+	+	+	+	—	—
Rhipidomys	+	+	+	+	—	—
Nectomys	+	+	+	+	—	—
Neacomys	+	+	+	—	—	—
Delomys	+	—	—	—	—	—
Wilfredomys	+	—	—	—	—	—
Aepeomys	—	+	—	—	—	—
Nesoryzomys	+	—	—	—	—	—
Phaenomys	+	—	—	—	—	—
Chilomys	—	+	—	—	—	—
Scolomys	—	+	—	—	—	—
Total number per unit	12	14	8	7	1	1
Number endemics	4	5	—	—	—	—

Note: + indicates presence; — indicates absence. See Table 16–7, Figs. 16–3 and 16–4 and Materials and Methods section for explanations of abbreviations.

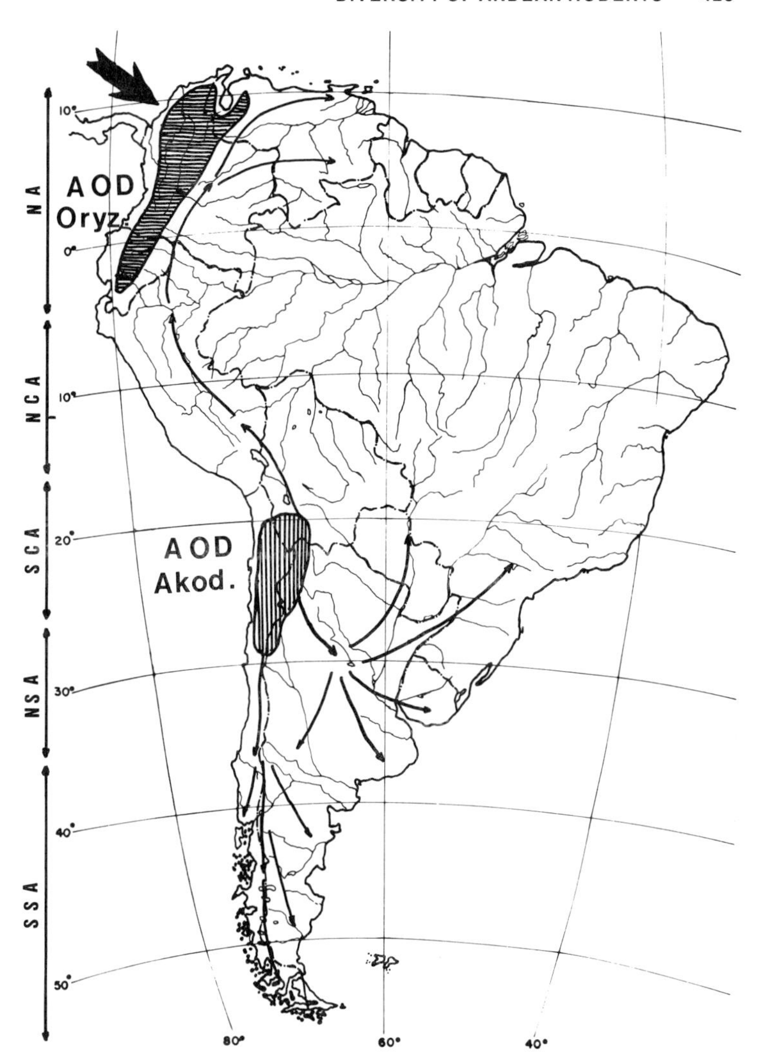

FIG. 16–3. Areas of original differentiation (AOD) of the tribes Oryzomyini (Oryz.) and Akodontini (Akod.) of South American cricetid rodents. Arrowed lines represent the suggested major lines of dispersal and secondary differentiation of the Akodontini, as explained in the text. Limits of the conventional Andean units are indicated on the left, as defined in Materials and Methods.

The Oryzomyini, the Ichthyomyini and Zygodontomys

Of the 435 localities in my analysis where oryzomyine species occur, 185, representing an f_T of 0.425, are located in the North Andean unit (see Materials and Methods section). This is the highest frequency of occurrence in a geographic unit of members of this taxon and is by far the highest frequency in all Andean units (Table 16–8). Additionally, the Northern Andean region is the area of common occurrence of the greatest number of genera and subgenera of the Oryzomyini, with 14 taxa of generic rank occurring together here, 5 of which are endemic to that region (36% endemism at the generic level) (Table 16–9). These data support the conclusion that the AOD of the Oryzomyini was in the northern Andes (Ecuador, Colombia, and Venezuela). This inferred oryzomyine AOD is equivalent to Müller's (1973) Northern Andean Dispersal Centre for the South American land vertebrates as a whole; it is also characterized by the presence of several other endemic mammalian genera, such as *Cryptotis*, *Nasuella*, and *Orolestes*. The placement of the oryzomyine AOD in the Andes of northwestern South America is in good agreement with the reasonable idea that the early sigmodontine (proto-oryzomyine) invading stock may have arrived on the coasts of northwestern South America by rafting from Central America (Fig. 16–3).

The northern Andes of Colombia, Ecuador, and Venezuela are known to have been an area of positive relief since the Paleocene, with phases of uplift occurring in the Eocene, the Miocene, and the Plio-Pleistocene (van der Hammen, 1961). Proto-oryzomyines coming by waif dispersal from more temperate latitudes in Central America probably dispersed rapidly to the emerged Early Miocene Andean highlands, finding appropriate conditions there. The subsequent uplifting of the Andes during a Miocene orogenic phase, which occurred at least in the Eastern Cordillera of Colombia and Venezuela, and the Western Cordillera of Colombia and Ecuador (van der Hammen, 1961; B. B. Simpson, 1975), surely increased the heterogeneity of the Tertiary Andean environments, triggering the differentiation of the original oryzomyine stock to adapt to different biotopes, mostly in montane forests. *Thomasomys*, *Aepeomys*, *Chilomys*, *Microryzomys*, and *Oryzomys* s.s. may have resulted from this process. Some of them adjusted more and more to high altitudes, and became preadapted to, or at least tolerant of, the higher elevations that resulted from the Plio-Pleistocene orogenic phase from which originated the present relief and the evolution of the páramo biome. Other oryzomyines diversified in adaptation to different biotopes at lower elevations to invade northeastern lowlands and eventually more southern regions. This is probably the origin of *Rhipidomys* (middle elevation to lowland, arboreal), *Oecomys* (lowland, arboreal), *Oligoryzomys* (mostly open lowland, granivorous [Meserve, 1981]), and *Nectomys* (forested lowland, semiaquatic). *Zygodontomys*, a genus of dubious classification among the sigmodontine tribes, may also represent an oryzomyine derivative that differentiated in the same AOD in adaptation to grassland in mountain valleys to eventually disperse in the savanna–grassland biomes of Colombia, Venezuela, and the Guyanas. The ichthyomyines are a small tribe likely to have originated in the same AOD from a *Thomasomys*-like ancestor. They are also typically Northern Andean in f_T distribution (Table 16–8), and two (*Neusticomys*, *Anotomys*) of their four genera are endemic to Northern Andean sites. Ichthyomyines probably differentiated in high mountain forests and became adapted to semiaquatic life in mountain streams and to an animal diet.

The Akodontini and the Scapteromyini

The akodont rodents can easily be interpreted, on morphological and chromosomal grounds, to be derived from oryzomyine ancestors. Taking into account molar teeth morphology and skull specializations, the akodonts are, in a sense, intermediate in grade of evolution between the oryzomyines and the phyllotines. Akodontines are mostly generalists in diet, with the exception of some predominantly insectivorous puna species and genera (*Oxymycterus*, *Lenoxus*, *Notiomys*, and, presumably, *Blarinomys* and the more vegetarian *Bolomys*), remarkable for their great versatility in habitat preferences. The great majority (80.3%) of the akodontine species belong to one of the three polytypic genera, *Akodon*, *Oxymycterus*, or *Bolomys*. Species of *Akodon* live in the puna, the páramos, in montane tropical and subtropical forests, in lowland subtropical forests, in grassy pampas, dry montane Andean valleys, semidesert Patagonian tablelands, and cold southern Andean forests. *Bolomys* occurs either at the highest altitudes on the altiplano, or in the chaco, pampean, and Brazilian lowlands. *Oxymycterus* has representatives in puna localities, subtropical mountain and lowland forests and in the grassy pampean steppes. However, several genera or subgenera are more restricted in distribution, for example, *Chroeomys*, found only in the puna, *Podoxymys*, of the high tepuis of the Guayana region, *Notiomys* of the Valdivian forest, and *Microxus* of the Andean heights.

The habitat and food niche versatility of most akodontines may have favored their dispersal. In fact, they are broadly dispersed in South America, their representatives occurring in almost all regions, with the exception of the Amazon Basin (Tables 16–8, 16–9). However, and as already noted, they are mostly Andean in distribution. In the Andes they are sparsely distributed, being rarer only in Northern Andean and North-Central Andean regions (Table 16–8). This is an indication that their AOD is not likely to belong to any of those Andean units. But as regards the three remaining partitions, the f_T analysis does not provide a particularly evident area of diversity concentration, as is the case with the Oryzomyini and the Phyllotini (discussed later). Although the South-Central Andes show a lower frequency (0.147) of akodontine occurrences than the North-Southern Andes (0.207) or the South-Southern Andes (0.200), the differences are certainly not impressive. If we look at the areas of greater overlap of genera (Table 16–10), the Non-Andean area shows the greater one, but this is in fact a misleading effect of the pooling in our analysis of the different non-Andean regions. Within the Andes proper the SCA is the unit of greatest occurrence of akodontine genera and subgenera, but the difference is not great as regards the NSA. Taking the distribution of genera and the f_T data in the SCA plus the NSA units (Table 16–8, Table 16–10) together, we have nine genera and a frequency of akodontine localities occurrence of 0.354. We have also three endemic subgenera, *Chroeomys*, *Lenoxus*, and

TABLE 16–10. Distribution of the akodontine genera and subgenera in Andean conventional units and in non-Andean localities.

Genera and subgenera	Non-Andean	NA	NCA	SCA	NSA	SSA
Akodon s.s.	+	+	+	+	+	+
Deltamys	+	—	—	—	—	—
Hypsimys	—	—	—	—	+	—
Chroeomys	—	—	+	+	—	—
Abrothrix	—	—	—	—	+	+
Oxymycterus	+	—	+	+	—	—
Juscelinomys	+	—	—	—	—	—
Bolomys	+	—	+	+	+	—
Chelemys	—	—	—	—	+	+
Microxus	—	+	—	+	—	—
Notiomys	—	—	—	—	—	+
Blarinomys	+	—	—	—	—	—
Podoxymys	+	—	—	—	—	—
Lenoxus	—	—	—	+	—	—
Total number per unit	7	2	4	6	5	4
Number endemics	4	—	—	1	1	1

Note: + indicates presence; — indicates absence. See Table 16–7, Figs. 16–3 and 16–4 and Materials and Methods section for explanations of abbreviations.

Hypsimys, in the two combined units. This suggests, as advanced elsewhere (Bianchi et al., 1971), that the akodontines probably differentiated in this general region, at intermediate zone between the SCA and the NSA regions being their more likely AOD (Fig. 16–3). This AOD would geographically correspond with the southern half of Müller's (1973) Puna Dispersal Centre.

It must be taken into account, however, that the South-Southern Andean unit is also relatively rich in akodontines, both in f_T distribution and in number of genera (Tables 16–8, 16–10). This region also has an endemic genus, *Notiomys*, and a large number (11) of endemic species. But it is evident that it cannot compete with the southern puna region as the main AOD of the akodontines. The observed richness in genera, species, and geographic occurrences of akodontines in the SSA unit might be interpreted, at the most, as an indication that it may be a secondary AOD for this tribe.

The ancestral akodontine may have been a generalized *Akodon*-like form of oreal origin, which colonized the area of the puna from the north in Late Miocene or Early Pliocene times, before the altiplano reached considerable heights, which started to occur in the Middle Pliocene (Ahlfeld, 1970). This ancestral stock may have found adequate conditions in the southern protopuna, and from this placement local differentiation may have started, triggered by the changing Andean environments, resulting in the local origination of different genera and subgenera. *Akodon* s.s. may have been the original taxon to differentiate, as inferred from the necessary time to account for its extensive polytypism and from its plesiomorphism as compared to the remaining members of the tribe. A nonspecialized *Akodon* with $2n = 52$ chromosomes (which is most probably the primitive karyotype for the tribe; Reig, unpublished data), similar to *Akodon andinus*, may have been sufficiently generalist to have lived in different habitats, and to have settled either in montane forests of the eastern slopes of the rising Andes or in the dry high mountain valleys and open semidesert heights. *Bolomys* may have differentiated early also, as indicated both by its extremely derived karyotype of $2n = 34$, and by its presence as a fossil in Lower Pliocene (Montehermosan) strata (Reig, 1978). *Chroeomys* and *Hypsimys* represent well-differentiated *Akodon* derivatives that adapted to arid heights and remained as endemics on the uplifting altiplano and high Andean valleys. *Oxymycterus* may have originated locally from *Akodon* in adaptation to an animal diet, probably in more humid habitats, then secondarily adapted to puna habitats, and finally spread into the eastern lowlands. *Lenoxus* is certainly a well-differentiated *Oxymycterus* derivative that quite probably evolved in the same AOD.

To explain the present distribution of all the members of the Akodontini, it is necessary to assume that most of their taxa originated in what is now the southern puna region, and that they migrated in various directions and experienced further differentiation in other areas. Three main directions of dispersal may be assumed. A dispersal to the south of a generalized *Akodon* through the Andean highlands was connected with the differentiation of *Abrothrix*, *Chelemys*, and *Notiomys*, which are endemic either to the NSA, or the SSA units, or both. *Abrothrix* certainly originated in the Pliocene, as indicated by its presence in Upper Pliocene (Chapadmalalan) strata (Reig, 1978). The presence of fossil *Abrothrix* in the Upper Pliocene and Lower Pleistocene of the pampas region, and of extant species of *Akodon* in the Patagonian tablelands, indicates that this southern Andean dispersal flow also secondarily spread into eastern steppes. A dispersal to the north from the original akodontine AOD, following the eastern forest fringe of the Andes, is necessary to explain the present distribution of *Akodon orophilus*, *A. mollis*, *A. tolimae* and *A. urichi*, as well as of *Microxus* and *Podoxymys*. One may question why, given the presence of most of these taxa in the northern Andes (where they may have originated directly from the oryzomyines), it is postulated that the Northern Andean akodontines migrated to the north from a South-Central Andean region. The answer is that the chromosome evidence is highly suggestive of the derived condition of the aforementioned species of *Akodon*, with *A. orophilus* ($2n = 26$), *A. mollis* ($2n = 22$), and *A. urichi* ($2n = 18$) indicating a northward direction of decrease in karyotypic number (chromosomal data from Gardner and Patton [1976] and from Reig, unpublished results). The third main direction of dispersal, from the southern puna AOD directly toward the eastern lowlands, is necessary to explain the present distribution of lowland representatives of *Akodon* s.s. *Oxymycterus*, and *Bolomys*, as well as that of the exclusively lowland *Akodon* subgenus *Deltamys*, and of the genera *Blarinomys* and *Juscelinomys*. It is significant to point out here that *A.* (*Akodon*) *boliviensis*, a typical inhabitant of the present puna, but also found at lower altitudes in sites surrounding the puna, has $2n = 40$ chromosomes. Most *Akodon* species of the lowlands of central Argentina show either identical $2n = 40$ karyotypes (*A. varius*), or very similar $2n = 42–43$ (*A. molinae*), or else $2n = 38$ (*A. azarae*, *A. dolores*) ones which are easily derived from the former (Bianchi et al., 1971). This may be taken as a suggestive indication that the Argentine forms originated in ancestors from the puna, an idea further supported by the fact that both *A. boliviensis* and the puna species *A. albiventer* (including *berlepschii* as a subspecies) have a $2n = 40$ karyotype (Gardner and Patton, 1976).

The Scapteromyines are close to the akodontines in molar pattern, and they are likely to have derived directly from an akodontine stock that invaded the eastern lowlands, to become adapted to semiaquatic habitats (*Scapteromys*, *Bibimys*?) and to semiburrowing (*Kunsia*) forms of life. In any case, they are typically non-Andean rodents and they played no role in the history of high Andean rodent faunas.

The Phyllotini and Punomys

Baskin (1978) recently proposed that the Phyllotines have derived directly from Miocene forms of the southern United States related through *Bensonomys* to a *Copemys* ancestor. I have claimed elsewhere (Reig, 1980), that this hypothesis is quite unlikely. *Calomys* (which Baskin alleged differs from *Bensonomys* only at the subgeneric level) is a typical "complex-penis cricetid" related to the remaining Sigmodontinae both in morphology (Hershkovitz, 1962) and in karyology (Gardner and Patton, 1976; Pearson and Patton, 1976) whereas *Copemys* is not only the probable ancestor of the fossil *Bensonomys*, but also the direct ancestor of *Peromyscus* (James, 1963; Clark, Dawson, and Wood, 1964). *Peromyscus* is a typical "simple-penis cricetid" related at the subfamily level to the remaining North American Neotominae. The similarities in molar structure that Baskin observed between *Calomys* and *Bensonomys* are therefore more likely to be due to convergence. Moreover, they are less impressive than supposed by Baskin and are contradicted by other molar features not taken into account by this author (Reig, 1980).

A more parsimonious hypothesis, which is also in better agreement with the available evidence, is that phyllotines differentiated in South America directly from the oryzomyines through the akodontines (Reig, 1980). The evidence does not seem conclusive to decide between the two alternatives. *Zygodontomys* was supposed to be an intermediate taxon between akodontines and phyllotines, but it seems better considered now as an independent offshoot of the Oryzomyini, probably unconnected with the ancestry of either akodontines or phyllotines. The akodontines represent an intermediate grade between oryzomyines and phyllotines in the evolution of molar teeth and in increasing adaptation to open, treeless biomes, this being the main reason to postulate that phyllotines may have been direct descendants of akodontines. Nevertheless, there are no objections to supposing that akodontines and phyllotines radiated from oryzomyines independently. It can be argued that this alternative is in better agreement with the chromosomal evidence. However, the latter is not necessarily conclusive in this respect.

Calomys is probably the most primitive living phyllotine, showing comparatively brachyodont and crested molar teeth. Species of this genus inhabit grasslands, scrublands, and forest fringes. *Calomys* is likely to be very close to the hypothetical phyllotine ancestor that departed from sylvan oryzomyines or subsylvan akodontines. *Calomys sorellus*, a widespread and common species of the puna, has been considered to be the most generalized member of this genus (Hershkovitz, 1962), which is also in agreement with its primitive karyotype (Pearson and Patton, 1976). This may be a clue in the search for the center of original differentiation of the phyllotines.

In fact, the phyllotines are the more clearly Andean Sigmodontinae, showing 80.9% of their occurrences in the Andes (Table 16–7). The f_T analysis also suggests a South-Central Andean AOD (f_T in SCA = 0.387), the frequency of occurrences in this geographic unit being strikingly greater than in the other units (Table 16–8). The data on range overlap of genera also support this suggestion, as the SCA unit shows 10 phyllotine genera (conventionally including *Punomys*), 3 of which (*Galenomys*, *Chinchillula*, *Punomys*) are endemic to the SCA, and two others (*Neotomys*, *Andinomys*) are puna endemics shared either with the NCA or the NSA units, respectively (Table 16–11). The conclusion seems to be fully warranted, therefore, that the Andes of southern Peru, Bolivia, and northern Chile and Argentina are the AOD of the Phyllotini (Fig. 16–4). This area matches the present distribution of the puna biome and coincides with Müller's (1973) Puna Dispersal Centre. However, it is almost certain that phyllotines differentiated in that region before the altiplano reached its present height.

A phyllotine ancestor may have arrived in this area by the Middle or Upper Miocene, at a time when the altiplano experienced an important phase of uplift (Lohmann, 1970), but before its main uplift in Pliocene and Pleistocene times (B. B. Simpson, 1975). An at least Upper Miocene time of arrival is indicated by the presence of *Auliscomys* in Montehermosan strata (now dated as Early Pliocene; see Marshall et al., 1979) of the pampean region, certainly derived from puna forms (Reig, 1978). The hypothetical phyllotine ancestor diversified in semiarid to arid protopuna, probably heterogeneous, environments toward increasingly pastoral life forms. This diversification resulted in the origination in the protopuna region of the more generalized genera

TABLE 16–11. Distribution of the phyllotine genera in Andean conventional units and in non-Andean localities.

Genera and subgenera	Non-Andean	NA	NCA	SCA	NSA	SSA
Phyllotis	+	+	+	+	+	+
Calomys	+	—	+	+	+	—
Euneomys	—	—	—	—	+	+
Auliscomys	—	—	+	+	—	—
Graomys	+	—	—	+	+	—
Andalgalomys	+	—	—	—	+	—
Eligmodontia	+	—	+	+	+	+
Pseudoryzomys	+	—	—	—	—	—
Galenomys	—	—	—	+	—	—
Reithrodon	+	—	—	—	+	+
Chinchillula	—	—	—	+	—	—
Neotomys	—	—	+	+	—	—
Andinomys	—	—	—	+	+	—
Irenomys	—	—	—	—	—	+
Punomys [a]	—	—	—	+	—	—
Total number per unit	7	1	5	10	8	5
Number endemics	—	—	—	3	—	1

[a] *Punomys* is not a typical phyllotine. It is placed here for convenience.

Note: + indicates presence; — indicates absence. See Table 16–7, Figs. 16–3 and 16–4, and Materials and Methods section for explanations of abbreviations.

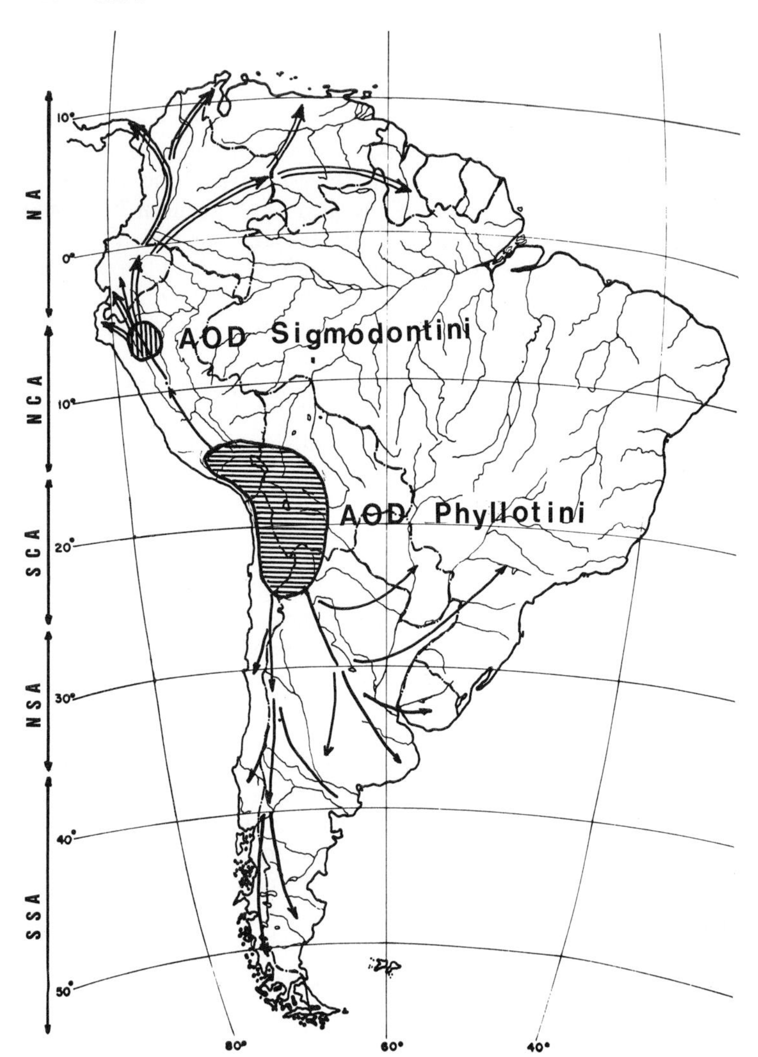

FIG. 16–4. Areas of original differentiation (AOD) of the tribes Phyllotini and Sigmodontini of South American cricetid rodents. Double arrowed lines represent the suggested directions of dispersal of the genus *Sigmodon* in South America. Simple arrowed lines represent the suggested lines of dispersal and secondary differentiation of the Phyllotini as explained in the text. Limits of the conventional Andean units are indicated on the left, as defined in Materials and Methods.

Calomys and *Phyllotis*, as well as in the early diversified pastoral genus *Neotomys*. This is in agreement with the chromosomal phylogeny presented by Pearson and Patton (1976, fig. 5), which also postulates that *Chinchillula* and *Andinomys* split early from a *Phyllotis osilae*-like ancestor. These two genera remained in their place of origin and became increasingly adapted to an herbivorous diet and to drier biotopes, to adjust later to the high altitudes of the puna after the Plio-Pleistocene uplift. The more pastoral genus *Auliscomys* may have originated as a lateral offshoot of the *Phyllotis* cladogenesis evolving also in situ to speciate later as a typical puna form. Another event of local differentiation resulted in *Galenomys*, which also eventually became typical of the puna. *Graomys* probably diverged in the protopuna from a common stem with *Auliscomys* and remained highly versatile ecologically.

Thus, the pre-Pliocene protopuna may have witnessed the local differentiation of *Calomys*, *Phyllotis*, *Graomys*, *Auliscomys*, *Andinomys*, *Neotomys*, *Galenomys*, and *Chinchillula*. The last four genera eventually became puna specialists after the altiplano reached its present heights, thanks to preadaptations they had gained in semidesert protopuna habitats. It is interesting to note that those genera are now monotypic puna endemics. *Calomys*, *Phyllotis*, *Graomys*, and *Auliscomys* remained more eurytopic and generalized in ecology and experienced active speciation and dispersal. Some of the species of *Calomys*, *Phyllotis*, and *Auliscomys* later evolved forms welladapted to the puna, and probably experienced further speciation during the Pleistocene in the puna proper in connection with expansion–retraction evironmental cycles in the Quaternary. *Graomys* remained at lower altitudes and only secondarily evolved adaptations to the puna. The four more generalized genera originated species that dispersed in three main directions. A northward dispersal included only *Phyllotis*, which reached the Andes of Ecuador, where it is now represented by *P. andium* and *P. haggardi*, the latter becoming eventually the only phyllotine that adapted to the southern páramos. Another dispersal line brought *P. darwini* and *Auliscomys* to the southern Andes of Argentina and Chile, and resulted in further local differentiation of *Phyllotis* in the south and in *Auliscomys micropus* [= *Phyllotis* (*Loxodontomys*) *micropus*]. *Irenomys* may have evolved from members of this dispersal event. There is evidence that forms involved in the southern Andean dispersal expanded eastward to the lowland Patagonian steppes and pampas, as demonstrated by the present distribution of *Auliscomys micropus* and *Phyllotis bonaerensis* (see Reig, 1978, for recognition of this form as a full species) in pampean "sierras" of Buenos Aires Province. Additionally, by the presence of the extinct *Auliscomys formosus* in the Montehermosan of the southeastern pampas region (Reig, 1978). It is of interest to remember here that as typical a southern Andean akodontine as *Abrothrix* is also represented, as a fossil, in Plio-Pleistocene strata of southeastern Buenos Aires Province.

But more important dispersal of phyllotines from an original protopuna AOD was southeasterly in direction and involved local differentiation in monte, chaco, pampa, and northern Patagonian regions. Species of *Calomys* invaded the eastern lowlands and differentiated in a complex array of species in Bolivia, Paraguay, east-central Brazil, Argentina, and Uruguay. The intriguing presence of *C. hummelincki* in Venezuela may indicate a relict of a northward dispersal of the genus through a savanna corridor east of the Andes. From a *Calomys*-like ancestor probably evolved *Eligmodontia* in the protopuna, and further dispersed into the monte biome and the Patagonian steppes. The chacoan *Pseudoryzomys* may also have evolved from a *Calomys*-like ancestor that differentiated in semiarid lowlands of eastern Bolivia and Paraguay. *Graomys* also expanded from its protopuna center of origin to the monte, Patagonia, dry pampas, and the chaco. *Graomys dorae* (Reig, 1978) indicates the presence of the genus in the pampas already in the Upper Pliocene. *Andalgalomys* is apparently more similar to the *Calomys-Eligmodontia* complex (Williams and Mares, 1978), and its $2n = 60$ karyotype suggests a direct descendance from *Calomys*. Its present distribution suggests that it originated in peri-Andean dry lowlands, and that it segregated into different species in the chaco and the monte. The presence of the living species *Reithrodon auritus* in the Chapadmalalan (Reig, 1978) suggests that the genus may have evolved in Lower Pliocene times. Recent distribution may be interpreted as indicating either an origin in the southern Andean phyllotine dispersal group, or directly in a protopuna ancestor that invaded the monte lowlands. The ecological specialization and morphological distinction of this genus and the lack of intermediate forms as regards other phyllotine genera, both in morphology and in karyotype, make any further comment on its origin seem like guesswork. Finally, *Euneomys* is also likely to have derived, probably from *Auliscomys*, in the process of the southeasterly dispersal of the phyllotines, diversifying into different species in adaptation to life in dry montane valleys of the North-Southern Andean unit and extending southward to southern Patagonia.

The origins of the peculiar and distinctive puna genus *Punomys* are intriguing. In molar structure it is more primitive than any of the phyllotines, but in skull characters and external appearance it is fairly similar to some of the phyllotine genera, though more specialized. If the phyllotines are considered to be the direct descendants of the Oryzomyini, *Punomys* may represent an early descendant of a protophyllotine stock, which specialized to high elevations from the beginning. We need more comparative information, including chromosomal data, to further advance our understanding of origins and history of this typically puna genus.

The Probable Geographic Origin and Further Evolution of the Sigmodontini

After withdrawing *Reithrodon* and *Neotomys* from the sigmodont group of rodents (Gardner and Patton, 1976, Pearson and Patton, 1976), the tribe Sigmodontini remains reduced to *Sigmodon* (including *Sigmomys* as a junior synonym) and *Holochilus*, the cotton rats and the marsh rats, respectively. The two genera comprise herbivorous inhabitants of open country, although *Sigmodon* may reach high Andean altitudes. *Sigmodon* is mainly a grassland genus, whereas *Holochilus* has palustrine preferences and a certain degree of adaptation to semi-aquatic life. Both may invade cultivated fields, where they can attain high densities.

Even though the Sigmodontini are now in South America a predominantly non-Andean group (f_T in non-Andean regions = 0.794; see Table 16–7), there are strong indications that they first evolved in the North-Central Andes from a phyllotine ancestor. *Neotomys* is intermediate between phyllotines and sigmodont rodents both in morphology and in chromosomes (Gardner and Patton, 1976), and its presence in the living fauna of the northern puna is suggestive of an origin of the genus *Sigmodon* in the pre-Pliocene Andes of northern Peru. The occurrence of *S. peruanus* (in which I include *simonsi*, *chonensis*, *peruanus*, and *puna* as subspecies) in the lowlands and mountain slopes west of the Andes of Ecuador and north-central Peru, and of *S. inopinatus* in the high Andes of Ecuador, is in good agreement with this hypothesis. Thus, *Sigmodon* is likely to have originated from a *Neotomys*-like ancestor in north-central Peru, and to have further dispersed westward to the western coastal lowland region of the north-central Peru and Ecuador, and northward through the valleys of the Andes of Ecuador up to the proto-Magdalena Valley of Colombia, and eventually to have spread from the latter to the lowlands of Colombia and Venezuela (Fig. 16–4). *Sigmodon inopinatus*, found isolated on the páramo of Chimborazo (Anthony, 1924), is likely to have arrived there by the uplifting of the páramos during Plio-Pleistocene times. As a consequence of its northward spread, *Sigmodon* most probably entered Central America by overland dispersal once the Panamanian land bridge was established, some 5.5 million years ago (Reig, 1981c), and differentiated in the arid and mesic biomes of North America into several species (Zimmerman, 1970). The presence of *S. medius* in the Early Blancan of North America (Gidley, 1922) makes it necessary to place the origin of *Sigmodon* at an earlier date, probably in Late Hemphillian-Huayquerian times. The relationships between the living North and South American species of *Sigmodon* are still obscure. The very primitive, $2n = 82$ karyotype (Reig and Barros, unpublished) of *S. alstoni* suggests that its carrier is a remnant of the early *Sigmodon* differentiation, but the knowledge of the chromosomes of *S. peruanus* and *S. inopinatus* would be critical for this evaluation, and their karyotypes are unknown so far. (Note that several living North American species of *Sigmodon* have derived, low-number karyotypes [Zimmerman, 1970].)

Holochilus is likely to have originated from a *Neotomys*-like ancestor that descended to the lowlands of eastern Peru, where it is now represented by a $2n = 56$ form (Gardner and Patton, 1976) that I tentatively refer with full species status to *H. amazonicus*. It is possible to think of this species as the one from which originated the other living species of *Holochilus*, from a western Amazonian center of dispersal, with different forms adapted to different grassland-savanna habitats in Colombia, Venezuela, the Guianas, eastern Brazil, central Argentina, Paraguay, and Uruguay.

Overview of the History of the High Andean Sigmodontine Rodents

I hope to have convinced the reader that the distributional and systematic facts compel one to conclude that the high Andean biomes played the major role as the area where sigmodontine rodents evolved the main features of their present diversity, whereas the lowland biomes contributed only secondarily to the process of their differentiation.

This is quite a reasonable conclusion, additionally, if we consider that the incorporation of cricetid mice into South America must have taken place in the northwestern part of the continent at a time coinciding roughly with a major episode of Andean uplift, i.e., the second phase of the Andean orogeny (Groeber, 1951). The first populations that differentiated in South America were deeply influenced by the uplifting mountains: they thus gradually evolved to adapt to the different habitats that resulted from the changing topographies and climates.

Moreover, the Andes are certainly a region of great spatial and temporal heterogeneity, where the differences in relief and altitude determine a great diversity of habitats, which moreover changed during the different episodes of uplifting. This situation is ideal to promote a relatively speedy and extensive differentiation such as the one observed in the living South American sigmodontines.

In the case of caviomorphs, their previous arrival in South America when the Andes were just beginning to take shape, together with their most probable place of implantation far east from the Andean region, cooperated to make, first the humid lowland forest biomes, and later the mesic to arid lowland biomes, the main region of their original differentiation. However, when the Andes reached a more significant height during Miocene times, the evolution of caviomorphs became clearly influenced by the orogenic changes and their climatic consequences. Some of the caviomorph taxa further differentiated in connection with the uprising cordilleras. This happened particularly in the South-Central and the North-Southern Andean regions.

The contrasts in the history of sigmodontine and caviomorph rodents as far as the role of the Andes in their cladogenetic processes is concerned are certainly responsible for the primarily autochthonous Andean character and greater diversity of the present high Andean sigmodontines, as opposed to the allochthony and smaller species numbers of the high Andean caviomorphs. In other words, the sigmodontines are more abundant in the two high Andean biomes because they are a primarily Andean group, and the caviomorphs are less abundant in the puna and in fact rare in the páramos, because they are only secondarily Andean.

The previous analysis also demonstrates, as in the case of caviomorphs, a sharp distinction between the history of the páramo and that of the puna sigmodontine faunas, as exemplified by the recorded differences in the taxonomic composition of the two faunas. Thus, the páramo sigmodontine fauna is made up mainly of oryzomyines, because it resulted from in situ gradual adaptation to the páramos of oryzomyine taxa that originated and evolved in the same general area when the páramos developed as a result of an increasing orogenic uplift.

Similarly, the puna sigmodontine fauna mainly comprises phyllotine and akodontine species because these two tribes evolved their major differentiation in protopuna environments of the South-Central and North-Southern Andes. An important portion of their genera and species became gradually adapted to the uplifting puna in Late Tertiary and Pleistocene times.

The minor representation of phyllotines, akodontines, and sigmodontines of protopuna origin in the páramos, and the even less significant presence of a few marginal oryzomyines of Northern Andean origin in the puna, is well explained by inter-Andean dispersal. But these dispersal patterns cannot jeopardize the compelling conclusion that the observed differences in sigmodontine composition between the puna and the páramo faunas are the consequence of the parellel in situ diversification and adaptation of different phylogenetic units, a process Sukachev (1958) called "phylocoenogenesis."

THE PROBABLE HISTORY OF THE SOUTH AMERICAN AND ANDEAN SCIURID RODENTS

The living South American sciurids belong to the subfamily of diurnal squirrels, Sciurinae, and, among them, solely to the tribe Sciurini of arboreal squirrels (Moore, 1959). The Sciurinae are an almost cosmopolitan taxon, and the Palearctic region is currently regarded as its original center of differentiation (Moore, 1961). Sciurines surely populated South America from tropical North America (Patterson and Pascual, 1972). There are suggestions that the migration of sciurines into South America took place in two different episodes.

Moore (1959, 1961) grouped the species of South American squirrels in four different genera, namely, *Guerlinguetus* (including *Hadrosciurus* as a subgenus), *Syntheosciurus* (including *Mesosciurus*), *Microsciurus* (including *Leptosciurus* and *Simosciurus*), and *Sciurillus*. *Guerlinguetus* occurs in the Amazon Basin, and *Sciurillus* in the Amazonian and Guianan region; neither has representatives outside South America. By contrast, *Syntheosciurus* and *Microsciurus* live in South, North, and in Central America, and in South America have representatives both in lowland forests and in Andean forests. Hershkovitz (1972) suggested that *Sciurillus* and the Amazonian species that Moore referred to *Guerlinguetus* had an ancestor that entered South America possibly in the Middle Tertiary, "before the Andes arose as a barrier to the eastward spread of oversea migrants" (Hershkovitz, 1972, p. 353). The common occurrence of species referred to *Syntheosciurus* (comprising "*Sciurus*" *granatensis*) and of species of *Microsciurus* in Central America and northwestern South America is interpreted by Hershkovitz as an indication that *Syntheosciurus* represents a late overland migration, whereas *Microsciurus*, which also reached upper Amazonia, may belong to the same migration phase as *Sciurillus* and *Guerlinguetus* and later reinvaded Central America. Hershkovitz disagrees with the previous conclusions of Moore (1959, 1961), who had rejected the idea of an early arrival of *Sciurillus* and had contended that all South American squirrels differentiated in Central America, which served as a "staging area" in which sciurids of Nearctic origin adapted to tropical conditions and differentiated before invading South America after the Panamanian land bridge was established. The concept of Central America as a staging region for the differentiation of typical South American taxa was also invoked to explain the origin of the diversity of South American cricetids (Patterson and Pascual, 1972). I have tried to demonstrate, however, that this view is untenable in the case of those rodents. In any event, it seems to me that there are no reasons to postulate that taxa that occur in a given region differentiated in an area where they are absent and where they have no immediate ancestors. Needless to say, there are taxa, now endemic to a given region, that have been demonstrated to be autochthonous of another, as in the case of the South American

Lamini (llamas, vicuñas, etc.; Webb, 1974), to cite just one example. But in those cases the presence of a well-documented fossil record makes the explanation mandatory, which is not necessarily the case for the origin of the diversity of the South American Sigmodontinae nor, I believe, that of the origin of such typically South American sciurids as *Guerlinguetus* and *Sciurillus*. Not having firsthand experience with sciurid rodents, I find it moot to further develop an argument in favor of one or the other of the two alternative views on the antiquity of South American sciurids. All I can say is that the idea of two episodes of migration of the South American Sciurinae, an earlier one for the ancestors of *Sciurillus* and *Guerlinguetus*, and a more recent one for *Syntheosciurus* and *Microsciurus*, is plausible (see Reig, 1981c) and seems to be in good agreement with the present evidence of the degree of differentiation and distribution of the living forms.

But in any case, the history of Andean sciurids is limited to *Microsciurus* and *Syntheosciurus*, the only two genera with representatives now living in Andean localities, and the idea that they migrated into South America by overland dispersal after the Panamanian land bridge emerged is more compelling than in the case of *Guerlinguetus* and *Sciurillus*. Species of *Microsciurus* are now present in Central America, and in South America they are distributed mostly along the western slopes of the Andes, reaching southward to northern Argentina as indicated by *M. (Leptosciurus) argentinus*. In peripuna habitats they reach up to 2700 m, but in the northern Andes they never reach higher than 2200 m. Quite probably, the South American species entered the continent by the Panamanian land bridge in Plio-Pleistocene times, mostly spreading in low- to middle-elevation mountain forests, but never becoming fully adapted to higher altitudes. Some populations spread from the Andes eastward, reaching the lowlands of the upper Amazon and of eastern Bolivia. *Syntheosciurus* is now restricted to Northern Andean localities and to lowlands surrounding the Andes of Ecuador, Colombia, and Venezuela, being frequent at altitudes above 2200 m and reaching the páramo in some places. Central American species of the genus (*S. brochus*, *S. poasensis*) now live in the cordillera of southern Central America, which has been suggested as the center of their local differentiation (Moore, 1961, p. 15). The montane distribution of these Central American species may be taken as an indication that species of this genus invading South America through the Panama land bridge may have been preadapted to high altitudes. It thus would have been easier for them to make their way in the uprising northern Andes, where they spead to high Andean forests, eventually becoming tolerant of páramo heights.

GENERAL DISCUSSION AND CONCLUSIONS

It should be clear from the data presented in this chapter that the puna and páramo rodent faunas evolved mostly as separate units and that the striking differences in their taxonomic composition are the result of different evolutionary histories of their component faunal elements. It also seems mandatory to conclude that those different histories are rooted long ago in Tertiary times. A complex series of events of dispersal, local differentiation, preadaptation, and further adjustment to increasing altitudes is necessary to explain the present diversity patterns of the two rodent faunas.

The puna rodent fauna is mainly the result of adjustment to arid conditions at high altitudes of phyllotine and akodontine cricetids that probably originated in the area of the present puna from northern Andean oryzomyine ancestors, before the uplift of the altiplano to its present heights. The phyllotines and akodontines may have evolved preadaptations to aridity in protopuna environments during Miocene and Early Pliocene times. These two cricetid components were supplemented by the important contribution of caviomorphs (octodontids, abrocomids, chinchillids, caviids), which originated during Miocene times in non-Andean lowlands, and became preadapted to aridity in peri-Andean regions and middle-elevation Andean valleys, to further adjust to arid conditions of high altitudes along with the continuing elevation of the altiplano. Additionally, a few oryzomyine cricetids of northern Andean origin expanded their range southward to become differentiated at the species or subspecies level as components of the puna biome.

The páramo rodent fauna is dominated by oryzomyine cricetids. They are most likely to have originated in northern Andean mountain forest biomes from a protosigmodontine ancestor that migrated to South America by overwater dispersal in the earlier Miocene. The Oryzomyini gradually differentiated in the Miocene with the uplift of the Andes. Some of them became tolerant of humid high altitudes and were able to adjust to the páramo heights that resulted from the final uplift of the Andes in Plio-Pleistocene times. Another component element of the present páramo rodent fauna is the akodontine cricetids, which probably differentiated in a protopuna region from where they dispersed northward through the Andes to evolve forms adapted to the high-altitude northern Andean forests prior to the Upper Pliocene. From there they reached the páramos with the subsequent uprising of the cordilleras. Additionally, a few endemic ichthyomyine cricetids evolved in the Northern Andes from oryzomyine ancestors and eventually adjusted to the páramo heights. The páramo representation of the sigmodontine cricetid rodents is completed with one species of Phyllotini (*Phyllotis haggardi*) and one of Sigmodontini (*Sigmodon inopinatus*), which are both also of central Andean origin and show marginal distribution in the southern páramos of Ecuador. Two caviomorph species are also present in a few páramos. One is the agoutid *Stictomys taczanowski*, a species that originated in Northern Andean forests from lowland ancestors of the Amazonian forests. The other is the caviid *Cavia porcellus*, an occasional inhabitant of a few páramos that became secondarily adjusted to them from populations adapted to high Andean forests that had their ancestry in the open lowlands east of the Andes. Finally, a few populations of the squirrel species *Syntheosciurus granatensis* occasionally occur in a few páramos. They are descendants of Central American populations that invaded the Northern Andean forests after the Panamanian land bridge was established, in Plio-Pleistocene times.

The puna rodent fauna has many more genera and species and also shows a much greater degree of endemism than the páramo rodent fauna. These intriguing facts compel one to look for an explanation. They can be interpreted either (a) as an indication that the puna biome is older than the páramo biome (the historical-effect hypothesis), or (b) as an indication that taxonomic diversification in the altiplano was favored by a greater environmental heterogeneity determining narrower niches and greater species packing (the ecological-effect hypothesis), or else (c) as a result of the area effect of a larger and much more continuous extension of the puna biome, or (d) as a combination of (a), (b), and (c).

I believe the area-effect hypothesis (c) to be unlikely, because the puna surface, even though larger and more continuous than the smaller and discontinuous páramos, is rather uniform, and extensive environmental uniformity is not precisely a condition to trigger taxonomic differentiation. The same argument may hold against the ecological-effect hypothesis (b). But narrower niches and greater species packing may result from long-term competition, an idea that would support the historical-effect hypothesis (a).

In discussions about the bearing of historical effects on the diversity of taxa of the puna and páramo biota, the climatic and vegetational changes that occurred during the Pleistocene are commonly invoked as the main factors of taxonomic differentiation under the operation of isolation–dispersal cycles determined by the glacial–interglacial oscillations that, as is well known, occurred during the Quaternary (Haffer, 1969, 1970, 1974; van der Hammen, 1972; B. B. Simpson, 1975).

That Pleistocene changes have been an active factor in speciation of birds and plants in the high Andes seems to be established beyond any serious doubt. The problem is whether most of the speciation occurred during the Pleistocene in birds, plants, or in other organisms, and whether allopatric speciation (which is a prerequisite for speciation through cycles of isolation and dispersal) is applicable as an exclusive mode of species formation for all sorts of organisms.

As regards the first question, I believe that the proponents of the refuge theory went too far in trying to explain the patterns of species diversity found in the present biota as having been exclusively or mostly determined by Pleistocene events. Actually, it is a truism that speciation rates may be of different tempos for different kinds of organisms and that they can even vary within a given major taxon. For instance, remains inseparable from living species of *Bufo* have been found in the Paleocene of Brazil (Estes and Reig, 1973), and the still living *Bufo marinus* was found in the Miocene of Colombia (Estes and Wasserzug, 1963). Similarly, I have not been able to separate the Pliocene fossil remains of the phyllotine *Reithrodon* from the living species *R. auritus* (Reig, 1978). These few examples suffice to demonstrate that a certain number of species of the living South American biota may have originated in pre-Pleistocene times, and even far back in the Tertiary. Besides, to claim that Pleistocene fluctuations can explain the whole pattern of the present diversity of a given major taxon, e.g., most birds of Amazonia (see Haffer, 1974, p. 345), implies that evolution was much slower in pre-Pleistocene times, which is in full disagreement with the history depicted by the fossil record for most of the animal and plant taxa. It seems evident that contraction and expansion of suitable habitats during the climatic cycles of the Pleistocene played an important role in triggering speciation in birds (Haffer, 1974), lizards (Vanzolini and Williams, 1970), and some groups of plants (B. B. Simpson, 1975). But it is equally evident to me that to understand the pattern of species diversity of those and of other taxa, due allowance must be given to speciation events that took place in Tertiary times. This seems to be particularly evident in the case of high Andean rodents, for which I attempted to demonstrate compelling reasons to assume the occurrence of Tertiary speciation and major diversification events.

A detailed analysis of speciation patterns in high Andean rodents is probably premature as it would require a better understanding of the still poorly known systematic status of many species. But the evidence at hand overrides the idea that solely Pleistocene events acted as the causative agency of the present diversity of rodents of the puna and páramos. On the contrary, the observed contrast in the rodent faunas of the two biomes, and the reconstruction of their probable histories, obliges one to assume that not only the cyclic environmental changes of the Quaternary, but also the changing history of the Andes since the Miocene onward, must be taken into account to depict reasonably the process of their origin and differentiation. If this conclusion is granted, the recorded differences in diversity and endemism between the puna and páramo rodent faunas might be ascribed to a longer historical effect in the former than in the latter. This conclusion must be taken as an indication that the present physical and vegetational characteristics of the puna originated earlier than the corresponding features of the páramos, an inference that seems to match the known geologic evidence and that is worthy of confrontation with the data provided by other taxa (i.e., birds, frogs, insects) of the puna and páramo biota.

As regards the allopatric model of speciation, which would imply a principal effect of the Pleistocene climatic oscillations to explain the origin of the species diversity in rodents of high Andean biomes, there are cogent reasons to maintain that this model cannot be taken as the exclusive, or even the dominant, mechanism of speciation in the case of these organisms. On the contrary, rodents are becoming a frequent example of speciation through chromosomal changes in conditions of parapatry (Nevo et al., 1974; Reig et al., 1980; Reig, 1981a,b). This mechanism of species differentiation does not necessarily require the agency of long periods of isolation for the process to result in full species separation. Therefore, the frequent occurrence of chromosomal speciation in rodents does not make it necessary to invoke the refuge theory as a main explanation of the high Andean rodent species diversity, but it is quite possible that the cycles of contraction and expansion of suitable habitats during the Pleistocene played an additional role in promoting rodent speciation in high Andean biomes. The importance of this role can be assessed only after gaining a better knowledge of the systematics, the detailed distribution, and the habitat preferences of the rodent species that now inhabit the puna and the páramos.

Certainly, we are still far from having a fully satisfactory explanation of all factors, and an understanding of their relative importance, which acted to shape the present patterns of species diversity of high Andean rodents. I have tried to put together the facts known at present, and to develop the explanations that, in my belief, are in keeping with the different sorts of data and with the available body of theory. However, even if my explanations appear warranted by the present state of knowledge, they remain tentative. Rodent evolution is a field that requires further elucidation of facts and results from new and detailed systematic, ecological, and biogeographic studies. Therefore, this chapter purports to provoke thought and to stimulate further research more than to present definite answers to the problems analyzed.

ACKNOWLEDGMENTS

The data and interpretations presented here are the result of many years of field, museum, and laboratory work, and of theoretical elaboration. Many people collaborated with me in the factual work, and several colleagues contributed their criticisms and suggestions to develop and clarify my ideas. Begging the indulgence of those who collaborated with me on one or another aspect and who are unwittingly overlooked here, I will acknowledge the people who most closely collaborated with me in the factual work and who more significantly contributed to the shaping of my ideas. Among the former, I am deeply indebted to Marisol Aguilera, María Alicia Barros, Antonio Pérez-Zapata, Galileo J. Scaglia, Orlando A. Scaglia, Alfredo Vitullo, and Myriam Useche. Among the latter, I benefited from discussions with and comments by Moritz Benado, Néstor O. Bianchi, Rosendo Pascual, James L. Patton, Francis Petter, Oliver P. Pearson, Vincent Sarich, George G. Simpson, Angel Spotorno, Nikolay N. Vorontzov, Robert Voss, and François Vuilleumier. I must also mention the stimulus I received from Maximina Monasterio and François Vuilleumier in inviting me to contribute this chapter to the present volume. This essay must be considered an unplanned by-product of research projects sponsored and financed, in temporal sequence, by the John Simon Guggenheim Memorial Foundation, the Smithsonian Institution, the Secretaría de Ciencia y Técnica de la República Argentina (SECYT), the Consejo de Desarrollo Cientifico y Humanístico de la Universidad de Los Andes, the Consejo Nacional de Investigaciones Científicas y Tecnológicas de Venezuela (CONICYT), the Decananato de Investigaciones de la Universidad Simón Bolívar, and the Consejo Nacional de Investigaciones Científicas y Técnicas de Argentina (CONICET).

REFERENCES

Ahlfeld, F. 1970. Zur Tektonik des andinen Bolivien. *Geol. Rundsch.* 59: 1124–1140.

Allen, J. A. 1916. List of mammals collected in Colombia by the American Museum of Natural History expedition, 1910–1915. *Bull. Amer. Mus. Nat. Hist.* 35: 191–238.

Anthony, H. E. 1924. Preliminary report on Ecuadorian mammals. No. 4. *Amer. Mus. Novitates* 114: 1–6.

Barros, M. A., and O. A. Reig. 1979. Doble polimorfismo robertsoniano en *Microxus bogotensis* (Rodentia, Cricetidae) del Páramo de Mucubají (Mérida, Venezuela). *Acta Cient. Venez.* 30 (supp. 1): 106.

Baskin, J. A. 1978. *Bensonomys*, *Calomys*, and the origin of the phyllotine group of neotropical cricetines (Rodentia, Cricetidae). *J. Mammal.* 59: 125–135.

Berggren, W. A., and C. D. Hollister. 1974. Paleogeography, paleobiogeography and the history of circulation in the Atlantic Ocean. Studies in paleo-oceanography, ed. W. W. Hay. *Soc. Econ. Paleont. Mineral.*, spec. publ., 20: 126–186.

Bianchi, N. O., O. A. Reig, O. J. Molina, and N. F. Dulout. 1971. Cytogenetics of the South American akodont rodents (Cricetidae). I. A progress report of Argentinian and Venezuelan forms. *Evolution* 25: 724–736.

Brum, N. 1965. Investigaciones citogenéticas sobre algunas especies de Cricetinae (Rodentia) del Uruguay. *Anais Segundo Congr. Latino-Americano Zool.* 2: 315–320.

Brum-Zorrilla, N., N. Lafuente, and P. Kiblisky. 1972. Cytogenetic studies in the cricetid rodent *Scrapteromys tumidus* (Rodentia-Cricetidae). *Experientia* 28: 1373.

Cabrera, A. 1953. Los roedores argentinos de la familia Caviidae. *Fac. Agron. Vet. Univ. Buenos Aires, Escuela de Veterinaria Publicación* no. 6: 1–93.

———. 1961. Catálogo de los mamíferos de América del Sur. *Rev. Mus. Argentino Cien. Nat. "Bernardino Rivadavia," Cien. Zool.* 4: xxii, 309–732, frontispiece.

Carleton, M. D. 1980. Phylogenetic relationships in neotomine-peromyscine rodents (Muroidea) and reappraisal of the dichotomy within New World Cricetinae. *Misc. Publ. Mus. Zool. Univ. Michigan*, no. 157: 1–146.

Chapman, F. M. 1921. The distribution of bird-life in the Urubamba Valley of Peru. *Bull. U. S. Nat. Mus.* 117: 1–138.

Clark, J. B., M. R. Dawson, and A. E. Wood. 1964. Fossil mammals from the lower Pliocene of Fish Lake Valley, Nevada. *Bull. Mus. Comp. Zool. Harv.* 131: 27–63.

Cox, C. B. 1974. Vertebrate palaeodistributional patterns and continental drift. *J. Biogeog.* 1: 75–94.

Croizat, L., G. Nelson, and D. E. Rosen. 1974. Centers of origin and related concepts. *Syst. Zool.* 23: 265–287.

Cuatrecasas, J. 1957. A sketch of the vegetation of the north-andean province. *Proc. 8th. Pac. Sci. Congr.* 9: 167–173.

Dorst, J. 1958. Contribution à l'étude écologique des rongeurs des hauts plateaux du Pérou méridional. *Mammalia* (Paris) 22: 547–565.

Ellerman, J. R., 1941. *The families and genera of living rodents*. Vol. 2, *Family Muridae*. London: British Museum of Natural History.

Estes, R., and O. A. Reig. 1973. The early fossil record of frogs. A review of the evidence. In *Evolutionary biology of the anurans*, ed. J. L. Vial. Columbia: Univ. of Missouri Press, pp. 11–63.

Estes, R., and R. Wasserzug. 1963. A Miocene toad from Colombia, South America. *Breviora Mus. Comp. Zool.* 193: 1–13.

Gallardo, M. 1979. Las especies chilenas de *Ctenomys* (Rodentia, Octodontidae). I. Estabilidad cariotípica. *Arch. Biol. Med. Exper.* 12: 71–82.

Gardner, A. L., and J. L. Patton. 1976. Karyotypic variation in oryzomyine rodents (Cricetinae) with comments on chromosomal evolution in the Neotropical cricetine complex. *Occ. Papers Mus. Zool. Louisiana State Univ.*, no. 49: 1–48.

Gentile de Fronza, T. 1970. Cariotipo de *Scapteromys tumidus aquaticus* (Rodentia, Cricetidae). *Physis* 30: 343.

George, W., and B. J. Weir. 1974. Hystricomorph chromosomes. *Symp. Zool. Soc. London* 34: 79–108.

Gidley, J. W. 1922. Preliminary report of fossil vertebrates of the San Pedro Valley, Arizona, with descriptions of new species of Rodentia and Lagomorpha. *Prof. Pap. United States Geol. Surv.*, no. 131: 119–130.

Groeber, P. 1951. La Alta Cordillera entre las latitudes 34° y 29°30′ S. *Rev. Inst. Nac. Invest. Cien. Nat. y Museo Argentino de Cien. Nat. "Bernardino Rivadavia," Cien. Geol.* 1: 235–252.

Gyldenstolpe, N. 1932. A manual of Neotropical sigmodont rodents. *Kungl. Svensk. Vetenskapsakad. Handl.* 11: 1–164.

Haffer, J. 1969. Speciation in Amazonian forest birds. *Science* 165: 131–137.

———. 1970. Geologic-climatic history and zoogeographic significance of the Urabá region in northwestern Colombia. *Caldasia* 10: 603–636.

———. 1974. Avian speciation in tropical South America. *Publ. Nuttall Ornithol. Club* 14: 1–390.

———. 1977. Pleistocene speciation in Amazonian birds. *Amazoniana* 6: 161–191.

Handley, C. O. 1976. Mammals of the Smithsonian-Venezuelan Project. *Brigham Young Univ. Sci. Bull., Biol. Ser.* 20: 1–89.

Harrington, H. J. 1956. An explanation of the geological map of South America. *Mem. Geol. Soc. Amer.* 65: xii–xviii.

Hershkovitz, P. 1944. A systematic review of the Neotropical water rats of the genus *Nectomys* (Cricetinae). *Misc. Publ. Mus. Zool. Univ. Michigan*, no. 58. 1–101.

———. 1948. Mammals of northern Colombia. Preliminary report no. 2: Spiny rats (Echimyidae), with supplemental notes on related forms. *Proc. U. S. Nat. Mus.* 97: 125–140.

———. 1955. South American marsh rats, genus *Holochilus*, with a summary of sigmodont rodents. *Fieldiana, Zool.* 37: 639–687.

———. 1960. Mammals of northern Colombia. Preliminary report no. 8: Arboreal rice rats, a systematic revision of the subgenus *Oecomys*, genus *Oryzomys*. *Proc. U. S. Nat. Mus.* 110: 513–568.

———. 1962. Evolution of Neotropical cricetine rodents (Muridae) with special reference to the phyllotine group. *Fieldiana, Zool.* 46: 1–525.

———. 1966a. South American swamp and fossorial rats of the scapteromyine group (Cricetinae, Muridae), with comments on the gland penis in murid taxonomy. *Z. Säugetierk.* 31: 81–149.

———. 1966b. Mice, land bridges and Latin American faunal interchange. In *Ectoparasites of Panama*, ed. R. L. Wenzel and V. J. Tipton. Chicago: Field Museum of Natural History, pp. 725–751.

———. 1972. The recent mammals of the Neotropical Region: A zoogeographic and ecological review. In *Evolution, mammals and southern continents*, ed. A. Keast, F. C. Erk, and B. Glass. Albany: State Univ. of New York Press, pp. 311–431.

Hoffstetter, R. 1975. El origen de los Caviomorpha y el problema de los Hystricognathi (Rodentia). *Actas 1° Congr. Argentino Paleont. y Bioestrat.* (Tucumán, Argentina, 1974) 2:

———. 1976. Histoire des mammifères et dérive des continents. *La Recherche* 7: 124–138.

Hoffstetter, R., and R. Lavocat. 1970. Découverte dans le Déséadien de Bolivie de genres pentalophodontes appuyant les affinités africaines des rongeurs caviomorphes. *C.R. Acad. Sci.* Paris (D) 271: 172–175.

Honacki, J. S., K. E. Kinnan, and J. W. Koeppl. 1982. *Mammal species of the world.* Lawrence, Kansas: Allen Press.

Husson, A. M. 1978. *The mammals of Surinam.* Leyden: Brill.

James, G. T. 1963. Paleontology and nonmarine stratigraphy of the Cuyama Valley badlands, California. *Univ. Calif. Publ. Geol. Sci.* 45: 1–145.

Jenks, W. F. 1956. Peru. *Mem. Geol. Soc. Amer.* 65: 213–247.

Kratochvíl, J., L. Rodríguez, and V. Barus. 1978. Capromyinae (Rodentia) of Cuba I. *Acta Sc. Nat. Brno* 12: 1–60.

Lavocat, R. 1971. Affinités systematiques des caviomorphes et des phiomorphes et origine africaine des caviomorphes. *Anais Acad. Bras. Cienc.*, supp. 43: 515–522.

———. 1973. Les rongeurs du Miocene d'Afrique orientale. *Mém. et Trav. Inst. Montpellier, E.P.H.E.* 2: 1–284.

———. 1974. What is an hystricomorph? *Symp. Zool. Soc. London* 34: 7–20.

Lohmann, H. H. 1970. Outline of tectonic history of Bolivian Andes. *Bull. Amer. Assoc. Petr. Geol.* 54: 735–757.

Mares, M. A. 1975. South American mammal zoogeography: Evidence from convergent evolution in desert rodents. *Proc. Nat. Acad. Sci. U.S.A.* 72: 1702–1706.

———. 1980. Convergent evolution among desert rodents: A global perspective. *Bull. Carnegie Mus. Nat. Hist.* 16: 1–51.

Marshall, L. G. 1979. A model for paleobiogeography of South American cricetine rodents. *Paleobiology* 5: 126–132.

Marshall, L. G., R. F. Butler, R. E. Drake, G. H. Curtis, and R. H. Tedford. 1979. Calibration of the Great American Interchange. *Science* 204: 272–279.

Martin, L. D. 1980. The early evolution of the Cricetidae in North America. *Univ. Kansas Paleont. Contr.* 102: 1–42.

Massoia, E. 1979. Descripción de un género y especie nuevos: *Bibimys torresi* (Mammalia-Rodentia-Cricetidae-Sigmodontinae-Scapteromyini). *Physis*, sec. C, 38: 1–7.

———. 1980. El estado sistemático de cuatro especies de cricétidos sudamericanos y comentarios sobre otras congenéricas (Mammalia-Rodentia). *Ameghiniana* 17: 280–287.

Meserve, P. L. 1981. Trophic relationships among small mammals in a Chilean semiarid thorn scrub community. *J. Mammal.* 62: 304–314.

Moojen, J. 1948. Speciation in the Brazilian spiny rats (genus *Proechimys*, family Echimyidae). *Univ. Kansas Mus. Nat. Hist.* publ. 1: 301–406.

———. 1952. *Os roedores do Brasil.* Rio de Janeiro: Min. Educ. e Saúde, Inst. Nac. do Livro, Bibl. Cien. Brasil. Ser. A, II: 1–214.

Moore, J. C. 1959. Relationships among living squirrels of the Sciurinae. *Bull. Amer. Mus. Nat. Hist.* 113: 1–71.

———. 1961. The spread of existing diurnal squirrels across the Bering and Panamanian land bridges. *Amer. Mus. Novitates* 2044: 1–26.

Müller, P. 1973. *The dispersal centres of terrestrial vertebrates in the Neotropical realm.* The Hague: Junk.

Musser, G. G., and A. L. Gardner. 1974. A new species of the ichthyomyine *Daptomys* from Perú. *Amer. Mus. Novitates* 2537: 1–23.

Nevo, E., Y. J. Kim, C. Shaw, and C. S. Thaeler, Jr. 1974. Genetic variation, selection and speciation in *Thomomys talpoides* pocket gophers. *Evolution* 28: 1–23.

Ojasti, J. 1972. Revisión preliminar de los picures o agutíes de Venezuela (Rodentia, Dasyproctidae). *Mem. Soc. Cien. Nat. La Salle* no. 93, vol. 32: 159–204.

Oliveira Pinto, O. M. de. 1931. Ensaio sobre a fauna de Sciurideos do Brasil consoante sua representação nas collec c ̃es do Museo Paulista. *Rev. Mus. Paulista* 17: 263–321.

Osgood, W. H. 1943a. A new genus of rodents from Perú. *J. Mammal.* 24: 369–371.

———. 1943b. The mammals of Chile. *Field Mus. Nat. Hist. Zool. Ser. Publ.* 542: 1–268.

Pascual, R. 1967. Los roedores Octodontoidea (Caviomorpha) de la formación Arroyo Chasicó (Plioceno inferior) de la Provincia de Buenos Aires. *Rev. Mus. La Plata* (n.s.) *Sec. Paleont.* 5: 259–282.

Pascual, R., J. Pisano, and E. J. Ortega. 1965. Un nuevo Octodontidae (Rodentia, Caviomorpha) de la formación Epecuén (Plioceno medio) de Hidalgo (Provincia de La Pampa). Consideraciones sobre los Ctenomyinae Reig, 1958, y la morfología de sus molariformes. *Ameghiniana* 4: 19–29.

Patterson, B., and R. Pascual. 1968. New echimyid rodents from the Oligocene of Patagonia, and a synopsis of the family. *Breviora Mus. Comp. Zool.* 301: 1–14.

———. 1972. The fossil mammal fauna of South America. In *Evolution, mammals and southern continents*, ed. A. Keast, F. C. Erk, and B. Glass. Albany: New York State Univ. Press, pp. 247–309.

Patterson, B., and A. E. Wood. 1982. Rodents of the Deseadan Oligocene of Bolivia and the relationships of the Caviomorpha. *Bull. Mus. Comp. Zool.* 149: 371–543.

Pearson, O. P. 1951. Mammals in the highlands of southern Perú. *Bull. Mus. Comp. Zool.* 106: 117–174.

———. 1958. A taxonomic revision of the rodent genus Phyllotis. *Univ. Calif. Publ. Zool.* 54: 391–496.

———. 1982. Distribución de pequeños mamíferos en el altiplano y los desiertos de Perú. *Actas VIII Congr. Latinoamer. Zool., Mérida, Venezuela*: 263–284.

Pearson, O. P., and J. L. Patton. 1976. Relationships among South American phyllotine rodents based on chromosome analysis. *J. Mammal.* 57: 339–350.

Petter, F. 1973. Les noms de genre *Cercomys*, *Nelomys*, *Trichomys* et *Proechimys*. *Mammalia* (Paris) 37: 422–426.

Pielou, E. C. 1979. *Biogeography*. New York: Wiley.

Pine, R. H., S. D. Miller, and M. L. Schamberger. 1979. Contributions to the mammalogy of Chile. *Mammalia* (Paris) 43: 339–376.

Raven, P. H., and D. I. Axelrod. 1975. History of the flora and fauna of Latin America. *Amer. Scient.* 63: 420–429.

Reig, O. A. 1975. Diversidad, historia evolutiva y dispersión de los roedores cricétidos sudamericanos. *Acta Cient. Venez.* 26 (supp. 1): 7.

———. 1977. A proposed unified nomenclature for the enamelled components of the molar teeth of the Cricetidae (Rodentia). *J. Zool., Lond.* 181: 227–241.

———. 1978. Roedores cricétidos del Plioceno superior de la Provincia de Buenos Aires (Argentina). *Publ. Mus. Munic. Cienc. Nat. "Lorenzo Scaglia"* 2: 164–190.

———. 1979. Proposiciones para una solución al problema de la realidad de las especies biológicas. *Rev. Venez. Filos.* 11: 79–106.

———. 1980. A new fossil genus of South American cricetid rodents allied to *Wiedomys*, with an assessment of the Sigmodontinae. *J. Zool., Lond.* 192: 257–281.

———. 1981a. Breve reseña del estado actual de la teoría de la especiación. In *Ecología y genética de la especiación animal*, O. A. Reig, ed., pp. 11–42. Caracas: Editorial Equinoccio, Universidad Simón Bolivar.

———. 1981b. Modelos de especiación cromosómica en las casiraguas (genus *Proechimys*) de Venezuela. In *Ecología y genética de la especiación animal*, O. A. Reig, ed., pp. 149–190. Caracas: Editorial Equinoccio, Universidad Simón Bolivar.

———. 1981c. Teoría del origen y desarrollo de la fauna de mamíferos de América del Sur. *Monogr. Naturae, Mus. Munic. Cienc. Nat. "Lorenzo Scaglia"* 1: 1–161.

Reig, O. A., and P. Kiblisky. 1969. Chromosome multiformity in the genus *Ctenomys* (Rodentia, Octodontidae). A progress report. *Chromosoma* (Berlin) 28: 211–244.

Reig, O. A., M, A. Barros, M. Useche, M. Aguilera, and O. Linares. 1979a. The chromosomes of spiny rats, *Proechimys trinitatis*, from Trinidad and eastern Venezuela (Rodentia, Echimyidae). *Genetica* 51: 153–158.

Reig, O. A., M. Tranier, and M. A. Barros. 1979b. Sur l'identification chromosomique de *Proechimys guyannensis* (E. Geoffroy) 1803, et de *Proechimys cuvieri* Petter 1978 (Rodentia, Echimyidae). *Mammalia* (Paris) 43: 35–39.

Reig, O. A., Aguilera, M. A. Barros, and M. Useche. 1980. Chromosomal speciation in a Rassenkreis of Venezuelan spiny rats (genus *Proechimys*; Rodentia: Echimyidae). Animal Genetics and Evolution, ed. N. N. Vorontzov and J. M. van Brink, *Genetica* 53–54: 291–312.

Reinig, W. F. 1950. Chorologische Voraussetzungen für die Analyse von Formenkreisen. In *Syllegomena Biologica, Festschrift Otto Kleinschmidt*, A. von Jordans and F. Peus, eds., pp. 346–378. Wittenberg: Ziemsen.

Rosen, D. E. 1975. The vicariance model of Caribbean biogeography. *Syst. Zool.* 24: 431–464.

Sarich, V. M., and J. E. Cronin. 1980. South American mammal molecular systematics, evolutionary clocks, and continental drift. In *Origin of New World monkeys and continental drift*, B. Chiarelli and R. Ciochon, eds., pp. 399–421. New York: Plenum Press.

Sarmiento, G. 1976. Evolution of the arid vegetation in tropical America. In *Evolution of desert biota*, D. S. Goodall, ed., pp. 69–99. Austin: Univ. of Texas Press.

Savage, J. M. 1974. The isthmian link and the evolution of Neotropical mammals. *Contr. Sci., Nat. Hist. Mus. Los Angeles County*. 260: 1–51.

Shamel, H. J. 1931. *Akodon chacoensis*, a new cricetid rodent from Argentina. *J. Wash. Acad. Sci.* 21: 427–429.

Simpson, B. B. 1975. Pleistocene changes in the flora of the high tropical Andes. *Paleobiology* 1: 273–294.

Simpson, G. G. 1950. History of the fauna of Latin America. *Amer. Scient.* 38: 361–389.

———. 1969. South American mammals. In *Biogeography and ecology of South America*, ed. E. J. Fittkau et al. The Hague: Junk, pp. 879–909.

———. 1980. *Splendid isolation. The curious history of South American mammals.* New Haven: Yale Univ. Press.

Solbrig, O. T. 1976. The origin and floristic affinities of the South American temperate desert and semidesert regions. In *Evolution of desert biota*, D. W. Goodall, ed., pp. 7–49. Austin: Univ. of Texas Press.

Soukup, J. 1961. Materiales para el catálogo de los mamíferos peruanos. *Biota* 3: 238–276.

Spotorno, A. E. 1976. Análisis taxonómico de tres especies altiplánicas del género *Phyllotis* (Rodentia, Cricetidae). *Anal. Mus. Hist. Nat. Chile* 9: 141–161.

Taseer Hussain, S., H. de Bruijn, and J. M. Leinders.

1978. Middle Eocene rodents from the Kala Chitta Range (Punjab, Pakistan) (III). *Proc. Konink. Nederl. Akad. Wetenshappen* (b) 81: 101–112.

Tate, G. H. H. 1932a. The taxonomic history of the genus *Reithrodon* Waterhouse (Cricetidae). *Amer. Mus. Novitates* 529: 1–4.

———. 1932b. The taxonomic history of the South American cricetid genera *Euneomys* (subgenera *Euneomys* and *Galenomys*), *Auliscomys*, *Chelemiscus*, *Chinchillula*, *Phyllotis*, *Paralomys*, *Graomys*, *Eligmodontia* and *Hesperomys*. *Amer. Mus. Novitates* 541: 1–21.

———. 1932c. The taxonomic history of the Neotropical cricetid genera *Holochilus*, *Nectomys*, *Scapteromys*, *Megalomys*, *Tylomys* and *Ototylomys*. *Amer. Mus. Novitates* 562: 1–19.

———. 1932d. The taxonomic history of the South and Central American cricetid rodents of the genus *Oryzomys*. Part 1: subgenus *Oryzomys*. *Amer. Mus. Novitates* 579: 1–18.

———. 1932e. The taxonomic history of the South and Central American cricetid rodents of the genus *Oryzomys*. Part 2: subgenera *Oligoryzomys*, *Thallomyscus*, and *Melanomys*. *Amer. Mus. Novitates* 580: 1–17.

———. 1932f. The taxonomic history of the South and Central American oryzomine genera of rodents (excluding *Oryzomys*): *Nesoryzomys*, *Zygodontomys*, *Chilomys*, *Delomys*, *Phaenomys*, *Rhagomys*, *Rhipidomys*, *Nyctomys*, *Oecomys*, *Thomasomys*, *Inomys*, *Aepeomys*, *Neacomys* and *Scolomys*. *Amer. Mus. Novitates* 581: 1–28.

———. 1932g. The taxonomic history of the South and Central American akodont rodent genera: *Thalpomys*, *Deltamys*, *Thaptomys*, *Hypsimys*, *Bolomys*, *Chroemys*, *Abrothrix*, *Scotinomys*, *Akodon* (*Chalcomys* and *Akodon*), *Microxus*, *Podoxymys*, *Lenoxus*, *Oxymycterus*, *Notiomys*, and *Blarinomys*, *Amer. Mus. Novitates* 582: 1–32.

———. 1932h. The taxonomic history of certain South and Central American cricetid Rodentia: *Neotomys*, with remarks upon its relationships; the cotton rats (*Sigmodon* and *Sigmomys*); and the "fish-eating" rats (*Ichthyomys*, *Anotomys*, *Rheomys*, *Neusticomys*, and *Daptomys*). *Amer. Mus. Novitates* 583: 1–10.

———. 1935. The taxonomy of the genera of Neotropical hystricoid rodents. *Bull. Amer. Mus. Nat. Hist.* 68: 295–447.

———. 1939. The mammals of the Guiana region. *Bull. Amer. Mus. Nat. Hist.* 76: 151–229.

Thomas, O. 1917. Preliminary diagnoses of new mammals obtained by the Yale-National Geographic Society Peruvian Expeditions. *Smith. Misc. Coll.* 68: 1–3.

———. 1920. On mammals from near Tinogasta, Catamarca, collected by Sr. Budin. *Ann. Mag. Nat. Hist.* (9) 6: 116–120.

———. 1926. The Godman-Thomas expedition to Peru. I. On mammals collected by Mr. R. W. Hendee near Lake Junín. *Ann. Mag. Nat. Hist.* (9) 17: 313–318.

Tranier, M. 1976. Nouvelles données sur l'évolution non parallèle du caryotype et de la morphologie chez les Phyllotinés (Rongeurs, Cricétidés). *C. R. Acad. Sci. Paris* 283, ser. D: 1201–1203.

van der Hammen, T. 1961. Late Cretaceous and Tertiary stratigraphy and tectogenesis of the Colombian Andes. *Geol. Mijnbouw* 40: 181–188.

———. 1972. Historia de la vegetación y el medio ambiente del norte sudamericano. *Mem. Symp. I. Congr. Latin. -Amer. Mex. Bot., Mexico*: 119–134.

Vanzolini, P. E., and E. E. Williams. 1970. South American anoles: The geographic differentiation and evolution of the *Anolis chrysolepis* species group (Sauria, Iguanidae). *Arq. Zool., S. Paulo* 19: 1–298.

Venegas, W. 1974. Variación cariotípica en *Phyllotis micropus micropus* Waterhouse (Rodentia, Cricetidae). *Bol. Soc. Biol. Concepción* 48: 69–76.

Villarroel, C. A. 1975. Dos nuevos Ctenomyinae (Caviomorpha, Rodentia) en los estratos de la formación Umala (Plioceno superior) de Vizcachani (Prov. Aroma. Dpto. La Paz. Bolivia). *Actas. Prim. Congr. Argent. Paleont. Bioestrat., Tucumán*, 12–16 August 1974, vol.1: 495–503.

Vorontzov, N. N. 1967. *Evoliutzia pishtchevaritelnoi sistemy gryzunov (mysheobraznye)* (Evolution of the digestive system of Muroid rodents). Novosibirsk: Izdatel'stvo "Nauka" Sib. Otd. Akad. Nauk SSSR.

Voss, R. S., and A. V. Linzey. 1981. Comparative gross morphology of male accessory glands among Neotropical Muridae (Mammalia: Rodentia) with comments on systematic implications. *Misc. Publ. Mus. Zool. Univ. Michigan*, no. 159: 1–41.

Vuilleumier, F. 1969. Pleistocene speciation in birds living in the high tropical Andes. *Nature* (London) 223: 1179–1180.

———. 1975. Zoogeography. In *Avian biology*, Vol. 5, D. S. Farner and J. R. King, eds., pp. 421–496. New York: Academic Press.

———. 1978. Qu'est-ce que la biogéographie? *C. R. Soc. Biogéogr. Paris* 475: 41–66.

———. 1979. Comparación y evolución de las comunidades de aves de páramo y puna. In *El medio ambiente páramo*, M. L. Salgado-Labouriau, ed., pp. 181–205. Caracas: Edic. Centro de Estudios Avanzados, IVIC.

Vuilleumier, F., and D. Simberloff. 1980. Ecology versus history as determinants of patchy and insular distributions in high Andean birds. In *Evolutionary Biology*, Vol. 12, M. K. Hecht, W. C. Steere, and B. Wallace, eds., pp. 236–379. New York: Plenum.

Walker, L. I., A. E. Spotorno, and R. Fernández-Donoso. 1979. Conservation of whole arms during chromosomal divergence of phyllotine rodents. *Cytogenet. Cell Genet.* 24: 209–216.

Webb, S. D. 1974. Pleistocene llamas of Florida, with a brief review of the Lamini. In *Pleistocene mammals of Florida*, S. D. Webb, ed., pp. 170–213. Gainesville: Univ. of Florida Press.

Weberbauer, A. 1936. Phytogeography of the Peruvian Andes. *Field Mus. Nat. Hist., Bot. Ser.* 13: 13–81.

Wenzel, R. L., and V. J. Tipton. 1966. Some relationships between mammal hosts and their ecotoparasites. In *Ectoparasites of Panama*, R. L. Wenzel and V. J. Tipton, eds., pp. 677–723. Chicago: Field Museum of Natural History.

Wetzel, R. M., and J. W. Lovett. 1974. A collection of mammals from the Chaco of Paraguay. *Univ. Connecticut Occ. Pap. Biol. Sci.*, ser. 2: 203–216.

Williams, D. F., and M. A. Mares. 1978. A new genus and species of phyllotine rodent (Mammalia: Muridae) from northwestern Argentina. *Ann. Carnegie Mus.* 47: 193–221.

Wood, A. E. 1974. The evolution of the Old World and New World Hystricomorphs. *Symp. Zool. Soc. London* 34: 21–60.

———.1975. The problems of the Hystricognathous rodents. *Papers in Paleont.* 12: 75–80.

Wood, A. E., and B. Patterson. 1959. The rodents of the Deseadan Oligocene of Patagonia and the beginning of South American rodent evolution. *Bull. Mus. Comp. Zool. Harvard* 120: 279–428.

Woods, C. A., and E. B. Howland. 1979. Adaptive radiation of capromyid rodents: Anatomy of the masticatory apparatus. *J. Mammal.* 60: 95–116.

Ximenez, A., A. Langguth, and R. Praderi. 1972. Lista sistemática de los mamíferos del Uruguay. *An. Mus. Nac. Hist. Nat. Montevideo* (2) 7, no. 5: 1–49.

Yonenaga, Y. 1975. Karyotypes and chromosome polymorphism in Brazilian rodents. *Caryologia* 28: 269–286.

Zimmerman, E. G. 1970. Karyology, systematics and chromosomal evolution in the rodent genus *Sigmodon*. *Publ. Mus. Michigan State Univ. Biol.*, ser. 4: 389–454.

V

ORIGINS OF SELECTED FLORAS AND FAUNAS

Webster's Ninth New Collegiate Dictionary defines the word *origin* as follows: "1: ancestry, parentage; 2a: rise, beginning, or derivation from a source; b: the point at which something begins or rises or from which it derives." In studying the origins of a modern fauna or flora, or of part of a modern flora or fauna, similarly, the biologist must determine relationships of parentage and/or ancestry (in a taxonomic or phylogenetic sense), and determine or assess the source of the flora or fauna, and how it derived from that source (in a geographic sense as well as a taxonomic one).

In this book, floras or faunas are delimited by environmental factors within a given geographic area, and so we speak, for instance, of the afroalpine flora, or of the avifauna of the páramo and the puna. In other words, generally speaking, each flora and fauna is defined ecologically by the boundaries of the high tropical biomes found above the upper limit of continuous forest. Flora or fauna then simply means the sum of species of a given taxon (or group of taxa) living and reproducing in a given ecologically delimited area. The flora or fauna so defined is then usually analyzed in terms of components, often called floral or faunal elements, or at times floral or faunal groups (as distinct from "taxonomic groups," which can be a group of related or unrelated species, or a group of genera). The units of analysis in the analytical process of subdividing the total flora or fauna into elements are usually taxonomic, most often at the species or genus level.

A number of questions are asked, and sometimes answered, in studies of the origins of a modern fauna or flora. The ultimate question, or series of questions, is of the form: Where does a given flora (fauna) come from? But this question really is: What are the floral (faunal) elements of a given flora (fauna), and where does each come from? Mayr (1965) analyzed the concept of fauna, and defined the term thus: "A fauna consists of the kinds of animals found in an area as a result of the history of the area and its present ecological conditions" (p. 484). He added (p. 484):

> One might even include in the definition a reference to the isolation that is a prerequisite for the development of a truly distinctive fauna. Or one might stress the fact that a fauna consists of faunal elements which differ in origin, age, and adaptation. Perhaps we need more faunal analyses before a satisfactory definition can be attempted.

Udvardy (1969, p. 282) further discussed the concept of fauna, and did so also in terms of elements:

> *Fauna* may be understood to have three distinct meanings: (1) the species list of the total faunation of any geographic entity, delimited naturally (topographically) or arbitrarily, e.g., by political boundaries; (2) the species list of a single major taxon (order, class, etc.) on a geographic entity (also called avifauna, malacofauna, entomofauna, and so on); (3) the species list of an ecological unit (a community, biome, and the like).

Similar remarks can be made about the concept of flora in these two quotations by simply transposing the word flora for fauna. In the present book, the chapters of Part V thus seem to be a combination of all three of the meanings defined by Udvardy, and seem to follow the general ideas put forth by Mayr.

It is worth noting here that biogeographers who place themselves in the so-called vicariance school, and who use cladistic methods of taxonomic analysis in order to reconstruct phylogenetic histories, would probably disapprove of the procedures outlined by Mayr and Udvardy. Nelson and Platnick (1981), for instance, discuss in detail the methods used by such biologists. We do not want to enter here into a philosophical assessment of the relative merits of one or the other of these views, for such is not the goal of this book. The different chapters in part V provide the reader with a panorama of ideas concerning floral or faunal origins, whose range touches in Chapter 19 on the cladistic methodology advocated by Nelson and Platnick (1981).

Once more, as in previous parts of this book, a hierarchical approach is evident. Faunal (or floral) elements, as well as endemic taxa, are usually classified into different levels including the species, the genus, and the family. The problem of the ecological characteristics of an area considered to be a source of floral or faunal elements is repeatedly approached. The question of the relative amount of active or passive dispersal or, on the contrary, the relative amount of geographic stability, of given elements is one of major concern in faunal or floral reconstructions. When geographic stability is presumed, then the next question becomes one of adaptability to changing conditions. The significance of range discontinuities is another pattern that recurs in part V. Finally, and ultimately in some sense, all faunal or floral origin analysis rests on sound taxonomic knowledge, and hence an evaluation of taxonomic resemblances between and among floras or faunas. This is perhaps where methodologies in today's biogeography are in greatest contrast, since it appears that cladists do not admit any other form of analysis but theirs.

Of the several patterns that emerge from the different chapters of part V, perhaps the most conspicuous is that of the generally low endemism at the genus level, but relatively high endemism at the species level (at least in some lineages), in the floras and faunas of the high tropical montane ecosystems. The second pattern is the relative differences between different taxa (e.g., Lepidoptera versus Aves) in the range of physiological adaptations to the high tropical environment. Thus, whereas birds do not seem, on the whole, to "need" special physiological mechanisms to be able to survive at high tropical altitudes, butterflies apparently do. If this is the case, then taxonomic constraints exist that might restrict the possibility, for a given animal or plant group, to either successfully invade the high tropical environments from elsewhere, especially from low altitudes, or to suceed there by progressive and adaptive modifications in situ.

Part V of this book thus comes back to some of the problems evoked in part II and shows us that a study of adaptations in the present-day components of the biota cannot be undertaken in intellectual isolation, and can only be nourished by the knowledge of the historical antecedents of these components. The development of some components of the biota was studied historically on the basis of fossil evidence in part III, in which the procedure was essentially to go from the past toward the present. In part V, by contrast, attempts are made to retrace the past history of the modern elements, in a procedure, or rather in a series of procedures, that attempt to reconstruct the course of history by matching inferences from the present with information on the past (see, e.g., Vuilleumier, 1980, for a critical discussion of these procedures in speciation studies).

REFERENCES

Mayr, E. 1965. What is a fauna? *Zool. Jb. Syst.* 92: 473–486.

Nelson, G., and N. Platnick. 1981. *Systematics and biogeography. Cladistics and vicariance.* New York: Columbia Univ. Press.

Udvardy, M. D. F. 1969. *l ynamic zoogeography.* New York: Van Nostrand Reinhold.

Vuilleumier, F. 1980. Reconstructing the course of speciation. *Acta XVII Congr. Int. Ornithol., Berlin 1978,* Vol. II: 1296–1301.

17

Origins of the Afroalpine Flora

OLOV HEDBERG

The major parts of tropical East Africa and Ethiopia consist of vast plateaus at altitudes of 1000–2000 m in East Africa and 1000–3000 m in Ethiopia. These plateaus are dissected in the south by two longitudinal rift valleys, the eastern one of which continues northward through Ethiopia. Along these rifts are scattered a number of isolated high mountains, mainly of volcanic origin, several of which reach altitudes between 4000 and 6000 m. The highest summits in East Africa are the Virunga volcanoes and Ruwenzori along the western rift; Elgon, Aberdare, Mt. Kenya, Kilimanjaro, and Mt. Meru along the eastern rift (cf. Hedberg, 1951). In Ethiopia at least 10 mountains reach altitudes above 4000 m, notably the Semien, Gughé and Balé mountains.

These high mountains display vegetation zonation reflecting gradual climatic changes with increasing altitude. In East Africa most of this zonation, comprising a montane forest belt, an ericaceous belt, and an afroalpine belt (Hedberg, 1951) is still relatively intact (Fig. 17–1). In Ethiopia, however, only fragments of the montane forest belt remain, and the ericaceous and afroal-

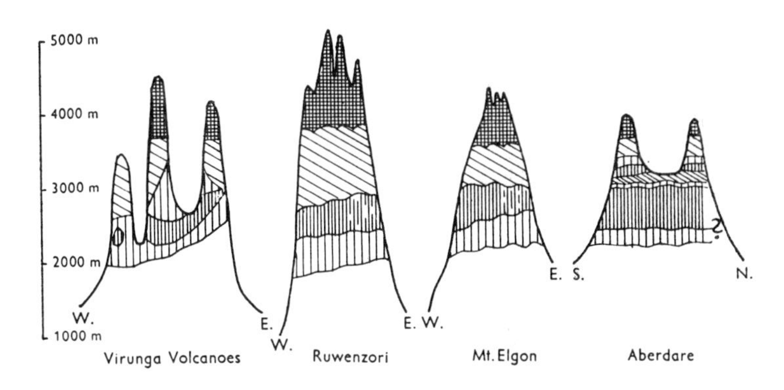

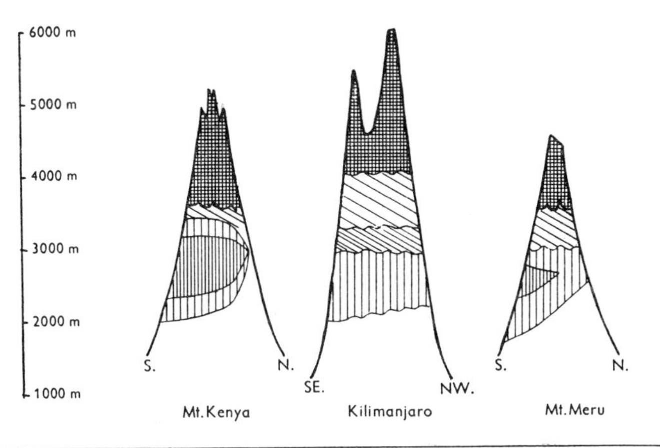

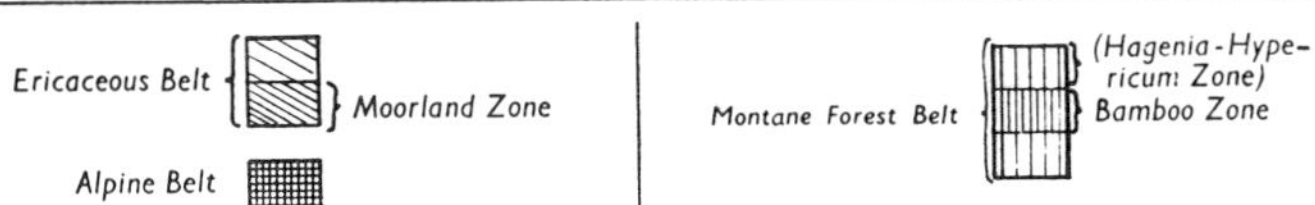

FIG. 17–1. Schematic profiles showing the vegetation zonation of the seven highest mountains of East Africa. The wettest side of each mountain is on the left; the letters at their base show the orientation. From Hedberg (1951).

pine belts have been intensely degraded through burning, firewood collecting, grazing, and agriculture (Scott, 1952, 1958; Hedberg, 1971, 1978).

The montane forest belt extends between 1700–2300 m and 3000–3300 m, and is distinguished by the occurrence of a great number of hardwood trees and some conifers, containing sometimes a marked zone of bamboo (*Arundinaria alpina*). The ericaceous belt occurs between 3000–3300 m and 3550–4100 m, being dominated by trees and shrubs of *Erica arborea* and a couple of *Philippia* species, which in many areas have been crippled by fire. The afroalpine belt, finally, starts above the ericaceous belt (that is, above the upper limit of more or less continuous ericaceous (tree or shrub) vegetation) and extends to the highest summits.

The annual mean temperatures of the afroalpine belt resemble those of a high-latitude alpine belt or of the Arctic region. But as expertly described by Troll (1943, 1944, 1948, 1955, 1959, and other papers), there are fundamental differences between the climates of tropical-alpine environments and those of extratropical high mountains and the Arctic. The former have important diurnal variations but insignificant seasonal variations in temperature—in effect they have "winter every night and summer every day" (Hedberg, 1957; cf. Rauh, 1977; see Chap. 2). The harsh conditions evidently impose a strong climatic stress acting as a selective filter (Larcher, 1980), which allows only a limited number of species to survive.

The flora of this milieu, the afroalpine flora (Hauman, 1933), differs so much from the flora at lower levels on the same mountains and in surrounding uplands that it has been treated as a separate floristic region (Hauman, 1955; Monod, 1957; Hedberg, 1965; White, 1970). Its delimitation from the less elevated afromontane flora (White, 1970, 1978) is not easy, however. Few afroalpine species are entirely restricted to the afroalpine belt; most of them occur also in the ericaceous belt and many even in the montane forest belt (Fig. 17–2; Hedberg, 1957, 1969).

The afroalpine flora of Ethiopia shows broad similarities to that of tropical East Africa. Because giant senecios and most shrubby alchemillas are absent from Ethiopia, whereas the genera *Rosa*, *Saxifraga*, and *Primula* are present, the Ethiopian afroalpine flora has been accorded the rank of a subregion (Pichi-Sermolli in Exell, 1955, p. 491; Pichi-Sermolli, 1957, p. 90; Hedberg, 1961b, p. 916; 1965, p. 519). I have

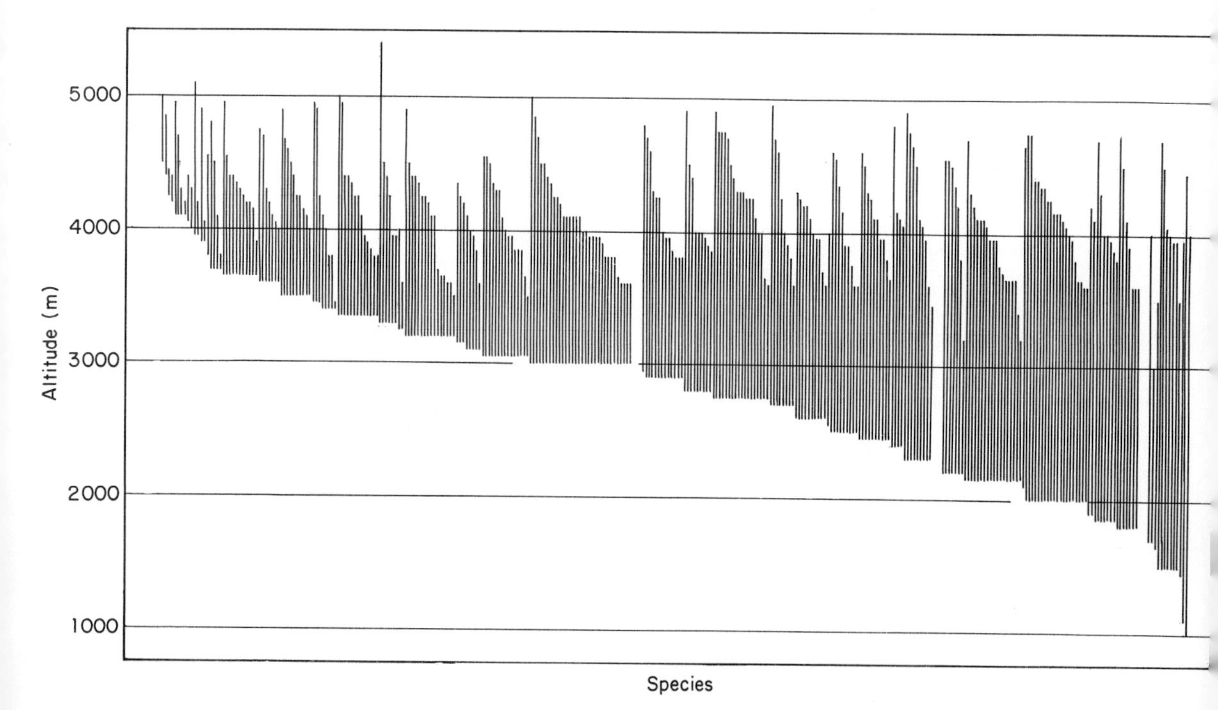

FIG. 17–2. Variability in altitudinal distribution displayed by vascular plant species of the afroalpine flora. Each species regularly occurring in the afroalpine belt is represented by a vertical line indicating its known altitudinal range. The ranges have been sorted according to decreasing values of, first, their lower altitudinal limit and, second, their upper altitudinal limit. From Hedberg (1969); altitudinal data from Hedberg (1957).

included in the afroalpine flora all species of vascular plants that to my knowledge have been found in the afroalpine belt on one or more of the high mountains of East Africa and Ethiopia.

A third subregion, an "austral domain" of the afroalpine floristic region, has been proposed by Killick (1978, p. 517) for the flora of the "austro-afroalpine belt" of Coetze (1967), synonymous with the alpine belt of Killick (1978, p. 545), which extends from 2860 to 3480 m in the Drakensberg area (South Africa). The vascular flora of this domain, explored by Killick (1963, 1978) and Jacot Guillarmod (1971), has only about 20 species in common with the "Central" and "Ethiopian" domains of the afroalpine floristic region, being otherwise mainly dominated by species with South African affinities. Climatically, however, the austro-afroalpine belt of the Drakensberg region is not tropical-alpine—at a latitude of 29° S it is exposed to a distinctly seasonal climate (Killick, 1978).

The occurrence of a certain number of species in common with the afroalpine floristic region of East Africa and Ethiopia does not in my opinion justify the inclusion of the austral domain in the afroalpine floristic region. The criterion stated earlier for admittance of a plant species into the afroalpine flora is that it occur in the afroalpine belt of one or more mountains of East Africa and Ethiopia. Appreciable numbers of the afroalpines enumerated in Hedberg (1957) occur also in mountain areas of West Africa (47 species), Arabia (20), Madagascar (13), and Eurasia (24). I therefore propose to treat the flora of the austro-afroalpine belt as an essentially independent floristic region. It offers fascinating phytogeographic and ecological problems, but they are largely different from those encountered in equatorial Africa (White, 1978, pp. 501, 508). A similar example is provided by the flora of Jebel Marra in the Sudan, which harbors 21 afroalpine species on its upper parts but, as stated by Wickens (1976, p. 46), "There is no suggestion of an Afro-alpine Region on Jebel Marra, merely the representation of an Afro-alpine element in the montane flora." The same argument can be applied to the afroalpine representatives on the upper part of Cameroon Mt. and the Bamenda Highlands (Keay, 1955).

In this chapter I will focus on the afroalpine flora of East Africa and to some extent that of Ethiopia. The former area has been reasonably well explored botanically (Hedberg, 1957, 1964b), but the latter is still insufficiently known (Scott, 1952, 1955, 1958; Gillet, 1955; Pichi-Sermolli, 1957; Hedberg, 1971, and in press).

ECOLOGICAL ADAPTATIONS AND LIFE FORMS

As stated earlier the selective filter imposed by the harsh afroalpine environment has allowed only a limited number of vascular plant species to survive in it—so few that White (1978, p. 466) classified the area as an archipelago-like "region of extreme floristic impoverishment." Some of the most important limiting factors are the frequent and intense night frosts, solifluction phenomena on open soil, and the intense insolation provoking rapid heating in the morning (Hedberg, 1964a,b). Most of the specialized morphological types occurring in the afroalpine flora (Fig. 17–3) can be distributed among five tropical-alpine life forms, providing temperature insulation to the most sensitive parts of the plant and hence facilitating the maintenance of their water balance and securing their sexual reproduction (Fig. 17–4; Hedberg, 1964a,b). It is striking that exactly the same life forms have evolved in parallel fashion though largely in different genera and families, under similar climatic conditions in other parts of the world, notably in the tropical Andes (Meyer, 1900; Cuatrecasas, 1934, 1968; Hauman, 1935; Cotton, 1944; Troll, 1948, 1959; Jeannel, 1950; Rauh, 1977; Hedberg and Hedberg, 1979).

The morphologically strongly specialized taxa or ecotypes occurring in the afroalpine belt may in many cases have been derived through successive adaptations from less specialized afromontane taxa or populations occurring at lower levels on the same mountains. In several cases clinal variation with altitude has been described, for example, in leaf size in *Crassula granvikii* (Hedberg, 1957, fig. 24), size of capitula in *Cineraria grandiflora* (Hedberg, 1957, fig. 48), leaf pubescence in *Helichrysum stuhlmannii* (Hedberg, 1957, p. 344), and specimen height in *Dipsacus pinnatifidus* (Hedberg and Hedberg, 1977a, fig. 2). A gradual transition from erect to decumbent or cushion-forming growth toward higher altitudes occurs, among others, in *Alchemilla subnivalis* (Hedberg, 1957, p. 286), *Helichrysum newii* (Hedberg, 1957, p. 343), and *H. gofense* (Hedberg, 1975, pl. 1). A good example of the autochthonous origin of an afroalpine taxon from an afromontane ancestor is provided by *Bartsia abyssinica* var. *petitiana*, which has evidently been derived from the less specialized var. *abyssinica* occurring at lower levels in the same area (Hedberg et al., 1980). Replacement of an afromontane taxon at lower level by a closely related afroalpine species at high altitude is exemplified

FIG. 17–3. Afroalpine vegetation at about 4100 m in the Teleki Valley, Mt. Kenya. In the center a flowering specimen of *Lobelia telekii*, further away nonflowering *Dendrosenecio keniodendron*, and in the foreground tussocks of *Festuca pilgeri* and shrubs of *Alchemilla argyrophylla.* Photo by O. Hedberg.

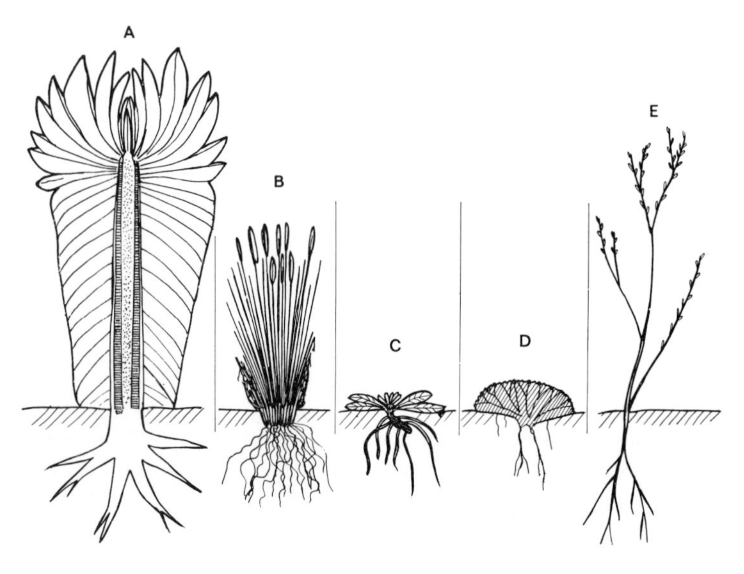

FIG. 17–4. The five most important phanerogamic life forms of the afroalpine belt. A: giant rosette plant; B: tussock grass; C: acaulescent rosette plant; D: cushion plant; E: sclerophyllous shrub. From Hedberg (1964a).

by the *Wahlenbergia krebsii* group (Thulin, 1975, pp. 97f., maps 14 and 45), as well as by the genus *Dichrocephala* (Hedberg, 1957, p. 335). The most famous of all afroalpines, the giant senecios and giant lobelias, have also been found to provide the most conspicuous examples of adaptive trends in morphological features associated with increasing altitude (Mabberley, 1973, 1977, and Chap. 4, this volume).

ENDEMISM AND VICARIOUS TAXA

The afroalpine flora has many endemic taxa, in particular pairs or groups of vicarious (or geographically representative) taxa in many genera. The first and most remarkable cases were published for the giant senecios and giant lobelias by Fries and Fries (1922a,b). Numerous other instances of vicariism (geographic replacement) were described in later papers by the same and many other botanists. Fries and Fries (1948) gave a summary of the cases described for Mt. Kenya and Aberdare. Unfortunately the concept of vicarious species was sometimes pushed too far on insufficient evidence. Thorough investigations of the much greater material later available has resulted in the reduction of many vicarious species to synonymy (Hedberg, 1955, 1957). In spite of this reduction the extent of endemism in the afroalpine flora is quite impressive. This endemism may be studied at several different levels.

First, many species are restricted—as far as is known—to a single mountain, for example, *Calamagrostis hedbergii* to Mt. Kenya, *Poa kiliman-*

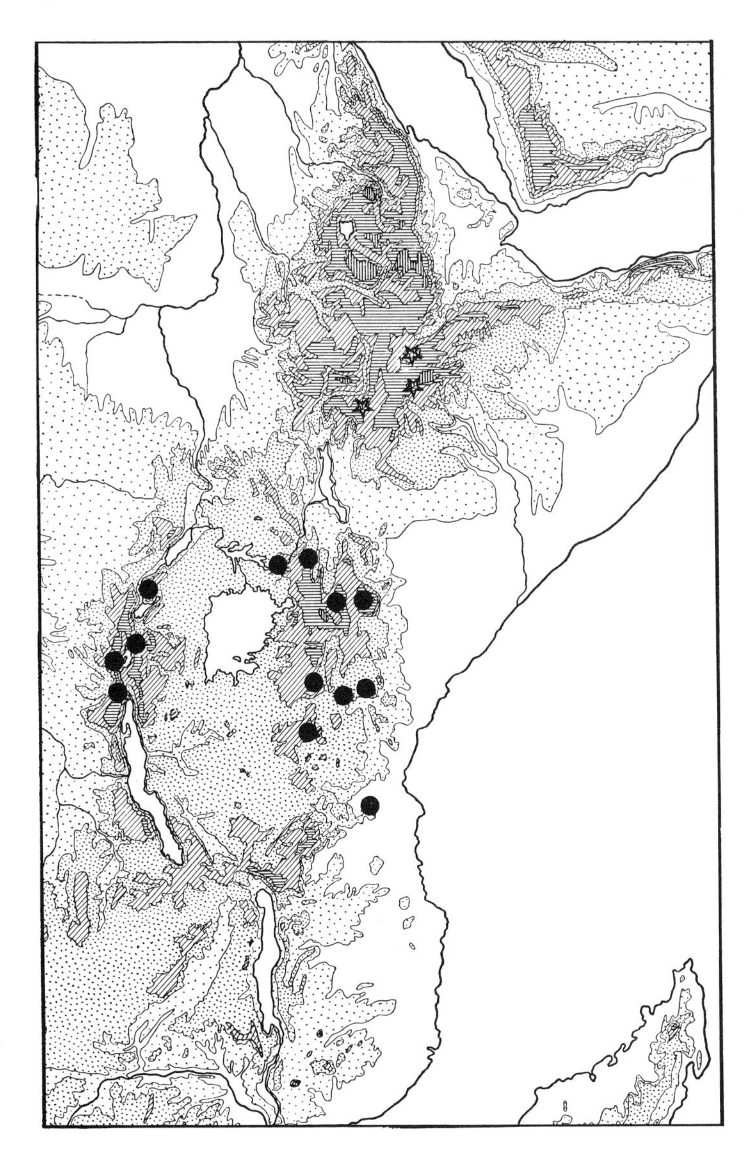

FIG. 17–5. Distribution of *Alchemilla johnstonii* Oliv. (*black circles*) and *A. haumanii* Rothm. (*open stars*). From Hedberg (1962a). For scale and altitudinal belts, see FIG. 17–7.

jarica to Kilimanjaro (Markgraf-Dannenberg, 1969), *Ranunculus cryptanthus* to Elgon, and *Helichrysum gloria-dei* to Aberdare (Hedberg, 1957). Some of these species may of course be discovered later on other mountains as well. *Myosotis keniensis*, for instance, once believed to be endemic to Mt. Kenya, has recently been discovered in southern Ethiopia (Hedberg, in press).

Second, one species may occur on a group of mountains, for instance, *Carpha eminii* on the western group in East Africa, or *Festuca pilgeri* on the eastern group. A pair of vicarious species is *Swertia macrosepala* on the western group and *S. volkensii* on the eastern group. Two pairs of vicarious species are *Lobelia wollastonii-L. lanuriensis* on the western group and *L. telekii-L. bambuseti* on the eastern group (Mabberley, 1974, p. 569).

On a third level, vicariism between tropical East Africa and Ethiopia is exemplified by the species pair *Alchemilla johnstonii-A. haumanii* (Fig. 17–5). But in most genera the differentiation between these two areas is not more pronounced than between the western and eastern groups of mountains within East Africa.

A fourth level of vicariism is represented by *Subularia monticola* (mapped in Hedberg, 1962a, fig. 5), forming an afroalpine vicariad of the boreal *S. aquatica* (mapped by Hultén, 1958, map 167; cf. Hedberg, 1957).

In some genera the patterns become more complicated because altitudinal adaptation occurs in combination with differentiation of geographically isolated populations. A good example is provided by *Stachys aculeolata* (Björnstad, Friis, and Thulin, 1971). On some mountains this species displays seemingly clearcut differentiation into two altitudinally and morphologically separate population systems, whereas other mountains harbor continuously varying population systems bridging the gaps between those first mentioned. In a case like this taxonomic segregation is impracticable.

Endemism in the afroalpine flora is high: about 80% of the species are endemic to the high mountains of East Africa and Ethiopia (Hedberg, 1961b). Endemism is also very high in the afromontane flora (White, 1978, p. 469). This pronounced endemism indicates that the afroalpine and afromontane floras have long been efficiently isolated from other high mountain or temperate floras (Hedberg, 1961b). A comparison of patterns of endemism between the two groups of mountains in East Africa and Ethiopia supports the hypothesis that the afroalpine enclaves have been efficiently isolated from each other for a considerable time (Hedberg, 1961b, 1969, 1970a; Mabberley, 1973).

As emphasized earlier (see section on Ecological Adaptations and Life Forms and Fig. 17–2) different afroalpine species show considerable variation in altitudinal range. It is noteworthy that the percentage of endemism is much higher among those species restricted to the upper parts of the mountains than among those also occurring lower down (Table 17–1; Hedberg, 1969). There is also a rough inverse correlation between intermountain distance and phytogeographic affinity. The differentiation is more marked between the

TABLE 17–1. Number of afroalpine vascular plant species in East Africa (some also occur in Ethiopia) reaching their lower altitudinal limit within each of four altitude intervals, and percentage of one-mountain endemics, one-mountain-group endemics, two-mountain-group endemics, and nonendemic species in each of these altitude groups.

Altitude intervals	Total number of species	One mountain only		One group of mountains		Both Groups of mountains		Occurring also outside East Africa and Ethiopia	
		Number	%	Number	%	Number	%	Number	%
Above 3000 m	101	65	64	27	27	6	6	3	3
			92	91%					
3000–2400 m	70	8	11	30	43	18	26	14	20
			38	54%					
2400–1800 m	56	1	2	13	23	20	36	22	39
			14	25%					
Below 1800 m	23	1	4	2	9	3	13	17	74
			3	13%					

eastern and western groups of mountains in East Africa than between the mountains within each group (Hedberg, 1961b, 1970a).

PHYTOGEOGRAPHIC ELEMENTS

Having briefly considered the distribution, adaptations, and endemism of the afroalpine flora, I now turn to the topic of this chapter, the geographic derivation of this flora. For an analysis of this kind it is customary to distribute the species concerned among floral elements. Because more than 80% of the species of the afroalpine flora are endemic to the high mountains of East Africa and Ethiopia, delimiting floral elements is evidently of limited usefulness. I have therefore resorted to a subdivision of the flora into *genetic floral elements*. According to White (1970, p. 55), a genetic floral element consists of those species "whose closest relatives, at the present time, occur in a particular area, which is not necessarily the place where they originated, and might have been occupied only relatively recently." For this analysis I used a reduced list of East African afroalpines. Weeds and accidental species as well as hybrids and other imperfectly known taxa have been removed, leaving 260 taxa.

In an earlier paper on the afroalpine flora (Hedberg, 1965) I recognized nine genetic floral elements: (1) an afroalpine element, (2) an afromontane element, (3) a South African element, (4) a Cape element, (5) a Southern Hemisphere temperate element, (6) a Northern Hemisphere temperate element, (7) a Mediterranean element, (8) a Himalayan element, and (9) a pantemperate element. Since it is difficult to distinguish between the afroalpine and afromontane elements, I later combined these two into the endemic afromontane element. Similarly, I found it desirable to combine the South African and the Cape elements into the South African element (see also Nordenstam, 1969, p. 70). The resulting classification is presented in Table 17–2.

Any classification of this sort is largely arbitrary. Thus the genus *Protea* is mostly South African, whereas the family Proteaceae is Southern Hemisphere temperate. *Callitriche stagnalis* is mostly a Northern Hemisphere temperate species, but the genus *Callitriche* is pantemperate. In an earlier paper (Hedberg, 1965) I stressed that the survey was tentative, and that many species might later have to be moved to other elements when their taxonomic relationships were better understood. An example of such a change is afforded by the genus *Keniochloa*, treated earlier as an afroalpine endemic, but now sunk into the South Asiatic genus *Colpodium* (Tzvelev, 1965) thus being instead a pantemperate genus. Similar changes have been made or are under way in other genera so the classification given here must be regarded as provisional. Keeping these difficulties in mind I now discuss these elements one by one.

The Endemic Afromontane Element

I have assigned to this element, first, monotypic genera that are entirely or almost entirely confined to the upper parts of the high East African and Ethiopian mountains. Examples are *Oreophyton* (Jonsell, 1976), *Haplosciadium*, and *Dianthoseris*. Second, I have assigned to this element species groups or genera whose distribution is centered on the same mountains. Examples include *Dendrosenecio* (Hauman ex Hedb.) Nordenstam (Fig. 17–6), *Lobelia* sect. *Rhynchopetalum* (Mabberley, 1974), and the suffrutescent species of *Alchemilla*. The list of taxa assigned to the endemic afromontane element is given in Appendix 17–A.

As discussed further under Rates of Differen-

TABLE 17–2. Genetic flora elements of the afroalpine flora.

Element	Number of taxa	Percent
Endemic afromontane element	82	32
South African element	25	10
Southern Hemisphere temperate element	6	2
Northern Hemisphere temperate element	34	13
Mediterranean element	18	7
Himalayan element	8	3
Pantemperate element	87	33
Total	260	100

Note: The endemic afromontane element combines the afroalpine element and the afromontane; the South African element combines the South African and the Cape elements.

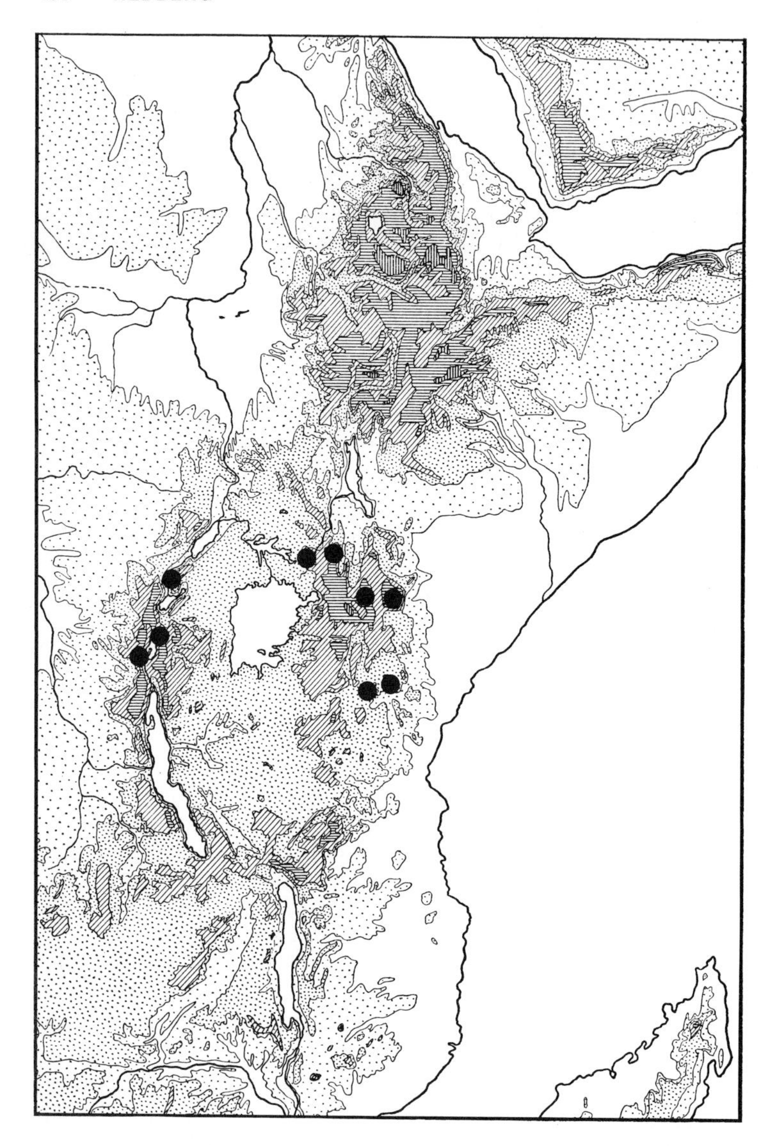

FIG. 17–6. Distribution of the genus *Dendrosenecio* (Hauman ex Hedb.) Nordenstam. From Hedberg (1964a). For scale and altitudinal belts, see FIG. 17–7.

tiation and Speciation and Genesis of the Afroalpine Flora, the taxa belonging to this element, or their ancestors, must evidently have existed in East Africa for a very long time, certainly much longer than the age of the mountains now harboring them (Hedberg, 1961b; Clayton, 1976). Their original derivation can rarely be established; in some cases there may be indications of transatlantic connections, for instance, in *Bartsia decurva*, which in vegetative parts displays a remarkable resemblance to some Andean species (Hedberg, 1957).

The South African Element

The South African element includes taxa that have (the majority of) their closest relatives in South Africa, including the Cape region. They include, among others, the afroalpine representatives of the genera *Pentaschistis*, *Kniphofia*, *Protea*, *Stoebe*, and *Euryops*. The example of *Zaluzianskya elgonensis* is mapped in Fig. 17–7 (Hedberg, 1970b). The distance separating that species from its closest relatives in South Africa seems far too great to have been bridged by suitable vegetation during climatic changes of the Pleistocene. Its extension northward to East Africa must in all probability be due to long-distance dispersal. Appendix 17–B lists the taxa included in the South African element. Afromontane representatives of this element that do not or just barely reach the afroalpine belt include *Clifforthia aequatorialis* R. E. Fr. and Th. Fr. jr. (Weimarck, 1934), *Juncus dregeanus* ssp. *bachiti* (Weimarck, 1946, p. 172), *Pelargonium whytei*,

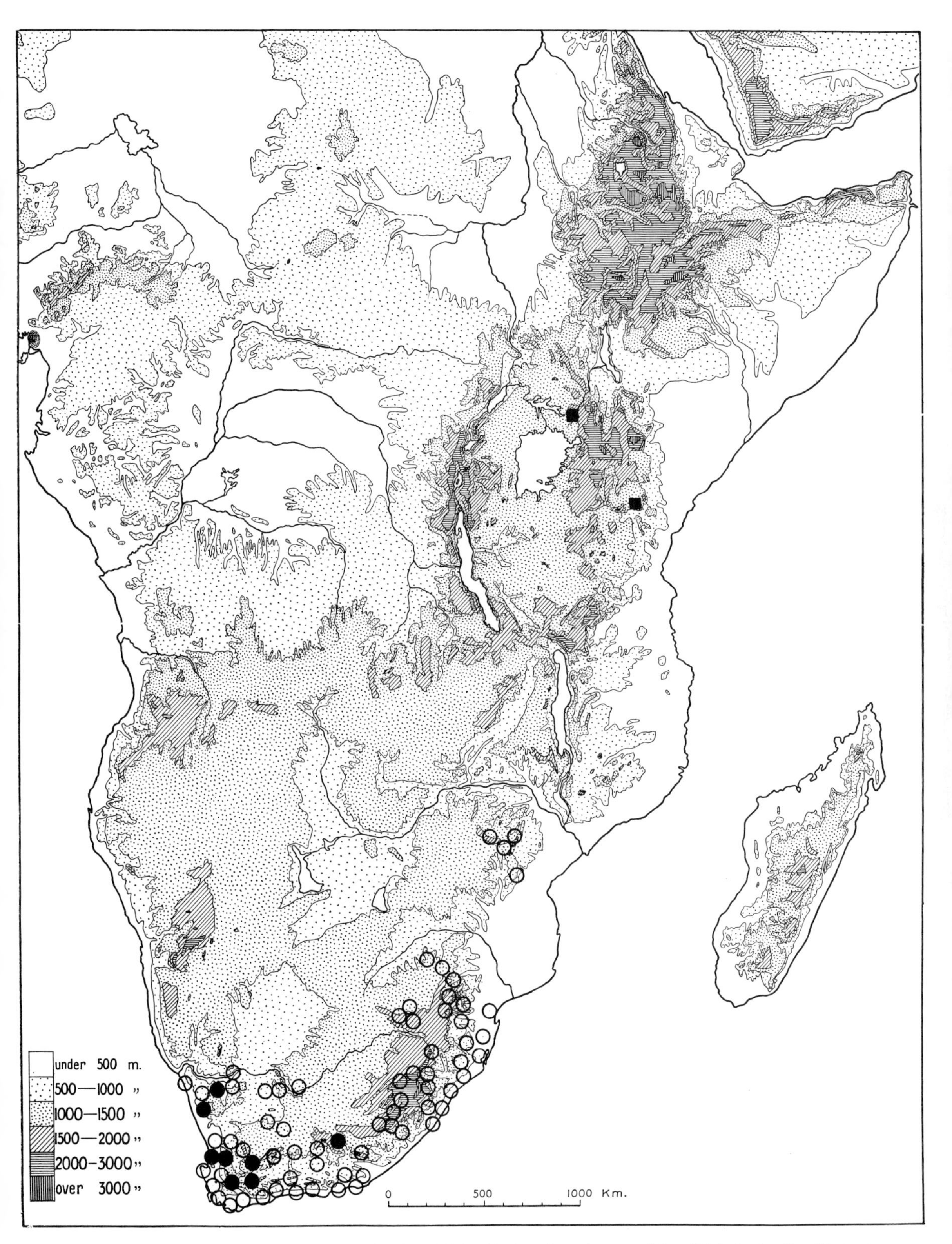

FIG. 17–7. Distribution of the genus *Zaluzianskya* F. W. Schmidt, based on the material available in the Kew Herbarium. The *black squares* mark localities for *Z. elgonensis* Hedb.; the *black circles* indicate localities for its closest allies *Z. gilioides* Schlecht. and *Z. pusilla* (Benth.) Walp.; the *open circles* represent localities for other species of the genus. Modified from Hedberg (1970b).

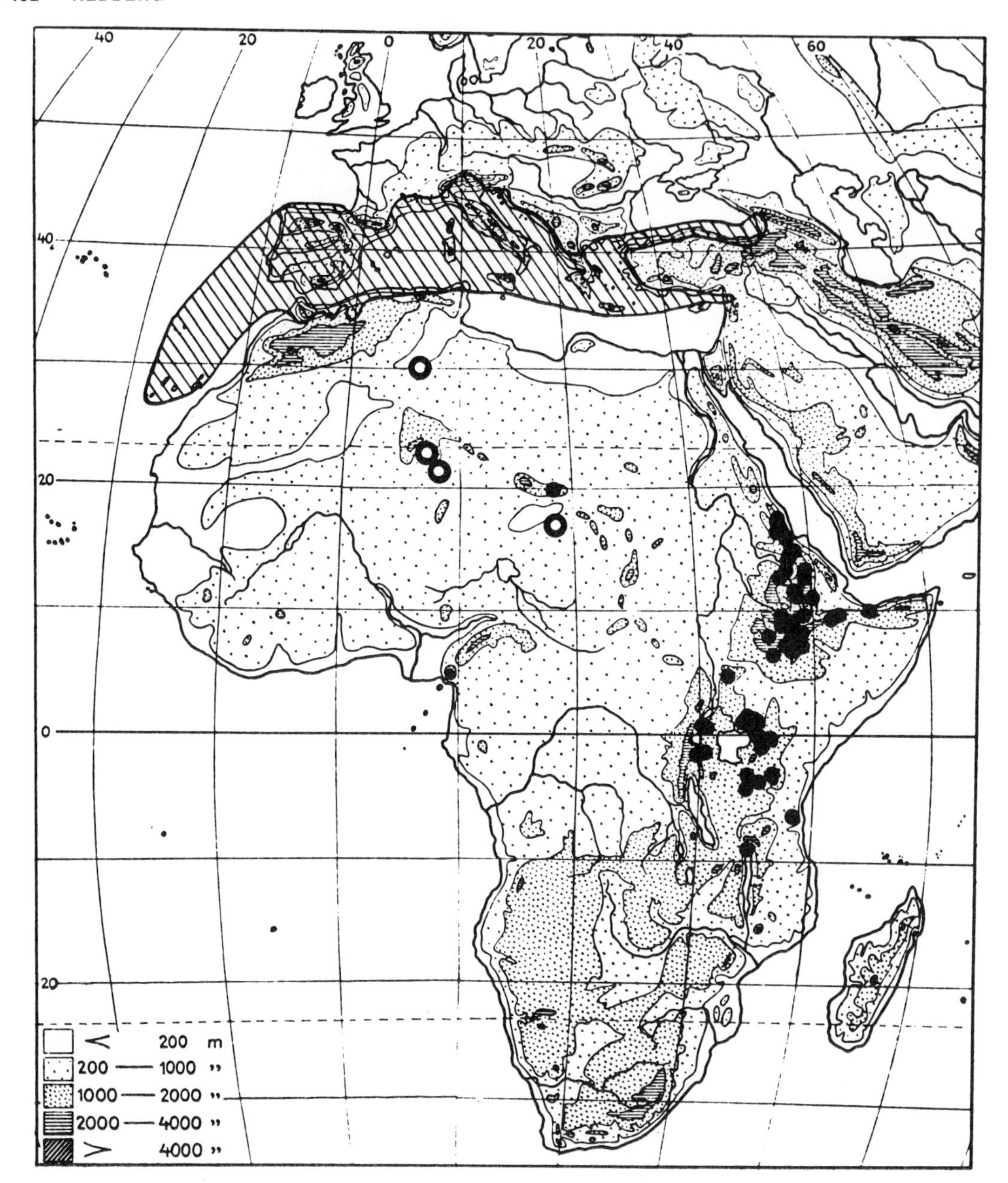

FIG. 17–8. Distribution of *Erica arborea* L. *Hatching* shows more or less continuous distribution; *black circles* indicate verified recent localities; *thick open rings* represent finds of subfossil pollen tetrads. From Hedberg (1961a).

Artemisia afra, and *Osteospermum volkensii* (Nordlindh, 1943).

The Southern Hemisphere Temperate Element

It is difficult to separate this element from the South African element. As mentioned earlier the genus *Protea* has its center in South Africa, but the family Proteaceae is definitely Southern Hemisphere temperate. Similarly the genus *Cotula* has more than 20 species in South Africa, but there are many more elsewhere, especially in the southern temperate region. Two species of this group, *Lycopodium saururus* and *Polypodium rigescens*, occur themselves in other Southern Hemisphere temperate areas, whereas *Carpha* and *Valeriana* are represented there by closely related taxa. The best known representative of this element outside Africa is *Nothofagus* (see later section on Rates of Differentiation and Speciation). The following afroalpine taxa belong here: *Lycopodium saururus* (Rauh, 1977, fig. 5), *Polypodium rigescens*, *Carpha eminii*, *Valeriana kilimandscharica* (Meyer, 1958), *Cotula cryptocephala*, and *C. abyssinica*.

The Northern Hemisphere Temperate Element

Some of the taxa belonging to this element have East African populations that are taxonomically indistinguishable from boreal ones. Examples are *Cardamine hirsuta*, *Arabidopsis thaliana*, *Arabis alpina* (Hedberg, 1962b), *Callitriche stagnalis*, and *Anthriscus sylvestris*. Closely related vicarious taxa occur, for instance, in *Subularia* (Hedberg, 1957, p. 272) and *Dipsacus* (Hedberg and Hedberg, 1977a). The taxa included in this element are listed in Appendix C.

The Mediterranean Element

The Mediterranean element comprises species that either occur themselves in the Mediterranean area or have the greatest number of their close relatives there. A good example of the former is *Erica arborea* (Fig. 17–8). Pollen studies have shown that this species and several other Mediterranean taxa occurred much farther south in the Sahara between 10,000 and 6,000 years ago than they do now, when they reached Ahaggar and Tibesti (Wickens, 1976, fig. 23). The southward migration of some Mediterranean species was facilitated at that time, but there cannot have been easy access eastward from Ahaggar to Ethiopia and East Africa, as evidenced by the fact that *E. arborea* is absent from Jebel Marra in the Sudan (Wickens, 1976, p. 80). Appendix D lists the taxa included in this element.

Related to this element are a number of afromontane taxa at lower altitudes (below 3000 m) with a Macaronesian-East African disjunction, for instance, in the genera *Canarina* (Hedberg, 1961a), *Dracaena* (Sunding, 1970, fig. 9), and *Aeonium*.

The Himalayan Element

To the Himalayan element belong taxa that either occur themselves on high mountains of central Asia, like *Satureja punctata* and *S. biflora*, or have their nearest known relatives in this area. It comprises the following taxa: *Colpodium chionogeiton* and *C. hedbergii* (Clayton, 1970), *Satureja punctata* (mapped for Africa by Wickens, 1976, map 157), *S. biflora*, *Dichrocephala alpina*, *Crepis scaposa* ssp. *afromontana*, and *C. suffruticosa*. An Ethiopian afroalpine species belonging to this element is *Primula verticillata* (map in Hedge and Wendelbo, 1978, fig. 12).

While the representation of this element in the afroalpine flora is rather poor in number of species, its influence at lower altitudes (below 3000 m) in the afromontane flora is much more impressive. Examples of the same species occurring in East Africa and the mountains of central Asia are *Persicaria nepalensis* (Meisn.) H. Gross, *Cerastium indicum* Wight and Arn., *Myrsine africana* L. (Fig. 17–9), *Sanicula elata* Ham. ex D. Don (Shan and Constance, 1951, fig. 46), and *Piloselloides hirsuta* (Forsk.) C. Jeffrey. An afromontane endemic with near relatives both in the Cape and in the Himalayan areas is *Polygonum afromontanum* Greenway (Hedberg, 1962a, p. 424).

The Pantemperate Element

The pantemperate element comprises species that either occur themselves on three or more continents and in both hemispheres, or whose closest relatives are so distributed. The element is to some extent a dumping place to which have been assigned species whose detailed taxonomic relationships are insufficiently known. Several of the taxa enumerated in Appendix E may have to be placed in other elements when the genera concerned are better understood, for instance, *Myosotis keniensis*, whose closest relative may well be the Southern Hemisphere *M. antarctica*. Attempts at phytogeographic analysis of genera like *Ranunculus* and *Senecio* are futile without thorough taxonomic revisions. A good example of a pantemperate species is *Deschampsia flexuosa*, which was mapped by Hultén (1958, map 323). Appendix E lists the taxa included in this element.

RATES OF DIFFERENTIATION AND SPECIATION

The occurrence, or absence, of taxonomically recognizable differentiation between geographically separated populations of the same species has often been given great weight in phytogeographic and evolutionary discussions. When the same species occurs in widely separated areas

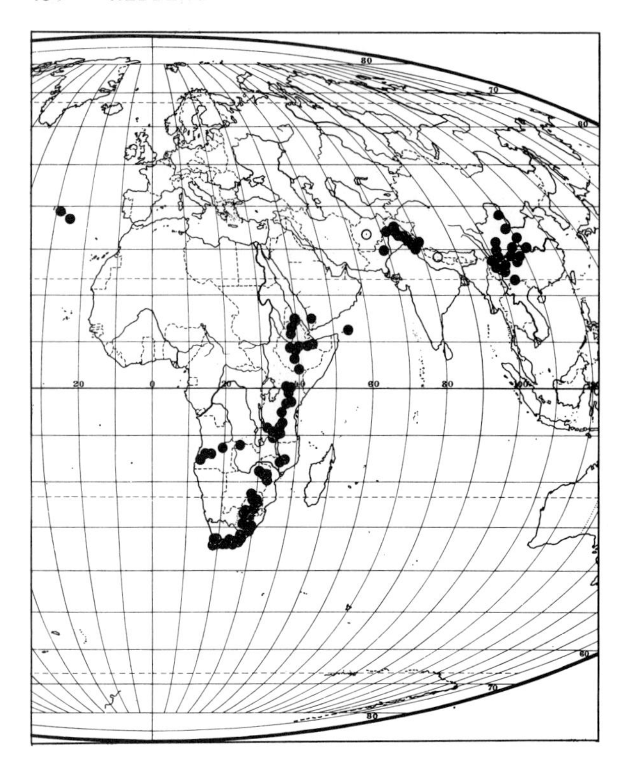

FIG. 17–9. Distribution of *Myrsine africana* L., based on herbarium material examined at BM, BR, EA, FI, K, P, S, UPS (abbreviations according to Holmgren and Keuken, 1974). *Open rings* indicate localities that could not be identified on maps.

without perceptible morphological differentiation this may either be interpreted as an example of recent migration, as assumed by Schnell (1977, p. 158) for *Sanicula elata* on the Cameroon Mt. and the East African mountains (and southern Asia), or as a result of slow evolutionary change. Before offering phytogeographic conclusions it may therefore be useful to discuss the problem of evolutionary rates.

The occurrence of the afroalpine flora on widely scattered mountain enclaves provides unusually good opportunities to compare the rates of evolutionary differentiation and speciation in different plant genera and families. Since for taxa occupying comparable altitudinal ranges the efficiency of geographic isolation may be considered to be comparable, the amount of observable differentiation between different mountain populations may be taken as a rough measure of the rate of evolutionary diversification (Hedberg, 1957, p. 376; cf. Wild, 1968, p. 215). Absence or low level of differentiation between isolates is taken to mean a slow rate of evolutionary change; marked differentiation indicates a rapid rate of change. From such comparison it seems that the rate of differentiation has been comparatively fast in genera like *Trifolium*, *Alchemilla*, *Philippa* and *Swertia*, in the giant lobelias, and in some genera of Compositae, particularly *Helichrysum*, *Dendrosenecio*, and *Senecio*. In representatives of the families Juncaceae, Carophyllaceae, Cruciferae, and Crassulaceae, by contrast, the rate of evolutionary change appears to have been much slower.

Apart from a few cases of man-made or man-induced amphidiploids there seem to be few known examples of dated speciation in plants. It appears considerably easier to find examples of lack of differentiation in spite of prolonged geographic isolation. *Arabis alpina* is one example. First, there is no recognizable differentiation between different mountain populations of East Africa and Ethiopia. Second, the populations of northern Europe and East Africa are clearly compatible (Hedberg, 1962b), although they can hardly have been in breeding contact since before the origin of the desert belt through the Sahara and Arabia. This means that an isolation of at least a few million years has resulted in no taxonomic differentiation. A similar example is provided by *Myrsine africana* (Fig. 17–9), whose widely separated populations show no recognizable morphological differentiation.

An even more striking example from another part of the world is found in the species pair *Platanus orientalis* (northeastern Mediterranean area) and *P. occidentalis* (eastern North America). Although morphologically quite distinct they are reproductively compatible (Stebbins, 1950)—after a period of isolation perhaps amounting to some 50 million years.

One of the most famous of all disjunct distributions occurs in the genus *Nothofagus*, mapped by Du Rietz (1949, fig. 13). Fossil *Nothofagus* is known from Cretaceous and Tertiary deposits in both South America and Australia, as well as from parts of Antarctica. The connections between these different areas appear to have been broken in the Oligocene (Thorne, 1978, p. 311), more than 25 million years ago. Although there are different *Nothofagus* species in the different regions, they have the same parasitic fungi (*Cyttaria*) and largely the same epiphytic lichen flora (Du Rietz, 1940, p. 252). In this case 25 million years has evidently been enough for the differentiation of new species but not for the segregation of new genera.

The preceding examples evidently suggest considerably slower rates of evolutionary differentiation and speciation in some groups than has often been assumed, indicating a long-term equilibrium between these populations and their environment (Stebbins, 1950). To what extent could such a slow rate be compatible with the amount of evolutionary divergence that has evidently taken place among the angiosperms since the Early Cretaceous? To investigate this apparent contradiction, let us assume that the average number of new taxa arising through differentiation from an old one would be two, and that the average time for each such process has been three million years. If no extinction occurs this would lead in 30 million years from one ancestral taxon to more than 1000 descendants, and in 60 million years to more than one million. In fact, the situation will of course be much more complicated because of large-scale extinction, of the number of daughter taxa being more than two in many cases, and of vast differences in evolutionary rates between groups. But our exercise certainly indicates that even with an average time of more than three million years for differentiation of one ancestral taxon into two daughter taxa the time available since the Early Cretaceous would have been amply sufficient for the origin of our present-day flora. Against this background it is easier to accept such wide disjunctions as those of *Sanicula elata* and *Myrsine africana* as representing ancient relict distributions.

There are of course other examples suggesting a considerably faster rate of differentiation. The six subspecies of *Lobelia deckenii* have differentiated presumably since the origin of their isolated mountain enclaves, perhaps in the last million years. The same thing applies to the species of the other fast-evolving taxa mentioned previously.

EVOLUTIONARY MECHANISMS IN THE AFROALPINE FLORA

Evolution is generally considered to depend on three main processes: mutation, recombination, and natural selection. Although all three of these certainly act wherever evolution is at work their relative importance might be expected to vary under different environmental conditions. The striking adaptations (see Chap. 4) and high degree of endemism in the afroalpine flora justify a discussion of the evolutionary processes in this particular environment.

"Because in any one generation the amount of variation contributed to a population by mutation is tiny compared to that brought about by recombination of pre-existing genetical differences, even a doubling or trebling of the mutation rate will have very little effect upon the amount of genetic variability available to the action of natural selection" (Stebbins, 1966, p. 29). But under special environmental circumstances the mutation rate may be increased to such an extent that it might contribute to an increased rate of adaptive evolution. One environmental factor capable of inducing large numbers of mutations is radiant heat shocks, as demonstrated experimentally by Pettersson (1961). Such heat shocks may have been of widespread occurrence in connection with volcanic eruptions and may therefore have played an important role during the period of mountain building in East Africa. The increased genetic variability induced in this way may have speeded up evolutionary adaptation under the harsh environmental conditions of the afroalpine belt.

According to Stebbins (1966, p. 31) the chief limiting factors on the supply of variability for the action of natural selection are the restrictions on gene exchange and recombination that are imposed by the mating structure of populations and the structural patterns of chromosomes. Species with a breeding system securing outcrossing may obviously be expected to have a greater capacity for recombination and hence for differentiation than less efficient outcrossers, or inbreeders. One of the most efficient outbreeding devices in the plant kingdom is the capitulum in the Compositae (Burtt, 1961), and it therefore stands to reason that this family displays a higher rate of differentiation and speciation than, for example, the largely inbreeding Cruciferae.

One important process increasing the amount of genetic variability available for selection is hybridization. Field observations indicating interspecific hybridization in the afroalpine flora have been reported by Hedberg (1957) for the genera *Alchemilla*, *Lobelia*, *Helichrysum*, *Senecio*, and *Dendrosenecio*, and by Mabberley (1973, 1974) for *Lobelia* and *Dendrosenecio*. A striking example of a hybrid swarm with beginning introgression in *Senecio* was reported by Hedberg (1955), and an example of introgression in *Bartsia* was studied by Zemede (unpublished B. Sc. thesis, Addis Ababa University). Hybridization and introgression may be much more widespread and important in the afroalpine flora than has been documented so far.

Another significant evolutionary process is polyploidy. In 200 afroalpine angiosperm species investigated the percentage of polyploid taxa was found to be 49 (Hedberg and Hedberg, 1977b). Out of 116 taxa for which more than one collection was studied, 23% were found to possess more than one euploid chromosome number. The high frequency of infraspecific polyploidy discovered in a fairly small sample suggests that this phenomenon may be much more widespread in the afroalpine flora.

Most of the relevant polyploids presumably are allopolyploids combining the gene pools of different populations, infraspecific taxa, or even species. Assessment of hybridization behind each case of polyploidy would be very time-consuming, but this matter need not concern us here. My main point is that since each infraspecific polyploid will be more or less completely reproductively isolated from its lower-ploid ancestor(s), infraspecific polyploidy is likely to be a common cause of barriers to interbreeding and hence an important agent of speciation in the afroalpine flora.

Directional natural selection under the influence of strong environmental stress must be largely responsible for the adaptive evolution in the afroalpine flora discussed briefly earlier and described in more detail in Hedberg (1964a, b). In some cases there are considerable environmental differences between corresponding habitats on different mountains (Hedberg, 1951, 1964b). The conspicuous differences in height and foliage between *Lobelia wollastonii* on the western mountains and *L. telekii* on the eastern ones (Hedberg, 1957, 1969) might be largely explained by such (micro)climatic differences (Hedberg, 1969). Directional natural selection must also be held responsible for much of the ecoclinal variation described earlier (see Ecological Adaptations and Life Forms).

In other cases, however, the morphological differences found between different mountain populations seem very difficult to explain in adaptive terms. This applies, for instance, to the degree of splitting of the corolla and the amount of pubescence of anthers, calyx, and bracts in the subspecies of *Lobelia deckenii* (Hedberg, 1957, p. 187; Mabberley, 1974, p. 574). An analogous case is found in leaf crenulation, corolla width, and fruit pubescence in *Valeriana kilimandscharica* (Hedberg, 1957, p. 330; Kokwaro, 1968, p. 4). Numerous similar examples are described in Hedberg (1957). Such cases are most easily explained as resulting from the Sewall Wright effect (Wright, 1951) acting on originally very small founder populations resulting from long-distance dispersal by chance occurrence (Hedberg, 1969).

GENESIS OF THE AFROALPINE FLORA

Wickens (1976, p. 74) has stated that "the upper parts of the mountains of Africa must have remained virtually unpopulated until conditions were suitable for an influx of temperate taxa, especially Mediterranean plants which are already adapted for a short day length and low temperature survival." I cannot subscribe to this view. Wickens's hypothesis implicitly assumes that high mountains in Africa are of more recent origin than those in Eurasia. I see no evidence for such a difference.

Franz (1979, p. 393) assumed that the marvelous morphological and ecophysiological adaptations displayed by the afroalpine flora had been achieved during the Pleistocene. I find this to be extremely unlikely. Long before the origin of those high East African and Ethiopian mountains, which now harbor an afroalpine flora, there has evidently existed in Africa a number of fairly high mountains inhabited by at least an afromontane flora. Several of these mountains are believed to represent residuals of a Jurassic planation, e.g., Mlanje, the Aberdares, and much of Ethiopia. Some of them certainly reached altitudes of well above 2000 m already before the Plio-Pleistocene uplift and should thus have been inhabitable by montane species for a very long time (Clayton, 1976). A considerable number of afromontane taxa, including the ancestors of the endemic afromontane flora element, are likely to have existed there at least since the Miocene. The upwarping in East Africa in the Late Pliocene folded the Miocene "African surface" up to 1600 m and accelerated the rifting, which resulted also in the block faulting of Ruwenzori. The accompanying volcanism created the majority of the present high mountains.

These young mountains have evidently remained isolated from each other since their origin (Hedberg, 1961b) and will have provided ideal habitats for colonization by afromontane and afroalpine taxa already occurring on the preexisting mountains. Pleistocene climatic changes have evidently modified their vegetation zonation as well as the general vegetation pattern of tropical Africa (Hamilton, 1976, 1982), so that intermountain contacts may have been established for the montane forest belt, but such contacts are unlikely to have been possible for the ericaceous and afroalpine belts (Hedberg, 1970a). Intermountain migration of afroalpine vascular plants has therefore probably occurred mainly by independent long-distant dispersal, i.e., by air flotation (cyclones) and by birds (internal, in feathers, or in mud on feet) (Hedberg, 1969; Clayton, 1976; Wickens, 1976). Circumstantial evidence supporting this view is found in the random distribution of many afroalpine taxa and in the composition of vicarious afroalpine plant communities, which appear to have been "synthesized" separately on each mountain (Hedberg, 1970a). The difficulties of intermountain dispersal in tropical Africa are probably the main factor accounting for the poverty of species in the afroalpine flora in comparison with the páramo flora of South America, where "mountain hopping" must have been facilitated by the greater continuity of mountain ranges. The Andes also provided a convenient avenue for the northward extension of the Southern Hemisphere temperate element into the tropics.

Reinforcement of the afromontane flora with taxa migrating from the north, and belonging to the Northern Hemisphere temperate, Mediterranean, Himalayan, or pantemperate elements would seem all but impossible under present climatic conditions. Their migration may have been easier in connection with the upwarping of the Red Sea hills during the Late Miocene (Wickens, 1976, p. 50) and may have been similarily facilitated by climatic changes during the Pleistocene. According to Wickens (1976, p. 70), "The eastern Mediterranean element could not have entered Africa before the Pliocene, with the Euro-Siberian and western Mediterranean element following during the Pleistocene." Such immigration is of course much easier to visualize for afromontane taxa now occurring down to fairly low altitudes than for more exclusively high-altitude afroalpines. Wickens postulated the immigration of the Himalayan element to date from the Miocene, when the southern Arabian pathway was open for tropical species, or perhaps from the Pliocene, but he found no evidence supporting a later migration. The present distribution of *Myrsine africana* (Fig. 17–9) may be a relict from a time when biological contacts between East Africa and India were less discontinuous. In the same way the present distribution patterns of *Canarina* (Hedberg, 1961a, fig. 19) and *Dracaena* (Sunding, 1970, fig. 9) seem to represent relicts of a moist-temperate flora with much wider distribution in what is now southern Europe and North Africa up to Late Tertiary times (see Sunding [1979] for literature references). It is interesting to note in this connection that the important northern temperate mountain genus *Primula* is represented in the afromontane flora by a single species only, considered by Wendelbo (1961) to belong to the oldest and most primitive part of the genus. An analogous case is provided by the genus *Saxifraga*, which is also represented by a single species in tropical Africa, occurring, like *Primula*, southward only to Ethiopia (Fig. 17–10). Presumably these genera barely managed to disperse across the Sahara and Arabia before they became too dry (Hedberg, 1969).

Reinforcement to the afroalpine flora from the south (by South African and Southern Hemisphere temperate elements) would also seem most improbable under present climatic conditions. The possibilities of intermountain migration presumably will have been increased through the upwarping along the edges of the Great Rift Valley system as well as through climatic changes during the Pleistocene, but there is no evidence of any large-scale Pleistocene intermountain dispersal of afroalpine vascular plants.

As emphasized earlier, intermountain dispersal of afroalpine vascular plants has probably occurred mainly through independent long-distance dispersal. Such dispersal, as a phenomenon, is unfortunately beyond observation or experimental study (Hedberg, 1969): one dispersal event per millennium might suffice to achieve the distribution pattern we find today. The nearest we can get to a good understanding of dispersal is to investigate diaspore morphology of the plants concerned and classify them into groups according to their probable mode of dispersal (Hedberg, 1970a). Of the afroalpines restricted to altitudes above 3000 m, 35% showed adaptation to air flotation, 13% to dispersal in fur or feathers of mammals or birds, 51% seemed likely to be dispersed in mud on feet of mammals or

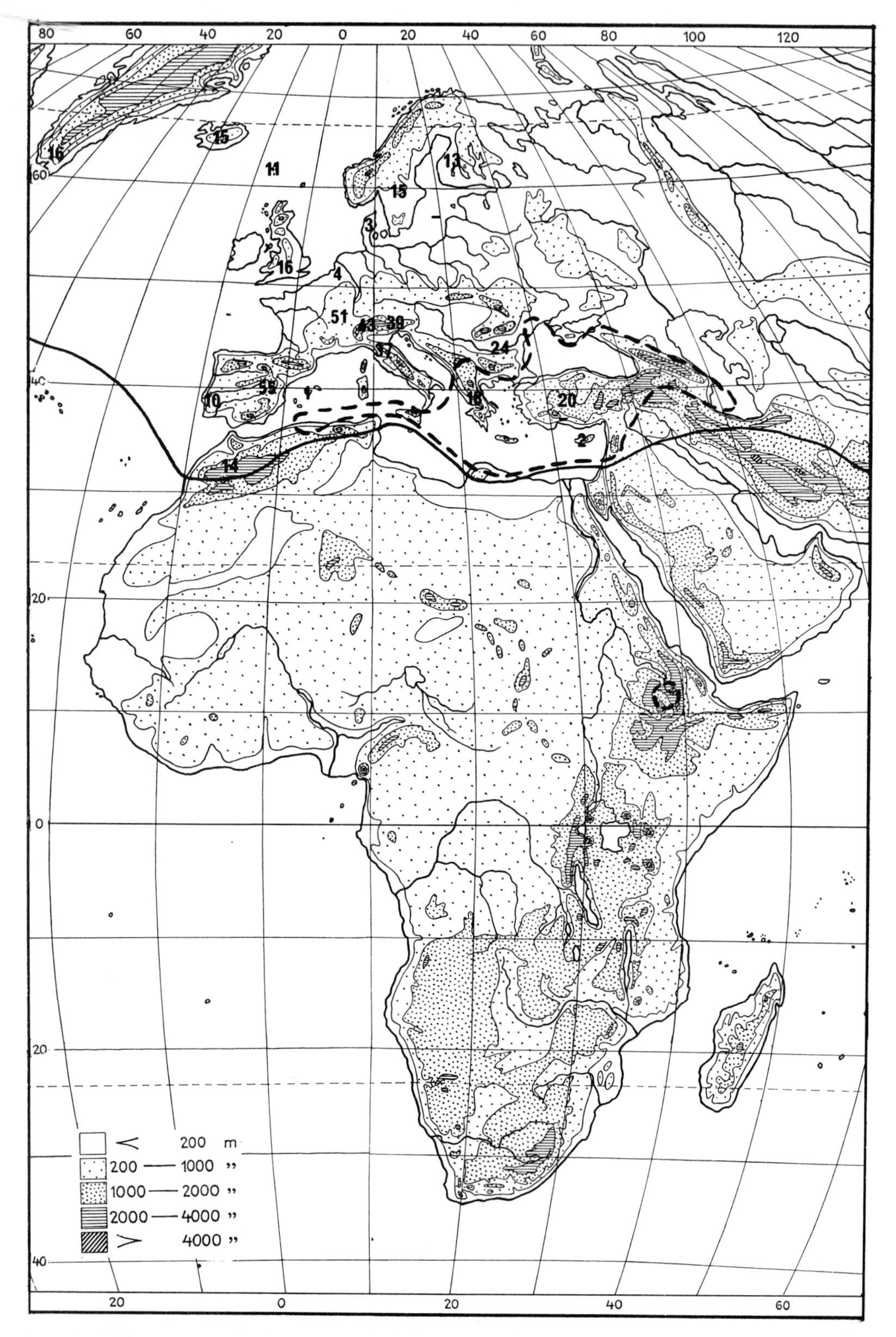

FIG. 17–10. Distribution of the genus *Saxifraga* in Africa, the Mediterranean area, and the rest of Europe and Greenland. The *figures* give the number of species known from each country. The *broken line* marks the distribution boundary of the section *Cymbalaria* Griseb., comprising also the Ethiopian species *S. hederifolia* Hochst. ex A. Rich. The *solid line* marks the southern boundary of the genus outside Ethiopia.

birds, and 1% were antropochorous. (See Hedberg [1970a] for a detailed discussion with some examples.) More important for intermountain dispersal than the aforementioned vectors may have been such catastrophic events as cyclones, which may easily transport diaspores that have no apparent adaptations for air flotation (Hedberg, 1969).

Successful migration of an afroalpine species from one mountain to another calls not only for intermountain transport of its diaspores but also for establishment in the new environment. A diaspore landing in closed vegetation usually stands a poor chance—establishment is much easier on open habitats like scree slopes or ground recently laid bare in front of glacier snouts. Even more important for the establishment of alien species are the vast areas of lava streams and ash deposits formed through volcanic activities in connection with mountain building. The recruitment of new members to the afroalpine flora of a volcanic mountain is much facilitated during periods of volcanic activity. Of course, lava flows and other types of volcanic activity destroy the vegetation, but when the lava has cooled down, new niches are at hand for diaspores to colonize (see, e.g., van Leeuwen, 1936; Skottsberg, 1941).

APPENDIX 17–A

Taxa included in the Endemic Afromontane element.

Cyperaceae	*Carex runssoroensis* var. *runssoroensis* and var. *aberdarensis*
	C. monostachya
	C. elgonensis
	C. simensis
Cruciferae (Brassicaceae)	*Oreophyton falcatum*
Rosaceae	*Alchemilla ellenbeckii*
	A. cyclophylla
	A. microbetula
	A. subnivalis
	A. elgonensis
	A. argyrophylla ssp. *argyrophylla* and ssp. *argyrophylloides*
	A. triphylla
	A. stuhlmannii
	A. johnstonii
Leguminosae	*Adenocarpus mannii*
	Trifolium cryptopodium var. *kilimanjaricum*
Violaceae	*Viola eminii*
Hypericaceae	*Hypericum revolutum* ssp. *keniense*
	H. bequaertii (for the latter two see Robson, 1979)
Umbelliferae (Apiaceae)	*Haplosciadium abyssinicum*
Ericaceae	*Philippia excelsa*
	P. johnstonii
	P. trimera
	P. keniensis ssp. *keniensis*, ssp. *meruensis* and ssp. elgonensis (Ross, 1957)
	Blaeria filago ssp. *filago* and ssp. *saxicola*
	B. johnstonii ssp. *johnstonii* and ssp. *keniensis*

APPENDIX 17–A CONTINUED.

Primulaceae	*Anagallis serpens* ssp. *meyeri-johannis*
Scrophulariaceae	*Celsia floccosa* Benth. *Veronica gunae* *V. glandulosa* *Bartsia abyssinica* var. *petitiana* (Hedberg et al., 1980) *B. longiflora* ssp. *longiflora* and ssp. *macrophylla* *B. decurva* (for the last three see Hedberg et al., 1979)
Rubiaceae	*Galium glaciale* *G. ruwenzoriense*
Campanulaceae	*Wahlenbergia krebsii* ssp. *arguta* (Thulin, 1975, p.63) *W. pusilla* (Thulin, 1975, p. 63)
Lobeliaceae	*Lobelia deckenii* ssp. *deckenii*, ssp. *bequaertii*, ssp. *burttii*, ssp. *keniensis*, ssp. *sattimae*, and ssp. *elgonensis* *L. telekii* *L. wollastonii* *L. lanuriensis* *L. bambuseti*
Compositae (Asteraceae)	*Helichrysum kilimanjari* *H. forskahlii* (J. F. Gmel.) Hilliard and Burtt 1980 *H. citrispinum* *H. newii* *H. brownei* *H. amblyphyllum* *H. stuhlmannii* *H. chionoides* *H. argyranthum* *H. meyeri-johannis* *H. ellipticifolium* *H. formosissimum* var. *formosissimum* and var. *volkensii* *H. guilelmii* *H. gloria-dei* *Coreopsis elgonensis* *Dendrosenecio johnstonii* ssp. *dalei*, ssp. *refractisquamatus*, ssp. *battiscombei*, ssp. *cheranganiensis*, ssp. elgonensis, ssp. *barbatipes*, ssp. *johnstonii*, and ssp. *cottonii* *D. keniodendron* *D. brassica* ssp. *brassica* and ssp. *brassiciformis* *Dianthoseris schimperi*

Note: Author's names are given only when the nomenclature deviates from Hedberg (1957). The nomenclature for the giant senecios follows Nordenstam (1978).

APPENDIX 17–B

Taxa included in the South African element.

Gramineae (Poaceae)	*Pentaschistis borussica* (incl. *P. meruensis* and *P. ruwenzoriensis*; Clayton, 1970)
	P. pictigluma (Steud.) Pilg. (Wickens, 1974)
	P. minor
Liliaceae	*Wurmbea hamiltonii* (Wendelbo, 1968)
	Kniphofia thomsonii
	K. snowdenii
Iridaceae	*Hesperantha petitiana* var. *volkensii*
	Dierama pendulum (mapped in Hedberg, 1962a, fig. 2)
	Gladiolus watsonioides
Orchidaceae	*Disa stairsii*
Santalaceae	*Thesium kilimandscharicum*
Crassulaceae	*Crassula schimperi* ssp. *phyturus* (Mildbr.) Fernandes (Fernandes, 1978)
	C. granvikii
Ericaceae	*Erica whyteana* ssp. *princeana*
Scrophulariaceae	*Zaluzianskya elgonensis*
	Hebenstretia dentata
Rubiaceae	*Anthospermum usambarense*
Compositae (Asteraceae)	*Helichrysum odoratissimum*
	H. splendidum
	Stoebe kilimandscharica
	Cineraria grandiflora
	Euryops dacrydioides
	E. brownei
	E. elgonensis
	Haplocarpha rueppellii

Note: Author's names are given only when the nomenclature deviates from Hedberg (1957).

APPENDIX 17–C

Taxa included in the Northern Hemisphere temperate element.

Gramineae	*Anthoxanthum nivale* (Hedberg, 1976)
Cyperaceae	*Carex bequaertii*
Caryophyllaceae	*Cerastium octandrum* var. *octandrum* (Wickens, 1976, map 37) and var. *adnivale* *C. afromontanum* *Sagina abyssinica* ssp. *aequinoctialis* *S. afroalpina*
Ranunculaceae	*Delphinium macrocentrum* *Anemone thomsonii*
Cruciferae	*Subularia monticola* (mapped in Hedberg, 1962a fig. 5C — the vicariad *S. aquatica* was mapped in Hultén, 1958, map 197) *Cardamine hirsuta* *C. obliqua* *Arabidopsis thaliana* *Arabis alpina* (mapped in Hultén, 1958, map 31; cf. Hedberg 1962b)
Crassulaceae	*Sedum churchillianum* *S. ruwenzoriense* *S. meyeri-johannis* *S. crassularia*
Leguminosae	*Astragalus atropilosulus*
Callitrichaceae	*Callitriche stagnalis*
Hypericaceae	*Hypericum kiboense* (Robson, 1977, p. 336) *H. afromontanum*
Umbelliferae	*Anthriscus sylvestris* *Caucalis melantha* (mapped by Wickens, 1976, map 99) *Pimpinella kilimandscharica* *Peucedanum friesiorum* *P. kerstenii* *Heracleum elgonense* *H. taylorii*
Rubiaceae	*Galium ossirwaënse*
Dipsacaceae	*Dipsacus pinnatifidus*
Compositae	*Carduus chamaecephalus* *C. keniensis* *C. ruwenzoriensis*

Note: Author's names are given only when the nomenclature deviates from Hedberg (1957).

APPENDIX 17–D

Taxa included in the Mediterranean element.

Gramineae	*Vulpia bromoides* (African distribution mapped by Wickens, 1976, map 208) *Aira caryophyllea* (Wickens, 1976, map 181)
Iridaceae	*Romulea cameroniana* Baker (Wickens, 1976, p. 158 and map 163) *R. congoensis* *R. keniensis*
Caryophyllaceae	*Silene burchellii* (Wickens, 1976, map 40)
Cruciferae	*Thlapsi alliaceum* *Barbarea intermedia*
Crassulaceae	*Umbilicus botryoides* (Wickens, 1976, map 32)
Leguminosae	*Trifolium acaule* *T. tembense* *T. multinerve* *T. elgonense*
Umbelliferae	*Ferula montis-elgonis*
Ericaceae	*Erica arborea* (Fig. 17–8)
Scrophulariaceae	*Sibthorpia europaea* (mapped by Hedberg, 1961a, fig. 20)
Dipsacaceae	*Scabiosa columbaria*
Compositae	*Anthemis tigreensis*

Note: Author's names are given only when the nomenclature deviates from Hedberg (1957).

APPENDIX 17–E.
Taxa Included in the Pantemperate Element.

Family	Taxa
Pteridophyta	*Polystichum magnificum* *Elaphoglossum subcinnamomeum* *Cystopteris fragilis* *Asplenium actinopteroides* *A. adiantum-nigrum* *A. kassneri* *A. uhligii*
Gramineae (Poaceae)	*Andropogon amethystinus* *Vulpia bromoides* *Koeleria capensis* *Festuca pilgeri* ssp. *pilgeri* and ssp. *supina* *F. abyssinica* (Wickens, 1976, map 189) *Poa ruwenzoriensis* *P. leptoclada* (Wickens, 1976, Map 201) *P. schimperiana* *P. kilimanjarica* (Hedb.) Markgraf-Dannenberg *Deschampsia angusta* *D. caespitosa* (Morton, 1961, map 10) *D. flexuosa* *Calamagrostis hedbergii* *Agrostis schimperiana* *A. sclerophylla* *A. gracilifolia* *A. taylorii* *A. quinqueseta* *A. kilimandscharica* *A. trachyphylla* *A. volkensii*
Cyperaceae	*Scirpus setaceus* *Isolepis kilimanjarica* *I. ruwenzoriensis* (Lye and Haines, 1974)
Eriocaulaceae	*Eriocaulon volkensii*
Juncaceae	*Luzula abyssinica* *L. campestris* var. *gracilis* *L. johnstonii*
Orchidaceae	*Habenaria eggelingii*
Urticaceae	*Parietaria debilis* (mapped for Africa by Wickens, 1976, map 90)
Portulacaceae	*Montia fontana*
Caryophyllaceae	*Stellaria sennii*
Ranunculaceae	*Ranunculus volkensii* *R. oreophytus* (mapped by Hedberg, 1962a, fig. 4) *R. stagnalis* *R. cryptanthus* *R. keniensis*
Geraniaceae	*Geranium vagans* *G. kilimandscharicum* *G. arabicum*
Euphorbiaceae	*Euphorbia wellbyi*
Balsaminaceae	*Impatiens phlyctidoceras* *I. tweediae*

Gentianaceae	*Swertia kilimandscharica*
	S. crassiuscula
	S. lugardae
	S. subnivalis
	S. uniflora
	S. volkensii
	S. macrosepala
Boraginaceae	*Myosotis abyssinica* (Wickens,1976, map 141)
	M. vestergrenii
	M. keniensis
Labiatae (Lamiaceae)	*Salvia merjamie*
	S. nilotica
	Satureja kilimandschari
	S. uhligii
	S. abyssinica var. *condensata*
Scrophulariaceae	*Limosella africana*
	L. macrantha
Lobeliaceae	*Lobelia lindblomii*
Compositae (Asteraceae)	*Conyza subscaposa*
	Crassocephalum ducis-aprutii
	Senecio roseiflorus
	S. mattirolii
	S. sotikensis
	S. snowdenii
	S. polyadenus
	S. purtschelleri
	S. rhammatophyllus
	S. telekii
	S. meyeri-johannis
	S. jacksonii ssp. *jacksonii,* ssp. *caryophyllus* and ssp. *sympodialis*
	S. keniophytum
	S. aequinoctialis
	S. subsessilis
	S. transmarinus var. *transmarinus* and var. *sycephyllus*
	S. scheinfurthii

Note: Author's names are given only when the nomenclature deviates from Hedberg (1957).

REFERENCES

Björnstad, I., I. Friis, and M. Thulin. 1971. A revision of the *Stachys aculeolata* group (Labiatae) in tropical Africa. *Norw. J. Bot.* 18: 121–137.

Burtt, B. L. 1961. Compositae and the study of functional evolution. *Transact. Bot. Soc. Edinb.* 39: 216–232.

Clayton, W. D. 1970. Gramineae (Part 1). In E. Milne-Redhead and R. M. Polhill, eds., *Flora of tropical East Africa*. London: Crown Agents for Oversea Governments and Administrations.

———. 1976. The chorology of African mountain grasses. *Kew Bull.* 31: 273–288.

Coetze, J. A. 1967. Pollen analytical studies in East and Southern Africa. In E. M. van Zinderen Bakker, ed., *Palaeoecology of Africa* 3: I–XII, 1–146.

Cotton, A. D. 1940. The megaphytic habit in tree Senecios and other genera. *Proc. Linn. Soc. Lond.* 156: 158–168.

Cuatrecasas, J. 1934. Observaciones geobotánicas en Colombia. *Trab. Mus. Nac. Cienc. Nat. Ser. Bot.* 27: 1–144.

———. 1968. Paramo vegetation and its life forms. *Colloquium Geogr.* 9: 163–186.

Du Rietz, G. E. 1940. Problems of bipolar plant distribution. *Acta Phytog. Suec.* 13: 215–282.

Exell, A. W. et al. 1955. Some aspects of the montane flora of tropical Africa. *Webbia* 11: 489–496.

Fernandes, R. B. 1978. Crassulaceae africanae novae vel minus cognitae. *Bol. Soc. Brot.*, ser. 2, 52: 165–220.

Franz, H. 1979. *Ökologie der Hochgebirge*. Stuttgart: Ulmer.

Fries, R. E., and T. C. E. Fries, 1922a. Ueber die Riesen-Senecionen der afrikanischen Hochgebirge. *Svensk bot. Tidskr.* 16: 321–340.

———. 1922b. Die Riesen-Lobelien Afrikas. *Svensk bot. Tidskr.* 16: 382–416.

———. 1948. Phytogeographical researches on Mt. Kenya and Mt. Aberdare, British East Africa. *K. Svenska Vetensk. Akad. Handl. III* (Stockholm) 25(5).

Gillett, J. B. 1955. The relation between the highland floras of Ethiopia and British East Africa. *Webbia* 11: 459–466.

Hamilton, A. C. 1976. The significance of patterns of distribution shown by forest plants and animals in tropical Africa for the reconstruction of upper Pleistocene palaeoenvironments: A review. In E. M. van Zinderen Bakker, ed., *Palaeoecology of Africa* 9: 63–97.

———. 1982. *Environmental history of East Africa: A Study of the Quaternary*. New York: Academic Press.

Hauman, L. 1933. Esquisse de la végétation des hautes altitudes sur le Ruwenzori. *Bull. Acad. Belg. Cl. Sci.*, ser. 5, 19: 602–616, 702–717, 900–917.

———. 1935. Les "Senecio" arborescents du Congo. *Rev. Zool. Bot. Afr.* 28: 1–76.

———. 1955. La "Région afroalpine" en phytogéographie centro-africaine. *Webbia* 11: 467–469.

Hedberg, I. 1976. A cytotaxonomic reconnaissance of tropical African *Anthoxanthum* L. (Gramineae). *Bot. Notiser* 129: 85–90.

Hedberg, I., and O. Hedberg. 1977a. The genus Dipsacus in tropical Africa. *Bot. Notiser* 129: 382–389.

———. 1977b. Chromosome numbers of afroalpine and afromontane angiosperms. *Bot. Notiser* 130: 1–24.

———. 1979. Tropical-alpine life-forms of vascular plants. *Oikos* 33: 297–307.

Hedberg, O. 1951. Vegetation belts of the East African mountains. *Svensk bot. Tidskr.* 45: 140–202.

———. 1955. Some taxonomic problems concerning the afroalpine flora. *Webbia* 11: 471–487.

———. 1957. Afroalpine vascular plants. A taxonomic revision. *Symb. bot. Upsal.* 15: 1–411.

———. 1961a. Monograph of the genus *Canarina* L. (Campanulaceae). *Svensk bot. Tidskr.* 55: 17–62.

———. 1961b. The phytogeographical position of the afroalpine flora. *Rec. Adv. Botany* (Toronto) 1: 914–919.

———. 1962a. Mountain plants from southern Ethiopia, collected by Dr. John Eriksson. *Ark. Bot.* (Stockholm), ser. 2, 4: 421–435.

———. 1962b. Intercontinental crosses in *Arabis alpina* L. *Caryologia*. 15: 252–260.

———. 1964a. Etudes écologiques de la flore afroalpine. *Bull Soc. Roy. Bot. Belgique* 97: 5–18.

———. 1964b. Features of afroalpine plant ecology. *Acta Phytogeogr. Suecica* (Uppsala) 49: 1–144.

———. 1965. Afroalpine flora elements. *Webbia* 19: 519–529.

———. 1969. Evolution and speciation in a tropical high mountain flora. *Biol. J. Linn. Soc. London* 1: 135–148.

———. 1970a. Evolution of the afroalpine flora. *Biotropica* 2: 16–23.

———. 1970b. The genus *Zaluzianskya* F. W. Schmidt (Scrophulariaceae) found in tropical East Africa. *Bot. Notiser* 123: 512–518.

———. 1971. The high mountain flora of the Galama mountain in Arussi province, Ethiopia. *Webbia* 26: 101–128.

———. 1973. Adaptive evolution in a tropical-alpine environment. In *Taxonomy and ecology*, V. N. Heywood, ed., Vol. 5, pp. 71–92. The Systematics Assoc.

———. 1975. Studies of adaptation and speciation in the afro-alpine flora of Ethiopia. *Boissiera* 24: 71–74.

———. 1977. Origin and differentiation of the afro-alpine flora. Bicentenary celebration of C. P. Thunberg's visit to Japan: 36–42. Tokyo: Royal Swedish Embassy and Botanical Society of Japan.

———. 1978. Nature in utilization and conservation of high mountains in Eastern Africa (Ethiopia to Lesotho). *The use of high mountains of the world*, pp. 42–54. A series of papers commissioned by IUCN and published by Dept. of Lands and Survey, Wellington, New Zealand.

———. In press. The high mountain flora of the Bale Mts. in S. Ethiopia. *Webbia*.

Hedberg, O., B. Ericson, A. Grill-Willén, A. Hunde, L. Källsten, O. Löfgren, T. Ruuth, and O. Ryding. 1979. The yellow-flowered species of *Bartsia* (Scrophulariaceae) in tropical Africa. *Norw. J. Bot.* 26: 1–9.

Hedberg, O., P.-E. Holmlund, R. L. A. Mahunnah, B. Mhoro, W. R. Mziray, and A.-C. Nordenhed. 1980. The *Bartsia abyssinica*-group (Scrophulariaceae) in tropical Africa. *Bot. Notiser* 113: 205–213.

Hedge, I. C., and P. Wendelbo, 1978. Patterns of distribution and endemism in Iran. *Notes Roy. Bot. Gard. Edinb.* 36: 441–464.

Holmgren, P. K., and W. Keuken. 1974. *Index Herbariorum* 1. *The Herbaria of the world*. 6th ed. Regnum Vegetabile 92.

Hultén, E. 1958. The amphi-atlantic plants and their phytogeographical connections. *K. Svenska Vet. Akad. Handl.*, ser. 4, 7: 1–340.

———. 1962. The circumpolar plants. I. Vascular cryptogams, conifers, monocotyledons. *K. Svenska Vet. Akad. Handl.*, ser. 4, 8: 1–275.

Jacot Guillarmod, A. 1971. *Flora of Lesotho*. Lehre: Cramer.

Jeannel, R. 1950. *Hautes montagnes d'Afrique*. Paris: *Publ. Mus. Nat. Hist. Nat.*, Supplement No. 1.

Jonsell, B. 1976. Some tropical Africa Cruciferae. Chromosome numbers and taxonomic comments. *Bot. Notiser* 129: 123–130.

Keay, E. W. J. 1955. Montane vegetation and flora of the British Cameroons. *Proc. Linn. Soc. Lond.* 165: 140–143.

Killick, D. J. B. 1963. An account of the plant ecology of the Cathedral Peak area of the Natal Drakensberg. *Mem. Bot. Surv. S. Afr.* 34: 1–178.

———. 1978. The afro-alpine region. In *Biogeography and ecology of Southern Africa*, M. J. A. Werger and A. C. van Bruggen, eds., pp. 515–560. The Hague: Junk.

Kokwaro, J. 1968. Valerianaceae. In *Flora of tropical East Africa*, E. Milne-Redhead and R. M. Polhill, eds. London: Crown Agents for Oversea Governments and Administrations.

Larcher, W. 1980. Klimastress im Gebirge—Adaptationstraining und Selektionsfilter für Pflanzen. *Rheinisch-Westfäl. Akad. Wiss. Vorträge* 291: 49–88.

Lye, K. A., and R. W. Haines. 1974. Studies in African Cyperaceae. XIII. New Taxa and Combinations in *Isolepis*. R. Br. *Bot. Notiser* 127: 522–526.

Mabberley, D. J. 1973. Evolution in the giant groundsels. *Kew Bull.* 28: 61–96.

———. 1974. The pachycaul lobelias of Africa and St. Helena. *Kew Bull.* 29: 535–584.

Markgraf-Dannenberg, I. 1969. Die systematische Zugehörigkeit von *Festuca kilimanjarica* Hedb. *Wildenowia* 5: 271–278.

Meyer, F. G. 1958. The genus *Valeriana* in East Tropical and South Africa. *J. Linn. Soc. Lond. Bot.* 55: 761–771.

Meyer, H. 1900. *Der Kilimandscharo, Reisen und Studien*. Berlin.

Monod, T. 1957. Les grandes division chorologiques de l'Afrique. CCTA/CSA (London), publ. no. 24: 1–147.

Morton, J. K. 1961. The upland floras of west tropical Africa—their composition, distribution and significance in relation to climate changes. *Comptes rendus de la IVe réunion plénière de l'AETFAT* (l'Association pour l'Etude Taxonomique de la Flore d'Afrique Tropicale), pp. 391–409. Lisbon.

Nordenstam, B. 1969. Phytogeography of the genus *Euryops* (Compositae). *Opera Botanica* 23: 1–77.

———. 1978. Taxonomic studies in the tribe Senecioneae (Compositae). *Opera Botanica* 44: 1–83.

Nordlindh, T. 1943. *Studies in the Calendulae. I. Monograph of the genera Dimorphoteca, Castalis, Osteospermum, Gibbaria and Chrysanthemoides*. Lund.

Pettersson, B. 1961. Mutagenic effects of radiant heat shock on phanerogamous plants. *Nature* 191: 1167–1169.

Pichi-Sermolli, R. E. G. 1957. Una carta geobotanica dell'Africa Orientale (Eritrea, Etiopia, Somalia). *Webbia* 13: 15–132.

Rauh W. 1977. Die Wuchs- und Lebensformen der tropischen Hochgebirgsregionen und der Subantarktis, ein Vergleich. In *Geoecological relations between the southern temperate zone and the tropical mountains*, C. Troll and W. Lauer, eds., *Erdwiss. Forsch.* 11: 62–92.

Robson, N. K. B. 1977. Studies in the genus *Hypericum* L. I. Infrageneric classification. *Bull. Brit. Mus. Nat. Hist. (Bot.)* 5: 291–355.

———. 1979. Parallel evolution in African and Mascarene Hypericum. *Kew Bull.* 33: 571–584.

Ross, R. 1957. Notes on Philippia. *Bull. Jard. Bot. de l'Etat* (Brussels) 27: 733–754.

Schnell, R. 1977. *La flore et la végétation de l'Afrique tropicale. 2. Introduction à la phytogégraphie des pays tropicaux*. Vol. 4. Paris: Gauthier-Villars.

Scott, H. 1952. Journey to the Gughé Highlands (Southern Ethiopia), 1948–49; biogeographical research at high altitudes. *Proc. Linn. Soc. Lond.* 163: 85–189.

———. 1955. Journey to the high Simien district, Northern Ethiopia, 1952–53. *Webbia* 11: 425–450.

———. 1958. Biogeographical research in High Simien (Northern Ethiopia), 1952–53. *Proc. Linn. Soc. Lond.* 170: 1–91.

Shan, R. H., and L. Constance. 1951. The genus *Sanicula* (Umbelliferae) in the Old World and the New. *Univ. Calif. Publ. Bot.* 25: 1–78.

Skottsberg, C. 1941. Plant succession on recent lava flows in the island of Hawaii. *Göteborgs Kungl. Vetenskaps- och Vitterhets-samhälles Handlingar* 6B: 1–32.

Stebbins, G. L. 1950. *Variation and evolution in plants*. New York: Columbia Univ. Press.

———. 1966. *Processes of organic evolution*. Englewood Cliffs, N.J.: Prentice-Hall.

Sunding, P. 1970. Elements in the flora of the Canary Islands and theories on the origin of their flora (in Norwegian; English summary). *Blyttia* 28: 229–259.

———. 1979. Origins of the Macaronesian flora. In *Plants and islands*, D. Bramwell, ed., pp. 13–40. London: Academic Press.

Thorne, R. F. 1978. Plate tectonics and angiosperm distribution. *Notes Roy. Bot. Gard. Edinb.* 36: 297–315.

Thulin, M. 1975. The genus *Wahlenbergia* s. lat. (Campanulacae) in tropical Africa and Madagascar. *Symb. Bot. Upsal.* 21: 1–123.

Troll, C. 1943. Die Frostwechselhäufigkeit in den Luft- und Bodenklimaten der Erde. *Meteorol. Zeit.* (Braunschweig) 60: 161–171.

———. 1944. Strukturböden, Solifluktion und Frostklimate der Erde. *Geol. Rundschau* 34: 545–694.

———. 1948. Der assymetrische Aufbau der Vegetationszonen und Vegetations-Stufen auf der Nord- und Südhalbkugel. *Ber. Geobot. Forschungsinst. Rübel* (Zurich) 1947: 46–83.

———. 1955. Der jahreszeitliche Ablauf des Naturgeschehens in den verschiedenen Klimagürteln der Erde. *Studium Generale* 8: 713–733.

———. 1959. Die tropischen Gebirge, ihre dreidimensionale klimatische und pflanzengeographische Zonierung. *Bonner Geogr. Abh.* 25: 1–93.

Tzvelev, N. N., and Z. V. Bolchovshiek. 1965. On the genus Zingeria P. Smith and closely related genera within the family Gramineae (Caryo-systematic investigation) (in Russian). *Bot. Journal* 50: 1317–1320.

van Leeuwen, W. M. D. 1936. Krakatau, 1883–1933. In K. W. Dammerman and H. J. Lam, eds., *Annales du Jardin Botanique de Buitenzorg*, Vols. XLVI–XLVII. Leiden: Brill.

Weimarck, H. 1934. *Monograph of the genus Cliffortzia*. Lund.

———. 1946. Studies in the Juncaceae with special reference to the species in Ethiopia and the Cape. *Svensk Bot. Tidskr.* 40: 141–178.

Wendelbo, P. 1961. Studies in Primulaceae. II. *Acta Univ. Bergen. Ser-Math. Nat.* 11.

———. 1968. *Wurmbea hamiltonii*, a new afroalpine species of Liliaceae. *Bot. Notiser* 121: 114–116.

White, F. 1970. Floristics and plant geography. In *The evergreen forests of Malawi*, J. D. Chapman and F. White, eds., pp. 38–77. Oxford: Commonwealth Forestry Institute.

White, F. 1978. The afromontane region. In *Biogeography and ecology of Southern Africa*, M. I. A. Werger, ed., pp. 463–513. The Hague: Junk.

Wickens, G. E. 1974. Studies in the Gramineae 25. The range of variation in *Pentaschistis pictigluma. Kew Bull.* 26: 41–44.

———. 1976. The flora of Jebel Marra (Sudan Republic) and its geographical affinities. *Kew Bull.*, add. ser. 5: 1–368.

Wild, H. 1968. Phytogeography in South Central Africa. *Kirkia* 6: 197–222.

Wright, S. 1951. The genetical structure of populations. *Ann. Eugen.* 15: 323–354.

Zemede, A. 1972. Introgressive hybridization in two species of *Bartsia*. Unpublished B.Sc. thesis, Addis Ababa University.

18

Origins and History of the Malesian High Mountain Flora

JEREMY M. B. SMITH

It would be inappropriate to discuss the origins of the Malesian high mountain flora without first describing briefly the ecological setting of that flora today, and in the past as far as we know it. In considering such aspects in the first sections of this chapter, reference will be made to some of the many publications on the mountains and their biotas, whose material can only be summarized here. Particular attention is given to four regional compilations: Gressitt (1982); Luping, Chin Wen and Dingley (1978); van Royen (1980); and van Steenis (1972).

Consideration in this chapter is focused particularly on the tropical-alpine flora, that assemblage of microtherm, light-demanding small plants that compose New Guinea's mountain grasslands and inhabit similar but less extensive environments on high mountains in other parts of Malesia. Many species and most genera are common to the floras of mountains in more than one island, but few occur in lowland habitats below about 1000 m. The tropical-alpine flora of Malesia can be thought of as a single flora with local variants, as was recognized in the detailed, pioneering work of van Steenis (1934–1936).

THE MOUNTAINS AND THEIR ENVIRONMENTS

The Malesian floristic region extends approximately from Burma to Fiji, including principally the multitude of tropical islands to the north and northwest of Australia, and to the south and southeast of China and Indochina. The high mountains of the region occur in Malaysia, Indonesia, the Philippines, and Papua New Guinea.

The highest mountains and the only real cordillera are on the island of New Guinea, the highest of all (with glaciers) being in the western part, Irian Jaya. Eight New Guinea mountains exceed 4000 m in height: Jaya (or Carstensz, 4884 m, with the nearby Idenberg), Trikora (or Wilhelmina, 4750 m), and Mandala (or Juliana, 4702 m) in Irian Jaya; and Wilhelm (4510 m), Giluwe (4368 m), Bangeta (4121 m), Antares (4120 m), and Victoria (4036 m) in Papua New Guinea.

The only other Malesian mountain of this height is Kinabalu (4101 m) in northern Borneo (see Fig. 18–1). However there are many more mountains in the region higher than 3000 m along the island arc from Sumatra to Flores, most of the others being in New Guinea and Sulawesi.

Geologically the mountains vary. Active volcanism occurs, especially along the Sumatra-Flores arc, in New Guinea, and in the Philippines. In New Guinea the volcanoes generally lie to the north and east of the high mountains, which themselves are of Cenozoic sedimentary rocks such as limestone (e.g., Jaya, Bangeta), or intrusive rocks such as gabbro (e.g., Wilhelm), or are extinct volcanoes (e.g., Giluwe). Volcanoes have contributed tephra to some of the high mountain environments in the recent geologic past in New Guinea, but these mostly have not experienced other recent volcanic disturbance. This is not so for many of the mountains in Sumatra (e.g., Kerinci), Java (e.g., Semeru) and farther east, where recent or continuing volcanic activity has made their upper reaches impoverished as ecosystems. Some extinct volcanoes on these islands support richer vegetation, as do nonvolcanic mountains, for example, in Borneo (e.g., Kinabalu, adamellite) and Sulawesi (e.g., Rantemario, metamorphosed sediments).

Soil development is affected by factors additional to age and composition of the substrate. In New Guinea mountains, perhaps because of low-intensity precipitation and/or inputs of tephra, soils generally are deep and fertile by comparison with other mountains in the region. By contrast Kinabalu, for example, has very little soil in its summit zone, which is mostly bare rock (Smith, 1977a). The summit ridge of

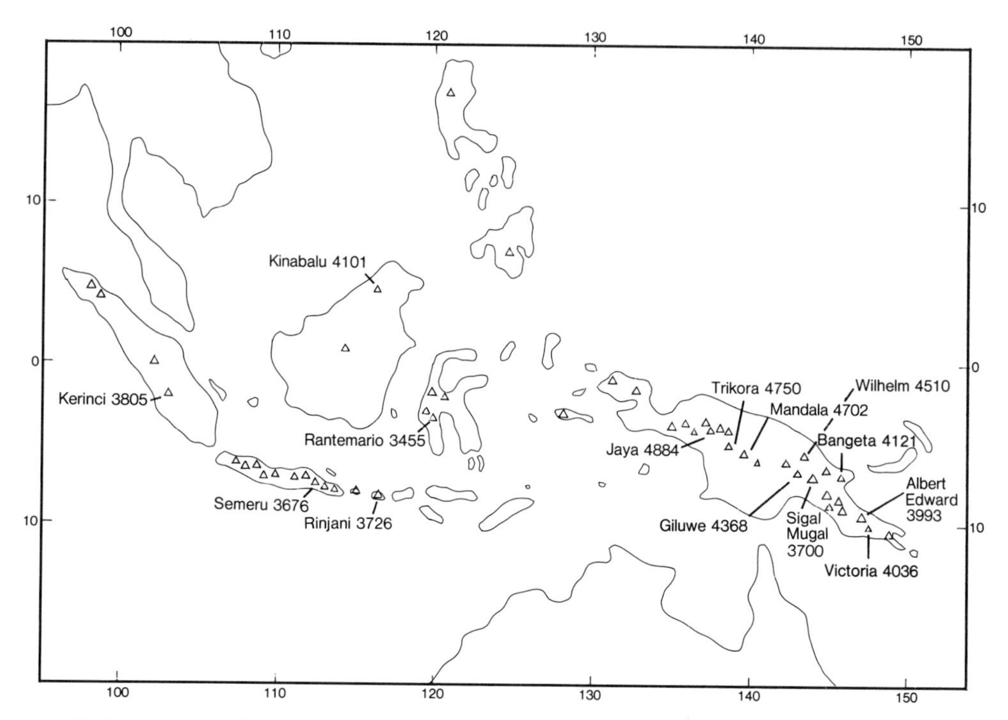

FIG. 18–1. Locations of Malesian mountains higher than 2900 m. Not all New Guinea mountains are shown.

Rantemario has a shallow, infertile soil. Where soils have been analyzed they are commonly acidic and rich in humus.

Most parts of the Malesian region have an ever-wet, equatorial climate, and the mountains reflect this. With increasing altitude temperatures fall, but precipitation may increase to a maximum at 3000–3500 m. Dry seasons occur annually but are not pronounced except from eastern Java to Timor, in south Sulawesi, and in southeast Papua New Guinea. Elsewhere, including on all the highest mountains, the climate is usually wet and cloudy throughout the year, though drier periods may occur occasionally, as during 1972 on Kinabalu (Lowry, Lee and Stone, 1973) and in eastern New Guinea (Brown and Powell, 1974). As the only cordillera in the region, in New Guinea, is oriented approximately east–west, major rainshadow effects do not occur to create dry mountain environments.

PALEOGEOGRAPHY

In early Mesozoic times, before about 140 million years ago, all the southern continents with India were thought to be contiguous and to form a vast southern continent, Gondwanaland. At about that time India, and later South America and Africa, rifted away and drifted toward their present positions. Australia (including southern New Guinea, New Caledonia, New Zealand, and perhaps Timor and some other islands or parts of islands now in eastern Indonesia), hitherto a part of Antarctica, broke away about 55 million years ago and moved north to collide with Asia, in the Malesian region, in the Miocene 10–15 million years ago. The bulk of Malesia, however, from Thailand to the Philippines, western Sulawesi and Flores, was never part of Gondwanaland, but belonged instead to a northern grouping of continents called Laurasia.

Malesia therefore consists of lands belonging, paleogeographically, to both south and north, and came into existence as a single region only 10–15 million years ago (see also Chap. 21). Most reconstructions show Australia and Asia separated by a wide expanse of ocean before then (e.g., Audley-Charles, Hurley and Smith, 1981) providing little opportunity for land plant migration between them. This dual origin of Malesia is reflected today in the Sunda and Sahul shelves, areas of shallow sea extending from Asia as far as Borneo and Bali, and from Australia to New Guinea, respectively. Between these shelves lies deeper ocean, with islands including the Philippines, Sulawesi, and Timor. There has probably never been a terrestrial connection across this area, which is called Wallacea.

This general picture (which lacks clarity in many of its details) has been used in explaining present distributions of vertebrate fauna and lowland flora (e.g., Whitmore, 1981). However, its relevance to interpreting the origins and migrations of tropical-alpine flora—except in apparently ruling out Mesozoic or Early Tertiary mountain migration corridors between Australia and Asia—is doubtful. One reason for this is that the Malesian tropical-alpine environment itself may not have existed before the Quaternary (the last 1.8 million years) or latest Tertiary.

The orogenies and volcanism that created today's Malesian mountains postdate the collision between Australia and Asia. It is unlikely that high mountains existed, at least continuously, in either the Australian or the Asian parts of what was to become Malesia during earlier times. Even if some high mountains were present in Cretaceous or Tertiary times, because world climates were warmer, they would have needed to be very high indeed to have provided a suitable habitat for the tropical-alpine flora. The high volcanic mountains in the Sumatra-Flores arc are geologically young. So is the granitic intrusion forming Kinabalu (Jacobson, 1978). In New Guinea, uplift began about 30 million years ago, but development of high mountains probably occurred during much later phases of orogeny and volcanism (Thompson, 1967; Bain, 1973). The coast north of Bangeta is rising today at a rate of 3.5 mm annually (Chappell, 1974).

During the Quaternary world climate was at most times cooler than at present, and much cooler than during the Tertiary. For most of the period climates were "glacial," with briefer "interglacial" intervals (Emiliani, 1972), probably having major effects on the extent of tropical-alpine environments in Malesia.

PAST AND PRESENT VEGETATION

Pre-Quaternary fossils of high mountain floras in Malesia are few. One record of interest concerns the arrival of the southern conifers *Phyllocladus* and *Podocarpus* in northern Borneo about two million years ago, after being absent from the pollen record of the preceding 28 million years (Muller, 1966).

The main fossil record for Malesian tropical-alpine vegetation comes from the Late Quaternary of New Guinea. A unified picture is starting to emerge, as reviewed by Hope (1980a; Fig. 18–2) and Walker and Hope (1982).

From before 50,000 years to about 15,000 years ago, climates were colder (and at least in western New Guinea probably drier) than at present. At the time of ice maximum (18,000–15,000 years ago), temperatures in the mountains may have been 7°C lower than today's, glaciers extended down to about 3200 m on many mountains (Löffler, 1982), and the upper forest limit in the intermontane valleys of the central highlands was depressed to an altitude of 2000–2250 m. Tropical-alpine vegetation at that time occupied an area of about 55,000 km^2 compared with only about 800 km^2 by 6000 years ago. By comparison with modern vegetation zonation the more shrubby and open upper mountain ("subalpine") forest (occurring above about 3000 m today) was very compressed, and may not have existed as such at all, whereas shrub-rich tropical-alpine grasslands were much more extensive, occupying an altitudinal belt about 1000 m deep (see Fig. 18–2). Outside New Guinea, Kinabalu is the only Malesian mountain known to have been glaciated during Late Quaternary times.

Between 15,000 and 8500 years ago climate warmed. Vegetation belts rose to slightly above their present levels and assumed their modern composition. Fires affected the mountain grasslands to different extents and over different periods, indicated by the presence and abundance in sediments of carbonized particles. Fires occurred, at variable frequencies, since 12,000 years ago on Albert Edward and even earlier at Kosipe (2000 m) nearby; from 10,800 to 1500 years ago on Jaya; since about 1000 years ago and especially in the last 250 years on Wilhelm; and hardly at all on Scorpio, which even today is remote from human populations. Fires increase the area and probably alter the composition of grasslands, and progressively destroy forests. They are considered to be almost all lit by human beings (Smith, 1975a), who are known to have been in highland New Guinea for at least 25,000 years.

The vegetation of only the most recent glacial period and the succeeding postglacial interval has so far been revealed by palynological and other studies. There is, however, no reason to suppose that the 25 or more earlier glacial and interglacial periods shown elsewhere did not have similar impacts in New Guinea to those of the most recent, or that they were not all reflected in other mountain environments in Malesia. For most of the past 1.8 million years, climates have probably been cooler, and tropical-alpine vegetation more extensive, than today, though the available record suggests no long-term stability in either (Walker and Flenley, 1979; Walker and Hope, 1982). This is of significance to consideration of origins, migrations, and evolution of the high mountain flora, as discussed later.

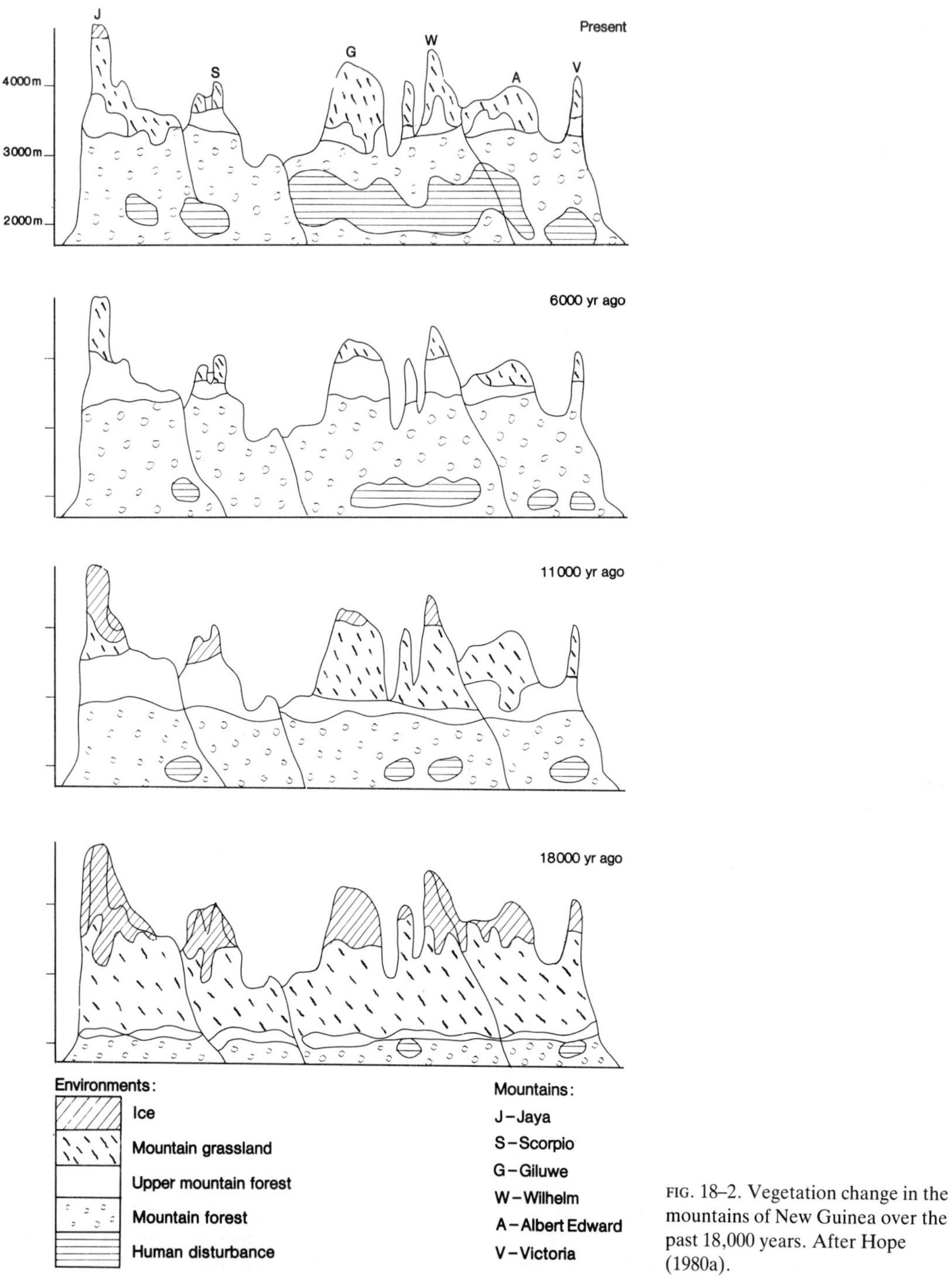

FIG. 18–2. Vegetation change in the mountains of New Guinea over the past 18,000 years. After Hope (1980a).

Present vegetation of Malesian high mountains can only briefly be described and exemplified here. The principal concern in this chapter is the tropical-alpine flora, but the distribution of this is closely dependent on the distribution and density of the forests, since tropical-alpine species cannot tolerate the shade imposed by dense forest cover. Tropical-alpine species therefore occur in the following situations: (1) above the climatic forest limit; (2) where stress (e.g., waterlogging, shallow soil) precludes forest growth at relatively high altitudes; (3) where such stress leads to a more open forest canopy, permitting shade-intolerant plants to form a field layer; and (4) where human or other disturbance to mountain forest creates open conditions.

The first of these categories is restricted to New Guinea, where grasslands and other nonforest

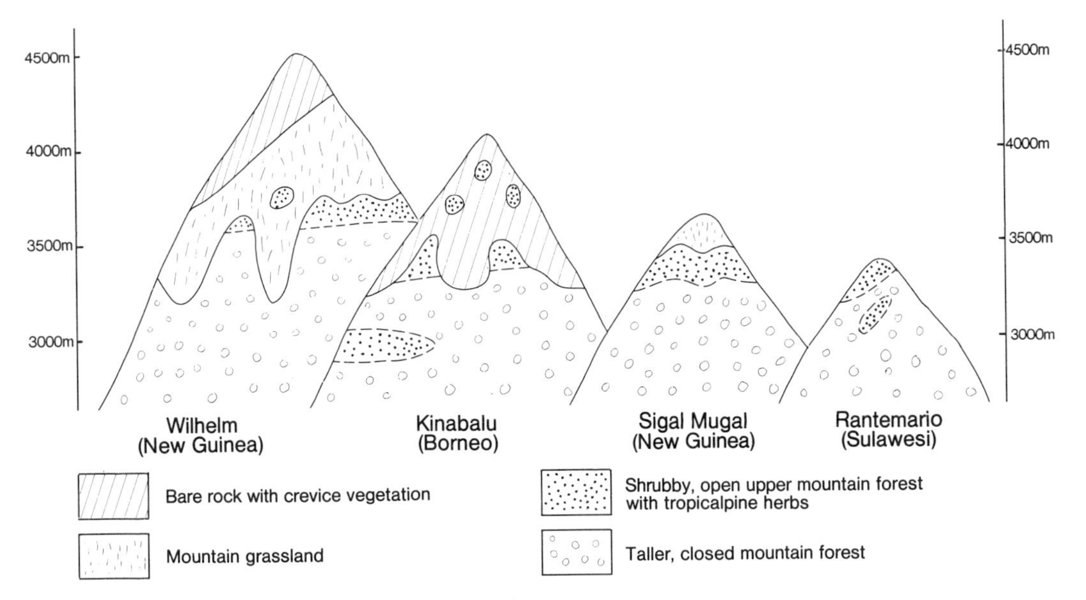

FIG. 18–3. Present vegetation of four Malesian mountains.

communities are extensive and varied (Wade and McVean, 1969; Hope, 1980b). The highest stands of forest on Kinabalu grow at 3950 m, only 150 m below the summit; there is little reason to suppose that they represent a climatic forest limit as throughout the summit zone about 3300 m, forest is restricted severely in distribution by lack of soil (Smith, 1980a). The highest forests with true trees in New Guinea grow at about 3930 m on Jaya, but with stands of tall shrubland (arguably also forests) found as high as 4170 m (Hope, 1976). On Wilhelm a diffuse upper forest limit is found in an unburned valley at 3760–3810 m (Smith, 1975a). On smaller mountains the forest limit is yet lower, for example, at 3430–3530 on Sigal Mugal (Smith, 1975a); this may not be entirely a climatic restriction. An edaphic factor is probably also involved in the exclusion of forest from the summits of tropical mountains that do not reach the altitude of the climatic forest limit (Grubb, 1971).

Even mountains that are almost entirely forested can support tropical-alpine flora if those forests are not too dense. On Rantemario (3455 m, Sulawesi), for example, the windward side of the summit ridge is clothed by an open heath, and the opposite slope by a shrubland 2–3 m tall, both including many of the same woody species that occur also in denser forests on lower slopes. Tropical-alpine herbaceous plants grow beneath these shrubs, and beneath the open shrubby forest of ridge crests elsewhere above 3000 m, as well as beside streams and pools, and on local rocky outcrops.

Human destruction of forests can greatly increase the area available to tropical-alpine plants. Fires have led to great increases in the area of grassland on most New Guinea mountains (Hope, 1980a; Smith, 1980b) and have had more local effects on mountains in other areas (van Steenis, 1972). Tropical-alpine species may grow far below their usual altitudes in areas of human as well as natural disturbance (Smith, 1975b); such areas also provide sites for invasion by alien species.

The present vegetation types of four mountains with which this author is somewhat familiar are compared in Fig. 18–3. Wilhelm and Kinabalu are contrasting examples of high mountains. Sigal Mugal is typical of many smaller mountains in New Guinea, in this case little disturbed by man. Rantemario exemplifies an even smaller mountain that, though almost completely forested, still supports a varied tropical-alpine flora.

GEOGRAPHIC ORIGINS OF THE FLORA

The forest floras of Malesia, including those of the high mountains, differ from the tropical-alpine flora in their geographic affinities and distribution, and probable history and origins. Although some taxa have apparently migrated into or through the region in the past, including some with their centers to the south (e.g., *Leptospermum*, *Podocarpus*, *Tasmannia*) or the north (e.g., *Photinia*, *Rhododendron*, *Schima*), a larger proportion has less distant affinities (at least at a close taxonomic level) and belongs more exclusively to the Malesian region or a part of it. Forests have a longer continuous history there than

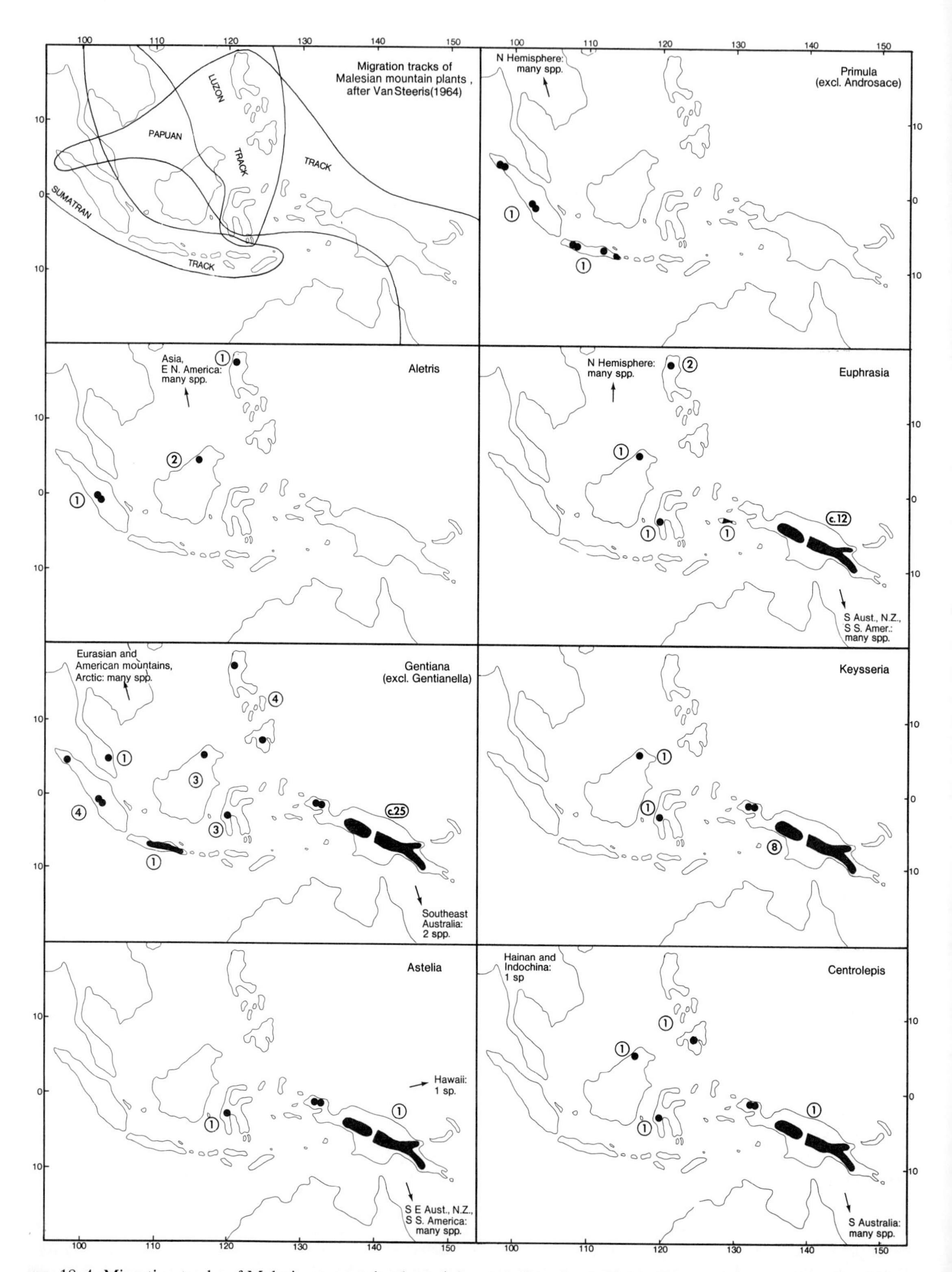

FIG. 18–4. Migration tracks of Malesian mountain plants (after van Steenis, 1964), and the distributions of seven tropical-alpine genera (numbers give indications of species diversity in different areas of Malesia).

do high mountain environments with tropical-alpine vegetation, and migrations that occurred probably mostly did so more slowly and at a more remote time in the past. It is a mistake to equate the histories of long-lived forest trees such as Fagaceae (Fagoideae) with those of herbs and small shrubs such as *Veronica-Hebe* simply on the grounds of similar present ranges, as does van Steenis (1979); their ecological strategies and migration potentials are so different that their histories are most unlikely to have been the same.

The notion of Malesia constituting a cradle of origin of many plant taxa, perhaps even of the angiosperms themselves (e.g., Takhtajan, 1969), needs modification in the light of paleogeographic reconstructions showing the region disunited until Late Tertiary times. Nevertheless many primitive and often geographically restricted taxa survive in Malesian forests, one example being *Trigonobalanus* which appears to retain many primitive characters of its family Fagaceae.

The tropical-alpine flora presents a different picture. Most of its genera and many of its species grow also in cool environments in one or other of the temperate zones, and few genera are endemic. The flora is unusual, not in its individual constituent taxa, but in their assemblage together in a phytogeographically diverse mixture.

Van Steenis (1934–36; 1964) has defined three principal tracks: the Sumatra (from the Himalaya), the Luzon (from China), and the New Guinea (from New Zealand and southeastern Australia). They are mapped in Fig. 18–4, which also includes examples of distributions that may be interpreted as variants of each track. The tracks generally follow the lines of mountains as we see them today, which is where the modern tropical-alpine flora survives. They are usually thought to indicate migration routes along which tropical-alpine plant taxa have spread, predominantly from temperate to Malesian localities. Van Steenis considered them also to indicate the locations of former cordilleras. Not all tropical-alpine plants can be allocated certainly to one of these tracks, as is clear from Fig. 18–4.

Most Malesian tropical-alpine taxa have close affinities with populations in northern or southern temperate zones (or both), and in most cases the extra-Malesian populations are the most diverse and presumably ancestral. To take the example of the Wilhelm tropical-alpine flora considered by Smith (1977b), of 107 native genera, only 10 are endemic to Malesia or nearly so; 11 are northern, 28 are southern, and 58 occur in both northern and southern temperate zones as well as in Malesia. Of 182 native species, 133 are endemic, 9 northern, 23 southern, and 17 widespread. (The tropical-alpine genera and species on Mt. Wilhelm are fully listed in Smith [1977b, Appendix], and need not be given again in this chapter.) It is widely agreed that the Malesian tropical-alpine flora consists mainly of a mixture of the descendants of immigrants from both north and south, with a rather small autochthonous element (at the generic level). What is not so fully agreed is when the migrations took place.

THE AGE OF THE TROPICAL-ALPINE FLORA

In his pioneering account of the composition and history of the high mountain flora of Malesia (1934–36), and subsequently (1964, 1972), van Steenis has argued that it can have achieved its present distribution only by progressive dispersal of entire communities along ancient cordilleras more extensive than today's mountains. He does not believe that establishment of new populations following dispersal over distances of several hundred kilometers has occurred to a significant degree, and he points to a lack of correlation between dispersal mechanisms and distributions, and to the ecological difficulty of such establishment in preexisting vegetation, as supporting evidence. He earlier thought (1964, 1967) that such cordilleras or chains of peaks, traversing Malesia between southern and northern temperate regions, must have been of Cretaceous age. More recently (1972) he has admitted the possibility of Miocene to Pliocene migrations, though still (1979) denying the probability of dispersal over long distances.

Cretaceous or Early Tertiary mountain links between Australasia and Asia are untenable on the basis of recent paleogeographic reconstructions, as outlined earlier. Late Tertiary links are not ruled out so easily, though independent evidence for such mountains is lacking. In response to the objection that modern Malesian mountains supporting tropical-alpine flora are geologically young, van Steenis (1967) has pointed out that young mountains (specifically Kinabalu) may have "inherited" parts of an old flora from nearby ancient mountains that have since been eroded to low altitudes.

A younger age was suggested for temperate elements in the Kinabalu flora by Holloway (1970), who pointed out that lowered montane vegetation zones during Pleistocene glacial periods would have been mutually closer than they are today, facilitating migration of taxa between them. Of greater importance may have been the increase in the number of mountains reaching toward or above the lowered climatic forest limit of such periods, and so acting as "stepping-stones" for migrating plants. Repeated climatic fluctuations during the Pleistocene may also have led to the creation of bare or poorly vegetated areas such as moraines and fluvioglacial fans that could have acted as competition-free seedbeds for distantly dispersed seeds, as well as causing local extinctions. Pleistocene migrations of Malesian mountain plants have been considered important by other authors, for example, Gupta (1972) and Raven and Axelrod (1972). The probable evolution of tropical-alpine species from lower-altitude ancestors during the Pleistocene has been discussed with reference to *Leptospermum* by Lee and Lowry (1980).

Some features of the Malesian tropical-alpine flora support the idea of geologically recent migration involving dispersal over long distances, such as the relatively low rate of species endemism and high proportion of polyploids. Again taking the Wilhelm flora as an example (Smith, 1977b), of 163 species only 2% are endemic to that mountain, 73% are endemic to Malesia, and 25% are not endemic even at that scale; the corresponding figures for the afroalpine flora of East African mountains are 64%, 33%, and 3% (Hedberg, 1969). This suggests geologically recent achievement of present distributions in Malesia, at least by comparison with the afroalpine flora, unless evolution is proceeding unusually slowly, which seems unlikely in a flora of relatively fast maturing species occurring in small and (in the geologically recent past) fluctuating populations.

A high level of polyploidy may also indicate recent immigration over long distances, because polyploidy may confer various characteristics appropriate to colonizing organisms, including self-compatibility, better vegetative reproduction and vigor, and broader ecological tolerance. Two thirds of 75 tropical-alpine species on Wilhelm investigated by Borgmann (1962) were polyploids. This compares with polyploidy levels of 45% and 49% in the tropical-alpine floras of East and West Africa, respectively (Morton, 1966).

In an ecological study of the Wilhelm flora, Smith (1977b; 1982) found differences between a small "ancient" element (generic endemics and endemic species in genera with distributions suggesting Gondwanaland derivation); a larger "peregrine" element (including most native species); and aliens, introduced to New Guinea by people. All the demonstrated differences showed the "ancient" plants, overall, to have to the least extent characteristics expected in a colonizing taxon (e.g., ability to invade disturbed sites, rapid growth, broad altitudinal range); the most recent immigrants, the aliens, showed such characteristics to the greatest degree. These differences were considered to support the idea that most members of the native tropical-alpine flora are descended from ancestors with only a short history in the region, having arrived probably during Quaternary glacial periods by seed dispersal over long distances. Of 11 characters assessed, only 2, both pertaining to dispersal, did not show differences between elements, and it was concluded that adaptations effective in promoting dispersal over distances of up to several kilometers are irrelevant to the rare but important dispersal events in which seeds are transported over much longer distances.

Such a recent "waif" origin for the Malesian flora provides a possible explanation for its relative lack of "typical" tropical-alpine life forms, by comparison with possibly older floras of East African mountains and the northern Andes. Ecological reasons may also be proposed, particularly relating to differences in radiation levels and moisture availability. Tussock (bunch) grasses and rosettes are common, as are species with pubescent shoots; but stem rosette plants are absent (except for tree ferns), and cushions (e.g., species of *Astelia*, *Centrolepis*, *Danthonia*, *Gaimardia*, *Oreobolus*) are not as large, well formed, or conspicuous as those of, for example, species of *Aciachne* and *Plantago* in the Andes, or of other genera in Tibet and in cool southern temperate mountain regions.

CONCLUSION

Present evidence suggests that the mountains of the Malesian region are young, and that before the Pleistocene little or no suitable environment for a tropical-alpine flora existed. This flora, having close affinities with various distant regions, is probably mainly derived from immigrants that established themselves on the mountains primarily during Pleistocene glacial periods, after occasional seed dispersal over long distances from other cool places, especially in mainland Asia and temperate Australasia.

ACKNOWLEDGMENT

I thank C. D. Ollier for providing paleogeographic information.

REFERENCES

Audley-Charles, M. G., A. M. Hurley, and A. G. Smith. 1981. Continental movements in the Mesozoic and Cenozoic. In *Wallace's line and plate tectonics*, T. C. Whitmore, ed., pp. 9–23. Oxford: Clarendon Press.

Bain, J. H. C. 1973. A summary of the main structural elements of Papua New Guinea. In *The western Pacific: Island arcs, marginal seas, geochemistry*, P. J. Coleman, ed., pp. 149–161. Perth: Univ. of Western Australia.

Borgmann, E. 1962. Anteil der Polyploiden in der Flora des Bismarck-gebirges von Ostneuguinea. *Zeitschrift für Botanik* 52: 118–172.

Brown, M., and J. M. Powell. 1974. Frost and drought in the highlands of Papua New Guinea. *J. Trop. Geog.* 38: 1–6.

Chappell, J. M. A. 1974. Geology of coral terraces, Huon Peninsula, New Guinea: A study of Quaternary tectonic movements and sea level changes. *Bull. Geol. Soc. America* 85: 553–570.

Emiliani, C. 1972. Quaternary hypsithermals. *Quatern. Res.* 2: 270–273.

Gressitt, J. L., ed. 1982. *Biogeography and ecology of New Guinea*. The Hague: Junk.

Grubb, P. J. 1971. Interpretation of the "Massenerhebung" effect on tropical mountains. *Nature* 229: 44–45.

Gupta, R. K. 1972. Boreal and arcto-alpine elements in the flora of western Himalayas. *Vegetatio* 24: 159–175.

Hedberg, O. 1969. Evolution and speciation in a tropical high mountain flora. *Biol. J. Linn. Soc.* 1: 135–148.

Holloway, J. D. 1970. The biogeographical analysis of a transect sample of the moth fauna of Mt Kinabalu, using numerical methods. *Biol. J. Linn. Soc.* 2: 259–286.

Hope, G. S. 1976. Vegetation. In *The equatorial glaciers of New Guinea*, G. S. Hope, J. A. Peterson, I. Allison, and U. Radok, eds., pp. 113–172. Rotterdam: Balkema.

———. 1980a. Historical influences on the New Guinea flora. In *The alpine flora of New Guinea*, vol. 1, P. van Royen, ed., pp. 223–248. Vaduz: Cramer.

———. 1980b. New Guinea mountain vegetation communities. In *The alpine flora of New Guinea*, vol. 1, P. van Royen, ed., pp. 153–222. Vaduz: Cramer.

Jacobson, G. 1978. Geology. In *Kinabalu: Summit of Borneo*, M. Luping, Chin Wen, and E. R. Dingley, eds., pp. 101–110. Kota Kinabalu: Sabah Society.

Lee, D. W. and J. B. Lowry. 1980. Plant speciation on tropical mountains: *Leptospermum* (Myrtaceae) on Mount Kinabalu, Borneo. *Bot. J. Linn. Soc.* 80: 223–242.

Lowry, J. B., D. W. Lee, and B. C. Stone. 1973. Effect of drought on Mount Kinabalu. *Malayan Nature J.* 26: 178–179.

Luping, M., Chin Wen, and E. R. Dingley, eds. 1978. *Kinabalu: Summit of Borneo*. Kota Kinabalu: Sabah Society.

Morton, J. K. 1966. The role of polyploidy in the evolution of a tropical flora. In *Chromosomes Today*, Vol. 1, C. D. Darlington and K. R. Lewis, eds., pp. 73–76. Edinburgh: Oliver & Boyd.

Muller, J. 1966. Montane pollen from the Tertiary of north-west Borneo. *Blumea* 14: 231–235.

Raven, P. H., and D. I. Axelrod. 1972. Plate tectonics and Australasian paleobiogeography. *Science* 176: 1379–1386.

Smith, J. M. B. 1975a. Mountain grasslands of New Guinea. *J. Biogeog.* 2: 27–44.

———. 1975b. Notes on the distributions of herbaceous angiosperm species in the mountains of New Guinea. *J. Biogeog.* 2: 87–101.

———. 1977a. An ecological comparison of two tropical high mountains. *J. Trop. Geog.* 44: 71–80.

———. 1977b. Origins and ecology of the tropicalpine flora of Mount Wilhelm, New Guinea. *Biol. J. Linn. Soc.* 9: 87–131.

———. 1980a. The vegetation of the summit zone of Mount Kinabalu. *New Phytologist* 84: 547–573.

———. 1980b. Ecology of the high mountains of New Guinea. In *The alpine flora of New Guinea*, Vol. 1, P. van Royen, ed., pp. 111–132. Vaduz: Cramer.

———. 1982. Origins of the tropicalpine flora. In *Biogeography and ecology of New Guinea*, J. L. Gressitt, ed., pp. 287–308. The Hague: Junk.

Takhtajan, A. 1969. *Flowering plants: Origin and dispersal*. Edinburgh: Oliver & Boyd.

Thompson, J. E. 1967. A geological history of eastern New Guinea. *J. Aust. Pet. Explor. Assoc.* 7: 83–93.

van Royen, P., ed. 1980. *The alpine flora of New Guinea*, Vol. 1. Vaduz: Cramer.

van Steenis, C. G. G. J. 1934–36. On the origin of the Malaysian mountain flora. *Bull. Jard. Bot. Buitenzorg* 13: 135–262; 13: 289–417; 14: 56–72.

———. 1964. Plant geography of the mountain flora of Mt Kinabalu. *Proc. Roy. Soc.* London, ser. B, 161: 7–38.

———. 1967. The age of the Kinabalu flora. *Malayan Nat. J.* 20: 39–43.

———. 1972. *The mountain flora of Java*. Leiden: Brill.

———. 1979. Plant geography of east Malesia. *Bot. J. Linn. Soc.* 79: 97–178.

Wade, L. K., and D. N. McVean. 1969. *Mt Wilhelm studies. 1. The alpine and subalpine vegetation*. Canberra: Australian National Univ.

Walker, D., and J. R. Flenley. 1979. Late Quaternary vegetational history of the Enga Province of upland Papua New Guinea. *Philos. Trans. Roy. Soc.* London, ser. B, 286: 265–344.

Walker, D., and G. S. Hope. 1982. Late Quaternary vegetation history. In *Biogeography and ecology of New Guinea*, J. L. Gressitt, ed., pp. 263–285. The Hague: Junk.

Whitmore, T. C., 1981. *Wallace's line and plate tectonics*. Oxford: Clarendon Press.

19

Origins of the High Andean Herpetological Fauna

JOHN D. LYNCH

The rich, nonpareil South American herpetofauna is composed of no fewer that 2215 species of amphibians and reptiles and accounts for one third of all amphibians and one fifth of all reptiles world wide (Duellman, 1979a). These 2215 species are partitioned into 58 family groups (including tribes). Although the Andes (1000 m and above) make up a relatively modest fraction (9% by planimetric measure) of the South American continent, 15 of the 26 amphibian family groups and 17 of the 32 reptilian family groups occur in the Andes (i.e., at least one endemic species found above 1000 m). The overpowering influence of the Andes on the herpetofauna of South America is seen in its proportion of endemic species (approximately 730 species). Nearly one half (465 of 1100) of the South American amphibians and one fourth (265 of 1115) of the South American reptiles are Andean endemics.

THE ANDEAN FAUNA

Generic endemism is poorly expressed in the Andes (Fig. 19–1). According to Duellman (1978a, p. 3), 86 of the 115 amphibian genera and 114 of the 203 reptilian genera found in South America are endemic to the continent. However, only 9 amphibian (all frogs) genera (*Alsodes*, *Atopophrynus*, *Batrachophrynus*, *Centrolene*, *Cryptobatrachus*, *Geobatrachus*, *Osornophryne*, *Phrynopus*, and *Telmatobius*) and 12 reptilian genera (*Macropholidus*, *Opipeuter*, *Phenacosaurus*, *Pholidobolus*, *Phymaturus*, *Polychroides*, *Proctoporus*, *Saphenophis*, *Stenocercus*, *Synophis*, *Umbrivaga*, and "*Urotheca*" *williamsi*) are Andean endemics. The relative significance of the endemism differs among these 21 genera because *Atopophrynus*, *Centrolene*, and *Geobatrachus* (frogs), *Macropholidus*, *Opipeuter*, *Phymaturus*, and *Polychroides* (lizards), and *Umbrivaga* and "*Urotheca*" (snakes) are monotypic wheras *Phrynopus* and *Telmatobius* (frogs) and *Stenocercus* (lizards) are modest-sized genera (14–30 species).

Among the more prominent (speciose) genera, *Eleutherodactylus* stands out (Table 19–1), accounting for more than 20% of the herpetological diversity in the Andes. Although large genera

FIG. 19–1. Numbers of endemic frog genera (1–11 species) found in various ecogeographic units of South America. Southeastern Brazil is distinctive for its high proportion of generic endemism, as is Patagonia. Stippling separates Patagonia, eastern Brazil, and the cerrado-chaco-pampas regions.

TABLE 19–1. Prominent amphibian and reptile genera encountered in the Andes (1000–5000m).

Genus	Suborder [a]	Andean species	Size of genus (no. of species)	Percent Andean	Percent Andean herpetofauna	Cumulative percent
Eleutherodactylus	F	158	425	37(70) [b]	21.4	21.4
Centrolenella	F	45	60	75	6.2	27.5
Colostethus	F	43	55	78	5.6	33.2
Atractus	S	38	70	54	5.2	38.4
Gastrotheca	F	32	35	91	4.4	42.7
Telmatobius	F	30	30	100	4.1	46.8
Liolaemus	L	28	47	60	3.8	50.7
Atelopus	F	27	36	75	3.7	54.4
Stenocercus	L	26	29	90	3.6	57.9
Bufo	F	20	186	11	2.7	60.7
Hyla	F	19	240	8(14) [c]	2.6	63.3
Anolis	L	18	200	9	2.5	65.8
Proctoporus	L	17	18	94	2.3	68.1
Bolitoglossa	U	14	62	22	1.9	70.0
Phrynopus	F	14	14	100	1.9	71.9
Caecilia	C	12	27	44	1.6	73.6
Dipsas	S	11	31	35	1.5	75.1
Bothrops	S	9	49	18	1.2	76.3
Dendrobates	F	9	36	25	1.1	77.4

[a] Suborders are identified as follows: F, frog; S, snake; L, lizard; U, salamander; C, caecilian.

[b] If only South American taxa are counted (225), 70% of *Eleutherodactylus* are Andean.

[c] If only South American taxa are counted (132), 14% of *Hyla* are Andean.

are normally rare, four are represented in the Andean herpetofauna. *Eleutherodactylus* is the largest vertebrate genus known (and, of course, the largest neotropical amphibian genus). Two other large frog genera, *Bufo* and *Hyla* (186 and 240 species, respectively), are poorly represented in the Andean fauna. The largest reptilian genus, *Anolis*, is likewise poorly represented in the Andes (Table 19–1).

The amphibian and reptile genera represented in the Andes fall into two groups (Figs. 19–2, 19–3). Most genera are poorly represented (fewer than 50% of the species of the genus are Andean), but some (see Fig. 19–3) are very well represented in the Andes. The importance of the Andes to the evolution of the larger genera (e.g., *Centrolenella*,[1] *Colostethus*, *Eleutherodactylus*, *Stenocercus*, and *Telmatobius*) is obvious when contrasted with such large genera as *Anolis*, *Bothrops*, *Bufo*, *Hyla*, *Leptotyphlops*, and *Rhadinaea* having poor (8%–14%) Andean representations. The two groups are most apparent in those genera having 33 or more species (Figs. 19–2, 19–3) where fewer than one third of the species, or more than two thirds, are Andean (Fig. 19–3). In smaller genera, the two modes merge.

The Andean herpetofauna may thus be restricted to those genera having appreciable Andean representations (Tables 19–2 and 19–3) and probably intimate evolutionary involvement with the Andes. The 15 frog genera include 9 ranging in altitude to the páramo or puna habitats above 3000 m. The 13 lizard and snake genera include 11 ranging into the high Andes.

The single most prominent physiographic element of the high Andean herpetofauna (although evident also at lower elevations) is the relatively low and xeric Huancabamba Depression in southern Ecuador and northern Peru. Of the 14 dominant frog genera, 9 occur essentially north of the depression (Table 19–2). Such large genera as *Centrolenella* and *Eleutherodactylus* are found in the Peruvian and Bolivian Andes but only as cloud forest elements. *Alsodes* and *Batrachophrynus* are southern elements as is *Telmatobius* although three species of *Telmatobius* invade the Andes of Ecuador (Trueb, 1979). *Gastrotheca* is distributed throughout the Andes (both northern and southern) but is represented by distinct species groups on opposite sides of the Huancabamba Depression (Duellman and Fritts, 1972; Duellman, 1974). *Phrynopus* likewise is distributed without regard to the Huancabamba

[1] Although the generic name *Centrolenella* is used throughout this chapter, some authors use the name *Hylopsis* for these leaf frogs (Lynch, 1981a; Lynch and Ruíz, 1983). The correct name is *Centrolenella*.

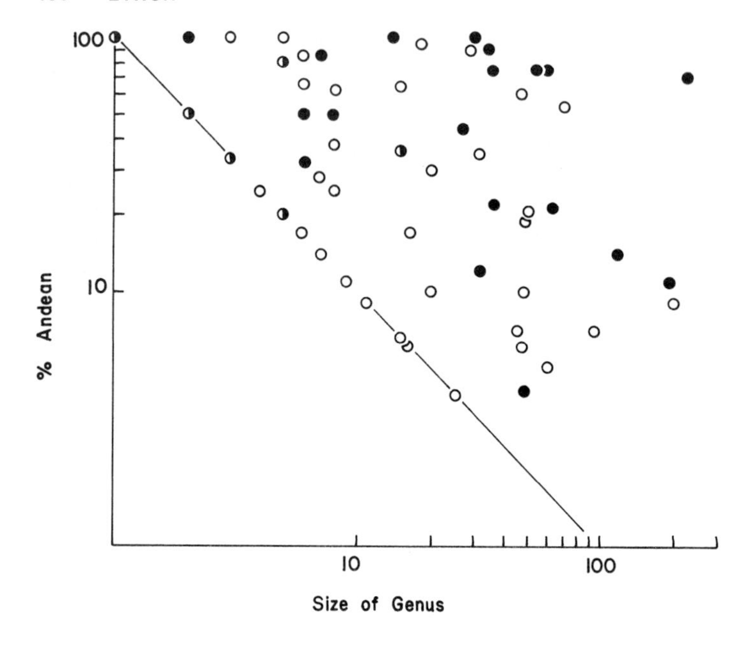

FIG. 19–2. Log-log plot of genus size (number of species) and the proportion of the species of the genus found in the Andes (1000–5000 m). Open circles are genera of squamate reptiles (lizards, snakes); solid circles are genera of amphibians (caecilians, frogs, salamanders); bicolor symbols represent points shared by one amphibian and one reptile genus. The diagonal line connects minimal representation points (the lower left one-half of the plot is a null set).

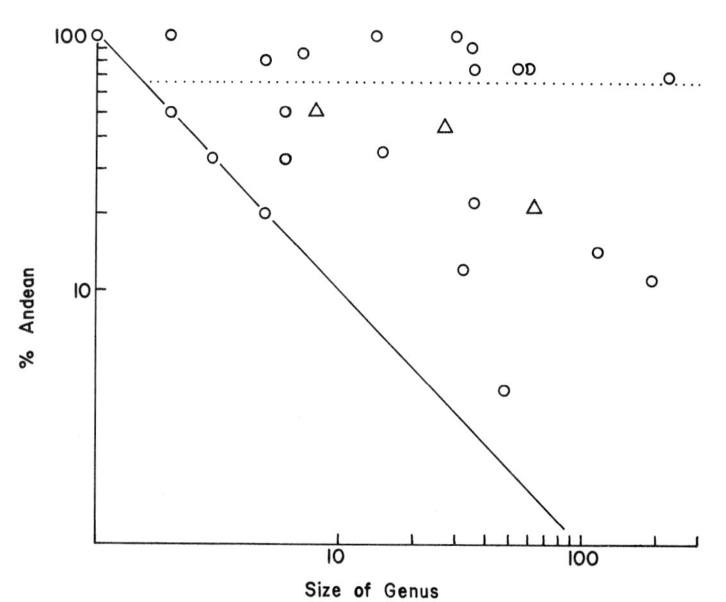

FIG. 19–3. Log-log plot of genus size (number of species) and the proportion of species of each amphibian genus found in the Andes (1000–5000 m). Triangles are caecilian and salamander genera; circles are frog genera. The solid line connects minimal representation points, as in FIG. 19–2. The dotted line represents the 65th percentile (points above the dotted lines are genera termed Andean in the text; points below the line are non-Andean genera). Also see Tables 19–1 and 19–2.

Depression but its species groups do not segregate according to the depression (Lynch, 1975).

Reptiles contrast with the northern distribution of Andean amphibians (Table 19–3). Two lizard and four snake genera are northern elements. Two monotypic lizard genera (*Macropholidus* and *Polychroides*) occur in the Huancabamba Depression. Six genera (five lizards, one snake) are southern elements and the lizard genus *Stenocercus* is essentially southern (Fritts, 1974), paralleling the distibution seen in *Telmatobius*. Three lizard genera (*Euspondylus*, *Prionodactylus*, and *Proctoporus*) and the snake genus *Atractus* occur both north and south of the depression. None of these genera is so well known as to allow inspection of species group distributions relative to the Huancabamba Depression.

The distinction between the distributions of amphibians and reptiles relative to the Huancabamba Depression is most evident in the high Andean elements (occurring above 3000 m). Five anuran genera (*Atelopus*, *Centrolenella*, *Colostethus*, *Eleutherodactylus*, and *Osornophryne*) are northern elements, two (*Batrachophrynus* and *Telmatobius*) are southern elements, and two (*Gastrotheca* and *Phrynopus*) are found in the northern and southern Andes, whereas among reptiles, three are northern elements (*Phenaco-*

TABLE 19–2. Andean genera of amphibians.

Genus	Number of species	Percent Andean	Altitude range (m)	Occurrence
Frogs				
Amphignathodon	1	100	1200–2010	Northern
Alsodes	7	85	0–2500	Southern
Atelopus	36	75	0–4500	Northern
Atopophrynus	1	100	2780	Northern
Batrachophrynus	2	100	4000–4100	Southern
Centrolene	1	100	1920–2150	Northern
Centrolenella	60	75	0–3400	Northern
Colostethus	55	78	0–3800	Northern
Eleutherodactylus [a]	225	70	0–4400	Northern
Gastrotheca	35	91	0–4600	Both
Geobatrachus	1	100	1550–2870	Northern
Hemiphractus	5	80	250–1910	Northern
Osornophryne	2	100	2700–3700	Northern
Phrynopus	14	100	2000–4100	Both
Telmatobius	30	100	1200–5000	Both/southern

[a] South American taxa only.

TABLE 19–3. Andean genera of reptiles.

Genus	Number of species	Percent Andean	Altitude range (m)	Occurrence
Lizards				
Cercosaura	1	100	1000–1600	Southern
Ctenoblepharis	6	83	1000–4600	Southern
Euspondylus	8	62	1100–2980	Both
Liolaemus	47	60	750–4880	Southern
Macropholidus	1	100	3100	Depression
Opipeuter	1	100	1000–3000	Southern
Phenacosaurus	3	100	1750–3750	Northern
Pholidobolus	5	100	1800–3960	Northern
Phymaturus	1	100	1200–3500	Southern
Polychroides	1	100	1000–1750	Depression
Prionodactylus	5	80	100–3020	Both
Proctoporus	18	94	1000–4080	Both
Stenocercus	29	90	1000–3890	Both/southern
Snakes				
Atractus	70	54	800–4000	Both
Saphenophis	5	100	300–3200	Northern
Synophis	3	100	450–2200	Northern
Tachymenis	6	67	70–4570	Southern
Umbrivaga	1	100	1000–1200	Northern
'Urotheca'	1	100	1400–2000	Northern

saurus, *Pholidobolus*, and *Saphenophis*), one (*Macropholidus*) occurs only in the depression, five are southern elements (*Ctenoblepharis*, *Liolaemus*, *Phymaturus*, *Stenocercus*, and *Tachymenis*), and two (*Atractus* and *Proctoporus*) are found in the northern and southern Andes.

These nine anuran, nine lizard, and three snake genera vary in terms of numbers of species found in high-altitude habitats and the degree to which the species are adapted to (or restricted to) high-altitudes habitats (Tables 19–4, 19–5). Some 79 species of frogs, 38 lizards, and 8 snakes of these 21 genera have appreciable proportions of their distribution areas in high-altitude habitats. To these we must add those species from non-Andean genera also penetrating high-altitude habitats. Three salamanders (*Bolitoglossa*) are found in páramos in Colombia and Venezuela, a microhylid toad (*Glossostoma*) is found in subpáramo habitats in southern Ecuador, and five toads (*Bufo*) occur in puna in Peru and Bolivia as do two species of the leptodactylid genus *Pleurodema* and one tree frog (*Hyla pulchella*). One lizard (*Anadia*) and one snake (*Leimandophis*) encroach on high-altitude habitats in Colombia and one viper (*Bothrops*) does so in Peru.

TABLE 19–4. Ecological distributions of northern Andean species (primarily after Duellman, 1979b).

Species	Altitudes (m)	Habitat [a]
Bolitoglossa adspersa	2500–3400	Páramo
B. hypacra	3610	Páramo
B. orestes	2000–3500	Sub [b], páramo
Atelopus carrikeri	2350–4400	Páramo
A. ebenoides	2660–3600	Sub, páramo
A. ignescens	2900–4500	Páramo
A. mucubajiensis	2900–3100	Páramo
A. oxyrhynchus	2010–3500	Cloud forest-páramo
Atopophrynus syntomopus	2780	Cloud forest-subpáramo
Centrolenella buckleyi	2100–3400	Cloud forest–páramo
Colostethus anthracinus	2500–3500	Sub, páramo
C. edwardsi	3000–3200	Páramo
Colostethus sp.	3800	Páramo
Colostethus sp.	2300–3500	Sub, páramo
Colostethus sp.	2100–3500	Sub, páramo
C. subpunctatus	2100–3300	Sub, páramo
C. vertebralis	2500–3200	Sub, páramo
Eleutherodactylus bogotensis	2900–3600	Páramo
E. buckleyi	2400–3700	Sub, páramo
E. chloronotus	2280–3440	Cloud forest-páramo
E. cryophilius	2800–3100	Subpáramo
E. curtipes	2750–4400	Páramo
E. elegans	2900–3800	Páramo
E. ginesi	2800–4000	Páramo
E. lancinii	2800–3000	Páramo
E. leoni	1960–3400	Cloud forest–páramo
E. leptolophus	3180	Subpáramo
E. lynchi	2460–3500	Sub, páramo
E. modipeplus	2560–3700	Sub, páramo
E. myersi	2900–3275	Páramo
E. nicefori	2850–3400	Páramo
E. obmutescens	3275–3400	Páramo
E. ocreatus	3500–4150	Páramo
E. orcesi	3160–3800	Páramo
E. peraticus	2850–3460	Sub, páramo
E. pycnodermis	2600–3400	Sub, páramo
E. racemus	3050–3570	Sub, páramo
E. riveti	2600–3400	Sub, páramo
E. simoterus	3200–3900	Páramo
E. supernatis	2540–3700	Sub, páramo
E. thymelensis	3310–4150	Páramo
E. trepidotus	3050–3650	Sub, páramo
Gastrotheca argenteovirens	2850–3300	Sub, páramo
G. helenae	3400	Páramo
G. riobambae	1800–4135	Sub, páramo
Glossotoma aequatoriale	2500–3615	Sub, páramo
Osornophryne bufoniformis	2700–3700	Páramo
O. percrassa	3750	Páramo
Phrynopus brunneus	3000–3200	Páramo
P. nanus	2640–3400	Páramo
P. peraccae	3100–3350	Subpáramo

TABLE 19–4 CONTINUED.

Species	Altitudes (m)	Habitat[a]
Phrynopus sp.	3200	Páramo
Telmatobius niger	2400–3500	Sub, páramo
Anadia bogotensis	2000–3600	Sub, páramo
Macropholidus ruthveni	3100	Cloud forest
Phenacosaurus heterodermus	1800–3750	Sub, páramo
P. orcesi	1750–3100	Sub, páramo
Pholidobolus macbrydei	2315–3960	Sub, páramo
Proctoporus laevis	3000	Subpáramo
P. unicolor	2800–3200	Sub, páramo
Stenocercus guentheri	2100–3890	Sub, páramo
S. ivitus	3100	Cloud forest
S. nubicola	3100	Cloud forest
Atractus nigriventris	3000	Páramo
A. trivittatus	3000	Páramo
A. variegatus	4000	Páramo
Leimandophis bimaculatus	1225–3250	Cloud forest-subpáramo
Saphenophis tristriatus	3200	Subpáramo

[a] These habitat types are described and illustrated by Duellman (1979b, pp. 373–79).

[b] 'sub' = subpáramo and is used whenever the taxon occurs in both subpáramo and páramo.

TABLE 19–5. Ecological distributions of southern Andean species (primarily after Duellman, 1979b).

Species	Altitudes (m)	Habitat
Batrachophrynus brachydactylus	4000–4100	Puna
B. macrostomus	4000–4100	Puna
Bufo cophites	2200–3500	Arid
B. flavolineatus	4000–4600	Puna
B. spinulosus	1000–5000	Arid-puna
B. trifolium	2900–4600	Arid-puna
Bufo sp.	3500–3750	Puna
Gastrotheca excubitor	3100–3550	Puna
G. griswoldi	3200–3800	Puna
G. marsupiata	2760–4360	Subpáramo-puna
G. peruana	2300–4600	Subpáramo-puna
Gastrotheca 'E'	3500–3800	Puna
Hyla pulchella	500–3300	Dry forest-subpáramo
Phrynopus cophites	3400–4100	Puna
P. laplacae	3400	Subpáramo
P. montium	3400	Subpáramo
P. peruanus	3600	Puna
P. peruvianus	2400–3700	Cloud forest-puna
P. simonsii	3050–3500	Puna
Pleurodema cinerea	2900–4100	Puna
P. marmorata	3000–5000	Puna
Telmatobius albiventris	3800–3825	Puna
T. atacamensis	3775	Puna
T. brevipes	3000	Arid
T. brevirostris	3000–3600	Puna
T. contrerasi	3050	Arid
T. crawfordi	4100–4200	Puna
T. culeus	2575–4150	Puna
T. halli	3000	Arid
T. hauthali	2400–4000	Arid
T. intermedius	3300	Arid
T. jelskii	2700–4500	Subpáramo-puna
T. juninensis	3650–3800	Puna

TABLE 19–5 CONTINUED.

Species	Altitudes (m)	Habitat
T. marmoratus	2500–5000	Puna
T. oxycephalus	1600–3800	Arid-puna
T. pefauri	3200	Arid
T. peruvianus	3000–3500	Arid
T. sanborni	4000	Puna
T. verrucosus	3800	Puna
T. sp.	3200–3500	Puna
Ctenoblepharis jamesi [a]	3600–4600	Arid
C. nigriceps	3000–3500	Arid
C. schmidti [a]	4000–4300	Arid
C. stolzmanni	1000–3800	Arid
Liolaemus alticolor	2860–4800	Arid-puna
L. altissimus	1400–3500	Subpáramo
L. mocquardi	3800–4300	Arid
L. multiformis	3650–4880	Arid-puna
L. ornatus	3300–4300	Puna
L. pantherinus	3500–4300	Puna
L. signifer	3600–4300	Puna
L. walkeri	3400–4100	Puna
Liolaemus sp. A	4300	Puna
Liolaemus sp. B	4080	Puna
Liolaemus sp. C	3820–4800	Puna
Phymaturus palluma	1200–3500	Arid
Proctoporus bolivianus	3300–4080	Puna
P. pachyurus	2900–3800	Puna
Stenocercus boettgeri	2900–3250	Dry forests
S. chrysopygus	2200–3500	Arid
S. empetrus	2650–3350	Arid
S. marmoratus	2750–3350	Arid
S. melanopygus	2700–3250	Arid
S. ornatissimus	2000–3400	Arid
Atractus nigricaudatus	3300	Subpáramo
A. pauciscutatus	2900–3000	Subpáramo
Bothrops andianus	2000–3300	Cloud forest-subpáramo
Tachymenis peruviana	70–4570	Arid-puna
T. tarmenis	3000	Puna

[a] Cei (1979a) suggested that both are species of *Liolaemus*.

Even casual inspection of Tables 19–4 and 19–5 reveals that most species are distributed across more than the high-altitude habitats. Many of these species are known from few localities only and little information is available about their ecological or geographic distributions. Nevertheless, the only genera endemic to high-altitude habitats are small genera (e.g., *Batrachophrynus* and *Osornophryne*). Those genera having several to many Andean species (first 15 in Table 19–1) are mostly poorly represented in high-altitude habitats (Table 19–6) because the bulk of the species are cloud forest species. Only *Phrynopus* and *Telmatobius* (see Chap. 14) have substantial representations in high-altitude habitats.

The southern Andean frog genera exhibit reproductive mode 1 (of Crump, 1974; Lynch, 1979a) in contrast to an array of more specialized reproductive modes in the northern Andean frog genera and in the two widespread Andean genera (Table 19–7). On the basis of reproductive biologies and their distributions within the Andes, it is tempting to suggest that in spite of their geographic distributions, *Gastrotheca* and *Phrynopus* should be termed northern Andean elements. Lynch (1979a) considered modes 6, 8, and 10 to be forest-dependent reproductive modes whereas modes 1 and 4 are less dependent on forest-mediated environments. Reproductive mode 1 is the most widely distributed among amphibians and accordingly the most primitive. Modes 6, 8 and 10 are clearly derived reproductive modes.

TABLE 19–6. High-altitude representations in largest Andean genera.

	Percent Andean [a]	Andean species (N)	High-altitude species (N)	Percent Andean in high altitudes
Andean genera				
Eleutherodactylus	70	158	25	16
Centrolenella	75	45	1	2
Colostethus	78	43	7	16
Atractus	54	38	5	13
Gastrotheca	91	32	8	25
Telmatobius	100	30	20	67
Liolaemus	60	28	11	39
Atelopus	75	27	5	18
Stenocercus	90	26	9	35
Proctoporus	94	17	4	24
Phrynopus	100	14	10	71
Non-Andean genera				
Bufo	11	20	5	25
Hyla	14	19	1	5
Anolis	9	18	0	0
Bolitoglossa	22	14	3	21

[a] See Table 19–1.

TABLE 19–7. Reproductive biologies of Andean frog genera.

Genus	Reproductive Mode [a]
Northern Genera	
Amphignathodon	(10) Direct development, in pouch on female's back
Atelopus	(1) Aquatic development, swift streams
Centrolene	Unknown, probably as in *Centrolenella*
Centrolenella	(4) Aquatic tadpoles, eggs deposited on leaves above streams
Colostethus	(6) Terrestrial eggs, aquatic larvae
Eleutherodactylus	(8) Direct development, terrestrial
Geobatrachus	(8) Direct development, terrestrial
Hemiphractus	(10) Direct development, on female's back
Osornophryne	Unknown, probably as in *Atelopus*
Widespread Genera	
Gastrotheca	(10) Partial or complete development in pouch on female's back
Phrynopus	(8) Direct development, terrestrial
Southern Genera	
Alsodes	(1) Aquatic tadpoles, eggs in ponds or streams
Batrachophrynus	(1) Aquatic tadpoles, eggs in ponds or streams
Telmatobius	(1) Aquatic tadpoles, eggs in ponds or streams

[a] Modes are numbered following Crump (1974) and Lynch (1979a).

Thus, the patterns of distributions for the Andean amphibians and reptiles are as follows:

1. The Andes harbor a diversity greater than might be predicted from their geographic contribution alone.
2. Fifteen amphibian genera (all frogs) and 19 reptile genera (13 lizards and 6 snakes) are strongly Andean in their distributions. These genera account for 16% of the endemic amphibian genera and 17% of the endemic reptile genera of South America.
3. Among the nine dominant frog genera, most are distributed north of the Huancabamba Depression, whereas among the eleven dominant reptile genera, most are distributed south of the depression.
4. Only two frog genera (aside from very small genera) have appreciable representations in high-altitude habitats. *Telmatobius* is a genus of the southern Andes and *Phrynopus* occurs both north and south of the Huancabamba Depression. With the exception of these two genera and the small genera *Batrachophrynus* and *Osornophryne*, the high-altitude herpetofauna appears invasive, probably from cloud forests.
5. Amphibians in the northern Andes exhibit derived reproductive modes whereas those of the southern Andes exhibit the primitive reproductive mode.

The contrasting distributions of amphibians (mostly northern Andes) and reptiles (mostly southern Andes) and of the distributions of reproductive modes within the amphibians probably reflect the general N→S decline in moisture through the Andean chain and thus agree with Schall and Pianka's (1978) general observations for frogs and for squamate reptiles that species densities for frogs are inversely correlated with sunfall but positively correlated with precipitation whereas for lizards the reverse obtains.

PHYLOGENETIC RELATIONSHIPS OF ANDEAN ELEMENTS

Ignoring those Andean elements that appear autochthonous, phylogenetic tracings extend from the Andes to at least three other Neotropical units. The plethodontid salamanders of the genus *Bolitoglossa* appear to have radiated primarily in Central America (Savage, 1966; Wake and Lynch, 1976). Wake and Lynch (1976) considered the highland species to be derived from lowland elements. *Batrachophrnyus* and *Telmatobius* are derived from Patagonian elements (Lynch, 1978; see Chap. 14). The Andean *Pleurodema* are also Patagonian derivatives (Duellman and Veloso, 1977) as is *Liolaemus* (Cei, 1979b). The third tracing (Fig. 19–4) extends from Patagonia to southeastern Brazil to the northern Andes (dendrobatid and eleutherodactyline frogs). That some such tendrils should be evident is expected from the assumption that lowlands represent the primitive areas and highlands represent the more derived areas. The Andes are a comparatively recent landform on the continent (Harrington, 1962; Jacobs, Bürgl, and Conley, 1963; Sonnenberg, 1963; James, 1973; Irving 1975; Simpson, 1979;). There exists fossil evidence to require the conclusion that many amphibian and reptilian groups were well established on the continent before most of the Andean uplift occurred (Estes, 1970; Báez and de Gasparini, 1979). These facts normally encourage the "conclusion" that the Andean fauna originated by means of dispersal of lowland groups onto the Andes.

Andean elevations in excess of 2000 m were achieved within the past five million years (Simpson, 1979). The youth of the Andes means that dispersal models are not necessarily the best models to explain their fauna. Janzen's (1967) theoretical exploration suggests that invasion of high montane regions by tropical lowland faunas should be difficult. Peters (1973), Lynch (1979b, 1981b), and Lynch and Duellman (1980) observed that closest relatives were found at comparable rather than at lower elevations in frogs of the genera *Atelopus* and *Eleutherodactylus*.

Duellman (1979b) emphasized the ecogeographic divisions in the Andean herpetofauna in lieu of discussing the origins of the fauna. The data he presented (also see Tables 19–2–19–5) clearly point to a northern Andean fauna and a southern Andean fauna divided by the Huancabamba Depression. The existence of the two faunas is suggestive of two very distinct evolutionary histories (see also Chap. 16 on rodents).

Furthermore, the two faunas differ in another significant way. If one abstracts the altitudinal ranges for each species from Duellman's compilations, one finds that in the northern Andes fewer species have broad altitudinal distributions and in the southern Andes more species have broad altitudinal distributions (Table 19–8). The percentages of species of frogs, lizards, and snakes having altitudinal ranges of less than 1000 m are greatest in the northern Andes (77.3, 73.0, and 82.1, respectively), intermediate in the Andes of Peru and Bolivia (72.7, 54.8, and 73.3 respectively), and lowest in the Andes of Argentina and

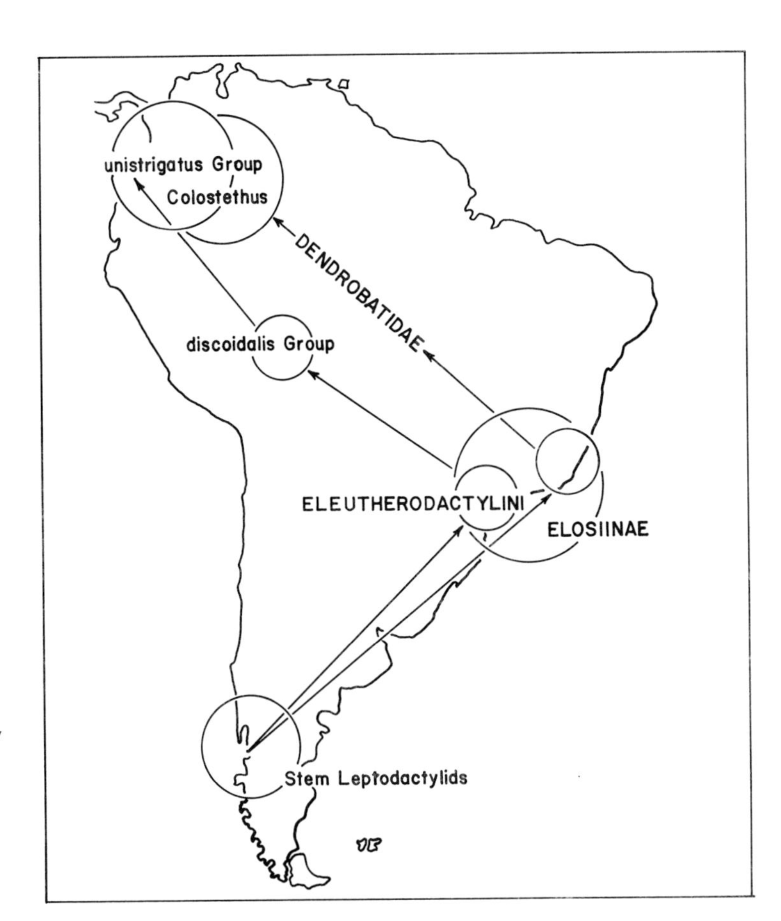

FIG. 19–4. A generalized track (terminology after Croizat et al., 1974) of leptodactyloid frogs. The cladistically most primitive elements are Patagonian with progressively derived elements in southeastern Brazil and the northern Andes. Groups are positioned geographically by plotting their centers of diversity. The major evolutionary diversification for each lineage is in the northern Andes (*Colostethus* and the *unistrigatus* group of *Eleutherodactylus*).

TABLE 19–8. Altitudinal ranges (differences between highest and lowest elevations of collection sites) of species of amphibians and reptiles in the Andes.

	Elevational range in meters					
	<500	501–1000	1001–1500	1501–2000	2001–3000	>3001
Northern Andes						
Caecilians (15)	13	2	0	0	0	0
Salamanders (12)	11	1	0	0	0	0
Frogs (286)	150	76	46	17	2	0
Lizards (63)	31	15	13	3	1	0
Snakes (84)	55	14	8	6	1	0
Bolivia-Peru						
Frogs (77)	43	13	9	7	4	1
Lizards (42)	9	14	8	6	5	0
Snakes (15)	11	0	2	0	1	1
Argentina-Chile						
Frogs (24)	14	1	1	3	4	1
Lizards (24)	6	12	3	3	4	0
Snakes (6)	0	1	0	1	3	1

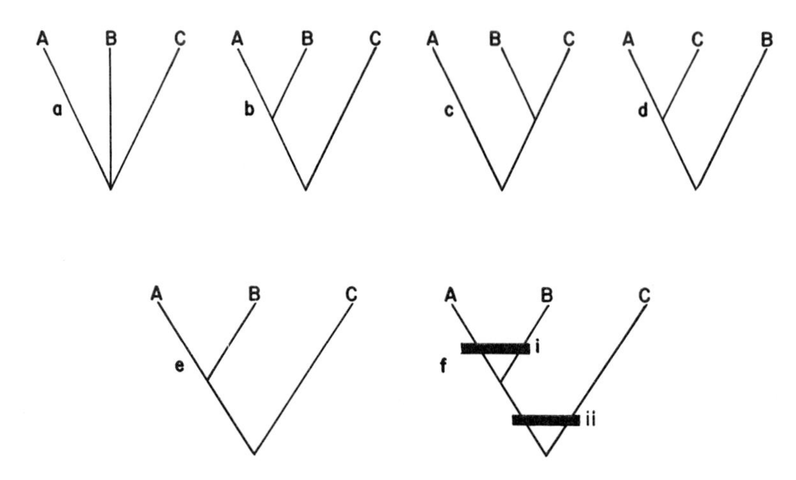

FIG. 19–5. Phylogenetic hypotheses. (a) Unrestrictive (and imprecise), states only that A, B, and C share a common ancestor (the group is a *presumed* monophyletic group without an hypothesized synapomorphy). Cladograms (b, c, d) are restrictive (precise), i.e., only one can be true (the other two must be false), and all three are possible under cladogram (a). (e) An implicit cladogram (as are a–d)—no data are advanced to provide support for the proposed monophyly of group ABC or for the subgroup AB. (f) An explicit cladogram with two hypothesized synapomorphies (synapomorphy i is an uniquely derived character shared by taxa A and B; synapomorphy ii is an uniquely derived character shared by taxa A, B, and C; each synapomorphy provides sufficient support for the two postulates of monophyly, A and B, and A, B, and C.

Chile (62.5, 64.3, and 16.7 respectively). The percentages of species having altitudinal ranges of more than 2000 m show an opposite trend: northern Andes (0.7 and 1.4, frogs and reptiles, respectively), central Andes (6.5 and 12.3, respectively), and southern Andes (20.8 and 23.5, respectively).

This curious fact, not noted by Duellman (1979b), is explicable by means of Janzen's (1967; p. 242) thesis. In temperate lowlands, the wide annual variation in climate requires organisms to have a broad physiological adaptiveness that serves as a preadaptation allowing invasion of a wide array of altitudes, whereas in tropical lowlands, the narrow annual variation in climate selects for narrow physiological adaptiveness. As a result, these lowland tropical species are able to select a narrower array of altitudes (in Janzen's words, mountain passes are [biologically] "higher" in the tropics than in the temperate zones).

The available phylogenetic hypotheses for Andean amphibians and reptiles are inadequate to resolve the choice between dispersal and vicariance modes of populating the Andes but they are not so rare as suggested by Duellman (1979b; p. 409). That most published accounts assume a dispersalist mode provides testimony to the persuasiveness of one school of biogeographers and to the general absence of precise phylogenetic hypotheses.

Most phylogenetic hypotheses for Andean amphibians and reptiles are in the form of unrestrictive hypotheses and are implicit rather than explicit. Imprecise (and unrestrictive) hypotheses provide the illusion of information (Fig. 19–5) because they allow any data to be accepted (the hypotheses are not falsifiable).

In spite of these shortcomings, three phylogenetic and distributional patterns are evident for the Andean herpetofauna: (1) exclusively Andean groups, (2) Andean groups having a few extra-Andean members, and (3) non-Andean groups having a few Andean members.

Exclusively Andean Groups

This is the most common phylogenetic-distributional pattern for the Andean herpetofauna. The most common herpetofaunal element of the northern Andes is the leptodactylid frog genus *Eleutherodactylus* (*unistrigatus* group; see Lynch, 1976). Several subgroups of the *unistrigatus* group have been recognized. The *curtipes* assembly consists of three subpáramo-páramo species (2400–4400 m) found from central Colombia to southern Ecuador (Lynch, 1981b). The *myersi* assembly consists of seven subpáramo-páramo species (2800–4150 m) found in the Mérida Andes, Cordillera Oriental of Colombia, and southern Colombia south to southern Ecuador (Lynch, 1981b). The *orcesi* assembly consists of five species found in páramo and subpáramo habitats (3000–4150 m) from central Ecuador north along the Cordillera Central of Colombia

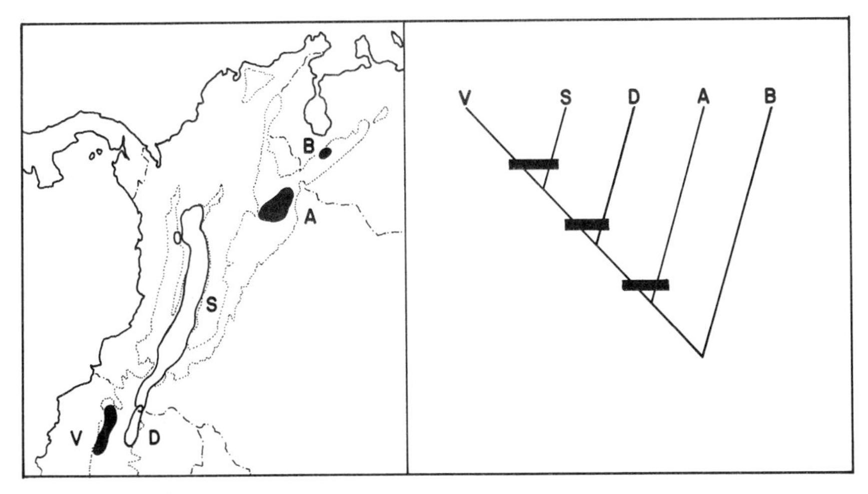

FIG. 19–6. *Left*: Distribution of cloud forest and subpáramo frogs of the *devillei* assembly (*unistrigatus* group) of *Eleutherodactylus* (A, *E. anolirex*; B, *E. briceni*; D, *E. devillei*; S, *E. supernatis*; V, *E. vertebralis*). *Right*: the cladogram published by Lynch (1983). For synapomorphies see Lynch (1983). The 1000 m contour (*dotted lines*) of the Andes is indicated for geographic reference.

(Lynch, 1980a, 1981b). The *unistrigatus* assembly consists of seven species found in cloud forests, subpáramo, and páramo habitats (1500–3700 m) on the Cordillera Oriental of Colombia and from southern Colombia to northern Peru. The *riveti* assembly consists of three species found in southern Ecuador (Lynch, 1979b). *E. balionotus* (2800 m) and *E. riveti* (2620–3420 m) are subpáramo-páramo species more closely related to one another than to the cloud forest *E. ruidus* (2317 m). Lynch and Duellman (1980) identified four subgroups of the *unistrigatus* group found in cloud forests from southern Colombia to northern Peru (*celator* assembly, three species, 1707–3050 m; *glandulosus* assembly, four species, 1710–2750 m; *nigrogriseus* assembly, three species, 1150–2835 m; and *pyrrhomerus* assembly, three species, 2350–3400 m).

One of the few explicit phylogenetic hypotheses available is that for the *surdus* assembly of the *unistrigatus* group (Lynch, 1980b). The five species occur in cloud forests (1660–2988 m) in southern Colombia and Ecuador. One pair of sister species occurs on opposite sides of the Ecuadorian Andes (*E. baryecuus* and *E. surdus*); another pair occurs in partial sympatry (*E. duellmani* and *E. sobetes*). These four species occur at slightly higher elevations than the cladistically most remote *E. pugnax*.

The *devillei* assembly of the *unistrigatus* group is another with an explicit phylogenetic hypothesis (Lynch, 1983). Five species are known (Fig. 19–6) and distributed in cloud forests and subpáramo habitats (1600–3300 m) in Colombia, Ecuador, and Venezuela. The cladistically most remote taxon (*E. briceni*) occurs in relatively low-elevation cloud forests near Mérida (around 1600 m). The other species usually occur at elevations above 2500m.

The *ignescens* group of *Atelopus* occurs in high cloud forests, subpáramo, and páramo habitats in Colombia, Ecuador, and Venezuela (Rivero, 1963, 1965, Cochran and Goin, 1970; Peters, 1973; Duellman, 1979b), but only unrestrictive and imprecise phylogenetic hypotheses are available. The allied toads of the genus *Osornophryne* occur in similar habitats (high cloud forests and subpáramos, 2700–3400 m) in northern Ecuador and on the Cordillera Central of Colombia (Ruíz and Hernández, 1967). Two dendrobatids from the Cordillera Oriental of Colombia (*Colostethus edwardsi* and *C. ruizi*) form a monophyletic group (Lynch, 1982) found in high cloud forests (*C. ruizi*) and páramo (*C. edwardsi*).

Tree frogs are poorly represented in the Andes with the exception of the pouch-brooding frogs of the genus *Gastrotheca*. Duellman (1974), Duellman and Fritts (1972), and Duellman and Pyles (1980) recognized three species groups of exclusively Andean forms. Unfortunately, only unrestrictive phylogenetic hypotheses are available for the *argenteovirens*, *marsupiata*, and *plumbea* groups of *Gastrotheca*. Duellman and Fritts (1972) reviewed the *marsupiata* group and recognized seven species distributed from northern Argentina to northern Peru; the species occur at elevations between 1500 (cloud forests) and 4600 m (puna). Duellman (1974) and Duellman and Pyles (1980) recognized seven species in the *plumbea* group. These frogs occur in cloud forests, subpáramo, and páramo (1600–4135 m) in southern Colombia, Ecuador, and northern Peru. This species group was considered derived from the *marsupiata* group (Duellman, 1974, p. 24); if so, the *marsupiata* group would not be monophyletic. The *argenteovirens* group has not been monographed but includes six species distributed in Colombia, eastern Panama, and Venezuela (Duellman, 1974). The most widely distributed species (*G. nicefori*) occurs in low-altitude forests (northern Colombia, eastern Panama, western Venezuela) and in cloud forests to elevations of at least 2000 m. Other species occur in cloud forests and subpáramo habitats (2000–3200 m).

The *larinopygion* group of *Hyla* is composed of no fewer than five species found in high cloud forests (2000–3200 m) of the Cordillera Central of Colombia and northern Ecuador with no apparent relatives (Duellman and Altig, 1978; Duellman and Berger, 1982; Ruíz and Lynch, 1982) or allied to the *bogotensis* group of *Hyla*.

The leptodactylid frog genus *Phrynopus* occurs in cloud forests, páramos, and puna from northern Colombia south to Bolivia (Lynch, 1975). Lynch recognized four species groups (one of which, the *peruanus* group, is not monophyletic). The *flavomaculatus* group consists of six species found in cloud forests, subpáramo, and páramo habitats (1659–3350 m) in the Cordillera Oriental (Colombia), Ecuador, and northern Peru. Geographically adjacent species are adjacent on the cladogram (northernmost species cladistically primitive, those in the vicinity of the Huancabamba Depression the most advanced). The grouping of *P. nanus* (Cordillera Oriental, Colombia, páramos) and *P. simonsi* (northern Peru, páramos) proposed by Lynch is in error (C. Ardila, pers. comm.). *Phrynopus nanus* is most closely related to undescribed species from the Cordillera Central of Colombia. The remaining six species occur in cloud forests, páramos, and puna in Bolivia and Peru. The cladistically primitive *P. simonsi* occurs at high elevations (as do most species of the group with the exception of the poorly known *P. wettsteini*).

The leptodactylid frog genus *Telmatobius* is presumably monophyletic and distributed through moderate- and high-elevation habitats from Ecuador south to Andean Argentina and Chile (Lynch, 1978; Duellman, 1979b; Trueb, 1979; see Chap. 14). Intrageneric relationships are poorly understood but the three species found in subpáramo and páramo habitats in Ecuador appear closely interrelated (Trueb, 1979). The two species of *Batrachoprynus* occur in high-altitude Peru, but in spite of their superficial similarity to some *Telmatobius* they are not closely related to *Telmatobius* (Lynch, 1978; but see Chap. 14).

Reptiles are not common in high Andean habitats. Nevertheless, additional examples of exclusively Andean groups are apparent for at least three putative monophyletic groups. The anoline lizard genus *Phenacosaurus* consists of three montane species (Lazell, 1969). *P. heterodermus* (1800–3750 m, Cordillera Oriental, C. Central, and C. Occidental of Colombia) and *P. nicefori* (1900–2340 m, Cordillera Oriental and Sierra de Perijá, Venezuela) are more closely allied to one another than to the third species, *P. orcesi* (1750–3100 m, northern Ecuador). Lazell (1969; p. 4) considered *Anolis jacare* most closely allied to *Phenacosaurus*. *A. jacare* occurs at elevations of 1600–1640 m in the Mérida Andes (Williams et al. 1970).

Myers (1973) proposed the genus *Saphenophis* for five Andean species found in cloud forests (1100–3200 m) in western Colombia and Ecuador. His phylogenetic hypothesis (implicit) placed one pair of species (*S. atahuallpae* and *S. sneideri*) as a branch distinct from the other three species. He considered *S. antioquiensis* (2650 m) a sister species to *S. tristriatus* (3200 m); this pair of species is related to *S. boursieri* (1100–2000 m) found on the western slopes of the Andes in southern Colombia and northern Ecuador. Myers also reported *S. boursieri* from the Amazon Basin of Ecuador but those records probably are the result of confounded locality data. The record from "Pastaza River, Canelos to Marañon" is suspect; another specimen (MCZ 19628, *E. necerus*) from the same locality is of a species found in cloud forests 1140 to 1500 m in western Ecuador. Myers suggested that speciation in *Saphenophis* was promoted by habitat fragmentation occurring with upward displacement of habitats during interglacials.

Montanucci (1973) reviewed the microteid lizard genus *Pholidobolus* and recognized five species distributed in cloud forests, semiarid grasslands of intermontane basins, and páramos (1800–4000 m) in Ecuador. The most generalized taxon is *P. affinis*, an inhabitant of the intermontane basins of central Ecuador (1800–3050 m). Montanucci (1973, p. 30) considered *Macropholidus ruthveni* a "highly specialized derivative of *Pholidobolus* stock"; *M. ruthveni* occurs in northern Peru (3100 m).

The many groups cited earlier provide evidence that a substantial fraction (here estimated at 80%) of the Andean herpetofauna is most closely related to *other* Andean species. This allows the conclusion that much of the Andean herpetofauna had originated in situ (see similar conclusion for high Andean birds in Chap. 23). If my estimate is of the correct order of magnitude, one could describe the Andean herpetofauna as essentially autochthonous with much of the species diversity resulting from adaptation and speciation as the Andes elevated during the Tertiary. However, the remaining phylogenetic-distributional patterns potentially dictate other conclusions. These patterns are especially significant because they have the greatest possibility of revealing the origins of the herpetofauna because each pattern is one in which some species are Andean and some are not.

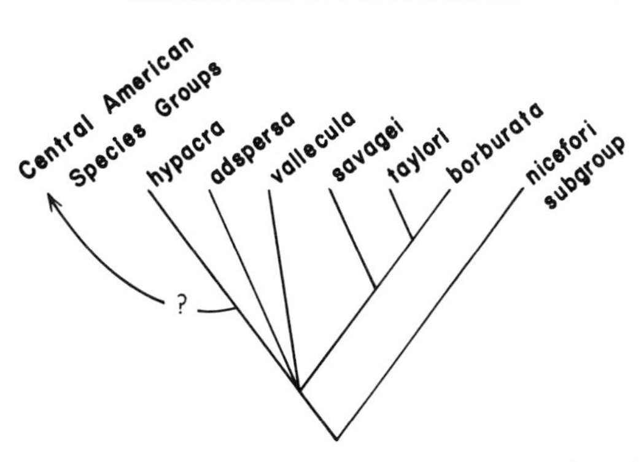

FIG. 19–7. Partially resolved cladogram for salamanders of the *adspersa* subgroup (*adspersa* group) of *Bolitoglossa*. *B. adspersa*, *B. hypacra*, and *B. vallecula* are high altitude forms.

Andean Groups with a Few Extra-Andean Members

Duellman (1979b) considered *Rhamphophryne* an Andean genus. Trueb (1971) and Izecksohn (1976) recognized six species, three of which occur at moderate altitiudes (1450–2670 m) on the Cordillera Central of Colombia. Another species occurs on the Serranía del Darién (1265–1480 m) in eastern Panama and adjacent Colombia, another in eastern Ecuador (100–700 m), and another in extreme eastern Brazil. In her review of the genus, Trueb (1971) did not provide a phylogenetic hypothesis except in suggesting that *R. macrorhina* and *R. rostrata* (two species from the Cordillera Central) were most like the ancestral stock of the genus. Thus she implies that an Andean genus has dispersed away from the Andes. However, her analysis was done without considering the geographically isolated *R. proboscideus* (Izecksohn, 1976).

Duellman (1972) reviewed the *Hyla bogotensis* group and recognized six allopatric species (*H. bogotensis* in the Cordillera Oriental of Colombia, 2500–2900 m; *H. platydactyla* in the Mérida Andes of Venezuela, 1600–2700 m; *H. colymba* in Costa Rica and Panama, 560–1410 m; *H. alytolylax* on the Pacific slopes of the Andes in Colombia and Ecuador, 800–1460 m; *H. phyllognatha* on the Amazonian slopes of the Andes in Ecuador and Peru, 610–1740 m; and *H. denticulata* on the Caribbean slopes of the Andes in northern Colombia, 1400–2400 m). Duellman and Altig (1978) added *H. torrenticola*, sympatric with *H. phyllognatha*, from the Amazonian slopes of the Andes of southern Colombia and adjacent Ecuador (1440–1490 m). Myers and Duellman (1982) assigned *H. palmeri* from Colombia, Ecuador, and Panama (100–920 m) to the group. Work on frogs from the Cordillera Oriental of Colombia has revealed at least one other species (Ruíz and Lynch, 1982).

The only available phylogenetic hypotheses are those published by Duellman (1972). *Hyla bogotensis* and *H. platydactyla* form a branch separate from the branch containing *H. alytolylax*, *H. colymba*, *H. denticulata*, *H. phyllognatta*, and, *fide* Duellman and Altig (1978), *H. torrenticola*. The phylogenetic hypothesis is one of an Andean group having a single derived taxon (*H. colymba*) in the lowlands.

The distribution of the *Bolitoglossa adspersa* group (subgroup *adspersa*) closely parallels that of the *Hyla bogotensis* group (Brame and Wake, 1963, 1972; Wake, Brame, and Myers, 1970). The only dendrogram of relationships available (Fig. 19–7) is that published by Brame and Wake (1963) but, if still valid, this is significant because it groups a species from the Coastal Range of Venezuela (*B. borburata*) and one from the Sierra Nevada de Santa Marta (*B. savagei*). This pair is in turn grouped with *B. adspersa* from the subpáramos and páramos of the Cordillera Oriental of Colombia, and with *B. vallecula*, a high cloud forest and subpáramo species from the Cordillera Central of Colombia. Wake et al. (1970; pp. 15–16) modified this implicit dendrogram only in considering *B. taylori* (Serranía de Darién) and *B. borburata* sister species.

Athough not included in their dendrogram, *B. hypacra*, another páramo-dwelling species, was grouped with the *adspersa* series by Brame and Wake (1963, p. 67); *B. hypacra*, however, is also somehow "closely related to Central American species" (Brame and Wake, 1963, pp. 59, 67). This series of species is more remotely (Brame and Wake, 1963, pp. 61–62) related to three species from low cloud forests of the Cordillera Oriental of Colombia (*B. capitana*, *B. nicefori*, and *B. pandi*).

The iguanid lizard genus *Stenocercus* is essentially Andean (sea level to 4000 m) but has one lowland species, *S. roseiventris* (Fritts, 1974). Of

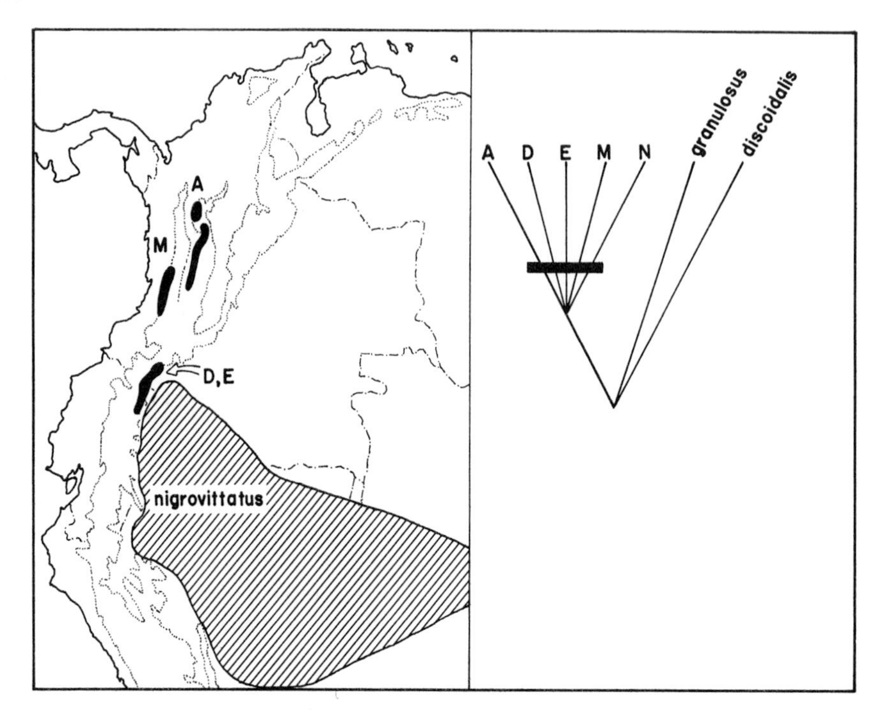

FIG. 19–8. *Left*: Distribution of the northern subgroup of the *discoidalis* group of *Eleutherodactylus*. *Right*: A partially resolved cladogram for the species group (the synapomorphy is broad vomerine odontophores). The northern species are identified by letters (A, undescribed species; D, *E. dolops*; E, *E. elassodiscus*; M, *E. mantipus*; N, *E. nigrovittatus*). *Dotted lines* indicate 1000 m contour.

the 19 species included in his phylogenetic hypothesis, only *S. carrioni*, *S. moestus*, and *S. roseiventris* are lowland lizards. In each case, the more primitive members of the cladogram are highland taxa (2000–3000 m) as are the sister species of each of the taxa found in low-elevation sites peripheral to the Andes.

These four cases are alike in that the lowland taxa are advanced rather than primitive. The next example presents a different pattern. The *Eleutherodactylus discoidalis* group consists of seven species. The group has not yet been analyzed phylogenetically (Lynch, in prep.) but the pattern emerging is of advanced taxa being Andean (Fig. 19–8). *E. discoidalis* and *E. granulosus* occur on the eastern slopes of the Andes in northern Argentina and Bolivia and southern Peru. The cladistically advanced taxa include one species in the Amazon Basin (*E. nigrovittatus*) and four Andean (cloud forests, subpáramo) species (*E. dolops*, *E. elassodiscus*, *E. mantipus*, and an undescribed species).

The *lacrimosus* assembly of the *unistrigatus* group of *Eleutherodactylus* (Lynch and Duellman, 1980; Lynch, 1980c) provides another example of an Andean group having a few extra-Andean members. Five species are known. *E. lacrimosus* occurs in the Amazon Basin (Lynch, 1980d) and is considered most closely related to *E. petersi*, found in cloud forests (1100–1920 m) of the Amazonian versant of the Andes in Ecuador. Lynch (1980c) considered *E. bromeliaceus* (Amazon versant, southern Andean Ecuador, 1707–2622 m) the nearest relative of *E. eremitus* (Pacific versant of Andes in Ecuador, 1540–2100 m). This is related to *E. mendax* (Bolivia and southern Peru, 200–2200 m).

Other subgroups of the *unistrigatus* group of *Eleutherodactylus* (for which only unrestrictive hypotheses are available), having Andean and lowland species, include the *acuminatus*, *crucifer*, *rubicundus*, and *trachyblepharis* assemblies (Lynch and Duellman, 1980).

The microteid lizard genus *Proctoporus* includes some 15 species distributed in the Andes from Bolivia to Venezuela and Trinidad (Peters and Donoso-Barros, 1970) at elevations between 1000 and 4080m. The *luctuosus* group consists of five species found in Colombia, Ecuador, Venezuela, and Trinidad at relatively low elevations in cloud forests (Uzzell, 1958). Uzzell (1958) did not comment on relationships except to point out that *P. achlyens* (Coastal Range of Venezuela) and *P. shrevei* (Trinidad) were a subgroup within the *luctuosus* group. Uzzell (1970) also recognized the *pachyurus* group for three species found in Bolivia and Peru (at high altitudes). He considered the *pachyurus* group related to *P. striatus*

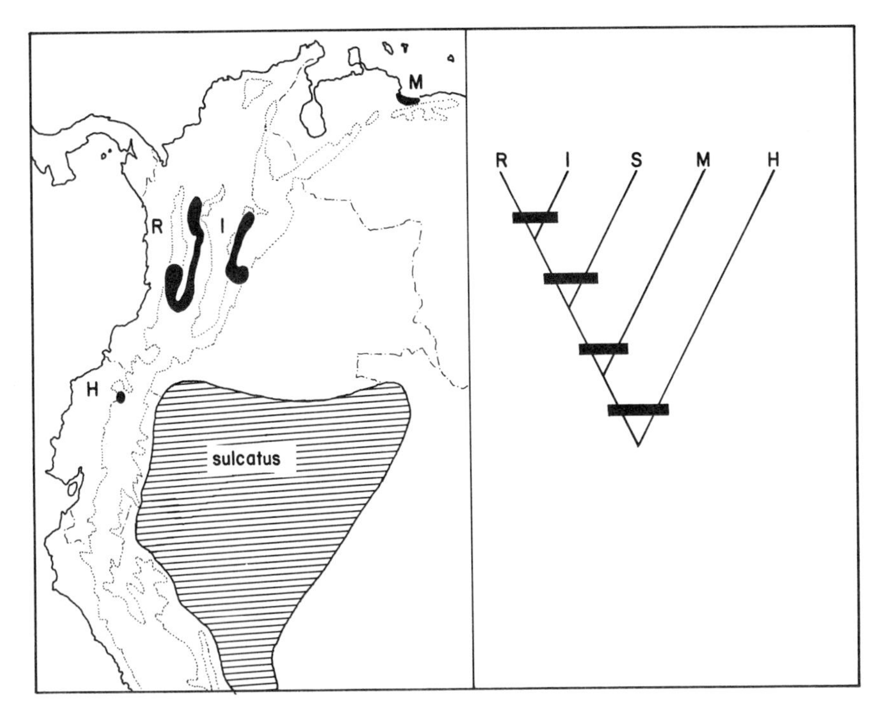

FIG. 19–9. *Left*: Distribution of frogs of the *sulcatus* group of *Eleutherodactylus* (H, *E. helonotus*, low cloud forests; I, *E. ingeri*, cloud forests; M, *E. mausii*, lowland; R, *E. ruizi*, cloud forests; sulcatus, *E. sulcatus*, rain forests). *Right*: A fully resolved cladogram of their relationships (for synapomorphies see Lynch, 1981c). Dotted lines indicate 1000 m contour of the Andes.

of Colombia. If Uzzell's (1958) groupings of the *luctuosus* group are right and if Mencher (1963, p. 84) is correct in viewing the Coastal Range of Venezuela as geologically independent of the Andes, the *shrevei* subgroup may have differentiated on these low extra-Andean ranges as the result of upward displacement of cloud forests in the Late Pleistocene.

Non-Andean Groups with a Few Andean Members

The *minutus* group of *Dendrobates* consists of at least 14 species of small dendrobatids distributed from western Panama to the Amazon Basin (Silverstone, 1975; Myers and Daly, 1976, 1980; Myers, 1982). Most species occur in the lowlands but an apparently monophyletic subgroup (*D. abditus*, *D. bombetes*, and *D. opisthomelas*) occurs in the Andes (670–2200 m) of Colombia and Ecuador (Myers and Daly, 1980, p. 20). The relationships among the species of the *opisthomelas* subgroup remain unknown but the most plausible hypothesis to explain the origins of this cloud forest group is fragmentation of habitat rather than independent dispersal-adaptation by lowland stocks (Myers and Daly, 1980, p. 22).

The *sulcatus* group of *Eleutherodactylus* consists of five species found in northwestern South America (Lynch, 1981c). One widespread species occurs in the Amazon Basin, a second in the lowlands of Venezuela, a third in cloud forests of western Ecuador, and the last two in cloud forests of Colombia (Fig. 19–9). The phylogenetic hypothesis for the group (Lynch, 1981c) holds that the most advanced species are those from the Colombian cloud forests. The origins of *E. ingeri* and *E. ruizi* could be independent dispersal-adaptation by lowland stocks but the most parsimonious explanation is that the ancestral distribution was fragmented by Andean orogeny and/or the developing aridity of the Cauca and Magdalena river valleys.

Duellman and Veloso (1977) reviewed the frog genus *Pleurodema* and recognized 12 taxa and two other, unnamed, populations. In general, *Pleurodema* species are lowland frogs of nonforested environments. The most prominent exceptions are *P. thaul*, a composite species found in austral forests and subhumid chaparral, from sea level to 1500 m (Duellman and Veloso, 1977; p. 22), and the two Andean species *P. cinerea* (2900–4100 m, northern Argentina to southern Peru) and *P. marmorata* (3000–5000 m, puna from central Peru to northeastern Chile and Central Bolivia). The cladograms published by Duellman and Veloso (1977, p. 27) are contrary to their "evolutionary scheme" (1977, p. 39). Their implicit cladogram (p. 39) reveals that the Andean taxa are not closely related and that *P. marmorata* is most closely allied to a series of lowland species from arid habitats in Brazil (*caatinga*) and central Argentina whereas *P. cinerea* is most closely allied to *P. borelli*, a species found in northern Argentina in subtropical environments (400–3000 m).

The Andean *Pleurodema* are phylogenetically independent; the independency allows an explanation of independent dispersal. The distributions of *P. borelli* and *P. cinerea* may represent fragmentation induced by the uplifting of the Andes.

The *spinulosus* group of *Bufo* (Cei, 1972) is distributed from low-elevation forests in Chile and coastal deserts of Chile and Peru to 5000 m in Peru. The group occurs in the Andes (and adjacent lowlands) south from the Huancabamba Depression (Cei, 1972; Duellman, 1979b). Unfortunately, no phylogenetic hypotheses are available for the species group. Cei (1972, p. 85) suggested that the group antedates the Oligocene uplifts of the southern Andes.

The iguanid lizard genus *Liolaemus* contains nearly 50 species (Peters and Donoso-Barros, 1970), 28 of which occur in the Andes south of 25° S. Most of the taxonomic attention paid to the genus has been at the level of distinguishing local populations, although Cei (1979b) considered the Patagonian species to be the most primitive. The genus is distributed in such a fashion as to tell us much about the evolution of the southern Andean herpetofauna. Especially of interest will be discovering whether the phylogenetic-distributional patterns of *Liolaemus* are congruent with those for *Pleurodema*, the *spinulosus* group of *Bufo*, the Patagonian telmatobiines, and the montane species groups of *Gastrotheca*.

DISCUSSION

A discussion of the origins of the Andean herpetofauna in the absence of precise phylogenetic hypotheses is a waste of time and paper (Ball, 1975). Many biogeographic expositions have been advanced and serve now only to cloud issues. Biogeographic as opposed to ecogeographic hypotheses involve history. The temporal component cannot be known a priori (except in the most general way) but must be deduced from the patterns evident in geography *and* relationships (Croizat, 1964; Croizat, Nelson, and Rosen, 1974; Rosen, 1975; Nelson and Platnick, 1981; Wiley, 1981). The contributions of ecogeography improve our descriptions but do not improve our analyses.

There are but two general hypotheses (termed scenarios later) to account for the origins of the Andean herpetofauna. The first makes little use of phylogenetic information and tends to "explain" distributional patterns in light of the most recent and most popular geological base. The second emphasizes phylogenetic and distributional patterns and tends to "explain" or recount a history. The two hypotheses might also be termed static versus dynamic or proximate versus ultimate (not in Mayr's [1961] sense, but in the sense of how much influence the Pleistocene perturbations are hypothesized to have had; the proximate versus ultimate distinction also confounds the explanations under the second general hypothesis).

Idiographic Explanations: The Static-Continent Hypothesis

Not all specific hypotheses grouped here are proposed by persons believing the Andes to be a sta-

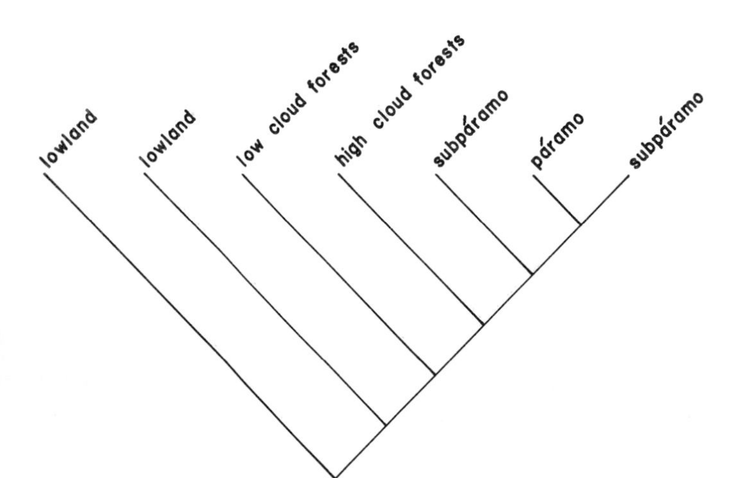

FIG. 19–10. Phylogenetic hypothesis for a group of organisms dispersing into the Andes. Species are represented by the environments they occupy. This phylogenetic hypothesis is not known for any group of amphibians or reptiles but is the fully resolved hypothesis expected if the high Andean fauna is invasive.

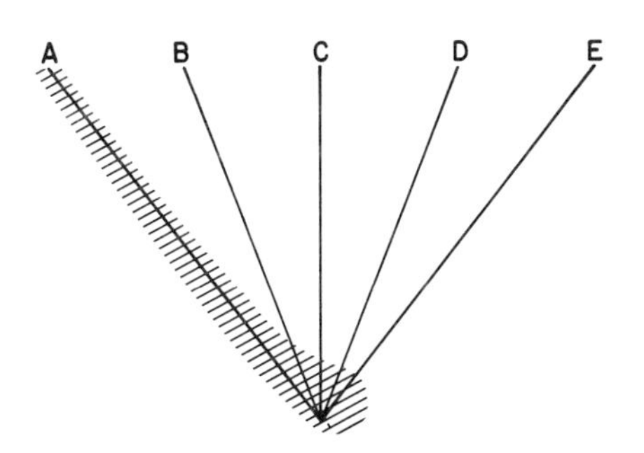

FIG. 19–11. Phylogenetic hypothesis for a group of organisms originating in the lowlands and spawning several high-altitude taxa (B–E). The high-altitude taxa are independent derivatives of a lowland ancestor (A). This hypothesis is unrestrictive (and unresolved) and not supported by data (see text). Lowland environments are denoted with hatching.

tic landform. In fact, most authors would concede that the Andes have experienced a complex history, some parts older and others more recent; but the explanations offered by these authors are explanations in which the mountains are presumed to antedate the organisms—i.e., the explanation is more proximate.

Under this general hypothesis, some member(s) of a lowland group disperse(s) (via simple dispersal) into vacant habitats at a higher altitude and adapt(s) to the more temperate conditions of the higher altitudinal stratum. Once established, subpopulations of the derived form invade higher elevational strata until a terminal taxon is established in the highest stratum available. The cladistic relationships of such a group would be as in Fig. 19–10. This static view also includes the subset wherein the Andes rise on the western edge of the continent, followed by invasion and adaptation-speciation from the lowlands.

Under this hypothesis, high-altitude forms, isolated geographically, would not be closely related (i.e., would not be sister species) but would be the result of independent invasions of habitat islands formed by the Andean topography. Alternatively, those species found in high-altitude habitats would all be derived independently from a lower-elevation ancestor (Fig. 19–11). During the Pleistocene, present distributions were established (or all the events outlined here occurred after the last glacial cycle).

That many groups of Andean amphibians and reptiles now have unrestrictive phylogenetic hypotheses would seem consistent with this idea. However, the unrestrictive hypotheses are not sufficient support because they are not subject to falsification (Nelson and Platnick, 1981; Wiley, 1981). These phylogenetic hypotheses are inadequate because they are not substantiated by synapomorphies.

This general hypothesis is termed idiographic because for each group of organisms, a unique explanation is offered—birds fly, frogs hop, and snakes crawl.

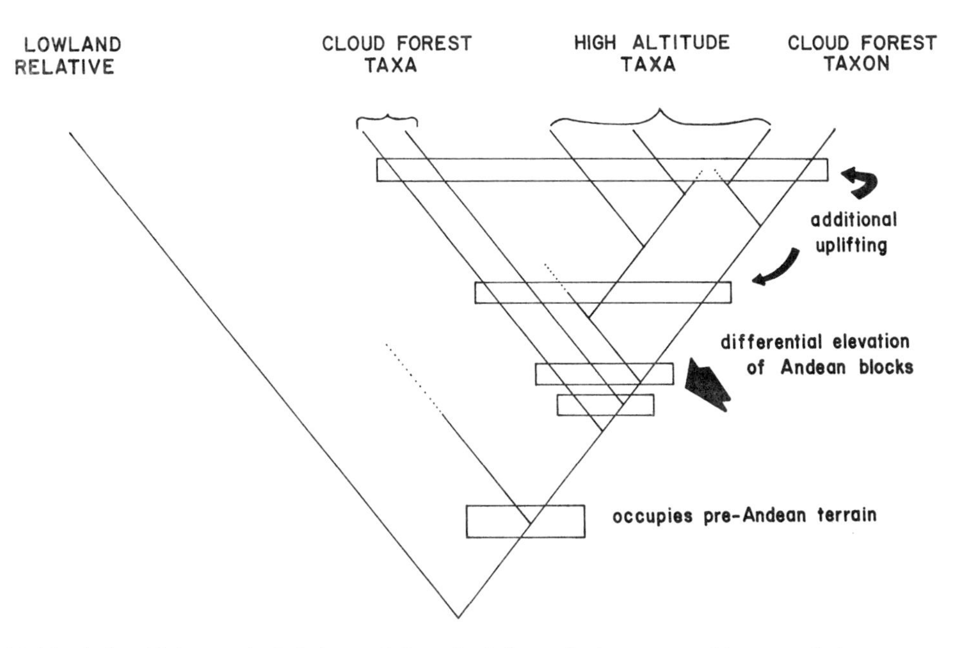

FIG. 19–12. Matrix (pre-Pleistocene) of phylogenetic hypothesis for an Andean group with some splitting events associated with orogenic activity. Dotted lines denote extinctions. Species are identified by the habitats occupied. Also see FIGS. 19–13 and 19–14 (Pleistocene impacts).

Synthetic Explanations: The Dynamic-Continent Hypothesis

Under this hypothesis, as soon as the pre-Andean landscape forms (series of low but slightly elevated hills and ridges), the resident organisms begin to differentiate. These slightly elevated areas would have different environments than those found in the adjacent wet or dry lowland environments. The Andean orogenies elevated various blocks of land to various altitudes carrying the resident herpetofaunas into more temperate environments. These uplifted faunas (which did not disperse) then must adapt or die out. The various blocks of upland areas are isolated from one another by erosion and/or inhospitable climates and/or time of elevation so as to produce a matrix (Fig. 19–12) of relationships (nonsimultaneous splitting of ancestral populations).

As the Andes achieve their present altitudes, a global crisis in climate develops. The changing availabilities of rainfall restrict or expand forested regions alternatively with nonforested regions. When the global climate cools and glaciation envelops the higher altitudes of the Andes, the habitat zones are depressed and adpressed (or compressed) against lower elevation zones. The depression of habitat may unite once-isolated areas and allow range expansion (complex "dispersal") of taxa into previously inaccessible geography (Fig. 19–13). With the coming of interglacials, upland displacement of habitats fragments contiguous distributions promoting speciation at high altitudes. Upward displacement also fragments the distributions of cloud-forest-dwelling organisms and speciation ensues in the isolated serranías peripheral to the Andes (Fig. 19–14).

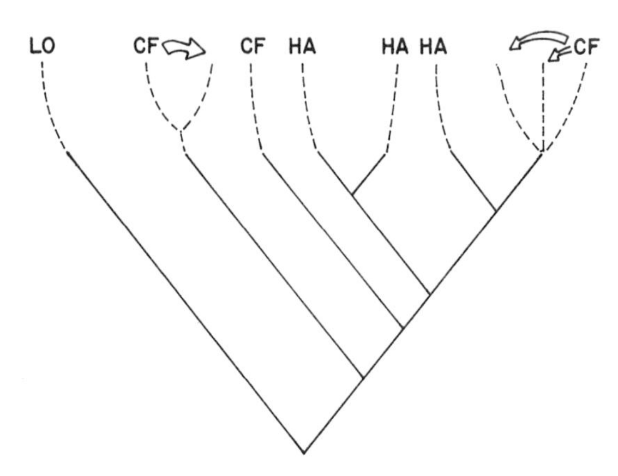

FIG. 19–13. Impact of habitat zone depression (during glaciations) on the phylogenetic hypothesis in FIG. 19–12. Dashed lines represent continuation of lineages (those splitting are dispersing into newly accessible habitat) during habitat depression. Taxa (and habitats) are as follows: LO, lowland; CF, cloud forest; HA, páramo and subpáramo.

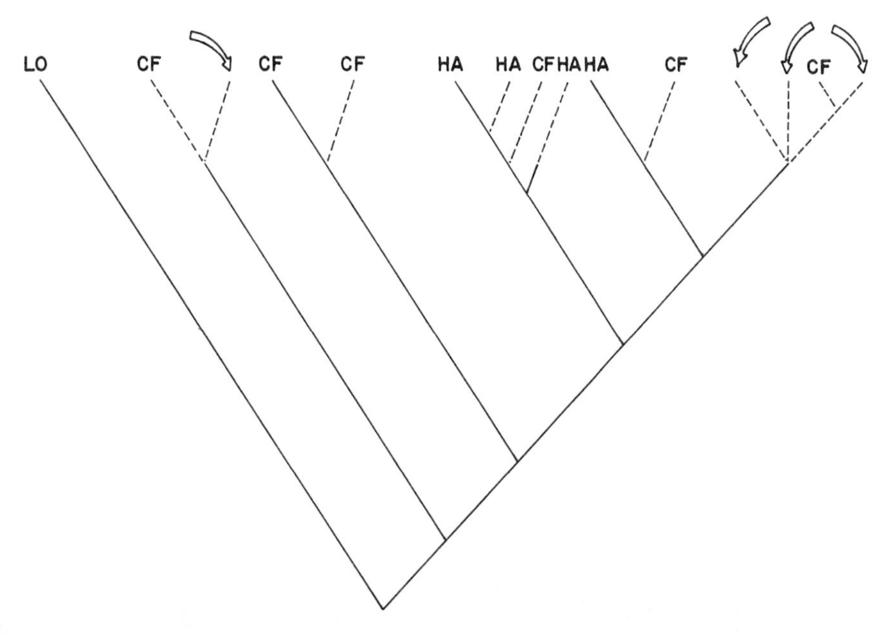

FIG. 19–14. Impact of upward displacement of habitat (during interglacials) superimposed on the phylogenetic hypothesis in FIGS. 19–12 and 19–13. Dashed lines represent new speciation events resulting from habitat fragmentation. Certain pairs of taxa (*arrows*), appear to have originated by simple dispersal into the lowlands (or onto outlying serranías).

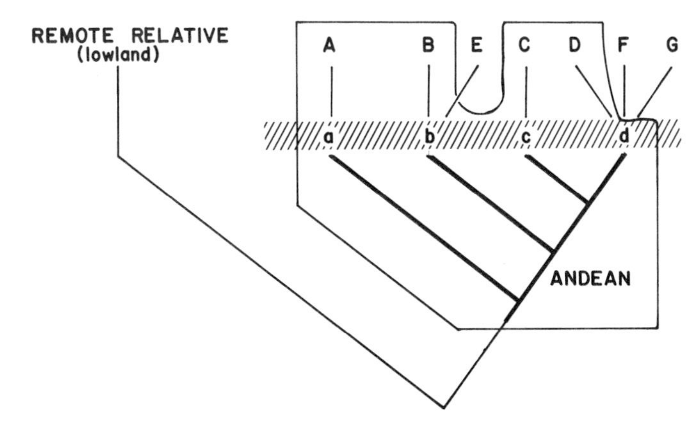

FIG. 19–15. Model of the evolution of an Andean group. The basic pattern of relationships (*heavy lines*) is produced by orogeny. Pleistocene depression-displacement phases are indicated by the horizontal bar (*hatching*). Taxa found in the Andes are enclosed in the box labeled "Andean." Taxa E, F, and G are Pleistocene derivatives occupying peripheral serranías or lowland tropical forests adjacent to the Andes. Pleistocene ancestors are identified with lower-case letters (a–d).

Resolution between Choices: Research Needs

At present, data are inadequate to resolve an explanation of the origins of the Andean herpetofauna. It is probably significant that none of the phylogenetic hypotheses available is of the form of Fig. 19–10 (idiographic explanation, dispersal into the Andes). Those organism groups now having unrestrictive phylogenetic hypotheses (Fig. 19–11) do not provide support for the dispersal hypothesis because the cladograms (as in Fig. 19–11) are founded on a *lack* of data. Dispersal provides no explanation because it is offered in lieu of a phylogenetic hypothesis. If a group actually evolved as in Fig. 19–11, there should be data to support such a hypothesis. The data would be in the form of one synapomorphy uniting all members of the group but no other synapomorphies uniting subsets (all taxa would be characterized by unique derived features—autapomorphies). Until such phylogenetic data are available, this hypothesis cannot be entertained rationally.

The other hypothesis is not contradicted by the available phylogenetic hypotheses. On the contrary, the phylogenetic hypotheses for some groups seem to demand that this hypothesis be true (*Bolitoglossa*, Fig. 19–7; *Eleutherodactylus*, Figs. 19–8, 19–9; *Phenacosaurus*; *Saphenophis*; *Hyla bogotensis* group; *Dendrobates minutus* group). A major problem is that not enough phylogenetic hypotheses are available. Most of those available are for groups found in the Andes north of the Huancabamba Depression. The paucity of phylogenetic constructs for groups in the southern Andes is another impediment to the corroboration of a general hypothesis. The complexity of the Andes, even so simple a complexity as northern versus southern, suggests that there may not be a single explanation. The proposal advanced here (Fig. 19–15) applies to the northern Andes but remains to be tested by developing phylogenetic hypotheses for other monophyletic groups found in the northern Andes: if the proposal is true, the new phylogenetic hypotheses will not be contrary to the proposal, but, if false, the new phylogenetic hypotheses will be incongruent with those identified here. The development of a hypothesis for the southern Andes must await proposals of phylogenetic hypotheses for groups present in the southern Andes.

ACKNOWLEDGMENTS

Four colleagues William E. Duellman, Jorge I. Hernández, Charles W. Myers, and Pedro M. Ruíz have generously shared their thoughts on Andean evolution with me. George Byers and Ed Wiley helped formalize my thinking in systematics. Marsha Lynch typed and retyped my manuscript and Jennifer Lynch aided in preparation of tables. My deepest thanks to all of them.

REFERENCES

Báez, A. M., and Z. B. de Gasparini. 1979. The South American herpetofauna: An evaluation of the fossil record. In *The South American Herpetofauna: Its origins, evolution, and dispersal*, W. E. Duellman, ed., pp. 29–54. Mus. Nat. Hist. Monograph 7. Lawrence: University of Kansas Press.

Ball, I. R. 1975. Nature and formulation of biogeographic hypotheses. *Syst. Zool.* 24: 407–430.

Brame, A. H., Jr., and D. B. Wake. 1963. The salamanders of South America. *Contrib. Sci. Los Angeles County Mus.* 69: 1–72.

———. 1972. New species of salamanders (genus *Bolitoglossa*) from Colombia, Ecuador and Panama. *Contrib. Sci. Nat. Hist. Mus. Los Angeles County* 219: 1–34.

Cei, J. M. 1972. *Bufo* of South America. In *Evolution in the genus Bufo*, W. F. Blair, ed., pp. 82–92. Austin: University of Texas Press.

———. 1979a. A reassessment of the genus *Ctenoblepharis* (Reptilia, Lacertilia, Iguanidae) with a description of a new subspecies of *Liolaemus multimaculatus* from western Argentina. *J. Herpetol.* 13: 297–302.

———. 1979b. The Patagonian herpetofauna. In *The South American herpetofauna: Its origin, evolution, and dispersal*, W. E. Duellman, ed., pp. 309–339. Mus. Nat. Hist. Monograph 7. Lawrence: University of Kansas Press.

Cochran, D. M., and C. J. Goin. 1970. Frogs of Colombia. *U.S. Natl. Mus. Bull.* 288: 1–655.

Croizat, L. 1964. *Space, time, form: The biological synthesis*. Caracas: Published by the author.

Croizat, L., G. Nelson, and D. E. Rosen. 1974. Centers of origin and related concepts. *Syst. Zool.* 23: 265–287.

Crump, M. L. 1974. Reproductive strategies in a tropical anuran community. *Univ. Kansas Mus. Nat. Hist. Misc. Publ.* 61: 1–68.

Duellman, W. E. 1972. A review of the neotropical frogs of the *Hyla bogotensis* group. *Occas. Pap. Mus. Nat. Hist. Univ. Kansas* 22: 1–31.

———. 1974. A systematic review of the marsupial frogs (Hylidae: *Gastrotheca*) of the Andes of Ecuador. *Occas. Pap. Mus. Nat. Hist. Univ. Kansas* 22: 1–27.

———. 1979a. The South American herpetofauna: A panoramic view. In *The South American herpetofauna: Its origin, evolution, and dispersal*, W. E. Duellman, ed., pp. 1–28. Mus. Nat. Hist. Monograph 7. Lawrence: University of Kansas Press.

———. 1979b. The herpetofauna of the Andes: Patterns of distribution, origin, differentiation, and present communities. In *The South American herpetofauna: Its origin, evolution, and dispersal*, W. E. Duellman, ed., pp. 371–459. Mus. Nat. Hist. Monograph 7. Lawrence: University of Kansas Press.

Duellman, W. E., and R. Altig. 1978. New species of treefrogs (family Hylidae) from the Andes of Colombia and Ecuador. *Herpetologica* 34: 177–185.

Duellman, E. W., and T. J. Berger. 1982. A new species of Andean treefrog (Hylidae). *Herpetologica* 38: 456–460.

Duellman, W. E., and T. H. Fritts. 1972. A taxonomic review of the southern Andean marsupial frogs (Hylidae: *Gastrotheca*). *Occas. Pap. Mus. Nat. Hist. Univ. Kansas* 9: 1–37.

Duellman, W. E., and R. A. Pyles. 1980. A new marsupial frog (Hylidae: *Gastrotheca*) from the Andes of Ecuador. *Occas. Pap. Mus. Nat. Hist. Univ. Kansas* 84: 1–13.

Duellman, W. E., and A. Veloso M. 1977. Phylogeny of *Pleurodema* (Anura: Leptodactylidae): A biogeographic model. *Occas. Pap. Mus. Nat. Hist. Univ. Kansas* 64: 1–46.

Estes, R. 1970. Origin of the Recent North American lower vertebrate fauna: An inquiry into the fossil record. *Forma et Functio* 3: 139–163.

Fritts, T. H. 1974. A multivariate evolutionary analysis of the Andean iguanid lizards of the genus *Stenocercus*. *San Diego Soc. Nat. Hist. Memoir* 7: 1–89.

Harrington, H. J. 1962. Paleogeographic development of South America. *Bull. Amer. Assoc. Petrol. Geol.* 46: 1773–1814.

Irving, E. M. 1975. Structural evolution of the northernmost Andes, Colombia. *Geol. Survey Profess. Pap.* 846: 1–47.

Izecksohn, E. 1976. O status sistemático de *Phryniscus proboscideus* Boulenger (Amphibia, Anura, Bufonidae). *Rev. Bras. Biol.* 36: 341–345.

Jacobs, E., H. Bürgl, and D. L. Conley. 1963. Backbone of Colombia. In *Backbone of the Americas*, ed. O. E. Childs and B. W. Beebe. *Amer. Assoc. Petrol. Geol., Mem.* 2: 62–72.

James, D. E. 1973. The evolution of the Andes. *Sci. Amer.* 229: 60–70.

Janzen, D. H. 1967. Why mountain passes are higher in the tropics. *Amer. Nat.* 101: 233–249.

Lazell, J. D., Jr. 1969. The genus *Phenacosaurus* (Sauria: Iguanidae). *Breviora* 325: 1–24.

Lynch, J. D. 1975. A review of the Andean leptodactylid frog genus *Phrynopus*. *Occas. Pap. Mus. Nat. Hist. Univ. Kansas* 35: 1–51.

———. 1976. The species groups of the South American frogs of the genus *Eleutherodactylus* (Leptodactylidae). *Occas. Pap. Mus. Nat. Hist. Univ. Kansas* 61: 1–24.

———. 1978. A reassessment of the telmatobiine leptodactylid frogs of Patagonia. *Occas. Pap. Mus. Nat. Hist. Univ. Kansas* 72: 1–57.

———. 1979a. The amphibians of the lowland tropical forests. In *The South American herpetofauna: Its origin, evolution, and dispersal*, W. E. Duellman, ed., pp. 189–215. Mus. Nat. Hist. Monograph 7. Lawrence: University of Kansas Press.

———. 1979b. Leptodactylid frogs of the genus *Eleutherodactylus* from the Andes of southern Ecuador. *Univ. Kansas Mus. Nat. Hist. Misc. Publ.* 66: 1–62.

———. 1980a. New species of *Eleutherodactylus* of Colombia (Amphibia: Leptodactylidae). I. Five new species from the Paramos of the Cordillera Central. *Caldasia* 13: 165–188.

———. 1980b. Two new species of earless frogs allied to *Eleutherodactylus surdus* (Leptodactylidae) from the Pacific slopes of the Ecuadorian Andes. *Proc. Biol. Soc. Washington* 93: 327–338.

———. 1980c. *Eleutherodactylus eremitus*, a new trans-Andean species of the *lacrimosus* assembly from Ecuador (Amphibia: Leptodactylidae). *Breviora* 462: 1–7.

———. 1980d. A taxonomic and distributional synopsis of the Amazonian frogs of the genus *Eleutherodactylus*. *Amer. Mus. Novitates* 2969: 1–24.

———. 1981a. The identity of *Hylopsis platycephala*

Werner, a centrolenid frog from northern Colombia. *J. Herpetol.* 15: 283–291.

———. 1981b. Leptodactylid frogs of the genus *Eleutherodactylus* in the Andes of northern Ecuador and adjacent Colombia. *Univ. Kansas Mus. Nat. Hist. Misc. Publ.* 72: 1–46.

———. 1981c. The systematic status of *Amblyphrynus ingeri* (Amphibia: Leptodactylidae) with the description of an allied species in western Colombia. *Caldasia* 13: 313–332.

———. 1982. Two new species of poison-dart frogs (*Colostethus*) from Colombia. *Herpetologica* 38: 366–374.

———. 1983. A new leptodactylid frog from the Cordillera Oriental of Colombia. In *Advances in herpetology and evolutionary biology*, A. G. J. Rhodin and K. Miyata, eds., pp. 52–57. Cambridge: Harvard Univ. Press.

Lynch, J. D., and W. E. Duellman. 1980. The *Eleutherodactylus* of the Amazonian slopes of the Ecuadorian Andes (Anura: Leptodactylidae). *Univ. Kansas Mus. Nat. Hist. Misc. Publ.* 69: 1–86.

Lynch, J. D., and P. M. Ruíz. 1983. New frogs of the genus *Eleutherodactylus* from the Andes of southern Colombia. *Trans. Kansas Acad. Sci.* 86: 99–112.

Mayr, E. 1961. Cause and effect in biology: Kinds of causes, predictability, and teleology are viewed by a practising biologist. *Science* 134: 1501–1506.

Mencher, E. 1963. Tectonic history of Venezuela. In *Backbone of the Americas*, ed. O. E. Childs and B. W. Beebe. *Amer. Assoc. Petrol. Geol., Mem* 2: 73–87.

Montanucci, R. R. 1973. Systematics and evolution of the Andean lizard genus *Pholidobolus* (Sauria: Teiidae). *Univ. Kansas Must. Nat. Hist. Misc. Publ.* 59: 1–52.

Myers, C. W. 1973. A new genus for Andean snakes related to *Lygophis boursieri* and a new species (Colubridae). *Amer. Mus. Novitates* 2522: 1–37.

———. 1982. Spotted poison frogs: Descriptions of three new *Dendrobates* from western Amazonia, and resurrection of a lost species from "Chiriqui." *Amer. Mus. Novitates* 2721: 1–23.

Myers, C. W., and J. W. Daly. 1976. A new species of poison frog (*Dendrobates*) from Andean Ecuador, including an analysis of its skin toxins. *Occas. Pap. Mus. Nat. Hist. Univ. Kansas* 59: 1–12.

———. 1980. Taxonomy and ecology of *Dendrobates bombetes*, a new Andean poison frog with new skin toxins. *Amer. Mus. Novitates* 2692: 1–23.

Myers, C. W., and W. E. Duellman. 1982. A new species of *Hyla* from Cerro Colorado, and other tree frog records and geographical notes from western Panama. *Amer. Mus. Novitates* 2752: 1–32.

Nelson, G., and N. Platnick. 1981. *Systematics and biogeography/cladistics and vicariance.* New York: Columbia University Press.

Peters, J. A. 1973. The frog genus *Atelopus* in Ecuador. *Smithsonian Contrib. Zool.* 145: 1–49.

Peters, J. A., and R. Donoso-Barros. 1970. Catalogue of the neotropical squamata. Part II. Lizards and amphisbaenians. *U.S. Natl. Mus. Bull.* 297: 1–293.

Rivero, J. A. 1963. Five new species of *Atelopus* from Colombia, with notes on other forms from Colombia and Ecuador. *Caribbean J. Sci.* 3: 103–124.

———. 1965. Notes on the Andean salientian (Amphibia) *Atelopus ignescens* (Cornalia). *Caribbean J. Sci.* 5: 137–140.

Rosen, D. E. 1975. A vicariance model of Caribbean biogeography. *Syst. Zool.* 24: 431–464.

Ruíz, P. M., and J. I. Hernández. 1976. *Osornophryne*, género nuevo de anfibios bufónidos de Colombia y Ecuador. *Caldasia* 11: 93–148.

Ruíz P. M., and J. D. Lynch. 1982. Dos nuevas especies de *Hyla* (Amphibia: Anura) de Colombia, con aportes al conocimiento de *Hyla bogotensis*. *Caldasia* 13: 647–671.

Savage, J. M. 1966. The origins and history of the Central American herpetofauna. *Copeia* 1966: 719–766.

Schall, J. J., and E. R. Pianka. 1978. Geographic trends in numbers of species. *Science* 201: 679–686.

Silverstone, P. A. 1975. A revision of the poison-arrow frogs of the genus *Dendrobates* Wagler. *Nat. Hist. Mus. Los Angeles County, Sci. Bull.* 21: 1–55.

Simpson, B. B. 1979. Quaternary biogeography of the high montane regions of South America. In *The South American herpetofauna: Its origin, evolution, and dispersal*, W. E. Duellman, ed., pp. 157–88. Mus. Nat. History Monograph 7. Lawrence: University of Kansas Press.

Sonnenberg, F. P. 1963. Bolivia and the Andes. In *Backbone of the Americas*, ed. O. E. Childs and B. W. Beebe. *Amer. Assoc. Petrol. Geol., Mem.* 2: 36–46.

Trueb, L. 1971. Phylogenetic relationships of certain neotropical toads with the description of a new genus (Anura: Bufonidae). *Contrib. Sci. Los Angeles County Mus.* 216: 1–40.

———. 1979. Leptodactylid frogs of the genus *Telmatobius* in Ecuador with the description of a new species. *Copeia* 1979: 714–733.

Uzzell, T. M. Jr. 1958. Teiid lizards related to *Proctoporus luctuosus*, with the description of a new species from Venezuela. *Occas. Pap. Mus. Nat. Hist. Univ. Michigan* 597: 1–15.

———. 1970. Teiid lizards of the genus Proctoporus from Bolivia and Peru. *Postilla* 142: 1–39.

Wake, D. B., and J. F. Lynch. 1976. The distribution, ecology, and evolutionary history of plethodontid salamanders in tropical America. *Nat. Hist. Mus. Los Angeles County, Sci. Bull.* 25: 1–65.

Wake, D. B., A. H. Brame, Jr., and C. W. Myers. 1970. *Bolitoglossa taylori*, a new salamander from cloud forest of the Serranía de Pierre, eastern Panama. *Amer. Mus. Novitates* 2430: 1–18.

Wiley, E. O. 1981. *Phylogenetics: The theory and practice of phylogenetic systematics*. New York: Wiley.

Williams, E. E., O. A. Reig, P. Kiblisky, and C. Rivero-Blanco. 1970. *Anolis jacare* Boulenger, a "solitary" anole from the Andes of Venezuela. *Breviora* 353: 1–15.

20

Origins of Lepidopteran Faunas in the High Tropical Andes

HENRI DESCIMON

When an entomologist from the northern temperate zone travels in the high Andes, in the páramos or the puna, it is generally not his first contact with a lepidopteran fauna. If he has already collected in the European alpine zone or in the American Rockies, he has at first a strange impression: butterflies *look* exactly the same, but they are not the same taxa! This black-veined white pierid with a hurried flight is not a *Synchloe*, but a *Tatochila*; this small fritillary, an *Yramea*, not a *Boloria*; this black, tricky, stone-resting satyrid is not an *Erebia* but a *Punapedaliodes* or a *Faunula*. However, this *Colias* is really a *Colias*. There is nonetheless a difference with Holarctic alpine faunas: species are not numerous and rather monotonous, and individuals are often scarce. By contrast, as soon as he goes down below timberline and enters montane forest, our entomologist finds butterflies are thriving and, if conditions are fair, he may see at least hundreds of individuals and dozens of species, a huge diversity in a very short time. Moreover, it appears at a glance that he is in another world, with a deep faunistic originality: here he actually enters South America. I will try in this chapter to bring some answers to the following questions:

1. Does the similarity between the species of the oreal Neotropics and those of the alpine zone of the Holarctic reflect real taxonomic affinities, pure convergences, or some more complex evolutionary phenomena (Fig. 20–1)?
2. Where do the faunistic elements of the high Andes originate?
3. What are the causes of the low faunistic diversity of the páramo-puna biome, when compared to other biomes at lower altitudes in the Neotropics or, to a lesser extent, to some Holarctic oreal biota?
4. Why is there such a clear-cut difference between oreal and montane forest faunas in the Andes?

I will also attempt to define distribution patterns, faunistic subregions, and dispersion centers and to compare them with the data obtained from other groups.

THE BUTTERFLIES OF THE HIGH ANDES: A PRESENTATION

Four families of Lepidoptera Rhopalocera have members in the fauna of the puna-páramo ecosystems: Pieridae, Satyridae, Nymphalidae, and Lycaenidae. The Hesperiidae are also included in this fauna, but they are so poorly studied that I will not take them into account.

The Pieridae

The Pieridae are represented in the high Andes by species belonging to two subfamilies: Pierinae, with some genera included in the tribe Pierini, and Coliadinae, with the genus *Colias* and, somewhat marginally, *Teriocolias*.

The Pierini are certainly the best known and probably the most interesting butterflies of this region. The work of Klots (1933), Field (1958), Herrera and Field (1959), Ackery (1975), Field and Herrera (1977), Shapiro (1977a, b, 1978, 1979a, b), and Shapiro and Torres (1978) have brought this group (a homogeneous subtribe according to Field, 1958) to a very high level of biotaxonomic knowledge. Most strikingly, they are confined to cool or even cold conditions, that is, either to the Andes or to the southern temperate tip of the continent. Maps of the distributions of the most important genera are presented in Figs. 20–2 to 20–4. They are compiled from the data of Herrera and Field (1959) and Field and Herrera (1977), simplified and completed from the author's observations and various more

FIG. 20–1. Convergence versus taxonomic relationship between pairs of high Andean Palearctic butterfly species from oreal biomes. For each pair, the Andean species is listed first. First pair: 1, *Tatochila sagittata* (Paucartambo, southern Peru); 2, *Metaporia leucodice* (central Himalaya). Second pair: 3, *Tatochila microdice macrodice* female (Sillustani, central Peru); 4, *Synchloe callidice* (Hautes-Alpes, France). Third pair: 5, *Phulia nymphula* (*up*, upperside, *un*, underside) (Cordillera Negra, central Peru); 6, *Baltia butleri* (Ladakh). Fourth pair: 7, *Colias vauthieri* (Argentina); 8, *Colias croceus* (France). Fifth pair: 9, *Colias lesbia dinora* female (Ecuador); 10, *Colias croceus* female (France). Sixth pair: 11, *Punapedaliodes albopunctata* (Ecuador); 12, *Erebia lefebvrei pyrenea* (Pyrénées-Orientales, France). Seventh pair: 13, *Yramea cora* (Abra Soqilacasa, central Peru); 14, *Boloria pales* (Hautes Alpes, France).

recent references. Eight genera are currently distinguished:

1. *Theochila*, monotypic and confined to lowland temperate regions;
2. *Tatochila*, comprising 14 species divided into five closely related groups (Herrera and Field, 1959); this genus covers the whole area under consideration here, from the Sierra Nevada de Santa Marta south to Tierra del Fuego (Fig. 20–2). It reaches its maximum diversity in the southern temperate tip of the continent. Some species display clear-cut subspecific variation.
3. *Hypsochila*, seven species, extending from southern Peru into Tierra del Fuego. The genus also reaches its maximum diversity south of the Tropic of Capricorn.
4. *Phulia*, four species, with a typically Andean distribution from the Cordillera Blanca area in Peru south to central Chile and west-central Argentina (Fig. 20–3); the core of species diversity is more northern than in previous genera.
5. *Infraphulia*, three species distributed at high elevations (3000–5000 m) from central Peru to northern Chile (Fig. 20–3).
6. *Pierphulia*, three species, with a similar, but a little more southern distribution (Fig. 20–4).
7. *Piercolias* (= *Andina*; = *Trifurcula*), three species, probably the butterflies living at the highest elevations in the Andes (Garlepp, 1892); the genus ranges from southern Peru to northern Chile (Fig. 20–4).

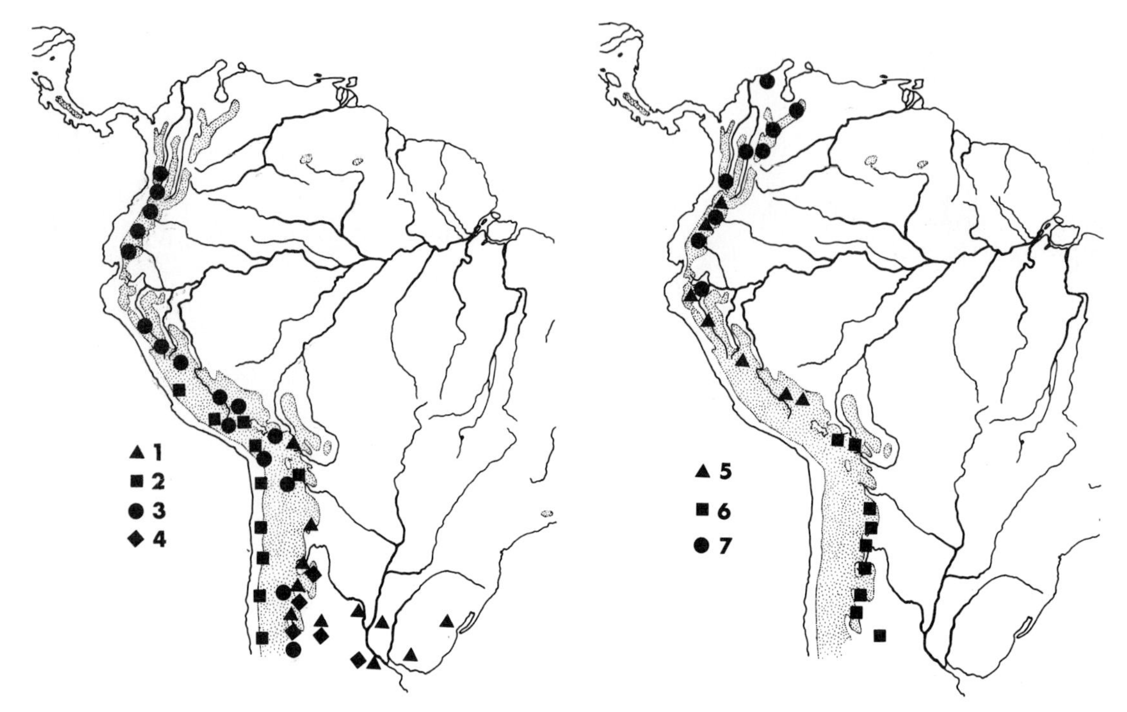

FIG. 20–2. Distribution of seven species of the genus *Tatochila*. 1: *T. autodice*; 2: *T. blanchardii*; 3: *T. microdice*; 4: *T. vanvolxemii*; 5: *T. sagittata*; 6: *T. stigmadice*; 7: *T. xanthodice*.

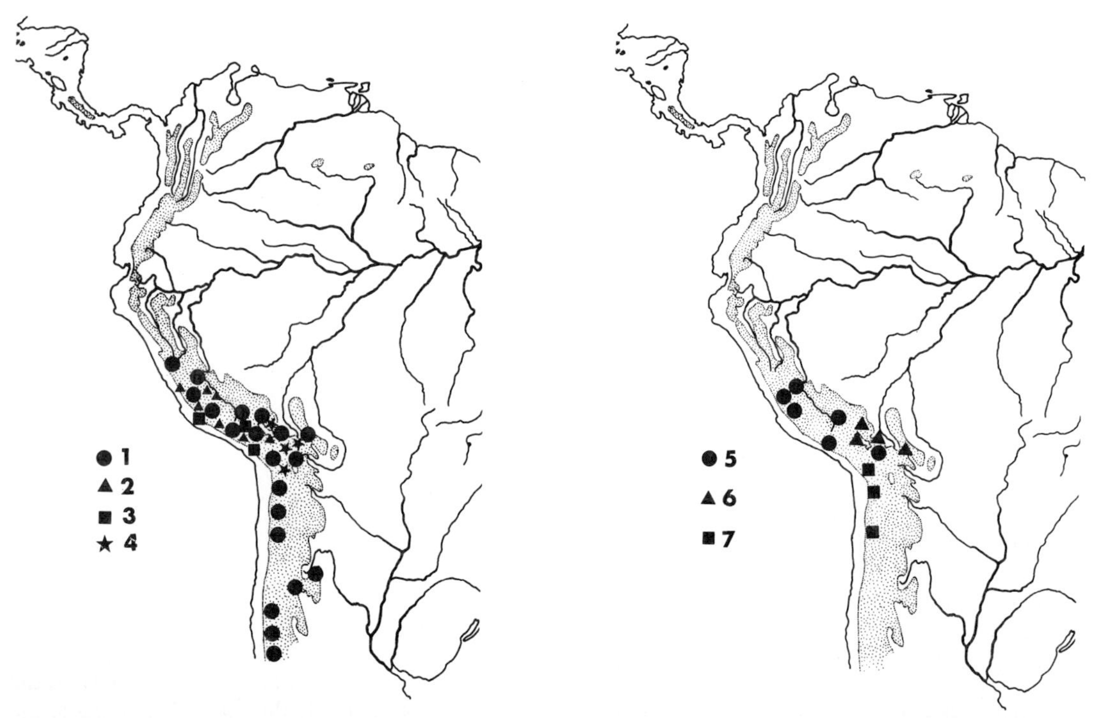

FIG. 20–3. Distribution of the genera *Phulia* and *Infraphulia*. 1: *P. nymphula*; 2: *P. garleppi*; 3: *P. nannophyes*; 4: *P. paranympha*; 5: *I. madeleina*; 6: *I. illimani*; 7: *I. ilyodes*.

FIG. 20–4. Distribution of the genera *Reliquia, Pierphulia* and *Piercolias*. 1: *R. santamarta*; 2: *Pierphulia rosea*; 3: *Pierphulia nysias*; 4: *Pierphulia isabella*; 5: *Piercolias coropunae*; 6: *Piercolias forsteri*; 7: *Piercolias huanaco*.

8. *Reliquia*, recently discovered by M. Adams (Ackery, 1975), monotypic and limited to the Sierra Nevada de Santa Marta (Fig. 20–4), where it occurs at high elevations.

These butterflies are small to medium (length of forewing between base and apex: 10–30 mm) and their habitus, rather homogeneous, is that of a Holarctic "Green-Veined White." The taxonomic structure of the complex is not yet well understood, and the grouping of genera varies according to the morphological criteria under consideration (Field, 1958). My own preliminary attempts, taking advantage of the work necessary to prepare this chapter, to trace a cladogram of the whole group of species have not been successful. Physiological data (presented later in the section on Adaptive Features) only result in additional sources of perplexity (Shapiro, 1978). *Theochila*, *Tatochila*, and *Hypsochila* appear to form a first group, with morphological, biogeographic, and ecological coherence, and the remaining genera another one, the position of *Reliquia* being rather uncertain, though it may be more related to the second group, which is made up of high-elevation dwellers. A striking feature is that the genera of the *Phulia* group clearly show some parallelism in the number (three or four) and distribution patterns of species.

South American Pierini have no relatives in the other parts of the Southern Hemisphere, but there is an undoubtedly closely related genus in the Himalaya, *Baltia*, with two species, akin to *Phulia* and *Piercolias* and showing the same ecological characters (Field, 1958): the similarity of the habitats is, according to F. and J. Michel (pers. comm.), extraordinary. The Pierini of the Nearctic and, with the exception of *Baltia*, those of the Palearctic, though clearly related to the Andean ones, are more distant. Since the work of Klots (1933), the South American species have often been considered to be derived from an ancestral form close to *Pieris* (*Synchloe*) *callidice*, an alpine species of the Palearctic and its Nearctic relatives *P. nelsoni*, *P. protodice*, and *P. occidentalis* (Forster, 1958; Mani, 1968), though the closeness of *Phulia* and *Piercolias* to *Baltia* had been stressed as early as 1924 by Röber (1926). Anyway, it must be remembered that the undoubted similarity in wing patterns and morphology may be due as much to convergence as to actual evolutionary relations. All these problems will be discussed further.

The foodplants of all these butterflies are probably Cruciferae and perhaps also the South American family Tropeolaceae (Shapiro, 1978). A striking fact is that, in many places in South America, all *Tatochila* species have adopted European cruciferous weeds as foodplants, to such an extent that it is impossible to determine their genuine original hosts (Shapiro, 1979a); this is also true for *Phulia nymphula* in the Cordillera Blanca in Peru (Descimon, unpublished).

The South American *Colias* (subfamily Coliadinae) represent also a very homogeneous and even more compact group of species. Though obviously belonging to this huge genus, which has undergone an explosive adaptive radiation in central Asia and a minor one in North America, the South American species possess some "family resemblances" which suggests a monophyletic origin, as do several peculiar features of the genitalia (Petersen, 1963). This author proposes to separate *C. ponteni* in a separate genus, *Protocolias*, and the two South American species he has studied (*C. vauthierii* and *C. "cunninghami"*) in a subgenus of *Colias*, *Mesocolias*. The study of other South American *Colias* genitalia (Descimon, unpublished) suggests rather that one monophyletic unit, with some primitive morphological features, is present in the continent; for reasons which will not be discussed here, it does not appear urgent to separate them from the bulk of the genus. South American *Colias* have been far less studied than pierine butterflies. Their treatment by Röber (1926) is unfortunately one of the worst that could be conceived: they are scattered with neither apparent nor internal logic among the North American species. The more recent comprehensive book by d'Abrera (1981), based only upon the British Museum collection, does not take into account the literature on the subject and fails to bring the necessary clarification to it. Two recent papers by Berger (1981, 1983), although extremely interesting, appear also more confusing than enlightening. One must rely only on scattered papers (Talbot, 1925; Ureta, 1937; Forster, 1955; Reissinger, 1972; Lamas, 1981). Actually, the only relevant treatment of the South American representatives of the genus that has been made available to the author is that by G. Lamas of the Museo de Historia Natural "Javier Prado" in Lima, unfortunately not yet published. With this help, combined with field data collected during several trips to Ecuador and Peru, studies on preimaginal stages of several species (Descimon, unpublished), and preliminary results of enzyme electrophoresis (in collaboration with H. Geiger), it is possible to

FIG. 20–5. Distribution of the genus *Colias* in Central and South America. 1: North American species; 2: *C. euxanthe*; 3: *C. lesbia*; 4: *C. flaveola*; 5: *C. dimera*; 6: *C. vauthieri*.

present here a tentative arrangement of the group. There are at least five Andean species (Fig. 20–5):

1. *C. dimera* Dbld. and Hew. is a northern Andean species (ranging from the Cordillera de Mérida to Chachapoyas). It is widespread from 2000 m to 3200–3500 m, and feeds on white clover (*Trifolium repens*); because this plant was imported by the Spaniards, the butterfly obviously has adapted to it recently. A wide range extension of the species, consecutive to the opening of artificial meadows in the inter-Andean valleys, is thus quite probable. Also worth noting is that until the present time, alfalfa was not used as a foodplant, while in captivity the caterpillar accepts many papilionaceous herbs, including various *Medicago* (Descimon, unpublished). Like *C. lesbia* and *C. euxanthe* its preimaginal stages are as reminiscent of Holarctic *Colias* as the adult ones; in fact, their pattern—green with fine setae and longitudinal stigmatal lines—is the generalized "primitive" one of Coliadinae (Descimon, to be published).

2. *C. vauthieri* Guér. is distributed from southern Argentina and Chile to Bolivia. It feeds also at the present time on white clover (Courtney, pers. comm.); the caterpillar is markedly different from those of other Andean *Colias* but recalls some Holarctic species (Descimon, unpublished). (Perhaps to be included here is *C. ponteni* Wall. = *C. imperialis* Btlr., not found again since its description in the mid-nineteenth century; its genuine geographic distribution is quite a mystery [Tierra del Fuego?].)

3. *C. lesbia* F. is widespread in southern South America (nominal form and hardly distinct subspecies). Largely polymorphic, this species has adapted to alfalfa since the conquest and has become a crop pest. Like many *Colias*, it is highly migratory (Williams, 1930). Absent, of course, from the lowlands of tropical South America, the species is found at medium elevations (2500–3500 m) in the Andes (subspecies *andina* Stgr., Bolivia to central Peru, subspecies *dinora* Kirby, northern Peru and Ecuador); in this region, it feeds on white clover. The subspecies *dinora*, the largest South American *Colias*, has been often mistakenly designated as "*cunninghami* Btlr." (e.g., in Hovanitz, 1945b), due to a mislabeling of plates in Seitz's book (Röber, 1926).

4. *C. euxanthe* Feld. is a central-Andean species, ranging from Bolivia to the high mountains of Ecuador. Generally speaking, it has a broad altitudinal range, from above 4800 m down to 3000 m and even lower, with marked variation, especially in size (smaller individuals in the higher reaches). It is not absolutely certain that *alticola* God. and Salv. from Ecuador actually belongs to this species. The larva feeds at upper elevations on *Astragalus* sp. (Descimon, unpublished) but, at lower ones, it has adapted to white clover. This gives it the opportunity of cohabiting with *C. lesbia* in artificial meadows. Here arises a puzzling problem: how many *Colias* species do exist in cultivated areas of middle elevation in Peru? The phenotypic diversity of a population flying in a single meadow is often confusing. It is tempting to interpret this diversity in terms of polyphenism and polymorphism and to state, as did Shapiro (1984), that only a single species is present. On the other hand, Perez Ramos (1982) considers that three species are found together in the Callejón de Huaylas area (*C. lesbia*, *C. alticola*, and *C. euxanthe*); similarly, Berger (1983) describes a new species, *C. verhulsti* from the Cuzco region, which also implies that three species occur there (*C. lesbia*, *C. verhulsti*, and *C. euxanthe*). Now, it seems obvious that the *C. alticola* figured by Perez Ramos in his work are perfectly intermediate between the *C. lesbia* and *C. euxanthe* of the same plate. My opinion, based largely on field observations, is that there exists a situation somewhat comparable to that found in the United States with two other *Colias* species: *C. eurytheme* and *C. philodice*. It is well known that these taxa, recently brought into contact by the extension of alfalfa fields, hybridize at a high rate, although persisting as separate species (Taylor, 1972).

C. lesbia and *C. euxanthe* would therefore represent two species that are imperfectly separated—semispecies, or, better, quasispecies (Bernardi, 1980)—and probably brought into coexistence by the action of man, as is frequently the case with quasispecies (Woodruff, 1973). It is easy to conceive that intraspecific polymorphism and polyphenism, combined with interspecific hybridization, would produce a number of variants that are difficult to interpret at first sight. Breeding and electrophoresis experiments that I have undertaken with H. Geiger have so far been insufficient to disentangle this situation.

5. *C. flaveola* Blanchard is distinguished by its suffused pattern, reminiscent of certain Holarctic *Colias* (e.g., *C. phicomone* in Europe, *C. cocandica* in central Asia—the closest phenotypically—or *C. behrii* in North America). The males of some populations lack the "androconia" observed in all other South American species, which has been interpreted by Berger (1983) as a sign of specific distinction. However, the "andro-

conia" is an unstable character in some *Colias* species (Verity, 1911) and some *flaveola* populations present a patch of minute size that could indicate a transitional situation. Thus, for the sake of simplicity, I will consider a single species. Some geographic variation is known. The populations of the western side of the Andes (*flaveola* sensu stricto from the Coquimbo region in Chile and *erika* Lamas from the Arequipa highlands in peru) are very pale, whereas those from the eastern side (*blameyi* Jörg. from Aconquija in Argentina, *weberbaueri* Strand from Bolivia and southern Peru, *mossi* Rotsch. from San Marcos, Peru) are often heavily suffused with black scales. This gives them a green appearance and is probably an adaptive character related to the intensity of the sun. A peculiar feature of this species is that it flies only during the rainy season (March-April, or December in the Arequipa region) (Forster, 1955; Lamas, 1981; Berger, 1983); other *Colias* fly throughout the year, with a peak during the dry season (Perez Ramos, 1982; Descimon, unpublished).

The adaptive radiation of Andean *Colias*, although much less spectacular than that of the Whites, has allowed them to colonize many habitats. Except for *C. electo* (a close relative of the highly migratory Western Palearctic *C. croceus*), which is common in the temperate parts of Africa south of the Sahara, there are no other *Colias* in the Southern Hemisphere. A clearly Nearctic species, without marked affinities with Andean ones, occurs in the mountains of southern Mexico and northern Guatemala but, as is the case for Pierini, no species is known from the southernmost regions of Central America and even the Sierra Nevada de Santa Marta appears to be devoid of *Colias*. Many representatives of the genus from the Holarctic region present a close phenotypic resemblance with the Andean ones; Verity (1911) already mentioned the almost perfect similarity between the central Asian *C. felderi* and *C. aurora* and the South American *C. lesbia*; the suite of resemblances between the various forms of *C. cocandica* (also from central Asia) and those of *C. flaveola* is no less striking. In North America, *C. hecla*, which feeds on *Astragalus*, may be compared also to Andean species; an arctic-alpine element (the only Rhopaloceran, along with *Boloria chariclea*, to have been able to colonize Greenland), it is widespread in the circumboreal regions and reaches southward to British Columbia and central Alberta. But perhaps the closest kin to Andean taxa is *C. meadii*: the only North American "androconia"-bearing species, it is found above the timberline in the Rocky Mountains from British Columbia to northern New Mexico. Its foodplants are alpine clovers; it is hard not to compare this fact with the greediness of Andean *Colias* rushing onto the "good old" Shamrock! The extent to which these resemblances are merely convergences is yet to be determined.

The Satyridae

The picture here is quite different. No group exhibits clear-cut boreal affinities and no adaptive radiation is observed in the puna-páramo biomes. This fact is all the more striking because in the immediately lower-lying biota of montane forests, the tribe Pronophilini has undergone a huge explosion of genera and species. Above the timberline, they remain present, but with a low diversity. Perhaps the most characteristic feature is that there is no single homogeneous group adapted to the oreal biomes, but rather a series of isolated species that have settled in this environment.

By far the best studied area is the northern Andean region, thanks to the work of Adams (1973, 1977) and Adams and Bernard (1977, 1979). In the páramos of the Sierra Nevada de Santa Marta is found one endemic species, belonging to a monotypic genus, *Paramo oculata* (Fig. 20–6), which belongs in turn to the "*Pedaliodes*" group of Pronophilini. In the Serranía de Valledupar, Adams and Bernard (1979) have discovered another endemic monotypic genus, *Dangond dangondi*, which lives in the marshy parts of the páramo; the only relative of this relict taxon is a species occurring in the Cordillera de Mérida: "*Pedaliodes*" *empetrus*. With (or close to) it is to be found another species, *Lymanopoda paramaera*, which is also characteristic of the upper montane forest.

Going southward, the data become ancient and vague, often unreliable, with imprecise indications as to the place of capture and its ecological characteristics; indeed, it is often impossible to determine whether a given species has been collected in páramo or in montane forest. The most reliable works are those of Fassl (1910, 1911, 1914a, b, 1915a, 1918), Godman and Salvin (1891), and Weymer and Maassen (1891). Most of the cited species belong to the genus *Lymanopoda* (or *Penrosada*?), which is also found in montane forest: *L. levana*, *L. gortina*, *L. tolima*. They are often white, possibly mimetic of *Tatochila*. *Altopedaliodes*, especially *A. reissi*, and *Pseudosteroma* species are also reported from high altitudes, probably in páramo.

In southern Ecuador, in the páramo of Cumbe,

FIG. 20–6. Distribution of high-altitude Pronophilinae. 1: *Paramo oculata*; 2: *Punapedaliodes albopunctata*; 3: *Punargentus lamna*.

between Guamote and Alausi, I have observed extensive colonies of *Punapedaliodes albopunctata*. The butterflies occurred in a dry, low, grassy páramo, physiognomically close to the puna, at an elevation of about 3500 m. This species, however, is found mainly farther south, into Bolivia (Forster, 1958) (Fig. 20–6). It was found to be very common in the neighborhood of Cajamarca (northern Peru) in July 1980 (F. Michel, pers. comm.). It is a local and patchy species, not observed everywhere though widespread, and it looks very much like an *Erebia*.

By contrast, *Punargentus lamna* is perhaps the commonest and most widespread butterfly of the puna (Fig. 20–6). Its distributional area seems to begin in the Cordilleras Blanca and Negra (unpublished data from Descimon and Mast de Maeght, 1971, and Michel, 1980). This genus includes six species, living mostly in the Bolivian, Chilean, and Argentine Andes in puna or alpine-like biota; it is related to *Argyrophorus* and *Cosmosatyrus* (Heimlich, 1963), both genera characteristic of the alpine zone of the southern temperate tip of South America. *P. lamna* is characterized by a silvery coloration that renders it most conspicuous and attractive when in flight (though to a lesser degree than the glorious Chilean *Argyrophorus argenteus*). We have observed it in practically every puna section along the road between Lima and Puno (A. and H. Descimon, unpublished) and it is also very common in Bolivia (Forster, 1958). *P. angusta* and *P. gustavi*, of uncertain taxonomic status, are less well known.

The taxonomic affinities of high Andean satyrids are quite clear: all belong to the characteristically neotropical Pronophilini tribe. The presence in South America of species of the genus *Erebia*, which belong to a different tribe, is a myth based on old, inaccurate systematics; it is a pity that it should have reappeared in recent publications (Fittkau, 1969; Shields and Dvorak, 1979). According to the basic work of Miller (1968), the Pronophilini belong to the mostly Holarctic subfamily Satyrinae and stand close to the fundamentally Palearctic Satyrini. They are less closely related to the South African Dirini, whereas the Hypocystini, from Australia and New Zealand, in spite of convergent features, are still more distant. However, some aspects of the work of Miller have been criticized by Adams and Bernard (1977), and a revision of the taxonomic structure of the family using Hennig's cladistic methods would perhaps give somewhat different results. It may be pointed out that the haunts of Satyrinae are generally open landscapes: Mediterranean and steppe-like grasslands ("*Satyrus* series") or alpine meadows ("*Oeneis* series"); however, some species live also in temperate forest biomes (*Minois*, some *Hipparchia*). The Pronophilini are dwellers of humid and cool forests (absent in lowland tropical forest) that can adapt secondarily to the dry conditions of the puna, where they are less diverse than in páramo ecotones.

The Nymphalidae

The nymphalids are poorly represented over the whole range of the Andes. Ubiquitous and migratory species such as *Vanessa virginiensis* and *Dione moneta* are very often met with in páramo (Adams, 1973) as well as in puna biomes (Descimon, unpublished). Their presence here is of little biogeographic significance. The only characteristic elements are the four species of the genus *Yramea* (Staudinger, 1894; Fassl, 1920; Forster, 1955; Hughes, 1956, 1958; Grey, 1957). These butterflies are restricted to southern Peru, Bolivia, and Argentina (Hayward, 1964). They resemble extraordinarily, both in their behavior and habitus, the Holarctic genus *Boloria* (de Lesse, 1966), to which they are closely related, according to Grey (1957).

The Lycaenidae

Although rather discreet, this family is well represented in the highest Andes. Unfortunately, other than cladistic revisions by Kurt Johnson (1981), which include a number of Andean lycaenid genera, the systematics of montane Neotropical lycaenids have not been revised for a long time. Their overwhelming diversity and heterogeneity would appear to preclude both a partial study such as that of the Andean species and a general revision that would necessitate several lives. Neotropical lycaenids will probably have to be slowly approached through the revision of assemblages of monophyletic groups as in several recent works by S. S. Nicolay. Eliot (1973) provided a very good arrangement of tribes, which he himself considered tentative. Concerning the Andes, other than including the general reference of taxa erected by Johnson (1981), one must follow the order of Draudt (1926), which is very defective but the only general one available.

The Andean species belong to two subfamilies, both cosmopolitan: the Theclinae and the Polyommatinae. All taxa of the former subfamily belong to the tribe Eumaeini, both Holarctic and Neotropical, and are clearly divided into five groups:

1. The "*Thecla*" *loxurina* group (number 29 of Draudt, 1926). It consists of *T. loxurina* (probably an aggregate of species whose range extends from Colombia to Bolivia), and *T. dissentanea* (from the Cuzco region). These species apparently belong to a Neotropical unit that is well represented in the lowlands.

2. The "*Thecla*" *culminicola* group. A rapid glance corroborates the long-neglected opinion of Staudinger (1894) that it might be connected to the Holarctic genus *Callophrys* and more precisely to its subgenus *Incisalia*, some species of which look similar. Johnson, however (1981) has shown that affinities are not directly to the Nearctic *Incisalia* but to sister groups in the Himalaya. In fact, the sister group of pine-feeding Nearctic *Incisalia* is in the Himalaya also, and the other forb-feeding taxa traditionally placed with Nearctic *Incisalia* (*fotis*, *augustinus*, *henrici*, *polios*, etc.) are actually the phylogenetic sister group of the taxa *Sandia macfarlandi* (southern United States and northern Mexico) and *S. xami* (United States and Mexico with a sister species in Guatemala). These two assemblages themselves are the sister group of a clade of Neotropical taxa that extend into Colombia and Venezuela but are not montane. The sister group of the "*Thecla*" *culminicola* group is the *Satsuma* (sensu lato) taxon of western China and Tibet, though it must be *Satsuma* sensu lato as divided by Johnson into several genera in a primitive-to-derived hierarchy. Of these, Nearctic *Incisalia* (*niphon*, *eryphon*, *lanoraieensis*) relate to sister groups in the Himalaya and China that are more primitive than the "*Thecla*" *culminicola* group—*Satsuma* affinities. *Satsuma* taxa to which "*Thecla*" *culminicola* are related have a huge diversity in Tibet, western China, Mongolia, and the Amur–Manchuria area, Johnson adding over 30 new species. Johnson does not place special emphasis on their high Andean relations, noting that more primitive stems of the cladograms including Himalayan, montane Nearctic, and Andean taxa include groups of much wider distribution and endemics in Guatemala. He would seem to prefer differential extinction to a more special relation uniquely Andean and Himalayan. According to Johnson's work, and a subsequent revision specifically of the Andean representatives, about 10 species are present in the Andes. These range from northern Chile to central Colombia. The species *culminicola* and *alatus* are hitherto most familiar to lepidopterists.

3. The "*Thecla*" *arria* group (number 30 of Draudt, 1926). In this group Draudt clustered 12 taxa that occur at high Andean altitudes and that superficially resemble members of the *culminicola* group in apparent mottled and suffused underwing patterns. Johnson (1981) has shown that this resemblance is not indicative of phylogenetic relationship. His work suggests that members of the *arria* group may themselves have an

eastern Palearctic sister group, particularly the genus *Panchala* of Asia. Other than Johnson's comments separating the *arria* group from those clustered with *culminicola* little work has been done on the former since their original descriptions.

4. *Eiseliana* (Toledo, 1978). Toledo named the genus *Eiseliana* from a single Chilean species *koehleri* occurring at very high altitudes. Johnson in a subsequent review has documented that *Eiseliana* is a monophyletic assemblage containing at least six Andean species and also includes the name *punona* Clench, formerly placed in *Thecla*. Johnson suggests that *Eiseliana* may be a Neotropical sister group of *Strymon* sensu lato based on comparison of the genitalia of both sexes.

5. *Cyanophrys* (sensu lato). Johnson has added over 40 new species to the knowledge of this familiar Neotropical group of lycaenids. He has also divided them into several sister genera in a primitive-to-derived hierarchy. An important aspect of his work is the documentation that species of this group are indeed Andean and that numerous Andean endemics occur, especially in northern Argentina. The altitudinal range of these species needs to be studied. The taxa that have many of the most primitive character states studied by Johnson in the lycaenids have eastern Palearctic sister groups. More derived Argentine Andean taxa in this group show a pattern of sister group occurrence eastward in the mountains of southeastern Brazil, and Andean endemics from central Bolivia northward have sister groups occurring in Amazonian Brazil and on the Guayana Shield. The most common pattern in three and four taxon statements in *Cyanophrys* sensu lato is that southeastern Brazil is plesiotypic with various areas of the Andes north of central Bolivia or the Guayana Shield as apotypic. Three and four taxon statements with Argentine Andean apotypic taxa seem to occur independently of the preceding pattern with their plesiotypic counterparts in southeastern Brazil.

In the Polyommatinae, the more numerous species all belong to the cosmopolitan tribe Polyommatini. Two sections may be distinguished:

1. The section *Leptotes*, pantropical, with the genus *Leptotes* (likely to be divided) well represented in the Neotropical lowlands and sharing two species in the high Andean fauna: *L. andicola*, found by Whymper in the páramos of Ecuador, and *L. callanga*, described from Cuzco.

2. The section *Polyommatus*, cosmopolitan. South American species are well characterized by a peculiar feature of the male genitalia, the sagum, and are confined to montane and temperate biota. A partial revision, at the genus level, has been provided by Nabokov (1945). Three genera are known from the high Andes: *Itylos*, *Parachilades*, and *Paralycaeides*, which include, in an unrevised state, nine species, mostly from central Peru to Bolivia, northern Chile, and north-western Argentina. They are often met with at considerable elevations.

ADAPTIVE FEATURES OF BUTTERFLIES IN THE ANDEAN OREAL BIOME

Adaptive traits are likely to be observed in all aspects of the morphology and physiology of the puna-páramo insects. However, it is mainly in fields immediately accessible to our observations that such peculiarities have been pointed out, i.e., coloration, developmental cycle, and behavior.

Adaptive Colorations

Adaptive colorations concern chiefly the wings, which make up the major part of the surface of the body.

Melanism

Melanism is a complex phenomenon, which is part of several different ecophysiological strategies (Descimon and Renon, 1975; Descimon, 1980a). In high-altitude butterflies, two types of melanism can be considered: "thermoregulatory" melanism, and melanism linked to cuticular waterproofing. The former results in an improved heating of the insect hemolymph, which circulates actively in the wings; this phenomenon was observed long ago in alpine insects and Lepidoptera (Heer, 1836). To avoid oversimplification, it must be thought of as reflecting an equilibrium between contradictory selective pressures, with overheating being avoided by behavioral adaptations such as posture and flight activity. Polyphenism, i.e., differences in habitus from one brood to the next, may also be encountered as an adaptation to seasonal climates (Shapiro, 1976): the brood that flies under cool conditions is darker and absorbs more solar radiation than the warm-season brood (Watt, 1969); this reaction is triggered by photoperiod and, to a lesser extent, by temperature. It is very widespread in the Pieridae of temperate regions and may, in some cases, be present in a latent form that may be triggered under artificial conditions of photoperiod and temperature (Shapiro, 1975). Polyphenism exists also in tropical species, where it may occasionally reach such an extreme form that the dry and wet season broods have in some cases been mistaken for different species (Mell, 1931). In this latter case, its adaptive significance is not well understood. It may be added that, according to Field (1951), the coliadine butterfly *Teriocolias zelia* shows a marked polymorphism between "dry" and "wet" season broods that involves not only both pteridine and melanin pigments at the same time, but the wing pattern as well (Forster, 1955). This species flies close to, if not within, the limits of the oreal zone: its haunts are the upper parts of the dry, nonforest biota of the inter-Andean valleys, at elevations of 3000–4000 m. Seasonal variation has been observed close to the limit of the lowland tropical zone, where seasonal differences are already clear-cut.

In Andean butterflies, adaptive melanism appears to be widespread. In Pieridae, it expresses itself through the enhancement of preexisting melanic patterns, both on the upper- and underside of the wings. The wing pattern of Pieridae is quite degenerate (Schwanwitsch, 1956), which severely limits blackening opportunities by stressing preexistent patterns. This might account for the striking likeness between the Andean *Phulia*, the alpine *Synchloe*, and the Himalayan *Baltia*: since the basic pattern-coding information is the same in the whole Pierinae subfamily, it may be expected to be similarly revealed by similar selective pressures. However, the characteristic shape of marginal markings, among many other features peculiar to the Andean Pierini (with the exception of *Reliquia*) and *Baltia*, argues against the hypothesis of pure convergence. In pierine butterflies, the extension of black areas is generally limited, especially in males. Do they possess any other thermoregulatory mechanism? It may simply be recalled that the "background" pigments, pteridines, absorb markedly (though to a lesser extent than melanin) infrared and ultraviolet radiations, which may provide a sufficient temperature increase for these animals. The deepest, almost pure black pierids (*Pereute*, *Archonias*) live in lowland regions and their color is determined by mimicry with *Heliconius*.

Melanism is widespread in *Colias*, where it is the component of a large polymorphism. In *C. flaveola weberbaueri*, Fassl (1915b) has described a very dark morph from the vicinity of La Paz. According to Forster (1955), its frequency changes from one locality to another. *C. lesbia* and *C. euxanthe* are also very polymorphic in coloration, but both the extent (pattern) and the intensity (suffusion) of melanin pigmentation, as well as of pteridines, are involved. It seems that these polymorphisms are subject to clinal variation, with the highest proportion of melanics being found at the highest altitudes.

Nymphalids (*Yramea*) are not especially melanic when compared to lowland species of Neotropical fritillaries (e.g., *Euptoieta*). On the contrary, satyrids *are* melanic, and extremely so, most of the Pronophilini species characteristic of oreal biota, for instance (the ones belonging to the genera *Punapedaliodes*, *Paramo*, *Dangond*, "*Pedaliodes*"), being completely black or almost so. It should, however, be kept in mind that melanism is by no means restricted to the inhabitants of the highest regions: many Pronophilini of the montane forest are also deeply melanic. *Punargentus lamna*, a most characteristic species of the puna, has wings covered with silvery splashes, whereas some *Lymanopoda* are white. Using the Palearctic genus *Melanargia*, we have shown (Descimon and Renon, 1975) that melanization has several, often contradictory, functions: not only does it improve the absorption of solar radiation, but it also protects against desiccation. While if melanization is excessive, it might result in overheating and cause the cuticula to reach its "critical melting point" (that of the wax layer), which results in a dramatic increase in water loss. There is a behavioral safeguard against this hazard; I have sometimes observed an alpine all-black *Erebia* (such as *E. lefebvrei* or *E. pluto*) die in a few minutes from desiccation after being laid anesthesized on the very rock on which it had rested previously.

If the ecophysiological factors that act for or against melanism in satyrids are complex, many convergences between oreal and montane species from very different regions may nonetheless be observed. The resemblance of some South American genera (*Cosmosatyrus*, *Faunula*, *Manerebia*, *Punapedaliodes*) to *Erebia* is quite obvious (de Lesse, 1966). This well-known Holarctic genus, which possesses many species, all melanic, also has a very wide ecological range within its mainly montane habitat. Convergence also may be observed with some alpine species of New Zealand, such as *Percnodaimon pluto* (long confused with *Erebia pluto* from the Alps) and *Erebiola butleri*. These latter species are considered by Miller (1968) as belonging to the tribe Hypocystini of the Satyrinae, not too close to either the Pronophilini or the Erebiini.

In lycaenids, *Itylos*, *Parachilades*, *Paralycaeides*, and "*Callophrys*" *culminicola* are indeed melanic. Only the ventral face is involved, a fact that is likely to be in relation with their behavior.

C. culminicola is fond of resting on sun-drenched rocks, its wings closed, and may even occasionally lie down on one side (F. Michel, pers. comm.), in convergence with some Palearctic species (*Tomares mauretanicus*, Barragué, 1954; *Agriades pyrenaica*, Descimon, unpublished). This melanism cannot, however, be assigned to mere thermoregulation, since it usually consists in some shades of gray with broken lines and rocklike stereomorphic mottled patterns that strongly suggest a concealing function, as does the intense polymorphism that affects this coloration.

Camouflage patterns

In montane forests, an impressive number of butterflies bear very striking leaflike designs on their ventral face, exposed at rest. Both coloration and shape are often associated, enhancing the effect of a leaf with its ribs, stalk, mold spots, holes, and water droplets; a remarkable example has been studied in detail by Schwanwitsch (1940). In cloud forest, colors are often rusty, with silvery spots, thus providing the insects with the very appearance of the rotten leaves seen in abundance in this extremely wet environment (Fassl, 1913). As soon as one enters open, high-altitude biomes, such patterns disappear suddenly, to be replaced in some species either by grass-tuft-looking patterns or by mottled, often stereomorphic, grayish designs. This is, of course, in correlation with resting sites: grassy vegetation or barren rocks. Both types of graphism also have been observed in Asian steppe butterflies by Schwanwitsch (1943, 1946). "Jack-of-all-trades" patterns, hard to describe but extremely efficient for concealment (at least from my own butterfly hunter experience!), are also met with. In Pieridae, *Tatochila* and *Phulia* illustrate the grass-blades theme with green or brownish streaks along the veins, which consist in a "pointillist" mixture of yellow scales (from pteridines) and bluish black scales (from melanin and thin layers). This is another element of resemblance with Northern Hemisphere species such as *Pieris napi* and *Synchloe callidice*. *Piercolias* are quite different; their underside recalls, obviously by pure convergence, that of a *Colias*; since they live in places from which vegetation is nearly absent, grass homotypy would have no significance here. In Satyridae, *Punargentus lamna* exhibits a striped pattern of brownish and silver that is quite efficient for concealment in ambient vegetation. More generally, most satyrids display mottled graphisms with stereomorphic effects on a melanic background, quite comparable to those described by Schwanwitsch (1943) in rock-resting Palearctic species; these are very good imitations of the surface of a stone with its reliefs. Deflectory eyespots may also be superimposed on these designs. In northern Peru, in puna vegetation near Abra Porculla, I observed birds (unidentified passerines, somewhat larklike) attack, and even catch satyrids, especially *Punargentus lamna*. This observation leaves no doubt about the functional significance of these protective devices.

Polymorphism

Polymorphism in female *Colias* should be mentioned here, since it is especially striking in the Andes (Hovanitz, 1945b; Descimon, 1980a). To summarize this very complex and not yet completely understood problem, most *Colias* species are usually vividly colored (yellow or orange) by pteridines. The females, however, are dimorphic, a variable proportion of them being whitish and called "*Alba*." This character is determined by an autosomal, sex-conditioned dominant gene. The whitish phenotype is merely due to a decreased synthesis of pteridines (Descimon, 1966; Watt, 1973). The metabolic significance of this point is complex. In light of the latest available results (Descimon, 1979 and to be published), white and colored females appear to diverge early in the metabolic chains. In "*Alba*," glycin (the initial precursor common to pteridines and purines) is directed toward protein synthesis rather than toward heterocycles as in "colored" females. Graham, Watt, and Gall (1980) have shown that the females of the former phenotype retain more larva-derived resources for somatic maintenance and for reproduction and, under certain conditions, mature their eggs faster than do "colored" females. On the contrary, they attract the males less and mate less frequently (Graham et al., 1980). It may be added that the frequency of both morphs varies largely from year to year; "*Alba*" is disadvantaged under warm conditions and during phases of high population density, which may be related to a higher sensitivity to viral diseases (Descimon, 1980).

In Andean *Colias*, most species possess the "*Alba*" morph, in particular *C. dimera*, *C. lesbia*, *C. vauthieri*, and *C. euxanthe*. In *C. dimera* (and, to a lesser extent, *C. lesbia dinora*), Hovanitz (1945b) has described a cline of "*Alba*" frequency, positively correlated with altitude; in the same species, breeding experiments have shown that "*Alba*" is severely handicapped at high temperatures (Descimon, 1980a).

These data are very interesting from an evolutionary point of view; they show that the same polymorphism, highly characteristic of *Colias* (Remington, 1954), has been conserved during the colonization of South America and the subsequent adaptive radiation that has given rise to the various species, almost all dimorphic. Of further interest is that the only nondimorphic, all-"*Alba*" species, *C. flaveola*, has also a whitish male and assumes a striking convergence toward the Holarctic group of *C. phicomone*, which is monomorphic in the same way.

Physiology of Life-Cycle Regulation

Under temperate climates, most butterflies overwinter in a metabolically resting form (involving diapause or at least arrested growth). The physiology of diapause and of its triggering is extremely well known in insects. This reaction is very generally released by the photoperiod and, to a lesser degree, by temperature. In the tropics, the evidence for a diapausing stage is rather meager (Shapiro, 1978). However, there are some indications that many tropical Andean butterflies emerge according to well-defined broods (Forster, 1955; Shapiro and Torres, 1978; Descimon, unpublished observations). In *Reliquia santamarta*, for example, there are three broods, two during the "long dry season" and one during the "short dry season" of the Sierra Nevada de Santa Marta (Shapiro and Torres, 1978). Most species of the Peruvian puna fly only during the dry season (July-November) in one or several broods; this is the case, for instance, in *Punargentus lamna* (F. Koenig, pers. comm.). In Bolivia, Forster (1962) states that "a whole series of species, especially those that live in the high valleys of the Cordillera, during the cold dry period fly in the sunshine low over the ground in protected places. During the summer rainy season insect life at these heights is presumably nearly impossible because of the heavy snowfall." In the vicinity of Quito, on the other hand, *C. dimera* seems to have broods all year round (Descimon, unpublished observation), which might be accounted for by the very weak seasonality of this equatorial region.

Even in temperate regions species such as *Colias croceus* (western Europe) and *C. eurytheme* (United States) lack diapause and merely grow very slowly under cool conditions—and die under cold ones; they originate from regions with a mediterranean climate. It is not surprising therefore that a certain diversity should be met with in tropical highland butterflies also.

The little we know about the physiology of life-cycle regulation and about the polyphenism it governs is already very significant; all the results we have in hand are due to a long series of experiments undertaken by Shapiro (1975, 1976, 1977a, b, 1978, 1979a, b, 1980a, b) and concern exclusively the family Pierini. The members of this group that live in temperate Holarctic regions are generally plurivoltine with a conditional, photoperiodically released diapause. A marked polyphenism is associated with this plurivoltinism. Some univoltine species are able to disclose latent reactions of plurivoltinism and polyphenism when submitted to suitable artificial rearing conditions (Shapiro, 1975, 1976). The picture is quite different in Andean Pierini. Some species of *Tatochila* from South American temperate regions (*T. vanvolxemii*, *T. mercedis*) present a clearcut polyphenism, but recent experiments by Shapiro (1980a) have shown that its determinism is not the same as that of Holarctic Pierini. In tropical species (*T. xanthodice*, *R. santamarta*) there is no polyphenism at all (Shapiro, 1978). The most obvious explanation of these facts is that Neotropical Pierini originated from a monophenic ancestor; the seasonal variation observed in species from southern South America would result from convergence (Shapiro, 1980a). Interesting elements of comparison are provided by the Palearctic alpine species *Synchloe callidice* (Descimon, unpublished results). It is normally single-brooded but, in some localities of the southern Alps, a partial, occasionally abundant second brood may be observed. It does not present any marked polyphenism, a fact that has been confirmed by rearing it under controlled conditions. So, in this case, the situation is similar to that observed in Andean Pierini. The phylogenetic implications of these data will be discussed later.

Physiology of Reproduction

It may be mentioned that *Infraphulia ilyodes* from Chile is reputedly viviparous (Field and Herrera, 1977); it would be interesting to know whether the phenomenon extends to other species.

Behavioral Adaptations

Flight

The main adverse factor in the high Andes might be strong winds, as Forster (1955) observed on the altiplano of Bolivia. I have observed butterflies flying against a strong wind on the páramo of Cumbe (southern Ecuador) in January, and on the altiplano near Putina (southern Peru) in August. Also, it must be stressed that butterflies are able to fly normally only when the sun is out (hence the importance of melanism). Most Andean Rhopalocera are strong fliers, especially *Hypsochila* (Ureta, 1955) and *Infraphulia illimani* (Field and Herrera, 1977), but they remain generally close to the ground. However, *Pierphulia nysias* (Garlepp, 1892) and *Infraphulia ilyodes ilyodes* (Field and Herrera, 1977) have a low and weak flight, remain in sheltered places, and enjoy the rare moments when the winds abate somewhat. It seems thus that at least two flight strategies may be used. Shapiro (1979b) has described the striking "hilltopping" behavior of *Reliquia santamarta*: the males of this species go up and down on the slopes of the peaks in a constant repetitive pattern; the purpose of this behavior is apparently to search for virgin females (after mating, the latter go down to more sheltered sites for egg laying). Such behavior is often met with in mountain butterflies (Shields, 1967; Scott, 1970). *Synchloe callidice*, the related Palearctic species, behaves in the same manner, as reported as early as 1932 by Rondou for Pyrenean populations: *S. callidice* "soars up to the highest summits from which it disappears to dash down into the chasms and whirl into the precipices." *R. santamarta* also displays an extreme sensitivity to the weakest sunbeam. Precisely the same behavior has been described by Petersen (1954) in the montane species *Pieris bryoniae* from Sweden, another country where sun is scarce.

When compared to Pieridae, Satyridae do not appear really weak, but rather episodic fliers: *Punapedaliodes albomaculata* and *Punargentus lamna*, for instance, spend most of their time at rest and fly only for short periods, when the wind is less fierce, or after disturbance or for courtship. Moreover, their skipping flight is highly characteristic of this family; *Lymanopoda* species that mimic *Tatochila* may easily be recognized thanks to this detail. A very important ecoethological feature peculiar to some satyrids is that they are almost the only Rhopalocera to fly in dull weather, when the sun is veiled; this is especially true of montane Pronophilini, which sometimes may have an almost normal activity under these conditions. Such an adaptation is obviously favorable in cloud forest and might also account for this group's being well represented in humid páramo.

Many species of Pieridae, Satyridae, and Lycaenidae, although they "dislike" excessive wind, use it to escape aggressors. In doing so, they face the wind direction, but do not fight it and use its speed, controlling only the height of their flight; they will land by going suddenly down to the ground, after a "jump" from 10 to 50 m, or even more. They share such a behavior with European alpine Lepidoptera. It appears to be both energy saving and a very efficient means of escaping predation (especially by the collector!).

Choice of Habitat and Resting Habits

The choice of its surroundings is usually characteristic of a species. In the puna biome, most species favor boulders and rocks, which offer a "heat oasis in a cold desert" (Weberbauer, 1945); marshy basins, whose vegetation provides some protection against wind, are also frequented, generally by other species. I have also observed, in the Cordillera de Mérida, that *Polylepis sericea* sparse woods showed no faunistic originality when compared to the neighboring vegetation, but a greater wealth of individuals; the same fact has been observed by F. Michel (pers. comm.) in the Cordillera Blanca in Peru in woods of other species of *Polylepis*. Large areas are often crossed without seeing any butterflies, and the reasons for this are not always obvious.

Most resting habits may be interpreted in terms of thermoregulation and concealment. Clench (1966) has attempted to analyze heating behavior in butterflies, by distinguishing two possible types of orientation with respect to the sun: "lateral heating," in which one of the ventral sides of the wings is exposed to the sun, and "dorsal heating," in which the dorsal face of the four open wings is exposed. Although some aspects of Clench's observations have been criticized by Watt (1969), this distinction seems relevant in most cases. A behavior implying both positions successively has been described in *Reliquia* by Shapiro and Torres (1978) and Shapiro (1979b); dorsal heating is mentioned as the usual behavior in *Phulia* by Field and Herrera (1977) and has indeed often been observed by the author and F. Michel. Both postures may also be noted in satyrids, though lateral heating is more frequent.

Resting places are not always the same for males and females, as is the case, for instance, in *Reliquia*, the females of which dwell several hundred meters below the males (Shapiro, 1979b). In *Punargentus lamna*, females stay mainly in the grass and males on the rocks. This is obviously related to their reproductive behavior: satyrids are Gramineae feeders and females remain close to their egg-laying place, which provides them also with shelter and concealment; males, much more active, search for the hottest places, which results in an elevation of their body temperature. Sometimes, the butterflies settle on the rocks during the day but hide in the vegetation at night; some also find shelter under stones or boulders—a fact observed also in scree-haunting *Erebia*, such as *E. lefebvrei* (Pyrenees), *E. pluto*, and *E. scipio* (Alps).

Generally speaking, all the aforementioned features are highly convergent with those observed in species from European, Asiatic, North American, and New Zealand mountains.

FAUNAL STRATIFICATION ACCORDING TO ELEVATION

It is interesting to compare the butterfly faunas of puna-páramo biomes with those of arboreal and nonforest (sensu Müller, 1973) communities. At the family-subfamily level and on a qualitative basis, the comparison is very easy and the results are obvious. From Table 20–1, it appears that seven families (Papilionidae, Danaidae, Ithomiidae, Morphidae, Acraeidae, Libythaeidae, Riodinidae, and even Heliconiidae are only marginally present) and six subfamilies (Dismorphiinae, Brassolinae, Haeterinae, Biinae, Apaturinae, Charaxidinae) are completely missing in these biomes, although most of them are very abundant in lowland forest. At this taxonomic level, lowland and montane forests have strong affinities, whereas nonforest lowland and oreal biomes stand closer together than the oreal and montane forest biomes. What these rather surprising results point to is the existence of a discontinuity between the faunas of forest and open landscapes, a discontinuity that probably reflects a deep divergence in adaptive strategies that occurred early in the process of evolution of butterflies (note that similar observations were made by Moreau [1966] on birds). Moreover, families

TABLE 20–1. Relative abundance of butterfly species at the family-subfamily level in the major biomes of the Neotropical region.

	Lowland forest	Montane forest	Nonforest	Oreal biome
Papilionidae	+++	+	++	
Pieridae				
Dismorphiinae [a]	+	+++	+	
Coliadinae	++	++	++	++
Pierinae	+	+++	++	+++
Danaidae	++	++	+	
Ithomiidae	+++	+++	(+)	
Satyridae				
Brassolinae [b]	+++	++		
Haeterinae [b]	+++	+		
Biinae	++	+		
Satyrinae	++	+++	+	++
Morphidae [b]	+++	++	(+)	
Acraeidae	++	+++		
Heliconiidae [b]	+++	+++	+	(+)
Nymphalidae				
Apaturinae	+++	++		
Charaxidinae	+++	+++	+	
Nymphalinae	+++	+++	++	+
Libythaeidae	+	+		
Lycaenidae				
Theclinae	+++	++	+	++
Polyommatinae	+	++	+	+++
Riodinidae [a]	+++	++	+	

+, represented by few taxa; ++, numerous species; +++, very numerous species; (+), marginal occurrence.

[a] (Sub)family almost restricted to the Neotropics.

[b] (Sub)family endemic to the Neotropics.

TABLE 20–2. Number of butterfly species according to elevation and ecological zonation in the Andes. A. Southern Peruvian Andes east to Cuzco. B. Colombian Cordillera Central.

	Ecological zonation and elevation					
A[a]	Submacrothermic rain forest (500–1000 m)	Lower montane rain forest (1000–1750 m)	Upper montane rain forest (1750–2500 m)	Mountane cloud forest (2500–3500 m)	Lower puna (3500–4000 m)	Upper puna (4000–4500 m)
Papilionidae	24	7	1			
Pieridae	22	14	5	1	3	5
Danaidae	5	1				
Ithomiidae	12[b]	2[b]	1			
Satyridae	25	12	16	26	1	1
Morphidae	6	1		1		
Heliconiidae	8	2	2	1	(1)	
Acraeidae	2	3	2	1		
Nymphalidae	46	28	14	3	1	
Charaxidae	19	4	1	1		
Riodinidae	15[b]	3				
Lycaenidae	8[b]	3[b]	5	2	1	4
Total	192	80	47	36	7	10

	Elevation							
B[c]	500–1000 m	1000–1500 m	1500–2000 m	2000–2500 m	2500–3000 m	3000–3500 m	3500–4000 m	4000–4500 m
Papilionidae	2	5	3					
Pieridae	9	15	17	2	1	6	5	2
Danaidae	4	4	3					
Ithomiidae	1	3	4	1				
Satyridae	6	16	11	21	22	15	11	3
Morphidae		1			1			
Heliconiidae	7	8	5	2	1	1		
Acraeidae	1	5	5	2				
Nymphalidae	27	30	28	7	3	3		
Charaxidae	5	3	7	1				
Riodinidae	1	4	10					
Lycaenidae	1	2	4	2	1	1	1	
Total	64	96	97	38	29	26	17	5

[a] Data pooled from two transects: Ollantaytambo-Quillabamba and Urcos-Paucartambo-Kosnipata. Unpublished data from A. and H. Descimon and J. Mast de Maeght (1971, 1973).

[b] Species number is probably strongly underestimated.

[c] Because of the lack of precise ecological data, species numbers are given according to elevation only. Data from Fassl (1911).

and subfamilies that are restricted to, or best developed in, the Neotropics are missing in the upper Andes.

Unfortunately, our incomplete knowledge of the systematics of many groups and the incredible diversity of lowland faunas (Ebert, 1969) make it very difficult to paint an exact, quantitative picture of the fauna of lowland regions, and leads to severe underestimation. This is especially true of Theclinae (which are canopy dwellers) and Riodinidae (which are extremely localized and have a very whimsical flight period).

Only two typical transects will be considered here, one in the northern Andean region (Fassl, 1911, reinterpreted) and another in the puna region (H. Descimon, J. Mast de Maeght, unpublished data) (Table 20–2). It is clear that lower montane (rain)forest possesses the most diverse fauna; in upper montane (cloud)forest, Pronophilini become overwhelmingly abundant. All this diversity (and, often, abundance) vanishes suddenly on entering the oreal biome.

The only work to offer modern ecological data is the very careful study by Adams (1973) of the butterflies of the Sierra Nevada de Santa Marta. This region, which displays a unique variety of biomes and habitats for so small an area (roughly 10,000 km^2), is exceptionally favorable to a comprehensive study. Eight types of habitats, ranging from sea level up to 5000 m were investigated. Páramo is by far the poorest zone as far as Rhopalocera are concerned (six species). By all methods of numerical analysis used (Jaccard's, Cole's, and Cook's coefficients), it stands far apart from other habitats; the only species common to this zone and some others are migrants and vagrants. In this particular region, the richest zone was found to be low montane gallery forest (214 species), followed by tropical evergreen dry forest (206 species) and montane gallery forest (178 species).

More generally, it should be stressed that almost no species of the immediately underlying biomes (montane cloud forest generally) enter the oreal biome. This point has behavioral implications; at the limit of the two ecosystems, butterflies that have ventured a few meters into the open landscape are often observed flying headlong back.

Even inside the oreal biome, stratification is often clear-cut. Most *Tatochila* and some *Colias* (*C. dimera*, *C. lesbia*) fly in the lowest parts, and even go down to open cultivated areas in zones that actually lie at montane forest elevations. Members of the genus *Teriocolias* also fly at low altitudes, in hot and dry places. At medium heights, around 4000 m, most of the fauna is present, whereas from 4500 m upward, it is essentially pierids of the genera *Phulia*, *Pierphulia*, and *Piercolias* that are observed.

AREAL PATTERNS

Distributional data are scattered in numerous publications of unequal value (Weymer and Maassen, 1891; Godman and Salvin, 1891; Garlepp, 1892; Staudinger, 1894; Fassl, 1910, 1911, 1914a, b, 1915a, 1918, 1920; Dyar, 1913; Campos, 1926; Ureta, 1941, 1955; Zischka, 1947; Forster, 1955, 1958; Schröder, 1955; Hughes, 1956, 1958; de Lesse, 1966; Hayward, 1967, 1973; Heimlich, 1972; Adams, 1973). Some additional data were collected by the author, his wife, and J. Mast de Maeght during travels in Peru in July-August 1971 and 1973 and by the author alone in Ecuador (January 1970 and 1975) and Venezuela (December 1977). F. Michel kindly communicated part of the results of his collectings in Peru in July-August 1980. Some serious gaps remain in our knowledge of the ranges of many species. It is a pity that no report is available for most parts of the páramo islands of Colombia, for instance in the Nevado del Cocuy region. Also, as has already been stated, only the Pierini and (in part), the Satyridae have yet been subjected to a thorough taxonomic revision. However, and keeping in mind that some omissions and (hopefully) minor errors are likely to be present in the data collected, the information available is already sufficient to draw up a comprehensive picture of the overall distribution of butterflies in the high Andes (for a better comprehension of dispersal patterns, data about the southern temperate region must of course also be included).

From Table 20–3 and Figs. 20–2 to 20–4, it appears clear that the division between a "North Andean Center" and a "Puna Center" (to follow the terminology of Müller, 1973) applies just as well to butterflies. Of these two regions, the puna is by far the richest. In the Pieridae, the whole genus *Phulia*, as well as *Pierphulia*, *Hypsochila*, *Infraphulia*, and *Piercolias* are restricted to this center and also to the very rich and clearly related Patagonian center (Figs. 20–3, 20–4). In *Tatochila*, only three species (out of more than 14) penetrate in the North Andean Center; two of them occur in the form of "peripheral" subspecies, which is an indication of recent arrival (Fig. 20–2). Only *Reliquia santamarta* is peculiar to the fairly isolated Sierra Nevada de Santa Marta, hardly a dependency on the main nucleus of páramos (Fig. 20–4). In *Colias*, the southern

TABLE 20–3. Distribution patterns of butterflies in the oreal zone along the Andes.

Species	SNSM	CM	Col	Ec	N-Pe	C-Pe	S-Pe	Bol	Arg	Chi	Pat	TdF	S-TdF
Tatochila													
theodice t.									+	+			
t. gymnodice										+	+	+	
t. staudingeri													+
autodice								+	+				
blanchardii b.										+			
b. ernestae						+	+	+	+	+			
sterodice s.									+	+	+	+	
s. fueguensis													+
s. macrodice							+	+	+	+			
s. arctodice			+	+	+	+							
vanvolxemi									+				
mercedis									+	+			
inversa							+		+				
homeodice						+							
orthodice							+	+	+				
sagittata			+	+	+	+							
stigmadice								+	+				
pyrrhoma						+	+	+					
xanthodice x.		+	+	+	+	+							
x. paramosa	+												
distincta						+	+		+				
Hypsochila													
argyrodice										+	+	+	+
microdice											+		
huemul										+	+		
galactodice										+			
wagenknechti w.									+	+			
w. sulphurodice							+	+	+	+			
penai										+			
Phulia													
nymphula n.						+	+	+	+	+			
n. nympha								+					
paranympha p.								+					
p. ernesta								+					
nannophyes							+						
garleppi						+	+	+					

TABLE 20–3 CONTINUED.

Species	SNSM	CM	Col	Ec	N-Pe	C-Pe	S-Pe	Bol	Arg	Chi	Pat	TdF	S-TdF
Infraphulia													
illimani							+	+					
madeleina						+	+	+					
ilyodes										+			
Pierphulia													
nysias n.								+					
n. nysiella								+					
rosea r.										+			
r. maria							+						
r. annamariea						+	+	+					
isabella										+			
Piercolias													
huanaco								+					
forsteri								+					
coropunae							+						
Colias													
dimera		+	+	+	+								
lesbia											+		
l. pyrrhothea								+	+	+			
l. andina						+	+	+					
l. dinora				+	+								
vauthieri								+	+	+			
v. cunninghami											+		
ponteni													+?
euxanthe						+	+	+					
e. stuebeli					+								
e. coeneni					+								
e. alticola				+	+								
flaveola										+			
f. erika							+						
f. blameyi									+				
f. weberbaueri						+	+	+					
f. mossi					+								
Paramo oculata	+												
Pedaliodes empetrus		+											

Punapedaliodes albopunctata				+	+	+	+	+					
Punargentus lamna						+	+	+					
Yramea													
sobrina							+	+	+				
modesta									+	+			
lathonioides									+	+	+	+	
cytheris									+	+		+	+
'Thecla													
loxurina			+	+	+	+	+	+					
dissentanea							+						
culminicola			+	+	+	+	+	+					
alatus					+								
Leptotes													
andicola				+									
callanga							+						
Itylos													
moza								+					
pelorias								+					
pacis							+						
koa							+	+					
vapa							+	+					
ludicra								+					
speciosa							+	+					
Paralycaeides													
inconspicua							+						
Parachilades													
titicaca							+	+					
Total species number	2	3	6	10	13	18	32	35	21	23	8	5	5

Note: Abbreviations: SNSM, Sierra Nevada de Santa Marta; CM, Cordillera de Mérida; Col, Colombian Cordilleras; Ec, Ecuadorian Andes; N-Pe, northern Peru north to Abra de Porculla and south to Callejón de Huaylas; C-Pe, central Peru from Cordilleras Negra and Blanca to left side of Rio Apur ímac; S-Pe, southern Peru; Bol, Bolivia; Arg, Andean region of Argentina; Chi, northern and central Chile; Pat, Patagonia and southern Chile; TdF, Tierra del Fuego (north); S-TdF, southern Tierra del Fuego. N. B. The list may be considered complete for Pieridae, but is surely not so for other families.

region of the northern Andean center, especially Ecuador, is also the richest in species (Fig. 20–5). The genus is represented by a single species (*C. dimera*) in the Cordillera de Mérida, and has not been found in the Sierra Nevada de Santa Marta. Thus we observe a striking parallelism between the distribution of butterflies and that of birds, as analyzed in the pioneer work of Vuilleumier (1970).

The Satyridae present a different pattern: broadly speaking and based on the scanty data we have in hand, in the North Andean Center, one or two highly endemic species occur per massif; in the Puna Center, there are two widespread species, *Punapedaliodes albomaculata* (if a single species), entering the Ecuadorian region, and *Punargentus lamna*, stopping more to the south (Fig. 20–6). The southern region is also the richest in lycaenids and includes the only characteristic representative of the Nymphalidae (*Yramea*).

The affinities of the Puna Center with temperate montane and even lowland regions of Chile and Argentina are obvious, especially in Pieridae, where there is a definite faunal continuity.

No cases of clearcut vicariance beyond the subspecific level (e.g., *Tatochila sterodice*) may be observed between the two high Andean centers. Only *Tatochila sagittata* and *T. xanthodice* represent an area that is centered around northern Peru and overlaps both southern and northern tips of the two regions.

Both centers appear to be nonhomologous and their faunistic differences may not result from simple disjunction phenomena. The most conspicuous pattern of species distribution is a system of patches of various areas but definitely centered around the southern Peru–Bolivia region. From this true "dispersion center" arose peripheral subspeciation toward both north and south, such as in *Tatochila sterodice*. Between the Puna and Patagonian Centers, more complex phenomena are to be observed and there is often a "chain" of more or less overlapping species that suggest segregations and allopatric speciation, especially in the *Phulia* complex of genera.

At the level of resolution available here, it is not possible to discern the isolating effect of geographic barriers such as that analyzed by Vuilleumier (1977; 1984) in Andean birds. Even in the most distinctive barrier mentioned by Vuilleumier (1977, 1984), the northern Peruvian depression (Abra Porculla), we have observed (Descimon and Mast de Maeght, unpublished data) that many faunal elements that are characteristic of Ecuador (e.g., *Colias dimera*) still fly in the Chachapoyas Massif, southeast of the Marañon. Many puna elements reach the Callejón de Huaylas Mountains, but not the region of Cajamarca and here there is no obvious geographic cut. On the contrary, the puna species *Punapedaliodes albopunctata* goes north to central Ecuador and stops there. Now, the Chachapoyas-Balsas Massif has a physiognomy of páramo, the Ecuadorian Cumbe Plateau a puna one. In these cases, the relative importance of ecological and historico-geographic factors remains to be determined.

In any case, the northern Andean region actually does not appear, as far as butterflies are concerned, to be a dispersal center (sensu Müller, 1973), but rather a cluster of scattered areas.

DISCUSSION

Around 65–70 species of Lepidoptera Rhopalocera are residents of the oreal ecosystems (páramo, puna) in the Andes. Very likely, further studies of the Satyridae and Lycaenidae will increase slightly the number of specific taxa, but it seems improbable, however, that this number will ever reach 100. Moreover, these species belong to a few groups only; the family Pieridae possesses nearly half this total. Is such a low number of taxa due to the harshness of climatic conditions that prevail in the temperate as well as in the tropical Andes? It must be remarked that the only great range comparable to the Andes, the Himalaya, is far richer and presents an impressive explosion of a great many lepidopteran taxa.

The impression—subjective, of course—that is felt by a naturalist looking at the rhopaloceran fauna of the Andes is one of "unsaturation": many ecological niches appear "empty", in particular many foodplants remain without insects (but I must say that very few, if any, "generalists" penetrate in these biomes). Many times, wandering in the Great Andes, I stopped to look at a peculiar-looking biotope, in which I guessed there "surely" were special—and interesting, perhaps new!—butterflies. And there were none.

Such a feeling is obviously evocative of oceanic islands. However, the species richness of such a huge "island" as the high Andes is so low that one may be allowed to think that it has not yet reached an equilibrium. This non-equilibrium state would be easy to explain by considering the great isolation and the very young age of this range. This point will be discussed thoroughly later in this chapter. Here it is relevant to say that I do not have complete inventories of species, even in the better-known family Pieridae, especially in some important areas of Colombia. Thus the data that would allow one to test the hypothesis of equilibrium by an application of the MacArthur and Wilson (1967) model are not yet available. This is clearly a topic for further research.

We may set the problem in terms of evolution: the uplift of the Andes has created the possibility of a new, original ecosystem, or series of ecosystems, where there is a place, as almost everywhere, for butterflies. Adaptation to this biome requires, as we have seen, a considerable ecophysiological effort. The taxa that may contribute colonists to this new island (or even continent) must be located in three regions: (1) tropical South America (possibly also having affinities with African, Holarctic, or tropical Indo-Asiatic faunas); (2) the southern temperate tip of South America (this origin will be pointed out by affinities with temperate South Africa, Australia, and New Zealand, such as with plants like *Nothofagus*, Proteaceae, Podocarpaceae); (3) the Holarctic, naturally through North America.

The first type of origin is that most obviously represented by the lycaenids of the genus *Leptotes* and of the "*Thecla*" *loxurina* group. The heliconine *Dione moneta*, an occasional dweller of the oreal biomes, may also be reckoned among this group and, in disagreement with Forster (1955), I think that *Teriocolias* species also belong to it. But the Pronophilini are by far the most apt candidates for such a colonization in satyrids (and even among Neotropical butterflies in general), since they are localized exclusively in the zone lying immediately below montane forest. Nonetheless, the number of species having become adapted to open upper montane biomes is quite low. Two types of taxa adapted to oreal habitats seem to have arisen: (1) in the northern Andean region, isolated monotypic genera, from a different strain in each massif. Rather likely, this results from local evolution. We may point out that the páramo, wet and often with rather high vegetation, resembles montane forest more than does the puna. (2) In this latter community, only two widespread species occur and they are likely to be derived from the more southern temperate fauna; here, a more complex turn seems to have been taken, since the Pronophilini from temperate South America seem to be derived from a formerly tropical stock. At least two steps are thus involved and may have rendered the adaptive performance less difficult.

We have already seen that the affinities of the Pronophilini are Holarctic. But do the Satyrini come from South America or Pronophilini from the Holarctic? The question is not as devoid of sense as one might think, since Miller (1968) considers, with some support, that South America has played a major role in the evolution of satyrids: the most primitive subfamilies of this family, Brassolinae and Haeterinae, are endemic to the Neotropics and are tropical forest dwellers. Thus the family arose very likely in "Ameraustralia" at the end of the Cretaceous; according to Miller (1968), the stem of the Satyrinae differentiated during the Early Tertiary in Eurasia and reinvaded the Neotropics during the Eocene or Miocene. This scheme is very likely, but Miller's genealogical tree of Satyrids is not a true cladogram; it is mainly based on now outdated paleogeographic data. It would be desirable to apply here the method of Rosen (1978), using congruence between taxonomic and geologic cladograms. Perhaps then the statement of Adams and Bernard (1977) that Pronophilini are basically a Neotropical group would be confirmed. Nonetheless, we will consider for the time being that these butterflies are of medium old Holarctic origin.

Why did so few neotropical taxa invade the oreal biome? It seems that adaptation of Lepidoptera to warm tropical forest involves ecophysiological features that are contradictory to subsequent adaptation to cold, dry, windy conditions. We have seen that, behaviorally, even the Pronophilini from montane forest actively shun the open landscapes of páramo and puna; this behavioral reaction is perhaps not the least factor precluding from the start an adaptation to the oreal biome. The affinity, on the family-subfamily level, that has been mentioned between semiarid and puna-páramo biomes, has a similar significance: the change of ecophysiological strategy is a hard evolutionary step.

The second type of origin, "Gondwanian" or "*Nothofagus*-like," is very easy to deal with: it is nonexistent. No Andean butterflies reveal an affinity with Australian or New Zealand groups. Furthermore, the depauperate lepidopteran fauna of South American *Nothofagus* forests, as well as of the Patagonian steppe, is devoid of such faunistic elements. Forster (1955) cites as belonging to such a biogeographic group the pierid genera *Teriocolias* and *Mathania*. However, the former genus is very closely related to *Eurema*, which is mainly linked to semiarid biomes and whose center of origin is, to my mind, the southern tropical tip of North America. Thus *Teriocolias* may rather be reckoned as a Neotropical pierid stock adapting, very partially, to the highlands, since it is rather a dry montane landscape dweller than a puna or páramo inhabitant.

The Holarctic stock is the overwhelming faunistic element in the high Andes; it is the richest in species, and the one which has undergone the most striking adaptive radiation, and this very readily, as if it had found its perfect home in these rugged barren landscapes. Pierini, *Colias*, and, in some likelihood, the *Itylos* complex and the genus *Yramea* derive from Holarctic taxa. This assemblage or "coenocron" (Reig, 1968) represents up to 70% of the Andean butterfly fauna, counting the Pronophilini as Neotropical in spite of their undoubtedly Holarctic affinities. However, I must state that many taxa from the Northern Hemisphere montane fauna are absent; to give but a striking example, there are no Parnassiinae in the Andes, a fact that every lepidopterist painfully resents!

If the origin of the taxa is clear, everything else about them is not, as we will see. The Pierini, being the best studied group, will be used as an example in the following biogeographic analysis. The main problem is, how and when did this tribe enter the Andes? Biogeography has always been a very controversial field. The practice of basing paleogeographic assumptions on distributional patterns is not yet lost. For instance, such a keen and serious ecologist and taxonomist as Adams (1973) has dug out the old hypothesis of Todd and Carriker (1922), claiming the existence of an ancient high mountain range to explain the butterfly fauna of the Sierra Nevada de Santa Marta. Now, such an hypothesis has been authoritatively ruled out, especially, in a field closely related to ours, by the geologist–ornithologist Haffer (1974). I try here to consider geologic and paleogeographic data as the "independent variable" and to use as much as possible the rule of parsimony (Rosen, 1978).

Perhaps the two touchstones of all biogeographic explanations are the tempo of evolution (especially in groups without fossils) and dispersal modes. For the first point, nearly all controversies may be reduced to "tertiarist," or even "secondarist," as opposed to "quaternarist," hypotheses. For the second, the main opposition lies between the upholders of continuous dispersal and the supporters of chance long-distance dispersal.

The origin of Lepidoptera is likely to be ancient enough (Shields, 1976). The Rhopalocera appeared probably during the Cretaceous (Shields and Dvorak, 1979) and family differentiation arose at the end of that epoch (Zeuner, 1962; Nekrutenko, 1965) the divergence of subfamilies and tribes occurred mainly during the Early and mid-Tertiary. At the genus and species level, the speed of differentiation is likely to be more irregular and to depend largely on the opportunities for speciation and adaptive radiation, or for phyletic evolution. Thus it is conceivable that specific and even generic differentiations took place during the Pleistocene, but they may also be older.

Dispersal of butterflies is highly variable. It has been demonstrated that, depending on species, the populations may be either exceedingly sedentary or quite vagile (Ehrlich and Raven, 1969). However, field experiments have shown that an "empty" region may be colonized by strays, even only one single fertilized female (Descimon, 1976). It is obvious that long-distance transport by wind is severely limited by the relative resistance of butterflies to starvation and water loss. Successful high-altitude transport will meet with a behavioral obstacle, since under these conditions butterflies fly restlessly until they are exhausted. Johnson (1969) has stressed that arctic species, adapted to live in very precarious and hard habitats, are probably most apt to resist the hard trial of long-distance travel. Moreover, the migratory tendencies of some *Colias* are well known; Darwin even observed an enormous swarm of *C. lesbia* off the Argentine coast up to 200 m in the air and coming with a storm. The importance of migration in the distribution of some *Pieris* in California has also been demonstrated by Shapiro (1974). However, the extreme poverty in Lepidoptera of remote oceanic islands suggests that, beyond a distance of the order of 1000 km, the survival of a healthy, able-to-lay female propagule is improbable.

As a paleogeographic scheme, I will use here the simplest and most austere pattern: whatever the Tertiary or Secondary history of South America, the final uplift of the Andes was very late and it is most unlikely that areas reaching the altitude of the oreal biomes existed much before the Pleistocene (Gansser, 1973; James, 1973; Haffer, 1974; see also van der Hammen and Cleef, Chap. 7, and Hoffstetter, Chap. 9, this volume). The contrary has been argued by Todd and Carriker (1922) for the northern Andes and by Croizat (1958) for the puna region, but these assumptions have received absolutely no geologic support. There is no evidence for a continuous mountain ridge having existed in Central America at any epoch. "All of the Andean Cordillera . . . is very young; in fact, it was raised above sea level only after the end of the Cretaceous and elevations above 2000 m were achieved only within the last 2 to 5 million years. The faunas and floras of these high elevations could not have migrated into high elevation habitats nor begun to differentiate until the later Pliocene or Pleistocene" (Simpson, 1979).

The only factor that might have decreased the

insularity of the Andes is thus the descent of ecological zones during Pleistocene glaciations. This descent has surely been deeply marked, at least during glacial maxima: van der Hammen (1974) reports a depression of the temperature of 11° C in some regions of the Colombian Andes during the Riss glaciation (see also Chap. 7 and 8). Moreover, the climate was also at some times very arid and a relatively important extension of puna-like conditions is likely. However, this by no means implies that a continuous bridge of oreal conditions ever existed between the North American ranges and the Andes, but the gaps must have been seriously reduced. We are thus drawn to an explanatory mechanism parallel to that evoked by Haffer (1974) for birds, combining long-distance dispersal and climatic reduction of the gaps. The strong north winds that probably blew during these periods are a possible factor of enhanced dispersal of insects of arctic origin. Anyway, such a type of colonization has almost surely taken the form of foundation by a single fertilized female, with the now classic genetic effects of the founder principle of Mayr (1963) and the more recently developed implication of the genetic revolution (Templeton, 1979).

It also may be considered that dispersal of higher groups occurred through species adapted to low- or medium-elevation open biomes; examples are provided by *Pieris protodice* or *Colias eurytheme*, which are widespread as far south as central Mexico. For such species, at least during glacial times, a nearly continuous form of dispersal might be possible. The most favorable period for this invasion was the peniglacial dry and cold phases (Fairbridge, 1972), when open landscapes were surely more extensive. The genetic features of such a colonization are quite different from those of the founder principle of the previous case. However, this hypothesis must be advanced cautiously, since it implies close taxonomic relationships between the extant colonizing generalists and the stem of Andean species; the evidence of such closeness is not too convinving in most cases, except perhaps in some lycaenids (Johnson, 1981). According to various authors (e.g., Haffer, 1974), the first (Mindel) glaciation is hard to detect in many parts of the Andes, especially the Sierra Nevada de Santa Marta, likely due to the insufficient altitude of these mountains during the Early Pleistocene. On the contrary, the Riss and Würm phases have left traces everywhere (a very complete synthesis of paleoclimatic data of the Quaternary in the Andes is provided by Simpson, 1979). Moreover, Vuilleumier (1969) and Vuilleumier and Simberloff (1980) keenly stressed that the effects of glaciations were somewhat opposed in the northern and southern Andes: in the former, glacial phases produced a coalescence of the páramo islands; in the latter, they fragmented the puna area through the advance of glacial tongues and the formation of lakes.

Considering now that the Andean island has been colonized from North America, according to the most obvious and already classic aspects of MacArthur and Wilson's theories, the number of species must decrease from north to south. However, exactly the contrary is observed, especially with taxa characteristically of northern origin—the only taxa that are not impoverished are those of Neotropical origin! Thus, the history has undergone several ebb and flow phases.

A supplementary difficulty is brought by the absence of any member of oreal Andean (or North American) faunas in the high mountains of Mexico (Munroe, 1963) or the Talamanca Páramo region of Costa Rica. Here, a purely MacArthurian explanation is feasible: these regions may serve as stepping-stones, but their small area will produce a rapid species turnover, hence an extinction of the remnants of the invasion. Naturally, this argument makes no sense if we admit that the colonization has taken place through semilowland generalists, which are still present in Mexico.

Let us now try to reconstitute tentatively the history of some groups. The easiest to begin with is the genus *Colias*. South American species are very probably monophyletic; the differentiation between species is not very accentuated. They may be recent colonists, having arrived, for instance, at the beginning of the Würm glaciation, and may ultimately issue from a single female stray, probably close to the arctic *C. hecla* or *C. concandica*. In spite of this heavy genetic bottleneck effect, most potentialities have been kept in the new strain, e.g., "*Alba*" polymophism, capability for polyphenism, and migratory tendencies; the latter are particularly well developed in *C. lesbia* (Williams, 1930). Subsequent speciation would have been caused by repeated foundings or momentary isolation caused by the minor fluctuations of the Würm and post-Würm phases. There are no reasons to doubt that speciation might not be as fast in butterflies as in birds, where it is considered as taking around 10,000–20,000 years in some cases (Moreau, 1966; Vuilleumier, 1969; Haffer, 1974). It seems that the only mode of speciation known in Lepidoptera is the allopatric one (Guillaumin and Descimon, 1976; Descimon, 1980a); this fact provides one more element of resemblance between butterflies and birds, where allopatric speciation is also the only one known (Simpson and Haffer, 1978).

However, they may well have arrived somewhat earlier. Indeed, South American *Colias* present some primitive features that they share in common with tropical and subtropical American genera such as *Phoebis*, *Anteos*, *Zerene* (actually considered as a subgenus of *Colias*). Their closest relatives amongst Holarctic *Colias* are to be found in the Palearctic region, while the "androconia"-bearing group, to which they belong, is very poorly represented in North America. Further studies may therefore disclose an early divergence of the Andean lineage.

Things are much more complicated with the Pierini. First, the evolutionary differentiation has been more accentuated, reaching the generic level. Second, both morphologically and physiologically, the closest relatives of Andean species are not the extant ones from North America but the Palearctic *Baltia* and *Synchloe callidice*. Thus it might be more satisfactory to admit that colonization has occurred at least during the Riss glaciation. A few years ago, a rather simple hypothesis might have been put forward: the ancestor of Andean Whites derived from a species close to the stem of the North American *P. protodice* and *P. occidentalis* (more or less the classical Klots-Forster-Mani scheme, from which Shapiro [1978, 1980a, b] has removed himself somewhat). During the Riss glaciation, cold and dry conditions have prevailed at least at some time even in the northern Andes, which moreover offered more continuous habitats. Accordingly, the newcomers adapted at the start to puna-like conditions (if they were not already preadapted to them) rather than to páramo-like ones. During the Riss-Würm interglacial phase, the butterflies colonized the southern regions and were mostly eliminated (with the exception of *Reliquia*) from the northern Andean region, which was insular and less favorable to pierid butterflies. During the Würm, a new series of migrations occurred, with insular splitting of the Puna Center and reinvasion of the northern Andes. It can be hypothesized that, at this time, colonists related to *Phulia* invaded North America, Alaska, and eastern Asia, yielding the *Baltia* strain. Subsequent extinction of this pulsation (the accurate term of Jeannel, 1942), occurred readily, except in the most extreme parts of its range (centrifugal segregation; Jeannel, 1942); it is not impossible that competition with the polyphenic, better-adapted "modern" species of Pierini could have taken some part in this extinction. A likely period for the pulsation of the *Phulia-Baltia* stock would be a dry-sunny-cold phase, such as the well-known tardiglacial of the Palearctic. Naturally, the genus- and species-level radiation of the group in the Andes would be the result of a series of allopatric speciation events, followed by phases of ecological partitioning during sympatry.

In light of the most recent data, this scheme needs to be reconsidered in part. Cold and dry, periglacial conditions are known in circumpolar boreal regions at the end of the Tertiary (6–10 million years B.P.) (Herman, 1970). It is moreover well established that Jeannel's (1942) "angarian" region is (probably with Scandinavia and Alaska) the source where, for a long time, cold temperate climate prepared and preadapted faunal elements to invade the Holarctic owing to the progressive Pliocene-Pleistocene cooling (Gross, 1961). Here might have arisen the primeval differentiation of the stem common to *Phulia* and *Baltia*, that afterward invaded both the Himalaya and the Pan-American cordilleras during the beginning of glacial times (or even before, as a medium elevation open landscape generalist). We meet the former scenario here. In this case, centrifugal segregation is most typical and we must also admit the extinction of the *Phulia* lineage in all parts of the world, except the Himalaya and the Andes. Such a phenomenon is by no means exceptional; the examples of camelids or horses come readily to mind.

Are the Andean Pierini monophyletic? Until now, nobody cast any doubts on this assumption. In the present state of affairs, it would be possible to suggest that their apparent homogeneity lies in their archaism; the *Phulia* lineage, related to *Baltia*. may have penetrated in South America separately from the *Tatochila* one, more akin to *Synchloe*. This would be the finishing stroke to the "classical" hypothesis. The part of certainty in this problem is always becoming thinner, which agrees with a recent opinion of Shapiro (1981, pers. comm.): "I have gotten less radical the more I know." The results recently obtained by Johnson (1981) after a cladistic analysis of some groups of lycaenids also suggest an extremely complex history involving many separations and extinctions at various times.

Would it be possible to propose a "tertiarist" or even "secondarist" alternative hypothesis? Shields and Dvorak (1979) attempted to do so by comparing the *Phulia-Baltia* pattern with Maekawa's (1965) theory of "trans-paleo-equatorial" distribution. This author explains the distributional characters of some plant taxa, such as *Coriaria* and Lardizabalaceae, by the course of the equator during the Jurassic. It is evident that such an already quite debatable theory is irrelevant, since the Rhopalocera presumably did not exist at that time and extant genera certainly did not. In fact, the flaw in this kind of interpretation is that there is absolutely no geologic evidence that mountains reaching the elevation of the oreal biome existed in South America prior to the very late Tertiary.

It is possible that, when more information is available about *Yramea* and the *Itylos* complex, their history might be found to be parallel with that of the Pierini.

The data obtained with Lepidoptera may be compared to those from other taxa. In Coleoptera there is some parallelism: a good number of genera show marked affinities with the Holarctic fauna (Mani, 1968), but there is a somewhat higher proportion of elements from lowland stock and of taxa of Gondwanian origin; for instance, the genus *Trechisibus*, widespread and well diversified in the Andes from Chile north into Panama, exhibits indubitable relations with Australo-Tasmanian genera (Jeannel, 1942). The dipteran fauna of the family Tipulidae clearly originated from a Holarctic stock (Alexander, 1962), and the same is true of acridians (Descamps, pers. comm.).

Among vertebrates, a great variation is observed, owing to the ecological valency of the groups. In fishes, the data provided by Géry (1969) afford absolutely no point of comparison with butterflies, and the same is true for amphibians and reptiles (Duellman, 1979): they evolved from a lowland stock and the last two are distributed clearly around the two classical dispersion centers: northern Andean and puna (see also Chap. 19). Birds, on the other hand, present a striking similarity with butterflies: the proportion of Holarctic elements is high (Dorst, 1967; Vuilleumier, Chap. 23, this volume); there is, however, a significant percentage of species of Patagonian origin (Müller, 1973), as stated earlier by Chapman (e.g., 1921). Attention has already been drawn to the parallel impoverishment of bird and butterfly faunas from the core of the puna center northward to the south of Ecuador. For mammals, I will only mention some similarity between the adaptive radiation of Cricetidae (G. G. Simpson, 1969; see Reig, Chap. 16, this volume) and that of Pierini. The latter show also, at first glance, an obvious ecological and biogeographic parallelism with the camelids. As early as 1950, Stirton suggested that the existence of extensive open landscapes during cold-arid glacial phases enhanced the possibilities of dispersion for steppe species such as horses and camels (see also Hoffstetter, Chap. 9, this volume).

Surely the analysis of plant distribution and, above all, of palynological data provides the most interesting elements of comparison. Plants have a predominant importance for Lepidoptera, as nectar sources for adults and as foodplants for larvae; they are often the actual link between the habitats and the Lepidoptera that live in them. Most butterflies are oligophagous and are intimately related to plants through a long period of coevolution (Ehrlich and Raven, 1965). Of course, some "infidelities" are always possible and are especially evident in arctic-alpine species, e.g., *Colias* and *Boloria*, where the genuine foodplant families are forsaken for Ericacae, which happen to be numerous. Generally speaking, a lepidopteran species cannot colonize a region lacking a population of a foodplant that can attract laying females. In fact, the distributional area of a foodplant is almost always larger than that of its insect enemy. So, we may state that plants must precede their hosts. Unfortunately, the genuine foodplants of most Andean butterflies are unknown and distributional correlations are impossible. Even in the case of *Reliquia santamarta*, Shapiro and Torres have established only that the foodplant is one of the several endemic crucifer species, some of which belong to the holarctic genus *Draba*.

The floristic analysis of plant origins and distribution shows that they form a much more diverse sample than Rhopalocera in the Andes. This is not surprising if we consider their large species number and their ecological diversity. According to B. B. Simpson (1979), the types of origins that may be considered are the same as those I have taken into account for butterflies. North American elements are abundant; their arrival is recent and has been progressive (van der Hammen, 1974). Truly southern temperate taxa are also well represented (e.g., *Azorella*, *Acaena*), and there is a very important truly Neotropical stock that derives from lowland floras; among it, the Compositae have many taxa, which have undergone explosive adaptive radiation (*Espeletia*, sensu lato for instance; see Chap. 11). The appearance of plant species, as revealed by palynology, is often very sudden in a given locality.

Generally speaking, when compared with other groups, Lepidoptera do not appear the best equipped for rapid adaptation to a new environment. Perhaps because they are geologically older than them, vertebrate poikilotherms have been able to exploit much better the opportunity offered by the uplift of the Andes. Actually, the group that recalls butterflies is birds, in this aspect as in most others; in both groups, the rate of colonization from long distances appears to be greater than that of adaptation of local taxa to a different environment.

Coming back to the hyperdisjunct puzzling pattern *Phulia-Baltia*, it may be interesting to seek similar examples in other groups. The distribution of the genus *Callipogon* (Coleoptera) in the Andes and Vladivostock region (Jeannel, 1942) is somewhat similar. The dipteran *Tipula phalangioides* is reported by Alexander (1962) from the Andes of Ecuador and from the Sikkim Himalaya; since it belongs to a Holarctic group, this reinforces the second scenario. In plants, if we exclude the species tied to other (especially forest) biomes, the most striking case concerns two closely related herbaceous Bignoniaceae: the Andean *Argylia* and the Himalayan *Incarvillea*. Raven and Axelrod (1974) seem to be very embarrassed by this disjunction and do not propose any explanation; according to A. H. Gentry (pers. comm.), the resemblance is actually due to convergence. The genus *Hypochaeris* (Compositae) also deserves attention: it is well represented in the Andes and in the Palearctics but absent from North America. In opposition with Pierini and *Tipula*, Bignoniaceae, very probably Compositae and *Callipogon* originated from South America. Thus the common feature of these distributions is the absence of taxa from North America; it is very possible that this continent has had numerous extinctions during Quaternary times.

It also may be interesting to compare the Andean lepidopteran faunas with those from other tropical or subtropical high mountains. All other massifs, with the exception of the Himalayas, are poorer. In Africa, the Ethiopia highlands have but a few endemic subspecies of lowland widespread Palearctic species, such as *Pieris brassicae brassicoides*, *Colias croceus electo*, *Pontia daplidice aethiops*, and *Lycaena phlaeas pseudophlaeas* (Carpenter, 1935). The situation is, in some respects, a "reduction" of the Andean one. The other African mountains show very few, if any, endemic elements: most species encountered in the oreal zone are lowland species, mainly accidentals or vagrants. *C. croceus electo* is, however, present on Kilimanjaro only (this migratory species reaches South Africa, which contradicts somewhat the earlier statement that *Colias* is absent from the Southern Hemisphere, except South America). *Cupido aequatorialis*, a lycaenid found on practically all the massifs, and perhaps some nymphalid fritillaries of the genus *Issoria* represent the only species peculiar to the high montane biomes of Africa. The situation is the same in New Guinea where the oreal biome is devoid of endemic taxa (Toxopeus, 1950); this range is considered to be even younger than the Andes. In the Himalaya, however, the picture is quite different. There one finds proliferation of endemic species or subspecies, with complex adaptive radiations in many genera, such as *Parnassius*, *Colias*, *Aporia*, *Pieris*, *Oeneis*, and various fritillary genera. Most of them are obviously Palearctic; a few, such as some species of *Lethe*, belong to tropical Indo-Australian groups (Mani, 1968). However, the oreal zone of the more southern tropical massifs is often depauperate in butterfly species. (For intertropical and Tibetan comparisons in birds, see Chap. 6, which reveals some differences with the butterfly patterns. See in particular the discussion of the species-area effect.)

The poverty in butterflies of the African and New Guinean oreal regions is easily explained by the youth, the small size, and the isolation of these montane islands. However, some other groups, plants for instance, show explosive proliferation of endemic species. These examples and that of the Himalaya confirm that (1) oreal biomes in the tropics are inhospitable to butterflies; for local lowland (mainly forest) fauna, adaptation to such a different environment is an extremely difficult evolutionary performance; for the temperate fauna, there is also some hindrance, such as the lack of clear-cut seasonality; and (2) the source of taxa preadapted to high montane life is the Holarctic. I thus agree with Müller (1973) that the limit of the Holarctic, as far as the oreal fauna is concerned, is not in southern Mexico, as Hoffman argued, but rather in Tierra del Fuego.

CONCLUSIONS

In summary, it is clear that the Neotropical and southern temperate regions contributed little (or nothing) to the oreal butterfly fauna of the Andes. Its affinities lie instead with the Holarctic realm. But colonization was not simultaneous, and several faunal strata of invaders (G. G. Simpson, 1965) can be identified on the basis of their present state of evolutionary differentiation (assuming that a single species in each group made the step, that the rates of diversification were much the same, and that diversity was not drastically reduced by subsequent extinctions in some groups). The oldest assemblage is the Pronophilini, which have reached considerable diversification in montane forest ("strong" genus level); their separation from the Holarctic satyrine stock may be dated back at least 5–10 million years B.P. Later, and subsequent to the extension of páramos and puna habitats following the final uplift of the Andes, some of them entered these high-altitude biomes, either directly or *via* southern South American regions. The history of the lycaenids of the "*Thecla*" *culminicola* group and of the genus *Cyanophrys* may be somewhat parallel to that of the Pronophilini, though in some cases they may be of more recent origin. All other groups are fundamentally associated with open, cool, or cold biomes and must have evolved some 5–10 million years ago in an already cold (and probably dry) region that is likely to have been located in the circumpolar arctic zone. The Pierini and the *Itylos* complex constitute the next stratum ("weak" genus level). They must have arrived during a cold-dry period, through long-distance dispersal, most likely by using the Mexican and Central American mountain ranges as stepping-stones, from which they ultimately disappeared. It is in the southernmost part of the Andes that their major adaptive radiation took place. Their time of arrival remains uncertain. Geologic and paleoclimatological data argue for a relatively recent time—the Riss glaciation. But an earlier colonization by less specialized ancestral forms dwelling in medium low, open habitats (and having left no trace in North America) cannot be excluded. *Colias* and, possibly, *Yramea* make up the last stratum (species level of diversification). Their most likely time of arrival is the Würm, although this date may be too recent. They have numerous relatives in the Palearctic and the Nearctic.

Is there a test that might allow one to "falsify" this scenario and hence give it fuller scientific status (sensu Popper), in spite of the fact that decisive fossil finds are extremely unlikely for Lepidoptera? Recent advances in molecular biology suggest that one test would be quite feasible, and might be even relatively easy, unless meeting with difficulties specific to Lepidoptera, which seems most unlikely. Owing to their high rate of evolution and the ease of their analysis by restriction endonucleases, comparison of mitochondrial DNAs (Brown, George, and Wilson, 1976) allows dated cladograms to be constructed over precisely the time span of interest here. All that would be needed in our case is that living material from the Andes and a few Palearctic, Nearctic, and Himalayan species be made available. As a result, both the times T_1 at which the Andean stems separated from their nearest living relatives in the Holarctic and T_2 at which living Andean species diverged among themselves could be estimated. These times would not date the faunal exchanges themselves, but would at least set upper and lower limits to their timing.

A more classical approach, which is being carried out, is the one provided by the electrophoretic study of enzymes. It may be somewhat difficult to calibrate the time scale of the cladograms so obtained, but the actual affinities of Andean butterflies with Holarctic ones will be worked out to a large extent. This result will hopefully allow one to choose among the hypotheses proposed here, or else to put forward new ones.

ACKNOWLEDGMENTS

Cordial thanks are due Arthur M. Shapiro for much information, advice, discussion, and suggestions for improvement of the manuscript. I am also very much indebted to François Michel for having communicated his observations and for having discussed extensively the form and the substance of this chapter. Kurt Johnson kindly helped me with the text on the Lycaenidae, providing me with a novel insight to the taxonomy of this group; not all of his results could be incorporated into the discussion. In the field I received efficient assistance from Maximina Monasterio and Guillermo Sarmiento (Mérida) and N. Venedictoff (Quito). Travel to Venezuela in 1977 was supported by a grant from the Centre National de la Recherche Scientifique (R.C.P. 317, "Polymorphisme, spéciation, mimétisme"). I am most grateful to François Vuilleumier for having overcome my reluctance to write the present chapter. During the last phase of this work, the friendly welcome and efficient help of G. Lamas, in the Javier Prado Museum of Lima, was appreciated.

POSTSCRIPT

Between the writing of this chapter and its publication many weaknesses and inconsistencies have become apparent to the author, and new papers have been published that deserve citation. Among the most noteworthy, only two will be mentioned here. The first is an extensive ecological study of a lepidopteran taxocene in the Cordillera Blanca in Peru (Perez Ramos, 1982), which brings to light many interesting features concerning especially altitudinal distribution and phenology. The second is a biogeographic zonation of Peru by Lamas (1982), whose conclusions about the oreal zone agree quite well with the ideas presented here.

Writing the present chapter has made the author a more active participant in research on Andean butterflies. More Andean Pierinae have been bred in captivity. This has permitted me to study their biology and the morphology of preimaginal stages, in particular of *Phulia* and *Infraphulia* (Courtney, Descimon, and Shapiro, in preparation). In addition to interesting behavioral features, these observations underline the primitiveness of Andean pierids. The same is true of enzyme electrophoresis, which is bringing some novel insights to the subject (Geiger, Descimon, and Shapiro, in preparation). The *Tatochila* and *Phulia* generic complexes are indeed quite distinct, as foreseen in this chapter. *Tatochila* butterflies are related to *Ascia* and to the Palearctic genera *Aporia-Metaporia*, which present a large adaptive radiation in central Asia. The resemblance between the pupa of *Tatochila* and that of the European *Aporia crataegi* struck me as soon as I bred them; the resemblance in wing pattern between *Tatochila* and Asiatic *Aporia-Metaporia* also appears very striking, but only after attention is drawn to it by electrophoresis. The probable kinship with *Ascia* is unexpected and very interesting; this Neotropical genus (which has lost the *Tatochila-Metaporia* pattern) is a lowland generalist of extreme migratory habit, the "dreamed of" candidate to represent the ancestor that crossed Central America to reach the Andes. Things are not as clear for *Phulia* and related genera; they appear to be closer to the Holarctic *Pontia* (*P. callidice*, *P. occidentalis* and others). F. Michel has recently bred *Baltia* from Ladakh; the preimaginal stages strongly recall those of *Phulia*. It must be noted that the pupa of *Phulia* resembles strongly those of *Tatochila* and *Aporia*. Is this a primitive feature shared by both lineages?

In any event, although the Klots-Mani scheme of invasion of the Andes by Holarctic Pierini has been seriously modified, it remains the most likely one in its basic principle. There is no serious reason to postulate, as have Herrera and Covarrubias (1983), that *Phulia* and *Tatochila* are of southern origin.

With the same techniques, it has been possible to see that South American *Colias* shows marked affinities with the "androconia"-bearing species (such as the European *C. croceus*), whereas they are more distant from other species clusters; thus they differentiated during the beginning of the major adaptive radiation of the genus, that is, probably long before the Pleistocene. Finally, all data suggests that a single massive arrival of Holarctic elements took place, rather early, before the beginning of the Pleistocene, during a cold-arid phase, with in particular the camelids. The Coenocron passed rapidly through the northern Andes and underwent adaptive radiations more southwards, from where it expanded more recently in the reverse direction.

Concerning the colonization of the Andes, the idea that is perhaps becoming the clearest is that it is unnecessary to postulate spectacular geoclimatic events to explain it. Migratory lowland-open landscape species, of which *Ascia monuste* and *Colias lesbia* give us some idea, are good candidates to be the first colonists, provided they had potentialities for adapting to highlands. In this case, the most significant distinction for adaptive strategies in butterflies is not between cold and warm, dry and moist, or sunny and cloudy, but between forest and nonforest biomes.

REFERENCES

Ackery, P. H. 1975. A new pierine genus and species with notes on the genus *Tatochila* (Lepidoptera: Pieridae). *Bull. Allyn Mus.* 30: 1–9.

Adams, M. J. 1973. Ecological zonation and the butterflies of the Sierra Nevada de Santa Marta, Colombia. *J. Nat. Hist.* (London) 7: 699–718.

———. 1977. Trapped in a Colombian Sierra. *Geogr. Mag.* 49: 250–254.

Adams, M. J., and G. I. Bernard. 1977. Pronophiline butterflies (Satyridae) of the Sierra Nevada de Santa Marta, Colombia. *Syst. Entomol.* 2: 263–281.

———. 1979. Pronophiline butterflies of the Serrania de Valledupar, Colombia-Venezuela border. *Syst. Entomol.* 4: 95–118.

Alexander, C. P. 1962. Beiträge zur Kenntnis der Insektanfauna Boliviens XVII. Diptera II. The craneflies (Tipulidae: Diptera). *Veroff. Zool. Staatsammlung* (Munich) 7: 9–159.

Barragué, G. 1954. Contribution à une faune des Lépidoptères Rhopalocères des environs d'Alger. *Bull. Soc. Hist. Nat. Afrique Nord* 45: 179–188.

Berger, L. A. 1981. Observations sur quelques espèces de *Colias* (Lep. Pieridae). *Lambillionea* 81: 28–31.

———. 1983. Notes sur les *Colias* néotropicaux (Lépidoptères Pieridae). *Lambillionea* 83: 4–13.

Bernardi, G. 1980. Les catégories taxonomiques de la systématique évolutive. In *Les problèmes de l'espèce dans le règne animal*, vol. 3. Eds. C. Bocquet, J. Génermont, and M. Lamotte, pp. 373–425. Paris: Soc. Zool. Fr.

Brown, W. M., M. George, Jr., and A. C. Wilson. 1976. Rapid evolution of animal mitochondrial DNA. *Proc. Natl. Acad. Sci.* 76: 1967–1971.

Campos, F. 1926. Contribución al estudio de los insectos del callejón interandino. *Rev. Coleg. Nac. V. Rocafuerte* 8: 1–40.

Carpenter, G. D. Hale. 1935. The Rhopalocera of Abyssinia: A faunistic study. *Trans. R. Ent. Soc. London.* 83: 313–448.

Chapman, F. M. 1921. The distribution of bird life in the Urubamba Valley of Peru. *U. S. National Museum Bull.* 117: 1–138.

Clench, H. K. 1966. Behavioral thermoregulation in butterflies. *Ecology* 47: 1021–1034.

Croizat, L. 1958. *Panbiogeography*, Vol. I. *The New World.* Caracas: Published by the author.

d'Abrera, B. 1981. *Butterflies of the neotropical region*,

part I. *Papilionidae and Pieridae*. Melbourne: Landsdowne.

Descimon, H. 1966. Variations quantitatives des ptérines de *Colias croceus* (Fourcroy) et de son mutant *helice* (Hbn) (Lepidoptera Pieridae) et leur signification dans la biosynthèse des ptérines. *C. R. Acad. Sci. Paris* 262: 390–393.

———. 1976. L'acclimatation des Lépidoptères: Un essai d'expérimentation en biogéographie. *Alexanor* 9: 195–204.

———. 1979. Pteridine biosynthesis and nitrogen metabolism in the butterfly *Colias croceus* and its "*Alba*" mutant. In *Chemistry and Biology of Pteridines*, R. L. Kisliuk and G. M. Brown, eds., pp. 93–98. New York: Elsevier.

———. 1980a. Polymorphisme de la coloration et stratégies adaptatives chez les Lépidoptères. In *Recherches d'écologie théorique: Les stratégies adaptatives*, R. Barbault, P. Blandin, and J. A. Meyer, eds., pp. 53–76. Paris: Maloine.

———. 1980b. *Heodes tityrus tityrus* Poda et *H. tityrus subalpina* Speyer (Lycaenidae): Un problème de spéciation en milieu alpin. *Nota Lepid.* 2: 123–125.

Descimon, H., and C. Renon. 1975. Mélanisme et facteurs climatiques. II. Corrélation entre la mélanisation et certains facteurs climatiques chez *Melanargia galathea* (Lepidoptera Satyridae) en France. *Arch. Zool. Exp. Gén.* 112: 437–468.

Dorst, J. (1967). Considérations zoogéographiques et écologiques sur les oiseaux des hautes Andes. In *Biologie de l'Amérique australe*, Vol III, C. Delamare-Deboutteville and E. Rapoport, eds., pp. 471–504. Paris: Centre National de la Recherche Scientifique.

Draudt, M. 1926. *Thecla*. In *Les Macrolépidoptères du Globe*. Vol. V, *Diurnes américains*, A. Seitz, ed., pp. 739–831. French ed. Paris: Le Moult.

Duellman, W. E. 1979. The herpetofauna of the Andes: Patterns of distribution, origin, differentiation, and present communities. In *The South American herpetofauna: Its origin, evolution, and dispersal*, W. E. Duellman, ed., pp. 371–459. Mus. Nat. Hist. Monograph 7. Lawrence: University of Kansas Press.

Dyar, H. G. 1913. Results of the Yale Peruvian Expedition of 1911. Lepidoptera. *Proc. U. S. Natl. Mus.* 44: 279–324.

Ebert, H. 1969. On the frequency of butterflies in eastern Brazil, with a list of the butterfly fauna of Poços de Caldas, Minas Gerais. *J. Lep. Soc.* 23 (Suppl. 3): 1–48.

Ehrlich, P. R. and P. H. Raven. 1965. Butterflies and plants: A study in coevolution. *Evolution* 18: 586–608.

———. 1969. The differentiation of populations. *Science* 165: 1228–1232.

Eliot, J. N. 1973. The higher classification of the Lycaenidae (Lepidoptera): A tentative arrangement. *Bull. Br. Mus. (Nat. Hist.)* 28: 373–506.

Fairbridge, R. W. 1972. Climatology of a glacial cycle. *Quat. Res.* (New York) 2: 283–302.

Fassl, A. H. 1910. Tropische Reisen. II. Ueber den Quindiúpass. *Entomol. Ztschr.* 24: 113–138.

———. 1911. Die vertikale Verbreitung der Lepidopteren in der Columbischen Central-Cordillera. *Fauna exot.* 1: 25–26, 29–30.

———. 1913. Neue Preponen aus Bolivien. *Entomol. Rundsch.* 30: 43–44.

———. 1914a. Tropische Reisen. V. Das obere Caucaltal und die Westkordillere. *Entomol. Rundsch.* 31: 35–38, 42–46, 50–52, 57–58.

———. 1914b. Tropische Reisen. VI. Die Hochkordillere von Bogota. *Entomol. Rundsch.* 31: 97–100, 104–105, 108–110, 115–116.

———. 1915a. Die vertikale Verbreitung der Lepidopteren in der Columbischen West-Cordillera. *Entomol. Rundsch.* 32: 9–12.

———. 1915b. Neue Pieriden aus Süd-Amerika. *Deutsch. Entomol. Ztschr. "Iris"* 29: 176–181.

———. 1918. Die vertikale Verbreitung der Lepidopteren in der Colombischen Ost-Cordillera. *Entomol. Rundsch.* 35: 1–4, 30–31, 44, 48–50.

———. 1920. Meine Bolivia-Reise. *Entomol. Rundsch.* 37: 10–11, 15–18, 22–23, 25–27, 29–30, 34–35, 41–43, 45–48.

Field, W. D. 1951. A revision of *Eurema* Hübner subgenus *Teriocolias* Röber (Lepidoptera Pieridae). *Acta. zool. Lilloana* 9: 359–373.

———. 1958. A redefinition of the butterfly genera *Tatochila*, *Phulia*, *Piercolias*, and *Baltia*, with descriptions of related genera and subgenera. *Proc. U. S. Natl. Mus.* 108: 103–131.

Field, W. D., and G. Herrera. 1977. The pierid butterflies of the genera *Hypsochila* Ureta, *Phulia* Herrich-Schäffer, *Infraphulia* Field, *Pierphulia* Field, and *Piercolias* Staudinger. *Smithson. Contr. Zool.* 232: 1–64.

Fittkau, E. J. 1969. The fauna of South America. In *Biogeography and ecology in South America*, pt. II, E. J. Fittkau, J. Illies, H. Klinge, G. H. Schwabe, and H. Sioli, eds., pp. 624–658. The Hague: Junk.

Forster, W. 1955. Beiträge zur Kenntnis der Insektenfauna Boliviens. I. Enleitung. Lepidoptera I. *Veröff. Zool. Staatsamml.* (Munich). 3: 81–160.

———. 1958. Beiträge zur Kenntnis der Insektenfauna Boliviens XIX. Lepidoptera III. Satyridae. *Veröff. Zool. Staatsamml.* (Munich) 8: 51–188.

———. 1962. Presidential address to the seventh pacific slope meeting of the Lepidopterists' Society. *J. Lep. Soc.* 15: 116–120.

Gansser, A. 1973. Facts and theories of the Andes. *J. Geol. Soc. London.* 129: 93–131.

Garlepp, G. 1892. Brief aus Bolivien. *Deutsch. Entomol. Ztschr., "Iris"* 5: 272, 273–276.

Géry, J. 1969. The fresh-water fishes of South America. In *Biogeography and ecology in South America*, E. J. Fittkau, J. Illies, H. Klinge, G. H. Schwabe, and H. Sioli, eds., pt. II, pp. 828–848. The Hague: Junk.

Godman, F. D. and O. Salvin. 1891. *Lepidoptera Rhopalocera*. In *Supplementary appendix to travels amongst the Great Andes of the Equator* (E. Whymper). London: John Murray.

Grey, L. P. 1957. Warren's argynnid classification (Nymphalidae). *Lepidopt. News* 11: 171–176.

Gross, F. J. 1961. Zur Evolution europäischer Lepidopteren. *Verh. Dtsch. Zool. Ges. Saarbrüken.* 1961: 461–478.

Guillaumin, M., and H. Descimon. 1976. La notion d'espèce chez les Lépidoptères. In *Les problémes de l'espèce dans le règne animal*, C. Bocquet, J. Génermont, and M. Lamotte, eds., Paris: Soc. Zool. Fr., Mém. no. 38. pp. 129–201

Haffer, J. 1974. *Avian speciation in tropical South America*. Nuttall Ornithological Club, Publ. no. 14.

Graham, S. M., W. B. Watt, and L. F. Gall. 1980. Metabolic resource allocation vs. mating attractiveness: Adaptive pressures on the "alba" polymorphism of *Colias* butterflies. *Proc. Natl. Acad. Sci. U.S.A.* 77: 3615–3619.

Hayward, K. J. 1964. Insecta Lepidoptera Rhopalocera, Vol. 3. In *Genera et Species animalium argentinorum*, A. Willink, ed., Bonariae, Argentina.

———. 1967. Lista de los tipos de insectos y otros invertebrados conservados en el Instituto Miguel Lillo. *Acta Zool. Lilloana* 22: 337-352.

———. 1973. Catálogo de los ropaloceros aregentinos. *Opera Lilloana* 23: 1–318.

Heimlich, W. 1963. Eine neue Satyride aus Chile. *Entomol. Zeitschr.* 69: 173–179.

———. 1972. Satyridae der südlichen Neotropis und Subantarktis (Lepidoptera Satyridae). *Beitr. Entomol.* 22: 149–197.

Herman, Y. 1970. Arctic Paleo-Oceanography in late Cenozoic Times. *Science* 164: 474–477.

Herrera, G. J., and R. Covarrubias. 1983. Distribución biogeográfica del grupo *Tatocheila* (sic)-*Phulia* (Lepidoptera: Pieridae). *Res. Comm. Sci. IX Congr. Latinoamer. Zool. Arequipa-Peru.* 9–15 Oct., 1983. p. 213.

Herrera, G. J., W. D. Field. 1959. A revision of the butterfly genera *Theschila* and *Tatochila* (Lepidoptera: Pieridae). *Proc. U. S. Natl. Mus.* 108: 467–514.

Hoffman, C. C. 1940. Catálogo sistemático y zoogeográfico de los Lepidopteros mexicanos. Primera parte. Papilionoidea. *An. Inst. Biol. México* 11: 639–739.

Hovanitz, W. 1945a. Comparisons of some Andean butterfly faunas. *Caldasia* 3: 301–336.

———. 1945b. Distribution of *Colias* in the equatorial Andes. *Caldasia* 3: 283–300.

Hughes, R. A. 1956. Notes on the butterflies of the Arequipa District of southwest Peru. *The Entomol.* 89: 248–251.

———. 1958. Butterfly collecting in the High Andes of southern Peru, March 1956. *The Entomol.* 91: 1–8.

James, D. E. 1973. The evolution of the Andes. *Sci. Amer.* 229: 60–79.

Jeannel, R. 1942. *La genèse des faunes terrestres*. Paris: Presses Universitaires de France.

———. 1967. Biogéographie de l'Amérique australe. In *Biologie de l'Amérique australe*, C. Delamare-Deboutteville and E. Rapoport, eds., pp. 401–460. Paris: Centre National de la Recherche Scientifique.

Johnson, C. G. 1969. *Migration and dispersal of insects by flight*. London: Methuen.

Johnson, K. 1981. Revision of the Callophryina of the world with phylogenetic and biogeographic analyses (Lepidoptera: Lycaenidae). Ph.D. thesis. New York: City University.

Klots, A. D. 1933. A generic revision of the Pieridae (Lepidoptera), together with a study of the male genitalia. *Entomol. Amer.* 12: 139–242.

Lamas, G. 1977. A preliminary check-list of the Butterflies (Lepidoptera) of Peru West of the Andes. *Rev. Cien. Univ. Mayor. San Marcos*, Lima 70: 59–77.

———. 1981. Notes on Peruvian butterflies (Lepidoptera). VI. Twelve new Pieridae. *Rev. Ci. Univ. Nac. Mayor San Marcos* 73: 44–53.

———. 1982. A preliminary zoogeographical division of Peru, based on butterfly distributions (Lepidoptera, Papilionoidea). In *Biological diversification in the Tropics*, G. T. Prance, ed., pp. 336–357. New York: Columbia University Press.

Lesse, H. de 1966. Impressions d'un Lépidoptériste en Amérique du Sud. *Alexanor* 4: 171–178, 225–232, 242–252.

———. 1967. Les nombres de chromosomes chez les Lépidoptères néotropicaux. *Ann. Soc. Entomol. Fr., N. S.* 3: 67–136.

MacArthur, R. H., and E. O. Wilson. 1967. *The theory of island biogeography*. Princeton: Princeton Univ. Press.

Maekawa, F. (1965). Floristic relation of the Andes to eastern Asia with special reference to trans-paleoequatorial distribution. *J. Fac. Sci. Tokyo.* 3, *Botany*, 9: 161–195.

Mani, M. S. 1968. *Ecology and biogeography of high altitude insects*. The Hague: Junk.

Mayr, E. 1963. *Animal species and evolution*. Cambridge, Mass.: Harvard Univ. Press.

Mell, R. 1931. Die Trockenzeitform als Hemmungserscheinung (Diagora nigrivena Leech) als Trockenzeitform von Hestina assimilis (L.). *Biol. Zbl.* 51: 187–194.

Miller, L. D. 1968. The higher classification, phylogeny and zoogeography of the Satyridae (Lepidoptera). *Mem. Amer. Entomol. Soc.* 24: 1–174.

Moreau, R. E. 1966. *The bird faunas of Africa and its islands*. New York: Academic Press.

Müller, P. 1973. *The dispersal centres of terrestrial vertebrates in the neotropical realm*. The Hague: Junk.

Munroe, E. 1963. Characteristics and history of the North American fauna: Lepidoptera. *Proc. 16th Int. Congr. Zool.* 4: 21–47.

Nabokov, V. 1945. Notes on Neotropical Plebejiinae (Lycaenidae, Lepidoptera). *Psyche* 52: 1–61.

Nekrutenko, Y. P. 1965. Tertiary Nymphalid butterflies and some phylogenetic aspects of systematic Lepidopterology. *J. Res. Lepid.* 4: 149–158.

Perez Ramos, J. E. 1982. Ecología de las mariposas diurnas (Lepidoptera, Rhopalocera) del sector Llanganuco, Parque Nacional Huascarán, Ancash, Perú. Lima: Universidad Particular Ricardo Palma.

Petersen, B. 1954. Egg-laying and habitat selection in some *Pieris* species. *Entomol. Tidschr.* 75: 194–203.

———. 1963. The male genitalia of some *Colias* species. *J. Res. Lepidopt.* 1: 135–156.

Raven, P. H., D. I. Axelrod. 1974. Angiosperm biogeography and past continental movements. *Ann. Mo. Bot. Gard.* 61: 539–673.

Reig, O. 1968. Peuplement en vertébrés Tétrapodes de l'Amérique du Sud. In *Biologie de l'Amérique Australe*. Vol. IV, C. Delamare-Deboutteville and C. Rapoport, eds., pp. 215–260. Paris: Centre National de la Recherche Scientifique.

Reissinger, E. 1972. Eine neue *Colias*-Unterart aus Perú (Lepidoptera Pieridae). *Atalanta* B, 4: 60–64.

Remington, C. L. 1954. The genetics of *Colias* (Lepidoptera). *Adv. Genet.* 6: 403–450.

Röber, J. 1926. Pieridae. In *Les Macrolépidoptères du Globe*, Vol. V, *Diurnes américains*, A. Seitz, ed., pp. 53–111. French ed. Paris: Le Moult.

Rondou, J. P. 1932. Catologue des Lépidoptères des Pyrénées. *Ann. Soc. entomol. Fr.* 101: 165–244.

Rosen, D. E. 1978. Vicariant patterns and historical explanation in biogeography. *Syst. Zool.* 27: 159–188.

Schröder, H. 1955. Eine Falter-Ausbeute aus dem westlichen Bolivien (Ins. Lepid. Rhopal.) *Senck. biol.* 36: 329–338.

Schwanwitsch, B. N. 1940. On a remarkable dead leaf imitation in *Zaretes*, a genus of Nymphalid butterflies. *J. Zool. Mosk.* 19: 14–25.

———. 1943. Stereomorphism in the cryptic colour patterns of Rhopalocera. *Zool. J. Mosk.* 22: 323–339.

———. 1946. Imitation of a plant in the cryptom of *Satyrus huebneri* Feld. (Lepidoptera). *Trav. Soc. Natur. Leningrad.* 69: 223–228.

———. 1956. Wing-pattern of Pierid butterflies (Lepidoptera, Pieridae). *Rev. Entomol. U.S.S.R.* 35: 285–301.

Scott, A. A. 1970. Hilltopping as a mating mechanism to aid survival of low-density species. *J. Res. Lepidopt.* 7: 191–204.

Shapiro, A. M. 1974. Altitudinal migration of Central California butterflies. *J. Res. Lepidopt.* 13: 157–161.

———. 1975. Development and phenotypic responses of uni- and bivoltine *Pieris napi* (Lepidoptera Pieridae) in California. *Trans. R. Entomol. Soc. London.* 127: 65–71.

———. 1976. Seasonal polyphenism. *Evol. Biol.* 9: 259–333.

———. 1977a. Evidence for obligate monophenism in *Reliquia santamarta*, a neotropical-alpine Pierine butterfly (Lepidoptera: Pieridae). *Psyche* 84: 183–190.

———. 1977b. The life-history of *Tatochila xanthodice*, a montane Neotropical butterfly. *J. N.Y. Entomol. Soc.* 86: 51–55.

———. 1978. Developmental and phenotypic responses to photoperiod and temperature in an equatorial montane butterfly, *Tatochila xanthodice* (Lepidoptera: Pieridae). *Biotropica* 10: 297–301.

———. 1979a. The life histories of the *autodice* and *sterodice* species-groups of *Tatochila* (Lepidoptera: Pieridae). *J. N.Y. Entomol. Soc.* 87: 236–255.

———. 1979b. Notes on the behaviour and ecology of *Reliquia santamarta*, an alpine butterfly (Lepidoptera: Pieridae) from the Sierra Nevada de Santa Marta, Colombia, with comparisons to Nearctic alpine Pierini. *Stud. Neotr. Fauna Envir.* 14: 161–170.

———. 1980a. Convergence in pierine polyphenisms (Lepidoptera). *J. Nat. Hist.* 14: 781–802.

———. 1980b. Physiological and developmental responses to photoperiod and temperate as data in phylogenetic and biogeographic inference. *Syst. Zool.* 29: 335–341.

———. 1984. The genetics of seasonal polyphenism and the evolution of "general purpose genotypes" in butterflies. In *Population biology and evolution*, K. Wöhrmann and V. Loeschke, eds., pp. 16–30. Berlin, Heidelberg: Springer-Verlag.

Shapiro, A. M., and Torres, N. 1978. Notas sobre la biología de dos mariposas Pieridae de grandes alturas de Colombia (Lepidoptera: Pieridae). *Cespedesia* 7: 7–23.

Shields, O. 1967. Hilltopping. *J. Res. Lepidopt.* 6: 69–178.

———. 1976. Fossil butterflies and the evolution of Lepidoptera. *J. Res. Lepidopt.* 15: 132–143.

Shields, O., and S. K. Dvorak. 1979. Butterfly distribution and continental drift between the Americas, the Caribbean and Africa. *J. Nat. Hist.* 13: 221–250.

Simpson, B. B. 1975. Pleistocene changes in the flora of the high tropical Andes. *Paleobiology* 1: 273–294.

———. 1979. Quaternary biogeography of the high montane regions of South America. In *The South American herpetofauna: Its origin, evolution, and dispersal*, W. E. Duellman, ed., *Univ. Kansas Mus. Nat. Hist. Monogr.* 7: 157–188. Lawrence: Univ. of Kansas Press.

Simpson, B. B., and J. Haffer. 1978. Speciation patterns in the Amazonian forest biota. *Ann. Rev. Ecol. Syst.* 9: 497–518.

Simpson, G. G. 1965. *The geography of evolution.* New York: Capricorn Books.

———. 1969. South American mammals. In *Biogeography and ecology in South America*, part II, E. J. Fittkau. J. Illies, H. Klinge, G. H. Schwabe, and H. Sioli, eds., pp. 879–909. The Hague: Junk.

Staudinger, O. 1894. Hochandine Lepidopteren. *Dtsch. Entomol. Ztschr. "Iris"* 7: 43–100.

Stirton, R. A. 1950. Late Cenozoic avenue of dispersal for terrestrial animals between North America and South America. *Bull. Geol. Soc. Amer.* 61: 1541–1542.

Talbot, G. 1925. Forms of *Colias* from Peru. *Proc. Entomol. Soc. London.* 1925–26: XVI.

Taylor, O. R. 1972. Random vs. non-random mating in the sulfur butterflies, *Colias eurytheme* and *Colias philodice* (Lepidoptera: Pieridae). *Evolution* 26: 344–356.

Templeton, A. R. 1979. Genetics of colonization and establishment of exotic species. In *Genetics in relation to insect management*, M. A. Hoy and J. McKelvey, Jr. eds., pp. 41–49. Working papers, the Rockefeller Foundation.

Todd, C. T., and M. A. Carriker. 1922. The birds of the Santa Marta region of Colombia: A study in altitudinal distribution. *Ann. Carnegie Mus.* 14: 1–611.

Toledo, Z. D. A. 1978. Contribución al conocimiento de los Lepidopteros argentinos VI. *Eiseliana* nuevo género de Lycaenidae (Theclinae, Stimoni). *Acta Zoológica Lilloana* 33: 80–84.

Toxopeus, L. J. 1950. Geological principles of species evolution in New Guinea. *Proc. 8th Int. Congr. Entomol.*: 508–522.

Ureta, E. 1937. Lepidopteros de Chile. *Rev. Chil. Hist. Nat.* 40: 343–380.

———. 1941. Lepidopteros ropaloceros de Bolivia. *Bol. Mus. Nac. Hist. Nat.* Chile 19: 31–41.

———. 1955. Nuevas especies de Pieridae (Lep. Rhopalocera) de Chile y Argentina. *Bol. Mus. Nac. Hist. Nat.* Chile 26: 57–61.

van der Hammen, T. 1974. The Pleistocene changes of

vegetation and climate in tropical South America. *J. Biogeog.* 1: 3–26.

Verity, R. 1911. *Rhopalocera Palearctica.* Firenze (Italia): R. Verity Publ.

Vuilleumier, F. 1969. Pleistocene speciation in birds living in the high Andes. *Nature* (London) 223: 1179–1180.

———. 1970. Insular biogeography in continental regions. I. The northern Andes of South America. *Amer. Natur.* 104: 373–388.

———. 1977. Barrières écogéographiques permettant la spéciation des Oiseaux des hautes Andes. In *Biogéographie et Evolution en Amérique tropicale*, H. Descimon, ed., pp. 29–51. Paris: Publ. Lab. Zool. Ecole Normale Supérieure.

———. 1984. Zoogeography of Andean birds: Two major barriers; and speciation and taxonomy of the *Diglossa carbonaria* superspecies. *Nat. Geogr. Soc. Res. Reports* 16: 713–731.

Vuilleumier, F., and D. Simberloff. 1980. Ecology versus history as determinates of patchy and insular distributions in high Andean birds. In *Evolutionary Biology*, Vol. 12. M. K. Hecht, W. C. Steere, and B. Wallace, eds., pp. 235–379. New York: Plenum Press.

Watt, W. B. 1969. Adaptive significance of pigment polymorphism in *Colias* butterflies. II. Thermoregulation and photoperiodically controlled melanin variation in *Colias eurytheme. Proc. Natl. Acad. Sci. U.S.A.* 63: 767–774.

———. 1973. Adaptative significance of pigment polymorphisms in *Colias* butterflies. III. Progress in the study of the "*Alba*" variant. *Evolution* 27: 537–548.

Weberbauer, A. 1945. *El mundo vegetal de los Andes peruanos.* Lima: Minist. Agricultura.

Weeks, A. G., Jr. 1905. *Illustrations of diurnal Lepidoptera with descriptions*, part I. Boston: The University Press.

Weymer, G., and P. Massen. 1891. Lepidopteren gesammelt auf einer Reise durch Colombia, Ecuador, Peru, Brasilien, Argentinien und Bolivien in den Jahren 1868–77 von Alphons Stübel. In *Reisen in Süd-Amerika,* W. Reiss and A. Stübel.

Williams, C. B. 1930. *The migration of butterflies.* London: Oliver and Boyd.

Woodruff, D. S. 1973. Natural hybridization and hybrid zones. *Syst. Zool.* 22: 213–217.

Zeuner, F. E. 1962. Notes on the evolution of the Rhopalocera (Lep.). *Proc. 11th Congr. Entomol.* 1: 310–313.

Zischka, R. 1947. Catálogo de los Insectos de Bolivia. Primera contribución: Rhopalocera. *Fol. Univ.* (Cochabamba, Bolivia) 1: 27–36.

———. 1948. Catálogo de los insectos de Bolivia. Contribución número tres. *Fol. Univ.* 2: 3–5.

———. 1950. Catálogo de los insectos de Bolivia. Contribución número ocho. *Fol. Univ.* 4: 51–56.

———. 1951. Los piéridos bolivianos. *Fol. Univ.* 5: 7–32.

21

Origins of Lepidopteran Faunas in High Mountains of the Indo-Australian Tropics

JEREMY D. HOLLOWAY

The Indo-Australian tropics in the strict sense extend from the southeast Himalaya and India to northern Australia and Polynesia. The rest of the Himalayan range is only just outside the tropics and is included in this discussion. For, apart from the Himalaya and the Tibetan Plateau, most high mountains of the area are geologically extremely young, and only began to attain their present altitudes a few million years ago in the latter half of the Miocene and the Pliocene. Possible exceptions to this youth may be southeast New Guinea, the Philippines, Malaya, and western Borneo; mountains in Borneo may have been higher in the past and are now reduced by erosion.

The southern and eastern ranges of the Himalaya are much younger than the rest but, with the older Tibetan Plateau, the whole Himalayan complex is likely to have supported a montane flora and fauna over a long period of geologic time and, together with centers of diversity in higher latitudes, might be predicted to have acted as a source area for the biotas of the mainly much more recent mountains of Malesia (here defined as including Malaysia, the Philippines and Indonesia as far east as Sulawesi and Timor) and Melanesia (the Moluccas to Fiji and New Caledonia). (For a definition of the Malesian floristic region, see Chap. 18.)

Evolution and radiation of *purely* montane groups within Malesia and Melanesia are likely to have been much more restricted although, as will be seen, a significant proportion of the high-altitude biota has been derived through adaptation of taxa characteristic of lower altitudes, more especially in Melanesia. An overview of general geologic history and lowland biogeography of Malesia and Melanesia in relation to Asia and Australia is consequently essential to the understanding of the high-altitude biotas. The first few sections of this chapter will review the geology, phytogeography, and vegetation. The major section on the Geography of Montane Lepidoptera, will be devoted to all montane taxa, including the montane forest ones. The other chapter on Lepidoptera in this volume (Chap. 20) considered chiefly the taxa occurring above timberline. My focus is therefore somewhat broader than the other chapters, except Chapter 22 on African birds.

Lepidoptera, through their often specific larval food requirements, are closely associated with the flora of an area. Their distribution therefore might be expected to reflect phytogeographic patterns to some extent, though this does not hold in New Caledonia and New Zealand (Holloway, 1979; J. S. Dugdale, pers. comm.).

Extensive collections of Lepidoptera from high altitudes in the area are uncommon, but, fortunately, have been made relatively strategically. The rich collections of the British Museum (Natural History) contain much material from mountains in Borneo, Sumatra, Java, Luzon, Sulawesi, Ceram, Buru, and New Guinea. Material from these areas lacks precise ecological information with the exception of that collected by the author in Borneo. Ecologically based surveys have been made of Fiji, the New Hebrides (now Vanuatu), and New Caledonia.

Much collecting and taxonomic work remains to be done, so the treatment presented here is tentative and will certainly have to be modified in the future. Nevertheless a general picture is beginning to emerge. My approach is narrative and, for reasons outlined elsewhere (Holloway, 1982a), dispersalist.

GEOGRAPHY

The Indo-Australian tropics, unlike the tropical regions of Africa and South America, have a majority of areas of steep, dissected, high relief. The undulating, ancient peneplains of moderate altitude of Africa and South America have counterparts only in central India and northern Australia. The rest of the area is characterized by a young, mountainous landscape, punctuated by a few very low lying, often swampy flat areas (southern New Guinea, western Sumatra, and in Kampuchea and Thailand). The cratonic peneplains of central India and northern Australia have a monsoonal climate, or one with a marked dry season, leading to the development of savanna. Central India has probably been relatively arid throughout the Tertiary but the arid zone of Australia is more recent and the fringe of rain forest habitats in the very north and northeast was probably more extensive in the past (literature reviewed by Holloway, 1979). These dry or seasonally dry areas are not associated biogeographically to any extent with the tropical mountains and need not be considered in this chapter. Only southern China, especially in the warmer past, remains an area where lowland rain forest habitats may have been extensive and predominant over a long period of time.

Climatically one may distinguish an equatorial area extending from the lands on the Sunda Shelf (Sundaland: Malaya, Sumatra, Borneo and [though only in part climatically] Java) through northern Sulawesi and the Moluccas to New Guinea and northeastern Melanesia where rainfall is high and distributed relatively evenly through the year. Elsewhere within the tropics the rainfall is more seasonal, usually with a definite regular dry season.

The present-day distribution of land above 1000 m (Fig. 21–1) largely reflects the geologic structure of the area. The highlands of Burma, briefly interrupted at the Isthmus of Kra, continue in the mountains of Malaya (geologically older and somewhat distinct), Sumatra, Java, and the Lesser Sunda chain east to Timor. From southeast China a line of mountains extends from Taiwan through the Philippines and continues after a break east through the North Moluccas to New Guinea or south along the Minahassa Peninsula to Sulawesi. From Luzon the mountains of Palawan offer a connection to those of northern Borneo. The central range of New Guinea is the largest montane area in the Indo-Australian tropics after the Himalaya. It forms part of a "montane archipelago" that continues westward with the mountains of the Vogelkop in western New Guinea, of Ceram and Buru, and of Sulawesi and Borneo. In terms of geologic history this last group of montane areas is more disjunct and varied than the others, but has been brought into juxtaposition relatively recently by a continuing, massive, westward thrust along the Sorong Fault through the center of New Guinea.

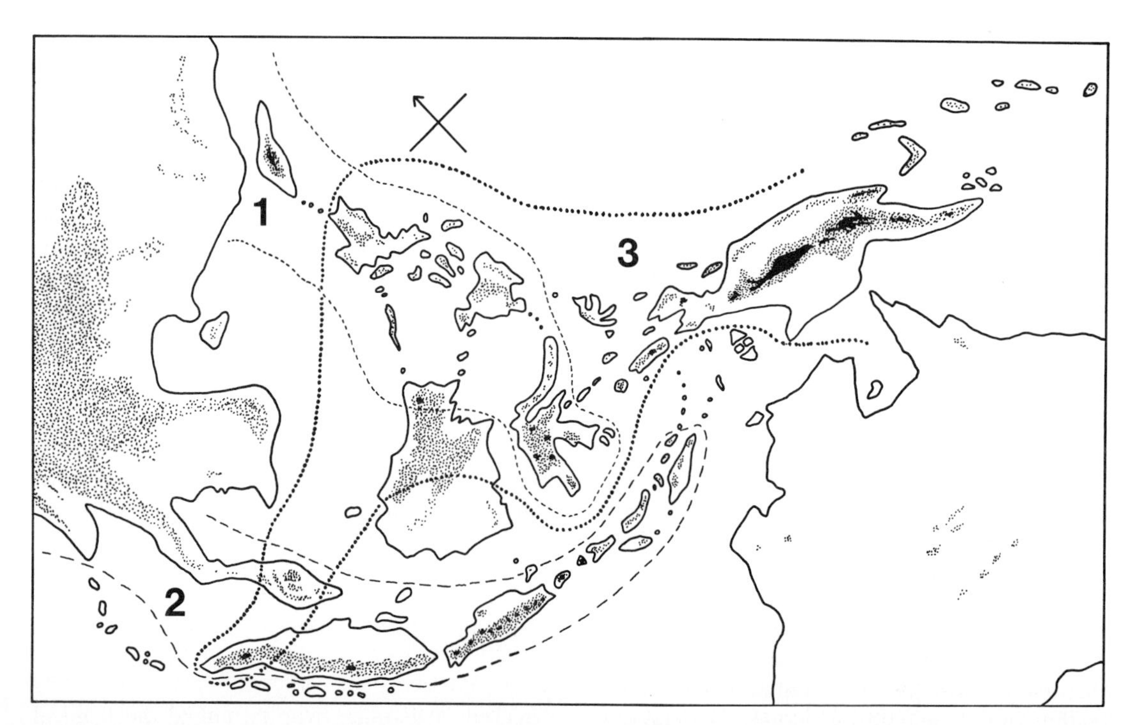

FIG. 21–1. Present-day geography of the Indo-Australian tropics, showing the trends of plant distribution described by van Steenis (1964). Areas where land over 1000 m is frequent are stippled and land over 3000 m is shown in black. 1: Luzon trend; 2: Sumatran trend; 3: Papuan trend.

The mountains of Sulawesi and Borneo, between the Philippine and Sumatran chains, might be expected to show some degree of biological interaction with them as well as with more easterly areas. Sulawesi in particular appears today as a sort of node for the three chains of mountains and might be expected to feature strongly in any widespread dispersal of montane organisms.

There is very little land over 3000 m (Fig. 21–1) in the 5000 km between the Himalaya and New Guinea. Several very isolated peaks reach above this altitude in the Sumatran chain (Leuser and Kerintji in Sumatra, nine or so, often still volcanic, in Java, and Rindjani in Lombok). The altitude of 3000 m is exceeded only by Kinabalu (4101) in Borneo where other mountains are very much lower, but reached nowhere in the Philippines, although a number of peaks in northern Luzon (Central Cordillera) and Mindanao (Mounts Apo and Ragang) approach it. Five peaks in central Sulawesi and one in central Ceram exceed 3000 m. (Many parts of central Taiwan are above 3000 m, but are strictly outside the tropics.)

Permanent snow occurs only in the Himalaya, Taiwan and New Guinea, areas which also have a forest limit or tree line. Frost has been recorded from the summit slopes of Mt. Kinabalu (Smith, 1980). Frost pockets in valleys, creating an "inverted tree line", occur on G. Leuser in Sumatra (J. Dransfield, pers. comm.). Elsewhere temperature, but not necessarily precipitation, declines uniformly with altitude, though perhaps variably from mountain to mountain (Flenley, 1979). More persistent cloud belts occur on many peaks from 1000 m upward, perhaps most commonly at 2000 m on major ranges in the equatorial zone, leading to the development of mossy cloud forest.

GEOLOGIC HISTORY

Indo-Australian tropical lands, with the exception of peninsular India and northern Australia, are all situated on or near the boundaries and zones of interaction of the Indian Ocean, Pacific, and Southeast Asian tectonic plates. The highly complex structure of the area must be interpreted in terms of the phenomena of island arc and marginal basin formation that are associated with zones where the oceanic parts of the major tectonic plates interact with each other or with continental crust.

The plate tectonic history of the area was reviewed in detail by Holloway (1979), so only a summary is presented in this section, with more detailed consideration of the geology of the lands on the Sunda Shelf.

From the Late Cretaceous to −30 my (million years ago) the portion of the Indian Ocean Plate west of Ninety-east Ridge moved northward from a southerly spreading ridge, bringing peninsular India into collision with Asia. This led to uplift of the Tibetan Plateau initially and, through the later Tertiary, to that of the southern Himalayan ranges with an east–west axis. A lateral component of the collision caused uplift of the north–south ranges of northern Southeast Asia, mainly in Burma.

After about −30 my the eastern portion of the Indian Ocean Plate, bearing most of Australasia, fused with the western portion and shared its northerly motion, gradually bringing Australia to interact with island arcs evolving in the Indonesian archipelago. Prior to this the biological isolation between Asia and Australasia was probably almost complete.

The earlier history of the eastern portion may have harmonized more with that of the Pacific Plate, which had a slower, more northerly motion than today. The formation of the North Tasman Sea was from −80 my to −65 my, separating New Caledonia and New Zealand from Australia to their current relative positions.

From −50 my to −30 my the South Fiji Basin was formed. The spreading axis involved may have extended round the north of Australia to continue in another axis detected in the Indian Ocean *east* of Ninety-east Ridge that was active over a similar period. Archipelagos may have begun to evolve along the northern and southern margins of this hypothetical spreading zone: the Outer Melanesian Arc, consisting of the North Moluccas, northern New Guinea, the Bismarcks, Solomons, New Hebrides, Fiji, and Tonga, and the Inner Melanesian Arc, more fragmented but perhaps with more substantial and older land areas, consisting of the northern part of New Zealand, New Caledonia, possibly Rennell Island, southeast and central New Guinea, and perhaps fragments of land farther west now associated with Ceram, Buru, Kei, Tenimber, and Timor. The Outer Arc may have had a continuation north to the Philippines.

At −30 my the motion of the Pacific Plate changed from northerly to westerly. The main active spreading axis in Australasia was thereafter not in the Fiji Basin but between Australia and Antarctica. The extensive Fiji Basin came under compression, the relative motion of the Pacific causing left lateral shear along its northern margin, the Outer Arc. The northerly motion of Australasia was accommodated by subduction south of the Outer Arc from the Moluccas to the New Hebrides, a process that continues today. These events led to progressive uplift in both arcs, their convergence in New Guinea resulting in the uplift of the massive central range.

In addition, the evolution of Sulawesi as a land area and of the Lesser Sunda chain (the Banda Arcs), together with the westward thrusting along the Sorong Fault in central New Guinea (south of the Outer Arc, now, from the Solomons westward, part of the westerly moving Pacific Plate), and uplift of associated fragments such as Ceram and Buru led to increasing biological contact between areas on the Sunda Shelf and tropical Australasia.

Geologic events connected with Sulawesi and the Sunda Shelf during this period are complex and not yet fully elucidated.

The Sunda Shelf is partially underlain by a basement of Paleozoic and Mesozoic rocks, which occurs in the Malay Peninsula and has its eastern extremity in western Sarawak and western Kalimantan, e.g., the Schwaner Mts.; it has been in its present latitude since at least the middle of the Cretaceous (Haile, McElhinny and McDougall, 1977). The rest of the Sunda Shelf has accreted to this basement during the Tertiary through development of small peripheral basins and island arc systems in the course of subduction of major plates to the east, south, and west of the basement. The complex of arcs composing the Philippines probably also developed during this period. Mountain ranges referable to these peripheral arc systems in Java and Sumatra were mostly uplifted in the Miocene and onward. The submerged areas of the Sunda Shelf between Singapore, Borneo, Sumatra, and Java are underlain by shallow , peneplained basement that has been low lying throughout the Tertiary, but provided sediments for the peripheral basins to the southwest from the Oligocene to the Miocene. In the Pliocene this area was probably low-lying, swampy, with local marine transgressions from the northeast (N. H. Holloway, pers. comm.).

Sulawesi on the one hand and the Banda Arcs on the other are probably two double-arc systems that have extended and developed at the margin of expanding back-arc basins developing southeast of the Sunda Shelf. The Banda Arcs are thought to have evolved by collision between the submarine continental margin of northern Australia-New Guinea and such newly formed volcanic arcs, producing the eastern part of the Lesser Sunda chain, Kei, Tenimber, Ceram, Buru, and fragments such as Misool (Carter, Audley-Charles, and Barber, 1976; Audley-Charles et al., 1979; Norvick, 1979). Thus Timor, Ceram, and Misool are underlain by fragments of the Australian craton. The main period of orogenesis in the Banda Arcs was in the Late Miocene and Pliocene, a period that probably also saw uplift in parts of Sulawesi. The northern part of the Banda and Sulawesi arc systems has been influenced by shear along the Sorong Fault system of New Guinea with the result that these areas are being thrust westward toward Borneo, leading to the flexure observed in both arcs today.

The history of the mountain systems in the area can be inferred to some extent from this train of events. The hypothesis of an old range of mountains from Asia to Australia, continuous in the past but now sundered and fragmented, favored by some phytogeographers (e.g., van Steenis, 1964) can be discounted on geologic grounds. The distribution patterns they sought to explain by this hypothesis are most readily understood in terms of modern geography (see also Chap. 18).

Mountainous land has probably existed north of India since the Late Cretaceous: the Tibetan Plateau is much older than the Himalayan chain and the mountains of Assam, Burma, and western China. High ground has existed in the Malay Peninsula since the Paleozoic and in western Borneo since the Mesozoic (N. H. Holloway, pers. comm.). Arnhem Land and the Kimberley Plateau in northern, cratonic Australia may have been more lofty in the past. Parts of the Philippines may also have been montane through the Tertiary and southeastern New Guinea may have had a longer subaerial and possibly montane history than the rest of Melanesia, comparable with that of New Caledonia.

All other major mountainous areas of Indonesia and Melanesia have been uplifted since the middle of the Miocene or later. East of the Sunda Shelf this orogeny was the result of the collision of the Melanesian Arcs in New Guinea and massive westward shearing along the Sorong Fault, caused by westward motion of the Pacific Plate relative to the Indian Ocean Plate.

This tentative overall picture provides a basis for discussion of the biogeographic patterns.

PHYTOGEOGRAPHY

The discussion in this section is drawn mainly from the reviews by Flenley (1979) and Holloway (1979), which contain many references.

The geologic history of the Indo-Australian tropics indicates that the floras and vegetation of Australasia and Southeast Asia developed largely in isolation from each other until the Miocene. The Early Tertiary pollen record from the then much warmer New Zealand includes only a trickle of tropical Asian or Malesian colonists.

At the beginning of the Miocene, tropical vegetation in the Melanesian Arcs was characterized by a mixture of old austral elements in southern families (Araucariaceae, Podocarpaceae, Casuarinaceae, Cunoniaceae, Epacridaceae, Monimiaceae, Proteaceae, and Winteraceae), and others with an ancient southern history (Myrtaceae), together with elements resulting from early radiations of plants of Malesian origin (Araliaceae, Lauraceae, Loranthaceae, Meliaceae, Palmae, Pittosporaceae, Rubiaceae, Rutaceae, Sapindaceae, and Sapotaceae), as well as Malesian elements of the Myrtaceae.

The geologic history suggests that this Melanesian Arc vegetation would have been found in young landscapes of steep relief in mainly maritime conditions except in tropical parts of northern Australia. Montane vegetation zones tend to occur at lower altitudes on small peaks near the sea than on large mountain ranges inland (the *Massenerhebung* effect; Flenley, 1979), so the Melanesian vegetation may have been preadapted to montane conditions. Such a hypothesis of "insular montane" preadaptation of the Melanesian flora could explain its subsequent history in interaction with the Malesian flora.

The historical development of the Southeast Asian flora has been reviewed by Flenley (1979). Angiosperms had largely replaced gymnosperms as dominant elements in the flora by the end of the Cretaceous. In the Early Tertiary flora the Dipterocarpaceae, Fagaceae, Leguminosae, Moraceae, and Palmae were probably common; these families are still amongst the most important ones today. The presence of Fagaceae might be taken to indicate the presence of highlands, in view of their current montane ecology in the tropics.

The geologic events of the Miocene permitted components of the vegetation of Asia and Melanesia to intermingle to a much greater extent than previously. Perhaps lowland elements were exchanged earlier than high-altitude ones, the isolating distances being less. On the other hand the insular nature of montane habitats may have favored the evolution of better dispersal powers in their flora (Smith, 1977).

The Melanesian flora, or elements of it, has been successful at spreading westward into Malesia only in the upper montane zone or in habitats with similar, poor, acid soils in the lowlands such as heath forest or forest on ultrabasic igneous rock. It has reached Sulawesi and Borneo, but also extends weakly to the Philippines, Malaya, and Sumatra.

Vegetation Zonation

Primary control of the development of altitudinal zonation is probably temperature (Flenley, 1979) and associated climatic factors involving precipitation, but the type of relief and its effect on drainage and soil formation are probably also important. Diurnal variations in leaf temperature at high altitudes may lead to physiological drought, and prevalent foggy conditions (in "cloud" forest) may impair transpiration. Thus montane vegetation is usually sclerophyllous.

Vegetation on Southeast Asian mountains can be placed into three forest types: lowland rain forest (up to 1000–1500 m), dominated by trees of the Dipterocarpaceae; lower montane forest (a narrow, ill-defined zone at 1000–2000 m), with Fagaceae such as *Quercus*, *Castanopsis*, and *Lithocarpus* abundant and the Guttiferae, Lauraceae, and Myrtaceae important; and upper montane forest (upward of 1500–2000 m), often mossy in zones of frequent cloud cover, with Fagaceae declining with altitude and dominant groups being conifers, Myrtaceae, and Ericaceae, the last becoming more important in high altitude forest of low stature. At very high altitudes on Mt. Kinabalu in Borneo various open habitats with herbaceous communities occur, though still within the forest limit (Smith, 1980).

In New Guinea the Malesian lowland dipterocarp zone is well developed and the fagaceous lower montane zone is also well represented. Immediately above the latter, *Nothofagus*, also fagaceous but with a southern history, dominates a more typically Melanesian zone. At higher altitudes still the Melanesian flora generally predominates, though Malesian elements such as Ericaceae do occur. At and above the forest limit herbaceous species become increasingly important, these being derived primarily from centers in temperate latitudes or the Himalaya; these habitats have been described by Smith (1975), and the history of this flora is discussed in Chapter 18.

Farther east in Melanesia (excluding New Caledonia, where the vegetation has unique characteristics associated with the antiquity of the flora and the extensive ultrabasic rock) the fagaceous zone is absent, lowland forest of primarily Malesian character but lacking dipterocarps giving way at about 1000 m to a forest dominated by conifers, Myrtaceae, and other elements of Melanesian character.

Plant Distribution Patterns

Van Steenis (1964) identified three major distribution trends (called tracks in Chap. 18) in montane plants in Malesia (Fig. 21–1): the Papuan trend, extension of Melanesian elements westward, primarily to Sulawesi and Borneo; the Luzon trend, extension of temperate and subtropical Chinese elements to northern Borneo and Sulawesi via Taiwan and the Philippines; the Sumatran trend, extension of Himalayan elements to Malaya, Sumatra, Java, and the Lesser Sundas. The trends follow the modern distribution of montane land very closely, given that extension of the Papuan trend taxa to the Philippines, Malaya, and Sumatra is weak. Van Steenis suggested these trends were relicts of less disjunct distributions established in the Late Cretaceous and Early Tertiary along more continuous mountain ranges, now considerably eroded and fragmented, because he assumed that long-distance dispersal of the plants concerned was not possible. Geographic, geologic, and ecological considerations (Holloway, 1970, 1978; Smith, 1977) do not support this hypothesis and indicate that these trends are of more recent origin (Pliocene or later) involving long-distance dispersal, perhaps most frequent during periods of glaciation and depression of vegetation zones in the Pleistocene (see Chap. 18). Palynological evidence (Muller, 1966) is consistent with a more recent date for these trends, as discussed later.

Most of the important Malesian montane plant genera, such as *Rhododendron*, *Vaccinium*, and *Quercus*, are centered in the Himalaya, whence presumably they dispersed south as montane land in Malesia became uplifted and more frequent. In addition, it is likely that the old mountains of western Borneo and Malaya also would have played a role, albeit perhaps secondary, in the development and radiation of montane floristic elements in Malesia. Many Himalayan montane genera have secondary centers of diversity in the region, for example the fagaceous genus *Castanopsis* of the lower montane zone (Corner, 1978). Primary centers of diversity occur in other, largely montane genera. Perhaps the most striking of these is the pitcher plant genus *Nepenthes*, mainly montane, but also occurring in comparable lowland areas with acid, nutrient-poor soils such as heath forest. Almost half the known species of *Nepenthes* are recorded from Borneo, most being endemic. Figs, though not typically montane, attain their greatest diversity in Borneo (Corner, 1978).

The fossil record (Muller, 1966) indicates that prior to the Pliocene arrival of conifers of Australasian origin in montane Borneo, the northern temperate and montane genera *Pinus*, *Picea*, *Tsuga*, and *Alnus* grew there in the Oligocene and Miocene. This suggests that montane areas of significant extent and stature occurred in Borneo at that time. Of the conifers, only *Pinus* occurs on the Sunda Shelf today, growing in the higher ranges of northern and central Sumatra; the genus also grows in the Central Cordillera of Luzon.

Raven (1973) has suggested that many of the alpine and subalpine plant genera of New Zealand, often including large monophyletic groups of endemic species, did not arrive in New Zealand until the Pliocene, colonizing by dispersal across the Tasman Sea from Australia. Most belong to obligate temperate or montane genera with their major development in the Northern Hemisphere. An hypothesis of recent northern origin for such genera presupposes a combination of good dispersal powers and the presence of a chain of suitable montane stepping-stones through the Malesian tropics. A plausible route to Australia in the Pliocene might have been via Taiwan, Luzon, Borneo, Sulawesi, Ceram, and New Guinea—in effect the tropical localities for the genus *Euphrasia* as mapped by van Steenis (1971).

The Glaciations

The Pleistocene glaciations were probably accompanied by great changes in the distribution of vegetation types in the Indo-Australian tropics. The summit of Kinabalu (Myers, 1978) and much of the central range of New Guinea (Flenley, 1979) were extensively glaciated, and other high peaks in the area also could have been affected. During the time of the last glaciation there are indications of a general temperature drop in tropical montane areas, perhaps as much as 10° C. This was probably associated with a generally drier climate, though not serious enough to affect the viability of lowland rain forest (Flenley 1979). At the same time a fall in sea level of about 100 m exposed most of the Sunda Shelf and also the Sahul Shelf between Australia and New Guinea, as well as many smaller areas now covered by shallow seas.

The effects of these changes on the vegetation are as yet incompletely understood, investigated mainly by means of pollen analyses from stratified deposits of known age (Flenley, 1979, reviewing his own extensive work in this field; Chaps. 7 and 8). The forest limit in equatorial areas was generally depressed by about 1000–1500 m, occurring then at about 2000 m. The upper limits of lowland forest may have been depressed by less, perhaps only by 500 m. Even so, this would mean that a significant portion, perhaps a third, of the *present* land area of Malesia was under montane vegetation of some sort.

The exposed continental shelf areas would no doubt have become forested, but perhaps by no more than swamp or heath forests of low diversity as occur in flat alluvial basins or on raised terraces of unconsolidated sediment in the lowlands of Malesia today. Thus it does not necessarily follow that the rich lowland dipterocarp forest would have become more extensive during periods of low sea level; the reverse could equally well have occurred with the depression of the upper limit of this forest type.

The Pleistocene increases in areas of montane vegetation would naturally have been associated with reduction in the distances between them. These recent most optimum conditions for spread and radiation of montane organisms are likely to have produced most of the patterns of specific and even generic distribution observed today; patterns of distribution and radiation of greater antiquity are likely to have been obscured or blurred, making them hard to recognize, particularly in mobile and evolutionarily plastic organisms such as Lepidoptera.

THE GEOGRAPHY OF MONTANE LEPIDOPTERA

The discussion of Lepidoptera distribution will be restricted to the superfamilies mentioned later in this section. Collection and investigation of the "microlepidoptera" in the Indo-Australian tropics have been much more haphazard, but reference should be made to the work of Diakonoff (1952–55, 1967) in New Guinea and the Philippines.

The taxonomic treatment of the macrolepidoptera has generally been more detailed and somewhat more reliable, but it is still far from perfect and biogeographic inferences made from the modern classification should be regarded with extreme caution. A preliminary program of genitalia dissection was undertaken in two geometrid subfamilies (Ennominae, Larentiinae) that contribute a major portion of the tropical montane species, in order to check the validity of generic assignments, particularly of species groups endemic to New Guinea. The results often did not support the current classification and are therefore presented in some detail to assist future workers.

In every montane area of the Indo-Australian tropics the lepidopteran fauna can be divided into two components. In the first belong species derived from montane centers well outside the area concerned. In the second component belong those species derived, by a process of adaptation, from taxa in the surrounding lowlands. The adaptation process has obviously predominated in the primary source area for tropical montane Lepidoptera, the Himalaya.

The Role of the Himalaya in Indo-Australian Lepidoptera Geography

In the Himalaya (including here the Tibetan Plateau as well as the southern ranges and those of western China and northern Burma) plants and animals were able to adapt over a long period of time, from the Early Cretaceous onward, to new ecological conditions as the mountain ranges were uplifted. Within the Himalaya there has been some segregation of centers of speciation and radiation, the products of these centers dispersing to interact in intervening localities (Holloway, 1969, 1974) . Though moths have yet to be sampled along an altitude transect in the Himalaya, it is likely that most Malesian montane moths are drawn from below the forest limit there rather than from the alpine zone.

Butterflies in the alpine zone of the Himalaya belong almost entirely to predominantly Palearctic genera centered in the western Himalaya, Chitral, and Pamirs where previous adaptation to aridity and seasonality, and to consequently more open habitats, may have led to a greater facility for colonization of habitats of rather inhospitable nature above the tree line.

Middle-altitude Himalayan butterflies belong mainly to genera centered in the northeastern Himalaya or western China that also extend strongly into the Malesian tropics. Within the eastern Himalaya can be identified a "distributional continuum" from genera centered in the Assamese region of the northeastern Himalaya and northern Burma, with a strong representation of lowland and lower montane species in Malesia, to those with more northerly centers in western China, contributing montane species to Malesia but with greater representation in the Palearctic, particularly the eastern Palearctic of China, Japan, and Korea.

Thus the alpine zone of the Himalaya is biogeographically fairly distinct from zones below the tree line, species from western- and eastern-centered genera being segregated from each other with altitude in intervening areas such as the Himalaya of Nepal (Holloway, 1974). Insects of the alpine zone encounter conditions of seasonality with winter temperatures below freezing; rarefaction of the atmosphere at such altitudes leads to specific problems of aridity and insolation. The ecology of insects in the Himalayan alpine zone has been discussed in detail by Mani (1968).

Tropical-alpine habitats in Malesia are to be found to any great extent only in New Guinea. Conditions of cold, atmospheric rarefaction and consequent aridity tend to be more uniform through the year, presenting somewhat different problems to the high-altitude biota in terms of adaptation. General hardiness and tolerance of such conditions would be expected to be more important than development of seasonal diapause in insects. Seasonal fluctuations in the Lepidoptera fauna at somewhat lower altitudes (2000–3000 m) are not evident in New Guinea (Hebert, 1980).

There is little information on the Lepidoptera of the New Guinea tropical-alpine zone. Diakonoff (1952–55) indicated that any specialized fauna was probably very small, perhaps absent from some mountains. Most of the species that do occur tend to be consistent in their characteristics with the youth of high elevations in New Guinea. Ten species of microlepidoptera out of 44 thought by Diakonoff to be characteristic of the tropical-alpine zone belong to eight cosmopolitan or circumtropical genera, and 11 belong to eight endemic genera, two of which (*Emmetrophysis*, a gelechiid, and *Orochion*, a lyonetiid) are endemic to the zone.

The remaining 23 species are endemic and belong to 14 genera of more restricted distribution. Seven of these genera were restricted in New Guinea to the tropical-alpine zone. Two, *Aeolostoma* (Tortricidae) and *Antiopala* (Oecophoridae), are otherwise known only from Australia, and one, *Harmologa* (Tortricidae), is a New Zealand genus. Of the other four, three are otherwise confined to Europe and North America (*Cosmiotes*, Elachistidae; *Ochromolopis*, Epermeniidae; *Eidophasia*, Plutellidae) and the fourth (*Argyresthia*, Yponomeutidae) is virtually restricted to these areas but has one species in Australia among a few other stragglers. There is thus a small purely Holarctic element in the New Guinea tropical-alpine Lepidoptera.

Trends of Dispersal within the Tropics

Most of the truly montane Lepidoptera of the lands on the Sunda Shelf, of the Philippines, and of Sulawesi are shared with the Himalaya or are endemic taxa with close relatives in the Himalaya. Perhaps two thirds of the species characteristic of the four highest altitudinal elements on Kinabalu recognized by Holloway (1970) are of this nature. Many Himalayan species have dispersed widely through the Malesian mountains, such as *Euplexia albovittata* Moore (Table 21–1); other examples will be given later.

Correlation with Montane Plant Trends

More localized dispersals often reflect the trends noted for plants by van Steenis. Several noctuid

TABLE 21–1. Representation of various genera in the mountains of the Indo-Australian tropics and elsewhere, and the distribution of *Euplexia albovittata*. None of the taxa have been recorded from the Lesser Sunda Islands.

	Loxofidonia	*Xanthorhoe*	*Chloroclystis*	*Synegia*	*Carea (Calymera)*	*Euplexia*	*E. albovittata*
Himalaya	3	6+	3	3	12	15	+
China	?	?	?	4	2	?	+
Taiwan	1	3+	?	3	?	?	+
Malaya	—	1	6	3	26	1	+
Sumatra	1	1	—?	5	30	1	+
Java	1	3	3	—	9	1	+
Borneo	—	2	10	10	53	9	+
Bali	1	—	—	—	—	—	+
Sulawesi	1	3	4	8	10	2	+
Ceram and Buru	—	3	3	5	6	1	+
New Guinea	3	1	4	21	6	13	—
Bismarcks	—	—	—	—?	—	1	—
Australia	—	4+	?	1	1	—	—
New Zealand	—	Many	Many	—	—	—	—

species of the temperate and tropical montane subfamily Pantheinae and species of the Thyatiridae have extended along the Sumatran trend from the Himalaya, thence north to Sulawesi in one instance.

A number of Palearctic and Chinese species have reached the mountains of Luzon from Taiwan but only two instances involve Mt. Kinabalu in Borneo. The species are larentiine geometrids belonging to predominantly Palearctic and Himalayan genera. *Cosmorhoe* is centered in the eastern Himalaya. *C. chalybearia* Moore extends from the northeastern Himalaya to Taiwan. The related species *C. dispar* Warren and *C. moroessa* Prout fly in the mountains of Luzon and Mindoro and on Kinabalu, respectively.

The second genus is *Dysstroma*, now included by some authors in *Chloroclysta*. It is centered in the mountains of western China and is strongly represented in the Holarctic region. A trio of very similar species is distributed along the Luzon trend, with *D. fumata* Bastelberger in Taiwan, *D. heydemanni* Prout in montane Luzon, and *D. pendleburyi* Prout common at 2100 m on Mt. Kinabalu. Related species also fly in Java (*D. cuneifera* Warren) and Sumatra (*D. ceprona* Swinhoe).

Several species are exclusive to the mountains of Borneo and Luzon, such as the geometrid *Idaea themeropis* West and the noctuid "*Hypopteridia*" *luminosa* Wileman and West; others, such as the geometrids *Spaniocentra apatella* West, *Collix mesopora* Prout, and *Hydrelia flexilinea* Warren occur also in Sulawesi.

Species ranging from the Himalaya to New Guinea are either distributed along the Sumatran trend to Java, Bali, and Lombok, thence to Sulawesi, Ceram, Buru, and New Guinea (as in the *Loxofidonia* species, all closely related, illustrated in Table 21–1), or along the Luzon trend to Borneo, thence to Sulawesi and on as before. Even the species distributed widely through the mountains on both trends are usually unrecorded from the Lesser Sunda Islands or northern Moluccas, but this impression may be altered when these areas have been better collected. Thus Himalayan species, when they colonized New Guinea, appear to have dispersed along the Papuan trend from west to east.

About 10 Geometridae are shared between Borneo and the mountainous areas of the Papuan trend. They include three *Anisodes* (Sterrhinae) and *Ruttellerona lithina* Warren (Ennominae), the rest belonging to the subfamily Larentiinae. Many of these may have originated in New Guinea and dispersed westward, such as "*Horisme*" *labeculata* Prout, shared with New Guinea, where numerous related species occur, and Sulawesi; two further species are endemic to Borneo though with relatives in New Guinea. Three *Micromia* species are common to Borneo and New Guinea. Most *Micromia* are found in the mountains of New Guinea but the genus requires revision. Three *Sauris* species, *S. arfakensis* Joicey and Talbot and the related pair of *S. erecta* Warren and *S. ceramica* Rothschild, are shared between Borneo and the southern Moluccas, with *ceramica* also in Sulawesi and *arfakensis* in New Guinea. These three species may also have dispersed westwards. *S. erecta* is now placed in *Tympanota* by Dugdale (1980), a genus centered in New Guinea and strongly represented along the Papuan trend; *S. ceramica* is probably also a *Tympanota*.

Transtropical Genera

Two groups of Larentiinae appear to correspond generally in their distribution with the northern temperate, tropical montane, and New Zealand alpine plant genera discussed by Raven (1973), and might therefore also be suggested to represent a Pliocene or later dispersal across the equator from north to south.

The first group is best referred to the genus *Xanthorhoe*, here defined by the possession in the male genitalia of a prominent, curved, setose club, or calcar, arising centrally between the bases of the valves. The valves generally consist of a ventral, weakly sclerotized apical portion, with the dorsal costal margin more sclerotized and often produced into some sort of projection. The males generally have bipectinate antennae and, in common with a number of other larentiine genera such as *Piercia*, a pair of eversible corematous sacs between the seventh and eighth abdominal segments.

Dugdale (1971) placed New Zealand and Australian calcar-bearing larentiines in the genus *Helastia* as these did not bear very close resemblance to the type species of *Xanthorhoe*, the Palearctic *X. montanata* Denis and Schiffermuller. *X. montanata* is morphologically somewhat distinct from its Palearctic congeners, which resemble *Helastia* more, but it is perhaps most informative at this stage of our knowledge of larentiine taxonomy to group all the calcar-bearing species under *Xanthorhoe*.

The southern *Xanthorhoe* prove not to be so isolated biogeographically as Dugdale (1971) originally considered. Several montane species are recorded from a number of localities through the Indo-Australian tropics (Table 21–1). A group of species with greenish forewings and a rounded valve apex and reduced costal projection to the male genitalia is centered in the northeastern Himalaya (e.g., *X. curcumata* Moore and *X. curcumoides* Prout, the latter possibly occurring also on Kinabalu [Holloway, 1976]), with species also in Taiwan. The group is represented by *X. fissiferula* Prout in Sumatra and *X. hyphagna* Prout in Java. *X. saturata* Guenée extends from the Himalaya to Taiwan, whence presumably it dispersed to the mountains of Luzon. Two stout species with green markings on the forewings from Ceram, *X. hedyphaes* Prout and *X. pratti* Prout, are not closely related to other green species though they are definitely members of *Xanthorhoe*.

Two species of uncertain affinity, *X. ludifica* Warren and *X. nubilosa* Warren, occur in the mountains of Java and two new and distinctive, but as yet undescribed species were taken in Sulawesi by Sutton and Rees.

Associated mainly with the Papuan trend of mountains is a series of species with rather similar pinkish gray wings and sexual dimorphism, though only *dissociata* Warren from Luzon and *mesilauensis* Holloway from Kinabalu are confirmed as closely related by the structure of their male genitalia. The other species are *X. everetti* Warren from Sulawesi, *X. callisthenes* Prout from Ceram, and (not dissected) *X. simplicata* Prout from Buru.

Few calcar-bearing species have so far been identified in New Guinea. One is *X. vulgaris* Rothschild, a slender, pale montane species, possibly allied to *X. finitima* Walker from New Caledonia. *X. pallida* Rothschild from New Guinea may also be related, but has not been dissected.

A number of New Guinea species usually attributed to *Xanthorhoe* transpire on dissection to be unrelated, having no calcar and complex valve structure. These include: a related trio, *albiapicata* Warren, *fulvinotata* Warren, and *bifulvata* Warren; *coeruleata* Warren with asymmetric male genitalia; and a group of five, large, satiny brown species, *monastica* Warren, *interrufata* Warren, *cerasina* Warren, *succerasina* Prout, and *lucirivata* Warren. The relationships of these groups of species are obscure; all are montane.

Another Australasian group of species ill placed in *Xanthorhoe* appears related to the Himalayan *sordidata* Moore, and may therefore provide another example of a transtropical montane dispersal. The generic name *Visiana* is based on *sordidata*, which also occurs in the mountains of Sumatra, Java, and Sumbawa and on Kinabalu, and so this name could be applied to the whole group. In the Australasian species the male genitalia are less robust, more slender than in *sordidata*, though all lack a calcar and have the valve apex produced into a spine. The group includes the eastern Australian *brujata* Guenée, *vinosa* Warren from New Guinea and Buru, *hyperctenista* Prout from the Vulcan Islands, and an unnamed species from Sulawesi.

Further *Xanthorhoe* species from the area may be found to have been misplaced in other genera. For example, the genus *Loxofidonia* is based on a North American calcar-bearing species and at present includes a homogeneous group of montane Indo-Australian tropical species (Table 21–1) that would perhaps be better placed in *Xanthorhoe*. Three, *buda* Swinhoe, *obfuscata* Warren, and *bareconia* Warren, fly in the northeastern Himalaya, the last also occurring in the mountains of Sumatra, Java, and Bali. In Sulawesi and western New Guinea there occurs *sigmata* Prout, and two more species, *lipernes* Prout and *plumbilinea* Warren, occur farther east in New Guinea. There are single species in the mountains of Sri Lanka and Taiwan.

The extent of *Xanthorhoe* in Australia is not yet fully ascertained as so few of the Australian species attributed to the genus have been dissected. Those recognized so far include *X. strumosata* Guenée, *X. subidaria* Guenée, *X. centroneura* Meyrick, and *X. sodaliata* Walker. The last extends to Norfolk Island, has a very close relative in New Zealand, and is either vagrant to or resident in New Caledonia (Holloway, 1977, 1979). The group is therefore mobile.

Dugdale (1971) attributed 20 New Zealand species to *Helastia*, here included broadly with *Xanthorhoe*. Most are moss feeders though some feed on herbaceous plants like the majority of their Palearctic relatives (J. S. Dugdale, pers. comm.). The hypothesis that these species radiated recently in New Zealand, following derivation via Australia across the mountains of the tropics from the north, seems plausible.

The *Xanthorhoe* group is also distributed through high-altitude localities in the African tropics (e.g., Fletcher, 1958) to temperate southern Africa.

The second group of Larentiinae is based in the broad sense on the genus *Chloroclystis* in the tribe Eupitheciini. It can also be defined on the basis of characters of the male genitalia: the lack of an uncus; an hourglass-shaped juxta, the waist associated with prominent ventral labides from the transtilla; a spined ventral manica pad, often bearing two much larger "horns"; prominent cornuti in the aedeagus vesica; broad valves, rounded apically, with a broad costa and well-defined saccular margin, the latter lobed basally whence arise several long hairlike setae; a pair of projections or octavals on the eighth sternite, well developed and separated. Such genitalia have been illustrated for British species by Pierce (1914), for Bornean ones by Holloway (1976), and for New Zealand ones by Dugdale (1971), the last attributed to the genus *Pasiphila*.

Many Oriental and Australasian tropical montane species generally attributed to *Chloroclystis* will eventually have to be excluded; the group requires revision.

Genuine *Chloroclystis* are widespread but not diverse in the Palearctic region and Himalaya, have a center of diversity in Borneo, and are moderately represented also in Sulawesi, Ceram, and New Guinea (Table 21–1). There are numerous species in New Zealand, referred by Dugdale to *Pasiphila*. Australian members of the group have not yet been identified, though *C. testulata* Guenée, recorded from Australia, New Zealand, and Norfolk Island, has several characteristics in common with the group. It is distinguished in the male genitalia by a setose papilla central to the valve. Its presence on Norfolk Island is testimony of its mobility. *C. nina* Robinson, one of the few definitely montane Fijian species (Robinson, 1975), is very closely related to *testulata*.

Some species of the heterogeneous and primarily New Guinea genus *Micromia* have a number of characteristics in common with *Chloroclystis* in their male genitalia, such as the lack of an uncus, the general shape and setation of the valves, and the prominent octavals (*latistriga* Prout and *fletcheri* Holloway, both common to Borneo and New Guinea). They are distinguished by highly complex coremata at the base of the valve and modifications to the male hindwing involving scent scales. One of the Kinabalu *Chloroclystis*, *C. layanga* Holloway, also has the male hindwing slightly modified with scent scales.

Here again the hypothesis of transtropical derivation of the New Zealand representatives of the group in recent times is plausible, but the pattern is not as clear as for the *Xanthorhoe* group. The center of diversity in Borneo and the possibly related and divergent *Micromia* species in New Guinea might be indicative of a somewhat longer history for the group in the mountains of the Indo-Australian tropics.

Other examples of apparently recent spread of northern temperate taxa into Australasia include the noctuid genus *Diarsia* and two butterfly groups. The *Vanessa* group of the Nymphalidae contains a number of highly mobile species, so it is not surprising to find several species in the mountains of Java, Luzon, and Sulawesi and two species in Australia and New Zealand (Leestmans, 1978).

More surprising is the occurrence of three *Lycaena* species in New Zealand, well separated from what appear to be their closest relatives in eastern Asia. The discovery of two species of a new lycaenine genus (*Melanolycaena*) at 2000–3000 m in the mountains of New Guinea may not in itself indicate the route of access of *Lycaena* to New Zealand, as the New Guinea genus has most in common morphologically with the *Heliophorus* section of the subfamily, characteristic of tropical and subtropical montane forest (Sibatani, 1974). The discovery does at least give some indication of the dispersal ability of these Lycaenidae. But Sibatani did not entirely rule out the possibility that the New Zealand and New Guinea lycaenines are monophyletic, representing a remnant of a now largely distinct old lycaenine stock, a branch of which may have invaded Australasia from Southeast Asia during the Pleistocene.

Himalayan Genera Extending to New Guinea

Several other northern temperate or Himalayan genera would appear to have spread through the mountains of Malesia to radiate in New Guinea, and no doubt others will become apparent when the taxonomy and affinities of endemic montane groups in New Guinea have been studied further. Four examples are worth mentioning.

The larentiine genus *Piercia*, contains several Himalayan and Palearctic species as well as a number of species in montane tropical and southern temperate Africa (Fletcher, 1958). Key characters of the male genitalia are a sclerotized and produced costa to the valve and a pair of prominent slender hooks arising from an extension of the transtilla. Fletcher (1958, fig. 183) has illustrated these characters. The group shows some relationship with *Xanthorhoe* in the valve costal projection and in having eversible coremata between segments seven and eight in the male abdomen, but it does not have a prominent calcar. The monobasic Himalayan species *Apithecia viridata* Moore is also related to *Piercia*. Related, perhaps synonymous genera are *Xenoclystia* Warren (*X. nigroviridata* Warren [northeastern Himalaya] and *X. delectans* Warren [New Guinea] dissected) and *Desmoclystia* (*D. unipuncta* Warren and *D. falsidica* Warren [both New Guinea] dissected). *Xenoclystia* has a species extending from the northeastern Himalaya to Burma, 2 more in Burma and 2 in New Guinea, and *Desmoclystia* has 18 species in the mountains of New Guinea. The whole group requires revision.

Taxonomic revision is also needed to clarify the picture concerning the ennomine geometrid genus *Myrioblephara*. This contains a number of Himalayan and Chinese species, one (*simplaria*

Swinhoe) ranging from the Himalaya via Java, Borneo, and the Philippines to Ceram and New Guinea, another (*flexilinea* Warren) with similar range but excluding the Himalaya and including Sulawesi, 42 species endemic to New Guinea, and 3 in the hills of Queensland. The genitalia indicate affinities with the Holarctic genus *Aethalura*, perhaps even synonymy.

The third group, the mainly tropical pierid butterfly genus *Delias*, ranges from the Himalaya and China to Australia and New Caledonia. A very high proportion of the species are montane, flying between 1000 m and 3000 m (Talbot, 1928–37). Talbot, in his revision of the genus, recognized a number of species groupings. Several of these range widely through the Indo-Australian tropics or are predominantly Malesian, but others range from the Moluccas to the Solomons only and the largest number of groups is found in New Guinea only, each with several species. These last two distribution patterns may reflect rather dramatically the segregation of the two Melanesian Arcs discussed earlier, but no attempt has been made yet to test the validity of Talbot's groupings. The genus feeds almost exclusively on the plant family Loranthaceae, already noted as an Early Tertiary Malesian invader of Australasia.

The last example, like the *Piercia* group, is of a genus that has also dispersed in the mountains of Africa. The hadenine noctuid genus *Apospasta*, as currently defined, has two species in the Himalaya, one of which extends to Java and Luzon, together with a number in the mountains of tropical Africa (Fletcher, 1961) and one on the island of Reunion. The species referred by Holloway (1976) to the genus *Hypopteridia* have male genitalia that indicate close affinity with *Apospasta* with which perhaps they should be combined. One species, *nyei* Holloway, is endemic to Borneo, another, *luminosa* Wileman and West, is common to the mountains of Borneo and Luzon, and a third, *stigma* Joicey and Talbot, flies in the mountains of New Guinea.

A Southern Temperate Genus Extending Northward

There is only one definite instance among the "macrolepidoptera" superfamilies where a southern temperate-centered genus has contributed species to the mountains of Malesia. The larentiine geometrid genus *Poecilasthena* is centered in southern Australia with a few species in New Zealand. Recorded foodplants are *Leptospermum* (McFarland, 1979) and *Styphelia* (J. S. Dugdale, pers. comm.), both belonging to families, Myrtaceae and Epacridaceae respectively, with a long history of radiation in Australasia. Both genera have spread westward into the Malesian tropics, mainly in the mountains. *Poecilasthena* would appear to have followed its foodplants westward. Two species occur over a range of altitude in New Caledonia, one shared with Fiji and Samoa. Single species occur in the mountains of Queensland, New Guinea, Sulawesi, Borneo, Luzon, Malaya, and Burma.

The Composition of the Montane Faunas

A large number of the species and genera mentioned in the previous section are Geometridae, particularly of the subfamily Larentiinae. A family-by-family survey of the Lepidoptera reveals great disparity of representation in Indo-Australian montane habitats. The following account makes particular reference to Borneo as detailed information from other localities is mainly lacking.

Zygaenoidea and Cossoidea

The superfamilies Zygaenoidea and Cossoidea have few montane species. In the former the subfamily Chalcosiinae are virtually restricted to the lowlands; the Zygaeninae include the diverse Palearctic genus *Zygaena*, which is characteristic of open habitats, often fairly dry, but does not feature in the forested mountains of the tropics. In the Cossoidea the only truly montane member of the Limacodidae in Borneo is *Thosea kinabalua* Holloway, related to Philippines species. Among the Cossidae only those genera that extend weakly to the Palearctic include tropical montane species, namely, *Cossus* and *Zeuzera*. In the Metarbelidae a distinctive species, *Indarbela kinabalua* Holloway, is found only in the mountains of Borneo. Other members of the family are restricted to the lowlands.

Bombycoidea

The Bombycoidea are also poorly represented with a handful of montane Sphingidae, a few Lasiocampidae, mainly of the predominantly lowland genus *Trabala*, and the *Loepa katinka* Westwood group and some *Antheraea* and *Cricula* species of the Saturniidae. The Eupterotidae, predominantly tropical and apparently flying mainly in the rain forest understory, include the genus *Melanothrix*, which contains several montane species as well as lowland ones, the latter often frequenting heath forest habitats. *Mela-*

nothrix has its center of species richness in Borneo. *Tagora weberi* Holloway is a species of upper montane forest in Borneo.

Noctuoidea

The superfamily Noctuoidea has greater representation at high altitudes, but not all families participate to the same extent. The Lymantriidae are predominantly lowland (Holloway, in prep.) but a number of *Calliteara*, *Lymantria* and *Euproctis* species are characteristically montane; all three genera include a few Palearctic species.

The Notodontidae are well represented, with a number of Himalayan elements such as species of the genera *Fentonia* and *Suzukiana* flying in the mountains of Sundaland. The most characteristic montane notodontid genus is *Quadricalcarifera*, which, though it contains a few lowland species, attains its greatest diversity in the mountains of the Himalaya, Sundaland, and New Guinea. The genus *Cascera* contains 12 species that would appear to have radiated within the mountains of New Guinea, only one being recorded elsewhere, from Queensland. A few smaller notodontid genera are also endemic to, and mainly montane in, New Guinea. The Notodontidae provide a significant component of the Palearctic fauna also.

The primarily tropical Agaristidae, day-flying, are rarely recorded at high altitudes. The Arctiidae and Ctenuchidae are perhaps richest in species in the lowlands but also have many exclusively montane species. In Borneo the montane Arctiidae include a number of *Eilema* species, all four Bornean *Agylla*, several *Spilosoma* (a large genus extending into both northern and southern temperate regions in the area but most diverse in the tropics), and montane species of predominantly lowland tropical genera such as *Nyctemera*, *Chionaema*, and *Asura*. *Asota kinabaluensis* Rothschild is a denizen of upper montane forest in Borneo, morphologically somewhat distinct from its several lowland congeners. The Bornean Ctenuchidae include a number of *Callitomis* species apparently endemic to individual mountain peaks, such as *C. trifascia* Holloway from Kinabalu.

There is great disparity in representation among the subfamilies of the Noctuidae and, with a few exceptions, the ones with high montane representation are also those predominating in the Palearctic. This correlation generally holds throughout the macrolepidopteran families. Departures, such as in many butterfly families, the geometrid subfamily Sterrhinae, and the noctuid subfamilies Hadeninae, Heliothinae, and Plusiinae, where high representation in the Palearctic is not associated with high representation in tropical mountains, would appear to be related to general adaptation to open habitats and herbaceous or graminaceous larval feeding habits.

The trifine subfamilies Noctuinae, Hadeninae, and Acronictinae predominate at higher altitudes in Malesia and also in the Palearctic fauna. The Heliothinae are open- and disturbed-habitat specialists of tropical and warm temperate latitudes and do not contribute to montane faunas. The primarily Palearctic Cuculliinae are not represented but these, with the exception of the genus *Cucullia* itself, are predominantly vernal and autumnal (as adults) specialists that may be unable to tolerate the seasonally uniform conditions characteristic of tropical mountains. A similar blend of trifine subfamilies is predominant at high altitudes within tropical Africa (Fletcher, 1961).

The Noctuinae are represented by the genera *Agrotis*, *Xestia*, and *Diarsia*. *Agrotis* contains a number of pest species (open-habitat specialists) and also occurs at lower altitudes. Though centered in the Northern Hemisphere, the genus has been successful in colonizing parts of Australasia and the Pacific, a number of localized species occurring there together with more widespread ones. *Agrotis kinabaluensis* Holloway is a high-altitude Bornean relative of the Palearctic *A. cinerea* Schiffermuller; *A. magnipunctata* Prout is also related, flying in the mountains of Ceram and Buru.

Xestia and *Diarsia* have their centers of species richness in the mountains of western China and include numerous Holarctic species. In the former genus races of the widespread Palearctic *X. c-nigrum* Linnaeus or closely related species occur in the mountains of Sri Lanka, Sundaland, Luzon, and Sulawesi. *Diarsia* has been more successful at penetrating the tropics. Three species range from the Himalaya to the mountains of Sundaland, namely, *D. nigrosigna* Moore, *D. stictica* Poujade, and *D. ochracea* Walker. *D. flavostigma* Holloway (Borneo and Sumatra) is part of a complex of species represented also in Sulawesi and New Guinea and by *D. intermixta* Guenée in Australasia. The last has colonized both New Zealand and Norfolk Island, presumably from Australia. Three morphologically isolated species, *D. banksi* Holloway, *D. barlowi* Holloway, and *D. serrata* Holloway, are endemic to Kinabalu; such endemism has not been recorded for *Diarsia* on other Malesian mountains.

The Hadeninae include species of the *Mythimna-Leucania* complex that are characteristic of open grassland habitats in the tropics and in temperate latitudes. In Malesia these species are found most commonly in second growth areas with *Imperata* grass but a few do occur at higher altitudes in association with natural vegetation. More typically montane hadenines in Malesia are Himalayan species such as *Heliophobus dissectus* Walker (known from Kinabalu and Ceram in Malesia) and *Pseudaletia albicosta* Moore (known from the Philippines, Borneo, Java, and Flores in Malesia), the *Apospasta* species mentioned earlier in this chapter, and the Bornean endemic *Hadena kaburonga* Holloway, a species of uncertain affinity. The tropical lowland genus *Elusa* needs further examination and may prove to be incorrectly assigned to the Hadeninae.

The Acronictinae are a heterogeneous assemblage of all other trifines and are strongly represented at all altitudes. Typical temperate genera such as *Acronicta* and *Trachea* are represented in the Malesian mountains. The genus *Euplexia* (as treated by Holloway, 1967) is of great interest because of its centers of diversity in Borneo and New Guinea; it is discussed in more detail in the next section.

The quadrifine subfamilies Acontiinae, Euteliinae, Ophiderinae, and Hypeninae are primarily tropical and lowland. The Plusiinae are a small subfamily, well represented everywhere except at extreme latitudes, often migratory and, like the *Mythimna* complex, mainly open-habitat specialists. There are rather few rain forest species in Malesia, and only one, *Plusia confusa* Moore, is definitely montane, extending from the Himalaya to the mountains of Sundaland and the Philippines.

The Catocalinae have a greater range of altitude than the previous quadrifine subfamilies, both as a group and with regard to individual species. The *Sypna* group of genera is perhaps most characteristic of middle altitudes. The Pantheinae are a small subfamily, almost entirely montane in Malesia, and their species are derived mainly from Himalayan centers; the subfamily is well represented in the eastern Palearctic.

The three remaining subfamilies in the Quadrifinae are predominantly tropical, but are well represented at medium altitudes up to about 2000 m on average, where the Ericaceae become an important component of the vegetation. In the Stictopterinae and Sarrothripinae numerous species occur from the lowlands to this altitude and other species are restricted to montane habitats. Species of the subfamily Chloephorinae tend to be more localized. The *Didigua* and *Calymera* sections (B and C + D, respectively, in Holloway, 1976) of the *Carea* complex are diverse and fairly evenly represented at all altitudes up to about 2000 m in Borneo; the center of species richness for both sections is in Borneo. The group will be discussed further in the next section.

Geometroidea

Similar distinctions may be made among geometroid families. The temperate and Himalayan Thyatiridae include a handful of species from the mountains of Sundaland derived from Himalayan centers; in New Guinea the genus *Habrona* contains five montane species and there is a sixth in Buru.

The Uraniidae are a mainly lowland, tropical family but the related Epiplemidae, also mainly tropical, have a number of montane species in the genus *Epiplema*. *Epiplema* is particularly diverse in New Guinea and other parts of Melanesia.

The Drepanidae, generally a forest family, are moderately represented in temperate forests, particularly in eastern Asia and the Himalaya, but reach their greatest diversity in the tropics. In Borneo they are most diverse in the lowlands but have several exclusively montane species usually shared with the Himalaya or derived from Himalayan-centered genera (for example two *Paralbara* species, *Nordstroemia duplicata* Warren and *Macrocilix maia* Leech).

The Geometridae, having more reliably defined subfamilies than the Noctuidae and smaller flight range leading to more localized distributional patterns, are perhaps the ideal large macrolepidopteran family for biogeographic studies. The subfamilies segregate to a large degree with altitude in the tropics (e.g., Fig. 21–2, for Borneo), with the Geometrinae mainly in the lowlands, Ennominae covering a wide range but perhaps most diverse at middle altitudes, and Larentiinae predominating at high altitudes. The Oenochrominae and Sterrhinae, much smaller subfamilies, are moderately represented at all altitudes up to about 2000 m in Borneo. It is interesting to note that in eastern Melanesia (Fiji, New Caledonia), farthest away from Malesian source areas, the Larentiinae are strongly represented in lowland habitats. This might be indicative of a less than exclusively montane history in Melanesia as a whole in the past.

Proportions of the various subfamilies at high altitudes in Borneo are very similar to those at high latitudes. Proportions in the British fauna are also shown in Fig. 21–2. They correlate well with proportions in the Bornean fauna at around 2500 m.

Within the Sterrhinae the lowland species are drawn mainly from the genera *Idaea* and *Scopula*. These include a majority of open-habitat specia-

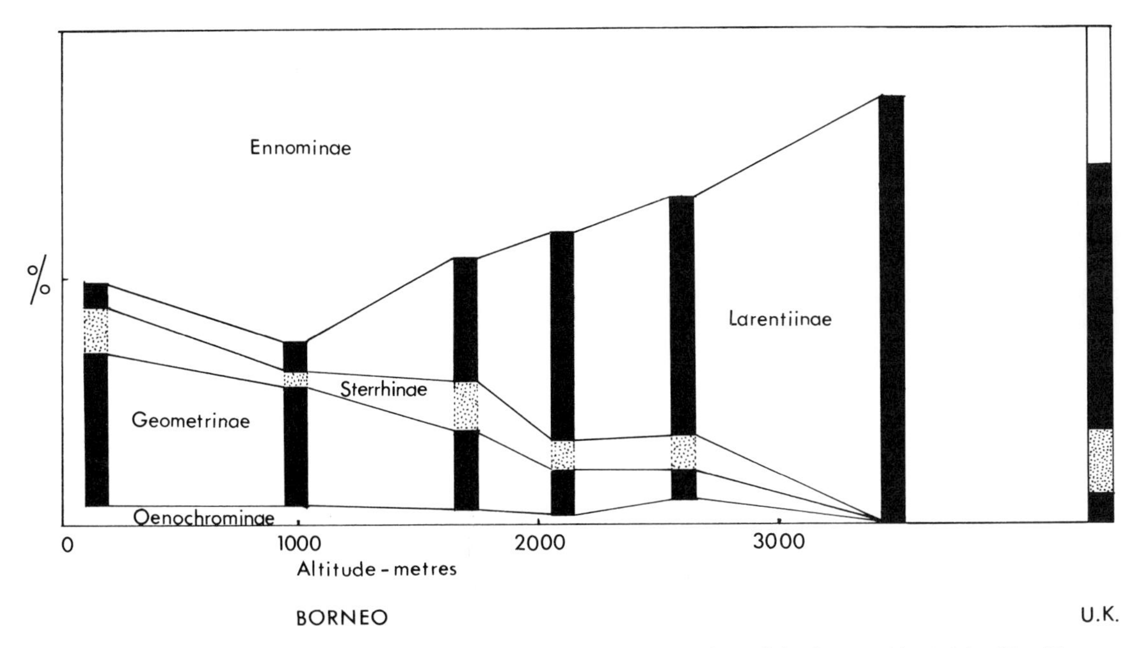

FIG. 21–2. Composite altitudinal transect from Borneo, showing the proportions of the Geometridae subfamilies. The two lowest measures are from mixed dipterocarp forest and lower montane forest on the Mulu transect, the rest from Kinabalu. Proportions in the British fauna are illustrated on the right for comparison.

lists and are diverse also in temperate latitudes. They contain few tropical montane species. The genera *Chrysocraspeda* and *Ptochophyle* also contain many lowland species but are restricted to the tropics and are both rich in species in Borneo. Sterrhinae are unusually abundant in lowland forest on limestone in Borneo. The reason for this is as yet obscure but the extreme roughness and relief of the terrain have resulted in a more open forest structure, in places interspersed with cliff habitats. The montane Sterrhinae are mainly drawn from *Anisodes* and related genera. Species of these genera are found at low altitudes but the greatest diversity is encountered in the upper montane forests. Over half the sterrhine species at about 2000 m on Kinabalu are drawn from these genera. The *Anisodes* group is extremely diverse throughout the Indo-Australian tropics but not significant in temperate latitudes, though the related genus *Cyclophora* has diversified in northern temperate regions, the larvae being arboreal feeders. The *Anisodes* group is also rarely found away from forests in the tropics. Many *Anisodes* species have dispersed within the tropical mountain systems, such as the three spanning the Papuan distribution trend.

Many of the montane Larentiinae have already been discussed in the previous section. Representation of temperate groups is strong but variable. For example, in the Eupitheciini, *Eupithecia* predominates in the Palearctic whereas it is relatively unimportant in the tropics where *Chloroclystis*, *Gymnoscelis*, and a number of smaller related genera thrive. There are also a number of almost exclusively tropical montane groups, including genera such as *Collix*, *Sauris*, the *Phthonoloba* and *Hypocometa* complex and a few Melanesian groups. These are relevant to the discussion of evolution and radiation of montane groups within the Malesian tropics.

Radiation of Montane Groups within the Tropics

I have already mentioned the Himalayan centers of species richness, montane taxa evolving in conjunction with the uplift of the mountain ranges by a slow process of adaptation. The western centers, arising from a possibly more seasonal, arid, northerly lowland condition, have contributed mainly to the radiation of species in the Palearctic region. The eastern centers, developing in an area where lowland habitats were more humid, aseasonal, and tropical, have contributed species mainly to montane habitats in the Indo-Australian tropics and only secondarily to the Palearctic. This contribution to the montane tropics has been extensive, augmented by events in the Pleistocene, and always may have eclipsed most evolution or radiation of montane taxa within the more adjacent parts of Malesia.

Three questions will be examined in this section. First, is there any evidence of the development of a montane fauna related to the continuous history of montane land in Borneo suggested by geologic and palynological evidence? Second, can any groups of Lepidoptera be recognized that might be associated with the old Melanesian flora, spreading with it into the Malesian tropics? This question also involves the radiation of montane genera within New Guinea. Last, can the montane faunas of Fiji and New Caledonia, most isolated from the dispersal of Himalayan and Malesian taxa, provide any clues to the processes of evolution of montane species in Melanesia?

Borneo

Montane habitats in Sumatra and Java are more extensive and higher on average today than they are in Borneo; Sulawesi has an equivalent extent of montane habitat that is also, on average, higher and may have existed for several million years. In the absence of any historical evidence to the contrary, these areas might be expected to have a montane fauna equivalent to, or richer than, that of Borneo, perhaps with the evolution of localized groups. The evidence would appear to be contradictory: groups more or less endemic are frequently encountered in Borneo, yet no definite ones have been identified for Sumatra and Java, though this pattern must be tested by further collecting in the latter localities.

A number of the Bornean taxa are single, taxonomically rather isolated species such as *Hadena kamburonga* Holloway, *Asota kinabaluensis* Rothschild, *Indarbela kinabalua* Holloway, and the three *Diarsia* species mentioned earlier.

Others are small monophyletic groups endemic to the mountains of Borneo. Several may occur in the genus *Euplexia* discussed later in this section, but the clearest examples are larentiines of the tribe Trichopterygini (Dugdale, 1980). They are defined by their slender builds with elongate abdomens that, in the male, usually bear on the first segment ventrally, a posteriorly directed ampulla or more complex structure. The hindwings are reduced in size relative to other members of the subfamily, modified in the male with raised scales and a pouch. They are most characteristic of the Indo-Australian tropics and are predominantly montane, though occurring at low altitudes in eastern Melanesia. Typical genera are *Sauris*, *Episteira*, *Hypocometa*, and *Phthonoloba*; the genus *Tatosoma* occurs in New Zealand, and Australasian members of the tribe were discussed by Dugdale (1980), who also presented a checklist of Indo-Australian species of the *Sauris* group. The group is represented in Madagascar by *Aposteira* and in the mountains of tropical Africa by *Protosteira* and species attributed to *Episteira*. Numerous genera occur in South America, mainly in temperate latitudes.

The endemic Bornean groups occur one in each of *Hypocometa* and *Phthonoloba*, both rather isolated morphologically within their genera (Holloway, 1976). The first includes the duplex of *H. leptomita* Prout, an upper montane species found on several ranges of mountains, and *H. titanis* Prout, a much larger species restricted to the summit zone of Kinabalu, which it shares with one of the distinctive *Diarsia*, *Poecilasthena nubivaga* Prout, and a *Chloroclystis* species (Holloway, 1970). Three interrelated species of *Phthonoloba* fly in Borneo, two of which, *P. stigmalephora* Prout and *P. caliginosa* Holloway, are known only from 2000–2500 m on Kinabalu. The third, *P. lutosa* Holloway, described from Kinabalu, occurs at a slightly lower altitude than the other two, but has recently been recorded from the upper montane forest of G. Mulu in northern Sarawak and Bukit Retak in Brunei.

The genus *Goniopteroloba* also possesses the abdominal structure mentioned earlier, but the forewing fasciation is simpler, the margin bifalcate; it too is mainly montane. A few species occur in the Himalaya, Sumatra, Malaya, Philippines, and Sulawesi, but the highest concentration, five, is in Borneo. Diversity of *Sauris* in Borneo is relatively high.

Other genera that include a high proportion of montane species and exhibit high diversity in Borneo are the eupterotid *Melanothrix*, the larentiine *Chloroclystis* (Table 21–1), the ennomine *Synegia* and the noctuid genera *Carea* (Chloephorinae) and *Euplexia* (Acronictinae). The first two have been discussed earlier. The last three will be discussed next.

Synegia needs revision and the species group referred to here may eventually have to be placed with *Eugnesia*. Those species illustrated and described by Holloway (1976) may be regarded as typical of it. Table 21–1 illustrates the representation of species of this general facies type throughout the Indo-Australian tropics, drawn from both *Synegia* and *Eugnesia* and, as far as dissected, with male genitalia akin to those of *E. correspondens* Warren, the type species of *Eugnesia*. The Oriental center of diversity is in Borneo, Sulawesi is well endowed, and there has been also considerable speciation in New Guinea. Specific affinities within the genus have yet to be elucidated, but Borneo has five endemics (one with a relative in Luzon) and two shared only with montane Malaya. *Synegia eumeleata* Walker would appear to be a heath forest species in Borneo and ranges from the Himalaya to New Guinea. *S. imitaria* Walker is found most frequently in lowland forest but occurs occasionally in the upper montane forest; it ranges from the Himalaya to Sundaland and is known from Sri Lanka. The remaining species are montane, with the greatest diversity at about 2000 m.

The section *Calymera* (C + D of Holloway, 1976) of *Carea* is more emphatically centered on the Sunda Shelf, with the greatest number of species in Borneo (Table 21–1). The section has not radiated in New Guinea. In Borneo the species appear to be rather localized with regard to forest type or altitude, One quarter are lowland, one third upper montane, and the rest lower montane. Here would appear to be a case of radiation and adaptation within a genus to a range of altitudes in a tropical locality, and a detailed investigation of its systematics would be interesting.

Euplexia presents yet a different pattern (Table 21–1). It has distinct centers in the Himalaya, in Borneo, and in New Guinea. All the species are of upper montane forest in Borneo. The genus is here defined as including species with valve structure of the male genitalia as illustrated by Holloway (1976) for Bornean species, though the genus and its relation to *Trachea* need clarification. Only the Bornean species have been dissected to any degree.

One section of *Euplexia* is widespread, consisting of species with black and white forewings related to *E. albovittata* Moore. A number of species described from the mountains of Sundaland, particularly Sumatra, will probably prove on dissection to be variants of *albovittata*, but on Kinabalu there is a genuine duplex with the endemic *E. styx* Holloway flying at altitudes above those frequented by *albovittata*. *E. albovittata* or relatives range eastward from the Himalaya as far as Ceram. Allied species are also found on tropical African mountains (e.g., Fletcher, 1961). A number of temperate southeastern Australian species referred to *Euplexia*, such as *E. iorrhoa* Meyrick, share the characters of the tropical species (L. Hill, pers. comm.), so this would appear to be another example of a transtropical group.

A group of nine large, dark brown species in the northeastern Himalaya have no apparent relatives in Malesia, but three others may well be related to the smaller, greener species of Borneo and New Guinea. All the Bornean species of this group are endemic except one shared with Java. The Malayan species is related to the largest Bornean group recognized by Holloway (1976). Those species in Sulawesi and Ceram would appear to have affinities with the New Guinea center.

Another genus that should be investigated in the context of Bornean radiation is the ennomine *Diplurodes*. Numerous species occur in Borneo at all except very high altitudes, with the greatest concentration in lower montane forest.

Borneo would therefore seem to have served at least as a minor center for the evolution of montane Lepidoptera. Further evidence may become apparent when the taxonomy of the fauna has been elucidated to a much greater degree than at present. The sampling program of a wide range of habitats and mountains in Sabah, Sarawak, and Brunei undertaken by the author and colleagues should facilitate this work.

Luzon in the Philippines is notable for a strong Palearctic influence at high altitudes (Diakonoff, 1967), but apart from three endemic species in the ennomine genus *Arctoscelia* (the only other congener, from Borneo, is probably unrelated), no endemic groups of species are apparent.

Sulawesi

Sulawesi, so much the crossroads for montane dispersals and exhibiting significant endemism in its lowland butterfly fauna (Holloway and Jardine, 1968), is still undercollected so it is not possible to state categorically whether or not it has acted as a center for the evolution and radiation of montane Lepidoptera. Indications so far are that it has not to any great extent. The collection made by Sutton and Rees contained several *Arycanda* and *Milionia* (Ennominae), *Euproctis* (Lymantriidae), and two *Euplexia* of Papuan character. *Synegia* and *Chloroclystis* were moderately represented and the collection included a few ennomines of uncertain affinity but possibly old endemics. The most bizarre of these species is one where the male has a tailed hindwing, and a flap, almost a jugum, from the posterior margin of the forewing covering a patch of scent scales on the hindwing. Long projections from the sixth and seventh sternites and a small one on the eighth complete a weird and unique assemblage of characters presumably connected with scent dispersal in courtship.

New Guinea

Numerous large, endemic species groups are known from the mountains of New Guinea. Several have already been discussed in this chapter, such as the relatives of *Piercia*, the *Micromia* group, *Delias*, *Euplexia*, and *Synegia* species, species of *Quadricalcarifera* and *Cascera* in the Notodontidae, and species of *Habrona* in the Thyatiridae. Many of these are of northern or Malesian derivation and may have radiated relatively recently. Other genera of Malesian origin, however, such as *Delias* with its association with Loranthaceae, may have had a longer history in Melanesia. There may also be typically Australasian groups that have always been associated with the early Gondwanan angiosperm flora. Neither these nor the old radiations of Malesian origin in Melanesia would be expected to be exclusively montane but, as with the Melanesian flora, today could be largely limited to montane habitats by competition from lowland Malesian elements of more recent arrival.

To elucidate this situation as far as possible I made a preliminary and, of necessity, rather cursory survey of the larentiine and ennomine genera with large numbers of species recorded from the mountains of New Guinea. The results of this survey mainly suggest areas of interest for future, more thorough taxonomic work, preferably with the accumulation of more precise field data on the habitat preferences and host plants of the species concerned.

The majority of the endemic larentiine genera are restricted to high altitudes in New Guinea. For many of them there are no clear pointers as to their biogeographic affinities: the small groups mentioned earlier as misplaced in *Xanthorhoe*; *Spectrobasis* (8 species); *Lasioedma* (2 species, plus one included at present in *Xanthorhoe*); and *Sterrhochaeta* (30 species). The male genitalia in these genera usually bear structures alien to Malesian and Palearctic genera as far as these are known, and could represent an old Melanesian tropical element.

The genus *Chaetolopha* is in need of revision. The species *incurvata* Moore (northeastern Himalaya) and *rubicunda* Swinhoe (mountains of Sundaland) are misassigned and bear no relationship to Melanesian and Australian taxa. The type species, *C. oxyntis* Meyrick, flies in southeastern Australia together with two congeners. These have markedly different facies from the group of seven species found in the mountains of New Guinea, one of which (*minutipennis* Warren) extends to the mountains of New Ireland, Ceram, and Buru. Dissection of members of the Australian and New Guinea groups has revealed that they share characters such as a very broad saccus, large hooked projections from the transtilla, and coremata between segments five and four on the abdomen, but that they also have a number of characters distinguishing one from the other, such as reduction of the uncus in the New Guinea group and possession of prominent lateral "wings" of fused scales on the tegumen by *oxyntis*. The significance of such genitalia characters in terms of synapomorphy (shared unique derivation) and therefore possible monophyly will only become apparent when the genitalia and other structures of the Larentiinae generally and those of Australian larentiines in particular are better known. On present evidence the genus *Chaetolopha* is a candidate for consideration as an old Australasian group. A possible parallel may be seen in the butterfly genus *Hypocysta* where a similar dichotomy into Papuan and temperate Australian groups occurs.

A stronger case can be made for another group of larentiines as an old Australian radiation, in this case restricted to tropical rain forest habitats, some montane. The group is defined by several characters of the male genitalia not found in northern groups, and it also lacks eversible coremata on the abdomen though long hairs may be present on a Y-shaped thickening of the membrane between the genitalia and the eighth sternite. The most singular feature of the male genitalia is a pair of anteriorly directed broad hooks that project from the dorsal extremities of the vinculum; the valves are very rounded with characteristic molding and projections on the costa. The genitalia of a member genus, *Papuanticlea*, were illustrated by Holloway (1979).

Genera included in this group of larentiines are *Crasilogia*, with five species at about 2000 m in New Guinea, and *Papuanticlea*, with a lowland rain forest species in New Caledonia, one from Queensland, probably montane, and three from moderate altitudes in New Guinea, one also recorded from montane Ceram. Two further Queensland species, possibly largely montane, must also be added, with the exclusion at present of their congeners. One is *Crasilogia gressitti* Holloway, the sister species of *C. fumipennis* Warren, and the other is *Heterochasta conglobata* Walker. *Polyclysta gonycrota* Prout from lowland rain forest in Fiji is misplaced, having coremata between segments four and five and basal hooks from the male valve, suggesting some relationship with *Chaetolopha*. The distribution of the *Crasilogia* group extends over the Inner Melanesian Arc and the New Caledonian species indicates that the group is not exclusively montane. The *Crasilogia* group has been studied cladistically by Holloway (1984b).

The last major New Guinea larentiine group to be discussed is at present included in the genus *Horisme*. *Horisme* is based on the Palearctic *H. tersata* Denis and Schiffermuller, but includes a cosmopolitan and heterogeneous array of species. About sixteen species from New Guinea such as *disrupta* Warren, *aeolotis* Prout, and *albimedia* Warren, together with 3 from montane Borneo and 1 from Borneo, Sulawesi, and New Guinea, form a homogeneous group with regard to wing pattern and characters of male genitalia (illustrated by Holloway, 1976). They lack abdominal coremata and have rounded valves with a submarginal fringe of very long setae. The species are obviously misplaced in *Horisme* and might even be close to the *Crasilogia* group. All are montane, and the Bornean species have their closest relatives in, or are shared with, New Guinea. This is therefore a possible case of an old Melanesian genus penetrating westward in recent times into the mountains of Malesia.

The Larentiinae are predominantly montane but many genera in the Ennominae endemic to New Guinea or centered there, such as *Heterodisca* (4 species), *Orthotmeta* (4 species), *Paradromulia* (19 species), and *Milionia* (numerous species, mentioned later in this section), include species from all altitudes. Strictly montane species in these genera have presumably evolved by processes of adaptation and speciation within New Guinea as parts of its land area were uplifted and united. Unfortunately we know too little about the local distributions or affinities of the species concerned to hazard a guess as to what those processes may have been, though some indications of this nature are emerging with regard to the Bornean fauna.

These processes might be considered as a more localized counterpart (perhaps restricted in history to the lands now forming central New Guinea) of those involved in the radiation of other genera more widely distributed through the Indo-Australian tropics. *Synegia*, primarily montane, containing both localized and widely distributed groups, might be ideal for investigation in this context. The genus *Ruttellerona*, possibly related to *Paradromulia* and to the monotypic New Caledonian genus *Eunnoumeana*, is similarly interesting. *Ruttellerona* at present includes the following species: *cessaria* Walker (Himalaya to New Guinea, lowland to montane); *harmonica* Hampson (Sri Lanka); *stigmaticosta* Prout (Sumatra, montane); *kalisa* Prout (Java, montane); *obsequens* Prout (Ceram, Buru, montane); *lithina* Warren (Borneo, Sulawesi to New Guinea, montane); *scotozonea* Hampson (Christmas Island, Indian Ocean); *indiligens* Prout (New Guinea, montane); *ochreocosta* Bethune-Baker (New Guinea, Bismarcks, mainly montane but also on Feni Island); *presbytica* Robinson (Fiji, mainly lowland forests).

A detailed systematic treatment of the brightly colored or otherwise striking ennomine species placed in genera such as *Eucharidema*, *Ctimene*, and *Milionia* could lead to interesting biogeographic discoveries. These genera are only weakly represented in Malesia but have radiated extensively, often with montane species (e.g., eight *Eucharidema* in New Guinea), from the Moluccas eastward. They could reveal much about the history of the Melanesian biota, both with reference to the hypothesis of converging Melanesian Arcs and the development of montane habitat. Several other families have given rise to groups of brightly colored species centered in New Guinea, such as the Lithosiinae, and this might give a clue to past conditions. Are the bright colors indicative of a change to diurnal habit and if so, why did this occur?

The 4 *Milionia* species present in Borneo today are mainly diurnal and all primarily montane. They may well feed on conifers such as *Dacrydium* and *Podocarpus*. One of these species, *M. basalis* Walker, feeds naturally as a larva on conifers of southern origin in the mountains of Malaya, but has recently colonized podocarp and *Pinus* plantations in the lowlands (Tho, 1978). It may feed on *Pinus* naturally in the Himalayan and Japanese parts of its range. In New Guinea a similar case has been documented (Wylie, 1974). *M. isodoxa* Prout probably feeds naturally on *Araucaria hunsteinii*, a dominant of montane forest, but has recently become a pest in *A. cunninghami* plantations and been recorded from other *Araucaria* and planted *Pinus*.

Dietary restriction of *Milionia* to conifers, if confirmed by further records, could indicate a long history of coevolution with southern conifer genera in Melanesia, as well as recent invasion with the Melanesian flora into the mountains of Southeast Asia, with potential for a switch to *Pinus* and penetration farther north. In general, transference from a conifer-feeding to an angiosperm-feeding habit or vice versa in Lepidoptera is unusual, though specificity to particular conifer families or genera is much weaker than it is as regards the angiosperms (Holloway and Hebert, 1979).

Arycanda is weakly represented in Sundaland with species ranging individually from the lowlands to montane habitats, but it has, like the genera mentioned earlier, radiated most strongly farther east and includes several montane species in New Guinea. It would also repay systematic investigation.

The genus *Casbia* is diverse throughout Australia and is also well represented in Melanesia, especially New Guinea and New Caledonia, though few of the species in New Guinea can strictly be considered montane. Numerous larval host records are from *Alphitonia* and other genera in the Rhamnaceae (Robinson, 1975; Holloway, 1979; McFarland, 1979). This may prove to be another instance of coevolution that could have interesting biogeographic ramificatons.

There is a group of ennomine taxa, mostly within endemic genera, that are more strictly montane within New Guinea. They are all of relatively slender build with forewings more elongate and less triangular than in most other Ennominae. They have generally similar wing pattern, dark, with blocks of color rather than the more usual ennomine fine fasciation. All have a small blisterlike structure, or fovea, at the base of the forewing just posterior to the cell in the male. The fovea and general characteristics of the male genitalia are common to a large number of ennomine genera throughout the world, such as *Alcis* and *Hyposidra*. General convergence in build and facies through adaptation to a similar mode of life cannot be ruled out, but the group should definitely be examined in more detail in case it represents an old radiation within New Guinea.

The genera involved are *Tolmera* (10 species), *Microtome* (5 species), several species at present placed in *Paralcis* (e.g., *coerulescens* Warren and *intertexans* Prout), and 7 misplaced in *Cleora* (*discipuncta* Joicey and Talbot; *fenestrata* Prout; *hoplogaster* Prout, and two undescribed in New Guinea; *argicerauna* Prout in montane Buru; and a further undescribed species in the mountains of Sulawesi). The Sulawesi ennomines mentioned earlier are generally similar, as is *Arctoscelia epelys* Prout from montane Borneo, wrongly associated with its current Luzon congeners that lack a fovea. Generally similar species occur in New Zealand in the diverse genus *Pseudocoremia* (one species also on Norfolk Island; Holloway, 1977) and the monotypic *Challastra*, but this resemblance may again be convergent. If not, then the group may provide another instance of an old Australasian radiation.

Paralcidia, with similar build and facies to the *Tolmera* group, has no fovea in the male; it contains seven species endemic to the mountains of New Guinea. Its affinities are not evident, though more thorough comparison with the Luzon *Arctoscelia* and the genus *Ischalis* in New Zealand would test the possibility that certain general resemblances in the male genitalia are more than fortuitous.

An ennomine complex that may parallel the *Crasilogia* group of larentiines in its distribution over lands of the Inner Melanesian Arc includes all species at present attributed to the genera *Polyacme* and *Anisographe*. *Polyacme* includes one species in New Caledonia and the Loyalty Islands, characteristic of lowland rain forest, and four endemic to New Guinea and taken most frequently at 2000 m. *Anisographe* includes a Queensland species, a montane New Guinea species, and one, flying up to 2000 m, ranging from Queensland through New Guinea to Buru. These two groups were analyzed cladistically by Holloway (1984b).

Fiji and New Caledonia

Fiji and New Caledonia are fairly extreme outliers on the Outer and Inner Melanesian Arcs, respectively. Both appear to have had a longer subaerial history than most lands in the area, particularly New Caledonia, which may date back well into the Mesozoic (Holloway, 1979). In both areas the Larentiinae are found commonly at low altitudes as well as in the mountains.

Fiji has a small group of eight species more or less restricted to mossy, upper montane forest (Robinson, 1975). Two are of probable Australian derivation: *Chloroclystis nina* Robinson is endemic and *Poecilasthena leucydra* Prout is shared with New Caledonia, where it is more a lowland species. One is in a noctuid genus in which widespread Indo-Australian tropical species predominate, though the Fijian species, *Sasunaga tomaniiviensis* Robinson, is endemic. The remaining species, all endemic, belong either to small species groups restricted to Fiji or are morphologically isolated within their genus, suggestive of a long history within the Fijian biota (Holloway, 1979). They include one hypenine noctuid, three larentiines, and one ennomine.

Thus it may be that in Melanesia, throughout its biological history, montane species have evolved predominantly through the adaptation of lowland taxa to the perhaps more stringent conditions at high altitudes, rather than through invasion by taxa from temperate or montane centers outside Melanesia. This latter mode of acquiring montane species has predominated in the mountains of Malesia and has recently become important in New Guinea; in areas well isolated from such temperate centers, however, such as Fiji, it has been of minor importance only.

New Caledonia appears to have a much smaller strictly montane fauna than Fiji (Holloway, 1979), but again the older elements predominate. The oenochromine *Adeixis montana* Holloway belongs to a trio of species that have radiated within the island. The larentiine *Caledasthena montana* Holloway is monotypic, endemic and of uncertain affinity. *Chlorocoma octoplagiata* Holloway, a geometrine, is tentatively assigned to a genus with numerous species in temperate parts of Australia, but is probably one of the older members of the New Caledonian fauna. Only *Chloroclystis lunifera* Holloway is closely related to an Australian species (known to migrate infrequently to Norfolk Island; Holloway, 1977); it is not strictly montane, only somewhat commoner at high altitudes.

Diversity

Any discussion of the evolution of tropical Lepidoptera is hampered by great gaps in our knowledge of the life histories of the species concerned. The macrolepidoptera are generally defoliators, often with very specific host plant requirements. For example, the *Ocinara* group of genera in the Bombycidae appears to be restricted to *Ficus* (Dierl, 1978) and hence might not be expected to feature strongly in montane biotas. Among the ennomines potentially coevolutionary trends have already been noted for *Milionia* and *Casbia*. However, even inklings of trends such as this are exceptional in our knowledge. Only when information of this nature is much more plentiful will it be possible to piece together the processes involved in the evolution of montane Lepidoptera in the tropics (see also Chap. 20). For instance, it is impossible to tell in most cases whether a species is restricted to montane habitats by virtue of foodplant specialization, physiological specialization with regard to temperature or other environmental factors, or through competition.

In cases of competition congeners may segregate with regard to altitude. This can only be observed readily if a quantitative sampling program with regard to habitat and altitude is undertaken. Interesting features have emerged during the course of such work in Borneo, some of which have already been mentioned here or elsewhere (Holloway, 1970, 1978 and 1984a).

Observations on general diversity at different altitudes are as yet few and indicate little more than a general decline with altitude (Holloway, 1977, 1978, and 1984a [Borneo]; Hebert, 1980 [New Guinea]). Variation is hard to explain and correlation with floristic diversity, despite the degree of specificity of larval food requirements, can be surprisingly poor. Anomalous low lepidopteran diversity in habitats with very high floristic diversity in New Caledonia was tentatively related by Holloway (1979) to a possible high level of toxicity in the vegetation produced by tolerance or even accumulation by the plants of the unusual metal ions in the particular soils on which this vegetation grows.

Moth diversity in Bornean lowland forests also shows little correlation with floristic (tree) diversity or relative area of habitat type, a situation discussed in greater detail by Holloway (in press).

Therefore, interpretation of the observation that the highest geometrid diversity encountered in Borneo so far is at 1000 m (lower montane forest on both the shales of G. Mulu and the limestone of G. Api) must be speculative and tempered with caution (Holloway, 1984a). At this altitude the subfamily Ennominae is at its highest proportional representation. Diversity approaching this level was also encountered at 1930 m on Kinabalu (at 1000 m the slopes are almost entirely deforested by man), much higher than that recorded at similar altitudes on Mulu where similar species were nevertheless involved. In this latter instance the effect of area may well be operative as habitats at that altitude are very much more extensive on Kinabalu.

A number of geometrid species could be regarded as characteristic of lower montane forest on Mulu and Api, though this has yet to be confirmed by numerical analysis of the data. The high diversity observed is therefore more than just a reflection of the overlap of the lowland fauna with that of the upper montane forests above.

There is little information on the relative palatability to Lepidoptera larvae of the lowland Dipterocarpaceae as compared with the Fagaceae, Lauraceae, and Myrtaceae of the lower montane zone, though, in general, lowland rain forest trees contain more chemical defenses against insect attack than do temperate ones; certainly the Fagales are very important as larval hosts in the Holarctic (Holloway and Hebert, 1979). On the other hand, a diversity of chemical defenses in the lowland forests might have been expected to produce a diversity of coevolutionary lines among defoliating Lepidoptera. Only with much more information on the foodplant range of the species concerned will this problem be open to investigation.

The high diversity at 1000 m might also have a recent historical explanation if, as suggested earlier, depression of vegetation zones during the glaciations led to a predominance of montane habitats and similar lowland heath forest in Sundaland, coupled with greater access to centers of diversity in the Himalaya. Lowland dipterocarp forest at such times might have been less extensive, its reduction having caused extinction of a previously richer fauna, a richness that has not developed again subsequently.

Duplex Species

Study of the ecology of species duplexes and complexes may be relevant to understanding the evolution of tropical montane faunas. The Borneo sampling program is providing much information on the ecological segregation of closely related species pairs. Often these species replace each other with altitude with varying degrees of overlap. As taxonomic work on the fauna progresses so more of such duplexes are coming to light (Holloway, 1970, 1973, 1978, and 1982b, pp. 185, 191, 226).

In the majority of cases the member of the pair favoring higher altitudes also has the most restricted geographic distribution. Such pairs can be lowland versus montane as in the *Eurema* (Pieridae) duplexes discussed by Holloway (1973) or both montane as in the moth examples noted by Holloway (1970). In general, the greater the altitude at which the duplex features, the more likely is the higher member of the pair to be endemic to Borneo. In extreme cases, such as in the *Hypocometa*, *Fascellina*, *Apophyga*, and *Garaeus* duplexes mentioned by Holloway, the higher member is only known from Kinabalu; in most cases for some reason the higher species is also very much the larger.

The geographic restriction of the higher altitude member of the duplex prompts the explanation of a double invasion of a mountain area peripheral to the usual range of the species, each invasion triggered by an event such as the depression of montane zones in the Pleistocene which facilitated dispersal and establishment. The descendant population of the first invasion must diverge so as to be reproductively isolated from the source population prior to the second invasion, and be so adapted to local conditions that it can compete successfully at higher altitudes with the population established by the second invasion even if the latter is more successful overall as a result of the larger gene pool of the source population in the central part of its range.

Another type of duplex in Borneo has the two member species separated by intervening areas of apparently unsuitable habitat. The cases detected so far involve lowland heath forest on the one hand and lower or upper montane forest on the other, one species of the duplex flying in each habitat. Both heath and montane forest types grow on peaty, acid soils poor in nutrients and have certain floristic elements in common. Indeed, the lower montane forest on G. Mulu was described by Hanbury-Tenison and Jermy (1979) as ecotonal between mixed dipterocarp forest and kerangas (wet heath forest).

As well as members of a duplex being split between these two habitat types, there are several instances where populations of the same species occur in each habitat. Two *Calliteara* species (Lymantriidae) are common in the lower montane forest of the limestone G. Api and also in lowland dry heath forest dominated by *Gymnostoma* (Casuarinaceae) in Brunei, but are absent from richer lowland forests, wet heath forest, or the less well drained lower montane forest of the Mulu range. They belong to a species group with a Papuan trend distribution. A number of species were found to be common to kerangas in the lowlands and Mulu lower montane forest, but were not taken on Api or in dry heath forest. Hydrology would therefore appear to be important.

Several duplex pairs are discussed by Holloway (1984a) but the most interesting case is where both species, ennomines of the genus *Peratophyga*, are known only from Borneo, with no close relatives. *P. sobrina* Prout was described from Kinabalu, where it is relatively common between 1500 m and 2000 m. It was taken at similar altitudes on G. Mulu. A related species was discovered in the kerangas forests in the lowlands of the G. Mulu National Park and was later found to fly also in the dry heath forest of lowland Brunei. It is obviously not possible from this to identify the habitat of the original ancestral population for this duplex, but the phenomenon prompts two questions. Does adaptation to lowland heath forest by Lepidoptera serve as preadaptation for colonization of tropical montane habitats? Second, was exchange between heath forest and montane forest habitats greatly facilitated during the glaciations when depression of the montane zones may have brought them into greater proximity or indeed contiguity?

Further indirect support for the similarity of these two types of habitat is provided by the observation that it is into precisely these habitats that Melanesian floristic elements have been able to penetrate in their relatively recent spread westward.

CONCLUSIONS AND PROSPECT

The Indo-Australian montane Lepidoptera show general distribution patterns that reflect in many ways those of the flora. Only the emphasis is different, particularly in the degree to which Melanesian elements have recently penetrated the mountains of Sundaland, much less in the Lepidoptera than in the flora. This difference in emphasis may be taken in itself to indicate that the evolution of the Indo-Australian montane biota has been primarily by dispersal events rather than through the vicariant fragmentation of a uniform biota spanning the tropics along previously more continuous ranges of mountains as has been suggested by some biogeographers. The overall picture is still relatively obscure compared to that available for the flora. It will be clarified only by further fieldwork, preferably the quantitative sampling of altitude transects coupled with attempts to obtain host plant data, backed by taxonomic investigation of some of the interesting taxonomic groups noted in this chapter. But the destruction of forest habitats in the area is proceeding at an alarming rate, so this prospect may well prove to be a pipedream.

ACKNOWLEDGEMENTS

I would like to thank the Trustees and Keeper of Entomology at the British Museum (Natural History) for facilities provided. I am very grateful to Professor M. G. Audley-Charles, Mr. J. S. Dugdale, Dr. J. R. Flenley, and Mr. D. S. Fletcher for their helpful comments on a draft of this chapter. Dr S. L. Sutton kindly made available to me at very short notice the collection of Lepidoptera made by Dr. C. J. Rees and himself while members of the Operation Drake project. My brother, Mr. N. H. Holloway, directed my attention to recent literature on the geology of the Sunda Shelf and added comments from his own experience; my thanks are due to him also.

REFERENCES

Audley-Charles, M. G., D. J. Carter, M. S. Norvick, and S. Tjokrosapoetro. 1979. Reinterpretation of the geology of Seram: Implications for the Banda Arcs and northern Australia. *J. Geol. Soc. Lond.* 136: 547–568.

Carter, D. J., M. G. Audley-Charles, and A. J. Barber. 1976. Stratigraphical analysis of island arc-continental margin collision in eastern Indonesia. *J. Geol. Soc. Lond.* 132: 179–198.

Corner, E. J. H. 1978. Plant life. In *Kinabalu, summit of Borneo*, M. Luping, Chin Wen and E. R. Dingley, eds., pp. 112–178. Kota Kinabalu: Sabah Society.

Diakonoff, A. 1952–1955. Microlepidoptera of New Guinea; results of the third Archbold expedition. *Verh. K. Ned. Akad. Wet.* 2d ser., 49: parts 1,3, and 4; 50: parts 1 and 3.

———. 1967. Microlepidoptera of the Philippine Islands. *Bull. U.S. Nat. Mus.* No. 257: i–vii, 1–484.

Dierl, W. 1978. Revision der orientalischen Bombycidae (Lepidoptera). Teil 1: Die *Ocinara*-Gruppe. *Spixiana* 1: 225–268.

Dugdale, J. S. 1971. Entomology of the Aucklands and other islands south of New Zealand: Lepidoptera, excluding non-crambine Pyralidae. *Pacific Insects Monograph* 27: 55–172.

———. 1980. Australian Trichopterygini (Lepidoptera: Geometridae) with descriptions of eight new taxa. *Aust. J. Zool.* 28: 301–340.

Flenley, J. R. 1979. *The equatorial rain forest: A geological history*. London: Butterworths.

Fletcher, D. S. 1958. Geometridae. In *Ruwenzori expedition 1952*, Vol. 1, pp. 77–176. London: British Museum (Natural History).

———. 1961. Noctuidae. In *Ruwenzori expedition 1952*, Vol. 1, pp. 177–323. London: British Museum (Natural History).

Haile, N. S., M. W. McElhinny, and I. McDougall. 1977. Palaeomagnetic data and radiometric ages from the Cretaceous of West Kalimantan (Borneo), and their significance in interpreting regional structure. *J. Geol. Soc. Lond.* 133: 133–144.

Hanbury-Tenison, A. R., and A. C. Jermy. 1979. The R. G. S. Expedition to Gunong Mulu, Sarawak. *Geogr. J.* 145: 175–191.

Hebert, P. D. N. 1980. Moth communities in montane Papua New Guinea. *J. Anim. Ecol.* 49: 593–602.

Holloway, J. D. 1969. A numerical investigation of the biogeography of the butterfly fauna of India, and its relation to continental drift. *Biol. J. Linn. Soc.* 1: 373–385.

———. 1970. The biogeographical analysis of a transect sample of the moth fauna of Mt. Kinabalu, Sabah, using numerical methods. *Biol. J. Linn. Soc.* 2: 259–286.

———. 1973. The taxonomy of four groups of butterflies (Lepidoptera) in relation to general patterns of butterfly distribution in the Indo-Australian area. *Trans. R. Ent. Soc. Lond.* 125: 125–176.

———. 1974. The biogeography of Indian butterflies. In *Ecology and biogeography in India*, M. S. Mani, ed. pp. 473–499. The Hague: Junk.

———. 1976. *Moths of Borneo with special reference to Mt. Kinabalu.* Kuala Lumpur: Malayan Nature Society.

———. 1977. *The Lepidoptera of Norfolk Island, their biogeography and ecology.* Series Entomologica 13. The Hague: Junk.

———. 1978. The butterflies and moths. In *Kinabalu, summit of Borneo*, M. Luping, Chin Wen, and E. R. Dingley, eds. pp. 254–263. Kota Kinabalu: Sabah Society.

———. 1979. *A survey of the Lepidoptera, biogeography and ecology of New Caledonia.* Series Entomologica 15. The Hague: Junk.

———. 1982a. Mobile organisms in a geologically complex area: Lepidoptera in the Indo-Australian tropics. *Zool. J. Linn. Soc.* 76: 353–373.

———. 1982b. Taxonomic appendix. In *An introduction to the moths of South East Asia* (H. S. Barlow), pp. 174–271. Kuala Lumpur: H. S. Barlow.

———. 1984a. The larger moths of the Gunung Mulu National Park; a preliminary assessment of their distribution, ecology, and potential as biological indicators. In *Gunung Mulu National Park, Sarawak*, part II, A. C. Jermy and K. P. Kavanagh, eds, *Sarawak Mus. J.* n. s., 30 (51), Special Issue 2: pp. 149–190.

———. 1984b. Lepidoptera and the Melanesian Arcs. In *Biogeography of the tropical Pacific*, P. H. Raven, F. J. Radovsky, and S. H. Sohmer, eds., Honolulu: Association of Systematics Collections and Bishop Museum.

———. in press. Moths in tropical rain forest. In *Tropical rain forest ecosystems, structure and function*, H. Lieth and H. J. A. Werger, eds., *Ecosystems of the World*, 14B. Amsterdam: Elsevier.

Holloway, J. D. and P. D. N. Hebert. 1979. Ecological and taxonomic trends in macrolepidopteran host plant selection. *Biol. J. Linn. Soc.* 11: 229–251.

Holloway, J. D. and N. Jardine. 1968. Two approaches to zoogeography: A study based on the distributions of butterflies, birds and bats in the Indo-Australian area. *Proc. Linn. Soc. Lond.* 179: 153–188.

Leestmans, R. 1978. Problèmes de spéciation dans le genre *Vanessa*. *Linn. Belg.* 5: 130–156.

Mani, M. S. 1968. *Ecology and biogeography of high altitude insects*. Series Entomologica 4. The Hague: Junk.

McFarland, N. 1979. Annotated list of larval foodplant records for 280 species of Australian moths. *J. Lepid. Soc.* 33: Supplement.

Muller, J. 1966. Montane pollen from the Tertiary of N. W. Borneo. *Blumea* 14: 231–235.

Myers, L. C. 1978. Geomorphology. In *Kinabalu, summit of Borneo*, M. Luping, Chin Wen, and E. R. Dingley, eds., pp. 91–94. Kota Kinabalu: Sabah Society.

Norvick, M. S. 1979. The tectonic history of the Banda Arcs, eastern Indonesia: A review. *J. Geol. Soc. Lond.* 136: 519–527.

Pierce, F. N. 1914. *The genitalia of the group Geometridae of the Lepidoptera of the British Islands*. Liverpool: Northern.

Raven, P. H. 1973. Evolution of subalpine and alpine plant groups in New Zealand. *N. Z. J. Bot.* 11: 177–200.

Robinson, G. S. 1975. *Macrolepidoptera of Fiji and Rotuma, a taxonomic and geographic study*. Faringdon: E. W. Classey.

Sibatani, A. 1974. A new genus for two new species of Lycaeninae (*s. str.*) (Lepidoptera: Lycaenidae) from Papua New Guinea. *J. Aust. Ent. Soc.* 13: 95–110.

Smith, J. M. B. 1975. Mountain grasslands of New Guinea. *J. Biogeogr.* 2: 27–44.

———. 1977. Origins and ecology of the tropicalpine flora of Mount Wilhelm, New Guinea. *Biol. J. Linn. Soc.* 9: 87–131.

———. 1980. The vegetation of the summit zone of Mt. Kinabalu. *New Phytol.* 84: 547–573.

Talbot, G. 1928–1937. *Monograph of the pierine genus Delias*. London: John Bale, Danielsson and British Museum (Natural History).

Tho Yow Pong. 1978. A day-flying moth. *Nature Malaysiana* 3: 40–43.

van Steenis, C. G. G. J. 1964. Plant geography of the mountain flora of Mt. Kinabalu. *Proc. R. Soc. Lond.*, ser. B, 161: 7–38.

———. 1971. *Nothofagus*, key genus of plant geography, in time and space, living and fossil, ecology and phylogeny. *Blumea*, 19: 65–98.

Wylie, F. R. 1974. The distribution and life history of *Milionia isodoxa* Prout (Lepidoptera, Geometridae), a pest of planted hoop pine in Papua New Guinea. *Bull. Ent. Res.* 63: 641–659.

22

Origins of the High-Altitude Avifaunas of Tropical Africa

R. J. DOWSETT

A chain of mountains runs for some 5000 km through the eastern half of Africa, from the Ethiopian Plateau in the north to coastal areas of South Africa. Much of this chain is centered along rift valleys. Two isolated areas farther west—in Angola and Cameroon—are closely related to the eastern mountains, both faunistically (Moreau, 1966) and floristically (White, 1981), notwithstanding their considerable geographic separation.

Most of these mountains are sufficiently high—with low temperature and high precipitation—to carry evergreen forest. Although many mountains have an ericaceous belt, fewer also have a higher belt of afroalpine vegetation. Afroalpine moorland usually occurs only above about 3500 m; its avifauna has been discussed in detail by Moreau (1966) (see also Chap. 6). The present chapter is concerned mainly with the afromontane forest avifauna, which totals some 160 stenotopic species. Thus the focus in this chapter is at slightly lower elevations than in most other chapters in this volume. Nonforest birds of open situations such as grassland and afroalpine moorland amount to 62 species in all (Table 22–1.) The evolution and origins of the afroalpine vegetation are discussed elsewhere in this book (Hedberg, Chap. 17).

These high-altitude environments have seen the evolution of a special avifauna. Above a certain altitude on most high mountains there is a montane life zone whose dominant species of birds are absent at lower levels. This transition usually seems well marked, as human activity and environment have now limited the montane vegetation to the highest levels, and the lower slopes of mountains may be clothed in deciduous woodland or in cultivation. But in reality the transition is gradual, from an avifauna in which lowland bird species are dominant to another avifauna in which few lowland species occur, and forms adapted to life at high altitudes predominate.

Regrettably, in few parts of Africa is there still an unbroken progression from lowland to montane evergreen forest, notable exceptions being the Impenetrable Forest in western Uganda (Keith et al., 1969) and some of the mountains of eastern Zaire (Prigogine, 1974). Away from the equator the vertical range over which evergreen forest extends is much less, rarely exceeding 1500 m in the Tanganyika-Nyasa and Southeastern groups (Fig. 22–1) for example. In most heavily populated areas mountains may have no more than thin strips of riparian forest along streams reaching to lower altitudes, but even here can be seen the gradual progression from a lowland to a montane avifauna. Species appear and others drop out every few hundred meters gained, a phenomenon so striking on mountains elsewhere in the tropics, as in New Guinea (Diamond, 1972) and Peru (Terborgh, 1971).

THE MONTANE AVIFAUNA

The altitude at which a change from lowland to montane zone occurs is about 1500–1800 m on most tropical mountains, but there is a cline of decreasing altitude as one moves south of the equator toward and beyond the Tropic of Capricorn, some forests containing birds that are typically montane elsewhere reaching sea level in coastal South Africa. With due allowance for these changes, it is possible to recognize a montane avifauna, containing some 222 species (Appendix 22–A) of the 1500 or so breeding birds recorded from the Afrotropical region by Moreau (1966, pp. 84–87).

The island of Madagascar is excluded from consideration in this chapter, as it now contains no strictly montane avifauna (Benson, Colebrook-Robjent, and Williams, 1976–77). The Gulf of Guinea islands are also excluded, as the

TABLE 22–1. Summary of the stenotopic montane avifauna of the Afrotropical region.

Family	Number of genera [a]		Number of species	
	Forest [b]	Nonforest	Forest [b]	Nonforest
Threskiornithidae		2 (1)		2
Anatidae		1 (1)		1
Accipitridae	2	1	2	1
Phasianidae	1	2	5	5
Rallidae		2 (1)		3
Charadriidae		1		1
Columbidae	3 (1)	1	3	1
Psittacidae	2 (2)		2	
Musophagidae	1 (1)		5	
Cuculidae	1 (1)		1	
Tytonidae	1		1	
Strigidae	1	1	1	1
Caprimulgidae		1		1
Apodidae	1 (1)		2	
Trogonidae	1 (1)		1	
Meropidae	1		1	
Bucerotidae	1 (1)		1	
Capitonidae	2 (2)		3	
Indicatoridae	1		1	
Picidae	2 (2)		3	
Eurylaimidae	1 (1)		1	
Alaudidae		1		1
Hirundinidae		2 (1)		3
Oriolidae	1		3	
Corvidae		3 (1)		3
Paridae	1	1	1	1
Timaliidae	5 (3)		7	
Campephagidae	1		2	
Pycnonotidae	2 (2)		7	
Turdidae	8 (7)	4 (2)	22	5
Sylviidae	13 (9)	2	31	9
Muscicapidae	4 (3)		7	
Motacillidae		2 (1)		4
Malaconotidae	2 (2)		11	
Prionopidae	1 (1)		1	
Sturnidae	3 (3)	1 (1)	6	1
Promeropidae		1 (1)		2
Nectariniidae	2	1	11	6
Zosteropidae	2 (1)		2	
Ploceidae	1	2 (1)	5	3
Estrildidae	4 (4)		7	
Fringillidae	2 (1)	1	4	8
Total	74 (49)	33 (11)	160	62
	99 [c] (60)		222	

[a] Figures in parentheses denote number of genera endemic to the afrotropical region.

[b] Numbers of forest genera and species include those species that are not confined to this habitat (classed F/NF in Appendix A).

[c] In some genera, one species may occur in forest and another in nonforest; these genera are listed twice, thus the grand total of genera (99) is not simply the addition of 74 forest and 33 nonforest genera.

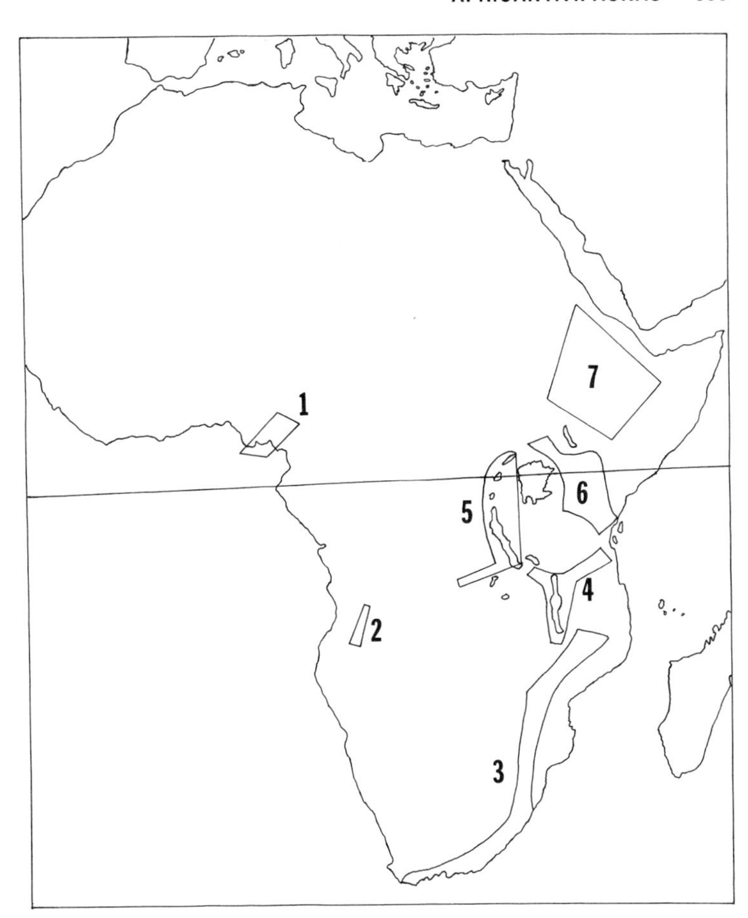

FIG. 22–1. The Afrotropical montane groups. 1: Cameroon; 2: Angola; 3: Southeastern; 4: Tanganyika-Nyasa; 5: East Congo; 6: Kenya; 7: Ethiopia.

factors influencing evolution in the few montane birds there (discussed by Moreau, 1966, pp. 318–26) differ from those acting on mainland avifaunas. However, some of the Cameroon Mt. species do occur on Fernando Po, while the uncertain occurrence of *Schoutedenapus myoptilus* in the Cameroon group (Appendix 22–A) is based on the fact that this easily overlooked swift is known from all other montane groups and from Fernando Po (White, 1965).

The composition of the stenotopic montane avifauna as listed in Appendix A, and summarized in Table 22–1, differs to some extent from the analyses of Moreau (1966, pp. 84–87) and of Hall and Moreau (1970, pp. 382–83). The composition of the non-forest montane avifauna of Appendix 22–A also differs slightly from that of the afroalpine avifauna listed in Appendix 6–A. The main criterion used here is that the species should be wholly confined as a breeding bird to high altitude, or that this is at least the case in two of the seven Afrotropical montane groups (discussed later in more detail). This latter proviso allows inclusion of several species that behave as strictly high-altitude birds in a substantial part, but not all, of their ranges. In addition to lowland forest species reaching their upper limits on mountains, other lowland species occur opportunistically, too many to be discussed here. In all areas for which adequate data exist, the preponderance of the montane breeding avifauna is stenotopic, for example, 68% of the 63 forest-associated species on the high Nyika Plateau in northern Malawi are confined to high altitudes (pers. obs.).

I have generally followed the specific nomenclature and taxonomic viewpoints of Hall and Moreau (1970) and of Snow (1978), but have diverged in a few examples. Justification is made in Appendix 22–B for these changes, but for some the exact course to be followed will remain a matter of opinion, pending further field research. This is particularly so in the case of allospecies, although for some controversial examples it has been possible to present comparisons of vocalizations, sometimes supported by tape playback experiments (Dowsett and Dowsett-Lemaire, 1980). For allopatric taxa that do appear to be specifically distinct, the superspecies concept (e.g., Hall and Moreau, 1970) is useful in ensuring that common origins are not lost sight of.

At the family level I follow Benson et al. (1971), that is to say the Wetmore classification with the recognition of a few additional families, most notably the Malaconotidae.

THE MONTANE GROUPS

Those Afrotropical mountains that possess a stenotopic montane avifauna were divided into seven groups by Moreau (1966, pp. 196–97), although he gave formal names only to six, excluding the montane forests south of the Zambezi River (Moreau, 1966, p. 190). Dowsett (1971) suggested the formal recognition of a seventh group, the Southeastern, in which he included the mountains of southern Malawi, east of the rift. As stressed by Moreau (1966, p. 197), such groups are necessarily an oversimplification, but I retain here those I recognized earlier (Dowsett, 1971) as analysis of distribution patterns within them throws useful light on the possible origins of montane species.

Using these groupings (as shown in Fig. 22–1), I have determined the percentages of forest species shared by pairs of groups, taking the distributional data from Appendix 22–A. Analysis is confined to forest species, as suitable habitat is found on a great number of mountains, in all groups, whereas not all have montane grassland or afroalpine vegetation. Results are presented in Table 22–2.

Although several formulas exist for various coefficients of association, I have found the best simple assessment to involve the proportion of the smaller avifauna that is shared by both, or Simpson's (1960) index. This is because the most frequent biases in comparing the avifaunas of individual mountains are the results either of insufficient exploration or of the effect of pauperization in small, satellite habitats. Consequently, a mountain that has a small avifauna, but that nevertheless shares every species with a larger, richer, or better known neighbor, must be considered very closely related to that neighbor. Conversely, endemism is considered a sign of distinctiveness, although it must be borne in mind that areas containing endemics may be refugia as much as earlier centers of evolution in the absence of knowledge of extinctions elsewhere. It is desirable that intermontane "island" avifaunal relationships be assessed as objectively as possible, and I will elsewhere be undertaking multivariate analyses for the better-known African mountain forests, similar to studies of páramo-puna birds in the Andes by Vuilleumier and Simberloff (1980).

Table 22–2 shows that the percentages of stenotopic forest species shared between pairs of mountain groups varies from only 21% (Cameroon and Ethiopia) to 89% (Angola and East Congo). Before the significance of these figures can be appreciated, it is necessary to know the intragroup variation. I have determined it for 40 or more mountains whose avifaunas I have studied thoroughly in the Tanganyika-Nyasa and Southeastern groups with F. Dowsett-Lemaire (unpublished data). Neighboring mountains in these areas all share at least 75% of stenotopic high-altitude forest species, and this proportion is not altered significantly if analyses cover the larger samples available for all forest-breeding birds (not just those that are strictly montane). This proportion is reduced a little where widely separated mountains are compared, but even the most northerly satellite in the Tanganyika-Nyasa group (Ufipa, southwestern Tanzania) shares 75% of montane species (and 62% of all forest birds) with perhaps the most southerly (Malawi

TABLE 22–2. Percentage of stenotopic montane forest avifaunas shared between groups [a].

Groups	1 (48) [b]	2 (18)	3 (51)	4 (64)	5 (92)	6 (83)	7 (34)
1. Cameroon (37%) [c]	—	67	27	35	60	50	21
2. Angola (6%)		—	56	72	89	78	28
3. Southeastern (16%)			—	78	51	78	35
4. Tanganyika-Nyasa (8%)				—	62	87	53
5. East Congo (29%)					—	71	65
6. Kenya (5%)						—	74
7. Ethiopia (26%)							—

[a] Results are presented as the percentage of the smaller avifauna shared with a larger neighbor.

[b] Figures in parentheses are the numbers of forest species recorded from each group.

[c] Figures in parentheses are the percentages of forest species endemic to each group (Table 22–3).

Hills, southern Malawi). Consequently, it may be assumed for the purposes of intergroup comparisons that pairs of groups sharing at least 70% of stenotopic montane forest species are likely to be closely related.

It is clear from Table 22–2 that (with one notable exception) the percentage of avifauna shared is lowest when the two groups are separated geographically by at least one other group (Fig. 22–1). The exception is the lack of any significant relationship between the Cameroon and Ethiopia groups, both of which have an above average proportion of endemic species (Table 22–2). A similar pattern of distribution is apparent when individual species are considered. As shown by Dowsett (1980a) and Moreau (1966), species common to two separate groups will always be found in any intervening one, for example, all the Tanganyika-Nyasa species common to the Cameroon and Kenya groups occur also in the intervening East Congo group.

Three of the seven montane groups are so isolated that movements by forest birds from any one to a neighbor are unlikely to have been possible for a long time. These are the Ethiopia, Cameroon, and Angola groups. Table 22–2 confirms that the Ethiopia group has rather little in common with any but the two groups that are nearest to it (Kenya and East Congo), and the same is true of the Cameroon group and its nearest neighbors (Angola and East Congo). However, it is striking from the analysis in Table 22–2 that the Angola group shares an extremely high proportion (89%) of its avifauna (which is admittedly small, only 24 stenotopic forest species) with the East Congo group. A former link between these two areas receives some support from the studies of forest tree distribution made by White (1981), discussed further at the end of this chapter. Unlike the Cameroon and Ethiopia groups, which have 37% and 26% endemics, respectively, the montane forests of Angola are almost completely lacking in endemic birds (6%). Were the Angola and East Congo groups in closer proximity, rather than some 2000 km apart, they possibly could not be considered separate (Dowsett, 1980a).

The other groups in the eastern half of the continent, from the mountains of southern Sudan down to the coastal forests of South Africa, form much more of a continuum, and it is less easy to determine which natural boundaries were of most influence in the immediate past. Moreau (1966) and Dowsett (1971) determined the boundaries of the Kenya, East Congo, and Tanganyika-Nyasa groups rather subjectively. I was myself influenced by the potential importance of valleys as effective barriers to highland bird dispersal, as discussed in part by Benson, Irwin, and White (1962). The possible boundaries of montane groups can now be considered rather more objectively, with the data from Table 22–2 in mind. Useful maps of each area were presented by Moreau (1966), topography being of special importance in suggesting which groups of mountains may have been in most recent close proximity with each other, prior to the fragmentation and contraction of forest habitats during the relative desiccation in recent millennia (Moreau, 1966; White, 1981). I discuss three groups in greater detail below.

Kenya Group

In the north, this group is clearly separated from the Ethiopia group by low-lying desert country from Lake Turkana northward. Similarly an extensive belt of lowland in the Nile valleys of central Uganda, between Lakes Victoria and Albert, separates the mountains of the Kenya group from those of the East Congo. But to the south the situation is less clear.

Moreau (1966, pp. 205–207) excluded from the Kenya group montane forests from the Taita Hills southward. These he included in the Tanganyika-Nyasa group, together with the rich avifauna of the Usambaras in northeastern Tanzania. On the other hand, Dowsett (1971) included all the mountains from the Usambaras northward in the Kenya group. I was influenced not only by the high level of endemism in the Usambaras, but also by the existence of a lowland gap, perhaps long established, between the Usambaras and their southern neighbors the Ulugurus (an equally striking center of endemism). The endemic species of these two areas have recently been discussed by Stuart (1981). Endemism need not in itself mean that an area has been long isolated, nor that it is not related to a neighboring area lacking endemics. Rather more instructive when examining possible intermontane relationships are the species shared. Table 22–2 shows that the Kenya and Tanganyika-Nyasa groups share 87% of their stenotopic forest avifaunas, when the Usambaras are included in the former group. If the Usambaras are removed and placed in the southern group, then the two have only 71% of species in common. The Usambaras form the northern limit in eastern Africa of eight montane forest species (*Alethe fuelleborni*, *Sheppardia sharpei*, *Modulatrix stictigula*, *M. orostruthus*, *Apalis moreaui*, *Orthotomus metopias*, *Laniarius fuelleborni*, and

Anthreptes rubritorques). On the other hand, it is the southern limit of four montane species (*Tauraco hartlaubi*, *Merops oreobates*, *Nectarinia reichenowi*, and *Spermophaga ruficapilla*), as well as of the eastern subspecies (*akleyorum*) of *Bostrychia olivacea*. These patterns of distribution do not seem to place the Usambaras firmly in either the Kenya group or the Tanganyika-Nyasa. As pointed out by Stuart and Turner (1980), I was wrong to state that the Usambaras would seem to have no connection with the Tanganyika-Nyasa group (Dowsett, 1971), but for the moment no firm decision is possible. White (1981) has examined some of the botanical evidence for afromontane migratory tracks, or areas of origin, and these kinds of data will need to be taken into account, when available, in any further consideration of bird distribution patterns.

East Congo Group

The northern boundary of this group is well marked, the peripheral highlands being the Lendu Plateau, west of Lake Albert. The group consists of the mountains on either side of the rift valley containing Lakes Albert, Edward, Kivu, and Tanganyika. The southern limit is less obvious. Moreau (1966, pp. 204–205) considered the Ufipa Plateau in southwestern Tanzania to be in the Tanganyika-Nyasa group (although not marking it as such on his accompanying map), pointing out that its avifauna differed considerably from that of Kungwe-Mahari to the north. The map presented by Dowsett (1971, p. 322) shows the Ufipa in the East Congo group, but this is in error. The Ufipa forests have been explored in recent years by D. C. Moyer, R. Stjernstedt, and the author, and most of the birds present are species typical of the Tanganyika-Nyasa group. The proportion of species shared between the Ufipa and Kungwe-Mahari, the nearest East Congo mountain east of the rift, is only 35%. The relationship is closer (68%) between Ufipa and Marungu, the most southerly of the East Congo group west of the rift, and its nearest neighbor.

There can be little doubt of Kungwe-Mahari's affinity with the East Congo group, all but one of its stenotopic montane species (*Phylloscopus ruficapilla*, a typical species of the Tanganyika-Nyasa group) reappearing in the Itombwe highlands (Prigogine, 1971). In spite of containing such nonmontane Congo elements as *Andropadus latirostris* and *Alethe poliocephala*, the forest avifauna of Ufipa is closely related to that of its nearest Tanganyika-Nyasa neighbor, Mt. Rungwe (85%). Marungu has a large number of species typical of the East Congo group (Dowsett and Prigogine, 1974).

Southeastern Group

This group, as proposed by Dowsett (1971), has its northern elements east of the Lake Malawi-Shire River rift in Malawi and adjacent Mozambique, and runs south to the coastal forests of South Africa. This subjective assessment is confirmed by comparison of the avifauna of Mt. Dedza (the most southerly outlier of the Tanganyika-Nyasa group) to that of its nearest neighbor east of the rift, Mangochi (only 44% of species shared). On the other hand, Dedza shares 94% with its largest northern neighbor, the South Viphya Mts., and Mangochi 88% with its largest southern neighbor, Mulanje. Clearly the effectiveness of the Shire rift as a barrier between montane groups is significant and probably long established. Although forests occur west of the rift as far south as the Malawi Hills, most are impoverished and thus of uncertain affinities, although the Malawi Hills themselves may be more closely related to forests to the east, such as Mt. Thyolo (Cholo) (85% of forest species shared, and 100% of the few montane ones).

Within the Southeastern group there is a close relationship between neighboring mountains (e.g., 88% of species shared between Mulanje and Gorongoza), in spite of the greater isolation of some elements in this group. There is a southward cline of decreasing species richness, many species being absent south of the Zambezi River, where montane forests occur at much lower altitudes. Indeed, the fact that all but one of the 20 endemic species in the Southeastern group (Table 22–3), including 7 forest species, are absent north of the Zambezi suggests that this perhaps ought to be the boundary between the groups, or the Limpopo River. Certainly the proportions of shared species throughout most of Malawi are in

TABLE 22–3. Numbers of forest and non-forest species endemic to each montane group.

Groups	Forest	Nonforest	Total
1. Cameroon	18	1	19
2. Angola	1	0	1
3. Southeastern	8	12	20
4. Tanganyika-Nyasa	5	3	8
5. East Congo	27	1	28
6. Kenya	4	4	8
7. Ethiopia	9	18	27
Total	72	39	111

excess of 70%, and the Shire River is a barrier to relatively few species.

It is clear from the analysis of these data that all seven groups of the Afrotropical region are interrelated to varying degrees, with shared avifaunal elements. Two forest birds occur in all groups (*Columba arquatrix* and *Alcippe abyssinica*), and to these should probably be added *Schoutedenapus myoptilus*, which may well occur in the mainland Cameroon group (as discussed earlier). The small babbler *Alcippe* presents a remarkable contrast to the two nonpasserines, which have extensive movements: it is very habitat-bound and sedentary, although its nearest present-day relatives are in the Oriental region. An originally continuous distribution is, as mentioned earlier, a feature in all the Afrotropical montane groups. This is true also of the forest birds that are found in mountains in the Arabian peninsula (*Streptopelia lugens*, *Turdus olivaceus* and *Phylloscopus umbrovirens*), all occurring in the Ethiopia group, from which they presumably originated (Moreau, 1966, p. 209).

ENDEMISM IN THE MONTANE GROUPS

Species that are today endemic to a single montane group, at times to a single mountain, may have originated in that area or may merely have found refuge there. Such species may once have had more extensive ranges, and have become locally extinct in some parts, or they always may have been rare. Because we are looking at a jigsaw puzzle from which many pieces have been lost, the ornithological picture is far from being evidence of the true patterns of changes in the montane environment. Nevertheless, study of montane endemism does provide clues to the origins of such birds, and this is best done by considering group endemism, rather than just those possibly relict forms that are today confined to single mountains.

Cameroon Group

The 19 species endemic to this group of mountains (Table 22–3) are predominantly forest birds, and their distributions may have been more extensive in the past when lowland forest still prevailed over much of western Africa. Apart from isolated Mt. Cameroon itself, which reaches over 4000 m, montane forest does not go much above 2000 m altitude in this area (Moreau, 1966, pp. 197–201).

Two species of Sylviidae are in monotypic genera, and without obvious close relatives, namely *Urolais epichlora* and *Poliolais lopezi*. The latter has been placed in *Camaroptera*, but it may also be close to the Asian tailorbirds of the genus *Orthotomus* (Fry, 1976; Hall and Moreau, 1970), although further study of this group is required. Several nonmontane forest birds in West Africa have probable affinities with species in Asia (Olson, 1979). Possible relationships with the rich Neotropical avifauna should also be sought, in view of the greater age now ascribed by some authors to some modern species and the essentially pantropical distribution of several of the families occurring in the Afrotropical region (Cracraft, 1973).

The genus *Speirops* (Zosteropidae) is endemic to the Cameroon group and nearby Gulf of Guinea islands. It exhibits the tendency to enlarged size often present in island birds, and probably evolved locally from *Zosterops* stock (Moreau, 1966).

Among the remaining endemics, two of the most striking are *Lioptilus gilberti* (Timaliidae) and *Malaconotus kupeensis* (Malaconotidae). The babbler is now considered to be a member of a superspecies that also occurs in the East Congo group, but the affinities of the bush-shrike are less clear, although it may well prove to be another of the very localized species placed in the *Malaconotus blanchoti* superspecies by Hall and Moreau (1970), which occur on mountains widely scattered across Africa. The family Malaconotidae is confined to the Afrotropical region (with the exception of a species marginally reaching North Africa). On the other hand, the Timaliidae have radiated most extensively in the Oriental region, and although *Lioptilus* is a purely African genus, taxonomic ties should be sought among Asian species.

The 14 endemics left have all been placed in superspecies or species groups with taxa occurring outside the Cameroon group, by Hall and Moreau (1970) and by Snow (1978). Three are allied with nonmontane species: *Tauraco bannermani* (with *T. erythrolophus* of the Angola Plateau), *Laniarius atroflavus* (with the over-large assemblage of crimson-breasted bush-shrikes named the "*L. barbarus* superspecies"), *Phyllastrephus poensis* (allied to bulbuls of nonmontane distribution in western and eastern Africa). In addition, *Nectarinia ursulae* is perhaps most closely related to the widespread *N. venusta*, but more evidence is needed to substantiate this claim.

The montane relatives of Cameroon endemics occur in the East Congo, Kenya, and Tanganyika-Nyasa groups, especially the first, which is more likely than the others to have had closer geographic ties with the Cameroon group in the past. *Francolinus camerunensis* is placed in the large, and perhaps rather heterogeneous superspecies that contains several francolins endemic to other groups, but its closest relative would seem to be *F. nobilis* of the East Congo group. Similarly, *Ploceus baglafecht* and *P. bertrandi* should perhaps not be placed in the same superspecies (they are sympatric in part of their range), nor may the Cameroon endemic *P. bannermani* be related to *P. bertrandi*, which, interestingly, does not occur in the intervening East Congo group.

Cossypha isabellae seems to be related to the montane populations of *C. bocagei*, but not to *C. polioptera*, which is mistakenly placed in the same species group by Hall and Moreau (1970). Dowsett and Dowsett-Lemaire (1980) stressed that these latter two species are not closely related; their morphological similarities are the result of convergence.

Two of the Cameroon endemics have in the past been considered conspecific with allies elsewhere (*Andropadus montanus* with *A. masukuensis*, *Nesocharis shelleyi* with *N. ansorgei*), but Hall and Moreau (1970) thought that genetic separation from the East Congo group had been long enough for them to be specifically separable. However, this remains to be tested in the field, for example, with vocal playback experiments.

Angola Group

The one montane endemic (*Francolinus swierstrai*) is a forest francolin, placed in the heterogeneous *F. erckelii* superspecies by Snow (1978). As with *F. camerunensis*, its closest relative is likely to be *F. nobilis* of the East Congo group. *Melaenornis brunnea* was treated by Moreau (1966) as a species endemic to the Angola group, but it is no more than a well-marked race of *M. chocolatina*, the latter widespread in the eastern African montane groups.

Southeastern Group

Only one of the 20 species endemic to this montane group occurs north of the Zambezi, namely, *Alethe choloensis*. The species within this small genus are all allopatric and appear to be very closely interrelated. Interestingly, *A. choloensis* separates a recently discovered population of *A. fuelleborni* in Mozambique from the main populations of that species in the Tanganyika-Nyasa and (marginally) Kenya groups.

South of the Zambezi, most of the mountain endemics are well-marked species of nonforest situations (as, too, in the Ethiopia group). Indeed, the one nonpasserine among them (the ibis *Geronticus calvus*) is placed in a superspecies with *G. eremita*, a bird of the southern Palearctic that winters within the Ethiopian highlands.

Few other species have any close relationship with birds outside the Southeastern group. The timaliid *Lioptilus nigricapilla* is a member of a genus that occurs elsewhere in the Afrotropical region, but it is not known to have any especially close affinities with its congeners. *Bradypterus barratti* is no longer considered conspecific with *B. mariae*; instead its nearest relative may be *B. cinnamomeus*, which ranges allopatrically northward from the extreme north of this group. *Apalis chirindensis* is related to, but specifically distinct from, the more widespread (and not solely montane) *A. melanocephala*. *Mirafra ruddi* has been considered conspecific with *archeri* of Somalia (Hall and Moreau, 1970), but I would consider them specifically distinct, pending field investigation. Hall and Moreau (1970) have placed *Anthus crenatus* in a superspecies with *A. lineiventris*, but there is little evidence for their close affinity. Indeed the latter may prove to be related to Palearctic populations of *A. trivialis*. These same authors have argued that *Nectarinia violacea* forms a superspecies with *N. olivacea*, but they are dissimilar in several respects. The endemic South African canaries *Serinus* are also well marked, but two may have relatives that are widespread in the montane groups farther north in Africa. *S. scotops* may be related to members of the *S. citrinelloides* species group, and there seem to be good reasons for regarding *S. leucoptera* as closely allied to *S. burtoni*.

The remaining 10 species are not known to have any close relationship to extralimital species, but further investigation for possible links would be worthwhile. Two of the canaries (*Serinus totta* and *S. symondsi*) do form a superspecies between themselves, and indeed may be conspecific. Two species are in an endemic family (Promeropidae), which is perhaps most closely related to the Sturnidae (Sibley and Ahlquist, 1974). *Chaetops frenatus* is probably in an endemic monotypic genus, if it proves to be unrelated to *Achaetops pycnopygius* as is suggested by Clancey (1980), who places the former in the Turdidae, the latter in the Sylviidae.

Tanganyika-Nyasa Group

This group is very poor in endemics. The bushshrike *Malaconotus alius* is known from only the

Ulugurus in Tanzania (Stuart and Jensen, 1981). It is of uncertain affinities, being included by Hall and Moreau (1970) in the heterogeneous superspecies headed by *M. blanchoti*, itself a widespread savanna species. *Nectarinia rufipennis*, a recently described species known only from the Uzungwa Mts. in southeastern Tanzania (Stuart and Jensen, 1981), is a double-collared sunbird, but so distinctive that its relationships are uncertain. *Dessonornis lowei* occurs only in Tanzania, and is most closely related to *D. montana*, which is endemic to the Usambaras. Indeed, some authors have considered the two to be conspecific.

The warblers *Apalis chapini*, *Cisticola nigriloris*, and *Bathmocercus winifredae* are constituents of superspecies whose members range from the Kenya montane group westward to West Africa. *B. winifredae* has an extremely limited and isolated distribution, being known only from the Ulugurus and Ukagurus in northeastern Tanzania (Britton, 1980).

The remaining two species have limited ranges in southwestern Tanzania and northern Malawi. *Cisticola njombe* was placed tentatively by Hall and Moreau (1970) in a superspecies with the widespread savanna *C. chiniana*, but recent observations by F. Dowsett-Lemaire (pers. comm.) suggest that this is unlikely to be correct, and its relationships remain uncertain. *Euplectes psammocromius* is best considered specifically distinct from *E. hartlaubi*, but may be closely related to both that species and *E. progne*.

East Congo Group

The mountains in this group provide some of the most intriguing problems of avian relationships, and also some of the best evidence for extralimital origins. These latter (discussed further later) include the owl *Phodilus prigoginei* (known from a single specimen from Itombwe; Prigogine, 1973) and the broadbill *Pseudocalyptomena graueri* (a monotypic genus), whose relatives are undoubtedly Asian and whose position in Africa is unique. The warbler *Hemitesia naumanni* is also in a monotypic genus, but it may be more closely related to the Oriental *Tesia* than to any Afrotropical species (Hall and Moreau, 1970). The status of *Graueria vittata* is much more uncertain, and although it may have affinities with the African genus *Camaroptera*, as suggested by Hall and Moreau (1970), the possibility that its origins too are in the Oriental region should not be overlooked.

As with the preceding species, the great majority of the remaining endemics are forest birds. Most are members of montane superspecies of wider distribution, some (such as *Francolinus nobilis* and *Nectarinia alinae*) having allies in the distant Cameroon group as well as the nearer Kenya and Tanganyika-Nyasa groups. The relationships of others are much less clear. *Prionops alberti*, for example, is a well-marked species, although Hall and Moreau (1970) have suggested that it may be related to *P. plumata*, a widespread bird of lowland savannas. Hall and Moreau (1970) stressed the uncertain affinities of the weaver *Ploceus alienus*.

The East Congo group provides examples of the extent of radiation in some genera. No fewer than four species of sunbird *Nectarinia* are endemic to the group, all members of superspecies of wider distribution (and all with elements in the Kenya group). The genus *Cryptospiza* comprises four very similar species, all of which have been found together on some mountains in the group. Ecological differences and the pattern of evolution within the genus are unknown. The genus *Apalis* is also well represented in this group. There are some forms whose status as species requires investigation. Montane and lowland subspecies of *A. binotata* occur almost sympatrically (Hall and Moreau, 1970) and the presence of any genetic isolating mechanisms between them should be tested. Dowsett and Dowsett-Lemaire (1980) have shown that *A. cinerea* and the mainly midaltitude *alticola* are in reality conspecific; similar treatment is here accorded *A. argentea* and *eidos*, following Hall and Moreau (1970), and *A. porphyrolaema* and *kaboboensis*, although the latter pair are given specific status by Prigogine (1978b). *A. ruwenzorii* and *A. pulchra* have often been considered conspecific, but Hall and Moreau (1970) have suggested reasons for keeping them separate. These various problems involving *Apalis* have been studied mainly via their morphology, but studies of their specific interaction (especially response to song) are needed before any degree of certainty about their status as species can be achieved.

Kenya Group

In marked contrast to the high level of endemism in the East Congo group, namely 28 of 106 (26%) of the stenotopic montane forest and nonforest avifaunas (Table 22–3; Appendix A), endemic species in the Kenya group amount to only 8% in total (8 of 105). As in the Cameroon group, the endemics include a francolin (*Francolinus jacksoni*) and a tauraco (*Tauraco hartlaubi*), both closely related to elements in other groups. The six other endemics include only two more forest birds, the robin *Dessonornis montana* (endemic to the Usambaras) and the starling *Cinnyricinclus femoralis*. The latter forms a superspecies with another forest starling, the allopatric *C. sharpii* (though, contra Hall and Moreau [1970] not with the widespread savanna congener *C. leucogaster*).

The four nonforest endemics all have close relatives elsewhere. *Macronyx sharpei* is a member of a superspecies with *M. flavicollis* of the Ethiopia group, and *Cisticola hunteri* is a member of the same superspecies as *C. nigriloris* of the Tanganyika-Nyasa group. Incidentally, *C. hunteri* can be added to the list of East African endemics presented by Turner (1977), following the recommendations regarding its specific status made by Dowsett and Dowsett-Lemaire (1980) and others. Another warbler, *C. aberdare*, has only recently been separated from a more widespread lowland relative *C. robusta* (Traylor, 1967). Finally, the affinities of *Euplectes jacksoni* remain unclear, although its closest relatives may well include the sympatric *E. progne* and its ecological counterpart in the Tanganyika-Nyasa group, *E. psammocromius*.

Ethiopia Group

What is striking about the montane avifauna of this group is the very high level of endemism (27 of 65 species or 42% in total), and the impoverished forest element. Indeed, whereas 26% of the limited forest avifauna is endemic, the level is 58% in nonforest birds. As stated elsewhere, this situation is mirrored (though less strongly) by that in the group at the other end of Africa, the Southeastern.

Three of the endemics are placed in monotypic genera (a goose, *Cyanochen cyanopterus*; a babbler, *Parophasma galinieri*; and a species of uncertain affinities, *Zavattariornis stresemanni*). *Zavattariornis* remains one of the great enigmas of African ornithology, probably a member of the Corvidae, but possibly of the Sturnidae (Hall and Moreau, 1970). As with other taxa of uncertain status in the northern half of the Afrotropical region, the possibility should be borne in mind that its closest relatives may be found extralimitally, perhaps in the Oriental region.

Several of the other species in the Ethiopia group have no obviously close relatives elsewhere, such as the rail *Rougetius rougetii* (sometimes placed in the genus *Rallus*), the plover *Vanellus melanocephalus* and the swallow *Hirundo megaensis* (the latter, as with *Zavattariornis*, included here, although not occurring at such high altitudes as most of the montane avifauna).

Situated as it is, it might be expected that the Ethiopia group would contain a good number of species of clearly Palearctic origin, absent from the montane groups farther south. But the only such species appears to be the chough *Pyrrhocorax pyrrhocorax*. Instead, some well-differentiated endemics do have their closest relatives farther south within the Afrotropical region, even though these may not be obvious. Such examples are the ibis *Bostrychia carunculata*, the parrot *Poicephalus flavifrons*, the lovebird *Agapornis taranta*, the tauraco *Tauraco ruspolii* and the woodpecker *Dendropicos abyssinicus*—for all these genera are confined to the Afrotropical region.

Even species whose genera do have representatives outside Africa appear to have their closest affinities within the continent, for example, the oriole *Oriolus monacha*, the tit *Parus leuconotus* and the canaries *Serinus nigriceps* and *S. tristriatus* (Hall and Moreau, 1970). The recently described *S. ankoberensis* is said by Ash (1979) to be most closely related to *S. menachensis* of southern Arabia, but the latter is in turn a member of a group of species within the Afrotropical region.

WELL-MARKED MONTANE RACES AND INCIPIENT SPECIES

The preceding discussion made it clear that it is not always obvious, on present evidence, whether isolated taxa have reached the level of species. Because they have a bearing on the effects of montane isolation and speciation, it is useful to examine the distribution of some well-marked montane subspecies, which are considered incipient species by some authors. Table 22–4 summarizes these data. White's (1960–65) findings have been followed, as his work is the only modern evaluation of the taxonomy of the whole Afrotropical region. Complete agreement with these findings is unlikely to be achieved by other taxonomic reviews, but they provide a useful basis for discussion.

Containing, as they do, only species that are nonmontane for much of their range, these data are not concerned with the patterns of subspeciation within stenotopic montane species, which were reviewed briefly by Dowsett (1980a).

Table 22–4 contains a total of 50 lowland species with montane races. (Many lowland species occur in highland areas without exhibiting any morphological differentiation.) The Kenya and East Congo groups contain a large proportion of the montane forms of these species, whereas the Angola and Southeastern groups contain hardly

TABLE 22–4. Lowland species with montane races in the Afrotropical groups.

Family	No. of species with montane races	No. of such species in each group [a]						
		1	2	3	4	5	6	7
Threskiornithidae	1					1		
Anatidae	1						1	
Accipitridae	1						1	1
Phasianidae	3				1	2	2	
Columbidae	1			1	1	1		
Psittacidae	1						1	
Strigidae	1						1	
Meropidae	1						1	1
Phoeniculidae	1	1				1	1	
Capitonidae	1				1			
Picidae	1	1						
Eurylaimidae	1					1	1	
Alaudidae	2				1	1		
Hirundinidae	2	2			1	1	1	2
Pycnonotidae	3				1	1	3	
Turdidae	4	1	1		1	2	1	2
Sylviidae	5	2		1		3	3	1
Muscicapidae	3	2			1	2	1	1
Motacillidae	3	2	1		1	3	2	1
Malaconotidae	2	1				1	2	
Laniidae	1				1			
Zosteropidae	1			1	1	1	1	
Ploceidae	6	1				1	4	2
Estrildidae	3					2	3	
Fringillidae	1		1					
Total	50	13	3	3	11	23	30	12
(%)	(100)	(26)	(6)	(6)	(22)	(46)	(60)	(24)

[a] 1, Cameroon; 2, Angola; 3, Southeastern; 4, Tanganyika-Nyasa; 5, East Congo; 6, Kenya; 7, Ethiopia.

any. Although this is not shown in the table, exactly half of the 50 species are found in their montane form in only one montane group.

In most cases these montane races are probably only a reflection of environmental changes and are unlikely to lead to the formation of genetically distinct new species. Even in some of the most isolated examples (e.g.,the ibis *Bostrychia olivacea akleyorum* in the Kenya montane group) there has been little divergence from the parent form. However, some recognizable taxa have been suggested as incipient species, and these are worthy of attention. *Columbidae*. As mentioned in Appendix 22–B, the montane form of *Columba delegorguei* does not seem to warrant specific recognition (Dowsett and Dowsett-Lemaire, 1980).

Strigidae. Snow (1978) has pointed out that *Bubo poensis vosseleri* may be a good species (as others have also suggested), but that too little is known of this rare owl for a firm conclusion to be drawn.

Alaudidae. Hall and Moreau (1970) have suggested that the highland form of *Mirafra africana* in southwestern Tanzania and northern Malawi might be an incipient species. Dowsett (1972b) has shown that two subspecies are involved in these isolated populations (*nigrescens* and *nyikae*); there is as yet no evidence that either is specifically distinct, and given the great amount of racial variation within this family, not too much weight should be given to morphological divergence.

Hirundinidae. Hall and Moreau (1970) have suggested that the small, basically West African race (*domicella*) of *Hirundo daurica* might be an incipient species. The Afrotropical highland forms of this species seem most distinct from the small western Palearctic taxa (F. Dowsett-Lemaire, pers. comm.), and the possibility that there is an essentially afromontane species should be investigated.

Turdidae. Hall and Moreau (1970) have mentioned that there may be good reasons for recognizing the montane forms of *Oenanthe bottae* (Ethiopia group) as specifically distinct from *heuglini*, which was the course followed by Mackworth-Praed and Grant (1960).

Muscicapidae. The small, gray flycatchers of the genus *Muscicapa* are especially troublesome taxonomically. Although *itombwensis* is here treated as a form of *M. olivascens* (following Hall and Moreau, 1970), it may well prove to be specifically distinct, when better known. This is less likely to be true of the various montane races of *M. adusta*.

Motacillidae. As mentioned in Appendix 22–B, *Anthus latistriatus* is here treated as a race of *A. novaeseelandiae*, though without any specially great conviction. Other forms of this pipit have also given rise to questions concerning their specific status, as too have montane populations of *A. similis* (Dowsett and Dowsett-Lemaire, 1980). This difficult genus is in need of detailed field research before conclusions can be reached with confidence. Hall and Moreau (1970) have also suggested that the isolated population of *Motacilla capensis* in the East Congo and Kenya groups may be an incipient species, but this is less likely on present evidence.

Laniidae. Hall and Moreau (1970) have claimed that *Lanius collaris marwitzi* (a white-browed form in the highlands of southwestern Tanzania) appears not to interbreed with neighboring populations. The behavior and habitat of *marwitzi* are the same as those of other montane populations of the species. White-browed individuals resembling this form appear elsewhere at times, as in the population of *L. collaris* on the montane Nyika Plateau in northern Malawi (pers. obs.). It seems they are likely to interbreed, and for the moment I would consider the two no more than subspecifically distinct.

Zosteropidae. The recognition of *Zosterops poliogastra* as a good species in the mountains of Kenya and Ethiopia (Hall and Moreau, 1970) has clarified the confused taxonomy of *Z. senegalensis* somewhat. There remain in that species several richly colored montane forms, parapatrically distributed with lowland races, that is to say in contact but ecologically segregated (Hall and Moreau, 1970). Some of these may prove to be genetically discrete, which was the view taken by Mackworth-Praed and Grant (1960), but further investigation is needed in the field.

Ploceidae. No fewer than five species have montane forms within the Kenya and Ethiopia groups. Of these, the very isolated and ecologically segregated *Ploceus olivaceiceps nicolli* should perhaps be accorded specific recognition (S. N. Stuart, pers. comm.).

Estrildidae. The distinctive and isolated *Spermophaga ruficapilla cana* of the Usambaras in northeastern Tanzania should be studied to determine its genetic relationship to other populations of the species.

INTRA-AFRICAN ORIGINS OF THE MONTANE AVIFAUNA

The foregoing sections have examined patterns of distribution and endemism within the afromontane archipelago. A major feature that arises is the high proportion of the stenotopic avifauna that is shared between several montane groups. These groups are artificial associations in so far as most of their avifaunas are not significantly different in a way that can be tested statistically. With intragroup variation in the order of 75%–90% of species shared by pairs of montane forest "islands," it is clear that many groups exhibit the same degree of similarity, notwithstanding the much greater isolation involved. As a result, it is difficult to pinpoint the possible origins of relatively widespread species.

Endemism is high in the forest avifaunas of the Cameroon, East Congo, and Ethiopia groups (Table 22–2), although the total number of forest birds in this last group is small. The Southeastern and Ethiopia groups have a high proportion of endemic nonforest species. But it is not clear if present-day concentrations of endemic taxa represent centers of evolution or refugia for birds whose distributions during the last pluvial period were more extensive.

The history of the afromontane "islands" is uncertain. Moreau (1966), on the evidence then available, concluded that montane species now confined to isolated and small areas would have occurred much more widely during the last glaciation, when cool forest conditions would have existed at much lower altitudes than now. This period was considered to have lasted for at least 50,000 years before 18,000 years B. P. Moreau's (1963, 1966) interpretation of paleoclimatic data was criticized by Livingstone (1975), who considered from paleobotanical investigations that prior to 12,000 B.P. the climate was drier, not colder, and that forest trees were missing from high altitudes. The suggestion that there may thus have been a considerable spread of forest conditions in Africa after 18,000 B.P. (Livingstone, 1975; Hamilton, 1976) prompted Diamond and Hamilton (1980) to conclude that montane bird species that today show disjunct distributions "have probably either flown from one area to the other or else once occurred in intervening lowland forest." This interpretation of often widely discontinuous distribution patterns (Appendix 22–A) would mean these birds were either very mobile or previously showed a much wider tolerance of altitudinal range. In fact, what evidence there is suggests strongly that most forest bird species are today essentially sedentary (Dowsett,

1980a, b; Dowsett and Dowsett-Lemaire, MS), and striking genetic changes need to be postulated if Diamond and Hamilton (1980) are correct. Some authors have rightly stressed the speed at which speciation can occur (Milstein, 1979), but it cannot be assumed that the short time during which racial differences have become noticeable in some populations of introduced *Passer domesticus* (Johnston and Selander, 1964) applies to other, native and/or more genetically stable species. The fact that relatively little subspecific variation is apparent in most African montane birds (Dowsett, 1980a, b) suggests there may be much genetic conservatism in tropical forest birds.

It also seems unlikely that many forest species that are now strictly high-altitude birds would have occurred in lowland forest in the past, although it is true that in some parts of the afromontane archipelago a small proportion of species today migrate to lower levels in the nonbreeding season (see chap. 23 for similar observations on páramo-puna birds in the Andes). Such migrants amount to about a third of the montane forest avifauna in an area such as Mulanje Mountain (southern Malawi), where lowland forest is still accessible (pers. obs.). None of these breeds at low altitude, and only part of the population moves down—in the best studied species, the robin *Pogonocichla stellata*, the individuals that are resident at high altitude are territorial males (Dowsett, 1982). These movements appear to be prompted by seasonal food shortages at high levels during the cold, dry season months, and seem to lend no support to Diamond and Hamilton's (1980) theories. The Afrotropical lowland forests today have a very large and ecologically complicated avifauna, as discussed by Moreau (1966), which is likely to be long established, any theory to the contrary requiring very convincing evidence.

White (1981) has criticized the paleobotanical conclusions of Livingstone (1975) as often being inconsistent with the evidence available. No firm dating of past changes in forest distribution is possible at present, but it does appear probable that the postglaciation contraction of forests postulated by Moreau (1966) and White (1981) is essentially correct, whatever its timing. Suggestions that montane forest is as extensive now as it ever has been (Diamond, 1981) are surely wrong. Indeed, the drastic effects of fire and also of felling—influences that are growing as human population densities increase—have greatly reduced the extent of forest within the last few centuries.

White (1981) suggested that present-day distribution patterns among forest trees point to there being "migratory tracks" along which species spread in the past. The ornithological evidence presented in this chapter shows that distributions of montane forest birds were essentially continuous (although now disjunct) between groups, and a similar pattern is evident in intragroup distributions of many species (Dowsett, 1980a). The checkerboard patterns presented elsewhere in the tropics, as in New Guinea (Diamond, 1972), are largely lacking, and in the Andes Vuilleumier and Simberloff (1980) also found no firm evidence of such distributions.

As with forest trees (White, 1981), there is a strong suggestion from the ornithological data that the Angola group of montane forests was closely connected to forests to the east, as in the East Congo group (Table 22–2). This may, as postulated by Moreau (1966, pp. 210–211), point to a link through intervening northern Zambia, where higher rainfall and the reduced effects of fire in the past could have meant that present-day midaltitude riparian forests were far more extensive (Dowsett, 1980a).

White (1981) has stressed that forest trees present no evidence of any recent link between the Cameroon and East Congo-Kenya groups, and it shares only 50% and 60% of forest birds with these two groups, respectively (Table 22–2). However, three birds from Cameroon do reappear in only the East Congo group—*Sheppardia roberti*, *Elminia albiventris* and *Laniarius poensis*—as is true also of montane populations of *Cossypha bocagei*. Many others occur more widely (Appendix 22–A), not all of them also in the impoverished Angola group, and it remains probable from the ornithological evidence that some connection between Cameroon and East African mountains did exist in the past. Further correlations between the distribution patterns of montane vegetation, birds and other classes should be investigated.

The existence of "migratory tracks" need not mean that there were long-distance movements by trees or birds in the past, merely that montane forests were much closer to each other then (owing to climatic changes), and so some inter-forest movement was possible on occasion. This might even have been possible for the most sedentary species of birds where montane islands were linked by riparian forest through which juvenile or other nonresident individuals could pass.

The patterns suggested by the data reviewed here are of fragmentation of the ranges of montane birds at some time in the past, dating from a period when precipitation became less and temperatures higher (whenever that may have been). To an important extent the gaps between montane forests are also the results of more recent human activity. Within and between the montane

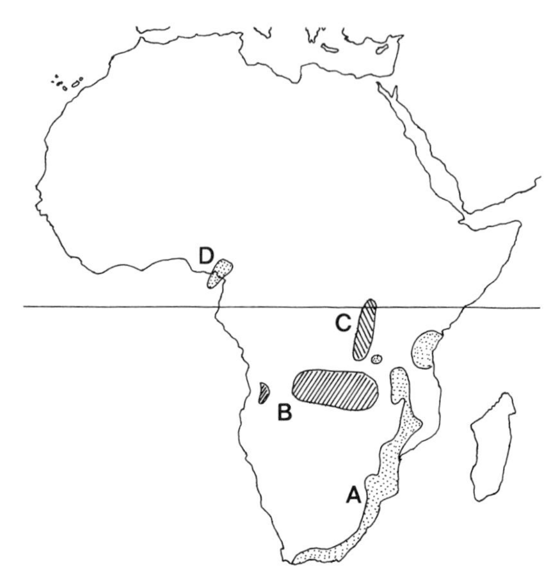

FIG. 22–2. Distribution of the *Phylloscopus ruficapilla* superspecies, an example of widespread allopatry. A: *P. ruficapilla*; B: *P. laurae;* C: *P. laetus*; D: *P. herberti.*

groups, species distribution was originally mostly continuous. A typical example is shown by the *Phylloscopus ruficapilla* superspecies (Fig. 22–2). There are many such examples of closely related species occurring allopatrically over a wide range, and further evidence for the generally limited extent of speciation suggested by this is afforded by the very low ratio of species to genera (about 2:1) in the African montane avifauna (Table 22–1) (1.39 for afroalpine birds according to Chap. 6). On many mountains most forest genera are represented by single species, or by very few species clearly segregated ecologically. Limited morphological variation, limited speciation, and what was originally a continuous distribution, lacking any checkerboard patchiness, are perhaps a reflection of the stability of tropical forest environments over long periods of time.

EXTRALIMITAL ORIGINS OF THE AFROMONTANE AVIFAUNAS

This chapter has so far considered species distribution patterns within Africa, which (in the absence of paleobotanical, paleoclimatic, or other comparative data) allows for few firm conclusions to be drawn regarding the origins of these species. However, some families and genera are not endemic to Africa (Table 22–1), and some of these provide interesting pointers to possible origins outside the Afrotropical region.

Threskiornithidae. Of the two species of ibis, *Bostrychia carunculata* certainly has an immediately Afrotropical origin (the genus being endemic to the region). The genus *Geronticus* comprises two species, one endemic to South Africa, the other in the southern Palearctic. It is not clear to which extralimital genera the two Afrotropical ones are most closely related, but comparison with the Oriental genus *Pseudibis* might be revealing.

Anatidae. Although some species of duck are typically montane (e.g., *Anas undulata*), the only stenotopic montane species in the family is the very distinctive *Cyanochen cyanopterus* of the Ethiopia group. Snow (1978) suggests that the genus is distantly related to the South American genus *Chloephaga*, which comprises five species (see Chap. 6, this volume).

Accipitridae. The monotypic genus *Gypaetus* is widespread in Eurasia as well as in the Afrotropical montane groups, and may well have originated in the Palearctic. *Accipiter* is a cosmopolitan genus, and in addition to the one montane species there are a further six to eight species (depending on taxonomy) in lowland forest and savanna in Africa. *A. rufiventris* may have originated from immigration of *A. nisus* from the Palearctic, although such migrants no longer occur commonly in the Afrotropical region. They are similar in appearance, and are even considered conspecific by Snow (1978). There is a striking parallel in *Buteo oreophilus*, with a montane distribution in Africa and a probable Palearctic origin in *B. buteo*, populations of which still migrate to Africa. *B. brachypterus* of Madagascar appears less closely related to either of these two species (Snow, 1978; Dowsett, pers. obs.).

Phasianidae. Of the 10 stenotopic montane species in this family, 9 are of the genus *Francolinus*, which is mainly Afrotropical in terms of both geographical distribution and radiation. The richest area in species and endemics is the most northerly montane group, Ethiopia. Palearctic populations of the quail *Coturnix coturnix* migrate to Africa and doubtless are the origins of Afrotropical breeding populations, although contra Snow (1978) there is no evidence that these migrants now reach South Africa, where breeding birds are not only montane.

Rallidae. Two of the montane stenotopic species are in the genus *Sarothrura*, which is confined to the African continent and Madagascar. Although Ripley (1977) has merged this genus with the New World *Coturnicops*, and Olsen (1973) has stressed their close relationship, it seems best to retain the Afrotropical genus *Sarothrura* for the time being (Dowsett and Dowsett-

Lemaire, 1980), pending better knowledge of the species most like some *Coturnicops*, the enigmatic *S. ayresi* of the Ethiopia montane group and southern Africa. The other montane representative of this family is also Ethiopian, *Rougetius rougetii*—a monotypic genus of uncertain affinities (Snow, 1978).

Charadriidae. The cosmopolitan genus *Vanellus* includes 11 Afrotropical species, which show a good deal of radiation in some characters and have been placed in as many as 10 genera (Snow, 1978). The Ethiopian endemic *V. melanocephalus* seems to have no immediately close relative, although its affinities may be with *V. tectus*, a species of the semiarid northern half of Africa.

Columbidae. Two species of *Columba* are predominantly montane, of which the widespread *C. arquatrix* (one of only two stenotopic nonpasserines to occur in all seven montane groups) is closely related to *C. hodgsonii* of the Himalaya (Snow, 1978). *C. albitorques* of the Ethiopia group is less easy to place, but may be fairly closely related to the essentially Palearctic *C. livia* species group (Snow, 1978). *Streptopelia lugens* also has obvious links with the Palearctic *S. turtur* species group; although most similar in appearance to eastern *orientalis* (considered by Snow to be specifically distinct from *S. turtur*), it is perhaps relevant that the western Palearctic populations migrate to the northern half of Africa, and from them may have come the parent stock of *S. lugens* (and *hypopyrrha*).

The fourth montane columbid is the monotypic genus *Aplopelia*, purely Afrotropical in distribution but, like the three species discussed earlier, present in the most northerly group (Ethiopia), and hence prima facie likely to have had an extralimital origin. However, no acceptable candidate for a close relationship has yet been put forward.

Psittacidae. The two Ethiopia group endemics (*Poicephalus flavifrons* and *Agapornis taranta*) both have immediate relatives in species groups within Africa. *Poicephalus* is a purely Afrotropical genus, but although *Agapornis* is too (occurring on Madagascar as well as the African mainland), it is likely to be closely related to *Loriculus* of the Oriental region (Snow, 1978).

Musophagidae. This family is endemic to Africa, and four of the five stenotopic species (excluding *Tauraco leucotis*) are members of a fairly closely interrelated complex. In view of egg-white protein and other evidence for an affinity between the families Musophagidae and Cuculidae (Sibley and Ahlquist, 1972), the superficial similarities between *Tauraco* and some members of the Malagasy genus *Coua* (notably *C. caerulea*) (pers. obs.) are interesting.

Cuculidae. There is only one montane species (*Cercococcyx montanus*), and even that occurs in some lowland localities, as in Malawi (Benson and Benson, 1977), and is closely related to a lowland congener, *C. olivinus*. This Afrotropical genus may be most closely related to *Cuculus* species of the Palearctic or Oriental regions.

Tytonidae. Perhaps the most intriguing montane endemic, and certainly the least known, is the single specimen of *Phodilus prigoginei* from the East Congo group. Its closest relative, and indeed its only congener, is *P. badius* of Southeast Asia, which ranges from India eastward. The two are closely related, but specifically distinct (Prigogine, 1973).

Strigidae. The two montane owls in this family have probably a Palearctic origin. *Bubo capensis* is closely related to *B. bubo* of the Palearctic, which occurs as far into Africa as the southern Sahara. Although the genus *Bubo* has speciated more widely in Africa than on any other continent, the essentially montane distribution of northern populations of *B. capensis* does lend support to the idea of an immediate Eurasian origin.

Asio abyssinicus is clearly closely related to the Holarctic *A. otus* superspecies, but pace Snow (1978) it seems best to continue to grant it specific recognition (as, too, the Malagasy *A. madagascariensis*). The other African member of the genus, the widespread *A. capensis*, likewise has a Palearctic origin.

Caprimulgidae. The single montane nightjar is a member of the cosmopolitan genus *Caprimulgus*. *C. poliocephalus* is certainly most closely related to another Afrotropical species, *C. pectoralis*, and although Snow (1978) has not done so, they might even be considered members of a superspecies, as their breeding ranges are allopatric.

Apodidae. The single stenotopic genus *Schoutedenapus* probably contains two species, which are narrowly sympatric. Their taxonomic position is unclear (Snow, 1978).

Trogonidae. *Apaloderma vittatum* is fairly closely related to the two other members of this Afrotropical genus, even though it has at times been placed in a separate genus. Snow (1978) suggests that *Apaloderma* has a monophyletic origin, but detailed comparisons with Neotropical and Oriental genera (in both of which regions the family is richly represented) would be of interest.

Meropidae. *Merops oreobates* has not been allied with any extralimital member of the genus, but its relationships within Africa are unclear, as discussed by Snow (1978). It is superficially not unlike another member of the subgenus *Melitto-*

phagus (which is purely Afrotropical), *M. pusillus*. I have here followed Snow (1978) in considering *lafresnayii* a form of the essentially lowland *M. variegatus*, but with some reservation, in view of the criticism of this course by Devillers (1977). Further investigation is desirable, but it may well be that *lafresnayii* is best given specific status, in which case it would be a second stenotopic montane species.

Bucerotidae. Although this family occurs also in the Oriental and Australasian regions, *Bycanistes* is an Afrotropical genus with no recent extralimital relationship, a superficial similarity to the Oriental *Anthracoceros* being considered the result of convergence (Snow, 1978). Relationships within the genus *Bycanistes* are unclear, and *B. brevis* occurs at times sympatrically with two of the other four members of the genus, and may not be strictly montane.

Capitonidae. *Stactolaema (Buccanodon) olivacea* belongs to an Afrotropical genus, which has no known relative in the Oriental region. The undesirability of including this species in the genus *Pogoniulus* has been mentioned in Appendix B. The two montane species in the latter genus are members of the same superspecies. Snow (1978) considers that all the Afrotropical barbets are more closely interrelated, despite their great degree of radiation, than any is to an extralimital taxon.

Indicatoridae. The smaller members of this mainly Afrotropical family present systematic problems on a par with those of the small, grey *Muscicapa* flycatchers, mentioned elsewhere. Until species limits are determined with any degree of certainty, relationships and origins will remain unclear. *I. pumilio* seems to be essentially montane, and is perhaps most closely related to the lowland forest *I. willcocksi*, but clearly much further information is needed (Prigogine, 1978c).

Picidae. The three stenotopic montane woodpeckers belong to two genera, each confined to the Afrotropical region. Two are members of wide-ranging superspecies (*Dendropicos abyssinicus* and *D. griseocephalus*), whereas *Campethera tullbergi* has no close relative. Snow (1978) has suggested that these two genera are unrelated to any Oriental or Palearctic woodpeckers, but may be allied to the American genus *Colaptes*, and Dowsett-Lemaire (1983) has pointed to shared features of *D. griseocephalus* and species of South American *Melanerpes*.

Eurylaimidae. The monotypic genus *Pseudocalyptomena* (*P. graueri*) of the East Congo group is another example of a taxon whose nearest allies are in the Oriental region, although no particular genus has been proposed as an especially close relation.

Alaudidae. The South African endemic *Mirafra ruddi* (excluding *archeri* of Somalia) is of uncertain affinities. It is a bird of the highveld, not strictly montane, and so may not warrant inclusion in Appendix A.

Hirundinidae. Two members of the cosmopolitan genus *Hirundo* are stenotopic Afrotropical montane endemics (*H. atrocaerulea* and *H. megaensis*). The relationships of both are by no means certain, although Hall and Moreau (1970) have allied them with, respectively, *H. nigrorufa* and *H. dimidiata*. *Psalidoprocne* is an Afrotropical genus, and *P. fuliginosa* of the Cameroon montane group is a member of a typically confusing superspecies (with *P. albiceps*, which is to a great extent also montane in distribution). Hall and Moreau (1970) consider unlikely any relationship between *Psalidoprocne* and *Stelgidopteryx* of the Americas.

Oriolidae. Two of the three montane *Oriolus* species are members of the same superspecies (*O. monacha* and *O. percivali*), whereas *O. chlorocephalus* is much more distinctive, with no obviously close ally. Although the genus *Oriolus* occurs widely in the Oriental region, with several similar blackheaded species, no close relationships have been determined.

Corvidae. The three genera all occur in the Ethiopia group. *Pyrrhocorax pyrrhocorax* is obviously of Palearctic origin, being scarcely separable from Eurasian populations of the species. *Corvus crassirostris* appears to be closely related to the widespread (and often montane) *C. albicollis*, perhaps less closely to the sympatric *C. rhipidurus*, which has been placed in the same superspecies by Hall and Moreau (1970). The extent to which these species are related to the Holarctic *C. corax* is undecided. *Zavattariornis stresemanni* is of very uncertain affinities, perhaps not even placed in the correct family, and the possibility of relationship with some Oriental taxon should be investigated.

Paridae. Although the genus *Parus* is widespread in the Afrotropical as well as Holarctic and Oriental regions, it seems certain that Palearctic birds were the predecessors of African ones. The two montane species (*P. fasciiventer* of the East Congo group and *P. leuconotus* of Ethiopia) are both members of superspecies within Africa, together with species of more widespread distribution. *P. fasciiventer* has perhaps the closer links with the Palearctic, through the *P. ater* superspecies and *P. major*.

Timaliidae. Radiation within this family is

nowhere near as great in Africa as in Asia, and consequently most authorities have suggested an Oriental origin for the family. *Alcippe abyssinica* and *Trichastoma pyrrhopterum* are both congeneric with many Oriental forms, although the latter is one of seven species of *Trichastoma* in Africa, all closely interrelated. An eighth species (*T. poliothorax*) has been placed in the monotypic genus *Kakamega* (Mann, Burton, and Lennerstedt, 1978). *Parophasma galinieri* of the Ethiopia group may well prove to be congeneric with *Lioptilus* (Hall and Moreau, 1970). The montane species of *Lioptilus* include the strikingly distinctive *L. gilberti* of the Cameroon group, clearly rather divergent, and originally placed in its own genus *Kupeornis*. Links should be sought between *Lioptilus* and the plethora of babblers in the Oriental region.

Campephagidae. The genus *Coracina* is most profusely represented in the Oriental and Australasian regions. It is there that the Afrotropical species may be expected to have had their origins, especially as there are also members of the genus in the Indian Ocean islands. *C. caesia* seems closely related to the more widespread, lowland *C. pectoralis*, but the immediate affinities of the other species, *C. graueri* of the East Congo group, are less obvious.

Pycnonotidae. Neither of the two montane genera in the Afrotropical region (*Andropadus* and *Phyllastrephus*) has any immediate relative outside the region, unless the extreme view is taken that *Andropadus* is congeneric with *Pycnonotus*. However, it is now believed that several bulbuls on Madagascar are correctly placed in *Phyllastrephus* (Benson et al., 1976–77; pers. obs.), and it would be interesting to know if allies can be identified in the Oriental region (the lowland forest *Criniger* being common to both regions).

The four *Andropadus* species form two related pairs, *A. montanus-masukuensis* and *A. tephrolaemus-milanjensis*, generally allopatric, whereas *Phyllastrephus flavostriatus*, *P. poensis*, and *P. placidus* are each members of separate superspecies.

Turdidae. No Afrotropical species ranges far outside the region, but the genera *Turdus* (four montane species), *Monticola* (one species), and *Cercomela* (one species) occur outside, and the first two probably have Palearctic origins.

Ripley (1952) proposed to enlarge the essentially Palearctic genus *Erithacus* with the addition, among others, of members of the Afrotropical genus *Sheppardia*. This would produce an unacceptably heterogeneous assemblage, but the possible external origin of the small akalats should not be forgotten. Hall and Moreau's (1970) suggestion that *Pogonocichla stellata* may be related to the morphologically similar *Tarsiger chrysaeus* of the Himalaya needs to be assessed by comparative study.

Sylviidae. Most of the montane genera are essentially Afrotropical, such as the complex *Apalis* (a forest offshoot of *Prinia*?) and *Cisticola* groups. However, other genera occur extralimitally, whereas some may have allies elsewhere, and most of these suggest origins in the Oriental region.

Whether or not the African species of *Phylloscopus* (four of them montane) are better placed in the genus *Seicercus*, their origin is clearly Oriental or Palearctic. The genus *Prinia* is also shared with the Oriental region (its one montane Afrotropical species being the isolated *P. robertsi* of the Southeastern group). The subjective impression given by the genus *Bradypterus* is that it could profitably be associated with such Palearctic genera as *Cettia* and *Locustella*, perhaps via *Bebrornis* of some Indian Ocean islands, and the few Himalayan species now placed in *Bradypterus*. *Artisornis metopias*, with a very limited distribution in eastern Africa, should probably be considered congeneric with the Asian tailor-birds of the genus *Orthotomus* (Hall and Moreau, 1970; Fry, 1976), and the affinities of the monotypic *Poliolais lopezi* are perhaps with the Afrotropical *Camaroptera*, itself possibly a tailor-bird. *Hemitesia neumanni* is also the only member of its genus, and may be closely related to *Tesia* of the Oriental region.

Muscicapidae. This is another family in which there are evident suggestions of Asian origins (the Monarchinae and Rhipidurinae being here given the status of subfamilies). Dowsett and Stjernstedt (1973) pointed out the similarities between Afrotropical *Trochocercus* and the Australasian and Oriental *Rhipidura* species, echoing, unknown to them at the time, Mayr and Moynihan (1946). As discussed further by Dowsett and Dowsett-Lemaire (1980), the possibility of a close relationship between these genera requires detailed consideration. Perhaps the African genus *Platysteira* will also be found to be allied to some Oriental monarchine flycatcher, in view of the presence of the related *Pseudobias wardi* on Madagascar. *Terpsiphone* is already accepted as having populations in Asia, Africa, and the Indian Ocean islands. Some members of the Afrotropical genus *Batis* are closely related to *Platysteira*.

Apart from the monarchine flycatchers, only

Muscicapa lendu may possibly have an extralimital ally, and the systematics of these small, gray flycatchers are too uncertain at present for this to be assessed with any certainty. The genus *Melaenornis* (with two montane species) is purely Afrotropical, but its possible relationship to such other African genera as *Sigelus* (Hall and Moreau, 1970) and *Bradornis* (pers. obs.) remains unsettled.

Motacillidae. *Anthus* is a cosmopolitan genus, and it is not known to which species the two South African endemics (*A. crenatus* and *A. chloris*) are mostly closely related. On the other hand, *Macronyx* is a purely Afrotropical genus, quite unrelated to the New World *Sturnella* (family Icteridae) in spite of many convergent characters shared (Thomson, 1964) (see also Chap. 6). *M. sharpei* and *M. flavicollis* of the Kenya and Ethiopia groups, respectively, are typical members of a genus in which radiation has been very limited.

Malaconotidae. Many authorities now accept the bush-shrikes as a family distinct from the widespread Laniidae. It is essentially an Afrotropical family, and the montane species (*Laniarius* spp. and *Malaconotus* spp.) present many problems relating to the processes of speciation within the continent, but no pointers to their ultimate origins.

Prionopidae. Prionopidae are another Afrotropical family presumed to be closely related to the true shrikes Laniidae, although some striking points of similarity are found among the Malagasy family Vangidae (pers. obs.), including cooperative breeding (Grimes, 1976). Within *Prionops* there has been relatively little radiation, and the one montane species (the localized *P. alberti* of the East Congo group) is unexceptional.

Sturnidae. The three montane genera are essentially Afrotropical, although one species of *Onychognathus* occurs marginally in the Palearctic. Except that *O. walleri* and *O. tenuirostris* are widely sympatric and not closely related, the species of *Onychognathus*, *Poeoptera*, and *Cinnyricinclus* in the African montane groups present similar patterns of allopatric, superspecies distribution. *Poeoptera* is perhaps most closely related to *Onychognathus* (Hall and Moreau, 1970), but any extralimital allies of the three genera are unclear.

Promeropidae. This is an endemic Afrotropical family of rather uncertain affinities. At one time thought to be an offshoot of the Australasian honey-eaters (family Meliphagidae), the Promeropidae have more recently been suggested to be probably closest to the Sturnidae (Sibley and Ahlquist, 1974).

Nectariniidae. Both genera, *Anthreptes* (1 montane species) and *Nectarinia* (16 species), are mainly Afrotropical, with some representatives in the Oriental region. For none of the species concerned is any extralimital relationship obvious, and the intra-African patterns of distribution have already been discussed here.

Zosteropidae. Another family whose distribution is widespread in the Old World tropics, the white-eyes have radiated rather widely in Africa (witness the evolution of the distinctive endemic genus *Speirops* in the Cameroon group and nearby Gulf of Guinea islands, and the complicated patterns of speciation evident in *Zosterops*, described by Hall and Moreau, 1970). Any relationships outside Africa are unclear, although the family is represented in the Indian Ocean islands, as is usual when families are shared between the Afrotropical and Oriental regions.

Ploceidae. The genus *Ploceus* (six montane species) is basically Afrotropical (with a few Oriental species), and *Euplectes* (two species) entirely so. Speciation within *Ploceus* is great, but whether or not this implies an Afrotropical origin for all the species concerned is unclear.

Estrildidae. The four montane genera (*Nesocharis*, *Cryptospiza*, *Clytospiza* and *Estrilda*) are all Afrotropical endemics, although *Estrilda* may be related to *Amandava*, which also occurs in the Oriental region. Within Africa, all these montane species have close relatives of lowland disposition, with the notable exception of *Cryptospiza*, whose four species are all montane and in places broadly sympatric.

Fringillidae. *Linurgus* is a monotypic genus of no obviously close affinity with anything else, although Hall and Moreau (1970) suggest it may be nearest *Carduelis spinoides* of the Himalaya. Most of the African montane canaries *Serinus* are in the two extreme groups, Ethiopia and the Southeastern (South Africa). *S. canicollis* may be related to *S. serinus* of the western Palearctic (Hall and Moreau, [1970] place them in the same superspecies), but with the exception of *S. ankoberensis* (allied to the Arabian *S. menachensis*), no other species has any extralimital affinities. Indeed, the great majority of *Serinus* species are Afrotropical.

CONCLUSION

The preceding review has undoubtedly raised more questions than can immediately be answered. This is understandable, in view of our very incomplete knowledge of the birds and their past environments. What must be emphasized is that there is no ornithological *evidence* regarding the origins of the Afrotropical montane avifaunas, merely pointers.

The relationships of the African avifaunas have been reviewed by Hall and Moreau (1970) and Snow (1978), but until similar analyses are undertaken for other zoogeographic regions, it is difficult to identify potential extralimital relationships. Even within Africa our knowledge of interspecific affinities is very superficial and often based merely on subjective assessments of morphological characters. The specific status of many taxa (isolated and parapatric) cannot be properly determined without comparative studies of genetic characters (e.g., blood, egg-white, or feather proteins) or possible behavioral isolating mechanisms (e.g., vocalizations). Until such data are available, assessments of status and origin often must remain a matter of personal opinion.

Moreover, knowledge of the timing of speciation in African birds is remarkably incomplete, and it is unfortunate that scientists in other disciplines have looked to ornithologists for firm evidence of past environmental changes (Hamilton, 1976). Consequently, the search for extralimital allies for members of the Afrotropical montane avifauna should not be influenced by preconceived ideas of timing. For example, the fact that the Arabian peninsula has long been devoid of forest should not invalidate possible relationships between Oriental and African species. Indeed, there are several suggestions of Himalayan species in particular having African allies, and others may appear with further knowledge. It is not necessary to postulate relatively recent migration to account, say, for the presence of the forest owl *Phodilus prigoginei* in the East Congo group. The fact is that such species are today probably very sedentary, although genetic changes may have occurred in the past (as in some populations of *Passer domesticus*). Genetic conservatism, and the fact that environmental and competitive pressures might be expected to have differing effects on different populations, could account for a situation in which many forest species in Africa may have existed essentially unchanged for many millennia.

Table 22–5 summarizes the tentative data relating to extralimital affinities discussed in this chapter. About 8% of species are shared with the Palearctic region, although none with the Oriental (contrary to the situation with such lowland species as *Salpornis spilonota* and *Anthus similis*). Nevertheless, some relationships with Oriental species are close, as in the genus *Alcippe* (Fig. 22–3). Other Afrotropical representatives are far less closely related to their Oriental allies (e.g., *Pseudocalyptomena graueri*, and such non-montane species as *Afropavo congensis* and

TABLE 22–5. The extralimital relationships of the Afrotropical montane avifauna.

Family	Total species	Number of species with certain (or possible) allies in other regions [a]		
		Palearctic	Oriental	Neotropical
Threskiornithidae	2	1		
Anatidae	1			(1)
Accipitridae	3	3		
Phasianidae	10	1		
Columbidae	4	1(1)	1	
Psittacidae	2		(1)	
Tytonidae	1		1	
Strigidae	2	2		
Eurylaimidae	1		1	
Corvidae	3	1		
Paridae	2	(1)		
Timaliidae	7		2	
Campephagidae	2		2	
Turdidae	27	5(1)		
Sylviidae	40	4	4(3)	
Muscicapidae	7		(2)	
Zosteropidae	2		1	
Fringillidae	12	(1)	(1)	
24 other families (Table 1)	94	0	0	0
Total	222	18(4)	12(7)	(1)

[a] Figures in parentheses denote unproven alliances.

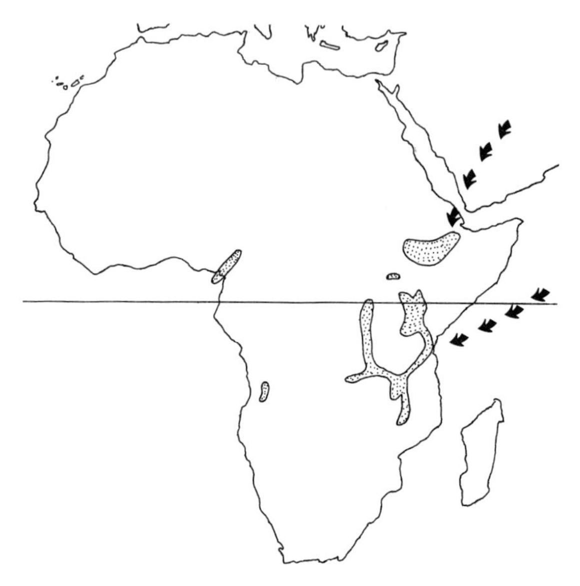

FIG. 22–3. Distribution of *Alcippe abyssinica*, and possible routes of arrival of the genus in Africa.

others discussed by Olson [1979]). It may be supposed these have been long isolated, although the lack of firm knowledge of the rate at which speciation has progressed needs to be reemphasized.

Croizat (1958) has suggested that many intercontinental distributions in the Southern Hemisphere are the result of events before the breakup of Gondwanaland. His contention that distributions such as those of *Rhipidura* flycatchers cannot be explained by migration (on account of their weak flying abilities) is open to question. Apart from the possibility of significant genetic change in this respect, too little is known of the dispersing ability of tropical forest birds, although Dowsett and Dowsett-Lemaire (MS) have shown that in Malawi a few montane forest birds are capable of interforest movements, but that most are highly sedentary. Thus it seems possible that some anomalous distribution patterns might make more sense if they were known to have originated in the very distant past. Some of the more acceptable theories have been proposed by Cracraft (1973), although he stresses the need for caution in interpreting distributions or migrations of recent genera in terms of continental drift. Brodkorb (1971) has emphasized that probably all living species of birds originated during the Quaternary. Haffer (1974) also suggested that in Amazonia there was repeated speciation in a large number of Quaternary forest refugia. It is possible that the origins of many afromontane bird species are much older than hitherto supposed, and ornithologists must look to paleontologists for firm evidence of past environmental changes and their timing, against which to set the picture of avian evolution, as well as seeking more distant relationships in other zoogeographic regions.

ACKNOWLEDGEMENTS

My studies of African montane avifaunas have been supported by the National Museums Board of Zambia, the Wildlife Conservation Society of Zambia, the National Geographic Society, the South African Nature Foundation, and the Frank M. Chapman Fund of the American Museum of Natural History.

APPENDIX 22–A

The stenotopic avifauna of the high mountains of the Afrotropical region.

		Occurrence on montane group [b]						
Species (by family)	Habitat [a]	1	2	3	4	5	6	7
Threskiornithidae								
Geronticus calvus	NF			+				
Bostrychia carunculata	NF							+
Anatidae								
Cyanochen cyanopterus	NF							+
Accipitridae								
Gypaetus barbatus	NF			+		+	+	+
Accipiter rufiventris	F/NF			+	+	+	+	+
Buteo oreophilus	F			+	+	+	+	+
Phasianidae								
Francolinus erckelii	NF							+
F. castaneicollis	F							+
F. jacksoni	F						+	
F. nobilis	F					+		

Species (by family)	Habitat [a]	Occurrence on montane group [b]						
		1	2	3	4	5	6	7
F. camerunensis	F	+						
F. swierstrai	F		+					
F. harwoodi	NF							+
F. psilolaemus	NF						+	+
F. levaillantii	NF		+	+	+	+	+	
Coturnix coturnix	NF			+	+	+	+	+
Rallidae								
Sarothrura ayresi	NF			+				+
S. affinis	NF			+	+		+	
Rougetius rougetii	NF							+
Charadriidae								
Vanellus melanocephalus	NF							+
Columbidae								
Columba albitorques	NF							+
C. arquatrix	F	+	+	+	+	+	+	+
Streptopelia lugens	F/NF				+	+	+	+
Apolopelia larvata	F	+		+	+	+	+	+
Psittacidae								
Poicephalus flavifrons	F							+
Agapornis taranta	F/NF							+
Musophagidae								
Tauraco leucotis	F							+
T. ruspolii	F							+
T. hartlaubi	F						+	
T. bannermani	F	+						
T. johnstoni	F					+		
Cuculidae								
Cercococcyx montanus	F			+	+	+	+	
Tytonidae								
Phodilus prigoginei	F					+		
Strigidae								
Bubo capensis	NF			+	+		+	+
Asio abyssinicus	F					+	+	+
Caprimulgidae								
Caprimulgus poliocephalus	NF		+		+	+	+	+
Apodidae								
Schoutedenapus myoptilus	F/NF	+	+	+	+	+	+	+
S. schoutedeni	F/NF?					+		
Trogonidae								
Apaloderma vittatum	F	+	+	+	+	+	+	
Meropidae								
Merops oreobates	F					+	+	+
Bucerotidae								
Bycanistes brevis	F			+	+		+	+
Capitonidae								
Stactolaema olivacea	F			+	+		+	
Pogoniulus leucomystax	F				+		+	
P. coryphaeus	F	+	+			+		
Indicatoridae								
Indicator pumilio	F					+	+	
Picidae								
Campethera tullbergi	F	+				+	+	
Dendropicos abyssinicus	F/NF							+
D. griseocephalus	F		+	+	+	+	+	
Eurylaimidae								
Pseudocalyptomena graueri	F					+		
Alaudidae								
Mirafra ruddi	NF			+				

APPENDIX 22–A CONTINUED.

Species (by family)	Habitat [a]	Occurrence on montane group [b]						
		1	2	3	4	5	6	7
Hirundinidae								
Hirundo atrocaerulea	NF			+	+	(+)	(+)	
H. megaensis	NF							+
Psalidoprocne fuliginosa	NF	+						
Oriolidae								
Oriolus percivali	F					+	+	
O. monacha	F							+
O. chlorocephalus	F			+	+		+	
Corvidae								
Pyrrhocorax pyrrhocorax	NF							+
Corvus crassirostris	NF							+
Zavattariornis stresemanni	NF							+
Paridae								
Parus fasciiventer	F					+		
P. leuconotus	F/NF							+
Timaliidae								
Trichastoma pyrrhopterum	F				+	+	+	
Kakamega poliothorax	F	+				+	+	
Alcippe abyssinica	F	+	+	+	+	+	+	+
Lioptilus rufocinctus	F					+		
L. gilberti	F	+						
L. nigricapillus	F			+				
Parophasma galinieri	F							+
Campephagidae								
Coracina caesia	F	+		+	+	+	+	+
C. graueri	F					+		
Pycnonotidae								
Andropadus tephrolaemus	F	+		+	+	+	+	
A. milanjensis	F			+	+		+	
A. montanus	F	+						
A. masukuensis	F				+	+	+	
Phyllastrephus placidus	F			+	+	+	+	
P. flavostriatus	F	+		+	+	+	+	
P. poensis	F	+						
Turdidae								
Cercomela sordida	NF						+	+
Monticola explorator	NF			+				
Myrmecocichla melaena	NF							+
M. semirufa	NF							+
Chaetops frenatus	NF			+				
Pogonocichla stellata	F			+	+	+	+	
Swynnertonia swynnertoni	F			+	+			
Alethe poliophrys	F					+		
A. fuelleborni	F			+	+		+	
A. choloensis	F			+				
Dessonornis montana	F						+	
D. lowei	F				+			
D. archeri	F					+		
Cossypha anomala	F			+	+		+	
C. semirufa	F						+	+
C. caffra	F/NF			+	+	+	+	
C. bocagei	F	+				+		
C. isabellae	F	+						
Sheppardia aequatorialis	F					+	+	
S. sharpei	F				+		+	
S. roberti	F	+				+		
Modulatrix stictigula	F				+		+	
M. orostruthus	F			+			+	

Species (by family)	Habitat [a]	Occurrence on montane group [b]						
		1	2	3	4	5	6	7
Turdus olivaceus	F/NF	+		+	+	+	+	+
T. gurneyi	F	+	+	+	+	+	+	
T. piaggiae	F					+	+	+
T. fischeri	F			+		+	+	
Sylviidae								
Bradypterus graueri	NF					+		
B. cinnamomeus	F/NF	+		+	+	+	+	+
B. victorini	F/NF			+				
B. barratti	F			+				
B. mariae	F	+	+	+	+	+	+	
B. sylvaticus	F/NF			+				
Chloropeta similis	F/NF				+	+	+	
Phylloscopus ruficapilla	F			+	+	+	+	
P. laetus	F					+		
P. herberti	F	+						
P. umbrovirens	F				+	+	+	+
Cisticola ayresii	NF		+	+	+	+	+	
C. aberdare	NF						+	
C. lais	NF		+	+	+		+	
C. njombe	NF				+			
C. bodessa	NF						+	+
C. hunteri	NF						+	
C. chubbi	NF	+				+	+	
C. nigriloris	NF				+			
Prinia robertsi	F/NF			+				
Urolais epichlora	F/NF	+						
Apalis jacksoni	F	+				+	+	
A. chariessa	F			+	+		(+)	
A. thoracica	F			+	+		+	
A. cinerea	F	+	+		+	+	+	
A. argentea	F					+		
A. porphyrolaema	F					+	+	
A. chapini	F				+			
A. bamendae	F	+						
A. chirindensis	F			+				
A. pulchra	F	+				+	+	
A. ruwenzorii	F					+		
A. moreaui	F			+			+	
Orthotomus metopias	F			+	+		+	
Poliolais lopezi	F	+						
Bathmocercus winifredae	F				+			
Sylvietta leucophrys	F					+	+	
Graueria vittata	F					+		
Hemitesia naumanni	F					+		
Parisoma lugens	F/NF				+	+	+	+
Muscicapidae								
Muscicapa lendu	F					+		
Melaenornis ardesiaca	F					+		
M. chocolatina	F		+		+	+	+	+
Batis capensis	F			+	+		+	
B. diops	F					+		
Elminia albonotata	F			+	+	+	+	
E. albiventris	F	+				+		
Motacillidae								
Anthus crenatus	NF			+				
A. chloris	NF			+				
Macronyx flavicollis	NF							+
M. sharpei	NF						+	

APPENDIX 22–A CONTINUED.

Species (by family)	Habitat [a]	Occurrence on montane group [b]						
		1	2	3	4	5	6	7
Malaconotidae								
Laniarius atroflavus	F	+						
L. poensis	F	+				+		
L. fuelleborni	F				+		+	
Malaconotus multicolor	F			+	+	+	+	
M. olivaceus	F			+	+			
M. monteiri	F	+	+				(+)	
M. lagdeni	F					+		
M. gladiator	F	+						
M. alius	F				+			
M. kupeensis	F	+						
M. dohertyi	F					+	+	
Prionopidae								
Prionops alberti	F					+		
Sturnidae								
Poeoptera stuhlmanni	F					+	+	+
P. kenricki	F				+		+	
Onychognathus tenuirostris	F/NF				+	+	+	+
O. walleri	F	+			+	+	+	
O. albirostris	NF							+
Cinnyricinclus sharpii	F				+	+	+	+
C. femoralis	F						+	
Promeropidae								
Promerops cafer	NF			+				
P. gurneyi	NF			+				
Nectariniidae								
Anthreptes rubritorques	F				+		+	
Nectarinia ursulae	F	+						
N. afra	NF		+	+	+	+		
N. preussi	F	+				+	+	
N. regia	F					+		
N. rockefelleri	F					+		
N. mediocris	F			+	+	+	+	
N. tacazze	NF						+	+
N. purpureiventris	F					+		
N. kilimensis	F/NF		+	+	+	+	+	
N. reichenowi	NF					+	+	
N. famosa	NF			+	+	+	+	+
N. johnstoni	NF				+	+	+	
N. violacea	NF			+				
N. oritis	F	+						
N. alinae	F					+		
N. rufipennis	F				+			
Zosteropidae								
Zosterops poliogastra	F						+	+
Speirops lugubris	F	+						
Ploceidae								
Ploceus baglafecht	NF	+			+	+	+	+
P. bertrandi	F/NF			+	+			
P. melanogaster	F	+				+	+	
P. alienus	F					+		
P. insignis	F	+	+			+	+	
P. bannermani	F	+						
Euplectes psammocromius	NF				+			
E. jacksoni	NF						+	
Estrildidae								
Nesocharis shelleyi	F	+						

Species (by family)	Habitat [a]	Occurrence on montane group [b]						
		1	2	3	4	5	6	7
Cryptospiza reichenovii	F	+	+	+	+	+	+	
C. salvadorii	F					+	+	+
C. shelleyi	F					+		
C. jacksoni	F					+		
Clytospiza cinereovinacea	F		+			+		
Estrilda melanotis	F/NF		+	+	+	+	+	+
Fringillidae								
Serinus canicollis	NF		+	+	+	+	+	+
S. nigriceps	NF							+
S. totta	NF			+				
S. symondsi	NF			+				
S. citrinelloides	F/NF			+	+	+	+	+
S. scotops	F			+				
S. tristriatus	NF							+
S. ankoberensis	NF							+
S. burtoni	F	+	+		+	+	+	
S. striolatus	NF				+	+	+	+
S. leucopterus	NF			+				
Linurgus olivaceus	F	+			+	+	+	
Total (222 species)		51	24	75	81	106	105	65

(+) Probably not present as a breeding species.

[a] Habitat: F, evergreen forest; NF, nonforest.

[b] 1, Cameroon; 2, Angola; 3, Southeastern; 4, Tanganyika-Nyasa; 5, East Congo; 6, Kenya; 7, Ethiopia.

APPENDIX 22–B

Notes on the systematics of the Afrotropical montane avifaunas and on their composition.

Accipitridae. *Accipiter rufiventris* is included in Appendix A as it is strictly montane over much of its range. Several records from southern and southcentral Africa at low altitudes (e.g., Snow, 1978) are in fact referable to *A. ovampensis* (Irwin, Benson, and Steyn, 1983). Pace Snow (1978), I prefer to treat *A. rufiventris* as a species distinct from *A. nisus* of the Palearctic, though both are usefully placed in the same superspecies (Dowsett and Dowsett-Lemaire, 1980; Irwin et al., 1983), and probably have a common origin. I use the name *Buteo oreophilus*, rather than *B. tachardus* as proposed by Brooke (1974), following James and Wattell (1983).

Phasianidae. *Coturnix coturnix* is included in Appendix A, although it is not confined to montane areas in the southern half of its African range. Some Palearctic birds migrate to Africa, and in the past this was probably the origin of Afrotropical breeding populations.

Rallidae. *Sarothrura ayresi* is included in Appendix A, although recent reports from Zimbabwe and South Africa suggest that this little-known species may not be strictly montane in its southern populations (Hopkinson and Masterson, 1977; Mendelsohn, Sinclair, and Tarboton, 1983).

Columbidae. The *Columba iriditorques* superspecies of Snow (1978) is considered to comprise but a single species, *C. delegorguei* (Dowsett and Dowsett-Lemaire, 1980), whose mainly lowland distribution precludes it from inclusion in Appendix A. I also prefer to consider *C. sjostedti* and *C. thomensis* conspecific with *C. arquatrix*. There appear to be good reasons for considering *Streptopelia hypopyrrha* a species distinct from *S. lugens* (Snow, 1978), but the former is omitted here as it does not seem to occur in the Cameroon montane group proper.

Strigidae. I prefer to continue to recognize *Asio abyssinicus* as a species distinct from the Palearctic *A. otus* (cf. Snow, 1978).

Apodidae. I have followed Snow (1978) in recognizing two species in the genus *Schoutedenapus* and include both in Appendix A, although little is known of the altitudinal distribution of *S. schoutedeni* (Prigogine, 1971). My own experience in Malawi suggests that *S. myoptilus* may nest in forest trees, rather than on cliffs as hitherto supposed, and this may possibly be true also of *S. schoutedeni*.

Capitonidae. I cannot accept Snow's (1978) placement of *Buccanodon olivaceum* in the genus *Pogoniulus*, and tentatively follow Short and Horne (1980) in using *Stactolaema*. Although Clancey (1980) gives specific status (as *Cryptolybia woodwardi*) to isolated populations in Zululand and southern Tanzania, my own experience is that the voices of all are identical and the plumage differences seem insufficient to warrant specific separation. This species is included in Appendix 22–A, although much of its distribution is at low altitudes.

Alaudidae. *Mirafra ruddi* is included in Appendix 22–A as a South African endemic, the isolated *archeri* in Somalia perhaps being best treated as a separate species, as tentatively suggested by Hall and Moreau (1970).

Hirundinidae. *Hirundo atrocaerulea* is unique among Afrotropical montane birds in being a long-distance migrant, breeding only in groups 3 and 4.

Oriolidae. I follow Prigogine (1978a) in recognizing the montane *Oriolus percivali* as specifically distinct from *O. larvatus*.

Corvidae. *Zavattariornis stresemanni* (of doubtful affinities) is included in Appendix A, although neither it nor the other southern Ethiopian endemic *Hirundo megaensis* may be strictly montane.

Timaliidae. *Trichastoma poliothorax* is here placed in the genus *Kakamega*, created for it by Mann et al. (1978).

Pycnonotidae. The montane *Phyllastrephus placidus* is treated as a species distinct from *P. cabanisi*, following Dowsett (1972a), but I have merged *P. poliocephalus* with *P. flavostriatus*, pending further investigation of possible species-isolating mechanisms in this group (Dowsett and Dowsett-Lemaire, 1980). *P. orostruthus* is placed in the family Turdidae (genus *Modulatrix*), following Benson and Irwin (1975).

Turdidae. I follow Irwin and Clancey (1974) in placing *Pogonocichla swynnertoni* in the monotypic genus *Swynnertonia*, my limited field experience of this species suggesting it might more correctly be placed near the akalat genus *Sheppardia*. *Dessonornis (Alethe) anomala* is placed in the genus *Cossypha* (Dowsett and Dowsett-Lemaire, 1980). The genus *Dessonornis* is tentatively retained for *D. archeri*, *D. lowei* and *D. montana*; however, they may all be distinctive from the type species *D. humeralis*. *Turdus fischeri* is included in Appendix 22–A as its known and putative breeding forests are all within montane groups, records from coastal Kenya for example being of altitudinal migrants in the nonbreeding season (Britton and Rathbun, 1978). *T. fischeri* has only recently been discovered in montane group 5, on Upemba, southern Zaire (Lippens and Wille, 1976), and the only montane record farther north is a single specimen from the Imatongs, southern Sudan (Nikolaus, 1982).

Sylviidae. The *Bradypterus barratti-cinnamomeus* complex is resolved into three species (all wholly or mainly montane), namely, *B. cinnamomeus*, *B. barratti*, and *B. mariae* (Dowsett and Stjernstedt, 1979; Dowsett and Dowsett-Lemaire, 1980). The highland *Cisticola bodessa* is specifically distinct from *C. chiniana* (Dowsett-Lemaire and Dowsett, 1978, and references therein). Following Dowsett and Dowsett-Lemaire (1980), the *C. hunteri* superspecies is considered to comprise three semispecies (with specific status given to *C. nigriloris*), and *Apalis cinerea* and *A. alticola* are treated as conspecific. *Apalis chapini* and *A. bamendae* are best considered specifically distinct from *A. porphyrolaema* (Dowsett and Dowsett-Lemaire, 1980). *A. chariessa* is included in group 6 in Appendix 22–A, but its Kenyan population may now be extinct.

Muscicapidae. *Muscicapa lendu* is tentatively treated as a good species (Prigogine, 1978b), but the relationships of these difficult small, gray flycatchers are in need of much further field research. The *Batis capensis* superspecies also present problems; following Dowsett and Dowsett-Lemaire (1980), *B. capensis* and *B. mixta* are treated as conspecific, while the more isolated *B. diops* is for the moment given specific status. Two species of *Trochocercus* are here placed in the genus *Elminia* (Dowsett and Stjernstedt, 1973; Dowsett and Dowsett-Lemaire, 1980) as behaviorally they appear to be distinct from *Trochocercus* sensu stricto, the type species of which is *T. cyanomelas* (and not *T. nitens* as suggested by Dowsett and Dowsett-Lemaire [1980]).

Motacillidae. Although Prigogine (1971) gives specific status to *Anthus latistriatus*, I prefer for the moment to continue to treat this montane taxon as a subspecies of *A. novaeseelandiae*, pending a complete and thorough review of its montane populations.

Malaconotidae. I follow Hall and Moreau (1970) in treating *Laniarius poensis* and *L. fuelleborni* as separate species, but with some misgivings, as on present evidence they might equally be considered conspecific. The genus *Malaconotus* presents much confusion, and the arrangement adopted here cannot be considered final.

M. multicolor and *M. nigrifrons* are treated as conspecific (Dowsett and Dowsett-Lemaire, 1980), and are included in Appendix A on account of the essentially montane distribution of some members of both taxa. I tentatively follow Hall and Moreau (1970) in granting specific status to the very little known *M. monteiri*, but I do not accept the occurrence of the species in western Kenya, the supposed specimen having been lost (Britton, 1980). *M. lagdeni* may not be entirely montane, there being a specimen from lowland West Africa (Hall and Moreau, 1970).

Promeropidae. Sibley and Ahlquist (1974) reviewed the checkered history of the genus *Promerops* and placed it in its own family, closest to the Sturnidae. The two isolated taxa are best each given specific status.

Nectariniidae. Pending comparative field study, I prefer to treat as conspecific with the allopatric *Nectarinia mediocris* the two forms *loveridgei* and *moreaui* of northeastern Tanzania, as well as *prigoginei* of southeastern Zaire. Specific status has been claimed for the first two by Stuart and van der Willigen (1980) and for *prigoginei* by Benson and Prigogine (1981). Although Clancey and Irwin (1978) may be correct in separating some montane populations of *N. afra* from those in South Africa, further evidence is needed to confirm this, pending which I treat all as conspecific (Dowsett and Dowsett-Lemaire, 1980). *N. rufipennis* is a recently discovered species (Jensen, 1983; Stuart and Jensen, 1981; S. N. Stuart, pers. comm.).

Ploceidae. *Euplectes psammocromius* is here treated as specifically distinct from *E. hartlaubi*, as recommended by Dowsett and Dowsett-Lemaire (1980). *E. psammocromius* is the ecological counterpart of *E. progne* in many respects, and perhaps more closely related to that species than usually supposed.

Estrildidae. I would prefer to follow Prigogine (1971) and use the name *Clytospiza cinereovinacea*, rather than place the species in *Euschistospiza*, as done with reservations by Hall and Moreau (1970).

Fringillidae. To the montane species of *Serinus* considered by Hall and Moreau (1970) is added *S. ankoberensis* described from Ethiopia by Ash (1979). I follow Hall and Moreau (1970) in considering the status of *S.* "*flavigula*" to be uncertain, although evidence presented by Erard (1974) does suggest it may be specifically distinct. Treated here as conspecific with *S. burtoni* is the well-marked form *melanochrous*, now known to have a wider range than hitherto supposed (Stuart and Jensen, 1981); it may in fact be an incipient species (Britton, 1980).

REFERENCES

Ash, J. S. 1979. A new species of serin from Ethiopia. *Ibis* 121: 1–7.

Benson, C. W., and F. M. Benson. 1977. *The birds of Malawi*. Limbe, Malawi: Montfort Press.

Benson, C. W., and M. P. S. Irwin. 1975. The systematic position of *Phyllastrephus orostruthus* and *Phyllastrephus xanthophrys*, two species incorrectly placed in the family Pycnonotidae (Aves). *Arnoldia (Rhodesia)* 7(17): 1–10.

Benson, C. W., and A. Prigogine. 1981. The status of *Nectarinia afra prigoginei* (Macdonald). *Gerfaut* 71: 47–57.

Benson, C. W., M. P. S. Irwin, and C. M. N. White. 1962. The significance of valleys as avian zoogeographical barriers. *Ann. Cape Prov. Mus.* 2: 155–189.

Benson, C. W., R. K. Brooke, R. J. Dowsett, and M. P. S. Irwin. 1971. *The birds of Zambia*. London: Collins.

Benson, C. W., J. F. R. Colebrook-Robjent, and A. Williams. 1976–77. Contribution à l'ornithologie de Madagascar. *L'Oiseau R. F. O.* 46: 103–34, 209–42, 367–86; 47: 41–64, 167–191.

Britton, P. L., ed. 1980. *Birds of East Africa*. Nairobi: East African Natural History Society.

Britton, P. L., and G. B. Rathbun. 1978. Two migratory thrushes and the African Pitta in coastal Kenya. *Scopus* (Nairobi) 2: 11–17.

Brodkorb, P. 1971. Origin and evolution of birds. In *Avian biology*, Vol. 1, D. S. Farner and J. R. King, eds., pp. 19–55. New York: Academic Press.

Brooke, R. K. 1974. *Buteo tachardus* Andrew Smith 1830. *Bull. Brit. Orn. Club* 94: 59–62.

Clancey, P. A., ed. 1980. *S.A.O.S. checklist of Southern African birds*. Johannesburg: Southern African Ornithological Society.

Clancey, P. A., and M. P. S. Irwin. 1978. Species limits in the *Nectarinia afra/N. chalybea* complex of African double-collared Sunbirds. *Durban Mus. Novitates* 11: 333–351.

Cracraft, J. 1973. Continental drift, paleoclimatology, and the evolution and biogeography of birds. *J. Zool. Lond.* 169: 455–545.

Croizat, L. 1958. *Panbiogeography*, Vol. IIa. Caracas: Published by the author.

Devillers, P. 1977. Projet de nomenclature française des oiseaux du monde. 5. Trogonidés aux Picidés. *Gerfaut* 67: 469–489.

Diamond, A. W. 1981. Reserves as oceanic islands: Lessons for conserving some East African montane forests. *Afr. J. Ecol.* 19: 21–25.

Diamond, A. W., and A. C. Hamilton. 1980. The distribution of forest passerine birds and Quaternary climatic change in tropical Africa. *J. Zool. Lond.* 191: 379–402.

Diamond, J. M. 1972. *Avifauna of the eastern highlands of New Guinea*. Cambridge, Mass.: Nuttall Orn. Club.

Dowsett, R. J. 1971. The avifauna of the Makutu Plateau, Zambia. *Rev. Zool. Bot. Afr.* 84: 312–333.

———. 1972a. Is the bulbul *Phyllastrephus placidus* a good species? *Bull. Brit. Orn. Club* 92: 132–138.

———. 1972b. Races of the lark *Mirafra africana* in the

Tanganyika-Nyasa montane group. *Bull. Brit. Orn. Club* 92: 156–159.

———. 1980a. Extinctions and colonisations in the African montane island avifaunas. *Proc. IV Pan-Afr. Orn. Congr. (Seychelles)*: 185–197.

———. 1980b. Post-Pleistocene changes in the distributions of African montane forest birds. *Act. XVII Congr. Int. Orn. (Berlin)*: 787–792.

———. 1982. The population dynamics and seasonal dispersal of the Starred Robin *Pogonocichla stellata*. M.Sc. thesis, University of Natal, Pietermaritzburg.

Dowsett, R. J., and F. Dowsett-Lemaire. 1980. The systematic status of some Zambian birds. *Gerfaut* 70: 151–199.

———. MS. Homing ability and territorial replacement in some African forest birds.

Dowsett, R. J., and A. Prigogine. 1974. The avifauna of the Marungu Highlands. In *Hydrobiological survey of the Lake Bangweulu-Luapula River basin*, J. J. Symoens, ed., 19, pp. 1–67. Brussels: Cercle hydrobiologique de Bruxelles.

Dowsett, R. J., and R. Stjernstedt. 1973. The birds of the Mafinga Mountains. *Puku* 7: 107–123.

———. 1979. The *Bradypterus cinnamomeus-mariae* complex in Central Africa. *Bull. Brit. Orn. Club 99*: 86–94.

Dowsett-Lemaire, F. 1983. Studies of a breeding population of Olive Woodpeckers, *Dendropicos griseocephalus*, in montane forests of southcentral Africa. *Gerfaut* 73: 221–237.

Dowsett-Lemaire, F., and R. J. Dowsett. 1978. The Boran Cisticola *Cisticola bodessa* in Kenya, and its possible affinities. *Scopus* (*Kenya*) 2: 29–33.

Erard, C. 1974. Taxonomie des serins à gorge jaune d'Ethiopie. *L'Oiseau R.F.O.* 44: 308–323.

Fry, C. H. 1976. On the systematics of African and Asian tailor-birds (Sylviinae). *Arnoldia (Rhodesia)* 8(6): 1–15.

Grimes, L. G. 1976. The occurrence of cooperative breeding behaviour in African birds. *Ostrich* 47: 1–15.

Haffer, J. 1974. Avian speciation in tropical South America. *Pub. Nuttall Orn. Club*, no. 14: 1–390.

Hall, B. P., and R. E. Moreau. 1970. *An atlas of speciation in African passerine birds*. London: British Museum (Nat. Hist.).

Hamilton, A. 1976. The significance of patterns of distribution shown by forest plants and animals in tropical Africa for the reconstruction of upper Pleistocene palaeoenvironments: A review. In *Palaeoecology of Africa, the surrounding islands and Antarctica*, E. M. van Zinderen Bakker, ed., 9, pp. 63–97. Cape Town: Balkema.

Hopkinson, G., and A. N. B. Masterson. 1977. On the occurrence near Salisbury of the white-winged flufftail. *Honeyguide* (Rhodesia) 91: 25, 27–28.

Irwin, M. P. S., and P. A. Clancey. 1974. A re-appraisal of the generic relationships of some African forest-dwelling robins (Aves: Turdidae). *Arnoldia (Rhodesia)* 6(34): 1–19.

Irwin, M. P. S., C. W. Benson, and P. Steyn. 1983. The identification of the Ovambo and Red-breasted Sparrow Hawks in South Central Africa. *Honeyguide* (Zimbabwe) 111–112: 28–44.

James, A. H. and J. Wattell. 1983. *Buteo tachardus* Smith, 1830, preoccupied by *Buteo tachardus* Vieillot, 1823. *Ostrich* 54: 126.

Jensen, F. P. 1983. A new species of sunbird from Tanzania. *Ibis* 125: 447–449.

Johnston, R. F., and R. K. Selander. 1964. House sparrows: Rapid evolution of races in North America. *Science* 144: 548–550.

Keith, S., A. Twomey, H. Friedmann, and J. Williams. 1969. The avifauna of the Impenetrable Forest, Uganda. *Amer. Mus. Novitates* 2389: 1–41.

Lippens, L., and H. Wille. 1976. *Les oiseaux du Zaire*. Tielt, Belgium: Lannoo.

Livingstone. D. A. 1975. Late Quaternary climatic change in Africa. *Ann. Rev. Ecol. Syst.* 6: 249–280.

Mackworth-Praed, C. W., and C. H. B. Grant, 1960. *Birds of eastern and northeastern Africa*, vol. 2. London: Longmans.

Mann, C. F., P. J. K. Burton, and I. Lennerstedt. 1978. A reappraisal of the systematic position of *Trichastoma poliothorax* (Timaliinae, Muscicapidae). *Bull. Brit. Orn Club* 98: 131–140.

Mayr, E., and M. Moynihan. 1946. Evolution in the *Rhipidura rufifrons* group. *Amer. Mus. Novitates* 1321: 1–21.

Mendelsohn, J. M., J. C. Sinclair, and W. R. Tarboton. 1983. Flushing flufftails out of vleis. *Bokmakierie* (Johannesburg), 35: 9–11.

Milstein, P. le S. 1979. The evolutionary significance of wild hybridization in South African highveld ducks. *Ostrich*, supp. 13: 1–48.

Moreau, R. E. 1963. Vicissitudes of the African biomes in the Late Pleistocene. *Proc. Zool. Soc. Lond.* 141: 395–421.

———. 1966. *The bird faunas of Africa and its islands*. London and New York: Academic Press.

Nikolaus, G. 1982. A new race of the spotted ground thrush *Turdus fischeri* from South Sudan. *Bull. Brit. Orn Club* 102: 45–47.

Olson, S. L. 1973. A classification of the Rallidae. *Wilson Bull.* 85: 381–416.

———. 1979. *Picathartes*—another West African forest relict with probable Asian affinities. *Bull. Brit. Orn. Club* 99: 112–113.

Prigogine, A. 1971. Les oiseaux de l'Itombwe et de son hinterland. I. *Ann. Mus. Roy. Afr. Centr.* 185: 1–298.

———. 1973. Le statut de *Phodilus prigoginei* Schouteden. *Gerfaut* 63: 177–185.

———. 1974. Contribution à l'étude de la distribution verticale des oiseaux orophiles. *Gerfaut* 64: 75–88.

———. 1978a. Le statut du Loriot de Percival, *Oriolus percivali*, et son hybridation avec *Oriolus larvatus* dans l'est africain. *Gerfaut* 68: 253–320.

——— 1978b. Les oiseaux de l'Itombwe et de son hinterland. II. *Ann. Mus. Roy. Afr. Centr.* 223: 1–134.

———. 1978c. Note sur les petits indicateurs de la forêt de Kakamega. *Gerfaut* 68: 87–89.

Ripley, S. D. 1952. The thrushes. *Postilla* 13: 1–48.

———. 1977. *Rails of the world*. Toronto: M. F. Feheley.

Short, L. L., and J. F. M. Horne. 1980. Vocal and other behaviour of the green barbet in Kenya. *Ostrich* 51: 219–229.

Sibley, C. G., and J. E. Ahlquist. 1972. A comparative

study of the egg-white proteins of non-passerine birds. *Peabody Mus. Nat. Hist. Bull.* 39: 1–276.

———. 1974. The relationships of the African sugarbirds (*Promerops*). *Ostrich* 45: 22–30.

Simpson, G. G. 1960. Notes on the measurement of faunal resemblance. *Amer. J. Sci.* 258–A: 300–311.

Snow, D. W. ed. 1978. *An atlas of speciation in African non-passerine birds*. London: British Museum (Nat. Hist.).

Stuart, S. N. 1981. A comparison of the avifaunas of seven east African forest islands. *Afr. J. Ecol.* 19: 133–151.

Stuart, S. N., and F. P. Jensen. 1981. Further range extensions and other notable records of forest birds from Tanzania. *Scopus* (Kenya) 5: 106–115.

Stuart, S. N., and D. A. Turner. 1980. Some range extensions and other notable records of forest birds from eastern and northeastern Tanzania. *Scopus* (Kenya) 4: 36–41.

Stuart, S. N., and T. A. van der Willigen. 1980. Is Moreau's sunbird *Nectarinia moreaui* a hybrid species? *Scopus* (Kenya) 4: 56–58.

Terborgh, J. 1971. Distribution on environmental gradients: Theory and a preliminary interpretation of distributional patterns in the avifauna of the Cordillera Vilcabamba, Peru. *Ecology* 52: 23–40.

Thomson, A. L., ed. 1964. *A new dictionary of birds*. London: Academic Press.

Traylor, M. A. 1967. *Cisticola aberdare* a good species. *Bull. Brit. Orn. Club* 87: 137–141.

Turner, D. A. 1977. Status and distribution of the East African endemic species. *Scopus* (Kenya) 1: 2–11.

Vuilleumier, F., and D. Simberloff. 1980. Ecology versus history as determinants of patchy and insular distributions in high Andean birds. In *Evolutionary Biology*, Vol. 12, M. K. Hecht, W. C. Steere, and B. Wallace, eds, pp. 235–379. New York: Plenum.

White, C. M. N. 1960. A checklist of the Ethiopian Muscicapidae (Sylviinae). Part 1. *Occ. Pap. Natn. Mus. Sth. Rhod.* 24B: 399–430.

———. 1961. *A revised checklist of African broadbills, pittas, larks, swallows, wagtails and pipits*. Lusaka, Zambia: Government Printer.

———. 1962a. *A revised checklist of African shrikes, orioles, drongos, starlings, crows, waxwings, cuckoo-shrikes, bulbuls, accentors, thrushes and babblers*. Lusaka, Zambia: Government Printer.

———. 1962b. A checklist of the Ethiopian Muscicapidae (Sylviinae). Parts II and III. *Occ. Pap. Natn. Mus. Sth. Rhod.* 26B: 653–694, 695–738.

———. 1963. *A revised checklist of African flycatchers, tits, tree creepers, sunbirds, white-eyes, honey eaters, buntings, finches, weavers and waxbills*. Lusaka, Zambia: Government Printer.

———. 1965. *A revised checklist of African non-passerine birds*. Lusaka, Zambia: Government Printer.

White, F. 1981. The history of the afromontane archipelago and the scientific need for its conservation. *Afr. J. Ecol.* 19: 33–54.

23

Origins of the Tropical Avifaunas of the High Andes

FRANÇOIS VUILLEUMIER

What are the origins of the birds living in the páramo and puna of the high tropical Andes? Chapman (1917, 1921, 1926) was probably the first author to give an answer to this question, after analyses of the distribution of páramo birds in Colombia and Ecuador and of puna birds in Peru. He believed that páramo and puna birds were largely derived from southern South America, especially Patagonia. Other birds came from elsewhere in South America and from North America, whereas some taxa remained of uncertain origin. In Chapman's view, there was little exchange between the avifaunas of the páramo or puna and that of his "Temperate Zone" below. He wrote (1926, p. 116) that "the Paramo has taken so little from the arid Temperate, and the fact that both draw their life from their sea-level equivalent, distant over 2,000 miles, rather than from contiguous regions, is an indication that the South Temperate and Paramo of Chile are more like the same zones in Ecuador than the latter are like other plains in the same latitude."

Unfortunately, Chapman did not carry his analyses further. This was due in part to his confusion between vegetation types and life zones, and to his apparent belief that the avifauna in life zones somehow evolved as a "block." Furthermore, he never really defined the limits of his life zones south of Peru. What, for instance, could be his "Paramo Zone" of Chile? Finally, Chapman was hampered by the split generic and specific taxonomy of his time.

In an important paper, Dorst (1967, pp. 481–497) analyzed the páramo and puna avifauna as a single unit for the first time. He clearly defined four faunal elements (ubiquitous, Patagonian, boreal, and local), and discussed the distribution and origin of each in detail. More recently, Short (1971, 1972) and Fjeldså (1981a, b, 1982a, 1983a) mentioned the probable origins of selected high Andean taxa, Haffer (1970, 1974) analyzed the zoogeography of Andean birds, and I discussed the evolution of páramo and puna birds (Vuilleumier, 1969b, 1981; also Vuilleumier and Ewert, 1978; Vuilleumier and Simberloff, 1980).

This chapter does not attempt to update these earlier analyses, but offers instead a new synthesis, based to a large extent on new information from my own field and museum studies, and from the results of faunistic and taxonomic work carried out by many authors.

Before studying the origins of tropical high Andean birds one must first define the area of study, and then draw a list of the species recorded in that region. The status of each species (whether breeding resident, local migrant, long-distance migrant, or accidental) must be determined, and its ecological preferences and taxonomic relationships must be assessed. Armed with this information one should eventually be able to determine whether each species originated in situ or near its present area of distribution, or whether it is an immigrant from elsewhere, and if so, from where. The migrants and accidentals come from somewhere else. The difficult work, of course, is to identify the areas of origin of the resident species. In this search clues provided by the migrant and accidental species can be of help. In other papers on high Andean birds I have dealt only with the breeding avifauna. I include here nonbreeding species as well.

DEFINITIONS, DATA BASE, ASSUMPTIONS, AND METHODS

Vegetation Types

The páramo and puna biomes and their avian habitats have been described by Dorst (1955, 1957, 1967), M. Koepcke (1954), Vuilleumier and Ewert (1978), Bourlière (1957), Corley Smith (1969), Fleming (1973), Niethammer (1953), Pearson and Ralph (1978), and others. Ornithologists since Dorst and M. Koepcke have usually adopted the definitions of páramo and puna given by geographers, botanists, and plant ecologists, such as Troll (1959), Cuatrecasas (1968), Weberbauer (1945), and Cabrera (1968) (see also Chap. 16).

In this chapter, I compare the páramo-puna avifauna to avifaunas living in other biomes or vegetation types. In order to describe these vegetation types in terms of a single scheme I adopt here for convenience that of Hueck and Seibert (1972), which, although outdated in some respects, has the advantage of covering all of South America. Table 23–1 lists these vegetation types, which are mapped in Fig. 23–1 and briefly described here:

Páramo: Usually wet, dense, alpinelike vegetation in the tropical Andes of Venezuela, Colombia, Ecuador, and northern Peru. Includes *Polylepis* forests and woodlands (see further discussion later in this chapter).

Puna: Usually dry to arid, alpinelike vegetation in the tropical Andes of Peru (except northern part), Bolivia, northern Argentina and northern Chile, south to about 27° S. Includes *Polylepis* forests and woodlands and *Puya* stands (discussed later).

Andino-alpine: Alpinelike vegetation (grassland and scrub) in the temperate Andes of central and southern Chile and Argentina, from about 27° S to southernmost South America (Tierra del Fuego and neighboring islands).

Wet montane: Several kinds of moist, cool, forested environments from about 2000 m (variable locally) to upper limit of continuous forests at 3000–3800 m in the tropical Andes from Venezuela and Colombia southward to northwestern Argentina along the eastern slopes of the Andes, and to northern Peru along the western slopes. The wet montane as defined here corresponds largely to what Moynihan (1979, p. 7) called "humid cold tropical" because he rightly found Chapman's term "Temperate Zone" inappropriate to a tropical situation.

Dry montane: Dry to arid, warm to cool, nonforest environments (woodlands, cactus stands, thornscrub, matorral) in the tropical Andes, adjacent to the páramo, puna, and/or the wet montane, in dry intermontane basins, or along the dry Andean slopes of Peru, from about 2000 to 3500 m (variable locally).

Patagonian steppes: Dry, cool, grasslands or shrubsteppes of Patagonia, east of the Andes from about 39° S to Tierra del Fuego.

Nothofagus forests: Extremely wet, evergreen or partially deciduous forests along the Andes and west of the Patagonian steppes (Vuilleumier, 1985).

Two more vegetation types, briefly mentioned earlier, must be discussed further. (1) *Polylepis* woodlands occur along the high Andes from Venezuela and Colombia to Chile and Argentina (Simpson, Chap. 12, this volume). I consider these woodlands to be part of the páramo and puna. However, some *Polylepis* woodlands occur at the upper edge of the wet montane along the eastern Andes of Ecuador and Peru, at the upper edge of the dry montane along the western slopes of the Andes of Peru, and at the upper limit of dry intermontane basins in Bolivia. A number of bird species occur in the dry or in the wet montane and also in *Polylepis*, but not in the characteristically open puna habitats, such as grasslands or scrub. Although not included in the basic páramo-puna avifauna as described in a later section, these species are nevertheless analyzed in this chapter under the name *peripheral species*. (2) Stands of the giant Bromeliaceae *Puya raimondii*, described as an avian habitat in the puna by Dorst (1957), Rees and Roe (1980), and Vuilleumier (unpublished data), are included in the puna.

Units of Analysis

The basic unit of analysis employed in this chapter is the species and/or the allospecies, more rarely the genus. I adopted Mayr's (1963) definitions of a species and superspecies, and Amadon's (1966) suggestion to use square brackets to designate superspecies in lists. Superspecies (and their component members, the allospecies) are widely used in systematic ornithology (e.g., Mayr and Vuilleumier, 1983) and in zoogeography (e.g., Short, 1975; Vuilleumier, 1985).

Although some authors (e.g., Mayr and Short, 1970, p. 3) have used the concept of "zoogeographical species" for "superspecies . . . or individual species not belonging to a superspecies," I here use as units allospecies (not superspecies) and/or species that are not members of superspecies. Mayr and Short's (1970) procedure would, of course, result in a list with fewer species than mine. Allospecies represent the end points of the allopatric speciation process (Mayr, 1963). Thus they are evolutionary units that show how much

TABLE 23–1. Vegetation types of the high Andes and Patagonia. Numbers refer to Hueck and Seibert (1972).

Number	Vegetation types	General names	Nomenclature in this chapter (biomes)
83, 82 (part)	Grassland, scrubland	Andean high montane vegetation	Páramo
84, 82 (part)	Grassland, scrubland		Puna
82 (part)	Grassland, scrubland		Andino-alpine
22	Cloud forest	Andean wet montane forests	Wet montane
21	Ceja, upper yungas forests		
20	Nogal-pino forest		
40	Dry woodlands in intermontane basins	Andean dry montane woodlands and arid scrub	Dry montane
33 (part)	Dry woodlands, open scrub		
82 (part)	Semidesert (Pacific slopes)		
64–67	Grassland, shrubsteppe	Patagonian steppes and semideserts	Patagonian steppes
75, 76, 78–80	Rain forests	Southern forests	*Nothofagus* forests

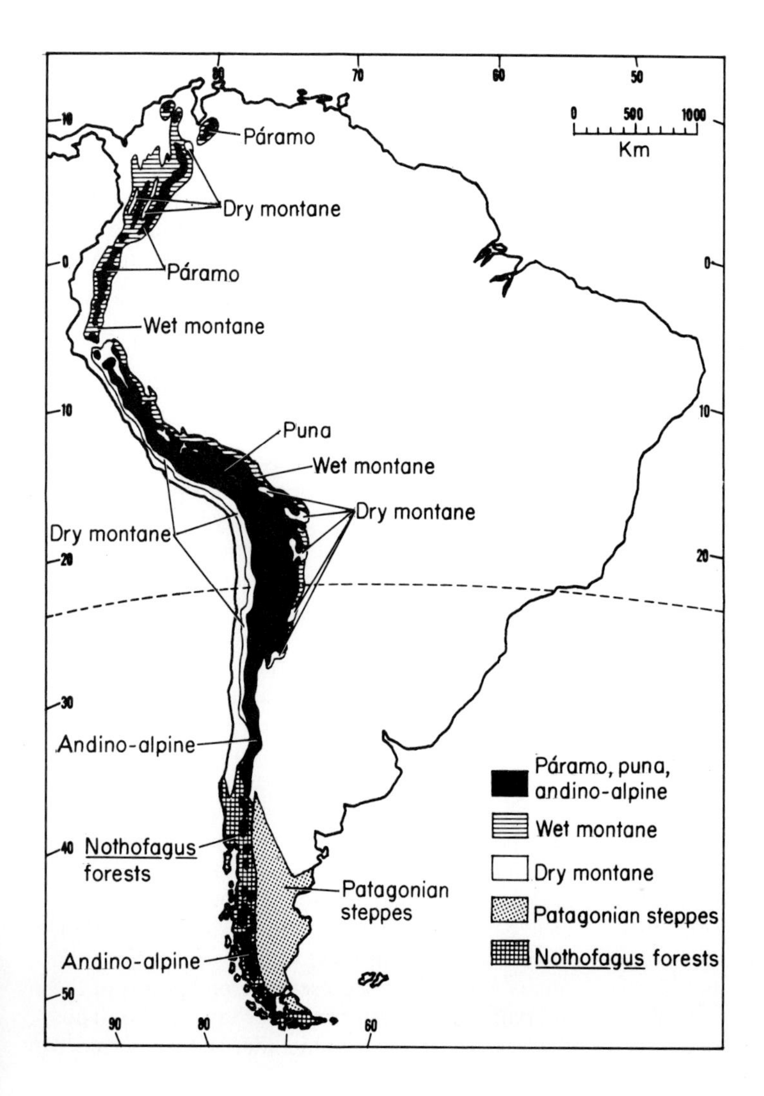

FIG. 23–1. Vegetation units (biomes) in the high Andes and in Patagonia. For further explanations see Table 23–1.

origination has taken place in situ through local speciation events. In a study of faunal origins such units of analysis are thus important.

The Avian Data Base

Data on taxonomy, geographic and altitudinal distribution, habitat preferences, seasonal movements, and breeding come either from data (many unpublished) gathered during my field trips to the Andes in 1964, 1965, 1967–68, 1975, 1978, 1980, and 1981, or from my museum studies in 1964–1971 and 1974–1984. I have had field experience with 85% of the 319 species in the total list of páramo-puna birds, and with 153 (92%) of the 166 species considered to form the basic breeding páramo-puna avifauna. Even though extensive, my own work is insufficient to form the data base for an analysis of faunal origins. It has been supplemented by the pertinent literature. Papers with taxonomic and distributional information are marked with an asterisk in the references.

I admit a total of 319 species of birds to the páramo-puna list. The decision to include—or exclude—a given species from the list was made after carefully considering the available evidence. A rather stringent approach was used in making such decisions. As a result, I excluded certain species from the páramo-puna list that other workers might have retained. Such differences in opinion are probably unavoidable when dealing with an ecologically defined fauna whose boundaries are not as sharp as those of a geographically defined one (political unit, or island).

In Table 23–2 the fauna is subdivided into components. Of the 248 species that breed in the páramo-puna, 82 (33%) are peripherally distributed in the páramo, the puna, or both. Among these 82 species are some that other authors might put in the "true" páramo-puna list of 166 species. Of the latter, 34 species (20%) are water birds, and 132 (80%) are land birds. Appendix 23–A lists the 166 páramo-puna species, and Appendix 23–B the 82 peripheral species.

Seventy species (22% of 319) do not breed in the páramo-puna. They migrate there from North America (30 species, 21 water birds), or from southern South America (5 land bird species) (Appendix 23–C), or else are accidentals from elsewhere in South America (35 species, 22 water birds) (Appendix 23–D). Finally, one introduced species (*Passer domesticus*) has been reported from the puna (Fjeldså, 1983b, p. 296).

The ecologically defined, but relatively insular, páramo-puna avifauna is compared in Table 23–3 with a geographically defined, and truly insular one of similar species richness, the Malagasy avifauna (Keith, 1980). Note that (1) the Malagasy

TABLE 23–2. The components of the páramo-puna avifauna.

	Waterbirds	Landbirds	Total
1. Breeding páramo-puna species			
1a. Endemic species	11	37	48
1b. Nearly endemic species	4	17	21
1c. Nonendemic species	19	78	97
Total 1	34	132	166
2. Peripherally breeding species	3	79	82
Total 1 + 2	37	211	248
3. Long-distance migrant species			
3a. Migrants from North America	21	9	30
3b. Migrants from South America	–	5	5
Total 3	21	14	35
4. Accidental species from South America	22	13	35
Total 3 + 4	43	27	70
5. Introduced species	–	1	1
Grand total	80	239	319

TABLE 23–3. Comparison of páramo-puna and Malagasy region avifaunas.

	Páramo-puna (185,000 km^2)		Malagasy[a] (365,000 km^2)
	Without peripheral species	With peripheral species	
Breeding species			
Endemics	48 (29%)	48 (19%)	180 (75%)
Nonendemics	118 (71%)	200 (81%)	60 (25%)
Total	166 (100%)	248 (100%)	240 (100%)
Nonbreeding species			
Migrants and accidentals	70	70	59
Grand total	236	318	299

[a] Data from Keith (1980).

region has about twice the surface area of the páramo-puna, and (2) the long-isolated Malagasy region has 75% endemism at the species level, as opposed to only 19%–29% in the continental and relatively young páramo-puna fauna.

Assumptions and Methods

I assume that the information obtained after studying the present distribution of bird species provides clues for the reconstruction of their past history. I am aware, of course, of the difficulties inherent in such an undertaking (Vuilleumier, 1980). Reasonable enough speculations about the origins of a given taxon can be formulated after analyzing its present-day distribution, but new data can most easily prove such conclusions to be wrong.

Human interference with natural vegetation has been substantial and has lasted a long time in the high Andes, and must have affected avian distributions there as elsewhere. Unfortunately, we are not in the enviable position of being able to draw maps of distribution that represent both pre- and post-disturbance, or at least different time intervals, as was done, for instance, for *Neophron percnopterus* in the Mediterranean Basin by Bergier and Cheylan (1980). Similarly, we cannot know yet in most (or perhaps even all) cases whether some high Andean distribution patterns are due to very recent and rapid changes, as can sometimes be illustrated clearly elsewhere (e.g., *Cisticola juncidis* in western Europe; Géroudet and Lévêque, 1976).

Methods in the present work are empiricial and comparative. As much as possible, I made comparisons in a numerical way, although I did not analyze the numerical data statistically. Chi-square tests can be performed on some of the tabulated results, but since we are dealing with trends, and with categories that are not always sharply separated, one may question the biological validity of such statistical tests. Thus, even though the results of my analyses are often expressed quantitatively, they are interpreted largely in qualitative terms. Perhaps this procedure will spur others on to carry out statistical tests after having defined appropriate null hypotheses.

I did not assume in my analysis that Chapman's or Dorst's theories of faunal origins needed to be tested. Rather, I worked on the premise that high Andean birds could have originated either in situ or elsewhere. I assumed that a species originated in situ if its closest relative(s) also occur(s) in the high Andes (see also Chap. 19). An extra-Andean origin was assumed when close relative(s) were distributed elsewhere, but this assumption was carefully weighted by other evidence, especially range continuity versus disjunction of the species inhabiting the high Andes.

TAXONOMIC COMPOSITION OF THE PÁRAMO-PUNA AVIFAUNA

Absent Versus Present Taxa

The following 12 families are absent from the páramo-puna, but are well represented in other biomes: Cracidae (wet montane), Phasianidae (wet montane; dry montane), Cuculidae (dry montane), Apodidae (wet montane; dry montane), Trogonidae (wet montane; dry montane), Ramphastidae (wet montane), Dendrocolaptidae (wet montane; dry montane), Cotingidae (wet montane; dry montane), Corvidae (wet montane; dry montane), Vireonidae (wet montane), Parulidae (wet montane), and Cardinalidae (wet montane; dry montane). These 12 families have radiated in forest or woodland habitats below the páramo-puna. They have not invaded more open habitats and successfully experienced a niche shift, or if they did, the experiment failed. Four of these families (Apodidae, Cotingidae, Corvidae, and Cardinalidae) occur peripherally at the edge of the páramo-puna, but do not actually live in páramo or puna habitats.

Six other families have only one or two species in the páramo-puna but are well represented in other Andean biomes: Psittacidae, Picidae, Formicariidae, Rhinocryptidae, Turdidae, and Icteridae. Finally, the Trochilidae, even though one of the numerically dominant families in the páramo-puna (see next section), have managed only a wedge there, so to speak, if I judge by the wealth of genera and species in the wet montane (including important radiations in *Aglaeactis*, *Coeligena*, *Eriocnemis*, *Metallura*, and *Heliangelus*).

Among the taxa present in the páramo-puna one can single out the water birds of open-water habitats (lakes, lagoons, marshes, salt flats), the ground foragers (which are not necessarily members of families tied to forest habitat, e.g., the Tinamidae, which in South America occur from lowland rain forests to the highest puna grasslands), the aerial–ground foragers (which elsewhere are rather restricted to trees: Tyrannidae), and the shrub or grass foragers (a guild classification of páramo-puna birds is presented in Chap. 6). Some of the most specialized páramo-puna taxa occupying these foraging substrates and the high-altitude *Polylepis* trees are:

Water: the flightless grebe, *Rollandia microptera* (Fjeldså, 1981a).
Ground: the terrestrial flicker, *Colaptes rupicola* (Short, 1971).
Aerial–ground: ground flycatchers, *Muscisaxicola* spp. (Vuilleumier, 1971; Traylor and Fitzpatrick, 1982).
Shrub–grass: ground furnariids, *Asthenes* spp. (Dorst, 1963).
Trees (*Polylepis*): nuthatch-like conebill, *Oreomanes* (Vuilleumier, 1984).

The ecological adaptations of páramo-puna birds have been discussed by Dorst (1955, 1967), M. Koepcke (1954), and Carpenter (1976). High Andean birds seem to have relatively few physiological adaptations to high altitudes (Dorst, 1972; Carpenter, 1975, 1976; Carey and Morton, 1976; Berger, 1978). For further details see Wolf and Gill, Chapter 5, this volume. The question of how physiological adaptations may be related to success or failure in colonization of the high Andes is outside the scope of this chapter.

Numerically Dominant Families

Five of the 36 families (14%) in the páramo-puna are numerically dominant (have more than six species) and are listed in Table 23–4. With the exception of the Trochilidae, the species-genus ratio is higher in these families than in all páramo-puna species combined (Table 23–4). The ratio for the Furnariidae is much larger than the mean for the five families, and is twice as large as the ratio for the entire avifauna. These figures suggest speciation events and radiations in situ. However, a comparison with the species-genus ratios for the Andes and Patagonia as a whole does not show large differences, except for the Anatidae and the Furnariidae. The páramo-puna thus has higher diversities of Anatidae and Furnariidae, but only slightly higher (or equal) diversity of Tyrannidae. The Andes-Patagonia are more diverse in Trochilidae and Emberizidae. Thus not all the numerically dominant families in

TABLE 23–4. Numerically dominant families (six species or more) in breeding páramo-puna avifauna.

Family	Number of genera	Number of species and allospecies	Species-genus ratio in páramo-puna	Species-genus ratio in Andean-Patagonian avifauna (exclusive of páramo-puna)
Anatidae	3	7	2.33	1.67
Trochilidae	6	11	1.83	2.20
Furnariidae	7	28	4.00	2.46
Tyrannidae	7	17	2.43	2.23
Emberizidae	6	18	3.00	3.33
(Mean)	(6)	(16)	(2.72)	(2.38)
Entire páramo-puna avifauna	84	166	1.98	

TABLE 23–5. Numerical distribution of Furnariidae, Tyrannidae, and Emberizidae (numerically dominant in páramo-puna avifauna) in Andean-Patagonian biomes and in South America.

	Páramo-puna (N = 166; 100%)	Andino-alpine (N = 48; 100%)	Wet montane (N = 636; 100%)	Dry montane (N = 290; 100%	Patagonian steppes (N = 121; 100%)	South America (N = 2700; 100%)
Furnariidae						
Number of species	28	7	47	32	16	210
(Percent)	(17)	(15)	(7)	(11)	(13)	(8)
Tyrannidae						
Number of species	17	6	78	31	15	310
(Percent)	(10)	(13)	(12)	(11)	(12)	(11)
Emberizidae						
Number of species	18	8	30	43	8	140
(Percent)	(11)	(17)	(5)	(15)	(7)	(6)
Total number	63	21	155	106	39	660
(Percent)	(38)	(44)	(24)	37	(32)	(24)

the páramo-puna owe this dominance to a greater potential for speciation.

Table 23–5 compares the three richest páramo-puna families in different faunas. The Furnariidae are better represented in the high Andes (páramo-puna and andino-alpine) and in the Patagonian steppes than elsewhere, especially the wet montane. Clearly, they thrive in open habitats. Perhaps the Furnariidae have been open-country birds for a very long time along the Andean-Patagonian axis.

The Tyrannidae are about equally represented in all the faunas; hence very open habitats (steppes) may have been occupied about equally extensively as closed habitats (forests) by birds of this family (see also Traylor and Fitzpatrick, 1982). Even though some of the most specialized of all flycatchers (the ground-tyrants, *Muscisaxicola*) have clearly evolved in the páramo-puna, this radiation is not accompanied there by a numerical diversification of the Tyrannidae as a whole. The Tyrannidae are basically ecologically ubiquitous, so it is not surprising that they should have done well in the high Andes. The point is that they have not diversified more in the páramo-puna than in other Andean biomes, in spite of their numerical dominance in the páramo-puna.

The Emberizidae are most diverse in the andino-alpine and in the dry montane, but only moderately so in the páramo-puna. They are least diverse in the wet montane (same representation as for South America as a whole). This pattern differs from those of the other two families and suggests that different taxonomic sets of Emberizidae occur in different faunas, perhaps as a result of different histories of radiation in different vegetation types. The high representation of the Emberizidae in the dry montane is noteworthy, since it involves a respectable 43 species.

GEOGRAPHIC DISTRIBUTION OF PÁRAMO-PUNA TAXA

Genera

Table 23–6 summarizes the distribution of the 84 genera (including 166 breeding species) in the páramo-puna and compares it to the distribution of genera in the *Nothofagus* forests. Many páramo-puna genera are Pan-American or cosmopolitan in distribution (total 46%). This matches Chapman's (1921, p. 40) figure of 47% for Peruvian puna genera that he called of "general distribution," but not his (1926, p. 113) data for Ecuadorian páramo birds, since only 25% of his 28 genera belong to his categories North and South Temperate and cosmopolitan pooled.

Many of the remaining genera in Table 23–6 are Andean-Patagonian (33%) and/or southern South American (39% pooling the two categories). For Peruvian puna genera, Chapman (1921) had a total of 24 of 58 genera (41%), a very similar percentage, of his categories, South Temperate Origin, Argentine Origin, and Temperate and Puna Zones combined. He had a much higher figure (15 of 28 genera or 54%) of páramo birds allocated to his South Temperate (Chapman, 1926).

Whereas Chapman (1921) considered 12% of 58 genera (in Peru) to be restricted to the puna,

TABLE 23–6. Geographic distribution of páramo-puna genera (breeding avifauna, N = 166 species, 84 genera) and of *Nothofagus* forest genera (land birds, N = 46 species, 40 genera).

Distribution	Páramo-puna avifauna				*Nothofagus* avifauna [a]			
	Number		Percent		Number		Percent	
Endemic	4		5		4		10	
Andean-Patagonian	29	33	34	39	6	11	15	27.5
Southern South American	4		5		5		12.5	
Widespread South American	8		10		5		12.5	
Pan-American	11	39	13	46	8	20	20	50
Cosmopolitan	28		33		12		30	
Total	84		100		40		100	

[a] Data from Vuilleumier (1985).

and (1926) 21% of 28 genera (in Ecuador) to be restricted to the páramo, Table 23–6 shows only 5% endemism for the páramo-puna genera. My data agree reasonably well with Chapman's figures for Ecuadorian páramo and Peruvian puna birds insofar as both data sets reveal that a relatively high percentage of genera in the páramo-puna are distributed along the Andean-Patagonian axis and/or have wider distributions still. There is little agreement, however, about the restricted or endemic genera. This difference is surely due in part to the emphasis on a rather narrow genus in Chapman's time versus a broader genus today. But it is also due to the fact that Chapman did not analyze the páramo and puna avifauna as a whole and that some of the genera he considered "restricted" to the puna were also listed by him for páramo (or vice versa).

Whether my results and figures are in fact comparable with Chapman's is actually uncertain, irrespective of the differences in generic concept between his time and today. He apparently confused "origin" and "distribution." For instance, he stated (1926, p. 112) that the four genera "*Fulica, Erismatura, Asio*, and *Cistothorus*, . . . are both North and South Temperate in their affinities," but on page 113 he tabulated only two genera under "North and South Temperate." Are three of these four genera of page 112 included in the five cosmopolitan genera of his table on page 113? It is unclear. What is clear, however, is that Chapman viewed the páramo and puna faunas as composed of a rather large proportion of restricted taxa and a large proportion of southern temperate and widely distributed taxa. He also clearly concluded that faunal exchange was negligible between his Paramo Zone or Puna Zone and neighboring Temperate Zone.

On the basis of my analysis I conclude that there is very little generic endemism, that one third of the genera are Andean-Patagonian in distribution, one third are cosmopolitan, about 13% Pan-American, and 10% widespread in South America. Note that the figures for the *Nothofagus* forest avifauna (which is geographically an Andean-Patagonian fauna) are quite similar, although higher proportions of genera are southern South American (12.5% v. 6%) and endemic (10% v. 5%).

Species

Table 23–7 was constructed like Table 23–6, but one additional distributional category was included (Andean). The two main categories are species of Andean distribution (35%) and endemics (29%). Together, they make up 64% of páramo-puna birds. Since by definition endemics are also Andean in their distribution, the majority (nearly two thirds) of páramo-puna birds are thus locally distributed. Another 14% of the species are distributed along the Andean-Patagonian axis. Pooling the first three categories one obtains 78% of the species. *Nothofagus* forest species are differently distributed, since proportionately more are endemic, and fewer are either Andean or Andean-Patagonian. Pooling these two categories yields only 58.5%. Proportionately more *Nothofagus* species are widely distributed (southern South American, widespread South American, Pan-American, or cosmopolitan): 41.5% pooling these four categories, versus only 22% for páramo-puna species.

Chapman (1926) categorized 5 of 33 species (15%) as "peculiar to Ecuador," but this category is not necessarily comparable to my "endemic" category. In 1921, Chapman attributed 46 of 82 species or 56% to the category "restricted to the Puna Zone," a much higher

TABLE 23–7 Geographic distribution of páramo-puna species (breeding avifauna, $N = 166$ species) and of *Nothofagus* forest species (land birds, $N = 46$ species).

Distribution	Páramo-puna avifauna				*Nothofagus* avifauna [a]	
	Number		Percent		Number	Percent
Endemic	48		29		19	41
Andean	58	81	35	49	8	17.5
Andean-Patagonian	23		14			
Southern South American	14		8.5		8	17.5
Widespread South American	9		5.5		6	13
Pan-American	10		6		5	11
Cosmopolitan	4		2		–	–
Total	166		100		46	100

[a] Data from Vuilleumier (1985).

figure than mine. Twenty-six species (32%) were placed in his three pooled categories: South Temperate Zone, Chile, and Temperate and Puna. This probably compares with 22.5% of the species in my analysis, a lower figure.

Again, some of the differences between Chapman's analysis and mine can be attributed to differences in taxonomic concepts (a broader, polytypic species concept today), and to the fact that he did not deal with the páramo and puna avifaunas as a whole.

The results of my analysis clearly show that the majority (78%) of páramo-puna species are distributed along the Andean-Patagonian axis and that only a few have much wider distributional ranges. By contrast, the majority (56%) of the genera are widely distributed, yet a large percentage are Andean-Patagonian (39%). These results suggest that the fauna as a whole, although ultimately of varied origins, has had much evolution in situ, an hypothesis that will be examined further in later sections.

ORIGINS OF SELECTED GENERA AND SPECIES

Several high Andean genera and species have been examined during systematic and/or zoogeographic studies by other authors, whose conclusions were as follows:

Rollandia: southern origin (Fjeldså, 1981a, pp. 227–228).

Podiceps [*nigricollis*] *taczanowskii*: recently speciated locally from *P. occipitalis* (Fjeldså, 1981b, pp. 239–241).

Podiceps [*nigricollis*] *occipitalis*: ?southern origin (Fjeldså, 1981a, p. 226).

Chloephaga: related to montane African *Cyanochen* (e.g., Snow, 1980, p. 74); see Dorst and Vuilleumier, Chapter 6, this volume.

Laterallus jamaicensis: local glacial relict (Fjeldså, 1983a, p. 281).

Fulica [?*atra*] *ardesiaca*: originated during geographic isolation in situ (Fjeldså, 1982a, pp. 16–19).

Vanellus [*chilensis*] *resplendens*: "*chilensis* and *resplendens* are closely related to one another and represent an invasion of South America [from the Old World] separate from *cayanus* . . . These species are certainly closest to *vanellus*" (Bock, 1958, p. 64).

Charadrius [*falklandicus*] *alticola*: ?southern (Patagonian) origin (Bock, 1958, pp. 72–73).

Oreopholus: ?related to Eurasian *Eudromias morinellus* (Bock, 1958, p. 79).

Phegornis: "perhaps allied to *Aechmorhynchus* and *Prosobonia*" (Bock, 1958, p. 83), two genera found in the southwest Pacific; "monotypic genus in the Charadriinae" of the Charadriidae (Zusi and Jehl, 1970, p. 768), whereas *Aechmorhynchus* and *Prosobonia* are considered congeneric members of the Scolopacidae (Zusi and Jehl, 1970).

Colaptes rupicola: from (?southern) lowland stock (Short, 1972).

Oreomanes: ?local (i.e., Andean) evolution from *Conirostrum*-like stock (judging from data on hybridization in Schulenberg, 1985).

Very diverse origins can be claimed for this sample of 12 genera and species: in situ evolution in perhaps four cases (33%) southern origin (?Patagonian) in about four instances (33%), African origins or affinities in two instances (17%), Eurasian affinity in perhaps one case (8%), and one unknown origin (8%). I will analyze the fauna more quantitatively in later sections.

TABLE 23–8. Faunal resemblance among breeding avifaunas (species level).

(Number of species in each avifauna)	Páramo (62)	Puna (148)	Andino-alpine (48)	Wet montane (636)	Dry montane (290)	Patagonian steppes (121)
A. Simpson's index [(C/N) × 100]						
Páramo	—	71%	33%	48%	53%	27%
Puna	44	—	71%	21%	42%	37%
Andino-alpine	16	34	—	19%	44%	44%
Wet montane	30	31	9	—	26%	10%
Dry montane	33	62	21	75	—	27%
Patagonian steppes	17	45	21	12	33	—
B. Jaccard's index [C(N₁ + N₂ − C) × 100]						
Páramo	-	27%	17%	5%	10%	10%
Puna		-	21%	4%	17%	20%
Andino-alpine			-	1%	7%	14%
Wet montane				-	9%	2%
Dry montane					-	8%
Patagonian steppes						-

SPECIES-LEVEL FAUNAL RESEMBLANCES

Table 23–8 shows the faunal resemblances among the breeding avifaunas of six vegetation types. Above and to the right of the diagonal in Table 23–8A are the indexes of faunal resemblance based on Simpson's (1960) index $(C/N) \times 100$, where C = number of species shared in common between two faunas, and N = number of species in the smaller of the two faunas. The figures below and to the left of the diagonal in Table 23–8A are the numbers of species shared by each pair of faunas.

The highest indexes of faunal resemblance are those between the páramo and the puna, and between the puna and the andino-alpine (both about 71%). All the other indexes are lower (less than 53%, down to 10%). The low (33%) resemblance between the páramo and the andino-alpine is noteworthy.

Of interest are the following, relatively high indexes: 53% between the páramo and the dry montane, and 48% between the páramo and the wet montane. These adjacent faunas thus have more faunal exchange than Chapman (1926) concluded, judging from Simpson's indexes. The puna, like the páramo, shares more species with the dry montane than with the wet montane, but the absolute values are less—42% and 21%, respectively. The puna shares more species with the dry montane (42%) than with the Patagonian steppes (37%), but the difference is small and may not be biologically significant. The páramo, however, shares only 28% of its species with the Patagonian steppes.

These figures agree only in part with Chapman's (1921) conclusion that the puna zone species are "largely" of "South Temperate Zone" origin, especially since his view is based on his allocating 22 of 82 species, or only 27%, to his South Temperate Zone. The figures of Table 23–8A, moreover, do not agree with Chapman's (1926) view that 67% (22 of 33 species) of his páramo birds of Ecuador are of southern origin. Instead, the data of Table 23–8A suggest faunal interchanges between páramo, wet montane, and dry montane, and between puna and dry montane. Resemblance of only 71% (and not 85%–90%) between páramo and puna species suggests furthermore much in situ differentiation or origination in the páramo and the puna, respectively, or else much differential immigration to the páramo and to the puna, or, more likely, both. Exchanges with the Patagonian steppes would appear to be relatively minor, but they could be masked the local evolution occurring after the exchanges took place.

Table 23–8B shows the faunal resemblances using Jaccard's index $[C/(N_1 + N_2 - C) \times 100]$, where C = number of species shared in common between two faunas, and N_1 and N_2 the number of species in each of the two faunas. This index is commonly used by biogeographers (e.g., Flessa, 1981). As is readily apparent from an examination of Table 23–9, the results obtained with Simpson's or with Jaccard's indexes vary. The most important differences are marked in boldface

TABLE 23–9. Faunal resemblances of pairs of avifaunas, comparing Simpson's and Jaccard's indexes (see Table 23–8).

	Jaccard's value	Index rank	Simpson's value	Index rank
Páramo–puna	27%	1	71%	1
Puna–Andino-alpine	21%	2	71%	1
Puna–Patagonian steppes	**20%**	**3**	**37%**	**8**
Páramo–Andino-alpine	**17%**	**4**	**33%**	**9**
Puna–Dry montane	17%	4	42%	7
Andino-alpine–Patagonian steppes	14%	6	44%	5
Páramo–Dry montane	**10%**	**7**	**53%**	**3**
Páramo–Patagonian steppes	10%	7	27%	10
Wet montane–Dry montane	9%	9	26%	12
Dry montane–Patagonian steppes	8%	10	27%	10
Andino-alpine–Dry montane	**7%**	**11**	**44%**	**5**
Páramo–Wet montane	**5%**	**12**	**48%**	**4**
Puna–Wet montane	4%	13	21%	13
Wet montane–Patagonian steppes	2%	14	10%	15
Andino-alpine–Wet montane	1%	15	19%	14

in the right-hand columns of Table 23–9. According to Jaccard's indexes the páramo and Patagonian steppes or andino-alpine are much alike, whereas the páramo and dry montane or wet montane share very few species in common. These results agree much more with Chapman's earlier results. Part of the differences in the results using these two indexes must have to do with sampling effects, since some of the faunas being compared are so uneven in species numbers. It is usually agreed that Simpson's index is less biased when comparing uneven samples (see Chap. 22).

ORIGINS OF ALTITUDINAL AND LATITUDINAL RANGE DISJUNCTIONS

Many taxa of páramo-puna birds (52 cases for 166 species, or 31%) exhibit clear-cut range disjunctions, with one taxon (allospecies, species, or subspecies) or one population in the páramo and/or puna, and another (or several others) elsewhere in South America, or even farther away. There are 23 instances of range disjunctions within superspecies. The disjunct taxa (one or more high Andean, the other or others extra-Andean) are here considered allospecies. I counted 20 instances of range disjunctions within species or allospecies where the disjunct taxa are considered subspecies. Finally, I detected 9 instances of range disjunctions within species that were not accompanied by subspecific recognition of the disjunct isolates.

Table 23–10 gives a numerical summary of the range disjunctions, arranged into six categories. A high percentage of all instances (48%) and of superspecies (43.5%), and the majority of cases for species (40% and 78%), involve disjunctions between the high Andes and the tropical lowlands, usually the eastern lowlands of central South America. Distinctly smaller proportions (9%–25%) involve high Andean and Patagonian (i.e., temperate lowland) disjunctions. However, another 11%–30% of the species show disjunctions involving the high Andes and both Patagonia and the tropical lowlands. If the two categories of range disjunctions involving Patagonia are pooled, then 24%–55% of the species analyzed show high Andean-Patagonian disjunctions. There are nevertheless more high Andean-tropical lowland disjunctions. These are clearly, then, the two most important patterns of range disjunctions at both the superspecies and species levels. Note that an Andean-North American disjunction pattern affects 43.5% of the superspecies but none of the species.

The range disjunctions summarized in Table 23–10 could be either (a) the result of a progressively widening break in a formerly continuous range from the tropical lowlands up to the high Andes, or from Patagonia to the high Andes, or (b) the result of dynamic populational changes in distribution. The latter would involve, for instance, the colonization of the páramo or puna from a tropical lowland area, and the successful establishment of an isolate in the high Andes. Both phenomena are at work here.

Examples of range disjunctions probably due to an extrinsic break (category a) include:

Tinamotis [*pentlandii*] (Andes-Patagonia), and *Asthenes* [*wyatti*] (Andes-Patagonia), among the superspecies (Fig. 23–2).

TABLE 23–10. Range disjunctions in the páramo-puna avifauna ($N = 166$ species).

Category of range disjunction	Superspecies (disjunct allospecies)		Species (disjunct subspecies)		Species (disjunct populations)		Total	
	Number	(Percent)	Number	(Percent)	Number	(Percent)	Number	(Percent)
High Andes–Patagonia	2	(9)	5	(25)	1	(11)	8	(15.5)
High Andes–Tropical lowlands	10	(43.5)	8	(40)	7	(78)	25	(48)
High Andes–Tropical lowlands–Patagonia	–	–	6	(30)	1	(11)	7	(13.5)
High Andes–southeastern Brazil	1	(4)	–	–	–	–	1	(2)
High Andes–Pantepui	–	–	1	(5)	–	–	1	(2)
High Andes–North America	10	(43.5)	–	–	–	–	10	(19)
Number of instances	23	(100)	20	(100)	9	(100)	52	(100)

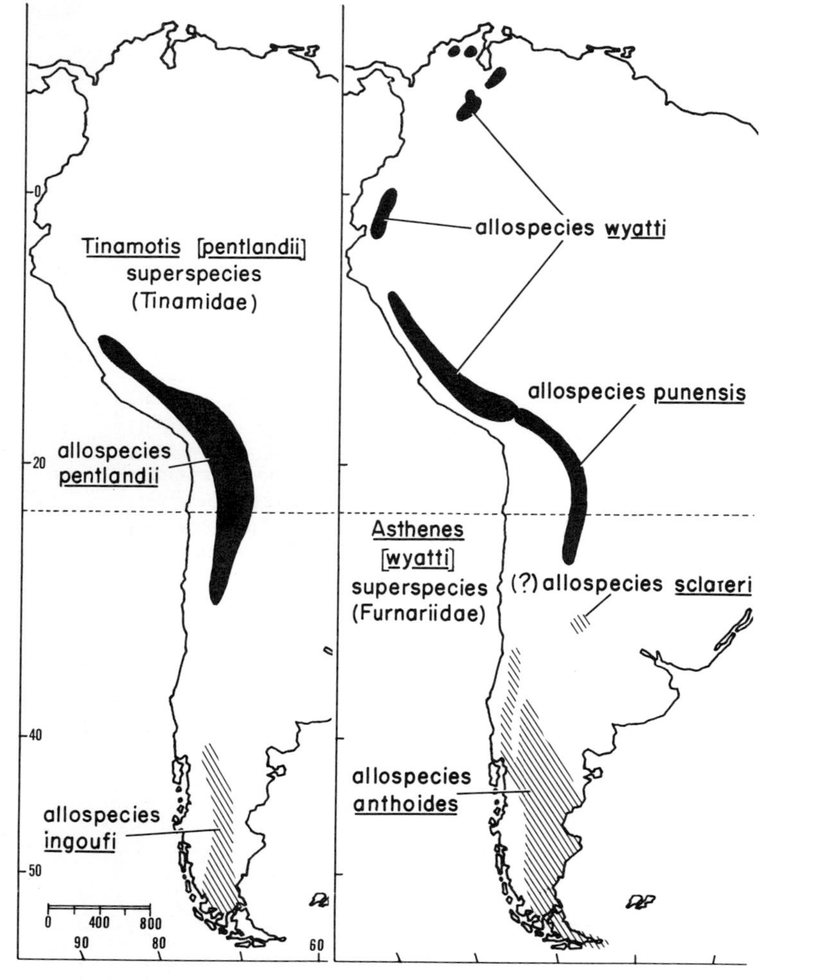

FIG. 23–2. Geographic distribution of the *Tinamotis* [*pentlandii*] superspecies (*left*) and the *Asthenes* [*wyatti*] superspecies (*right*). Distributional area of páramo-puna taxa in black, extra-Andean taxa shaded. Distribution gaps are believed to be due to extrinsic factors.

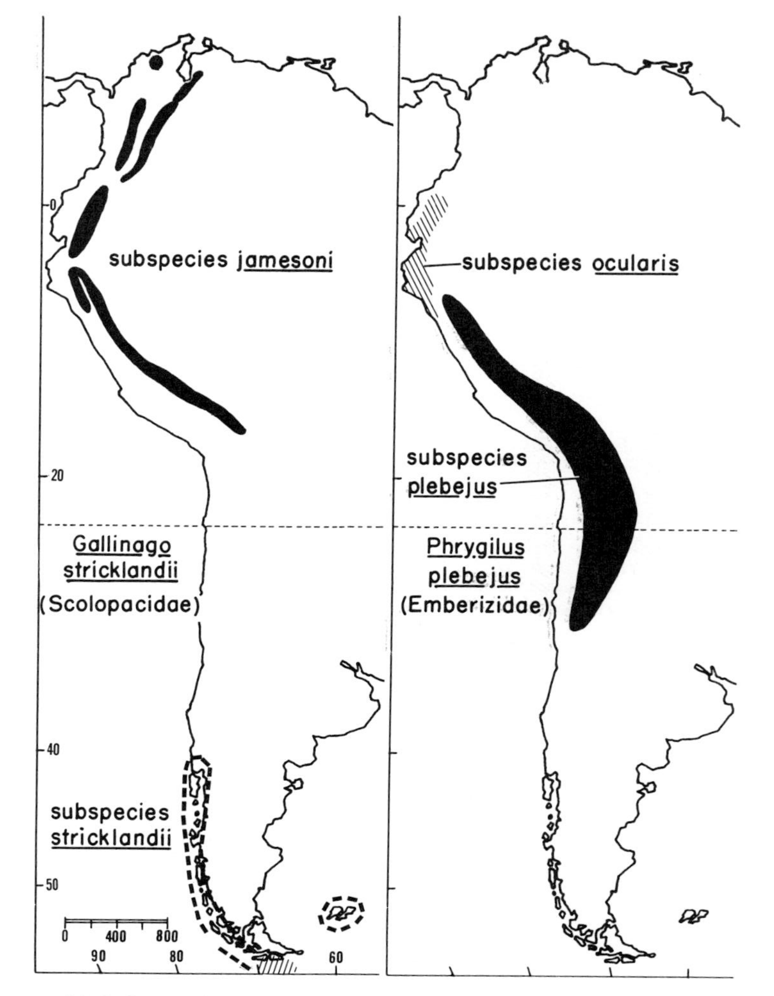

FIG. 23–3. Geographic distribution of *Gallinago stricklandii* (*left*) and *Phrygilus plebejus* (*right*). Páramo- puna taxon or population in black, extra-Andean or Andean and lowland taxon or population shaded. Gap within *Gallinago stricklandii* is believed to be extrinsic; northern population of *P. plebejus* may have invaded the lowlands from the highlands.

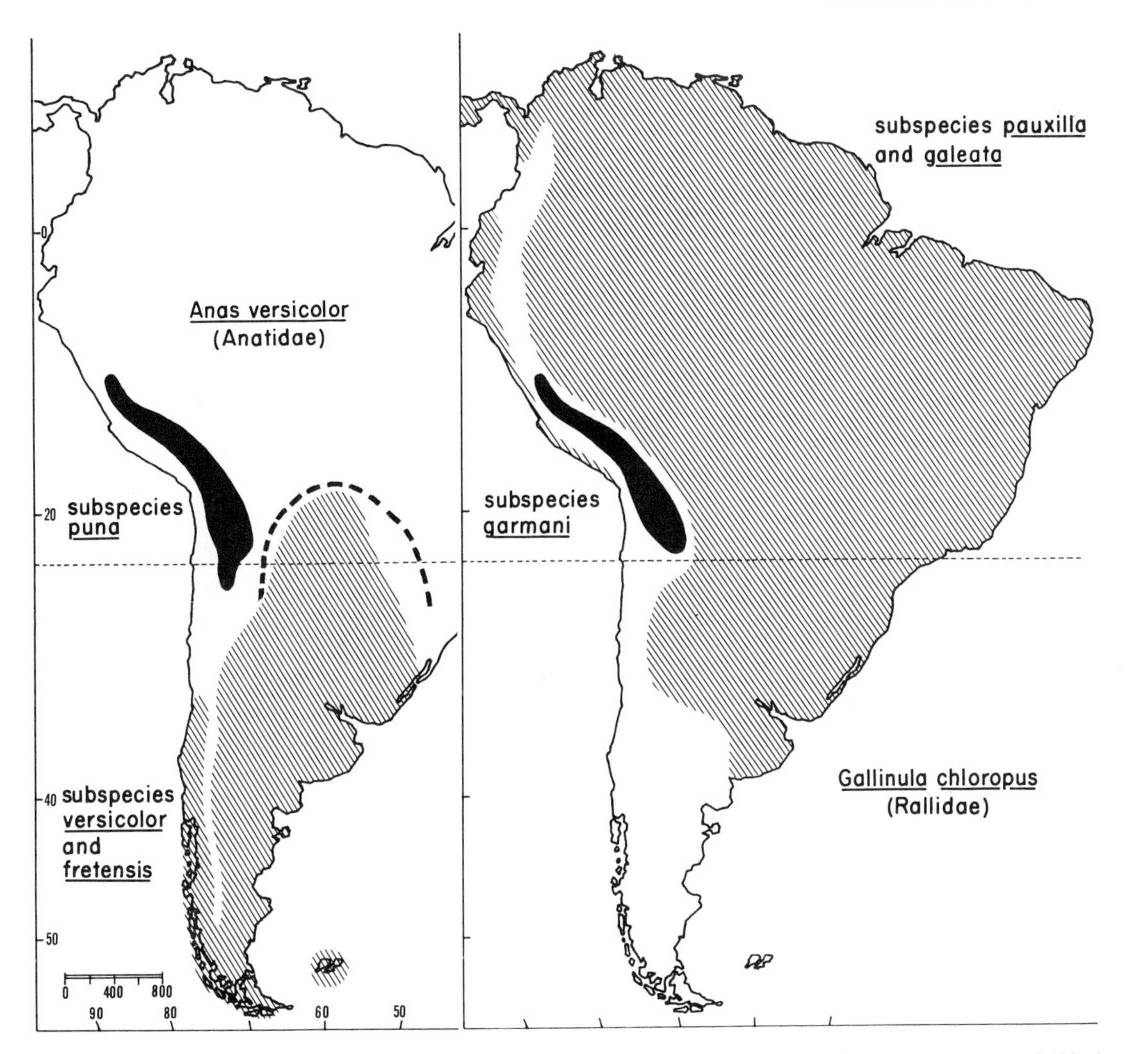

FIG. 23–4. Geographic distribution of *Anas versicolor* (*left*) and *Gallinula chloropus* (*right*). Páramo-puna taxa in black, extra-Andean ones shaded. Páramo-puna taxa may have originated from an invasion from the lowlands.

Gallinago stricklandii (Andes-Patagonia), and *Phrygilus plebejus* (Andes-lowlands), among the species (Fig. 23–3).

Examples of disjunctions that are probably the result of dynamic range expansions (category b) include:

Anas versicolor (Fig. 23–4) and *Agriornis microptera* (map in Vuilleumier, 1971), both probably from Patagonia or southern South America to the high Andes.

Nycticorax nycticorax, *Laterallus jamaicensis*, and *Gallinula chloropus* (Fig. 23–4) from the tropical lowlands to the high Andes.

Podiceps [*nigricollis*] *occipitalis*, *Oreopholus ruficollis*, *Lessonia* [*rufa*], and *Phrygilus alaudinus* (Fig. 23–5), perhaps from the high Andes to Patagonia or to the tropical lowlands (although the reverse immigration could be just about equally likely).

A number of range disjunctions are very difficult to analyze simply in terms of a break in the range versus active colonization, probably because they involved both phenomena at different times in the past. Such complex cases may include *Theristicus* [?*caudatus*], *Buteo* [?*albicaudatus*], *Recurvirostra* [*avosetta*], and *Zenaida* [*macroura*] among the superspecies, and *Phoenicopterus chilensis*, *Anas cyanoptera*, *Thinocorus rumicivorus*, and *Anthus hellmayri* among the species.

The analysis in this section shows that there is altitudinal and/or latitudinal speciation or incipient speciation in many high Andean birds, as

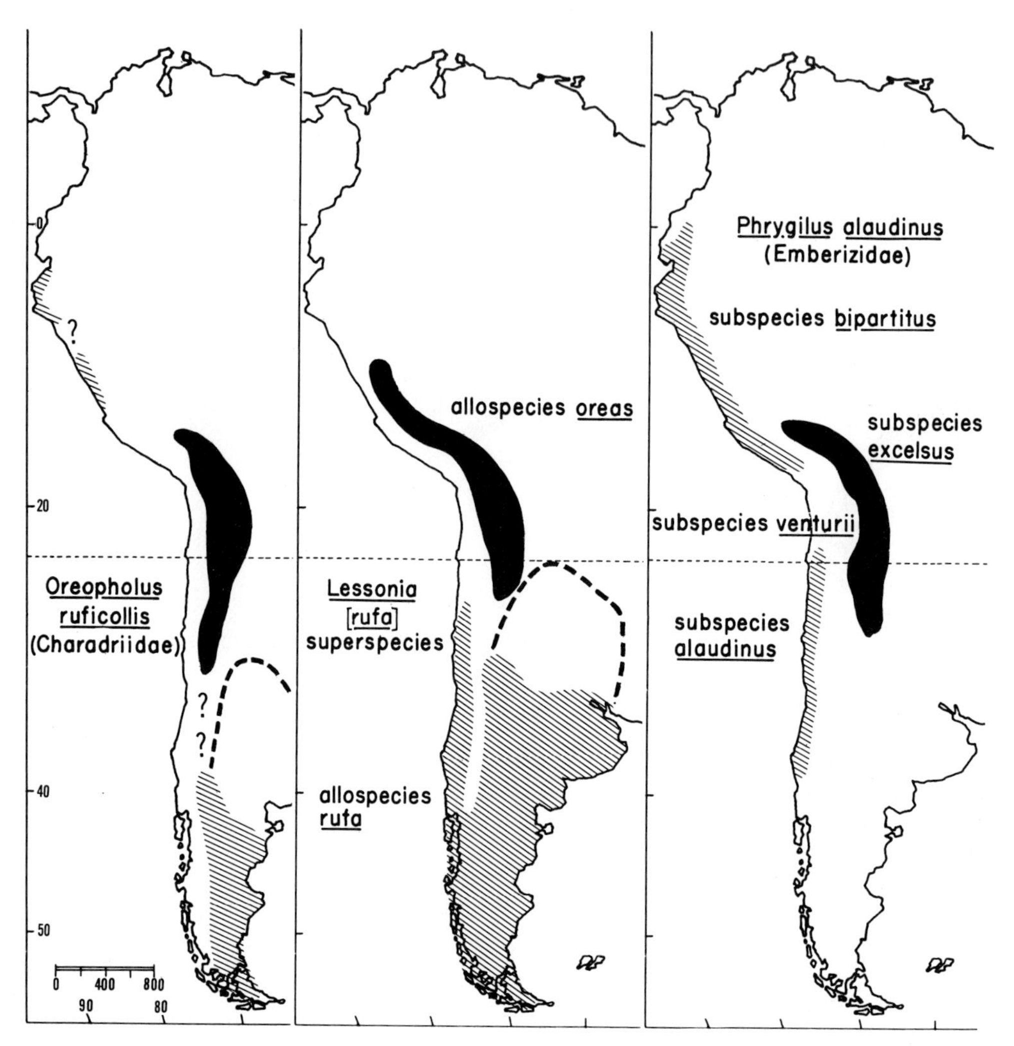

FIG. 23–5. Geographic distribution of *Oreopholus ruficollis* (*left*), the *Lessonia* [*rufa*] superspecies (*center*), and *Phrygilus alaudinus* (*right*). Páramo-puna taxa or populations in black, extra-Andean ones or extra-páramo-puna ones shaded. Lowland populations (northern ones of *Oreopholus*; southern South American ones of *Lessonia*; western slope and lowland ones of *Phrygilus*) may have originated from the highlands.

reflected in the large number of cases of complex range disjunctions, a topic on which much more work is needed.

THE ORIGINS OF ENDEMIC TAXA

Definition of an Endemic Taxon

A taxon (genus, species, subspecies) is defined as endemic if it is restricted in its ecogeographic distribution to the páramo and/or to the puna biomes (Table 23–1, Fig. 23–1). However, páramo or puna cannot, of course, always be defined unequivocally themselves, especially perhaps at the lower edges of these vegetation types in relatively dry areas, or in places that have been disturbed by man. Thus to define an endemic taxon may be a little difficult, and some workers may not recognize as endemic a taxon so admitted by others. In order to take this problem into account, I have included the category "nearly endemic."

A nearly endemic species occurs chiefly in the páramo and/or puna, but its distribution may include neighboring areas with a vegetation that is somewhat intermediate between pure páramo or pure puna and either the wet montane or the dry montane.

The páramo-puna avifauna has no endemic

TABLE 23–11. Comparisons of levels of endemism in the páramo-puna and the *Nothofagus* forest avifaunas.

Levels of endemism	Páramo-puna avifauna		*Nothofagus* avifauna [a]	
	Number	(Percent)	Number	(Percent)
Genus				
Endemic genera	4	(5)	4	(10)
Total genera	84	(100)	40	(100)
Species				
Endemic species and allospecies	48	(29)	19	(41)
Nonendemic species with endemic subspecies	31	(19)	5	(11)
Nonendemic species without endemic subspecies	87	(52)	22	(48)
Total species	166	(100)	46	(100)

[a] Data from Vuilleumier (1985).

family, only about 4 endemic genera, 48 endemic species, 21 nearly endemic species, and at least 31 endemic subspecies in nonendemic species. Table 23–11 compares these rates of endemism with those in the *Nothofagus* forest avifauna of Patagonia.

Endemism is clearly higher at the genus and species levels in the isolated *Nothofagus* forest avifauna than in the páramo-puna fauna. These comparisons suggest that the páramo-puna fauna is not very isolated at present and/or has not been very isolated in the past.

Endemic Genera

I do not know the identity of the six genera "restricted to Ecuador" mentioned by Chapman (1926) for his Paramo Zone of that country. For his Puna Zone of Peru he (1921) listed *Gymnopelia* (now = *Metriopelia*, a nonendemic genus), *Metriopelia*, *Ptiloscelys* (= *Vanellus*, nonendemic), *Pterophanes* (not included by me in the páramo-puna fauna), *Patagona* (nonendemic), *Oreotrochilus* (nonendemic, discussed later), and *Pseudochloris* (= *Sicalis*, nonendemic). Curiously, in his work on Ecuador, Chapman (1926) listed *Metriopelia*, *Ptiloscelys*, *Pterophanes*, and *Oreotrochilus* for the páramo, but did not mention that he had already cited them for the puna in 1921. Moreover, Chapman (1926) stated that the genera peculiar to his Paramo Zone included the three hummingbirds *Oreotrochilus*, *Chalcostigma*, and *Pterophanes*, but did not indicate that the first and third were also given by him in 1921 as "Genus Puna Zone."

In this chapter I recognize as endemic genera for the páramo-puna *Oxypogon* (páramo, Trochilidae), *Idiopsar* (puna, Emberizidae), *Xenodacnis* (páramo and puna, ?Thraupidae), and *Oreomanes* (páramo and puna, ?Thraupidae) (Fig. 23–6). One could argue that *Xenodacnis* and *Oreomanes*, both specialized to *Polylepis* woodlands, are not in fact truly part of the páramo-puna avifauna, since they do not live in the characteristic grassland and scrub of the puna. Should this view be adopted, then only two endemic genera must be retained.

Oxypogon is closely related to (and could conceivably be merged with) the genus *Chalcostigma* found in páramo, puna, and wet montane and its origin is most likely in situ. *Idiopsar* is closely related to (but probably not congeneric with) the Andean-Patagonian *Phrygilus-Diuca* complex and, similarly, has a probable in situ origin. *Xenodacnis* is an isolated genus whose origin is uncertain (?in situ). *Oreomanes* probably originated in situ from a "thraupid-coerebid" (conebill-like) stock in the upper wet montane.

I do not consider *Phoenicoparrus* as a genus distinct from *Phoenicopterus* (contra Kahl [1979b] for instance). Workers who keep *Phoenicoparrus* would thereby gain an endemic genus for the puna, with a local origin probably a rather long time ago. I do not consider *Oreotrochilus* (map in Carpenter, 1976) to be endemic to the páramo-puna because one species (*O. adela*) occurs almost entirely in the dry montane, while another (*O.* [*estella*] *leucopleurus*) occurs in the andino-alpine as well as the puna (e.g., Navas, 1965). A taxon as striking and isolated as *Phegornis* (Bock, 1958; Zusi and Jehl, 1970) is not included among the endemic genera because it occurs not only in the puna but also in the andino-alpine.

Irrespective of how restrictively an endemic genus is defined, however, there are very few of them, 2–5 out of 84 (2%–6%) in the páramo-puna fauna. The low generic endemism and the absence of family-level endemism clearly suggest

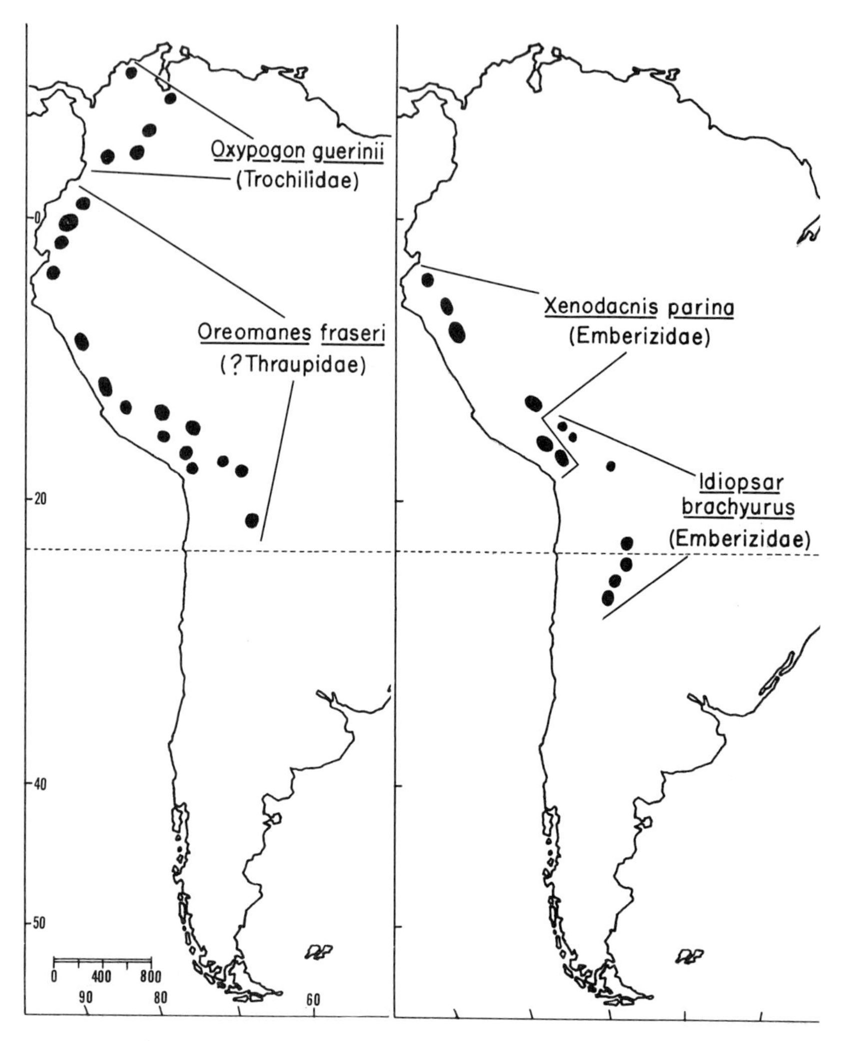

FIG. 23–6. Patchy geographic distribution of the four genera considered endemic to the páramo and puna biomes. Even though further work might reduce some of the gaps, the distribution will remain patchy in all four taxa.

recency of the fauna, or lack of isolation during its history, or both. The situation in the continental (yet island-like) páramo-puna avifauna is unlike that in a truly insular avifauna, like that of the Malagasy region. Keith (1980, p. 99) wrote that, with "40 endemic genera, one endemic subfamily, and five endemic families, . . . many of these birds, even at the species level, are so different from anything found outside the Malagasy Region that it is almost impossible to guess at their origins, while others afford us some clues."

Endemic Species

A total of 48 species are endemic, including the 4 species in the endemic monotypic genera. Of the remaining 44 species, 6 belong to 3 endemic superspecies: *Asthenes* [*flammulata*] (2 endemic allospecies) (see map in Vuilleumier, 1968), *Upucerthia* [*validirostris*] (2 endemic allospecies) (map in Vaurie, 1980), and *Phrygilus* [*erythronotus*] (2 endemic allospecies, see Fig. 23–7). These three superspecies show, first, local origination of six new allospecies in situ, and second, local origin of each stock. *Asthenes* [*flammulata*] is closely related to *A. urubambensis* (see The Origins of Peripheral Species), and may be from the wet montane or local in origins. *Upucerthia* [*validirostris*] is related to *U. albigula* of the puna and dry montane, and hence the whole complex is probably of local origin. *Phrygilus* [*erythronotus*] is related to other species of *Phrygilus* and to *Diuca*, a complex with an Andean-Patagonian

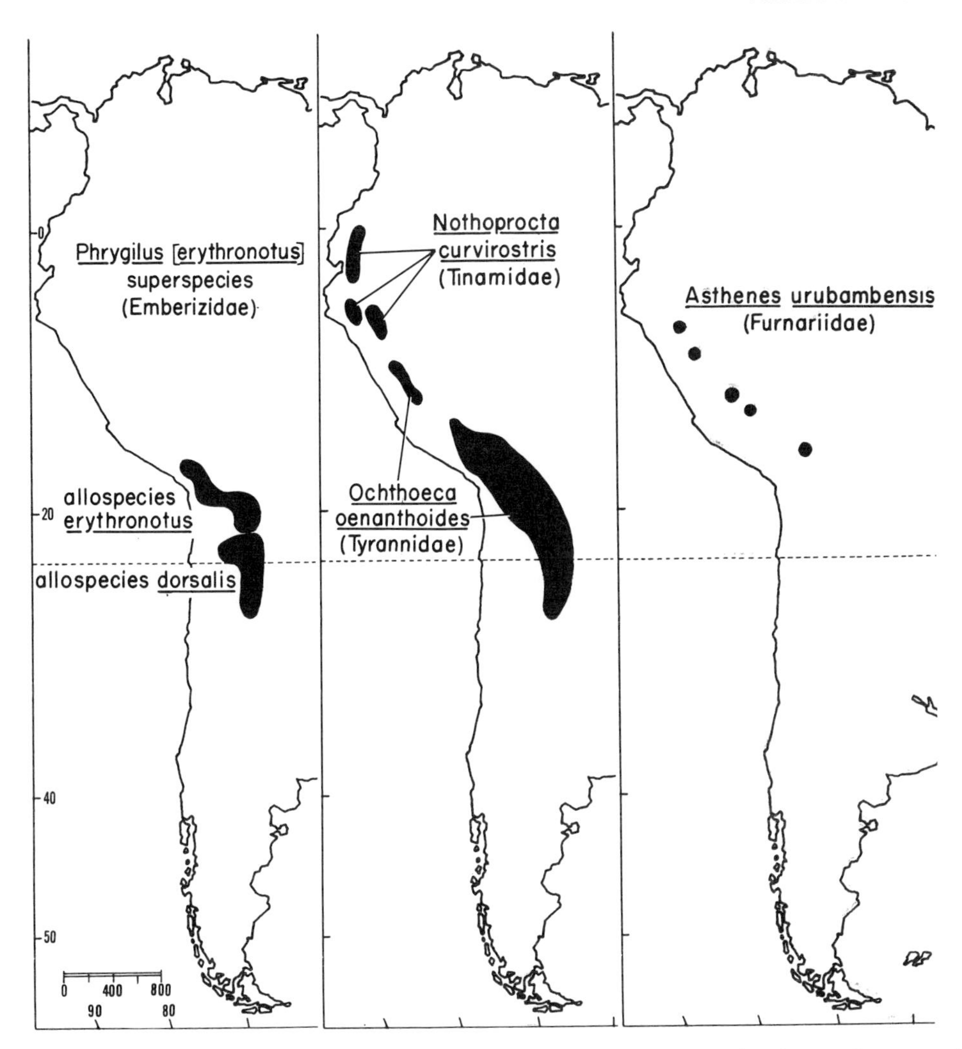

FIG. 23–7. Geographic distribution of a superspecies endemic to the puna, the *Phrygilus* [*erythronotus*] superspecies (*left*), and of two nearly endemic species, *Nothoprocta curvirostris* (páramo and neighboring "cultivation steppe") and *Ochthoeca oenanthoides* (puna and upper dry montane) (both *center*). Note that with the exception of the spilling of each species into another vegetation unit below páramo or puna, the distribution of these two species otherwise is very similar to those of endemics. *Right*: Patchy distribution of *Asthenes urubambensis* at the edge of the wet montane and in peripheral *Polylepis* woodlands along the eastern Andes of Peru and Bolivia.

pattern of distribution, and can be considered to have originated in situ.

The remaining 38 endemic species are diverse. They include land birds and water birds, passerines and nonpasserines, and a variety of trophic and substrate categories. The largest numbers of endemic species are found in the Trochilidae (4 species), Furnariidae (10), Tyrannidae (5), and Emberizidae (3). With the exception of the Anatidae, which lack endemic species, the best represented families among the endemics are also the numerically dominant ones.

Table 23–12 (column 1) gives the presumed geographic origins of the 48 endemic species, based on the taxonomic affinities and the geographic distribution of their nearest relatives. Whereas 17% have unknown origins, a substantial proportion (44%) are presumed to have originated in situ, close to where they live today. Twelve species (25%) have either Andean-Patagonian, or Patagonian origins. Another 7 species (14%) could have originated outside of the Andes, either elsewhere in South America, or in North America, or else even farther away.

TABLE 23–12. Presumed origins of páramo-puna bird species.

Origins	Endemic species		Nearly endemic species		Non-endemic species		Total		Peripheral species		Grand total	
	Number	(Percent)	Number	(Percent)	Number	(Percent)	Number	(Percent)	Number	(Percent)	Number	(Percent)
Local (high Andes)	21	(44)	7	(33)	34	(35)	62	(37.5)	57	(70)	119	(48)
Andean or Patagonian	9	(19)	3	(14)	23	(24)	35	(21)	5	(6)	40	(16)
Patagonian	3	(6)	1	(5)	3	(3)	7	(4)	1	(1)	8	(3)
South American	1	(2)	3	(14)	14	(14)	18	(11)	9	(11)	27	(11)
N. or S. American	2	(4)	1	(5)	17	(18)	20	(12)	6	(7)	26	(10.5)
Cosmopolitan	4	(8)	2	(10)	4	(4)	10	(6)	—	—	10	(4)
Unknown	8	(17)	4	(19)	2	(2)	14	(8.5)	4	(5)	18	(7.5)
Total	48	(100)	21	(100)	97	(100)	166	(100)	82	(100)	248	(100)

The 8 species of unknown origins are *Fulica gigantea*, *F. cornuta*, *Leptasthenura andicola*, *Geositta tenuirostris*, *Anthus bogotensis*, *Sicalis lutea*, *S. uropygialis*, and *Carduelis atrata*.

Nearly Endemic Species

These are basically páramo-puna species that spill over slightly into another vegetation type. *Nothura curvirostris*, for instance, is a páramo species also found marginally in the wet montane and locally in the dry montane ("cultivation steppes") (Fig. 23–7, center). *Ochthoeca oenanthoides* is a puna species that also occurs in the adjacent dry montane at high altitudes (Fig. 23–7). Table 23–12 (column 2) analyzes the presumed origins of the 21 nearly endemic species. One third of them may have originated in situ, and about 29% may have originated outside the Andes or Patagonia.

Endemic Subspecies in Nonendemic Species

At least 31 of the 97 nonendemic species have endemic subspecies. Of the 31 species, at least 2 (6%) have an endemic subspecies in the páramo-puna that some workers consider a species: *Pterocnemia pennata tarapacensis* (map in Plenge, 1982) and *Anas versicolor puna* (Fig. 23–4). In terms of faunal origins, it is noteworthy that 20 of the 31 species (65%) with endemic subspecies have disjunct subspecies and hence give us clues about possible pathways of geographic origin (see Origins of Altitudinal and Latitudinal Range Disjunctions).

THE ORIGINS OF NONENDEMIC SPECIES

Table 23–12 (column 3) summarizes the analysis of presumed origins for the 97 nonendemic species. About 35% of these species originated in situ, and another 27% originated along the Andean-Patagonian axis. Most of the remaining species (about 36%) may have originated outside the Andes (South America, North America or farther away), presumably by immigration.

THE ORIGINS OF PERIPHERAL SPECIES

A total of 82 species breed in habitats immediately adjacent to the páramo-puna, such as the upper edge of cloud forest or ceja (wet montane), or the upper reaches of dry montane vegetation. I allocated these species either to the wet montane or to the dry montane (see Appendix 23–B).

Fifteen families with 36 species live in the dry montane: Tinamidae (3 species), Columbidae (1), Psittacidae (1), Strigidae (1), *Apodidae* (1), Trochilidae (4), Picidae (1), Furnariidae (8), Tyrannidae (3), *Cotingidae* (1), *Phytotomidae* (1), Icteridae (1), Emberizidae (7), *Cardinalidae* (1), and Thraupidae (2). The four families in italics are absent from the páramo-puna. Note the large number of peripheral species of Furnariidae and Emberizidae in the dry montane: 15 of 36 species (42%).

There are 46 species in 15 families in the wet montane: Anatidae (1), Scolopacidae (1), Psittacidae (2), Trochilidae (10), Picidae (2), Furnariidae (10), Rhinocryptidae (2), Tyrannidae (4), *Cotingidae* (1), Hirundinidae (1), *Corvidae* (1), *Cinclidae* (2), Troglodytidae (1), Emberizidae (3) and Thraupidae (5). Again, families absent in the páramo-puna are in italics. Note the large numbers of Trochilidae, Furnariidae, and Thraupidae in the wet montane: 25 of 46 (54%).

Table 23–12 (column 5) shows the presumed origins of the 82 peripheral species. The great majority of them (70%) probably originated in situ, whereas only about 18% originated along the Andean-Patagonian axis, and another 18% probably originated outside the Andes and/or Patagonia.

THE ROLE OF *POLYLEPIS* WOODLANDS AND *PUYA* STANDS IN THE ORIGINS OF PÁRAMO-PUNA BIRDS

Many species of birds occur and breed in *Polylepis* woodlands (Vuilleumier, unpublished data) and a smaller number do so in *Puya raimondii* stands (Dorst, 1957; Rees and Roe, 1980; Vuilleumier, unpublished data). Two endemic monotypic genera, *Xenodacnis* and *Oreomanes*, are found apparently exclusively in *Polylepis*. Table 23–13 gives the numbers of species found in *Polylepis*, in *Puya*, and in both, for all components of the páramo-puna fauna, including the nonbreeding species (migrants and accidentals, cited as "other" in Table 23–13).

Polylepis and *Puya* are clearly important habitats for páramo-puna birds. *Polylepis* woodlands have proportionately more nearly endemic and nonendemic species than endemic species. *Puya* stands are about equally important for all three categories of breeding species. Note that a rather

TABLE 23–13. Occurrence of páramo-puna species in *Polylepis* woodlands and *Puya* stands.

Species and allospecies	*Polylepis*		*Puya*		Both	
	Number	(Percent)	Number	(Percent)	Number	(Percent)
Endemic ($N = 48$)	14	(29)	15	(31)	8	(17)
Nearly endemic ($N = 21$)	11	(52)	6	(29)	7	(33)
Nonendemic ($N = 97$)	45	(46)	27	(28)	26	(27)
Total ($N = 166$)	70	(42)	48	(29)	41	(25)
Peripheral ($N = 82$)	66	(80)	4	(5)	3	(4)
Other ($N = 71$)	2	(3)	–	–	–	–
Grand total ($N = 319$)	138	(43)	52	(16)	44	(14)

large proportion of these birds occurs in both *Polylepis* and *Puya* (17%–33%).

Of the 21 endemic species that occur in *Polylepis* and *Puya*, 16 (76%) belong to four of the five numerically dominant families discussed earlier: Trochilidae (2 species), Furnariidae (7), Tyrannidae (4), and Emberizidae (3). Of the nearly endemic or nonendemic species, several are specialized ecologically to *Polylepis* and/or *Puya*, although not restricted to these habitats. They include *Chalcostigma stanleyi*, *Leptasthenura yanacensis*, *Xolmis rufipennis*, *Agriornis andicola*, and *Carduelis crassirostris*. *Anairetes alpinus* and *Ampelion stresemanni*, here included among the peripheral species, could also be placed in the nonendemic species that are specialized to *Polylepis*.

As can be seen from Table 23–13, *Polylepis* woodlands are very important ecologically for peripheral species, since 80% of them are found there. Far fewer (5%) occur in *Puya*. Although many *Polylepis* woodlands occur right within the páramo and/or the puna, other woodlands are peripheral to the páramo-puna.

Of the 36 peripheral species in the dry montane, 28 (78%) are found in *Polylepis* woodlands, and 4 (11%) in *Puya* stands, a total of 32 or 89%. The four species in *Puya* stands are *Leptasthenura [pileata] pileata* (pers. obs.), *Phacellodomus striaticeps* (Dorst, 1957, p. 598), *Incaspiza personata* (pers. obs.), and *Thraupis bonariensis* (Venero and Brokaw, 1980).

Of the 46 peripheral species in the wet montane, 38 (83%) are found in *Polylepis* woodlands. Notable is the fact that all 10 species of Trochilidae are found in *Polylepis* (in the genera *Lafresnaya*, *Aglaeactis*, *Coeligena*, *Heliangelus*, *Eriocnemis*, *Metallura*, and *Chalcostigma*). Other peripheral species commonly found in *Polylepis* woodlands at the edge of the wet montane and the páramo-puna include the woodpecker *Piculus rivolii*, several Furnariidae (especially in the genera *Margarornis*, *Synallaxis*, and *Cranioleuca*), three Tyrannidae (*Mecocerculus leucophrys* and two species of *Ochthoeca*), the jay *Cyanolyca turcosa*, and Thraupidae of the genera *Hemispingus*, *Anisognathus*, and *Tangara*.

A large number of families, genera, and species absent from the grasslands and scrub of the páramo-puna thus inhabit the periphery of this biome. Most of these species (66 of 82 or 80%) live in the *Polylepis* woodlands found at the edge of the páramo-puna, but not, as a rule, in the isolated *Polylepis* woodlands occurring as ecological islands well within the grassland and the scrub. One thus has the impression that a series of taxa are poised at the edge of the páramo-puna and are ready to invade it, given the right circumstances. If some *Polylepis* woodlands at the edge of the puna were to become strongly isolated and "trapped" inside the puna, for instance, then substantial faunal enrichment of the puna avifauna could follow. This would be not so much a process of colonization from the outside as one of entrapment. This faunal enrichment could conceivably occur during a period of cooling and drying of the climate, resulting in a lowering of the timberline and consequently in the greater isolation of some of these *Polylepis* woods, now located at the lower edge of the páramo-puna biome.

To give an example, a species like *Asthenes urubambensis*, now living (patchy distribution) in the wet montane and in adjacent *Polylepis* (Fig. 23–7), might become a part of the puna fauna. If *A. urubambensis* is a secondary arrival in the wet montane (because its closest relatives are puna inhabitants), its shifting to the puna would be a sort of recapture. Careful field study of habitat preferences of peripheral species, especially those found in *Polylepis*, coupled with detailed taxonomic study, might provide a clue about the origins of habitat shifts within genera whose species live primarily in open (grassland, scrub) habitats. Such a study might be especially rewarding in Venezuela, where the work of Arnal (1983) has laid the foundation for all subsequent work on *Polylepis* and associated fauna. Future work on habitat shifts in the high Andes will have to take into account Martin's (1982) analysis of shifts from forest to scrub in Corsican birds.

PÁRAMO-PUNA TAXA COLONIZING OTHER VEGETATION TYPES

If the greatest diversity of species in a genus is assumed to occur approximately where that genus originated, and if the ecologically peripheral species represent "colonization" of a new, or potentially new, habitat, then several páramo-puna genera can be considered colonists. A few hypothetical examples are given below.

1. The grebe genus *Rollandia* has perhaps only recently colonized southern South America from the high Andes, although Fjeldså (1981a, p. 227) has written: "An early northwards spread of a primitive (White-tufted-like) *Rollandia* sp. to the Titicaca area may have taken place in early Pleistocene."
2. The miner genus *Geositta* has probably invaded the lowlands several times, including perhaps *G. poeciloptera* of Brazil. The desert lowlands of western Peru have three species of *Geositta* (*G. maritima*, *G. peruviana*, and *G. cunicularia*) that could be colonists from the high Andes, especially the latter two (map in Koepcke, 1965).
3. The genus *Cinclodes* has probably invaded the maritime habitat of the Peruvian and Chilean coasts (and also extreme southern South America?) from the high Andes (Fig. 23–8).
4. *Muscisaxicola fluviatilis* appears to be a col-

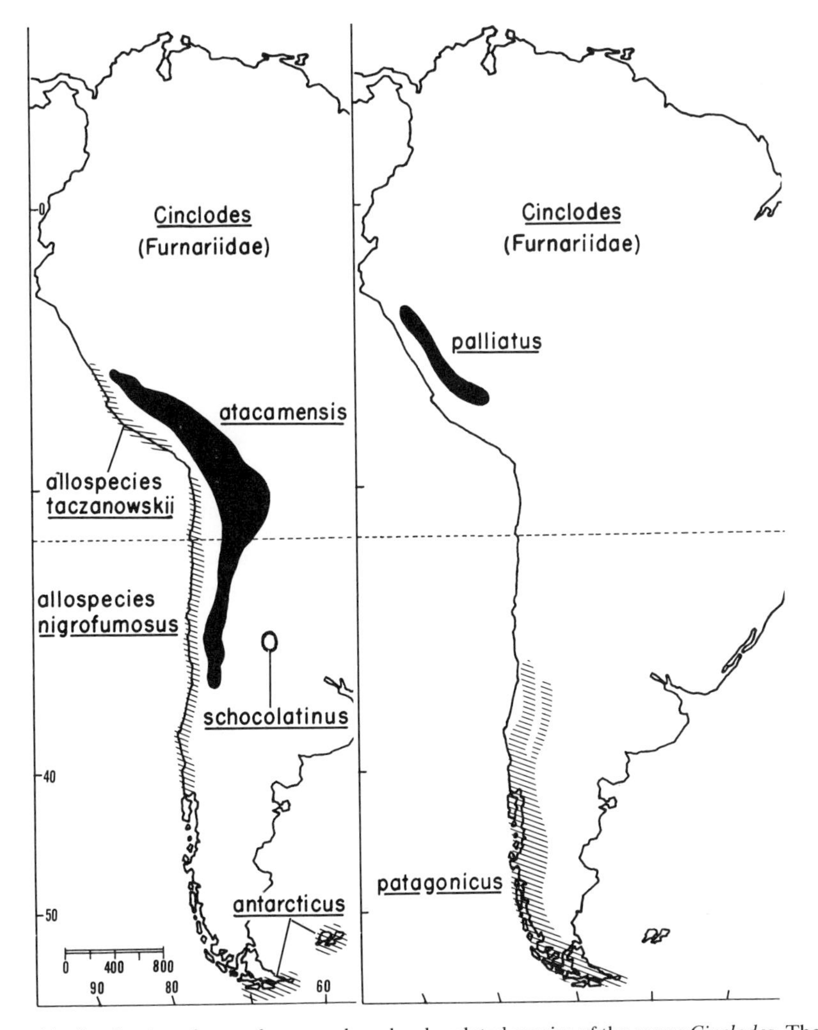

FIG. 23–8. Geographic distribution of several more or less closely related species of the genus *Cinclodes*. The coastal ("maritime") species *taczanowskii, nigrofumosus*, and *antarcticus* (*left*) and the Patagonian species *patagonicus* (*right*) may be derived from a highland stock, perhaps one ancestral to species like *palliatus* (*right*) or *atacamensis,* with its peripheral isolate *schocolatinus* (*left*), a relict from a period when distribution was more continuous (see also FIG. 23–2 [*right*] with *A.* [*wyatti*] *punensis* and similarly isolated *A.* [*wyatti*] *sclateri*).

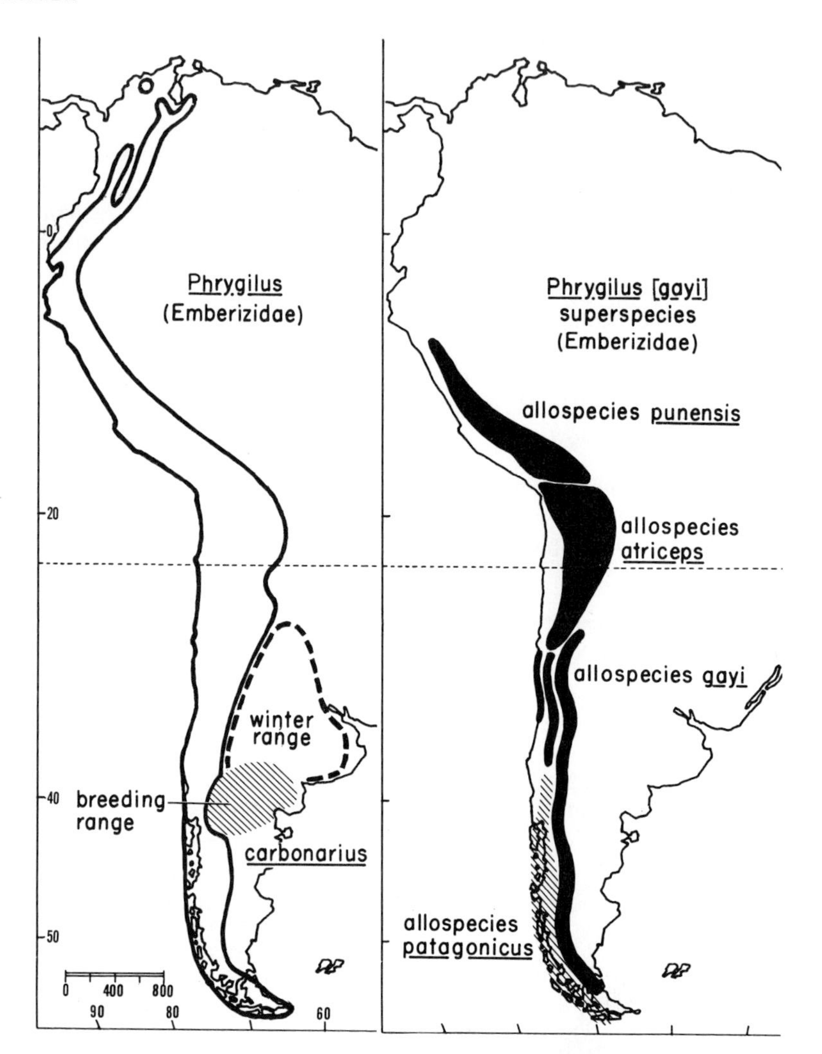

FIG. 23–9. Geographic distribution of *Phrygilus*. *Left*: *P. carbonarius* of northern Patagonia (*shaded*; lowlands) is believed to be a derivative of highland (Andean-Patagonian) stock, the present area of distribution of all other species in the genus (*circled area*). *Right*: the *P.* [*gayi*] *patagonicus* allospecies, the only taxon in the genus to live in forest (*Nothofagus* forests of Patagonia), probably derived from [*gayi*]-like stock living in open montane and steppelike vegetation.

onist of high Andean origin to the wet lowlands and foothills of the eastern Andes (Vuilleumier, 1971).

5. If *Muscigralla* is really an offshoot of proto-*Muscisaxicola* stock (Vuilleumier, 1971), it could be the result of a colonization from the Andes down to the western lowlands.
6. *Phrygilus carbonarius* could represent the result of a colonization of northern Patagonia from the Andes (Fig. 23–9).
7. *P.* [*gayi*] *patagonicus* is most certainly a colonist of *Nothofagus* forests from open Andean habitats (Fig. 23–9).

The more or less regular occurrence of puna-breeding species, either along the coast of Peru (Hughes, 1970, 1979, 1980a; Pearson and Plenge, 1974; Graves, 1981) or in the eastern lowlands of Ecuador (e.g., Norton, 1965), suggests that movements of populations of some species breeding at high altitudes, down to the tropical lowlands, take place as a result of populational or environmental pressures. The subsequent establishment of breeding populations at low altitudes, as a result of such movements, cannot be ruled out, although to my knowledge it has not been documented yet.

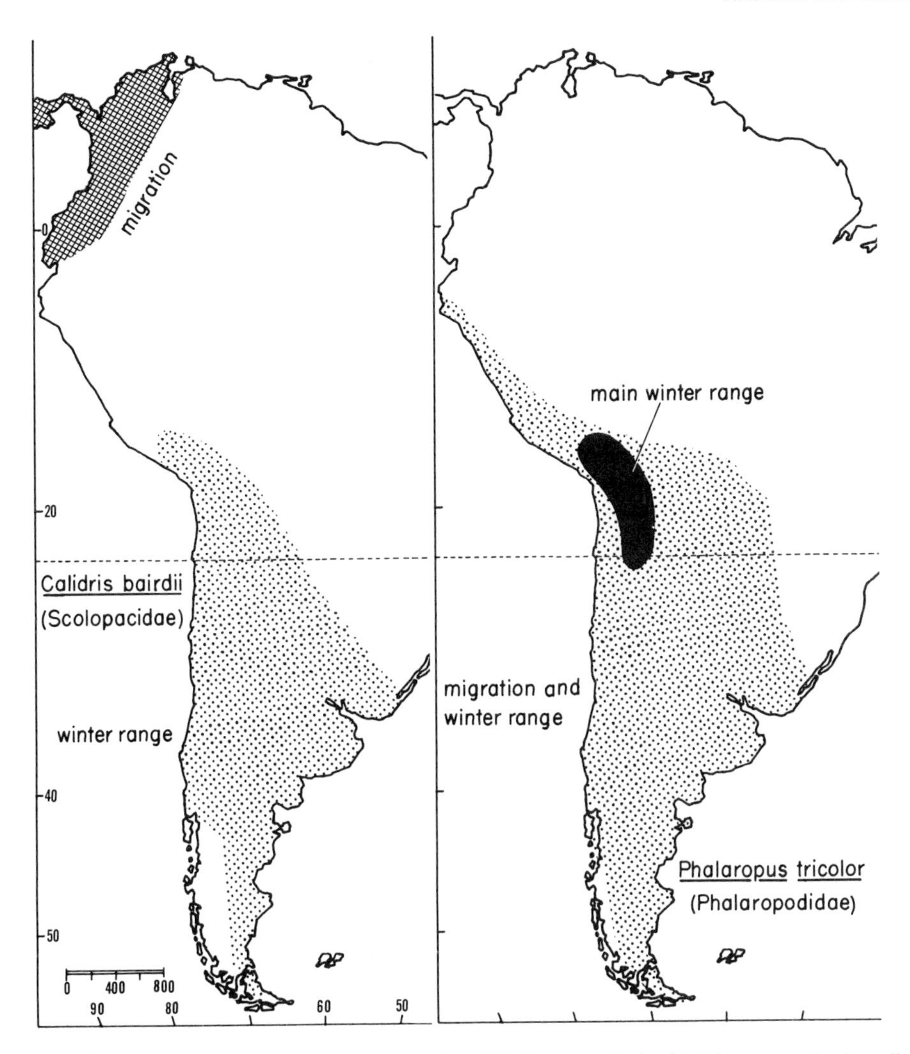

FIG. 23–10. Geographic distribution of *Calidris bairdii* (*left*) and *Phalaropus tricolor* (*right*), two species breeding in North America.

THE ORIGINS OF MIGRANT AND ACCIDENTAL SPECIES

Migrants from North America

Of the 30 species of North American migrants only about 12 (40%) can be considered to occur regularly in the páramo-puna, whereas the others are occasional only (Appendix 23–C). Of the 12 regular species, 9 are water birds, and 2 of these 9 spend much time and/or migrate in large numbers in the páramo-puna (*Calidris bairdii*, Jehl, 1979; and *Phalaropus tricolor*; Fig. 23–10). One of the North American migrant land birds, *Falco peregrinus*, also has breeding populations (subspecies *cassini*) in southern South America, and others (subspecies?) locally in the Andes (Gochfeld, 1977; see also Ellis and Glinski, 1980; Jenny, Ortiz, and Arnold, 1981). Such an amphitropical pattern is somewhat reminiscent of that of *Ciconia ciconia* in the Palearctic and in southern Africa (see, e.g., Snow, 1980, pp. 74–75). The probability of a North American migrant establishing itself as a breeding species in the páramo-puna seems relatively remote, but is not totally unlikely. Recently 17 nests of *Hirundo rustica* have been found in Albufera, Mar Chiquita, Buenos Aires Province, Argentina (M. Nores, pers. comm.). Until now this species was thought to be strictly a Northern Hemisphere migrant to South America. It seems possible that the population in Mar Chiquita is a colonization by migrant stock.

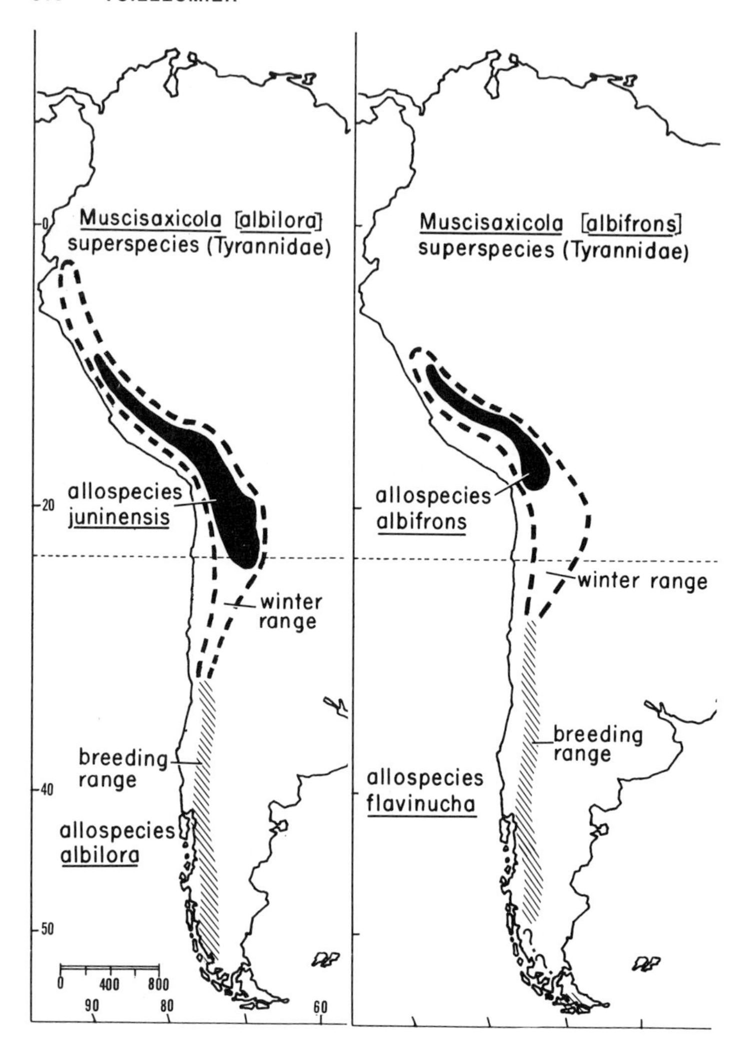

FIG. 23–11. Geographic distribution of two superspecies of *Muscisaxicola*, each having a component allospecies resident in the puna (*black*) and another, breeding in the andino-alpine (*shaded*), migrating northward to the puna and even (*albilora, left*) to the páramo.

Gallinago [*gallinago*] *gallinago*, for instance, breeds both in North and in South America. North American birds migrate to the páramos. A breeding allospecies, *andina*, is resident in the puna, and other taxa are either resident or partial migrants in the lowlands of tropical and temperate South America. The relationships and origins of the various allospecies and subspecies of the *Gallinago* [*gallinago*] superspecies (or species complex) in the Americas are in great need of detailed analysis.

Migrants from Southern South America

Only four species of *Muscisaxicola* seem to migrate regularly northward from their southern South American breeding range to the puna, and one even to the páramo: *M. capistrata*, *M.* [*albilora*] *albilora* (to páramo), *M.* [*albifrons*] *flavinucha*, and *M. frontalis*. These birds are essentially andino-alpine in their breeding distribution. Note that two of them belong to superspecies in which the other allospecies is a resident puna representative: andino-alpine *albilora* and puna *juninensis*, andino-alpine *flavinucha* and puna *albifrons* (Fig. 23–11).

In other páramo-puna–andino-alpine species partial migrations exist, and in several instances the exact breeding and nonbreeding ranges of some of the species with a resident puna population have not yet been satisfactorily determined. Table 23–14 shows the origins of migrants to the páramo-puna.

Accidentals from South America

Thirty-five species have been recorded as accidentals in the páramo-puna (Appendix 23–D).

TABLE 23–14. Origins of migrants and accidentals to páramo and puna.

Origins (in % of each faunal component)	Faunal component		
	Migrants (*N*=35)	Accidentals (*N*=35)	Total (*N*=70)
Local (high Andes)	—	26	13
Andes or Patagonia	—	—	—
Patagonia	14	3	8.5
South America (other than Andes and Patagonia)	—	71	35.5
North America	86	—	43
Cosmopolitan	—	—	—
Unknown	—	—	—
Total	100	100	100

Twenty-two of them (63%) are water birds. Six are accidental to the páramo, 24 to the puna, and 5 to both. The origins of these strays can be determined relatively easily (see Table 23–14).

The majority (25, or 71%) are tropical lowland birds, 9 (26%) live in the wet or the dry montane, and 1 (3%) breeds in southern South America. Most of the lowlands strays (17 of 25 or 68%; or 48% of the total number of accidentals) come from the eastern tropical lowlands. Of the 9 montane strays, 5 (56% of montane strays, 14% of total) occur in both the wet and the dry montane.

Accidentals are taxonomically quite diverse (grebes, anhinga, herons, storks, ducks, rails, jacana, tern, skimmer, pigeon, parakeet, nightjar, swifts, flycatcher, swallows, grosbeak, conebill and flowerpiercer). Some, but not all of these birds, are good fliers. The jacana, *Jacana spinosa* (Vuilleumier, 1981), is of interest, as is also the eastern tropical lowland species of ground-tyrant, *Muscisaxicola fluviatilis*, recorded occasionally in the puna (Vuilleumier, 1971; Olrog, 1979; Traylor, 1979a). The possibility of a population of *M. fluviatilis* breeding locally in the puna is not to be discarded.

The number of accidentals (35 out of a total of 319 species, or 11%) is relatively high, and will certainly grow as more ornithologists study birds in the páramo-puna. The likelihood of a flock of accidental birds establishing itself in the páramo-puna after having originated in the tropical lowlands is rather low, and depends, of course, on the birds finding the right kind of habitat in which to breed. H.-W. Koepcke (1963, pp. 403–404) mentioned the influx of species like *Ardea alba* in rather large numbers on the shores of high Andean lakes during periods of high water in the Amazonian lowlands. Natural fluctuations in some crucial environmental variable might thus "force" some birds upward in the Andes, just as high Andean species can be "forced" down by adverse conditions (severe snow storms) very high up (O'Neill and Parker, 1978).

DISCUSSION AND CONCLUSIONS

Table 23–12 summarizes the origins of páramo and puna bird species. In column 4 (total), the "true" páramo-puna avifauna of 166 species (Appendix A) is represented, pooling all three categories referred to in discussions of endemic species, nearly endemic species, and nonendemic species. Column 6 (grand total) includes all 248 species comprising the páramo and puna avifauna, including the peripheral species.

Many species of tropical high Andean birds appear to have originated in situ, in other words either within the páramo and puna biomes, or very close to the present location of these biomes: 62 of 166 species (37.5%) and 119 of the grand total of 248 species (48%) are of local origin. This conclusion was reached after an analysis of the present distributional area of the nearest relative(s) of each species, and after taking into account patterns of speciation as revealed by the presence of superspecies and as shown by range disjunctions within and between superspecies and species. Much local evolution has evidently occurred within the páramo-puna biomes.

The next group of species that stand out after those of local origin are those that originated either in the Andes or in Patagonia. For the purposes of this discussion they can be pooled and called Patagonian: thus 42 of 166 species (or 25%), and 48 of the grand total of 248 species (19%) are Patagonian in origin.

Of the remaining species, about 11% probably originated in South America outside the Andean-Patagonian axis, 10.5%–12% either in North or in South America. Finally, 4%–6% are cosmopolitan, and 7.5%–8.5% are of unknown origins. If the two categories "South American" and "North or South American" are pooled, on the assumption that these species are likely to have originated in extra-Andean South America, then 21.4%–23% of páramo-puna birds are South American.

Summarizing the data presented in this chapter and the speculations leading to the construction of Table 23–12, the origins of tropical high Andean birds can be said to have three main loci: the high Andes themselves (37.5%–48% of the species), Patagonia (19%–25%), and South America (21.5%–23%). Long ago, Chapman emphasized the contribution of Patagonia to the origin of tropical high Andean birds. But Chapman's data, although incomplete and somewhat difficult to interpret or to duplicate, for reasons given earlier in this chapter, also revealed other sources of origin. Thus qualitatively at least, my results confirm some of Chapman's results. But more detailed comparisons between the present data and Chapman's are futile. Whereas Chapman only discussed páramo birds in Ecuador, and puna species in part of Peru, I have dealt with the entire páramo-puna avifauna.

Chapman did not find much faunal similarity between páramo or puna birds and birds found in the dry or the wet montane. The results of my analyses of faunal resemblances suggest that faunal exchanges have taken place between the páramo-puna and the tropical montane biomes surrounding them at slightly lower altitudes. My analysis furthermore reveals the presence of a large pool of species living at the edge of the páramo-puna in peripheral situations, and occupying the biogeographically important *Polylepis* woodlands (and also to a lesser extent the stands of giant *Puya raimondii*). These species offer indirect evidence on the possible faunal interchanges between páramo-puna and surrounding vegetation types.

Finally, my analysis shows that a number of range disjunctions involving populations of species in the páramo-puna and other populations of the same species in the tropical lowlands are due to dynamic, probably relatively recent, range expansions. Most of these range expansions occurred from the lowlands up to the highlands, but some of them were in the other direction.

The present analysis was carried out chiefly at the species level. Genus-level analyses should also be undertaken. The detailed taxonomic analysis of interspecific relationships is only in its beginning. Further morphological studies must be carried out, supplemented with biochemical studies such as enzyme electrophoresis (for a beginning, see Braun and Parker [1985] on *Synallaxis*). The problem of the ecological occupation of space and the utilization of ecological resources by birds during colonization and after successful implantation in the high Andes needs to be studied critically. What trophic zones were available to colonists when the high Andes first became available for occupation? How did adaptive radiation proceed in the now most specialized páramo-puna groups? In some groups no radiation took place, for instance the flickers *Colaptes* (Short, 1971). The fossil and subfossil material present in archaeological sites (Wheeler, pers. comm.) may reveal the presence of taxa that are no longer living in the high Andes, or may show trends of morphological variation (see Wing, Chap. 10, this volume). The work of Matthiesen (pers. comm.) on avian remains from archaeological sites in and around the Sahara shows the potential of this approach in the Andes, with its rich archaeological source of documents.

SUMMARY

This chapter presents the first attempt to analyze the origins of all bird species found in the páramo and puna biomes of the high tropical Andes. Of the 319 species of birds that have been found in the páramo and/or the puna, 166 can be considered the core breeding group. Another 82 species breed at the periphery. Together they make up the breeding páramo-puna avifauna of 248 species. The remainder of the species are either migrants from North America (30 species), migrants from South America (5 species), accidentals from South America (35 species), or introduced (1 species). No family of bird is endemic and only about 4 genera (5%) are endemic to the páramo-puna. A much larger percentage of species are endemic (48 of 166 or 29%). Another 21 species (13%) are "nearly endemic." Thus species-level endemism is relatively high. Much species-level origination in the high tropical Andean avifauna occurred in situ. Other main geographic sources are Patagonia and non-Andean South America. Secondary sources are extra-South American. There is evidence of continued emigration today and of some outflow from the high Andes as well. Exchanges with the faunas of neighboring tropical montane biomes at lower altitudes have been or are relatively important.

ACKNOWLEDGMENTS

I thank my friends Maximina Monasterio for her encouragement with this work, help with field studies, and for critically reading the manuscript, and Osvaldo A. Reig for opportunities to discuss my work with him and other South

American colleagues. Detailed acknowledgments for much of the help received during fieldwork and for financial assistance during my research on Andean birds can be found in Vuilleumier and Simberloff (1980). I wish to acknowledge the work of the late Rodolphe Meyer de Schauensee, who, together with the late Eugene Eisenmann, produced the seminal checklist of South American birds that today serves as the basis for all biogeographic syntheses. I would like to acknowledge also the field work of the staff of the Museum of Zoology at Louisiana State University, thanks to whom much new and important distributional information has been gathered in the last 20 years. I thank Manuel Nores, who gave me important information about the breeding of *Hirundo rustica* in Argentina. The manuscript was typed by Mary Ardagna and the illustrations were drafted by Juan Barberis, both of whom I warmly thank for their collaboration.

APPENDIX 23–A

List of 166 species of birds breeding in the páramo and puna biomes. Square brackets indicate superspecies. For data on taxonomy, distribution, ecological preferences, and breeding, see references marked with an asterisk. E, endemic; Near E, nearly endemic; Non-E, nonendemic species. For Near E and Non-E species: A, Andean distribution; AP, Andean-Patagonian distribution; SSA, southern South American distribution; WSA, widespread South American distribution; PA, Pan-American distribution; C, cosmopolitan distribution.

Rheidae
1. *Pterocnemia pennata* (Non-E); AP)

Tinamidae
2. *Nothura ornata* (E)
3. *Nothura curvirostris* (Near E; A)
4. *Nothura darwinii* (Non-E; AP)
5. *Tinamotis [pentlandii] pentlandii* (E)

Podicipedidae
6. *Rollandia rolland* (Non-E; SSA)
7. *Rollandia microptera* (E)
8. *Podiceps [nigricollis] occipitalis* (Non-E; AP)
9. *Podiceps [nigricollis] taczanowskii* (E)

Phalacrocoracidae
10. *Phalacrocorax [auritus] olivaceus* (Non-E; WSA)

Ardeidae
11. *Nycticorax [nycticorax] nycticorax* (Non-E; C)

Threskiornithidae
12. *Plegadis [?falcinellus] ridgwayi* (E)
13. *Theristicus [?caudatus] melanopis* (Non-E; AP)

Phoenicopteridae
14. *Phoenicopterus chilensis* (Non-E; SSA)
15. *Phoenicopterus andinus* (E)
16. *Phoenicopterus jamesi* (E)

Cathartidae
17. *Cathartes aura* (Non-E; PA)
18. *Vultur gryphus* (Non-E; AP)

Accipitridae
19. *Circus [cyaneus] cinereus* (Non-E; WSA)
20. *Geranoaetus melanoleucus* (Non-E; AP)
21. *Buteo [?albicaudatus] polyosoma* (Non-E AP)
22. *Buteo [?albicaudatus] poecilochrous* (E)

Falconidae
23. *Polyborus [megalopterus] carunculatus* (Non-E; A)
24. *Polyborus [megalopterus] megalopterus* (Non-E; A)
25. *Falco [sparverius] sparverius* (Non-E; PA)
26. *Falco femoralis* (Non-E; PA)

Anatidae
27. *Chloephaga melanoptera* (Near E; A)
28. *Anas [crecca] flavirostris* (Non-E; SSA)
29. *Anas specularioides* (Non-E; AP)
30. *Anas [acuta] georgica* (Non-E; SSA)
31. *Anas versicolor* (Non-E; SSA)
32. *Anas cyanoptera* (Non-E; PA)
33. *Oxyura jamaicensis* (Non-E; PA)

Rallidae
34. *Rallus [nigricans] sanguinolentus* (Non-E; SSA)
35. *Laterallus jamaicensis* (Non-E; PA)
36. *Gallinula chloropus* (Non-E; C)
37. *Fulica [?atra] ardesiaca* (Near E; A)
38. *Fulica gigantea* (E)
39. *Fulica cornuta* (E)

Charadriidae
40. *Vanellus [chilensis] resplendens* (Near E; A)
41. *Charadrius [falklandicus] alticola* (E)
42. *Oreopholus ruficollis* (Non-E; AP)
43. *Phegornis mitchellii* (Non-E; A)

Scolopacidae
44. *Gallinago stricklandii* (Non-E; A)
45. *Gallinago nobilis* (Near E; A)
46. *Gallinago [gallinago] andina* (E)

Recurvirostridae
47. *Recurvirostra [avosetta] andina* (E)
48. *Himantopus [himantopus] melanurus* (Non-E; SSA)

Thinocoridae
49. *Attagis [malouinus] gayi* (Non-E; A)
50. *Thinocorus rumicivorus* (Non-E; AP)
51. *Thinocorus orbignyianus* (Non-E; AP)

Laridae
52. *Larus serranus* (Near E; A)

Columbidae

53. *Zenaida [macroura] auriculata* (Non-E; WSA)
54. *Metriopelia ceciliae* (Non-E; A)
55. *Metriopelia morenoi* (Non-E; A)
56. *Metriopelia aymara* (Near E; A)
57. *Metriopelia melanoptera* (Non-E; AP)

Psittacidae

58. *Bolborhynchus aurifrons* (Non-E; A)

Tytonidae

59. *Tyto [alba] alba* (Non-E; C)

Strigidae

60. *Bubo [bubo] virginianus* (Non-E; PA)
61. *Athene cunicularia* (Non-E; PA)
62. *Asio flammeus* (Non-E; C)

Caprimulgidae

63. *Caprimulgus longirostris* (Non-E; WSA)

Trochilidae

64. *Colibri coruscans* (Non-E; WSA)
65. *Oreotrochilus [estella] chimborazo* (E)
66. *Oreotrochilus [estella] estella* (Near E; A)
67. *Oreotrochilus [estella] leucopleurus* (Non-E; A)
68. *Oreotrochilus melanogaster* (E)
69. *Patagona gigas* (Non-E; A)
70. *Ramphomicron dorsale* (Near E; A)
71. *Chalcostigma olivaceum* (E)
72. *Chalcostigma stanleyi* (Near E; A)
73. *Chalcostigma heteropogon* (E)
74. *Oxypogon guerinii* (E)

Picidae

75. *Colaptes rupicola* (Near E; A)

Furnariidae

76. *Schizoeaca [fuliginosa] coryi* (Near E; A)
77. *Schizoeaca [fuliginosa] fuliginosa* (Near E; A)
78. *Phleocryptes melanops* (Non-E; SSA)
79. *Leptasthenura yanacensis* (Near E; A)
80. *Leptasthenura aegithaloides* (Non-E; SSA)
81. *Leptasthenura andicola* (E)
82. *Asthenes modesta* (Non-E; AP)
83. *Asthenes [dorbignyi] dorbignyi* (Non-E; A)
84. *Asthenes humilis* (E)
85. *Asthenes [wyatti] wyatti* (E)
86. *Asthenes [wyatti] punensis* (E)
87. *Asthenes [flammulata] flammulata* (E)
88. *Asthenes [flammulata] maculicauda* (E)
89. *Upucerthia [validirostris] jelskii* (E)
90. *Upucerthia [validirostris] validirostris* (E)
91. *Upucerthia albigula* (Non-E; A)
92. *Upucerthia ruficauda* (Non-E; AP)
93. *Upucerthia andaecola* (E)
94. *Upucerthia serrana* (Non-E; A)
95. *Upucerthia dumetaria* (Non-E; AP)
96. *Cinclodes [fuscus] fuscus* (Non-E; AP)
97. *Cinclodes excelsior* (E)
98. *Cinclodes atacamensis* (Near E; A)
99. *Cinclodes palliatus* (E)
100. *Geositta punensis* (E)
101. *Geositta [cunicularia] cunicularia* (Non-E; AP)
102. *Geositta saxicolina* (E)
103. *Geositta tenuirostris* (E)

Formicariidae

104. *Grallaria andicola* (Near E; A)
105. *Grallaria quitensis* (Non-E; A)

Rhinocryptidae

106. *Teledromas fuscus* (Non-E; AP)

Tyrannidae

107. *Anairetes parulus* (Non-E; AP)
108. *Tachuris rubrigastra* (Non-E; SSA)
109. *Ochthoeca fumicolor* (Non-E; A)
110. *Ochthoeca oenanthoides* (Near E; A)
111. *Ochthoeca leucophrys* (Non-E; A)
112. *Xolmis erythropygius* (Near E; A)
113. *Xolmis rufipennis* (Non-E; A)
114. *Agriornis montana* (Non-E; AP)
115. *Agriornis andicola* (E)
116. *Agriornis microptera* (Non-E; AP)
117. *Muscisaxicola maculirostris* (Non-E; AP)
118. *Muscisaxicola rufivertex* (Non-E; A)
119. *Muscisaxicola [albilora] juninensis* (E)
120. *Muscisaxicola [alpina] alpina* (E)
121. *Muscisaxicola [alpina] cinerea* (Non-E; A)
122. *Muscisaxicola [albifrons] albifrons* (E)
123. *Lessonia [rufa] oreas* (E)

Hirundinidae

124. *Notiochelidon murina* (Non-E; A)
125. *Petrochelidon andecola* (Near E; A)

Troglodytidae

126. *Cistothorus [platensis] platensis* (Non-E; PA)
127. *Cistothorus [platensis] meridae* (E)
128. *Cistothorus [platensis] apolinari* (Non-E; A)
129. *Troglodytes aedon* (Non-E; AP)

Mimidae

130. *Mimus [triurus] dorsalis* (Non-E; A)

Turdidae

131. *Turdus chiguanco* (Non-E; A)
132. *Turdus fuscater* (Non-E; A)

Motacillidae

133. *Anthus [furcatus] furcatus* (Non-E; SSA)
134. *Anthus hellmayri* (Non-E; SSA)
135. *Anthus correndera* (Non-E; SSA)
136. *Anthus bogotensis* (E)

Icteridae

137. *Agelaius thilius* (Non-E; SSA)
138. *Sturnella magna* (Non-E; PA)

Emberizidae

139. *Zonotrichia capensis* (Non-E; WSA)
140. *Phrygilus [gayi] punensis* (Non-E; A)
141. *Phrygilus [gayi] atriceps* (Non-E; A)
142. *Phrygilus [gayi] gayi* (Non-E; AP)

143. *Phrygilus fruticeti* (Non-E; A)
144. *Phrygilus unicolor* (Non-E; A)
145. *Phrygilus [erythronotus] erythronotus* (E)
146. *Phrygilus [erythronotus] dorsalis* (E)
147. *Phrygilus plebejus* (Non-E; A)
148. *Phrygilus alaudinus* (Non-E; A)
149. *Diuca [diuca] speculifera* (E)
150. *Idiopsar brachyurus* (E)
151. *Sicalis lutea* (E)
152. *Sicalis uropygialis* (E)
153. *Sicalis olivascens* (Non-E; A)
154. *Catamenia analis* (Non-E; A)
155. *Catamenia inornata* (Near E; A)
156. *Catamenia homochroa* (Non-E; WSA)

Thraupidae

157. *Xenodacnis parina* (E)
158. *Oreomanes fraseri* (E)
159. *Diglossa [carbonaria] gloriosa* (Non-E; A)
160. *Diglossa [carbonaria] humeralis* (Non-E; A)
161. *Diglossa [carbonaria] carbonaria* (Non-E; A)

Carduelidae

162. *Carduelis spinescens* (Near E; A)
163. *Carduelis crassirostris* (Non-E; A)
164. *Carduelis [magellanica] magellanica* (Non-E; WSA)
165. *Carduelis atrata* (E)
166. *Carduelis uropygialis* (Non-E; A)

APPENDIX 23–B

List of 82 species of birds living peripherally to the páramo and puna biomes. Square brackets indicate superspecies. For data on taxonomy, distribution, ecological preferences, and breeding, see references marked with an asterisk. DM, dry montane; WM, wet montane.

Tinamidae

1. *Nothura taczanowskii* (DM)
2. *Nothura kalinowskii* (DM)
3. *Nothura [?perdicaria] pentlandii* (DM)

Anatidae

4. *Merganetta armata* (WM; locally also DM)

Scolopacidae

5. *Gallinago imperialis* (WM)

Columbidae

6. *Columba maculosa* (DM)

Psittacidae

7. *Bolborhynchus [?orbygnesius] ferrugineifrons* (WM)
8. *Bolborhynchus [?orbygnesius] orbygnesius* (WM)
9. *Bolborhynchus aymara* (DM)

Strigidae

10. *Otus roboratus* (DM)

Apodidae

11. *Aeronautes andecolus* (DM)

Trochilidae

12. *Lafresnaya lafresnayi* (WM)
13. *Oreotrochilus adela* (DM)
14. *Aglaeactis cupripennis* (WM)
15. *Coeligena iris* (WM)
16. *Heliangelus viola* (WM)
17. *Eriocnemis vestitus* (WM)
18. *Eriocnemis cupreoventris* (WM)
19. *Eriocnemis mosquera* (WM)
20. *Lesbia nuna* (DM)
21. *Sappho sparganura* (DM)
22. *Metallura phoebe* (WM)
23. *Metallura williami* (WM)
24. *Metallura tyrianthina* (DM)
25. *Chalcostigma herrani* (WM)

Picidae

26. *Melanerpes cactorum* (DM)
27. *Veniliornis fumigatus* (WM)
28. *Piculus rivolii* (WM)

Furnariidae

29. *Margarornis squamiger* (WM)
30. *Synallaxis azarae* (WM)
31. *Cranioleuca antisiensis* (WM)
32. *Cranioleuca albicapilla* (WM)
33. *Schizoeaca [fuliginosa] perijana* (WM)
34. *Schizoeaca [fuliginosa] palpebralis* (WM)
35. *Schizoeaca [fuliginosa] helleri* (WM)
36. *Schizoeaca [fuliginosa] harterti* (WM)
37. *Leptasthenura striata* (DM)
38. *Leptasthenura [pileata] pileata* (DM)
39. *Leptasthenura [pileata] xenothorax* (WM)
40. *Asthenes [pudibunda] pudibunda* (DM)
41. *Asthenes [pudibunda] ottonis* (DM)
42. *Asthenes [pudibunda] heterura* (DM)
43. *Asthenes [dorbignyi] berlepschi* (DM)
44. *Asthenes urubambensis* (WM)
45. *Phacellodomus striaticeps* (DM)
46. *Geositta rufipennis* (DM)

Rhinocryptidae

47. *Scytalopus superciliaris* (WM)
48. *Scytalopus magellanicus* (WM)

Tyrannidae

49. *Mecocerculus leucophrys* (WM)
50. *Anairetes alpinus* (WM)
51. *Anairetes [?reguloides] nigrocristatus* (DM)
52. *Anairetes flavirostris* (DM)
53. *Ochthoeca pulchella* (WM)
54. *Ochthoeca rufipectoralis* (WM)
55. *Xolmis striaticollis* (DM)

Cotingidae

56. *Ampelion rubrocristatus* (WM)
57. *Ampelion stresemanni* (DM)

Phytotomidae
58. *Phytotoma* [*rutila*] *rutila* (DM)
Hirundinidae
59. *Notiochelidon cyanoleuca* (WM; locally also DM)
Corvidae
60. *Cyanolyca turcosa* (WM)
Cinclidae
61. *Cinclus* [*leucocephalus*] *leucocephalus* (WM)
62. *Cinclus* [*leucocephalus*] *schultzii* (WM)
Troglodytidae
63. *Troglodytes solstitialis* (WM)
Icteridae
64. *Molothrus badius* (DM)
Emberizidae
65. *Incaspiza pulchra* (DM)
66. *Incaspiza personata* (DM)
67. *Poospiza alticola* (DM)
68. *Poospiza hypochondria* (DM)
69. *Poospiza* [*garleppi*] *garleppi* (DM)
70. *Poospiza* [*garleppi*] *baeri* (DM)
71. *Atlapetes citrinella* (WM)
72. *Atlapetes schistaceus* (WM)
73. *Atlapetes nationi* (DM)
74. *Atlapetes rufigenis* (WM)
Cardinalidae
75. *Saltator* [*aurantiirostris*] *aurantiirostris* (DM)
Thraupidae
76. *Hemispingus superciliaris* (WM)
77. *Thraupis bonariensis* (DM)
78. *Anisognathus lacrymosus* (WM)
79. *Tangara vassorii* (WM)
80. *Conirostrum tamarugense* (DM)
81. *Conirostrum sitticolor* (WM)
82. *Conirostrum ferrugineiventre* (WM)

APPENDIX 23–C

List of species of migrants from North and South America to the páramo and puna biomes. RMNA, regular migrant from North America; OMNA, occasional migrant from North America; OMSA, occasional migrant from South America; RMSA, regular migrant from southern South America. Square brackets indicate superspecies.

Pandionidae
1. *Pandion haliaetus* (OMNA)
Falconidae
2. *Falco* [*peregrinus*] *peregrinus* (RMNA)
Anatidae
3. *Anas discors* (RMNA)
Rallidae
4. *Porzana carolina* (OMNA)
Charadriidae
5. *Charadrius vociferus* (OMNA)
6. *Pluvialis dominica* (OMNA)
7. *Pluvialis squatarola* (OMNA)
Scolopacidae
8. *Arenaria* [*interpres*] *interpres* (OMNA)
9. *Gallinago* [*gallinago*] *gallinago* (RMNA)
10. *Numenius phaeopus* (RMNA)
11. *Bartramia longicauda* (OMNA)
12. *Catoptrophorus semipalmatus* (OMNA)
13. *Tringa* [*incana*] *incana* (OMNA)
14. *Tringa* [*hypoleucos*] *macularia* (RMNA)
15. *Tringa* [*ochropus*] *solitaria* (RMNA)
16. *Tringa* [*nebularia*] *melanoleuca* (RMNA)
17. *Tringa flavipes* (RMNA)
18. *Calidris bairdii* (RMNA)
19. *Calidris melanotos* (OMNA)
20. *Calidris alba* (OMNA)
21. *Micropalama himantopus* (OMNA)
Phalaropodidae
22. *Phalaropus tricolor* (RMNA)
Laridae
23. *Larus pipixcan* (OMNA)
Cuculidae
24. *Coccyzus* [*americanus*] *americanus* (OMNA)
Tyrannidae
25. *Muscisaxicola macloviana* (OMSA)
26. *Muscisaxicola capistrata* (RMSA)
27. *Muscisaxicola* [*albilora*] *albilora* (RMSA)
28. *Muscisaxicola* [*albifrons*] *flavinucha* (RMSA)
29. *Muscisaxicola frontalis* (RMSA)
Hirundinidae
30. *Progne* [*subis*] *subis* (OMNA)
31. *Riparia riparia* (OMNA)
32. *Hirundo* [*rustica*] *rustica* (RMNA)
33. *Petrochelidon pyrrhonota* (RMNA)
Turdidae
34. *Catharus ustulatus* (OMNA)
Icteridae
35. *Dolichonyx oryzivorus* (OMNA)

APPENDIX 23–D

List of South American species that are accidental to the páramo and puna biomes. Square brackets indicate superspecies.

Podicipedidae
1. *Tachybaptus dominicus*
2. *Podilymbus* [*podiceps*] *podiceps*

Anhingidae
3. *Anhinga* [*anhinga*] *anhinga*
Ardeidae
4. *Ardea* [*cinerea*] *cocoi*
5. *Ardea alba*
6. *Egretta ibis*
7. *Egretta caerulea*
8. *Egretta* [*garzetta*] *thula*
9. *Ardeola striata*
Ciconiidae
10. *Mycteria americana*
11. *Jabiru mycteria*
Anatidae
12. *Dendrocygna* [*bicolor*] *bicolor*
13. *Dendrocygna viduata*
14. *Anas* [*bahamensis*] *bahamensis*
15. *Anas platalea*
16. *Netta erythrophthalma*
Rallidae
17. *Rallus* [*nigricans*] *nigricans*
18. *Rallus semiplumbeus*
19. *Porphyrula martinica*
Jacanidae
20. *Jacana spinosa*
Laridae
21. *Sterna simplex*
22. *Rynchops* [*niger*] *niger*
Columbidae
23. *Columba* [*fasciata*] *fasciata*
Psittacidae
24. *Aratinga wagleri*
Caprimulgidae
25. *Uropsalis segmentata*
Apodidae
26. *Streptoprocne zonaris*
27. *Cypseloides* [*rutilus*] *rutilus*
Tyrannidae
28. *Muscisaxicola fluviatilis*
Hirundinidae
29. *Tachycineta* [*leucorrhoa*] *leucopyga*
30. *Progne* [*subis*] *modesta*
31. *Stelgidopteryx ruficollis*
32. *Petrochelidon fulva*
Cardinalidae
33. *Pheucticus* [*chrysopeplus*] *chrysopeplus*
Thraupidae
34. *Conirostrum cinereum*
35. *Diglossa* [*baritula*] *sittoides*

REFERENCES

Asterisk indicates works containing taxonomic and distributional information.

Amadon, D. 1966. The superspecies concept. *Syst. Zool.* 15: 245–249.

*———. 1982. A revision of the sub-buteonine hawks (Accipitridae, Aves). *Amer. Mus. Novit.* 2741: 1–20.

Arnal, H. 1983. Estudio ecológico del bosque altiandino de *Polylepis sericea* en la Cordillera de Mérida. Tesis de Grado, Fac. Ciencias, Univ. Andes. Mérida, Venezuela.

*Behn, F., and G. Millie. 1959. Beitrag zur Kenntnis des Rüssel-blässhuhns (*Fulica cornuta* Bonaparte). *J. f. Ornithol.* 100: 119–131.

Berger, M. 1978. Ventilation in the humming birds *Colibri coruscans* during altitude hovering. In *Respiratory function in birds, adult and embryonic*, J Piiper, ed., pp. 85–88. New York: Springer-Verlag.

Bergier, P., and G. Cheylan. 1980. Statut, succès de reproduction et alimentation du Vautour percnoptère *Neophron percnopterus* en France méditerranéenne. *Alauda* 48: 75–97.

*Berlioz. J. 1974. Considérations sur le peuplement alticole des Colibris (Trochilidés) en Colombie. *C. R. Séances Soc. Biogéogr.* 50 (434–439): 16–19.

*Blake, E. R. 1977. *Manual of neotropical birds*, Vol. 1. *Spheniscidae (penguins) to Laridae (gulls and allies)*. Chicago: Univ. of Chicago Press.

*———. 1979. Order Tinamiformes. In *Check-list of birds of the world*, E. Mayr and G. W. Cottrell, eds., Vol. I, 2d ed. pp. 12–47. Cambridge, Mass: Mus. Comp. Zool.

*Bock, W. J. 1958. A generic review of the plovers (Charadriidae: Aves). *Bull. Mus. Comp. Zool.* 118: 27–97.

*Bond, J., and R. Meyer de Schauensee. 1942. The birds of Bolivia. Part I. *Proc. Acad. Nat. Sci. Philadelphia* 94: 307–391.

*———. 1943. The birds of Bolivia. Part II. *Proc. Acad. Nat. Sci. Philadelphia* 95: 167–221.

Bourlière, F. 1957. Un curieux biotope d'altitude: les paramos des Andes de Colombie. *Terre et Vie* 104: 297–304.

Braun, J. and T. A. Parker III. 1985. Molecular, morphological, and behavioral evidence concerning the taxonomic relationships of "*Synallaxis*" *gularis* and other synallaxines. In *Neotropical ornithology*, P. A. Buckley, M. S. Foster, E. S. Morton, R. S. Ridgely, and F. G. Buckley, eds. Amer. Ornithol. Union Monogr. No. 36: 333–346.

*Brokaw, H. P. 1976. Birds of Pampa Galeras, Peru. *Delmarva Ornithologist* 11: 26–30.

*Butler, T. Y. 1979. *The birds of Ecuador and the Galapagos Archipelago.* Portsmouth, N. H.: The Ramphastos Agency, P. O. Box 1091.

*Cabot, J., and P. Serrano. 1982. La comunidad de aves. In *Vegetación y fauna de la Reserva Nacional Altoandina "Eduardo Avaroa," Prov. Sud Lipez, Potosí, Bolivia.* In H. Alzerreca A., ed., pp. 40–60. La Paz, Bolivia: Oficina de Asesoria Agropecuario del INFOL.

Cabrera, A. L. 1968. Ecología vegetal de la puna. *Colloq. Geogr.* 9: 91–116.

Carey, C., and M. L. Morton. 1976. Aspects of circulatory physiology of montane and lowland birds. *Comp. Biochem. Physiol.* 54A: 61–74.

Carpenter, F. L. 1975. Bird hematocrits: Effects of high

altitude and strength of flight. *Comp. Biochem. Physiol.* 50A: 415–417.

———. 1976. Ecology and evolution of an Andean hummingbird (*Oreotrochilus estella*). *Univ. Calif. Publ. Zool.* 106: 1–74.

Chapman, F. M. 1917. The distribution of bird-life in Colombia; a contribution to a biological survey of South America. *Bull. Amer. Mus. Nat. Hist.* 36: i–x, 1–729.

———. 1921. The distribution of bird life in the Urubamba Valley of Peru. *Bull. U.S. National Mus.* 117: 1–138.

———. 1926. The distribution of bird-life in Ecuador. A contribution to the study of the origin of Andean bird-life. *Bull. Amer. Mus. Nat. Hist.* 55: i–xiii, 1–784.

———. 1927. The variations and distribution of *Saltator aurantiirostris*. *Amer. Mus. Novit.* 26: 1–19.

———. 1940. The post-glacial history of *Zonotrichia capensis*. *Bull. Amer. Mus. Nat. Hist.* 77: 381–438.

*Contreras, J. R. 1980. Furnariidae argentinos. I. Nuevos datos sobre *Thripophaga modesta navasi* y algunas consideraciones sobre *Thripophaga modesta* en la Argentina. *Historia Nat.* (Mendoza) 1: 49–68.

Corley Smith, G. T. 1969. A high altitude hummingbird on the volcano Cotopaxi. *Ibis* 111: 17–22.

Cuatrecasas, J. 1968. Paramo vegetation and its life forms. *Colloq. Geogr.* 9: 163–186.

Diamond, J. M. 1973. Distributional ecology of New Guinea birds. *Science* 179: 759–769.

Dorst, J. 1955. Recherches écologiques sur les oiseaux des hauts plateaux péruviens. *Travaux Inst. Français Études Andines* 5: 83–140.

———. 1957. The *Puya* stands of the Peruvian high plateaux as a bird habitat. *Ibis* 99: 594–599.

———. 1963. Note sur la nidification et le comportement acoustique du jeune *Asthenes wyatti punensis* (Furnariidés) au Pérou. *L'Oiseau et la Rev. Franç. Ornithol.* 33: 1–6.

———. 1967. Considérations zoogéographiques et écologiques sur les oiseaux des hautes Andes. In *Biologie de l'Amérique Australe*, C. Delamare-Deboutteville and E. Rapoport, eds., Vol. III, pp. 471–504. Paris: C.N.R.S.

———. 1972. Poids relatif du coeur chez quelques oiseaux des hautes Andes du Pérou. *L'Oiseau et la Rev. Franç. Ornithol.* 42: 66–73.

*Dorst, J., and J.-L. Mougin. 1979. Order Pelecaniformes. In *Check-list of birds of the world*, E. Mayr and G. W. Cottrell, eds., Vol. I, 2nd ed., pp. 155–193. Cambridge, Mass.: Mus. Comp. Zool.

Ellis, D. H., and R. L. Glinski. 1980. Some unusual records for the Peregrine and Pallid Falcons in South America. *Condor* 82: 350–351.

*Fitzpatrick, J. W. 1973. Speciation in the genus *Ochthoeca* (Aves: Tyrannidae). *Breviora, Mus. Comp. Zool.* 402: 1–13.

Fjeldså, J. 1981a. Comparative ecology of Peruvian grebes—a study of the mechanisms of evolution of ecological isolation. *Vidensk. Meddr. Dansk Naturh. Foren.* 143: 125–249.

———. 1981b. *Podiceps taczanowskii* (Aves, Podicipedidae), the endemic grebe of Lake Junin, Peru. A review. *Steenstrupia, Zool. Mus. Univ. Copenhagen* 7: 237–259.

———. 1981. Biological notes on the giant coot *Fulica gigantea*. *Ibis* 123: 423–437.

*———. 1982a. Biology and systematic relations of the Andean coot "*Fulica americana ardesiaca*" (Aves, Rallidae). *Steenstrupia, Zool. Mus. Univ. Copenhagen* 8: 1–21.

———. 1982b. Some behaviour patterns of four closely related grebes, *Podiceps nigricollis*, *P. gallardoi*, *P. occipitalis* and *P. taczanowskii*, with reflections on phylogeny and adaptive aspects of the evolution of displays. *Dansk Orn. Foren. Tiddskr.* 76: 37–68.

*———. 1983a. A Black Rail from Junin, central Peru: *Laterallus jamaicensis tuerosi* ssp. n. (Aves, Rallidae). *Steenstrupia, Zool. Mus. Univ. Copenhagen* 8: 277–282.

———. 1983b. Vertebrates of the Junin area, central Peru. *Steenstrupia, Zool. Mus. Univ. Copenhagen* 8: 285–298.

*———. 1983c. Geographic variation in the Andean coot *Fulica ardesiaca*. *Bull. Brit. Ornithol. Club* 103: 18–22.

*Fleming, R. S. 1973. Bird communities above treeline, a comparison of temperate and equatorial mountain faunas. Ph.D. thesis, Univ. of Washington, Seattle.

Flessa, K. W. 1981. The regulation of mammalian faunal similarity among the continents. *J. Biogeog.: 8: 427–437.* 427–437.

*Forshaw, J. M., and W. T. Cooper. 1973. *Parrots of the world.* Melbourne: Landsdowne Press.

Fry, C. H. 1980. An analysis of the avifauna of African northern tropical woodlands. *Proc. IV Pan-African Ornithol. Congr.:* 77-83.

Géroudet, P., and R. Lévêque. 1976. Une vague expansive de la Cisticole jusqu'en Europe centrale. *Nos Oiseaux* 33: 241–256.

*Gill, F. B. 1964. The shield color and relationships of certain Andean coots. *Condor* 66: 209–211.

Gochfeld, M. 1977. Peregrine falcon sightings in eastern Peru. *Condor* 79: 391–392.

*Goodwin, D. 1970. *Pigeons and doves of the world.* 2nd ed. London: Trustees Brit. Mus. (Nat. Hist.).

*Graves, G. R. 1981. New Charadriiform records from coastal Peru. *Gerfaut* 71: 75–79.

Griscom, L. 1950. Distribution and origin of the birds of Mexico. *Bull. Mus. Comp. Zool.* 103: 341–382.

Haffer, J. 1968. Ueber die Entstehung der nordlichen Anden und das vermutliche Alter columbianischer Vogelarten. *J. f. Ornithol.* 109: 67–69.

———. 1970. Entstehung und Ausbreitung nord-Andiner Bergvögel. *Zool. Jahrb. Syst.* 97: 301–337.

———. 1974. Avian speciation in tropical South America. *Publ. Nuttall Ornithol. Club* 14: 1–390.

*Hall, B. P. 1961. The taxonomy and identification of pipits (genus Anthus). *Bull. Brit. Mus. (Nat. Hist.)* 75: 243–289.

Hall, B. P., and R. E. Moreau. 1962. A study of the rare birds of Africa. *Bull. Brit. Mus. (Nat. Hist.)* 8: 313–378.

*Harris, M. P. 1981. The waterbirds of Lake Junin, central Peru. *Wildfowl* 32: 137–145.

*Hellack, J. J. 1976. Phenetic variation in the avian subfamily Cardinalinae. *Occ. Papers Mus. Nat. Hist. Univ. Kansas* 57: 1–22.

*Hilty, S. L., and W. L. Brown. 1983. Range extensions of Colombian birds as indicated by the M. A. Carriker Jr. collection at the National Museum of Natural History, Smithsonian Institution. *Bull. Brit. Ornithol. Club* 103: 5–17.

Howell, T. R. 1969. Avian distribution in Central America. *Auk* 86: 293–326.

*Hoy, G. 1965. Wiederentdeckung von *Asthenes sclateri* (Cabanis) in der Sierra de Cordoba (with note by E. Stresemann). *J. f. Ornithol.* 106: 204–207.

———. 1980a. Notas nidobiológicas del noroeste argentino. II. *Physis, Buenos Aires*, sec. C, 39: 63–66.

———. 1980b. Nota sobre la nidobiología de la "Dormilona cenicienta" *Muscisaxicola cinerea argentina* Hellmayr (Aves, Tyrannidae). *Historia Nat.* (Mendoza) 1: 180.

Hueck, K., and P. Seibert. 1972. *Vegetationskarte von Südamerika.* Stuttgart: Gustav Fischer.

Hughes, R. A. 1979. Notes on the birds of the Mollendo district, southwest Peru. *Ibis* 112: 229–241.

———. 1979. Notes on Charadriiforms of the south coast of Peru. In *Shorebirds in Marine Environments*, ed. F. A. Pitelka, Studies Avian Biol., no. 2: 49–53.

———. 1980a. Additional puna zone bird species on the coast of Peru. *Condor* 82: 475.

———. 1980b. Midwinter breeding by some birds in the high Andes of southern Peru. *Condor* 82: 229.

*Hurlbert, S. H., and J. O. Keith. 1979. Distribution and spatial patterning of flamingos in the Andean altiplano. *Auk* 96: 328–342.

Jehl, J. R., Jr. 1968. Relationships in the Charadrii (Shorebirds): A taxonomic study based on color patterns of the downy young. *San Diego Soc. Nat. Hist., Mem.* 3.

———. 1975. *Pluvianellus socialis*: Biology, ecology, and relationships of an enigmatic Patagonian shorebird. *Trans. San Diego Soc. Nat. Hist.* 18: 25–74.

———. 1979. The autumnal migration of Baird's sandpiper. *Shorebirds in Marine Environments*, ed. F. A. Pitelka, Studies Avian Biol., no. 2: 55–68.

Jenny, J. P., F. Ortiz, and M. A. Arnold. 1981. First nesting record of the peregrine falcon in Ecuador. *Condor* 83: 387.

*Johansen, H. 1969. Nordamerikanische Zugvögel in der südhälfte Südamerikas. *Bonn. Zool. Beitr.* 20: 182–190.

*Johnsgard, P. A. 1979. Order Anseriformes. In *Check-list of birds of the world*, E. Mayr and G. W. Cottrell, eds., Vol. I, 2d ed., pp. 425–506. Cambridge, Mass.: Mus. Comp. Zool.

*Johnson, A. W. 1965a. The horned coot *Fulica cornuta* Bonaparte. *Bull. Brit. Ornithol. Club* 85: 84–88.

*———. 1965b. *The Birds of Chile and adjacent regions of Argentina, Bolivia and Peru*, Vol. I. Buenos Aires: Platt Establecimientos Gráficos.

*———. 1967. *The Birds of Chile and adjacent regions of Argentina, Bolivia and Peru*, Vol. II. Buenos Aires: Platt Establecimientos Gráficos.

*———. 1972. *Supplement to The Birds of Chile and adjacent regions of Argentina, Bolivia and Peru.* Buenos Aires: Platt Establecimientos Gráficos.

*Johnson, A. W., F. Behn, and W. R. Millie. 1958. The South American flamingos. *Condor* 60: 289–299.

*Kahl, M. P. 1975. Distribution and numbers—a summary. In *Flamingos*, J. Kear and N. Duplaix-Hall, eds., pp. 93–192. Berkhamsted: T. and A. D. Poyser.

*———. 1979a. Family Ciconiidae. In *Check-list of birds of the world*, E. Mayr and G. W. Cottrell, eds., Vol. I, 2d ed., pp. 245–252. Cambridge, Mass.: Mus. Comp. Zool.

*———. 1976. Order Phoenicopteriformes. In *Check-list of birds of the world*, E. Mayr and G. W. Cottrell, eds., Vol. I, 2d ed., pp. 269–271. Cambridge, Mass.: Mus. Comp. Zool.

Keith, S. 1980. Origins of the avifauna of the Malagasy region. *Proc. IV Pan-African Ornithol. Congr.*: 99–108.

Koepcke, H.-W. 1963. Probleme des Vogelzuges in Peru. *Proc. XIII Intern. Ornithol. Congr.*: 396–411.

Koepcke, M. 1954. Corte ecológico transversal en los Andes del Perú central con especial consideracion de las Aves. Parte 1, Costa, vertientes occidentales y region altoandina. *Mem. Must. Hist. Nat. "Javier Prado"* (Lima) 3: 1–119.

*———. 1958. Die Vögel des Waldes von Zarate. *Bonn. Zool. Beitr.* 9: 130–193.

*———. 1961. Birds of the western slope of the Andes of Peru. *Amer. Mus. Novit.* 2028: 1–31.

———. 1965. Zur Kenntnis einiger Furnariiden (Aves) der Küste und des westlichen Andenabhanges Peru. *Beitr. Neotr. Fauna* 4: 150–173.

*———. 1970. *The birds of the Department of Lima, Peru* rev. and enl. Wynnewood, P.: Livingston.

*Maclean, G. L. 1969. A study of seedsnipe in southern South America. *Living Bird* 8: 33–80.

Martin, J.-L. 1982. L'infiltration des oiseaux forestiers dans les milieux buissonnants de Corse. *Rev. Ecol. (Terre Vie)* 36: 397–419.

Mayr, E. 1963. *Animal species and evolution.* Cambridge, Mass.: Belknap Press of Harvard Univ. Press.

*———. 1979. Order Struthioniformes. In *Check-list of birds of the world*, E. Mayr and G. W. Cottrell, eds., Vol. I, 2d ed., pp. 3–11. Cambridge, Mass.: Mus. Comp. Zool.

Mayr, E., and W. H. Phelps, Jr. 1967. The origin of the bird fauna of the South Venezuelan highlands. *Bull. Amer. Mus. Nat. Hist.* 136: 269–328.

*Mayr, E., and L. L. Short. 1970. Species taxa of North American birds. A contribution to comparative systematics. *Publ. Nuttall Ornithol. Club* 9: 1–127.

Mayr, E., and F. Vuilleumier. 1983. New species of birds described from 1966 to 1975. *J. f. Ornithol.* 124: 217–232.

*McFarlane, R. W. 1975a. Notes on the giant coot (*Fulica gigantea*). *Condor* 77: 324–327.

*———. 1975b. The status of certain birds in northern Chile. *Bull. Internat. Comm. Bird Preserv.* 12: 300–309.

*Meyer de Schauensee, R. 1966. *The species of birds of South America and their distribution.* Narberth, P.: Livingston.

*———. 1982. *A guide to the birds of South America.* 2nd rev. ed. Intercollegiate Press (no city given).

*Meyer de Schauensee, R., and W. H. Phelps, Jr. 1978. *A guide to the birds of Venezuela.* Princeton: Princeton Univ. Press.

*Morrison, A. 1939a. The birds of the Department of Huancavelica, Peru. *Ibis* 1939: 453–486.

*———. 1939b. Notes on the birds of Lake Junin, central Peru. *Ibis* 1939: 643–654.

*Moynihan, M. 1959. A revision of the family Laridae (Aves). *Amer. Mus. Novit.* 1928: 1–42.

———. 1979. Geographic variation in social behavior and in adaptations to competition among Andean birds. *Publ. Nuttall Ornithol. Club* 18: 1–162.

Navas, J. R. 1965. Nuevos aportes para *Oreotrochilus leucopleurus. Hornero* 10: 283–295.

*Navas, J. R. and J.A. Bo. 1982. La posición taxionomica de *Thripohaga sclateri* y *T. Punensis* (Aves, Furnariidae). *Com. Mus. Argentino Cienc. Nat. "Bernardino Rivadavia," Zoología* 4:85–93.

*Niethammer, G. 1953. Zur Vogelwelt Boliviens. *Bonn. Zool. Beitr.* 4: 195–303.

*——— 1956. Zur Vogelwelt Boliviens (Teil II: Passeres). *Bonn. Zool. Beitr.* 7: 74–150.

Norton, D. W. 1965. Notes on some non-passerine birds from eastern Ecuador. *Breviora, Mus. Comp. Zool.* 230: 1–11.

*Norton, W. J. E. 1975. Notes on the birds of the Sierra Nevada de Santa Marta, Colombia. *Bull. Brit. Ornithol. Club* 95: 109–115.

*Olivares, A. 1959. Aves migratorias en Colombia. *Rev. Acad. Colombiana Cienc. Exactas, Fis. Nat.* 10: 1–72.

*———. 1969. *Aves de Cundinamarca.* Bogotá: Univ. Nacional Colombia, Dirección de Divulgación Cultural y Publicaciones.

*———. 1973. Aves de la Sierra Nevada del Cocuy, Colombia. *Rev. Acad. Colombiana Cienc. Exactas, Fis. Nat.* 14: 39–48.

*Olrog, C. C. 1947. Breves notas sobre la avifauna del Aconquija. *Acta Zool. Lilloana* 7: 139–159.

———. 1962. Observaciones sobre becasinas neotropicales (Aves, Charadriiformes, Scolopacidae). *Neotrópica* 8: 111–114.

*———. 1979. Nueva lista de la avifauna argentina. *Opera Lilloana* 27: 1–324.

*O'Neill, J. P., and T. A. Parker III. 1976. New subspecies of *Schizoeaca fuliginosa* and *Uromyias agraphia* from Peru. *Bull. Brit. Ornithol. Club* 96: 136–141.

*———. 1978. Responses of birds to a snowstorm in the Andes of southern Peru. *Wilson Bull.* 90: 446–449.

*Ortiz-Crespo, F. I., and R. Bleiweiss. 1982. The northern limit of the hummingbird genus *Oreotrochilus* in South America. *Auk* 99: 376–378.

*Parker, T. A. III. 1981. Distribution and biology of the White-cheeked Cotinga *Zaratornis stresemanni*, a high Andean frugivore. *Bull. Brit. Ornithol. Club* 101: 256–265.

*Parker, T. A. III, S. A. Parker, and M. A. Plenge. 1982. *An annotated checklist of Peruvian birds.* Vermillion, S. D.: Buteo Books.

*Payne, R. B. 1979. Family Ardeidae. In *Check-list of birds of the world*, E. Mayr and G. W. Cottrell, eds., Vol. I, 2d ed., pp. 193–244. Cambridge, Mass.: Mus. Comp. Zool.

*Paynter, R. A., Jr. 1970a. Subfamily Emberizinae. In *Check-list of birds of the world*, P. A. Paynter, ed., Vol. XIII, pp. 3–124. Cambridge, Mass.: Mus. Comp. Zool.

*———. Subfamily Cardinalinae. In *Check-list of birds of the world*, R. A. Paynter, Jr. ed., Vol. XIII, pp. 216–245. Cambridge, Mass.: Mus. Comp. Zool.

*———. 1972. Biology and evolution of the *Atlapetes schistaceus* species-group (Aves: Emberizinae). *Bull. Mus. Comp. Zool.* 143: 297–320.

*———. 1978. Biology and evolution of the avian genus *Atlapetes* (Emberizinae). *Bull. Mus. Comp. Zool.* 148: 323–369.

Pearson, D. L., and M. A. Plenge. 1974. Puna bird species on the coast of Peru. *Auk* 91: 626–631.

Pearson, O. P., and C. P. Ralph. 1978. The diversity and abundance of vertebrates along an altitudinal gradient in Peru. *Mem. Mus. Hist. Nat. "Javier Prado"* 18: 1–97.

*Peña, L. E. 1961. Results of research in the Antofagasta Ranges of Chile and Bolivia. I. Birds. *Postilla* 49: 1–42.

*———. 1962. Notes on South American flamingos. *Postilla* 69: 1–8.

*Peters, J. L., and J. A. Griswold, Jr. 1943. Birds of the Harvard Peruvian expedition. *Bull. Mus. Comp. Zool.* 92: 281–327.

*Phelps, W. H., Jr. 1966. Contribución al análisis de los elementos que componen la avifauna subtropical de las cordilleras de la Costa Norte de Venezuela. *Bol. Acad. Cienc. Fis. Mat. Nat.* (Caracas) 26: 14–43.

*Philippi, R. A. 1964. Catálogo de las Aves Chilenas con su distribución geográfica. *Invest. Zool. Chilenas* 11: 1–179.

*Plenge, M. A. 1974. Notes on some birds in west central Peru. *Condor* 76: 326–330.

*———. 1982. The distribution of the lesser rhea *Pterocnemia pennata* in southern Peru and northern Chile. *Ibis* 124: 168–172.

Rees, W. E., and N. A. Roe. 1980. *Puya raimondii* (Pitcairnioideae, Bromeliaceae) and birds: An hypothesis on nutrient relationships. *Canadian J. Bot.* 58: 1262–1268.

*Remsen, J. V., Jr., and R. S. Ridgely. 1980. Additions to the avifauna of Bolivia. *Condor* 82: 69–75.

*Remsen, J. V., Jr., T. A. Parker, III, and R. S. Ridgely. 1982. Natural history notes on some poorly known Bolivian birds. *Gerfaut* 72: 77–87.

*Ribera, M. O., and W. Hanagarth. 1982. Aves de la region altoandina de la Reserva Nacional de Ulla-Ulla. *Ecología en Bolivia* 1: 35–45.

*Ridgely, R. S. 1980. Notes on some rare or previously unrecorded birds in Ecuador. *Amer. Birds* 34: 242–248.

*Ripley, S. D. 1957. Notes on the horned coot, *Fulica cornuta* Bonaparte. *Postilla* 30: 1–8.

*Roe, N. A., and W. E. Rees. 1979. Notes on the puna avifauna of Azangaro Province, Department of Puno, southern Peru. *Auk* 96: 475–482.

*Roig, V. R., and J. R. Contreras. 1975. Aportes ecológicos para la biogeografía de la Provincia de Mendoza. *Ecosur, Argentina* 2: 185–217.

*Rottman, J., and R. Kuschel. 1970. Observaciones ornitológicas en las Provincias de Antofagasta y Tarapacá. *Boletín Ornitológico* (Santiago) 2: 1–8.

*Schmidt-Marloh, D., and K.-L. Schuchmann. 1980. Zur Biologie des Blauen Veilchenohr-Kolibris (*Colibri coruscans*). *Bonn. Zool. Beitr.* 31: 61–77.

Schulenberg, T. S. 1985. An intergeneric hybrid conebill (*Conirostrum* × *Oreomanes*) from Peru. In *Neotropical ornithology*, P. A. Buckley, M. S. Foster, E. S. Morton, R. S. Ridgely, and F. G. Buckly, eds., Amer. Ornithol. Union Monogr. No. 36: 390–395.

*Serrano, P., and J. Cabot. 1982. *Las aves de la Reserva Nacional de Ulla Ulla, con comentarios sobre la abundancia y distribución de las especies.* La Paz, Bolivia: INFOL, Estudios Especializados EE–42: 1–16.

Short, L. L. 1971. The evolution of terrestrial woodpeckers. *Amer. Mus. Novit.* 2467: 1–23.

———. 1972. Systematics and behavior of South American flickers (Aves, *Colaptes*). *Bull. Amer. Mus. Nat. Hist.* 149: 1–110.

*———. 1975. A zoogeographic analysis of the South American chaco avifauna. *Bull. Amer. Mus. Nat. Hist.* 154: 163–352.

*———. 1982. *Woodpeckers of the world.* Greenville, Del.: Delaware Mus. Nat. Hist., Monograph series no. 4.

*Short, L. L., and J. J. Morony, Jr. 1968. Notes on some birds of central Peru. *Bull. Brit. Ornithol. Club* 89: 112–115.

*Sibley, C. G., K. W. Corbin, and J. E. Ahlquist. 1968. The relationships of the seed-snipe (Thinocoridae) as indicated by their egg white proteins and hemoglobins. *Bonn. Zool. Beitr.* 19: 235–248.

*Sick, H. 1968. Vogelwanderungen im kontinentalen Südamerika. *Vogelwarte* 24: 217–243.

Simpson, G. G. 1960. Notes on the measurement of faunal resemblance. *Amer. J. Sci.* 258–A: 300–311.

*Smyth, J. A. 1971. Observaciones ornitológicas en la región del Lago Titicaca, Peru-Bolivia. *Revista Univ. Nac. Técnica del Altiplano, Puno* 3: 76–99.

Snow, D. W. 1980. The affinities of African non-passerine birds to the Oriental and Palaearctic avifaunas. *Proc. IV Pan-African Ornithol. Congr.*: 71–76.

*———. 1983. The use of *Espeletia* by paramo hummingbirds in the eastern Andes of Colombia. *Bull. Brit. Ornithol. Club* 103: 89–94.

*Steinbacher, J. 1979. Family Threskiornithidae. In *Check-list of birds of the world*, E. Mayr and G. W. Cottrell, eds., Vol. I, 2d., pp. 254–268. Cambridge, Mass.: Mus. Comp. Zool.

*Storer, R. W. 1970. Subfamily Thraupinae. In *Check-list of birds of the world*, R. A. Paynter, Jr., ed., Vol. XIII, pp. 246–408. Cambridge, Mass.: Mus. Comp. Zool.

———. 1979. Order Podicipediformes. In *Check-list of birds of the world*, E. Mayr and G. W. Cottrell. eds., Vol. I, 2d ed., pp. 140–155. Cambridge, Mass.: Mus. Comp. Zool.

*Stresemann, E., and D. Amadon. 1979. Order Falconiformes. In *Check-list of the birds of the world*, E. Mayr and G. W. Cottrell, Vol. I, 2d ed., pp. 271–425. Cambridge, Mass.: Mus. Comp. Zool.

*Swales, B. H. 1926. *Idiopsar brachyurus* in Argentina. *Auk* 43: 547–548.

Terborgh, J. 1971. Distribution on environmental gradients: Theory and a preliminary interpretation of distributional patterns in the avifauna of the Cordillera Vilcabamba, Peru. *Ecology* 52: 23–40.

Terborgh, J., and J. S. Weske. 1975. The role of competition in the distribution of Andean birds. *Ecology* 56: 562–576.

*Todd, W. E. C., and M. A. Carriker, Jr. 1922. The birds of the Santa Marta region of Colombia: A study in altitudinal distribution. *Ann. Carnegie Mus.* 14: 1–611.

*Traylor, M. A., Jr. 1977. A classification of the tyrant flycatchers (Tyrannidae). *Bull. Mus. Comp. Zool.* 148: 129–184.

*———. 1979a. Family Tyrannidae. In *Check-list of birds of the world*, M. A. Traylor, Jr. ed., Vol. VIII, pp. 1–245. Cambridge, Mass.: Mus. Comp. Zool.

*———. 1979b. Family Phytotomidae. In *Check-list of birds of the world*, M. A. Traylor, Jr., ed., Vol. VIII, pp. 309–310. Cambridge, Mass.: Mus. Comp. Zool.

Traylor, M. A., Jr., and J. W. Fitzpatrick. 1982. A survey of the tyrant flycatchers. *Living Bird* 19: 7–50.

Troll, C. 1959. Die tropischen Gebirge. *Bonn. Geogr. Abh.* 25: 1–93.

*Tuck, L. M. 1972. The snipes: A study of the genus *Capella*. *Can. Wildl. Service Mon.* ser. 5: 1–429.

*Vaurie, C. 1962. A systematic study of the red-backed hawks of South America. *Condor* 64: 277–290.

*———. 1980. Taxonomy and geographical distribution of the Furnariidae (Aves, Passeriformes). *Bull. Amer. Mus. Nat. Hist.* 166: 1–357.

*Vaurie, C., J. S. Weske, and J. W. Terborgh. 1972. Taxonomy of *Schizoeaca fuliginosa* (Furnariidae), with description of two new subspecies. *Bull. Brit. Ornithol. Club* 92: 142–144.

*Venero, G. J. L., and H. P. Brokaw, 1980. Ornitofauna de Pampa Galeras, Ayacucho, Peru. *Publ. Mus. Hist. Nat. "Javier Prado" Zool.* (Lima) ser. A, 26: 1–32.

Vuilleumier, F. 1967. Speciation in high Andean birds. Ph.D. thesis, Harvard Univ., Cambridge, Mass.

*———. 1968. Population structure of the *Asthenes flammulata* superspecies (Aves: Furnariidae). *Breviora* 297: 1–21.

———. 1969a. Systematics and evolution in *Diglossa* (Aves, Coerebidae). *Amer. Mus. Novit.* 2381: 1–44.

———. 1969b. Pleistocene speciation in birds living in the high Andes. *Nature* 223; 1179–1180.

*———. 1969c. Field notes on some birds from the Bolivian Andes. *Ibis* 111: 599–608.

———. 1970a. Insular biogeography in continental regions. I. The northern Andes of South America. *Amer. Natur.* 104: 373–388.

———. 1970b. Generic relations and speciation patterns in the caracaras (Aves: Falconidae). *Breviora, Mus. Comp. Zool.* 355: 1–29.

———. 1971b. Generic relationships and speciation patterns in *Ochthoeca*, *Myiotheretes*, *Xolmis*, *Neoxolmis*, *Agriornis*, and *Muscisaxicola*. *Bull. Mus. Comp. Zool.* 141: 181–232.

———. 1981. The origin of high Andean birds. *Natural Hist.* 90: 50–57.

———. 1984. Patchy distribution and systematics of *Oreomanes fraseri* (Aves, ?Thraupidae) in Andean *Polylepis*. *Amer. Mus. Novit.* 2777: 1–17.

———. 1985. The forest birds of Patagonia: Ecological geography, speciation, endemism, and faunal history. In *Neotropical ornithology*, P. A. Buckley, M. S. Foster, E. S. Morton, R. S. Ridgely, and F. G. Buckley, eds., Amer. Ornithol. Union Monogr. No. 36: 255–304.

*Vuilleumier, F., and D. Ewert. 1978. The distribution of birds in Venezuelan páramos. *Bull. Amer. Mus. Nat. Hist.* 162: 47–90.

Vuilleumier, F., and D. Simberloff. 1980. Ecology versus history as determinants of patchy and insular distributions in high Andean birds. *Evol. Biol.* 12: 235–379.

Weberbauer, A. 1945. *El mundo vegetal de los Andes peruanos*. Lima: Minist. Agricultura.

Winterbottom, J. M. 1965. Avifaunal relationships between the Neotropical and Ethiopian regions. *Hornero* 10: 209–214.

*Zimmer, J. T. 1938. Notes on migrations of South American birds. *Auk* 55: 405–410.

*———. 1935–55. Studies of Peruvian birds, nos. 16, 19, 26, 32, 35, 43, 44, 48, 60, 61, 62, 65, 66. *Amer. Mus. Novit.* 757, 860, 930, 1044, 1095, 1193, 1203, 1262, 1513, 1540, 1595, 1649, 1723.

*Zusi, R. L., and J. R. Jehl. Jr. 1970. The systematic relationships of *Aechmorhynchus*, *Prosobonia*, and *Phegornis* (Charadriiformes: Charadrii). *Auk* 87: 760–780.

Author Index

Subject Index

Taxonomic Index

The taxonomic appendixes have not been indexed.